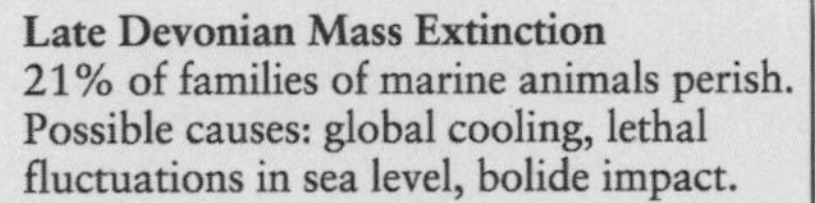

Late Permian Mass Extinction
57% of families of marine animals perish. Possible causes: severe climatic change resulting from assembly of Pangea, global increase in volcanic activity.

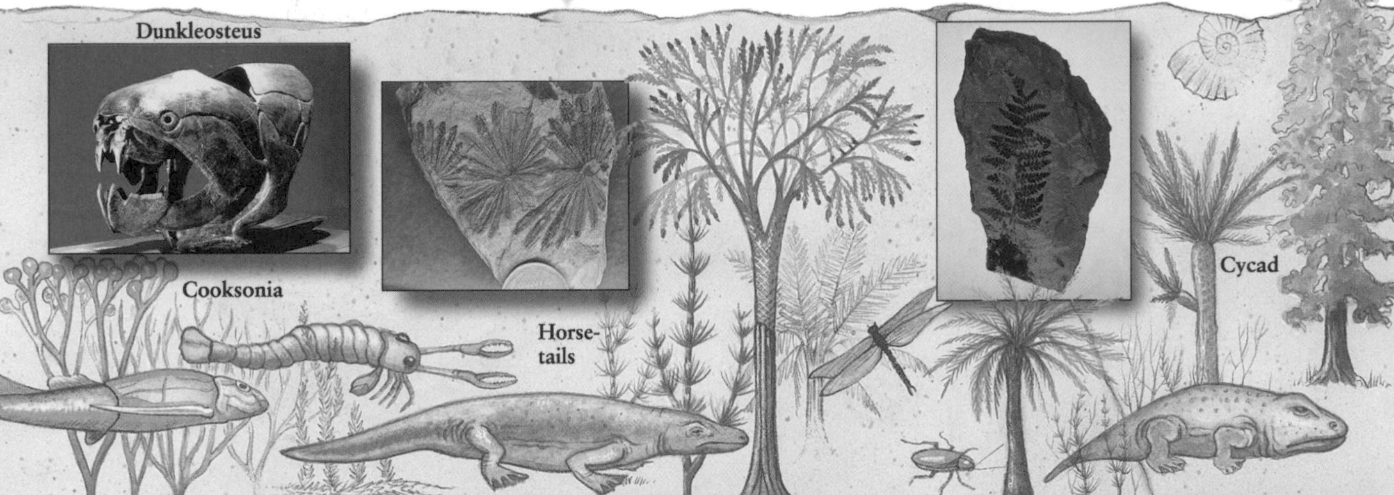

(continued on inside back cover)

Paleozoic Invertebrates:
articulate brachiopods, bryozoa, rugose and tabulate corals, stromatoporoids, cephalopods, crinoids, graptolites, ostracodes, and new species of trilobites.

GEOLOGIC TIME SCALE BOOKMARK

Use the attached time scale as a learning aid and to mark your place as you proceed through geologic time.

MARK YOUR PLACE!

Time Units of The Geologic Time Scale

(Numbers are absolute dates in millions of years before the present.)

Eon	Era	Period		Epoch	Date at base
Phanerozoic Eon (*Phaneros* = *"evident"*; *Zoo* = *"life"*)	Cenozoic Era	Quaternary		Holocene	
				Pleistocene	1.75
		Tertiary	Neogene	Pilocene	5.3
				Miocene	23.5
			Paleogene	Oligocene	33.7
				Eocene	53
				Paleocene	65
	Mesozoic Era	Cretaceous			135
		Jurassic			203
		Triassic			250
	Paleozoic Era	Permian			295
		Carboniferous	Pennsylvanian		320
			Mississippian		355
		Devonian			410
		Silurian			435
		Ordovician			500
		Cambrian			540
Proterozoic Eon	Neo-proterozoic				1000 ← 1 billion years ago
	Meso-proterozoic			Precambrian comprises about 87% of the geologic time scale	1600
	Paleo-proterozoic				2500
Archean Eon	Late				3000
	Middle				3400
	Early				3960
Hadean	No record			Origin of Earth 4.6 billion years ago	

See other side for more information.

MARK YOUR PLACE... IN GEOLOGIC TIME!

This bookmark depicting the geologic time scale units is provided with every purchase of

Levin: THE EARTH THROUGH TIME, 7/e.

Here is an additional aid corresponding to the geologic period scale shown throughout the text

GEOLOGIC PERIODS

Q	Quaternary
T	Tertiary
K	Cretaceous
J	Jurassic
Tr	Triassic
Pr	Permian
P	Pennsylvanian
M	Mississippian
D	Devonian
S	Silurian
O	Ordovician
Є	Cambrian
Pre-Є	Precambrian

HIGHLIGHTS OF GEOLOGIC EONS

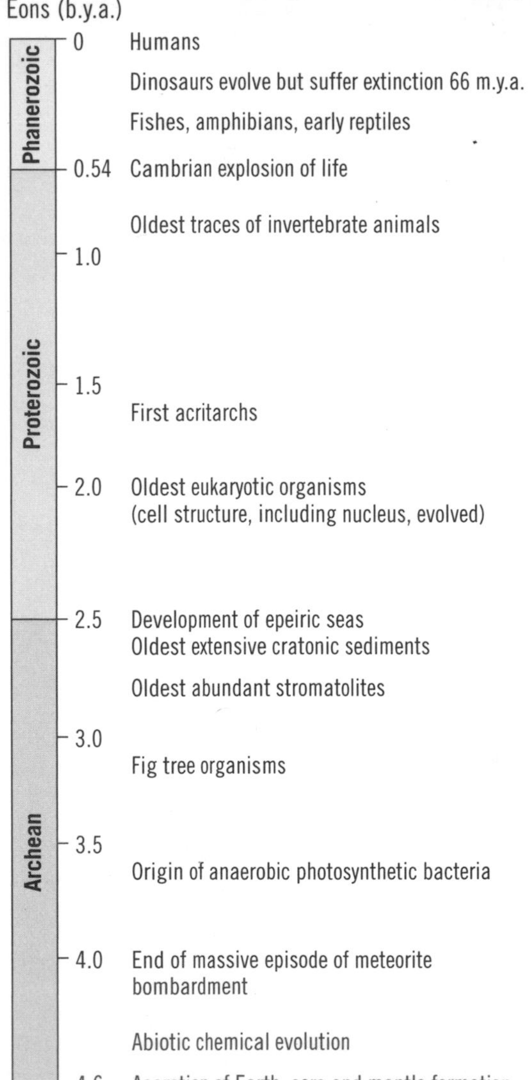

See other side for more geologic time units

SEVENTH EDITION

The Earth Through Time

To see where we might be going,
we must understand where we have been.

Robert Tamarkin, 1993

SEVENTH EDITION

The Earth Through Time

HAROLD L. LEVIN

Professor of Geology
Washington University,
St. Louis, Missouri

JOHN WILEY & SONS, INC.

ACQUISITIONS EDITOR Ryan Flahive

ASSISTANT EDITOR Denise Powell

EDITORIAL ASSISTANT Aliyah Vinikoor

MARKETING MANAGER Kevin Molloy

SENIOR PRODUCTION EDITOR Elizabeth Swain

TEXT AND COVER DESIGNER Madelyn Lesure

ILLUSTRATION EDITOR Anna Melhorn

PHOTO EDITOR Jennifer MacMillan

PHOTO RESEARCHER Amy Dunleavy

Cover: Triassic Welsh Desert with Prosauropod and Small Theropods. *(Copyright J. Sibbick)*

Unless otherwise noted, uncredited photos come from the collection of Harold Levin.

This book was set in 10/12 Janson by UG / GGS Information Services, Inc. and printed and bound by Von Hoffmann Press. The cover was printed by Von Hoffmann Press.

This book is printed on acid free paper. 

Harold L. Levin
The Earth Through Time

ISBN 0-470-00020-1

Printed in the United States of America

10 9 8 7 6 5 4 3 2 1

For Kay,
and for our grandchildren,
Noah, Lillie, Eli, Mollie, Natalie, Emily, Caitlyn, and Candis.

May they have the wisdom to treat the Earth kindly.

ABOUT THE AUTHOR

Harold ("Hal") Levin began his career as a petroleum geologist in 1956 after receiving bachelor's and master's degrees from the University of Missouri and a doctorate from Washington University. His fondness for teaching brought him back to Washington University in 1962, where he is currently professor of geology and paleontology in the Department of Earth and Planetary Sciences. His writing efforts include authorship of seven editions of *The Earth Through Time*; four editions of *Contemporary Physical Geology*; *Essentials of Earth Science*; and co-authorship of *Earth: Past and Present*, as well as eight editions of *Laboratory Studies in Historical Geology*; *Life Through Time*; and, most recently, *Ancient Invertebrates and Their Living Relatives.*

For his courses in physical geology, historical geology, paleontology, sedimentology, and stratigraphy, Hal has received several awards for excellence in teaching. The accompanying photograph was taken during a lecture on life of the Cenozoic Era. The horse skull serves to illustrate changes in the teeth and jaws of grazing animals in response to the spread of prairies and savannahs during the Miocene and subsequent epochs.

The history of the Earth can now be traced back through about 4 billion years, approaching the time when our planet had gathered most of its mass from a rotating cloud of dust, gases, and meteorites. From that time to the present, the Earth has undergone constant change. Ocean floors have expanded and moved, continents have broken apart, sea floors have been thrust skyward to form mountain ranges, and once-peaceful landscapes have been disrupted by earthquakes and volcanoes. For about 3 billion years life has existed on this dynamic planetary surface. Fossil remains of these ancient life forms attest to their success or failure in coping with perpetually changing environments. The record of how the Earth and its inhabitants have changed through time provides a fascinating story, and certainly deserves a place in the liberal education of students today.

While deciding on a course to fulfill the science requirement for a degree, students may question how learning about past geologic events and ancient life relate to the complex problems that presently confront society. Many of these problems are indeed unique to the present day, and are related to the activities of our species. We are a cause for ozone destruction, pollution of the atmosphere and water bodies, depletion of mineral resources, loss of animal and plant species, and a host of other problems. The Earth, however, experienced many catastrophic events long before it had human inhabitants. Repeatedly in the geologic past, land masses have been torn apart, the composition of the atmosphere has been altered by extensive volcanic activity, seas have alternately flooded and drained lands, huge meteorites have bombarded the Earth, and episodes of climatic cooling have resulted in great ice ages. These events triggered episodes of mass extinctions in the past. Understanding the relationship between these events and life of the past will help us cope with problems we face now and in the future.

Determining the cause of past geologic events is a complex task. As an example, one might infer that the demise of a particular group of ancient animals resulted from cooling of the planet. This leads one to ask *why* cooling occurred. Was it the result of changes in the composition of the atmosphere, changes in the amount of radiation received from the sun, the formation of wind-diverting mountain ranges, changes in the Earth's axis of rotation, or shifts in the location of lands and seas? The answers come only after study of the reciprocal actions of all the Earth's systems: the solid Earth, the atmosphere, and the biosphere. This integrated approach applies well both to the solution of contemporary problems and to an understanding of happenings in the remote geologic past. It is an approach embodied in this text.

As in earlier editions, this revision of *The Earth Through Time* is designed to meet the requirements of the undergraduate student who has had little or no previous acquaintance with geology. However, for those exploring the possibility of an academic major in geology, the text will provide the initial background needed for advanced courses. Information about geologic materials and processes are included in *The Earth Through Time* so that the book can be used either for a single, self-contained first course, or for the second course in a two-semester sequence of Physical and Historical Geology.

THE SEVENTH EDITION

The results of recent spacecraft explorations of neighboring planets have provided many clues to conditions on the Earth during its formative stages. Therefore, a chapter bearing the title "The Primordial Earth: Hadean and Archean Eons" has been prepared in order to encorporate a more thorough discussion of the Hadean. Because many inferences about the Hadean are based on knowledge of our planetary neighbors, a brief review of the characteristics of planets in our solar system is included. Sections describing the crystallization of magmas, metamorphic facies, sequence stratigraphy, and analysis of ancient environments have been revised and expanded. New information about how the microbial life of harsh environments relates to the origin of life is provided. The revision also includes a discussion of the proposed Precambrian "Snowball Earth." It is a topic currently receiving considerable attention in the professional literature. A new section in Chapter 1 describes the contributions of pioneer North American geologists.

In order to make the text a more effective and engaging instrument for learning, this seventh edition has many new illustrations and accounts of recent geologic research. Explanations have been refashioned so as to emphasize cause-and-effect relationships generated in the geologic past by the Earth's systems. A **Summary** at the end of each chapter provides a concise review of important concepts. So that students can self-assess their understanding of text information, **Questions for Review and Discussion** are provided for each chapter, and relevant internet **web sites** can be accessed at www.wiley.com/college/levin. Many new terms have been added to the **Glossary**. In the appendices, the **Classification of Living Things** has been extensively revised so as to better reflect recent discoveries relating to evolutionary relationships.

As in any book dealing with history, the chronological sequence of events forms a framework for discussion. To help students correlate geologic events to their time of occurrence, a removable **Geologic Time Scale Bookmark** is provided. As a further aid, small time scales containing an age indicator accompany many illustrations of rock formations and extinct forms of life. Where appropriate, figure captions include a question that draws the student's attention to the illustration or serves to emphasize a concept. Finally, when the need arises to check the age or correlation of rock formations, the student can refer to the **Formation Correlation Charts** in the appendices.

The **Enrichment** boxes and the **Geology of National Parks and Monuments** boxes of previous editions have been retained. Boxes new to this edition include "You Are the Geologist (Solve This Problem)," "The Geosynclinal Hypothesis Displaced by Plate Tectonics," "Banded Iron Formations: Civilization's Indispensible Treasure," "Can We Bring Back the Dinosaurs?," "Dinosaur National Monument," and "Grand Staircase–Escalante National Monument."

ANCILLARIES

The seventh edition of *The Earth Through Time* is accompanied by an extensive set of supporting materials. These include:

- A **Student Study Guide** prepared by Harry Wagner of Victoria College. Designed to enhance student's understanding of the text, the Guide includes Chapter Overviews, Learning Objectives for each chapter, Questions for Review, Key Terms, illustrations, and maps.
- An **Instructor's Manual and Test Bank** has been prepared by David T. King Jr., of Auburn University. To facilitate use of the text by instructors, the manual includes chapter outlines, questions that can be incorporated into examinations, and answers to questions in the textbook.
- **Computerized Test Banks** are available for Windows and Macintosh users.
- A set of full-color **Overhead Transparencies** is provided for use in the laboratory or lecture hall, as well as a set of 35-mm slides of illustrations and photographs.
- The **Earth Sciences Instructor's Resource** is a web site that provides numerous images from this and other John Wiley & Sons geology texts. It can be accessed at www.wiley.com/college/levin.

John Wiley & Sons, Inc. may provide complementary instructional aids and supplementary packages to those adopters qualified under our adoption policy. Pelase contact your sales representative for more information.

ACKNOWLEDGMENTS

Many of the changes in this edition are the result of incisive comments and suggestions of the diligent group of reviewers listed below:

Loren Babcock, *Ohio State University*
Merrill Foster, *Bradley University*
Fred Heck, *Ferris State University*
Stephen Mojzis, *University of Colorado*
Shuhai Xiao, *Tulane University*

I would also like to extend my thanks to the following earth scientists who contributed as reviewers to the previous editions of the book:

Dennis Allen, *University of South Carolina–Aiken*
Wiliam Ausich, *Ohio State University*
David R. Berry, *California State Polytechnic University*
William Berry, *University of California–Berkeley*
Michael Bikerman, *University of Pittsburgh*
Mark Camp, *University of Toledo*
Roger J. Cuffey, *University of Pennsylvania*
James H. Darrell, *Georgia Southern University*
Larry E. Davis, *Washington State University*
William H. Easton, *University of Southern California*
George F. Engelmann, *University of Nebraska at Omaha*
Stanley Fagerlin, *Southwest Missouri State College*
Cindy Ganes, *University of Saskatchewan*
Pamela Gore, *DeKalb College*
John Gosse, *University of Kansas*
Jack Hall, *University of North Carolina at Wilmington*
Vicki Harder, *Santa Teresa, New Mexico*
John A. Howe, *Bowling Green State University*
Warren Huff, *University of Cincinnati*
John R. Huntsman, *University of North Carolina*
Allen Johnson, *West Chester University*
Gary D. Johnson, *University of South Dakota*
William M. Jordan, *Millersville University of Pennsylvania*
Roger Kaesler, *University of Kansas*
Ernst Kastning, *Radford University*

David T. King, Jr., *Auburn University*
Larry Knox, *Tennessee Technical University*
Peter Kresan, *University of Arizona*
Ralph L. Langenheim, Jr., *University of Illinois*
W. Brit Leatham, *California State University–San Bernardino*
Peter Leavens, *University of Delaware*
Joseph Lintz, *University of Nevada at Reno*
Daniel L. Lumsden, *Memphis State University*
Donald Marchand, *Old Dominion University*
William H. Mathews, III, *University of British Columbia*
Dewey McLean, *Virginia Polytechnic Institute*
Eldridge Moores, *University of California–Davis*
Peter Nielsen, *Keene State College*
Cathryn Newton, *Syracuse University*
Shannon O'Dunn, *Grossmont College*
Donald E. Owen, *Lamar University*
William Parker, *Florida State University*
John Pope, *Miami University*
Jennifer Smith Prouty, *Corpus Christi State University*
G. J. Retallack, *University of Oregon*
Thomas Roberts, *University of Kentucky*
Barun K. Sen Gupta, *Louisiana State University*
Thomas W. Small, *Frostburg University*
Leonard W. Soroka, *St. Cloud University*
Don Steinker, *Bowling Green State University*
Calvin H. Stevens, *San Jose State University*
James Stevens, *Lamar University*
David Thomas, *Washtenaw Community College*
Harry Wagner, *Victoria College*
Kenneth Van Dellen, *Macomb Community College*
Carl Vondra, *Iowa State University*
Peter Whaley, *Murray State University*
Lisa White, *San Francisco State University*
William Zinmeister, *Purdue University*
Thomas T. Zwick, *Eastern Montana College*

Finally, my deep appreciation goes to my wife, Kay, who has cheerfully endured my preoccupation with preparing this new edition.

BRIEF CONTENTS

CONTENTS

1

Fossil ammonite cephalopod extracted from Jurassic age limestone at Lyme Regis on England's south coast. Ammonites are extinct mollusks distantly related to the living chambered nautilus. Mary Anning (1799–1847), known as the mother of paleontology, collected scores of these fossils from this locality. (Copyright Sinclair Stammers/ Science Photo Library/Photo Researchers.)

CHAPTER 1

Introduction to Earth History

The face of places and their forms decay; And that is solid earth that once was sea; Seas, in their turn, retreating from the shore, Make solid land, what ocean was before.
Ovid, *Metamorphoses, XV*

OUTLINE

We live on the third planet from the Sun. Our planet was formed 4.6 billion years ago and since that time has circled the Sun like a small spacecraft observing a rather average star. Somewhere between 300,000 and 150,000 years ago, a species of primate named *Homo sapiens* evolved on planet Earth. Unlike earlier animals, these creatures with oversized brains and nimble fingers asked questions about themselves and their surroundings. Their questioning has continued to the present day. How was the Earth formed? Why do earthquakes occur? What lies beneath the lands we live on and beneath the floor of the ocean? Even ancient people sought answers for these questions. In frail wooden ships they probed the limits of the known world, fearing that they might tumble from its edge. Their descendants came to know the planet as an imperfect sphere, and they began an examination of every obscure recess of its surface. In harsher regions, exploration proceeded slowly. It has been only within the last 100 years that humans have penetrated the deep interior of Antarctica. Today, except for a few areas of great cold or dense forest, the continents are well charted. New frontiers for exploration now lie beneath the oceans and outward into space.

GEOLOGY

Physical and Historical Components of Geology

Geology is the study of planet Earth. It is concerned with the materials of which the planet is made, the physical and chemical processes that act on these materials, and the history of the Earth and its inhabitants.

Geologists concern themselves with an exceptional variety of scientific tasks and therefore must employ knowledge from diverse fields. Some examine the composition and texture of meteorites and Moon rocks. With magnifiers and computers, others scrutinize photographs of planets to understand the origin of the features that characterize their surfaces. Still others are busily unraveling the structure of mountain ranges, attempting to predict the occurrence of earthquakes and volcanic eruptions, or studying the behavior of glaciers, streams, or underground water. Large numbers of geologists search for fossil fuels and the metallic ores vital to our standard of living. They worry, as do you and I, about the fate of humans in a world of diminishing resources. To do their work, geologists draw on the knowledge of astronomy, physics, chemistry, mathematics, and biology. For example, the petroleum geologist must understand the physics of moving fluids, the chemistry of hydrocarbons, and the biology of the fossils used to trace subsurface rock layers. Because geo-logy incorporates information from so many other scientific disciplines, it can be termed an eclectic science. The term *eclectic* is useful in describing a body of selected information drawn from a variety of sources. All sciences are eclectic to some degree, but geology is decidedly so.

For convenience of study, the body of knowledge called geology can be divided into **physical geology** and **historical geology**. The origin, classification, and composition of earth materials, as well as the varied processes that occur on the surface and in the deep interior of the Earth, are the usual subjects of physical geology. Historical geology addresses the Earth's origin, evolution, changes in the distribution of lands and seas, growth and destruction of mountains, succession of animals and plants through time, and developmental history of the solar system. The historical geologist examines planetary materials and structures to discover how they came into existence. He or she works with the tangible *results* of past events and must work backward in time to discover the cause of those events.

The Scientific Method in Geology

Geologists employ the same procedures used by scientists in other disciplines. Those procedures are rather formally referred to as the *scientific method*. The scientific method is merely a scheme for finding answers to questions and solutions to problems. It is not a fixed series of steps that researchers strictly and consciously follow. A scientific investigation often begins with the formulation of *questions*, proceeds to the collection of *observations* or data, and is followed by the development of an *explanation* or hypothesis. Further observations, tests, and scrutiny by other scientists serve to validate or invalidate the hypothesis.

As an example of scientific methodology, consider the work of geologists Anita Harris, Jack Epstein, and Leonard Harris. While working in the Appalachian Mountains for the United States Geologic Survey, these scientists observed that a group of tiny fossils called conodont elements (Fig. 1-1) differed in color from pale yellow to black. Conodont elements are the microscopic hard parts of organisms that lived on Earth from about 520 to 200 million years ago. They are composed of apatite, the same mineral of which bone is made, and are abundant in many localities and in all kinds of sedimentary rocks.

The geologists asked the question: "Why do conodont elements of the same age but from different parts of a region have different coloration?" They then set about obtaining the data that might provide an answer. In the laboratory they selected pale yellow conodont elements from a sedimentary rock that had never been deeply buried. These fossils were then heated to temperatures from 300°C to 600°C in 50° steps over a period of 10 to 50 days. They observed that the conodont elements changed in color through five phases from pale yellow to black. On further heating, so as to approximate conditions that change sedimentary rocks into metamorphic rocks (rocks which are altered by heat and pressure), black fossils changed sequentially to gray, milky white, and finally crystal clear.

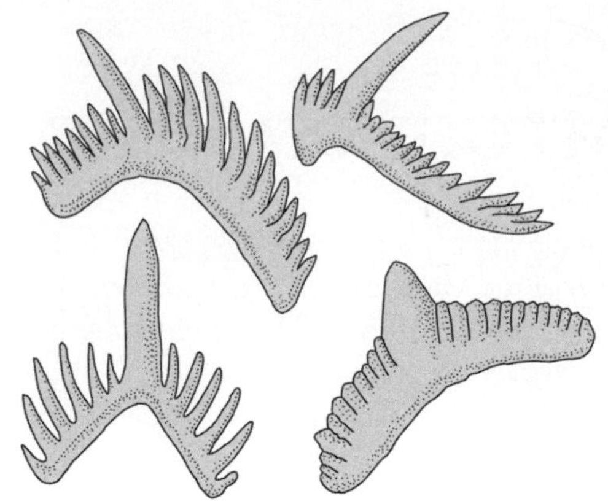

FIGURE 1-1 **Conodont elements** (magnified about 50 times).

Next, the colors produced experimentally were compared with conodont elements collected in the Appalachian Mountains from rocks that had been subjected to various depths and durations of burial. When plotted on a map, the data showed that the light-colored fossils occurred in rocks of the western Appalachians, where burial was least, and the dark-colored fossils were in rocks along the eastern side of the Appalachians, where burial was greatest. The original question could now be answered with a hypothesis stating that conodont element color alteration is caused by depth of burial and consequent increase in temperature. The study had practical value as well. Petroleum geologists learned that rocks containing black to clear conodont elements are less likely to yield commercial quantities of oil.

An Introduction to Plate Tectonics

A significant number of topics in both physical and historical geology are related to a grand unifying concept termed **plate tectonics**. "Tectonics" refers to large-scale deformation of rocks that compose the Earth's outer layers. The term "plate" is given to a large slab-like segment of the Earth's **lithosphere**. The lithosphere is the rigid outer 100 km or so of the Earth that includes the **crust** as well as the uppermost part of the **mantle** (Fig. 1-2).

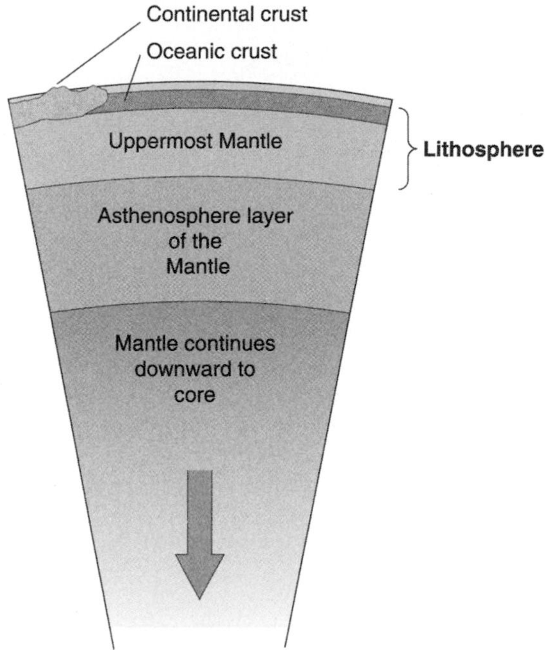

FIGURE 1-2 The lithosphere is the outer shell of the Earth that lies above the asthenosphere and comprises both the crust and an uppermost layer of the mantle.

On the Earth's surface there are seven large lithospheric plates and about 20 smaller plates. The plates rest on a partially molten layer of the mantle called the **asthenosphere**. The plates are in constant movement, probably because of heat-driven convectional plastic flow in the asthenosphere. Lithospheric plates must have margins or boundaries. Where two or more plates move apart from one another, the plate margins are termed **divergent boundaries. Convergent boundaries** occur where plates converge, and **transform boundaries** occur where they slide past one another. You will meet with all the above terms again in Chapter 5, where plate tectonics will be examined in more detail. Until then, this brief introduction will be useful.

THE FOUNDERS OF HISTORICAL GEOLOGY

Historical geology is a venerable science. Its beginnings can be traced to the time of classical Greece. Like other sciences, progress in historical geology has been based on the continuous accumulation of knowledge by past generations of workers. They have provided the foundations of geology upon which modern theories and precepts depend. A partial list of early contributors to our understanding of the origin and history of the Earth would include Nicolaus Steno, Abraham Gottlob Werner, James Hutton, William Smith, Georges Leopold Cuvier, Alexander Brongniart, Sir Charles Lyell, and Charles Darwin.

Nicolaus Steno (Niels Stensen)

Niels Stensen (1638–1687) was a Danish physician who was widely recognized for his studies in anatomy. Unable to secure a teaching position in Copenhagen's medical school, he settled in Florence, Italy. There he latinized his name to Nicolaus Steno and became physician to the Grand Duke of Tuscany. Since the duke was a generous employer, Steno had ample time to tramp across the countryside, visit quarries, and examine strata. His investigations of sedimentary rocks led him to formulate such basic principles of historical geology as superposition, original horizontality, and original lateral continuity.

The **principle of superposition** states that in any sequence of undisturbed strata, the oldest layer is at the bottom, and successively higher layers are successively

FIGURE 1-3 Steeply dipping strata grandly exposed in the Himalayan Mountains. It is often difficult to recognize the original tops of beds in strongly deformed sequences such as this. (*Courtesy of D. Bhattacharyya.*)

FIGURE 1-4 Superpositional sequence of undisturbed, horizontal, mainly Permian strata. Canyonlands National Park, Utah. The Colorado River is in the foreground. (*Photo by P. L. Kresan.*) ? *Along this bend in the river, where would one find the youngest stratum?**

younger. It is a rather obvious axiom. Yet Steno, on the basis of his observations of strata in northern Italy, was the first to explain the concept formally. The fact that it is self-evident does not diminish the principle's importance in deciphering Earth history. Furthermore, the superpositional relationship of strata is not always apparent in regions where layers have been steeply tilted or overturned (Fig. 1-3). In such instances, the geologist must examine the strata for clues useful in recognizing their uppermost layer. The way fossils lie in the rock and the evidence of mudcracks and ripple marks are particularly useful clues when one is trying to determine which way was up at the time of deposition.

The observation that strata are often tilted led Steno to his **principle of original horizontality**. He reasoned that most sedimentary particles settle from fluids under the influence of gravity. The sediment then must have been deposited in layers that were nearly horizontal and parallel to the surface on which they were accumulating (Fig. 1-4). Hence, steeply inclined strata indicate an episode of crustal disturbance after the time of deposition (Fig. 1-5).

The **principle of original lateral continuity** was the third of Steno's stratigraphic axioms. It pertains to the fact that, as originally deposited, strata extend in all directions until they terminate by thinning at the margin of the basin, end abruptly against some former barrier to deposition, or grade laterally into a different kind of sediment (Fig. 1-6). This observation is significant in that whenever one observes the exposed cross-section of strata in a cliff or valley wall, one should recognize that the strata, as originally deposited, should continue laterally for a distance that can be determined by field work and drilling. If lateral continuity is not observed and the lack of continuity is not related to one of the reasons given above, then the cause may be displacement of strata by faulting or erosional loss of strata. When geologists stand on a sandstone ledge at one side of a canyon, it is the principle of original lateral continuity that leads them to seek out the same

*Answers to the questions within the figure legends can be found in the *Student Study Guide* accompanying this text.

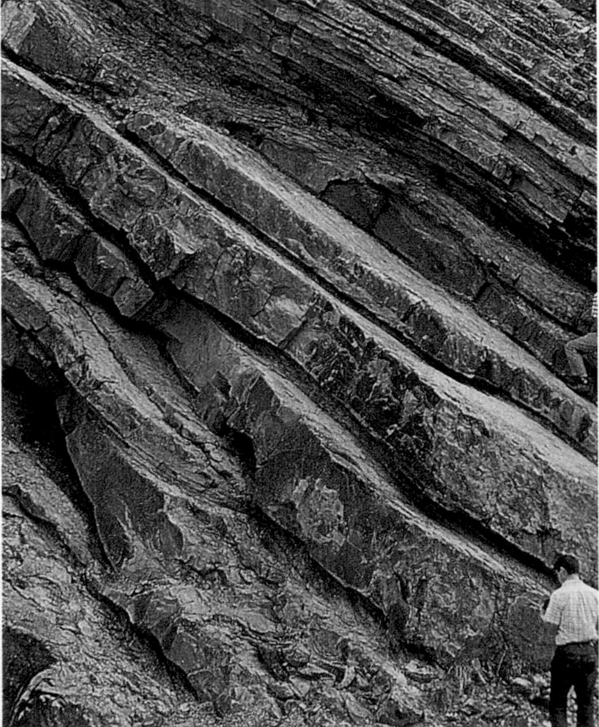

FIGURE 1-5 Steeply dipping shale and sandstone strata. Ouachita Mountains of Oklahoma.

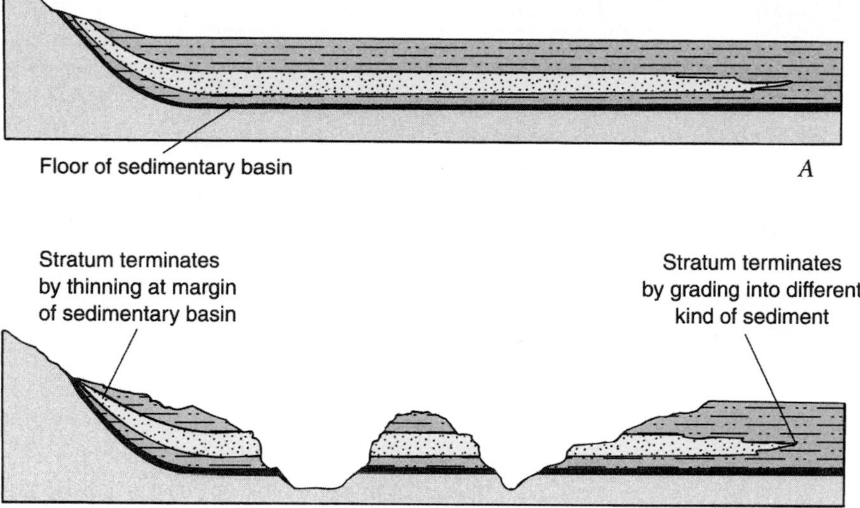

FIGURE 1-6 Illustration of original lateral continuity. Cross-section *A* shows a sandstone stratum deposited within a low-lying area or sedimentary basin that received sediment eroded from surrounding uplands. Cross-section *B* shows the same area after erosion has exposed the sandstone on hillsides. As indicated, the stratum continues horizontally until termination results from nondeposition at the margin of a basin or other obstruction or by gradation into different sediment. Strata may terminate abruptly as a result of erosion or faulting but were nevertheless *originally* laterally continuous. ? *What conditions in the environment of deposition may have caused the sandstone layer (stippled) to "pinch out" on the right side of section B?*

ledge of sandstone on the far canyon wall and then to realize that the two exposures were once continuous.

Today, we recognize that Steno's principles are basic to the geologic specialty known as **stratigraphy**, which is the study of layered rocks, including their texture, composition, arrangement, and correlation from place to place. Because stratigraphy enables geologists to place events as recorded in rocks in their correct sequence, it is the key to the history of the Earth.

Interpreters of the Geologic Succession

The stratigraphic principles formulated by Steno in the 17th century were rediscovered several decades later by other European scientists. Among the most prominent of these early geologists were John Strachey (1671–1743), Giovanni Arduino (1714–1795), Johann G. Lehmann (1719–1767), Georg Füchsel (1722–1776), and Peter Simon Pallas (1741–1811). John Strachey is best remembered for his use of the principles of superposition and original lateral continuity in deciphering the stratigraphic succession of coal-bearing formations in Somerset and Northumberland, England. He clearly illustrated the sequence of formations encountered at the surface and in mines and described the manner in which horizontal strata rested upon the eroded edges of inclined older formations. Years later this type of stratigraphic relationship would be termed *unconformable*.

Whereas John Strachey was particularly interested in a local stratigraphic succession, other naturalists developed a broader, more global view of the geologic succession. In Italy, Giovanni Arduino classified mountains according to the most abundant type of rock that composed them. He defined Primary mountains as those constructed of crystalline rocks of the kinds later to be named igneous and metamorphic. Arduino recognized that rocks of the Primary group were likely to be the oldest in a mountain system and were usually exposed along the central axis of ranges. Secondary mountains were constructed of layered, well-consolidated, fossiliferous rocks. Such rocks were later to be named sedimentary. Arduino's Tertiary designation was reserved for unconsolidated gravel, sand, and clay beds as well as lava flows.

Classifications similar to that of Arduino also appeared in the works of the German scientists Lehmann and Füchsel. These men were not rocking-chair theorists. Both were excellent field geologists. Füchsel worked chiefly in the mountains of Thuringia, whereas his contemporary Lehmann examined the rocks of the Harz and Erz Gebirge (Gebirge is the German word for mountains). They prepared excellent summaries of the stratigraphic succession in these mountains and further developed a remarkably perceptive understanding of some of the events involved in the making of mountain ranges.

This insight into the history of mountains was improved as a result of the work of a tireless field geologist named Peter Simon Pallas. Under the patronage of Catherine II of Russia, Pallas traveled across the whole of Asia and made careful studies of the Ural and Altai Mountains. He recognized the threefold division of mountains formulated by his predecessors. In addition, Pallas was able to construct a general geologic history of the Urals, and he provided a lucid description of how the rock assemblages change as one travels from the center to the flanks of mountain systems.

Abraham Gottlob Werner

One of the most influential geologists working in Europe near the close of the 18th century was Professor Abraham Gottlob Werner (1749–1817). Werner's eloquent and enthusiastic lectures at the Freiberg Mining Academy in Saxony transformed that school into an international center for geologic studies. Werner was a competent mineralogist, and many geologists of his day used his scheme for the identification of minerals and ores. He is not, however, remembered as much for his contributions to mineralogy as he is for his interpretation of the geologic history of the Earth. The cornerstone of that interpretation was his insistence that all rocks of the Earth's crust were deposited or precipitated from a great ocean that once enveloped the entire planet. Today we know that some rocks (the group called sedimentary) are indeed often of marine origin. Others, however, are decidedly not formed in water. Because they believed that all rocks had formed in the ocean, Werner and his many followers became known as *neptunists* (after Neptune, the Roman god of the sea).

Werner envisioned his universal ocean in the earliest stage of Earth history as a hot, steamy body saturated with all the dissolved minerals needed to form the rocks of his oldest division. He called these *Primitive Rocks*, or *Urgebirge*. Most of these rocks formed the cores of mountain ranges and would later come to be known as igneous and metamorphic.

In the second stage of the Wernerian interpretation of Earth history, the basin floor of the primitive ocean subsided and the waters filling the basin cooled. The ocean came to resemble the ocean of today. Werner told his students that this change was marked by the deposition of fossil-bearing, well-consolidated, stratified, and often structurally disturbed rocks that lie above the Urgebirge. These he designated Transition Rocks and suggested that they were deposited when the Earth had passed from an uninhabitable to an inhabitable condition. The fossils proved the planet had become suitable for life. Today, we recognize these rocks as part of Europe's predominantly sedimentary Paleozoic sequence of strata.

Above the Transition Rocks, Werner noted the occurrence of flat-lying sandstones, shales, coal beds, very fossiliferous limestones, and occasional layers of a black rock later determined to be basalt. These basalt layers were actually old lava flows. For all of these rocks lying above the Transition Rocks, Werner employed Johann Lehmann's term *Flötzgebirge*. A final term, Alluvium, was used for the unconsolidated sand, gravel, and clay that rested on the Flötzgebirge.

Although initially received with great interest and enthusiasm, Werner's ideas were soon to draw criticism. His theory failed to explain what had become of the immense volume of water that once covered the Earth to a depth so great that all continents were submerged. An even greater problem was his insistence that basaltic lava layers such as those in the Flötzgebirge were deposited in precisely the same manner as the enclosing limestones and shales. With visible, indisputable field evidence, geologists such as J. F. D'Aubisson de Voisins (1769–1832) in France clearly demonstrated the volcanic origin of these basaltic layers. Geologists with this opposing view came to be known as plutonists (after Pluto, the Roman god of the Underworld). According to the plutonists, fire rather than water was the key to the origin of igneous rocks. James Hutton of Scotland was a prominent plutonist who clearly stated that rocks such as basalt and granite "formed in the bowels of the Earth of melted matter poured into rents and openings of the strata."

James Hutton

James Hutton (1726–1797), an Edinburgh physician and geologist, is remembered not only as a staunch opponent of neptunism but also for his penetrating comprehension of how geologic processes alter the Earth's surface. For Hutton (Fig. 1-7), the Earth was a dynamic, ever-changing place in which new rocks, lands, and mountains arise continuously as a balance against their destruction by erosion and weathering. He took a cyclic view of our planet, as opposed to Werner's more static concept of an Earth that had changed very little from its beginning down to the present time. In addition, Hutton believed that "the past history of our globe must be explained by what can be seen to be happening now." This simple yet powerful idea was later to be named uniformitarianism by William Whewell. Charles Lyell (1797–1875) became the principal advocate and interpreter of uniformitarianism. We will speak of this great man again in the pages ahead.

Perhaps because it is so general a concept, uniformitarianism has been reinterpreted and altered in a variety of ways by scientists and theologians from Hutton's generation down to our own. Some of today's ideas

FIGURE 1-7 Portrait of James Hutton, the Scottish physician, farmer, and geologist. Hutton recognized that a study of present processes, such as weathering, erosion, the deposition of sediment, and volcanism, provided the means of understanding ancient rocks. The idea was later established as the principle of uniformitarianism. (*National Portrait Gallery of Scotland, Edinburgh, Scotland.*)

about what uniformitarianism implies would seem strange to Hutton himself. If the term uniformitarianism is to be used in geology (or any science), one must clearly understand what is uniform. The answer is that the physical and chemical laws that govern nature are uniform. Hence, the history of the Earth may be deciphered in terms of present observations on the assumption that natural laws are invariant with time. These so-called natural laws are merely the accumulation of all our observational and experimental knowledge. They permit us to predict the conditions under which water becomes ice, the behavior of a volcanic gas when it is expelled at the Earth's surface, or the effect of gravity on a grain of sand settling to the ocean floor. Uniform natural laws govern geologic processes such as weathering, erosion, transport of sediment by streams, movement of glaciers, and movement of water into wells.

Hutton's use of what later was termed uniformitarianism was simple and logical. By observing geologic processes in operation around him, he was able to infer the origin of particular features he discovered in rocks. When he witnessed ripple marks being produced by wave action along a coast, he was able to state that an ancient rock bearing similar markings was once a sandy deposit of some equally ancient shore. And if that rock now lay far inland from a coast, he recognized the existence of a sea that covered areas where Scottish sheep now grazed.

Hutton's method of interpreting rock exposures by observing present-day processes was given the catchy phrase "the present is the key to the past" by Sir Archibald Geike (1835–1924), a Scot with a brilliant career of discovery and experimentation in geology. The methodology implied in the phrase works very well for solving many geologic problems, but it must be remembered that the geologic past was sometimes quite unlike the present. For example, before the Earth had evolved an atmosphere like that existing today, different chemical reactions would have been prevalent during weathering of rocks. Life originated in the time of that primordial atmosphere under conditions that have no present-day counterpart. As a process in altering the Earth's surface, meteorite bombardment was once far more important than it has been for the past 3 billion years or so. Many times in the geologic past, continents have stood higher above the oceans, and this higher elevation resulted in higher rates of erosion and harsher climatic conditions, compared with intervening periods when the lands were low and partially covered with inland seas. Similarly, at one time or another in the geologic past, volcanism was more frequent than at present.

Nevertheless, ancient volcanoes disgorged gases and deposited lava and ash just as present-day volcanoes do. Modern glaciers are more limited in area than those of the recent geologic past, yet they form erosional and depositional features that resemble those of their more ancient counterparts. All of this suggests that present events do indeed give us clues to the past, but we must be constantly aware that in the past, the rates of change and intensity of processes often varied from those to which we are accustomed today and that some events of long ago simply do not have a modern analogue.

In order to emphasize the importance of natural laws over processes in the concept of uniformity, many geologists prefer to use the term *actualism* as a replacement for *uniformitarianism*. **Actualism** is the principle that natural laws governing both past and present processes on Earth have been the same. Hutton's friend John Playfair never suggested the term *actualism* but provided an eloquent statement of it when, in 1802, he wrote the following lines:

> Amid all the revolutions of the globe the economy of Nature has been uniform, and her laws are the only thing that have resisted the general movement. The rivers and rocks, the seas and the continents have changed in all their parts; but the laws that describe those changes, and the rules to which they are subject, have remained invariably the same.

The 18th-century concept of uniformitarianism was not the only contribution James Hutton made to geology. In his *Theory of the Earth*, published in 1785, he brought together many of the formerly separate thoughts of the naturalists who preceded him. He showed that rocks recorded events that had occurred over immense periods of time and that the Earth had experienced many episodes of upheaval, separated by quieter times of denudation and sedimentation. In his own words, there had been a "succession of former worlds." Hutton saw a world of cycles in which water sculpted the surface of the Earth and carried the erosional detritus from the land into the sea. The sediment of the sea was compacted into stratified rocks, and then by the action of enormous forces the layers were cast up to form new lands. In this endless process, Hutton found "no vestige of a beginning, no prospect of an end." No longer could geologists compress all of Earth history into the short span suggested by the Old Testament.

At Siccar Point on the North Sea coast of Scotland, Hutton came across exposures of rock where steeply inclined older strata had been beveled by erosion and covered by flat-lying younger layers (Figs. 1-8*A* and *B*). It was clear to Hutton that the older sequence was not only tilted but also partly removed by erosion before the younger rocks were deposited. The erosional surface meant that there was a time gap or hiatus in the rock record. In 1805, Robert Jameson named this relationship an **unconformity**. More specifically, Hutton's famous rock exposure was an *angular unconformity* because the lower beds were tilted at an angle to the upper beds. This and other unconformities provided Hutton with evidence for periods of denudation in his "succession of worlds." Although he did not use the word *unconformity*, he was the first to understand and explain the significance of this feature.

During most of his career, Hutton's published reports attracted only modest attention, and a good part of that attention came from opponents who preferred to follow the views of Abraham Gottlob Werner. To remedy this situation, British scientists who appreciated the value of Hutton's ideas convinced his friend John Playfair, a professor of mathematics and natural philosophy, to publish a summary of and commentary

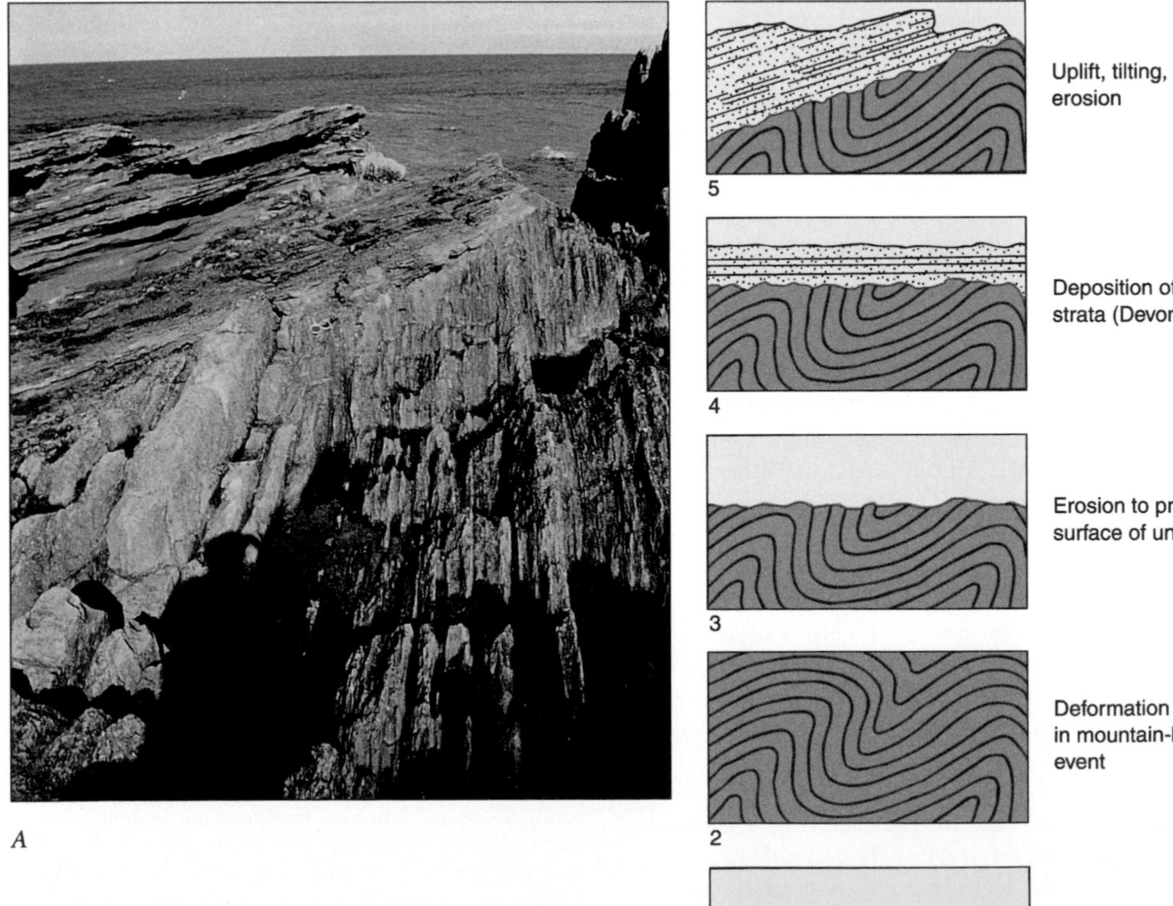

FIGURE 1-8 Angular unconformity at Siccar Point, eastern Scotland. (*A*) It was here that James Hutton first realized the historical significance of an unconformity. The drawings (*B*) indicate the sequence of events documented in this famous exposure. ? *Which of Steno's laws are illustrated in this rock exposure? (Photograph courtesy of E. H. Hay, De Anza College.)*

on *Theory of the Earth*. The work by Playfair was published in 1802 under the title *Illustrations of the Huttonian Theory of the Earth*. Whereas Hutton's writing was often complex and difficult to follow, Playfair's text was easy to read, unburdened by lengthy quotations from foreign sources, and highly persuasive. Indeed, subsequent geologists of the 19th century based much of their understanding of Hutton's ideas not on their reading of Hutton's original publications, but on the lucid, intelligent, and convincing phrases of John Playfair.

Hutton died 5 years before the publication of *Illustrations*. Throughout his life he had been absorbed in the investigation of the Earth. He was seen frequently in the field, scrutinizing every rock exposure he happened upon, and he soon became so familiar with certain strata that he was able to recognize them at different localities. What he was unable to do well, however, was determine whether dissimilar-looking strata were roughly equivalent in age. He had not discovered how to correlate beds that did not have a similar composition and texture (lithology). This problem was soon to be resolved by William Smith (1769–1839).

William Smith

William Smith was an English surveyor and engineer who devoted 24 years to the task of tracing out the strata of England and representing them on a map. Small wonder that he acquired the nickname "Strata Smith." He was employed to locate routes of canals, to design drainage for marshes, and to restore springs. In the course of this work, he independently came to understand the principles of stratigraphy, for they were of immediate use to him. By knowing that different types of stratified rocks occur in a definite sequence and that they can be identified by their lithology, the soils they form, and the fossils they contain, he was able to predict the kinds and thicknesses of rock that would have to be excavated in future engineering projects. His use of fossils was particularly significant. Prior to Smith's time, collectors rarely noted the precise beds from which fossils were taken. Smith, on the other hand, carefully recorded the occurrence of fossils and quickly became aware that certain rock units could be identified by the particular assemblages of fossils they contained. He used this knowledge first to trace strata over relatively short distances and then to extend over great distances his "correlations" to strata of the same age but of different lithology. Ultimately, this knowledge led to the formulation of the **principle of biologic succession**. This principle stipulates that the life forms of each age in the Earth's long history were unique for particular periods, that the fossil remains of life permit geologists to recognize contemporaneous deposits around the world, and that fossils could be used to assemble the scattered fragments of the record into a chronologic sequence.

Smith did not know why each unit of rock had a particular fauna. This was 60 years before the publication of Charles Darwin's *On the Origin of Species*. Today, we recognize that different kinds of animals and plants succeed one another in time because life has evolved continuously. Because of this continuous change, or evolution, only rocks formed during the same age contain similar assemblages of fossils.

News of Smith's success as a surveyor spread widely, and he was called to all parts of England for consultation. On his many trips, he kept careful records of the types of rocks he saw and the fossils they contained. Armed with his notes and observations, in 1815 he prepared a geologic map of England and Wales that is substantially accurate even today. In the 1830s, Smith was declared the "father of English geology."

Georges Cuvier and Alexandre Brongniart

The use of fossils for the correlation and recognition of formations was not exclusively William Smith's discovery. At the same time that Smith was making his observations in England, two scientists across the English Channel in France were diligently advancing the study of fossils. They were Baron Georges Léopold Cuvier (1769–1832) and his close associate Alexander Brongniart (1770–1847). Cuvier was an expert in comparative anatomy, and with this knowledge he became the most respected vertebrate paleontologist of his day. Brongniart was a naturalist who worked not only on fossil vertebrates but on plants and minerals as well. Together these men established the foundations of vertebrate paleontology. They validated Smith's findings that fossils display a definite succession of types within a sequence of strata and that this succession remains more or less constant wherever found. They also noticed that certain large groupings of strata were often separated by unconformities. As one would pass from one group of strata across the unconformity into the overlying group, a dramatic change in the kinds of animals preserved as fossils was apparent. From this observation, the two French scientists concluded that the history of life was marked by frightful catastrophes involving sudden violent flooding of the continents and abrupt crustal upheavals of stupendous magnitude. The last of these catastrophic episodes was considered to be the Noachian Deluge. Cuvier and Brongniart believed that each catastrophe resulted in the total extinction of life and was then followed by the appearance of new animals and plants. Cuvier did not speculate on how each of the many new species originated. Many geologists of the time, including the eminent Charles Lyell, held that geologic history was a uniform and gradual progression and could not accept Cuvier's concept of catastrophism. Thus began a catastrophism-versus-uniformitarianism controversy that rivaled the earlier neptunist-plutonist debates in scope and passion. Uniformitarianists

FIGURE 1-9 Sir Charles Lyell. (*Courtesy of Geological Society of London.*)

argued that many seemingly abrupt changes in fossil faunas were caused by missing strata or other imperfections in the geologic record. Other apparent breaks in the fossil record were actually not sudden, and the ancestors of each animal group could be found as fossils in underlying beds. Cuvier's idea that there were successive origins of life after each catastrophe is not supported by the fossil record, although rampant volcanism, asteroid impact, or the onslaught of harsh climatic conditions have caused mass extinctions at various times in the geologic past.

Charles Lyell

In the early 19th century, the English geologist Sir Charles Lyell (Fig. 1-9) authored the classic work *Principles of Geology*. This work both amplified the ideas of Hutton and presented the most important geologic concepts of the day. The first volume of this work was printed in 1830. It grew to five volumes and became immensely important in the Great Britain of Queen Victoria. In these volumes one can recognize Lyell's skill in explaining and synthesizing the geologic findings of contemporary geologists. As his friend Andrew C. Ramsey remarked, "We collect the data, and Lyell teaches us the meaning of them." Lyell's *Principles* became the indispensable handbook of every English geologist. In it are amplified many of the principles expressed earlier by Hutton regarding the recognition of the relative ages of rock bodies. For example, Lyell discusses the general principle that a geologic feature that cuts across or penetrates another body of rock must be younger than the rock mass penetrated (Fig. 1-10). In other words, the feature that is cut is older than the feature that crosses it. This generalization, called the **principle of cross-cutting relationships**, applies not only to rock bodies but also to geologic structures such as faults and unconformities. Thus, in Figure 1-11, fault *B* is younger than stratigraphic sequence *D*; the intrusion of igneous rock *C* is younger than the fault because it cuts across it; and by superposition, rock sequence *E* is youngest of all.

Another generalization to be found in Lyell's *Principles* relates to **inclusions**. Lyell logically discerned that fragments within larger rock masses are older than the rock masses in which they are enclosed. Thus, whenever two rock masses are in contact, the one containing pieces of the other will be the younger of the two.

In Figure 1-12*A*, the pebbles of granite (a coarse-grained igneous rock) within the sandstone tell us that

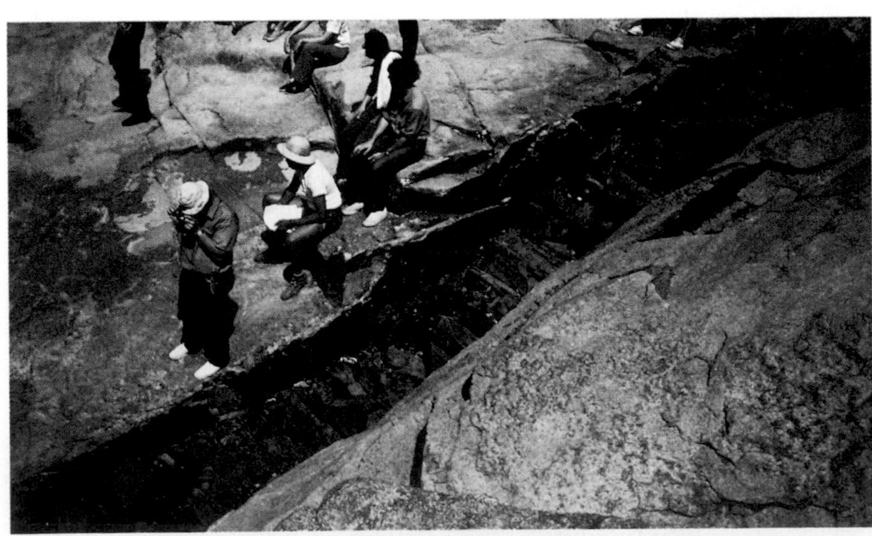

FIGURE 1-10 An example of the principle of cross-cutting relationships. The dark rock (basalt) was once molten. While in the molten state, it penetrated the surrounding light-colored rock that was already present and therefore older.

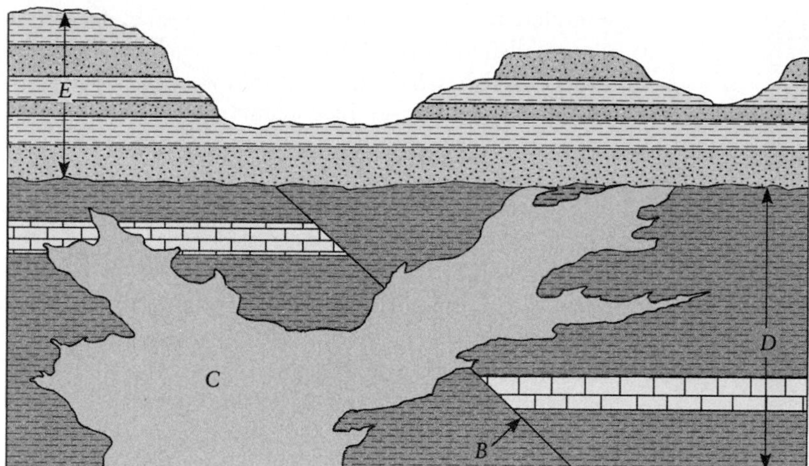

FIGURE 1-11 An example of how the sequence of geologic events can be determined from cross-cutting relationships and superposition. From first to last, the sequence indicated in the cross-section is first deposition of *D*, then faulting to produce fault *B*, then intrusion of igneous rock mass *C*, and finally erosion followed by deposition of *E*. Strata labeled *D* are oldest, and strata labeled *E* are youngest. *How do you know that the intrusion of C occurred after the formation of the fault?*

the granite is older and that the eroded granite fragments were incorporated into the sandstone. In Figure 1-12*B*, the granite was intruded as a melt into the sandstone. Because there are sandstone inclusions in the granite, the granite must be the younger of the two units.

Charles Darwin

As noted earlier, William Smith and some of his contemporaries were able to recognize that strata were often characterized by particular fossils and that there was a general progression toward more modern-looking assemblages of shells in higher, and thus younger, strata. It was Charles Darwin (1809–1882) who provided a general theory that would account for the changes seen in the fossil record.

As a young man (Fig. 1-13), Darwin had acquired an impressive knowledge of both biology and geology. That knowledge was the basis for his securing an unpaid position as a naturalist aboard the *H.M.S. Beagle*, bound for a 5-year mapping expedition around the world. On his return from the voyage in 1836, Darwin had assembled volumes of notes in support of his theory of evolution of organisms by natural selection. His theory was based on a logical system of observations and conclusions. He observed that all living things tend to increase their numbers at prodigious rates. Yet in spite of their reproductive potential to do so, no one group of organisms has been able to overwhelm the Earth's surface. In fact, the actual size of any given population remains fairly constant over long periods of time. Because of this, Darwin concluded that not all the individuals produced in any generation can survive. In addition, Darwin recognized that individuals of the same kind differ from one another in various morphologic and physiologic features. From this and the previous observation, he concluded that those individuals with variations most favorable in the existing environment would have the best chance of surviving and transmitting their favorable traits to the next generation. Darwin had no knowledge of genetics and

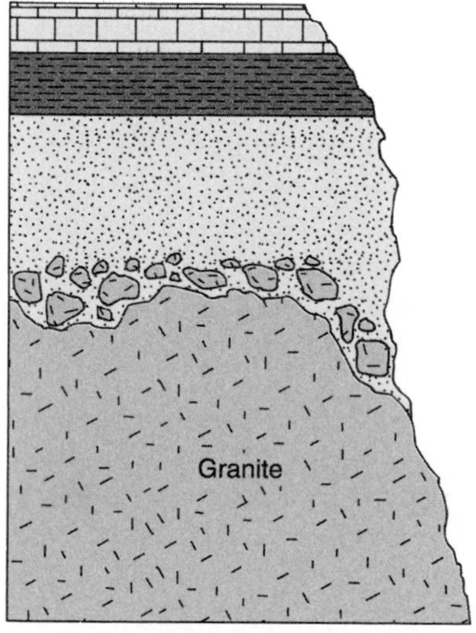

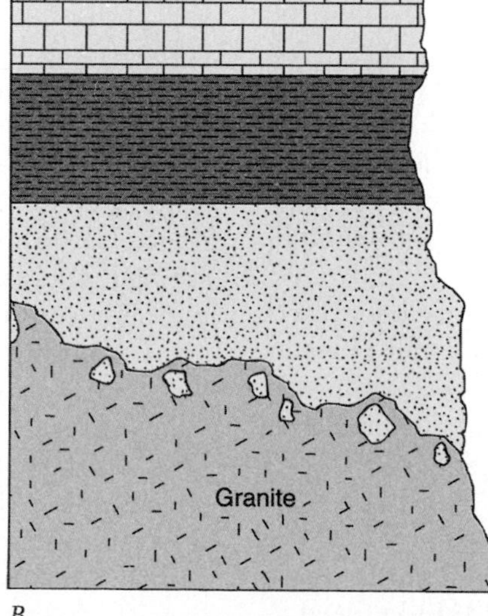

FIGURE 1-12 (*A*)Granite inclusions in sandstone indicate that granite is the older unit. (*B*) Inclusions of sandstone in granite indicate that sandstone is the older unit. *If the granite in (A) was found to be 150 million years old, and the shale above the sandstone 100 million years old, what can be stated about the age of the sandstone?*

FIGURE 1-13 Charles Darwin as a young man. This portrait was made shortly after Darwin returned to England from his voyage around the world on the *H.M.S. Beagle*. Observations made during this voyage helped him formulate the concept of evolution by natural selection. (*Bridgeman/Art Resource, NY.*)

therefore did not know the cause of the variation that was so important to his theory. Gregor Mendel's 1865 report of experiments in heredity had escaped his attention. In the decades following Darwin's death, geneticists clearly established that the variability essential to Darwin's theory of natural selection is derived from new gene combinations that occur during reproduction and from genetic mutation.

Possibly because he was reluctant to face the controversy that his theory would provoke, Darwin did not publish his findings on his return to England. He did, however, confide in Lyell and the great botanist Joseph Hooker. These friends urged him to publish quickly before someone else anticipated his discoveries. Yet Darwin continued to procrastinate. Then, in 1858, a comparatively unknown young naturalist named Alfred Russel Wallace sent Darwin a manuscript for review that contained the basic concepts of natural selection. Wallace had conceived of natural selection while on a biologic expedition to Indonesia. The idea came to him while he was suffering with malaria shortly after he read an essay on overpopulation by Thomas Malthus.

Understandably disturbed by Wallace's letter, Darwin sought the help of his friends Lyell and Hooker. Recognizing the importance of giving Darwin the credit he deserved for his discovery and the long years of assembling supporting evidence, the two scientists arranged for a presentation of Darwin's work and Wallace's paper before the Linnaean Society. Thus, the theory of natural selection was credited simultaneously to both scientists. Darwin now worked at top speed to complete his famous *On the Origin of Species*. The book was published in 1859. In that volume, Darwin hoped to accomplish two things. The first was to convince the world that evolution had occurred. Organisms had evolved or changed throughout geologic time. The second was to propose a mechanism for evolution. That mechanism was natural selection. Darwin's success in achieving his objectives can be measured by the fact that, within a decade, organic evolution had become the guiding principle in all paleontologic and biologic research. His book changed the way people viewed the world, and for this reason it has been described as one of the greatest books of all time.

Darwin died at his home in Down, England, in 1882. By that time, geologists everywhere were using their knowledge of evolution, biologic succession, superposition, cross-cutting relationships, and inclusions to decipher Earth history.

Geologists in the New World

It was inevitable that the success of European geologists during the 19th century would motivate scientists across the Atlantic to begin their own explorations. The rocks of the New World soon resounded with the blows of pioneer geologists armed with geology picks and firearms (the latter to counter the menace of wild animals and hostile Indians). Prominent among these early explorers of American geology was **William Maclure** (1763–1840). Maclure was a Scotsman who visited the United States in 1797 and decided to stay. He traveled by horseback across the Allegheny Mountains 50 times while examining their petrology and stratigraphy. In 1809, Maclure published the first geologic map of the United States in his *Observations of the Geology of the United States, Explanation of a Geologic Map*.

Before becoming a geologist, **Amos Eaton** (1776–1842) worked 3 years as a lawyer, worked 9 as a land agent, and then was sent to prison for 5 years for a crime he had not committed. After receiving a pardon from prison, he studied geology under the distinguished Yale professor Benjamin Silliman. In 1818 he published his *Index to the Geology of the Northern States*. Eaton founded the Rensselaer Institute (then called the Rensselaer School). He was an exceptional teacher, insisting that his students do "hands on" geology involving laboratory and field work. Such methods were unusual in an age when courses of study consisted only of listening to professors read from carefully prepared manuscripts.

Louis Agassiz (1807–1873) was another emigrant from Europe. He was born in Switzerland and came to the United States in 1846. After his formal education in Zurich, Heidelberg, and Munich, he began a comprehensive study of fishes that was to be the basis of his work entitled *Fossil Fishes*. For a decade before coming to America, he studied glaciers in the Alps and promoted the then-unheard-of (but valid) theory that immense ice sheets once covered much of North America and Eurasia. In his 1840 work entitled *Studies of Glaciers*, he wrote:

> The surface of Europe, adorned before by tropical vegetation and inhabited by troops of large elephants, enormous hippopotami, and gigantic carnivora, was suddenly buried under a vast mantle of ice, covering alike plains, lakes, seas, and plateaus.

Agassiz found ample evidence in America for his ice age theory. Along the shores of Lake Superior he showed his students bedrock bearing the aligned scratches (glacial striations) made by rocks locked in the base of the advancing ice, as well as huge boulders transported by the ice from distant northern terrains. Agassiz had become an extremely well known American scientist by 1859. He continued to publish in his first discipline, paleontology, and was the founder of the Harvard Museum of Comparative Zoology.

James Hall (1807–1873) frequently corresponded with Agassiz. Hall had been educated at the Rensselaer Institute, where he subsequently became a professor of chemistry. He was a brilliant geologist and paleontologist who became the director of New York's first geologic survey. Geologic mapping in New York revealed fossiliferous sequences over 40,000 feet thick. On the basis of fossils, Hall knew these rocks were deposited in shallow water. Thus, he correctly reasoned that the sea floor had subsided concurrently with deposition, but that subsequently mountains were raised from what were once marine basins. Hall gained fame the world over, not only for his knowledge of stratigraphy but for eight volumes of *The Paleontology of New York*.

James D. Dana (1813–1895) was a contemporary of Hall who became a professor at Yale University. His *Manual of Geology*, *Textbook of Geology*, and *System of Mineralogy* were among the most important texts of his time. Dana referred to Hall's elongated basins as *geosynclinals* (later shortened to *geosynclines*) but disagreed that the subsidence of these basins was caused by the ever-increasing load of sediment. Instead, he proposed that thick sequences of sediment accumulated where crustal movements had already made the basins.

Inevitably, the focus of geologic work in the United States would shift westward, often in the course of surveys of the Western Territories mandated by Congress. Prominent in these expeditions was **Ferdinand V. Hayden** (1829–1887). Known to Indians as "he who picks up rocks running," Hayden (Fig. 1-14) conducted surveys of the Badlands of South Dakota and the Black Hills and examined the geology along the Missouri, Yellowstone, Gallatin, and Madison Rivers. He was influential in convincing Congress to pass a bill authorizing the establishment of Yellowstone National Park, the oldest national park in the United States. Hayden became director of the United States Geological and Geographical Survey of the Territories. From 1879 to 1886 he was a leading geologist with the United States Geologic Survey.

John Powell (1834–1902) was the commander of an artillery battery during the Civil War. At the Battle of Shiloh, a rifle ball struck his right arm. Surgery on the wound was poorly done. A second operation was required, which reduced Powell's forearm to a mere

FIGURE 1-14 The Hayden field party in the summer of 1870, near Red Buttes, Wyoming. Geologist Ferdinand Hayden is the bearded man seated at the center of the table. (*Courtesy of the United States Geological Survey.*)

stump. The loss, however, never impeded his efficiency and endurance as a field geologist. He rose quickly to become the director of several geological and geographic surveys of the West as well as director of the United States Geologic Survey. His greatest feat was a journey by boat through the Grand Canyon of the Colorado River in the summer of 1869.

John Powell was the second director of the United States Geological Survey, succeeding **Clarence King** (1841–1901). King had studied under James Dana at Yale, where he became particularly interested in mineral resources. He was appointed by Congress to plan and direct the expedition for the geological survey of the 40th Parallel. King's *Systematic Geology* describes much of the topography and stratigraphy encountered during the survey.

Although the primary mission of the expeditions conducted by Hayden, Powell, King, and others was mapping and surveying, the crews were also to report on bedrock geology, biology, archaeology, and paleontology. With regard to paleontology, it quickly became apparent that treasure troves of dinosaur bones and bones of giant mammals were to be found in some of the Western formations. Two paleontologists who were particularly proficient in exploiting these bony treasures were **Othniel C. Marsh** (1831–1899) and **Edwin D. Cope** (1850–1897). Marsh had received his paleontologic training in Europe. He became the first professor of paleontology at Yale University and later founded the Peabody Museum of Natural History (named for his affluent uncle, George Peabody). Cope was a wealthy Quaker who became a protégé of Joseph Leidy, a highly regarded professor of anatomy at the University of Pennsylvania. Both Marsh and Cope were men of great endurance and not adverse to field work, but to hasten the task of preparing and describing the abundance of fossils being discovered in the West, they employed professional collectors. These stalwart bone hunters traveled great distances through pathless wilderness searching for fossils, while at the same time keeping a watchful eye for menacing Indians. They excavated fossils from quarries they had dug with pick and shovel, prepared them for transport, and shipped them back to Marsh in New Haven and Cope in Philadelphia. Unfortunately, there was no cooperation between Marsh and Cope. Both men were egotistical, inclined to jealousy, and competitive. A bitter feud developed between them as each tried to surpass the other in naming and describing newly discovered vertebrates. Unnecessary mistakes were made because of their haste to be the first to publish a description of a newly discovered fossil beast. Nevertheless, the "great dinosaur rush" led by Marsh and Cope had its benefits. It provided thousands of specimens for study and museum exhibits, motivated research worldwide, enhanced our understanding of life during the Mesozoic and Cenozoic, provided evidence for evolution, and established paleontology as a dynamic science imbued with a spirit of discovery.

TIME AND GEOLOGY

Many people are intrigued by the great age of rocks and fossils. Geology instructors are aware of this interest, for they are often asked the age of rock and mineral specimens brought to them by students and amateur collectors. When told that the samples are tens or even hundreds of millions of years old, the collectors are often pleased but also perplexed. "How can this person know the age of this specimen by just looking at it?" they wonder. If they insist on knowing the answer to that question, they may next receive a short discourse on the subject of geologic time. It is explained that the rock exposures from which the specimens were obtained had long ago been organized into a standard chronologic sequence based largely on superposition, evolution as indicated by fossils, and actual rock ages in years, obtained from the study of radioactive elements in the rock. The geologist's initial estimate of age is based on experience. He or she may have spent a few hours kneeling at those same collecting localities and had a background of information to draw on. Thus, at least sometimes, a geologist can recognize particular rocks as being of a certain age. The science that permits accomplishment of this feat is called geochronology. It is a science that began over 4 centuries ago when Nicholaus Steno described how the position of strata in a sequence could be used to show the relative geologic age of the layers. As described earlier, this simple but important idea was expanded and refined much later by William Smith and some of his contemporaries. These practical geologists showed how it was also possible to correlate strata. Outcrop by outcrop, the rock sequences with their contained fossils were pieced together, one above the other, until a standard geologic time scale based on relative ages had been constructed.

There are two different frames of reference when dealing with geologic time. The work of William Smith and his contemporaries was based on the concept of **relative geologic dating**. It involved placing geologic events and the rocks representing those events in the order in which they occurred, without reference to actual time or dates measured in years. Relative geologic time tells us which event preceded or followed another event or which rock mass was older or younger, relative to others. In contrast, **actual geologic dating** expresses, in years, the actual age of rocks or geologic events, as usually determined by the decay of radioactive elements.

The Standard Geologic Time Scale

The early geologists had no way of knowing how many time units would be represented in the completed geologic time scale. Nor could they know which fossils would be useful in correlation or which new strata might be discovered at a future time in some distant corner of the globe. Consequently, the time scale grew piecemeal, in an unsystematic manner. Units were named as they were discovered and studied. Sometimes the name for a unit was borrowed from local geography, from a mountain range in which rocks of a particular age were well exposed, or from an ancient tribe of Welshmen; sometimes the name was suggested by the kind of rocks that predominated.

DIVISIONS IN THE GEOLOGIC TIME SCALE Geologists have proposed the term *eon* for the largest divisions of the geologic time scale. In chronologic succession, the eons of geologic time are the **Hadean, Archean, Proterozoic**, and **Phanerozoic** (see geologic time scale, Fig. 1-15). The beginning of the Archean corresponds approximately to the ages of the oldest known rocks on Earth. Although not universally used, the term *Hadean* refers to that period of time for which we have no rock record, which began with the origin of the planet 4.6 billion years ago. The Proterozoic Eon refers to the time interval from 2500 to 544 million years ago.

The rocks of the Archean and Proterozoic are informally referred to as Precambrian. The antiquity of Precambrian rocks was recognized in the mid-1700s by Johann G. Lehman, a professor of mineralogy in Berlin, who referred to them as the "Primary Series." One frequently finds this term in the writing of French and Italian geologists who were contemporaries of Lehman. In 1833, the term appeared again when Lyell used it in his formation of a surprisingly modern geologic time scale. Lyell and his predecessors recognized these "primary" rocks by their crystalline character and took their uppermost boundary to be an unconformity that separated them from the overlying-and therefore younger-fossiliferous strata.

The remainder of geologic time is included in the **Phanerozoic Eon**. As a result of careful study of the superposition of rock bodies accompanied by correlations based on the abundant fossil record of the Phanerozoic, geologists have divided it into three major subdivisions, termed eras. The oldest is the **Paleozoic Era**, which we now know lasted about 300 million years. Following the Paleozoic is the **Mesozoic Era**, which continued for about 179 million years. The **Cenozoic Era**, in which we are now living, began about 65 million years ago.

As shown in the geologic time scale, eras are divided into shorter time units called **periods**, and periods can in turn be divided into **epochs**. Eras, periods, epochs, and divisions of epochs, called **ages**, all represent intangible increments of time. They are **geochronologic units**. The actual rocks formed or deposited during a specific time interval are called **chronostratigraphic units**. Table 1-1 indicates the chronostratigraphic units that correspond to geochronologic units. A **system** refers to all of the actual rock units of a given period, whereas a **series** is the chronostratigraphic equivalent of an epoch, and a **stage** represents the tangible rock record of an age. As an example of the way these terms are used, one might correctly speak of climatic changes during the Cambrian Period as indicated by fossils found in the rocks of the Cambrian System.

RECOGNITION OF GEOCHRONOLOGIC UNITS Geochronologic units bear the same names as the chronostratigraphic units to which they correspond. Thus, we may speak of the Jurassic System (a rock unit) or the Jurassic Period (a time unit) according to whether we are referring to the rocks themselves or to the time during which they accumulated. Geochronologic terms have come into use as a matter of convenience. Their definition is necessarily dependent on the existence of tangible chronostratigraphic units. The steps leading to the recognition of chronostratigraphic units began with the use of superposition in establishing relative age relationships. Local sections of strata were used by early geologists to recognize beds of successively different age and, thereby, to record successive evolutionary changes in fauna and flora. (The order and nature of these evolutionary changes could be determined because stratigraphically higher layers are successively younger.) Once the faunal and floral succession was deciphered, fossils provided an additional tool for establishing the order and nature of recorded geologic events. They could also be used for correlation, so that strata at one locality could be related to the strata of other localities. No single place on Earth contains a complete sequence of strata from all geologic ages. Hence, correlation to standard sections of many widely distributed local sections was necessary in constructing the geologic time scale (Fig. 1-16). Clearly, the time scale was not conceived as a coherent whole but rather evolved piece by piece as a result of the individual studies of many geologists. Indeed, for some units at the series and stage level, the process continues even today. The fact that the time scale developed in piecemeal fashion is apparent when one reviews its growth and development.

THE GEOLOGIC SYSTEMS

The Cambrian System The rocks of the **Cambrian System** take their name from *Cambria*, the Latin name for Wales. Exposures of strata in Wales (Fig. 1-17) provide a standard section with which rocks elsewhere in Europe and on other continents can be correlated. The standard section in Wales is named Cambrian by definition. All other sections deposited during the same time as the rocks in Wales are recognized as Cambrian by comparison to the standard section.

Adam Sedgwick, a highly regarded professor of geology at Cambridge (Fig. 1-18), named the Cambrian

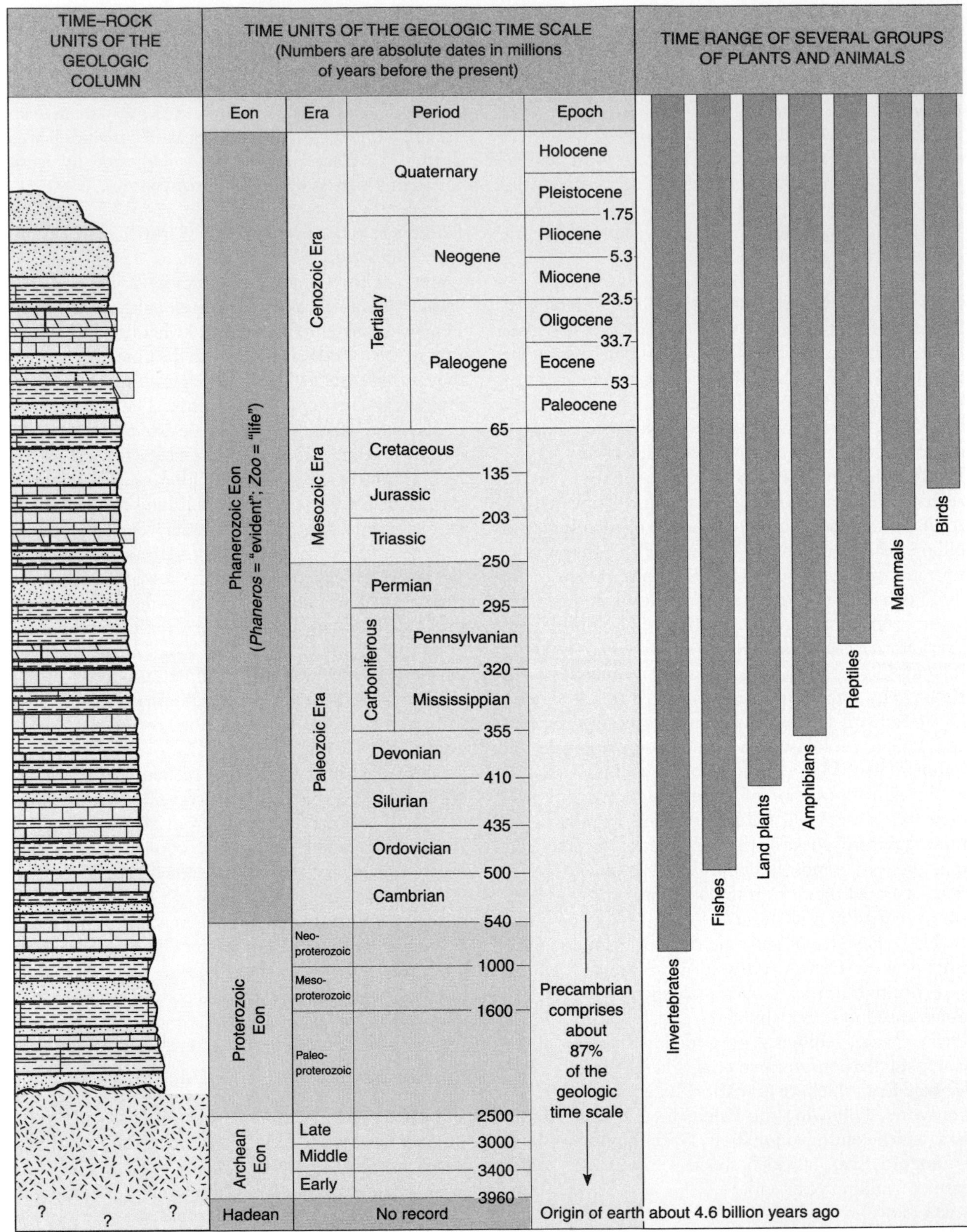

FIGURE 1-15 Geologic Time Scale. The age for the base of each division is in accordance with recommendations of the International Commission on Stratigraphy for the year 2000.

TABLE 1-1 Hierarchy of Geochronologic and Chronostratigraphic Terms

Geochronologic Divisions	Equlivalent Chronostratigraphic Divisions
Era	Erathem (rarely used)
Period	System
Epoch	Series
Age	Stage
Chron	Zone (Chronozone)

in the 1830s for outcrops of poorly fossiliferous dark siltstones and sandstones. The area in northern Wales that Sedgwick studied was noted for its complexity, yet he was able to unravel its geologic history on the basis of spatial relationships and lithology.

The Ordovician and Silurian Systems At about the same time that Sedgwick was laboring with outcrops that were to become the standard section for the Cambrian System, his former student, Sir Roderick Impey Murchison, had begun studies of fossiliferous strata in the hills of southern Wales. Murchison named these rocks the **Silurian System**, taking the name from early inhabitants of western England and Wales known as the *Silures*. In 1835, Murchison and Sedgwick presented a paper, *On the Silurian and Cambrian Systems, Exhibiting the Order in Which the Older Sedimentary Strata Succeed Each Other in England and Wales*. With this publication, the two geologists initiated the development of the early Paleozoic time scale. In the years that followed, a controversy arose between the two men that was to sever their friendship. Because Sedgwick had not described fossils distinctive of the Cambrian, the unit could not be recognized in other countries. Murchison argued, therefore, that the Cambrian was not a valid system. During the 1850s, he maintained that all fossiliferous strata above the "Primary Series" (the old name for Precambrian) and below the Old Red Sandstone (of Devonian age) belonged within the Silurian System. Sedgwick, of course, disagreed, but his opinion that the Cambrian was a valid system did not receive wide support until

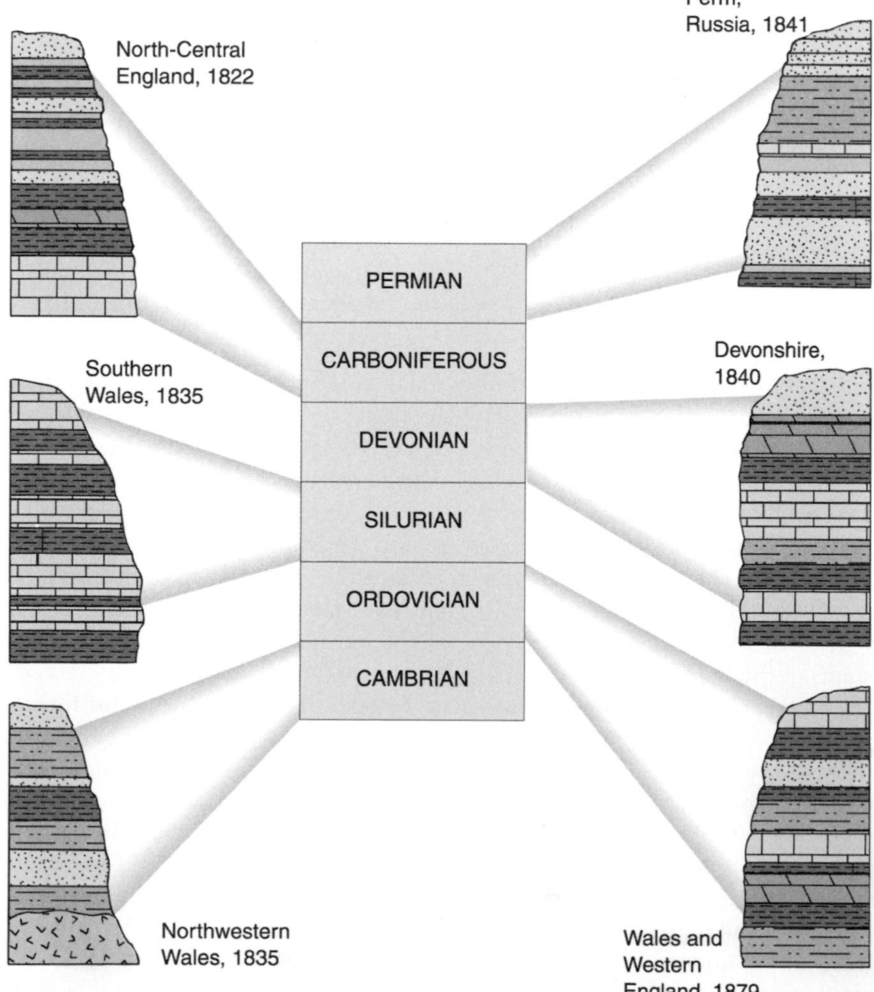

FIGURE 1-16 The standard geologic time scale for the Paleozoic and other eras developed without benefit of a grand plan. Instead, it developed by the compilation of "type sections" for each of the systems. ? *What criteria at Devonshire demonstrated that these strata were younger than those in Wales?*

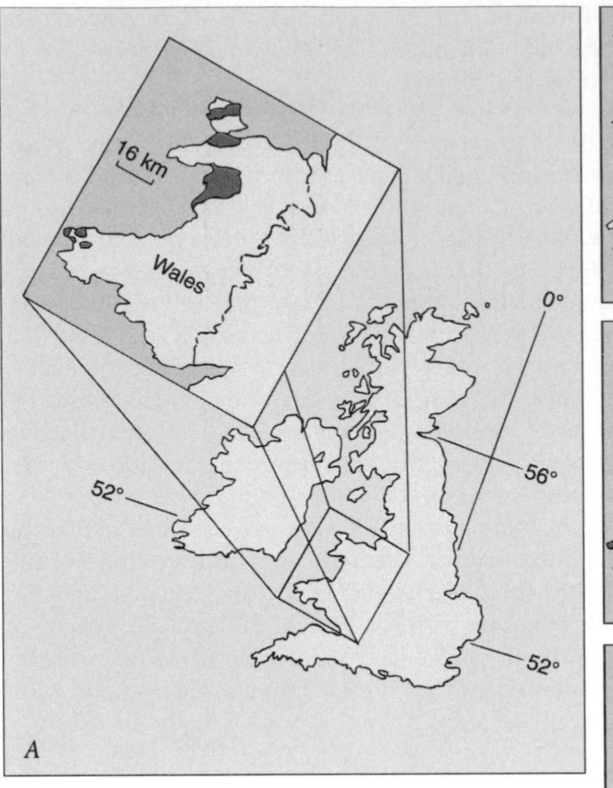

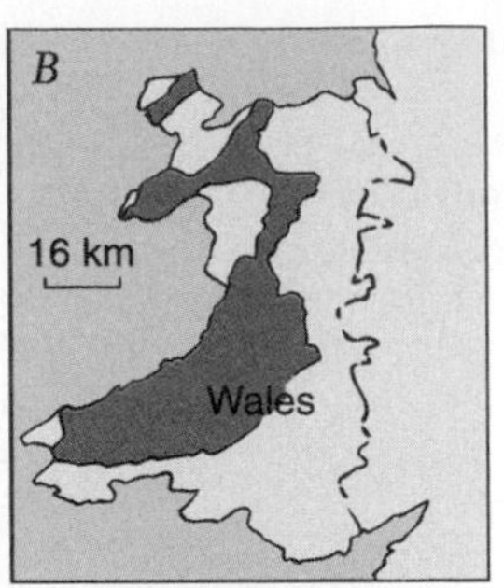

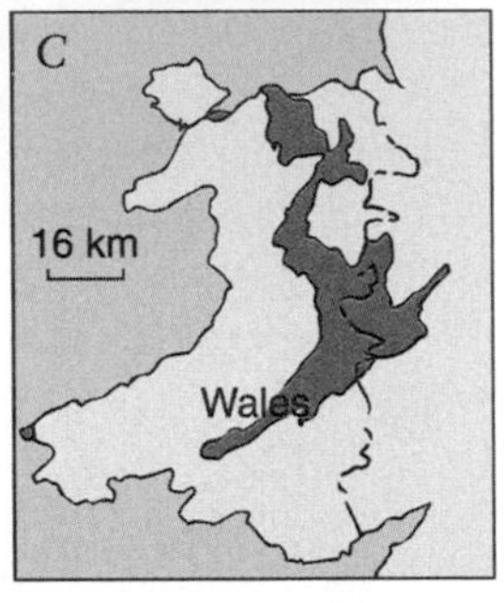

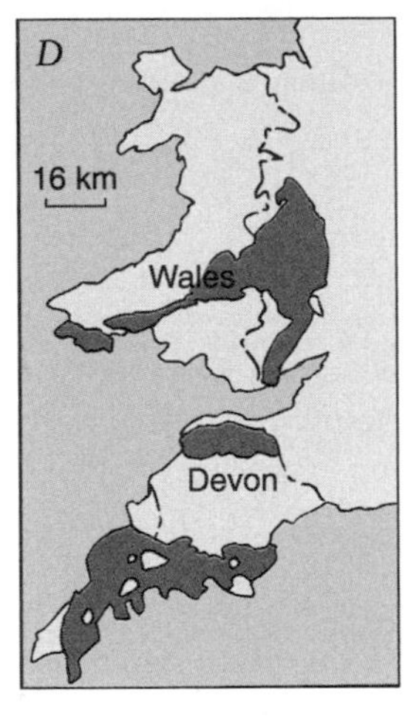

FIGURE 1-17 Outcrop areas for strata of the Cambrian (*A*), Ordovician (*B*), Silurian (*C*), and Devonian (*D*) systems in Great Britain.

fossils were described from the upper part of the sequence. The fossils proved to be similar to faunas in Europe and North America. Hence, the Cambrian did meet the test of recognition outside England. Using these fossils as a basis for reinterpretation, the English geologist Charles Lapworth proposed combining the upper part of Sedgwick's Cambrian and the lower part of Murchison's Silurian into a new system. In 1879, he named the system *Ordovician* after the *Ordovices*, an early Celtic tribe. The first three systems of the Paleozoic were thus established (Fig. 1-19).

The Devonian System The Devonian System was proposed for outcrops near Devonshire, England (Fig. 1-16), by Sedgwick and Murchison in 1839 (prior to the year of their bitter debate). They based their proposal on the fact that the rocks in question lay beneath the previously recognized Carboniferous System and contained a fauna that was different from that of the underlying Silurian and overlying Carboniferous. In their interpretation of the distinctive nature of the fauna, they were aided by the studies of William Lonsdale, a retired army officer who had become a self-taught specialist on fossil corals. Further evidence that the new unit was a valid one came when Murchison and Sedgwick were able to recognize it in the Rhineland region of Europe. The Devonian rocks of Devonshire were also found to be chronologically equivalent to the widely known *Old Red Sandstone* of Scotland and Wales.

The Carboniferous System The term **Carboniferous System** was coined in 1822 by the English geologists William Conybeare and William Phillips to designate strata that included beds of coal in north-central England. Subsequently, it became convenient in Europe and Britain to divide the system into a Lower Carboniferous and Upper Carboniferous—the latter containing most of the workable coal seams. Two systems in North America, the **Mississippian** and **Pennsylvanian**, are broadly comparable to these subdivisions.

FIGURE 1-18 Adam Sedgwick, one of the foremost geologists of the 19th century. A professor of geology at Cambridge University, Sedgwick is best remembered for deciphering the highly deformed system of rocks in northwestern Wales that he defined as the Cambrian System. He also founded the geologic museum at Cambridge that bears his name. (*Courtesy of the Cambridge Museum, Cambridge, England.*)

They are not precisely equivalent because detailed intercontinental correlations reveal that the boundary between the Mississippian and Pennsylvanian in North America is somewhat younger than the boundary between the Lower Carboniferous and Late Carboniferous in Europe. The American geologist Alexander Winchell formally proposed the name Mississippian in 1870 for the dominantly calcareous Lower Carboniferous strata that are extensively exposed in the upper Mississippi River drainage region. In 1891, Henry S. Williams provided the name Pennsylvanian for the coal-bearing Upper Carboniferous System.

The Permian System The **Permian System** takes its name from Permia, an ancient kingdom between the Urals and the Volga. In 1840 and 1841, Murchison, in company with the French paleontologist Edouard de Verneuil and several Russian geologists, traveled extensively across western Russia. To his delight, Murchison found he was able to recognize Silurian, Devonian, and Carboniferous rocks by the fossils they contained. As a result, he became even more convinced that groups of fossil organisms succeed one another in a definite and determinable order. Murchison established the new Permian System for rocks that overlay the Carboniferous System and contained fossils similar to those in German strata (the Zechstein beds), which had the same stratigraphic position as the Magnesian Limestone in England. Field studies had previously shown that the Magnesian Limestone rested on Carboniferous strata. Thus Murchison was able to include the Magnesian Limestone within the Permian by correlation. The fossils of the new system differed from those of the Carboniferous below and the Triassic above. In a letter to the Society of Naturalists of Moscow dated October 8, 1841, Murchison stated, "The Carboniferous System is surmounted, to the east of the Volga, by a vast series of marls, schists, limestones, sandstones, and conglomerates to which I propose to give the name 'Permian System.'" Murchison's establishment of the Permian System provides a fine example of the logic employed by early geologists in putting together the pieces of the standard time scale.

The Triassic System The influence of British geologists in providing names for the system of the Paleozoic is by now obvious. However, their presence is not as evident in the development of Mesozoic nomenclature. The **Triassic System**, for example, was applied in 1834 by a German geologist named Frederich von Alberti. The term refers to a threefold division of rocks of this age in

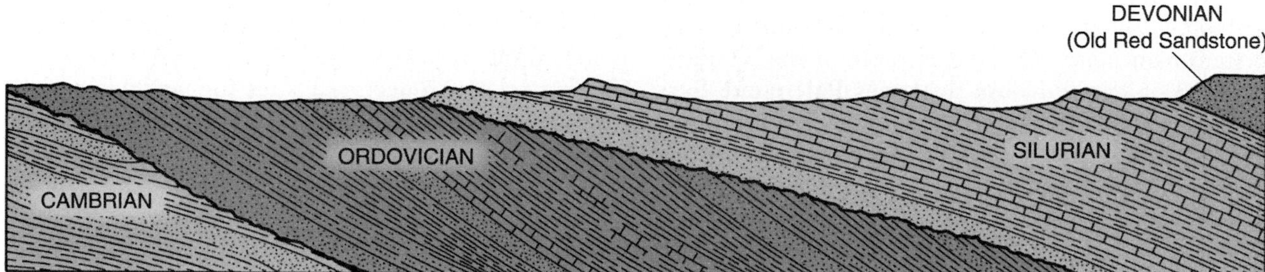

FIGURE 1-19 Generalized geologic cross-section for the Silurian-type region. Unconformities separate the Ordovician from the Cambrian and Silurian systems. Silurian strata are inclined toward the east, with more resistant rocks forming escarpments that face toward the west.

Germany. However, because the German strata in the type area are poorly fossiliferous, the standard of reference has been shifted to richly fossiliferous marine strata in the Alps.

The Jurassic System Another German scientist, Alexander von Humboldt, proposed the term **Jurassic** for strata of the Jura Mountains between France and Switzerland. However, in 1795, when he used the term, the concept of systems had not been developed. As a result, the Jurassic was redefined as a valid geologic system in 1839 by Leopold von Buch.

The Cretaceous System During the same year that Conybeare and Phillips were defining the Carboniferous, a Belgian geologist named Omalius d'Halloy proposed the term **Cretaceous** (from the Latin Creta, meaning "chalk") for rock outcrops in France, Belgium, and Holland. Although chalk beds are prevalent in some Cretaceous exposures, the system is actually recognized on the basis of fossils. Indeed, some thick sections of Cretaceous rocks contain no chalks whatsoever.

The Tertiary System The name **Tertiary** leads us back to the time when geology was just beginning as a science. Giovanni Arduino suggested a classification with four major divisions: Primary, Secondary, Tertiary, and Quaternary. The Tertiary was derived from his 1759 description of unconsolidated *montes tertiarii* sediments at the foot of the Italian Alps. Later, the Tertiary was more precisely defined, and standard sections for series of the Tertiary were established in France. The **Eocene, Miocene**, and **"Older" Pliocene**, for example, were proposed by Lyell in 1832 on the basis of the proportions of species of living marine invertebrates in the fossil fauna. He later used the name **Pleistocene** for a unit he had formerly called the Newer Pliocene. By Lyell's definition, 3 percent of the fossil fauna of the Eocene still live, whereas the Miocene contained 17 percent, and the Pliocene contained 50 to 67 percent. The term **Oligocene** was proposed by August von Beyrich in 1854, and the term **Paleocene** was proposed 20 years later by Wilhelm Schimper. Other system names are also used in place of the Tertiary. Many geologists now use the terms **Paleogene System** (for the Paleocene, Eocene, and Oligocene) and **Neogene System** (for the Miocene and Pliocene).

The Quaternary System In 1829, the French geologist Jules Desnoyers proposed the term **Quaternary** for certain sediments and volcanics exposed in northern France. Although these deposits contained few fossils, Desnoyers was convinced on the basis of field studies that they were younger than Tertiary rocks. In the decade following Desnoyer's establishment of the Quaternary, the unit was further divided into an older **Pleistocene Series**, composed primarily of deposits formed during the glacial ages, and the younger **Holocene Series**.

This brief review describing how geologists drew up a table of geologic time clearly shows a lack of any grand and coherent design. These geologic pioneers were influenced by conspicuous changes in assemblages of fossils from one sequence of strata to another. In many places in Europe they found that such changes frequently occurred above and below an unconformity. The success of their methods is apparent from the fact that, by and large, the systems have persisted and found wide use even to the present day.

Quantitative Geologic Time

EARLY ATTEMPTS AT QUANTITATIVE GEOCHRONOLOGY

Since the dawn of civilization, people have been curious about the age of the Earth. In addition, we have not been satisfied in being able to state the *relative* geologic age of a rock or fossil. Human curiosity demands that we know *actual* age in years. One early, but unscientific, attempt at quantitative dating was conducted in 1658 by James Ussher, Archbishop of Armagh and Primate of All England. After an analysis of solar and lunar cycles in the Julian Calendar that were calibrated against dates and events recorded in the Old Testament, Ussher stated that the Earth was created on October 23 in the year 4004 B.C. A refinement was added by Sir John Lightfoot, Vice Chancellor of Cambridge, who placed the precise hour at nine o'clock in the morning. The archbishop's date was inserted in many versions of the Bible and became widely accepted. However, geologists working during the 19th century showed that the archbishop's date could not be supported by objective scientific observation. They understood that if one were to discover the actual age of the Earth or of particular rock bodies, they would have to concentrate on natural processes that continue at a constant rate and that also leave some sort of tangible record in the rocks. Evolution is one such process, and Lyell recognized this. By comparing the amount of evolution exhibited by marine mollusks in the various series of the Tertiary System with the amount that had occurred since the beginning of the Pleistocene Ice Age, Lyell estimated that 80 million years had elapsed since the beginning of the Cenozoic. He came astonishingly close to the mark. However, for older sequences, estimates based on rates of evolution were difficult, not only because of missing parts in the fossil record but also because rates of evolution for many taxa were not well understood.

In another attempt, geologists reasoned that if rates of deposition could be determined for sedimentary rocks, they might be able to estimate the time required for deposition of a given thickness of strata. Similar reasoning suggested that one could estimate total elapsed geologic time by dividing the average thickness of sediment transported annually to the oceans into the total thickness of sedimentary rock that had ever been deposited in the past. Unfortunately, such estimates did not adequately account for past differences in rates of sedimentation or losses to the total stratigraphic section during episodes of erosion. Also, some very ancient sediments were no longer recognizable, having been converted to igneous and metamorphic rocks in the course of mountain building. Estimates of the Earth's total age based on sedimentation rates ranged from as little as a million to over a billion years.

Yet another scheme for approximating the Earth's age was proposed in 1715 by Sir Edmund Halley (1656–1742), whose name we associate with the famous comet. Halley surmised that the ocean formed soon after the origin of the planet and therefore would be only slightly younger than the age of the solid Earth. He reasoned that the original ocean was not salty and that subsequently salt derived from the weathering of rocks was brought to the sea by streams. Thus, if one knew the total amount of salt dissolved in the ocean and the amount added each year, it might be possible to calculate the ocean's age. In 1899, the Irish geologist John Joly attempted the calculation. From information provided by gauges placed at the mouths of streams, Joly was able to estimate the annual increment of salt to the oceans. Then, knowing the salinity of ocean water and the approximate volume of water, he calculated the amount of salt already held in solution in the oceans. An estimate of the age of the ocean was obtained by dividing the total salt in the ocean by the rate of salt added each year. Beginning with essentially nonsaline oceans, it would have taken about 90 million years for the oceans to reach their present salinity, according to Joly. The figure, however, was off the mark by a factor of 50, largely because there was no way to account accurately for recycled salt and salt incorporated into clay minerals deposited on the sea floors. Vast quantities of salt once in the sea had become extensive evaporite deposits on land; some of the salt being carried back to the sea had been dissolved, not from primary rocks but from eroding marine strata on the continents. Even though in error, Joly's calculations clearly supported those geologists who insisted on an age for the Earth far in excess of a few million years. The belief in the Earth's immense antiquity was also supported by Darwin, Huxley, and other evolutionary biologists, who saw the need for time in the hundreds of millions of years to accomplish the organic evolution apparent in the fossil record.

The opinion of the geologists and biologists that the Earth was immensely old was soon to be challenged by the physicists. Spearheading this attack was Lord William Thomson Kelvin, considered by many to be the outstanding physicist of the 19th century. Kelvin calculated the age of the Earth on the assumption that it had cooled from a molten state and that the rate of cooling followed ordinary laws of heat conduction and radiation. Kelvin estimated the number of years it would have taken the Earth to cool from a hot mass to its present condition. His assertions regarding the age of the Earth varied over 2 decades of debate, but in his later years he confidently believed that 24 to 40 million years was a reasonable age for the Earth. The biologists and geologists found Kelvin's estimates difficult to accept. But how could they do battle against his elegant mathematics when they were themselves armed only with inaccurate dating schemes and geologic intuition? For those geologists unwilling to capitulate, however, new discoveries showed their beliefs to be correct and Kelvin's to be unavoidably wrong.

A more correct answer to the question "How old is the Earth?" was provided only after the discovery of radioactivity, a phenomenon unknown to Kelvin during his active years. With the detection of natural radioactivity by Henri Becquerel in 1896, followed by the isolation of radium by Marie and Pierre Curie 2 years later, the world became aware that the Earth had its own built-in source of heat. It was not inexorably cooling at a steady and predictable rate, as Kelvin had suggested.

RADIOISOTOPIC METHODS FOR DATING ROCKS The solid Earth is composed of minerals and rocks. **Minerals** are solid, naturally occurring inorganic materials having a definite composition or range of compositions and usually possessing a uniform internal crystal structure. That uniform structure is derived from an orderly internal arrangement of the atoms that combine to make minerals. **Rocks** are solid, cohesive aggregates of the grains of one or more minerals occurring naturally in large quantities. To understand better the way atoms in rocks and minerals can reveal the numerical age of geologic events, a brief review of the nature of atoms is useful.

Atoms are the smallest particles of matter that can exist as an element. An individual atom consists of an extremely minute but heavy nucleus surrounded by rapidly moving negatively charged **electrons**. The electrons are relatively farther apart than are the planets surrounding our Sun; consequently, the atom consists primarily of empty space. However, electrons move so rapidly around the nucleus that they effectively fill the space within their orbits, giving volume to the atom and repelling other atoms that may approach.

In the nucleus of the atom are closely compacted particles called **protons**, which carry a unit charge of positive electricity equal to the unit charge of negative electricity carried by the electron. Associated with the protons in the nucleus are electrically neutral particles having the same mass as protons. These are called **neutrons**. Modern atomic physics has made us aware of still other particles in the nucleus. For our understanding of the atom, however, knowledge of protons and neutrons is sufficient. The number of protons in the nucleus of an atom establishes its number of positive charges and is called its **atomic number**. Each chemical element is composed of atoms having a particular atomic number. Thus, every element has a different number of protons in its nucleus. There are 90 naturally occurring elements that range in atomic number from 1 (for one proton) in hydrogen to 92 (for 92 protons) in uranium (Table 1-2 and Appendix D). The **mass** of an atom is approximately equal to the sum of the masses of its protons and neutrons. (The mass of electrons is so small that it need not be considered.) Carbon-12 is used as the standard for comparison of mass. By setting the atomic mass of carbon at 12, the atomic mass of hydrogen, which is the lightest of elements, is just a bit greater than 1 (1.008, to be precise). The nearest whole number to the total number of protons and neutrons in an element constitutes its **mass number**. Some atoms of the same substance have different mass numbers. Such variants are called **isotopes**. Isotopes are two or more varieties of the same element that have the same atomic number and chemical properties but differ in mass numbers because they have a varying number of neutrons in the nucleus. By convention, the mass number is noted as a superscript preceding the chemical symbol of an element, and the atomic number is placed beneath it as a subscript. Thus, $^{20}_{40}Ca$ is translated as the element calcium having an atomic number of 20 and mass number of 40 (Table 1-2).

Radioactivity The **radioactivity** discovered by Becquerel was observed in elements such as uranium and thorium, which are unstable and break down or decay to form other elements or other isotopes of the same element. Any individual uranium atom, for example, will eventually decay to lead if given a sufficient length of time. To understand what is meant by "decay," let us consider what happens to a radioactive element such as uranium-238 ($^{238}_{92}U$). Uranium-238 has a mass number of 238. The "238" represents the sum of the weights of the atom's protons and neutrons (each proton and neutron having a mass of 1). Uranium has an atomic number (number of protons) of 92. Such atoms with specific atomic number and weight are sometimes termed **nuclides**. Sooner or later (and entirely spontaneously), the uranium-238 atom will fire off a particle from the nucleus called an **alpha particle**. Alpha particles are positively charged ions of helium. They have an atomic weight of 4 and an atomic number of 2. Thus, when the alpha particle is emitted, the new atom will now have an atomic weight of 234 and an atomic number of 90. The new

TABLE 1-2 Number of Protons and Neutrons and Atomic Mass of Some Geologically Important Elements

Element and Symbol	Atomic Number (Number of Protons in Nucleus)	Number of Neutrons in Nucleus	Mass Number
Hydrogen (H)	1	0	1
Helium (He)	2	2	4
Carbon-12 (C)*	6	6	12
Carbon-14 (C)	6	8	14
Oxygen (O)	8	8	16
Sodium (Na)	11	12	23
Magnesium (Mg)	12	13	25
Aluminum (Al)	13	14	27
Silicon (Si)	14	14	28
Chlorine-35 (Cl)*	17	18	35
Chlorine-37 (Cl)	17	20	37
Potassium (K)	19	20	39
Calcium (Ca)	20	20	40
Iron (Fe)	26	30	56
Barium (Ba)	56	82	138
Lead-208 (Pb)*	82	126	208
Lead-206 (Pb)	82	124	206
Radium (Ra)	88	138	226
Uranium-238 (U)	92	146	238

*When two isotopes of an element are given, the most abundant is starred.

atom, which is formed from another by radioactive decay, is called a **daughter element**. From the decay of the parent nuclide, uranium-238, the daughter nuclide, thorium-234, is obtained (Fig. 1-20). A shorthand equation for this change is written:

$$^{238}_{92}\mathrm{U} \rightarrow {}^{234}_{90}\mathrm{Th} + {}^{4}_{2}\mathrm{He}$$

This change is not, however, the end of the process, for the nucleus of thorium-234 ($^{234}_{90}\mathrm{Th}$) is not stable. It eventually emits a beta particle (an electron discharged from the nucleus when a neutron splits into a proton and an electron). There is now an extra proton in the nucleus but no loss of atomic weight because electrons are essentially weightless. Thus, from $^{234}_{92}\mathrm{Th}$ the daughter element $^{234}_{91}\mathrm{Pa}$ (protactinium) is formed. In this case, the atomic number has been increased by 1. In other instances, a beta particle may be captured by the nucleus, where it combines with a proton to form a neutron. The loss of the proton would decrease the atomic number by 1.

Another kind of emission in the radioactive decay process is called gamma radiation. It consists of a form of invisible electromagnetic waves having even shorter wavelengths than X-rays.

The rate of decay of radioactive isotopes is uniform and is not affected by changes in pressure, temperature, or the chemical environment. Therefore, once a quantity of radioactive nuclides has been incorporated into a growing mineral crystal, that quantity will begin to decay at a steady rate, with a definite percentage of the radiogenic atoms undergoing decay in each increment of time. Each radioactive isotope has a particular mode of decay and a unique decay rate. As time passes, the quantity of the original or parent nuclide diminishes, and the number of the newly formed, or daughter, atoms increases, thereby indicating how much time has elapsed since the clock began its timekeeping. The beginning, or "time zero," for any mineral containing radioactive nuclides would be the moment when the radioactive parent atoms became part of a mineral from which daughter elements could not escape. The retention of daughter elements is essential, for they must be counted to determine the original quantity of the parent nuclide.

The determination of the ratio of parent to daughter nuclides is usually accomplished with the use of a mass spectrometer, an analytic instrument capable of measuring the atomic masses of elements and isotopes of elements. In the mass spectrometer, samples of elements are vaporized in an evacuated chamber, where they are bombarded by a stream of electrons. This bombardment knocks electrons off the atoms, leaving them positively charged. A stream of these positively charged ions is deflected as it passes between plates that bear opposite charges of electricity. The degree of deflection depends on the charge-to-mass ratio. In general, the heavier the ion, the less it will be deflected (Fig. 1-21).

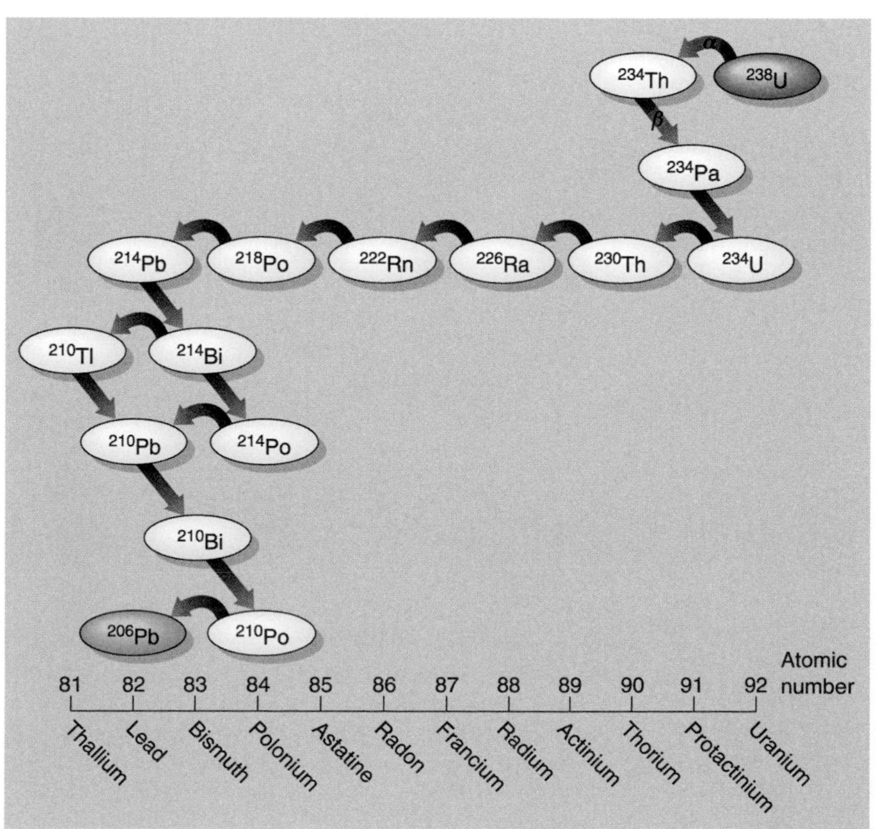

FIGURE 1-20 Radioactive decay series of uranium-238 (^{238}U) to lead-206 (^{206}Pb).

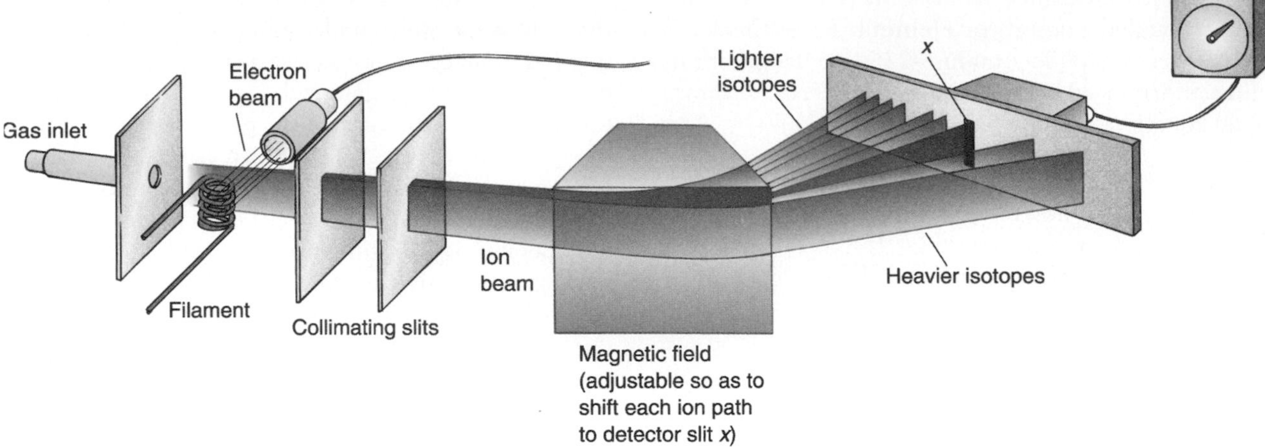

FIGURE 1-21 Schematic drawing of a mass spectrometer. In this type of spectrometer, the intensity of each ion beam is measured electrically (rather than recorded photographically) to permit determination of the isotopic abundances required for radiometric dating.

Of the three major families of rocks, the igneous clan is by far the best for isotopic dating. Fresh samples of igneous rocks are less likely to have experienced loss of daughter products, which must be accounted for in the age determination. Igneous rocks can provide a valid date for the time that a silicate melt containing radioactive elements solidified.

In contrast to igneous rocks, the minerals of sediments can be weathered and leached of radioactive components, and age determinations are far more prone to error. In addition, the age of a detrital grain in a sedimentary rock does not give an age of the sedimentary rock but only of the parent rock that was eroded much earlier.

Dates obtained from metamorphic rocks may also require special care in interpretation. The age of a particular mineral may record the time the rock first formed or any one of a number of subsequent metamorphic recrystallizations.

Once an age has been determined for a particular rock unit, it is often possible to use that data to approximate the age of adjacent rocks. For example, in Figure 1-22, a geology student on a field trip is showing the location of a thin, brown-colored layer of altered volcanic ash called **bentonite**. Zircon crystals within the altered ash yield uranium-to-lead ratios indicating the ash is 453.7 million years old. Thus, strata below the ash are older than 453.7 million years, and those above are younger.

In Figure 1-23, a shale bed lying below a lava flow that is 110 million years old and above another flow that is 180 million years old must be between 110 and 180 million years old. Similarly, as shown in Figure 1-24, the age of a shale deposited on the erosional surface of a 490-million-year-old granite mass and covered by a 450-million-year-old lava flow must be between 450 and 490 million years old. The fossils in that shale might then be used to assign the shale to a particular geologic system or series. Then, by correlation, the quantitative age determination obtained at the initial locality (Fig. 1-24, section *A*) could be assigned to formations at other locations (Fig. 1-24, section *B*).

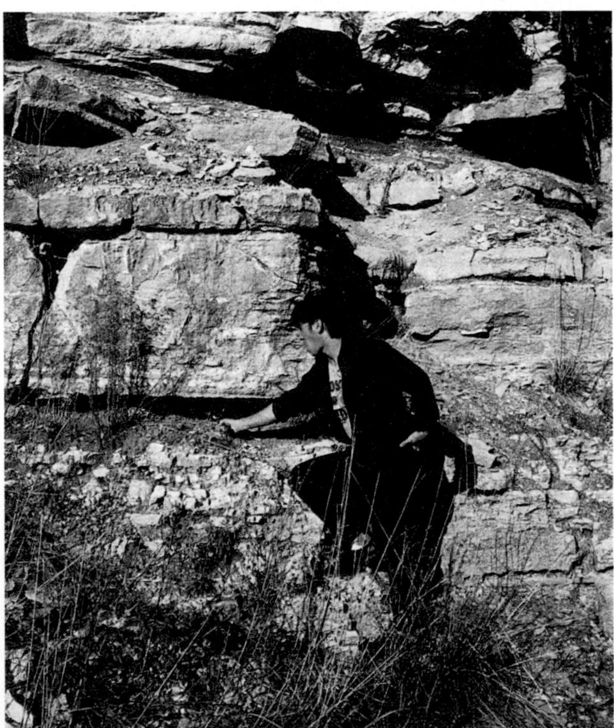

FIGURE 1-22 Thin layer of altered volcanic ash (dark brown) between layers of Ordovician marine limestones. ? *By means of a high-resolution uranium-lead method, the ash was found to be 453.7 million years old. What can now be stated about the age of the stratum beneath the ash?*

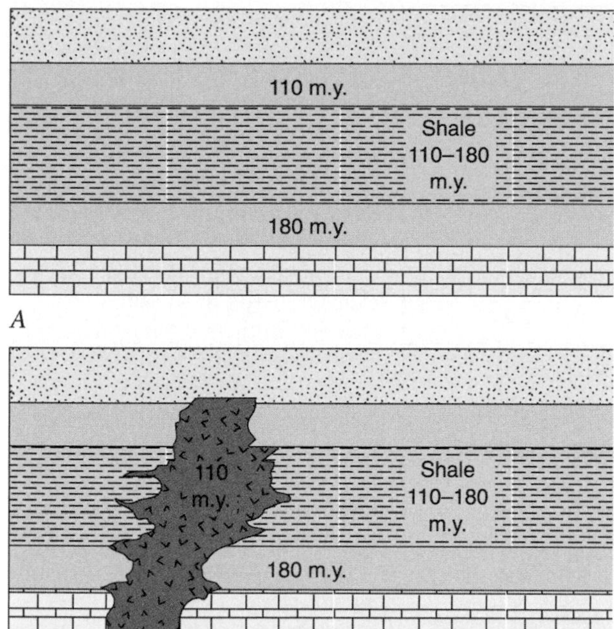

FIGURE 1-23 Igneous rocks that have provided absolute radiogenic ages can often be used to date sedimentary layers. (*A*) The shale is bracketed by two lava flows. (*B*) The shale lies above the older flow and is intruded by a younger igneous body. (Note: m.y. = million years.)

Half-Life One cannot predict with certainty the moment of disintegration for any individual radioactive atom in a mineral. We do know that it would take an infinitely long time for all of the atoms in a quantity of radioactive elements to be entirely transformed to stable daughter products. Experimenters have also shown that there are more disintegrations per increment of time in the early stages than in later stages (Fig. 1-25), and one can statistically forecast what percentage of a large population of atoms will decay in a certain amount of time.

Because of these features of radioactivity, it is convenient to consider the time needed for half of the original quantity of atoms to decay. This span of time is termed the **half-life**. Thus, at the end of the time constituting one half-life, half of the original quantity of radioactive element still has not undergone decay. After another half-life, half of what was left remains, or 1/4 of the original quantity. After a third half-life, only 1/8 would remain, and so on.

Every radioactive nuclide has its own unique half-life. Uranium-235, for example, has a half-life of 704 million years. Thus, if a sample contains 50 percent of the original amount of uranium-235 and 50 percent of its daughter product, lead-207, then that sample is 704 million years old. If the analyses indicate 25 percent of uranium-235 and 75 percent of lead-207,

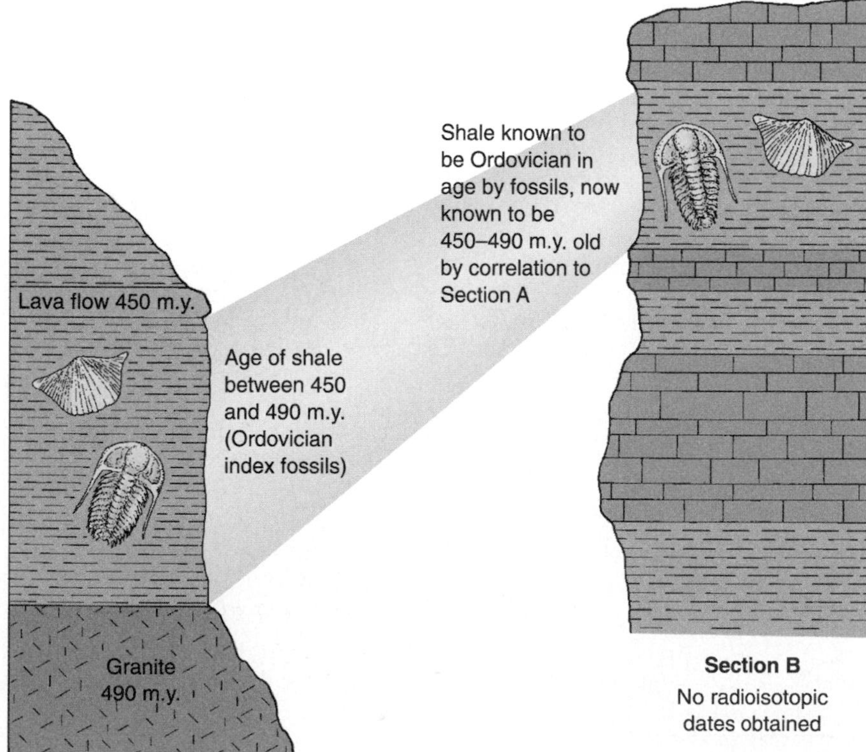

FIGURE 1-24 The actual age of rocks that cannot be dated isotopically can sometimes be ascertained by correlation.

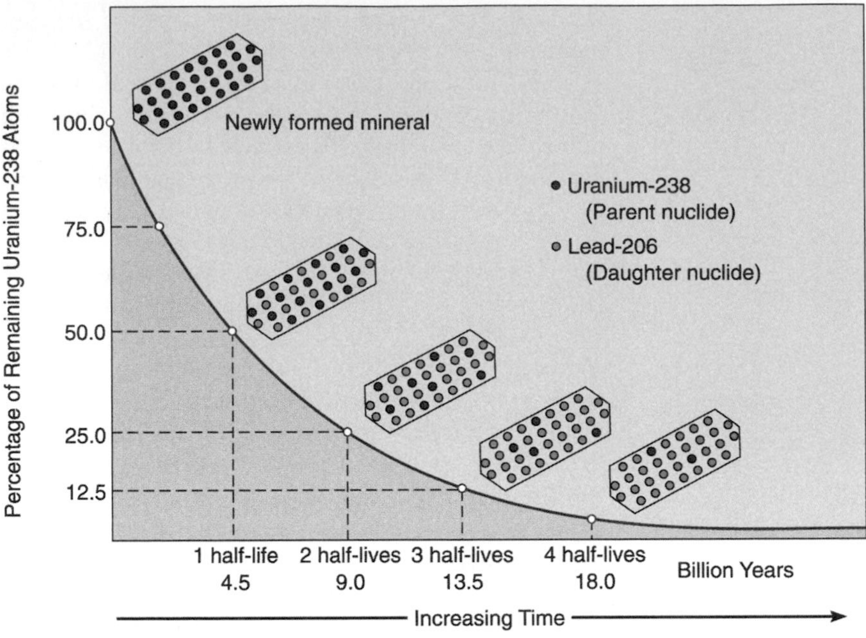

FIGURE 1-25 Rate of radioactive decay of uranium-238 to lead-206 During each half-life, half of the remaining amount of the radioactive element decays to its daughter element. In this simplified diagram, only the parent and daughter nuclides are shown, and the assumption is made that there was no contamination by daughter nuclides at the time the mineral formed. *If you were to draw a graph showing how many grains of sand passed through an hourglass each minute, how would the graph differ from the one depicted here?*

two half-lives would have elapsed, and the sample would be 1408 million years old.

THE PRINCIPAL GEOLOGIC TIMEKEEPERS At one time, there were many more radioactive nuclides present on Earth than there are now. Many of these had short half-lives and have long since decayed to undetectable quantities. Fortunately for those interested in dating the Earth's most ancient rocks, there remain a few long-lived radioactive nuclides. The most useful of these are uranium-238, uranium-235, rubidium-87, and potassium-40 (Table 1-3). There are also a few short-lived radioactive elements that are used for dating more recent events. Carbon-14 is an example of such a short-lived isotope. There are also short-lived nuclides that represent segments of a uranium or thorium decay series.

Uranium-Lead Methods Dating methods involving lead require the presence of radioactive nuclides of uranium or thorium that were incorporated into a rock when the rock originated. To determine the age of a sample of mineral or rock, one must know the original number of parent nuclides as well as the number remaining at the present time. The original number of parent atoms should be equal to the sum of the present

TABLE 1-3 Some of the More Useful Nuclides for Radioisotopic Dating

Parent Nuclide*	Half-Life†	Daughter Nuclide	Source Materials
Carbon-14	5730 years	Nitrogen-14	Organic matter
Uranium-238	4.5 billion years	Lead-296	Zircon, uraninite, pitchblende
Uranium-235	704 million years	Lead-207	
Thorium-232	14 billion years	Lead-208	
Rubidium-87	48.8 billion years	Strontium-87	Potassium mica, potassium feldspar, biotite, glauconite, whole metamorphic or igneous rock
Potassium-40	1251 million years (1.251 billion years)	Argon-40 (and calcium-40)‡	Muscovite, biotite, hornblende, whole volcanic rock, glauconite, and potassium feldspar†‡

**Nuclide* is a convenient term for any particular atom (recognized by its particular combination of neutrons and protons).

†Half-life data from Steiger, R. H., and Jäger, E. 1977. Subcomission on geochronology: Convention on the use of decay constants in geo- and cosmochronology, *Earth and Planetary Science Letters* 36:359–362.

‡Although potassium-40 decays to argon-40 and calcium-40, only argon is used in the dating method because most minerals contain considerable calcium-40, even before decay has begun.

number of parent atoms and daughter atoms. The assumptions are made that the system has remained closed, so neither parent nor daughter atoms have ever been added or removed from the sample except by decay, and that no daughter atoms were present in the system when it formed. The presence, for example, of original lead in the mineral would cause the radiometric age to exceed the true age. Fortunately, geochemists are able to recognize original lead and make the needed corrections.

As we have seen, different isotopes decay at different rates. Geochronologists take advantage of this fact by simultaneously analyzing two or three isotope pairs as a means to cross-check ages and detect errors. For example, if the $^{235}U/^{207}Pb$ radiometric ages and the $^{238}U/^{206}Pb$ ages from the same sample agree, then one can confidently assume that the age determination is valid.

Isotopic ages that depend on uranium-lead ratios may also be checked against ages derived from lead-207 to lead-206. Because the half-life of uranium-235 is much less than the half-life of uranium-238, the ratio of lead-207 (produced by the decay of uranium-235) to lead-206 will change regularly with age and can be used as a radioactive timekeeper (Fig. 1-26). This is called a lead-lead age, as opposed to a uranium-lead age.

Another uranium isotope, uranium-234, is often incorporated into the calcium carbonate skeletons of reef-building corals. Uranium-234 has a very short half-life and decays rapidly to thorium-230. The $^{234}U/^{230}Th$ dating method provides very reliable ages for reef corals that range in age from only a few thousand years to about 300,000 years. The ages, however, are accurate only if unaltered skeletal material is used in the analysis.

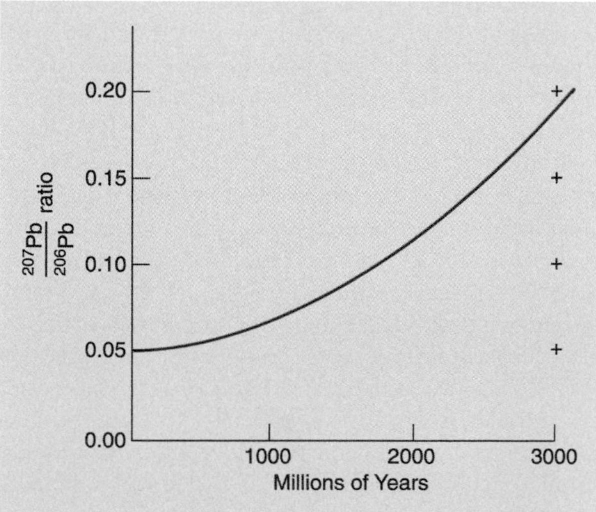

FIGURE 1-26 Graph showing how the ratio of lead-207 to lead-206 can be used as a measure of age.
What would be the age of a rock having a $^{207}Pb/^{206}Pb$ ratio of 0.15?

The Potassium-Argon Method Potassium and argon are another radioactive pair widely used for dating rocks. By means of electron capture (causing a proton to be transformed into a neutron), about 11 percent of the potassium-40 in a mineral decays to argon-40, which may then be retained within the parent mineral. The remaining potassium-40 decays to calcium-40 (by emission of a beta particle). The decay of potassium-40 to calcium-40 (by emission of a beta particle) is not used because the calcium-40 formed during radioactive disintegration cannot be distinguished from calcium-40 that may have originally existed in the rock. The advantage of using argon is that it is inert; that is, it does not combine chemically with other elements. Argon-40 found in a mineral is very likely to have originated there following the decay of adjacent potassium-40 atoms in the mineral. Also, potassium-40 is a constituent of many common minerals. However, like all isotopic dating methods, potassium-argon is not without its limitations. A sample will yield a valid age only if none of the argon has leaked out of the mineral being analyzed. Leakage may indeed occur if the rock has experienced temperatures above about 125°C. In specific localities, the ages of rocks dated by this method reflect the last episode of heating rather than the time of origin of the rock itself.

The half-life of potassium-40 is 1251 million years (1.251 billion years). As illustrated in Figure 1-27, if the ratio of potassium-40 to daughter products is found to be 1 to 1, then the age of the sample is 1251 million years (1.251 billion years). If the ratio is 1 to 3, then yet another half-life has elapsed, and the rock would have an isotopic age of two half-lives, or 2502 million years (2.502 billion years).

The Rubidium-Strontium Method The dating method based on the disintegration by beta decay of rubidium-87 to strontium-87 can sometimes be used as a check on potassium-argon dates because rubidium and potassium are often found in the same minerals. The rubidium-strontium scheme has a further advantage in that the strontium daughter nuclide is not diffused by relatively mild heating events, as is the case with argon.

In the rubidium-strontium method, a number of samples are collected from the rock body to be dated. With the aid of the mass spectrometer, the amounts of radioactive rubidium-87, its daughter product strontium-87, and strontium-86 are calculated for each sample. Strontium-86 is an isotope not derived from radioactive decay. A graph is then prepared in which the $^{87}Rb/^{86}Sr$ ratio in each sample is plotted against the $^{87}Sr/^{86}Sr$ ratio (Fig. 1-28). From the points on the graph, a straight line that is termed an isochron

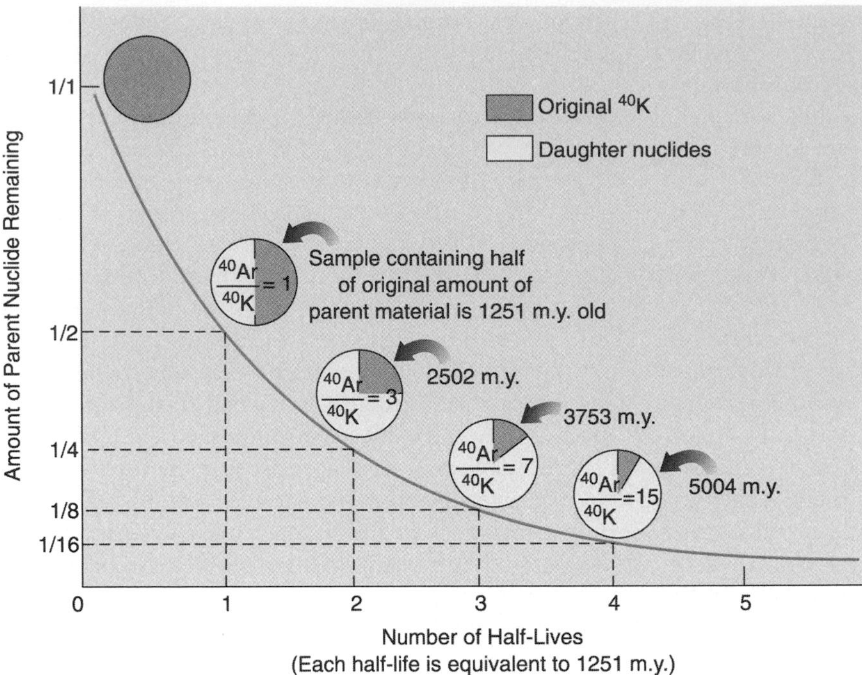

FIGURE 1-27 Decay curve for potassium-40.

is constructed. The slope of the isochron results from the fact that, with the passage of time, there is continuous decay of rubidium-87, which causes the rubidium-87/strontium-86 ratio to decrease. Conversely, the strontium-87/strontium-86 ratio increases as strontium-87 is produced by the decay of rubidium-87. The older the rocks being investigated, the more the original isotope ratios will have been changed and the greater will be the inclination of the isochron. The slope of the isochron permits a computation of the age of the rock.

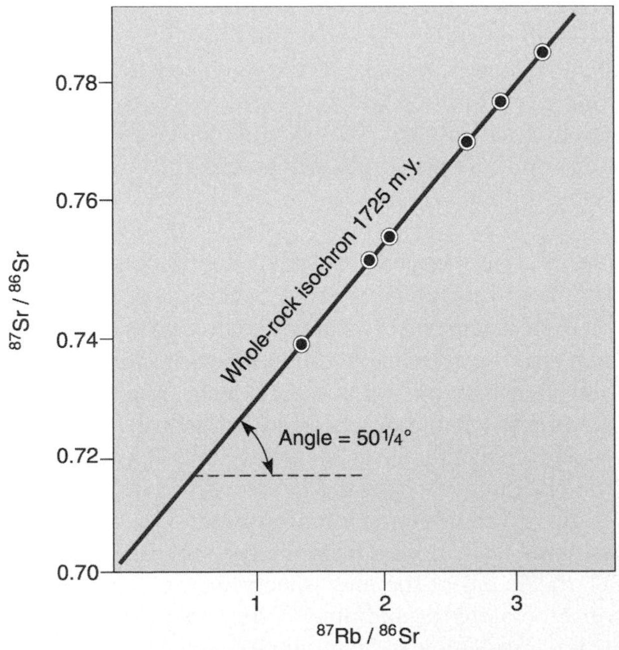

FIGURE 1-28 Whole-rock rubidium-strontium isochron for a set of samples of a Precambrian granite body exposed near Sudbury, Ontario. *(Modified from Krogh, T. E. et al. 1968. Carnegie Institute Washington Year Book 66:530.)*

The rubidium-strontium and potassium-argon methods need not always depend on the collection of discrete mineral grains containing the required isotopes. Sometimes the rock under investigation is so finely crystalline and the critical minerals so tiny and dispersed that it is difficult or impossible to obtain a suitable collection of minerals. In such instances, large samples of the entire rock may be used for age determination. This method is called **whole-rock analysis**. It is useful not only for fine-grained rocks but also for rocks in which the yield of useful isotopes from mineral separates is too low for analysis. Whole-rock analysis has also been useful in determining the age of rocks that have been so severely metamorphosed that the potassium-argon or rubidium-strontium radiometric clocks of individual minerals have been reset. In such cases, the age obtained from the minerals would be that of the episode of metamorphism, not the total age of the rock itself. The required isotopes and their decay products, however, may have merely moved to nearby locations within the same rock body, and therefore analyses of large chunks of the whole rock may provide valid radiometric age determinations.

The Carbon-14 Method Techniques for age determination based on content of radiocarbon were first devised by W. F. Libby and his associates at the University of

Chicago in 1947. The method is an indispensable aid to archaeologic research and is useful in deciphering very recent events in geologic history. Because of the short half-life of carbon-14—a mere 5730 years—organic substances older than about 50,000 years contain very little carbon-14. New techniques, however, allow geologists to extend the method's usefulness back to almost 100,000 years.

Unlike uranium-238, carbon-14 is created continuously in the Earth's upper atmosphere. The story of its origin begins with cosmic rays, which are extremely high-energy particles (mostly protons) that bombard the Earth continuously. Such particles strike atoms in the upper atmosphere and split their nuclei into small particles, among which are neutrons. Carbon-14 is formed when a neutron strikes an atom of nitrogen-14. As a result of the collision, the nitrogen atom emits a proton, captures a neutron, and becomes carbon-14 (Fig. 1-29). Radioactive carbon is being created by this process at the rate of about two atoms per second for every square centimeter of the Earth's surface. The newly created carbon-14 combines quickly with oxygen to form CO_2, which is then distributed by wind and water currents around the globe. It soon finds its way into photosynthetic plants because they utilize carbon dioxide from the atmosphere to build tissues. Plants containing carbon-14 are ingested by animals, and the isotope becomes a part of their tissue as well.

Eventually, carbon-14 decays back to nitrogen-14 by the emission of a beta particle. A plant removing CO_2 from the atmosphere should receive a share of carbon-14 proportional to that in the atmosphere. A state of equilibrium is reached in which the gain in newly produced carbon-14 is balanced by the decay loss. The rate of production of carbon-14 has varied somewhat over the past several thousand years (Fig. 1-30). As a result, corrections in age calculations must be made. Such corrections are derived from analyses of standards such as wood samples, whose exact age is known from tree ring counts.

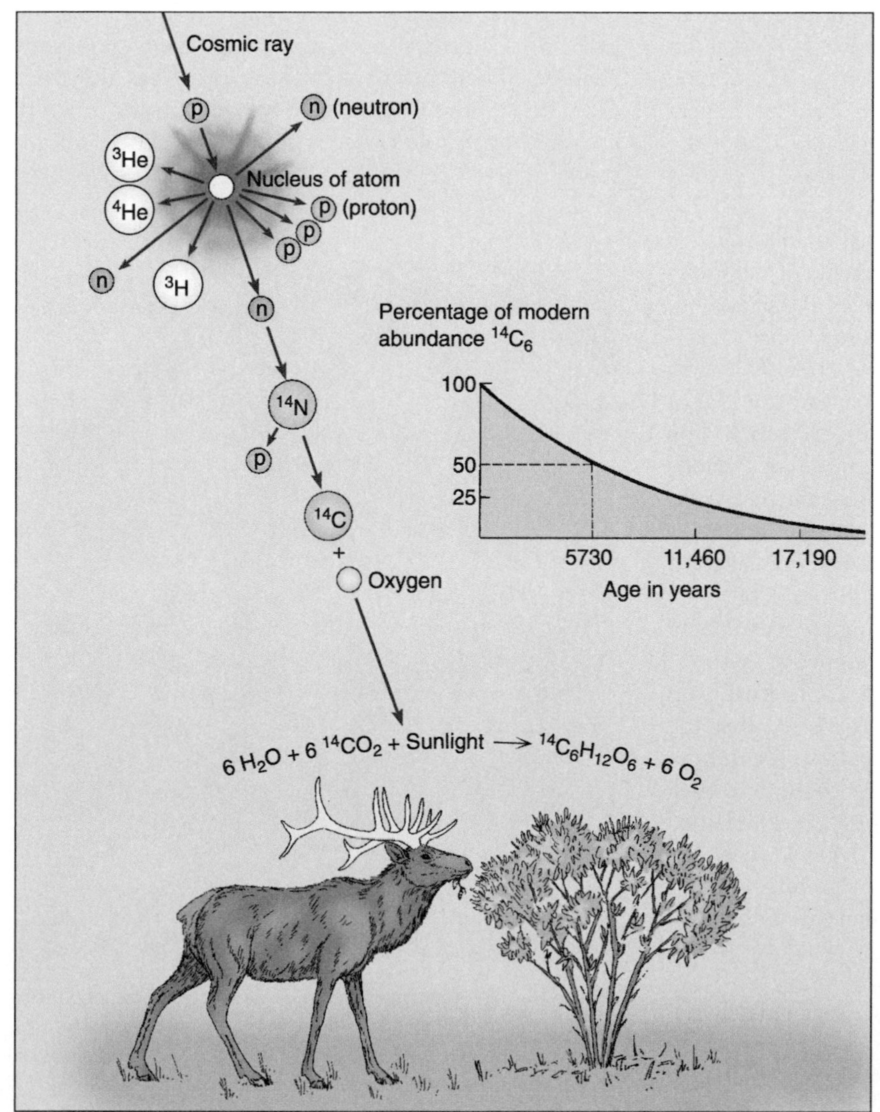

FIGURE 1-29 Carbon-14 is formed from nitrogen in the atmosphere. It combines with oxygen to form radioactive carbon dioxide and is then incorporated into all living things.

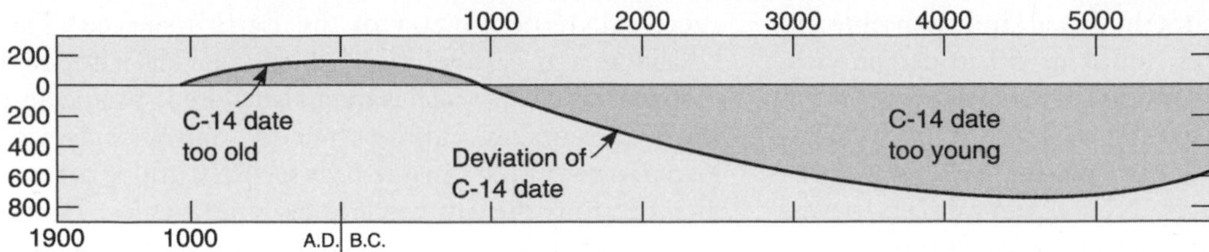

FIGURE 1-30 Deviation of carbon-14 ages to true ages from the present back to about 5000 B.C. Data are obtained from analysis of bristle cone pines from the western United States. Calculations of carbon-14 deviations are based on half-life of 5730 years. (*Adapted from Ralph, E. K., Michael, H. N., and Han, M. C. 1973. Radiocarbon dates and reality. MASCA Newsletter 9:1.*)

The age of some ancient bits of organic material is not determined from the ratio of parent to daughter nuclides, as is done with previously discussed dating schemes. Rather, the age is estimated from the ratio of carbon-14 to all other carbon in the sample. After an animal or plant dies, there can be no further replacement of carbon from atmospheric CO_2, and the amount of carbon-14 already present in the once-living organism begins to diminish in accordance with the rate of carbon-14 decay. Thus, if the carbon-14 fraction of the total carbon in a piece of pine tree buried in volcanic ash were found to be about 25 percent of the quantity in living pines, then the age of the wood (and the volcanic activity) would be two half-lives of 5730 years each, or 11,460 years.

The carbon-14 technique had considerable value to geologists studying the most recent events of the Pleistocene Ice Age. Prior to the development of the method, the age of sediments deposited by the last advance of continental glaciers was surmised to be about 25,000 years. Radiocarbon dates of a layer of peat beneath the glacial sediments provided an age of only 11,400 years. The method has also been found useful in studies of groundwater migration and in dating the geologically recent uppermost layer of sediment on the sea floors. Carbon-14 analysis of tissue from the baby mammoth depicted in Figure 4-7 indicates that the animal died 27,000 years ago. The age of giant ground sloth dung recovered from a cave near Las Vegas indicated the presence of these ancient beasts in Nevada 10,500 years ago. Charcoal from the famous Lascaux cave in France revealed that the artists who drew pictures of mammoths, woolly rhinoceroses, bison, and reindeer on the walls of the cave lived about 15,000 years ago. In archaeology, dates obtained by the carbon-14 method overturned many cherished concepts by demonstrating that the beginnings of agriculture and urbanization occurred much earlier than had formerly been thought.

Nuclear Fission Track Timekeepers Nuclear particle fission tracks were discovered in the early 1960s, when scientists using the electron microscope were able to examine the areas around presumed locations of radioactive particles that were embedded in mica. Closer examination showed that the tracks were really small tunnels—like bullet holes—that were produced when high-energy particles of the nucleus of uranium were fired off in the course of spontaneous fission (spontaneous fragmentation of an atom into two or more lighter atoms and nuclear particles). The particles speed through the orderly rows of atoms in the crystal, tearing away electrons from atoms located along the path of trajectory and rendering them positively charged. Their mutual repulsion produces the track (Fig. 1-31). The tracks are only a few atoms in width and are impossible to see without an electron microscope. Therefore, the sample is immersed for a short

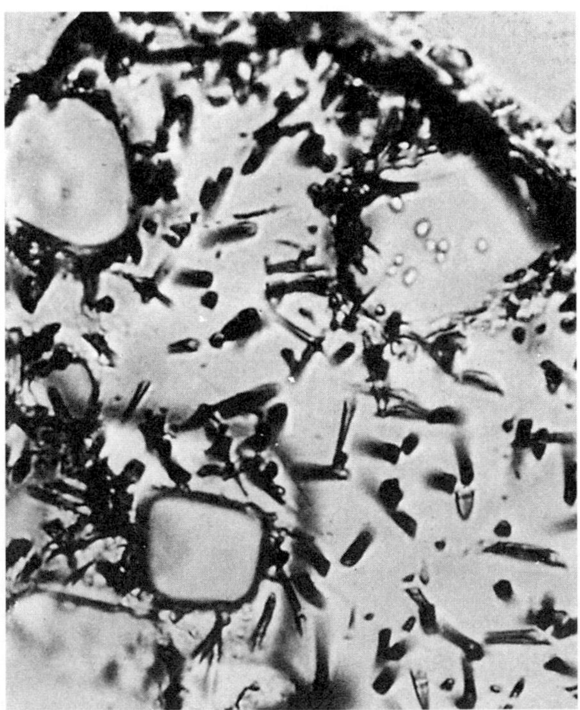

FIGURE 1-31 Fission tracks. These tracks were produced by plutonium-244 in a melilite crystal that was extracted from a meteorite. The small crystals are spinel inclusions. Melilite is a calcium-magnesium aluminosilicate. (*Photograph courtesy of F. Podosek, Department of Earth and Planetary Sciences, Washington University, St. Louis, MO.*)

period of time in a suitable solution (acid or alkali), which rushes up into the tubes, enlarging the track tunnel so that it can be seen with an ordinary microscope.

The natural rate of track production by uranium atoms is very slow and uniform. For this reason, the tracks can be used to determine the number of years that have elapsed since the uranium-bearing mineral solidified. One first determines the number of uranium atoms that have already disintegrated. This number is obtained with the aid of a microscope by counting the etched tracks. Next, one must find the original number of uranium atoms. This quantity can be determined by bombarding the sample with neutrons in a reactor and thereby causing the remaining uranium to undergo fission. A second count of tracks reveals the original quantity of uranium. Finally, one must know the spontaneous fission decay rate for uranium-238. This information is determined by counting the tracks in a piece of uranium-bearing synthetic glass of known date of manufacture.

Fission track dating is of particular interest to geochronologists because it can be used to date specimens only a few centuries old as well as to date rocks billions of years in age. The method helps to date the period between 50,000 and 1 million years ago; a period for which neither carbon-14 nor potassium-argon methods are suitable. As with all radiometric techniques, however, there can be problems. If rocks have been subjected to high temperatures, tracks may heal and fade away.

THE AGE OF THE EARTH

Anyone interested in the total age of the Earth must decide what event constitutes its "birth." Most geologists assume that "year 1" commenced as soon as the Earth had collected most of its present mass and had developed a solid crust. Unfortunately, rocks that date from those earliest years have not been found on the Earth. They have long since been altered and converted to other rocks by various geologic processes. The oldest materials known are grains of the mineral zircon taken from a sandstone in western Australia. The zircon grains are 4.1 to 4.2 billion years old. The zircons were probably eroded from nearby granitic rocks and deposited, along with quartz and other detrital grains, by rivers. Other very old rocks on Earth include 3.7-billion-year-old granites of southwestern Greenland, metamorphic rocks of about the same age from Minnesota, and 3.96-billion-year-old rocks from the Northwest territories of Canada (north of Yellowknife, Canada).

Meteorites, which many consider to be remnants of a shattered planet or asteroid that originally formed at about the same time as the Earth, have provided uranium-lead and rubidium-strontium ages of about 4.6 billion years. From such data, and from estimates of how long it would take radioactive decay to produce the quantities of various lead isotopes now found on the Earth, geochronologists feel that the 4.6-billion-year age for the Earth can be accepted with confidence. Evidence substantiating this conclusion comes from returned moon rocks. The ages of these rocks range from 3.3 to about 4.6 billion years. The older age determinations are derived from rocks collected on the lunar highland, which may represent the original lunar crust. Certainly, the moons and planets of our solar system originated as a result of the same cosmic processes and at about the same time.

SUMMARY

Simply stated, geology is the study of all naturally occurring processes, phenomena, and materials of the past and present Earth. Historical geology is that branch of the science concerned particularly with decoding the rock and fossil record of the Earth's long history. In the past few decades, advances in technology have added immensely to the store of geologic knowledge. Interpretation of the new data requires an exceptional understanding not only of geology but also of physics, chemistry, mathematics, and biology. However, the historical inferences that are drawn from the data are frequently derived from the fundamental geologic and paleontologic concepts introduced by Nicolaus Steno (superposition), James Hutton (uniformitarianism), Sir Charles Lyell (cross-cutting relationships, inclusion), William Smith (correlation), Charles Darwin (organic evolution), and others. These scientists formulated the principles by which geologists determine the relative age of rock outcrop, its history of deposition and deformation, and its spatial and chronologic relationship to strata in other regions of the Earth. American geologists such as William Maclure, Amos Eaton, James Hall, Ferdinand Hayden, and John Powell used the precepts of their European colleagues and predecessors in their studies of the geology of America, while Othniel Marsh and Edwin Cope revealed the rich fossil record of vertebrates in the American West.

In the early stages of its development, geology was totally dependent on relative dating of events. Hutton helped scientists visualize the enormous amount of time needed to accomplish the events indicated in sequences of strata, and the geologists who followed him pieced together the many local stratigraphic sections by using fossils and superposition. A scale of relative geologic time gradually emerged. Initial attempts to decide what the rock succession meant in terms of years were made by estimating the amount of salt in the ocean, the average rate of deposition of sediment, and the rate of cooling of the Earth. However, these early schemes did little more than suggest that the planet was at least tens of millions of years old and that the traditional concept of a 6000-year-old Earth did not agree with what could be observed geologically.

An adequate means of measuring geologic time was achieved only after the discovery of radioactivity at about the turn of the 20th century. Scientists found that the rate of decay by radioactivity of certain elements is constant and can be measured and that the proportion of parent and daughter elements can be used to reveal how long they had been pres- ent in a rock. Over the years, continuing efforts by investigators as well as improvements in instrumentation (particularly of the mass spectrometer) have provided many thousands of age determinations. Frequently, these numerical dates have shed light on difficult geologic problems, provided a way to determine rates of movement of crustal rocks, and permitted geologists to date mountain building or to determine the time of volcanic eruptions. In a few highly important regions, isotopic dates have been related to particular fossiliferous strata and have thereby helped to quantify the geologic time scale and to permit estimation of rates of organic evolution. Isotopic dating has also changed the way humans view their place in the totality of time.

The isotopic transformations most widely used in determining absolute ages are uranium-238 to lead-206, uranium-235 to lead-207, thorium-232 to lead-208, potassium-40 to argon-40, rubidium-87 to strontium-87, and carbon-14 to nitrogen-14. Methods involving uranium-lead ratios are of importance in dating the Earth's oldest rocks. The short-lived carbon-14 isotope that is created by cosmic ray bombardment of the atmosphere provides a means to date the most recent events in Earth history. For rocks of intermediate age, schemes involving potassium-argon ratios, those utilizing intermediate elements in decay series, or those employing fission tracks are most useful. A figure of 4.6 billion years for the Earth's total age is now supported by ages based on meteorites and on lead ratios from terrestrial samples.

Improvements in numerical geochronology are being made daily and will provide further calibration of the standard geologic time scale in the future. Some of the time boundaries in the scale, such as that between the Cretaceous and Tertiary systems, are already well validated. Others, such as the boundary between the Paleozoic and Mesozoic, require additional refinement. Additional efforts to incorporate isotopic ages into sections of sedimentary rocks are among the continuing tasks of historical geologists. The usual methods for determining the age of strata involve the dating of intrusions that penetrate these sediments or the dating of interbedded volcanic layers. Less frequently, strata can be dated by means of radioactive isotopes incorporated within sedimentary minerals that formed in place at the time of sedimentation. At present, the best numerical age estimates indicate that Paleozoic sedimentation began about 540 million years ago, the Mesozoic Era began about 250 million years ago, and the Cenozoic commenced about 65 million years ago.

QUESTIONS FOR REVIEW AND DISCUSSION

1. Describe the general steps used by geologists and other scientists in their attempt to solve particular problems or explain natural phenomena.

2. Discuss the principles that Steno, Lyell, and Smith formulated for the development of the geologic time scale.

3. What interpretation did Steno make on observing rock layers that were not horizontal? What could be said of a sequence of strata having oldest beds at the top and younger at the bottom?

4. Explain the difference between a geochronologic term (sometimes called a time term) and a chronostratigraphic term (sometimes called a time-rock term).

5. What features in a rock layer might a geologist seek in attempting to determine the top from the bottom of strata that had been forced into vertical alignment or overturned during mountain building?

6. What is meant by uniformitarianism?

7. Why is the concept of half-life necessary? (Why not use whole life?)

8. Pebbles of a black rock called basalt occur in a sedimentary rock composed of those pebbles. The sedimentary rock is called a conglomerate. The pebbles yield an isotopic age of 300 million years. Is the conglomerate younger or older than the pebbles?

9. How do isotopes of a given element differ from one another in regard to number of protons and neutrons in the nucleus?

10. How are dating methods involving decay of radioactive elements unlike methods for determining elapsed time that involve the funneling of sand through an hourglass?

11. What would be the effect on the isotopic age of a zircon crystal being dated by the potassium-argon method if a small amount of argon-40 escaped from the crystal?

12. State the age of a sample of prehistoric mummified human skin that contains 12.5 percent of the original amount of carbon-14.

13. How do fission tracks originate? What geologic events might destroy fission tracks in a mineral crystal?

READINGS

Adams, F. D. 1938. *The Birth and Development of the Geological Sciences*. Baltimore: Williams & Wilkins Co. (Reprinted by Dover Press, 1954.)

Albritton, C. C., Jr. 1980. *The Abyss of Time*. San Francisco: Freeman, Cooper & Co.

Berry, W. B. N. 1987. *Growth of a Prehistoric Time Scale*. San Francisco: W. H. Freeman & Co.

Dean, D. R. 1992. *James Hutton and the History of Geology*. Ithaca, NY: Cornell University Press.

Eicher, D. L. 1976. *Geologic Time*, 2nd ed. Englewood Cliffs, NJ: Prentice-Hall, Inc.

Eiseley, L. C. 1959. Charles Lyell, *Sci. Am.* 201(8): 98–106. (Offprint No. 846. San Francisco: W. H. Freeman & Co.)

Gohau, G. 1990. *A History of Geology*. New Brunswick, NJ: Rutgers University Press.

Hallam, A. 1983. *Great Geologic Controversies*. Oxford, England: Oxford University Press.

Jaffe, M. 2000. *The Gilded Dinosaur: the Fossil War Between E. D. Cope and O. C. Marsh and the Rise of American Science*. New York: Crown Publishers.

Speakman, C. 1982. *Adam Sedgwick, Geologist and Dalesman, 1785–1873*. Cambridge, England: Trinity College.

Winchester, S. 2001. *The Map that Changed the World: William Smith and the Birth of Modern Geology*. New York: HarperCollins.

York, D., and Farquhar, R. M. 1972. *The Earth's Age and Geochronology*. Oxford, England: Oxford University Press.

WEB SITES

The Earth Through Time Student Companion Web Site (www.wiley.com/college/levin) has online resources to help you expand your understanding of the topics in this chapter. Visit the Web Site to access the following:

1. Illustrated course notes covering key concepts in each chapter;
2. Online quizzes that provide immediate feedback;
3. Links to chapter-specific topics on the web;
4. Science news updates relating to recent developments in Historical Geology;
5. Web inquiry activities for further exploration;
6. A glossary of terms;
7. A Student Union with links to topics such as study skills, writing and grammar, and citing electronic information.

2

Pearly white, translucent crystals of selenite gypsum ($CaSO_4 \cdot 2H_2O$). Lechuguilla Cave, New Mexico. (Photograph by D. Bunnell.)

CHAPTER 2

Earth Materials: A Physical Geology Refresher

What stuff 'tis made of, whereof it is born, I am to learn.
Shakespeare, *The Merchant of Venice*

OUTLINE

Some readers of this book will have taken a prior course in physical geology. These students will have been introduced to rocks and minerals and may use this chapter as a review. For other readers, a basic knowledge of earth materials sets the stage for an understanding of the processes that have shaped our planet. Minerals and rocks (and the fossils within rocks) are the indispensible documents from which the history of the Earth is deciphered.

MINERALS

Minerals are important to all of us. They are the raw materials needed to manufacture the products that sustain our modern technologic society. For geologists, the search for new mineral deposits has always been a key objective. Minerals, however, are also of interest to earth scientists because of the evidence they provide about past events. Many minerals form only within a narrow range of physical conditions and therefore can be used to diagnose the pressures and temperatures involved in the formation of mountain ranges and volcanoes. Some minerals develop exclusively in ocean water and provide evidence of the incursion of seas across former areas of dry land. Others form under conditions of excessive aridity and are used to locate the arid tropical belts of long ago. The magnetic properties of certain minerals provide clues to drifting continents, the widening of oceans, and changes in the planet's magnetic poles. As a further aid to deciphering Earth history, some minerals contain radioactive elements that permit us to determine the ages of rocks and structures formed at particular times in the geologic past.

Common Rock-Forming Minerals

Minerals are naturally occurring solid, inorganic substances that have a definite chemical composition or range of compositions as well as distinctive properties that reflect their composition and regular internal atomic structure. Minerals can usually be identified by such physical properties as color, hardness, density, crystal form, and cleavage. Cleavage refers to the tendency of some minerals to break smoothly along certain planes of weakness.

TABLE 2-1 Abundances of Chemical Elements in the Earth's Crust*

Element and Symbol	Percentage by Weight	Percentage by Number of Atoms	Percentage by Volume
Oxygen (0)	46.6	62.6	93.8
Silicon (Si)	27.7	21.2	0.9
Aluminum (Al)	8.1	6.5	0.5
Iron (Fe)	5.0	1.9	0.4
Calcium (Ca)	3.6	1.9	1.0
Sodium (Na)	2.8	2.6	1.3
Potassium (K)	2.6	1.4	1.8
Magnesium (Mg)	2.1	1.9	0.3
All other elements	1.5	—	—
	100.0	100.0†	100.0†

*Based on Mason, B. 1966. *Principles of Geochemistry*. New York: John Wiley & Sons. Note the high percentage of oxygen in the Earth's crust.
†Includes only the first eight elements.

More than 3000 minerals have already been discovered and described, but most of these are rarely encountered. For our present purposes, it is important to consider only those minerals that compose the bulk of common rocks. Among these, the most important are the silicate minerals.

SILICATE MINERALS About 75 percent by weight of the Earth's crust is composed of the two elements oxygen and silicon (Table 2-1). These elements usually occur in combination with such abundant metals as aluminum, iron, calcium, sodium, potassium, and magnesium to form a group of minerals called the silicates. A single family of silicates, the feldspars, comprises about one-half of the material of the crust, and a single mineral species called quartz represents a sizable portion of the remainder.

Quartz The principal silicate minerals and some of their properties are listed in Table 2-2. The mineral **quartz** (SiO_2) is one of the most familiar and important of all the silicate minerals. It is common in many different families of rocks. In general appearance, quartz is a glassy, colorless, gray or white mineral. Quartz is relatively hard (see Mohs' scale in the footnote of Table 2-2) and will easily scratch glass. When quartz grows in an open cavity, it may form the hexagonal crystals (Fig. 2-1) that are prized by mineral collectors. More frequently, the crystal faces cannot be discerned because the orderly addition of atoms during crystal growth was interrupted by contact with other growing crystals and results in an aggregate that exhibits crystalline texture. Such common minerals as chert, flint, jasper, and agate are sedimentary varieties of quartz. Chert (Fig. 2-2) is a dense, hard, white mineral composed of sub-microcrystalline quartz that forms as a result of the precipitation of silicon dioxide by either biologic or chemical means. Flint is the popular name for a dark gray or black variety of chert in which the dark color results from inclusions of organic matter. Red or red-brown chert is called jasper. Agate is a submicroscopic variety of quartz characterized by bands of differing colors.

The origin of chert is a complex problem made even more difficult by the fact that different varieties are formed by somewhat different processes. Some cherts are replacements of earlier carbonate rocks. Others appear to have formed as a result of the solution and reprecipitation of silica from the siliceous skeletal remains of organisms. Small amounts of chert may also be precipitated directly from concentrated aqueous solutions.

The Feldspars The feldspars are the most abundant constituents of rocks, composing about 60 percent of the total weight of the Earth's crust. There are two major families of feldspars: the orthoclase, or potassium, group (Fig. 2-3), which comprises the potassium aluminosilicates; and the plagioclase group, which comprises the aluminosilicates of sodium and calcium (Fig. 2-4). Members of the plagioclase group exhibit a wide range in composition—from a calcium-rich end member called anorthite ($CaAl_2Si_2O_8$) to a sodium-rich end member called albite ($NaAlSi_3O_8$). Between these two extremes, plagioclase minerals containing both sodium and calcium occur. The substitution of sodium for calcium, however, is not random but is governed by the temperature and composition of the parent material. Thus, by examining the feldspar content of a once-molten rock, it is possible to infer the physical and chemical conditions under which it originated. Feldspars are nearly as hard as quartz and range in

TABLE 2-2 Common Rock-Forming Silicate Minerals

Silicate Mineral		Composition	Physical Properties
Quartz		Silicon dioxide (silica, SiO_2)	Hardness of 7 (on scale of 1 to 10)*; will not cleave (fractures unevenly); specific gravity: 2.65
Orthoclase feldspar group		Aluminosilicates of potassium	Hardness of 6.0–6.5; cleaves well in two directions; pink or white; specific gravity: 2.5–2.6
Plagioclase feldspar group		Aluminosilicates of sodium and calcium	Hardness of 6.0–6.5; cleaves well in two directions; white or gray; may show striations on cleavage planes; specific gravity: 2.6–2.7
Muscovite mica		Aluminosilicates of potassium with water	Hardness of 2–3; cleaves perfectly in one direction, yielding flexible, thin plates; colorless; transparent in thin sheets; specific gravity: 2.8–3.0
Biotite mica	Ferromagnesian minerals	Aluminosilicates of magnesium, iron, potassium, with water	Hardness of 2.5–3.0; cleaves perfectly in one direction, yielding flexible, thin plates; black to dark brown; specific gravity: 2.7–3.2
Pyroxene group		Silicates of aluminum, calcium, magnesium, and iron	Hardness of 5–6; cleaves in two directions at 87° and 93°; black to dark green; specific gravity: 3.1–3.5
Amphibole group		Silicates of aluminum, calcium, magnesium, and iron	Hardness of 5–6; cleaves in two directions at 56° and 124°; black to dark green; specific gravity: 3.0–3.3
Olivine		Silicates of magnesium and iron	Hardness of 6.5–7.0; light green; transparent to translucent; specific gravity: 3.2–3.6
Garnet group		Aluminosilicates of iron, calcium, magnesium, and manganese	Hardness of 6.5–7.5; uneven fracture; red, brown, or yellow; specific gravity: 3.5–4.3

*The scale of hardness used by geologists was formulated in 1822 by Frederich Mohs. Beginning with diamond as the hardest mineral, he arranged the following table:

10 Diamond	8 Topaz	6 Feldspar	4 Fluorite	2 Gypsum
9 Corundum	7 Quartz	5 Apatite	3 Calcite	1 Talc

color from white or pink to bluish gray. They have good cleavage in two directions, and the resulting flat, often rectangular surfaces are useful in identification. Those plagioclase feldspars with abundant sodium tend, along with potassium feldspars, to occur in silica-rich rocks such as granite. Calcium-rich plagioclases occur in such rocks as the Hawaiian lavas (called basalts).

Micas Mica is a family of silicate minerals easily recognized by its perfect and conspicuous cleavage (Fig. 2-5) along one directional plane. The two chief varieties are the colorless or pale-colored muscovite mica, which is a hydrous potassium aluminum silicate, and the dark-colored biotite mica, which also contains magnesium and iron. Both muscovite and biotite are common rock-forming silicates.

Hornblende Hornblende (Fig. 2-6) is a vitreous, black or very dark green mineral. It is the most common member of a larger family of minerals called amphiboles, which have generally similar properties. Because of its content of iron and magnesium, hornblende (along with biotite, augite, and olivine) is designated a ferromagnesian or mafic mineral. Crystals of hornblende tend to be long and narrow.

Augite Just as hornblende is only one member of a family of minerals called amphiboles, augite is an important member of the pyroxene family, in which many other mineral species also occur. Like hornblende, it is a ferromagnesian mineral and thus dark-colored. An augite crystal is typically rather stumpy in shape, with good cleavages developed along two planes that are nearly at right angles.

Olivine As you might guess from its name, this glassy-looking iron and magnesium silicate often has an olive green color. Olivine (Fig. 2-7) is present in dark rocks such as basalt. Along with pyroxene and minor calcium plagioclase, it is also an important component of the

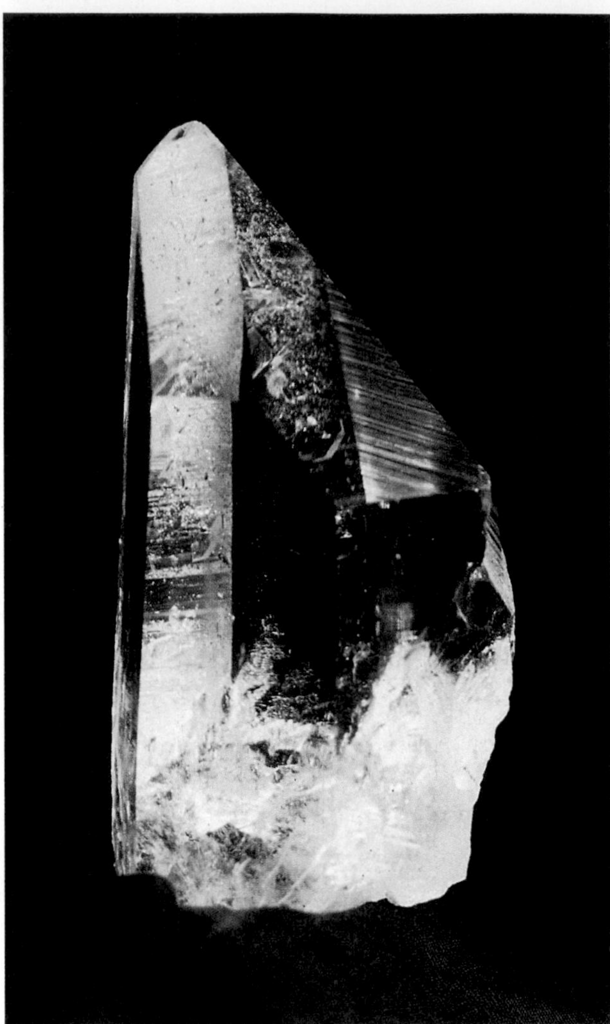

FIGURE 2-1 A large crystal of quartz. This specimen is 14 cm (about 5.5 inches) tall.

(*A*)

(*B*)

FIGURE 2-3 Orthoclase. (*A*) Crystals of orthoclase (potassium feldspar). (*B*) A cleavage fragment of orthoclase.

FIGURE 2-2 Chert. This variety (called novaculite) has a grainy texture that makes it useful as a grinding stone.

ultramafic rock called peridotite. Peridotite is a prominent rock type in the Earth's mantle, the rocky layer that lies beneath the crust.

Clay Minerals The word *clay* is so familiar to us that we may not appreciate its importance. Nearly seventy-five percent of the surface of continents is covered by clay minerals. They are the most abundant materials deposited in modern and ancient oceans. Shale, the most abundant of sedimentary rocks, is about 50 percent clay. Clay minerals are an essential component of agricultural soils, serving to bind soil particles together, and hold moisture and nutrients.

Clay minerals are silicates of hydrogen and aluminum with additions of magnesium, iron, and potassium. Their basic structure is similar to that of mica, but because individual flakes are extremely small (some

FIGURE 2-4 The variety of plagioclase feldspar known as labradorite. The mineral often displays beautiful blue and gold reflections as well as fine striations on cleavage planes. (*Courtesy of Wards Natural Science Establishment, Inc., Rochester, NY.*)

FIGURE 2-5 Muscovite mica, exhibiting its characteristic perfect cleavage in one plane. The blade of the screwdriver is shown here separating pieces of the mineral along its cleavage plane.

FIGURE 2-6 Hornblende. The black scaly flakes on the surface are biotite mica. Specimen is 6 cm long. *On the basis of its dark color, what metallic elements are likely to be present in this mineral?*

are comparable in size to some viruses). Their mica-like form can be seen only with the magnification provided by an electron microscope (Fig. 2-8). Unlike many of the silicate minerals already described, clay minerals form as a result of weathering of other aluminum silicate minerals such as feldspars. As a group, clays are known by their amorphous form, softness, low density, and ability to become plastic when wet.

Geologists recognize that there are two different uses for the term *clay*. As described above, a clay *mineral* is a hydrous aluminum silicate with mica-like form. However, the term *clay* can also be used to denote particles of sediment with a diameter of less than 1/256 mm.

NONSILICATE MINERALS Approximately 8 percent of the Earth's crust is composed of nonsilicate minerals. These include a host of carbonates, sulfides, sulfates, chlorides, and oxides. Among these groups, the carbonates, such as **calcite** and **dolomite**, are the most important. Calcite ($CaCO_3$), which is the main constituent of limestone and marble, forms in many ways. It is secreted as skeletal material by certain invertebrate animals, precipitated directly from sea water, or formed as dripstone in caverns. Calcite is

FIGURE 2-7 Olivine.

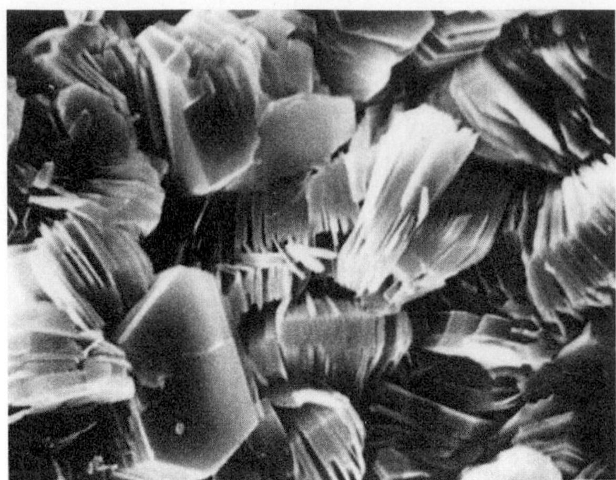

FIGURE 2-8 **Electron micrograph of the clay mineral kaolinite.** The flaky, stack-of-cards character of the clay crystals is a manifestation of their silicate sheet structure. (Magnified 32,000 times; *courtesy of Kevex Corporation.*)

FIGURE 2-10 **Halite.** (*Courtesy of Wards Natural Science Establishment, Inc., Rochester, NY.*) ? *How do these cleaved pieces of halite differ from the cleavage pieces of calcite in Figure 2-9?*

easily recognized by its rhombohedron-shaped cleaved fragments (Fig. 2-9) and by the fact that an application of dilute hydrochloric acid on its surface will cause effervescence (the bubbles are liberated carbon dioxide).

The term *dolomite* is used both for the carbonate mineral that has the chemical formula $CaMg(CO_3)_2$ and for the rock composed largely of that mineral. (The alternate term *dolostone* is also used occasionally to designate dolomite rock.) In the mineral dolomite, calcium and magnesium occur in approximately equal proportions.

Aragonite ($CaCO_3$) is another carbonate mineral that occurs in a different crystal form and more rarely than either calcite or dolomite. Most of us have seen it as the inner "mother-of-pearl" layer of clam shells. Because both calcite and aragonite are two different forms of the same compound, they are spoken of as *polymorphs* of calcium carbonate.

Other common nonsilicate minerals include common rock salt, or **halite** (NaCl), and **gypsum**. Halite is easily recognized by its salty taste and the fact that it crystallizes and cleaves to form cubes (Fig. 2-10). Gypsum is a soft, hydrous calcium sulfate ($CaSO_4 \cdot 2H_2O$). The variety of gypsum called satinspar has a fine, fibrous structure, whereas the variety known as selenite (Fig. 2-11) will split into thin plates. The finely crystalline, massive variety known as alabaster is used widely in carvings and sculpture because of its uniform texture and softness. Halite and the various gypsum minerals are sometimes referred to as evaporites be-

FIGURE 2-9 **Calcite, showing the characteristic rhombohedral shape of cleavage pieces.** (*Courtesy of Wards Natural Science Establishment, Inc., Rochester, NY.*)

FIGURE 2-11 **A cluster of selenite gypsum crystals.** (*Courtesy of Wards Natural Science Establishment, Inc., Rochester, NY.*)

cause they are often precipitated from bodies of water that have been subjected to intense evaporation.

If you were to make a list of the elements that compose the minerals described thus far, the list would be a surprisingly short one. Only eight elements make up the bulk of these minerals, and only the same eight are abundant in the Earth's crust as well. As shown in Table 2-1, these abundant elements are oxygen, silicon, aluminum, iron, calcium, sodium, potassium, and magnesium.

ROCKS

Rock Conversions

We have noted that minerals form under specific conditions of temperature and pressure. It is therefore possible to know the origin of the rocks containing those minerals. Geologists have agreed on a fundamental division of rocks into three great families according to difference of origin. Igneous rocks are those that have cooled from a molten state. Sedimentary rocks consist of materials formed from the weathered products of preexisting rocks that have been transported, deposited, and lithified. Metamorphic rocks are any that have been changed from previously existing rocks by the action of heat, pressure, and associated chemical activity. The changes may include a recrystallization of the previous minerals or growth of entirely new minerals.

It is important to remember that the rocks of any of the three major rock families are not immutable. The Earth's crust is dynamic and everchanging. Any sedimentary or metamorphic rock may be partially or completely melted to produce igneous rocks, and any previously existing rock of any category can be compressed and altered during mountain building to produce metamorphic rocks. The weathered and eroded residue of any family of rock can be observed today being transported to the sea for deposition and conversion into sedimentary rocks. These changes can be summarized in a schematic diagram that is designated the rock cycle (Fig. 2-12).

Although rocks are classified into groups that have had similar origin, the identification of rocks is not based on origin but on characteristics. For identification, it is necessary to know general appearance as well as mineral composition and physical properties. Texture (size, shape, and arrangement of constituent minerals) and mineral composition are essential for identification. Inferences regarding the origin of the rock are based on geologic observations and experimentation.

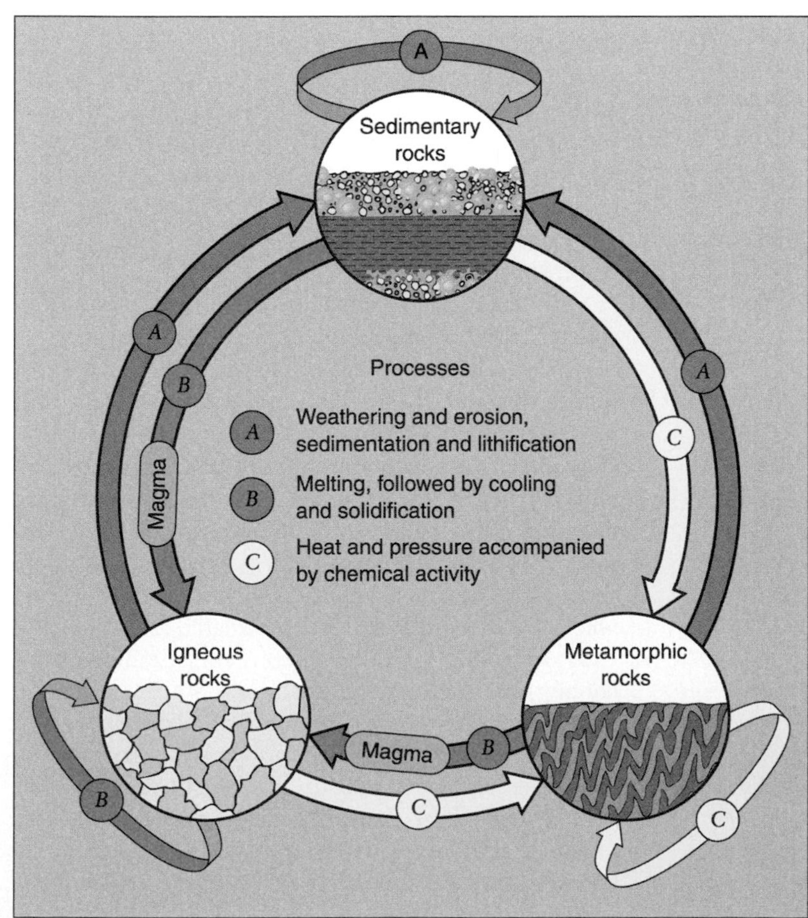

FIGURE 2-12 **Geologic processes act continuously on Earth to change one type of rock into another.**

Igneous Rocks

Igneous rocks constitute over 90 percent by volume of the Earth's crust, although their great abundance may go unnoticed because they are extensively covered by sedimentary rocks. Much of our mountain scenery is sculpted in igneous rocks. Recent volcanic activity (Fig. 2-13) provides an often spectacular reminder of the fiery processes that produce igneous rocks (the word *igneous* comes from the Latin *ignis*, meaning "fire"). It is an appropriate name for rocks that develop from cooling masses of molten material derived from exceptionally hot parts of the Earth's interior. **Magma** is the term used to describe this mixture of molten silicates and gases while it is still beneath the surface. If it should penetrate to the surface, it becomes **lava** (Fig. 2-14).

FIGURE 2-14 Front of a lava flow advancing over an older flow, Kilauea Volcano, Hawaii, February 1990. The advancing flow is breaking into a jagged, rough kind of lava termed *aa*. The underlying older flow exhibits the ropy texture of *pahoehoe* lava. (*Courtesy of J. Plaut.*)

COOLING HISTORY OF IGNEOUS ROCKS Igneous rocks that formed from magma that had penetrated into other rocks and solidified before reaching the surface are termed **intrusive** or **plutonic** igneous rocks. Very large masses of such rocks are sometimes called plutons. Their exposure at the Earth's surface results from crustal uplift and erosional removal of overlying rocks. In contrast, **extrusive** or **volcanic** igneous rocks form from melts that have reached the Earth's surface. This group includes rocks formed from lava erupted from volcanoes or lava that has welled out of fissures.

The grain size of igneous rocks is an index to their history of cooling (Fig. 2-15). Magmas deep within the Earth lose heat slowly and retain water. Water tends to inhibit formation of myriads of crystal nuclei. Thus, there is time and space for the growth of larger crystals around fewer nuclei. In typical intrusive rocks such as granite, diorite, and gabbro, the intergrown crystals are large enough to be seen readily without magnification. In contrast, extrusive igneous rocks have a finer texture in which crystals are too small to be seen with the unaided eye. The structure of such rocks reflects sudden chilling of molten silicates as they were ejected at the surface of the Earth. When a lava is extruded, there is not enough time for the growth of large crystals. Furthermore, water from the melt is quickly lost to the atmosphere, and without water, many tiny crystals form rather than fewer crystals that can grow to a larger size.

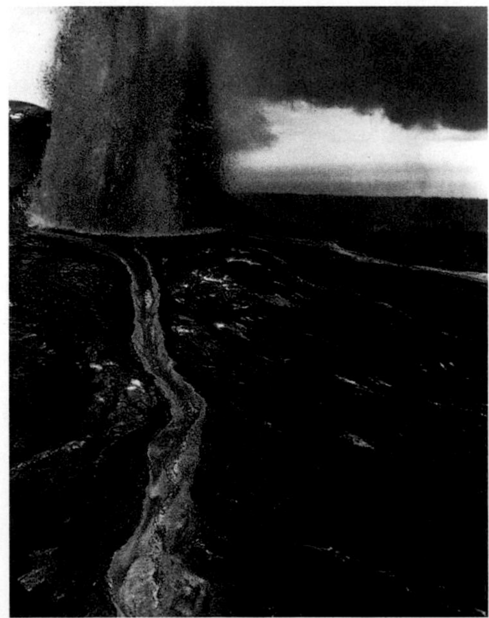

FIGURE 2-13 A stream of molten lava and a fire fountain pour from Pu'u'O'o vent, Hawaii, 1986. The lava solidifies to form an extrusive igneous rock. (*J. D. Griggs, U.S. Geological Survey.*)

One such extrusive rock is **basalt**, composed of ferromagnesian minerals and tiny rectangular grains of plagioclase feldspars. **Obsidian** is an extrusive rock that cooled so rapidly that there was insufficient time for crystallization; the melt therefore froze into a glass (Fig. 2-16). Obsidian forms from lavas that have lost most of their dissolved gases. If the lava still contains dissolved gases, bubbles can be released and form cavities called **vesicles**. **Pumice** is a vesicular glass. In a sense, it is actually a solidified silicate froth. **Tuff** is a volcanic rock composed of consolidated ash.

If coarsely crystalline igneous rocks indicate slow cooling and finely crystalline ones indicate rapid cooling, then what would be the cooling history of a rock with large crystals immersed in a very fine-grained matrix? Such rocks are said to have **porphyritic** texture (Fig. 2-15*C*). The large crystals (**phenocrysts**) were formed slowly, deep within the Earth, and were then swept upward and incorporated into the lava as it hardened at the surface.

(*A*)

(*B*)

(*C*)

COMPOSITION OF IGNEOUS ROCKS Both texture and composition are used in naming and classifying igneous rocks (Fig. 2-17). *Granite*, for example, refers to a coarsely textured rock composed largely of feldspar and quartz. Rhyolite has the same composition as granite but has a very fine-grained texture (Fig. 2-18).

The minerals in an igneous rock reflect the proportions of such elements as silicon, oxygen, aluminum, calcium, iron, magnesium, sodium, and potassium present in the magma from which the rock was formed. These eight elements combine in various ways to form feldspars, ferromagnesian minerals, micas, and quartz, which are the constituents of igneous rocks. If the magma is particularly rich in silica, for example, it is likely that quartz will be present in the rock formed by solidification (crystallization) of the melt. In general, intrusive igneous rocks are richer in silica than are extrusive rocks. This observation is related to the fact that greater amounts of silica in a magma increase its viscosity ("thickness," or resistance to flow). The thick liquid cannot readily work its way to the Earth's surface.

As noted earlier, granite (Fig. 2-18*A*) is a silica-rich, relatively light-colored intrusive rock composed primarily of potassium feldspar, quartz, sodium plagioclase, hornblende, and mica. It is derived from magmas so rich in silica that after all chemical linkages with metallic atoms are satisfied, enough silicon and oxygen still remain to form quartz grains. These form an interlocking network with feldspars and micas to form the crystalline texture of granite. Granodiorite is another quartz-bearing igneous rock in which plagioclase is the dominant feldspar mineral. Quartz-bearing rocks such as granite and granodiorite are loosely termed granitic rocks.

Basalt (Fig. 2-19) is a fine-grained extrusive rock derived from a low-silica melt rich in iron and magnesium. Because the silica percentage is low, quartz grains are rarely seen in basalt. Basaltic lavas have low viscosity and are able to flow for considerable distances across the Earth's surface before solidifying. Because of the presence of ferromagnesian minerals and gray calcium plagioclase, low-silica rocks such as basalt tend to be black, dark gray, or green.

CRYSTALLIZATION OF MAGMA Not all minerals form simultaneously during the crystallization of a magma. For example, if the magma originally had a basaltic

FIGURE 2-15 Differences in the texture of igneous rocks. (*A*) A coarse-grained intrusive rock called gabbro. (*B*) A porphyry with large phenocrysts of potassium feldspar. It is used here in the ornamental posts outside St. Paul's Cathedral in London. (*C*) A fine-grained igneous rock known as andesite. ? *What is the origin of this type of porphyritic texture? On the basis of color alone, what feldspar mineral is represented by the large phenocrysts?*

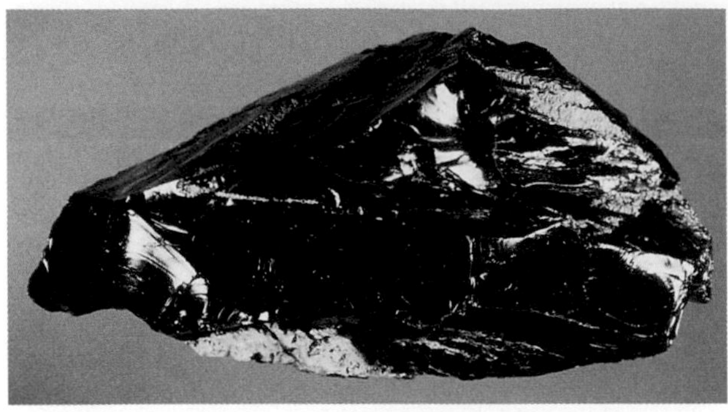

FIGURE 2-16 The volcanic glass obsidian. 🅰 *How does obsidian form?*

composition, olivine, pyroxene, and calcium feldspar will be among the earliest minerals to crystallize. Often these first-formed crystals are larger and more perfect than those that formed later because there was ample space for growth and an abundance of elements to incorporate in their crystal structure. The minerals that crystallize later must grow within the remaining spaces and thus tend to be smaller and less perfect. Also, minerals enclosed in other minerals must have formed before the enclosing mineral. When rocks are viewed microscopically, these observations permit one to determine the order in which specific minerals crystallized in a magma.

In the early 1900s an eminent geologist named N. L. Bowen studied the order in which minerals crystallize as a magma is slowly cooled. Bowen made artificial magmas by melting powdered rock samples in a

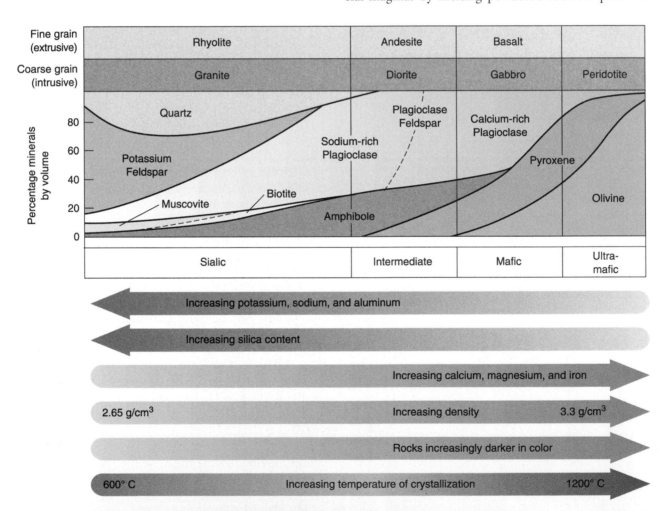

FIGURE 2-17 Mineral composition, texture, and other properties of common igneous rocks. The abundance of a particular mineral in an igneous rock can be estimated from the thickness of its colored area beneath the rock name. The scale on the left side can be used to estimate the percentage of each mineral. For example, granite at about the midpoint on the figure would be composed of about 26 percent quartz, 25 percent potassium feldspar, 31 percent sodium-rich plagioclase, 12 percent amphibole, 5 percent biotite, and 1 percent muscovite. 🅰 *What would be the composition of a midpoint basalt? What minerals might you expect to find in the porphyry shown in Figure 2-15B?*

(*A*)

(*B*)

FIGURE 2-18 **Granite and rhyolite.** (*A*) Granite, a coarse-grained intrusive rock that has the same composition as its intrusive, fine-grained equivalent, rhyolite (*B*). The lighter-colored grains in the granite are potassium feldspar and quartz, whereas the black minerals are hornblende and biotite. These minerals are also present in the rhyolite but cannot be seen with the unaided eye.

steel cylinder called a bomb. The bomb was heated until the powder melted and was then slowly cooled to the temperature and pressure selected for the experiment. That temperature and pressure was maintained for a period of a few months so that some minerals could crystallize within the melt. Thus, there would be a mixture of crystals and uncrystallized melt within the bomb. The bomb was then plunged into cold water or oil, causing the mixture to solidify. The melt solidified as glass, and within the glass the crystals were preserved. Bowen repeated the experiment many times, forming crystals at different temperatures before quick-cooling them. By identifying the crystals that formed at each temperature, he was able to determine the order of crystallization of minerals in a cooling magma. For a melt of basaltic composition, Bowen found that olivine would crystallize first and at the highest temperature. Pyroxene and calcium plagioclase

(*A*)

(*B*)

FIGURE 2-19 **Basalt.** (*A*) A hand specimen of basalt. (*B*) A thin section of basalt viewed with polarized light. Tabular crystals are plagioclase and brightly colored crystals are iron-magnesian silicates. Width of field of view is 2 mm.

would form next, followed by hornblende, biotite, and more sodium-rich feldspars. The order in which the minerals crystallized came to be called **Bowen's Reaction Series** (Fig. 2-20). It is called a *reaction series* because early formed crystals *react* with the melt to yield a new mineral further down in the series. For example, silica in the melt would react with olivine crystals by simultaneously dissolving and precipitating the dissolved components to form pyroxene, and pyroxene crystals would begin to grow in spaces between earlier formed crystals. At still lower temperatures, the melt would similarly react with pyroxene to form hornblende, and subsequently hornblende would react to form biotite. For the complete sequence to occur, however, the original magma must have sufficient magnesium, iron, and calcium. If the melt is rich in silica but deficient in these elements, the first-formed crystals might be biotite and sodic feldspars, followed by orthoclase, muscovite, and quartz. Thus, reaction is prevented if silica in the melt is exhausted in forming the early minerals. It is also prevented if minerals are removed from contact with the magma after they have crystallized.

On the right branch of Bowen's chart, one can trace the changes that occur as calcium-rich plagioclase reacts with the magma to produce varieties of plagioclase that are successively richer in sodium. Because the plagioclase minerals maintain the same basic crystal structure but change continuously in their content of calcium and sodium, the right side of the diagram is called the *continuous series*. The left side of the diagram depicts reactions that result in minerals of distinctly different structure. It is therefore called the *discontinuous series*. The three minerals at the base of the chart do not react with the melt. By the time they crystallize, little liquid remains, and the residues of silicon, aluminum, oxygen, and potassium join to form orthoclase, muscovite, and quartz.

One can see evidence of the reactions described above when igneous rocks are examined microscopically. One observes plagioclase crystals in which the innermost layers or zones are more calcium-rich and the outer zones more sodium-rich. Similarly, reaction rims of pyroxene can be observed around olivine, and rims of hornblende around pyroxene.

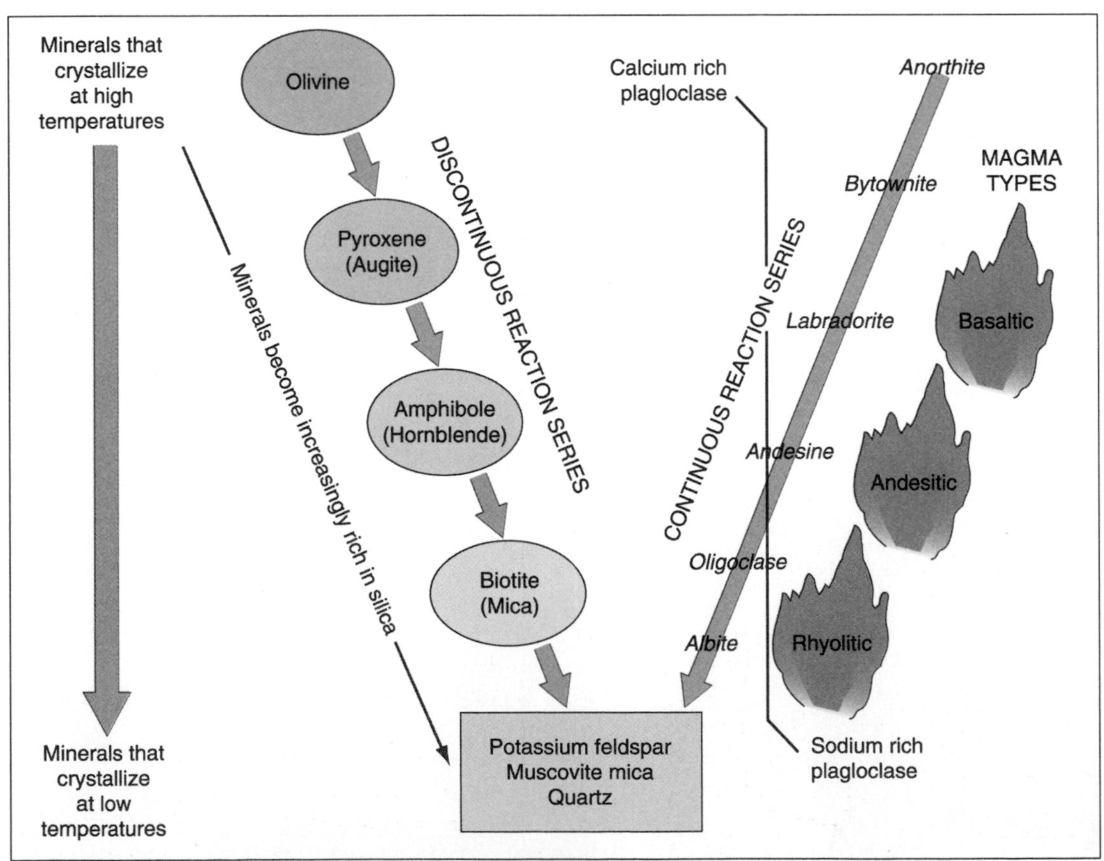

FIGURE 2-20 A depiction of Bowen's Reaction Series. Note that the earliest minerals to crystallize are olivine and calcium-rich plagioclase. As crystallization proceeds, each mineral reacts with the melt to form the mineral beneath it. ? *What minerals are likely to form when pyroxene crystals react with the remaining liquid magma? Are plagioclase crystals in a granite likely to be of the sodium-rich or calcium-rich variety?*

VOLCANIC ACTIVITY Although deciphering the history of a granitic mass is certainly intellectually stimulating, it is unlikely to evoke the feelings of awe and excitement one experiences when viewing volcanic activity. Volcanoes are basically vents in the Earth's surface through which hot gases and molten rock flow from the Earth's interior. The extrusions may be quiet or explosive. Quiet eruptions are exemplified by the Hawaiian volcanoes and are frequently characterized by truly enormous outpourings of low-viscosity lava (Fig. 2-21). The lava spreads widely to form the gentle slopes of a shield volcano. Explosive eruptions are caused by the sudden release of molten rock, driven upward by large pockets of compressed steam and gases. Explosive eruptions can literally blow the volcanic cone to bits. The explosive eruption of Krakatoa in 1883 was heard 5000 km away and was responsible for the death of 36,000 people.

Water is an important cause of explosive volcanism. It accounts for the violence of many Pacific rim volcanoes, such as Washington's Mount St. Helens. The water released by these volcanoes as steam and vapor is actually recycled sea water. At midocean ridges, sea water reacts with extruding basalts to form hydrous minerals. These water-bearing minerals are then conveyed by sea-floor spreading to subduction zones. As the crust descends into the subduction zone, water is released from the hydrous minerals. Pressures build in areas containing the trapped water until they become sufficient to cause explosive release.

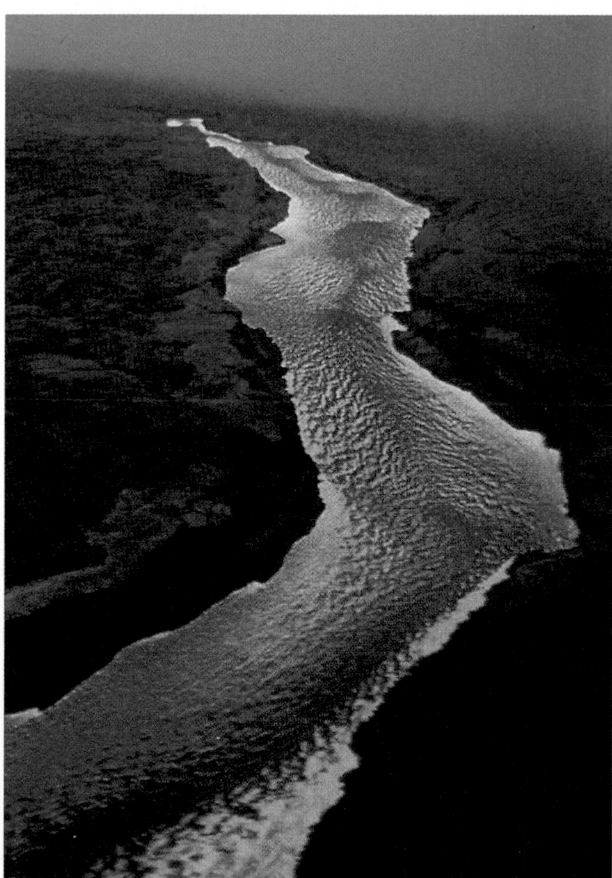

FIGURE 2-21 A river of lava formed during the 1984 dual eruption of the Mauna Loa and Kilauea volcanoes, Hawaii. (*Courtesy of I. Duncan.*)

There are, of course, all degrees of volcanic activity between quiet and violently explosive. Perhaps in response to changes in the composition of the melt, some volcanoes have even been known to shift from one type of activity to another. Volcanic activity also includes successive outpourings of lava from great fissures to form lava plateaus that extend over thousands of square kilometers. Such fiery floods produced the Columbia and Snake River Plateau as well as the Deccan Plateau of India.

By far, the most abundant kind of volcanic rock is basalt. It underlies the ocean basins, has been built into midoceanic ridges, and has accumulated sufficiently in such places as Hawaii and Iceland to have produced substantial land area.

How and at what depth did this great volume of basalt originate? To answer this question, it is necessary to refer briefly to a model of the Earth's interior that has been formulated by the study of earthquake waves. The model depicts the Earth's basaltic crust as a thin zone averaging about 6 kilometers thick and overlying the mantle of denser olivine- and pyroxene-rich rocks. The boundary between the crust and the mantle is recognized by an abrupt change in the velocity of earthquake waves as they travel downward into the Earth. For many years geologists believed that basaltic lavas originated from the lower part of the basaltic crust. However, several recent lines of evidence suggest that the basaltic lavas may have come from molten pockets of upper mantle material. For example, present-day volcanic activity is closely associated with deep earthquakes that occur within the mantle far beneath the crust. It is likely that fractures produced by these earthquakes could serve as passages for the escape of molten material to the surface. A detailed study of earthquake shocks from particular volcanic eruptions in Hawaii indicates that the erupting lavas were derived from pockets of molten material within the upper mantle at depths of about 100 kilometers. A weak plastic zone (called the "low-velocity zone") in the mantle appears to represent the level at which the lavas originated. The mechanism by which they developed is called partial melting. **Partial melting** (Fig. 2-22) is that general process by which a rock subjected to high temperature and pressure is partly melted and the liquid component is moved to another location. At the new location, the separated liquid may solidify into rocks that have a different composition from the parent mass. The word "partial" in the expression partial melting refers to the fact that some minerals melt at

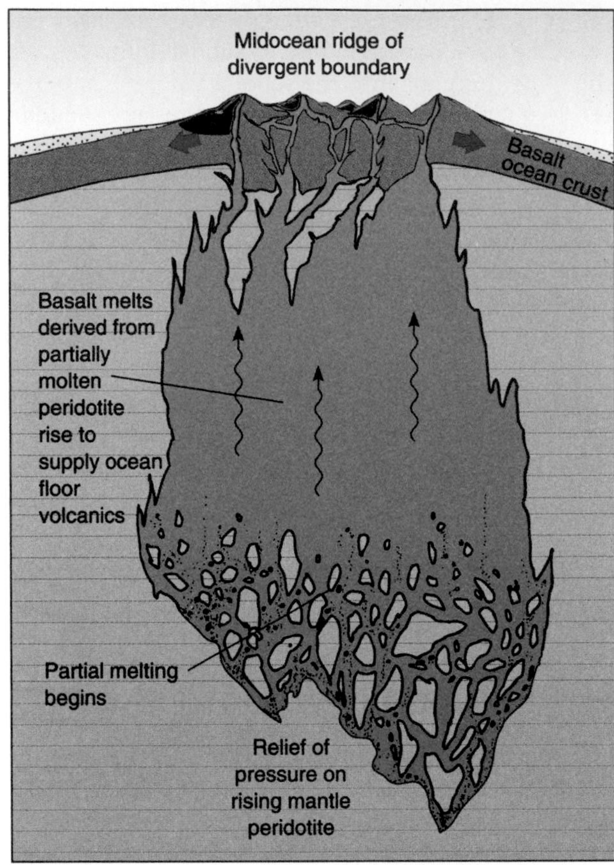

FIGURE 2-22 **A conceptual illustration of the process of partial melting of peridotite to form basalt.** With relief of pressure at divergent plate boundaries, rocks of the upper mantle begin to melt and release liquid that on solidification would form basalt. Basaltic lava extruded along divergent zones of tectonic plates is then moved laterally in the course of sea-floor spreading.

lower temperatures than do others, and so for a time the material being melted resembles a hot slush composed of liquid and still-solid crystals. The molten fraction is usually less dense than the solids from which it was derived and thus tends to separate from the parent mass and work its way toward the surface. In this way, melts of basaltic composition separated from denser rocks of the upper mantle and eventually made their way to the surface to form volcanoes.

Many complex and interrelated factors control where in the mantle partial melting may occur and even whether it will occur at all. Generally, heat in excess of 1500°C is required, but the precise temperature for melting is also influenced by pressure and the water content of the rock. As pressure increases, the temperature at which particular minerals melt also rises. Thus, a rock that would melt at 1000°C near the surface will not melt in deeper zones of higher pressure until it reaches far greater temperatures. Water has an effect opposite that of pressure, for its presence will allow a rock to start melting at lower temperatures and shallower depths than it would have otherwise. Laboratory experiments indicate that the melting of "dry" mantle rock can occur at depths of about 350 kilometers but that the presence of only a little water can cause partial melting and yield basaltic liquid from depths as shallow as 100 kilometers.

Not all lavas found at the Earth's surface are basaltic. Volcanoes of the more explosive type that are located at the edge of continents around the Pacific and in the Mediterranean extrude a lava called andesite. Andesite contains more silica than basalt does, and its lava is more viscous. This greater resistance to flow contributes to the gas containment that precedes explosive volcanic activity. Andesites are intermediate in silica content between the rocks of the continental crust and those of the oceanic crust of the Earth.

Andesitic rocks may originate in more than one way. Some emplacements result from originally basaltic magmas in which minerals such as olivine and pyroxene form early and settle out, thus leaving the remaining melt relatively richer in silica. This process is called **fractional crystallization** (Fig. 2-23). Other andesites may result from the melting of basaltic ocean crust and siliceous marine sediments as they descend into hot zones of the mantle. The water-rich and silica-enriched melts of andesitic composition are able to rise buoyantly and erupt along volcanic island arcs. We will examine this interesting idea more fully in Chapter 5, which deals with plate tectonics.

Sedimentary Rocks

Sedimentary rocks are rocks composed of consolidated sediment—particles and chemical compounds that are the product of weathering and erosion of any previously

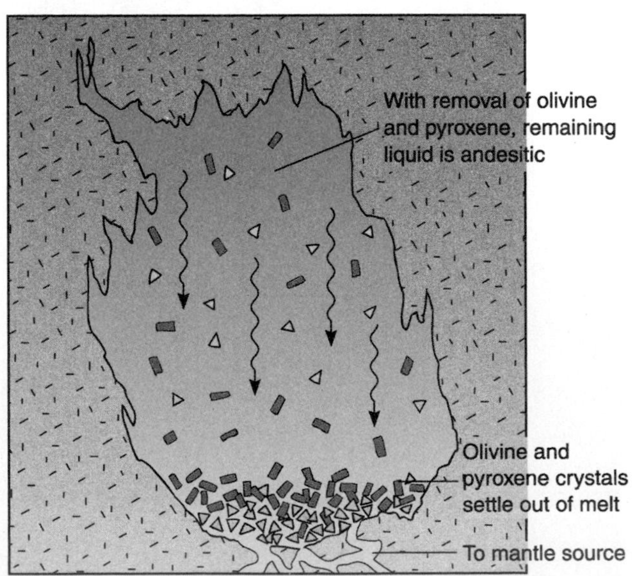

FIGURE 2-23 **An andesitic melt resulting from fractional crystallization of a basaltic magma.**

existing rock or soil. Components of sediments may range from large boulders to the molecules dissolved in water. Sediment is deposited through such agents as wind, water, ice, or mineral-secreting organisms. The loose sediment is converted to coherent solid rock by any of several processes: precipitation of a cementing material around individual grains, compaction, or crystallization. These processes constitute **lithification**.

The most obvious feature of sedimentary rocks is their occurrence in beds or layers called **strata**. Stratification is commonly the result of changes in the conditions of deposition that cause materials of a somewhat different nature to be deposited for a period of time. For example, the velocity of a stream might decrease, causing particles to settle out that might otherwise have stayed in suspension. In another situation, the kind of materials brought into a given depositional site by streams might change, and there would then be a corresponding change in composition of the accumulating layers.

Shale, sandstone, and carbonate rocks (such as limestone) constitute the most abundant sedimentary rocks. **Sandstones** are composed of grains of quartz, feldspar, and other particles that are cemented or otherwise consolidated. **Shale** consists largely of very fine particles of quartz and abundant clay. Shale is the most abundant of sedimentary rocks. The carbonates are rocks formed when carbon dioxide in water combines with oxides of calcium and magnesium.

DERIVATION OF SEDIMENTARY MATERIALS Sedimentary rocks must have originally come from the disintegration and decomposition of older rocks. Commonly, the older rocks are igneous; indeed, these were once the only rocks on Earth. It is therefore instructive to review the manner in which the common components of sedimentary rocks might be derived from an abundant kind of igneous rock, such as granodiorite (Fig. 2-24).

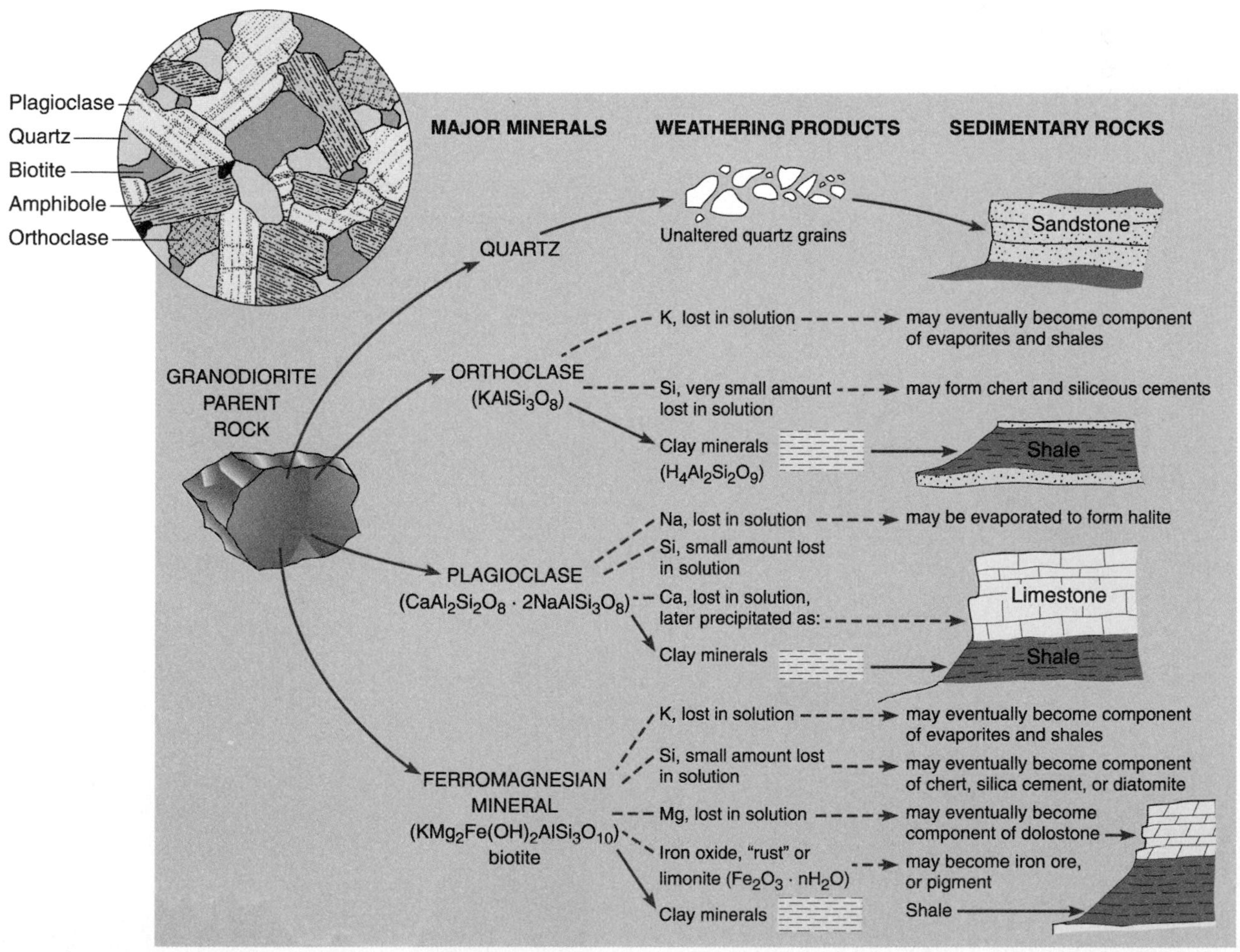

FIGURE 2-24 A conceptual diagram showing how the weathering of granitic rock yields quartz grains for quartz sandstone, clay for shale, and calcium for limestone. If weathering is not too severe, detrital grains of feldspar will also be included in sands and sandstones. For simplicity, minor mineral components are not included. ? *What is the most abundant, relatively insoluble product of the weathering of orthoclase and plagioclase? Which mineral is most stable (least likely to experience dissolution during chemical weathering)?*

Consider first the quartz in the granodiorite. Quartz will persist almost unchanged during weathering. It is one of the most chemically stable of all the common silicate materials. As the parent rock is gradually decomposed, quartz grains tend to be washed out and carried away to be deposited as sand that will one day become sandstone (Fig. 2-25).

The feldspars decay more readily than quartz. They are primarily aluminum silicates of potassium, sodium, and calcium. In the weathering process, the last three elements are largely dissolved and carried away by solutions (although some may remain in soils within clay minerals). Ultimately, they reach the sea, where they may stay in solution, or they may be precipitated as layers of sediment. Under suitable conditions, loss of CO_2 from ocean water can result in the precipitation of calcite. If large quantities of lake or sea water are evaporated, evaporites such as halite or gypsum may be formed. Of course, not all the feldspars and micas in the granodiorite necessarily decay. Some may persist as detrital grains that become incorporated into sandstones and other sediments.

The most voluminous product of the weathering of feldspars is clay. As described earlier, potassium, sodium, and calcium are largely dissolved during weathering. Aluminum and silicon are left behind to form the chief ingredients of clay, which then finds its way into the making of shales and claystones.

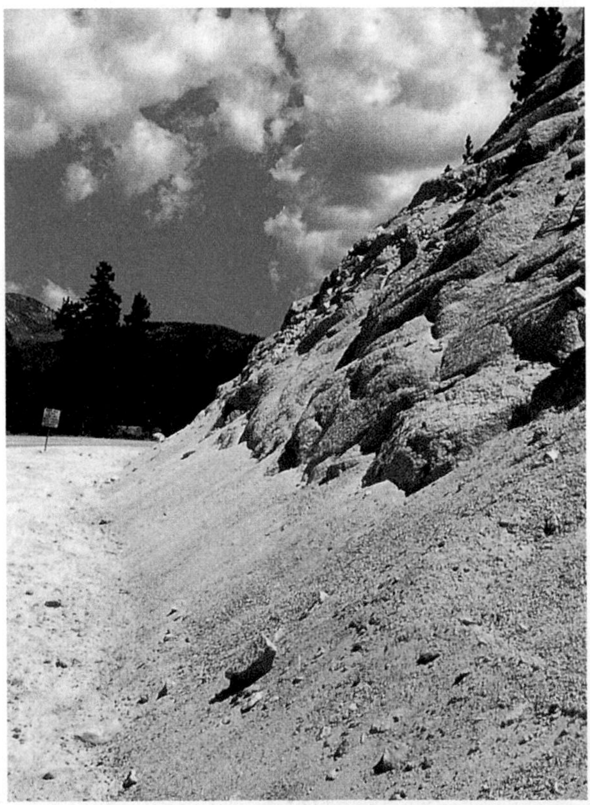

FIGURE 2-25 Exposure of weathered granodiorite, showing an accumulation of quartz grains and partially decomposed feldspar resulting from the disintegration of the parent rock. ? *In addition to the quartz and feldspar, what other weathering product is present in the debris beneath the granodiorite?*

Biotite is another mineral in the granodiorite source rock. Decomposition of biotite, which is a potassium, magnesium, and iron aluminosilicate, yields soluble potassium and magnesium carbonates, small amounts of soluble silica, iron oxides, and clay. The iron oxides serve to color many sedimentary rocks in shades of brown and red.

VARIETY AMONG SEDIMENTARY ROCKS Sedimentary rocks are classified according to their composition and texture. The term texture refers to the size and shape of the individual grains and to their arrangement in the rock. A rock that has a clastic texture is composed of grains and broken fragments (clasts) of pre-existing minerals, rocks, and fossils. Other sedimentary rocks are composed of a network of intergrown crystals and therefore have a crystalline texture.

Clastic Rocks. The fragments of pre-existing materials that compose clastic sedimentary rocks range in size from huge boulders to microscopic particles. Particle size is particularly useful in classifying these rocks, which include conglomerates, sandstones, siltstones, and shales. **Conglomerate** (Fig. 2-26) is composed of water-worn, rounded particles larger than 2 mm in diameter. The term **breccia** is reserved for clastic rocks composed of fragments that are angular (Fig. 2-27) but similar in size to those of conglomerates. In **sandstones**, grains range between 1/16 mm and 2 mm in diameter (Table 3-1). The varieties of sandstone are then subdivided mainly according to composition. **Siltstones** are finer than sandstones (1/16 mm to 1/256 mm). **Shales** may contain abundant clay minerals, which are flaky minerals that align

FIGURE 2-26 Conglomerate. The pebbles in this particular sample are chert in a carbonate matrix. The largest pebbles are about 2.5 cm in diameter. ? *How do conglomerates differ from breccias?*

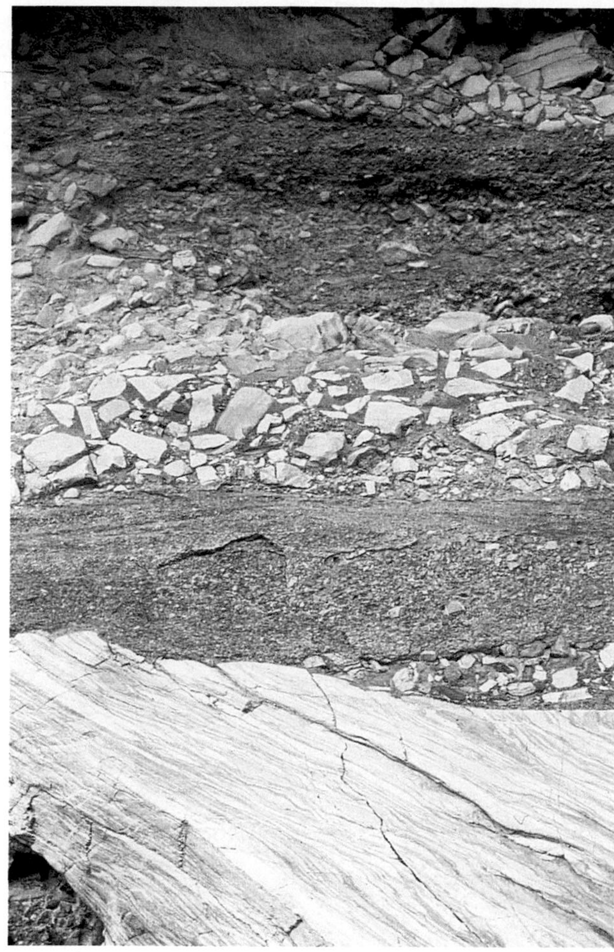

FIGURE 2-27 A breccia (bed across center) composed of angular fragments or clasts of Precambrian metamorphic rock (quartzite). Mosaic Canyon, Death Valley, California. (*Photograph by R.F. Dymek.*)

parallel to bedding planes. As a result, shales characteristically split into thin slabs parallel to bedding planes. This property is termed **fissility**. Rocks lacking fissility but composed of clay-sized particles are called claystones or mudstones.

Most clastic sedimentary rocks are composed of particles of minerals like quartz, feldspar, and mica. Because these are all silicate minerals, such rocks are termed **siliciclastics**. Sandstones, shales, and most conglomerates are siliciclastics. Approximately 75 percent of all sedimentary rocks are siliciclastics.

Carbonate Rocks. The minerals that compose carbonate rocks are primarily calcite, aragonite, and dolomite. Calcite ($CaCO_3$) and aragonite ($CaCO_3$) have the same chemical composition but differ in crystal form. (Aragonite is relatively unstable and is usually converted to calcite.) Calcite is the predominant mineral in limestone. Dolomite—$CaMg(CO_3)_2$—predominates in the rock dolomite. (Here both mineral and rock bear the same name.) Carbonate rocks usually also contain variable amounts of impurities, including iron oxide, clay, and particles of sand and silt swept into the depositional environment by currents.

Limestones The most abundant limestones are of marine origin and have formed as a result of precipitation of calcite or aragonite by organisms and the incorporation of skeletons of those organisms into sedimentary deposits. Inorganic precipitation of carbonate minerals may also form deposits of limestone. The importance of this process is questionable, however, because the precipitation is nearly always closely associated with photosynthetic and respiratory activities of organisms or with the release of tiny particles of aragonite upon the decay of green algae. Strictly speaking, it appears that very few marine limestones are the result of direct chemical precipitation.

After the calcium carbonate has accumulated, it becomes recrystallized or otherwise consolidated into indurated rock that may be variously colored-from white, through tints of brown, to gray. Limestones tend to be well stratified (Fig. 2-28), frequently contain nodules and inclusions of chert, and are often highly fossiliferous (containing fossils). The rock may range in texture from coarsely granular to very fine-grained and aphanitic.

FIGURE 2-28 Roadcuts exposing limestone strata are familiar to travelers on interstate highways throughout the Mississippi Valley region. In this roadcut near House Springs, Missouri, beds dip about 35°. The beds are the limb of a monocline (a fold with only one limb).

ENRICHMENT

Glass from Sand

Glass, a product formed by melting and cooling quartz sand and minor amounts of certain other common materials, is so familiar to us that we sometimes forget its importance. In fact, it would be difficult to imagine a world without glass. There would be no goblets, bottles, mirrors, or windows; no glass laboratory retorts, tubes, or beakers; no electric light bulbs or neon signs; no glass lenses for cameras, microscopes, telescopes, or motion picture projectors; and no fiber optics for use in viewing the inside of the human body. We see glass used everywhere in our buildings, electronic equipment, and medical devices. We not only see it, we see through it.

Glass has a long and colorful history. Natural glass, such as obsidian, was used by prehistoric humans to make arrowheads and knife blades. Human-made glass was manufactured by Egyptians as long ago as 3000 B.C. At Ur in Mesopotamia, glass beads nearly 4500 years old have been found in archaeologic excavations. The Greeks and Romans learned the art of making glass ornaments, bottles, vases, and trinkets that were prized throughout the ancient world. During the Middle Ages, Europeans produced beautiful stained-glass windows for cathedrals. Blood-red stained glass was made by adding copper compounds to the molten silica. The Europeans knew that cobalt gave the glass a rich, deep-blue color; manganese turned it purple; and antimony provided golden yellows. Iron oxides were added to color glass green or brown.

Today, the sands used in making glass have a silica content as high as 95.0 to 99.8 percent. Sources of such exceptionally pure sands are the Ordovician St. Peter Sandstone of Missouri and Illinois and the sandstone members of the Devonian Oriskany Formation of West Virginia and Pennsylvania. These clean, pure sandstones were deposited in shallow marine environments where wave action could remove clay impurities and concentrate grains of sand-sized quartz. Both sandstones are composed of grains that have been eroded from older sandstones, and this reworking has contributed to their extraordinary purity.

In the manufacture of window glass, the mixture prepared for melting consists of about 72 percent silica from quartz grains. Certain metallic oxides, such as soda (Na_2O) and lime (CaO), serve to lower the temperature required for melting. That temperature for soda-lime-silica glass is 600 to 700°C. Once molten, the batch (as it is called) is cooled, poured, rolled, or blown into the shapes desired. Compounds of boron and aluminum can be added to provide heat-resistant glass, as in Pyrex® cookware. Fine cut glass and so-called crystal are a silica-lead-soda glass known not only for its brilliance but for its musical tone as well. By varying the kinds and amounts of metallic oxides, glass can be produced for a great variety of special uses and products. There is little doubt that it will always be an essential part of life in this age of technology.

In general, limestones consist of one of a combination of textural components such as micrite, carbonate clasts, oölites, or carbonate spar. Micrite is exceptionally fine-grained carbonate mud. Carbonate clasts are sand- or gravel-sized pieces of carbonate. The most common clasts are either bioclasts (skeletal fragments of marine invertebrates) or oöids (Fig. 2-29), which are spherical grains formed by the precipitation of carbonate around a nucleus. Sparry carbonate is a clear, crystalline carbonate that is normally deposited between the clasts as a cement or has developed by replacement of calcite. These textural categories of limestone as seen through the microscope are shown in Figure 2-30. They permit classification of particular samples as micritic limestone, clastic limestone, oölitic limestone, or sparry (crystalline) limestone.

FIGURE 2-29 Oöids in an oölitic limestone. (*Copyright William E. Ferguson.*)

There are many varieties of limestone. Chalk is a soft, porous variety that is composed largely of calcareous, microscopic skeletal remains of marine planktonic (floating) animals and plants. Lithographic limestone is a dense, micritic limestone once widely used as an etching surface in printing illustrations. Some limestones consist almost entirely of the skeletal remains of reef corals and other frequently skeletonized marine invertebrates.

Dolomite As we have already noted, dolomite is a rock composed of the calcium-magnesium mineral of the same name. Like limestone, it occurs in extensive stratified sequences, and it is not easily distinguishable from limestone. The usual field test for distinguishing dolomite from limestone is to apply cold, dilute hydrochloric acid. Unlike limestone, which

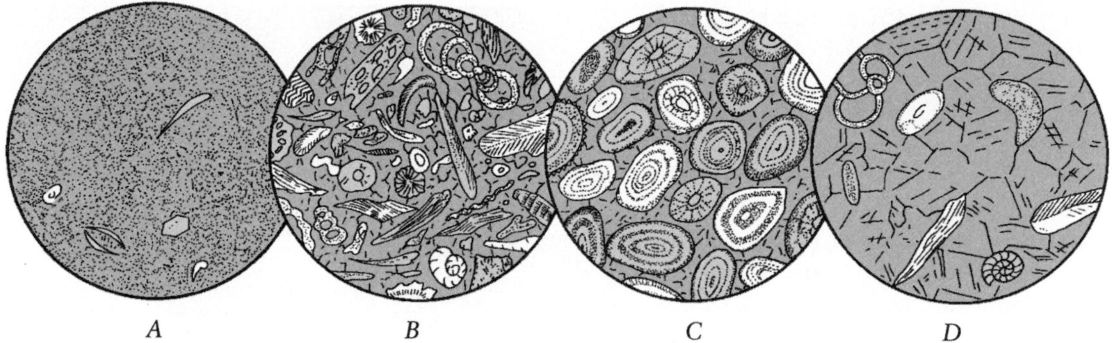

FIGURE 2-30 Textures of limestones as seen in thin section under a microscope. (*A*) Aphanitic limestone or micrite. (*B*) Bioclastic limestone with fine-grained sparry calcite as cement. (*C*) Oölitic limestone. (*D*) Sparry or crystalline limestone.

bubbles readily, dolomite effervesces only slightly, if at all.

The origin of dolomites is somewhat problematic. The mineral dolomite is not secreted by organisms during shell building. Direct precipitation from sea water does not normally occur today, except in a few environments where the sediment is steeped in abnormally saline water. Such an origin is not considered adequate to explain the thick sequences of dolomitic rock commonly found in the geologic record. The most widely believed theory for the origin of dolomites is that they result from partial replacement of calcium by magnesium in the original calcareous sediment. However, it is not known for how long or at what time in the history of the rock this dolomitization occurs.

Other Sedimentary Rocks

Chert We have previously mentioned a form of microcrystalline quartz called chert (SiO_2), noting its occurrence as nodules in limestones (Fig. 2-31). The origin of these nodules is still being debated among petrologists, although the majority believe that they form as replacements of carbonate sediment by silica dissolved in sea water trapped in the sediment. Some cherts occur in areally extensive layers and thus qualify as monomineralic rocks. These so-called bedded cherts are thought to have formed from the accumulation of the siliceous remains of diatoms and radiolaria and from subsequent reorganization of the silica into a microcrystalline quartz. Silica from the dissolution of volcanic ash is believed to enhance the process; indeed, many bedded cherts are found in association with ash beds and submarine lava flows.

Evaporites As indicated by their name, evaporites are chemically precipitated rocks that are formed as a result of evaporation of saline water bodies. Only about three percent of all sedimentary rocks consist of evaporites. Evaporite sequences of strata are composed chiefly of such minerals as gypsum, anhydrite, halite, and associated calcite and dolomite. Gypsum is a hydrous calcium sulfate, whereas anhydrite lacks water. Halite has the same composition as ordinary table salt. The conditions required for precipitation of thick sequences of marine evaporites include warm, relatively arid conditions and periodic addition of sea water to the evaporating marine basin. Evaporites,

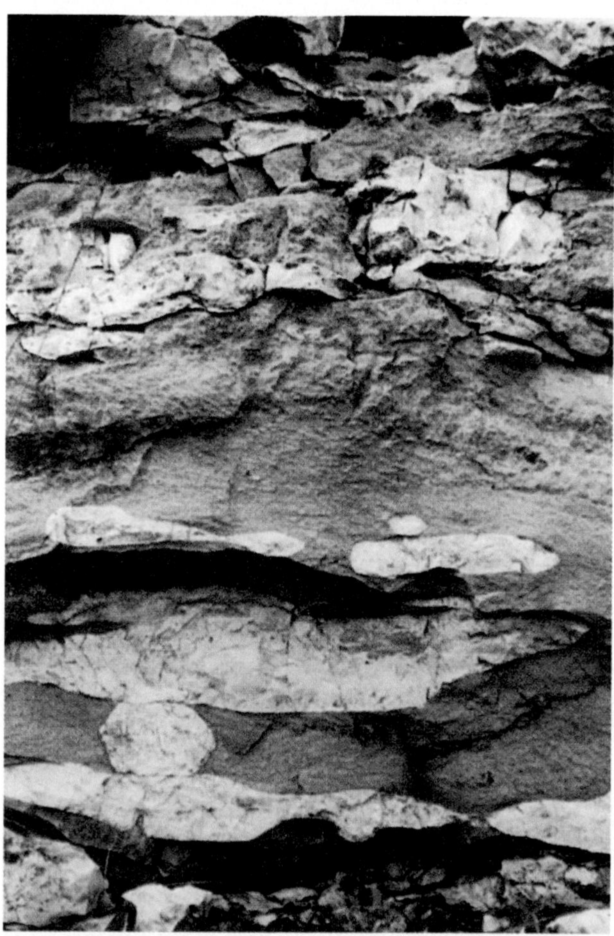

FIGURE 2-31 White nodular chert in tan limestone. Fern Glen Formation, west of St. Louis, Missouri.

however, are also precipitated from inland lakes in arid regions such as the Basin and Range Province of the United States.

Coal Coal is a black or dark-brown combustible rock formed during the biochemical and physical alteration of plant material that has accumulated in a swamp or marshlike environment. For coal to form, the plant matter must accumulate under water or be quickly buried so that it does not have access to air. If this were not the case, the carbon would be lost as oxygen combined with carbon from the plant matter to form carbon dioxide (and water).

Metamorphic Rocks

Sir Charles Lyell recognized that igneous or sedimentary rocks, if subjected to high temperature, pressure, and the chemical action of solutions and gases, can be altered to quite different kinds of rocks. Lyell employed the term *metamorphism* (from the Latin metamorphosis, meaning "change of form") to describe this process. It is still used today to describe alterations in rocks brought about by physical or chemical changes in the environment that are intermediate between those that result in igneous rocks and those that produce sedimentary rocks. Any previously existing rock may be converted to a metamorphic rock, and the changes primarily involve recrystallization of minerals in the rock while it remains in the solid state. In the process of recrystallization, the textural characteristics of the parent rock may be changed, while at the same time new minerals develop that are stable under the new conditions of pressure and temperature. New elements need not be introduced; instead, those that are already present are incorporated into different and often denser minerals. Variations in heat and pressure may result in different kinds of metamorphic rocks, even from the same parent material.

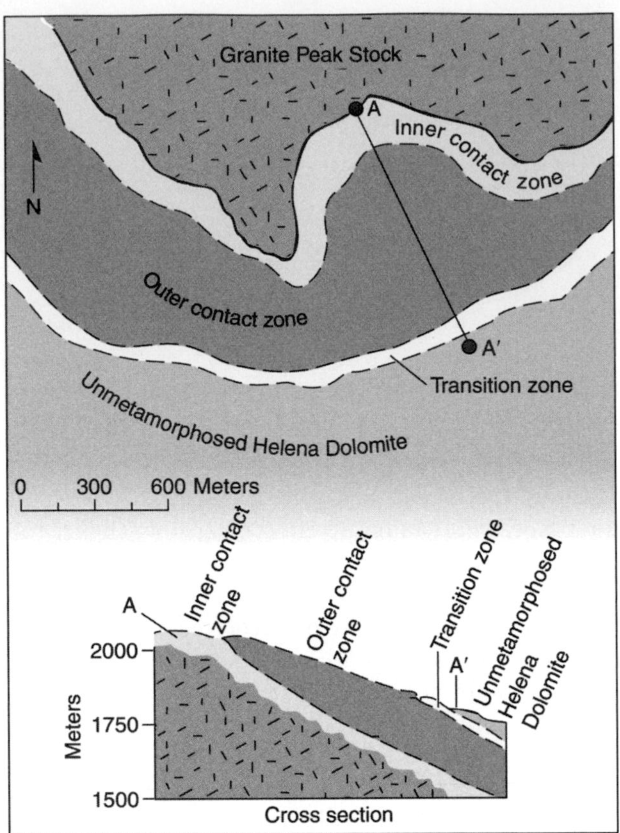

FIGURE 2-32 The Granite Peak aureole, located 12 miles east of Lincoln, Montana. An aureole is the zone of contact metamorphism surrounding an igneous intrusion. This metamorphic aureole is developed around a granitic stock that has intruded into dolomite country rock. The intensity of metamorphism diminished outward from the granite margin, and particular assemblages of metamorphic minerals occur within each contact metamorphic zone. (*Simplified from Melson, W. G. 1966. Am. Mineralogist 51:404, Fig. 1.*).

METAMORPHISM Alterations of rock immediately adjacent to igneous intrusions constitute **contact metamorphism** (Fig. 2-32). The changes that occur in the intruded rock are largely the result of high temperatures and the emanation of chemically active fluids that accompany igneous intrusions. Such factors as the size of the magmatic body, its composition and fluidity, and the nature of the intruded rock also influence the kind and degree of contact metamorphism. Important ore deposits are commonly situated in metamorphosed rock surrounding intrusives. Examples of such deposits include magnetite and copper ores in metamorphic zones around granite intrusives in the Urals, central Asia, the Appalachian Mountains, Utah, and New Mexico.

Regional or **dynamothermal metamorphism** is a type of rock alteration that is areally extensive and occurs under the conditions of great confining pressures and heat accompanying deep burial and mountain building. In a subsequent chapter, we will discuss how rocks deposited in crustal troughs adjacent to continents may be compressed into mountain systems and thus be regionally metamorphosed. Metamorphic index minerals known to form under specific temperature and pressure conditions are used to decipher the history of growth of these ancient mountainous regions, even when only the roots of the ranges remain. Figure 2-33 shows the temperatures at which certain metamorphic index minerals form in rock that is being subjected to heat and pressure. Note that the mineral chlorite forms at temperatures of 50°C to 300°C. The metamorphic index mineral garnet forms at higher temperatures and pressures, and sillimanite indicates the highest level of temperature and pressure.

KINDS OF METAMORPHIC ROCKS Because any rock can be metamorphosed in a number of different ways,

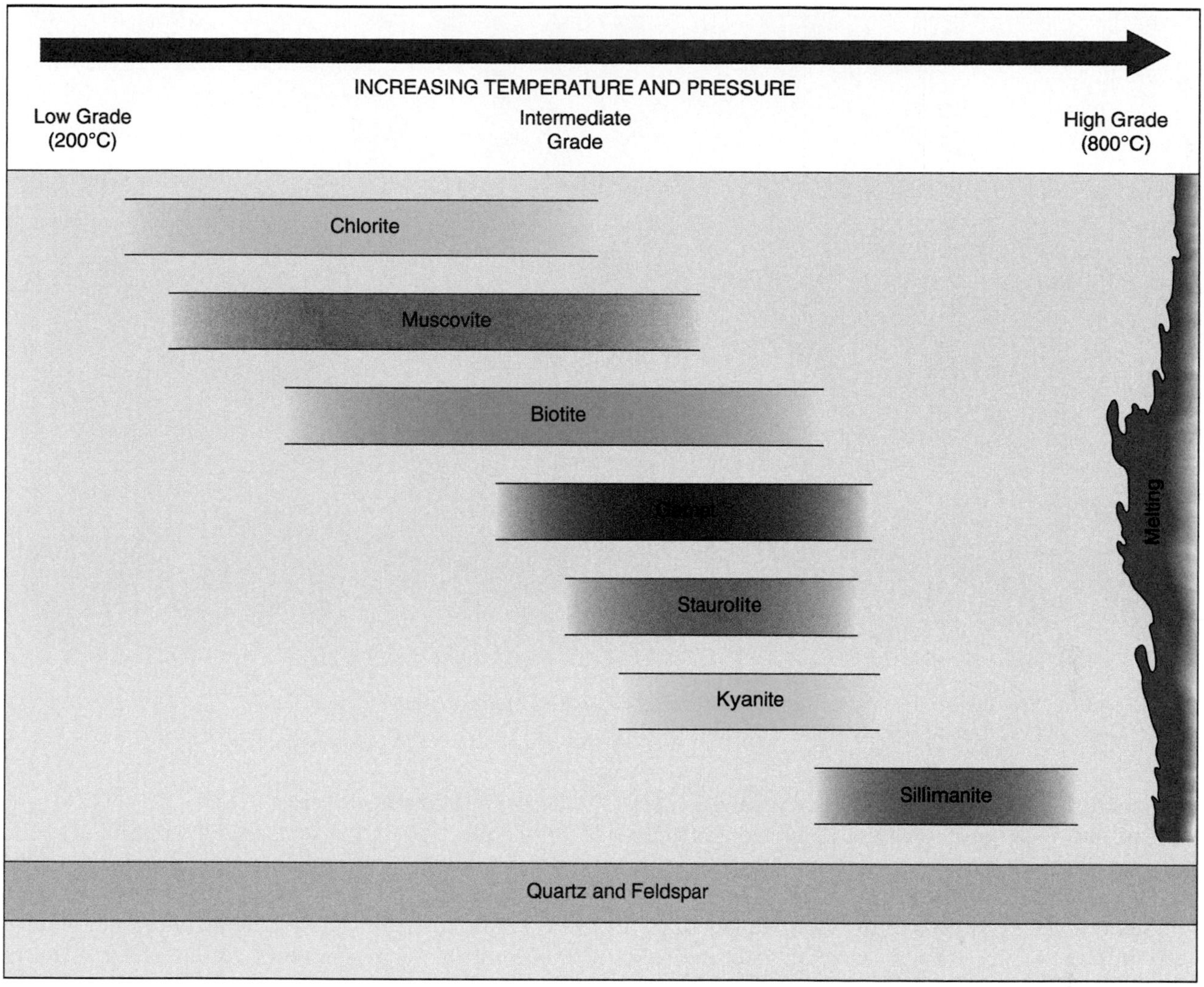

INCREASING TEMPERATURE AND PRESSURE
Low Grade (200°C)
Intermediate Grade
High Grade (800°C)
Chlorite
Muscovite
Biotite
Staurolite
Kyanite
Sillimanite
Melting
Quartz and Feldspar

FIGURE 2-33 Changes in minerals that will develop during the progressive metamorphism of shale. As the parent rock is subjected to low-grade metamorphism, chlorite and muscovite develop. The shale is metamorphosed to slate and then phyllite. With increasing temperature and pressure, muscovite will be joined by biotite, garnet, and staurolite. As temperatures approach the level of high-grade metamorphism, kyanite and sillimanite appear. Beyond 800A7C, the rock may be melted.

there are hundreds of different kinds of metamorphic rocks. However, we need only consider the most abundant kinds. It is convenient to divide metamorphic rocks into two groups based on the presence or absence of foliation. **Foliation** is laminated structure in a metamorphic rock resulting from a parallel alignment of platy or elongate mineral grains.

Foliated Metamorphic Rocks

Slate In slate, the foliation is microscopic and caused by the parallel alignment of minute flakes of silicates such as mica. The planes of foliation are quite smooth, and the rock may be split along these planes of "slaty cleavage." The planes of foliation may lie at any angle to the bedding in the parent rock, and this characteristic helps to differentiate slate from dense shale. Slate is derived from the regional metamorphism of shale. As heat and pressure are applied to the clay minerals in shale, they are converted to chlorite and mica, two platy minerals that contribute to the rock's fissility.

Phyllite The texture in phyllite is also very fine, although some grains of mica, chlorite, garnet, or quartz may be visible. Phyllite surfaces often develop a wrinkled aspect and are more lustrous than slate. Phyllite represents an intermediate degree of metamorphism between slate and schist. The parent rocks are commonly shale or slate.

Schist The platy or needlelike minerals in schist have grown sufficiently large to be visible to the unaided eye; the minerals tend to be segregated into distinct layers. Schists are named according to the most conspicuous mineral present. Thus, there are mica schists (Fig. 2-34), amphibole schists, chlorite schists,

FIGURE 2-34 **Steeply inclined mica schists splitting apart along foliation planes.** Great Smoky Mountain National Park, North Carolina-Tennessee.

and many others. Shales are the usual parent rocks for schists, although some schists are derived from fine-grained volcanic rocks.

Gneiss This is a coarse-grained (Fig. 2-35), evenly granular rock. Foliation results from segregation of minerals into bands rich in quartz, feldspar, biotite, or amphibole (Fig. 2-36). Foliation is coarse and appears less distinct than in schist. High-silica igneous rocks and sandstones are the usual parent rocks for gneisses.

FIGURE 2-35 **Coarse (gneissic) foliation developed in a quartz-feldspar-biotite gneiss.**

Nonfoliated Metamorphic Rocks

Marble A fine to coarsely crystalline rock, marble (Fig. 2-37) is composed of calcite or dolomite and therefore is relatively soft. (It can be scratched with steel.) Marble is derived from limestone or dolomite.

Quartzite A fine-grained, often sugary-textured rock, quartzite (Fig. 2-38) is composed of intergrown quartz and therefore is very hard. The rock will break through, rather than around, constituent grains, and may be any color. Quartzite is derived from quartz sandstone.

Greenstone A dark-green rock, greenstone has a texture so fine that mineral components, except for scattered larger crystals, cannot be seen without magnification. It is derived by the low-grade metamorphism of low-silica volcanic rocks such as basalt.

Hornfels This very hard, fine-grained rock is often studded with small crystals of mica and garnet that have no preferred orientation. Hornfels may form from shale or other fine-grained rocks that are intensely heated at their contact with intrusive igneous bodies.

THE HISTORICAL SIGNIFICANCE OF METAMORPHIC ROCKS We have noted that the conditions for metamorphism are developed in regions that have been subjected to intense compressional deformation. Such regions of the Earth's crust either now have, or once had, great mountain ranges. Thus, where large tracts of low-lying metamorphic terrane are exposed at the Earth's surface, geologists conclude that crustal uplift and long periods of erosion leveled the mountains.

FIGURE 2-36 An exposure of the Monson Gneiss of New Hampshire exhibiting plastic deformation resulting from intense pressure and heat. The dark bands are composed of hornblende, biotite, and plagioclase, whereas the lighter bands are rich in quartz and white feldspar. (*Courtesy of Robert Tucker.*)

Metamorphic rock exposures at many localities across the eastern half of Canada represent the truncated foundations of ancient mountain systems.

From studies of the mineralogic composition of metamorphic rocks, it is often possible for geologists to reconstruct the conditions under which the rocks were altered and then to make inferences about the directions of compressional forces, pressures, temperatures, and the nature of parent rocks. Investigators are aided in these studies by the knowledge that specific metamorphic minerals form and are stable within finite limits of temperature and pressure. Maps of metamorphic facies, or zones of rocks that formed under specific conditions, can be constructed. Commonly, such maps delineate broad bands of metamorphic rocks, each of which formed under sequentially more intense conditions of pressure and temperature.

Imagine a terrane that was once underlain by a thick sequence of calcareous shales, was subjected to compression to produce mountains, and then experienced loss of those mountains by erosion. One might then begin a traverse across the eroded surface on unmetamorphosed shales that were not involved in the mountain building. These shales would contain only unaltered sedimentary minerals. Progressing farther toward the area of most intense metamorphism, one might see that the shales had given way to slates bearing the green metamorphic mineral chlorite. Still farther along the traverse, schists containing intermediate-grade metamorphic minerals would appear. Finally, one might come upon coarsely foliated schists containing minerals that form only under high temperature and pressure. Although many characteristics of metamorphic rocks reveal the conditions under which they formed, they may also contain clues as to conditions on the Earth prior to metamorphism. For example, the discovery of a marble formed a billion years ago by metamorphism of a limestone indicates that the composition of the Earth's atmosphere and ocean 1 billion years ago was similar to that in which carbonates can be formed today.

FIGURE 2-37 Marble. On close examination, one can discern the lustrous cleavage surfaces of the calcite crystals. Width of specimen is about 9.0 cm

FIGURE 2-38 Hand specimen of an attractive green quartzite. ? *What simple physical test would convince you that this specimen was quartzite and not a green marble?*

SUMMARY

Rocks are the materials of which the Earth is composed. Rocks are themselves composed of minerals, and minerals, in turn, are constructed of chemical elements. The most common elements in rocks are oxygen, silicon, aluminum, iron, calcium, sodium, potassium, and magnesium. Rocks and minerals are the materials from which geologists make interpretations of ancient environments and past geologic happenings. Silicate minerals are the most important of the rock-forming minerals. Quartz, mica, feldspar, hornblende, augite, and olivine are silicates that were initially crystallized from molten rock. Clay, calcite, dolomite, gypsum, halite, and certain varieties of quartz (such as chert and flint) are formed by processes of weathering and precipitation or deposition at temperatures that prevail at the Earth's surface.

Rocks are aggregates of minerals. Igneous rocks are those that have cooled and solidified from a molten silicate body. If solidified beneath the Earth's surface, the igneous rocks are termed intrusive. Granite is an example. Igneous rocks, such as those formed from the lavas flowing from volcanoes, are termed extrusive. Basalt is an example. The texture of igneous rocks provides an index to their cooling history, in that coarse-grained varieties cooled slowly (and are mainly intrusive), whereas fine-grained varieties cooled rapidly (and are mainly extrusive).

Melting and crystallization of rock materials in the laboratory have provided an understanding of the causes of igneous rock diversity. Among the mechanisms that provide for a variety of rock types are fractional crystallization, as seen in Bowen's Reaction Series, and partial melting, a process by which rock is incompletely melted and molten fractions are removed.

Calcite is the main constituent of the carbonate rock limestone, and the mineral dolomite forms a carbonate rock that is also called dolomite. In water bodies in which intense evaporation occurs, evaporite minerals such as gypsum and halite may be precipitated. Clastic sedimentary rocks are composed of fragments of weathered rock that have been transported some distance from their place of origin. Transported pebbles and cobbles, for example, may be lithified to form the rock called conglomerate, sand grains may form sandstone, and particles of clay may accumulate to form shale.

Metamorphism refers to all the processes by which rocks of any kind undergo mineral and textural changes in the solid state in response to changing physical and chemical conditions. The agents of metamorphism include heat, pressure, and chemically active solutions. The basic kinds of metamorphism are contact and regional. Contact metamorphism occurs around the margins of bodies of molten rock. Unlike contact metamorphism, which may be relatively local, regional metamorphism occurs on a large (regional) scale and is usually associated with mountain building. The rocks in the regional belts are well foliated and divisible into distinct metamorphic zones or facies, characterized by minerals that formed in response to particular conditions of pressure and temperature.

QUESTIONS FOR REVIEW AND DISCUSSION

1. What is a mineral? What characteristics of a true mineral such as quartz or feldspar would not be present in a piece of glass?

2. What are the eight most abundant elements found in rocks and minerals?

3. Why are silicate minerals important in geology? Which silicates might one expect to find in granite? Which silicates occur in sedimentary rocks?

4. Which igneous rock best approximates the composition of the continental crust? The oceanic crust?

5. With regard to the origin of igneous rocks, describe what is meant by partial melting and fractional crystallization.

6. What inferences can be drawn from the color of igneous rocks? From the grain size of igneous rocks?

7. List the clastic sedimentary rocks in order of increasingly finer grain size.

8. What mineral groups discussed in this chapter are particularly common in sedimentary rocks?

9. List the foliated metamorphic rocks in order of increasingly coarser foliation.

10. If you were a Stone-Age (Paleolithic) human and had to choose between limestone and chert as the material for a spearhead, which would you select? Why?

11. Imagine that you are a geologist making a study of an area in which there are tabular layers of basalt lying between sedimentary strata. Some of the basalt layers represent lava that once flowed out upon the surface of the ground. Others were formed when basaltic magma was injected along sedimentary bedding surfaces. The former are called **lava flows** and the latter **sills**.

a. Which of the accompanying figures represents a lava flow?

b. Cite two lines of evidence that support your interpretation.

c. In which of the figures is the age of the bed above the basalt older than the basalt itself?

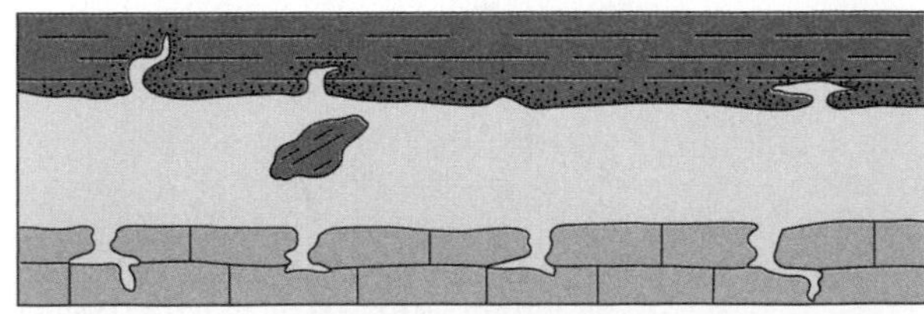

A

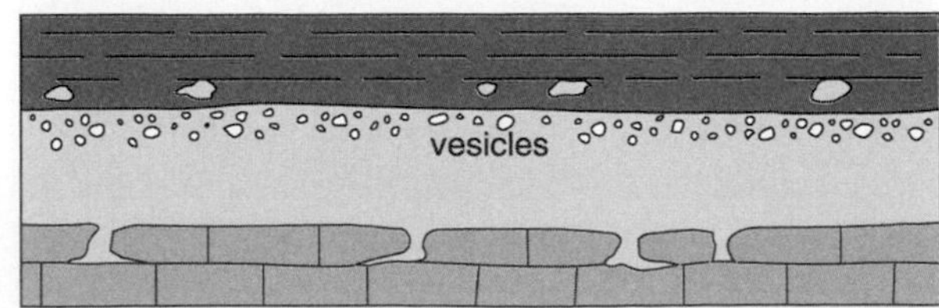

B

READINGS

Arem, J. 1991. *Rocks and Minerals*. Phoenix: Geoscience Press.

Blatt, H. and Tracy, R. J. 1996. *Petrology: Igneous, Sedimentary, and Metamorphic*, 2d ed., New York: W. H. Freeman.

Decker, R. W. and Decker, B. B. 1991. *Mountains of Fire: The Nature of Volcanoes*. New York: Cambridge University Press.

Fisher, R. V., Heiken, G., and Hulen, J. B. 1997. *Volcanoes: Crucibles of Change*. Princeton: Princeton University Press.

Pough, F. H. 1960. *A Field Guide to Rocks and Minerals*. Cambridge, MA: Houghton Mifflin.

Raymond, L. A. 1995. *Petrology*. Dubuque, Iowa: Wm. C. Brown.

Robinson, G. W. 1994. *Minerals: An Illustrated Exploration of the Dynamic World of Minerals and Their Properties*. New York: Simon & Schuster.

Sigurdson, H. (ed) 2000. *The Encyclopedia of Volcanoes*. New York: Academic Press.

WEB SITES

The Earth Through Time Student Companion Web Site (www.wiley.com/college/levin) has online resources to help you expand your understanding of the topics in this chapter. Visit the Web Site to access the following:

1. Illustrated course notes covering key concepts in each chapter;
2. Online quizzes that provide immediate feedback;
3. Links to chapter-specific topics on the web;
4. Science news updates relating to recent developments in Historical Geology;
5. Web inquiry activities for further exploration;
6. A glossary of terms;
7. A Student Union with links to topics such as study skills, writing and grammar, and citing electronic information.

3

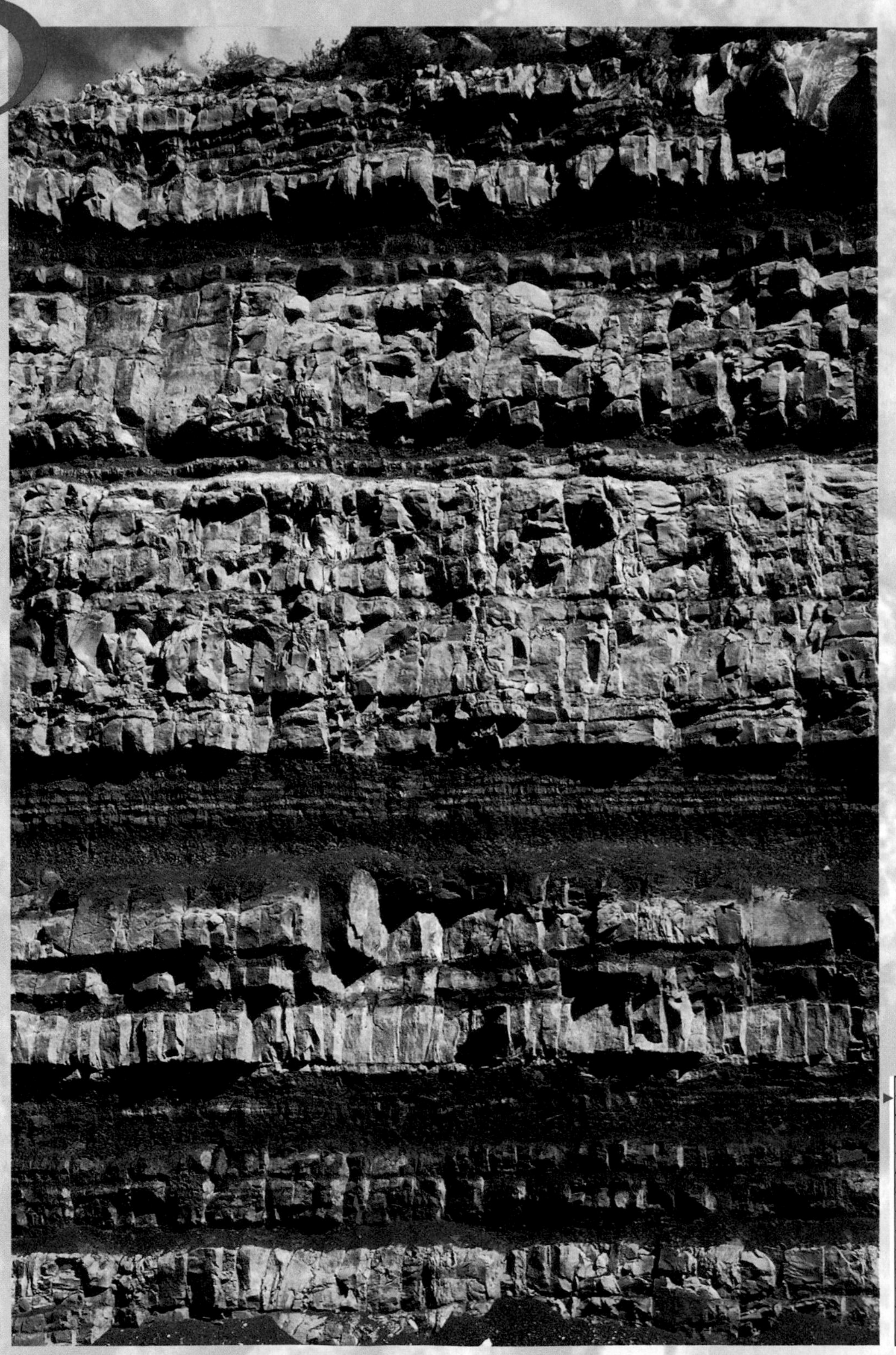

Sandstones and siltstones of the Uinta Formation near Park City, Utah. The Uinta Formation is Eocene in age. (Copyright William E. Ferguson.)

CHAPTER 3

The Sedimentary Archives

In the high mountains, I have seen shells. They are sometimes embedded in rocks. The rocks must have been earthy materials in days of old, and the shells must have lived in water. The low places are now elevated high, and the soft material turned into hard stone.
Chu-Hsi, A.D. 1200

OUTLINE

- THE TECTONIC SETTING
- ENVIRONMENTS OF DEPOSITION
  - The Marine Environment
  - Transitional Environments
  - Continental Environments
- COLOR OF SEDIMENTARY ROCKS
  - Black Coloration
  - Red Coloration
  - Red Beds
- TEXTURE OF CLASTIC SEDIMENTARY ROCKS
  - Size and Sorting of Clastic Grains
  - Shape of Clastic Grains
  - Arrangement of Clastic Grains
- INTERPRETING SEDIMENTARY ROCKS
  - Inferences from Sedimentary Structures
  - Geopetal Structures
  - Interpretation of Sands and Sandstones
  - Interpretation of Carbonate Rocks
  - Interpretation of Clays and Shales
- THE SEDIMENTARY ROCK RECORD
  - Rock Units
  - Facies
  - The Pervasive Effects of Sea-Level Changes
  - Correlation
  - Unconformities
- DEPICTING THE PAST
  - Geologic Columns and Cross-Sections
  - Geologic Maps
  - Paleogeographic Maps
  - Isopach Maps
  - Lithofacies Maps

Ever since the Earth has had an atmosphere and hydrosphere, sediments have been accumulating on its surface. The sediments, now formed into sedimentary rocks, contain features that tell us about the environment in which they were deposited. By interpreting these bits of evidence in successively higher strata, one can decipher the geologic history of a part of the Earth.

THE TECTONIC SETTING

Many factors determine the kind of sedimentary rock that will be formed in a particular area. Among these are the nature of the source area; the method of transport of sedimentary materials; the physical, chemical, and biologic processes operating in the place of deposition; the climate under which processes of weathering and erosion take place; and the changes that occur to sediment as it is being converted to solid rock. On a grander scale, the characteristics of an entire assemblage of rocks are also influenced by the tectonics of the region in which deposition takes place. The term *tectonics* refers to deformation or structural behavior of a large area of the Earth's lithosphere over a long period of time. For example, a region may be tectonically stable, broadly subsiding, or rising only gently. Other areas are tectonically active and are experiencing uplift and compressive forces that produce mountain ranges. Where a source area has recently experienced crustal compression and uplift, an abundance of coarse sediment (forming sandstones and conglomerates) derived from the rugged upland source area will be supplied to the basin. In the geologic past, such a tectonic setting has resulted in the accumulation of great "clastic wedges" of sediment that thinned and became finer away from the former mountainous source area. In other tectonic settings of the past, the source area has been stable and topographically more subdued, so that finer particles and dissolved solids became the most abundant components being carried by streams.

The tectonic setting influences not only the size of clastic particles being carried to sites of deposition but also the thickness of the accumulating deposit. For example, if a former marine basin of deposition had been provided with an ample supply of sediment and was experiencing tectonic subsidence (sinking), enormous thicknesses of sediments might accumulate. Over a century ago, James Hall (1811–1898), the eminent

American geologist mentioned in Chapter 1, recognized that the thick accumulations of shallow-water sedimentary rocks in the Appalachian region implied that crustal subsidence had accompanied deposition. His reasoning was quite straightforward. It was easy to visualize filling a basin that was 40,000 feet deep with 40,000 feet of sediment. However, where fossils indicated a basin never more than several hundred feet deep, the only way to get tens of thousands of feet of sediment into it would be to have subsidence occurring simultaneously with sedimentation.

In a marine basin of deposition that is stable or subsiding very slowly, the surface on which sedimentation is occurring is likely to remain within the zone of wave activity for a long time. Wave action and currents will wear, sort, and distribute the sediment into broad, blanketlike layers. If the supply of sediment is small, this type of sedimentation will continue indefinitely. Should the supply of sediment become too great for currents and waves to transport, however, the surface of sedimentation would rise above sea level, and deltas would form.

It is also important to consider the tectonic framework of entire continents as well as of particular areas. The principal tectonic elements of a continent are **cratons** and **orogenic belts** (Fig. 3-1). Cratons have two components: large areas of exposed ancient crystalline rocks called **shields**, as well as surrounding regions called **platforms**, in which these ancient rocks are covered by flat-lying or gently warped layers of sedimentary rocks. Cratons have been undisturbed by tectonic events since Precambrian time (about 540 million years ago). They comprise the stable interiors of continents.

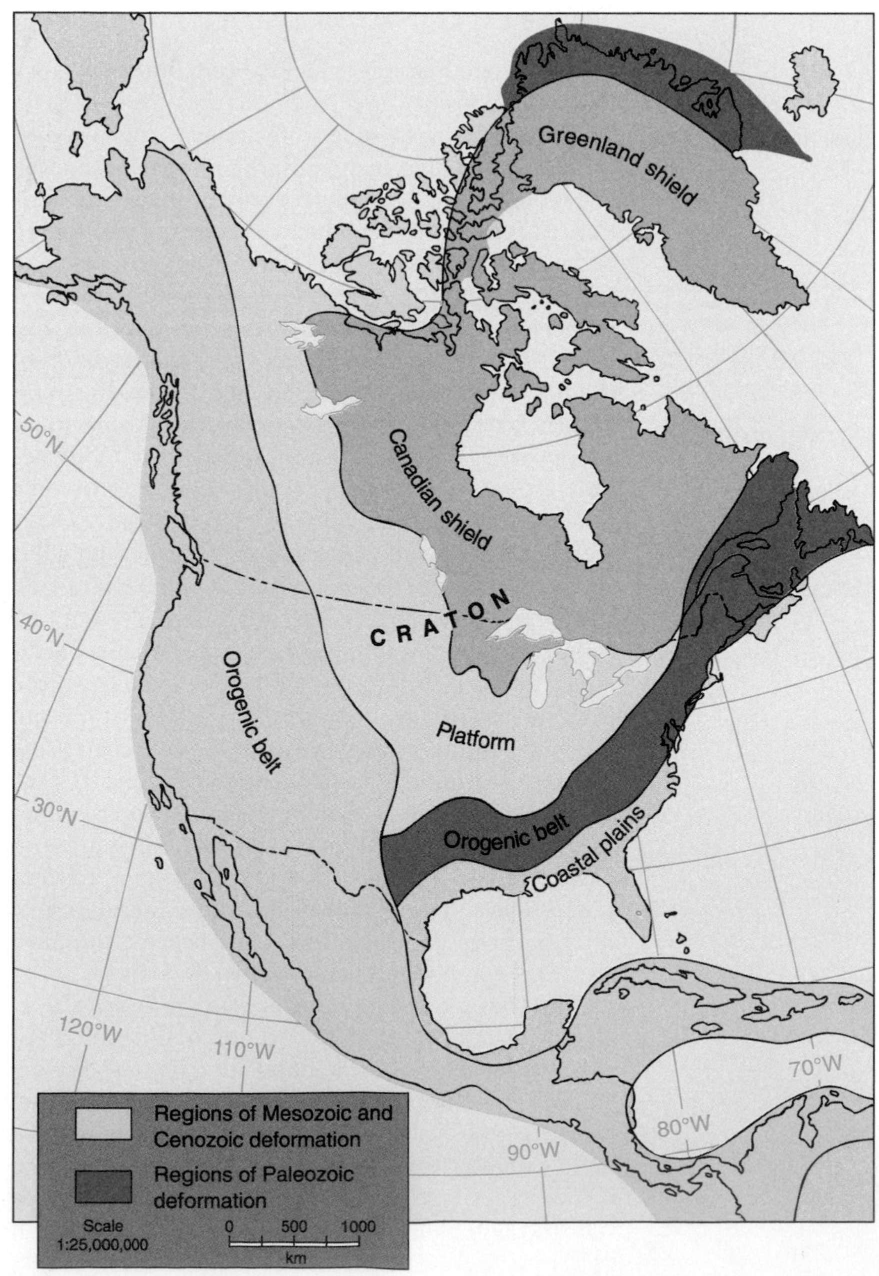

FIGURE 3-1 The craton and orogenic belts of North America.

Orogenic belts are elongated regions that border the craton and have been deformed by compressional forces since Precambrian time. Today, young orogenic belts are recognized by their high frequency of earthquakes and volcanic eruptions, whereas older belts are marked by severely deformed strata, crustal displacements, metamorphic terranes, and huge exposed bodies of intrusive igneous rocks.

The history of the Earth provides many examples of marginal orogenic belts that were once sites for the accumulation of great thicknesses of sediment. Such elongate tracts of sedimentation are found today along many continental margins. Following long episodes of deposition, the sediments along these tracts may be deformed as a result of an encounter with an oncoming tectonic plate, and a mountain range may form where there was once only a sedimentary basin. These events and their causes will be examined in Chapter 5.

The tectonic setting of deposition largely determines the nature of sedimentary deposits. Conversely, the kind of tectonic setting often can be inferred from a rock's textural and structural features and from its color, composition, and fossils. The tectonic and historical significance of some of these characteristics of sedimentary rocks will be described on the following pages.

ENVIRONMENTS OF DEPOSITION

Environment of deposition refers to all the physical, chemical, biologic, and geographic conditions under which sediments are deposited. Each environment of deposition is characterized by geographic and climatic conditions that modify or determine the properties of sediment that is deposited within it. Thus, the type of sediment becomes the key to the environment. Some sediments, such as chemical precipitates in water bodies, are solely the products of their environment of deposition. Their component minerals formed and were deposited at the same place. Other sediments consist of materials formed elsewhere and transported to the site of deposition. Geologists are keenly interested in the sediment of today's depositional environments because the features they find in the deposits of these modern areas can also be seen in ancient sedimentary rocks. Comparing present-day sedimentary deposits to old sedimentary rocks permits one to reconstruct conditions in various parts of the Earth as they were hundreds of millions of years ago. We apply, as did James Hutton over 200 years ago, the principle of uniformitarianism.

The Marine Environment

To facilitate discussion, the marine realm can be divided into shallow marine, deep marine, and continental slope environments. The shallow division includes the oceanic topographic regions called continental shelves. **Continental shelves** (Fig. 3-2) are nearly flat, smooth surfaces that fringe the continents in widths that range from only a few kilometers to about 300 kilometers and depths that range from low tide to about 200 meters. In a geologic sense, these shallow areas are not part of the oceanic crust but resemble the continents in their structure and composition. They are, in fact, the submerged edges of the continents and have a readily apparent continuity with the coastal plains. The outer boundaries of the shelves are defined by a marked increase in slope to greater depths. The smoothness of parts of the continental shelves seems to have been produced in part by the action of waves and currents during the last Ice Age. At that time, sea level was lowered at times by as much as 140 meters as a result of water being locked in glacial ice. Waves and currents sweeping across the shelves shifted sediment into low places and generally leveled the surface.

For geologists specializing in the study of sedimentary rocks, the continental shelves hold great interest. All of the sediment eroded from the continents and carried to the sea in streams must ultimately cross the shelves or be deposited on them. Many factors influence the kind of sediment deposited on the shelves, including the nature of the source rock on adjacent landmasses, the elevation of source areas, the distance from shore, and the presence of carbonate-secreting organisms. Because large grains are heavier than small particles, they tend to be deposited closer to shore. At the same time, shallow-water currents and wave action keep finer particles in suspension and carry them farther out to sea. For these reasons, clastic sediment deposited in the shallow marine environment tends to be coarser than material laid down in the deeper parts of the ocean. Sand, silt, and clay are common. Where there are few continent-derived sediments and the seas are relatively warm, lime muds of biochemical origin may be the predominant sediment. Coral reefs are also characteristic of warm, shallow seas. They remind us of the enormous biologic importance of the shallow marine environment. Over most of this realm, sunlight penetrates all the way to the sea floor. Algae and other forms of plant life proliferate. Here one finds the "pastures of the sea" on which, directly or indirectly, a multitude of swimming and bottom-dwelling animals are dependent (Fig. 3-3).

Those areas of the ocean floor that extend from the seaward edge of the continental shelves down to the ocean depths are named the **continental slopes**. Physiographic diagrams of the ocean floor are usually drawn with a large amount of vertical exaggeration, so that the continental slopes appear as steep escarpments. Actually, the inclination of the surface is only 3° to 6°. From the sharply defined upper boundary of the continental slope, the surface of the ocean floor drops to depths of 1400 to 3200 meters. At these depths the slope of the

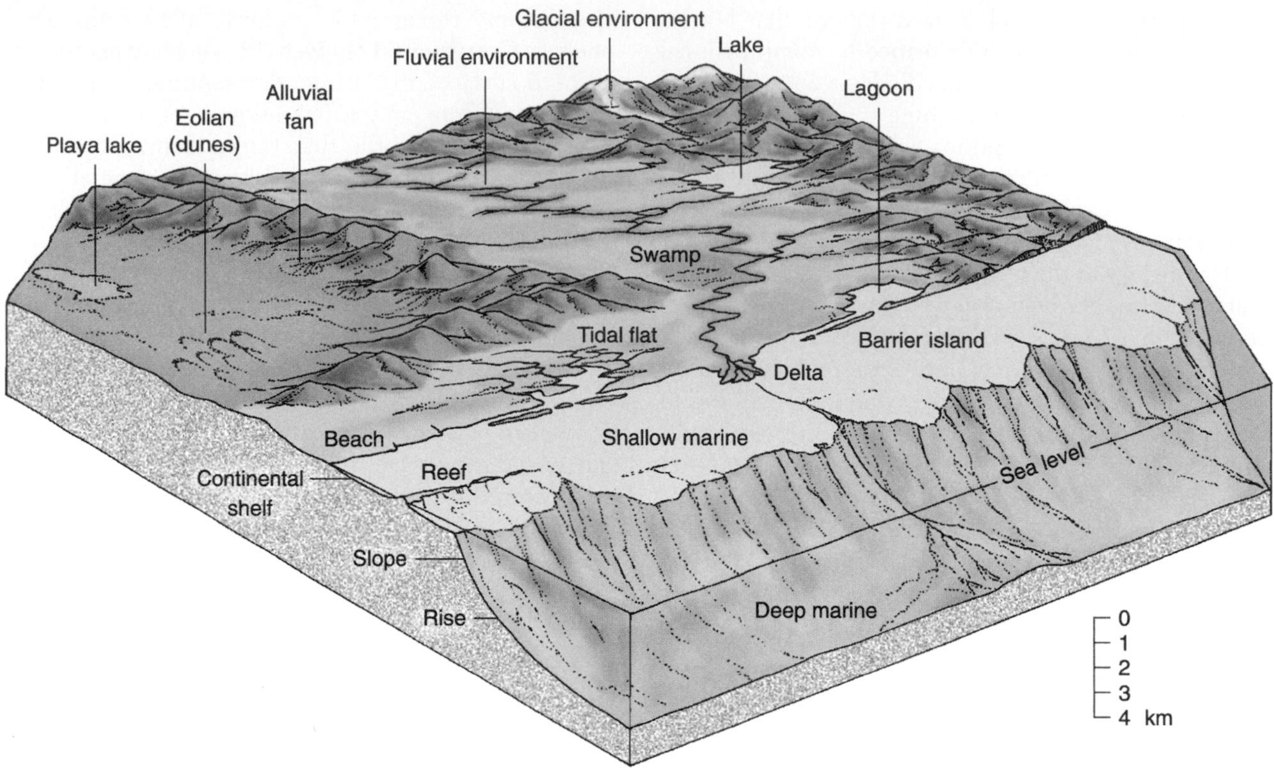

FIGURE 3-2 Diagram illustrating some marine, transitional, and continental environments of deposition. *What features of a shale formed in an ancient lake might be used to distinguish it from a shale formed in an ancient lagoon?*

ocean floor becomes gentler. The less-pronounced slopes comprise the **continental rises** (see Fig. 3-2).

Sediment deposited on slopes and rises is mostly fine sand, silt, and clay. These materials are often transported to sites of deposition by **turbidity currents**. Water in a turbidity current is denser than surrounding water because it is laden with suspended sediment. It therefore flows down the slope of the ocean floor beneath the surrounding clear water. Upon reaching more level areas, the current slows and drops its load of suspended particles. The deposits, called turbidites, may form submarine fans at the base of the continental slope (Fig. 3-4). In addition to turbidites, slope-and-rise deposits include fine clay that has slowly settled out of the water column and large masses of material that have slid or slumped down the slope under the influence of gravity.

In the deep marine environment far from the continents, only very fine clay, volcanic ash, and the calcareous or siliceous remains of microscopic organisms set-

FIGURE 3-3 Marine life flourishes in many areas of the well-lighted continental shelves, for light is required for the growth of marine plants, and plants are the basic components of the food chain that supports marine animals. *(Courtesy of L. E. Davis, Washington State University.)*

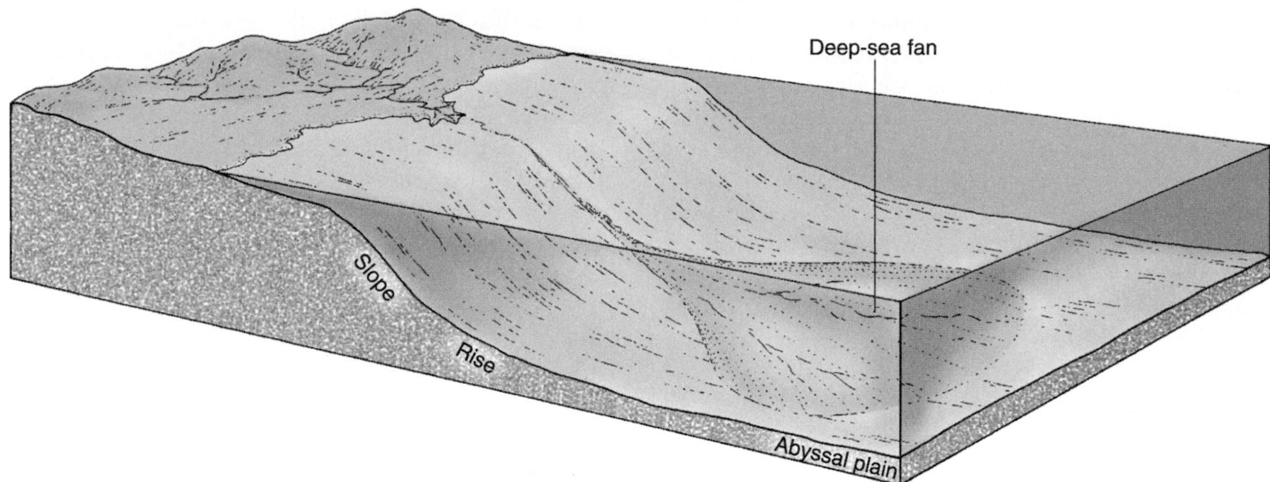

FIGURE 3-4 Deep-sea fan built of land-derived sediment emerging from the lower part of a submarine canyon. Such fans occur in association with large rivers, such as the Amazon, Congo, Ganges, and Indus. (Vertical exaggeration 200:1.) ? *Of the four major kinds of sandstones described in this chapter, which is most commonly associated with such deep-sea fan deposits?*

tle to the ocean floor. The exceptions are sporadic occurrences of coarser sediments that are carried down continental slopes into the deeper parts of the ocean by turbidity currents. Coarse sediments may also be dropped into deep water as they are released from melting icebergs.

Transitional Environments

The shoreline of a continent is the transitional zone between marine and nonmarine environments. Here one finds deltas and the familiar shoreline accumulations of sand or gravel that we call beaches. Mud-covered tidal flats that are alternately inundated and drained of water by tides are also found in the transitional zone.

Deltas are accumulations of sediment formed by the entrance of a stream into quiet water such as the ocean or a lake. The term was proposed about 25 centuries ago by Herodotus, who noted that the Nile Delta had the general shape of the Greek letter Δ. Not all deltas, however, have this shape, for every delta responds differently to depositional and erosional processes, which may act in opposition to one another.

The Mississippi Delta (Fig. 3-5*A*) is termed a *bird-foot delta* because its many divergent streams (distributaries) extend seaward to form a pattern that resembles the outstretched toes of a bird. In this great delta, the sediment supply has historically exceeded the reworking capabilities of ocean waves and currents in the northern Gulf of Mexico. The delta,

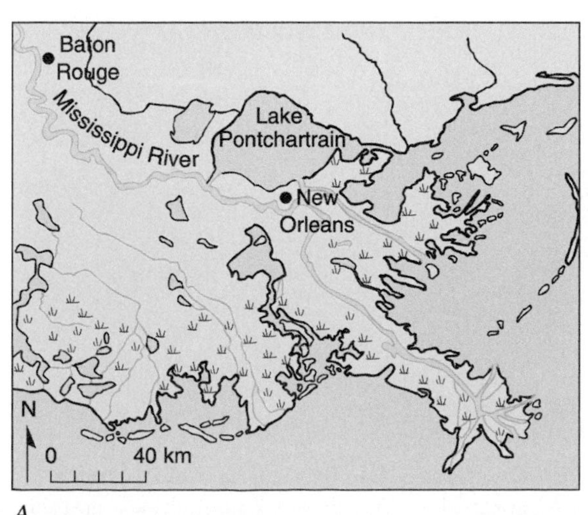

A

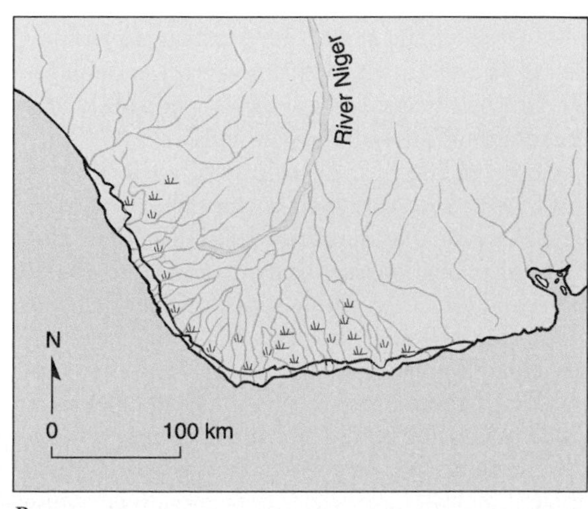

B

FIGURE 3-5 Comparison of the Mississippi (*A*) and Niger (*B*) deltas. ? *What changes or events might halt the progradation of these deltas?*

therefore, has been able to build seaward or **prograde**. Progradation occurs as lobelike accumulations of sand, silt, and clay are deposited at the mouths of one or more distributaries. After a lobe has formed, it will usually sink somewhat as its weight compresses the underlying material. One or more younger lobes may then accumulate over the first, producing cycles of deposits in which fine prodelta clays are successively overlain by delta front silts and sands. The vertical sequence of sediment for each successive cycle exhibits an upward progression of fine to coarse sediment, and each cycle may be capped by organic-rich marsh deposits.

Largely because dams constructed across tributaries of the Mississippi River have trapped sediment that might otherwise have contributed to delta expansion, the rate of growth of the Mississippi Delta has diminished since about 1950. In addition, the delta is sinking as a result of the tremendous weight of sediment on the crust and the withdrawal of underground water that supplies cities and towns along the margin of the Gulf of Mexico. The unfortunate result is a continuing loss of Louisiana's coastal wetlands and the rich harvest of seafood these wetlands provide.

In contrast to the Mississippi Delta, where the amount of sediment brought to the sea by streams exceeds the ability of the marine environment to rework and transport that sediment, the Niger Delta (Fig. 3-5*B*) on the west coast of Africa provides an example of a delta in which marine destructional processes are, to a far greater degree, in balance with the supply of sediment. The action of waves and longshore currents along the delta front is capable of reworking newly deposited sand, silt, and clay almost as fast as it is supplied. Over time, however, sedimentation has somewhat exceeded removal so that there has been concentric growth along the entire front of the delta. If sediment removal had exceeded sediment accumulation, there would be no delta at all. The Amazon is a river with no delta. The sediment load is not great enough to overwhelm the effect of tides, waves, and subsidence near the mouth of this largest river on Earth.

Some of the best exposures of ancient deltaic sequences can be observed in the Allegheny-Cumberland Plateau. Deltas in this region developed along the margins of shallow seas that covered western Pennsylvania and Ohio. An ample supply of sediment was provided by streams flowing from the eroding Appalachian highlands.

Deltaic sediments are rich in organic debris and include many alternating and intersecting bodies of permeable sands and impermeable clays. Because of this, ancient deltas have yielded tremendous volumes of coal, oil, and natural gas. Many of the once-prolific oil fields of the Texas and Louisiana Gulf Coast still pump oil from buried deltaic sedimentary sequences.

In addition to deltas, the transitional zone includes **barrier islands**, **lagoons** that lie between the barrier islands and the mainland, and **tidal flats**. Barrier islands are extensively developed along the Atlantic Coast of the United States (Fig. 3-6) as well as around Florida and the Gulf of Mexico. Sandy sediments predominate in this high-energy environment where sediment is winnowed and moved about by waves and longshore currents. Barrier islands are elongate in form and may attain lengths of over 100 kilometers. Most are less than a few kilometers in width. Coney Island off the coast of New York and Padre Island along the coast of Texas are examples familiar to vacationers. Bivalves, gastropods, echinoids, and crustaceans are prevalent in the barrier island environment.

The lagoons that lie behind barrier islands are protected from strong waves and currents. As a result, lagoonal deposits usually consist of fine-grained sediments. Lagoonal silts and clays tend to be densely burrowed by mollusks and worms, although species accomplishing this work are relatively few. Limited species diversity results from the water chemistry of

FIGURE 3-6 **Barrier island and lagoon on the south shore of Long Island.** Waves break along the seaward side of the barrier island, whereas conditions are relatively calm within the lagoon. (*Courtesy of G. R. Thompson and J. Turk.*)
? *Why are sediments coarser on the seaward side of the island?*

lagoons. Diversity is adversely affected in arid regions because of the limited tolerance of organisms to highly saline lagoonal water. Lagoons located in humid regions may also have lowered species diversity because of the influx of fresh water from streams or less saline water from connections to the open ocean.

Almost featureless, low-lying plains that are alternately inundated and drained by tides constitute tidal flats. These are commonly marshy areas in which fine-grained sediments predominate. Owing to their intermittent exposure, tidal flats are harsh environments for many organisms. Nevertheless, hardy species of mollusks, crustaceans, worms, and cyanobacteria thrive in many tidal flat environments.

Continental Environments

Continental environments of deposition include river floodplains, alluvial fans, lakes, glaciers, and eolian (wind) environments. The silt, sand, and clay found along the banks, bars, and floodplains of streams are familiar to most of us. In general, **stream deposits** develop as elongate bodies that reflect the course of the stream itself. Streams are highly complex systems affected by many interacting variables, including the quantity of water moving through the channel, the velocity of that moving water, the nature of the sediment being transported, and the shape of the stream channel. A change in any variable will cause a change in the sediment being deposited. Thus, sands, silts, and clays may grade abrubtly into one another in stream deposits. It is not an easy task to recognize ancient stream deposits. They tend to have few fossils (and no marine fossils). Traces of root systems in the clays of overbank deposits may provide evidence of stream deposits. Studies of sand and pebble grain orientation may indicate directional stream flow. Also useful is the observation that average particle size of sediment transported by a stream decreases downstream, whereas the rounding of grain corners (roundness) and approach to a sphere (sphericity) increase.

Stream-transported materials may accumulate quickly when a rapidly flowing river emerges from a mountainous area onto a flat plain. The result of the abrupt deposition is an **alluvial fan** (Fig. 3-7). Except for rare bones of vertebrates, spore and pollen grains, or fragmentary plant remains, ancient alluvial fan deposits lack fossils. They are recognized primarily on the basis of their lobate form and wedge-shaped cross-section.

Lakes are landlocked bodies of water. The sediments of lakes are termed **lacustrine deposits**. Somewhat quieter deposition occurs in lakes, which are ideal traps for sediment. Silt and clay are common lake sediments, although a variety of sediment is possible, depending on water depth, climate, and the character of the surrounding land areas. The chemistry of lake water is influenced by the environment. When traces of that chemistry are retained in ancient lake sediment, those traces may serve as indicators of former environmental conditions. The playa lakes of arid regions are shallow, temporary

FIGURE 3-7 Coalescing alluvial fans covering part of the floodplain, Gulf of Suez area, Egypt. The stream is unable to transport the huge amount of debris supplied to it and is dry during part of the year. (*Courtesy of D. Bhattacharyya.*) ? *If the mountains on the left side of the photograph are composed of granite, what silicate minerals are likely to be abundant in the sediments of the alluvial fans?*

(*A*)

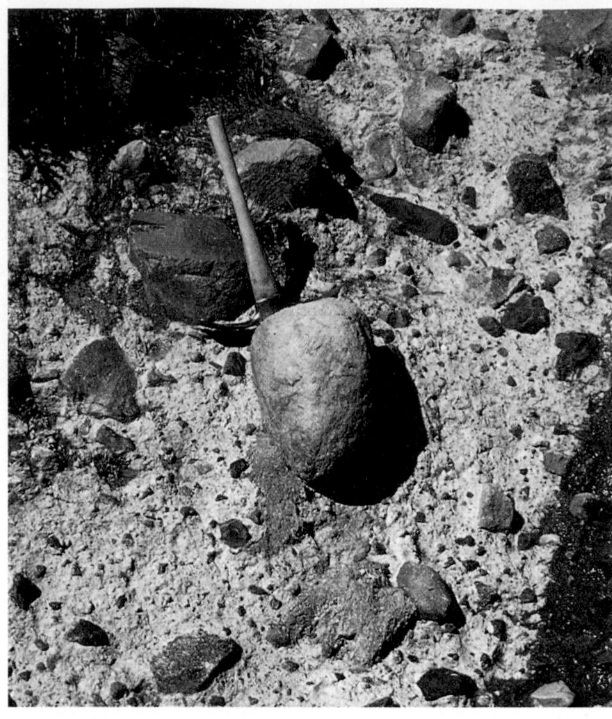

(*B*)

FIGURE 3-8 Glacial deposits. (*A*) Gravels in lateral and terminal moraines of a glacier on Baffin Island, Canada. (*B*) Till exposed on the flanks of Mt. Ranier, Ranier National Park. (*Photo by Steve Sheriff.*)

lakes that periodically become dry as a result of evaporation. Thus, evaporites characterize playa lake deposits. To identify lacustrine sediments, one looks for freshwater fossils, vertebrate tracks, and polygonal cracks (mud cracks) that may form when lake water levels are low and exposed mud dries.

The glacial environment may include a variety of other environments, including stream, lake, and even shallow marine. Glaciers have the ability to transport and deposit huge volumes and large fragments of rock detritus. Deposits are characteristically unsorted mixtures of boulders, gravel, sand, and clay (Fig. 3-8). Where such materials have been reworked by glacial meltwater, however, they become less chaotic and resemble stream deposits. One may even observe features resembling sediments deposited in deserts, as strong winds pick up fine glacial particles and deposit them in dunes.

In addition to moving ice and flowing water, wind can also erode, transport, and deposit sediment. Wind, however, is much more selective in the particle size it can transport. Air has only about 1/1000th the density of water and therefore can erode and transport only particles the size of sand or smaller. Environments where wind is an important agent of sediment transport and deposition are called **eolian environments** (Fig. 3-9). They are characterized by an abundance of sand and silt, little plant cover, and strong winds. These characteristics typify many desert regions.

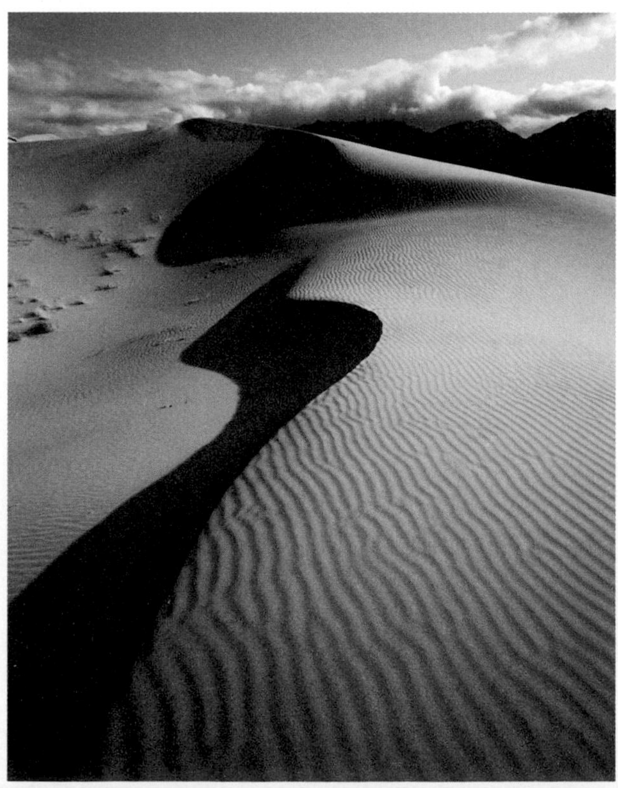

FIGURE 3-9 Dunes in Eureka Valley, California. (*Copyright Lee Rentz/Bruce Coleman, Inc.*)

COLOR OF SEDIMENTARY ROCKS

We have seen that color in igneous rocks can be used to indicate the approximate amount of ferromagnesian minerals present. Color in sedimentary rock can also provide useful clues to identification. For example, varieties of chert can be identified as flint if they are gray or black, or as jasper if they are red. Color is also useful in providing clues to the environment of deposition of sedimentary rocks. Of the sedimentary coloring agents, carbon and the oxides and hydroxides of iron are the most important.

Black Coloration

Black and dark-gray coloration in sedimentary rocks—especially shales—usually results from the presence of compounds containing organic carbon and iron. The occurrence of an amount of carbon sufficient to result in black coloration implies an abundance of organisms in or near the depositional areas as well as environmental circumstances that kept the remains of those organisms from being completely destroyed by oxidation or bacterial action. These circumstances are present in many marine, lake, and estuarine environments today. In a typical situation, the remains of organisms that lived in or near the depositional basin settle to the bottom and accumulate. In the quiet bottom environment, dissolved oxygen needed by aerobic bacteria to attack and break down organic matter may be lacking. There also may be insufficient oxygen for scavenging bottom dwellers that might feed on the debris. Thus, organic decay is limited to the slow and incomplete activity of anaerobic bacteria; consequently, incompletely decomposed material rich in black carbon tends to accumulate. In such an environment, iron combines with sulfur to form finely divided iron sulfide (pyrite, FeS_2), which further contributes to the blackish coloration. Such environments of deposition are likely to yield toxic solutions of hydrogen sulfide (H_2S). The lethal solutions rise to poison other organisms and thus contribute to the process of accumulation. Black sediments do not always form in restricted basins. They may develop in relatively open areas, provided the rate of accumulation of organic matter exceeds the ability of the environment to cause its decomposition.

Red Coloration

Hues of brown, red, and green often occur in sedimentary rocks as a result of their iron oxide content. Few, if any, sedimentary rocks are free of iron, and less than 0.1 percent of this metal can color a sediment a deep red. The iron pigments are not only ubiquitous in sediments but also difficult to remove in most natural solutions.

The iron present in sediment often occurs as either ferrous iron compounds or ferric iron compounds. Ferrous iron oxide (FeO) frequently occurs in oxygen-deficient environments. It is unstable and may slowly oxidize to form ferric iron oxide (Fe_2O_3). When oxygen is in short supply, ferric iron may be similarly reduced to ferrous iron. Ferric minerals such as hematite tend to color the rock red, brown, or purple, whereas the ferrous compounds impart hues of gray and green. Hydrous ferric oxide (limonite) is often yellow.

Red Beds

Strata colored in shades of red, brown, or purple by ferric iron are designated red beds by geologists. The compound Fe_20_3, which occurs as "rust" and the mineral hematite, provides the color. Oxidizing conditions required for the development of ferric compounds are more typical of nonmarine than marine environments; most red beds are floodplain, alluvial fan, or deltaic deposits. Some, however, are originally reddish sediment carried into the open sea. Electron microscopic studies of red beds forming today in Baja California indicate that the red coloration developed long after the sediment was deposited. After burial, the decay of clastic ferromagnesian minerals released iron that was oxidized by the oxygen in water circulating through the pore spaces. Thus, red coloration may be imparted in the subsurface and may be independent of climate. The paleoenvironmental interpretations of red beds should be based to a large degree on the associated rocks and sedimentary structures. Red beds interspersed with evaporite layers indicate warm and arid conditions.

Although red beds are more likely to represent nonmarine than marine deposition, occasionally one finds marine red beds interbedded with fossiliferous marine limestones. In such cases, the color may be inherited from red soils of nearby continental areas. Lands located in warm, humid climates often develop such red soils. When the soil particles arrive at the marine depositional site, they will retain their red coloration, provided there is insufficient organic matter present to reduce the ferric iron to the relatively soluble ferrous state. Otherwise, they will be converted to the gray or green colors of ferrous compounds.

In summary, sedimentary rocks of red coloration may be a product of the source materials, may have developed after burial as a result of a lengthy period of subsurface alteration, or may be the result of subaerial oxidation. Geologists are suspicious of the last possibility because most modern desert sediments are not red unless composed of sediment from nearby outcrops of older red beds.

TEXTURE OF CLASTIC SEDIMENTARY ROCKS

The size, shape, and arrangement of mineral and rock grains in a rock constitute its texture. In addition to the larger grains themselves, the textural appearance of a rock is influenced by the materials that hold the particles together. **Matrix** is bonding material that consists of finer clastic particles (often clay) that were deposited at the same time as the larger grains and that fill the spaces between them. **Cement**, on the other hand, is a chemical precipitate that crystallizes in the voids between grains following deposition. Silica (SiO_2) and calcium carbonate ($CaCO_3$) are common natural cements. Other cements include dolomite ($CaMg(CO_3)_2$), siderite ($FeCO_3$), hematite (Fe_2O_3), limonite ($2Fe_2O_3 3H_2O$), and gypsum ($CaSO_4 2H_2O$).

Texture can provide many clues to the history of a particular rock formation. In carbonate rocks, extremely fine-grained textures often indicate deposition in quiet water. Fine carbonate muds, which are the source sediment of such rocks, are not likely to settle to the bottom in turbulent water. Whole, unbroken fossil shells confirm the quiet-water interpretation. Limestones containing the worn and broken fragments of fossil shells are likely to be the products of reworking by wave action. They are turbulent shallow-water deposits.

Size and Sorting of Clastic Grains

Geologists universally use a scale of particle sizes known as the **Wentworth Scale** to categorize clastic sediments (Table 3-1). After disaggregation of a rock in the laboratory, the particles can be passed through a series of successively finer sieves, and the weight percentage of each size range in the rock can be determined. It is obvious that a stronger current of water (or wind) is required to move a large particle than to move a small one. Therefore, the size distribution of grains tells the geologist something about the turbulence and velocity of currents. It can also be an indicator of the mode and extent of transportation. If sand, silt, and clay are supplied by streams to a coastline, the turbulent nearshore waters will winnow out the finer particles, so that gradations from sandy nearshore deposits to offshore silty and clayey deposits frequently result (Fig. 3-10). Sandstones formed from such nearshore

TABLE 3-1 Size Range of Sedimentary Particles

Wentworth Scale (mm)	Fractional Equivalents (mm)	Particle Name
		Boulders
256		
128		Cobbles
64		
32		
16		Pebbles
8		
4		
		Granules
2		
		Very coarse sand
1.0		
		Coarse sand
0.5	1/2	
		Medium sand
0.25	1/4	
		Fine sand
0.125	1/8	Very fine sand
0.0625	1/16	
0.0313	1/32	
0.0156	1/64	
		Silt
0.0078	1/128	
0.0039	1/256	
		Clay

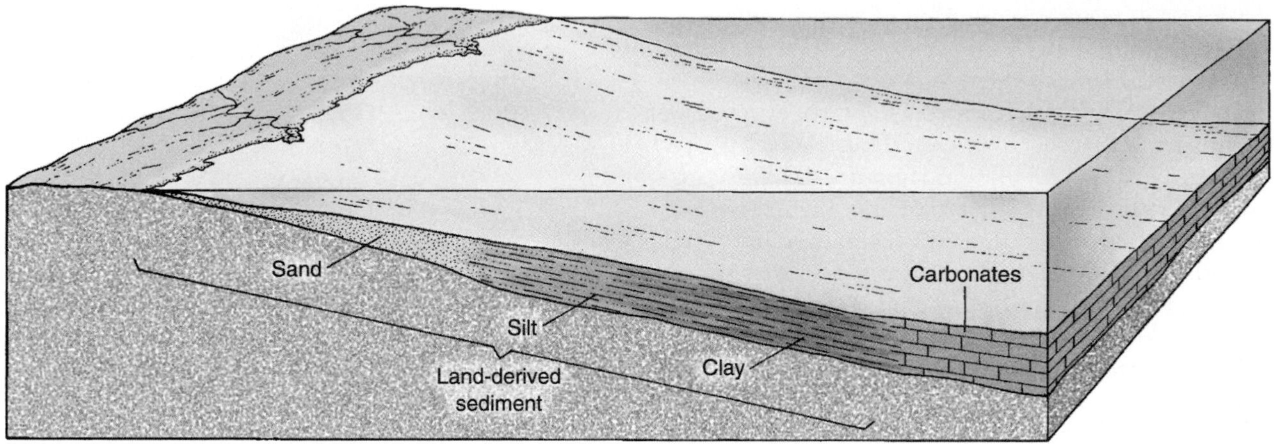

FIGURE 3-10 Idealized gradation of coarser nearshore sediments to finer offshore deposits.

sands may retain considerable porosity and provide void space for petroleum accumulations. For this reason, petroleum geologists draw maps showing the grain size of deeply buried ancient beaches and nearshore sandstone formations to determine areas of coarser and more permeable clastic rock.

One aspect of a clastic rock's texture that involves grain size is **sorting**. Sorting is an expression of the range of particle sizes deviating from the average size. Rocks composed of particles that are all about the same average size are said to be *well sorted* (Fig. 3-11), and those that include grains with a wide range of sizes are termed *poorly sorted*. Sorting provides clues to conditions of transportation and deposition. Wind, for example, winnows the dust particles from sand, producing grains that are all of about the same size. Wind also sorts the particles that it carries in suspension. Only rarely is the velocity of winds sufficient to carry grains larger than 0.2 millimeters. While carrying grains of that size, winds sweep finer particles into the higher regions of the atmosphere. When the wind subsides, well-sorted silt-sized particles drop and accumulate. In general, windblown deposits are better sorted than are deposits formed in an area of wave action, and wave-washed sediments are better sorted than are stream deposits. It must be kept in mind, however, that if a source sediment is already well sorted, the resulting deposit will be similarly well sorted. Provided they are not pervasively cemented, well-sorted sandstones have good porosity and permeability. They may, therefore, serve as reservoirs for petroleum and natural gas.

Poor sorting occurs when sediment is rapidly deposited without being selectively separated into sizes by currents (see Fig. 3-8*B*). Poorly sorted conglomerates and sandstones are deposited at the foot of mountains, where stream velocity is suddenly checked. Another example of a poorly sorted conglomerate is tillite, a rock deposited by glacial ice containing all particle sizes in a heterogeneous mixture.

Shape of Clastic Grains

The shape of particles in a clastic sedimentary rock can also be useful in determining its history. Shape can be

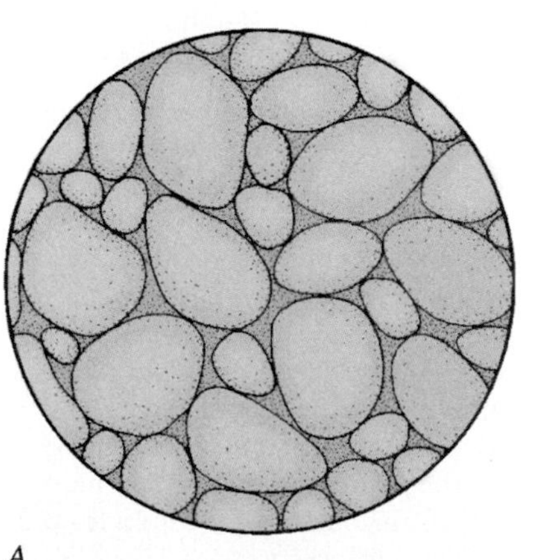

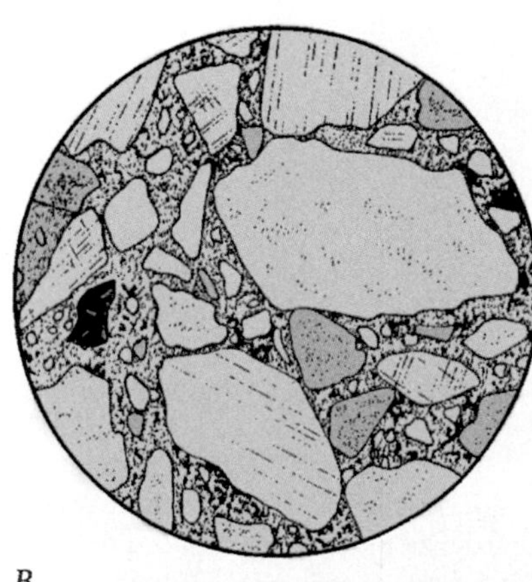

FIGURE 3-11 Sorting of grains in sandstones, as seen under the microscope, may range from good sorting (*A*) to poor sorting (*B*). (*A*) Quartz sandstone (light-tan grains) with carbonate (pink) cement. (*B*) A sandstone known as graywacke, composed of poorly sorted angular grains of quartz (light tan), feldspar (green), and rock fragments (orange). The graywacke lacks cement; spaces between grains are filled with a matrix of clay and silt. Width of fields is 1.5 mm.

? *Which of these sandstones can be considered an immature sandstone?*

described in terms of **rounding** of particle edges and **sphericity** (how closely the grain approaches the shape of a sphere; Figs. 3-12 and 3-13). A particle becomes rounded by having sharp corners and edges removed by impact with other particles. The relatively heavy impacts between pebbles and granules being transported by water cause rapid rounding. Lighter impacts occur between sand grains in water transport; the water provides a cushioning effect. The result is far slower rounding for sand grains. In conjunction with other evidence, the roundness of a particle can be used to infer the history of abrasion. It is a reflection of the distance the particle has traveled, the transporting medium, and the rigor of transport. It can also be used as evidence of recycling of older sediments.

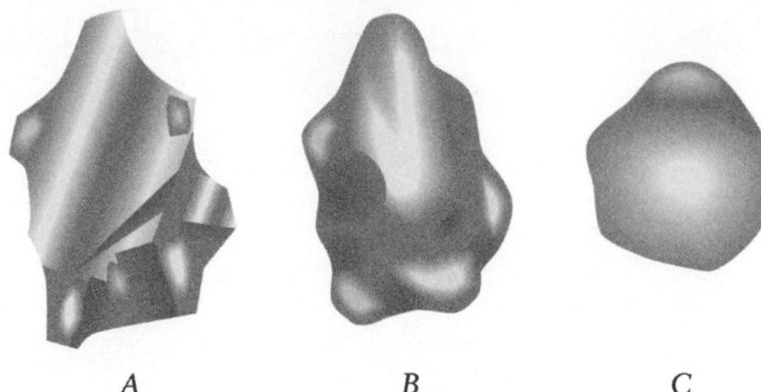

FIGURE 3-13 Shape of sediment particles. (*A*) An angular particle (all edges sharp). (*B*) A rounded grain that has little sphericity. (*C*) A well-rounded, highly spherical grain. Roundness refers to the smoothing of edges and corners, whereas sphericity measures the degree of approach of a particle to a sphere. ? *Although well-rounded, high-sphericity grains of quartz are common, feldspar grains are less likely to show good rounding and sphericity. What attribute of feldspar accounts for this difference?*

Arrangement of Clastic Grains

The third element in our definition of texture is the arrangement of the grains in the clastic rock. Geologists examine the rock to ascertain whether the grains are the same size and whether they are clustered into zones or heterogeneously mixed. These observations may help to determine whether the sediment had been winnowed and sorted by currents or had been dumped rapidly. Such factors as the medium of transport, surface of deposition, and direction and velocity of current control grain orientation. Geologists study grain orientation as a means of determining the direction of prevailing winds millions of years ago, the direction taken by ancient streams, or the trends of former longshore currents. In general, sand grains deposited in water currents acquire a preferred orientation in which the long axes of elongate grains are aligned parallel to the direction of flow. The preferred orientation of sand grains in a sandstone can be statistically analyzed in precisely oriented thin sections of sandstones (Fig. 3-14). For coarser sediments, studies of the preferred orientation of glacial and stream-deposited pebbles and cobbles indicate the direction of movement in glaciers and rivers that existed far back into the Precambrian. There is also a practical reason for studying grain orientation, for such information can provide clues to the subsurface location and trend of petroleum-bearing sandstone strata.

FIGURE 3-12 Well-rounded grains of quartz viewed under the microscope. From the St. Peter Formation, near Pacific, Missouri. Width of field is 1.85 mm.

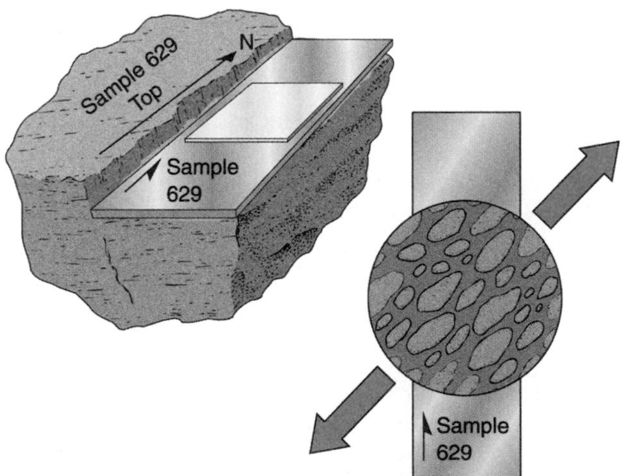

FIGURE 3-14 Grain orientation study. One method of studying grain orientation is to prepare an oriented thin section of a rock whose field orientation has been recorded. Grain orientations are then measured under a microscope equipped with a rotating stage. The angle of the long axis of each elongated grain from the north line is determined. From many individual measurements, a mean orientation is determined (in this example, about N 45°E or S 45°W) and its statistical significance is evaluated. A thin section cut perpendicular to bedding might reveal the tilt of the grains and might then be used to ascertain that the transportation medium flowed northeastward rather than southwestward.

ENRICHMENT

You Are the Geologist

Imagine you are a petroleum geologist working for a major oil company. Your company is informed that four parcels of continental shelf south of Louisiana are being offered for lease. You are informed, however, that the company has sufficient revenue to lease only one of the four parcels. You are asked to recommend one of the four parcels for lease. You have information that the area is underlain by the Gusher Sandstone, a formation which has yielded petroleum in adjacent area. The formation slopes gently toward the southeast. A few wells have been drilled into the Gusher Sandstone. Grain size analyses of Gusher Formation cores from these wells yielded the average grain size data on the Wentworth Scale (Table 3-1) shown on the accompanying map.

1. Based on the size analysis alone, which parcel of continental shelf would you recommend your company lease? Why?
2. What additional information would you hope to have in support of your recommendation?

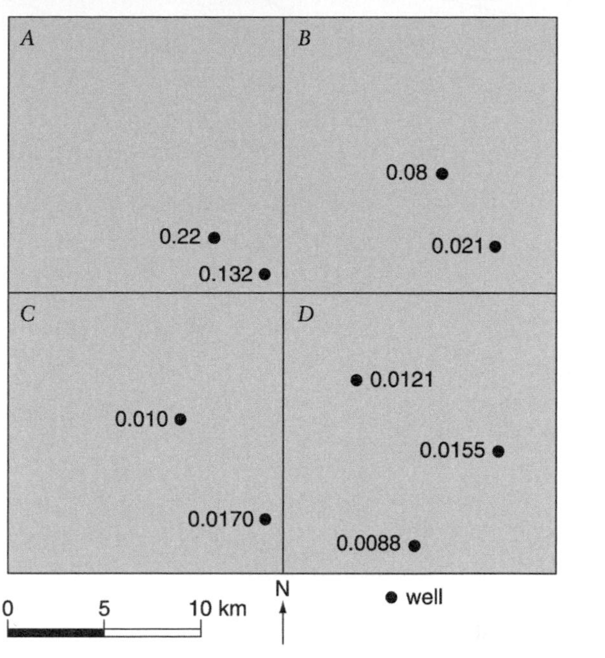

INTERPRETING SEDIMENTARY ROCKS

Inferences from Sedimentary Structures

Sedimentary structures are those larger features of sediments that are formed during or shortly after deposition and before lithification. Because particular sedimentary structures result from specific depositional processes, the structures are useful to geologists interested in reconstructing ancient environments. For example, **mud cracks** indicate drying after deposition. These conditions are common on valley flats, along the muddy margins of lakes, and in tidal zones. Mud cracks (Fig. 3-15) develop by shrinkage of mud or clay on drying and are most abundant in the subaerial environment. **Cross-bedding** (cross-stratification) is an

(*A*)

(*B*)

FIGURE 3-15 Modern and ancient mud cracks. (*A*) These modern mud cracks formed in soft clay around the margins of an evaporating pond. (*B*) Mud cracks and wave ripples (caused by wind blowing over a shallow lake) in mudstones of the Oneonta Formation, of Devonian age, near Unadilla, New York. Divisions on the scale are 1.0 cm. (*Modern mud cracks courtesy of L. E. Davis; ancient mud cracks courtesy of W. D. Sevon, Pennsylvania Bureau of Topography & Geological Survey.*) ? *How might the shape of a mud crack in a cross-section of an ancient mudstone be used to indicate the top of a stratum?*

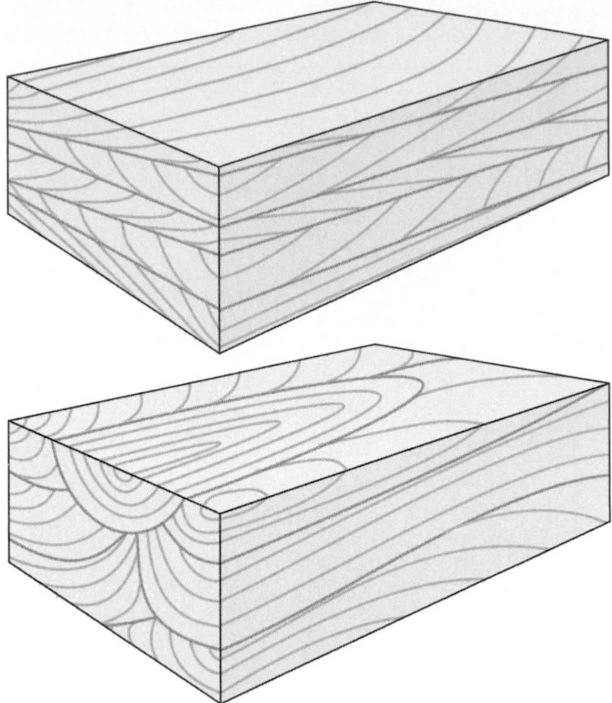

FIGURE 3-16 **Two types of cross-bedding.** The upper block shows tabular cross-bedding, as seen in beach deposits and dunes; the lower block represents trough cross-bedding, as often formed in river channels. The line at the base of each set of laminations represents a surface of erosion that truncates older sets below.

arrangement of beds or laminations in which one set of layers is inclined relative to the others (Figs. 3-16 and 3-17). The cross-bedding units can be formed by the advance of a delta (Fig. 3-18) or a dune (see Fig. 3-9). A depositional environment dominated by currents is inferred from cross-bedding. The currents may be wind or water. In either medium, the direction of the inclination of the sloping beds is a useful indicator of the direction taken by the current. By plotting these directions on maps, geologists have been able to determine the pattern of prevailing winds at various times in the geologic past.

Graded bedding consists of repeated beds, each of which has the coarsest grains at the base and successively finer grains nearer the top (Fig. 3-19). Although graded bedding may form simply as the result of faster settling of coarser, heavier grains in a sedimentary mix, it appears to be particularly characteristic of deposition by the turbidity currents discussed earlier. Turbidity currents are often triggered by submarine earthquakes and landslides that occur along steeply sloping regions of the sea floor. The forward part of the turbidity current contains coarser debris than does the tail. As a result, the sediment deposited at a given place on the sea bottom grades from coarse to fine as the "head" and then the "tail" of the current pass over it.

Ripple marks are commonly seen sedimentary features that developed along the surfaces of bedding planes (Fig. 3-20). *Symmetric ripple marks* are formed by the oscillatory motion of water beneath waves. *Asymmetric ripple marks* are formed by air or water currents and are useful in indicating the direction of movement of currents (Fig. 3-21). For example, ripple marks form at right angles to current directions; the steeper side of the asymmetric variety faces the direction in which the medium is flowing. Although some ripple marks have been found at great depths on the sea floor, these features occur more frequently in shallow-water areas and in streams.

FIGURE 3-17 **Tabular cross-bedding (cross-stratification) in the Mountain Lakes Formation, of Proterozoic age, Northwest Territories, Canada.** (*Courtesy of G. Ross.*) ? *Was the current that produced the cross-bedding flowing approximately from right to left or left to right?*

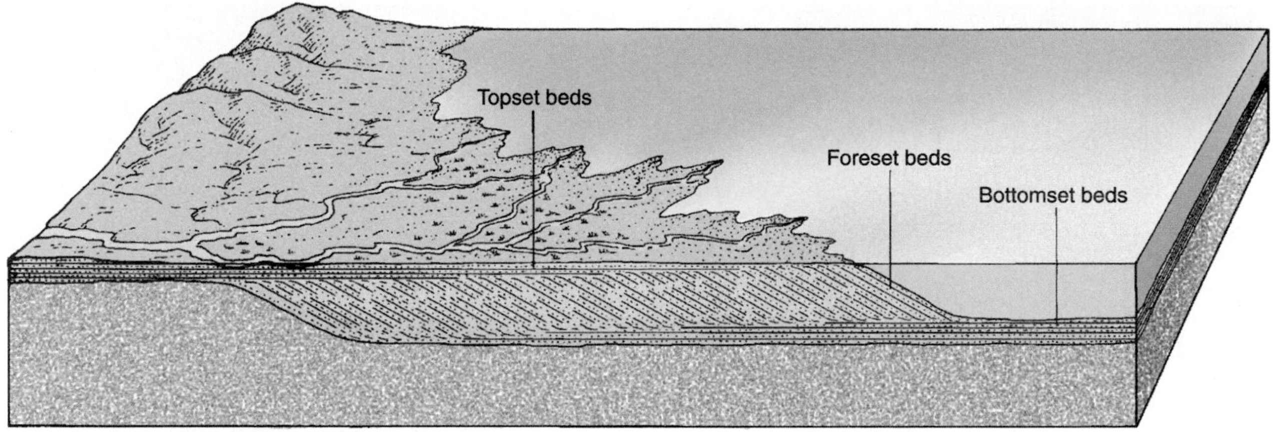

FIGURE 3-18 Cross-bedding in a delta. The succession of inclined foreset beds is deposited over bottomset beds that were laid down earlier. Topset beds are deposited by the stream above the foreset beds.

Geopetal Structures

The principle of superposition tells us that in undisturbed strata, the oldest bed is at the bottom and higher layers are successively younger. But what if the strata are deformed and overturned in such a way that the oldest beds are found at the top (Fig. 3-22*A*)? If overturning is not recognized, the geologic sequence of events, the kinds of folds, and features indicating the direction of sediment transport might be misinterpreted. For this reason, geologists working in areas of deformed strata (Fig. 1-1) carefully scrutinize rocks for indications of the original tops and bottoms of beds. Features providing such information are called **geopetal structures**. Among the more common geopetal structures are symmetric ripple marks (see Fig. 3-21*A*); included fragments; certain types of cross-bedding; graded bedding (see Fig. 3-19); mud cracks (see Fig. 3-15); bed surface markings, such as footprints, trails, and raindrop imprints; scour marks; fossils; and various biologically produced structures.

In the case of symmetric ripple marks (Fig. 3-22*B*), the sharp crests of the ripples normally identify the tops of beds. They point toward the younger beds. If fragments of the rippled rock are recycled by erosion and included in overlying strata, the overlying bed must be younger, and the interpretation is confirmed.

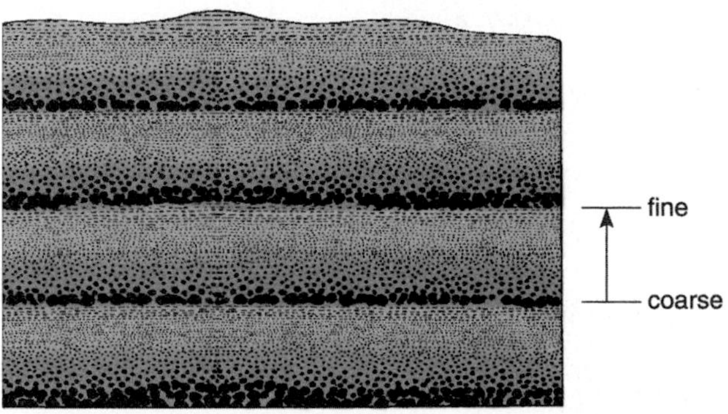

FIGURE 3-19 Graded bedding. *Of the four major kinds of sandstones, which most frequently displays graded bedding?*

In many kinds of cross-bedding, the cross-beds are concave upward, forming a small angle with beds below and a large angle with beds resting on their truncated upper edges (Fig. 3-22*C*). Geopetal interpretations based on cross-bedding, however, should be confirmed by other geopetal structures, as some cross-beds do not show the concave upward shape.

Graded bedding, in which grains are progressively finer from the bottom to the top of a bed, is another useful geopetal structure. As described earlier, graded beds are formed when fast-moving currents begin to slow, so that large particles are dropped first, followed by progressively finer grains.

Mud cracks are geopetal structures formed when mud dries, shrinks, and cracks. The cracks narrow downward, away from the top of the bed. Deposition above mud cracks would fill them, resulting in a corresponding pattern of ridges that identify the bottom of the overlying stratum.

As currents flow across beds of sand, they often erode various kinds of scour marks. An overlying layer of sediment may later fill these depressed markings, forming positive-relief casts in the covering bed (Fig. 3-22*D*). The casts are termed sole markings because they appear on the sole, or bottom, of the younger stratum.

Fossils of bottom-dwelling organisms such as corals may also be used to determine way-up, provided they have been buried in their natural, upright living positions. Some fossils that have been moved by currents may also be useful. For example, the curved shells of clams washed by currents may come to rest in a convex-upward position, as this is hydrodynamically most stable. Finally, many fossil organisms excavated and lived in burrows, such as the U-shaped burrows shown in

(A)

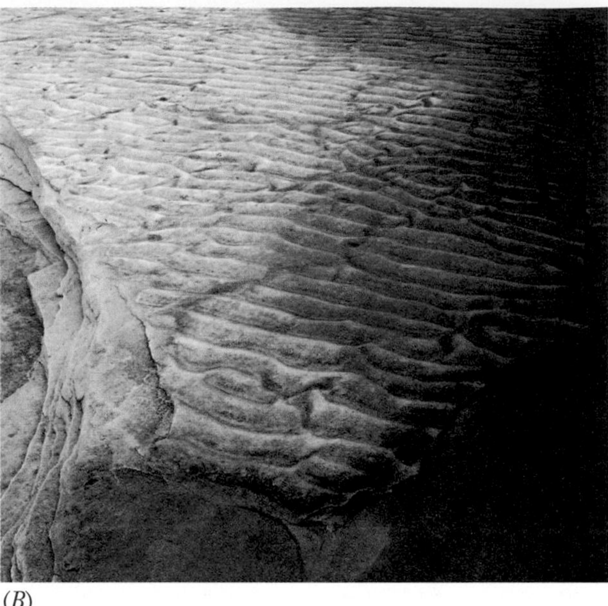
(B)

FIGURE 3-20 **Ripple marks.** (*A*) Ripple marks formed in sand along a modern beach. (*B*) Ancient ripple marks on a bedding surface of the Munising Formation, of Cambrian age, Mosquitoe Harbor, Pictured Rocks National Shoreline, Michigan. (*Copyright Robert P. Carr/ Bruce Coleman, Inc.*) ? *What do ripple marks in a marine sandstone indicate about the depth of water in which the sandstone was deposited?*

Figure 3-22*E*. Such structures are often excellent geopetal indicators.

Interpretation of Sands and Sandstones

Among clastic sedimentary rocks, sandstones have been studied in great detail and provide an extraordinary amount of information about conditions in and near the site of deposition. In particular, the mineral composition of sandstone grains can be used to identify source areas and to interpret what may have occurred prior to deposition. Often, by closely studying the grains, one can ascertain whether the source material was metamorphic, igneous, or sedimentary. The mineral content also provides a rough estimate of the amount of transport and erosion of the sand grains. Rigorous weathering and long transport tend to reduce the less-stable feldspars and ferromagnesian minerals to clay and iron compounds and tend to cause rounding and sorting of the remaining quartz grains. Hence, one can assume that a sandstone rich in these less-durable and angular components underwent relatively little transport and other forms of geologic duress. Such sediments are termed immature and are most frequently deposited close to their source areas. On the other hand, quartz can be used as an indicator of a sandstone's maturity; the higher the percentage of quartz, the greater the maturity.

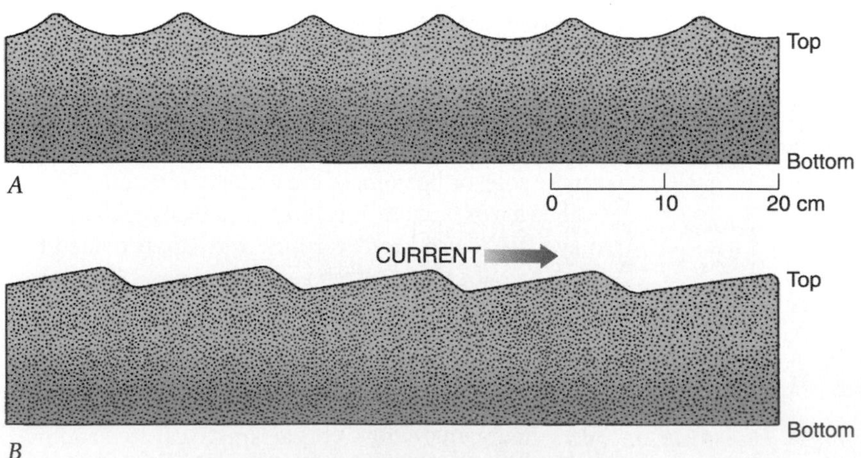

FIGURE 3-21 **Profiles of ripple marks.** (*A*) Symmetric ripples. (*B*) Asymmetric ripples.

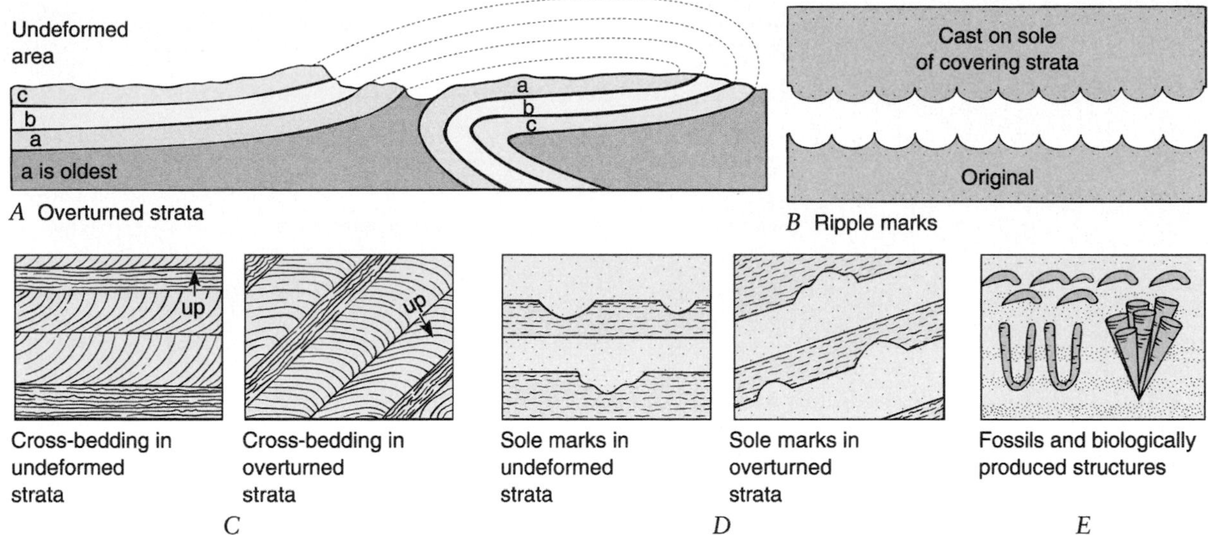

FIGURE 3-22 Various kinds of geopetal indicators. *In what regions of the United States might one encounter overturned strata?*

In addition to providing an indication of a rock's maturity, composition is an important factor in the classification of sandstones into quartz sandstone, arkose, graywacke, and lithic sandstone (sometimes termed subgraywacke; Fig. 3-23).

Quartz sandstones are characterized by a dominance of quartz with little or no feldspar, mica, or fine matrix. The quartz grains are well sorted and well rounded (see Fig. 3-12). They are most commonly held together by such cements as calcite and silica. Chemical cements such as these tend to be more characteristic of "clean" sandstones such as quartz sandstones and are not as prevalent in "dirtier" rocks containing clay. The presence of a dense, clayey matrix seems to retard the formation of chemical cement, perhaps because fine material fills pore openings where crystallization might occur.

Calcite cement may develop between the grains as a uniform, finely crystalline filling, or large crystals may form, and each may incorporate hundreds of quartz grains. Silica cement in quartz sandstone commonly develops as overgrowths on the original grain surfaces.

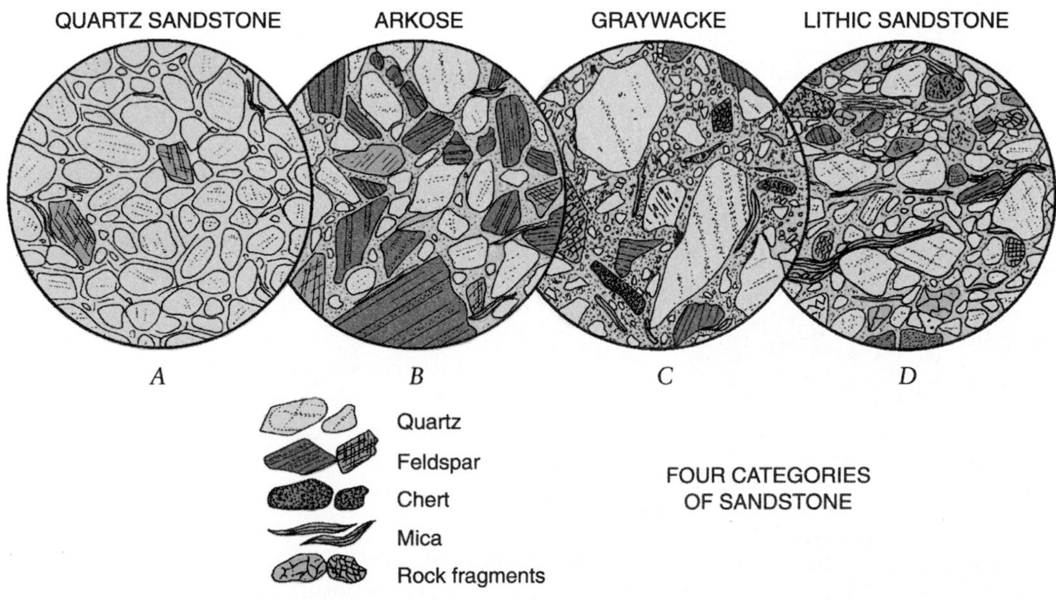

FIGURE 3-23 Four categories of sandstone as seen in thin section under the microscope. Diameter of field is about 4 mm.

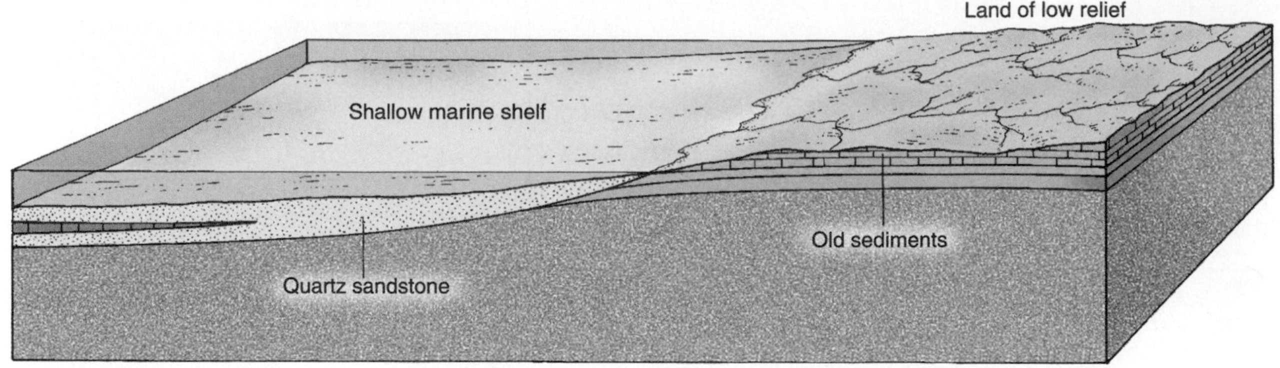

FIGURE 3-24 Idealized geologic conditions under which quartz sandstone may be deposited. There is little tectonic movement in this environment. Water depth is shallow, and the basin subsides very slowly. ? *What features formed by wave action would you expect to find in the marine sands of this environment?*

Quartz sandstones reflect deposition in stable, quiet, shallow-water environments, such as the ancient shallow seas that inundated large parts of low-lying continental regions in the geologic past or some parts of our modern continental shelves (Fig. 3-24). These sandstones, as well as clastic limestones, exhibit sedimentary features, such as cross-bedding and ripple marks, that permit one to infer shallow-water deposition.

Sandstones that are 25 percent or more feldspar (derived from erosion of a granitic source area) are called **arkoses** (Fig. 3-25). Quartz is the most abundant mineral, and the angular to subangular grains are bonded together by calcareous cement, clay minerals, or iron oxide. The presence of abundant feldspars and iron imparts a pinkish-gray color to many arkoses. In general, arkoses are coarse, moderately well-sorted sandstones. They may originate as basal sandstones derived from the erosion of a granitic coastal area experiencing an advance of the sea, or they may accumulate in fault troughs or low areas adjacent to granite mountains (Fig. 3-26).

FIGURE 3-25 Thin section of an arkose, viewed through a petrographic microscope. The clear grains are mostly quartz, whereas the grains that show stripes or a plaid pattern are feldspars. The matrix consists of kaolinite clay and fine particles of mica, quartz, and feldspar. (*Courtesy of the U.S. Geological Survey; photo by J. D. Vine.*) ? *How would you describe the sorting in this sandstone? Is it likely to have good permeability?*

Graywackes (from the German term wacken, meaning "waste" or "barren") are immature sandstones consisting of significant quantities of dark, very fine-grained material (Fig. 3-27*A*). Normally, this fine matrix consists of clay, chlorite, micas, and silt. There is little or no cement, and the sand-sized grains are separated by the finer matrix particles. Matrix constitutes approximately 30 percent of the rock, and the remaining coarser grains consist of quartz, feldspar, and rock particles. Graywacke has a dirty, "poured-in" appearance. The poor sorting, angularity of grains (Fig. 3-27*B*), and heterogeneous composition of graywackes indicate an unstable source and depositional area in which debris resulting from rapid erosion of highlands is transported quickly to subsiding basins. Graded bedding (see Fig. 3-19), interspersed layers of volcanic rocks, and chert (which may indirectly derive their silica from volcanic ash) further attest to dynamic conditions in the area of deposition. The inferred tectonic setting is dynamic and unstable, with deposition occurring offshore of an actively rising mountainous region (Fig. 3-28). Graywackes and associated shales and cherts may contain fossils of deep-water organisms, indicating deposition at great depth. Such shallow-water sedimentary structures as cross-bedding and ripple marks are rarely found.

Quartz sandstone, arkose, and graywacke are rather distinct kinds of sandstones. A sandstone that has a more transitional composition and texture is termed a **lithic sandstone** (subgraywacke). In lithic sandstones (see Fig. 3-23), feldspars are relatively scarce, whereas

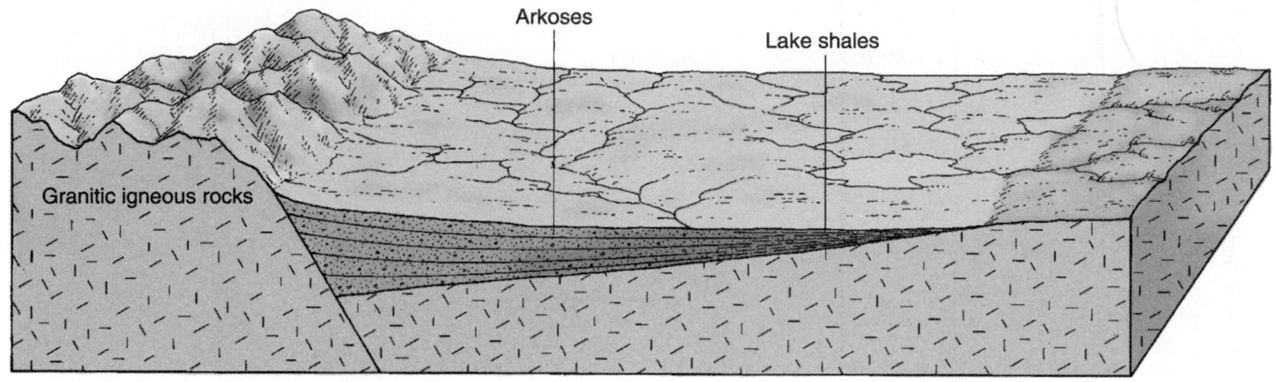

FIGURE 3-26 Geologic environment in which arkose may be deposited.

quartz, muscovite, chert, and rock fragments are abundant. There is a fine-grained detrital matrix that does not exceed 15 percent, and the remaining voids are filled with mineral cement or clay. Quartz grains are more rounded and abundant, sorting is better, and the quantity of matrix is lower in lithic sandstones than in graywackes.

The characteristic environments for lithic sandstones are deltaic coastal plains (Fig. 3-29), where lithic sandstones may be deposited in nearshore marine environments or swamps and marshes. Coal beds and micaceous shales are frequently associated with lithic sandstones.

Interpretation of Carbonate Rocks

Limestones are the most abundant of carbonate sedimentary rocks. Although limestone lake deposits do occur, most limestones originated in the seas. Nearly always, the formation of these marine limestones appears to have been either directly or indirectly associated with biologic processes. In some limestones, the importance of biology is obvious, for the bulk of the rock is composed of readily visible shells of mollusks and skeletal remains of corals and other marine organisms. In other limestones, skeletal remains are not present, but nevertheless the calcium carbonate ($CaCO_3$) that forms the bulk of the deposit was precipitated from sea water because of the life processes of organisms living in that water. For example, the relatively warm, clear ocean waters of tropical regions are usually slightly supersaturated with calcium carbonate. In this condition, only a slight increase in temperature, loss of dissolved carbon dioxide, or influx of supersaturated water containing calcium carbonate "seeds" can bring about the precipitation of tiny crystals of calcium carbonate. Organisms do not appreciably affect temperature, but through photosynthesis, myriad microscopic marine plants remove carbon dioxide from the water and thus may trigger the precipitation of calcium

(A)

(B)

FIGURE 3-27 Hand specimen (*A*) and thin section (*B*) of graywacke. The poorly sorted nature of graywacke and the angularity of its component grains are evident in the thin section as observed with the petrographic microscope and crossed polarizers. Width of field is 9.0 mm.

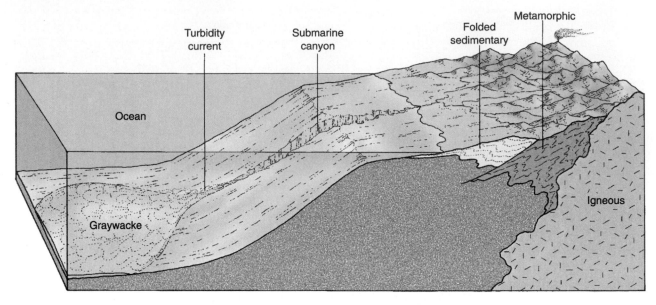

FIGURE 3-28 Tectonic setting in which graywacke is deposited. Frequently graywackes are transported by masses of water highly charged with suspended sediment. Because of the suspended matter, the mass is denser than surrounding water and moves along the sloping sea floor or down submarine canyons as a turbidity current. Graywacke sediment characteristically accumulates in deep-sea fans at the base of the continental slope.

carbonate. Bacterial decay may also enhance precipitation of calcium carbonate by generating ammonia and thereby changing the alkalinity of sea water. In either case, the precipitate can be considered an indirect product of organic processes. Carbonate sedimentation today is most rapid in shallow, clear-water tropical marine areas, such as the Bahama Banks east of Florida (Fig. 3-30). The carbonate sediments forming today in the Bahama Banks (Fig. 3-31) originate in more than one way. Some of them are derived from the death and dismemberment of calcareous algae, such as *Penicillus*, an organism that secretes tiny, needlelike crystals of calcium carbonate. The microscopic shells of other unicellular organisms also contribute to the carbonate buildup. In areas where tidal currents flow across the banks, oöids accumulate. As noted in the previous chapter, oöids are tiny spheres composed of calcium carbonate that are formed when particles roll back and forth on the sea floor and acquire concentric rings of carbonate (Fig. 3-32). Some of the sediment results from the precipitation of tiny crystals of lime from sea water that has been chemically altered by the biologic processes of marine plants. Coarser particles result from the abrasion of the shells of invertebrates or consist of fecal pellets produced by burrowing organisms.

The Bahama Banks are a modern example of a **carbonate platform**, defined as a broad, shallow marine structure composed largely of calcium carbonate that stands above the adjacent ocean floor on one or more

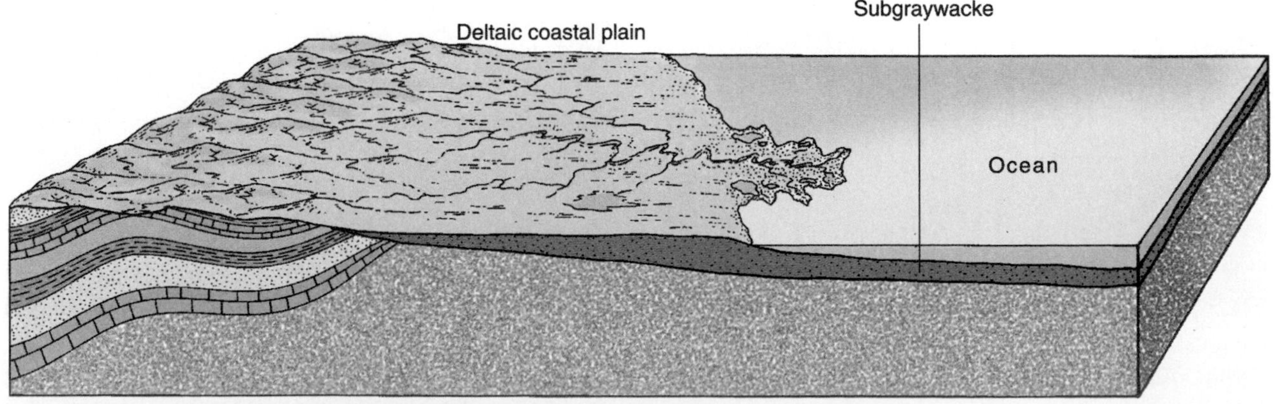

FIGURE 3-29 Deltaic environment in which lithic sandstones may be deposited.

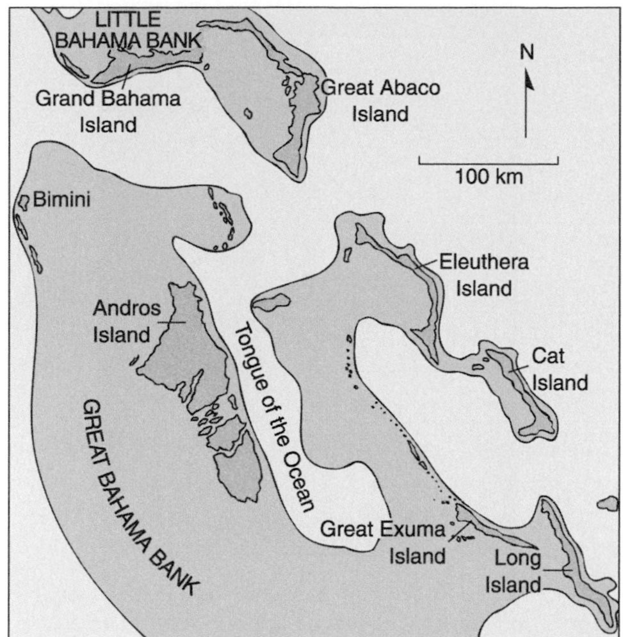

FIGURE 3-30 The Bahama Banks. The green areas have water depths of less than 180 m. Florida lies about 110 km west of Bimini.

sides. The Bahama platform is covered by less than 200 meters of water and is bordered by a steep slope into deep water on its eastern margin. At least 5000 meters of carbonate rock has been deposited on the Bahama Banks platform, indicating the structure has maintained itself by continuous production of carbonate sediment while simultaneously experiencing long-term subsidence.

FIGURE 3-31 Carbonate mud accumulating on the sea floor in the shallow warm waters of the Bahama Banks carbonate platform. Green algae of the genus *Penicillus* form the tuftlike growths in the background. These algae produce fine, needlelike crystallites of calcium carbonate (aragonite) that contribute to the production of carbonate sediment. Other algae, such as *Halimeda*, produce similar calcium carbonate particles. (*Courtesy of L. Walters.*)

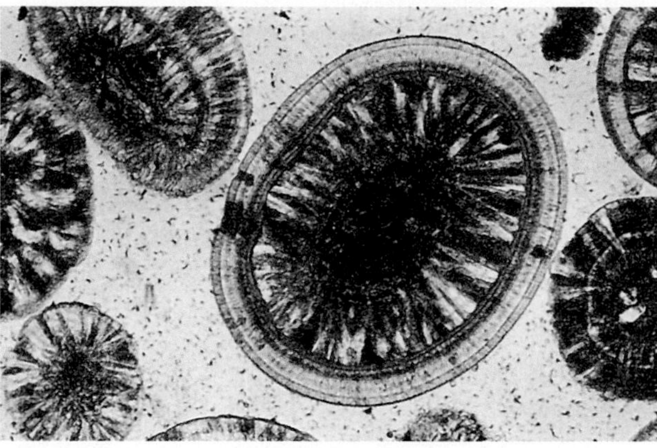

FIGURE 3-32 Thin section view of an oölitic limestone. The oöids are immersed in a cement of sparry (clear) calcite. The large oöid in the center has a maximum diameter of 0.74 cm. (*Courtesy of G. R. Thompson and J. Turk.*)

The requirement of warm, clear, shallow seas for the accumulation of modern carbonates seems to apply equally well to ancient deposits. Major sequences of ancient limestones are relatively free of clay and frequently contain an abundance of fossils of organisms that thrived in shallow, warm seas. Ancient carbonate rocks have developed in a variety of tectonic settings. Thick sections of limestones and dolomites have formed in ancient subsiding basins in western Texas, Alberta, and Michigan. In such areas, optimum conditions for carbonate sedimentation resulted in a rate of accumulation that approximately equaled subsidence. Thick deposits of limestones have also accumulated during the Paleozoic Era on carbonate platforms at the eastern margin of North America. During the geologic past, sea levels were typically higher than they are today and climates generally warmer. Thus, carbonate platforms were once more abundant and extensive.

One type of carbonate sedimentary rock that continues to perplex geologists is the magnesium-calcium carbonate rock dolomite. You will recall from Chapter 2 that dolomite is a rock composed primarily of the mineral dolomite, $CaMg(CO_3)_2$. Dolomite is believed to form when magnesium that has been concentrated in sea water replaces a portion of the calcium carbonate in previously deposited calcium carbonate sediment. Supporting this interpretation are fossil shells in dolomite strata that were originally composed of calcite but that subsequently have been changed to dolomite. Today, dolomite formation occurs in only a few areas, usually where evaporation of sea water is sufficiently intense to concentrate magnesium. Yet during parts of the Precambrian, Paleozoic, and Mesozoic eras, dolomites were extensively developed (Fig. 3-33). To account for these ancient deposits, it would appear that extensive areas of evaporative conditions would be required. Recently, geologist David Lumsden discovered a correlation between

FIGURE 3-33 **Dolomites of Triassic age exposed in the Italian Alps.** Dolomite was first recognized as a distinctive rock type by Guy de Dolomieu in 1791. The rock type was named dolomite in his honor by Richard Kirwan in 1794. (*Italian Tourist Board.*) ? *How do dolomites differ from limestones?*

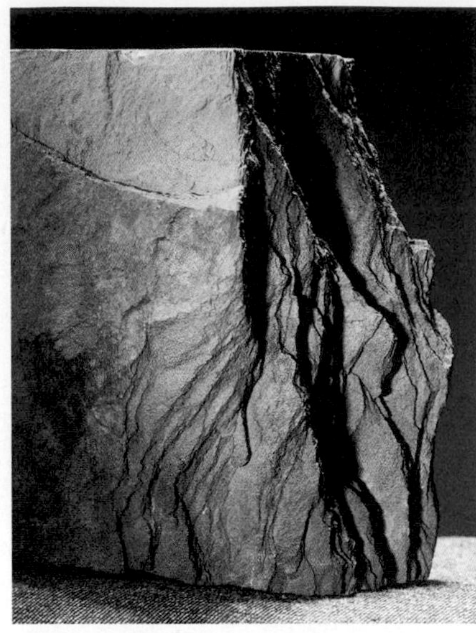

FIGURE 3-34 **Shale.** The dark color of this specimen results from its high content of carbon. Note the finely layered structure, which is termed fissility. Shale tends to split readily along the planes of fissility. (*Copyright Runk/Schoenberger/Grant Heilman.*)

ancient periods of high dolomite formation and episodes of high sea level. He suggests that during times of eustatic (worldwide) rise in sea level, broad, low-lying tracts of the continents were inundated by shallow seas. Where climatic conditions were favorable, these shallow seas provided the ideal evaporative environments needed to enrich sea water with magnesium. As the magnesium replaced part of the calcium in the calcium carbonate sediment that blanketed the sea floor, calcite was converted to dolomite. When the sea level subsequently fell, the magnesium enrichment process halted, and dolomite formation ceased.

Interpretation of Clays and Shales

Shale (Fig. 3-34) is a general term for a very fine-textured, fissile (capable of being split into thin layers) rock composed mainly of silt and clay-size particles or mixtures of the two. In general, the environmental significance of shales parallels that of the sandstones with which they are associated. Frequently, the silt-sized particles in shales are similar in composition and shape to sand grains in the associated sandstone beds. These silty components (Fig. 3-35) can be extracted for study by disaggregating the shale in water and repeatedly pouring off the muddy liquid, retaining the silt particles as a residue.

In shales associated with quartz sandstones, the silt fraction often consists predominantly of rounded quartz grains. Such quartz shales result from the reworking of older residual clays by transgressing shallow seas. Their association with thin, widespread limestones and quartz sandstones provides evidence of their deposition under stable tectonic conditions.

Feldspathic shales contain silt of at least 10 percent feldspar and tend to be rich in the clay mineral kaolinite. Feldspathic shales are common associates of arkoses and are presumed to have formed in a similar environment. Such shales are representative of the

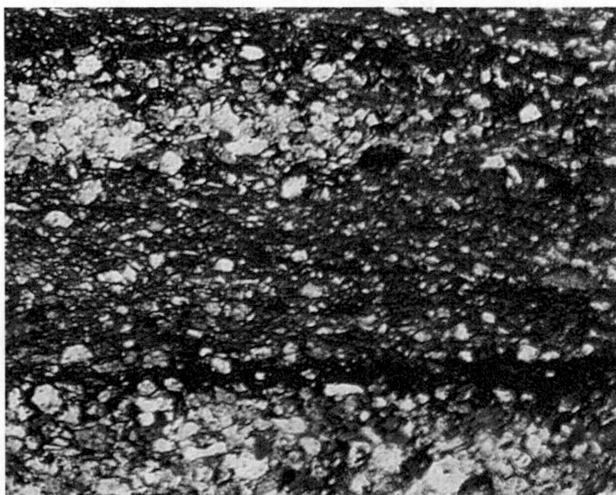

FIGURE 3-35 **Thin section of shale.** As indicated here, silt particles are abundant constituents of shale. The brown color results from organic matter that is mixed with the clay, which is the major constituent of shale. (*Courtesy of G. R. Thompson and J. Turk.*)

finer sediment winnowed from coarser detritus and deposited in quieter locations.

Chloritic shales are usually associated with graywackes. As implied by their name, flakes of the green silicate mineral chlorite are common among the silt-sized components. The less flaky particles tend to be angular. Fissility in these rocks is less well developed than in other shales. The clay and silt particles in chloritic shales are generally derived from mountainous, unstable source areas.

A shale type that is the approximate equivalent of a lithic sandstone is designated a micaceous shale. Mica flakes, quartz, and feldspar are all among its silt-sized components. Micaceous shales are deposited under conditions somewhat less stable than the environment for quartz shales. They are particularly characteristic of ancient deltaic deposits.

The clay minerals that occur in shales are complex hydrous aluminosilicates with constituent atoms arranged in silicate sheet structures. Kaolinites, smectites, and illites are the three major groups of clay minerals. Kaolinites are the purest and seem to have a preferred occurrence in terrestrial environments. Smectites may contain magnesium, calcium, or sodium or any combination of these three, whereas potassium is an essential constituent of illites. Illites are the predominant clay mineral in more ancient shales.

THE SEDIMENTARY ROCK RECORD

Rock Units

William Smith, the British surveyor mentioned in Chapter 1, demonstrated that distinctive bodies of strata could be traced over appreciable distances and therefore could be mapped. He produced an exceptionally fine geologic map of England and Wales in 1815. Smith's map, the first ever made of such high quality and accuracy, was accompanied by a comprehensive table of the rock units encountered in the area. Each of these units was given a particular name, such as the "Clunch Clay," the "Great Oölyte," or the "Cornbrash Limestone." Thus originated the concept of a fundamental unit in geology that was lithologically distinctive, that had recognizable contacts with other units both above and below, and that could be traced across the countryside from exposure to exposure (or in the subsurface from well to well). Smith referred to such a unit as a *stratum*, but today it is universally known as a **formation** (Fig. 3-36). A formation need not consist of a single rock type as long as the particular combination of rock types provides a distinctive aspect to the unit. Thus a formation may be composed entirely of beds of shale, or it may be a distinctive sequence of shale with interbeds of sandstone, evaporites, or limestone.

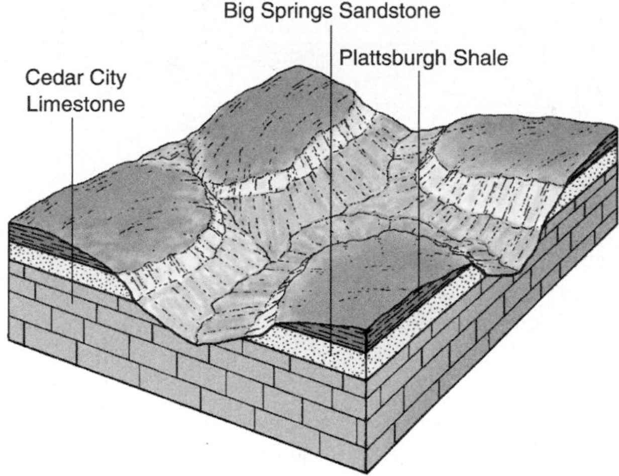

FIGURE 3-36 Formations. The diagram shows three formations. In practice, these formations would be formally named, often after a geographic location near which they are well exposed. For example, the three formations shown here might be designated the Cedar City Limestone, Big Springs Sandstone, and Plattsburgh Shale.

Formations and groupings or subdivisions of formations all constitute **rock units**. Rock units, also called **lithostratigraphic units**, are formally defined as bodies of rock identified by their distinctive lithologic and structural features without regard to time boundaries. They are *mappable* and are distinctly different from the time-rock or **chronostratigraphic units** defined in the previous chapter. Such features as texture, grain size, clastic or crystalline, color, composition, thickness, type of bedding, nature of organic remains, and appearance of the unit in surface exposures (or in the lithologic record of strata penetrated by wells) are all used to define a rock unit and recognize it in the field. Whereas a chronostratigraphic unit represents a body of rock deposited or emplaced during a specific interval of time, a rock unit such as a formation may or may not be the same age everywhere it is encountered. The nearshore sands deposited by a sea slowly advancing (transgressing) across a low coastal plain may form a single blanket of sand (perhaps later to be named the Oriskany Sandstone); however, that sand layer will be older where the sea began its advance and younger where the advance halted (Fig. 3-37).

NAMING ROCK UNITS Formations are given two names: first, a geographic name that refers to a locality where the formation is well exposed or where it was first described, and second, a rock name if the formation is primarily of one lithologic type. For example, the Kimmswick Limestone was first formally described in a professional publication in 1904 by Edward O. Ulrich. Ulrich named the formation for exposures near the small town of Kimmswick, Missouri. The Kimm-

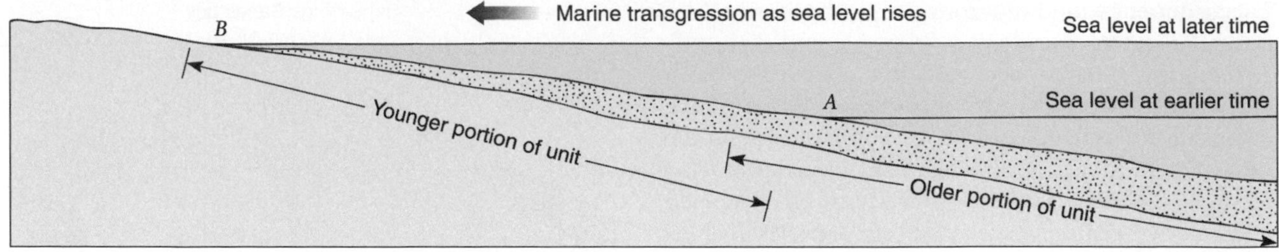

FIGURE 3-37 Diagram showing how the original deposits of a formation may vary in age from place to place.

swick Limestone is entirely composed of limestone, and therefore the rock name can follow the locality name. When formations are composed of several different kinds of rock, the locality name is simply followed by the term *Formation*, as in the Toroweap Formation of the Grand Canyon (Table 3-2).

There are other rock units in addition to formations. Distinctive smaller units within formations may be split out as **members**, and formations may be combined into larger units called **groups** because of related lithologic attributes (or their position between distinct stratigraphic breaks). For example, in Grand Canyon National Park, the Whitmore Wash, Thunder Springs, Mooney Falls, and Horseshoe Mesa rock units are *members* of the massive Redwall Limestone. In another part of the canyon wall, one finds the Tapeats Sandstone, Bright Angel Shale, and Muav Limestone combined to form a larger mappable rock unit known as the Tonto Group (Table 3-2).

Facies

The aforementioned rock terms provide for direct objective mapping of sedimentary beds as well as bodies of metamorphic and igneous rocks. If one is to make inferences about events recorded in rock units, it is useful to employ the term **facies**. A sedimentary facies refers to the characteristic aspects of a rock from which its environment of deposition can be inferred. For example, a body of rock might consist of a bioclastic limestone along one of its lateral margins and micritic limestone elsewhere. Geologists might then delineate a

TABLE 3-2 Rock Units of the Paleozoic Section in Grand Canyon National Park, Arizona

System	Group	Formation	Member
Permian		Kaibab	
		Toroweap	
	Hermit	Coconino	
		Hermit Shale	
Pennsylvanian		Supai	
Mississippian		Redwall Limestone	Horseshoe Mesa
			Mooney Falls
			Thunder Springs
			Whitmore Wash
Devonian		Temple Butte	
Cambrian	Tonto	Muav Limestone	
		Bright Angel Shale	
		Tapeats Sandstone	

"bioclastic limestone facies" and interpret it as a nearshore part of the rock body, whereas they might interpret the micritic limestone facies as a former offshore deposit. In this case, the distinguishing characteristics are lithologic (rather than biologic); therefore, the facies can be further designated as a **lithofacies**. In other cases, the rock unit may be lithologically uniform, but the fossil assemblages differ and permit recognition of different **biofacies** that reflect differences in the environment. A limestone unit, for example, might contain abundant fossils of shallow-water reef corals along its thinning edge and elsewhere be characterized by remains of deep-water sea urchins and snails. There would thus be two biofacies, one reflecting deeper water than the other. These could be designated the "coral" and the "echinoid-gastropod biofacies."

These examples illustrate that facies are clearly the products (sediment, shells of organisms) of particular environments of deposition. Today, as we travel across swamp, floodplain, and sea, we traverse different environments of deposition (Fig. 3-38). Each of these environments of deposition changes laterally into the adjacent environment of deposition, and each provides its present-day facies that likewise change to adjacent synchronous facies. Geologists record these facies changes on lithofacies and biofacies maps. Because these maps are based on time-rock units, they provide a view of different facies of essentially the same age. If one were able to make such maps of successively different times, it would become apparent that ancient facies have shifted their localities as the seas advanced or retreated or as environmental conditions changed.

Consider for a moment an arm of the sea slowly transgressing (advancing over) the land. The sediment deposited on the sea floor may ideally consist of a nearshore sand facies, an offshore mud facies, and a far-offshore carbonate facies. As the shoreline advances inland, the boundaries of these facies also shift in the same direction, thereby developing an **onlap sequence** (Fig. 3-39), in which coarser sediments are covered by finer ones. Should the sea subsequently begin a withdrawal (regression), the facies boundaries will again move in the same direction as the shoreline, creating as they do so an **offlap sequence** of beds (Fig. 3-40). In offlap situations, coarser nearshore sediment tends to lie above finer sediments. Also, because offlap units are deposited during marine regressions, recently deposited sediment is exposed to erosion, and part of the sedimentary sequence is lost. Study of sequential vertical changes in lithology, such as those represented by offlap and overlap relationships, is one method by which geologists recognize ancient advances and retreats of the seas and chart the positions of former shorelines.

Onlap and offlap patterns of sedimentation were recognized as early as 1894 by the German geologist Johannes Walther. Walther observed that the succession of facies occurring laterally is also seen in the vertical succession of facies. Thus, to find what facies are to be encountered laterally from a given locality, one need only examine the vertical sequence of beds at that locality. For example, section B in Figure 3-41 shows a typical "fining upward" succession of facies. Point X is in the nearshore silt facies and is overlain by finer shale and then limestone (hence the expression "fining upwards"). This same sequence of silt to shale to limestone is seen in moving westward to section A. Beneath point X is a coarse beach sand, which can be traced laterally (eastward) in section C. This relationship, in which the vertical succession of facies corresponds to the lateral succession, has been named **Walther's Principle**.

If the pattern of sediment spread seaward from shorelines always graded from nearshore sands to shales and carbonates, as depicted in Figure 3-39, predicting the locations of particular facies in ancient rocks would be comparatively easy. In reality, however, the task is usually more complex. For example,

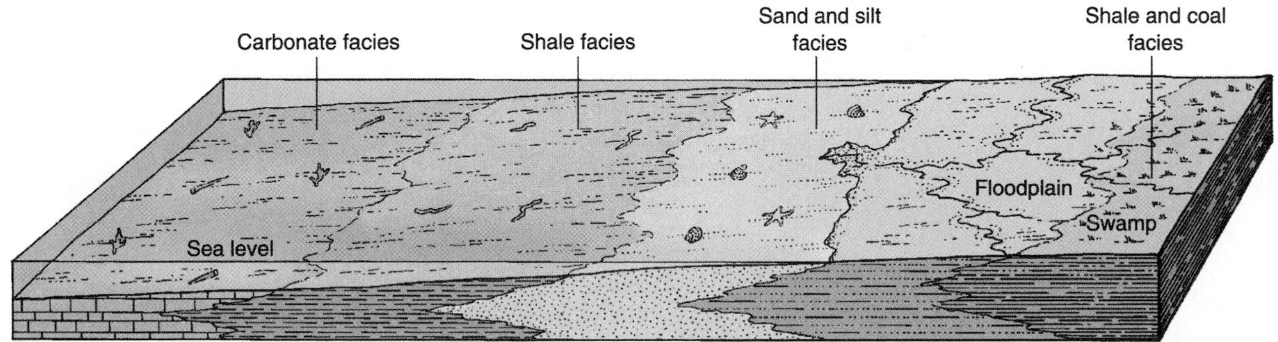

FIGURE 3-38 Sedimentary facies (lithofacies) developed in the sea adjacent to a land area. The upper surface of the diagram shows present-day facies, whereas the front face shows the shifting of facies through time. Notice that bottom-dwelling organisms also differ in environments having different bottom sediment and water depth.

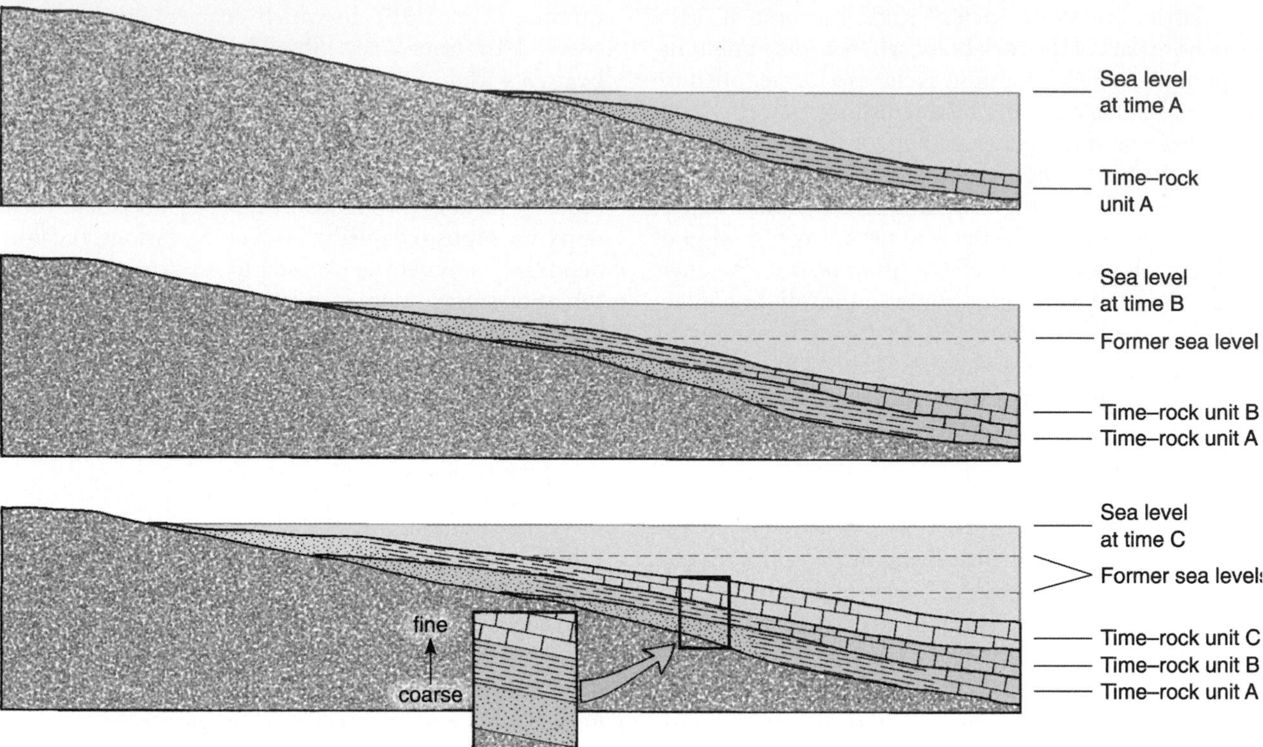

FIGURE 3-39 Sedimentation during a transgression produces an onlap relationship in which finer offshore lithofacies overlie coarser nearshore facies (see inset), nearshore facies are progressively displaced away from a marine point of reference, and older beds are protected from erosion by younger beds.

nearshore sandy facies are not present at all along some coastlines. This may occur because little sand is being brought to the coast by streams, vigorous wave and current action has carried the sand grains away, or possibly sand grains are trapped in submarine canyons farther up the coast. The nature of sedimentation along a coastline is also controlled by the direction of longshore currents, the location of the mouths of major streams that dump their sedimentary load into the sea, the amount of sediment supplied, the presence of barriers to dispersal of sediment, and whether the agent bringing the sediments to the sea is running water, wind, or glacial ice. Any of these factors complicate the study of facies, but they also provide fascinating problems for the geologist to solve.

The Pervasive Effects of Sea-Level Changes

Whenever a change in sea level occurs that is worldwide, the change is termed *eustatic*. Ice accumulating on the continents during an ice age causes lowering of sea level because much of the water making up the ice

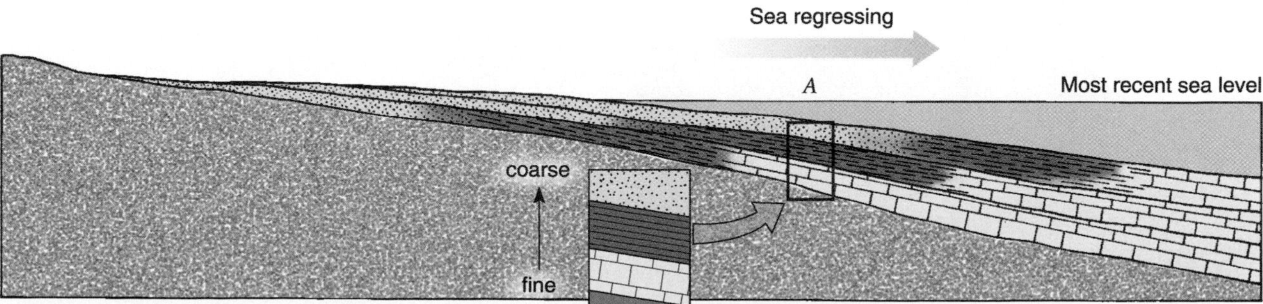

FIGURE 3-40 Sedimentation during a regression produces an offlap relationship in which coarser nearshore lithofacies overlie finer offshore lithofacies, as shown in *A*. The sandy nearshore facies is progressively displaced toward the marine point of reference. Older beds are subjected to erosion as the regression of the sea proceeds. As a result, offlap sequences are less commonly preserved than are onlap sequences.

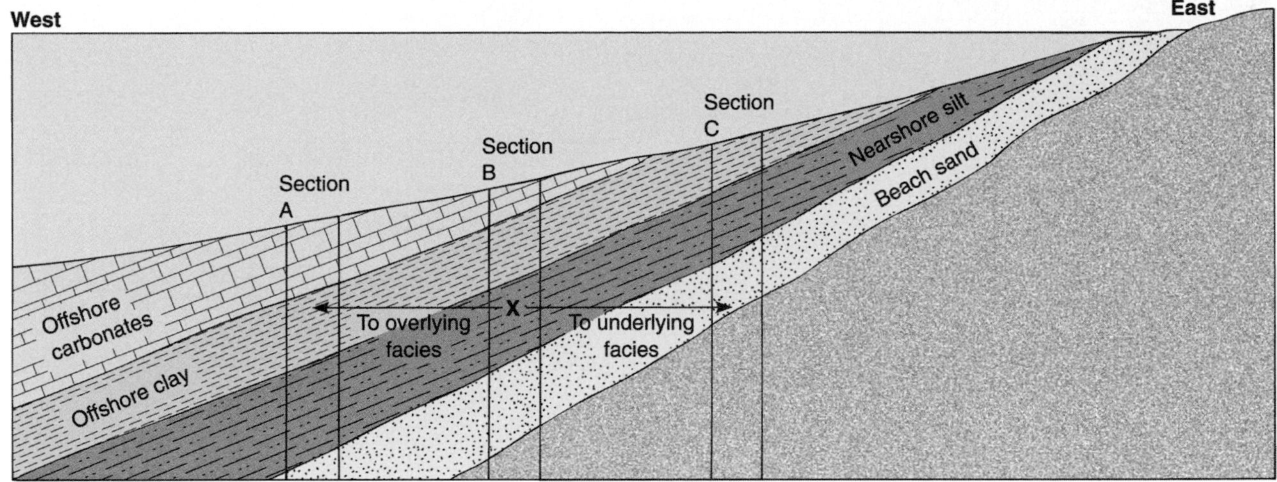

FIGURE 3-41 An illustration of Walther's Principle, which states that vertical facies changes correspond to lateral facies changes. (*After Brice, J. C., Levin, H. L., and Smith, M. S. 1993. Laboratory Studies in Earth History, 5th ed. Dubuque, IA: William C. Brown.*)

ultimately comes from the ocean. Conversely, during warmer episodes, water from melting ice flows back into the ocean, causing a rise in sea level and the landward advance of shorelines around the world. In addition to eustatic changes associated with ice ages, upwarping of the floor of an ocean basin or the development of extensive midoceanic ridges can result in a eustatic change in sea level. Midocean ridges are submarine mountain ranges extending for thousands of kilometers across the floors of the major ocean basins. They are composed of basaltic lavas derived from the mantle. Great rifts in the ocean floor widen to admit the basalt. As will be described in Chapter 5, the newly formed basaltic ocean floor moves laterally by a process called sea-floor spreading, and at a distance may plunge back into the mantle. The basaltic rocks formed along the midocean ridges are hot and thermally inflated. As a result, they displace a considerable volume of sea water and cause a worldwide rise in sea level. When the rate of extrusion of the basalts is rapid, there is likely to be a significant worldwide rise in sea level. Subsequent slower rates of extrusion and spreading would consequently result in a lowering of sea level. These changes profoundly influence the geologic history of continental and shelf areas, for they determine when these areas are inundated, when seas regressed, when there is deposition, and when there is erosion. The alternate advance and retreat of seas associated with events along midoceanic ridges permit geologists to recognize distinct packages or sequences of strata having erosional boundaries that reflect global cycles of sea-level fluctuations. The cycles can be plotted as sea level curves, such as the Vail sea level curves developed from seismic profiles across continental shelves by P. R. Vail (Fig. 3-42). If the Vail sea level curves detected in sediments of the continental shelves truly reflect eustatic changes in sea level, then they are global and permit worldwide correlation of the package of sediments representing each cycle. Thus, rather than correlating individual rock units, one correlates an entire *sequence* of beds. The term **sequence stratigraphy** has been given to this type of correlation. Sequence stratigraphy developed as a consequence of improved methods in oil exploration seismology. The superior technology provides cross-sectional images or **seismic profiles** of strata deep below the Earth's surface (Fig. 3-43).

Critics of Vail sea level curves contend that some of the cycles were the result of vertical movements of coastal regions. Without doubt, changes in the elevation of land areas bordering the ocean can cause effects similar to those resulting from eustatic change. A coastal tract may experience either tectonic uplift or tectonic subsidence. The former is likely to cause a retreat of the sea (offlap) from the rising land area, whereas subsidence might allow the sea to advance. In order to determine if a stratigraphic sequence deposited during the advance or the retreat of the sea validly indicates a eustatic change in sea level, one must show that the sequence can be correlated to similar sequences on other continents.

Precisely how far the sea will advance or retreat during a change in sea level is determined by the amount of change in sea level and the topography of the land. A low-lying, gently sloping terrain would have a much greater area inundated by a small increase in sea level than would a steeply sloping mountainous tract (Fig. 3-44).

Ultimately, the amount of inundation or regression along a coast must be related to the interaction between tectonic movements on the continents and eustatic sea-level changes. If the land area along a continental margin rises at the same time and amount as a eustatic rise in sea level, the tectonic change will cancel

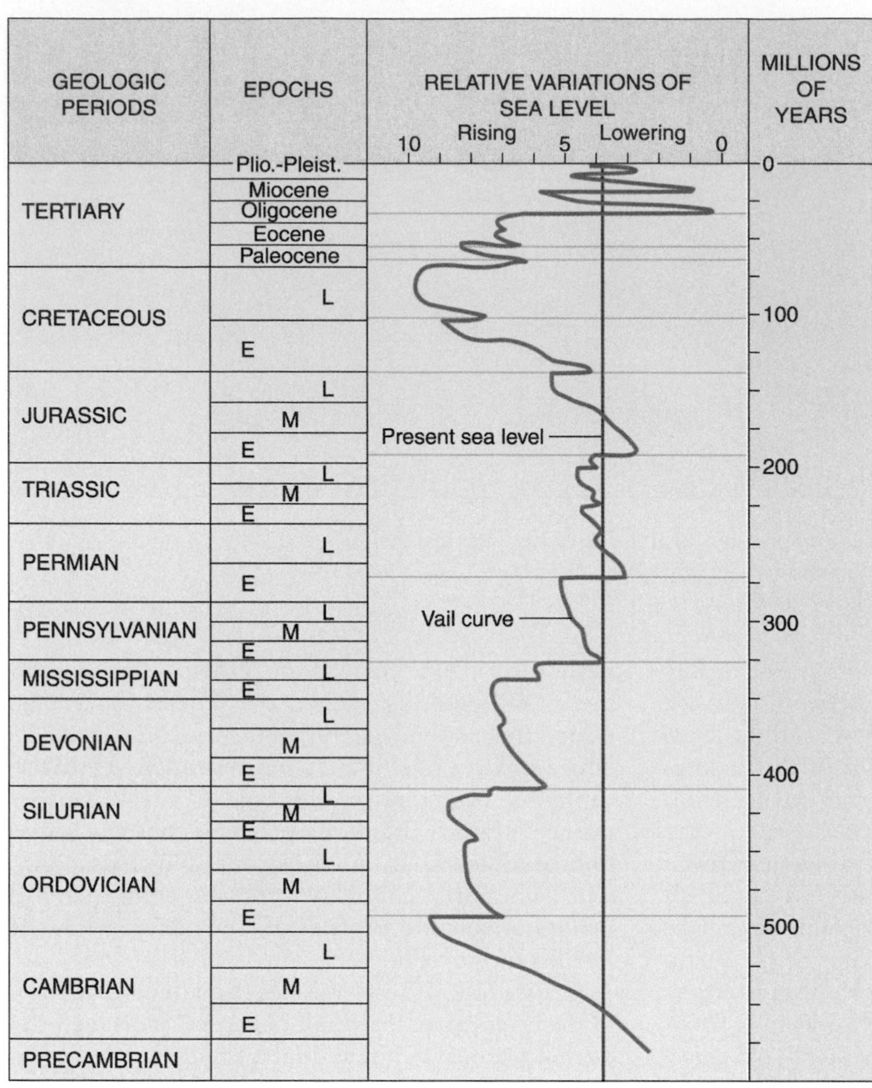

FIGURE 3-42 The Vail sea-level curve of major cycles of sea-level changes. The letters *E*, *M*, and *L* refer to Early, Middle, and Late. (*After Vail, P. R., et al. 1977. American Association of Petroleum Geologists Memoir 26.*) ? *How many major episodes of rising sea level occurred during the Paleozoic Era?*

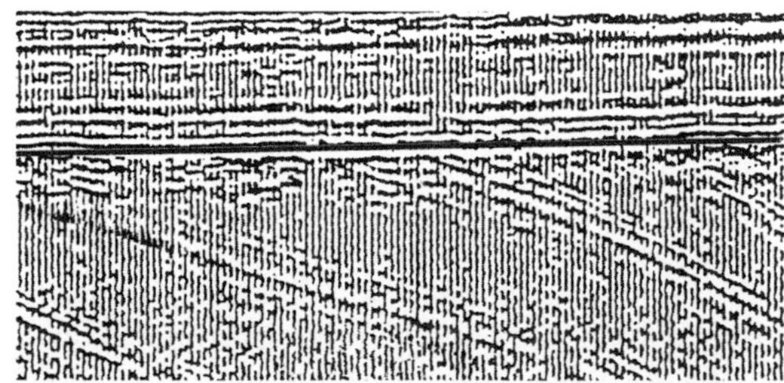

FIGURE 3-43 Seismic profile. The profile depicts tilted (dipping) strata that were beveled by erosion and subsequently covered by horizontal strata. To produce the profile, vibrations were generated into the Earth, either by detonating explosives in shallow drill holes or by hydraulic vibrators. The vibrations reflected off deep layers of rock and returned to the Earth's surface, where they were detected by receivers called geophones. Computers then processed the data received from the geophones to construct the profile.

out the eustatic one. The effect on shoreline displacement will be minimal. In contrast, if the land adjacent to the ocean is subsiding while the sea level rises, a greater marine advance can be expected.

Anyone examining a map depicting one of the world's greatest deltas would quickly recognize that tectonic and eustatic changes are not the only causes of a shift in shorelines. The rapid accumulation of sediment along a coast or at the mouths of rivers will cause land area to be built seaward by progradation.

In the geologic past, there have been repeated advances of seas into low-lying regions of continents. Many appear to be directly related to the Vail cycles just discussed. At times these marine transgressions covered as much as two-thirds of North America. The resulting inland seas are termed **epeiric**, meaning "a sea over a continent." In these epeiric seas were deposited the sedimentary rock record of much of the Paleozoic and Mesozoic eras. The advance and retreat of the epeiric seas were characteristically rather irregular, often interrupted by partial regressions, and ultimately followed

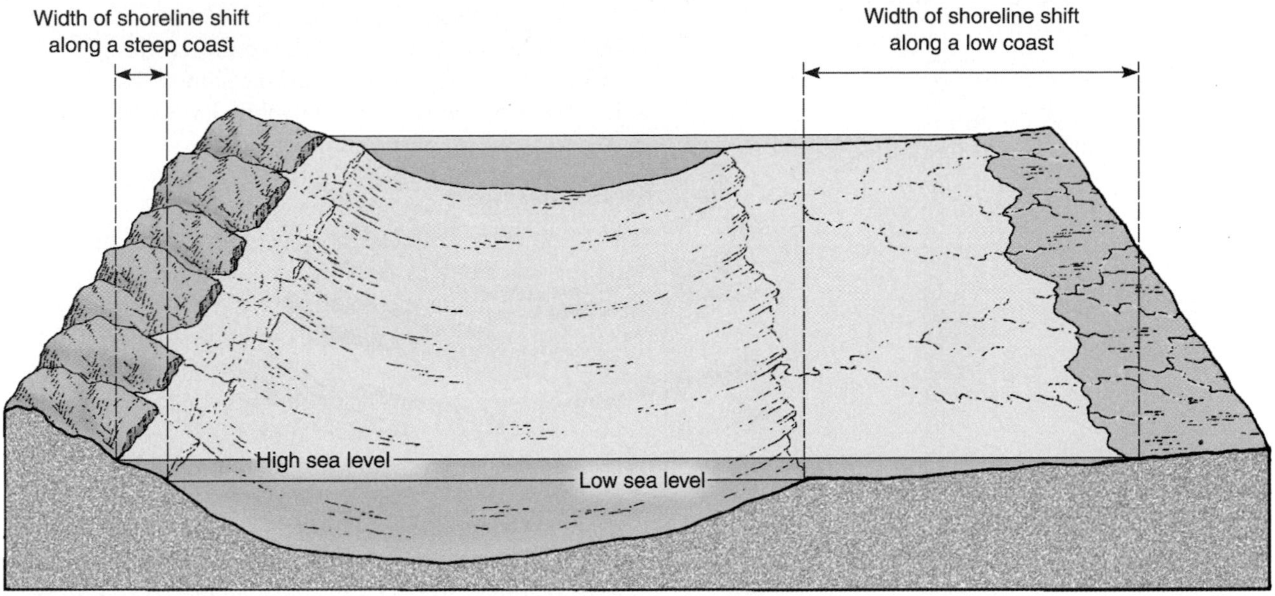

FIGURE 3-44 A rise or fall in sea level will affect a far greater area along a low coastline than along coastlines composed of highlands that rise steeply adjacent to the sea.

by gradual withdrawal back into the major ocean basins. Thus, one can recognize packages of strata separated by regional unconformities. The average rates of advance and retreat were in no sense catastrophic, for they rarely exceeded a few inches per century. However, they were sufficient to cause extensive inundation of the continents over the tens of millions of years encompassed by a geologic period.

Correlation

When examining an isolated exposure or rock in a road cut or the bank of a stream, a geologist is aware that the rock may continue laterally beneath the cover of soil and loose sediment and that the same stratum or rock body, or its equivalent, is likely to be found at other localities. The determination of the equivalence of bodies of rock in different localities is called **correlation** (Fig. 3-45). Correlation of both rock units and choro-stratigraphic units from locality to locality within and between continents or between bore holes drilled into the ocean floor is an important component of stratigraphy. Stratigraphy, however, is not confined to correlation alone. It includes all aspects of the study of rocks, including the conditions under which they originated, their mutual relationships, description, and identification. Because stratified rocks cover approximately three-fourths of the Earth's total land area and because strata contain our most readily interpreted clues to past events, stratigraphy forms the essential core of geologic history.

There are three principal kinds of correlation. The first, **lithostratigraphic**, attempts to correlate rock bodies on the basis of their lithology (composition, texture, color, and so on) and stratigraphic position. **Biostratigraphic** links units by similarity of fossil content. **Chronostratigraphic** correlation expresses equivalence in age as determined by fossils or radioactive dating. Patterns of transgression and regression (as in Vail cycles), distinctive chemical or isotopic characteristics, logs that reflect a unit's electrical or magnetic properties, or any combination of these and other attributes may be useful in correlation.

Because there is more than one meaning for the term, geologists are careful to indicate the kind of correlation used in solving a particular geologic problem. In some cases it is only necessary to trace the occurrence of a lithologically distinctive unit. Here, the age of that unit is not critical, and lithocorrelation will suffice. Other problems can be solved only through the chronocorrelations of rocks that are of the same age. Such correlations involve chronostratigraphic (time-rock) units and are of the utmost importance in geology. They are the basis for the geologic time scale and are essential in working out the geologic history of any region.

Lithocorrelation of strata from one locality to another may be accomplished in several ways. If the strata are well exposed at the Earth's surface, as in arid regions, where soil and plant cover is thin, then it may be possible to trace distinctive rock units for many kilometers across the countryside by actually walking along the exposed strata. In using this straightforward method of correlation, the geologist can sketch the contacts between units directly on topographic maps or aerial photographs. The notations can then be used in the construction of geologic maps. It is also possible to construct a map of the contacts between correlative units in the field by using aerial photographs or appropriate surveying instruments.

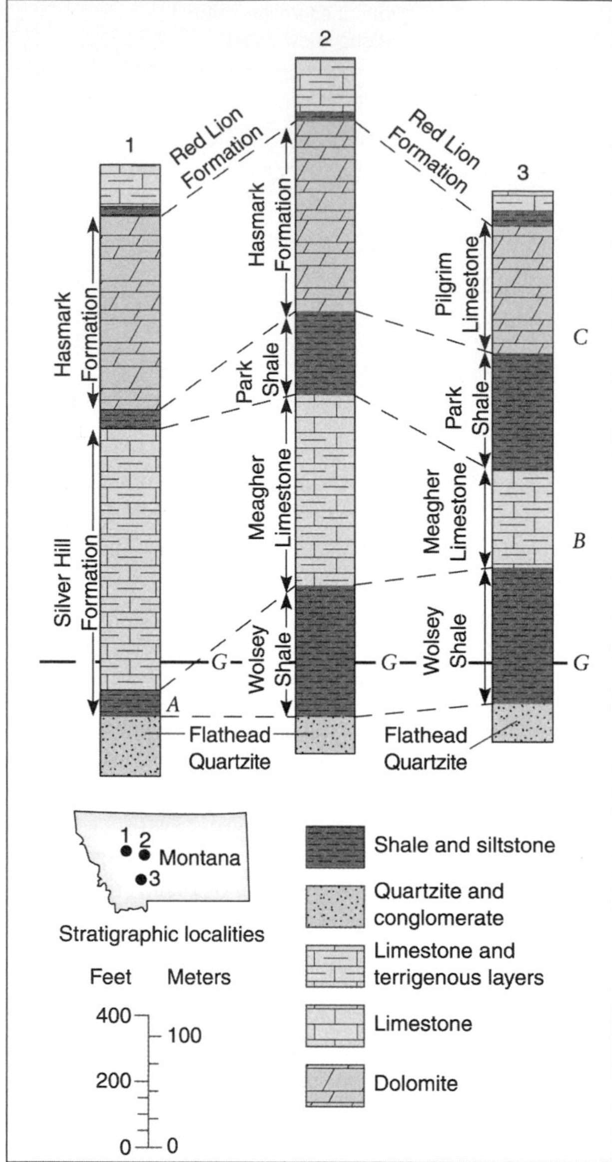

FIGURE 3-45 Correlation of lower Cambrian rock units in western Montana. The letters *C*, *B*, *G*, and *A* indicate the occurrences of trilobite index fossils *Cedaria*, *Bathyuriscus*, *Glossopleura*, and *Albertella*. (*Modified from Schmidt et al. 1994. U.S. Geological Survey Bulletin 2045.*)
? *Do the three lowermost formations in these three columnar sections indicate onlap or offlap?*

In areas where bedrock is covered by dense vegetation and a thick layer of soil, geologists must rely on intermittent exposures found along the sides of valleys, in stream beds, and in road cuts. Correlations are more difficult to make in these areas but can be facilitated by recognizing the similarity in position of the bed one is trying to correlate with other units in the total sequence of strata. A formation may have changed somewhat in appearance between two localities, but if it always lies above or below a distinctive stratum of consistent appearance, then the correlation of the problematic formation is confirmed (Fig. 3-46).

A simple illustration of how correlations are used to build a composite picture of the rock record is provided in Figure 3-47. A geologist working along the sea cliffs at location 1 recognizes a dense oölitic limestome (formation *F*) at the lip of the cliff. The limestone is underlain by formations *E* and *D*. Months later the geologist continues the survey in the canyon at location 2. Because of its distinctive character, the geologist recognizes the oölitic limestone in the canyon as the same formation seen earlier along the coast and makes this correlation. The formation below *F* in the canyon is somewhat more clayey than that at locality 1 but is inferred to be the same because it occurs right under the oölitic limestone. Working upward toward location 3, the geologist maps the sequence of formations from *G* to *K*. Questions still remain, however. What lies below the lowest formation thus far found? Perhaps years later an oil well, such as that at location 3, might provide the answer. Drilling reveals that formations *C*, *B*, and *A* lie beneath *D*. Petroleum geologists monitoring the drilling of the well would add to the correlations by matching all the formations penetrated by the drill to those found earlier in outcrop. In this way, piece by piece, a network of correlations across an entire region is built up.

For correlations of chronostratigraphic units, one cannot depend on similarities in lithology to establish equivalence. Rocks of similar appearance have been formed repeatedly over the long span of geologic time. Thus, there is the danger of correlating two apparently similar units that were deposited at quite different times. Fortunately, the use of fossils in correlation may help to prevent mismatching. Methods of correlation based on fossils (biostratigraphic correlation) are fully described in Chapter 4. They are based on the fact that animals and plants have undergone change through geologic time, and therefore the fossil remains of life are recognizably different in rocks of different ages. Conversely, rocks of the same age but from widely separated regions can be expected to contain similar assemblages of fossils.

Unfortunately, there are complications to these generalizations. For two strata to have similar fossils, they would have to have been deposited contemporaneously in rather similar environments. A sandstone formed on a river floodplain would have quite different fossils from one formed at the same time in a nearshore marine environment. How might one go about establishing that the floodplain deposit could be correlated to the marine deposit? In some cases this might be done by physically tracing out the beds along a cliff or valley side. Occasionally, one is able to find fossils that actually do occur in both deposits. Pollen grains, for example, could have been wafted by the wind into both environments. Possibly, both deposits occur directly above a distinctive, firmly correlated stratum, such as a layer of volcanic ash. Ash beds are particularly good time markers because they are deposited over a wide

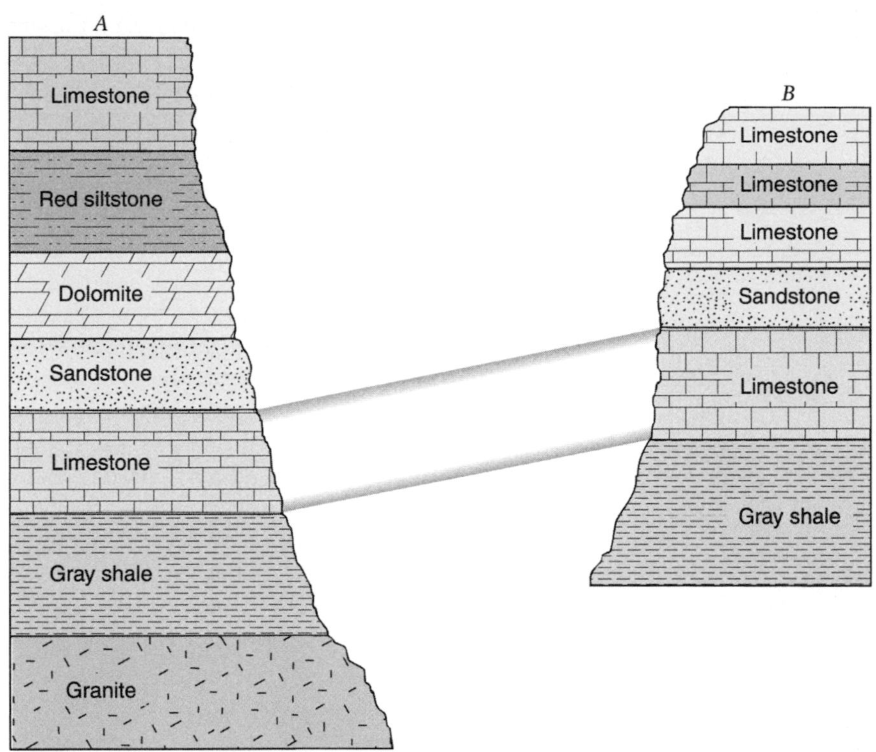

FIGURE 3-46 If the lithology of a rock is not sufficiently distinctive to permit its lithostratigraphic correlation from one locality to another, its position in relation to distinctive rock units above and below may aid in correlation. In the example shown here, the limestone unit at locality *A* can be correlated with the lowest of the four limestone units at locality *B* because of its position between the gray shale and the sandstone units.

area during a relatively brief interval of time. Such key beds are exceptionally useful in establishing the correlation of overlying strata. Finally, the geologist may be able to obtain the actual age of the strata using radioactive methods, and these values can then be used to establish the correlation.

Unconformities

Interpreting the geologic history of an area would be greatly facilitated if deposition were continuous over time and there were no erosional losses of sediment. Unfortunately, such an uninterrupted sequence of strata is rarely encountered. There are gaps in the geologic record where varying thicknesses of strata have been lost to erosion or where deposition did not occur for an interval of geologic time. As mentioned in Chapter 1, we call these breaks in stratigraphic continuity **unconformities**. Whether by erosion or nondeposition, the gap in the geologic record may encompass tens or even hundreds of millions of years of Earth history.

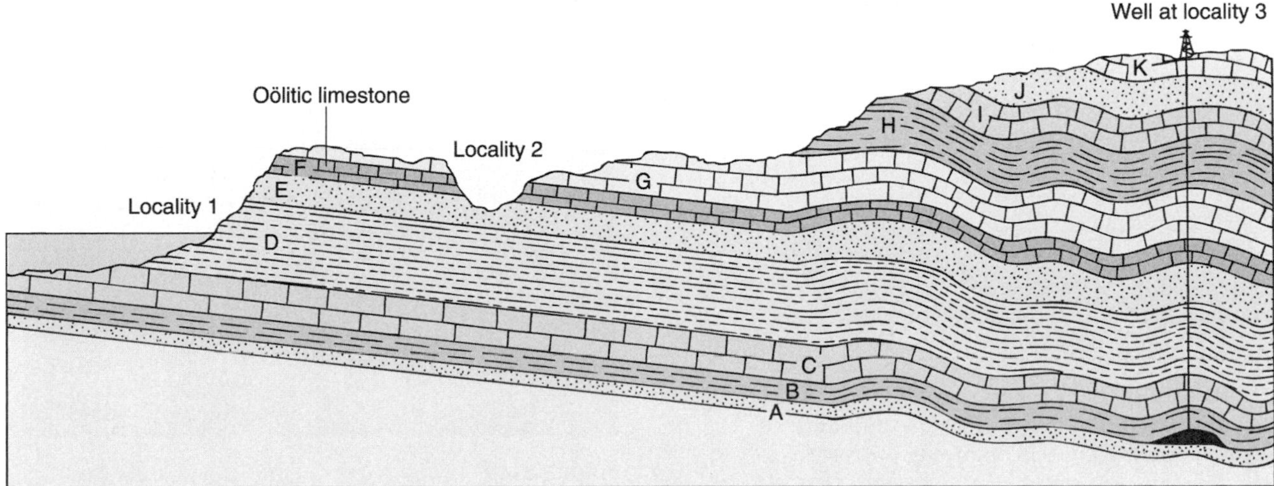

FIGURE 3-47 An understanding of the sequence of formations in an area usually begins with examination of surface rocks and correlation between isolated exposures. Study of samples from deep wells permits the geologist to expand the known sequence of formations and to verify the areal extent and thickness of both surface and subsurface formations.

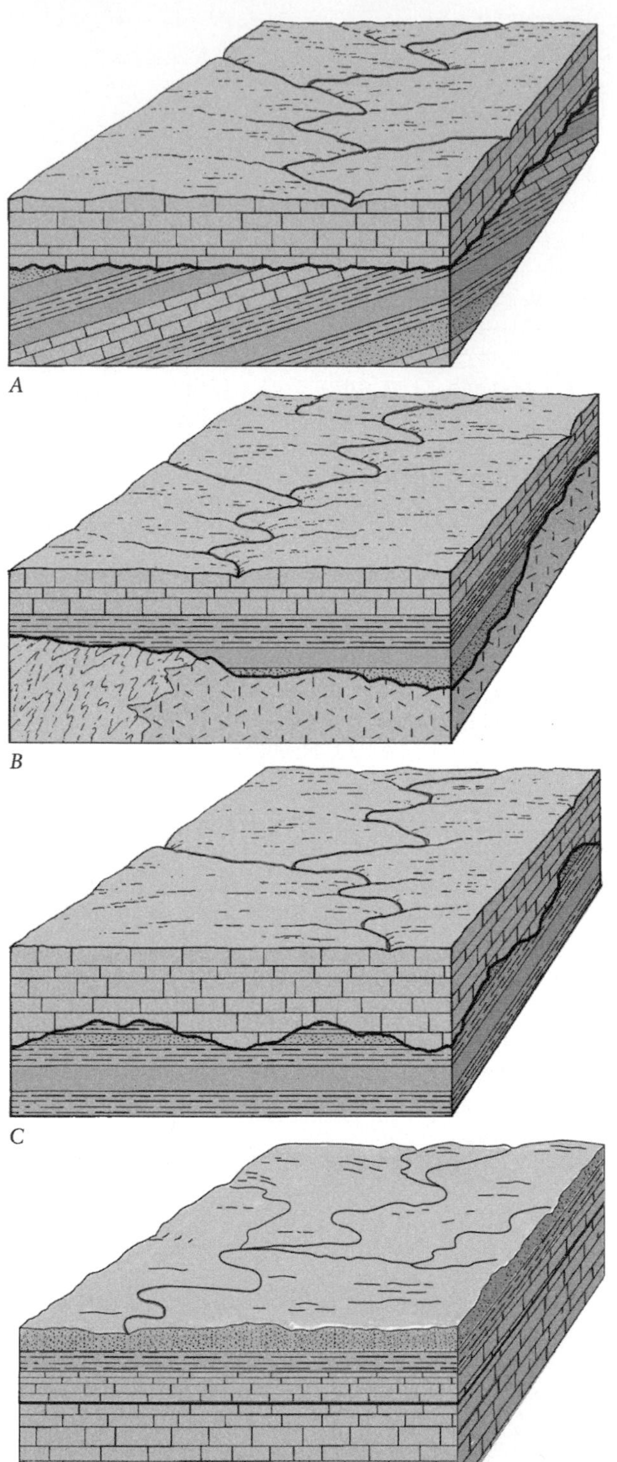

FIGURE 3-48 Four types of erosional unconformities. (*A*) Angular unconformity. (*B*) Nonconformity. (*C*) Disconformity. (*D*) Paraconformity.

The four major kinds of unconformities illustrated in Figure 3-48 differ with regard to the orientation of the rocks beneath the erosional surface. Of the types shown, the **angular unconformity** (Fig. 3-49 and Fig. 3-43) provides the most readily apparent evidence of crustal deformation. James Hutton recognized the significance of an angular unconformity when he observed this feature at Siccar Point on the Scottish coast of the North Sea (see Fig. 1-6).

Examples of unconformities are abundant on every continent. Some do not reflect the degree of deformation apparent in the strata at Siccar Point but rather document the simple withdrawal of the sea for a period of time, followed by another marine transgression. The result may be a **disconformity**, in which parallel strata are separated by an erosional surface. The withdrawal and advance of the sea may be caused by fluctuations in the volume of ocean water, but more commonly they are the result of crustal uplift and subsidence. **Nonconformities** are surfaces where stratified rocks rest on older intrusive igneous or metamorphic rocks (Fig. 3-50). In many nonconformities, crystalline rocks were emplaced deep within the roots of ancient mountain ranges that subsequently experienced repeated episodes of erosion and uplift. Eventu-

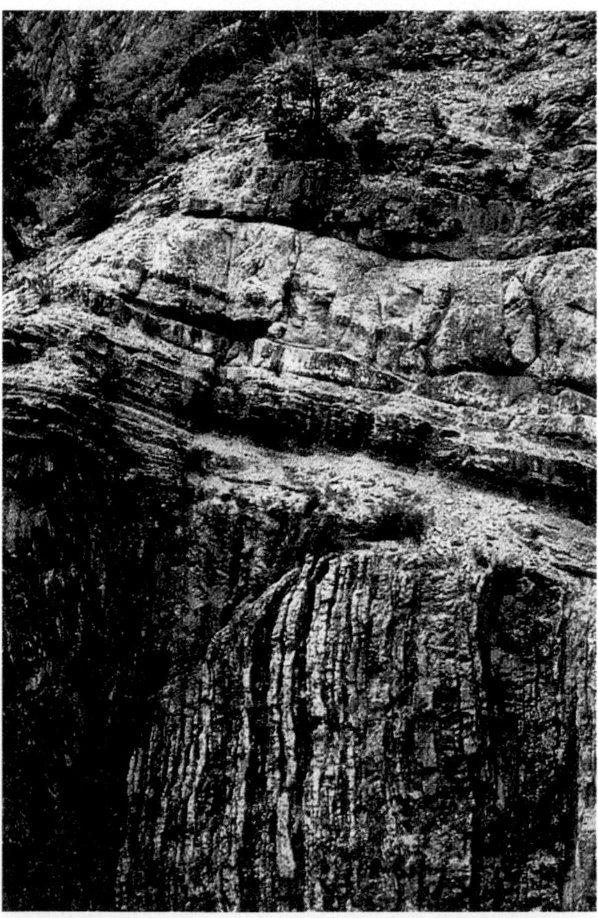

FIGURE 3-49 An angular unconformity separates vertical beds of the Precambrian (Proterozoic) Uncompahgre Sandstone from the overlying, nearly horizontal Devonian strata of the Elbert Formation, at Box Canyon Falls, Ouray, Colorado. (*Photograph by John H. Karachewski, from the Geological Society of America. 1987. Geology 15(5): Cover.*)

FIGURE 3-50 The erosional surface of this nonconformity is inclined at about 45° and separates Precambrian rhyolitic rock from overlying Upper Cambrian Bonneterre Dolomite. Lower and Middle Cambrian rocks are missing. Taum Sauk Mountain, southeastern Missouri. (*Courtesy of D. Bhattacharyya.*)

ally, the igneous and metamorphic core of the mountains lay exposed and provided the surface on which the younger strata were deposited.

Although unconformities represent a loss of geologic record, they are nevertheless also useful to geologists. Like lithostratigraphic units, they can be mapped and correlated. They often record episodes of terrestrial conditions that followed the withdrawal of seas. Where regionally extensive unconformities occur, they permit one to recognize distinct sequences or "packages" of strata of approximately equivalent age.

Depicting the Past

GEOLOGIC COLUMNS AND CROSS-SECTIONS To aid in the synthesis and interpretation of field observations of sedimentary and other rocks, geologists prepare a variety of maps, graphs, and charts designed to show relationships of rock bodies to one another, their thickness, the manner in which they are deformed, and their general composition. The most important graphic devices for communicating such information are columnar sections, cross-sections, and geologic maps.

Columnar sections, like those depicted in Figure 3-45, are made to show the vertical succession of rock units at a given location or for a specific region. They are used in correlation and in the construction of cross-sections. Cross-sections show the vertical dimension of a slice through the Earth's crust. Some cross-sections—namely, the stratigraphic type—emphasize the age or lithologic equivalence of the strata. The vertical measurements for such stratigraphic cross-sections are made from a horizontal line termed the *datum* that is drawn at the top of a definite rock unit or some other marker such as a particular fossil assemblage. The datum in Figure 3-45 is for the trilobite fossil *Glossopleura* (*G*).

Geologists construct stratigraphic cross-sections with the datum line horizontal. In the field, however, the marker for the datum line follows the configuration of the strata and may be variously inclined and folded. For this reason, stratigraphic sections do not validly show the tilt or position of beds relative to sea level.

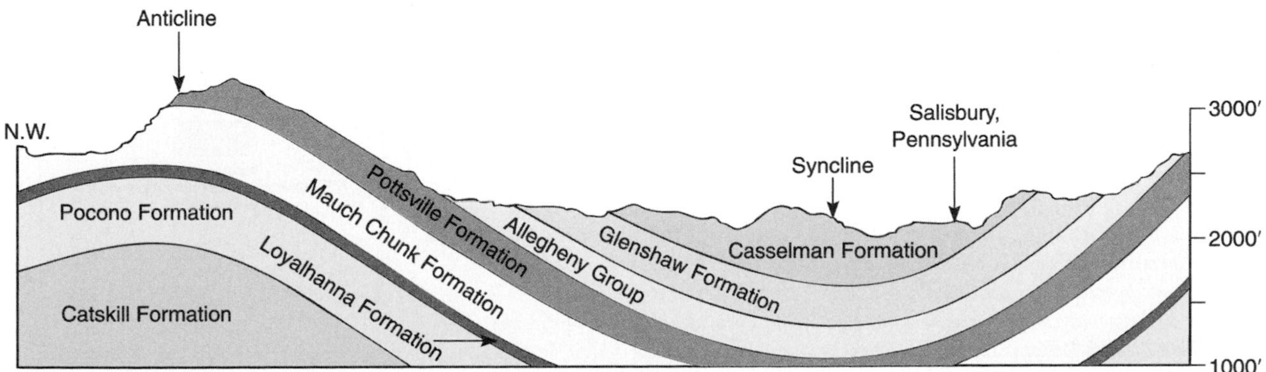

FIGURE 3-51 Geologic structural cross-section across Paleozoic rocks in the Appalachian Mountains, southeastern Pennsylvania, extending northwestward from the town of Salisbury.

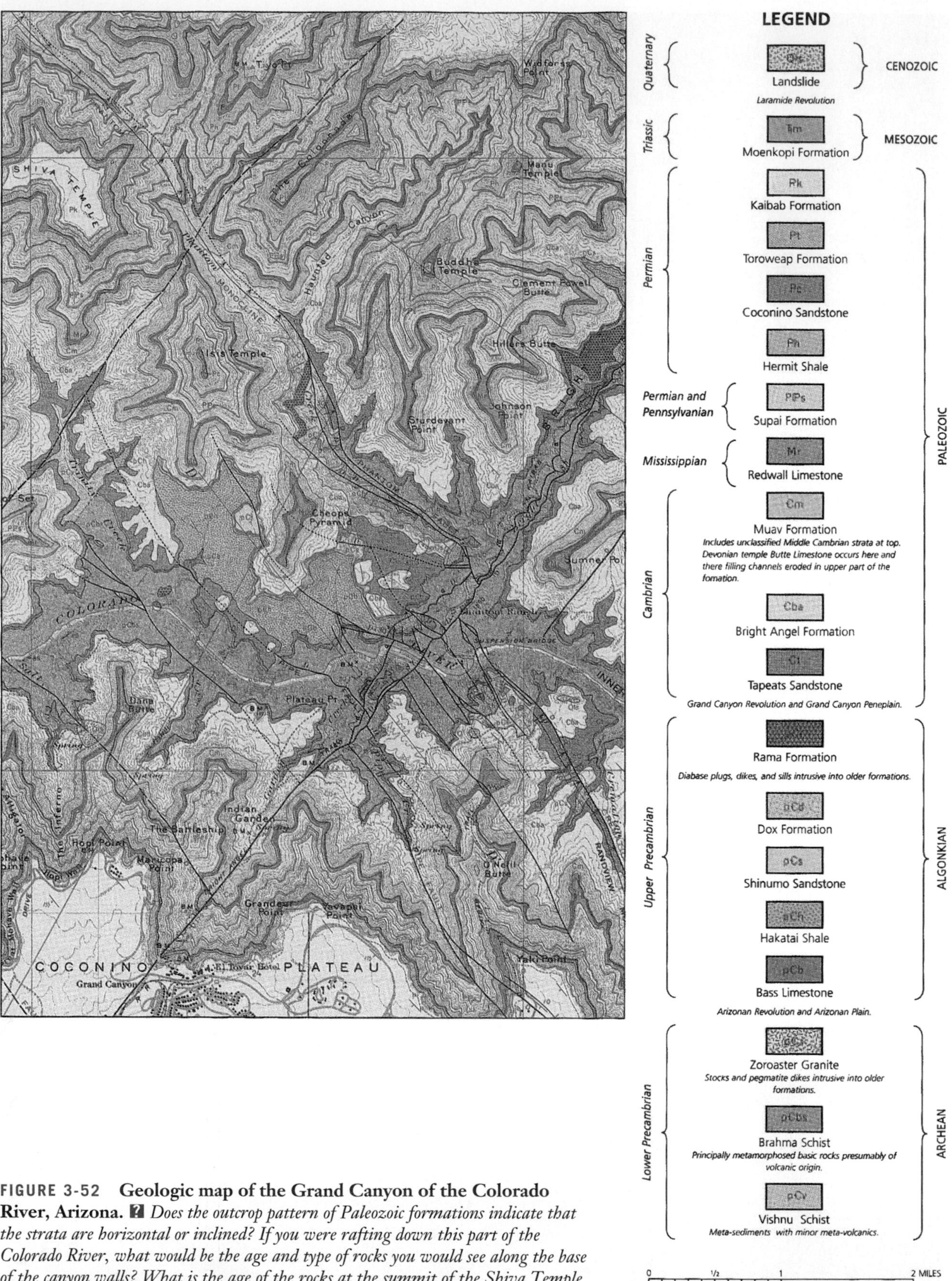

FIGURE 3-52 Geologic map of the Grand Canyon of the Colorado River, Arizona. ? *Does the outcrop pattern of Paleozoic formations indicate that the strata are horizontal or inclined? If you were rafting down this part of the Colorado River, what would be the age and type of rocks you would see along the base of the canyon walls? What is the age of the rocks at the summit of the Shiva Temple on the northwestern part of the map?*

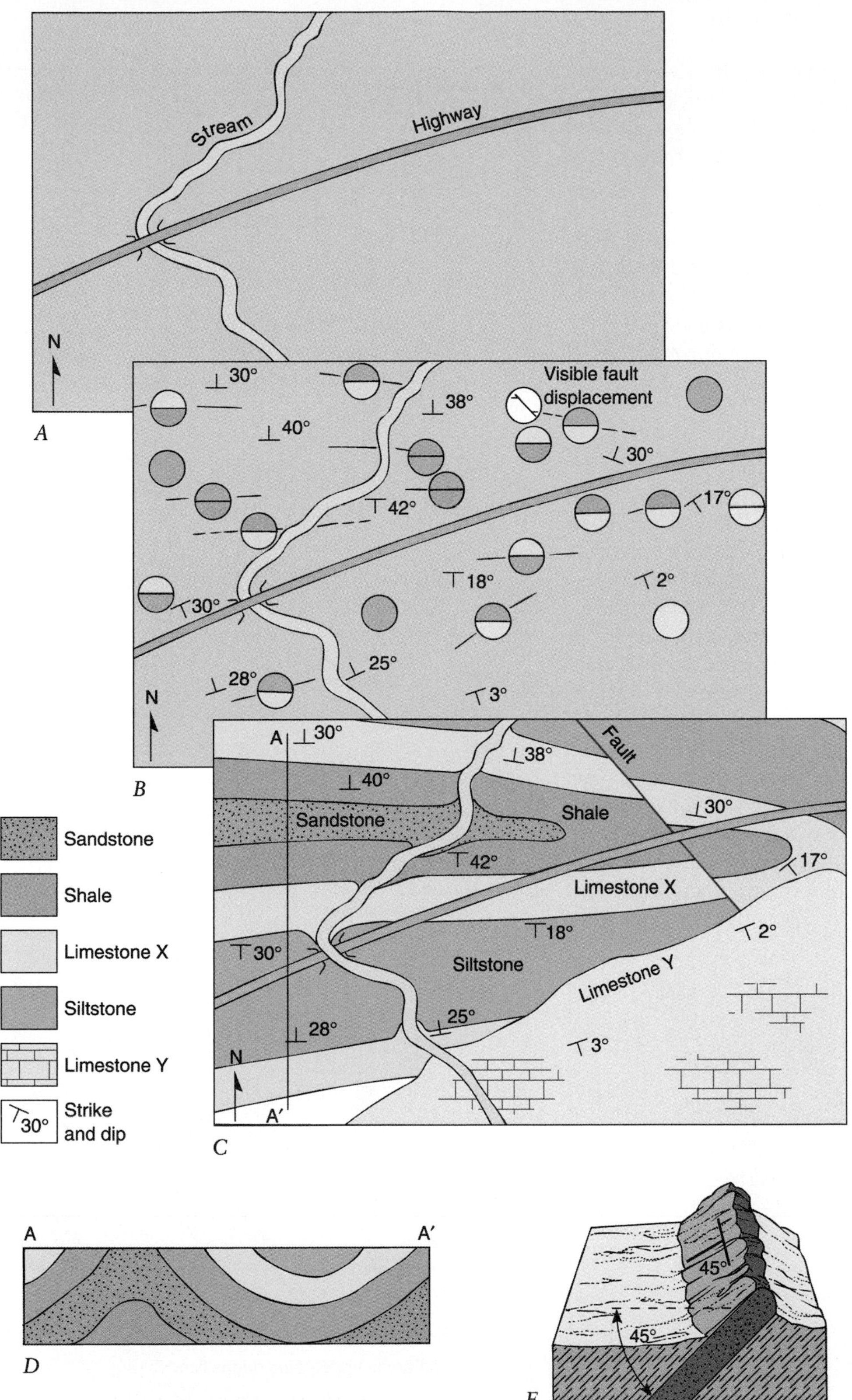

FIGURE 3-53 Steps in the preparation of a geologic map. (*A*) A suitable base map is selected. (*B*) The locations of rock exposures of the various formations are then plotted on the base map. Special attention is given to exposures that include contacts between formations; where they can be followed horizontally, they are traced onto the base map also. Strike (the compass direction of a line formed by the intersection of the surface of a bed and a horizontal plane) and dip (the angle an inclined stratum makes with the horizontal) are measured wherever possible and added to the data on the base map. After careful field study and synthesis of all the available information, formation boundaries are drawn to best fit the data. (*C*) On the completed map, color patterns are used to show the areal pattern of rocks beneath the cover of soil. (*D*) A cross-section is shown along line A-A9. (*E*) A block diagram illustrates strike and dip. ? *What is the oldest rock unit seen at the surface along A-A9? Where is the youngest unit located?*

Stratigraphic cross-sections are most effective in showing the way beds correlate and vary in thickness from exposure to exposure or well to well. To show the way beds are folded, faulted, or tilted, a structural cross-section can be prepared (Fig. 3-51). In the structural cross-section, the datum is a level line parallel to sea level, and the tops and bottoms of rock units are plotted according to their true elevations. If the vertical and horizontal scales are similar, the attitude of the beds will be correctly depicted. Many times, however, it is useful to have a larger vertical than horizontal scale to emphasize geologic features.

Geologic Maps

Geologic maps show the distribution of rocks of different kinds and ages that lie directly beneath the loose rock and soil covering most areas of the surface (Fig. 3-52). Assume for a moment that all of this loose material and the vegetation growing on it were miraculously removed from your home state, so that bedrock would be exposed everywhere. Imagine, further, that the surfaces of the formations now exposed were each painted a different color and photographed vertically from an airplane. Such a photograph would constitute a simple geologic map. In actual practice, a geologic map is prepared by locating contact lines between formations in the field and then plotting these contacts on a base map (Fig. 3-53*A* and *B*). Symbols are added to the colored areas to indicate formations and lithologic regions, mineral deposits, and structures such as folds

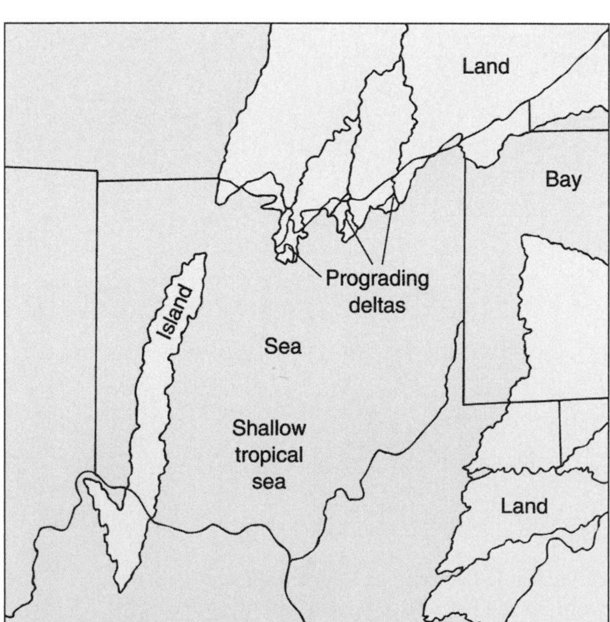

FIGURE 3-54 Paleogeographic map of Ohio and adjoining states during an early part of the Mississippian Period. The data for this study were obtained from outcrops and over 40,000 well records. (*After Pepper, J. F., de Witt, W. J., and Demarest, D. F. 1954. U.S. Geological Survey Professional Paper 259.*)

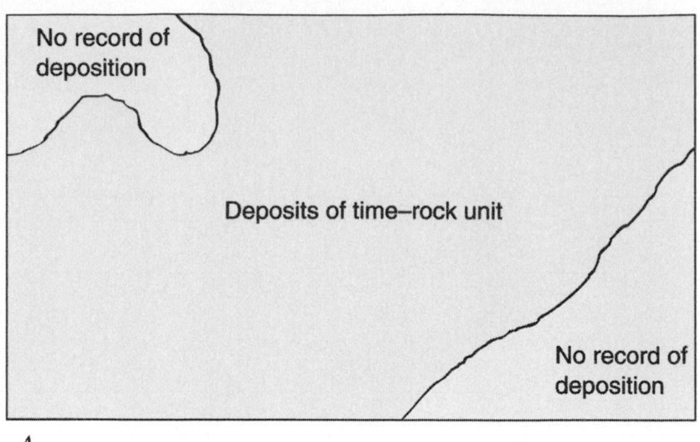

A

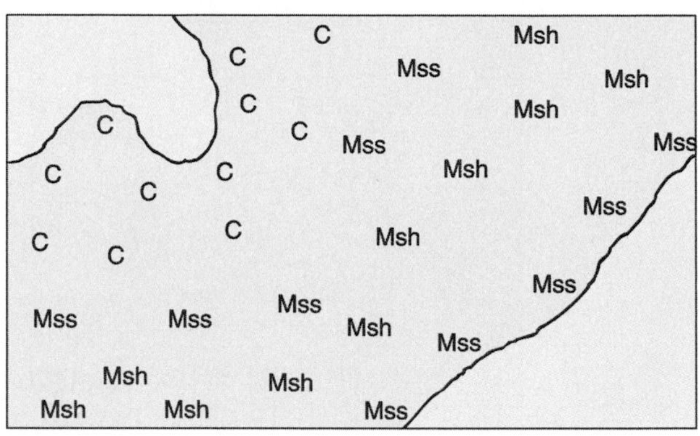

B

C: Continental sediments containing fossils of freshwater clams and land plants
Mss: Marine sandstone with fossils of marine invertebrates
Msh: Marine shale containing abundant fossils of marine microorganisms

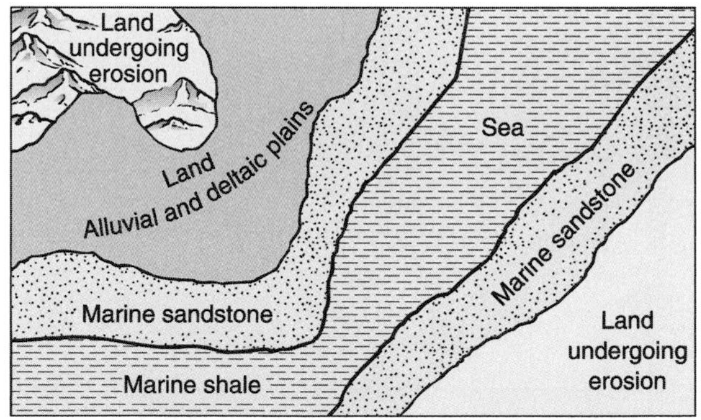

C

Land areas undergoing erosion
Land areas undergoing deposition
Marine areas undergoing deposition

FIGURE 3-55 Stages in the construction of a paleogeographic map. (*A*) Area of occurrence of a particular time-rock unit. (*B*) Plot of rock types within the time-rock unit. (*C*) Paleogeographic reconstruction.

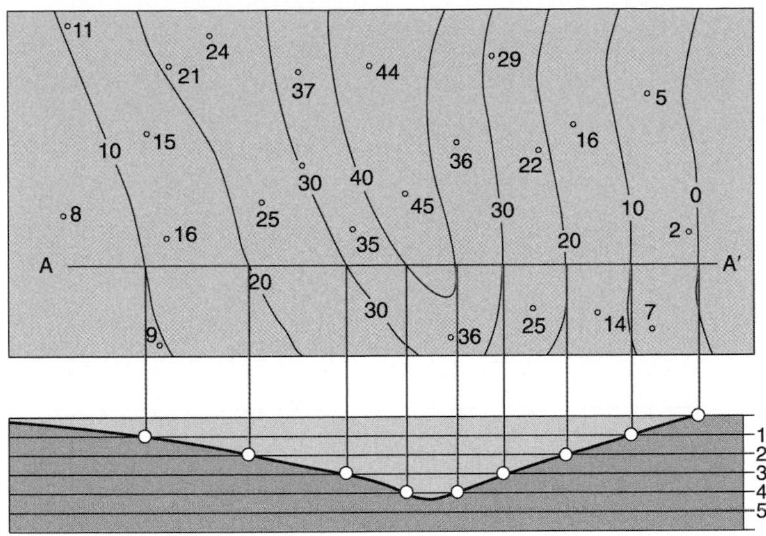

FIGURE 3-56 Diagram illustrating the construction of a simple isopach map in an area of undeformed strata.

and faults. Once the geologic map is completed, a geologist can tell a good deal about the geologic history of an area. The formations depicted represent sequential "pages" in the geologic record. From the simple geologic map shown in Figure 3-53*C*, the geologist is able to deduce that there was an ancient period of compressional folding, that the folds were subsequently faulted, and that an advance of the sea resulted in deposition of younger sedimentary layers unconformably above the more ancient folded strata.

Paleogeographic Maps

A map showing the geography of a region or area at some specific time in the geologic past is termed a **paleogeographic map**. Such maps are really interpretations based on paleontologic and geologic data. The majority of such maps show the distribution of ancient lands and seas (Fig. 3-54). Paleogeographic maps are, at best, of limited accuracy, because since seas advance and retreat endlessly through time, the line drawn at the sea's edge may represent an average of several shoreline positions. They are nevertheless useful for showing general geographic conditions within regions or continents. To prepare a paleogeographic map, one would plot all occurrences of rocks of a given time interval on a map and enclose the area of occurrence in boundary lines (Fig. 3-55*A* and *B*). Areas of nonoccurrence may be places of no deposition or places where deposits once existed but were subsequently eroded away. With the help of fossils, the nature of the sediments—that is, whether marine or nonmarine—is determined and plotted on the map. The final step is to complete the paleogeographic reconstruction (Fig. 3-55*C*).

Isopach Maps

Isopach maps are prepared by geologists to illustrate changes in the thickness of a formation or chronostratigraphic unit. The lines on an isopach map (Fig. 3-56) connect points at which the unit is of the same thickness. On a base map, the geologists plot the thickness of units as they are revealed by drilling or in measured surface sections. Isopach lines are then drawn to conform to the data points. Ordinarily, the upper surface of the unit being mapped is used as the horizontal plane or datum from which thickness measurements are made. An isopach map may be very useful in determining the size and shape of a depositional basin, the position of shorelines, and areas of uplift. Figure 3-57 is an isopach map of Upper Ordovician formations in Pennsylvania and adjoining states. The map indicates a

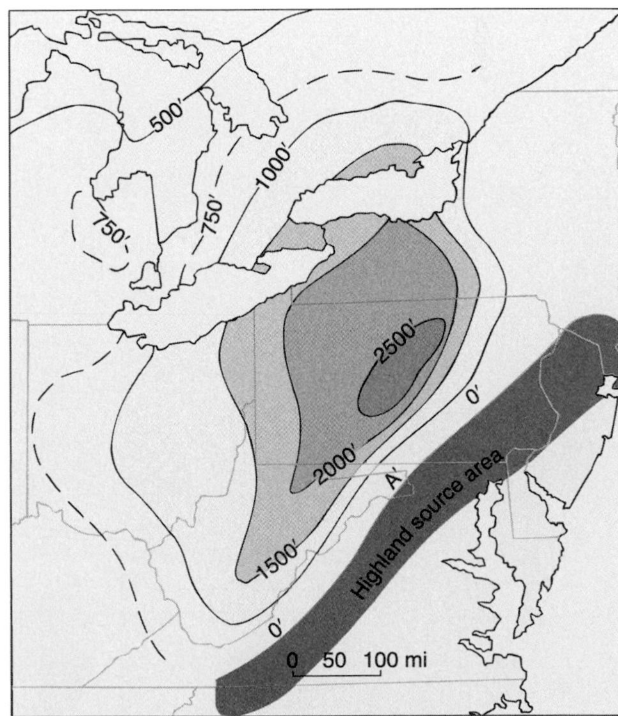

FIGURE 3-57 Isopach map of Upper Ordovician formations in Pennsylvania and adjoining states. (*After Kay, M. 1951. Geological Society of America Memoir No. 48.*)

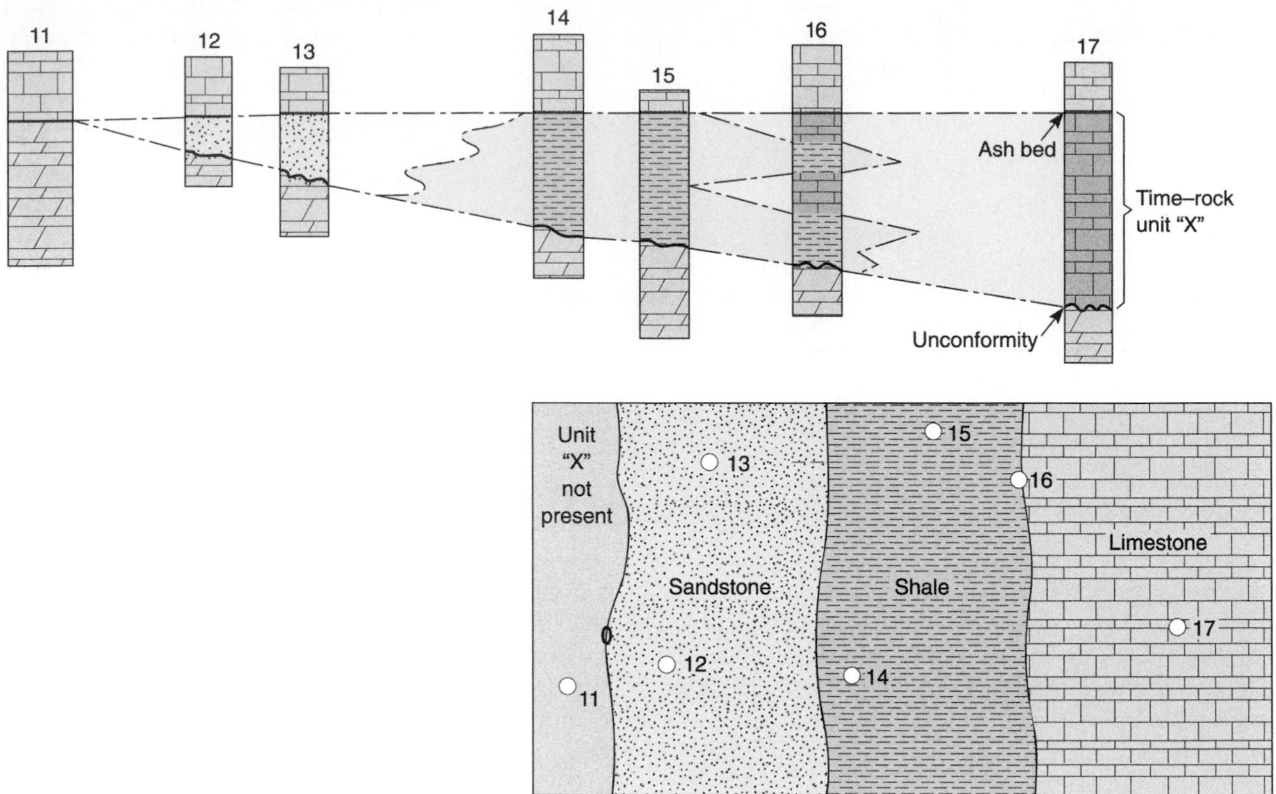

FIGURE 3-58 Diagram illustrating the preparation of a lithofacies map from a subsurface time-rock unit. Well locations are indicated by small circles. Lithostratigraphic correlation of rock units between wells is indicated by dashed lines. Because of the few control points, the exact position of lithofacies boundaries on this map is somewhat arbitrary. ? *Why is the ash bed a good datum for the cross-section?*

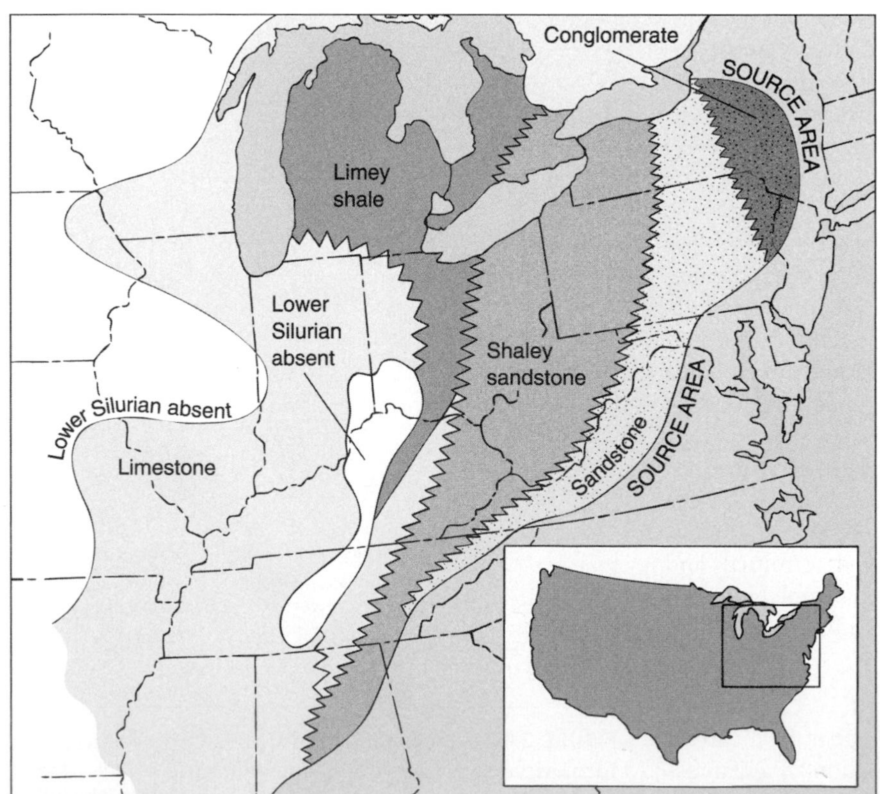

FIGURE 3-59 Lithofacies map of Lower Silurian rocks in the eastern United States. (*After Amsden, T. W. 1955. Bull Am Assoc Petrol Geol 39:60–74.*)

Grand Canyon National Park

In the year 1540, Hopi Indians directed a band of conquistadores to the rim of the Grand Canyon of the Colorado River. One can imagine their sense of wonder as they gazed at this stupendous natural spectacle. Awesome in magnitude and beauty, the great chasm was unlike anything Western humans had ever witnessed. It remains so today, an awesome monument to the erosive force of running water and gravitational downslope transfer of solid rock and weathered debris.

The historical geologist sees the layers of sandstone, shale, limestones, and lava flows exposed in the canyon walls as a great history book that reveals geologic events and changing life over a span of 2 billion years. Rocks of the first chapter are sands and muds that record the presence of a shallow water body in the region. Volcanoes erupted nearby, for the rocks are interbedded with layers of lava and volcanic ash. About 1.7 billion years ago, mountain building deformed and metamorphosed these sediments. In their altered state, they comprise the Vishnu Schist (Figs. 3-52 and 7-13), seen close up by rafters passing through the canyon's Inner Gorge. The Zoroaster Granite intrudes the Vishnu Schist and can also be seen in the deep clefts of the canyon. An interval of erosion followed the emplacement of the Zoroaster Granite. Then high-silica melts invaded joints and fractures and formed veins and igneous rock, which were later converted to light-colored gneisses. The region was mountainous at this time. Land plants had not evolved, and there was little to retard the forces of erosion. Eventually, the mountains were reduced to lowlands, and their cores can be seen in the canyon walls.

The next chapter in the Grand Canyon story is written in 3700 meters of sedimentary rocks and lava flows spread extensively over the Vishnu Schist. These are the rocks of the Grand Canyon Supergroup (Fig. 7-14). Following their deposition, the region was subjected to tensional forces that produced north-south trending fault-block mountains. The higher fault blocks became mountains. Debris eroded from these mountains filled intervening low areas where downfaulted blocks existed. The destructional forces of erosion gradually reduced the entire region to a low-lying terrain recorded in geologic history by "the great unconformity" that separates Precambrian from Paleozoic strata. Remnants of the once more extensive Grand Canyon Supergroup are nestled in remaining downfaulted blocks beneath the great unconformity.

Ascending the canyon, we reach rocks of the Paleozoic Era. The Paleozoic was a time of repeated inundation and regression of shallow seas. The first of the inundations laid down the nearshore Tapeats Sandstone (Fig. 8-10). These sands were followed by the Bright Angel Shale and Muav Limestone as the shoreline shifted eastward. The three formations comprise the Tonto Group. Their sequential change in lithology illustrates the way in which rock units may transgress time boundaries (Fig. 8-11).

As if pages in our history book had been ripped out, strata of Ordovician and Silurian age are not found in the Grand Canyon. They may have been deposited there, but if so, they have been lost to erosion. Thus, an unconformity (a gap in the stratigraphic record) caps the Cambrian sequence. Above that unconformity, carbonates of the Devonian Temple Butte Limestone were laid down in a shallow sea. Again the sea withdrew, only to return another time during the Mississippian Period. In this Mississippian sea, cherty carbonates of the Redwall Limestone were deposited (see accompanying photograph). Although freshly broken surfaces of the Redwall Limestone are gray, its weathered surface is stained red by iron oxide washed down the face of the precipice from red beds of the overlying Supai Group and Hermit Shale. The Redwall forms bold cliffs that front many of the canyon's promontories. It is richly fossiliferous with the remains of brachiopods, bryozoans, crinoids, and corals.

The withdrawal of the Redwall Sea is signaled by the presence of estuarian and tidal flat sediments that are part of the Surprise Canyon Formation near the top of the Redwall. Above these sediments of transitional envi-

View from the Nankoweap Indian Site of the Mississippian Redwall Limestone, Grand Canyon of the Colorado River, northwestern Arizona. The Redwall is actually a bluish gray limestone containing chert nodules. It takes its colorful name from a coating of red iron oxide stain derived from overlying strata. (*Copyright C. C. Lockwood/Earth Scenes.*)

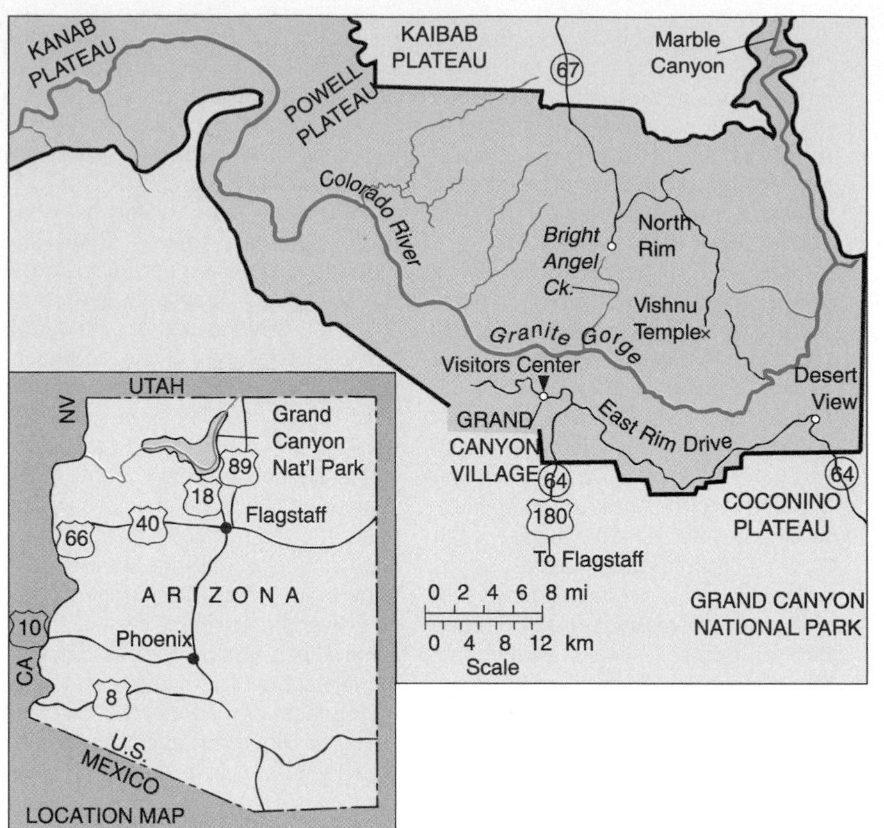

Location map.

ronments, one finds yet another erosional unconformity, and above that surface lie strata of the Pennsylvanian and early Permian Supai Group. The Buddha, Zoroaster, and other spectacular temples in the park are sculpted in the Supai. Nonmarine beds of the Supai exhibit tracks of amphibians, possibly those of reptiles, as well as imprints of ferns. In the overlying Hermit Shale, one can see evidence of nonmarine deposition in the formation's mud cracks and fossils of insects, conifers, and ferns. The floodplains and marshy tracts on which the sediments of the Hermit Shale were deposited were soon covered by migrating dunes of the Permian Coconino Sandstone. The nearly white Coconino sands are cross-bedded. The frosted, well-sorted, well-rounded grains reflect an origin as windblown sediment. Reptiles wandering across the dunes left their footprints in the sand.

Marine limestones and sandstones of the Toroweap Formation rest on the Coconino. They record the advance of a sea over the Coconino dune fields. Above the Toroweap are the bold vertical cliffs of the Kaibab Limestone. This thick and resistant formation forms the surface of the Kaibab Plateau north of the canyon and the Coconino Plateau on the south side. The Kaibab is the final rock unit of the Paleozoic in the Grand Canyon. During the Mesozoic, floodplain sands and silts of the Triassic Moenkopi Formation and gravels of the Shinarump Conglomerate were spread across the region. Near the Grand Canyon, however, all but a few remnants of these formations were swept away by erosion.

The region now known as the Colorado Plateau was tectonically uplifted late in the Cenozoic. That uplift resulted in steeper stream gradients and increased stream erosive powers. The Colorado River and its tributaries were able to deepen their channels at very rapid rates. Fifty years ago, most geologists estimated about 7 million years were required to erode the Grand Canyon. Recent measurements of rates of erosion along similar streams indicate that rivers having torrential flow can erode incredibly rapidly. Geologists involved in these studies now believe that erosion of the Grand Canyon and other deep gorges in eastern Arizona took only about 1.5 million years (a mere moment in geologic time). Thus, although the rocks of the canyon are very old, the great chasm itself is geologically young.

The Grand Canyon reached its present depth not only as a result of the erosive power of running water but also because of impact and abrasion by cobbles and gravels carried in the rushing currents. Yet if the river alone was the only mechanism for erosion, the canyon would have vertical walls. Weathering and gravitational mass movements of eroded debris downslope broadened the canyon to its present width of over 17 kilometers. The Colorado, like rivers everywhere, acts as a conveyor belt, carrying its own load as well as the debris supplied to it by slides, rockfalls, and other gravity-driven movements that we call mass wasting.

There is much more geology in the Grand Canyon than can be described in these few pages or that can be encompassed in a short visit. If you are fortunate enough to go there, stay the day and watch the sun go down over this magnificent colossus of canyons.

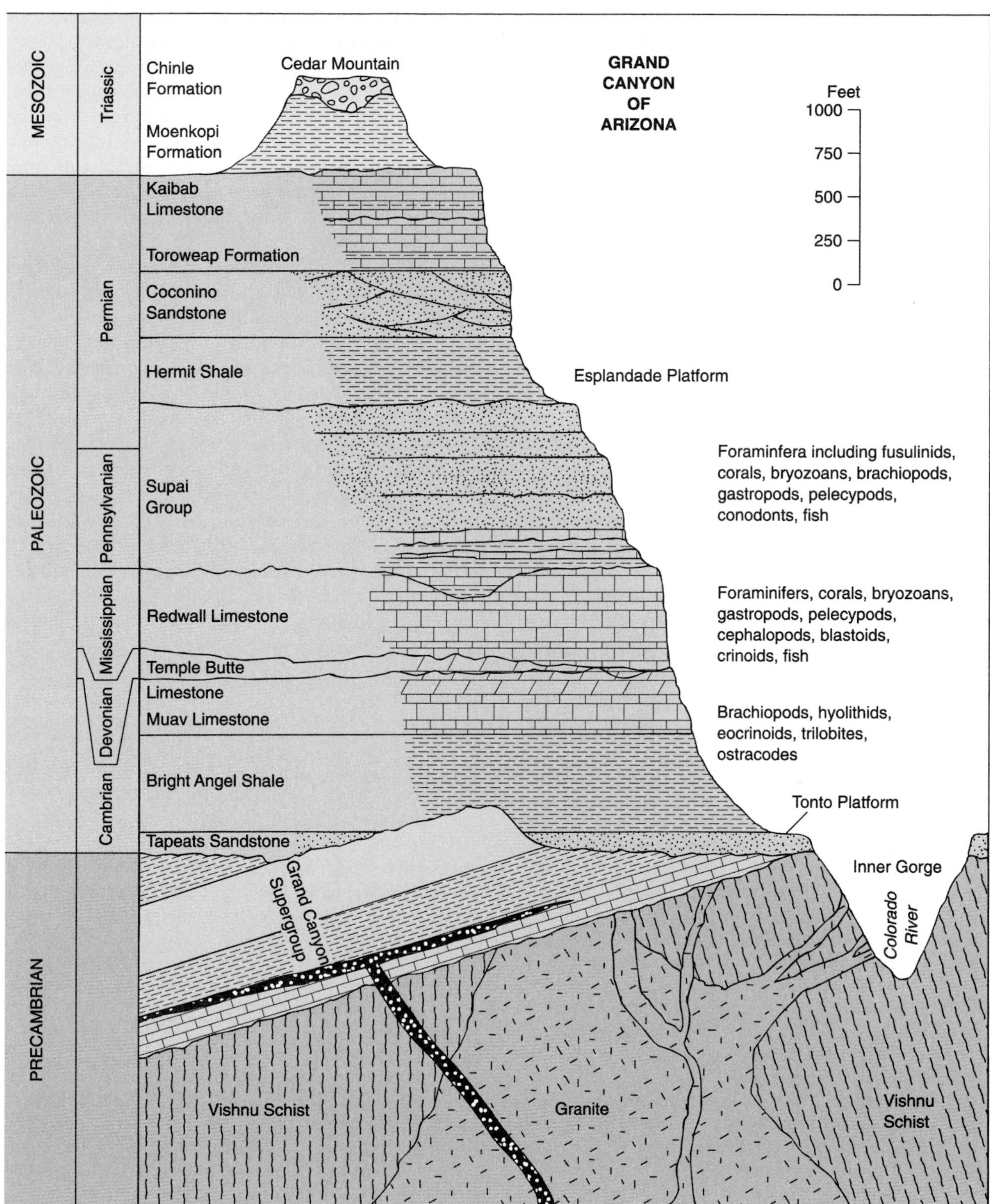

Generalized Geologic Column for Grand Canyon National Park. (*From McKee, E. D. 1982. The Supai Group of the Grand Canyon, U.S. Geological Survey Professional Paper 1173.*)

semicircular center of subsidence in southern New York and Pennsylvania in which over 2000 feet of sediment accumulated. The isopach pattern further indicates a highland source area to the southeast.

Lithofacies Maps

Maps constructed to show areal variations in facies can provide additional details and validity to paleogeographic interpretations. Such graphic representations are called **lithofacies maps**. Figure 3-58 is a hypothetical base map of an area subjected to exploratory drilling by oil companies. The logs for the wells are also shown. Geologists first correlate the formations. Then, assuming that the unconformity represents one time plane and the ash bed another, they define the time-rock unit as "X." Paleontologic study of the rocks between the time planes confirms the validity of the time-rock unit. Geologists may now prepare the lithofacies map. Time-rock unit X is missing at well number 11; this may be the result of its not being deposited there or, having been deposited, of its being eroded away. It is logical that the sandy facies was deposited adjacent to a north-south trending shoreline.

Figure 3-59 is a lithofacies map of rocks deposited over 400 million years ago in the eastern United States. From this map, one can infer the existence of a highland area that existed at that time along our eastern seaboard and that supplied the coarse clastics. Detrital sediments from the source area become fine, and the section thins as one proceeds westward from the source highlands. Finally, as far west as Indiana, the map indicates that only carbonate precipitates were laid down. The conglomerates were probably the deposits of great alluvial fans built out from the ancient mountain system.

The lithofacies maps just described provide a qualitative interpretation of areal changes in rock bodies. Quantitative lithofacies maps can also be constructed and are frequently used in the study of subsurface formations that are known primarily from well records. By means of contour lines, such maps show the areal distribution of some measurable characteristic of the unit being mapped. For example, contours may be drawn on the percentage of one lithologic component (such as clay) compared to the total unit or on the ratio of one rock type (such as sandstone) to the others within the unit. An isopach map is ordinarily the base map for any of the quantitative maps, since one must know the total stratigraphic thickness of the unit with which individual components are compared.

SUMMARY

Sedimentary rocks represent the material record of environments that once existed on the Earth's surface. For this reason, they are of great importance to the science of historical geology. All sedimentary rocks are formed by the accumulation and consolidation of the products of weathering derived from older rock masses, as well as by chemical precipitation and the accumulation of organic debris. Because of their mode of formation, the composition of sedimentary rocks provides information about source areas. Rock color can provide clues to the chemistry of the depositional medium. The materials of sedimentary rocks are often transported by wind, water, or ice, or they are carried in solution to be precipitated in a particular environment of deposition. The transporting medium imparts characteristics of texture or composition that can be used to reconstruct the depositional history and tectonic setting. Fossils in sedimentary rocks are splendid environmental indicators. They tell us if strata are marine or nonmarine, if the water was deep or shallow, or if the climate was cold or warm.

In the case of clastic rocks, the size, shape, and arrangement of grains can provide data about the energy of the transporting agent, the distance the grains had traveled, erosional recycling, and the degree to which movements of the Earth's crust had disturbed a basin of deposition. Similar kinds of information are elucidated by such primary sedimentary structures as graded bedding, cross-bedding, and current ripple marks. Sandstones are particularly useful in paleoenvironmental studies. Graywackes, arkoses, quartz sandstones, and lithic sandstones accumulate in particular paleogeographic and tectonic situations. It is the geologist's task to discover the details of those situations by examining the rocks.

Geologists usually divide successions of sedimentary rocks into rock units that are sufficiently distinctive in color, texture, or composition to be recognized easily and mapped. Such lithostratigraphic units are called formations and are not necessarily of the same age throughout their areal extent. A chronostratigraphic unit differs from a lithostratigraphic unit in that it is an assemblage of strata deposited within a particular interval of time. The Cambrian System, for example, is a chronostratigraphic unit including all the rocks deposited in the Cambrian Period.

Within any given chronostratigraphic unit, one may find rocks that vary in composition, texture, organic content, or other features from adjacent rocks. These rock bodies of distinctive appearance or aspect are called facies. Facies reflect deposition in a particular environmental setting. For example, along a coastline one may find nearshore sand facies that change seaward to shale facies and carbonate facies. As shorelines shift landward or seaward, facies shift accordingly, maintaining their association with a particular local set of environmental conditions.

Shoreline migrations may be the result of such factors as worldwide changes in sea level (eustatic changes), tectonic movements of continental borderlands, or progradation—the seaward advance of the coastline resulting from rapid deposition of sediment brought to the sea by rivers. Progradation is particularly evident in deltas.

That branch of geology that deals with the origin, composition, sequence, and correlation of stratified rocks is stratigraphy. Stratigraphic correlation involves determining the equivalence of strata in diverse locations by the use of lithology, fossil content, radioisotopic age, depositional cycles, isotopic characteristics, or any other distinctive physical or chemical characteristics of the units to be correlated. Lithostratigraphic correlation links units of similar lithology and stratigraphic position, whereas biostratigraphic correlation expresses similarity of fossil content and biostratigraphic position. Chronostratigraphic correlation links units of corresponding age.

The study of facies is of great importance in the development of reconstructions of conditions on Earth long ago. Geologists employ various graphic methods to record variations in facies and other attributes of sedimentary rocks. These methods include the preparation of lithofacies, biofacies, geologic, and isopach maps. If examined in chronologic sequence, such maps are useful not only in reconstructing ancient geography but also in providing a picture of the Earth's changing patterns of ancient lands and seas.

QUESTIONS FOR REVIEW AND DISCUSSION

1. What features in a sedimentary rock might indicate it was deposited in each of the following environments of deposition?

- **a.** Shallow marine environment
- **b.** Deep marine, continental rise environment
- **c.** Transitional, deltaic environment
- **d.** Continental, desert environment

2. Why are sandstones and siltstones of desert environments rarely black or gray?

3. How does matrix in a rock differ from cement? What are the most common kinds of cements found in sandstones? Which of these is most durable?

4. What would be the probable origin of a poorly sorted sandstone composed of angular grains in a 30 percent matrix of mud? What origin might you infer for a well-sorted sandstone that contains fossils of marine clams, is composed almost entirely of quartz, and has well-developed ripple marks?

5. What differences in texture and composition serve to distinguish between a mature and an immature sandstone?

6. In a columnar section of sedimentary rocks, a limestone is overlain by a shale, which in turn is overlain by sandstone. What might this coarsening upward sequence indicate with regard to the advance or retreat of a shoreline?

7. What conditions in the Bahama Banks carbonate platform result in the high production of calcium carbonate sediment?

8. What features of sedimentary rocks are useful in determining the direction of current of the depositing medium?

9. An isopach map shows an accumulation of 10,000 meters of sediments in a Paleozoic marine basin of deposition, yet the sedimentary rocks contain fossils indicating deposition in water no deeper than 200 meters. What has occurred in the basin of deposition?

10. In Figure 3-59, note the areas in which Lower Silurian rocks are absent. How do you account for their absence?

READINGS

Ager, D.V. 1973. *The Nature of the Stratigraphic Record*. London: Macmillan.

Boggs, S. Jr. 2001. *Principles of Sedimentology and Stratigraphy*, 3d ed. Englewood Cliffs, New Jersey: Prentice Hall.

Blatt, H., and Tracy, R. J. 1996. Sedimentary Rocks, In: *Petrology*, 2d ed. New York: W. H. Freeman.

Hsu, K. J. 1989. *Physical Principles of Sedimentology: A Readable Textbook for Beginners and Experts*. New York: Springer-Verlag.

Lemon, R. R. 1990. *Principles of Stratigraphy*. Columbus, OH: Merrill Publishing Co.

Mackenzie, F. T. 1998. *Our Changing Planet*. 2d ed. New Jersey: Prentice-Hall Inc.

McLane, M. 1995. *Sedimentology*. New York: Oxford University Press.

Pettijohn, F. J. and Potter, P. E. 1964. *Atlas and Glossary of Primary Sedimentary Structures*. New York: Springer Verlag.

Prothero, D. R., and Schwab, F. 1996. *Sedimentary Geology*. New York: W. H. Freeman.

Scholle, P. A., Bobout, D. G., and Moore, C. H. 1983. *Carbonate Depositional Environments*. Tulsa, OK: American Association of Petroleum Geologists.

Tucker, M. E. 1981. *Sedimentary Petrology*. New York: John Wiley & Sons.

WEB SITES

The Earth Through Time Student Companion Web Site (www.wiley.com/college/levin) has online resources to help you expand your understanding of the topics in this chapter. Visit the Web Site to access the following:

1. Illustrated course notes covering key concepts in each chapter;
2. Online quizzes that provide immediate feedback;
3. Links to chapter-specific topics on the web;
4. Science news updates relating to recent developments in Historical Geology;
5. Web inquiry activities for further exploration;
6. A glossary of terms;
7. A Student Union with links to topics such as study skills, writing and grammar, and citing electronic information.

4

An Ordovician age trilobite with eyes located on long stalks, which protruded out of sediment that at times covered the animal. The specimen is about 3 centimeters in length and it has been given the name Asaphia kowalewskii. *It was collected near St. Petersburg, Russia. (Copyright Sinclair Stammers/Science Photo Library/Photo Reseachers.)*

CHAPTER 4

The Fossil Record

Fossils have long been studied as great curiosities, collected with great pains, treasured with great care and at great expense, and shown and admired with as much pleasure as a child's rattle or a hobby horse is shown and admired by his playfellows, because it is pretty; and this has been done by thousands who have never paid the least regard to that wonderful order and regularity with which Nature has disposed of these singular productions, and assigned to each class its peculiar stratum.
William Smith, January 5, 1796

OUTLINE

The Earth is about 4.6 billion years old. Life has been present on it for at least 3.5 billion years. The science that seeks to understand all aspects of the succession of plants and animals over that great expanse of geologic time is called **paleontology**. It is a science based on the study of **fossils**, the remains or traces of ancient life. As paleontologists study these relics of former living organisms, they endeavor to learn how they lived and grew. What was the organism's original appearance? How did it interact with other organisms and its physical environment? What were its ancestors like, and what changes occurred in its descendants? Why did these changes occur?

FOSSILS

Preservation

When one considers the many ways by which organisms are completely destroyed after death, it is remarkable that fossils are as common as they are. Attack by scavengers and bacteria, chemical decay, and destruction by erosion and other geologic agencies make the odds against preservation very high. However, the chances of escaping complete destruction are vastly improved if the organism happens to have a mineralized skeleton and dies in a place where it can be quickly buried by sediment. Both of these conditions are often found on the ocean floors, where shelled invertebrates flourish and are covered by the continuous rain of sedimentary particles. Although most fossils are found in marine sedimentary rocks, they also are found in terrestrial deposits left by streams and lakes. On occasion, animals and plants have been preserved after becoming immersed in tar or quicksand, trapped in ice or lava flows, or engulfed by rapid falls of volcanic ash.

Paleontologists are interested in not only the discovery and interpretation of the evolution and habits of ancient organisms but also what occurred to them from the time they died to the time they reached the fossil state. The study of all that has happened to organisms from the moment of death to final preservation as a fossil is termed **taphonomy**. If the taphonomic history of an assemblage of fossils is not recognized, interpretation can be seriously flawed. Life on the deep sea floor, for example, might have consisted predominantly of worms and other soft-bodied animals. On death, it is

ENRICHMENT

Amber, the Golden Preservative

To many, the term fossil brings to mind remains of ancient organisms that have literally been turned to stone. In this chapter, however, a few rare instances of preservation of actual remains have been described. For such preservations, there is little superior to the organic substance known as amber (see Fig. 4-6). Insects are the usual organisms preserved in amber, and the bodies of insects, except for being desiccated, are preserved entirely. In addition to insects, spiders, crustaceans, snails, mammal hair, feathers, small lizards, and even a frog have been discovered in amber. By far the most abundant insects preserved in amber are flies, mosquitoes, and gnats of the Order Diptera. Paleontologists deplore the fact that during the late 19th century thousands of tons of raw amber were melted down for varnish, causing the loss of untold numbers of exquisitely preserved organisms.

Amber is a fossil resin produced by conifers. It is initially exuded from cracks and wounds in trees and, while still soft and sticky, traps and engulfs insects. The insect's struggle to escape can often be recognized in the swirl patterns around appendages. Later, through evaporation of the more volatile components, the soft resin hardens.

Although amber deposits are known in rocks ranging in age from Lower Cretaceous to the Holocene, the most famous are those found along the coast of the Baltic Sea, southwest of the Gulf of Riga, near the seaport of Kaliningrad (formerly Konigsberg). The conifer forests in this area were inundated during a marine transgression in early Oligocene time. Marine sediment buried the trees and their contained amber. Subsequently, chunks of amber eroded from the soft clays were distributed across adjacent areas by streams, wave action, and glaciers. Intrigued by their beauty, neolithic families are known to have gathered the rounded yellow and brown pieces of amber. Such prominent philosophers of antiquity as Aristotle and, later, Pliny and Tacitus described amber's physical and chemical properties. As a semiprecious gem, amber was transported along ancient trade routes from the Baltic region to the Mediterranean. Traders called amber "the gold from the north." Then as now, it was used in the manufacture of beads for jewelry. Amber beads, amulets, necklaces, and bracelets have been recovered from Etruscan tombs and from excavations in Mycenae, Egypt, and Rome. However, for modern women adorned in amber jewelry, a word of whimsical caution may be of interest. That person gazing intently at your necklace may be a paleontologist in search of an embalmed mosquito.

Reference

Poinar, G. O., Jr. 1992. *Life in Amber*. Stanford, CA: Stanford University Press.

likely that the bodies of those soft-bodied animals would be lost to scavengers and decay. Thus, a thriving community of ancient organisms would be unknown to geologists. In another scenario, currents may have swept the remains of shallow-water shelled animals into the deep marine environment. When the shelled animals have become fossils in a layer of sedimentary rock, that rock might be misinterpreted as a shallow- rather than deep-water deposit. The taphonomist examining the fossils would look for clues to such relocation, such as shells in non-living positions or with breakage caused by their being swept about. He would look for burrows and trails that might indicate the former presence of the soft-bodied animals, and he would attempt to determine losses to the fossil record from dissolution of shells and other forms of destruction.

The term *fossil* implies petrifaction, literally a transformation into stone. After the death of an organism, the soft tissue is ordinarily consumed by scavengers and bacteria. The empty shell of a snail or clam may be left behind. If it is sufficiently durable and resistant to dissolution, it may remain basically unchanged for a long period of time. Indeed, unaltered shells of marine invertebrates are known from deposits over 100 million years old. In many marine creatures, however, the skeleton is composed of a mineral variety of calcium carbonate called aragonite. Although aragonite has the same composition as the more familiar mineral known as calcite, it has a different crystal form, is relatively unstable, and in time changes to the more stable calcite.

Many other processes may alter the shell of a clam or snail and enhance its chances for preservation. Water containing dissolved silica, calcium carbonate, or iron may circulate through the enclosing sediment and be deposited in open spaces vacated by cells or in larger openings, such as marrow cavities and canals in bone once occupied by blood vessels and nerves. In such cases, the original composition of the bone or shell remains, but the fossil is made harder and more durable. This addition of a chemically precipitated substance into pore spaces is termed **permineralization** (Fig. 4-1).

Petrifaction may also involve a simultaneous exchange of the original substance of a dead plant or animal with mineral matter of a different composition. This process is termed **replacement** because solutions have dissolved the original material and replaced it with an equal volume of the new substance (Fig. 4-2). Replacement can be a marvelously precise process, so that details of shell ornamentation, tree

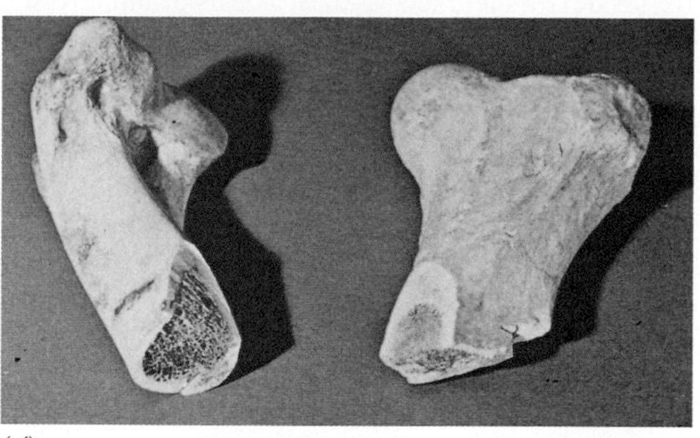

(*A*)

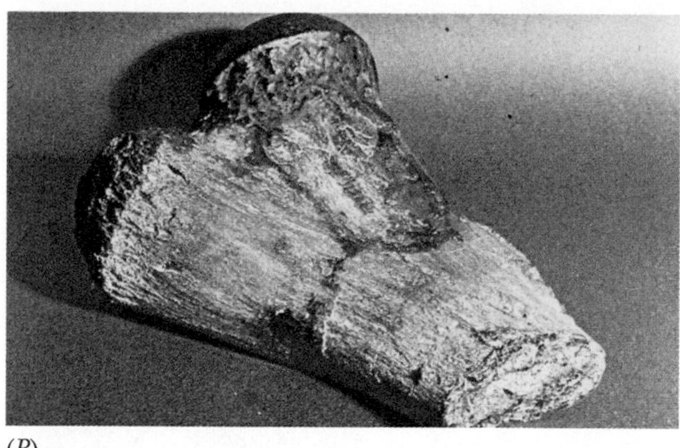

(*B*)

FIGURE 4-1 An example of permineralization. The bone fragments in the upper photograph (*A*) are part of a cow's femur (upper leg bone). They are unfossilized original bone in which one can readily see the central porous marrow cavity. The bone depicted in (*B*) is a permineralized dinosaur bone in which the porous space has been densely filled with mineral matter.

rings in wood, and delicate structures in bone are accurately preserved.

Another type of fossilization, known as **carbonization**, occurs when soft tissues are preserved as thin films of carbon. Leaves and tissue of soft-bodied organisms such as jellyfish or worms may accumulate, become buried and compressed, and lose their volatile constituents. The carbon often remains behind as a blackened silhouette (Fig. 4-3).

Fossils may also take the form of molds, imprints, or casts. Any organic structure may leave an impression of itself if it is pressed into a soft material and if that material is capable of retaining the imprint. Commonly among shell-bearing invertebrates, the shell is dissolved after burial and lithification, leaving a vacant **mold** bearing surface features of the original shell that are opposite of those on the shell itself (ridges are represented by grooves, knobs by depressions, and vice versa). If external features (growth lines, ornamentation) of the fossil are visible, the mold is an *external mold*. Conversely, the *internal mold* shows features of

FIGURE 4-2 The shell of an extinct marine organism known as an ammonoid. In this Jurassic fossil, the original calcium carbonate skeleton has been replaced with iron sulfide in the form of the mineral pyrite, known to some as "fool's gold." (*Copyright V. Fleming/Science Photo Library/Photo Researchers, Inc.*) ? *What other common minerals or compounds frequently occur as replacements?*

FIGURE 4-3 This fossil seed fern from rocks of Pennsylvanian age has been preserved by carbonization. The frond is approximately 27 centimeters in length.

FIGURE 4-4 An internal mold (or steinkern) formed by filling of the spiral cavity of an ancient marine snail.

the inside of a shell, such as muscle scars or supports for internal organs. Many invertebrate shells enclose a hollow space that may be left empty or filled with sediment (Fig. 4-4). The internal filling is called a *steinkern* (stone core). Finally, molds may be subsequently filled, forming **casts** that faithfully show the original form of the shell (Fig. 4-5).

Although it is certainly true that the possession of hard parts enhances the prospects of preservation, organisms having soft tissues and organs are also occasionally preserved. Insects and even small vertebrates have been found preserved in the hardened resin of conifers and certain other trees (Fig. 4-6 and box "Amber the Golden Preservative") X-ray examination of thin slabs of rock sometimes reveals the ghostly outlines of tentacles, digestive tracts, and visual organs of a variety of marine creatures. Soft parts, including skin, hair, and viscera of ice age mammoths, have been preserved in frozen soil (Fig. 4-7) or in tar oozing from oil seeps. A striking example of the preservation of soft parts was discovered in 1984 at Lindow Moss, England, when an excavating machine uncovered the body of a 2000-year-old human. Named Lindow Man

FIGURE 4-5 A cast (above) and mold (below) of a trilobite, formed in a nodule of calcareous shale. The trilobite (*Calliops*) is 4 centimeters long.

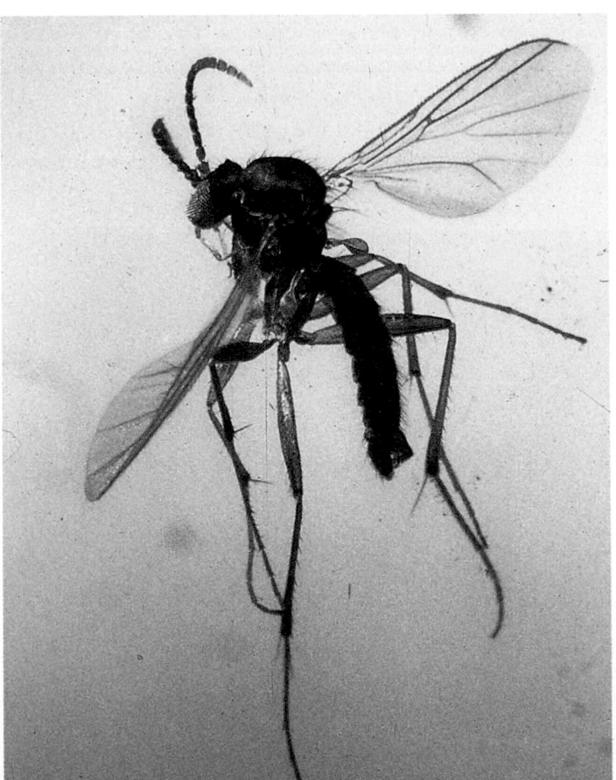

FIGURE 4-6 An Eocene insect preserved in amber. The insect is a member of the Order Diptera, which includes flies, mosquitoes, and gnats. Although preservation of insects is uncommon, it is likely that they approached modern levels of abundance and diversity early in the Cenozoic Era. (*Courtesy of W. Bruce Saunders, Bryn Mawr College.*)

FIGURE 4-7 In the summer of 1977, the carcass of this baby mammoth was dug from frozen soil (permafrost) in northeastern Siberia. The mammoth stood about 104 centimeters tall at the shoulders, was covered with reddish hair, and was judged to be only several months old at the time of death. Dating by the radiocarbon method indicates that death occurred 44,000 years ago. *(Photograph courtesy of K. Novikova, Biologopoczvennyj Institut, Vladivostok, former Soviet Union.)*

(Fig. 4-8), this unfortunate Briton had been ritually slaughtered (possibly by Druids), stripped, garroted, and then bled. The body was thrown in a bog, where the skin was preserved and the torso flattened by the weight of accumulating peat. (Only the upper part of the body was recovered; the lower part had been destroyed by the machine before the remains were detected.)

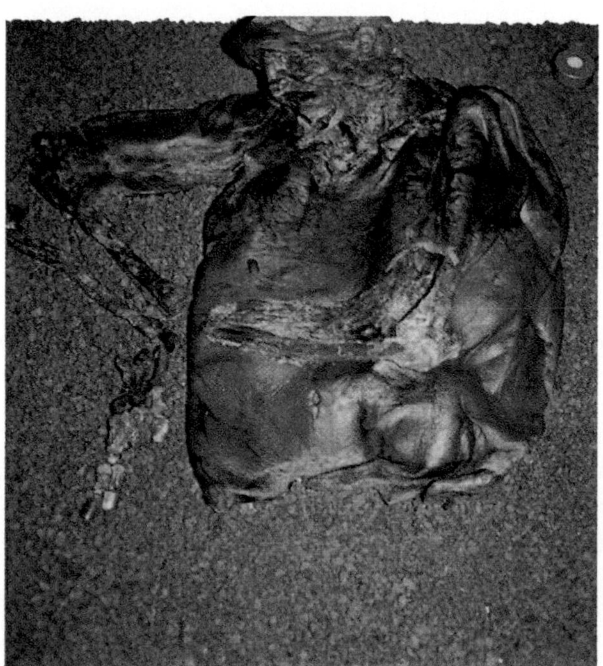

FIGURE 4-8 Preserved torso, arms, and head of the 2000-year-old Lindow Man. This example of preservation of soft tissue was found in a peat bog in 1984 at Lindow Moss, England. The lower half of the body was destroyed by the excavating machine. *(British Museum.)* **?** *The wet sediment of peat bogs is usually deficient in oxygen content. How might this have enhanced preservation of this soft tissue?*

The probability that actual remains of soft tissue will be preserved is improved if the organism dies in an environment of rapid deposition and little oxygen. Under such conditions, the destructive effects of bacteria are diminished. The Middle Eocene Messel Shale of Germany accumulated in such an environment. The shale was deposited in an oxygen-deficient lake where lethal gases sometimes bubbled up and killed animals. Their remains accumulated on the floor of the lake and were then covered by clay and silt. Among the superbly preserved Messel fossils are insects with iridescent exoskeletons, frogs with skin and blood vessels intact, and even entire small mammals with preserved fur and soft tissue.

Evidence of ancient life does not consist solely of petrifactions, molds, and casts. Sometimes the paleontologist is able to obtain clues to an animal's appearance and how it lived by examining tracks, trails (Fig. 4-9), burrows, and borings. Such markings are called **trace fossils**, and the study of trace fossils is termed **ichnology**. The tracks of an ancient vertebrate animal (Fig. 4-10) may indicate whether the animal that made them was bipedal (walked on two legs) or quadrupedal (walked on four legs), whether it was digitigrade (walked on toes) or plantigrade (walked on the flat of the foot), whether it had an elongate or a short body, whether it was lightly built or ponderous, and sometimes whether it was aquatic (with webbed toes) or possibly a flesh-eating predator (with sharp claws).

Trace fossils of invertebrate animals are more frequently found than are the traces of vertebrates, and

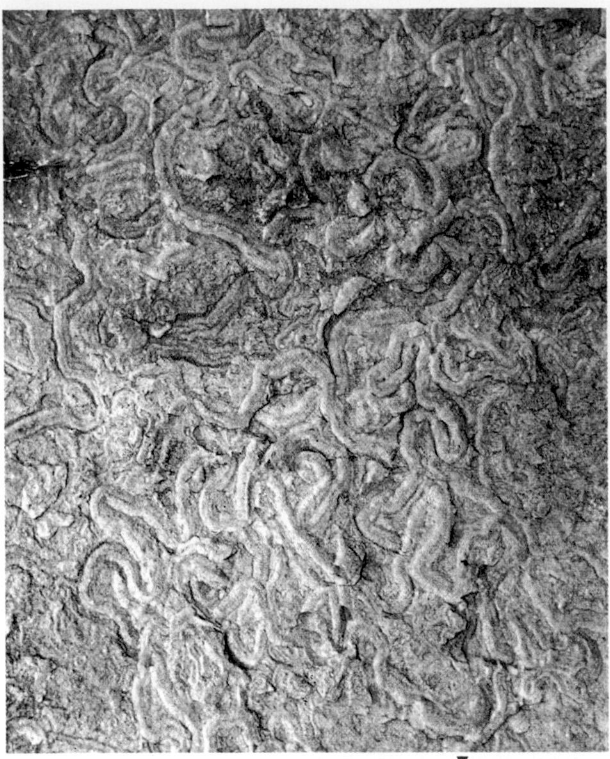

FIGURE 4-9 Trace fossils consisting of probable annelid worm trails in Upper Carboniferous siltstones that form the Cliffs of Moher along the southwest coast of Ireland. ? *Would this rock be useful for determining grain orientation and determination of direction of ancient ocean currents?*

FIGURE 4-10 Dinosaur trackways arranged to indicate the passage of a biped (identical three-toed imprints) whose tracks were crossed by a quadruped having larger rear than front feet (typical of many quadruped dinosaurs). Claw imprints on the biped suggest that it was a predator. ? *What indicates that the quadruped crossed the area after the biped?*

they are also useful indicators of the habits of ancient creatures. One can sometimes infer whether the trace-making invertebrate was crawling, resting, grazing, feeding, or simply living within a relatively permanent dwelling. For example, *crawling traces*, as might be expected, are linear and show directed movement (Fig. 4-11*A*). Shallow depressions that reflect the shape of the animal may be *resting traces* (Fig. 4-11*B*). Simple or U-shaped structures approximately perpendicular to bedding are often *dwelling traces* (Fig. 4-11*C*). *Grazing traces* (Fig. 4-11*D*) occur along bedding planes and are characterized by a systematic meandering or concentric and parallel patterns that represent the animal's effort to cover the area containing food in an efficient manner. The three-dimensional counterparts of grazing traces are called *feeding traces*. Feeding traces are made by animals that consume sediment for the organic nutrients within it. The traces consist of systems of branched or unbranched burrows, as shown in Figure 4-11.

The tracks, trails, and burrows of invertebrates have been generally similar throughout most of the Phanerozoic. We know that similar traces have been produced by different kinds of organisms, although often by members of the same taxonomic group. Because of their long geologic ranges, most trace fossils are unsuitable for biostratigraphic correlation. The value of these fossils lies in their use in recognizing the environment in which the trace-bearing sedimentary rock was deposited.

The Nature of the Fossil Record

The fossil record of life is incomplete. If it were a complete and total record, it would include information on all past forms of life for every increment of time and for every place on Earth. Clearly this is unattainable. Only a limited number of animals and plants have been preserved. Many that were preserved have never been exposed to our view by erosion or drilling. Still others have simply not yet been discovered.

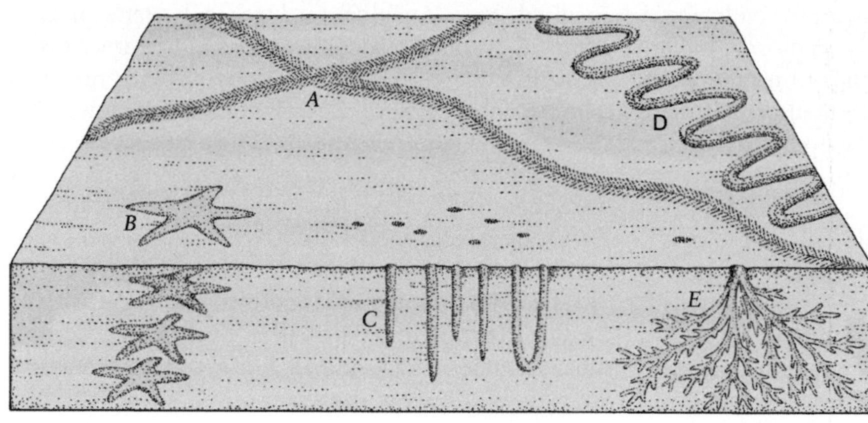

FIGURE 4-11 Traces that reflect animal behavior: (*A*) crawling traces, (*B*) resting traces, (*C*) dwelling traces, (*D*) grazing traces, and (*E*) feeding traces. ? *Which of these trace fossils might be useful as geopetal indicators?*

The record is more complete for marine life having hard external skeletons and for spore and pollen grains, which have highly resistant coverings. It is less complete for land life lacking bone or shell. Where burial is rapid and dead organisms are protected from scavengers and agents of decay, the probability of preservation improves. As we have noted, the ocean floor and low-lying land areas where deposition predominates are favorable places for fossil preservation. Many dinosaur remains, for example, have been discovered in the sands and clays deposited by streams that flowed across the low plains that lay east of the Rocky Mountains during the Cretaceous Period. There were certainly animals and plants living in nearby highland areas as well, but in such places erosion rather than deposition predominates, and fossils of local inhabitants are likely to have been destroyed.

Despite the many factors that prevent fossilization, the fossil record is remarkably comprehensive. On the basis of the tens of thousands of species known from fossils, paleontologists have been able to piece together a history of past life that is both accurate and verifiable.

THE RANK AND ORDER OF LIFE

The Linnaean System of Classification

Because of the large number of living and fossil animals and plants, random naming would be confusing and inefficient. Realizing this, the Swedish naturalist Carl von Linné (1707–1778), who is also known by his Latinized name Carolus Linnaeus, formulated a systematic method for naming animals and plants. As originally formulated, the Linnaean system uses morphology (the form and structure of organisms) as a basis for classification and employs what is known as binomial nomenclature at the species level. In this scheme, the first name is that of the genus, and it designates a group of animals or plants that appear to be related because of their general similarity. For example, *Felis* is the name of the genus in which cats are grouped. There are many kinds of cats, and therefore the second or trivial name denotes a morphologically distinct and restricted group belonging to the genus *Felis*. Thus, *Felis domesticus* is the common house cat, whereas *Felis leo* is the African lion and *Felis onca* is the jaguar. To illustrate how the common name for animals can lead to confusion, consider *Felis concolor*, a species that is also known as the cougar, mountain lion, puma, panther, brown tiger, and deer killer.

There must be consistency in the way organisms are formally named. For example, it is a rule that no two genera within a kingdom may have the same name. Further, names must be in Latin or Latinized and be printed, as they are above, in italics.

Concepts Involved in Classification

THE SPECIES The species is the fundamental unit in biologic classification. A **species** is a group of organisms that have structural, functional, and developmental similarities and that are able to interbreed and produce fertile offspring. Because members of one species do not breed with members of different species under natural conditions, species exist in reproductive isolation. Although individuals of a species are generally similar, they are not identical. They exhibit variation, and because of variation, the description of a single individual cannot include the range of variations present in a species. Many kinds of individual variation may exist, including differences between sexes or those associated with the development from juvenile to adult stages. If such differences are not recognized, one might err by applying two or more names to different members of the same species.

One might surmise that biologists have an advantage over paleontologists in deciding on the validity of a proposal to provide a species name to a group of similar animals. Biologists, after all, are able to work with living organisms and may be able to observe whether or not they can interbreed. Unfortunately, the breeding habits of over 80 percent of all living animals are not known. Thus, the recognition of species for biologists and paleontologists alike relies heavily on morphologic traits that are constant within the group. One must determine the range of variation within members of the proposed species and reach a well-reasoned decision about the kinds and levels of variation that can exist.

TAXONOMY The species is the basic unit in biologic classification, or **taxonomy**. In this system, the various categories of living things are arranged in a hierarchy that expresses levels of kinship. For example, a **genus** (pl. *genera*) is a group of species that have close ancestral relationships; a **family** is a group of related genera; an **order** is a group of related families; a **class** is a group of related orders; a **phylum** (pl. *phyla*) is a group of related classes; and a **kingdom** is a large group of related phyla. To use an example familiar to all, individual humans are members of the Kingdom Animalia, Phylum Chordata, Class Mammalia, Order Primates, Family Hominidae, Genus *Homo*, and species *sapiens*.

A traditional system of classification places all living things in six kingdoms (Fig. 4-12). Two of these are the **Archaeobacteria** and **Eubacteria**. These two kingdoms include microorganisms having nuclei that are not enclosed by a membrane. Archaeobacteria include methane-producing bacteria as well as bacteria capable of living under extreme conditions of temperature or salinity. Eubacteria are prokaryotes that live in water or soil or within larger organisms. The cyanobacteria (formerly called blue-green algae) are eubacteria. They were abundant on the Earth over 3 billion years ago and were responsible for critical changes in our planet's early atmosphere.

The remaining four kingdoms, are the **Protista**, **Fungi**, **Plantae**, and **Animalia**. The Protista include mostly single-celled, animal-like organisms that devour food for energy (**heterotrophs**), plant-like photosynthesizers (**autotrophs**), and decomposers (**saprophytes**). Fungi are multicellular eukaryotes, many of which are decomposers, absorbing nutrients from dead organisms or living as parasites on plants. Members of the Plantae are multicellular eukaryotes that typically live on land and undergo embryonic development. They are autotrophs. In contrast, members of the kingdom Animalia are multicellular heterotrophic organisms.

Although the six-kingdom system is widely used in biology textbooks, it has its shortcomings. It is based

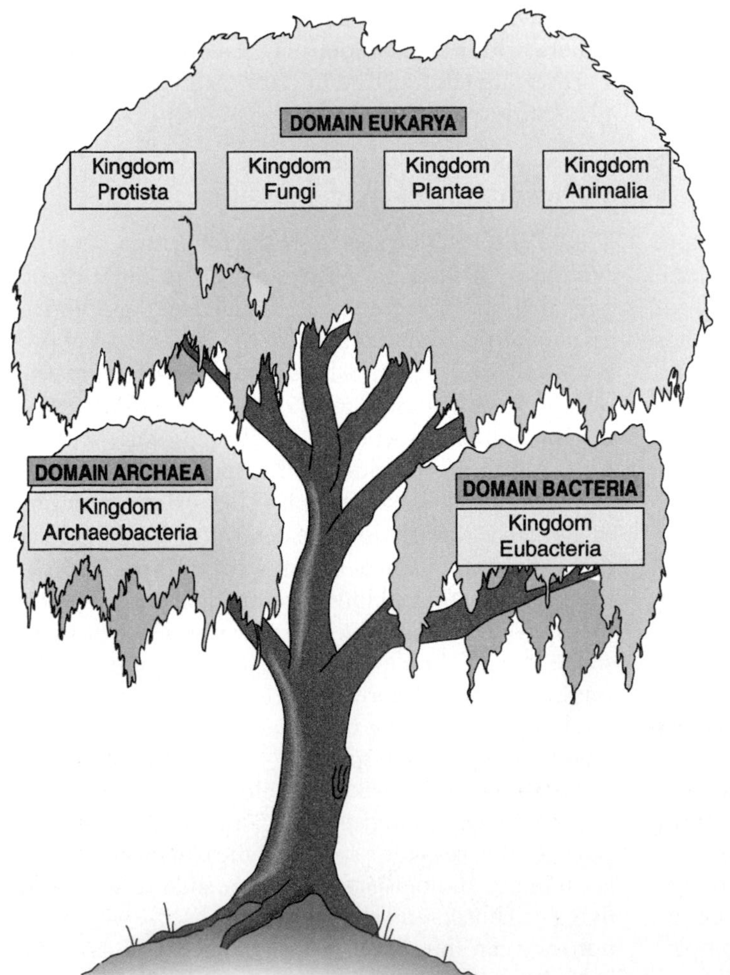

FIGURE 4-12 Six taxonomic kingdoms and the domains of which they are components.

largely on phenotypic traits, which are *observable* traits that arise from genetic processes. As we have seen, phenotypic traits have been essential to biologic classification and the study of evolutionary relationships. An even more convincing indicator of such relationships, however, is the structure of large molecules in the cell and in the sequence of amino acid components in such molecules as ribonucleic acid, or RNA. This molecular basis for phylogeny has been particularly instructive in determining affiliations among microbial organisms, although it has also provided a better understanding of the evolution of larger organisms. Molecular phylogeny indicates that superficially dissimilar groups such as plants, protists, animals, and fungi are actually closely related. To more validly show evolutionary relationships as revealed by molecular sequencing studies, it has been proposed that all life be divided into three great divisions termed **domains**. Microbiologist Carl Woese has named them the **Archaea**, **Bacteria**, and **Eukarya** (see Fig. 4-12). The domain Archaea (equivalent to the kingdom Archaeobacteria) includes methane-producing bacteria and an interesting group of heat-loving microbes called thermophiles that populate hydrothermal vent systems on the ocean floor (Fig. 4-13). Among the many organisms within the domain Bacteria are cyanobacteria, along with purple sulfur bacteria and various nonphotosynthetic groups of microbes. Fungi, plants, and animals, because of their molecular similarities, are placed within the domain Eukarya.

The subdivisions and common names for members of the five kingdoms are provided in Appendix A. Even a brief perusal of the listing indicates that biologic classification is more than a mere system for cataloging organisms. Because it is based on organic structures and embryologic development, it reflects the broad outlines of evolutionary relationships. In a sense, the classification is a blueprint for constructing the tree of life.

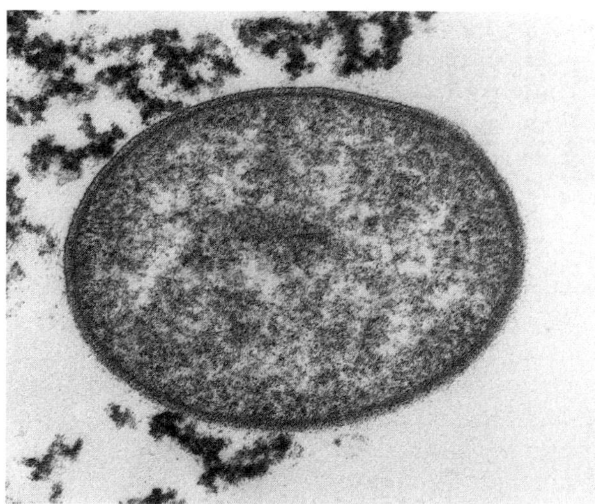

FIGURE 4-13 Transmission electron micrograph of an organism belonging to the domain Archaea. The microbe is a thermophile (tolerant of extreme temperatures). It was ejected from a thermal vent on the Pacific Ocean floor along the Gorda oceanic ridge (west of Oregon and Washington). The long axis of the cell is approximately one micron. (*Courtesy of Melanie Summit, Dept. of Earth & Planetary Sciences, Washington University*).

ORGANIC EVOLUTION

The Roman poet Ovid (43 B. C.–17 A. D.) once wrote that "there is nothing constant in the universe, all ebb and flow, and every shape that's born bears in its womb the seeds of change." These words are remarkably relevant when one considers the way life has changed through time, as revealed by the fossil record. At times that record indicates change may have been startlingly sudden, whereas at other times evidence shows a more gradual change. In either case, older forms have changed or evolved into newer forms to cope better with changes in their environment. This is not to say that the early, simpler forms of life are gone, for many persist along with their often more complex contemporaries.

The concept that organisms have changed through time is not a new one. Anaximander (611–547 B. C.) alluded to the idea 25 centuries ago. This famous Greek taught that life arose from mud warmed in the sun; that plants came first, then animals, and finally humans. Much later, during the Middle Ages, complete faith in the theologic doctrine of special creation of "species" effectively stifled imaginative thinking about evolution. It was not until the 18th century in Europe that intellectuals such as Jean Baptiste de Lamarck and George Buffon began to challenge the concept of special creation. Buffon was convinced that evolution had occurred, but he offered no explanation of its cause. He believed, however, that the environment was somehow involved in the process that resulted in evolution. Buffon also set down a careful definition of species, noting that species were separate entities that were not able to interbreed successfully with other dissimilar species. The first general theory of evolution to excite considerable public attention was developed by the celebrated naturalist Jean Baptiste de Lamarck in the period between 1801 and 1815. Lamarck's writings were followed a half century later by those of Charles Darwin and Alfred R. Wallace.

Lamarck's Theory of Evolution

The Lamarckian theory of evolution stipulated correctly that all species, including humans, are descended from other species. His theory, however, was based on the mistaken assumption that new structures in an organism appear because of need or "inner want" of the organisms, and that structures once acquired in this way during the lifespan of the organism are somehow inherited by later generations. In a similar but reverse fashion, little-used structures would disappear in succeeding generations. For example, Lamarck believed that snakes evolved from lizards that had a strong preference for crawling. Because of this inner need to have long, thin bodies, certain lizards developed such bodies, and because legs became less and less useful in crawling, these structures gradually disappeared. Lamarck's ideas were challenged almost immediately. There was no way to prove by experimentation that such a thing as "inner want" existed. More important, Lamarck's belief that characteristics acquired during the life span of an individual could then be inherited was tested and shown to be invalid. For example, a fair-skinned woman on the beach can be sunburned by exposure to the sun, but no amount of sunbathing will result in her first child being born with a sunburn. Circumcision of newborn males has been practiced for over 4000 years, but this *acquired* alteration does not appear in males born today. There is no way that somatic (or body) cells can pass characteristics over to reproductive cells and thereby on to the next generation. The Lamarckian concept of evolution based on use or disuse of organs was discredited.

Darwin's Theory of Natural Selection

Charles Darwin praised Lamarck for his courage in perceiving that the descendants of former creatures have undergone biologic change and are different from their ancestors. As described in Chapter 1, Darwin and his younger contemporary Alfred R. Wallace jointly proposed natural selection as an important mechanism of evolution. Darwin, with his years of careful data collection, and Wallace, by sudden insight, had discerned that competition for food, shelter, living space, and sexual partners among species with individual variations and surplus reproductive capacity will inevitably result in elimination of the less well-fitted and survival of those that are better fitted to their environments. In Darwin's own words,

> Can we doubt . . . that individuals having any advantage, however slight, over others, would have the best chance of surviving and of procreating their kind? On the other hand, we may feel sure that any variation in the least degree injurious would be rigidly destroyed. This preservation of favorable variations, I call Natural Selection.

Darwin's *On the Origin of Species by Means of Natural Selection* was published in 1859. It appeared at a time when at least part of the European intellectual atmosphere was more liberal and less satisfied with the theologic doctrine that every species had been independently created. Although the ideas of Darwin and Wallace continued to disturb the religious feelings of some, they nevertheless increasingly acquired adherents among 19th and 20th century scientists. Darwin did not consider his views impious. He saw a

> grandeur in this view of life with its several powers, having been originally breathed by the Creator into a few forms or into one; and that . . . from so simple a beginning endless forms most beautiful and most wonderful have been and are being evolved.

Mendelian Principles of Inheritance

Every theory has its strong and weak points. In Darwin's theory, the weakness lay in an inability to explain the cause of variability in a way that could be experimentally verified. The cause of at least a part of that variability was discovered in the course of elegant experiments on garden peas conducted by a Moravian monk named J. Gregor Mendel (1822–1884). Mendel discovered the basic principles of inheritance. His findings, printed in 1865 in an obscure journal, were unknown to Darwin and unheeded by the scientific community until 1900, when the article was rediscovered. Mendel described the mechanism by which traits are transmitted from adults to offspring. In his experiments with garden peas, he demonstrated that heredity in plants is determined by "character determiners" that divide in the pollen and ovules and are recombined in specific ways during fertilization. Mendel called these hereditary regulators "factors." They have since come to be known as genes.

As currently understood, genes are chemical units or segments of a nucleic acid, specifically, **deoxyribonucleic acid (DNA)**. As suggested by careful chemical and X-ray studies, the DNA molecule is conceived as two parallel strands twisted somewhat like the handrails of a spiral staircase (Fig. 4-14). The twisted strands are made up of phosphate and sugar compounds and are linked with cross-members composed of specific nitrogenous bases.

The importance of DNA is evident when we realize that it indirectly controls the production of proteins, the essential components of many basic structures and organs. Even the activities of organisms are regulated by specific catalytic proteins called enzymes. Without DNA and its products, there would be no life as we know it. Its ability to replicate itself precisely is the basis for heredity, and organic evolution ultimately depends on this remarkable molecule.

A **gene** is that part of the DNA molecule that is active in the transmission of hereditary traits. In nearly all organisms, genes are linked together to form larger units termed chromosomes, the central axes of which consists of a very long DNA molecule comprising hundreds of genes.

Reproduction and Cell Division

Reproduction of an organism may be sexual or asexual or may involve an alternation of sexual and asexual methods. All reproductive methods involve the division of cells. In sexual reproduction there is a union of reproductive or sex cells from separate individuals,

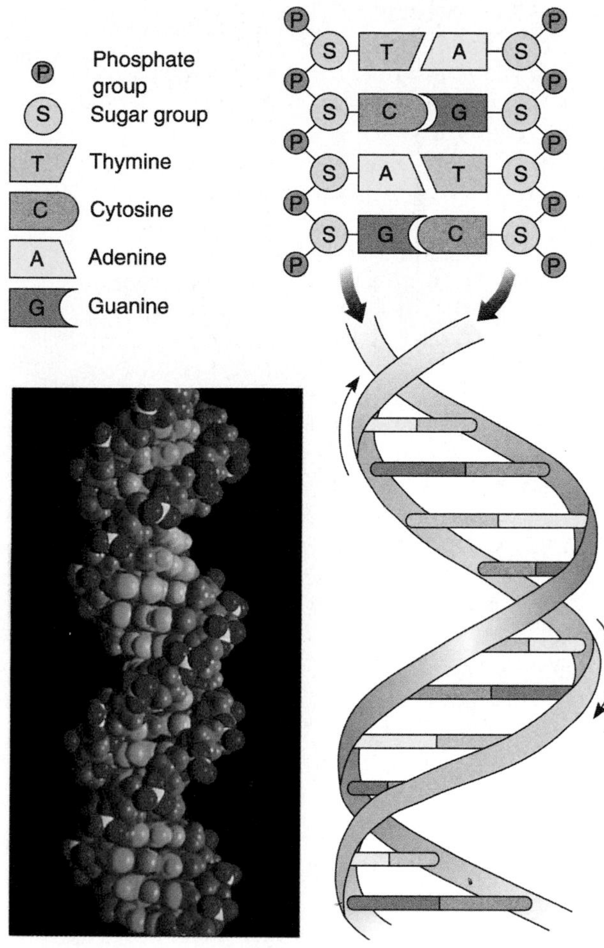

FIGURE 4-14 Representations of portions of the deoxyribonucleic acid (DNA) molecule. At the lower left is a computer reconstruction by R. J. Feldmann. The drawing depicts the twisted, double-stranded helix, with side rails composed of alternate sugar (deoxyribose) and phosphate molecules. Each rung of the twisted ladder is composed of one pair of nitrogenous bases. Of these, thymine links to adenine, and cytosine to guanine.

whereas in asexual reproduction cells do not unite. The more usual methods of asexual reproduction are binary fission, budding, and spore production. **Binary fission** is found in single-celled organisms, such as amebae, that divide to form genetically identical daughter organisms. **Budding** occurs in some unicellular as well as multicellular organisms. In this method, the parent organism simply sprouts a bulge or appendage that may either remain attached to the parent or separate and grow as an isolated individual. The third principal method involving asexual reproduction is the formation of tiny reproductive cells called spores by a parent organism, such as a seedless plant. The spores are formed by division of special spore parent cells and are shed by the parent organism (called the **sporophyte**) when they mature. When a spore settles onto moist soil, the floor of a water body, or some other suitable surface, it germinates and grows into a tiny plant (called a **gametophyte**), which, in turn, produces male and female sex cells. The union of these cells produces a new plant resembling the original sporophyte.

Organisms that produce asexually have the advantage of increasing their numbers rapidly whenever conditions are favorable. Some disadvantage exists, however, in that they are not able to develop as much variability among their offspring as do sexually reproducing organisms. To understand the reasons for that variability, we need to know some simple facts about chromosomes and their behavior during the development of reproductive cells and during fertilization. We note initially that the kind and number of chromosomes are constant for each species and differ between different species. Except in members of the Kingdom Monera (bacteria, including cyanobacteria), the chromosomes are located in the nucleus of the cells and occur in duplicate pairs. Thus, each chromosome has a homologous mate. In humans, for example, there are 46 chromosomes, or 23 pairs. Cells with paired homologous chromosomes are designated **diploid** cells. In all living things, new cells are being produced constantly to replace worn-out or injured cells and to permit growth. In asexual organisms, and in all the **somatic** or body cells of sexual organisms, the process of cell division that produces new diploid cells with exact replicas of the chromosomal components of the parent cells is called mitosis (Fig. 4-15).

In most organisms with sexual reproduction, a second type of division, called **meiosis**, takes place when **gametes** (egg cells or sperm cells) are formed. Meiosis may occur in unicellular organisms, but in multicellular forms it takes place only in reproductive organs (testes or ovaries). Meiosis consists of two quickly succeeding divisions so as to produce four final daughter cells, termed **haploid** because they do not have paired chromosomes. The haploid cells are the gametes, or reproductive cells. When two gametes meet during sexual reproduction, the sperm enters the egg to form a single cell, which can now be called the **fertilized egg**. Because two gametes have been combined into one cell, there is now a full complement of chromosomes and genes representing a mix from both parents. The fertilized egg now begins a process of growth by mitotic cell division that eventually leads to a complete organism.

The organism that develops from the union of haploid cells will already have some variability relative to its parents. There is, however, yet another aspect of meiosis that is important in producing variation in offspring. During the initial chromosome division while the chromosomes are still paired, they may break at corresponding places and exchange their severed segments in a process called **crossing over**. The result is an additional mixing of genes.

If one compares the effects of asexual reproduction to sexual reproduction, it is apparent that the

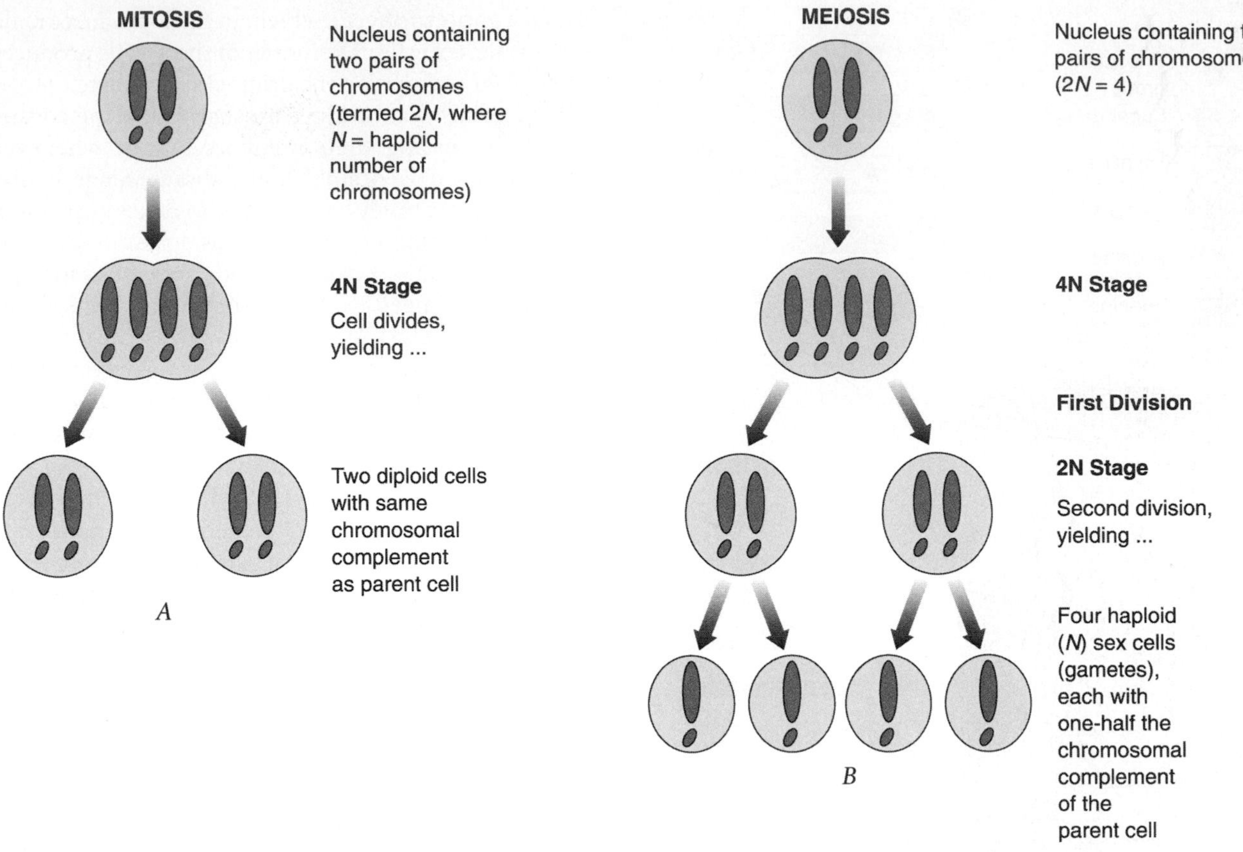

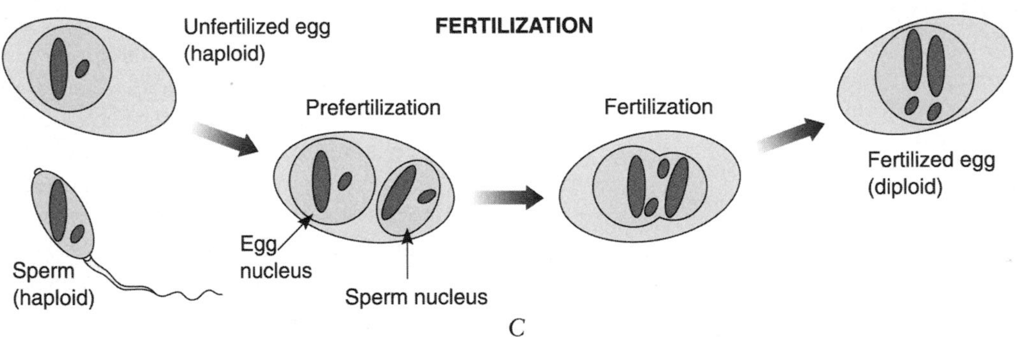

FIGURE 4-15 Greatly simplified summary comparison of the major features of (*A*) mitosis, (*B*) meiosis, and (*C*) fertilization.

latter fosters greater variation in offspring. Asexual organisms produce daughter cells that are identical to the parents. Unless there has been a fortuitous alteration of one or more genes (a mutation, as described in the next section), there is little or no change from generation to generation. In contrast, sexual reproduction provides variation by means of genetic recombination. In addition, sexually reproducing organisms accumulate considerable "invisible" variation in the form of recessive genes. As environmental changes occur, some of these recessive genes will produce traits in the animal or plant that have survival advantages, and new avenues of evolution will be opened.

Mutations

Some of the variation we see among individuals of the same species results from the mixing of genes that occurs during reproduction. However, if genes were never altered, then the number of variations they could produce would be limited. For organisms to evolve a truly new variation, a process is required that will change the genes themselves. That process is known as mutation. Mutations can be caused by ultraviolet light, cosmic and gamma rays, and chemicals, including certain drugs. They may also occur spontaneously without a specific causative agent. Mutations may occur in any cell, but their evolutionary impact is greater when they occur in the sex cells, for then succeeding generations will be affected.

To understand how mutations occur, it is useful to re-examine the configuration of DNA (see Fig. 4-13). The steps on the twisted DNA ladder are composed of two kinds of nitrogenous bases: *purines* and *pyrimidines*. Each step is made up of one purine joined to one pyrimidine. The purine called *adenine* is normally coupled with the pyrimidine called *thymine*. Similarly, the purine *guanine* joins with the pyrimidine *cytosine*. Thus, adenine-thymine and guanine-cytosine form base-pair steps of the ladder. Genes owe their specific characteristics to a particular order of these **base-pair** steps. If the order is disrupted, a mutation may result. For example, during cell division, when the twisted strands separate and then proceed to attract bases to rebuild the DNA molecule, there may be a mishap that results in the positioning of a guanine-cytosine base-pair in what should have been the location of an adenine-thymine base-pair. At that location, the gene is altered and may result in an inheritable change.

In summary, organic evolution may involve change from at least three different sources: mutations, gene recombinations, and natural selection. Mutations are the ultimate source of new and different genetic material. Recombination (mixing) spreads the new genetic material through the population and mixes the new with the old. Natural selection sorts out the multitude of varying traits, preserving those that, by chance, are best fitted to a particular environment.

Evolution in Populations

Evolution is a process of biologic change that occurs in populations. A **population** is a group of individuals of the same species that occupy a given area so that each individual has a chance to mate with members of the opposite sex within the group. Each of the individuals within the population has its own particular set of genes, and the sum of all of these within a breeding population constitutes the **gene pool** of that population. The gene pool is divided up in each generation and partitioned out to new offspring. Inevitably, in each successive generation, new mutations and genetic combinations occur that manifest themselves in the offspring. As natural selection comes into play, some of these will be passed on to the next generation in greater or lesser numbers than others, so that ultimately the gene pool is altered. Thus, evolution results from the impact of natural selection on the gene pool.

Speciation and Adaptive Radiation

The entire course of evolution depends on the origin of new species. The process by which this occurs is called **speciation**. A species is actually a collection of populations within which there is a free flow of genes. From this definition, it is implicit that there is no gene exchange between two different species. Barriers between different species keep the gene pools separate. These barriers may be reproductive barriers that prevent fertilization, mating, or development of offspring, or they may be geographic barriers, such as islands that separate and isolate land animals. Wherever reproductive or geographic barriers exist, various segments of the population become isolated for many generations. During this period, they are likely to accumulate enough genetic differences so that interbreeding between the segments is no longer possible. Once this has occurred, the altered isolated segments have become different species.

Once a new population has become established, pioneering segments of the population located around the fringes of the habitats may, like their parent species, undergo additional speciation. With innumerable successive speciations, diverse organisms characterized by diverse living strategies emerge. This branching of a population to produce descendants adapted to particular environments and living strategies is termed **adaptive radiation.**

The honey creepers of Hawaii provide an illustration of adaptive radiation. These birds, comprising many different species, are believed to have descended from a common ancestor. The most striking differences between the species occur in the sizes and shapes of the beaks, which are adaptively related to the kind of food on which the birds are dependent (Fig. 4-16). Some have stout beaks for crushing hard seeds, and others have beaks well adapted for seeking out insects in cracks and crevices or for sucking nectar. All of these variations in beak morphology are examples of **adaptations**, a word that means the acquisition of heritable characteristics that are advantageous to an individual and a population.

On a higher taxonomic level, one finds adaptive radiations among classes, orders, and families of organisms. All the orders of mammals, for example, had a common origin in an ancestral species that lived during the early Mesozoic. By adapting to different ways of life, the descendant forms came to diverge more and more from the ancestral stock, thereby providing today's rich diversity of mammalian plant eaters, flesh eaters, insect eaters, walkers, climbers, swimmers, and flyers.

Gradual or Punctuated Evolution

A question widely discussed by paleontologists is whether evolution proceeds gradually by means of an infinite number of subtle steps, as Darwin proposed, or if there are sudden sporadic advances during which most of the change in a lineage occurs over a relatively short span of time. Gradual progressive change is

FIGURE 4-16 The honey creepers of Hawaii provide a good example of adaptive radiation. When the ancestor of today's honey creepers first reached Hawaii, few birds were present. Succeeding generations diversified to occupy various ecologic niches. Their diversity is most apparent in the way their beaks have become adapted to different diets. Some have curved bills to extract nectar from tubular flowers, whereas others have short, sturdy beaks for cracking open seeds or pointed bills for seeking out insects in tree bark. *Himatione sanguinea*, at the lower right, has a relatively unspecialized beak and appears to be similar to the original honey creeper ancestor.

referred to as **phyletic gradualism**. *Phyletic* (from *phylogeny*) refers to evolutionary pathways. Thus, proponents of phyletic gradualism believe that change occurs by slow degrees along the evolutionary pathway of a lineage. Where striking evolutionary breaks occur, they are considered not true breaks at all but places where intermediate or transitional forms are missing because of imperfections in the geologic record. Darwin would insist that the transitional forms once existed but were not preserved, were destroyed, or have not yet been discovered.

Evidence of phyletic gradualism has been found in several groups of invertebrates, including lineages of well-preserved trilobites from the Ordovician of Wales, as well as among the marine planktonic organisms known as foraminifera. In the latter study, the fossil foraminifera were recovered from complete and uninterrupted well-dated sections of deep-sea cores. The fossils indicated a gradualistic line of descent that produced at least four species of the genus *Globorotalia* over a span of about 8 million years.

Paleontologists who oppose phyletic gradualism argue that the fossil record for the past 3 billion years contains many examples of new groups appearing suddenly in places where there is no evidence of any imperfections in the geologic record. Indeed, they find that the slow and stately advance of evolution can be documented only rarely and that evolutionary progress is more often sporadic. The term **punctuated equilibrium** was introduced in the 1970s by Stephen J. Gould and Niles Eldridge for evolution that progresses by sudden advances that "punctuate" long intervals of little change (termed **stasis**).

According to this concept, the punctuation or sudden morphologic change that interrupts equilibrium occurs at the periphery of the geographic area occupied by the population. These segments of the population are referred to as **peripheral isolates**. Gene flow is rapid among the individuals of peripheral isolates, and changes in morphology or physiology leading to speciation may occur within a short span of time. Species do not usually originate in the places where their parental stock exists but rather in boundary zones where variations can be tested against new environmental situations. Should the parent species suffer extinction or severely decline, the new species may move back into the parental domain, or they may expand into new territory. With adaptations that are a significant advantage, the new species are likely to enter a period of rapid evolution. This stage is usually followed by a period of moderate evolutionary rates and stability that lasts until the population begins to decline. Subsequently, yet another new group may begin its evolutionary expansion.

Phylogeny

The term phylogeny refers to the historical development of groups of organisms so as to depict descent from ancestors. The depiction is usually a diagram called a **phylogenetic tree**, such as shown in Figure

4-17. Branches on the tree are called **clades**. Most biologists and paleontologists today use one of two phylogenetic methods. The more traditional is called **stratophenetic phylogeny**. In the 1960s, a newer method termed **cladistic phylogeny** was introduced.

In stratophenetic phylogeny, organisms are arranged in treelike fashion, with the most recently evolved species or groups on the upper branches and older, ancestral species on the lower branches and trunk. Thus the tree includes the concept of change through time. The term *stratophenetic* is appropriate, for the tree relates the succession of life to the superpositional sequence of stratigraphic units. Stratophenetic phylogeny is dependent upon finding ancestral and descendent species in the fossil record. Unfortunately, this is not always possible. There are "missing links" where key forms are not found. As a result, parts of the tree may contain inferences about intermediary forms and ancestors. In some cases, cladistic phylogeny lessens the subjectivity inherent in stratophenic phylogeny.

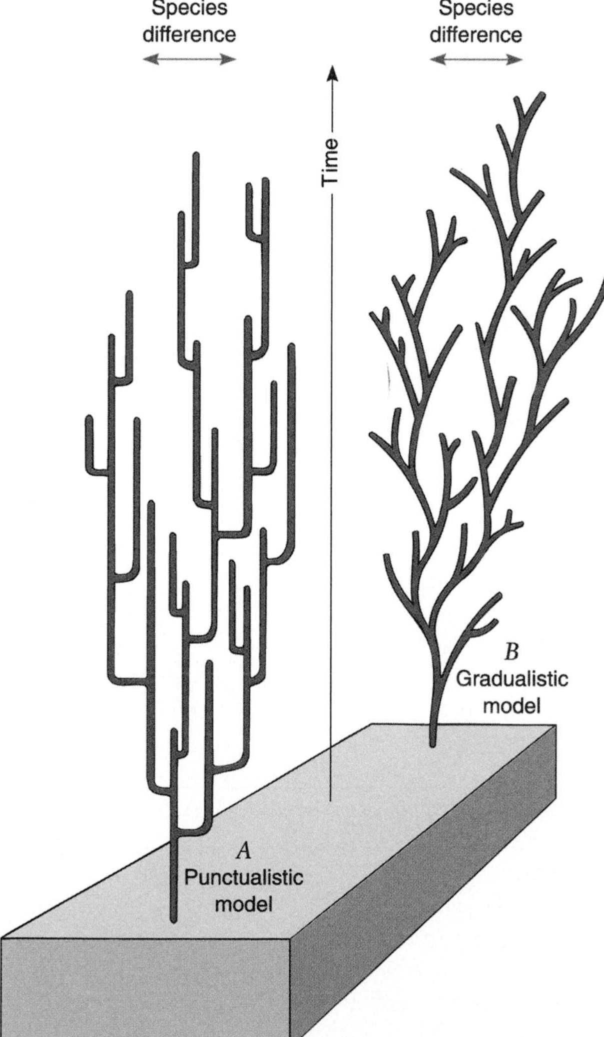

FIGURE 4-17 Phylogenetic trees depicting (*A*) the punctuated equilibrium model and (*B*) the gradualistic model of evolution. Morphologic change occurs in sideward directions. Time is depicted by the vertical direction. The short horizontal side branches of the punctuated equilibrium model depict sudden change, whereas the inclined branches of the gradualistic model suggest slow and uniform change through time.

In cladistic phylogeny, organisms are analyzed objectively on the basis of characteristics they share in order to determine their ancestor-descendant relationships. The analysis is depicted in a drawing called a **cladogram**. A cladogram shows closeness of relationship by the arrangement of groups. The shorter the links between groups, the closer the evolutionary relationship. In the very simple cladogram shown as Figure 4-18, the camel, tuna, and dolphin share the relatively primitive trait of a vertebral column. Although the dolphin and tuna superficially resemble one another, they are not placed close together because only the dolphin and camel share advanced characteristics, such as live birth, offspring nourished by milk from the female, and endothermy (warm-bloodedness). Unlike the stratophenetic phylogeny tree, a cladogram does not directly incorporate information about the time ranges of organisms. Its only purpose is to show how organisms are related.

Evidence of Evolution

PALEONTOLOGIC CLUES Early proponents of the concept of evolution had a difficult time convincing some of their 19th century contemporaries of the validity of Darwin's theory. One reason was that evolution is an almost imperceptibly slow process. The human life span is too short to witness evolutionary changes across generations of plants or animals. Fortunately, we can overcome this difficulty by examining the remains of organisms left in rocks of successively younger age. If life has evolved, the fossils preserved in

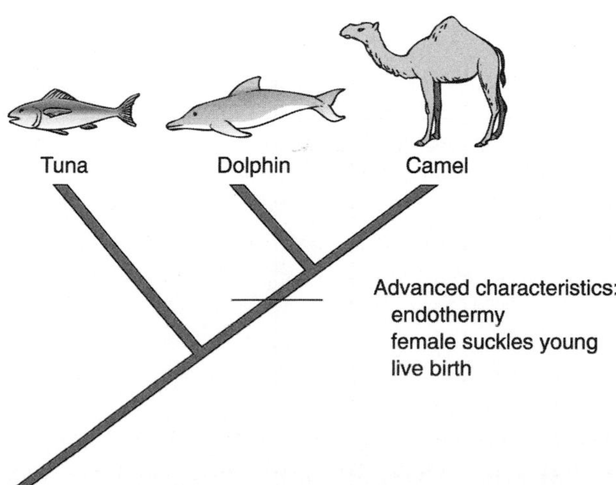

FIGURE 4-18 A simple cladogram. ? *Draw a cladogram depicting reptiles, mammals, amphibians, and birds. Where along the inclined line do the advanced characteristics of body hair, lungs, and claws appear?*

consecutive formations should exhibit those changes. Indeed, many examples are known of sequential morphologic changes among related creatures during successive intervals of geologic time. The most famous example is provided by Cenozoic fossil horses (Fig. 4-19). A small browsing animal from the upper Paleocene named *Radinskya* is presently considered the earliest relative of the horses. Unlike modern species, the earliest horses had four toes on their front feet and three on the rear. As one examines the many branching lineages of

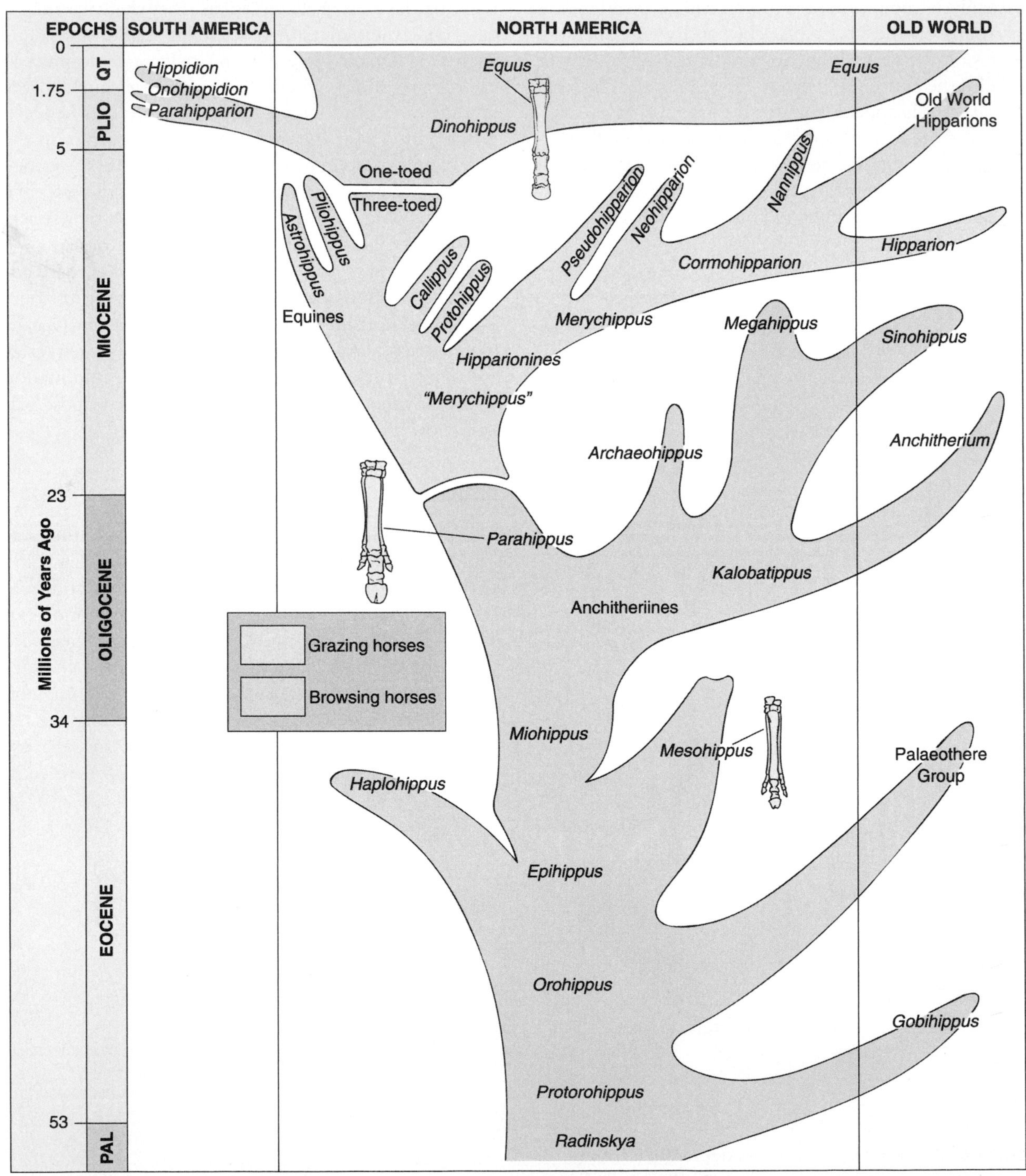

FIGURE 4-19 The very "bushy" phylogenetic tree of horses, showing evolutionary relationships between genera, the trend in toe reduction, and the transition from browsing to grazing horses. (*Based on Macfaddan, B. J. 1992. Fossil Horses: Systematics, Paleobiology, and Evolution of the Family Equidae. Cambridge: Cambridge University Press; and Prothero, D. R. and Schoch, R. M. 1994. Major Features of Vertebrate Evolution. Short Course in Paleontology No. 7. Knoxville, TN: University of Tennessee Press and The Paleontologic Society.*)

ENRICHMENT

Earbones Through the Ages

Humans and other terrestrial mammals hear because receptor cells in the inner ear generate electrical impulses in response to vibrations caused by sound waves. These waves travel through the external ear canal to the tympanic membrane or ear drum (see accompanying figure). The tympanic membrane vibrates in response, and the vibrations are transmitted through a chain of three small bones located in the middle ear to the sound organ or cochlea of the inner ear. The three bones (auditory ossicles) are hinged to one another in a manner that allows them to act like an amplifier, increasing the force of sound vibrations. The evolution of the ear bones and middle ear cavity are excellent examples of anatomic structures that served a particular function in ancestral animals but are changed in function and form in the descendant animals.

The first of the three auditory ossicles is the stapes. One can trace its transformation backward in time to the Silurian, when jaws evolved from cartilaginous or bony supports for gills. The bone that was to become the stapes was originally an upper element of one of these gill supports. In the evolution of fish with true jaws, the gill support just posterior to those that evolved into jaws became a supporting prop between the upper jaw and braincase. It is called the hyomandibular. With the advent of amphibians during the Devonian, the old fish hyomandibular was transformed into the stapes, where it functioned as a transmitter of vibrations from the tympanic membrane to the inner ear. The stapes persisted as the only auditory ossicle throughout the evolutionary history of both amphibians and reptiles. With the advent of mammals, however, two bones in reptiles that had served as the joint between the upper and lower jaws were changed in function to transmit vibrations. They became the incus and the malleus. Thus, over the past 400 million years or so, the bones that were to evolve into our present hearing apparatus served such diverse functions as supports for the gills, props for the braincase, and articulation for the jaws.

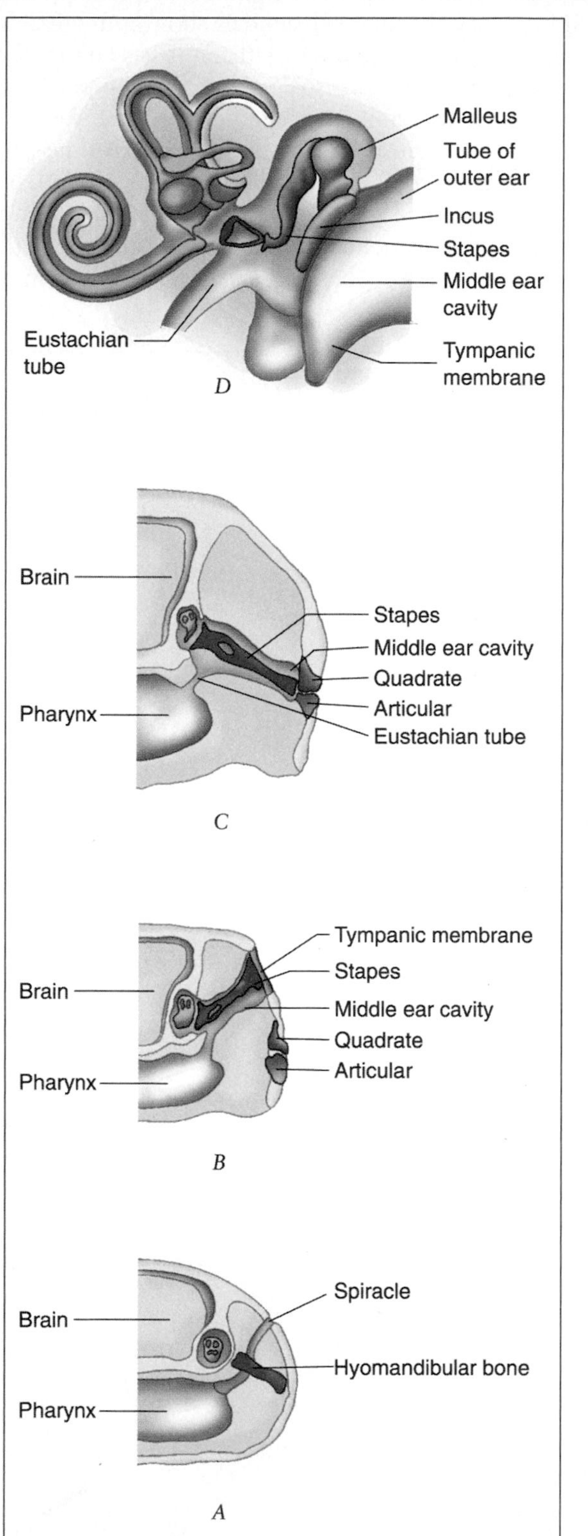

Evolution of auditory ossicles.
(*A*) Primitive fish with spiracle and hyomandibular bone, but without middle ear cavity. (*B*) Primitive amphibian with stapes (also called columella) derived from fish hyomandibular bone located within the former spiracular tunnel, which now functions as a middle ear cavity. (*C*) A reptile in which the stapes has shifted to a position near the quadrate. (*D*) Mammal in which the quadrate has been transformed into the incus, and the articular into the malleus. (*After Romer, A. S. 1941. Man and the Vertebrates. Chicago: University of Chicago Press.*)

the horse "family tree" in successively higher (hence, younger) formations, one is able to see the results of evolutionary change. The animals show an increase in size, a reduction of side toes with emphasis on the middle toe, an increase in the height and complexity of teeth, and a deepening and lengthening of the skull.

The change in horse dentition through time provides an interesting example of the important link between environment and organic evolution. There is paleontologic evidence that during the time that the horse family was evolving, grasslands were becoming increasingly widespread in North America and Eurasia. Grass is a rather harsh food for an animal that must feed on plants. The blades contain silica, and because grass grows close to the ground, it is usually coated with abrasive dust. Early members of the horse family lived in forested areas and fed on the less abrasive leaves of trees and shrubs. They were browsers and had the low-crowned teeth of leaf eaters. Later members of the horse family were affected by selection processes that favored variants better able to cope with the problem of tooth wear from eating grasses. Over

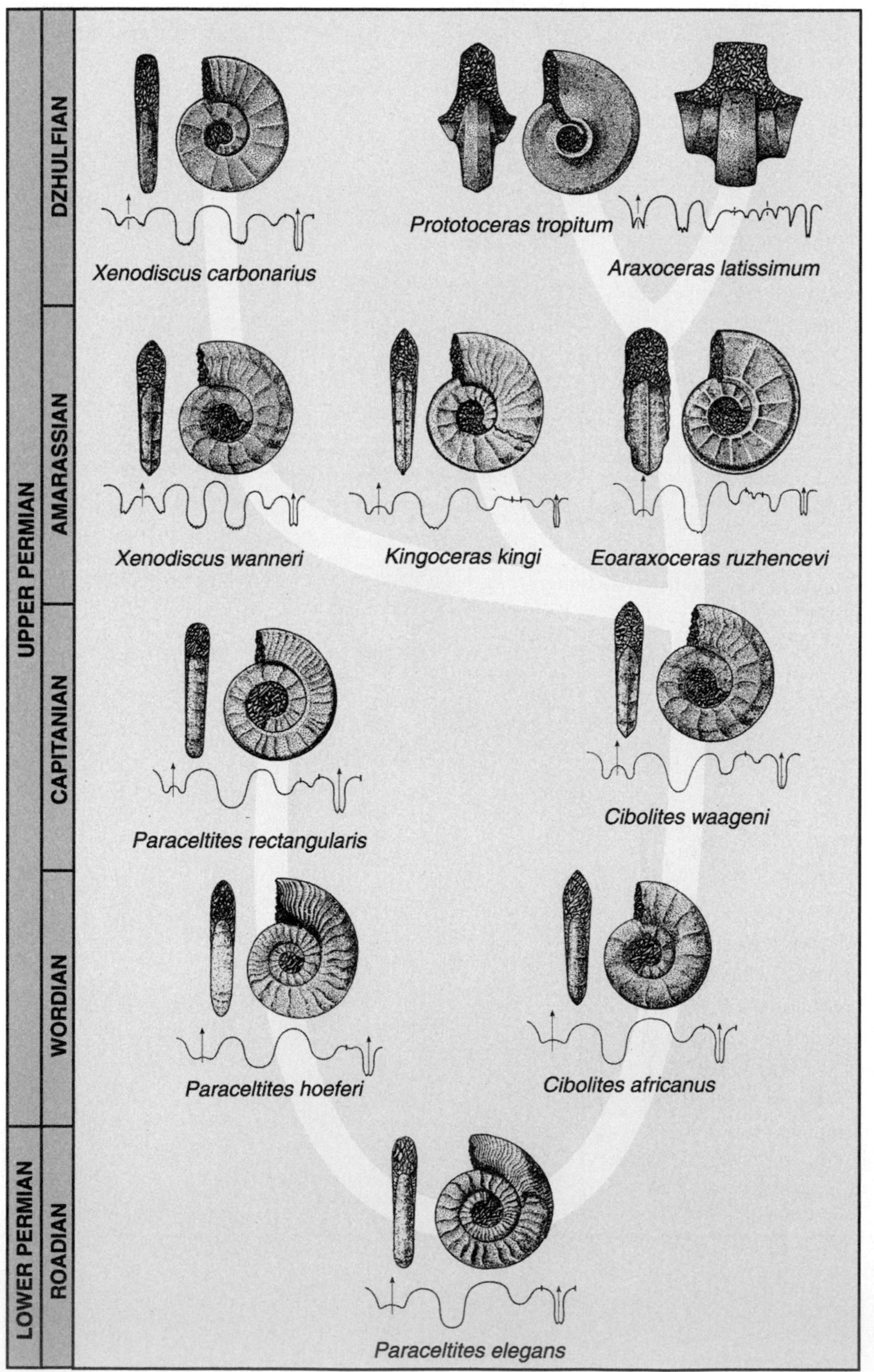

FIGURE 4-20 An example of progressive evolutionary change in a group of Permian ammonoid cephalopods. According to this interpretation, two evolutionary lineages originated from *Paraceltites elegans*, one terminating in *P. rectangularis* and a second producing *Cibolites waageni*. The latter was the ancestral stock for three additional lineages. The curved lines beneath each drawing are tracings of the suture lines (suture lines are formed where the edge of a chamber wall or septum joins the inner wall of the shell). The terms along the left border are the names of Permian stages. (*From Spinosa, C., Furnish, W. M., and Glenister, B. F. 1975. J. Paleontol. 49(2):239–283.*)

many generations, they had evolved the high-crowned teeth with complex patterns of enamel that characterize grazing animals. These horses were likely to have been well nourished and hence lived longer lives and were able to produce more progeny. Their evolution must also have affected the evolution of the predators that pursued them and may even have contributed to the evolution of species of grasses that were resistant to damage by grazing. Evolution is an intricate process driven by natural selection in which every animal and plant interacts with its neighbors and with the physical environment.

Although fossil remains of horses provide a fine illustration of paleontologic evidence of evolution, hundreds of other examples representing every major group of animals and plants would serve as well. The marine invertebrates known as cephalopods provide fine examples of progressive evolutionary changes (Fig. 4-20). Because of this, they are also exceptionally useful in correlation.

BIOLOGIC CLUES Supplemental to the paleontologic evidence favoring evolution are several persuasive arguments that are more directly biologic in nature. In studies of the comparative morphology of organisms, it is not at all unusual to find body parts that are evidently of similar origin, structure, and development, even though they may be adapted for different functions in various species. For example, in seed plants, leaves are found as petals, tendrils, and thorns. In four-limbed vertebrates, the bones of the limbs may vary in size and shape, but they are fundamentally similar and in similar relative positions in birds, horses, whales, and humans (Fig. 4-21). Such basically similar structures in superficially dissimilar organisms are referred to as **homologous**. The differences in homologous structures are the result of variations and adaptations to particular environmental conditions, but the similarities indicate genetic relationship and common ancestry.

As another line of evidence, biologists point to the existence in animals and plants of apparently useless and usually reduced structures that in other related species are well developed and functional. These **vestigial organs** are the "vestiges" of body parts that were utilized in earlier ancestral forms. Genetic material inherited from those ancestors still produces the vestigial organs, but the importance of these structures to the well-being of the organism has diminished with changes in environment and habits. Humans have over 100 of these vestigial structures, including the appendix, the ear muscles, and the coccyx (tail vertebrae). The vestigial pelvic bones in such different animals as the boa constructor and whale (Fig. 4-22) clearly suggest that they were evolved from four-legged animals. Proof of the whales' four-legged ancestry was obtained in 1994, when the well-preserved skeleton of an early Eocene whale was recovered from river sediments in Pakistan. The whale, named *Ambulocetus* ("walking swimming whale"), had front limbs developed as flippers, and long hind limbs with elongate toes for the support of webbed feet.

Biologists have provided further evidence of evolution through comparative studies of the embryos of vertebrate animals. In their early stages of development, embryos of fish, birds, and mammals are strikingly similar. It would seem that all these animals received basic sets of genes from remote common

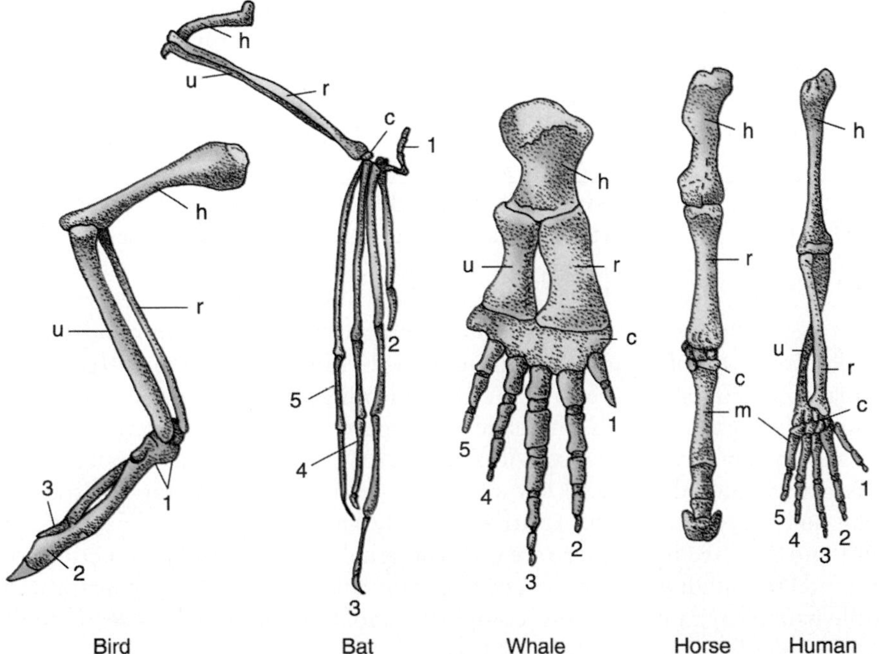

FIGURE 4-21 Skeleton of right forelimb of several vertebrates to show similarity of structure. Key: c, carpals; h, humerus; m, metacarpals; r, radius; u, ulna; 1–5, digits. Limbs are scaled to similar size for comparison. ? *Suggest a reason for the overlapping arrangement of the ulna and radius in the human forearm.*

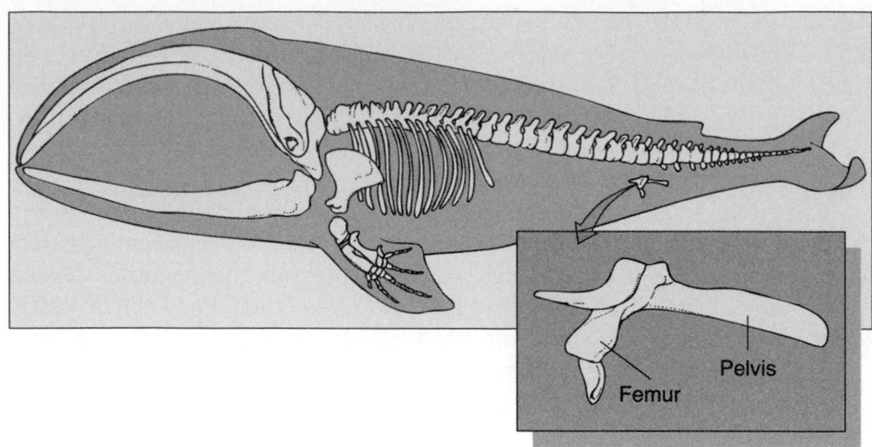

FIGURE 4-22 The pelvis and femur of a whale are vestigial organs. Vestigial organs are to be expected as animals evolve and adapt to different modes of life. Selective pressure for eliminating such organs is often weak, and so the vestige is retained for relatively long periods of time. In 1994, a fossil whale was exhumed from lower Eocene river sediments in Pakistan. Named Ambulocetus, the skeleton of this early whale included hind limbs with long toes for the support of webbed feet.

ancestors. These genes control embryologic development for a time. Later in the developmental process, other genes began to assume control and to cause each different species to develop in its own unique way.

Animals may reveal their evolutionary relationships not only in morphologic structures and embryology but in their molecular characteristics as well. In the DNA molecule, for example, molecular biologists are able to determine the sequence of the nucleotide base pairs (the "steps" on the DNA "ladder"), and when the sequences of different groups of animals or plants are compared, the degree to which they are related can be specified. If two groups are believed to be closely related on the basis of morphology, embryology, or the fossil record, they should have a greater percentage of DNA sequences in common than groups that are not closely related.

In addition to DNA, it has been found that digestive enzymes and hormone secretions are similar in related organisms. Proteins extracted from corresponding tissues of closely related animals often show striking similarities. The antigenic reactions of blood from various groups of humans have been found to be practically identical to such reactions in the blood of anthropoid apes.

FOSSILS AND STRATIGRAPHY

Establishing Age Equivalence of Strata with Fossils

One of William Smith's major contributions to geology was his recognition that individual strata contain definite assemblages of fossils. Because of the change in life through time, superposition, and the observation that once species have become extinct they do not reappear in later ages, fossils can be used to recognize the approximate age of a unit and its place in the stratigraphic column. (Such a method for judging the age of a unit would not be possible with inorganic characteristics of strata because they frequently recur in various parts of the geologic column.) Further, rocks formed during the same age in identical environments but diverse localities often contain similar fauna and flora if there has been an opportunity for genetic exchange. This permits geologists to match chronologically or correlate strata from place to place. It provides a means for establishing the age equivalence of strata in widely separated parts of the globe.

THE GEOLOGIC RANGE Before fossils could be used as indicators of age, it was first necessary to determine the relative ages of the major units of rock on the basis of superposition. Geologists began by working out the superpositional sequences locally, and then they added sections from other localities around the world, fitting the segments together into a summary or composite geologic column. The next step was to determine the fossil assemblage from each time-rock unit and to identify the various genera and species. This work began well over a century ago and is still very much in progress. Gradually, it became possible to recognize the oldest (or first appearance of) particular species as well as their youngest occurrence (last to appear) stratigraphically. The interval between the first and last appearance of a species constitutes its **geologic range**. Clearly, the geologic range of any ancient organism is not known *a priori* but is determined only by recording its occurrence in numerous stratigraphic sequences from hundreds of locations. Figure 4-23 illustrates one such study that records the geologic ranges of cephalopods (a class of mollusks having chambered shells) from a Cretaceous formation in Antarctica. Today, ranges are well known for some species and groups of species but relatively poorly known for others. Fortunately, there are now enough data so that isolated sections of strata in geologically unexplored areas can be located in the composite geologic column by use of their contained fossils.

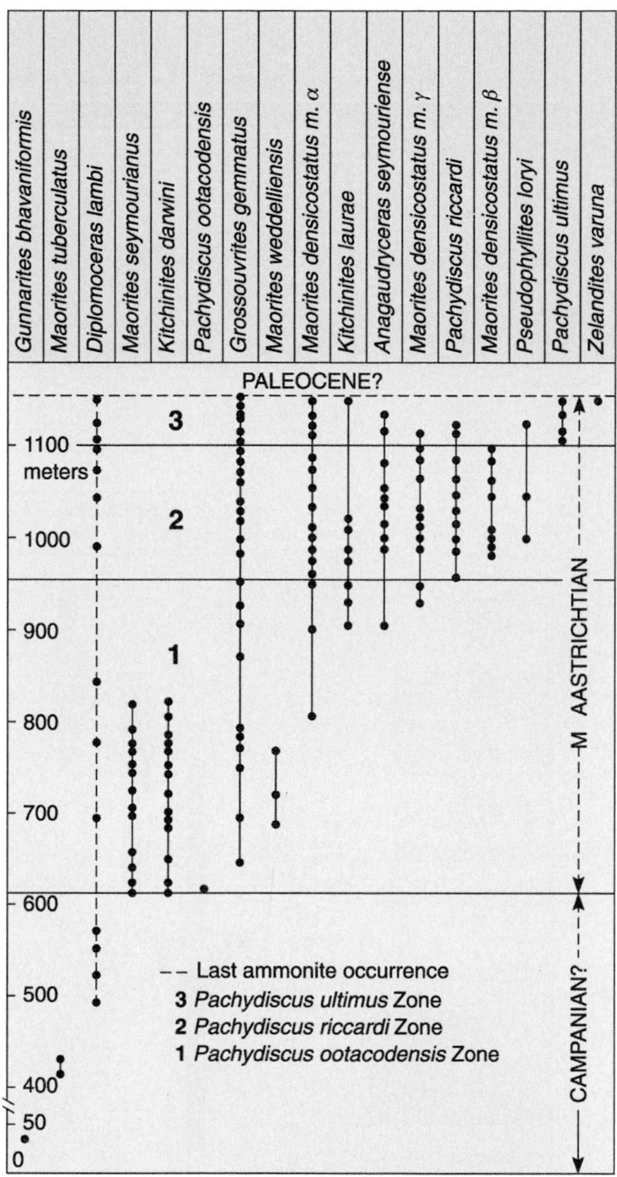

FIGURE 4-23 Example of a range chart showing the ranges of late Cretaceous ammonite cephalopods (chambered mollusks) from the Lopez de Bertodano Formation, Seymour Island, Antarctic Peninsula. The Campanian and Maastrichtian are stages of the Upper Cretaceous. Zonal boundaries are based on the first appearances of the species, after which each zone is named. (*From Macellari, E. E. 1986. J. Paleontol. Mem. 18, Part 2.*)

? *Draw a circle around the dot representing the first appearance of Pachydiscus ultimus, P. riccardi, and P. ootacodensis.*

IDENTIFICATION OF CHRONOSTRATIGRAPHIC UNITS The method for identifying time-stratigraphic units is illustrated in Figure 4-24. Geologists working in Region 1 come upon three time—rock systems of strata, designated O, D, and M. Perhaps years later in Region 2, they again find units O and D, but in addition, they recognize an older unit—namely, C—below (and hence older than) O. Finally, while working in Region 3, they find a "new" unit, S, sandwiched between units O and D. The section, complete insofar as can be known from the available evidence, consists of five units that decrease in age as one progresses upward from C to M. The geologists may then plot the ranges of fossil species found in C to M alongside the geologic column. If they should next find themselves in an unexplored region, they might experience difficulty in attempting to locate the position of this rock sequence in the standard column, especially if the lithologic traits of the rocks had changed. However, on discovering a bed containing species A, they are at least able to say that the rock sequence might be C, O, or S. Should they later find fossil species B in association with A, they might then state that the outcrop in the unexplored region correlates in time with unit S in Region 3. In this way, the strata around the world are incorporated into the total global stratigraphy.

PALEONTOLOGIC CORRELATION Of course, the preceding illustration is greatly simplified, and future discoveries may extend known fossil ranges. Geologists must always keep in mind that being unable to find key fossils might lead to erroneous interpretations. Fortunately, the chances of making such mistakes are diminished by the practice of using entire assemblages of fossils. Two or three million years from now, geologists might have difficulty in firmly establishing on fossil evidence that the North American opossum, the Australian wallaby, and the African aardvark lived during the same episode of geologic time. However, if they found fossils of *Homo sapiens* with each of these animals, it would indicate their contemporaneity. In this example, *Homo sapiens* can be considered the cosmopolitan species, for it is not restricted to any single geographic location within the terrestrial environment. The aardvark and wallaby are said to be *endemic species* in that they are confined to a particular area.

In the case of fossilized marine animals, cosmopolitan species have been especially useful in establishing the contemporaneity of strata, whereas endemic species are generally good indicators of the environment in which strata were deposited. Endemic species may slowly migrate from one locality to another. For example, the peculiar screw-shaped marine fossil bryozoan known as *Archimedes* (Fig. 4-25) was endemic to central North America during the Mississippian Period but migrated steadily westward, finally reaching Nevada by Pennsylvanian time and Russia by the Permian Period. Motile larvae of invertebrates that are attached to the sea floor as adults are often widely distributed by ocean currents, thus facilitating the migration of species.

One must be tentative in making correlations based on fossils. The validity of a correlation increases each time the sequence of faunal changes is found again at different locations around the world. Geologists must

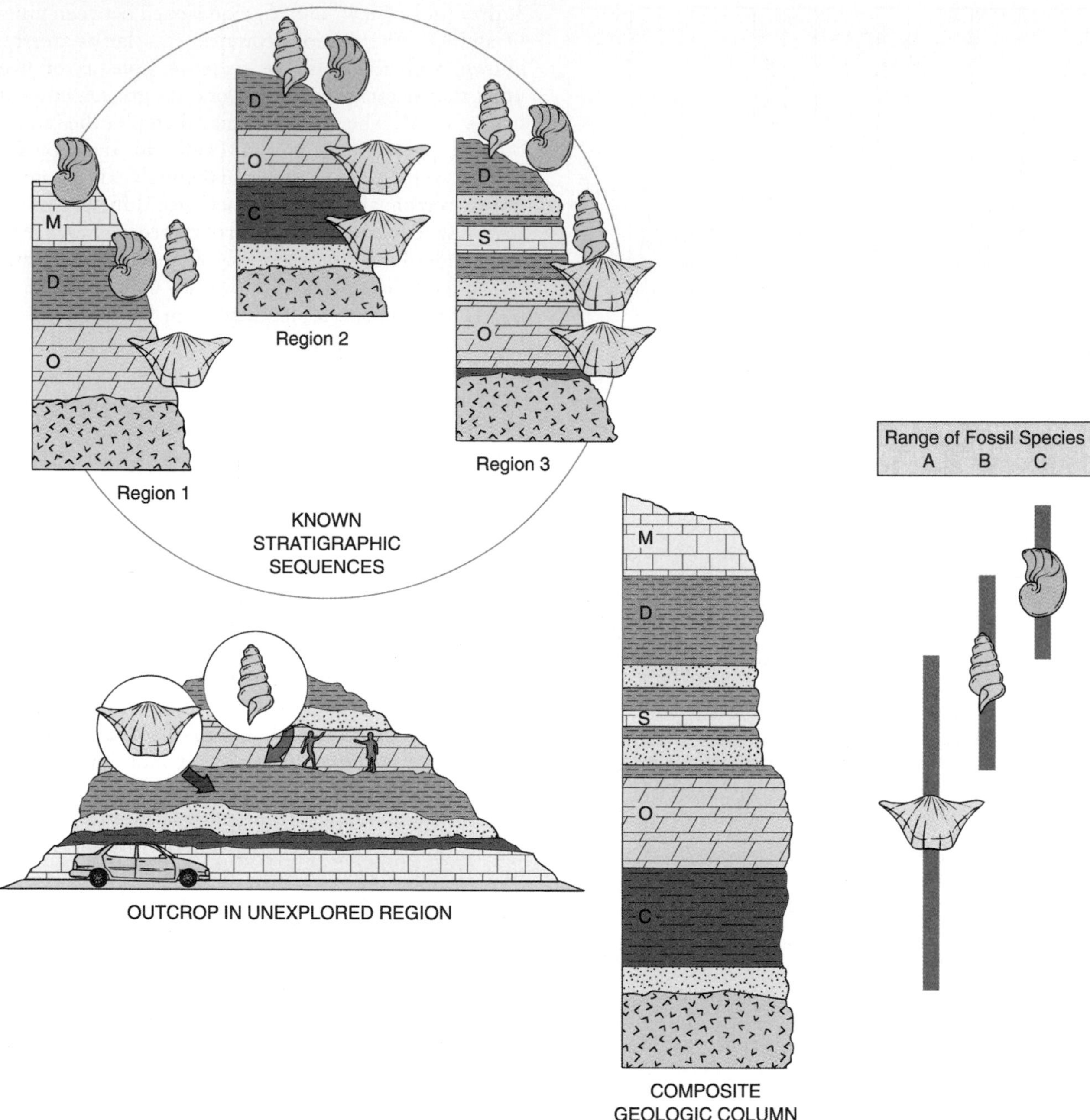

FIGURE 4-24 Use of geologic ranges of fossils to identify chronostratigraphic units.

also be aware that not all changes are caused by evolution but instead may indicate faunal migrations or shifts in flora that accompanied ancient environmental changes. In this regard, the sudden disappearance of a fossil need not mean that it became extinct but rather that it moved elsewhere. One might also note that the earliest appearance of a fossil in rocks of a given region might mean it had evolved there; however, it might also signify only that a pre-existing species had come into the new locality.

In addition to being on the alert for faunal changes that are related to shifting of ancient environments (as opposed to changes resulting from evolution), paleontologists must also be concerned about the possibility of **reworked fossils**. At various times in the geologic past, weathering and erosion have freed fossils from their host rock. Much as would be the case with any clastic particle, these fossils might then be reworked into younger beds, and the younger strata might then be mistakenly assigned to an older geologic time. Some fossils are particularly resistant to erosion and chemical decay and therefore are more susceptible to reworking. Among such resistant fossils are spore and pollen grains (Fig. 4-45) and conodonts (Fig. 1-1). The latter are tiny, toothlike structures composed of cal-

(A)

(B)

FIGURE 4-25 Two specimens of the guide fossil *Archimedes*. Specimen *A* shows only the screwlike axis of this bryozoan animal. In specimen *B*, one can see the fragile, lacy skeleton of the colony that is attached to the sharp, helical edge of the axis so as to form a netlike colony wound into an erect spiral. Specimens are 8 to 10 centimeters in height.

cium phosphate. Conodonts are the oldest remains of chordates in the fossil record. They occur in marine sedimentary rocks ranging in age from Late Proterozoic to Triassic and are particularly abundant in some Paleozoic formations.

INDEX FOSSILS Many fossils are rare and restricted to a few localities. Others are abundant, widely dispersed, and derived from organisms that lived during a relatively short span of geologic time. Such fossils are called **index** or **guide fossils** because they are especially useful in identifying time-rock units and correlating them from one area to another. A guide fossil with a short geologic range is clearly more useful than one with a long range. A fossil species that lived during the total duration of a geologic era would not be of much use in identifying the rocks of one of the subdivisions of lesser duration within that era. It is apparent that the rate of evolution is an important factor in the development of guide fossils. Simply stated, the rate of evolution is a measure of how much biologic change has occurred over a given interval of geologic time. Groups with rapid rates of evolution provide greater numbers of fossil species useful in stratigraphic correlation.

Although index fossils are of great convenience to geologists, correlations and interpretations based on assemblages of fossils are often more useful and less susceptible to error or uncertainties caused by undiscovered, reworked, or missing individual species.

BIOSTRATIGRAPHIC ZONES Geologists use the term **biozone** when describing a body of rock organized into stratigraphic units on the basis of its fossil content. Paleontologic biozones can vary in thickness or lithology and can be either local or global in lateral extent. Without formally naming them, we have already described the two major kinds of biostratigraphic zones: the range zone and the assemblage zone. A **range zone** is simply the rock body representing the total geologic life span of a distinct group of organisms. For example, in Figure 4-26 the *Assilina* range zone is marked by the first (lowest) occurrence of that genus at point A and its extinction at point B. Geologists may also designate **assemblage zones** selected on the basis of several coexisting taxa (the singular of *taxa* is *taxon*, a term which refers to a group of organisms that constitute a particular taxonomic category, such as species, genus, and family). Assemblage zones are named after an easily recognized and usually common member of the assemblage. In certain areas, however, even though the guide fossil for the assemblage zone may not be present, the other members permit recognition of the zone. Another kind of biostratigraphic zone is the **concurrent range zone**. It is recognized by the overlapping ranges of two or more taxa. For example, the interval between X and Y in Figure 4-26 might be designated the *Assilina-Heterostegina* concurrent range zone. Concurrent range zones often permit one to recognize the deposits representing smaller increments of time than might be provided by range zones.

Biozones are of fundamental importance in stratigraphy. They are the basic unit for all biostratigraphic classification and correlation. They are also the basis for time-stratigraphic terms because zones are aggregated into stages, stages into series, and series into systems. The boundaries of these units are usually zonal boundaries.

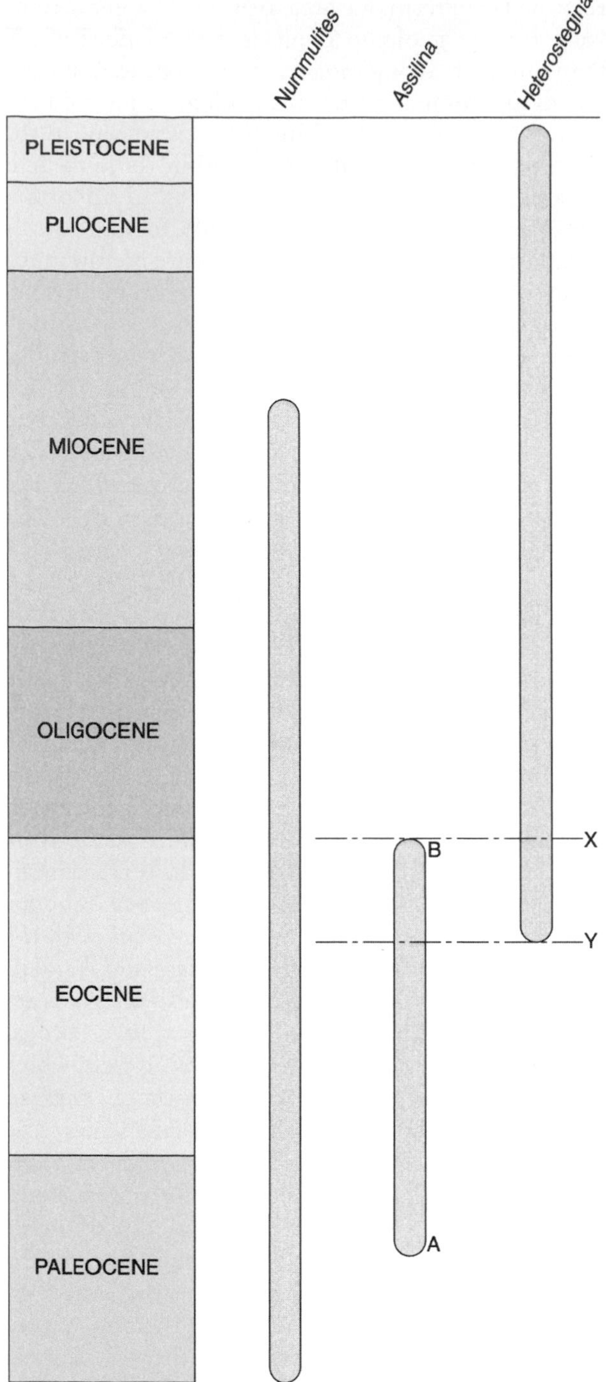

FIGURE 4-26 Geologic ranges of three genera of foraminifera. The interval between A and B is the total range of *Assilina*. The interval between X and Y could be designated the *Assilina–Heterostegina* concurrent range zone.

FOSSILS AS CLUES TO ANCIENT ENVIRONMENTS

Paleoecology

Although inert and mute, fossils are vestiges of once-lively animals and plants that nourished themselves, grew, reproduced, and interacted in countless ways with other organisms and their physical environments. The study of the interaction of ancient organisms with their environments is called **paleoecology**. Paleoecologists attempt to discover precisely where and how ancient creatures lived and what their habits and morphology reveal about the geography and climate of long ago. This scientific detective work is accomplished in various ways. One can, for example, compare species known only from fossils with living counterparts. The assumption is made that both living and fossil forms had approximately the same needs, habits, and tolerances. One can also examine the anatomy of the fossil and attempt to identify structures that were likely to have developed in response to particular biologic and physical conditions in the environment. As noted earlier, such modifications for living in a certain way and performing particular functions are called adaptations. An example of an adaptation is the broad, spiny valve of the brachiopod *Marginifera ornata* (Fig. 4-27) that served to support and anchor the animal on a sea bed composed of soft mud. In another example, reduction of the skeleton to mere spines in the Silurian trilobite *Deiphon* (Fig. 4-28) was probably an adaptation to provide buoyancy in the animal, which fed near the surface of the sea. Coiling in living and fossil cephalopods appears to be an adaptation designed to bring the center of buoyancy above the center of gravity and thereby permit these creatures to keep on an "even keel" while swimming. A list of additional examples of adaptations could be immense, for every creature is a result of hundreds of coordinated adaptations. Not all adaptations are morphologic. Biochemical and physiologic adaptations occur as well, although these are more difficult to recognize in fossils.

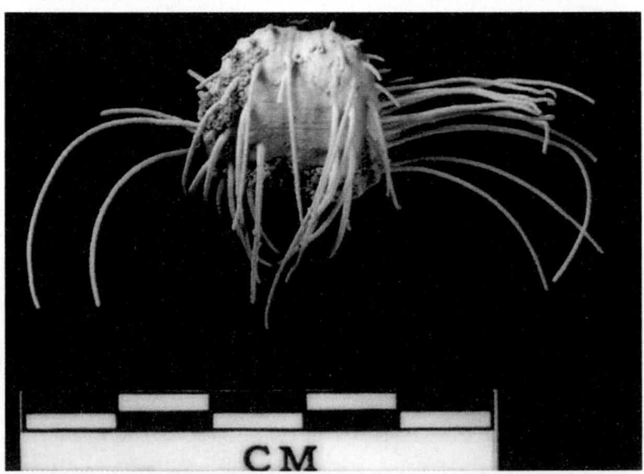

Pre-€ | € | O | S | D | M | P | Pr | Tr | J | K | T | Q

FIGURE 4-27 A fossil of the spinose brachiopod *Marginifera ornata*. The shell has been replaced by silica. The enclosing limestone has been dissolved so as to provide excellently preserved fossils with delicate spines intact. The animal rested on its ventral valve (shown here), and its spines provided anchorage. (*Photograph courtesy of R. E. Grant, U.S. Geological Survey.*)

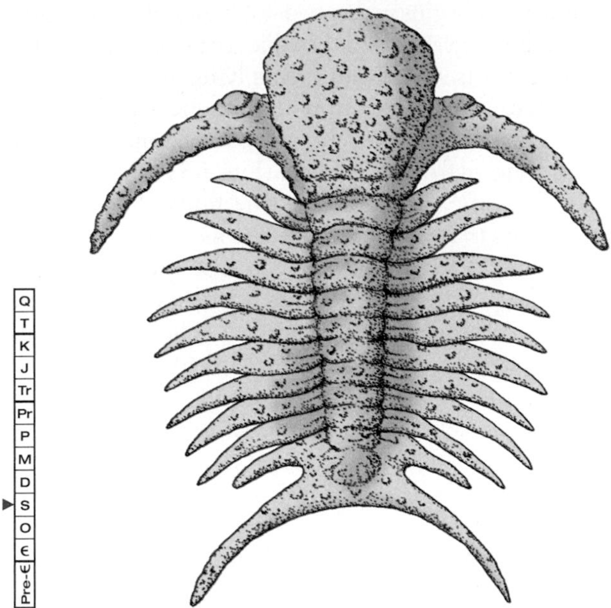

FIGURE 4-28 ***Deiphon* was a Silurian trilobite whose extreme spinosity suggests it was a swimmer and floater rather than a bottom dweller.** Spinosity increases surface area without adding weight, thus enhancing buoyancy. It is also possible that the swollen anterior structure (glabella) may have been filled with a liquid of low specific gravity, thus providing additional buoyancy. The length of the specimen is 27 millimeters.

The language of paleoecology is derived from **ecology**, the study of the present relationships between organisms and their environments. In ecologic studies, one tends to concentrate on the **ecosystem**, which is any selected part of the physical environment together with the animals and plants in it. An ecosystem may be as large as the Earth or as small as a garden pond. Paleoecologists are particularly interested in the ocean ecosystem because of the richness of the fossil record of marine life. The physical aspects of the ocean ecosystem include the water itself, its dissolved gases (especially carbon dioxide and oxygen), salts (phosphates, nitrates, chlorides, and carbonates of sodium, potassium, and calcium), various organic compounds, turbidity, pressure, light penetration, and temperature.

The biologic components of the ocean ecosystem are usually classified according to so-called **trophic** or feeding levels (Fig. 4-29). For example, **producer organisms** such as green plants manufacture compounds from simple inorganic substances by means of photosynthesis. These organisms in the sea are mostly very small (less than 0.1 millimeter in diameter) and include diatoms and other forms of algae. Producer organisms are eaten by **consumer organisms**, such as mollusks, crustaceans, and fish. The primary consumers commonly feed directly on unicellular plants and thus may be further designated as **herbivores**. The secondary consumers that eat the herbivores are called **carnivores**. Tertiary and even quaternary consumers feed on carnivores from lower trophic levels. The ecosystem also contains decomposers and transformers, including bacteria and fungi, which are able to break down the organic compounds in dead organisms and waste matter and produce simpler materials that can be used by the producers. Thus, in an ecosystem, the basic chemical components of life are continuously being recycled. The marine ecosystem also contains parasites, which feed on other organisms without necessarily killing them, and scavengers, which derive their nourishment from dead organisms.

Within any given ecosystem, one finds the specific environments or habitats where certain organisms live. In the ocean, the habitats range from the cold,

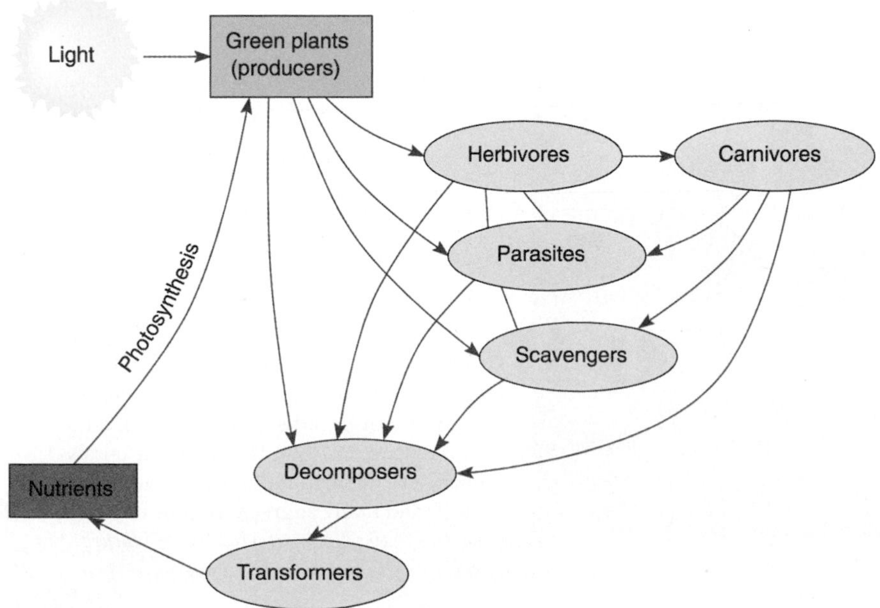

FIGURE 4-29 The movement of materials through an ecosystem. Components within ovals are consumers.

dark realm of the abyss to the warm, illuminated areas of tropical coral reefs. And within every habitat there are ecologic niches. Unlike the term **habitat**, which refers to the "address" or place where an organism lives, **niche** refers to the particular role played by the animal or plant as it interacts with all the physical, chemical, and biologic elements of its immediate environment. It is an organism's way of life. Within any given habitat there are many ways in which organisms can "earn a living." A coral reef (Fig. 4-30), for example, provides a niche for tiny tentacled coral animals that build the framework of the reef and feed as carnivores on smaller organisms. Certain marine snails occupy a niche along the surface of algal mats, where they sluggishly graze on films of algae. Behind the reef in areas of soft mud, plump lugworms consume the soupy sediment for its content of organic nutrients. Each creature has its characteristic ecologic niche. Each organism, however, is also a member of a natural assemblage of plants and animals living together in a **community**. One of the most interesting and challenging tasks of the paleontologist is to discover the interrelationships between members of ancient communities called paleocommunities. Which were the producers, herbivores, predators, and decomposers? What elements of the fauna might have been present but are not preserved? And what does the succession of paleocommunities, as seen in underlying and overlying strata, indicate about how the environment changed during the passage of geologic time?

The Marine Ecosystem

To facilitate the study of the ocean ecosystem, ecologists have developed a simple classification of marine environments. It begins with a two-fold division of the entire ocean into pelagic and benthic realms. The pelagic realm consists of the water mass lying above the ocean floor. It can be divided into a neritic zone, which overlies the continental shelves, and an oceanic zone, which extends seaward from the shelves (Fig. 4-31).

Within the pelagic realm one finds small animals and plants that float, drift, or feebly swim. These are **plankton**. **Phytoplankton** consist of plants and plantlike protists and include algae such as diatoms (Fig. 12-57) and coccolithophorids (Fig. 4-39). Planktonic animals constitute **zooplankton** and include protists such as radiolaria (Fig. 14-5), foraminifera (Fig. 4-32), certain tiny mollusks, small crustaceans, and the motile larva of many different families of invertebrates that live the adult stage of their life cycles on the sea floor.

The pelagic realm is also the home of **nekton**, or strongly swimming animals. Nekton are able to travel where they choose under their own power, and this is clearly advantageous. A swimming creature can search for its food and does not have to depend on food particles carried in chance currents. It can use its mobility to escape predators and can move to more favorable areas when conditions become difficult. The nekton is a diverse group that includes invertebrates such as shrimp, cephalopods, and certain extinct trilobites as well as vertebrates such as fish, whales, and marine turtles.

The second great division of the ocean ecosystem is the bottom or **benthic** realm. It begins with a narrow zone above high tide called the **supralittoral zone**. Relatively few kinds of marine plants and animals have adapted themselves to this harsh environment, where ocean spray provides vital moisture but where drying poses a constant danger. In addition, inhabitants of the

FIGURE 4-30 A Permian patch reef paleocommunity. In this diorama, one can recognize strings of bead sponges (top) and compact sponges (lower left), rugose corals (upper left), cephalopods (center and bottom center), and many aggregates of brachiopods. (*Courtesy of the National Museum of Natural History, Smithsonian Institution.*)

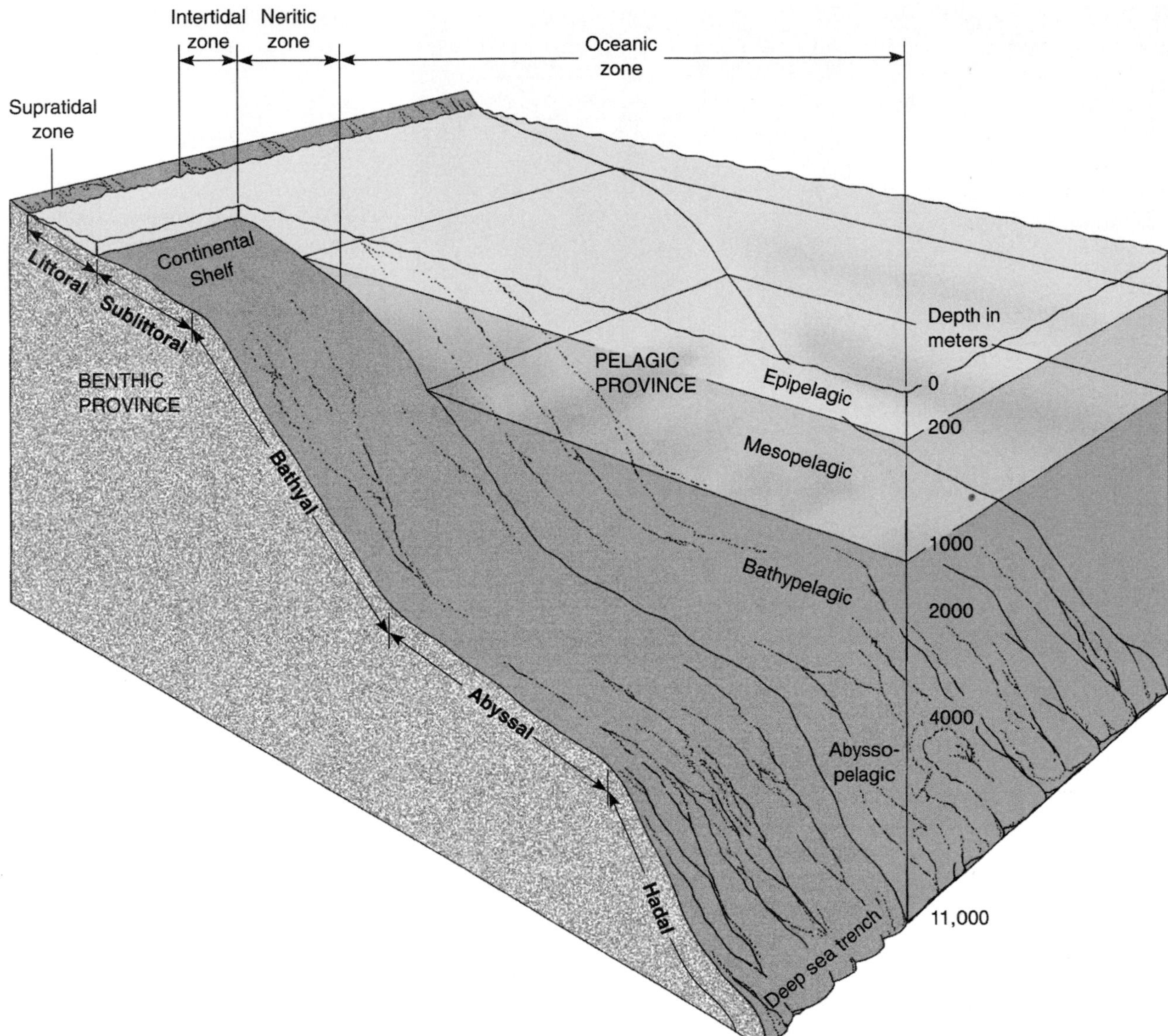

FIGURE 4-31 Classification of marine environments. (*After Hedgpeth, J. W., ed. 1957. Treatise on Marine Ecology and Paleoecology. Geological Society of America Memoirs 67(1):18.*)

supralittoral zone must be able to adjust to daily and seasonal temperature changes and the constant movement of sand as waves break along beaches.

Seaward of the supralittoral zone is the area between high and low tide. This is the **littoral zone**. Like those creatures of the supralittoral area, organisms in the littoral zone must be able to tolerate alternate wet and dry conditions. Some avoid the drying by burrowing into wet sand, whereas others have adapted themselves in various ways to retain body moisture when exposed to air.

Benthic animals and plants are most abundant seaward of the littoral zone in the continuously submerged **sublittoral zone** (Fig. 4-33). This zone extends from low tide levels down to the edge of the continental shelf (about 200 meters deep). Depending on the clarity of the water, light may penetrate to the sea floor in the sublittoral zone, although the base of light penetration is usually slightly less than 200 meters. Various kinds of algae thrive here as well as abundant protozoans, sponges, corals, worms, mollusks, crustaceans, and sea urchins. These animals are never subjected to desiccation. Their adaptations are mainly associated with food gathering and protection from predators. Some of these benthic animals live on top of the sediment that carpets the sea floor and are called **epifaunal**. Others, termed **infaunal**, burrow into the soft sediment or bore into harder substrates for food and protection (Fig. 4-34). Burrowers churn and mix sediment, a process known as **bioturbation**. Bioturbation may destroy original grain orientations in clastic sediments, rendering them useless for studies of

FIGURE 4-32 A living planktonic foraminifer similar to those that became abundant during the Cretaceous and Cenozoic. The fine rays are cytoplasm that is extruded through the pores of the tiny calcium carbonate shell. The shell is approximately 85 millimeters in diameter. (*M. Kage/Peter Arnold, Inc.*)
? *What functions do the pseudopodia (rays of cytoplasm) serve in this tiny animal?*

paleocurrent directions. Evidence of cyclic events may be obscured, and magnetic properties of the rock may be altered. On the other hand, bioturbation may provide evidence of the existence of organisms in strata where fossil remains such as shells are rare or absent.

Beyond the continental shelves, the benthic environment is subjected to low temperatures, little or no light, and high pressures. Without light, plants are unable to live at these depths. One encounters this **bathyal** environment from the edge of the shelf to a depth of about 4000 meters. Still deeper levels constitute the **abyssal** environment. The term **hadal** is reserved for the extreme depths found in oceanic trenches. As might be expected, animals are less abundant in the abyssal and hadal environments. Most hadal creatures are scavengers that depend on the slow fall of food from higher levels and predators that feed on the scavengers.

Physical Constraints Within the Ecosystem

THE CHEMISTRY OF SEA WATER The waters of the ocean are remarkably uniform in composition. An average sample of sea water is composed of about 35 parts per thousand of dissolved matter (Table 4-1). Nearly 86 percent of that dissolved matter consists of the elements sodium and chlorine. The sodium chloride, along with less common salts, gives the ocean its saltiness or **salinity**. Sea water with salinity less than about 30 to 40 parts per thousand is termed **brackish**. The waters of estuaries have lower-than-average salinity and are brackish because they receive fresh water from

FIGURE 4-33 Diorama depicting sublittoral zone organisms living on the floor of the sea covering the central United States during the Silurian Period. Members of this thriving paleocommunity include trilobites, horn and honeycomb corals, algae, branching bryozoan colonies, and small bivalved brachiopods. (*Courtesy of the Milwaukee Public Museum.*)
? *Draw an arrow to and label a representative of each of these groups.*

Pre-Є | Є | O | S | D | M | P | Pr | Tr | J | K | T | Q

FIGURE 4-34 Burrows made by infaunal organisms (probably worms) are preserved in the Northview Formation (Mississippian) of Missouri.

streams. The term **hypersaline** refers to water having higher than normal salinity. Hypersaline conditions may develop in enclosed areas like lagoons, bays, or isolated seas that are subjected to rapid evaporation and little addition of fresh water. Marine organisms are very sensitive to salinity. Relatively few species can tolerate salinity that is either abnormally high or abnormally low.

Some of the elements and compounds in sea water are in the form of dissolved gases, the most abundant of which are carbon dioxide and oxygen. Carbon dioxide is essential to the growth of marine plants. The amount of this gas often varies closely with the abundance of phytoplankton in any given region of the sea. By means of *photosynthesis*, plants replenish the supply of oxygen in sea water, providing this vital element for use by marine animals. By means of *respiration*, organisms then return carbon dioxide to plants (Fig. 4-35).

TABLE 4-1 Dissolved in Solids in Ocean Water

Chemical Constituent	Content (parts per thousand)
Calcium (Ca)	0.419
Magnesium (Mg)	1.304
Sodium (Na)	10.710
Potassium (K)	0.390
Bicarbonate ($HC0_3$)	0.146
Sulfate (SO_4)	2.690
Chloride (Cl)	19.350
Bromide (Br)	0.070
Total dissolved solids (salinity)	35.079

From U.S. Geological Survey publication, *Why is the Ocean Salty?*

The ocean has a great capacity for absorbing carbon dioxide from the air, and this helps to regulate the amount of this gas in the atmosphere. If the carbon dioxide content of the atmosphere rises, the rate at which it is dissolved in sea water also increases. Oxygen is also absorbed from the air above the oceans, but in lesser amounts than carbon dioxide. Because they are near the air-water interface, surface water tends to be richer in oxygen than deeper water. Also, phytoplankton flourish in the upper layers of water, where there is abundant light for photosynthesis. Beneath the near-surface layer, however, oxygen content decreases because of consumption of the element by animals and its use in reduction of animal organic waste.

Nitrogen is another vital element in ocean water. It is an indispensable component of proteins and nucleic acids. Nitrogen, however, is not abundant in a form usable by plants in photosynthesis. Most plants obtain their nitrogen from ammonia (NH_3) or nitrate (NO_3). These compounds are produced by certain bacteria (called nitrogen-fixing bacteria) as well as certain plants that are able to produce ammonia or nitrate.

No less important for life than nitrogen is the element phosphorus. Phosphorus is an essential component of DNA, RNA, and energy-rich molecules that participate in metabolism. Phosphorus derived from weathering of rocks forms compounds that are quite insoluble and are therefore deposited on the ocean floor. There is, however, a soluble form of phosphorus (called orthophosphate) that can be absorbed by algae and plants. These organisms then become sources of phosphorus for animals higher in the food chain.

Sulfur is another element in sea water that is of great importance to life. Elemental sulfur combines with oxygen to form sulfate, which is utilized by Bacteria and Archaea as a source of energy. Eurkaryotic organisms do not use sulfur as an energy source but incorporate it into proteins and certain other molecules.

MOVEMENTS OF OCEAN WATER The sea is in constant motion. Its movements provide for the transfer of nutrients and waste products essential for life. Among the many kinds of movement of ocean water are surface currents, tidal currents, currents induced by abnormal density, and currents produced by wave action along shorelines.

Surface currents are slow drifts of water set in motion by the Earth's prevailing winds. Energy is transferred from the moving air (wind) to the water. Because the transfer of energy occurs at the surface, the topmost level of water moves the fastest, and successively lower levels slower, until at about a depth of 100 meters, movement can hardly be detected.

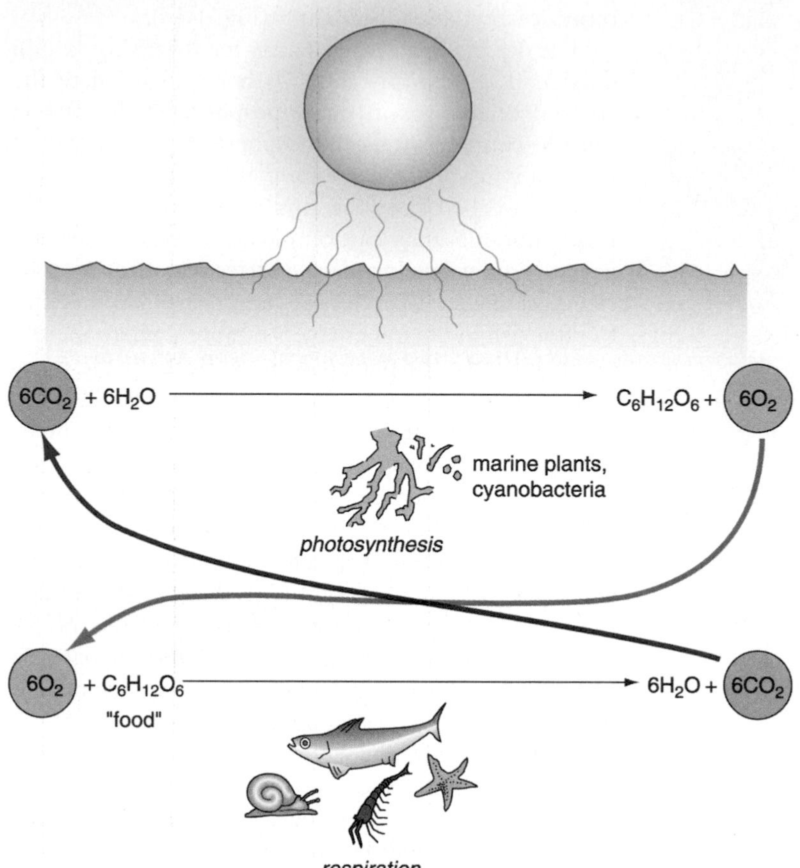

FIGURE 4-35 Interdependence of photosynthesis and respiration. ? *If there was sufficient sunlight to permit photosynthesis to proceed unchecked, what would become depleted?*

If surface currents derive their energy largely from wind, then one should be able to see some correspondence between the global current pattern and the pattern of atmospheric circulation. In general, such correlations do exist. In the low latitudes, winds blow from the east diagonally toward the equator, whereas to the north and south, winds blow from the west diagonally away from the equator. The result is large, rather elliptical, swirls of water circulation called **gyres**. Wind alone, however, is not the only factor that determines the global pattern of surface currents. Land masses cause currents to be deflected. Surface currents are also influenced by the rotation of the Earth. The Earth's rotation causes particles of matter in motion on the surface to be deflected toward the right in the Northern Hemisphere and toward the left in the Southern Hemisphere, regardless of the direction in which the particles are moving. This phenomenon has been named the **Coriolis effect**. Thus, gyres north of the equator rotate clockwise and those south of the equator rotate counterclockwise (Fig. 4-36).

Surface currents profoundly affect the climate and therefore the life of large regions of the Earth. As an example, the Gulf Stream and its extension as the North Atlantic Drift bring warm waters to a region well above the Arctic Circle. Without this current, Murmansk, located on the Arctic Circle, would not be an ice-free port, and the climate of the British Isles would be considerably harsher than at present. Those engaged in reconstructing the history of our planet attempt to plot the probable location of not only continents but also ancient surface currents. The distribution of temperature-sensitive fossil organisms can aid in such investigations, just as anomalous occurrences of fossils may be explained once the locations of ancient currents are determined.

Along the margins of continents where winds drive surface currents out to sea, deeper waters may rise to displace the surface waters. The process is called **upwelling**. Upwelling waters carry an abundance of nutrients, for they have been receiving organic detritus from above for long periods of time. Phytoplankton flourish and huge populations of fish occur in such areas.

Gravity also plays a role in mixing water masses in the ocean. Gravity pulls denser water downward, displacing lighter water upward. Temperature and salinity control the density of ocean water. For this reason, such movements are termed **thermohaline currents**.

Yet another factor involved in the motion of the ocean is tides. Tides are generated primarily by the grav-

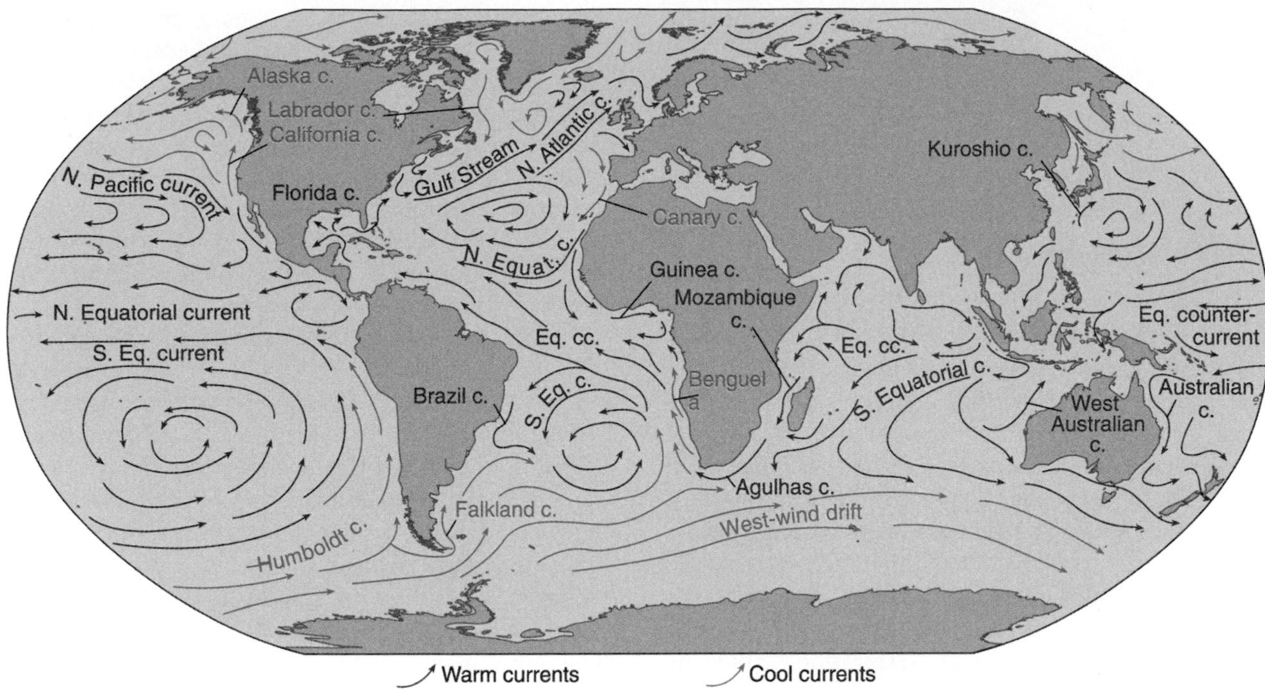

FIGURE 4-36 **Major ocean surface currents.** ? *In the Southern Hemisphere, why does the surface current turn to the left of the wind direction?*

itational attraction of the moon and, to a lesser degree, by the sun's gravitational pull (less because of the Sun's greater distance from the Earth). Plants and animals that live along most shorelines must have adaptations to survive, as their habitat is alternately flooded and drained twice in each span of approximately 24 hours.

Many other types of water movements occur in the ocean. Storm surges arise when strong winds drive water masses against the coast. Waves are a mechanism for mixing, as are nearshore currents of various kinds. Among these, nearshore drifts of water that parallel a shoreline (e.g., longshore currents) may bring nutrients to nearshore plants. Such currents, however, can also have a detrimental effect by stripping the coastal area of plankton or by making coastal waters so turbid that photosynthesis is inhibited.

WATER TEMPERATURE AND DEPTH The temperature of ocean water varies with both latitude and depth. It is also strongly affected by surface currents. Water is warmest near the equator, where it may reach a balmy 28°C. Near the poles, temperatures descend to -2°C. The ocean reacts only very slowly to changes in air temperature it therefore serves as a heat regulator for the atmosphere. Beginning at the surface of the ocean, where waters are generally the warmest, temperatures decrease slowly until reaching moderate depths, where there is a rapid decline in temperatures. The zone in which temperatures change rapidly with depth is called the **thermocline** (Fig. 4-37). Below the thermocline temperatures decrease very slowly, until at a depth of about 1500 meters the temperature remains constant at 1° to 3°C. Thus, there is a two-layer system, with warmer, less dense water lying above colder, more dense water. Animals that live below the thermocline have adaptations for coping with the darkness and cold. Like creatures of the hadal zone, they feed on detritus that slowly descends from higher levels.

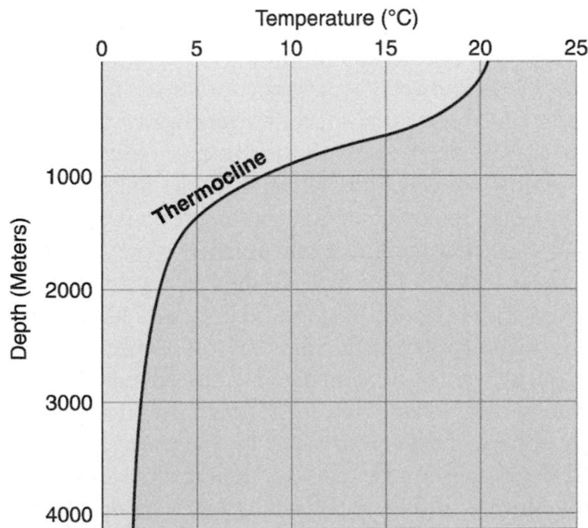

FIGURE 4-37 **The portion of the temperature-depth curve in the ocean that shows maximum change is the *thermocline*.**

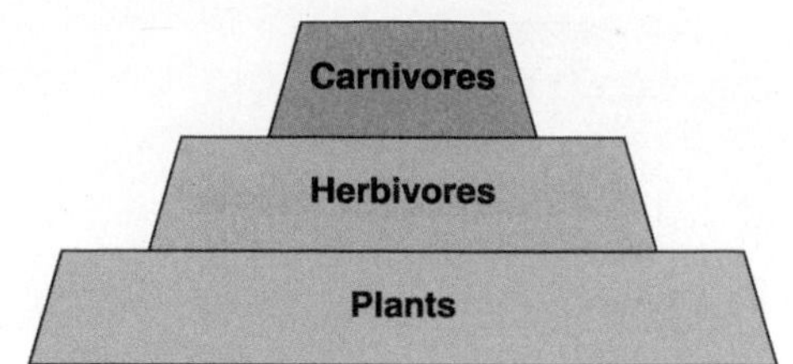

FIGURE 4-38 **Simple pyramid of ocean life.**

LIGHT Although some organisms can survive in the absence of light, the majority of life forms on Earth require sunlight to perform vital functions. Sunlight provides the energy required by organisms to perform vital functions. The near-surface layer of the ocean where there is sufficient light for photosynthesis is termed the **photic zone**. By means of photosynthesis, plants have the ability to convert sunlight into stored chemical energy. The energy is incorporated into the organic matter produced by the plant. It can be available when oxidized through respiration, either by plants themselves or by animals that feed on plants.

The amount of light that penetrates ocean water varies with the sun's angle, conditions in the atmosphere, conditions at the water surface, and the clarity of the water. As light travels through the water column, it becomes progressively dimmer until, at a depth of about 200 meters, darkness prevails. As a result, phytoplankton and marine plants that need sunlight to synthesize organic material live only in near-surface waters. The entire "pyramid of life" (Fig. 4-38) is dependent on these primary producers, for they support the ocean herbivores, which in turn support marine carnivores and omnivores.

Tropical waters have ample sunlight, and if nutrients are abundant, marine life flourishes. This is particularly true in both the shallow waters overlying the continental shelves and areas of upwelling. Tropical areas in the open ocean, however, may lack sufficient nutrients to support abundant life. Here, warmer waters above the thermocline do not mix with the nutrient-rich cold waters below, so recycling of nutrients does not occur.

In polar regions, surface waters are quite cold and there is no thermocline. Only organisms that can tolerate the extreme cold and diminished amount of sunlight inhabit these regions. Yet the polar waters have a positive effect on life of distant places. Cold, dense, nutrient-rich polar waters sink and become bottom currents that move slowly toward lower latitudes. In the northern hemisphere, the deep current originates near Greenland and flows southward. Chilled Antarctic bottom waters flow into the Indian and Pacific Ocean basins. As they move they begin to rise slowly in the open ocean and upwell rapidly near continental margins. As the bottom waters resurface, the nutrients they carry provide for bounteous biological activity at the surface.

ORGANIC DEEP SEA DEPOSITS Sediment covering the floor of the deep sea basins were first examined in the 1870's by scientists aboard the research vessel *H.M.S. Challenger*. Many of the samples dredged from the ocean floor were composed of microscopic shells and fragments of planktonic organisms. While still wet, the organic sediment had a slippery feel and were therefore named *oozes*. Depending on the composition of the shells, the oozes were further designated either **calcareous oozes** or **siliceous oozes**.

The main constituents of siliceous oozes are the tests (shells) of diatoms (see Fig. 14-2), which are algae, and tiny "protozoans" known as radiolaria (see Fig. 14-5). Siliceous oozes accumulate in colder, deeper oceanic zones where other sediments are lacking or in regions where an abundance of nutrients promotes high productivity of siliceous organisms. Calcareous oozes are composed mainly of the calcium carbonate coverings of coccolithophorids (Fig. 4-39) and foraminifera (Fig. 4-32), as well as the shells of tiny planktonic snails called pteropods (Fig. 4-40). In pteropods, the "foot" seen in ordinary snails is transformed into a pair of fins. If the tiny calcareous shells of coccolithophorids, foraminifera, and pteropods settle through a column of water no deeper than about 4000 meters, they accumulate to form calcareous ooze (Fig. 4-41). Colder waters below that depth hold more carbon dioxide in solution, which increases their acidity. Such water is corrosive to delicate calcium carbonate shells. At a depth of 4 to 5 kilometers, the supply of shells is approximately balanced by the amount being dissolved. As a result, calcareous oozes cannot accumulate. The water depth at which calcium carbonate is dissolved as fast as it falls from above is termed the **carbonate compensation depth**, or **CCD**. Not all calcareous material dissolves precisely at the CCD, for some shells can survive to greater depths because of their larger size or the presence of protective organic coatings.

Use of Fossils in Reconstructing Ancient Geography

The geographic distribution of present-day animals and plants is closely controlled by environment. Any given species has a definite range of conditions under which it can live and breed, and it is generally not found outside that range. Ancient organisms, of course, had similar restrictions on where they could survive. If we note the locations of fossil species of the same age on a map and correctly infer the environment in which they lived, we can produce a paleogeographic map for that particular time interval. One might begin by plotting on a simple base map the locations of fossils of marine organisms that lived at a particular time. This provides an idea of which areas were occupied by seas and

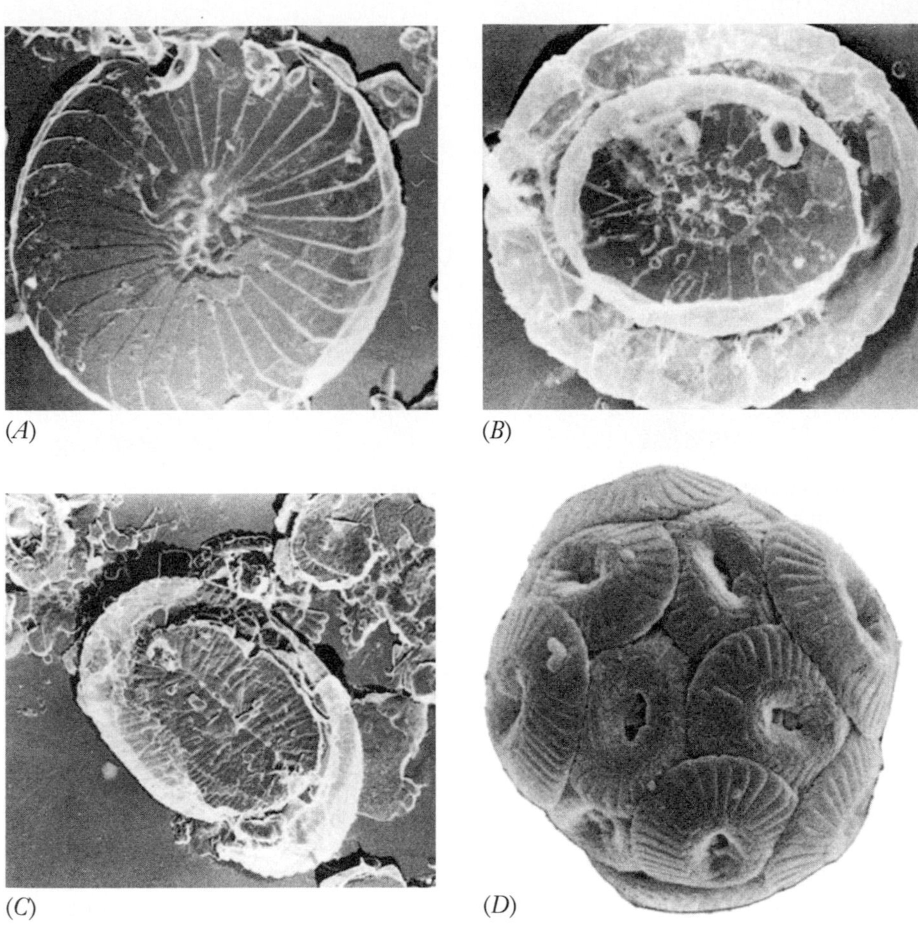

FIGURE 4-39 Coccoliths. (*A*) *Coccolithus leptoporous* (distal side). (*B*) *Coccolithus leptoporous* (proximal side). (*C*) *Helicosphaera carteri.* (*D*) Coccosphere of *Coccolithus pelagicus.* Coccoliths are about 18 to 22 microns in diameter.

might even suggest locations for ancient coastlines. Figure 4-42 shows major land and sea regions during the middle Carboniferous Period. The locations of marine protozoans called fusulinids are plotted as red circles. Notice that the Rocky Mountains were nonexistent, and their present location was occupied by a great north-south seaway.

Having obtained a fair idea of where seas and their shorelines existed, one might next give attention to evidence of the locations of land areas. The fossilized bones of land animals such as dinosaurs or mastodons would suggest a terrestrial environment, as would their preserved footprints. Fossil remains of land plants, including fossilized seeds and pollen, provide additional documentation of terrestrial paleoenvironments.

By an analysis of the fossils and the nature of the enclosing sediment, it is often possible to recognize deeper or shallower parts of the marine realm or to discern particular kinds of environments on land, such as ancient floodplains, prairies, deserts, and lakes. River deposits may yield the remains of freshwater clams and fossil leaves. A mingling of the fossils of land organisms and sea organisms might be the result of a stream entering the sea and perhaps building a delta in the process.

The migration and dispersal patterns of land animals, as indicated by the fossils, is one important indicator of former land connections as well as locations of mountainous or oceanic barriers that once existed between continents. Today, for example, the Bering Strait between North America and Asia prevents migration of land animals between the two continents. The approximately 80 kilometers between the two coastlines, however, is covered by less than 50 meters of water; this might lead one to suspect that Asia could once have been connected to North America. The fossil record shows that a land bridge did connect these two continents on several occasions during the Cenozoic Era. The earliest Cenozoic strata on both continents have a fossil fauna uncontaminated by foreign species. Somewhat younger rocks contain fossil

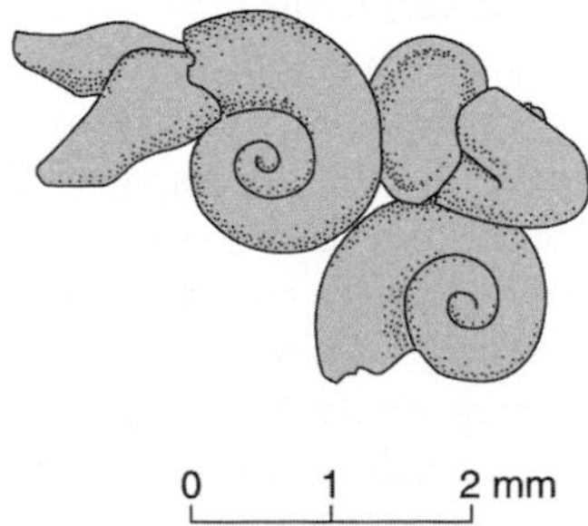

FIGURE 4-40 ***Limacina*, a tiny swimming marine snail or pteropod.** The foot is modified into a pair of winglike fins, shown at the left. At the right are two empty shells.

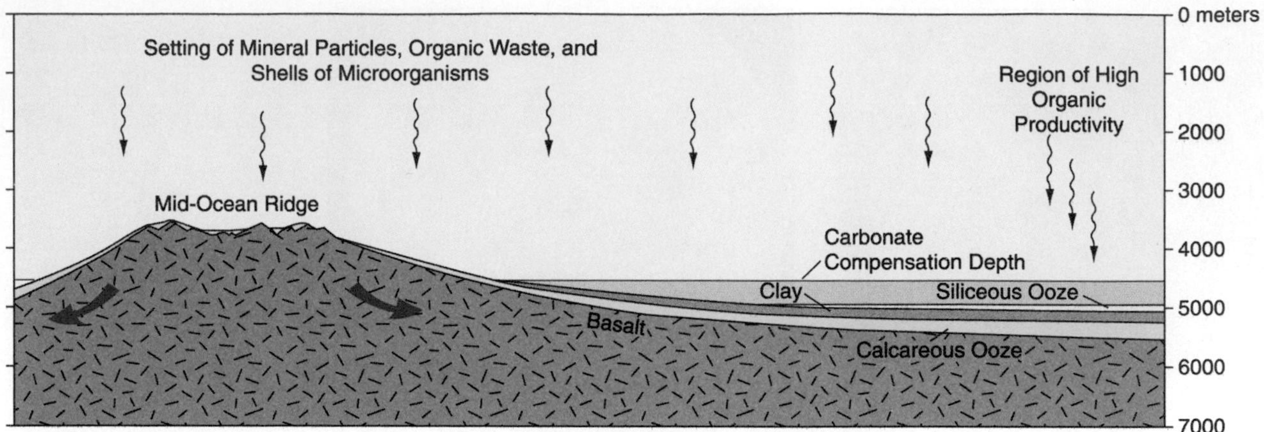

FIGURE 4-41 The ocean depth below which particles of calcium carbonate from microorganisms are dissolved as fast as they descend through the water column is termed the carbonate compensation depth, or CCD. The CCD varies at different locations in the ocean. In this conceptual drawing, note that calcium carbonate is able to accumulate along parts of the mid-oceanic ridge that are above the CCD. The accumulated layer is then carried away as lithospheric plates diverge from the ridge. When a given region of the ocean floor has reached depths below the CCD, calcium carbonate is no longer deposited on the ocean floor, whereas clay particles and siliceous remains of radiolaria and diatoms may accumulate.

remains of animals that heretofore had been found only on the opposite continent. During the Pleistocene Epoch, camels, horses, mammoths, and a rich variety of other land mammals migrated between North America and Eurasia across the Bering land bridge (Fig. 4-43). The bridge was still in place just 14,000 years ago when Stone Age hunters used the route to enter North America.

Another familiar example of how fossils aid in paleogeographic reconstructions is found in South America. Careful analyses of fossil remains indicate that, in the early Cenozoic Era, South America was isolated from North America; as a result, a uniquely South American fauna of mammals evolved over a period of 30 to 40 million years. The establishment of a land connection between the two Americas is recognized in strata only a few million years old (late Pliocene) by the appearance of a mixture of species formerly restricted to either North or South America.

In addition to deciphering the positions of shorelines or locations of land bridges, paleontologists can also provide data that help to locate the equator, parallels of latitude, and pole positions of long ago. It has been observed that in the higher latitudes of the globe, one is likely to find large numbers of individuals, but these are members of relatively few species. In contrast, equatorial regions tend to develop a large number of species, but with comparatively fewer individuals within each species. Stated differently, the variety or *species diversity* for most higher categories of plants and animals increases from the poles toward the equator. This is probably because relatively fewer species can adapt to the rigors of polar climates. Conversely, there is a stable input of solar energy at the equator, less duress caused by seasons, and a more stable food supply. Warmer areas place less stress on organisms and provide opportunity for continuous, uninterrupted evolution. Of course, in particular areas, generalizations about species diversity can be upset by local conditions.

Another way to locate former equatorial regions (and therefore also the polar regions that lie 90° of latitude to either side) is by plotting the locations of fossil coral reefs of a particular age on a world map. Nearly all living coral reefs (Fig. 4-44) lie within 30° of the equator. It is not an unreasonable assumption that the ancient reef corals had similar geographic preferences.

Use of Fossils in the Interpretation of Ancient Climatic Conditions

Among the many climatic factors that limit the distribution of organisms, climate (especially the temperature component of climate) is of great importance. There are many ways by which paleoecologists gain information about ancient climates. An analysis of fossil spore and pollen grains (Fig. 4-45) often provides evidence of past climatic conditions. Living organisms with known tolerances can be directly compared with fossil relatives. As noted earlier, corals thrive in regions where water temperatures rarely fall below 18°C, and it is likely that their ancient counterparts were similarly constrained. However, even when a close living analogue is not found, certain morphologic features can be useful in determining paleoclimatology. Plants with aerial roots, lack of annular rings, and large wood cell structure indicate tropical or subtropical climates.

Appalachian orogen

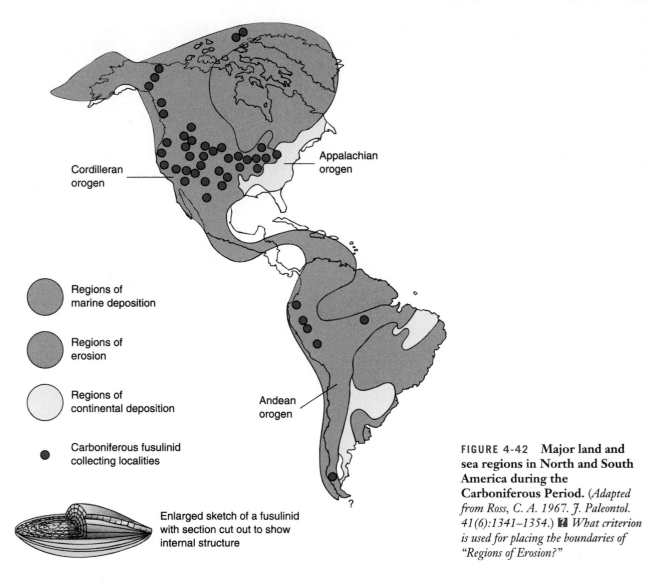

Cordilleran orogen

Regions of marine deposition

Regions of erosion

Regions of continental deposition

Andean orogen

Carboniferous fusulinid collecting localities

?

Enlarged sketch of a fusulinid with section cut out to show internal structure

FIGURE 4-42 Major land and sea regions in North and South America during the Carboniferous Period. (*Adapted from Ross, C. A. 1967. J. Paleontol. 41(6):1341–1354.*) ? *What criterion is used for placing the boundaries of "Regions of Erosion?"*

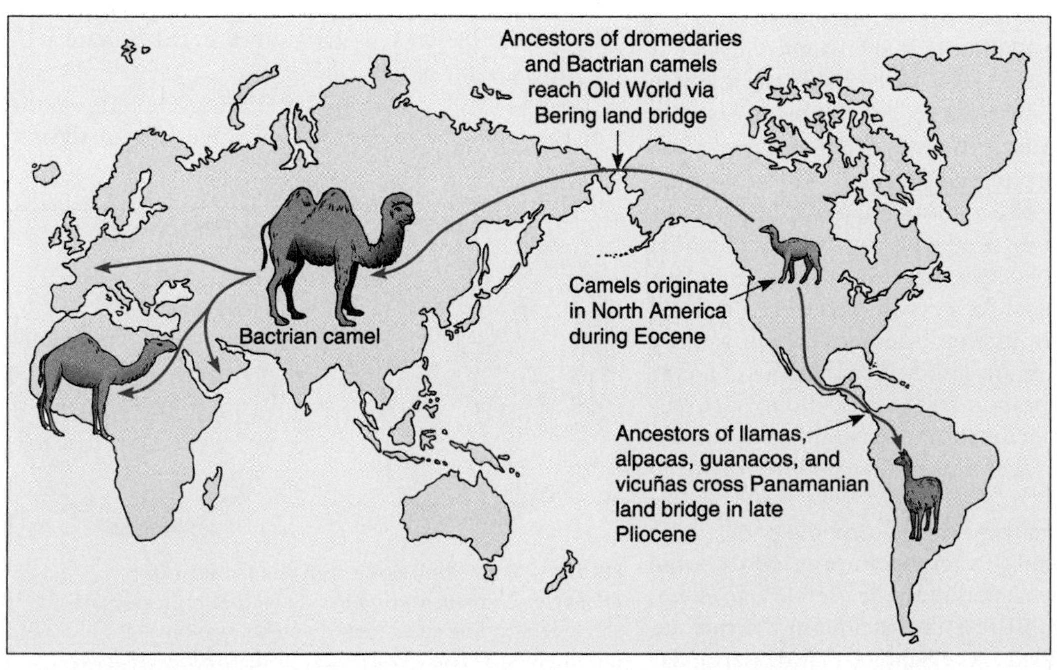

FIGURE 4-43 Intercontinental migrations of members of the camel family.

FIGURE 4-44 View of a modern coral reef. Coral reefs harbor an extraordinary diversity of organisms and, in some ways, rival tropical rain forests in their complexity. (*Copyright Jeffrey Rotman Photography.*)

(*A*)

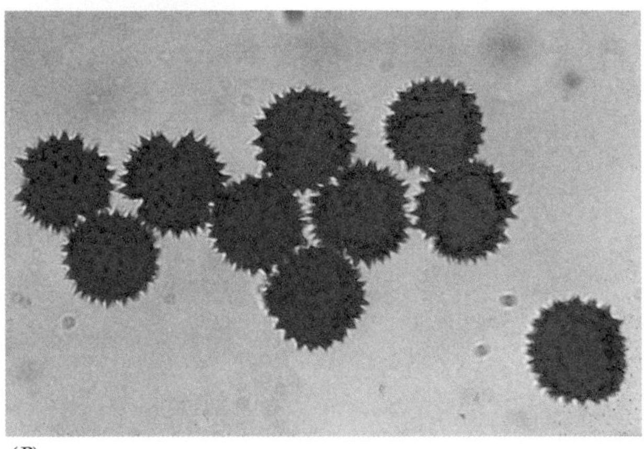

(*B*)

FIGURE 4-45 Pollen grains. (*A*) Pollen grain of a fir tree. Note the inflated bladders that make the pollen grain more buoyant. The specimen is about 125 microns in its longest dimension. (*B*) Pollen grains of ragweed. These grains are about 25 microns in diameter.

Marine mollusks (such as clams, oysters, and snails) with well-developed spinosity and thicker shells tend to occur in warmer regions of the oceans. In the case of particular species of foraminifera, variation in the average size or in the direction of coiling can provide clues to cooler or warmer conditions.

An example of environmentally induced changes in shell coiling directions is provided by the planktonic foraminifer *Globorotalia truncatulinoides* (Fig. 4-46). In the Pacific Ocean, left-coiling tests (shells) of these foraminifera have dominated in periods of relatively cold climate, whereas right-coiling tests predominated during warmer episodes. Another foraminifer, *Neogloboquadrina pachyderma*, exhibits similar changes in coiling directions, not only in the Pacific Ocean but in the Atlantic as well. Such reversals in coiling may occur quickly and over broad geographic areas. For this reason, they are exceptionally useful in correlation.

Aside from morphologic changes within species, entire assemblages of foraminifera are widely used in paleoecologic studies. Today, as in the past, benthic (bottom-dwelling) species have inhabited all major marine environments. As a result of the accumulated data on living foraminifera, inferences about salinity and water depths can be assigned to their fossil counterparts. Ancient foraminifera, for example, have frequently provided the means for recognition of ancient estuaries, coastal lagoons, and nearshore or deep oceanic deposits.

Sometimes, when the overall morphology of a fossil does not provide clues to temperature or climate, the compositions of the skeleton can be used. Magnesium, for example, can substitute for calcium in the calcium carbonate of shelled invertebrates. For particular groups of living marine invertebrates, it has been found that those living in warmer waters have higher magnesium values than do those residing in colder areas. This knowledge has been used to interpret the climate at the times fossil forms were living.

A widely used method for finding the temperature of the water of ancient seas involves the analysis of

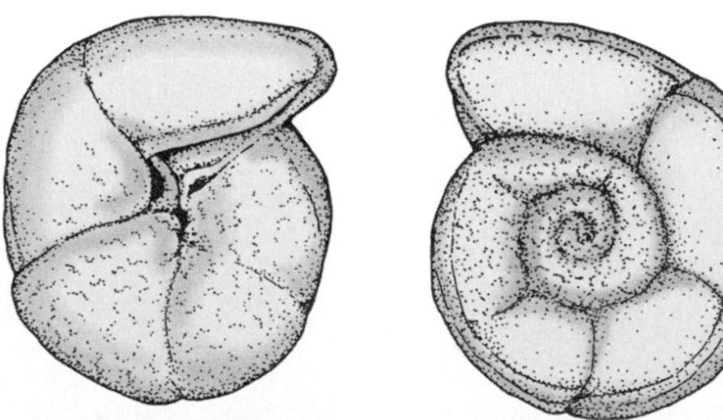

FIGURE 4-46 Shell coiling in the foraminifer *Globorotalia truncatulinoides*. Sketch of both sides of a single left-coiling specimen (diameter is about 0.9 millimeters). (*From Parker, F. L. 1962. Micropaleontology 8(2):219–254.*)

(A)

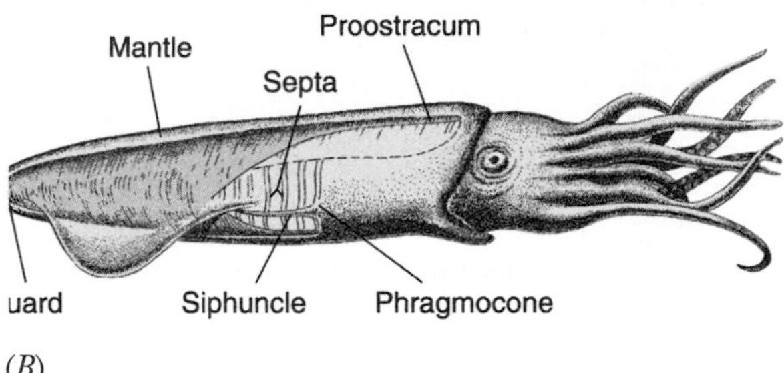

(B)

FIGURE 4-47 Belemnite. The usual belemnite fossil consists only of the solid part of the internal skeleton, called a guard. The guard here is 9.6 centimeters long. Belemnites are extinct relatives of squid and cuttlefish. Shown here is an interpretation of the appearance of a belemnite as it would appear if living.

oxygen isotopes in shell matter. When water evaporates from the ocean, there is a fractionation of the oxygen-16 and oxygen-18 in the water. (Oxygen-16, the usual form of oxygen, has an atomic mass of 16, whereas oxygen-18 has an atomic mass of 18.) Oxygen-16, being lighter, is preferentially removed, while the heavier oxygen-18 tends to remain behind. When the evaporated moisture is returned to Earth's surface as rain or snow, water containing the heavier isotope precipitates first, often near the coastlines, and flows quickly back into the ocean. Inland, the precipitation from the remaining water vapor is depleted in oxygen-18 relative to its initial quantity. If the interior of a continent is cold and contains growing glaciers, the glacial ice will lock up the lighter isotope, preventing its return to the ocean and thereby increasing the proportion of the heavier isotope in sea water. As this occurs, the calcium carbonate shells of marine invertebrates such as foraminifera will also be enriched in oxygen-18 and thereby reflect episodes of continental glaciation.

Even in the absence of ice ages, the oxygen isotope method may be useful as a paleotemperature indicator. As invertebrates extract oxygen from sea water to build their shells, a temperature-dependent fractionation of oxygen-18 occurs between the water and the secreted calcium carbonate in the shell matter. Provided there has been no alteration of the shell, it is possible to find the temperature of the water at the time of shell formation. In a famous early study, the oxygen isotope ratio in the calcium carbonate skeleton of a Jurassic belemnite indicated that the average temperature in which the animal lived was 17.6°C, plus or minus 6°C of seasonal variation. Subsequent studies on both Jurassic and Cretaceous belemnites (Fig. 4-47) provided confirmation of the inferred positions of the poles during the Mesozoic and indicated that tropical and semitropical conditions were far more widespread during late Mesozoic time than they have been in subsequent geologic time.

AN OVERVIEW OF THE HISTORY OF LIFE

Setting the Stage

As yet, no fossil evidence has been found of life on Earth during the planet's first billion years. The oldest direct indications of ancient life are remains of cyanobacteria discovered in rocks over 3.5 billion years old. These early fossils are usually considered evidence that the long evolutionary march had begun. However, they also stand at the end of another long and remarkable period during which living things presumably evolved from nonliving chemical compounds. We have no direct geologic evidence to tell us how and when the transition from nonliving to living occurred. What we do have are reasonable hypotheses supported by careful experimentation. Some of these hypotheses will be examined in Chapters 6 and 7, which deal with the Precambrian eons. It will suffice for this brief overview to note that life came into existence early in that great 4-billion-year span of time prior to the beginning of the Cambrian Period.

For most of Precambrian time, living things left only occasional traces. Here and there paleontologists have been rewarded with finds of bacteria, including filamentous cyanobacteria that formed extensive algal mats (Fig. 4-48). These microbial organisms produced important quantities of oxygen by photosynthesis. However, all life was at or below the unicellular level until about 1 billion years ago, when the world's first multicellular organisms left their trails and burrows in rocks that would later be called the Torrowangee Group of Australia. In rocks deposited about 0.7 billion years ago, fossil metazoans recognizable as worms, coelenterates (Fig. 4-49), and arthropods have been found—albeit rarely—in scattered spots around the world, including the United States, Australia, Canada, England, Russia, China, and South Africa. Thus, near the end of the Proterozoic, the stage was set for the evolution of a wide range of Paleozoic plants and animals.

FIGURE 4-48 The circular pods depicted in this mural of the margins of a Precambrian sea are calcareous algal structures called stromatolites. *(Courtesy of the National Museum of Natural History, Smithsonian Institution.)*

The Evolutionary History of Plants

Long before the first animal appeared on Earth, there were plants. Plants had their origin among unicellular aquatic protistids of the Precambrian. Among these, the green algae or chlorophytes are the most likely ancestors of vascular land plants. The first invasion of the land was made by plants that reproduced by means of spores (Fig. 4-50). From primitive forms that originated late in the Ordovician, vascular land plants expanded widely during the Devonian and Carboniferous, forming the bulk of the vegetation in the great coal forests. The most imposing of these spore bearers were the scale trees, whose close-set leaves left a pattern of scalelike scars on trunks and branches. The next group of plants to make their debut were pollen and seed producers. These so-called gymnosperms appeared in the Devonian and became widely dispersed during the Mesozoic. Plants that bore seeds and flowers, the angiosperms, had evolved by Cretaceous time. Plants with both seeds and flowers are called angiosperms. They are by far the most abundant plants today and include familiar trees such as oak, maple, sassafras, and birch as well as the grasses that became the primary food source for evolving Cenozoic mammals.

FIGURE 4-49 *Mawsonites*, a fossil found in the Pound Quartzite of Australia. The formation is late Proterozoic in age. Mawsonites is considered to be the mouth end of a jellyfish. *(Courtesy of B. N. Runnegar.)*

The Evolutionary History of Animals

Some general observations are possible after a brief scanning of the history of life following the Precambrian. One notes that the principal groups of invertebrates appear either very late in the Proterozoic or early in the Paleozoic (Fig. 4-51). Less advanced members of each phylum characterized the earlier geologic periods, whereas more advanced members came along later. Also, most of the principal phyla are still represented by animals today. In our review of fossil creatures, we are not startled by sudden appearances of bizarre or exotic animals and plants. Further, we are able to recognize periods of environmental adversity that caused extinctions. Such episodes were usually followed by much longer intervals of recovery and more or less orderly evolution. Several generalizations can be made about the history of life following the Precambrian. During the Cambrian, trilobites, brachiopods that lacked hinged shells, small cap-shaped mollusks, certain soft-bodied worms, chitin-shelled arthropods, and reef organisms called archaeocyathids predominated. Most of the Cambrian animals were deposit and suspension feeders. In subsequent periods of the Paleozoic, trilobites were joined by hinged brachiopods, nautiloids, crinoids, rugose corals, honeycomb corals, and twig-like bryozoans (Fig. 4-52). The Paleozoic was also the era when fishes, amphibians, and reptiles appeared and left behind a fascinating record of the conquest of the lands. Remains of more modern corals, diverse bivalves, sea urchins, and ammonoids characterize the marine strata of the Mesozoic Era. However, the Mesozoic is best known

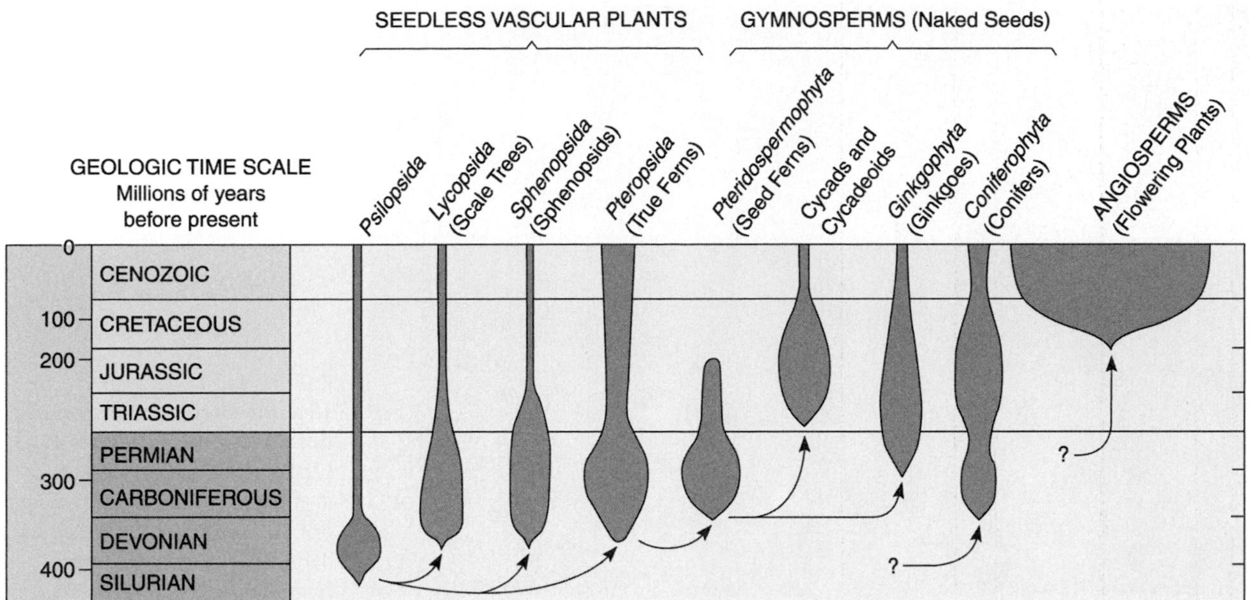

FIGURE 4-50 Geologic ranges, relative abundances, and evolutionary relationships of vascular land plants.

as the era when dinosaurs and their kin dominated the continents (Fig. 4-53). No less important than these big beasts, however, were the rat-sized primitive mammals and earliest birds that skittered about, perhaps unnoticed by the "thunder beasts."

The mollusks were particularly well represented in the marine invertebrate faunas of the Mesozoic, and they have continued in importance into the Cenozoic. However, no ammonoid cephalopods have occurred in this most recent era. Rather, rocks of the Cenozoic are recognized by distinctive families of protozoans (foraminifera) and a host of modern-looking snails, clams, sea urchins, barnacles, and encrusting bryozoa. Because the Cenozoic saw the expansion of

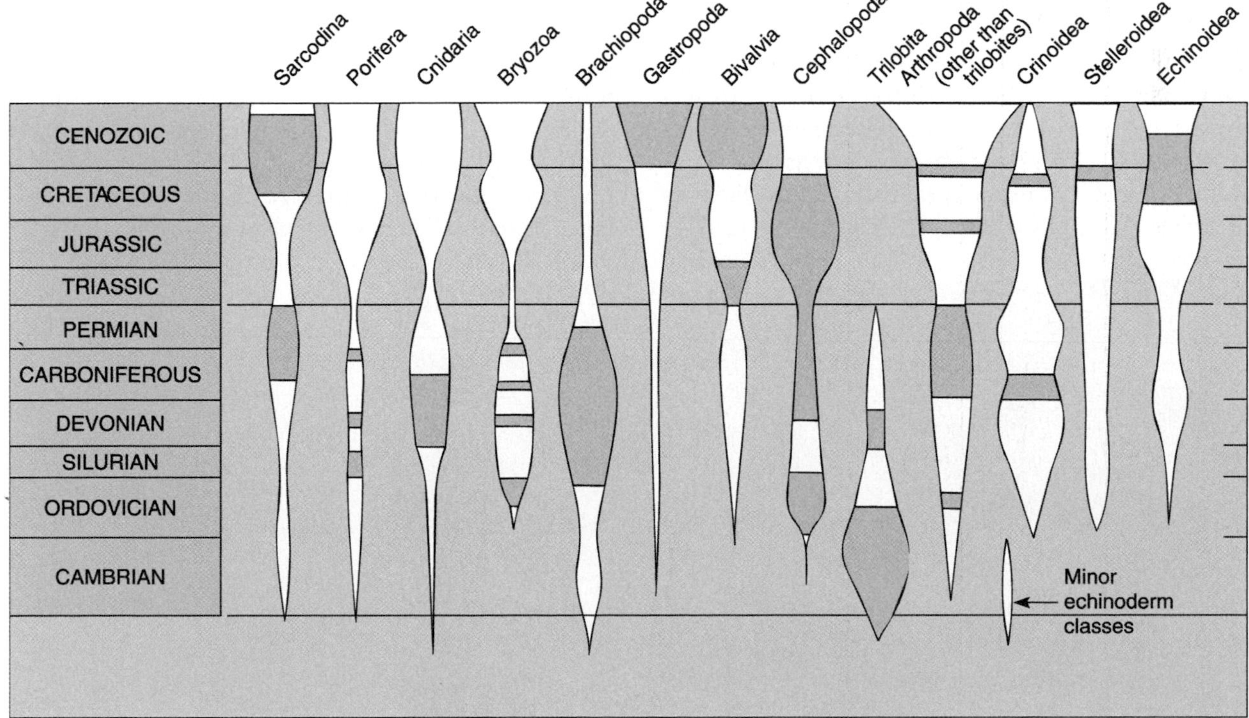

FIGURE 4-51 Geologic ranges and relative abundances of frequently fossilized categories of invertebrate animals. Width of range bands indicates relative abundance. Colored areas indicate where fossils of a particular category are widely used in zoning and correlation.

FIGURE 4-52 A large, straight-shelled cephalopod dominates this scene of an Ordovician sea floor. Also visible are corals, bryozoans, and crinoids. (*Courtesy of the National Museum of Natural History, Smithsonian Institution.*)

warm-blooded creatures such as ourselves, it is appropriately termed the "age of mammals."

Since that time over a half billion years ago, when most of the principal phyla of animals had become established, the history of life has not been a steady continuum. It has been marked by times of relatively sudden extinctions, affecting the entire Earth, and so devastating that they are termed **mass extinctions**. Five episodes of mass extinctions have been particularly catastrophic. These occurred at the end of the Ordovician, late in the Devonian, at the end of the Permian, in the late Triassic, and one that involved the demise of the dinosaurs and many less well known groups at the end of the Cretaceous Period. We will examine the cause and effects of these dramatic events in Chapters 10 and 12 and describe how each was followed by rapid evolutionary radiations into habitats and niches vacated by extinct predecessors.

SPECULATIONS ABOUT LIFE ON OTHER PLANETS

While reflecting upon the long and complex history of life on Earth, one might drift on to questions of whether or not other planets had—or now have—life as we know it. Is organic evolution a process that is unique to this planet? What properties of the Earth have made it a suitable place for the origin and evolution of life? Could these conditions exist elsewhere in the universe?

To begin with, the Earth is massive enough that its gravitational attraction is sufficient to retain an atmosphere. The temperature of most of the Earth's surface is low enough to provide an abundance of water in its liquid form. In addition, that temperature is suitable for chemical reactions required for life processes and is itself a consequence of the size of our sun and its distance from the Earth. Our sun-star also has a life span

FIGURE 4-53 Cretaceous carnivorous dinosaurs in combat. (*Courtesy of the American Museum of Natural History.*)

sufficiently long to permit time for the emergence and evolution of life. Finally, the Earth has always had all of the chemical elements required for life processes.

Life in Our Solar System

We may first consider the possibilities for extraterrestrial life here in our own solar system. All evidence to date indicates that because of either size or distance from the sun, conditions are currently too harsh on neighboring planets to permit the evolution of *higher* forms of life. However, scientists believe that Mars, Jupiter's moons Europa and Io, and Saturn's moon Titan all have properties that make them intriguing places to investigate for microbial life. Dry river channels, polar icecaps, and indications of ice stored beneath the surface show that life-sustaining water was abundant on Mars about 3.5 billion years ago. Europa has water slush or possibly even an ocean beneath its icy surface. Titan has a thick atmosphere of nitrogen molecules, and Io has conditions for life of the kind found here on Earth near volcanic vents. The Viking mission to Mars examined the Martian surface for indications of life but found nothing to suggest the existence of present or past organisms on the red planet. Recently, however, interest in the possibility of life on Mars was reawakened by reports of objects resembling microbes, or the chemical traces of microbes, within a Martian meteorite that had landed here on Earth.

How a piece of Mars got to our planet is a fascinating story in itself. It is a story that began about 16 million years ago when a large meteorite struck Mars with sufficient force to blast fragments of its rocky surface into space. For millions of years these chunks of rock traveled through the solar system. Some are still out there, continuously altering their trajectories in response to the pull of planets. The journey of the one in question, however, ended when it landed on an Antarctic ice field. It was found there in 1984 by scientists on a meteorite collection expedition. They named it ALH 84001.

But was this chunk of rock truly from Mars? If it was, it should contain unique chemical clues indicating its Martian derivation. In 1996, instruments aboard the Viking Lander had analyzed the atmosphere of Mars. ALH 84001 contained traces of gases identical to those identified in the Viking analysis. (Meteorites from the Moon or Venus lack these gases.)

Having confirmed the Martian origin of ALH 84001, the search for evidence of life within it began. One interesting discovery was tiny carbonate globules inside the meteorite. Provided the carbonates were not formed by rapid crystallization during the impact, then they may have been precipitated within the rock when it was part of the Martian surface about 3.6 billion years ago. At that time the planet was wetter and warmer than it is today.

The next task was to scrutinize the meteorite for fossils. Using a high-resolution scanning electron microscope, investigators found ultramicroscopic, sausage-shaped bodies similar in shape, but not size, to certain bacteria known here on Earth. The tiny structures lay within or near the carbonate globules and could not have gotten into them while lying on the Antarctic ice field. But were they actually fossils? Purely inorganic processes have been known to produce similar structures. Also, fossil bacteria found in Earth rocks are enormously larger than the sausage-shaped structures in ALH 84001. In addition, there was no evidence of cell walls, cell division, or cell growth in the structures.

From the above, it appears that confirmation of the former existence of life on Mars will require finding fossils of true cells: cells that are of the correct size and that show evidence of life processes. Actual samples of Martian rocks, if and when obtained, may provide the evidence.

Life in the Universe

The Earth is indeed biologically special in comparison with other planetary bodies in our solar system. However, it may not be so peculiar at all in the vast realms of the universe. Astronomers estimate that there are about 150 billion stars in our galaxy, and the number of galaxies now appears to be nearly limitless. Of those 150 billion stars, a not unreasonable estimate stipulates that at least 1 billion have planets with size and temperature conditions similar to the Earth's. These are potentially habitable planets. For the entire known universe (of which our galaxy is but a small part), reputable scientists have estimated that there are as many as 1020 planetary systems similar to our solar system. Such calculations indicate that it is probable that life does exist out there somewhere. Indeed, the universe may be rich in suitable habitats for life. In 1996, two such planets were found rotating around sunlike stars outside of our solar system. The first was discovered in the constellation Virgo, and the second in the constellation known as Ursa Major. The planet in the constellation Virgo lies at just the right distance from its sun-star for the liquid water essential for life to exist at its surface. The second planet may also contain liquid water, but only in its atmosphere.

What might be the nature of life if discovered on other planets? If we assume such life was formed from the same universal store of atoms and under physical conditions not too dissimilar to those that have existed on Earth, then we might very well recognize it as a living thing. But it is highly unlikely that duplicates of humans, cows, or butterflies exist on other planets. There are many variables in the evolutionary interactions of genetics, environment, and time involved in the making of a particular species. That the very same mutations, genetic recombinations, and environmental conditions producing a sparrow could occur in precisely the same sequential steps in time on a distant planet seems most improbable.

SUMMARY

Fossils are the remains or traces of life of the geologic past. Some processes that produce fossils include the precipitation of chemical substances into pore spaces (permineralization), molecular exchange for substances that were once part of the organism with inorganic substances (replacement), or compression of animals and plants to form a film of carbonized remains (carbonization). Molds and casts are additional types of fossils. Some fossils are indirect evidence of life. These include the tracks, trails, and burrows that comprise trace fossils. The fossil record of life is not complete. It is biased in favor of animals that possessed resistant parts or lived in a depositional environment where rapid burial occurred.

Fossils demonstrate that life has changed or evolved through time. Because of this, fossiliferous rock layers from different periods of geologic time can be recognized and correlated.

Our modern concept of evolution combines natural selection with hereditary mechanisms for change by recombining hereditary materials and mutations. The hereditary materials are chromosomes and genes located in the nuclei of cells. Genes are composed of DNA and are responsible for the total characteristics of an organism.

The evolutionary history of organisms can be depicted on a phylogenetic tree. In stratophenetic phylogeny the diagram is truly treelike, with descendants on upper branches and ancestors on the lower branches and trunk. Time is implied in the vertical dimension of the tree. Cladistic phylogeny produces a less treelike diagram called a cladogram, in which related organisms are placed in close proximity on the basis of characteristics they share that indicate a common ancestry.

The theory and practice of classifying and naming organisms are called taxonomy. Paleontologists, like biologists, use a binomial system of nomenclature in which each type of organism is assigned a two-part name designating genus and species. The taxonomic system used in classifying organisms is hierarchical, with levels including species, genus, family, order, class, phylum, kingdom, and domain. The three domains are the Bacteria, Archaea, and Eucarya. In the traditional five-kingdom system of classification, the kingdoms are the Archaeobacteria, Eubacteria, Protista, Fungi, Plantae, and Animalia.

Adaptation is a process by which populations change so as to become better fitted to their environments and strategy for survival. The changes may be morphologic, physiologic, or behavioral. When new habitats or other opportunities become available to a species, the new conditions may be rapidly exploited by descendant species; this results in diversification that we call adaptive radiation.

Darwin viewed evolution as a gradual change in lineages through time. This interpretation is called phyletic gradualism. Evolution may also entail sudden advances that punctuate periods of relative stability (stasis). The process is termed punctuated equilibrium and is thought to occur when new species arise suddenly from within isolated populations.

The area of paleontology that seeks to understand the relationships between ancient organisms and their environments is called paleoecology. On the basis of comparisons with present-day ecology, paleoecology provides information about the distribution of ancient lands and seas, ancient climates, depth of seas, barriers to migration, and the former location of continents. The paleoecology of marine areas is of particular importance to geologists because of the richness of the marine sedimentary and fossil record. To understand the ancient oceanic environment, knowledge of the chemical and physical properties of the ocean today is required. The ocean's chemical components, movements, temperature, and salinity control the nature and distribution of marine life.

Because fossils are useful in determining environments of deposition as well as in correlation, they are widely used in the search for fossil fuels. Microscopic fossils such as foraminifera, coccolithophorids, ostracodes, and conodonts are particularly useful in petroleum exploration, for they can often be recovered from well cores and drill cuttings.

Paleontologists are currently considering methods for detecting life on planetary bodies other than the Earth. Within our solar system, Mars and planetary satellites such as Europa and Titus appear to have physical and chemical conditions conducive to the origin of life. There is also a strong statistical probability of past or present life somewhere in the universe.

QUESTIONS FOR REVIEW AND DISCUSSION

1. What factors determine whether or not a particular fossil will be valuable as an indicator of the age and correlation of a stratum?

2. Fossil A occurs in rocks of Cambrian and Ordovician age. Fossil B occurs in rocks that range in age from early Ordovician through Permian. Fossil C is found in Mississippian through Permian strata.
   a. What is the maximum possible range of age for a stratum containing only fossil B?
   b. What is the maximum possible range for a stratum containing both A and B?
   c. Which is the better guide fossil, A or C?

3. A chronostratigraphic (time-rock) unit contains a different fossil assemblage at one locality than at another located 200 kilometers away. Suggest a possible cause of the dissimilarity.

4. In drilling for oil, geologists recover Devonian conodonts in a stratum known to be Permian in age. Explain how this may have occurred.

5. What kinds of vascular land plants existed during the Paleozoic Era?

6. What are the differences between the following?
   a. Mitosis and meiosis
   b. Haploid and diploid
   c. Gymnosperm and angiosperm
   d. Stratophenetic and cladistic phylogeny
   e. Domain and kingdom

7. What is meant by the term *adaptation*? Cite an example of adaptive radiation.

8. What evidence indicates that the meteorite ALH 84001 came from Mars?

9. What are peripheral isolates? Why do new species commonly arise from peripheral isolates?

10. How has the science of paleontology contributed to the following?

- **a.** Validation of the theory of organic evolution
- **b.** Recognition of geographic changes in the geologic past
- **c.** Recognition of climatic changes in the geologic past
- **d.** Recognition of the environment of deposition.

11. What are the contributions of Darwin and Mendel to our modern concept of organic evolution?

12. Using fossils for age correlation is dependent on a priori knowledge of their age ranges. How has this knowledge been obtained?

13. Distinguish between the concepts of phyletic gradualism and punctuated equilibrium. How did Charles Darwin account for the rapid or abrupt appearance of new species? Which, then, would be a more appropriate stratigraphic section to study for proof of punctuated equilibrium-a continuous set of cores from the floor of the ocean, or a section on the continent where there has been repeated episodes of uplift and erosion throughout geologic time?

14. What is the name of the process through which plants convert solar energy into stored chemical energy? What is the name of the process through which this stored energy is released by oxidation?

15. How does biological activity affect the amounts of oxygen and carbon dioxide in ocean water?

READINGS

Boardman, R. S., Cheetham, A. H., and Rowell, A. J. (eds.). 1987. *Fossil Invertebrates*. Oxford, England: Blackwell Scientific Publications.

Briggs, D. E. G., and Crowther, P. R. (eds.). 1990. *Paleobiology*. Oxford, England: Blackwell Scientific Publications.

Carroll, R. L. 1988. *Vertebrate Paleontology and Evolution*. Oxford, England: Blackwell Scientific Publications.

Clarkson, E. N. K. 1993. *Invertebrate Paleontology and Evolution*, 3rd ed. London: Geo. Allen & Unwin, Ltd.

Cowen, R. 1993. *History of Life*, 2nd ed. Palo Alto, CA: Blackwell Scientific Publications.

Darwin, C. 1859. *On the Origin of Species* (1963 ed.). Introduction by H. L. Carson. New York: Washington Square Press.

Doyle, P. 1996. *Understanding Fossils*. New York: John Wiley & Sons.

Eldredge, N., and Gould, S. J. 1972. Punctuated equilibria: An alternative to phyletic gradualism. In *Models in Paleobiology*. Schopf, T. J. M. (ed.) San Francisco: W. H. Freeman & Co.

Kemp, T. S. 1999. *Fossils and Evolution*. Oxford: Oxford University Press.

MacFadden, G. J. 1992. *Fossil Horses: Systematics, Paleobiology, and Evolution of the Family Equidae*. Cambridge, England: Cambridge University Press.

Schaal, S., and Ziegler, W. 1992. *Messel, An Insight into the History of Life and of the Earth*. Oxford, England: Clarendon Press.

Stewart, W. N. 1983. *Paleobotany and the Evolution of Plants*. Cambridge, England: Cambridge University Press.

Woese, C. R., Kandler, O., and Wheelis, M. L. 1990. Towards a natural system of organisms: Proposal for the domains Archaea, Bacteria, and Eukarya. *Proceedings of the National Academy of Sciences, U.S.A.* 87:4576–4579.

WEB SITES

The Earth Through Time Student Companion Web Site (www.wiley.com/college/levin) has online resources to help you expand your understanding of the topics in this chapter. Visit the Web Site to access the following:

1. Illustrated course notes covering key concepts in each chapter;
2. Online quizzes that provide immediate feedback;
3. Links to chapter-specific topics on the web;
4. Science news updates relating to recent developments in Historical Geology;
5. Web inquiry activities for further exploration;
6. A glossary of terms;
7. A Student Union with links to topics such as study skills, writing and grammar, and citing electronic information.

5

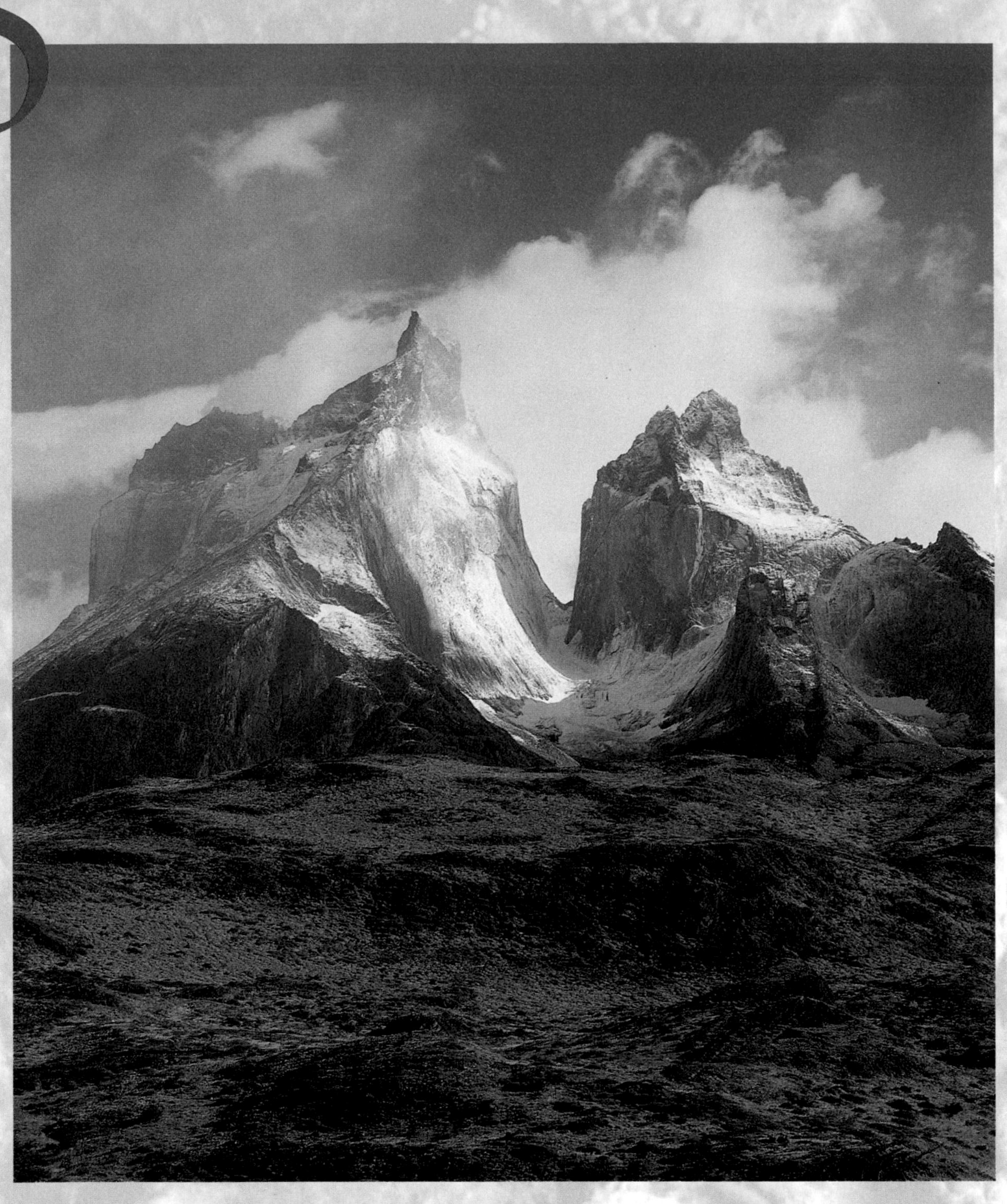

The Andes Mountains as seen from Pehoe Lake, Torres Del Paine National Park, Chile. The Andes formed at a convergent tectonic plate boundary when the westward moving South American plate converged on the Pacific Plate (Courtesy of T. Till.)

CHAPTER 5

Earth Structure and Plate Tectonics

The deep interior of the Earth is inaccessible, and no rays of light penetrate to let us see what is below the surface. But rays of another kind penetrate and carry with them their messages from the interior.
Inge Lehmann, 1959

OUTLINE

Before turning to the study of the geologic history of the Earth, it is necessary to understand the basic internal structure of our planet. In this age of artificial satellites and spacecraft, our attention is often directed toward the remarkable discoveries resulting from the exploration of outer space. We tend to forget that there is still much to learn about the "inner space" that lies beneath us. Our deepest wells penetrate only about 10 kilometers of the 6300 kilometers that separate us from the center of the planet. Only seismic waves generated by earthquakes permit us to "see" into the Earth's interior and to discern its boundaries and physical properties. Those properties bear directly on the forces that have deformed the Earth's crust, that trigger earthquakes, cause volcanic eruptions, and drive gargantuan plates of the planet's outermost layer from place to place around the globe. The formation of these plates, how they are formed, how they are destroyed, and how they interact are components of the process called plate tectonics.

SEISMIC WAVES

The "messages from the interior" mentioned in the epigram at the beginning of this chapter are, of course, seismic waves. Although devices that measure the Earth's gravity and magnetism are important sources of information about the Earth's interior, of even greater importance are instruments that record seismic waves. Seismic waves permit scientists to determine the location, thickness, and properties of the Earth's internal zones. They are generated when rock masses are suddenly disturbed, as when they break or rupture. Vibrations spread out in all directions from the source of the disturbance, traveling at different speeds through parts of the lithosphere, mantle, and core that differ in chemical composition and physical properties. The principal categories of these waves are **primary**, **secondary**, and **surface**. All three types of waves are recorded on an instrument called a **seismograph**. The record produced by the seismograph is termed a **seismogram**.

Primary waves, or **P-waves**, are the speediest of the three kinds of waves and therefore the first to arrive

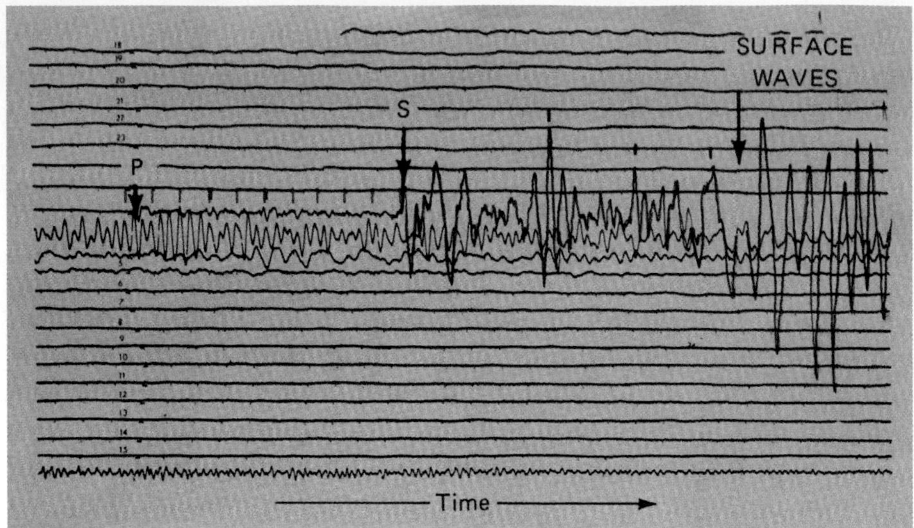

FIGURE 5-1 Record of a magnitude-6 earthquake that occurred in Turkey on March 28, 1964. The record was produced on a vertical component seismograph located in northwestern Canada. Time increases from left to right. Small tick marks represent 1-minute intervals. P means primary waves, S indicates secondary waves. Notice that the first S-waves arrive about 10 minutes after the arrival of the first P-waves. (*Courtesy of the Earth Physics Branch, Department of Energy, Mines, and Resources, Canada.*) ? *Judging from the magnitude of the vibrations recorded, which of the three kinds of seismic waves is likely to have caused the most damage?*

at a seismograph station after there has been an earthquake (Fig. 5-1). They travel through the upper crust of the Earth at speeds of 4 to 5 kilometers per second near the base of the crust, they speed along at 6 or 7 kilometers per second. In these primary waves, pulses of energy are transmitted as a succession of compressions and expansions that parallel the direction of propagation of the wave itself (Fig. 5-2). Thus, a given segment of rock set in motion during an earthquake is driven into its neighbor and bounces back. The neighbor strikes the next particle and rebounds, and subsequent particles continue the motion. Vibrational energy is an accordion-like "push-pull" movement that can be transmitted through solids, liquids, and gases. Of course, the speed of P-wave transmission will differ in materials of different density and elastic properties.

Secondary waves, which are also termed S-waves or transverse waves, travel 1 to 2 kilometers per second slower than do P-waves. Unlike the movement of P-waves, rock vibration in secondary waves is at right angles to the direction of propagation of the energy (Fig. 5-3). This type of wave is easily demonstrated by tying a length of rope to a hook and then shaking the free end. A series of undulations will develop in the rope and move toward the hook—that is, in the direction of propagation. Any given particle or point along the rope, however, will move up and down in a direction perpendicular to the direction of propagation. It is because of their more complex motion that S-waves travel more slowly than P-waves. They are the second group of oscillations to appear on the seismogram (see Fig. 5-1). Unlike P-waves, secondary waves will not pass through liquids or gases.

Both P- and S-waves are sometimes also termed **body waves** because they are able to penetrate deep into the interior or body of our planet. Body waves travel faster in rocks of greater elasticity, and their speeds therefore increase steadily as they move downward into more elastic zones of the Earth's interior and then decrease as they begin to make their ascent toward the Earth's surface. The change in velocity that occurs as body waves invade rocks of different elasticity results in a bending or refraction of the wave. The many small refractions cause the body waves to assume a curved travel path through the Earth.

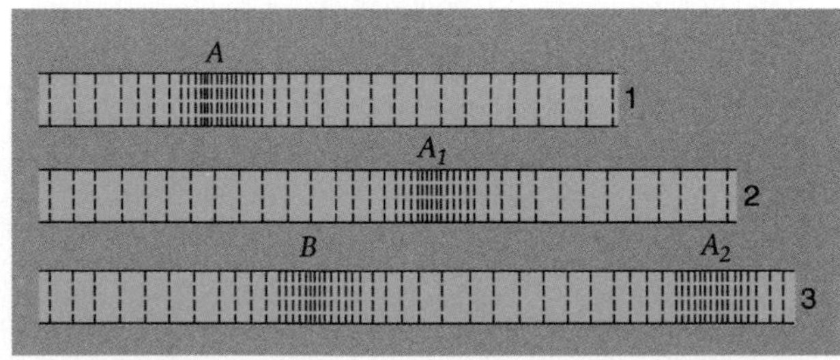

FIGURE 5-2 Movement of primary waves. In 1, the compression has moved the particles closer together at *A*. In 2, the zone of compression has moved to A_1. In 3, the zone of compression has moved to A_2 and a second compressional zone (*B*) has moved in from the left.

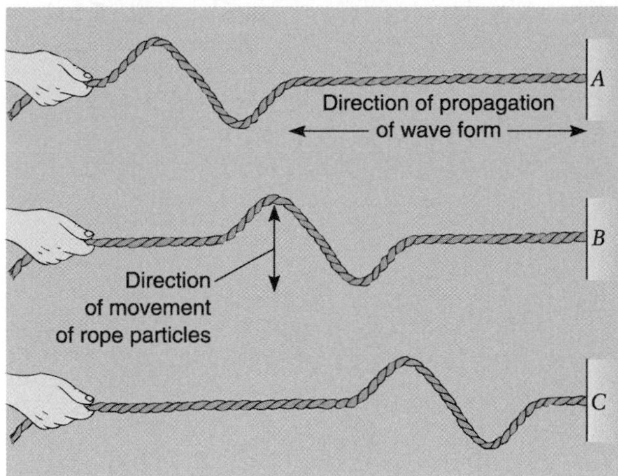

FIGURE 5-3 Analogy of propagation of S-waves by displacement of a rope. In rocks, as in this rope, particle movement is at right angles to the direction of propagation of the wave. *A*, *B*, and *C* show the displacement of the crest from left to right at successive increments of time.

Not only are body waves subjected to refraction, but they may also be partially reflected off the surface of a dense rock layer in much the same way as light is reflected off a polished surface. Many factors influence the behavior of body waves. An increase in the temperature of rocks through which body waves are traveling will cause a decrease in velocity, whereas an increase in confining pressure will cause a corresponding increase in wave velocity. In a fluid where no rigidity exists, S-waves cannot propagate and P-waves are markedly slowed.

Surface waves are large-motion waves that travel through the outer crust of the Earth. Their pattern of movement resembles that of waves caused when a pebble is tossed into the center of a pond. They develop whenever P- or S-waves disturb the surface of the Earth as they emerge from the interior. Surface waves are the last to arrive at a seismograph station. They are usually the primary cause of the destruction that can result from earthquakes affecting densely populated areas (Fig. 5-4). This destruction results because surface waves are channeled through the thin outer region of the Earth, and their energy is less rapidly dissipated into the large volumes of rock traversed by body waves.

THE DIVISIONS OF INNER SPACE

Most of what we know about the Earth's deep interior is derived from the interpretation of countless seismograms documenting recent and past earthquakes around the globe. From such studies, geophysicists have found evidence of the gradual change in rock properties with depth as well as the relatively abrupt boundaries between major internal zones. Boundaries where seismic waves experience an abrupt change in velocity or direction are called **discontinuities**. Two widely known breaks of this kind are named, after their discoverers, the Mohorovičić and Gutenberg discontinuities.

The discontinuity discovered by Mohorovičić (pronounced Mo-ho-ro-vitch-itz) was based on his observation that seismograph stations located about 150 kilometers from an earthquake received earthquake waves sooner than those nearer to the focus. Mohorovičić reasoned that below a depth of about 30 kilometers, there must be a zone having physical properties that permit earthquake waves to travel faster. That layer is now called the upper mantle.

In Figure 5-5, the shallow direct waves (*A*), although traveling at only about 6 kilometers per second, are the first to arrive at the closer seismograph station. At the farther station, however, the deeper but faster-traveling

FIGURE 5-4 Following the January 26, 2001 earthquake in Bhuj, India, a woman searches for belongings among the rubble that was once her home. (*Copyright AFP/Corbis*).

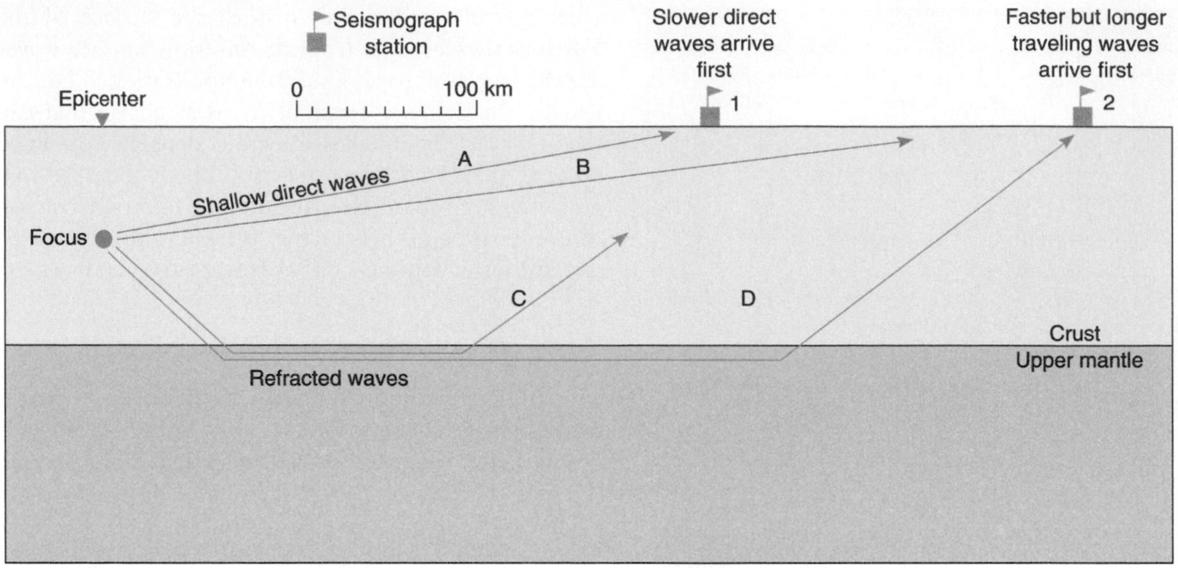

FIGURE 5-5 **Mohorovičić's conclusion about the location of the base of the Earth's crust was based on this interpretation of the travel paths of early and late-arriving seismic body waves.** (*Focus* is the true center of an earthquake and the point at which the disturbance originates. *Epicenter* is a point on the Earth's surface above the focus.)

wave (*D*) has caught up and moved ahead of the shallow direct wave and is therefore the first to arrive at seismograph station 2. The situation is analogous to what many of us do when we take a longer highway route to our destination than a more direct route through city streets because the greater speed on the highway allows us to arrive at the destination sooner.

The Mohorovičić discontinuity (or Moho, for short) lies at about 30 or 40 kilometers below the surface of the continents and at lesser depths beneath the ocean floors. It marks the boundary between the crust and the mantle.

The **Gutenberg discontinuity** is also called the CMB for core-mantle boundary. It is located nearly halfway to the center of the Earth at a depth of 2900 kilometers. Its location is marked by an abrupt decrease in P-wave velocities and the disappearance of S-waves, caused by a change in composition and a change from the solid to the liquid state that is characteristic of the outer core. The Gutenberg discontinuity marks the outer boundary of the Earth's core (Fig. 5-6).

The Core

INFERENCES FROM BODY WAVES As noted above, the boundary of the core was determined by the study of seismic waves. Seismology has also provided a means for discerning subdivisions of the core and deciphering some of its physical properties. For example, at a depth of 2900 kilometers (the core boundary), the S-waves cannot propagate, while at the same time P-wave velocity is drastically reduced from about 13.6 kilometers per second to 8.1 kilometers per second. Earlier, we noted that S-waves are unable to travel through fluids. Thus, if S-waves were to encounter a fluidlike region of the Earth's interior, they would be absorbed there and would not be able to continue. Geophysicists believe this is what happens to S-waves as they enter the outer core. As a result, the secondary waves generated on one side of the Earth fail to appear at seismograph stations on the opposite side of the Earth, and this observation is the principal evidence of an outer core that behaves as a fluid. The outer core barrier to S-waves results in an S-wave **shadow zone** on the side of the Earth opposite the earthquake focus (the exact rupture point of an earthquake). Within the shadow zone, which begins 105° from the earthquake's location, S-waves do not appear.

Unlike S-waves, primary waves are able to pass through liquids. They are, however, abruptly slowed and sharply refracted as they enter a fluid medium. Therefore, as primary seismic waves encounter the molten outer core of the Earth, their velocity is checked and they are refracted downward. The result is a **P-wave shadow zone** that extends from about 105° to 140° from the earthquake focus (Fig. 5-7). Beyond 140°, P-waves are so tardy in their arrival that they further validate the inference that they have passed through a liquid medium. At the upper boundary of the core, P-waves are also reflected back toward the Earth's surface. Such P-wave echos are clearly observed on seismograms.

The radius of the core is about 3500 kilometers. The inner core is solid and has a radius of about 1220 kilometers, which makes this inner core slightly larger than the Moon. Most geologists believe that the inner

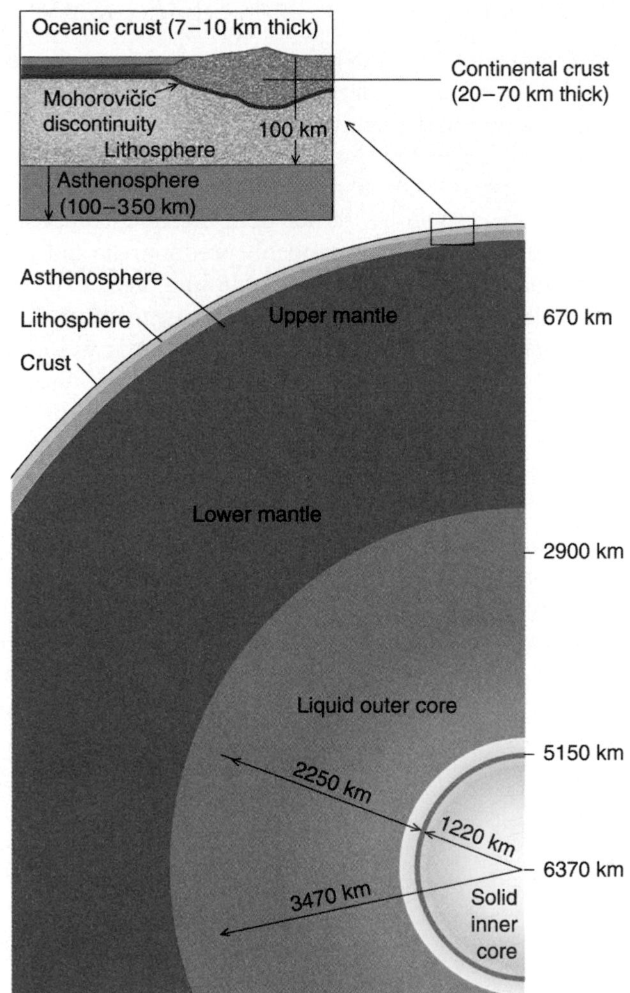

FIGURE 5-6 Interior of the Earth. (*Courtesy of G.R. Thompson and J. Turk*) ? *Label and draw a line on the diagram showing the location of the Gutenberg discontinuity, outer core, and inner core. In general, what are the differences in density between the inner core, outer core, mantle, and crust?*

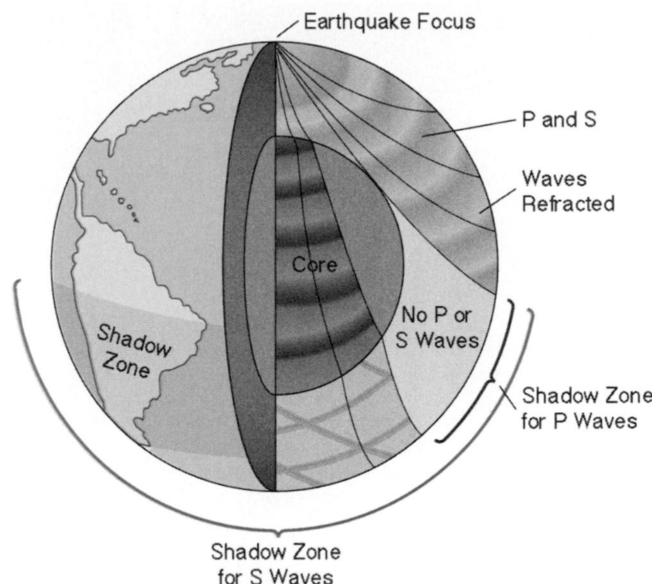

FIGURE 5-7 Refraction of seismic waves as they travel through the Earth. The travel paths bend gradually because of increasing velocity with depth. Abrupt bending or discontinuities occur at the boundaries of major zones. The S-wave shadow zone results from absorption of S-waves in the liquid outer core. P-waves entering the outer core are slowed and bent downward, giving rise to the P-wave shadow zone within which neither P- nor S-waves are received. ? *If an epicenter were located at the North Pole, seismographs in what large Argentinian city would not record secondary seismic waves?*

core has the same composition as the outer core and that it can exist only as a solid because of the enormous pressure at the center of the Earth.

Evidence of the existence of a solid inner core is derived from the study of hundreds of seismograms produced over several years. These studies showed that weak, late-arriving P-waves were somehow penetrating to stations that were within the P-wave shadow zone. Geophysicists recognized that this penetration could be explained by assuming that the inner core behaved seismically as if it were solid.

COMPOSITION OF THE CORE The Earth has an overall density of 5.5 g/cm^3, yet the average density of rocks at the surface is less than 3.0 g/cm^3. This difference indicates that materials of high density must exist in the deep interior of the planet to achieve the 5.5 g/cm^3 overall density. Calculations indicate that the rocks of the mantle have a density of about 4.5 g/cm^3, and that the average density of the core is about 10.7 g/cm^3. Under the extreme pressure conditions that exist in the region of the core, iron mixed with nickel would very likely have the required high density. Laboratory experiments, however, suggest that a highly pressurized iron-nickel alloy might be too dense and that minor amounts of such elements such as silicon, sulfur, carbon, or oxygen may also be present to "lighten" the core material.

Support for the theory that the core is composed of iron (85 percent) with lesser amounts of nickel has come from the study of meteorites. Many of these samples of solar system materials are iron meteorites (Fig. 5-8) that consist of metallic iron alloyed with a small percentage of nickel. Some geologists suspect that iron meteorites may be fragments from the core of a shattered planet. The presence of iron meteorites in our solar system suggests that the existence of an iron-nickel core for the Earth is plausible.

There is further evidence that the Earth may have a metallic core. Anyone who understands the functioning of a compass is aware that the Earth has a magnetic field. The planet itself behaves as if there was a great bar magnet embedded within it at a small angle to its

(*A*)

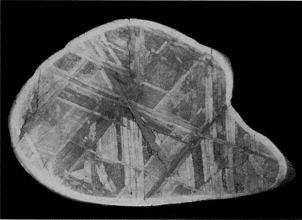

(*B*)

FIGURE 5-8 **Meteorites.** (*A*) is a meteorite that landed on the icy surface of Antarctica. (*B*) is a cut and polished iron meteorite from Mohave County, Arizona, showing the interesting pattern of interlocking crystals, called the Widmanstatten pattern. (*A, courtesy of NASA; B, from the Washington University collection.*) ? *Why are meteorite-collecting expeditions nearly always more successful in the Antarctic than on continents in warmer regions?*

rotational axis. A magnetic field is developed by the flow of electric charges and requires good electrical conductors. Silicate rocks, such as those in the mantle and lithosphere, do not conduct electricity very well, whereas metals such as iron and nickel are good conductors. Heat-driven convection in the outer core, coupled with movements induced by the Earth's spin, is thought to provide the necessary flow of electrons around the inner core that produces the magnetic field. Without a metallic core, this would not be possible.

The Mantle

MATERIALS OF THE MANTLE As was the case with the Earth's core, our understanding of the composition and structure of the mantle is based mainly on indirect evidence. As inferred from seismic wave data, the mantle's average density is about 4.5 g/cm^3, and it is believed to have a stony, rather than metallic, composition. Oxygen and silicon probably predominate and are accompanied by iron and magnesium as the most abundant metallic ions. The iron- and magnesium-rich rock peridotite (Fig. 5-9) approximates fairly well the kind of material inferred for the mantle. A peridotitic rock not only would be appropriate for the mantle's density but also is similar in composition to stony meteorites as well as volcanic rocks that are thought to have reached the Earth's surface from the upper part of the mantle itself. Such suspected mantle rocks are indeed rare. They are rich in olivine and pyroxenes and contain small amounts of certain minerals, including diamonds, that can form only under pressures greater than those characteristic of the crust.

The mantle is not uniform throughout but is composed of several concentric layers that differ in physical properties. One of these layers, named the **asthenos-**

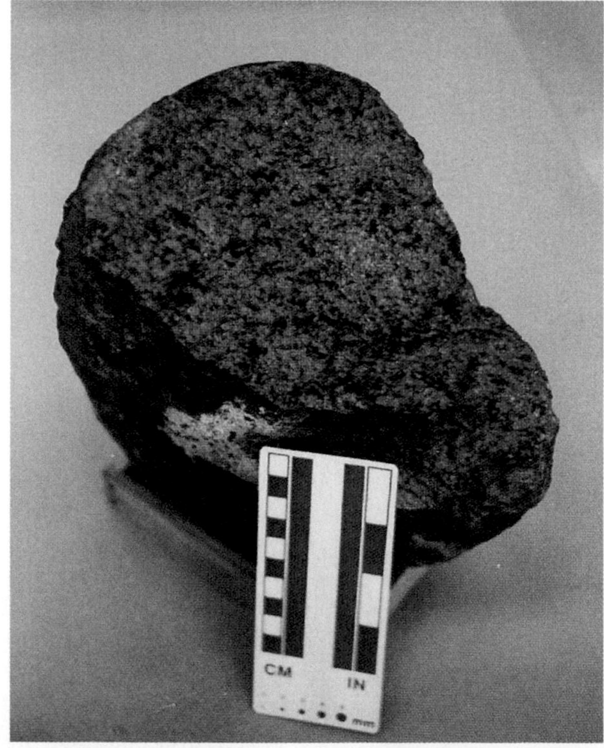

FIGURE 5-9 **Peridotite.** The green minerals are olivine and spinel, and the black grains are pyroxene. The specimen was collected at Kilbourne Hole maar, Dona Ana County, New Mexico. A maar is a crater formed during a violent volcanic explosion. The explosion brought chunks of mantle rock to the Earth's surface.

FIGURE 5-10 Pillow basalt that had erupted on the ocean floor of the Cayman Trough in the Carribbean. *(Courtesy of W. R. Normak and U.S. Geological Survey.)*

phere, is located high in the upper mantle (Fig. 5-6). Rock in the asthenosphere is at or near its melting point, and therefore magmas are generated in this zone. Seismic waves are slowed in the asthenosphere, not because of a decrease in density, but rather because they enter a region that is in a crystalline-liquid state. In such a "hot slush," perhaps up to 10 percent of the rock would consist of pockets and droplets of molten silicates. Such material would be capable of flow and would serve as a slippery, mobile layer for the overlying rigid layer known as the lithosphere. As illustrated in Figure 5-6, the lithosphere includes crust (oceanic and continental) and the uppermost part of the mantle that is above the asthenosphere. It is the dynamic enveloping shell of the Earth and is itself divided into lithospheric plates.

THE CRUST

The crust of the Earth is seismically defined as all of the solid Earth above the Mohorovičić discontinuity. It is the thin, brittle veneer that constitutes the continents and the floors of the oceans. The crust is not a homogeneous shell in which low places were filled with water to make oceans and higher places make continents. Rather, there are two distinct kinds of crust that, because of their distinctive compositions and physical properties, determine the very existence of separate continents and ocean basins.

The Oceanic Crust

Beneath the varied topography of the ocean floors lies an oceanic crust that is approximately 5 to 12 kilometers thick and has an average density of about 3.0 g/cm^3. Three layers of this oceanic crust can be recognized by seismic studies. On the upper surface is a thin layer of unconsolidated sediment that rests on the irregular surface of the igneous basement layer. The second layer consists of basalts that had been extruded under water (Fig. 5-10). The nature of the third and deepest layer of oceanic crust is not clear. Many suspect that it is metamorphosed basaltic mantle material that has become somewhat less dense by chemically combining with sea water. In rocks of the oceanic crust, we find a concentration of such common elements as iron, magnesium, and calcium. These elements are included in the plagioclase feldspars, amphiboles, and pyroxenes of basalts. The upper mantle is the ultimate source of the lavas that formed the oceanic crust.

The Continental Crust

PROPERTIES OF THE CONTINENTAL CRUST At the boundaries of the ocean basins, the Mohorovičić discontinuity plunges sharply beneath the thicker continental crust (Fig. 5-11). The depth of the Moho beneath the continents averages about 35 kilometers, although it may be considerably deeper or shallower in particular regions. The continental crust is not only thicker than its oceanic counterpart but also less dense, averaging about 2.7 g/cm^3. As a result, continents "float" higher on the denser mantle than the adjacent oceanic crustal segments. Somewhat like great stony icebergs, the roots of continents extend downward into the mantle.

The lower density of the continental crust results from its composition. Although it is referred to as granitic, it is really composed of a variety of rocks that approximate granite in bulk composition. Igneous

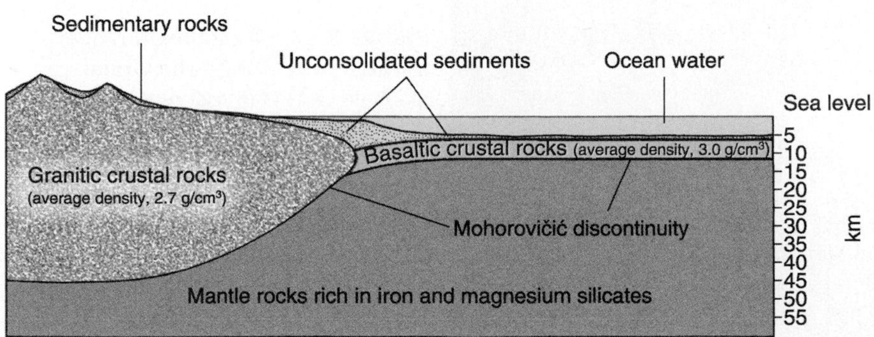

FIGURE 5-11 Generalized cross-section showing location of the Mohorovičić discontinuity.

FIGURE 5-12 Two normal faults in sandstones. The strata, exposed near Camberra, Australia, are Silurian in age. (*Copyright Fletcher & Baylis / Photo Researchers.*)

continental rocks are richer in silicon and potassium and poorer in iron, magnesium, and calcium than igneous upper mantle and oceanic rocks. In addition, extensive regions of the continents are blanketed by sedimentary rocks.

The condition of balance that exists among segments of the Earth's crust as they come into flotational equilibrium with denser mantle material is called isostasy. Because of isostasy, the continental crust stands higher than the denser oceanic crust. Isostasy also explains why mountain ranges experience repeated uplifts following erosional removal of great volumes of rock and soil. As the heavy burden of eroded material is carried away by streams, the ranges rise in much the same way that a boat rises in the water as its cargo is unloaded.

CRUSTAL STRUCTURES

Continents, mountains, and deep oceans appear to us to be everlasting. The human life span is far too short to permit a view of lands in upheaval, continents splitting and colliding, and ocean basins that expand and contract. People are allowed time only to witness an occasional manifestation of these changes, such as an earthquake or volcanic eruption. The effects of the changes, however, are recorded in rocks as they are deformed and broken. Among the effects of Earth movements are faults and folds, a brief discussion of which we will provide as background for the subsequent topic of plate tectonics.

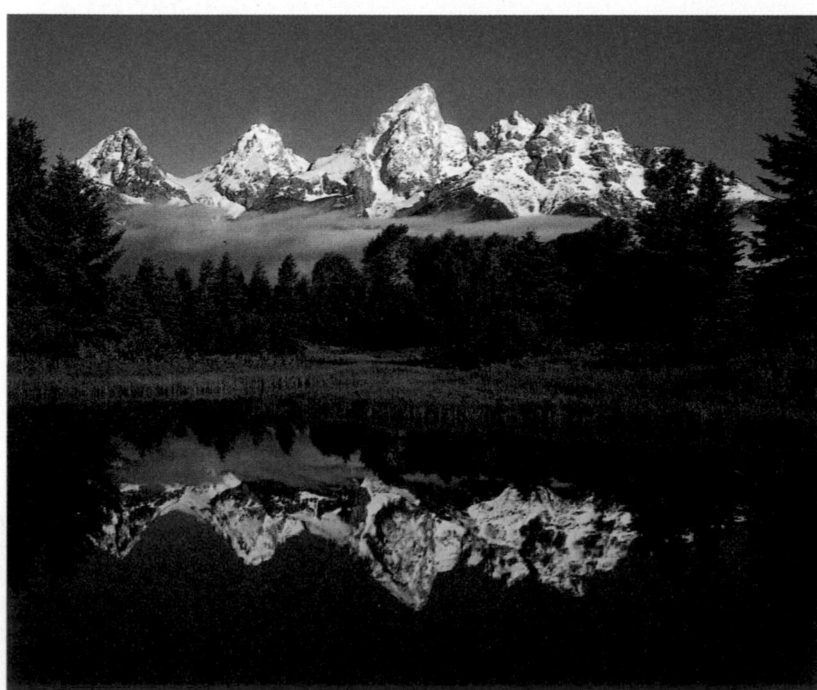

FIGURE 5-13 A mountain range formed by faulting—the Grand Tetons of Wyoming, viewed from Signal Mountain. The Grand Tetons were formed by normal faulting. The western block was uplifted to create the mountains, whereas the east block (represented by the foreground) dropped to form Jackson Hole, with its picturesque Jenny Lake. (*Copyright John Kieffer / Peter Arnold, Inc.*)

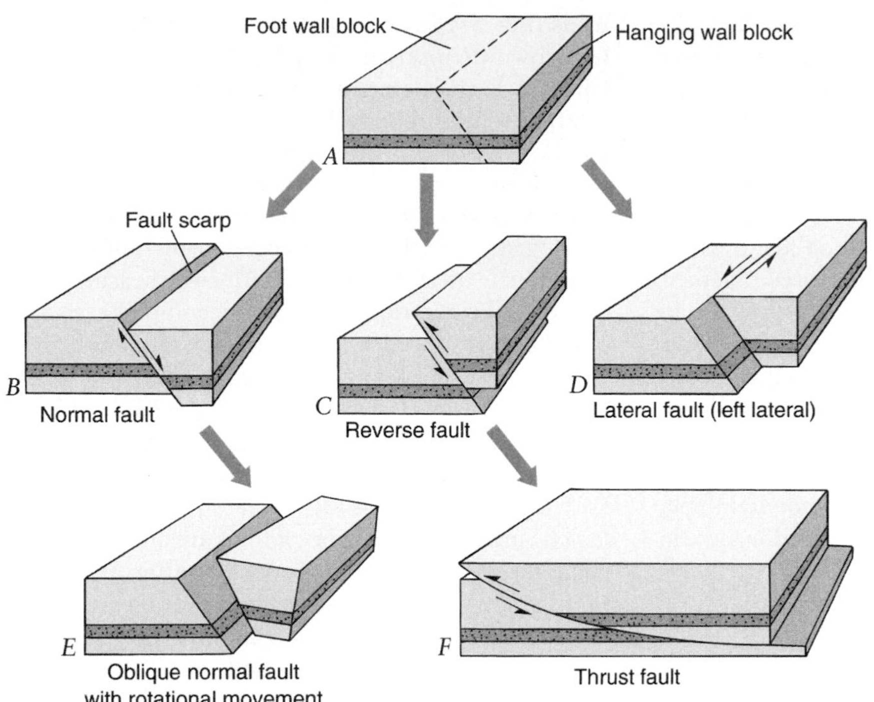

FIGURE 5-14 Types of faults. (*A*) The unfaulted block, with the position of the potential fault shown by the dashed line. In nature, movements along faults may vary in direction, as shown in (*E*). A thrust fault (*F*) is a type of reverse fault that is inclined at a low angle from the horizontal. ? *Which of these fault types are not the result of compressional forces?*

Faults

A **fault** is a break in the Earth's crustal rocks along which there has been displacement (Figs. 5-12 and 5-13). According to the relative direction of displacement of the rocks on either side of the plane of breakage, faults can be classified as **normal**, **reverse**, or **lateral** (Fig. 5-14). Lateral faults can be further designated as *right lateral* or *left lateral* if one looks across the fault zone to see if the opposite block has moved to the right or left. The San Andreas fault in California is a lateral fault of the right-lateral type (Fig. 5-15). Lateral faults are also called **strike-slip faults.**

FIGURE 5-15 Air photo mosaic in which the trace of the right-lateral strike-slip San Andreas fault is clearly visible. Notice the offset streams (*arrows*). San Luis Obispo County, California. (*Courtesy of U.S. Geological Survey.*)

In normal faults, the mass of rock that lies above the shear plane, called the **hanging wall**, appears to move downward relative to the opposite side or **foot wall.** Such faults occur where rocks are subjected to tensional forces-forces that tend to stretch the crust. On the other hand, reverse faults exhibit a hanging wall that has moved up relative to the foot wall. If we imagine ourselves holding two blocks cut from wood like those of Figure 5-15, we find that they must be shoved together, or compressed, to cause reverse faults. Thus, regions of the Earth's crust containing numerous reverse faults (and folded strata) are likely to have been compressed at some time in the geologic past. Reverse faults in which the shear zone is inclined only a few degrees from horizontal are termed **thrust faults**. Although examples of all kinds of faults may be found in any mountain belt, compressional faults are decidedly the most prevalent in the world's great mountain ranges.

Folds

No less important than faults as evidence of crustal instability are the bends in rock strata that are termed **folds** (Fig. 5-16). Sediments are, of course, originally laid down in approximately level layers. As noted in Chapter 1, the perceptive naturalist Nicolaus Steno referred to this observation as the "principle of original horizontality." Steno correctly inferred that if strata were originally horizontal, then folding of those beds was direct evidence of crustal movement. Geologists describe the orientation of folded or inclined beds or fault surfaces with two measurements called **strike** and **dip**. Strike is the compass direction of the line produced by the intersection of an inclined stratum (or other feature such as a fault plane) with a horizontal plane (Fig. 3-53). Dip is the angle of inclination of the tilted layer, also measured from the horizontal plane. The dip of the strata depicted in Figure 5-17 is 30°.

The principal categories of folds are anticlines, synclines, domes, basins, and monoclines (Fig. 5-18). **Anticlines** are upwardly folded or uparched layered rocks. On anticlines that have been truncated by erosion, the oldest strata are in the center and the youngest strata are on the flanks (Fig. 5-19). **Synclines** are downwardly folded rocks. When eroded, synclines have youngest beds in the center and oldest layers of rock on the flanks (Fig. 5-20). A special kind of fold called a **monocline** consists of a steplike bend in otherwise predominantly horizontal strata.

Folds tend to occur together, rather like a series of frozen wave crests and troughs. The erosion of anticlines and synclines characteristically produces a topographic pattern of ridges and valleys (Fig. 5-21). The

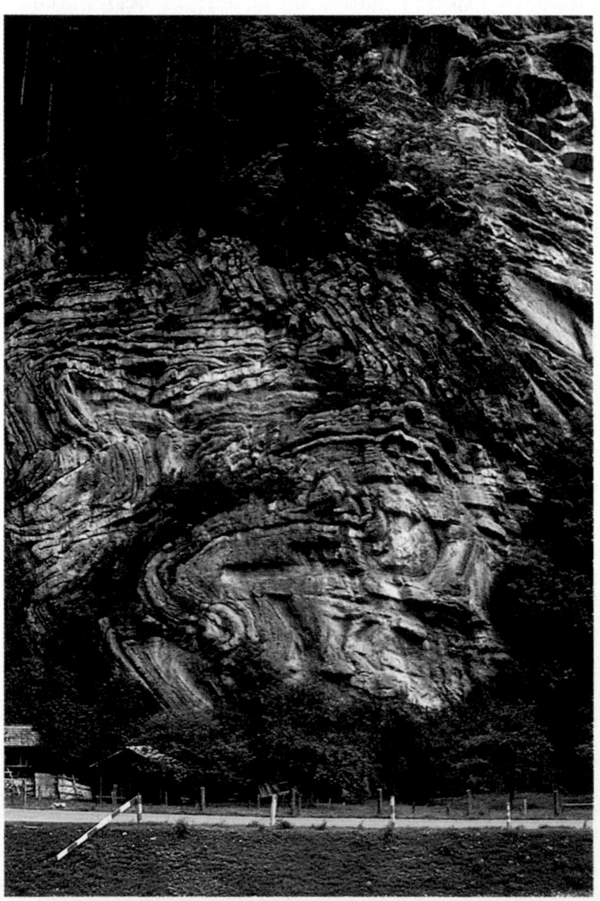

FIGURE 5-16 Severely folded strata exposed on the north side of the Rhone River valley, west of Sierre, Switzerland.

FIGURE 5-17 Geologist measuring the dip of strata in a roadcut.

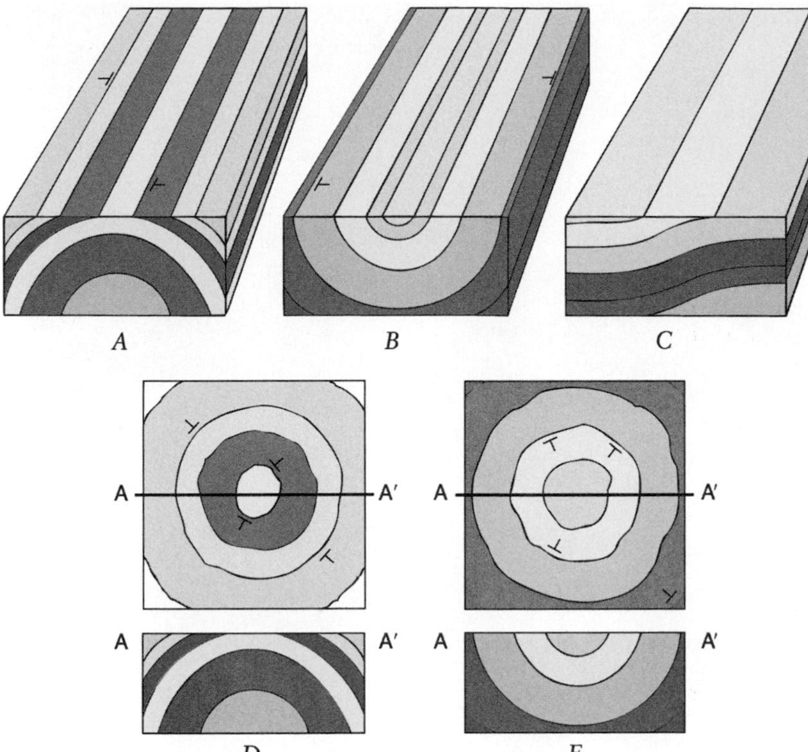

FIGURE 5-18 Types of folds. (*A*) Anticline. (*B*) Syncline. (*C*) Monocline. (*D*) Map view and cross-section of a dome. Notice that older strata are in the center of the outcrop pattern. (*E*) Map view and cross-section of a basin. Younger rocks occur in the central area of the outcrop pattern. ? *Find Missouri on the Bedrock Geology map of North America in the Appendices. Is the structure beneath southern Missouri a structural basin or dome? What other clearly domal structure can you discern in nearby states?*

ridges develop where resistant rocks project at the surface, whereas valleys develop along those parts of the fold underlain by more easily eroded rocks.

Domes and **basins** are similar to anticlines and synclines, except that they have elliptical to roughly circular outcrop patterns. A dome consists of uparched strata in which the beds dip (are inclined) in all directions away from the center. In contrast, a basin is a downwarp in which beds dip from all sides toward the center of the structure. The erosionally truncated beds of simple domes and basins may form a circular pattern of ridges and valleys. As is the case with anticlines and synclines, older beds are found in the center of truncated domes and younger strata at the center of basins (Fig. 5-18). Domes and basins vary in size from a few meters to hundreds of kilometers in diameter.

Although most folds (and particularly those found in mountain belts) are formed by compression, some may develop under a variety of other circumstances. For example, some relatively small folds may result from differential compaction as loose sediment is converted to sedimentary rock, from slumping of

FIGURE 5-19 Anticline exposed by Sun River Erosion, Teton County, Montana. (*Courtesy of R.R. Mudge and the U.S. Geological Survey.*)

FIGURE 5-20 Geologic structures often have no relationship to surface topography. Here a downfold in beds (syncline) lies beneath a high ridge. Sideling Hill road cut, 6 miles west of Hancock, Maryland, along U.S. Route 48, Ridge and Valley Province. The rocks are sandstones, minor conglomerates, and a few dark shales of the Mississippian Pocono Formation. ? *As the driver of the "18-wheeler" in the lower right area of the photograph continues toward the axial plane (center) of this structure, will he be passing increasingly younger or older strata?*

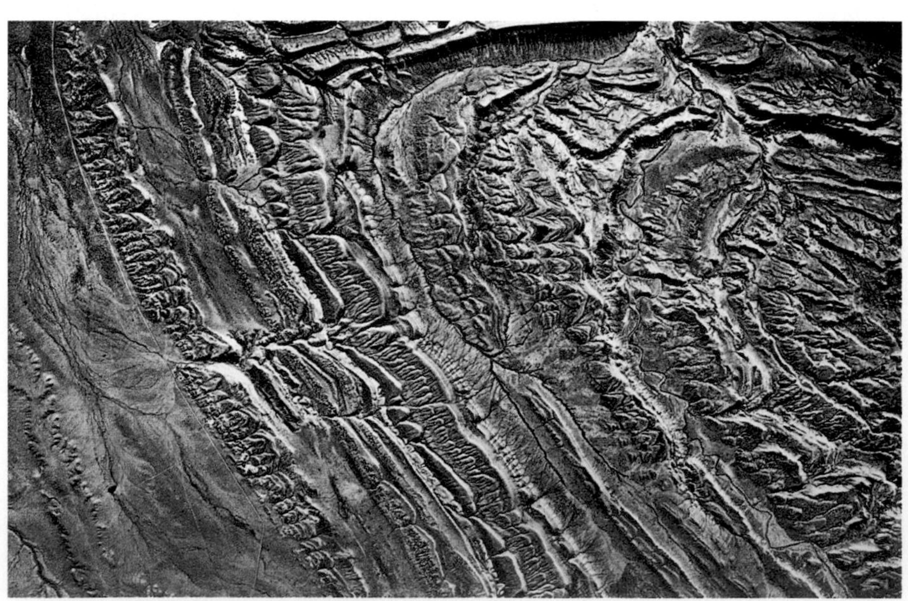

FIGURE 5-21 An anticlinal fold in Wyoming. Because of the sparseness of vegetation and soil cover, rock layers are clearly exposed, and differences in resistance to erosion cause layers to be etched into sharp relief. (*Courtesy of U.S. Geological Survey.*) ? *How has the geologic structure in this area affected drainage patterns?*

mounds of sediment at the time of its accumulation, from an upward protuberance of rock masses from below, from draping over uplifted fault blocks, and from crumpling as great blankets of strata slide over an older rock surface.

PLATE TECTONICS

Perhaps more than other scientists, geologists are accustomed to viewing the Earth in its entirety. It is their task to assemble the multitude of observations about the origin of mountains, the growth of continents, and the history of ocean basins into a coherent view of the whole Earth. Exciting and revolutionary discoveries made over the past 4 decades have provided that kind of integrated understanding of our planet. The current view of the Earth's dynamic geology, called **plate tectonics**, is based on the movements of plates of lithosphere driven by convection in the underlying mantle.

As with many scientific breakthroughs, the plate tectonics theory was not fully conceptualized until a foundation of relevant data had been assembled. In the years following World War II, research related to naval operations produced submarine detection devices that also proved useful in measuring the magnetic properties of the Earth's oceanic crust. In the mid-1940s to late 1950s, the need to monitor atomic explosions resulted in the establishment of a worldwide network of seismographs. This network provided precise information about the global pattern of earthquakes. The magnetic field over large portions of the sea floor was soon to be charted by the use of newly developed and delicate *magnetometers*. Other technologic advances ultimately permitted scientists to examine rock that had been dated by isotopic methods and then to determine the nature of the Earth's magnetic field at the time those rocks had formed. Geologically recent reversals of the magnetic field were soon detected, correlated, and dated. A massive federally funded program to map the bottom of the oceans was launched. Mapping was facilitated by improved echo-sounding devices capable of providing not only continuous topographic profiles of the sea floor (Fig. 5-22) but also images of the rock and sediment that lay beneath.

A new picture of the ocean floor began to emerge (Fig. 5-23). It was at once awesome, alien, and majestic. Great chasms, flat-topped submerged mountains, boundless abyssal plains, and globe-spanning volcanic ranges appeared on the new maps and begged an explanation. How did the volcanic midoceanic ridges and deep-sea trenches originate? Why were both so prone to earthquake activity? Why was the mid-Atlantic ridge so nicely centered and parallel to the coastlines of the continents on either side? As the topographic, magnetic, and geochronologic data accumulated, the relationship of these questions became apparent. An old theory called continental drift was re-examined, and the new, more encompassing theory of plate tectonics was formulated. It was an idea whose time had come.

Continental Drift

It requires only a brief examination of the world map to notice the remarkable parallelism of the continental shorelines on either side of the Atlantic Ocean. If the continents were pieces of a jigsaw puzzle, it would seem easy to fit the great "nose" of Brazil into the re-entrant of the African coastline. Similarly, Greenland might be inserted between North America and northwestern

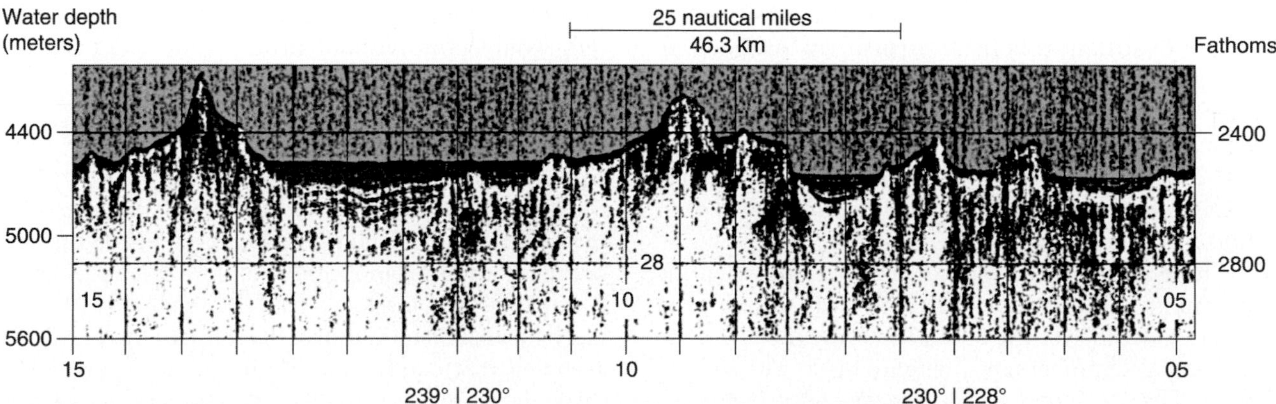

FIGURE 5-22 Continuously recorded profiles showing abyssal hills and abyssal plains along the edge of the Mid-Atlantic Ridge. *(From Hayes, D. E., and Pimm, A. C. 1972. Initial Reports of Deep Sea Drilling Project. National Science Foundation publication, vol. 14, pp. 341–376.)* *Note that the summits of the submarine volcanic mountains are at increasingly greater depth as one proceeds from left to right. What might be the reason for this?*

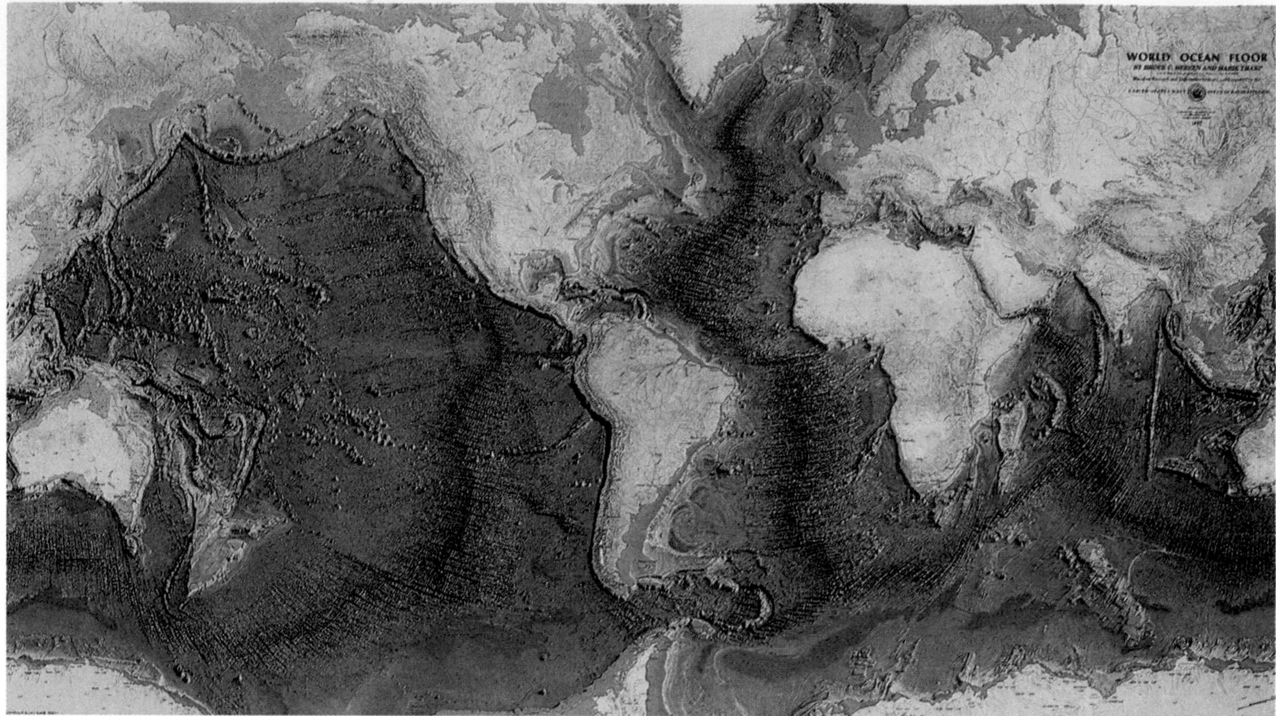

FIGURE 5-23 Midoceanic ridges, deep-sea trenches, and other features of the ocean floors. (*Copyright Marie Tharp.*)

Europe. It is not surprising, therefore, that earlier generations of map-gazers also noticed the fit and formulated theories involving the breakup of an ancient supercontinent.

In 1858 there appeared a work titled *La Création et ses Mystères Devoilés*. Its author, A. Snider, postulated that before the time of Noah and the biblical flood, there existed a great region of dry land. This antique land developed great cracks encrusted with volcanoes, and during the Great Deluge a portion separated at a north-south trending crack and drifted westward. Thus, North America came into existence.

Near the close of the 19th century, the Austrian scientist Eduard Suess became particularly intrigued by the many geologic similarities shared by India, Africa, and South America. He formulated a more complete theory of a supercontinent that he named Gondwanaland after a geologic province in east-central India. Today Suess's name for the supercontinent has been shortened to **Gondwana**.

The next serious effort to convince the scientific community of the validity of these ideas was made in the early decades of the 20th century by the energetic German geographer and meteorologist Alfred Wegener. His 1915 book, *Die Entstehung der Kontinente und Ozeane* ("The Origin of the Continents and Oceans"), is considered a milestone in the historical development of the concept of continental drift.

Wegener's hypothesis of continental drift was straightforward. Building on the earlier notions of Suess, he argued again for the existence in the past of a supercontinent that he dubbed **Pangea**. That portion of Pangea that was to separate and form North America and Eurasia came to be known as **Laurasia**, whereas the southern portion retained the earlier designation of Gondwanaland. According to Wegener's perception, Pangea was surrounded by a universal ocean named **Panthalassa**, into which the shifting continents moved when they began to split apart about 200 million years ago (Fig. 5-24). The fragments of Pangea were presumed to drift along like great stony rafts on the denser material below. In Wegener's view, the bulldozing forward edge of the slab would crumple and produce mountain ranges such as the Andes.

Wegener's theory of continental drift did not go unchallenged. Criticism was leveled chiefly against the concept that the continents were able to slide along through an oceanic crust. The eminent and scientifically formidable Sir Harold Jeffries calculated that the ocean floor was far too rigid to allow for the passage of continents, no matter what the imagined driving mechanism. It is now known that Jeffries was correct in asserting that continents cannot—and do not—plow through oceanic crust. Later it was determined that continents do move, but they move only as passive passengers on large rafts of lithosphere that glide over a comparatively soft and plastic upper layer of the Earth's mantle. Nevertheless, much of the evidence of drifting continents that Wegener and others had as-

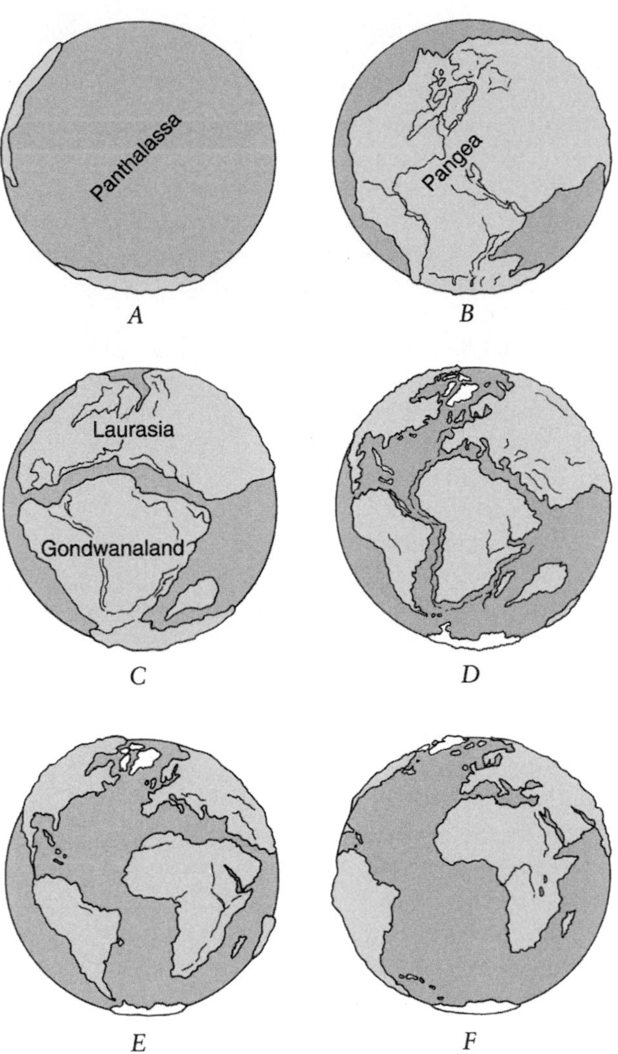

FIGURE 5-24 Break-Up of Pangea. (*A* and *B*) Alfred Wegener's view of the Earth as it existed 200 million years ago. At this time there was one ocean (Panthalassa) and one continent (Pangea). (*C*) By about 20 million years later, the supercontinent had begun to split into northern Laurasia and southern Gondwanaland. (*D*) Dismemberment continued, and (*E*) the Earth looked this way about 65 million years ago. (*F*) Further widening of the Atlantic and northward migration of India brought the Earth to its present state.

sembled was valid, and can be used to substantiate both the old and the newer concepts. We will consider this evidence.

Evidence of Continental Drift

The most convincing evidence of continental drift remains the geologic fit of many continents. Indeed, the correspondence is far too good to be fortuitous, even when one considers the expected modifications of shorelines resulting from erosion, deformation, or intrusions following the break-up of Pangea. A still closer match results when one fits the continents together to include the continental shelves, which are really only submerged portions of the continents. Such a computerized and error-tested fitting of continents was carried out by Sir Edward Bullard, J. E. Everett, and A. G. Smith of the University of Cambridge. The work showed a remarkable correspondence of continental margins (Fig. 5-25).

Another line of evidence favoring the drift theory involves sedimentologic criteria indicating similarity of climatic conditions for widely separated parts of the world that were once closely adjacent to one another. For example, in such now widely separated components of Gondwana as South America, southern Africa, India, Antarctica, Australia, and Tasmania, one finds glacially grooved rock surfaces and deposits of glacial rubble developed in the course of the late Paleozoic continental glaciation. The lithified deposits of poorly sorted clay, sand, cobbles, and boulders are called tillites. They, along with the grooves and scratches on rock surfaces that were apparently beneath the moving

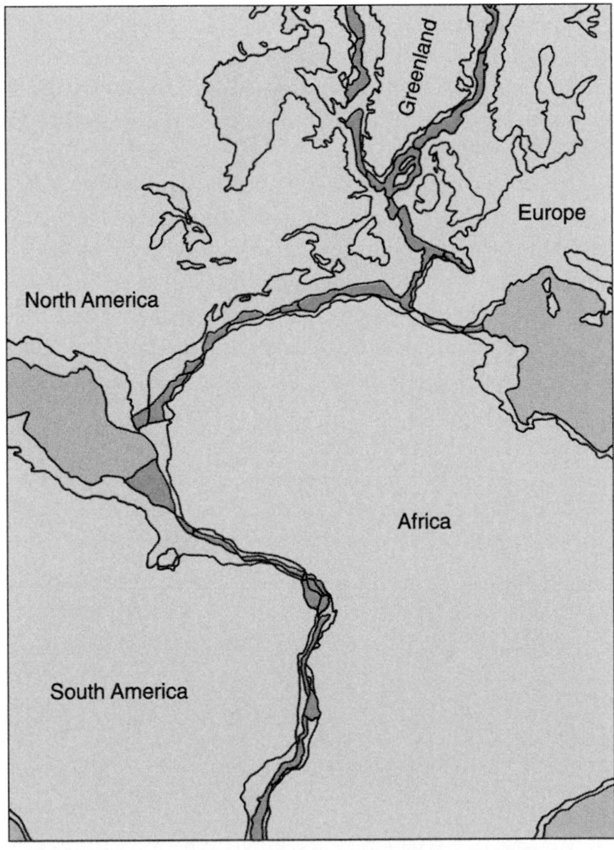

FIGURE 5-25 Fit of the continents, as determined by Sir Edward Bullard, J. E. Everett, and A. G. Smith. The fit was made along the continental slope (green color) at the 500-fathom contour line. Overlaps and gaps (shown in orange) are probably the result of deformations and sedimentation after rifting. (*Adapted from Bullard, E. C., et al. 1965. Philos. Trans. R. Soc. Lond. 258:41.*)

ice, attest to a great ice age that affected Gondwana at a time when it was as yet unfragmented and lying at or near the south polar region (Fig. 5-26). Furthermore, if the directions of the glacial grooves on the bedrock are plotted on a map, they indicate the center of ice accumulation and the directions in which it moved. Unless the southern continents are reassembled into Gondwana, this center would be located in the ocean, and great ice sheets do not develop centers of accumulation in the ocean. Hence, the existence of Gondwana seems plausible. In a few instances, oddly foreign boulders in the tillites of one continent are found in the deposits of another continent now located thousands of miles across the ocean. Petrologists are able to trace these boulders to parent rock masses in distant land masses.

There are additional clues to paleoclimatology that can be used to test the concept of moving land masses. Trees that have grown in tropical regions of the globe characteristically lack the annual rings resulting from seasonal variations in growth. Exceptionally thick coal seams containing fossil logs from such trees imply a tropical paleoclimate. The locations of such coal deposits should approximate an equatorial zone relative to the ancient pole position for that age. It is evident that such coal seams now being exploited in northern latitudes must have been moved to those locations from the equatorial zones along which the source vegetation accumulated.

Because of the decrease in solubility of calcium carbonate with rising temperatures, thick deposits of marine limestone also imply relatively warm climatic conditions. Arid conditions, such as those existing today on either side of the equatorial rain belt, can be recognized in ancient rocks by desert sandstones and evaporites. (Evaporites are chemical precipitates such as salt and gypsum that characteristically form when a body of water containing dissolved salts is evaporated.) Today such deposits are ordinarily formed in warm, arid regions located about 30° north or south of the equator. If one believes continents have always been where they are now, then it is difficult to explain the great Permian evaporite deposits found in northern Europe, the Urals, and the southwestern United States. The evaporites probably precipitated in warmer latitudes before Laurasia migrated northward.

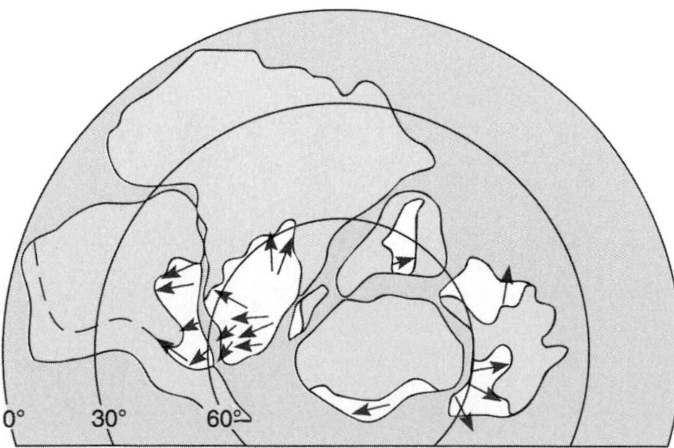

FIGURE 5-26 Reconstruction of Gondwana near the beginning of Permian time, showing the distribution of glacial deposits (shown in white). Arrows show direction of ice movement, as determined from glacial scratches on bedrock. (*Modified from Hamilton, W., and Kinsley, D. 1967. Geol. Soc. Am. Bull. 78:783–799.*)

A similar kind of evidence can be obtained by examining the locations of Permian reef deposits. Modern reef corals are restricted to a band around the Earth that is within 30° of the equator. Ancient reef deposits are now found far to the north of the latitudes at which they had originated.

At least some of the paleontologic support for continental drift was well known to Suess and Wegener. In the Gondwana strata overlying the tillites or glaciated surfaces, there can often be found nonmarine sedimentary rocks and coal beds containing a distinctive assemblage of fossil plants. Named after a prominent member of the assemblage, the plants are referred to as the ***Glossopteris* flora** (Fig. 5-27). Paleobotanists who have supported the idea of shifting continents have argued that it would be virtually impossible for this complex flora to have developed in identical ways on the southern continents as they are separated today. The seeds of *Glossopteris* were far too heavy to have been blown over such great distances of ocean by the wind.

Vertebrates also provide evidence of the existence of Pangea. A partial list of Gondwana vertebrates would include a small Permian aquatic reptile named *Mesosaurus* (Fig. 5-28), known only from localities in southern Africa and southern Brazil. As indicated by skeletal characteristics and the nature of the enclosing sediment, *Mesosaurus* inhabited freshwater bodies and used its long, sharp teeth to catch fish and small crustaceans. Another vertebrate indicator of adjoined continents is *Lystrosaurus*. This stocky, sheep-sized plant-eater left skeletal remains in Africa, India, Antarctica, Russia, and China, indicating these regions were once part of the supercontinent Pangea.

In strata somewhat younger than those yielding the remains of *Lystrosaurus*, paleontologists working in such widely separated localities as Africa, South America, China, and Russia have found the remains of a small carnivorous reptile named *Cynognathus*. In addition to its value in corroborating the existence of Pangea, *Cynognathus* is of special interest to those tracing the phylogeny of the vertebrates, for it possessed several skeletal characteristics indicating a relationship to the lineage that would ultimately lead to mammals. *Lystrosaurus* also possessed some mammal-like traits, but fewer than in *Cynognathus*.

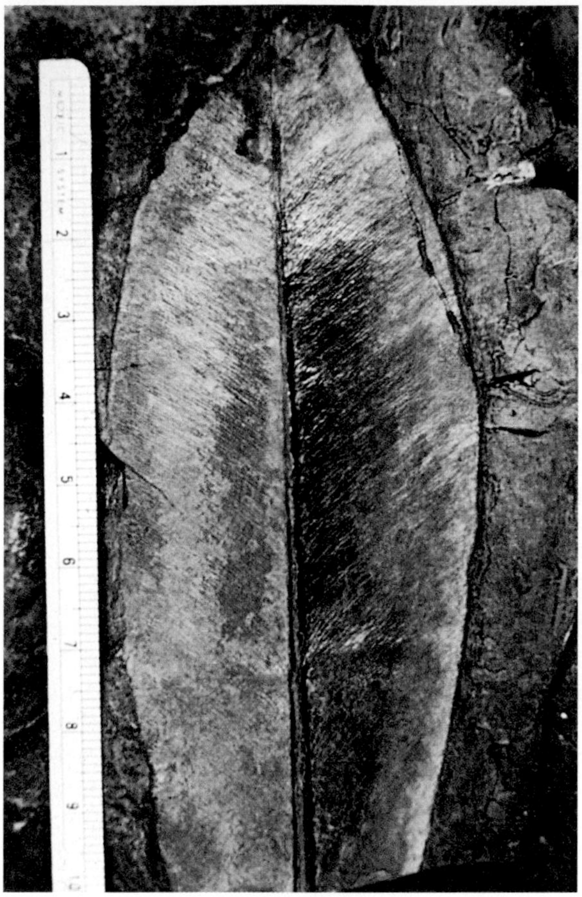

Before fragmentation of a supercontinent such as Laurasia, one would expect to find numerous similar plants and animals living at corresponding latitudes on either side of the line of future separation. Such similarities do occur in the fossil record of continents now separated by extensive oceanic tracts. Faunas of Silurian and Devonian fishes, for example, are comparable in such now-distant locations as Great Britain, Germany, Spitzbergen, eastern North America, and Quebec. These similarities, not only in fishes but also in amphibians, persist into Carboniferous rocks. In Permian and Mesozoic age rocks, one again finds striking similarities in the reptilian faunas of Europe and North America. The fossil evidence implies not only the former existence of continuous land connections but also a uniformity of environmental conditions between the elements of Laurasia that were once at approximately similar latitudinal zones.

When one turns to Cenozoic mammalian faunas, however, one finds the situation to be quite different. Distinctive faunal elements are evident in Australia, South America, and Africa. Apparently, as the continents

(A)

B

FIGURE 5-27 **Fossil *Glossopteris* leaf (*A*) associated with coal deposits and derived from glossopterid forests of Permian age. *(B) Glossopteris* tree. The leaf was found on Polarstar Peak, Ellsworth Land, Antarctica.** *(Courtesy of U.S. Geological Survey, J. M. Schopf, and C. J. Craddock.)*

FIGURE 5-28 **Fossil remains (above) and reconstruction (below) of the small freshwater reptile *Mesosaurus*, known only from Early Permian strata in a limited area along the coasts of Brazil and West Africa.**

became separated from each other by ocean barriers, genetic isolation resulted in morphologic divergence. The modern world's enormous biologic diversity is at least partially a result of evolutionary processes operating on more or less isolated continents. The faunas and floras during the periods prior to the break-up of Pangea were less diverse.

The character, sequence, age, and distribution of rock units have also been examined for insights into concepts of drift. One might presume that locations close to one another on the hypothetic supercontinent would have environmental similarities that would be indicated by the kinds of rocks deposited. As revealed in the correlation chart of southern continents (Table 5-1), there is such a similarity in the geologic sections of now widely separated land masses. The sections begin with glacial deposits such as the Dwyka Tillite of Africa and the Itararé Tillites of South America. These cobbly layers are overlain by nonmarine sandstones and shales containing the *Glossopteris* flora and layers of coal. The sequence may indicate a warming trend from cold glacial to more temperate climates. The next-higher group of strata shows evidence of deposition under terrestrial conditions with an abundance of alluvial, eolian, and stream deposits that contain fossils of mammal-like reptiles. The uppermost beds of the sequences include basalts and other volcanic rocks that are of similar age and composition in India and Africa but are slightly younger in South America. The absence of the reptile-bearing Triassic strata and the Jurassic volcanics in Australia is believed to be the result of Australia separating from Africa at an early date—probably early in the Permian—and long before the separation of Africa and South America. Australia must have retained just enough of a connection with Africa to permit the entry of dinosaurs and marsupials.

Yet another way to test the notion of a supercontinent is to see whether geologic structures, such as the trends of folds and faults, match up when now-distant continents are hypothetically juxtaposed. Such correlative trends do exist. Folds and faults are often difficult to date, although, if successful, one can establish the contemporaneity of a fault lineage or fold axis now on separated continents. One can also examine the outcrop patterns of Precambrian basement rocks and discern correlative boundaries between now widely separated continents. A folded geosynclinal sequence of Precambrian strata in central Gabon in Africa, for example, can be traced into the Bahia Province of Brazil. Also, isotopically dated Precambrian rocks of west Africa can be correlated with rocks of similar age in northeastern Brazil.

The Testimony of Paleomagnetism

Although Wegener was unable to provide a satisfactory explanation for what breaks continents apart and moves them across the face of the Earth, he had assembled a convincing body of evidence that such events occurred. Discoveries in the late 1950s and 1960s would provide further substantiating evidence. The new information came from the study of magnetism imparted to ancient rocks at the time of their formation and preserved down to the present time. To understand paleomagnetism, as it is called, it is necessary to digress for a moment and consider the general nature of the Earth's present magnetic field.

THE EARTH'S PRESENT MAGNETISM It is common knowledge that the Earth has a magnetic field. It is this field that causes the alignment of a compass needle. The origin of the magnetic field is still a question that

TABLE 5-1 Gondwana Correlations

System	Southern Brazil	South Africa	Penninsular India
Cretaceous	Basalt	Marine Sediments	Volcanics Marine Sediments
Jurassic	(Jurassic rocks not present)	Volcanics	Sandstone and shale Volcanics Sandstone and shale
Triassic	Sandstone and shale with reptiles	Sandstone and shale with reptiles	Sandstone and shale with reptiles
Permian	Shale and sandstones with coal and *Glossopteris* Shale with *Mesosaurus*	Shale and sandstones with coal and *Glossopteris* Shale with *Mesosaurus*	Shale and sandstones with coal and *Glossopteris* Shale
Carboniferous	Sandstone, shale, and coal with *Glossopteris* Tillite	Sandstone, shale, and coal with *Glossopteris* Tillite	Tillite

has not been fully resolved, but many geophysicists believe it is generated as the rotation of the Earth causes alignment of convective patterns in the liquid outer core. The magnetic lines of force resemble those that would be formed if there were an imaginary bar magnet extending through the Earth's interior. The long axis of the magnet would be the conceptual equivalent of the Earth's magnetic axis, and the ends would correspond to the north and south geomagnetic poles (Fig. 5-29). Although today the geomagnetic poles are located about 11° of latitude from the rotational axis, they slowly shift position. When averaged over several thousand years, the geomagnetic poles and the geographic poles do coincide. If we assume that this relationship has always held true, then by calculating ancient magnetic pole positions from paleomagnetism in rocks we have coincidentally located the Earth's geographic poles. It should be kept in mind, however, that such interpretations are based on the supposition that the rotational and magnetic poles have always been relatively close together. This seems a reasonable assumption based on the modern condition as well as on paleontologic studies that have shown inferred ancient climatic zones in plausible locations relative to ancient pole positions. Another assumption is that the Earth has always been dipolar. Paleomagnetic studies from around the world thus far support this supposition.

REMANENT MAGNETISM The magnetic information frozen into rocks may originate in several ways. Imagine for a moment the outpouring of lava from a volcano. As the lava begins to cool, magnetic iron oxide minerals form and align their polarity with the Earth's magnetic field. That alignment is then retained in the rock as its crystallization is completed. In a simple analogy, the magnetic orientations of the minerals responded as if they were tiny compass needles immersed in a viscous liquid. Because they are aligned parallel to the magnetic lines of force surrounding the Earth, they not only point toward the magnetic poles (**magnetic declination**) but also become increasingly more inclined from the horizontal as the poles are approached (Fig. 5-30). This **magnetic inclination**, when detected in paleomagnetic analysis, can be used to determine the latitude at which an igneous body containing magnetic minerals cooled and solidified.

Magnetism "frozen" into ancient rocks is called **remanent magnetism**. The type described previously, in which igneous rocks cool past the Curie temperature (also called the *Curie point*) of their magnetic minerals, is further classified as *thermoremanent magnetism*. The **Curie temperature** is simply that temperature above which a substance is no longer magnetic. A few minerals, the most important of which is magnetite (Fe_3O_4), have the property of acquiring remanent magnetism. Magnetite (Fig. 5-31) is widespread in varying amounts in virtually all rocks. The manner in which remanent magnetism is acquired is complicated but can be explained in a general way. Remanent magnetism in a mineral is ultimately due to the fact that some atoms and ions (charged atoms) have so-called magnetic moments; this means that they behave like tiny magnets. The magnetic moment of a single atom

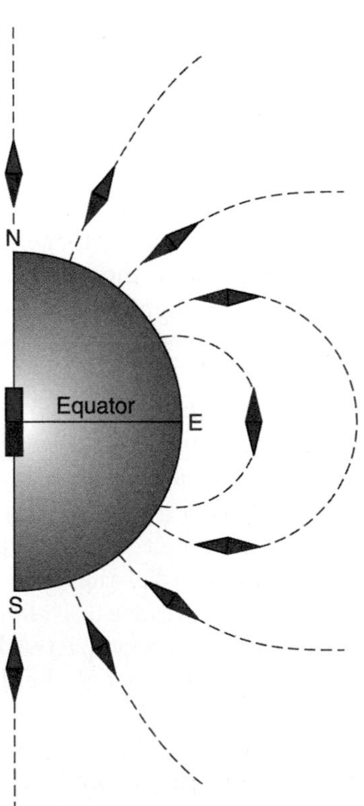

FIGURE 5-30 A freely suspended compass needle aligns itself in the direction of the Earth's magnetic field. The inclination of the needle will vary from horizontal at the equator to vertical at the poles.

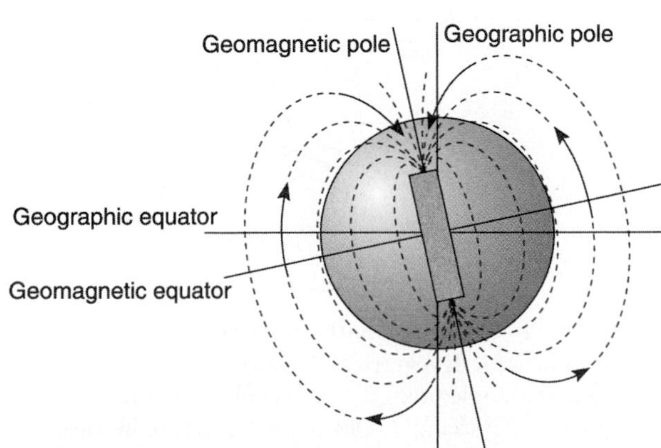

FIGURE 5-29 Magnetic lines of force for a simple dipole model of the Earth's magnetic field.

FIGURE 5-31 **Magnetite, an iron oxide mineral that acts as a natural magnet and is capable of picking up items composed of iron.** Magnetite possesses polarity like a compass needle. When suspended by a string, it will twist around until its south pole points to the Earth's North Pole. The paper clip has become magnetized and has tiny magnetite crystals attached. (*Copyright Runk / Shoenberger / Grant Heilman.*)

or ion is produced by the spin of its electrons. When magnetite takes on its remanent magnetism, the iron ions align themselves within the crystal lattice so that their magnetic moments are parallel. Most igneous rocks crystallize at temperatures in excess of 900°C. At these extreme temperatures, the atoms in the minerals have a large amount of energy and are being violently shaken. The vibrating, shaking atoms are unable to line up until more cooling occurs. Finally, when the Curie temperature is reached (578°C for magnetite), enough energy has been lost to allow the atoms to come into alignment. They will stay aligned at still lower temperatures because the Earth's field is not strong enough to alter the alignment already frozen into the minerals.

Igneous rocks are not the only kinds of Earth materials that can acquire remanent magnetism. In lakes and seas that receive sediments from the erosion of nearby land areas, detrital grains of magnetite settle slowly through the water and rotate so that their directions of magnetization parallel the Earth's magnetic field. They may continue to move into alignment while the sediment is still wet and uncompacted, but once the sediment is cemented or compacted, the **depositional remanent magnetism** is locked in.

Over the past 3 decades, geophysicists have been measuring and accumulating paleomagnetic data for all the major divisions of geologic time. Their results have been responsible for continuing interest in continental drift hypotheses. For example, when ancient pole positions were located on maps, it appeared that they were in different positions relative to a particular continent at different periods of time in the geologic past. Either the poles had moved relative to stationary continents or the poles had remained in fixed positions while the continents shifted about. If the poles were wandering and the continents "stayed put," then a geophysicist working on the paleomagnetism of Ordovician rocks in France, for example, should arrive at the same location for the Ordovician poles as a geophysicist doing similar work on Ordovician rocks from the United States. In short, the paleomagnetically determined pole positions for a particular age should be the same for all continents. On the other hand, if the continents had moved and the poles were fixed, then we should find that pole positions for a particular geologic time would be different for different continents. The data suggest that this latter situation is the more valid.

Another way to view the data of paleomagnetism is to examine what are called **apparent polar wandering curves**. The word *apparent* in the expression is necessary, for polar wandering curves alone might imply that the poles wander. As we have just noted, this is highly unlikely. The apparent polar wandering curves are merely lines on a map connecting ancient pole positions relative to a specific continent for various times during the geologic past. As shown in Figure 5-32, the curves for North America and Europe met in recent time at the present North Pole. This means that the

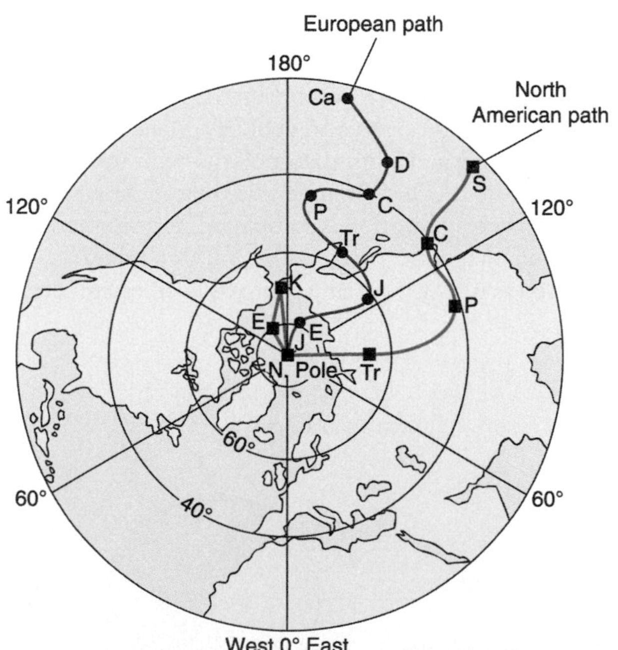

FIGURE 5-32 **Apparent paths of polar wandering for Europe and North America.** A scatter of points have been averaged to a single point for each geologic period: Ca, Cambrian; S, Silurian; D, Devonian; C, Carboniferous; P, Permian; Tr, Triassic; J, Jurassic; K, Cretaceous; E, Eocene. (*After Bott, M. H. P. 1971. The Interior of the Earth: Structure and Processes. New York: St. Martin's Press.*)

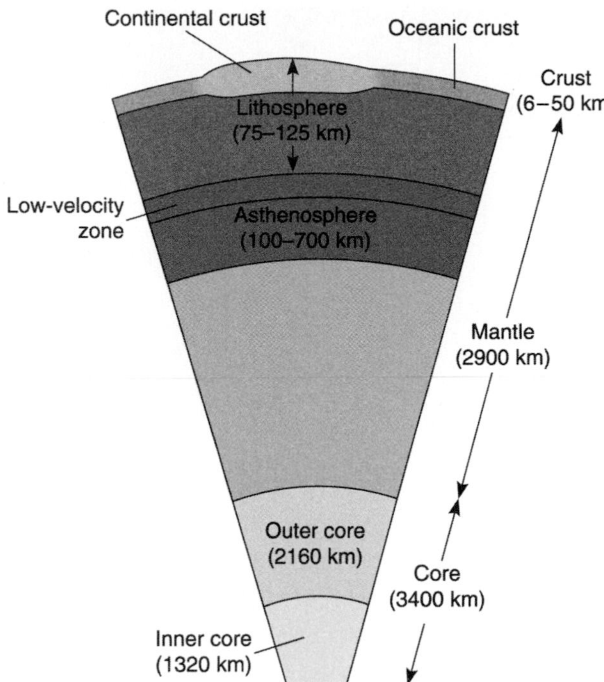

FIGURE 5-33 Divisions of the Earth's interior. (For clarity, the divisions have not been drawn to scale.)

paleomagnetic data from recently formed rocks from both continents indicate the same pole position. A plot of the more ancient poles results in two similarly shaped but increasingly divergent curves. If this divergence resulted from a drifting apart of Europe and North America, then one should be able to reverse the movements mentally and see if the curves do not come together. Indeed, the Paleozoic portions of the polar wandering curves could be brought into close accord if North America and its curve were to be slid eastward about 30° toward Europe. This sort of information gave new life to the old notion of continental drift and also supported the newer concept of tectonic plate movements.

The Basic Concept of Plate Tectonics

The basic elements of plate tectonics were described in the first chapter of this text. We noted that the **lithosphere**, or outer shell of the Earth (Fig. 5-33), is constructed of seven huge slabs and about 20 smaller plates that are squeezed in between them. The larger plates (Fig. 5-34) are approximately 75 to 125 kilometers thick. Movement of the plates causes them to

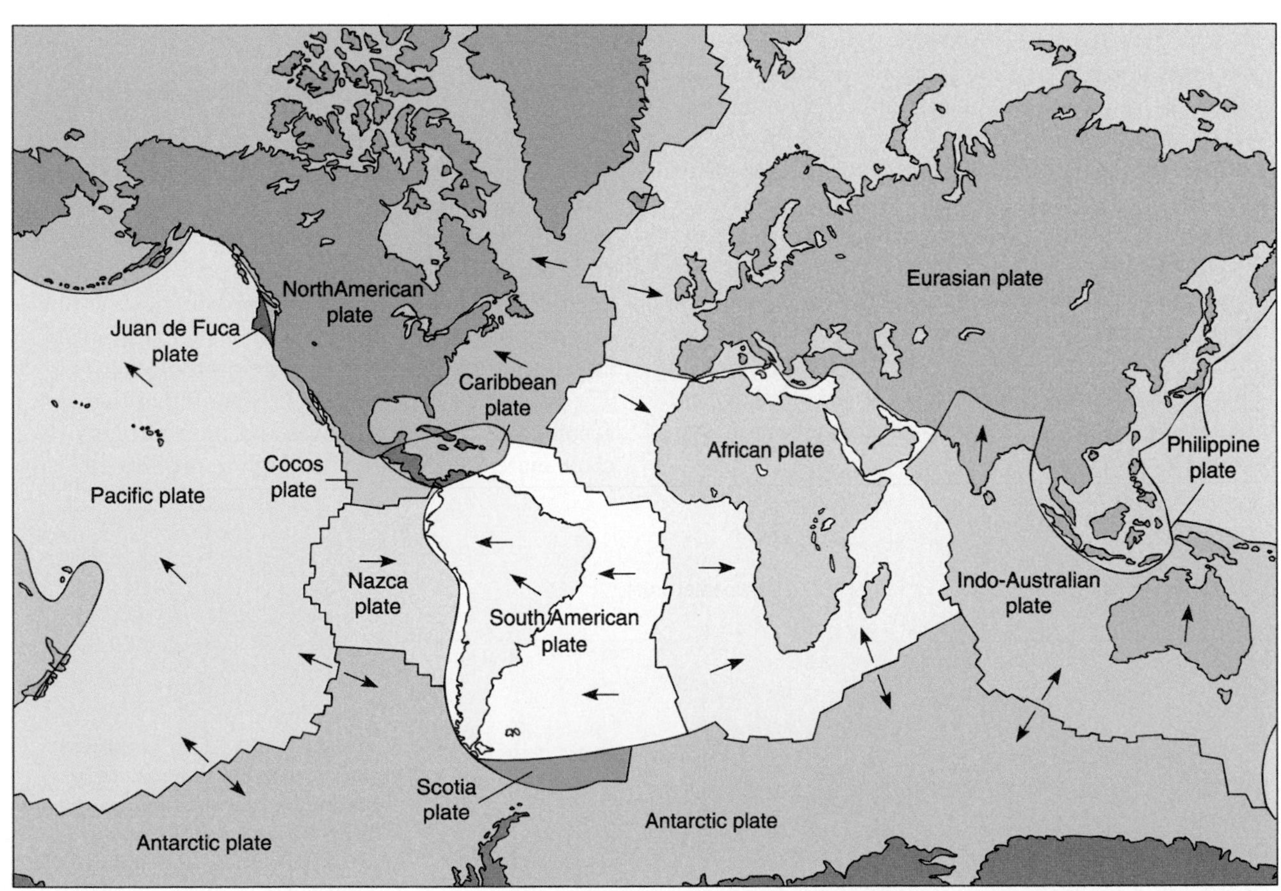

FIGURE 5-34 The Earth's major tectonic plates. *Arrows* indicate general direction of movement of plates.

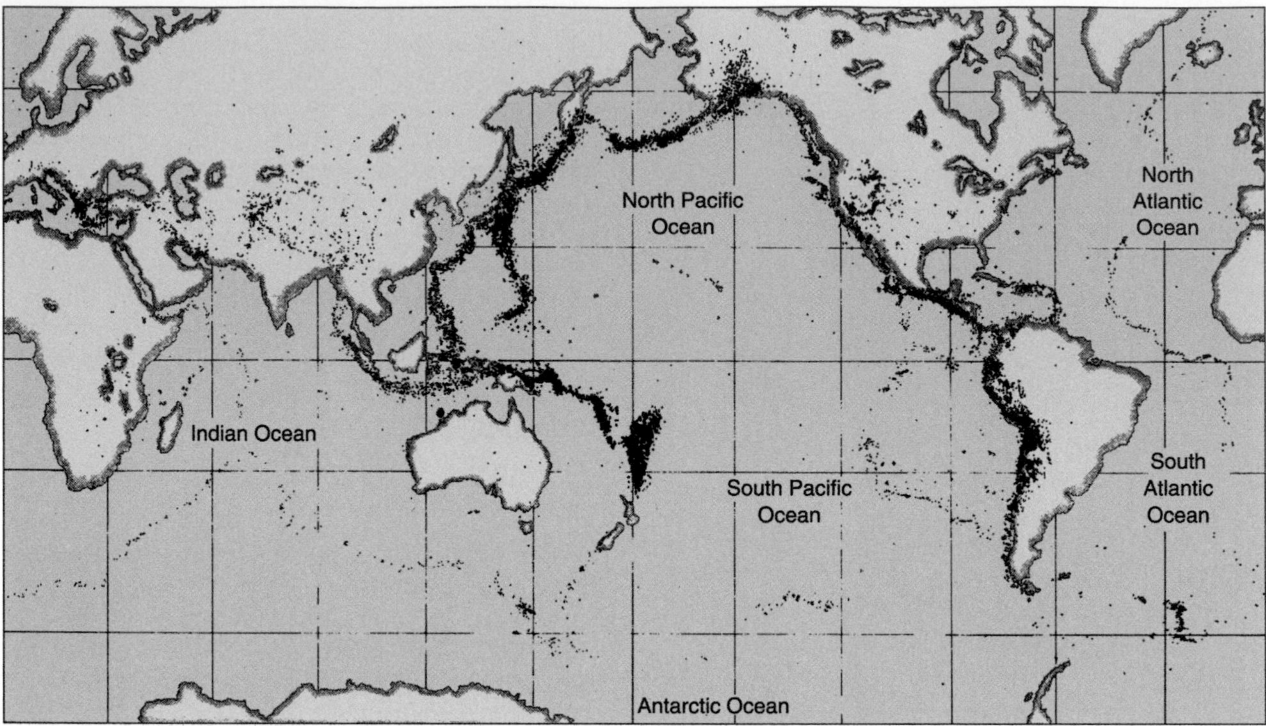

FIGURE 5-35 World distribution of earthquakes. Notice the major earthquake belt that encircles the Pacific and another that extends eastward from the Mediterranean toward the Himalayas and East Indies. The midoceanic ridges are also the site of many earthquakes.

converge, diverge, or slide past one another, and this results in frequent earthquakes along plate margins. When the locations of earthquakes are plotted on the world map, they clearly define the boundaries of tectonic plates (Fig. 5-35).

A part of a plate containing a continent would have the configuration shown in Figure 5-36. Lithospheric plates "float" on the weak, partially molten region of the upper mantle defined earlier as the **asthenosphere** (from the Greek *asthenos*, meaning "weak"). The asthenosphere is a region of rock plasticity and flowage detected on the basis of changes in seismic wave velocities.

Plate Boundaries and Sea-Floor Spreading

Central to the idea of plate tectonics is the differential movement of lithospheric plates. For example, plates move apart at **divergent plate boundaries**, which may manifest themselves as midoceanic ridges complete with tensional ("pull-apart") geologic structures. Indeed, the mid-Atlantic ridge approximates the line of separation between the Eurasian and African plate on one side and the American plate on the other (Fig. 5-37). As is to be expected, such a rending of the crust is accompanied by earthquakes and outpourings of volcanic materials that are piled high to produce the ridge

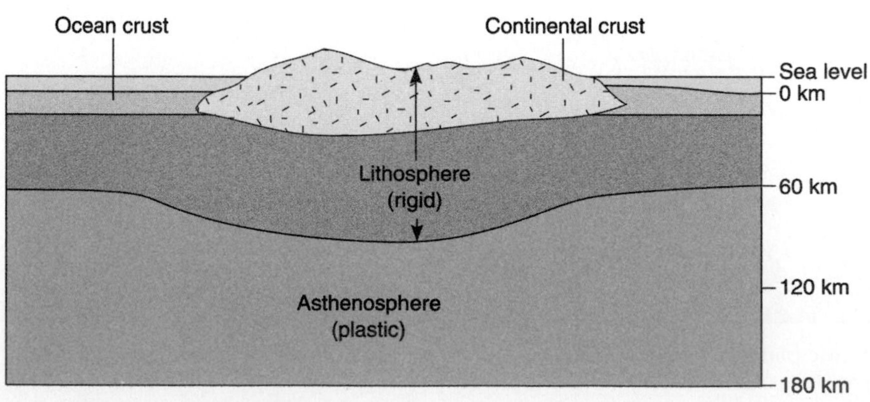

FIGURE 5-36 The outermost part of the Earth consists of a strong, relatively rigid lithosphere, which overlies a weak, plastic asthenosphere. The lithosphere is capped by a thin oceanic crust and a thicker continental crust.

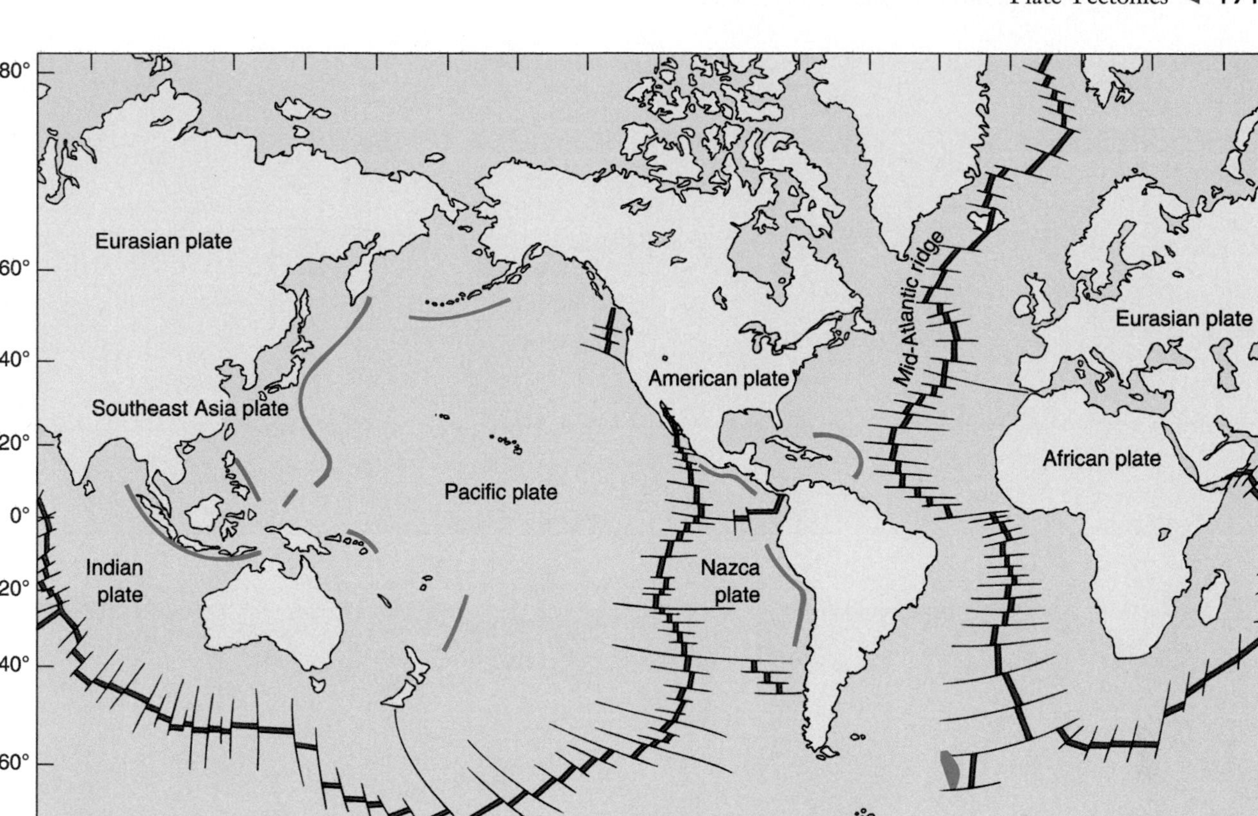

FIGURE 5-37 Locations of midoceanic ridges (red) and trenches (blue). Note fault zones offsetting the ridges. Dashed lines in Africa represent the East African rift zone. (*After Isacks, B. et al. 1968. J. Geophys. Res. 73:5855–5899. Copyright, The American Geophysical Union.*)
? *What are these faults called (see Fig. 5-39)? What does the location of Iceland relative to the mid-Atlantic ridge indicate about its origin?*

itself. The void between the separating plates is also filled with this molten rock, which rises from below the lithosphere and solidifies in the fissure. Thus, new crust (that is, *new sea floor*) is added to the **trailing edge** of each separating plate as it moves slowly away from the midoceanic ridge (Fig. 5-38). Zones of divergence may originate beneath continents, rupturing the overlying land mass and producing such rift features as the Red Sea, the Gulf of Aden, and the East African rift valleys. Because the trailing edge of a tectonic plate is not involved in encounters with other plates, it is also referred to as the plate's **passive margin**.

The process by which new oceanic crust is produced at mid-oceanic ridges and moved laterally away from the ridge is termed **sea-floor spreading**. Recognition of sea-floor spreading came as the result of several observations made by Harry H. Hess of Princeton University. During World War II, Hess commanded an attack transport in the Pacific. To determine water depth just offshore of enemy-held islands, the ship had been equipped with an echo sounder. When not employed for a military purpose, Hess used the device to produce topographic profiles of the ocean floor. The profiles showed midoceanic ridges, deep-sea trenches, and smaller features of the ocean floor. Among the latter were submerged mountains with flat, rather than conical, summits. These were later named **guyots**, after Arnold Guyot of Princeton University. Hess noticed that guyots located increasingly farther away from midoceanic ridges lie at increasingly greater depths. This led to the hypothesis that guyots were once volcanoes that were formed at the midoceanic ridges by upwelling lavas, that they subsequently were truncated by erosion at the ridge, and that they subsequently drifted away from the ridge while simultaneously sinking. Hess envisioned the ocean floors as continuously moving conveyor belts that moved from midoceanic ridges, where new floor was added by upwelling lavas, to deep-sea trenches, where the crust plunged downward to be consumed in the mantle. This spreading mechanism would account for not only the characteristics of guyots but also the relative thinness of the oceanic crust, the absence of oceanic crust much older than about 200 million years, and the absence of pre-Jurassic sediment above the basalt.

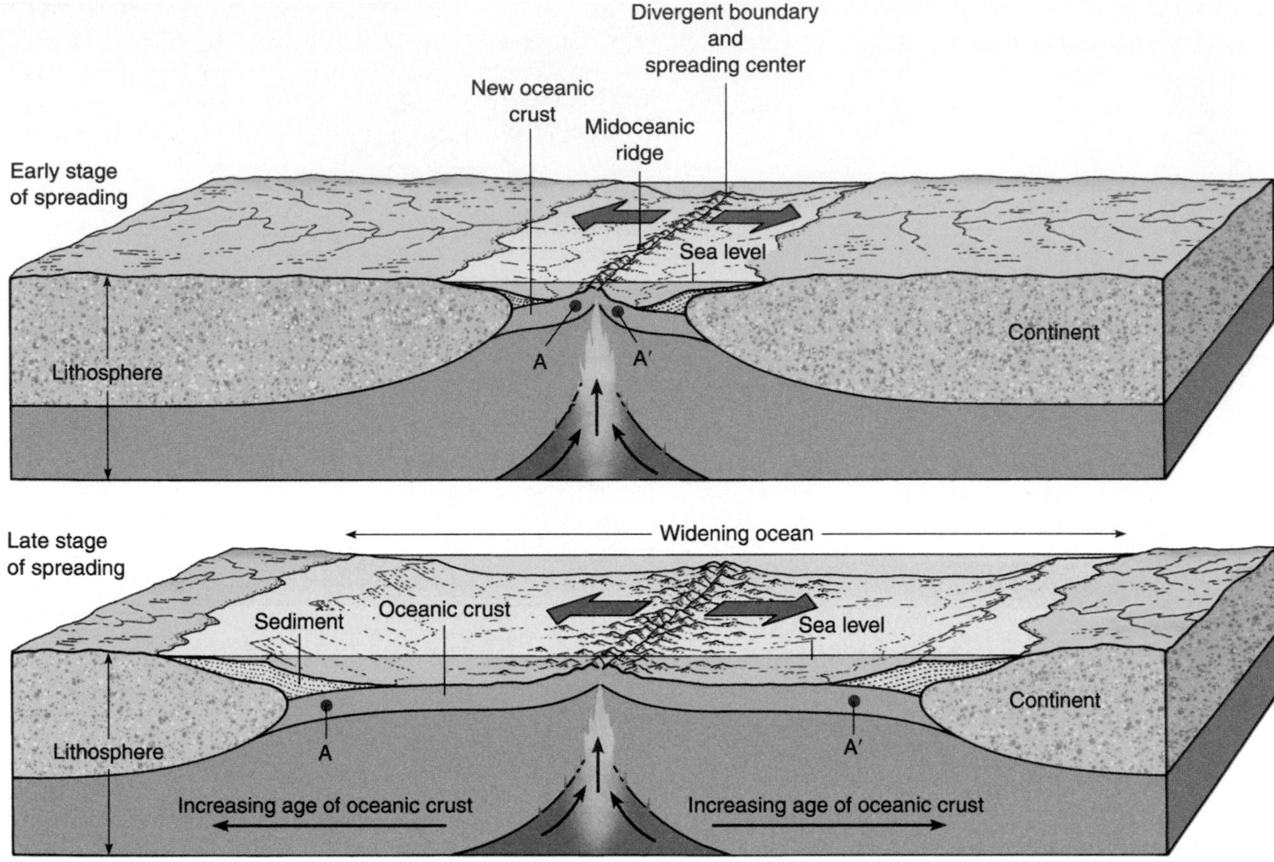

FIGURE 5-38 Sea-floor spreading. As the rift widens and newly formed crust moves away from the ridge axis, the crust stretches and fractures. Some blocks of fractured crust may drop while others remain elevated. Basaltic lavas rise to fill the lower elevations and the space between separating plates. As the lavas solidify, they become part of the trailing edge of a tectonic plate. **A** and **A′** are reference points.

The line along a midoceanic ridge where spreading begins is not straight or smoothly curving. Rather, it is offset by numerous faults. These features are called **transform faults** and are an expected consequence of horizontal spreading of the sea floor along the Earth's curved surface. The relative motions of transform faults are shown in Figure 5-39. The ridge acts as a spreading center that exists both to the north and to the south of the fault. The rate of relative movement on the opposite sides along fault segment X to X′ depends on the rate of extrusion of new crust at the ridge. Because of the spreading of the sea floor outward from the ridge, the relative movement between the offset ridge crests is opposite to that expected by ordinary fault

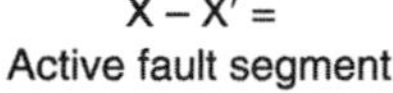

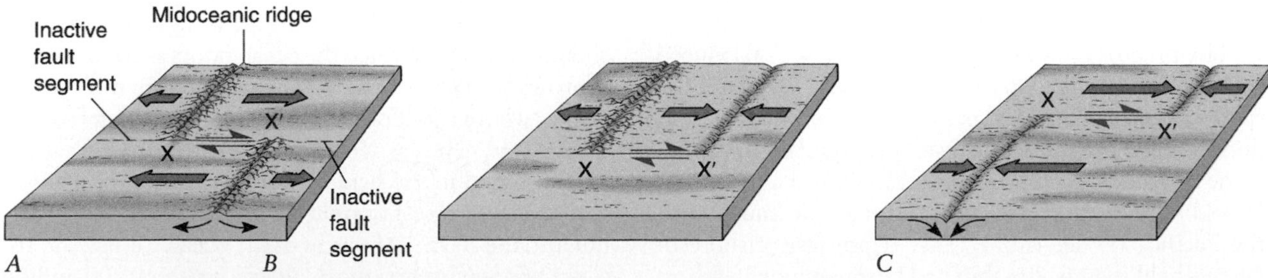

FIGURE 5-39 Three types of transform faults: (*A*) ridge-ridge transform, (*B*) ridge-trench, and (*C*) trench-trench.

movement. Thus, at first glance, the ridge-to-ridge transform fault (Fig. 5-39*A*) appears to be a left-lateral fault, but the actual movement along segment X-X′ is really right lateral. Notice that only the X-X′ segment shows movement of one side relative to the other. To the west of X and the east of X′, there is little or no relative movement.

If plates are receding from one another at one boundary, they may be expected to collide or slide past other plates at other boundaries. Thus, in addition to the divergent plate boundaries that occur along mid-oceanic ridges, there are **convergent** and **transform boundaries**. Convergent plate boundaries develop when two plates move toward one another and collide. As one might guess, these convergent junctions are characterized by a high frequency of earthquakes. In addition, they are the zones along which compressional mountain ranges or deep-sea trenches may develop (Fig. 5-40). The structural configuration of the convergent boundary is likely to vary according to the rate of spreading and whether the leading edges of the plates are composed of oceanic or continental crust. When the plates collide, one slab may slip and plunge below the other, producing what is called a **subduction zone**. The sediments and other rocks of this plunging plate are pulled downward (subducted), melted at depth, and, much later, rise to become incorporated into the materials of the upper mantle and crust. In some instances, the silicate melts provide the lavas for chains of volcanoes.

An example of a transform plate boundary (also called a shear boundary) is the well-known San Andreas fault in California. Along this great fault, the Pacific plate moves laterally against the American plate (Fig. 5-41). Transform plate boundaries are decidedly earthquake prone but are less likely to develop deep earthquakes or intense igneous activity. They are the active segments of transform faults along which no new surface is formed or old surface consumed.

Crustal Behavior at Plate Boundaries

We have noted that there are three basic kinds of plate boundaries: convergent, divergent, and transform (Plate 5-1). There are also three kinds of convergent boundaries that produce styles of deformation. The kind of deformation is determined primarily by the nature of the crust at the margins of the converging plates. If the leading edge of a plate is composed of continental crust and it collides with an opposing plate of similar composition, the result would be a folded mountain range having a core of granitic igneous rocks (see Fig. 5-40*A*). Because continental plate margins are too light and buoyant to be carried down into the asthenosphere, subduction does not occur in this type of continent-to-continent collision. Instead, the crust at

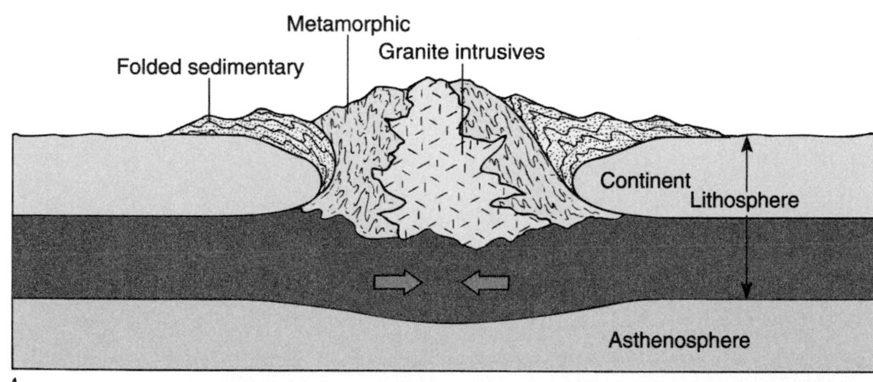

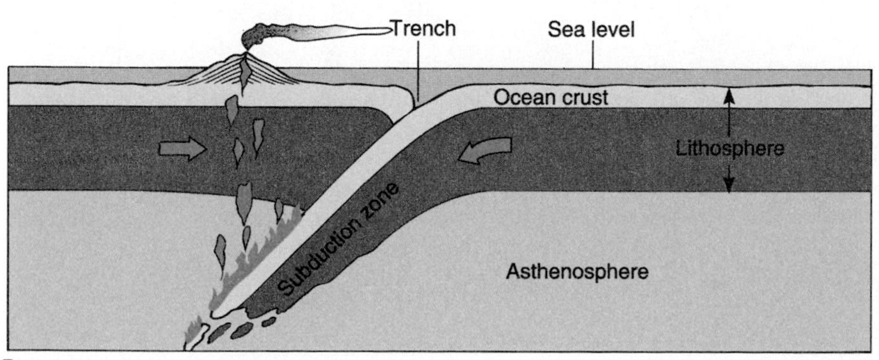

FIGURE 5-40 Two types of convergent plate boundaries. (*A*) Convergence of two plates, both bearing continents. (*B*) Convergence of two plates, both bearing oceanic crust. For convergence of an oceanic plate with a continental plate, see Figure 5-45.

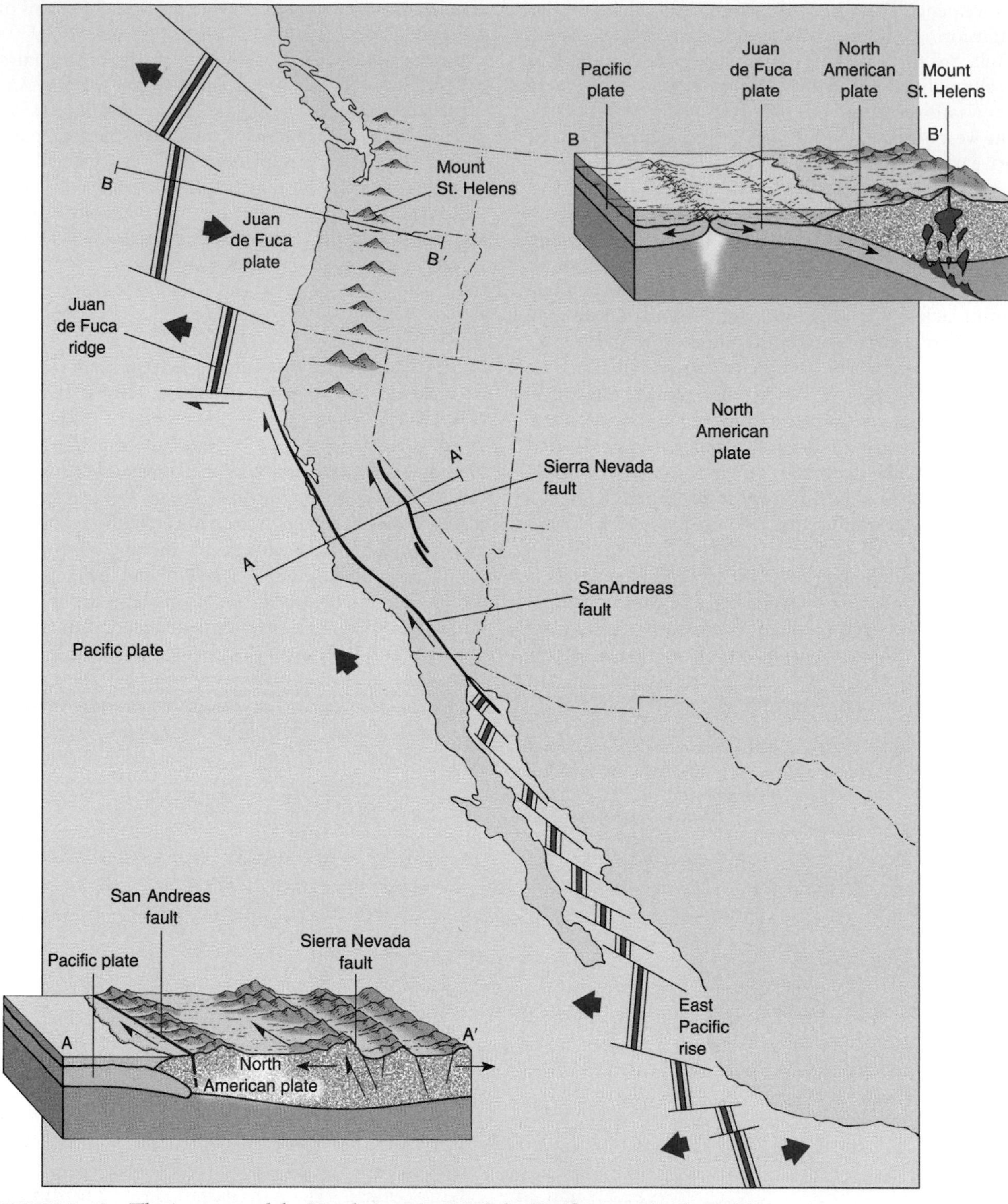

FIGURE 5-41 The juncture of the North American and the Pacific tectonic plates. The double lines are spreading centers. Note the trace of the San Andreas fault. To the north in Oregon and Washington, the small Juan de Fuca plate plunges beneath the North American continent to form the Cascades. (*Courtesy of U.S. Geological Survey*.) ? *Is the San Andreas fault a right- or left-lateral fault? Can it also be considered a transform fault*

FEATURES AND EVENTS AT THE BOUNDARIES OF TECTONIC PLATES

BOUNDARY	OPPOSING PLATE MARKINGS	GEOLOGIC EVENTS	TOPOGRAPHIC FEATURES	MODERN EXAMPLES
	Continent-Continent	Crustal deformation and mountain building	Mountains	Himalayas
CONVERGENT	Continent-Ocean	Subduction deformation, volcanism	Mountains ocean trenches	Andes
	Ocean-Ocean	Subduction deformation, volcanism	Island arcs ocean trenches	Western Aleutions
DIVERGENT	Continent-Continent	Continents fragment, magma rises beneath rift, volcanism	Rift valleys	East African Rift
	Ocean-Ocean	Sea floor spreading submarine volcanism	Mid-oceanic ridge	Mid-Atlantic Ridge
TRANSFORM	Continent-Continent	Crustal deformation	Crustal deformation along fault	San Andreas Fault
	Ocean-Ocean	Earthquakes	Offset of axes of mid-oceanic ridges	Offset, East Pacific Rise

Earthquakes

PLATE 5-1 Features and Events at the Boundaries of Tectonic Plates

the plate margins is deformed and may detach itself from deeper zones, and slabs of continental crust from one plate may ride up over the other. The zone of convergence between the two plates, recognized by the severity of folding, faulting, and intrusive activity, is termed the **suture zone**.

A dramatic and ongoing example of a collision between two continents is provided by the convergence of India and Eurasia (Fig. 5-42). The Himalayas are the spectacular result of that collision. The process of Himalayan mountain building began late in the Mesozoic. At that time, the leading (northern) edge of the Indian tectonic plate was being subducted beneath the southern margin of Eurasia (Fig. 5-43*A*). To the south of the subduction zone, the plate carried the continental mass of India. As the sea floor was being gobbled up at the subduction zone, the oceanic tract known as the **Tethys**, which separated Eurasia from northward-moving India, narrowed until the land masses collided (Fig. 5-43*B*). Sediments caught in the vise between the two continents were intensely deformed and pushed northward along great faults. Eventually, the northern border of India was firmly sutured against the buttress of Tibet (Fig. 5-43*C*). However, northward movement of the plate did not cease. Affected by continuing compression, slabs of crust were forced beneath Tibet along great thrust faults. Some of the Asian crust responded by breaking into lateral faults that are still actively transmitting the energy of this ongoing northerly convergence.

The amount of crustal shortening resulting from the India-Eurasia convergence has been estimated to be 1500 kilometers. Geologists can account for 500 to 1000 kilometers of that amount in the folds and thrust faults of the suture zone. The remaining 500 to 1000 kilometers was apparently dissipated along major east-west trending lateral faults in China and Mongolia (Fig. 5-44). Thus, plate convergence may continue long after initial closure by lateral release of plate movement along strike-slip faults (lateral faults) in the region peripheral to the suture zone. Today geologists are finding evidence of similar strike-slip faulting in older continent-continent convergences such as those that produced the Appalachian and Ural mountains.

The second kind of convergent plate boundary involves the meeting of two plates that both have oceanic

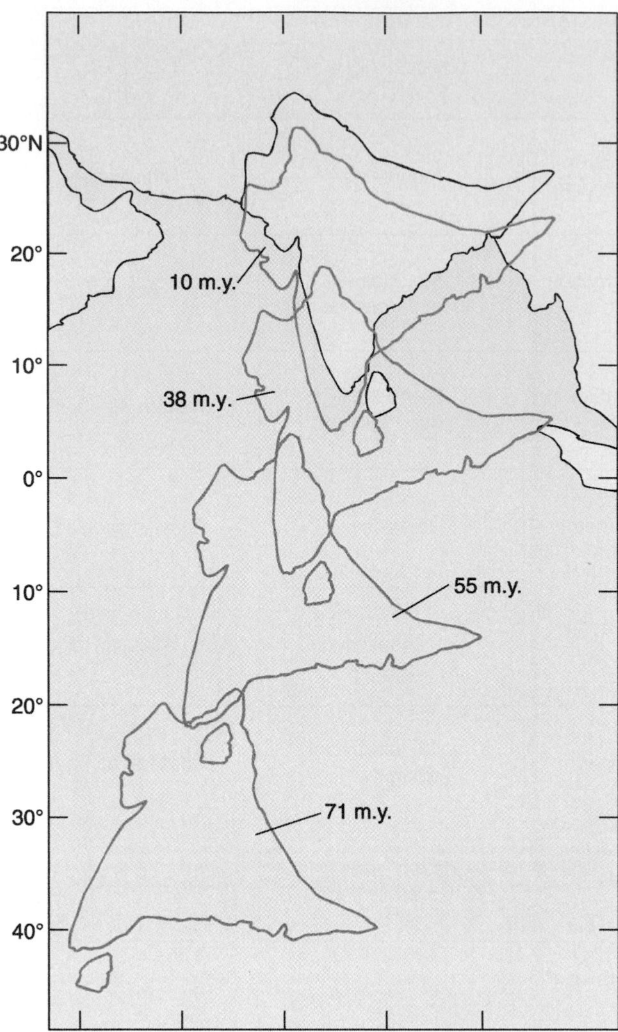

FIGURE 5-42 The northward migration of India relative to Eurasia, showing the position of the continent 71 million years (m.y.) ago (Cretaceous), 55 million years ago (early Eocene), 33 million years ago (Oligocene), and 10 million years ago (late Miocene). The configuration of the northern boundary of India is conjectural. *(From Molnar, P., and Tapponnier, P. 1975. Cenozoic tectonics of Asia: Effects of a continental collision. Science 189(4201):419–426. Copyright 1975, AAAS.)*

crust at their converging margins (see Fig. 5-40*B*). Although the rate of plate movement in an ocean-ocean convergence will affect the kinds of structures produced, it is likely that such locations will develop deep-sea trenches with bordering volcanic arcs such as those of the southwestern Pacific. Between the island arc and the deep-sea trench is an area called the **forearc.** The forearc includes an elongate, depressed **forearc basin** in which sediment and materials scraped off the descending plate may accumulate as an **accretionary prism.** Accretionary prisms tend to grow (by accretion) as more and more material is scraped off the descending plate and plastered onto the mass already in the prism.

Finally, there is the third possibility, which involves the collision of a continental (granitic) plate boundary with an oceanic (basaltic) one. The result of such a collision is a deep-sea trench located offshore from an associated range of mountains. Volcanic activity accompanying subduction would provide a compositional blend of granititic and basaltic lavas. The resulting rock is andesite, named for its prevalence in the Andes.

Regions of ocean-continent convergence (Fig. 5-45) are characterized by rather distinctive rock assemblages and geologic structure. As we have seen, the convergence of two lithospheric slabs results in subduction of the ocean plate, whereas the more buoyant continental plate maintains its position at the surface but experiences intense deformation, metamorphism, and melting. Accretionary prisms develop landward of the deep-sea trench, and great mountain ranges take form. At the same time, sediments and submarine volcanic rocks along the subduction zone are squeezed, sheared, and shoved into a gigantic, chaotic medley of complexly disturbed rocks within the accretionary wedge that are termed a **mélange**. Within the mélange one finds a distinctive assemblage of deep-sea sediments containing microfossils, submarine lavas, and serpentinized peridotite possibly derived from the upper mantle. Together, these rocks constitute an **ophiolite suite** (Fig. 5-46). The ophiolite suites are splinters of the oceanic plate that were scraped off the upper part of the descending plate and inserted into the accretionary wedge at the margin of the continent. Thus, ophiolites mark the zone of contact between colliding continental and oceanic plates. Another clue to the presence of such zones is a distinctive kind of metamorphic rock containing blue amphiboles. These rocks are called blue schists. They form at high pressures but relatively low temperatures. This rather unusual combination of high pressure and relatively low temperatures characterizes subduction zones where the relatively cool oceanic plate plunges rapidly into zones of high pressure.

Wilson Cycles

Plate tectonics has been in operation since late Archean time and possibly longer. Over this immense interval there have been many openings and closings of ocean basins. Indeed, plate tectonics controls the birth and death of ocean basins. The Pacific Ocean opened only about 300 million years ago. If rates and directions of sea floor spreading remain the same, it will be closed in another 200 million years. The opening of a new ocean basin along divergent zones, the expansion of the basin as sea floor spreading continues, and the ultimate closure of the basin as plates converge is termed a **Wilson Cycle**. The name honors J. Tuzo Wilson for his contributions to the theory of plate tectonics.

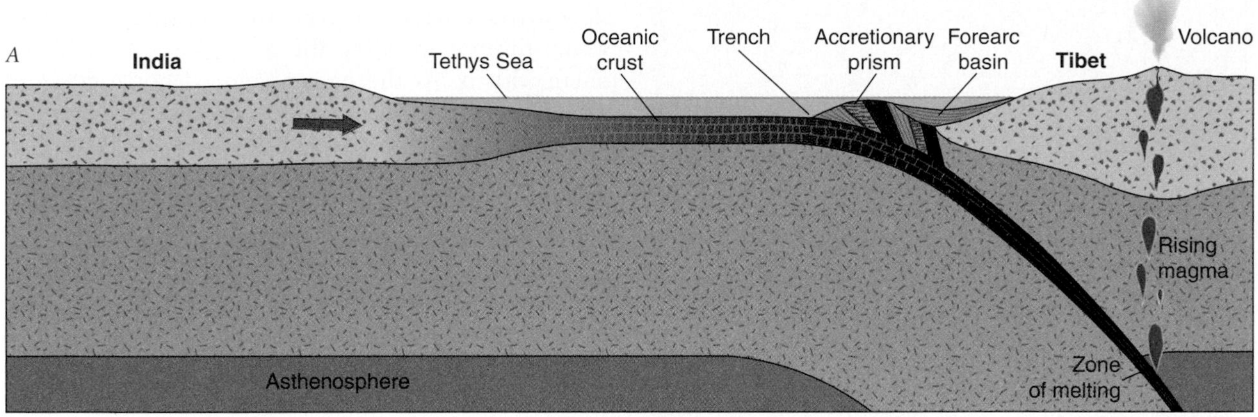

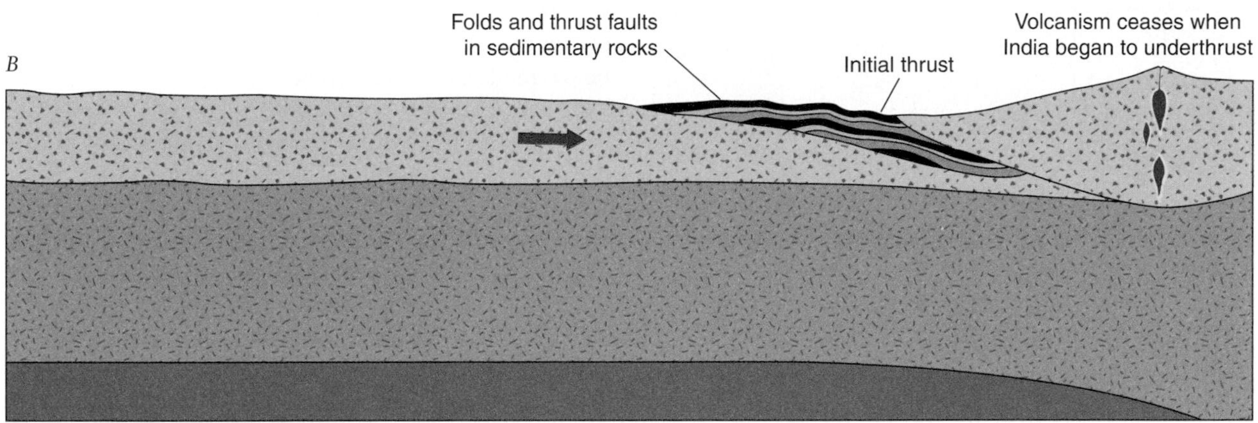

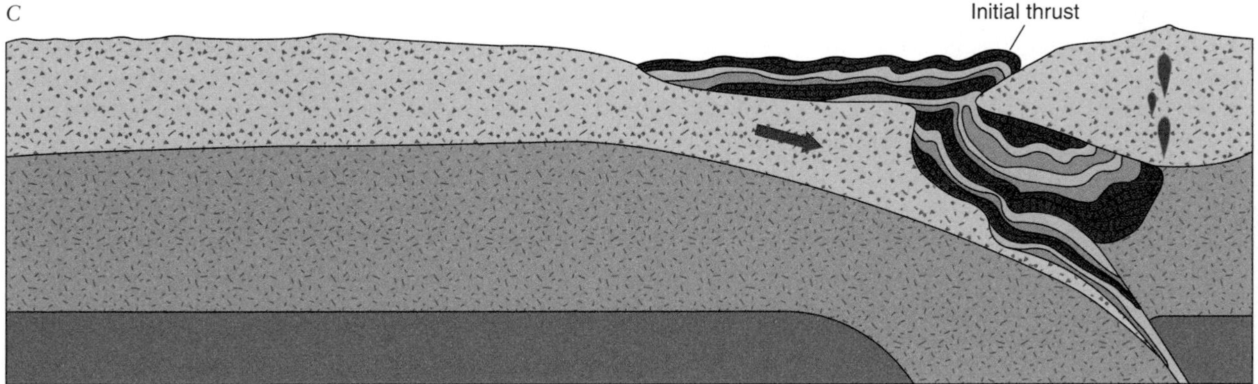

FIGURE 5-43 Cross-sections depicting how the Himalayas formed when the leading margin of the Indian plate subducted beneath the margin of Eurasia. (*A*) India moving northward as subduction zone develops at the southern margin of Eurasia. (*B*) By about 40 million years ago India had collided and the leading edge of India was thrust beneath southern Tibet. (*C*) Continued underthrusting and compression crushes Tibet and forms the high Himalayas. (*After Thompson, G.R., and Turk, J. 1997. Modern Physical Geology. Philadelphia: Saunders College Publishing.*)

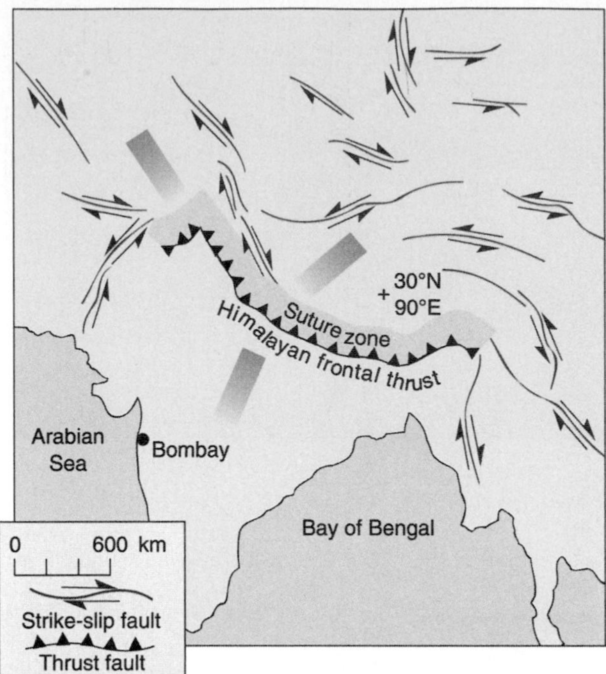

FIGURE 5-44 Major lateral (strike-slip) faults above and adjacent to the Indian suture zone. These faults may represent lateral transfer of some of the movement involved in the India-Eurasia convergence. Geologically recent movement along the faults implies that the plate movement involved in the convergence is still in progress. (*Fault locations from Molnar, P., and Tapponnier, P. 1975. Cenozoic tectonics of Asia: Effects of a continental collision. Science 189(4201):419–426. Copyright 1975, AAAS.*)

What Drives It All?

If we accept the fact that plates of lithosphere move across the surface of the globe, the next question concerns the cause of that movement. An early hypothesis was that the propelling mechanism consists of large thermal convection cells produced as mantle material is heated from below, expands, becomes less dense, and slowly rises. On encountering the lithosphere, the flow would diverge and drag the overlying slab of lithosphere with it. As the current moved horizontally, it would lose heat and become denser. Ultimately, it would encounter an opposing current, and both viscous streams would descend to be reheated and shunted toward a region of upwelling. Above the descending flow one would find subduction zones and deep-sea trenches. Midoceanic ridges would mark the location of the ascending flow.

Although convection-induced drag at the base of lithospheric plates may account for part of the forces needed to move tectonic plates, recent calculations indicate the force supplied by convection cells to the lithosphere may not be sufficient. Furthermore, the hypothesis does not take into account other forces acting on the plates. Many geologists now favor what can be termed the **ridge-push** and **slab-pull model.** Ridge-push forces arise from the fact that spreading centers, such as midoceanic ridges, stand high on the ocean floor and have low-density roots. Their elevation above adjacent regions of the ocean floor provides a tendency for the ridge material to slide downslope, thereby transmitting a push to the tectonic plate. At the same time, the mechanism of slab-pull operates at the subduction zones. There the subducting oceanic plate, being relatively cool and dense, sinks through the less-dense mantle, pulling the rest of the slab along as it does so (Fig. 5-47).

Thermal Plumes

Regardless of the mechanism for moving lithospheric plates, there is ample geophysical evidence of the existence of convection cells in the mantle. Mantle material circulates in great rolls and also rises from near the core-mantle boundary in a manner suggestive of a

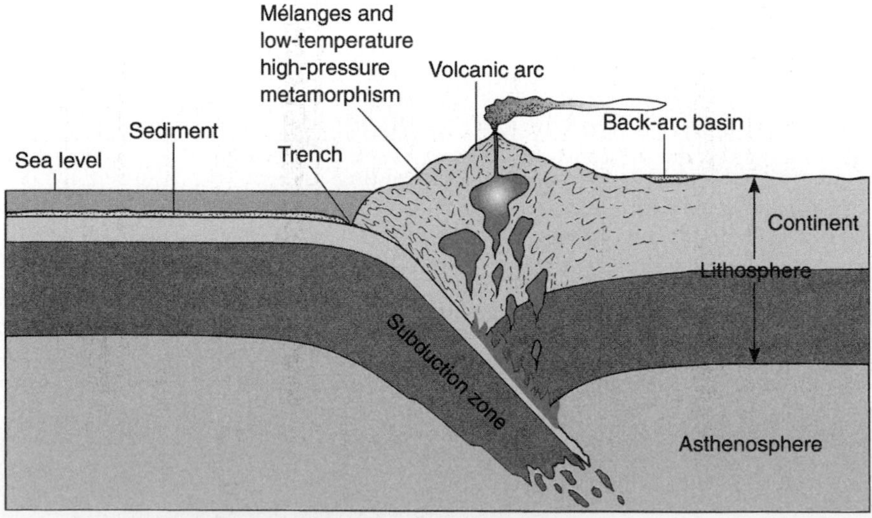

FIGURE 5-45 Convergence of the continental part of a plate with the oceanic part of another plate. The leading edge of the continental plate is crumpled, whereas the oceanic plate buckles downward, creating a trench offshore. The situation generalized here is similar to that off the west coast of South America, where the Nazca plate plunges beneath the South American plate.

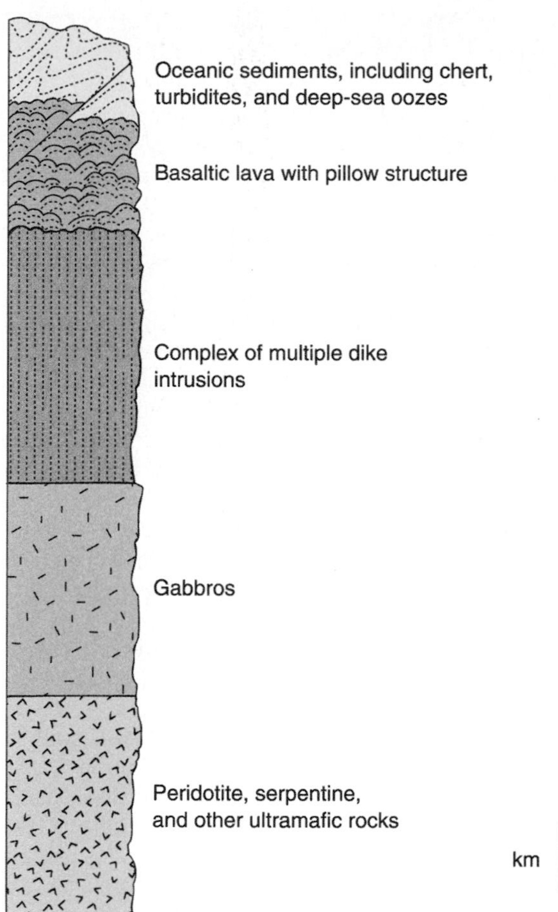

FIGURE 5-46 Idealized section of an ophiolite suite. Ophiolites are thought to be splinters of the ocean floor squeezed into the continental margin during plate convergence.

thundercloud. Such a configuration is termed a **thermal plume**. When a plume nears the lithosphere, it spreads laterally, doming the overlying plate and moving the rifted segments outward from the central area. As indicated in Figure 5-48, uplift results in a triple junction having three radiating fractures. As in the Afar triangle of Ethiopia, two of the fractures open to form narrow oceanic tracts (the Red Sea and Gulf of Aden), whereas the third becomes a fault-bound trough called a "failed arm" that fills with sediment. The failed arm is more formally known as an **aulacogen** (literally, "born as a furrow").

Tests of Plate Tectonics

PALEOMAGNETIC CLUES We have examined the various lines of evidence supporting Wegener's notion of continental drift. Nearly all of this evidence also supports the newer concepts embodied in the theory of plate tectonics. The earlier clues were based mostly on evidence found on land. The new theory, with its keystone concept of sea-floor spreading, was developed from evidence gleaned from the sea floor.

In the 1960s, when geophysicists such as Harry Hess began formulating ideas of sea-floor spreading, other scientists were puzzling over findings related to paleomagnetism. The new data were obtained from sensitive magnetometers that were being carried back and forth across the oceans by research vessels. These instruments were able to detect not only the Earth's main geomagnetic field but also local magnetic disturbances or *magnetic anomalies* frozen into the rocks along the sea floor. Maps produced from data obtained during traverses across the mid-Atlantic ridge exhibited mirror-image sets of bands of high- and low-field magnetic intensities (Fig. 5-49). In 1963, F. J. Vine, a research student at Cambridge University, and a senior colleague, Drummond Matthews, suggested that these variations in magnetic intensity were caused by reversals in the polarity of the Earth's magnetic field. The magnetometers towed behind the research vessels provided measurements that were the sum of the Earth's present magnetic field strength and the paleomagnetism frozen in the crustal rocks of the ocean floor. If the paleomagnetic polarity was opposite in sign to that of the Earth's present magnetic field, the sum would be

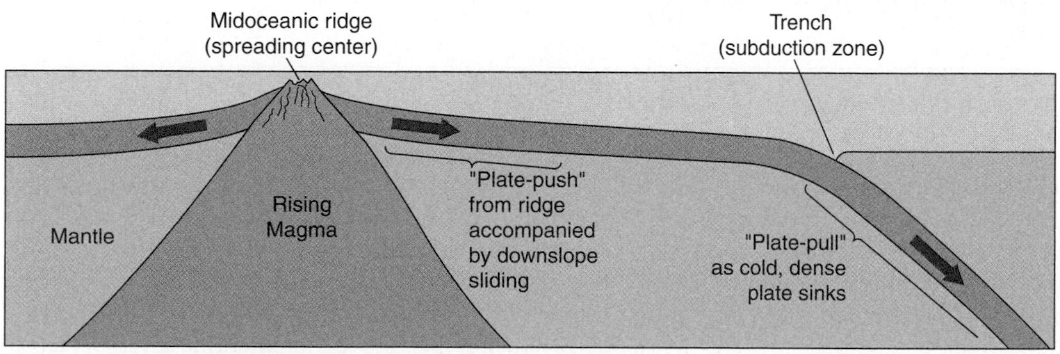

FIGURE 5-47 The push-pull mechanism of plate movement. Near the spreading center, the plate glides down the inclined surface on the asthenosphere, pushing the plate forward. At the subduction zone, the cool, dense plate sinks, pulling the rest of the plate downward with it.

ENRICHMENT

The Geosynclinal Hypothesis Displaced by Plate Tectonics

Long before geologists had acquired an understanding of plate tectonics, they struggled to find an explanation for the origin of mountains. Great mountain systems of the world are predominantly fold mountains that include ranges that parallel present or past continental boundaries. In their complex of anticlines, synclines, and thrust faults, they exhibit ample evidence of crustal compression. A hypothesis called the "geosynclinal hypothesis for mountain building" seemed to provide an explanation for how mountains develop. The hypothesis had its inception with the American geologist James Hall (1811–1898). As described in Chapter 1, Hall realized that the immense thickness of shallow water sedimentary layers in the Appalachians could not have accumulated unless the basin that received the sediment was sinking. Hall did not provide a name for the elongate subsiding basin that received the vast quantities of sediment. It was another famous American geologist, Yale professor James Dana (1813–1895), who suggested such tracts might be termed "geosynclinal." The term **geosyncline** was soon defined as a great elongate trough in the Earth's crust that, over a long span of geologic time, accumulates a huge thickness of sediment that may ultimately be compressed into a major system of mountains. In the 1930s, the German geologist Hans Stille suggested that a geosyncline would include two parallel tracts. The tract lying nearest the continent containing relatively shallow water sediments was designated the **miogeosyncline**, whereas the seaward, more rapidly subsiding, and dynamic tract could be termed the **eugeosyncline** (see accompanying figure). Graywacke sandstones, volcanic rocks, and cherts characterized the rocks of the eugeosyncline and indicated an adjacent volcanic source, perhaps in the form of a volcanic island arc.

The geosynclinal hypothesis postulated an early *depositional stage* during which the geosyncline received sediments from adjoining continents and volcanic island arcs. This was followed by an *orogenic stage* in which the geosyncline would be compressed and its content of sedimentary and volcanic rocks deformed, metamorphosed, and intruded by igneous melts. The orogenic stage would provide the many structural features typifying mountain systems around the world. It would be followed by a long episode of denudation and periodic uplift, constituting the *erosional stage* of geosynclinal history.

Although the geosynclinal hypothesis provided an explanation for most of the features found in mountain ranges, it failed to suggest a truly satisfactory mechanism for the origin of the geosynclinal trough and its subsequent compression. Concepts embodied in the hypothesis, however, facilitated an understanding of what occurs when tectonic plates converge. Science always stands ready to abandon hypotheses not supported by newer discoveries, and the geosynclinal explanation for the origin of mountain systems has now been replaced by modern concepts of plate tectonics.

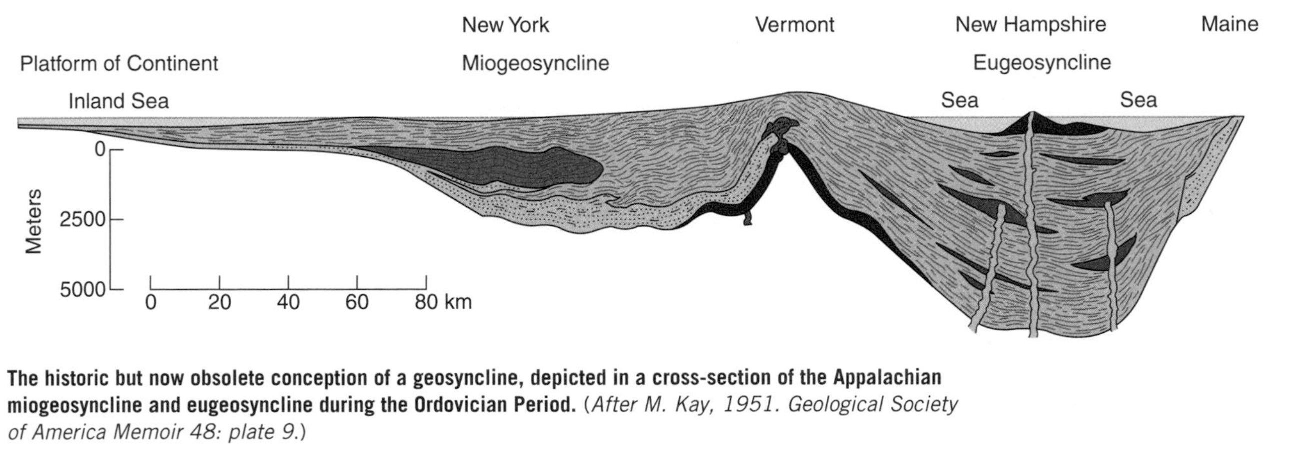

The historic but now obsolete conception of a geosyncline, depicted in a cross-section of the Appalachian miogeosyncline and eugeosyncline during the Ordovician Period. (*After M. Kay, 1951. Geological Society of America Memoir 48: plate 9.*)

less than the present magnetic field strength. This would indicate that the crust over which the ship was passing had reversed paleomagnetic polarity. Conversely, where the paleomagnetic polarity of the seafloor basalts was the same as that of the present magnetic field, the sum would be greater, and a normal polarity would be indicated.

Since 1963, geophysicists have learned that these irregularly occurring reversals of the Earth's magnetic field have not been infrequent. During the past 70 million years, the Earth's magnetic field has seen many episodes when the polarity was opposite to that of today (Fig. 5-50). These changes in polarity are incorporated into the remanent magnetism of the lavas that cooled into basalts at the midoceanic ridges. The lava acquires the magnetic polarity present at the time of extrusion and then moves out laterally, as has been previously described. If during that time the Earth's polarity was as it is today, the stripe is said to represent "normal" polarity. If the Earth's normal polarity reversed,

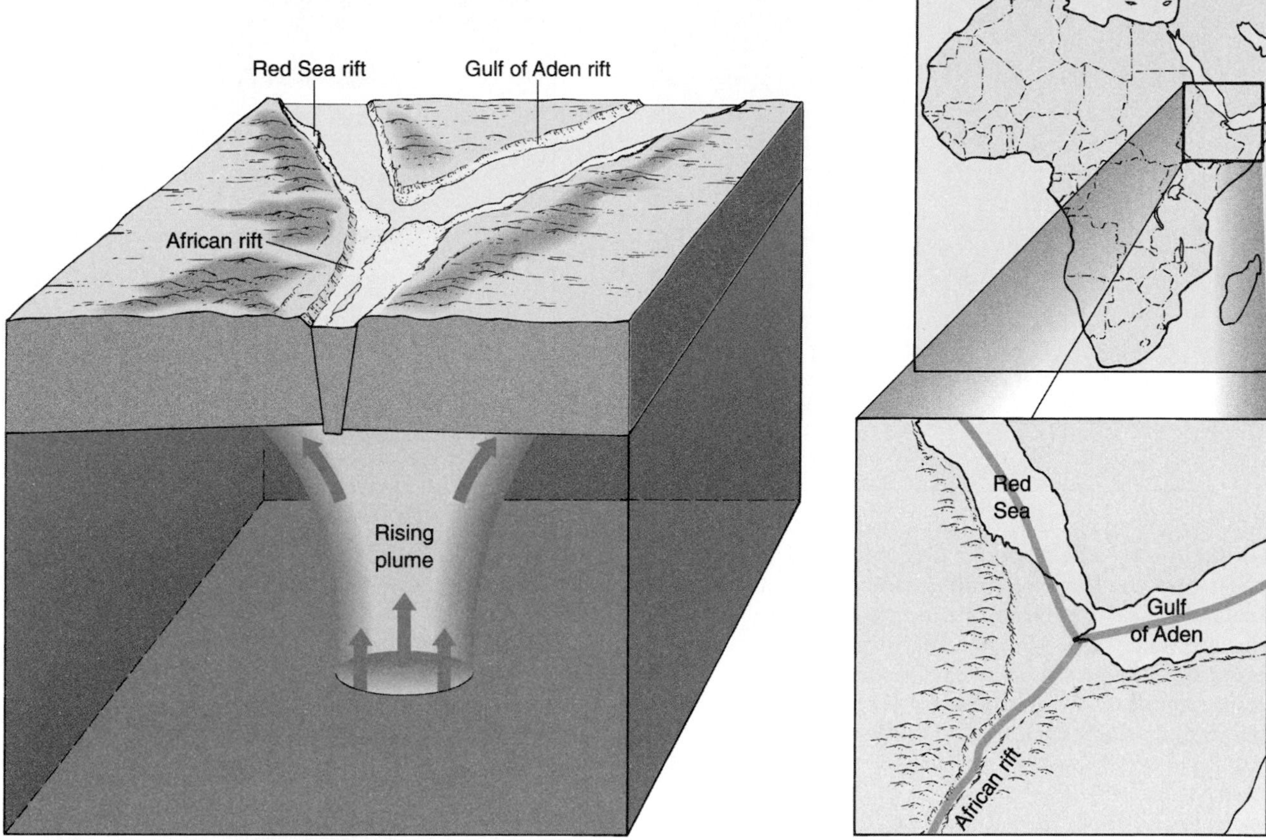

FIGURE 5-48 Rising plumes of hot mantle may cause severe rifts, often forming 120° angles with one another. An example is the Afar triangle, shown at the south end of the Red Sea on the small map of Africa. ? *Which of the three radiating arms is the aulacogen?*

the band of extruded lavas that followed behind the previous band would acquire "reverse" polarity as it cooled past the Curie temperature. As the process repeated itself through time, the result would be the symmetric, mirror-image patterns of normal and reverse stripes mentioned earlier (Fig. 5-51).

The stripes discovered by Vine and Matthews showed that Harry Hess was correct when he suggested in the early 1960s that sea-floor spreading occurs. At that time, however, there seemed to be no feasible way to determine the age of large areas of the ocean floor and thereby ascertain the rate of sea-floor movement. Cores of the basaltic sea floor are not suitable for radiometric dating because hot basalt is altered by contact with sea water. By lucky coincidence, a method for determining the age of the stripes was provided through the work of A. Cox, R. R. Doell, and G. B. Dalrymple in 1963. These geophysicists were investigating the magnetic properties of lava flows on the continents. They were able to accurately identify periodic reversals of the Earth's magnetic field imprinted within layers of basalt (Fig. 5-52). Many of the layers in these superpositional sequences could be dated by radiometric methods. With such data, the time sequence of magnetic reversals for the past 5 million years was determined. The final step was to correlate these dates obtained from land basalts to the rocks of the sea floor. Confirmation for the ages of some of the ocean basalts was obtained from the study of fossils in overlying sediments.

With knowledge of the age of particular normal or reverse magnetic stripes, it is possible to calculate rates of sea-floor spreading and sometimes to reconstruct the former positions of continents. Figure 5-53 illustrates how this can be done. For example, if one wishes to know the distance between the eastern coast of the United States and the northeastern coast of Africa about 81 million years ago, one brings together the traces of the two 81-million-year-old magnetic stripes, being careful to move the sea floor parallel to the transform faults, which indicate the direction of movement.

The velocity of plate movement is not uniform around the world. Plates that include large continents tend to move slowly. Their velocity relative to the underlying mantle rarely exceeds 2 centimeters per year. Plates that are largely devoid of large continents have average velocities of between 6 and 9 centimeters per

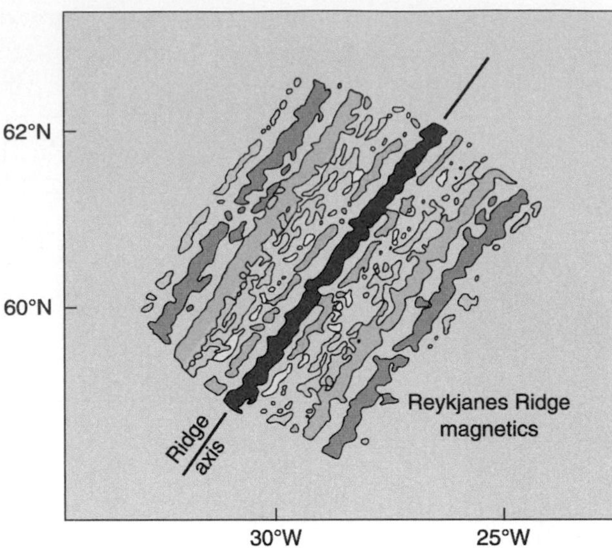

FIGURE 5-49 The magnetic field produced by the rocks over the Reykjanes midoceanic ridge southeast of Iceland. The colored bands are magnetically reversed and the intervening areas have the same magnetic orientation as exists today (termed "normal"). Note the symmetry with respect to the midoceanic ridge at the center. The ages of the stripes increase away from the ridge. (*After Heirtzler, J. R., et al. 1966. Magnetic anomalies over the Reykjanes ridge, Deep Sea Res., 13:427–443; and Vine, F. J. 1968. Magnetic anomalies associated with mid-ocean ridges, in Phinney, R. A. (ed). The History of the Earth's Crust. A Symposium. Princeton, NJ: Princeton University Press.*) **?** *By color, which of the magnetically reversed bonds represents the oldest rocks?*

year. Figure 5-54 provides a summary of the rates and directions of sea-floor spreading as determined by analyses of displacements along transform faults, magnetic anomalies, and other geophysical data.

Knowledge about magnetic reversals not only provides evidence of sea-floor spreading but also has permitted geologists to construct polarity time scales by which rocks exhibiting a distinctive pattern of reversals can be recognized as characteristic of a particular geologic period or epoch (See Fig 5-50). Because the reversals are simultaneous all around the world, they are valuable tools for chronostratigaphic correlation.

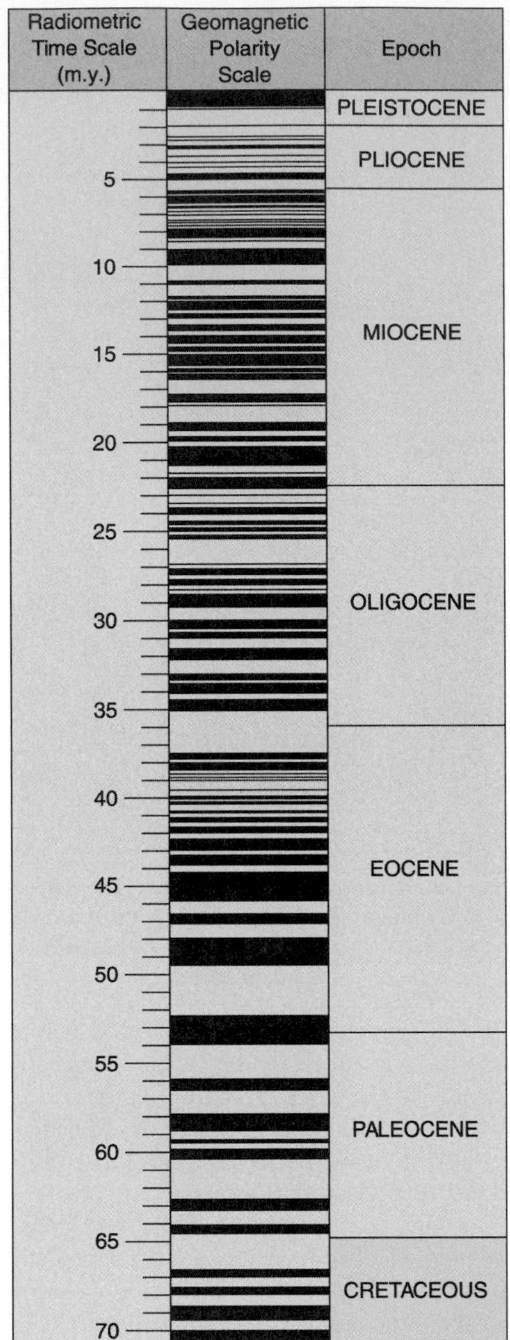

FIGURE 5-50 Reversals of the Earth's magnetic field during the past 70 million years (m.y.). Intervals shown in black indicate when the field was "normal," as today. (*Modified from Heirtzler, J. R., et al. 1968. J. Geophys. Res. 73:2119–2136. Copyright, The American Geophysical Union.*)

EVIDENCE FROM OCEANIC SEDIMENT If newly formed crust joins the ocean floor at spreading centers and then moves outward to either side, then the sediments, dated by the fossil planktonic organisms they contain, can be no older than the surface on which they came to rest. Near a midoceanic ridge, the sediments directly over basalt should be relatively young. Samples of the first sedimentary layer above the basalt farther away from the spreading center should be older (Fig. 5-55).

In a succession of cruises that began in 1969, the American drilling ship *Glomar Challenger* and subsequent cruises of the JOIDES *Resolution* (Fig. 5-56) collected ample evidence to prove that the previously stated conclusions on the age distribution of sediments are correct. The studies carried out by scientists associated with the drilling project confirmed earlier assumptions that there were no sediments older than about 200 million years on the sea floor, that the sediments on the sea floor were relatively thin, and that

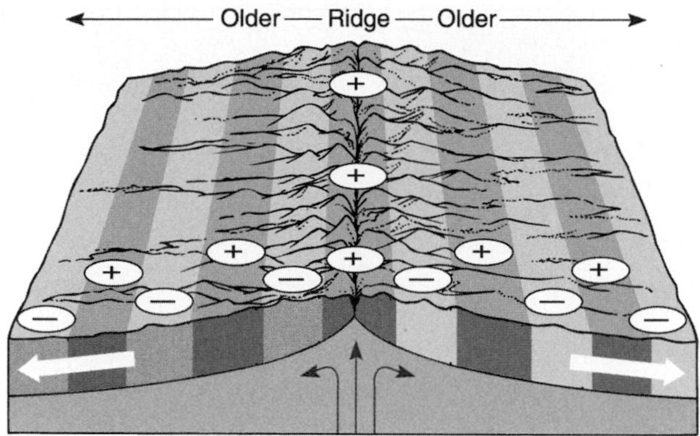

FIGURE 5-51 The normal (1) and reversed (2) magnetizations of the sea floor. Note the symmetry of the magnetizations with respect to the ridge. (*From McCormick, J. M., and Thiruvathukal, J. V. 1981. Elements of Oceanography, 2nd ed. Philadelphia: Saunders College Publishing.*)

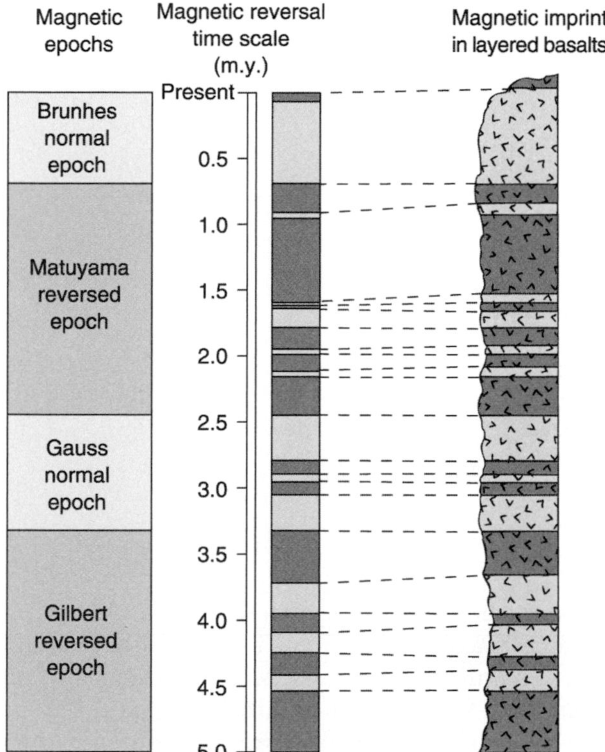

FIGURE 5-52 The imprint of normal (light tan) and reversed (dark tan) magnetic polarity in a composite section of layered basalts from many continental localities. The magnetic reversal time scale is shown in the center column (m.y. = million years), and at the left are the magnetic epochs, named after famous investigators of magnetic effects. The magnetic epochs reflect a predominance of one kind of polarity. Within the larger magnetic epochs are magnetic events of shorter duration. (*Modified from Cox, A. 1963. Geomagnetic reversals. Science 163:237–245.*)

they became thinnest closer to the midoceanic ridges. Sea floor spreading provided a logical reason for these observations. A given area of oceanic plate surface was simply not in existence long enough to accumulate a thick section of sediments over a long span of geologic time. The sea-floor clays and oozes were conveyed to subduction zones, dragged back down into the mantle, and (as suggested by rock compositional studies) resorbed. Thus, the entire world ocean is virtually swept free of deep-sea sediments every 200 to 300 million years.

As an example, sea-floor drilling and sampling have thus far been unable to locate sediments any older than 150 million years in the South Atlantic Ocean. This knowledge is related to the paleomagnetic calculations that suggest that South America and Africa have been moving apart from one another at a rate of about 4 centimeters per year. If this rate has been constant, then extrapolating back in time, one may estimate that the two continents should have been in contact about 150 million years ago during the Jurassic. After that time, the two fragments (Africa and South America) moved apart as oceanic basalts filled the gap left behind. In this way, only sediments younger than the data of fragmentation would be present, since there was not an oceanic basin of deposition in existence prior to that time. Similar reconstructions have been made from sea-floor samples and cores elsewhere around the word.

MEASUREMENTS FROM SPACE The space program has provided yet another method for measuring the movement of tectonic plates. This method uses laser devices that emit an intense beam of light of definite wavelength. The beam is capable of traveling across an enormous distance without becoming dispersed. During the *Apollo* mission, astronauts placed three clusters of special reflectors on the Moon. Laser beams from Earth are "fired" toward the Moon, and by recording the time required for the signal to travel to the Moon and return to Earth, one can determine the distance between the laser device and the lunar reflector. The method is accurate to within just 3 centimeters. Distances from the Moon to stations on two moving tectonic plates will, of course, change over a period of time, and such measurements can be used to calculate how fast continents are moving toward or away from one another.

Beaming lasers to artificial satellites from two widely spaced Earth stations, a technique called *satellite laser ranging* (SLR), can also be used to find the distance between two locations on Earth. The Lageos satellite, which circles the Earth at an altitude of 5800 kilometers, is currently being used to reflect laser pulses back to their source on Earth. By accurately measuring the time required for the laser pulse to travel to the satellite and return, the position of the

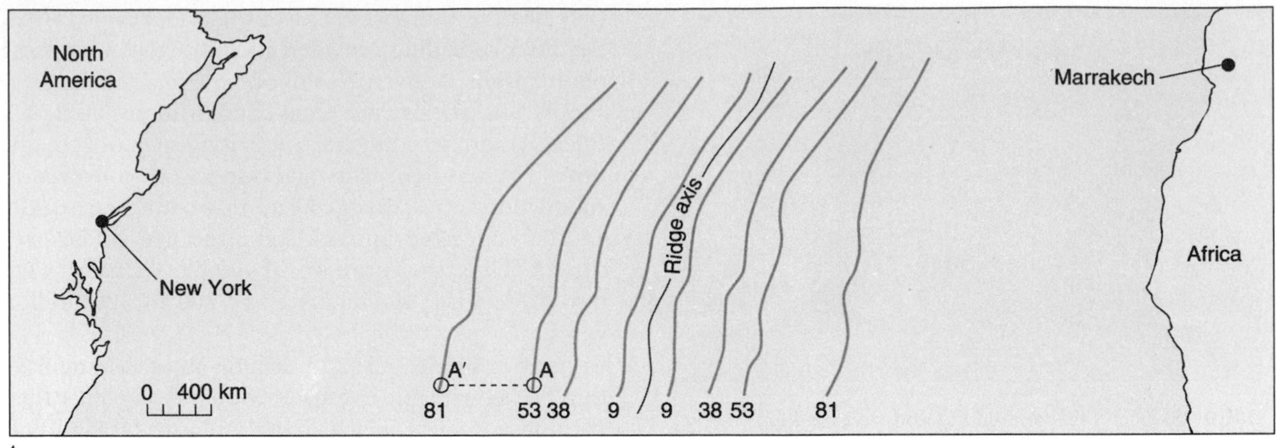

A

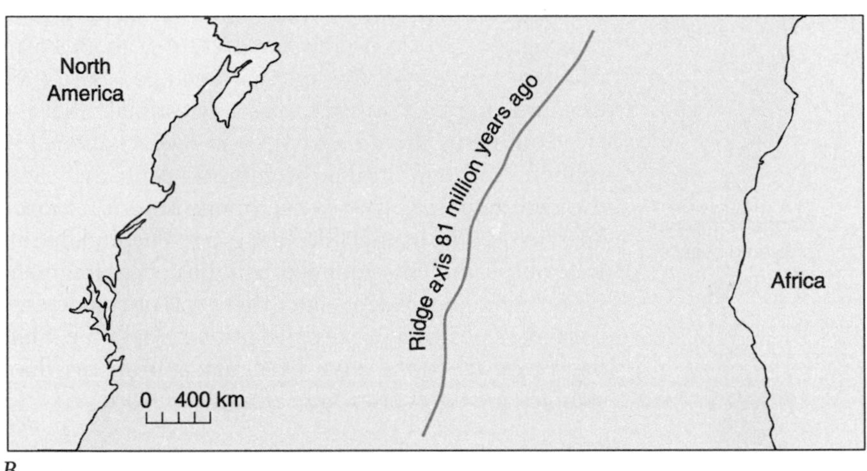

B

FIGURE 5-53 A method for determining the paleogeographic relations of continents. (*A*) The present North Atlantic with ages plotted for some of the magnetic stripes. To see the location of North America relative to Africa, say, 81 million years ago, the 81-million-year-old bands are brought together. (*B*) The result is the view of the much narrower North Atlantic of 81 million years ago. (*Modified from Pitman III, W. C., and Talwani, M. 1972. Geol. Soc. Am. Bull. 83:619–644.*) ? *In map A, how many kilometers did the ocean floor travel between 53 and 81 million years ago along the dashed line marked* **A** *and* **A′**?

earthbound station is precisely ascertained. A similar measurement made from a second station (perhaps on a different continent) fixes the distance between the two stations. Measurements are made again after a given lapse of time to determine the amount and rate of increase or decrease in distance between the two ground stations.

Another method that provides similar accuracy is called *very long baseline interferometry* (VLBI). In this method, several stations at widely spaced locations on Earth monitor the radio signals from distant quasars (very large redshift, intergalactic, luminous objects). Quasars are so far away from our planet (billions of light-years) that they serve as stationary reference points. The differences in the arrival times of radio signals received from the quasars determine the distance between the ground-based stations, and changes in those distances are then noted for given increments of time, as in the SLR method. Measurements based on monitoring stations in the United States and Europe indicate that the North American and European tectonic plates are moving apart at a rate of 1.9 centimeters per year and that Hawaii is drifting toward Japan at the relatively rapid rate of 8.3 centimeters per year. These determinations compare very well with measurements based on the spacing of magnetic bands on the floor of the Atlantic and Pacific oceans.

SEISMIC EVIDENCE The evidence of the actual movement of lithospheric plates appears to be fairly direct and susceptible to several methods of testing. It is more difficult to prove that plates of the lithosphere are

ENRICHMENT

Rates of Plate Movement

On examination of the rates at which tectonic plates move away from midoceanic ridges at various places around the Earth, it is immediately apparent that the rate of movement is greater at some places than at others. One reason for this relates to movement on a spherical surface. If the Earth were flat, as some viewed it before the voyages of Columbus, then all locations along a similarly flat plate would be moving away from a midoceanic ridge at about the same velocity. On a sphere, however, the plates are curved. They begin to separate at a location termed a pole of rotation and form an ever-widening split. The effect is somewhat like pulling apart the outer dry layer of an onion. The split in the skin would be widest where it was being pulled apart, and the gap would narrow to a point corresponding to the pole of rotation. On the Earth, plate movement away from the midoceanic ridge near the pole of rotation would be slower because of the lesser distance traversed over a given time interval. Conversely, far from the pole of rotation, plates diverge across a greater distance for the same time interval and hence have greater velocity. One can observe this effect in the North Atlantic, where spreading rates increase from about 0.75 cm per year near Iceland to over six times that amount near the equator. A corollary of this effect is that the width of new oceanic crust also increases with distance from the pole of rotation.

In addition to changes in plate velocity that are a consequence of the Earth's spherical shape, it appears that large expanses of continental crust incorporated into a plate can decrease its velocity. The Pacific and Nazca plates do not carry a heavy load of crust. These plates have higher velocities than the American and Eurasian plates, which are burdened with large continents.

dragged back down into the mantle and resorbed. The best indications that subduction actually does occur come from the study of earthquakes generated along the seismically active colliding margins of plates.

Well known among these studies are the investigations of Hugo Benioff, a seismologist and designer of improved seismographs. Benioff and subsequent investigators compiled records of earthquake foci that occurred along presumed plate boundaries. Charting the data showed that the deeper earthquake foci occur along a narrow zone that was tilted at an angle of 45° under the adjacent island arc or continent (Fig. 5-57).

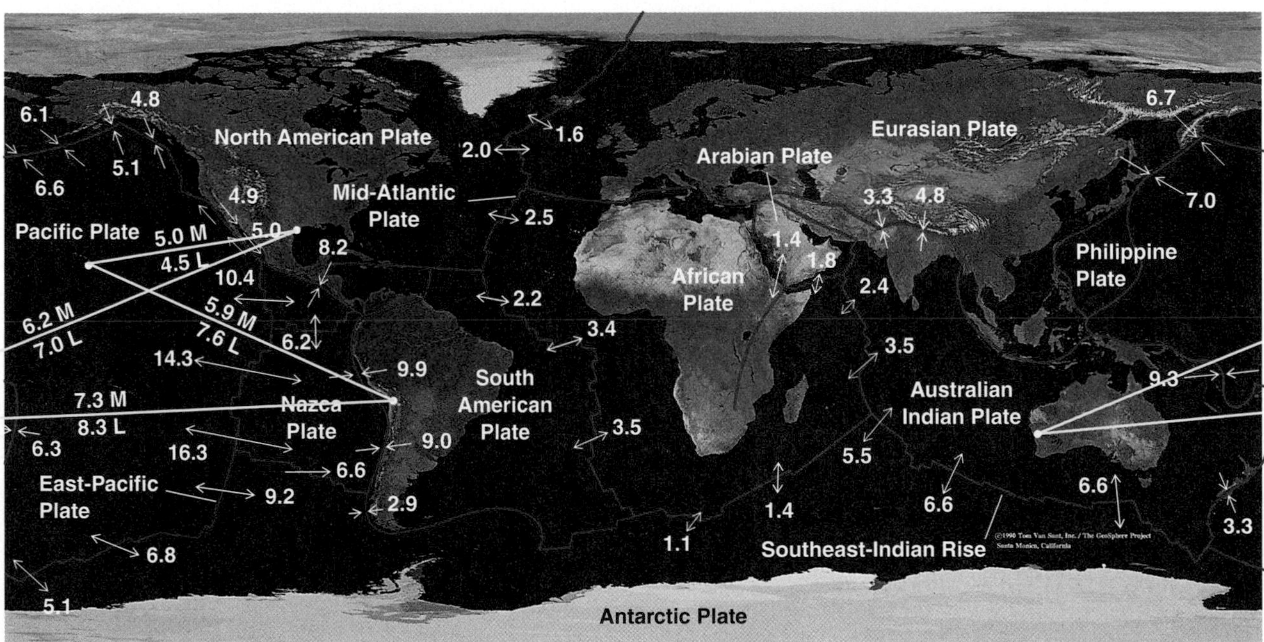

FIGURE 5-54 Plate velocities, in centimeters per year. Numbers along the midoceanic ridge system indicate the rates at which two plates are separating, based on magnetic reversal patterns on the sea floor. The arrows indicate the directions of plate motions. The yellow lines connect stations that measure present-day rates of plate motions with satellite laser ranging methods. The numbers followed by **L** are the present-day rates measured by laser. The numbers followed by **M** are the rates measured by magnetic reversal patterns. (Modified from NASA report, Geodynamics Branch, 1986. Tom Van Sant, Geosphere Project.) ? *What part of the world appears to have the fastest moving plates at present?*

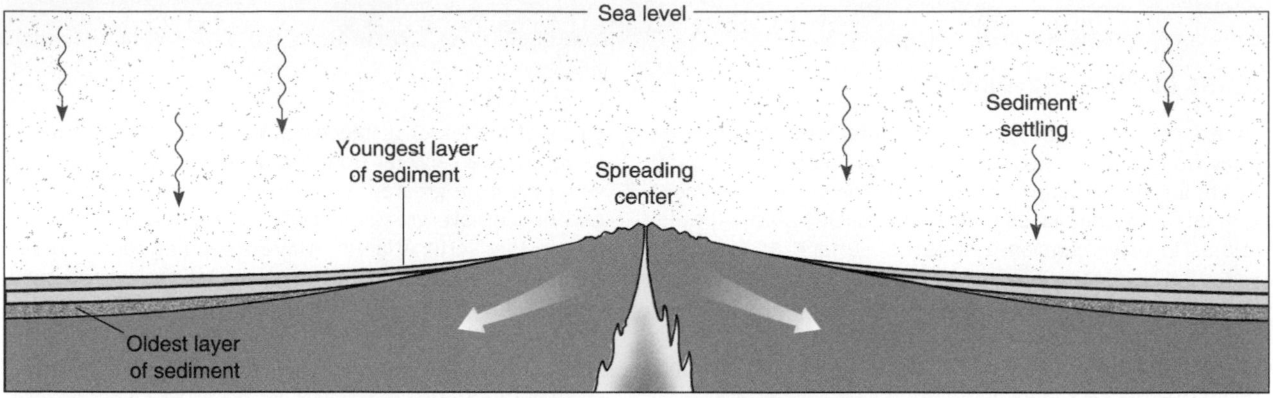

Sea level
Sediment settling
Youngest layer of sediment
Spreading center
Oldest layer of sediment

FIGURE 5-55 As a result of sea-floor spreading, sediments near the midoceanic ridges are youngest, and sediments directly above the basaltic crust but progressively farther away from the ridge are sequentially older.

This inclined seismic shear zone was traced to depths as great as 700 kilometers (435 miles). It was named the **Benioff seismic zone** after its discoverer. These inclined earthquake zones are believed to define the positions of the subducted plates where they plunge into the mantle beneath the overriding plate. The earthquakes are the result of fracturing and subsequent rupture as the cool, brittle lithospheric plate descends into the hot mantle.

In addition to the deep and intermediate focus earthquakes along the Benioff zone, there are intense and often destructive earthquakes at shallow depths where the edges of the two rigid plates press against one another. The 1964 Alaskan earthquake originated at this shallow level. Such quakes are thought to occur along the shear plane between the subducting ocean lithosphere and the continental lithosphere. Some may result from normal faulting as portions of the lithosphere are arched by the collision.

GRAVITY PLAYS A ROLE Gravity measured over deep-sea trenches is characteristically lower than that measured in areas adjacent to the trenches. Geophysicists refer to such phenomena as negative gravity anomalies. A **gravity anomaly** is the difference between the observed value of gravity at any point on the Earth and the computed theoretical value. **Negative gravity**

FIGURE 5-56 The JOIDES *Resolution*, successor to the *Glomar Challenger*. This oceanographic research vessel is designed for taking drill cores from the bottom of the ocean floor. The vessel is a floating oceanographic research center, with a seven-story laboratory stack that occupies 12,000 square feet. On-board facilities include laboratories for sedimentology, paleontology, geochemistry, and geophysics. The ship can suspend as much as 9000 meters of drill pipe to obtain core samples. (*Courtesy of Ocean Drilling Program, National Science Foundation.*)

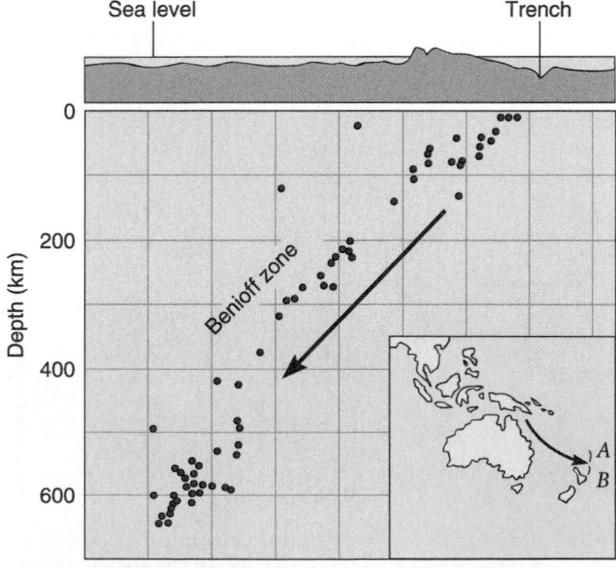

FIGURE 5-57 Vertical cross-section showing spatial distribution of earthquakes along a line perpendicular to the Tonga trench and volcanic island arc. The Tonga trench (*A* on inset map) lies just to the north of the Kermodec trench (*B*). (*Simplified from Isacks, B., Oliver, J., and Sykes, R. 1968. J. Geophys. Res. 73:5855.*)

anomalies occur where there is an excess of low-density rock beneath the surface. The very strong negative gravity anomalies extending along the margins of the deep-sea trenches can mean only that such belts are underlain by rocks much lighter than those at depth on either side. Because the zone of lower gravity values is narrow in most places, it is believed to mark trends where lighter rocks dip steeply into the denser mantle (Fig. 5-58). Geophysicists assume that the less-dense rocks of the negative zone must be held down by some force to prevent them from floating upward to a level appropriate to their density. That force might be provided by a descending convection current, or the negative gravity anomaly may reflect stretched and fractured oceanic crust undergoing slab pull.

Hot Spots

Anyone examining topographic maps (see Fig. 5-23) of the sea floor cannot help but notice the chains of volcanic islands and seamounts (submerged volcanoes). Because these volcanoes occur at great distances from plate margins, they must have originated differently from the volcanoes associated with midoceanic ridges or deep-sea trenches. Their striking alignment has been explained as being a consequence of sea-floor spreading. According to this notion, intraoceanic volcanoes develop over a "hot spot" in the asthenosphere. The hot spot is a manifestation of one of the deep plumes of upwelling mantle rock described earlier. Lava from the plume may work its way to the surface to erupt as a volcano on the sea floor. As the sea floor moves (at rates as high as 10 centimeters per year in the Pacific), volcanoes form over the hot spot and expire as they are conveyed away. New volcanoes form at the original location. The process may be repeated indefinitely, resulting in linear successions of volcanoes that may extend for thousands of kilometers in the direction that the sea floor has moved. The Hawaiian volcanic chain is believed to have formed in this manner from a single source of lava over which the Pacific plate has passed on a northwesterly course. In support of this concept are radioactive dates obtained from rocks of volcanoes that clearly indicate that those farthest from the source are the oldest.

At the western end of the Hawaiian Islands is a string of submerged peaks called the Emperor seamounts (Fig. 5-59). These submerged volcanoes trend in a more northerly direction than the Hawaiian Islands. However, both the seamounts and the islands are part of a single chain that has been bent as a result of a change in the direction of plate movement. Once again, such an interpretation is supported by age determinations. The oldest of the Hawaiian Islands (near the bend) was formed about 40 million years ago. The seamounts continue the age sequence toward the end of the chain, where the peaks are about 80 million years old. Thus, 40 million years ago the Pacific plate changed course and put a kink in the Hawaiian chain.

The ocean around Hawaii is not the only part of the globe that has hot spots. As indicated in Figure 5-60, hot spots are widely dispersed and occur beneath both continental and oceanic crust. Yellowstone National Park with its geysers and hot springs (Fig. 5-61) is over a hot spot that is in the interior of a continent.

Lost Continents and Alien Terranes

We are accustomed to thinking of continental crust in terms of large landmasses such as North America or Eurasia. There are, however, many relatively small patches of continental crust scattered about on the lithosphere. As long ago as 1915, Alfred Wegener described the Seychelles Bank (Fig. 5-62) in the Indian Ocean as a small continental fragment that had broken away from Africa. The higher parts of the Seychelles

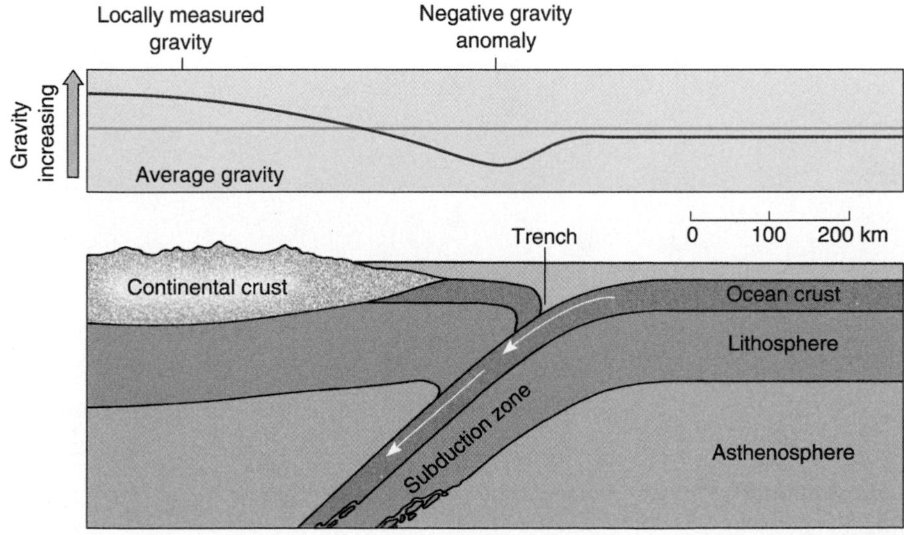

FIGURE 5-58 Diagram showing the variation in gravity over a deep-sea trench. The trench, subducting sediments, and relatively low-density rocks occupy space that would otherwise be filled with denser rocks. Hence, the force of gravity over the trench and subduction zone is weaker than over the Earth generally.

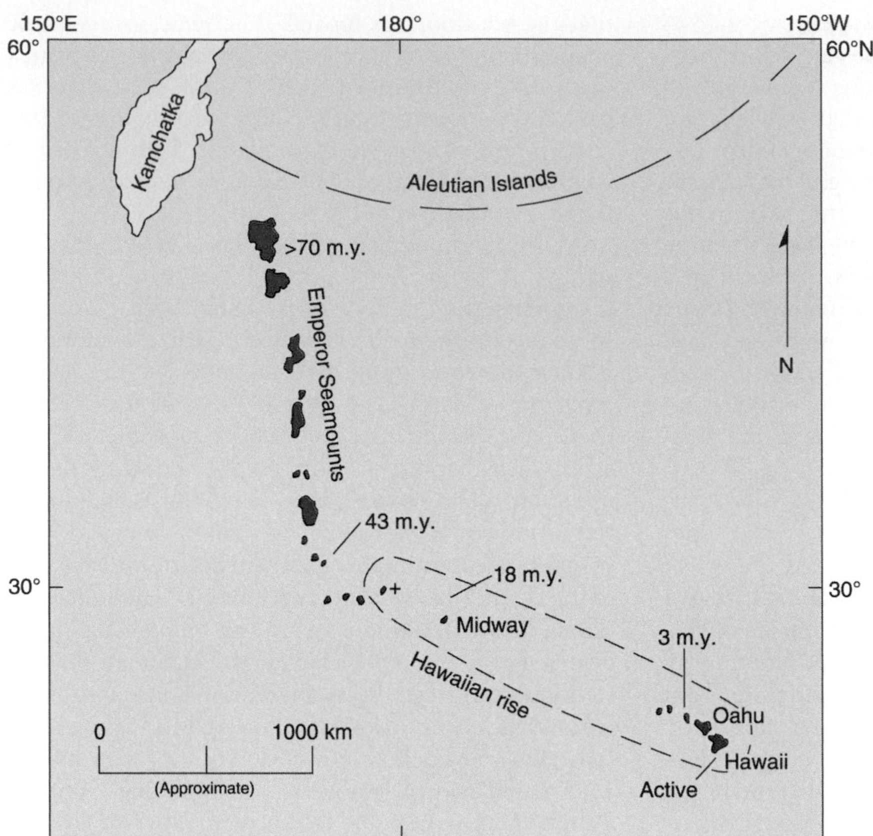

FIGURE 5-59 Bend in trend of the Hawaiian Island-Emperor Seamount chain was probably caused by change in direction of movement of the Pacific tectonic plate. *(From Watkins, J. S., Bottino, M. L., and Morisawa, M. 1975. Our Geological Environment. Philadelphia: W.B. Saunders Company.)*

Bank project above sea level as islands, but many other such small patches of continental crust are totally submerged. Geologists use the term **microcontinents** for these bits of continental crust that are surrounded by oceanic crust. They are recognized by their granitic composition, by the velocity with which compressional seismic waves traverse them (6.0 to 6.4 kilometers per second), by their general elevation above the oceanic crust, and by their comparatively quiet seismic nature.

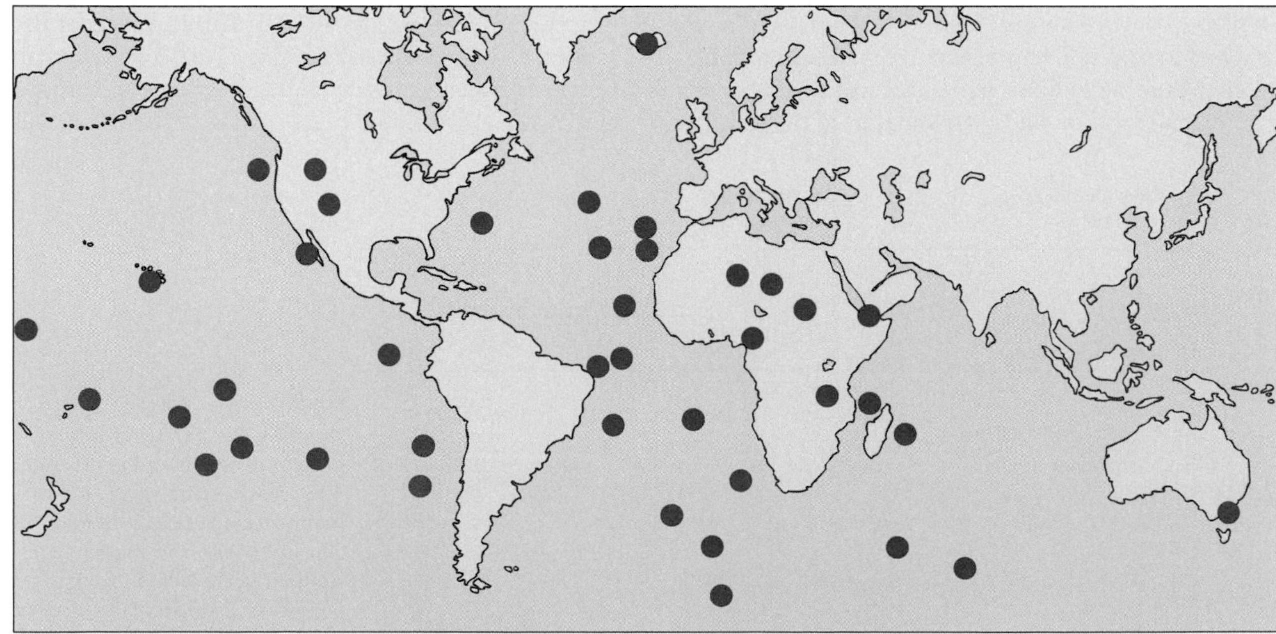

FIGURE 5-60 Locations of some of the major hot spots around the Earth. *(Courtesy of Tom Crough, Department of Geologic and Geophysical Sciences, Princeton University.)*

ENRICHMENT

The Missing Hawaiian Volcano

In this chapter we have described how the Hawaiian chain of volcanic islands originated by passage of the Pacific plate over a relatively stable hot spot. In addition to the general alignment of the Hawaiian Islands and Emperor seamounts, however, one can also observe that the subaerial volcanoes located along the southern segment of the chain occur sequentially along two lines, as shown in the accompanying figure. Along each line, the volcanic islands are fairly evenly spaced, except for a large gap in the western line between the islands Kahoolawe and Hualalai. Geologists have been concerned with this extra space, which seemed to disrupt an otherwise remarkably uniform pattern. Those concerns, however, were put to rest recently when new bathymetric data and geochemical analyses of lava dredged from the sea floor demonstrated the existence of a submerged volcano within what had been assumed was a gap in the chain. The missing volcano was about 70 kilometers long by about 30 kilometers wide, and it had a height of 4 kilometers above the abyssal sea floor. Geologists named the volcano Mahukona, after the closest settlement to the volcano on the island of Hawaii.

The question of the gap in the southern Hawaiian chain had been solved, but two questions remained. Why were the volcanoes in a paired sequence, and why did Mahukona become extinct before building itself upward to form an island? In laboratory experiments that model ascending hot spot plumes, the plumes bifurcate as they rise, forming two distinct branches. This may have occurred in the ascending plume that generated the volcanoes of the southern Hawaiian chain to produce two parallel lines of volcanoes about 35 kilometers apart. One of these volcanoes, Mahukona, did not reach sea level, possibly because the branch feeding the western line was smaller or became depleted, causing Mahukona to become dormant before it reached sea level. Other volcanoes on the western line, such as Lanai and Kahoolawe, although they reached sea level, tend to be smaller than volcanoes in the eastern line. The submerged volcano, Lo'ihi, has been described as "an island in the womb." Although its summit is 3000 feet below sea level, it rises 3 miles above the ocean floor. In about 50,000 years, Lo'ihi will be an additional island in the Hawaiian chain.

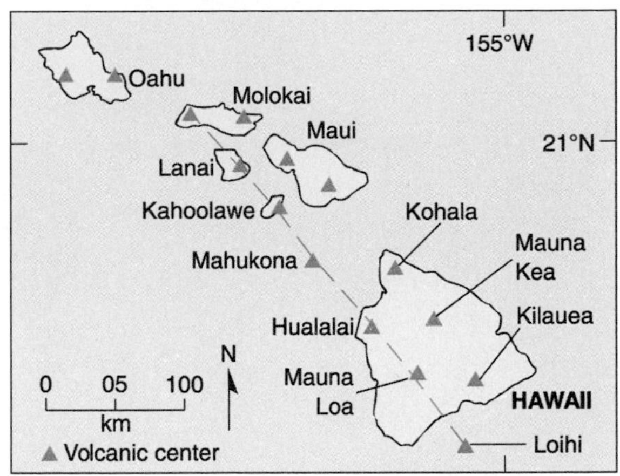

Southern Hawaiian Islands with paired sequence of volcanoes.

Reference

Garcia, M. O., Kurz, M. D., and Muenow, D. W., 1990. *Geology* 18:1111–1114.

It is apparent that microcontinents are small pieces of larger continents that have experienced fragmentation. As these smaller pieces of continental crust are moved along by sea-floor spreading, they may ultimately converge on the subduction zone at the margin of a large continent. Because they are composed of relatively low-density rock and hence are buoyant, they are a difficult bite for the subduction zone to swallow. Their buoyancy prevents their being carried down into the mantle and assimilated. Indeed, the small patch of crust may become incorporated into the crumpled margin of the larger continent as an exotic block, or so-called allochthonous terranes (also called suspect, exotic, or alien terranes).

It is interesting that geologists found evidence of allochthonous terranes long before the present theory for their origin was formulated. While mapping Precambrian rocks, they came across fault-bounded areas that were incongruous in structure, age, fossil content, lithology, and paleomagnetic orientation when compared to the surrounding rocks. It was as if these areas were small, self-contained, isolated geologic provinces.

Allochthonous terranes have been identified on every major land mass, with well-studied examples in the Appalachians, many parts of western North America, and Alaska (Fig. 5-63). A recently recognized allochthonous terrane in the Andes appears to have broken away from the southern margin of the United States and drifted across the ocean to the western margin of South America.

If splinters of continents can be transported by the spreading sea floor, so can thickened pieces of oceanic crust. Particularly in the Cordilleran mountain belt of North America, one finds allochthonous terranes that were apparently microplates of ocean crust containing volcanoes, seamounts, segments of island arcs, and other features of the ocean floor. All of these features were carried to the western margin of North America like passengers on a huge conveyor belt. As the plate that bore them plunged downward at the subduction

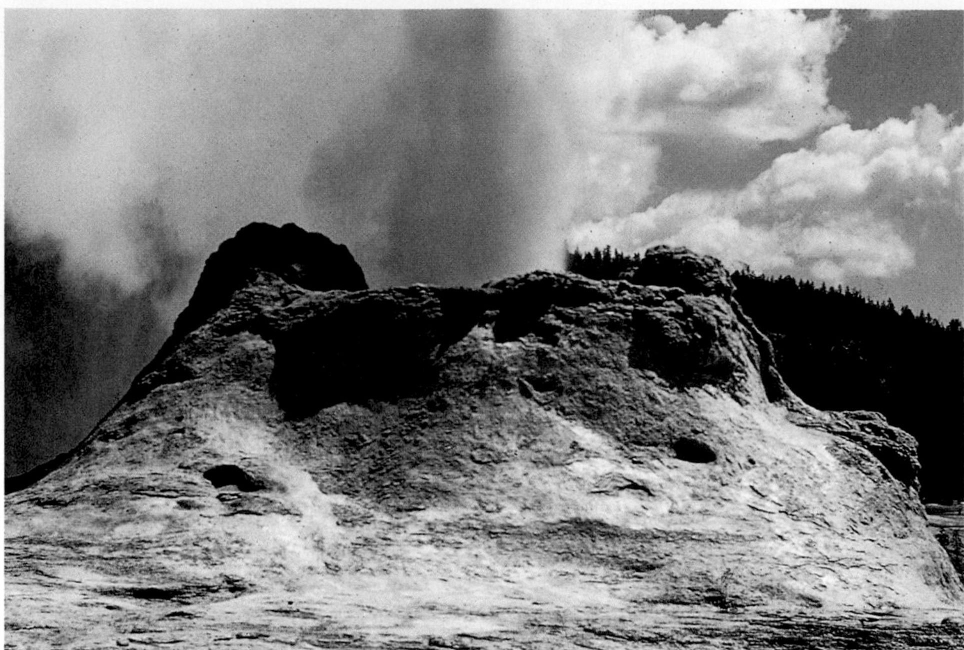

FIGURE 5-61 Mammoth Geyser, Yellowstone Park, Wyoming. Yellowstone is over a hot spot. Surface waters percolating through a system of deep fractures reach the hot rocks below and erupt in columns of hot water and steam.

zone located along the continental margin, volcanoes, seamounts, and the other features were scraped off the descending plate and plastered onto the continental margin as an accretionary prism. The result was a vast collage of accreted oceanic microplates interspersed with similarly transported microcontinents. It appears that about 50 such allochthonous or exotic terranes exist in the Cordillera, all lying west of the edge of the continent as it existed about 200 million years ago. Clearly this process of successive additions of oceanic and continental microplates can significantly increase the rate at which a continent is able to grow. It is also apparent that active continental margins such as those bordering the Pacific Ocean are likely to grow faster than passive margins because of microplate accretion.

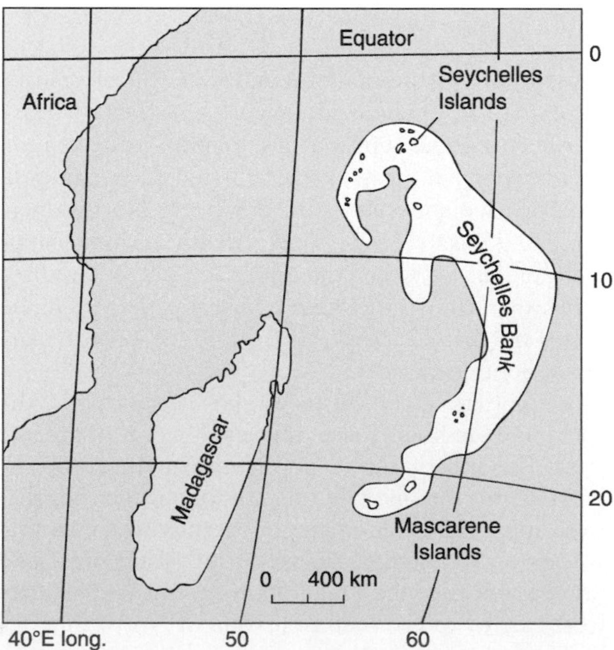

FIGURE 5-62 Location map of the Seychelles Bank.

Unraveling the orogenic history of mountain ranges containing multiple allochthonous terranes is an enormously complex task requiring the cooperation of geologists, geophysicists, and paleontologists. Such a collaboration can often yield dramatic results. Recently, paleontologists and geologists working in the Wallowa Mountains of Oregon discovered a massive coral reef of Triassic age. The reef, however, clearly did not belong where it was found. It rested on volcanic rocks whose paleomagnetic orientation indicated that they had solidified from lavas at a latitude considerably closer to the equator. Further investigation indicated that the fossil organisms in the reef were identical to those found in Triassic strata of the Austrian and German Alps. These strata were deposited prior to Alpine mountain building in a seaway called the Tethys, which extended from the present east coast of Japan into the Mediterranean region. The similarity between the fossils of Oregon and those of the Alps strongly suggests that the reefs now in Oregon actually grew around the margins of volcanic islands in the Tethys Sea and, in the subsequent 220 million years or so, were transported for thousands of kilometers before colliding with the North American continent.

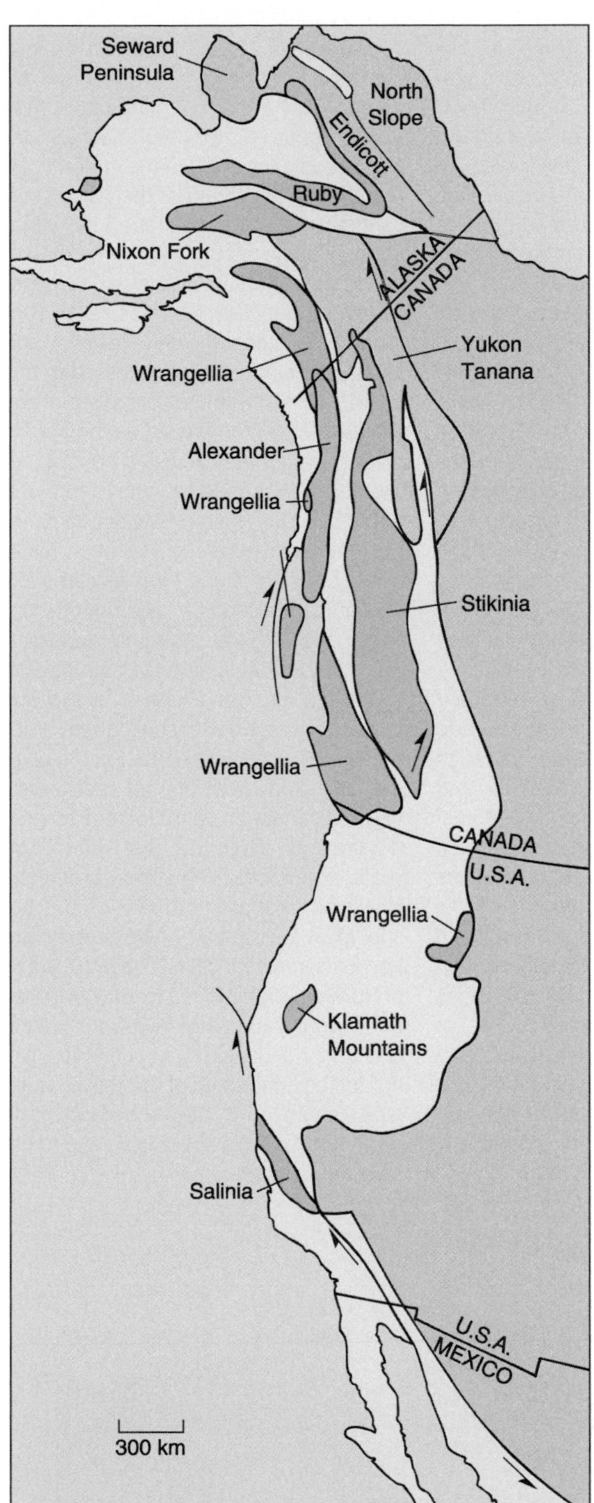

FIGURE 5-63 The larger allochthonous continental terranes in western North America that contain Paleozoic or older rocks. At some time prior to reaching their present locations, these terranes were continental fragments embedded in oceanic crust or microcontinents. Those colored green probably originated as parts of continents other than North America, whereas the pink blocks are possibly displaced parts of the North American continent. (*From Ben-Avraham, Z. 1981. Am. Sci. 69:298.*)

Plate Tectonics and Ore Deposits

In recent years, geologists have begun to see numerous correlations between patterns of mineral distribution and locations of present and former tectonic plate boundaries. Plate boundaries are likely sites for movements of hot, aqueous fluids that bear important metals in solutions. With changes in temperature or pressures, or on contact with reactive rocks, such hydrothermal solutions will precipitate ore minerals. The process may operate at both convergent and divergent plate boundaries. An example of hydrothermal mineralization at a divergent boundary is provided by the Red Sea, which has pools of hot and exceptionally salty water along its bottom. The pools are rich in iron, manganese, zinc, and copper. The brines have percolated upward through the young oceanic crust, dissolving the metals en route. Metals brought to the surface in this way in the past may have been conveyed in thin layers across immense tracts of the ocean floor by sea-floor spreading. Sediments containing metallic ions extracted from the sea water itself often enriched the accumulation. Ultimately, sea-floor spreading moved the sediments and their contained metals into collision with another plate, providing an opportunity for their inclusion as ore bodies within the deformed belt of the convergent plate margins.

LIFE ON A TRAVELING CRUST

The formation and disruption of supercontinents, the uplift of mountain systems along collisional plate boundaries, and the opening and closing of ocean basins produce changes in climate, oceanic circulations, the advance and retreat of inland seas, and even the chemistry of ocean waters. Such changes have a direct impact on life. They may enhance or inhibit biologic diversity, influence rates of evolution, and cause mass extinctions.

Pangea has not been the only supercontinent to exist on Earth. During the past 4.6 billion years of Earth history, there have been at least three supercontinents assembled. Pangea was one of these. Another, named **Rodinia**, was formed in the early Proterozoic, and the third, **Pannotia**, was assembled near the end of the Precambrian. Supercontinents afforded certain advantages to organisms. They facilitated migration and genetic interaction of terrestrial animals as well as nearshore marine faunas living around their circumventing coastal areas. Such faunas would not be confined to multiple, isolated locations. Thus, supercontinents provide favorable environments for the evolution of cosmopolitan faunas with a lower variety of animals than would be the case if there had been many smaller, isolated continents. During the Mesozoic Era, when continents were largely still in contact with one another, there were fewer major taxa of land

Hawaii Volcanoes National Park

If one wants to see geology in action, there are few places that surpass Hawaii Volcanoes National Park (Figure 1*A*). Within the park, the active volcanoes Mauna Loa and Kilauea periodically provide spectacular fiery fountains and rivers of incandescent lava as evidence that the primordial forces involved in the Earth's development are still at work. Mauna Loa attests to the prodigious ability of volcanoes as land builders. It is the most massive mountain on Earth, rising more than 6 miles (9.6 kilometers) above the floor of the Pacific Ocean.

Broad, gently sloping volcanoes such as Mauna Loa and the other volcanoes of the Hawaiian chain are termed shield volcanoes because their shape resembles that of the shields of ancient warriors. Their broadly convex shape results from repeated eruptions of highly fluid basaltic lava, emerging either from vents or along rift zones.

The Kilauea area of Hawaii Volcanoes National Park is an ideal place to view many of the characteristic features of shield volcanoes. One can easily explore this area along the 11-mile Crater Rim Drive that encircles Kilauea's summit caldera and craters (see accompanying photograph). Along this road, one passes through a lush rainforest environment, a rain shadow desert, lava flows, and well-marked stops and walks to view Sulphur Banks, Steam Vents, the Saggar Museum, Halema'uma'u Crater, Devastation Trail, and Kilauea Ili Crater. On the east side of the drive is the Thurston Lava Tube. Lava tubes begin as conduits through which lava flows. When lava breaks through the solidified crust of basalt and flows away, it leaves behind a now-vacated, cave-like lava tube. Along the trails leading from Crater Rim Drive, one can find good examples of pahoehoe and aa lava. Pahoehoe has a ropy appearance produced in the still-plastic surface scum of lava by the drag of more rapidly flowing lava below. In contrast, aa has a blocky, fragmented texture. It forms as the thicker upper layer of lava hardens and is carried in conveyor belt fashion to the front of the flow. There the brittle top layer is broken into jagged chunks that accumulate at the leading edge of the flow. If you were to walk on this jagged material, you might easily guess why Hawaiians call it aa.

Basalt is by far the most abundant rock in Hawaii Volcanoes National Park. With successive eruptions, Hawaiian basalts become poorer in silica and more enriched in sodium and potassium. The sodium- and potassium-rich alkalic lavas are more viscous. They are found largely near the summits of the shield volcanoes. As is common in many basalts, those of Hawaii contain the mineral olivine. Often one finds large crystals of olivine (phenocrysts) in the basalt, and these are cut and polished for jewelry. The gem name for olivine is peridot. Peridot rings and bracelets can be found in almost any Hawaiian shop catering to tourists.

Volcanoes of the Hawaiian Islands are *intraplate* volcanoes. They are located within a tectonic plate rather than along its more dynamic margins. Intraplate volcanoes are in the minority, for more than 75 percent of the Earth's volcanoes are distributed around the edges of the Pacific plate. They form the so-called ring of fire, which includes Washington's Mount St. Helens, Alaska's Katmai, Popacatepetl in Mexico, and Japan's Mount Fuji. As discussed in this chapter, the intraplate volcanoes of the Hawaiian chain originated as the Pacific plate moved slowly across a hot spot, or plume, of hot magma rising from the upper mantle. This process accounts for the general alignment of Hawaiian volcanoes. Early eruptions form a seamount, and after thousands of additional eruptions over several hundred thousand years, the volcanic edifice is built above sea level. The island formed in this way continues to grow with subsequent eruptions until movement of the Pacific plate carries it away from the hot spot, severing its connection to the parent magma (Figure 1*B*). Over the past 70 million years, the Pacific plate has carried the Hawaiian Islands northwest of the hot spot at an average rate of 10 centimeters per year. In recent years, the large island of Hawaii has continued to grow, with periodic contributions of lava from Mauna Loa and Kilauea. Hawaii, however, will not be the final island of the Hawaiian Island chain. To the southeast, the seamount Lo'ihi is building its way to the surface of the sea.

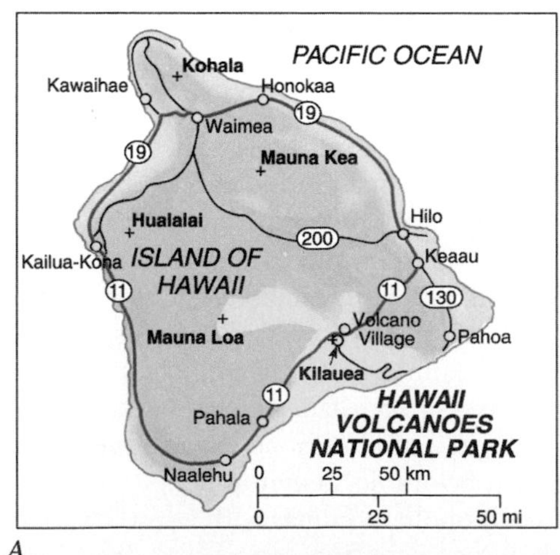

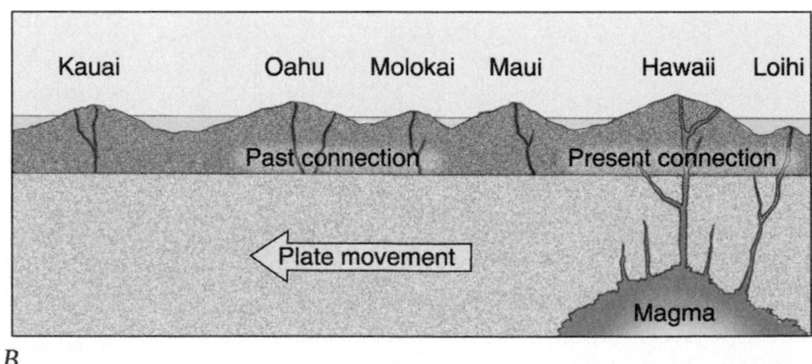

Hawaii volcanoes and plate movement. (*A*) Location map for Hawaii Volcanoes National Park. (*B*) Cross-section indicating origin of volcanoes of the Hawaiian Island chain as the Pacific plate passed over a hot spot. (*A and B courtesy of the U.S. National Park Service.*)

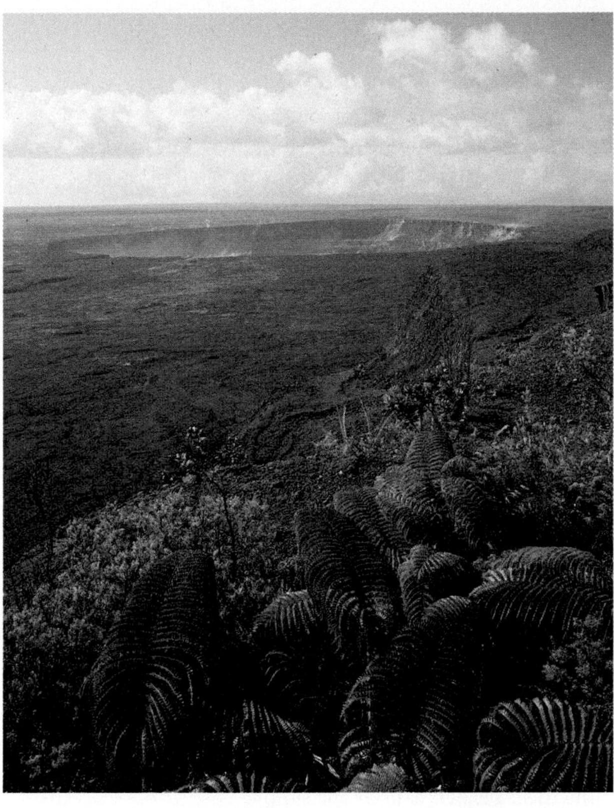

On the rim of the Kilauea caldera. A pit crater that last erupted in 1967–1968 is visible within the larger caldera. Ama'Uma' ferns grow in the foreground. (*Photograph by J. Gnass.*)

The figure accompanying the box entitled "The Missing Hawaiian Volcano" shows the location of volcanoes along the southern part of the Hawaiian Island chain. These locations suggest that the volcanoes occur in pairs that are about 35 kilometers apart. Thus, in Hawaii Volcanoes National Park, Mauna Loa has its mate in Kilauea. To account for the paired arrangement of volcanoes, geologists speculate that the plume supplying the volcanoes has two branches, with one branch providing lava to the western volcano and one to the eastern. Kohala, at the northern end of Hawaii, appears to lack a mate. Actually, however, the mate is present as the seamount Mahukona. Mahukona may have failed to become an island because the branch of the plume supplying the western line of volcanoes was smaller or became deplated.

For scientists, Mauna Loa and Kilauea provide a natural laboratory for on-site study of volcanism. At the Hawaii Volcano Observatory and the University of Hawaii, volcanologists record the periodicity of eruptions, changes in magnetic properties, and variations in the composition and temperature of gases emitted at fumaroles. They record the swelling or tilting of the land surface and carefully consider every tremor associated with subterranean movements of magma. These observations help in finding ways to forecast eruptions and in understanding the behavior of volcanoes everywhere around the world. What we learn from today's Hawaiian volcanoes can be used to interpret the volcanism of long ago and to predict what can be expected in the future.

vertebrates than in the succeeding Cenozoic, which is characterized by maximum continental fragmentation.

In addition to the correlation between continental fragmentation and increased biologic diversity, other events associated with plate tectonics have affected the history of life. Locally, organisms have suffered during supercontinent assembly because of the loss of barriers that protected them from competing faunas. Uplift accompanied by regression of inland seas may have destroyed habitats for marine creatures and produced more rigorous climates for animals on land. On the other hand, eustatic increases in sea level related to rapid sea-floor spreading and growth of midoceanic ridges would have expanded the fertile shallow marine realm onto the continents and produced a moderating effect on climate and seasonality. As an example of the effect of sea-floor spreading on life, one notes that during the Precambrian–Cambrian transition, lavas from a 10,000-kilometer-long midoceanic ridge may have displaced sufficient ocean water to cause a significant eustatic rise in sea level. The resulting Cambrian marine transgressions, occurring at a time of increased atmospheric oxygen, may have strongly contributed to the rapid expansion of the Earth's earliest multicellular animals.

SUMMARY

Knowledge of the Earth's deep interior is derived from the study of earthquake waves. Among the various kinds of earthquake or seismic waves are primary, secondary, and surface waves. Primary and secondary waves (also called body waves) pass deep within the Earth and therefore are the most instructive. Study of abrupt changes in the characteristics of seismic waves at different depths provides the basis for a threefold division of the Earth into a central core, a thick, overlying mantle, and a thin, enveloping crust. Sudden changes in earthquake wave velocities and angles of transmission are termed discontinuities.

The core of the Earth begins at a depth of 2900 kilometers. The Gutenberg Disconformity occurs at that depth. The core is divided into an inner and outer core. It is likely that the outer core is molten and that the entire core is composed of iron, with small amounts of nickel and possibly either silicon or sulfur. The primary evidence of the existence of a liquid outer core is the disappearance of S-waves at a depth of about 2900 kilometers (S-waves are not transmitted through liquids) and the Earth's magnetic field.

The mantle constitutes about 80 percent of the Earth's total volume. It is composed largely of iron and magnesium silicates such as pyroxene and olivine. That portion of the upper mantle that lies just below the lithosphere is called the asthenosphere. The asthenosphere serves as a weak, plastic layer on which the more rigid overlying layers move. As lithospheric plates move, they may break apart, collide, or slip past one another.

The seismic boundary that separates the mantle from the overlying crust is the Mohorovičić discontinuity. It lies far deeper under the continents than under the ocean basins. Thus, the continental crust is thicker than the oceanic crust. There are compositional and density differences as well. The continental crust has an overall granitic composition and is less dense than the oceanic crust, which is composed of basaltic rocks.

The crust of the Earth is not as as we sometimes think. We are reminded of this fact when earthquakes occur and when we observe strata deformed and broken by compressional and tensional forces. Geologic structures such as faults and folds are the result of such forces.

Faults are breaks in crustal rocks along which there has been a displacement of one side relative to the other. According to the directions of that movement, faults are classified as normal, reverse, thrust (low-angle reverse), or lateral (strike-slip). There are intermediate types as well. Normal faults result from tensional forces and are recognized by the apparent downward movement of the hanging wall relative to the footwall. In reverse faults, the hanging wall appears to move up relative to the footwall. A thrust fault is merely a variety of reverse fault characterized by a fault plane that has a low angle of inclination relative to a horizontal plane. A lateral (strike-slip) fault is one whose displacement has been predominantly parallel to the strike of the fault plane.

No less important than faults as evidence of the Earth's instability are folds, the principal categories of which are anticlines, domes, synclines, basins, and monoclines. Both anticlines and domes have uparched strata but differ in that domes are roughly symmetric, with beds dipping more or less equally away from some central point. Synclines and basins are composed of down-folded strata, with the basin being more or less circular and characterized by beds that dip inward toward the center. In monoclines (Fig. 5-18), strata dip for an indefinite length in one direction and then return to their former, usually horizontal, attitude.

The majority of geologic structures were either directly or indirectly produced by the movement of tectonic plates. According to the concepts embodied in plate tectonics, the crust of the Earth and part of the upper mantle compose a brittle shell called the lithosphere. The lithosphere is broken into a number of plates that ride over the asthenosphere. Heat from within the Earth may create convection currents in the mantle. Such currents may provide part of the force needed to move tectonic plates, but other forces may include the push on the lithospheric plate that is exerted by material sliding outward from the high-standing spreading center (midoceanic ridge) and the pull on the plate as the cold leading edge descends into a subduction zone.

Materials for the lithospheric plates originate along fracture zones typified by midoceanic ridges. Along these tensional features, basaltic lavas rise and become incorporated into the trailing edges of the plates, simultaneously taking on the magnetism of the Earth's field as they crystallize. In recent geologic time, this process has provided a record of geomagnetic reversals that in turn permit geophysicists to determine the rates of sea-floor spreading.

As long as they do not have continents on opposing edges, when plates collide, one of them may slide beneath the other and enter the mantle at a steep angle, to be remelted at depth. The zone of collision between plates may be marked by systems of volcanoes, deep-sea trenches, and great mountain ranges. The continental crustal segments of the lithosphere ride passively on the plates. They have a lower density than does oceanic crust and do not sink into subduction zones. This explains why continents are older than the ocean floors and why, when continents collide, they produce mountain ranges rather than trenches. Where continents straddle zones of divergent movements in the asthenosphere, the landmass may break and produce rift features such as the Red Sea and Gulf of Aden in the Afar triangle. These fracture zones may then widen further, resulting in the formation of new tracts of ocean floor.

Aside from the suturing of one large land mass onto another, a continent may grow in either of two ways, both involving mountain building. One process involves orogenic compression and metamorphism of sediments that had accumulated in marginal depositional basins. The second is by accretion of microcontinents and oceanic microplates bearing volcanoes, sea-floor plateaus, and seamounts. These bits and pieces of crust are carried by sea-floor spreading to marginal subduction zones, where they are accreted onto the edge of the continent as allochthonous terranes. The Cordillera of North America is largely composed of a complex collage of such terranes, each having its own distinctive geology that can often be traced to distant source regions.

Plate tectonic events and processes have had an important impact on the evolution and distribution of life on Earth. In general, fragmentation of a supercontinent has produced greater biologic diversity, whereas consolidation of landmasses into a supercontinent diminishes diversity.

QUESTIONS FOR REVIEW AND DISCUSSION

1. If one were able to drill a well from the North Pole to the center of the Earth, what internal zones would be penetrated?

2. What are the three major categories of seismic waves? Describe their characteristics.

3. What does the presence of an S-wave shadow zone indicate about the interior of the Earth?

4. What is a seismic discontinuity? Where are the Gutenberg and Mohorovičić discontinuities located?

5. How do anticlines (and domes) differ from synclines (and basins) with regard to the age relations of rocks exposed across the erosionally truncated surfaces of these structures?

6. What kind of rock approximates the average composition of the continental crust? The oceanic crust? The mantle?

7. What kind of body waves would be received on a seismograph located 180° from the epicenter of an earthquake?

8. What are the principal categories of faults? What kinds of faults might one find in regions subjected to great compressional forces? What kinds of faults result primarily from tension in the Earth's crust?

9. What is a gravity anomaly? What sort of gravity anomaly might one expect over a subduction zone? A midoceanic ridge?

10. What are folds? What are the principal kinds of folds?

11. Compile a list of items that Alfred Wegener might have used to convince a skeptic of the validity of his theory of continental drift.

12. Do midoceanic volcanoes have lavas of granitic or basaltic composition? What is the composition of lavas in mountain ranges adjacent to subduction zones, such as the Andes? Account for the differences in composition.

13. According to plate tectonics, how did the Himalaya Mountains form? The San Andreas fault? The Red Sea and Dead Sea?

14. According to plate tectonics, where is new material added to the sea floor, and where is older material consumed?

15. A moving tectonic plate must logically have a leading edge, a trailing edge, and sides. Where are the leading and trailing edges of the North American tectonic plate?

16. What is remanent magnetism? What is its origin? How is it used in finding ancient pole positions? How has remanent magnetism helped validate the concept of plate tectonics?

17. Why are the most ancient rocks on Earth found only on the continents, whereas only relatively younger rocks are found on the ocean floors?

18. What evidence would one seek to support an interpretation that a particular area in the Appalachians was an allochthonous terrane? How would an allochthonous terrane derived from a microcontinent differ from one derived from a volcanic island arc?

19. How has the detection of reversals in the Earth's magnetic field been used as support for the concept of sea-floor spreading?

20. How do the alignment and age distribution of volcanic islands in the Pacific Ocean provide evidence of sea-floor spreading?

READINGS

Allègre, C. 1988. *The behavior of the Earth*. Cambridge, MA: Harvard University Press.

Ben-Avraham, Z. 1981. The movement of the continents. *Am. Sci.* 69:291–299.

Bolt, B. A. 1982. *Inside the Earth*. San Francisco: W. H. Freeman & Co.

Condie, K. C. 1997. *Plate Tectonics and Crustal Evolution*. Newton, MA: Butterworth-Heinemann.

Erickson, J. 2001. *Plate Tectonics*, Revised ed., New York: Checkilometersark Books.

Kearey, P., and Vine, F. J. 1996. *Global Tectonics*, 2nd ed. Oxford, England: Blackwell Science Ltd.

Moores, E. M., and Twiss, R. J. 1995. *Tectonics*. New York: W. H. Freeman & Co.

Murphy, J.B. and Nance, R.D. 1992. Mountain belts and the supercontinent cycle. *Sci. Am.* (April), p. 84.

Valentine, J. W., and Moores, E. M. 1972. Plate tectonics and the history of life in the oceans. *Sci. Am.* 230:80–89.

Wegener, A. 1929 (1966 translations). *The Origin of the Continents and Oceans*. New York: Dover Publications.

WEB SITES

The Earth Through Time Student Companion Web Site (www.wiley.com/college/levin) has online resources to help you expand your understanding of the topics in this chapter. Visit the Web Site to access the following:

1. Illustrated course notes covering key concepts in each chapter;
2. Online quizzes that provide immediate feedback;
3. Links to chapter-specific topics on the web;
4. Science news updates relating to recent developments in Historical Geology;
5. Web inquiry activities for further exploration;
6. A glossary of terms;
7. A Student Union with links to topics such as study skills, writing and grammar, and citing electronic information.

6

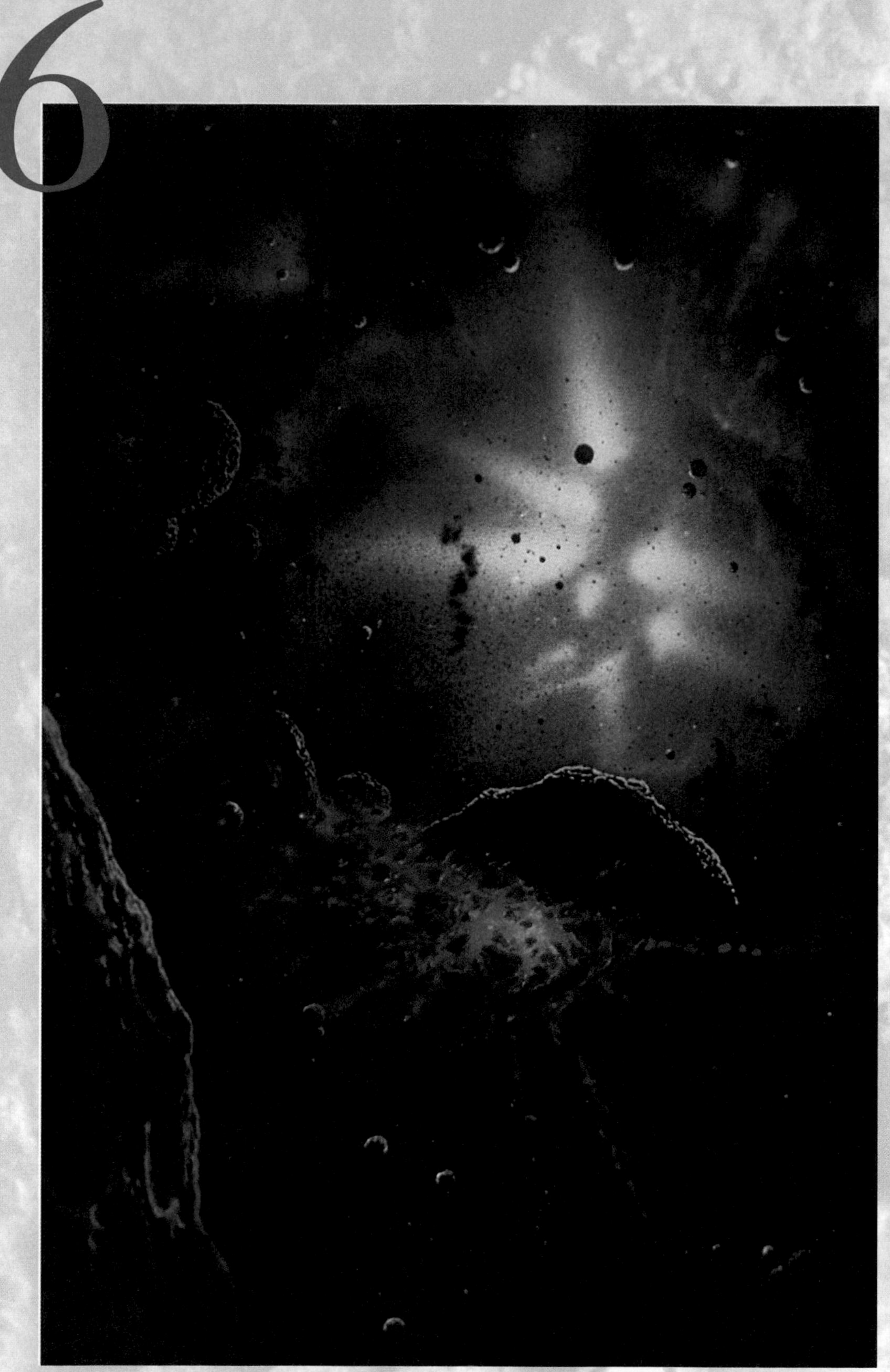

Collisions between large bodies that had aggregated in the solar nebula during the formative stages of the solar system. (Painting by Don Dixon.)

CHAPTER 6

The Primordial Earth: Hadean and Archean Eons

What was the Earth like in the very beginning? How does one get from a conglomerate ball of stardust to a planet with concentric structure—inner and outer core, silicate mantle, crust, oceans, and layered atmosphere?
Preston Cloud, *Oasis in Space*, W. W. Norton & Co., 1988

OUTLINE

Intense heat and meteor bombardment characterize the Earth's formative stages. The name *Hadean* for the planet's first eon appropriately suggests the hellish conditions existing at the dawn of Earth history. The Hadean began about 4.6 billion years ago. Much of what is known about the Hadean has been gleaned from meteorites and from explorations of the planets and their satellites in our solar system. The Archean Eon followed the Hadean (Fig. 6-1). Tentatively, the beginning of the Archean is placed at 3.96 billion years ago, the date provided by the Earth's oldest known expanse of crustal rocks, the Acasta Gneiss of northwestern Canada. There is, however, evidence of an even older crust. It consists of 4.4-billion-year-old detrital grains of zircon extracted from metamorphosed sediments in western Australia. Zircon most commonly crystallizes from a granitic melt, indicating the existence of granitic crust. Thus, the 4.4-billion-year-old zircons indicate that the beginning of the Archean may have been earlier than 3.96 billion years ago. Additional evidence of Hadean history must be gleaned from meteorites and from explorations of the moon and planets. Our solar system neighbors provide clues to the origin of the Earth's internal concentric structure, its radiation-shielding magnetosphere, its atmosphere, and its hydrosphere.

THE EARTH IN SPACE

The Earth is one of nine planets that revolve around a rather average star, our Sun. The Earth is an approximately spherical planet with a diameter of nearly 13,000 kilometers (8,000 miles), with a vast ocean that

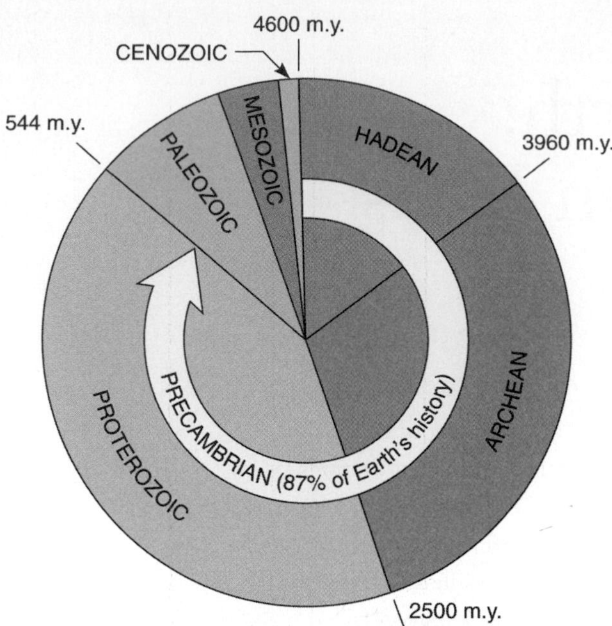

FIGURE 6-1 Proportions of geologic time encompassed by the Precambrian and its Hadean, Archean, and Proterozoic eons.

covers over 71 percent of its surface, and with an atmosphere composed prmarily of nitrogen (78 percent) and oxygen (21 percent). The average density of the solid Earth is 5.5 g/cm^3. Surface rocks, however, have densities of 2.5 to 3.0 g/cm^3. The disparity indicates that high-density material such as iron lies in the interior of the Earth. As described in the previous chapter, most of this high-density material is concentrated in the core, which has a diameter of 7000 kilometers (larger than the diameter of the planet Mercury). The core is surrounded by the mantle. The mantle extends from the base of the crust to the core-mantle boundary, which lies at a depth of 2900 kilometers.

In order of increasing distance from the Sun, the planets of our solar system are Mercury, Venus, Earth, Mars, Jupiter, Saturn, Uranus, Neptune, and Pluto. Pluto, however, may not be a true planet. It is considered by many astronomers to be a large satellite that escaped from Neptune. A belt of asteroids orbits the Sun in the region between the paths of Mars and Jupiter. This grouping of planets and asteroids around the Sun constitutes our **solar system** (Fig. 6-2). Certainly, ours is not the only such system in the universe. Planets of other systems are too distant to be detected directly, but we know that they are out there because of wobbling motions detected in distant stars. The wobbling is believed to be caused by the gravitational pull of orbiting, but not visible, planets.

Our solar system is a small part of a much larger aggregate of stars, planets, dust, and gases called a **galaxy**. Galaxies are also numerous in the universe, and their constituents are arranged into several different general forms: tightly packed elliptical galaxies, irregular galaxies, and the more familiar discoidal spiral galaxies with their glowing central bulge and great curving arms. The galaxy in which our solar system is located is called the Milky Way galaxy because as we look toward its dense central bulge, we see a great milky haze of light

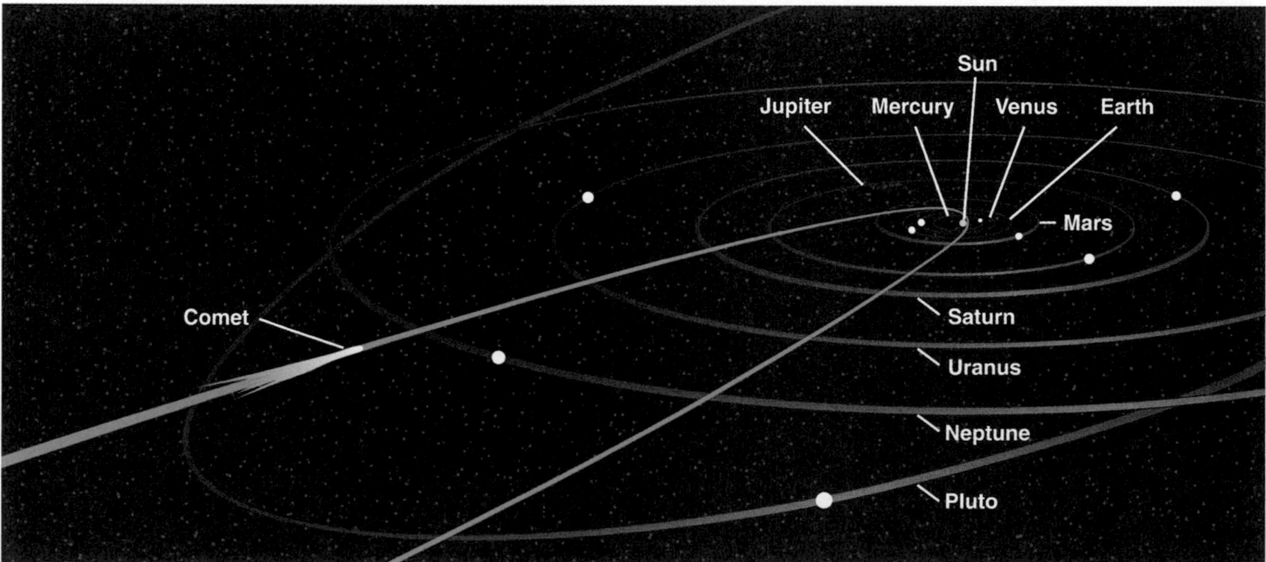

FIGURE 6-2 Schematic view of the solar system, showing orbits of the planets. The asteroid belt lies between the orbits of Mars and Jupiter. In 1997, another planetary body about 500 kilometers in diameter was discovered beyond Pluto, providing evidence that the solar system extends farther than was previously thought. **?** *In terms of size and density, how do the inner planets (Mercury, Venus, Earth, and Mars) differ from such outer planets as Jupiter, Saturn, Uranus, and Pluto? (See Table 6-1.)*

ENRICHMENT

The Origin of the Universe

Just as the origin of the Earth is dependent on the origin of the solar system, so also is the origin of the solar system linked to the formation of the universe. Theories for the origin of the universe must conform to an important astronomical observation called the **red shift**. To understand the red shift, one may recall that light, on passing through a glass prism, is separated into a band of differing colors—a spectrum. The spectrum of a star reveals not only the star's composition but also whether it is moving toward or away from the Earth and at what speed. For example, if a star is moving away from the Earth, wavelengths of the light from the star are shifted toward the red end of the spectrum. (If the star is moving toward the Earth, the shift would be toward the blue end.) In addition, the greater the velocity of separation, the wider the observed red shift. By 1914, the astronomer W. M. Slipher had found 12 galaxies that clearly exhibited red shift. He reasoned that most of the galaxies within his range of observation were moving away from each other at great speed. In 1929, Edwin Hubble made the further discovery that the red shift increased with increasing distances of the galaxies, indicating that the more distant the galaxy, the higher its receding velocity. Thus, red shift indicates that the universe is expanding.

If we were to make a movie of the galaxies moving apart as indicated by the red shift and then play back the film in reverse, it is clear that all galaxies would come together at a single location. This infinitely dense point would then explode. Astronomers have dubbed the great explosion the **big bang**. Calculations based on the amount of expansion that has already occurred in the universe indicate that the big bang occurred between 15 and 18 billion years ago. It marked the instantaneous creation of all matter in the present universe. Initial heat in the universe reached about 100 billion degrees Celsius. During the first second, the universe cooled to about 10 billion degrees (a thousand times hotter than the center of the Sun today). Atoms could not exist at such high temperatures, so only radiant energy, very light particles called neutrinos, and electrons existed. The formation of protons and neutrons quickly followed. After about 1.5 minutes, temperatures dropped to about 1 billion degrees Celsius. A few simple atomic nuclei formed by fusion (hydrogen and helium). During the subsequent million years or so, the universe continued to cool as it expanded. Atoms began to form when the temperature had fallen to a few thousand degrees. By about a billion years after the big bang, stars and galaxies had probably begun to form. In the interior of stars, matter was reheated, nuclear reactions ignited, and synthesis of heavier elements was initiated.

Some astronomers are uneasy about a cosmic evolution that starts with a big bang and ends with galaxies disappearing somewhere out in the farthest reaches of space. They support **steady-state** cosmology of a universe that will continue to expand forever. As expansion progresses, new matter, initially in the form of hydrogen, is formed from combinations of atomic particles in the space between galaxies at about the same rate that older material is receding. In this way, a uniform density of matter in the universe is maintained.

There is also the possibility of combining the big bang and steady-state theories into one theory that stipulates that for every big bang there is a big crunch. According to this so-called **oscillating universe** cosmology, after the great explosion that begins expansion, gravitational forces begin to prevail and cause matter to be drawn back to its place of origin. The expanding universe would thus become a contracting one. In the last stages of contraction, matter would be returning at enormous speed, compressing the returned particles into the infinitely dense mass that undergoes the next big bang. An estimated 100 billion years would be required to complete the cycle.

in the heavens. The Milky Way is a spiral-type galaxy (Fig. 6-3). It rotates slowly in space, completing one rotation about every 240 million years. Our Sun is located about two thirds of the distance (or 26,000 light-years) outward from the center of the galaxy to its edge. (A light-year is the distance light travels in a vacuum in 1 year; it amounts to 6×10^{12} miles.)

THE SUN

For many reasons, the Sun is the most important member of our solar system. It is the fountainhead and sustainer of advanced forms of life on Earth. Without the gravitational attraction of the Sun, the planets would wander off into space. Our sun is a modest star (in comparison to others in our galaxy). Nevertheless, it is about 1.5 million kilometers in diameter and contains 98.8 percent of the material in the solar system. The Sun is composed of the same elements as the Earth, but their proportions are quite different. About 73 percent of the Sun's mass is hydrogen. Another 25 percent is helium. The remaining 2 percent consists of heavier elements that exist as gases in the interior, where temperatures exceed 20 million degrees Celsius.

Although an enormous amount of solar energy is intercepted by the Earth, our planet is not roasted but is able to maintain a range of temperatures roughly between −50 and +50°C. The maintenance of this vital temperature range is made possible by three factors. First, because of rotation, the Earth receives energy from the Sun on one hemisphere, whereas it returns heat to space over its entire surface. Second, some of

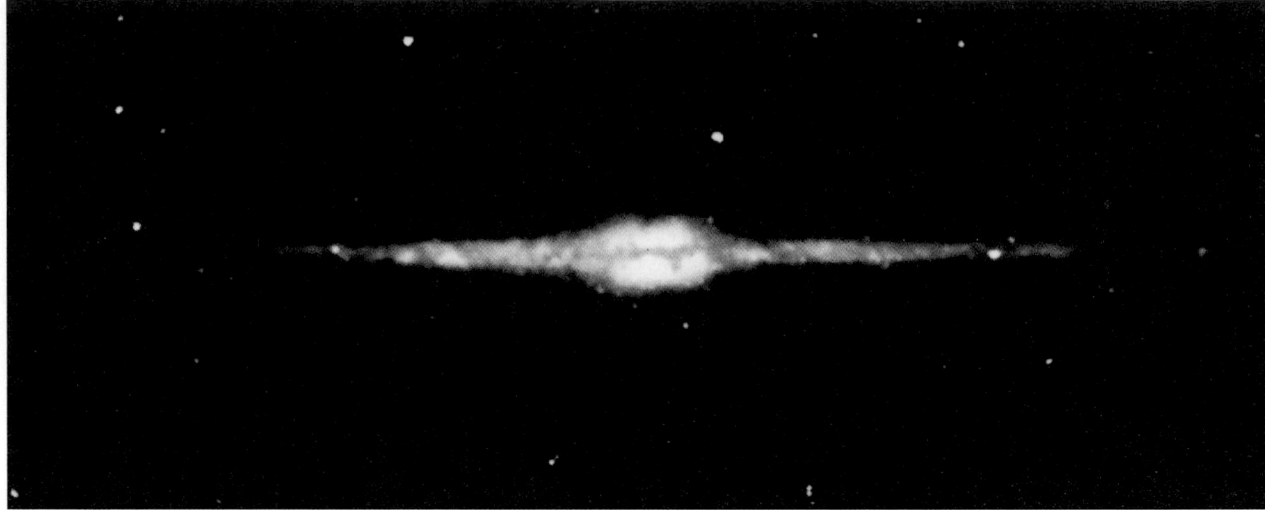

FIGURE 6-3 A view of the Milky Way Galaxy obtained by the diffuse infrared background experiment on NASA's *Cosmic Background Explorer Satellite* in 1990. When we look edge-on toward the center of the galaxy, so many stars are in our line of sight that we see a glow of light in a circular band that appears "milky" in its luminescence. (Courtesy of NASA.) ? *What would the galaxy look like if viewed from above?*

the incoming radiation is reflected off the atmosphere and clouds and is directed back into space without ever reaching ground level. Finally, a part of the intercepted radiation is absorbed by the atmosphere and radiated back into space without warming the Earth's surface.

The energy that maintains the Sun as a great glowing sphere of gases is derived from a continuous thermonuclear reaction called **fusion**. In the fusion process, hydrogen is changed to helium, and excess mass is converted to energy (Fig. 6-4). Each second, the Sun transmits an amount of energy equivalent to that which would result from the burning of 25 billion pounds of coal. This energy from the Sun is the ultimate force behind the many geologic processes that continuously change the Earth's surface. For example, the Sun's rays aid in the evaporation of surface waters, which in turn results in clouds that provide the precipitation required for erosion. Along with the Earth's rotation, the Sun's radiation results in winds and ocean currents. Some scientists believe that protracted variations in the heat received from the Sun may trigger episodes of continental glaciation or may reduce lush forests to barren wastelands. Although its effect is less than that of the moon, the Sun helps to move the tides.

THE MOON

As satellites go, our Moon is large relative to the size of its parent planet. It has a diameter of about one fourth that of the Earth. Its mean density of 3.3 g/cm^3 is similar to the density of the upper part of the Earth's mantle. The Moon rotates on its axis in exactly the same time that it takes to orbit the Earth. As a result of this synchronous rotation, we always see the same side of the Moon (Fig. 6-5). Images of the far side of the moon are transmitted to us from lunar orbiters. These images reveal a surface more densely cratered than the side of the Moon facing the Earth.

In 1610, Galileo Galilei made the first telescopic observations of the Moon's near side. With his primitive telescope, he was able to see the Moon's surface was "rough, full of cavities and prominences . . ." Galileo called the lighter-hued, craggy, and more reflective

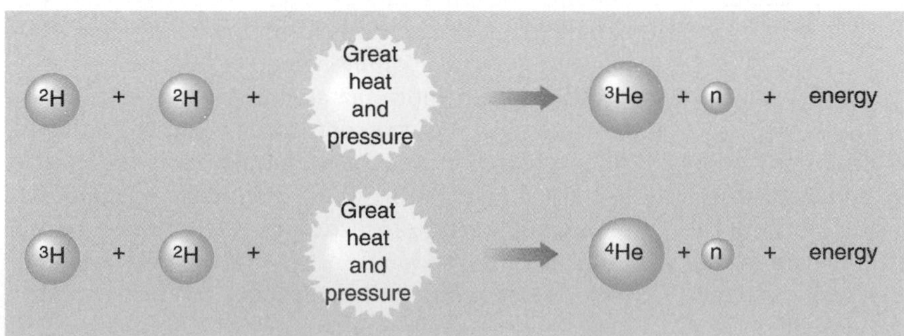

FIGURE 6-4 Examples of two fusion reactions. (n = neutron).

FIGURE 6-5 The Moon as photographed by the *Apollo 17* mission. The region at the right is part of the Moon's far side. It is the side never seen from the Earth. The dark regions are maria floored by basalt lava flows. The heavily cratered lighter regions are the lunar highlands, showing the effect of heavy meteoric bombardment. *(Courtesy of NASA.)* *What evidence suggests that, after the extrustion of mare basalts, meteors striking the moon were smaller and their arrival less frequent than in the Moon's earlier history?*

terrains **terrae**. Today, however, the term **lunar highlands** is preferred. Rocks collected from the moon as part of the Apollo Program have been isotopically dated. They reveal that the highlands originated more than 4.2 billion years ago. They are Hadean in age. The dense cratering of the highlands clearly indicates that our satellite, *as well as the Earth itself,* experienced a massive episode of meteoric bombardment 4.2 to 3.9 billion years ago. On Earth, because the original crater-bearing Hadean crust has been recycled into the mantle by plate tectonic processes, we no longer see Hadean craters. Plate tectonic processes have not been active on our satellite, so the Moon's cratered terrains are preserved.

In addition to the many "small spots" that we now recognize as craters, Galileo also observed the large dark areas that children imagine as facial features of "the man in the moon." The darker areas form the floors of immense basins. Galileo called these **maria** (seas) and suggested they might be filled with water. Today we know the dark areas have been flooded, not with water, but with dark basaltic lava. In many maria, the basalt is contained within basins that are hundreds of kilometers across and 10 to 20 kilometers deep. Isotopic dating of the mare basalts reveal that they were extruded from about 3.8 to 3.2 billion years ago. Because the mare basins are not heavily cratered, we know that meteor strikes had diminished for both the Moon and the Earth following the extrusion of the mare basalts.

Mare basalts (Fig. 6-6*A*) resemble basalts found on Earth. They are dark, finely crystalline rocks composed of calcium feldspar, pyroxene, olivine, and ilmenite. In contrast, rocks of the lunar highlands are coarsely crystalline. Calcium feldspar is the dominant mineral. They are termed anorthosites, a group of rocks related to gabbro.

Many planetary geologists believe that the moon originated when another body somewhat larger than Mars smashed into the Earth about 4.4 billion years ago. The debris from the impact went into orbit around the Earth, where it collected to form the Moon. Heat generated by the massive meteoric bombardment of the early Moon, as well as from gravitational compression of the debris, melted the Moon's outer zone, allowing lower-density alumino-silicates to rise to the surface and form a lunar crust. Higher density iron-rich minerals sank to deeper levels, where they solidified as a lunar

(*A*)

(*B*)

FIGURE 6-6 Moon rocks. (*A*) A+ lunar basalt collected by *Apollo 15* astronauts. (*B*) A rock from the lunar highlands. It is composed mainly of plagioclase (the white mineral) and olivine (yellow grains). *(Courtesy of NASA.)* *How were the spherical holes in the lunar basalt produced? What does this indicate about the rate of cooling of this rock?*

mantle. At the same time, heat from the decay of radioactive elements raised the temperature of the Moon's interior to the point where a small core may have developed. Finally, during the next several million years, lavas began to flow onto the surface of the satellite, filling the mare basins.

FORMATION OF THE SOLAR SYSTEM

Any account of the origin of the solar system is obliged to conform to certain basic characteristics of the system. These include dynamic constraints relating to the movements of planets, the composition and density of the planets, and their age. With regard to the dynamic constraints, we know that the planets all *revolve* around the Sun in the same direction and their paths are approximately in the plane of the Sun's rotation. Thus, the solar system is like a disk in shape. The direction taken by the planets as they revolve around the Sun is counterclockwise (called prograde) when viewed from a hypothetical point in space above the Sun's north pole. The direction of *rotation* of the planets on their axes is also counterclockwise, with the exception of Venus and Uranus. The tilt of the axes of rotation provides for seasonal change on planets. With few exceptions, satellites mimic the movements of the planets they orbit.

As indicated by their differing densities (Table 6-1) as well as by Earth, Moon, and meteorite samples, the planets differ in composition. Mercury, Venus, Earth, and Mars are small, dense, rocky, and rich in metals. They have mean densities of 5.4 g/cm^3, 5.2 g/cm^3, 5.5 g/cm^3, and 3.9 g/cm^3, respectively. Jupiter, Saturn, Uranus, and Neptune have lesser densities, of 1.3 g/cm^3, 0.7 g/cm^3, 1.3 g/cm^3, and 1.6 g/cm^3, respectively. They are termed the Jovian planets (see Fig. 6-5). Like the Sun, the composition of Jupiter and Saturn is dominated by hydrogen and helium.

There are mineral grains known to be 4.4 billion years old in sedimentary rocks found on Earth. The maximum age obtained from lunar samples and from meteorites is 4.6 billion years. It is a determination that agrees with theoretical calculations of the age of the Sun. Thus, we can place the birth of the solar system at about 4.6 billion years ago.

The Solar Nebula Hypothesis

When the dynamic, compositional, and age constraints are taken into account, they favor a hypothesis that the solar system was derived from a rotating cloud of dust particles and gases called the **solar nebula**. This basic concept was first suggested in 1755 by Immanuel Kant. About 40 years later, Pierre Laplace further developed Kant's original theory, and in recent years it has been improved with the rich flow of information emanating from the space program.

The modern theory begins with a cold, rarefied cloud of gases and dust particles. This initial material consisted of elements and the chemical compounds formed by combinations of elements. The elements forming the cloud were originally produced by nuclear reactions within stars or by great explosions that mark the death of stars. Hydrogen, helium, oxygen, silicon, and carbon were prominent elements in the cloud.

After the elements and their compounds had collected in a region of space, the dust cloud began to rotate in a counterclockwise direction, to contract, and to assume a discoidal shape. About 90 percent of the cloud's mass remained concentrated in the thicker central part. Then, while rotation and contraction were occurring, turbulence began to affect the cloud system. Chaotic swirling and churning movements were superimposed on the grander primary motion of the cloud. Each turbulent eddy was affected by adjacent eddies, and some served as collecting sites for matter from neighboring swirls. When the disk had shrunk to a size somewhat larger than the present solar system, its denser condition permitted condensation of concentrated knots of matter. Smaller particles merged to build chunks up to several meters in diameter, and

TABLE 6-1 The Planets

Name	Distance from Sun (AU)*	Revolution Period (yrs)	Diameter (km)	Mass (10^{23} kg)	Density (g/cm^3)
Mercury	0.39	0.24	4,878	3.3	5.4
Venus	0.72	0.62	12,102	48.7	5.3
Earth	1.00	1.00	12,756	59.8	5.5
Mars	1.52	1.88	6,787	6.4	3.9
Jupiter	5.20	11.86	142,984	18,991	1.3
Saturn	9.54	29.46	120,536	5,686	0.7
Uranus	19.18	84.07	51,118	866	1.2
Neptune	30.06	164.82	49,660	1,030	1.6
Pluto	39.44	248.60	2,200	0.01	2.1

*An AU (or astromonical unit) is the distance from the Earth to the Sun. (From Fraknoi, A, Morrison, D. and Wolff, S. 2001. Voyages Through the Universe. Philadelphia: Saunders College Publishers.)

these larger bodies swept up finer particles within their orbital paths. This process of accumulation of bits of matter around an initial mass is called **accretion**. Because the particles available for accretion were cold and essentially similar throughout the period of accretion, the process is often further described as **cold homogeneous accretion**. The large bodies formed from swarms of accreting materials can be termed **protoplanets**. The protoplanets were enormously larger than present-day planets. Each rotated somewhat like a miniature dust cloud, and each eventually swept away most of the debris in its orbital path and was able to revolve around the central mass without collision with other protoplanets (Fig. 6-7). The protoplanets' formation from the original dark, cold cloud required an estimated 10 million years.

While the protoplanets were in the process of condensation and accretion, material that had been pulled into the central region of the nebula condensed, shrank, and was heated to several million degrees by gravitational compression. The Sun was born. Later, when pressures and temperatures within the core of the Sun became sufficiently high, thermonuclear reactions began providing further energy by the nuclear fusion of hydrogen atoms into atoms of helium. (Today, the Sun converts about 596 million metric tons of hydrogen into 532 million metric tons of helium each second. The difference represents matter that has been converted to energy.) As the Sun began its thermal history, its radiation ionized surrounding gases in the nebula. These ionized gases in turn interacted with the Sun's magnetic lines of force, causing the magnetic field to adhere to the surrounding nebula and drag it around during rotation. At the same time, the nebula caused a drag on the magnetic field and slowed down the Sun's rotation.

The stream of radiation from the Sun, sometimes called solar wind, drove enormous quantities of the lighter elements and frozen gases outward into space. This solar force is what causes a comet's tail to show wavy streaming or to be bent away from the Sun. As would be expected, the planets nearest the Sun lost enormous quantities of lighter matter and thus have smaller masses but greater densities than the outer planets. They are composed primarily of rock and metal that could not be carried away by the solar wind. The Jovian planets were the least affected and retained their considerable volumes of hydrogen and helium surrounding their rocky inner cores.

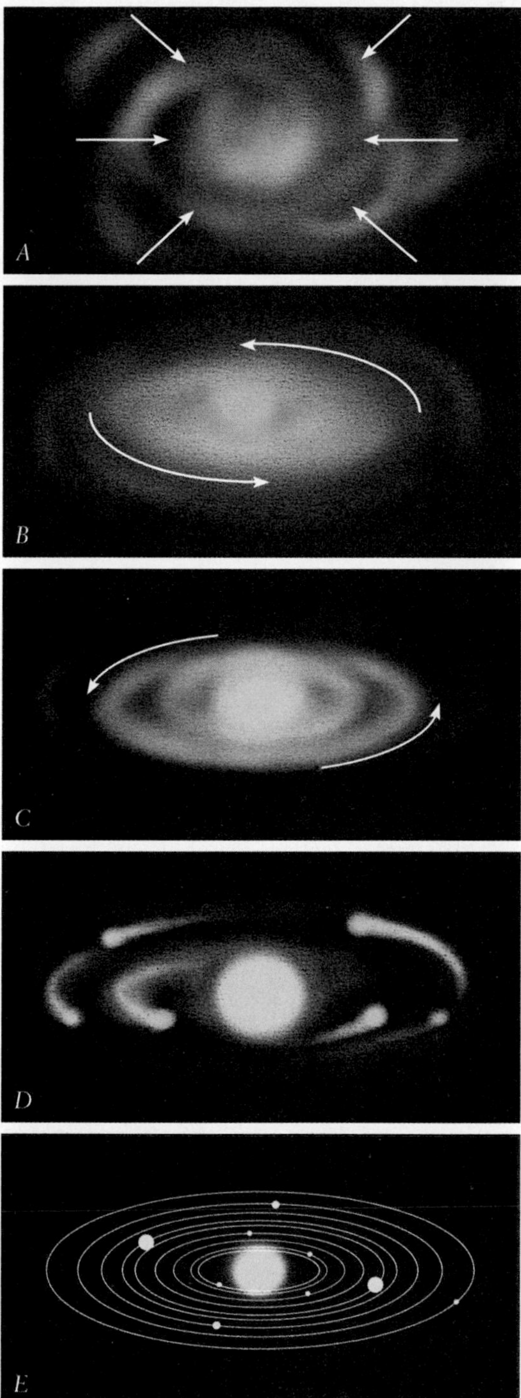

FIGURE 6-7 An illustration of the solar nebula hypothesis for the origin of the solar system. The solar nebula shrinks and forms a disk as it begins rotation. Much of the material is concentrated centrally, which ultimately becomes the sun. Solid particles in the disk condense as the nebula cools, and accretion of particles forms the planets, the asteroid belt, and satellites. (*From Fraknoi, A., Morrison, D., and Wolff, S. 1997. Voyages Through the Universe: Philadelphia, Saunders College Publishing.*)

Meteorites: Samples of the Solar System

Meteorites are interplanetary chunks of rock that have reached the Earth from space. They provide clues to the origin and Hadean history of the Earth, Moon, and planets. Meteorites that have impacted our planet are fragments of asteroids formed when they collided

FIGURE 6-8 Small carbonaceous chondrite. This meteorite fell near Murray, Kentucky, in 1950. Scale divisions are in millimeters. (*From the Washington University collection.*)

in space and were shattered. Such collisions would send chunks of rock into orbits that crossed the orbital paths of the Earth, Moon, and other planets in the solar system. Many meteorites bear compositional and textural evidence of such collisions. Some meteorites are chemically similar to lunar basalts and are believed to have come from our moon. Also of interest are some having similarities with Martian materials analyzed by the *Viking* lander. About 500 meteorites at least as large as a baseball survive passage through the Earth's atmosphere and crash into our planet each year. Weathering and erosion have erased most of the craters made by the impact of larger meteorites. Only about 70 partially preserved craters or clusters of craters remain.

Meteorites are named according to their composition as ordinary chondrites, carbonaceous chondrites, achondrites, and stony-irons. **Ordinary chondrites** are the most abundant. As revealed by uranium-lead, strontium-rubidium, and potassium argon dating methods, they are about 4.6 billion years old and thus constitute samples of Hadean materials. Many ordinary and carbonaceous chondrites contain spherical bodies called **chondrules**. The spherical shape of chondrules indicates they solidified from molten droplets splashed into space during collisions of objects swirling about in the solar nebula.

Carbonaceous chondrites (Fig. 6-8) contain about 5 percent organic compounds, including dozens of inorganically produced amino acids. Some of these are protein-building amino acids, and many contain the basic building blocks of DNA and RNA found in all living things. Thus, carbonaceous chondrites (as well as carbon-bearing asteroids and comets) not only provided much of the Earth's early carbon but may also have supplied the organic building blocks needed to construct life.

A small percentage of stony meteorites do not contain chondrules and are therefore termed **achondrites**. They are composed of angular fragments that indicate they are the products of collisions of larger bodies similar to basalts on Earth. Most have a composition resembling terrestrial basalt.

Iron meteorites (Fig. 6-9) are intergrowths of two varieties of iron-nickel alloy. The large size of the crystals indicates a long episode of slow crystallization within a parent body sufficiently large to provide insulation for the hot interior, followed by violent disruption of that body by a collision. Such larger bodies that really constitute minor planets are termed **asteroids**.

The least abundant of meteorites are the **stony-irons**. They are composed of silicate minerals and iron-nickel alloy in about equal proportions. The stony-irons probably originated near the boundary between the iron core and stony silicate mantle of an asteroid destroyed by collision.

FIGURE 6-9 A 15-ton iron meteorite. This large meteorite, named the Williamette Meteorite, was discovered in 1902 near Williamette, Oregon. It is on display at the American Museum of Natural History in New York City.

THE EARTH'S NEAREST NEIGHBORS

The first four planets in the solar system (Mercury, Venus, Earth, Mars) are called **terrestrial planets** because they are relatively small, rocky, and rich in metals (Fig. 6-10). The Jovian planets Jupiter, Saturn, Uranus, and Neptune are large and dominated by hydrogen and helium.

Mercury

Mercury is the smallest and swiftest of the inner planets. It revolves rapidly, making a complete journey around the sun every 88 Earth days. *Mariner 10* (launched in 1973) provided our first close look at Mercury. Photographs transmitted back to Earth revealed a planet with a moonlike densely cratered surface (Fig. 6-11) as well as smooth areas resembling maria. Great elongate scarps were discerned, and these are thought to be fracture zones formed by either tidal forces or shrinkage of the planet as it cooled. Mercury has a very thin atmosphere composed of sodium and much smaller amounts of helium, oxygen, potassium, and hydrogen. There is a weak magnetic field, a light-weight crust covered with dust, and an iron core. The existence of the core, inferred from Mercury's high density, suggests the planet underwent melting and differentiation. NASA plans to send another spacecraft back to Mercury. It will be called *Messenger* and will be launched in 2004 and go into a year-long orbit around the planet in 2009.

Venus

Although similar to the Earth in mass and volume (and therefore density and gravity), other characteristics of Venus differ from those of Earth. Venus has no oceans.

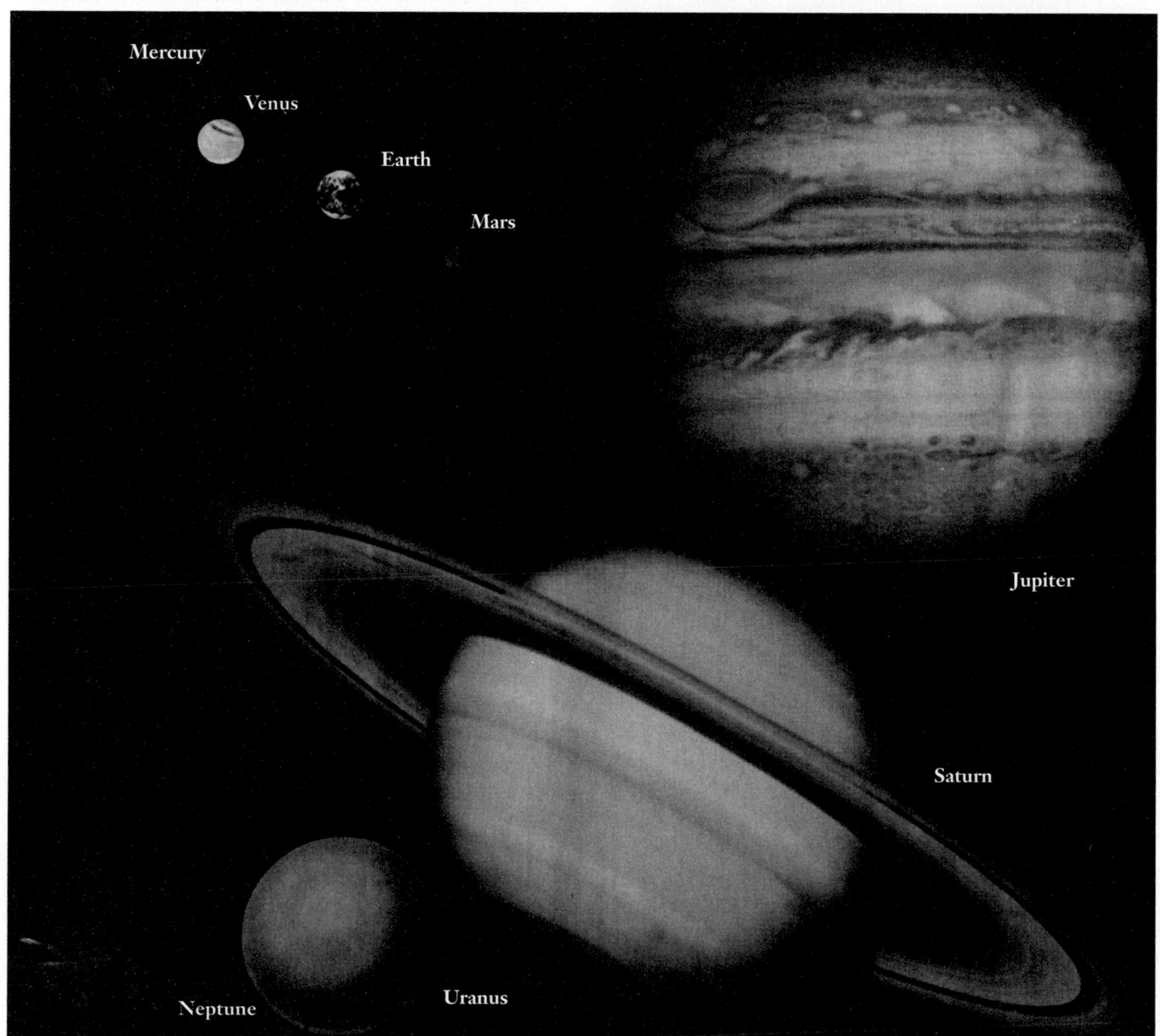

FIGURE 6-10 **The relative sizes of the planets from Mercury through Neptune.**

FIGURE 6-11 Photomosaic of Mercury, as photographed by the *Mariner* spacecraft. (*Courtesy of NASA.*)

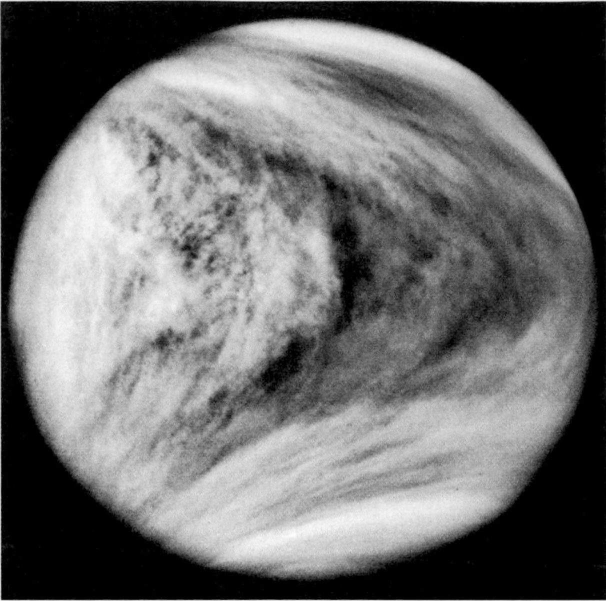

FIGURE 6-12 The Clouds of Venus, as photographed by the *Pioneer* spacecraft. In visible light nothing of the surface of Venus can be observed because of the dense cloud cover.

Venus makes one complete rotation each 243 days. Its direction of rotation is opposite to that of most other planets. One explanation for this is that during its formative stages it was struck at an angle by another huge body, causing it to rotate in the backward direction. From Venus's size and density, one can infer an interior rather like that of the Earth.

Atmospheric pressure at the surface is roughly equivalent to the pressure 1000 meters beneath the Earth's ocean. The Venusian atmosphere is 98 percent carbon dioxide. Without an ocean in which carbon dioxide can be combined with calcium and magnesium to form carbonates, there is no process to remove carbon dioxide from the atmosphere. Sunlight in the visible part of the spectrum passes through the atmosphere and heats the surface of the planet. The energy reflected off the surface, however, is mainly infrared radiation. Carbon dioxide and the other constituents of the atmosphere are largely opaque to infrared radiation. Radiation is trapped by this greenhouse effect and causes the temperature at the planet's surface to reach a searing 470°C (hot enough to melt lead). The lesson to be learned from Venus is that here on Earth we must be careful to guard against increases in the carbon dioxide content of our atmosphere.

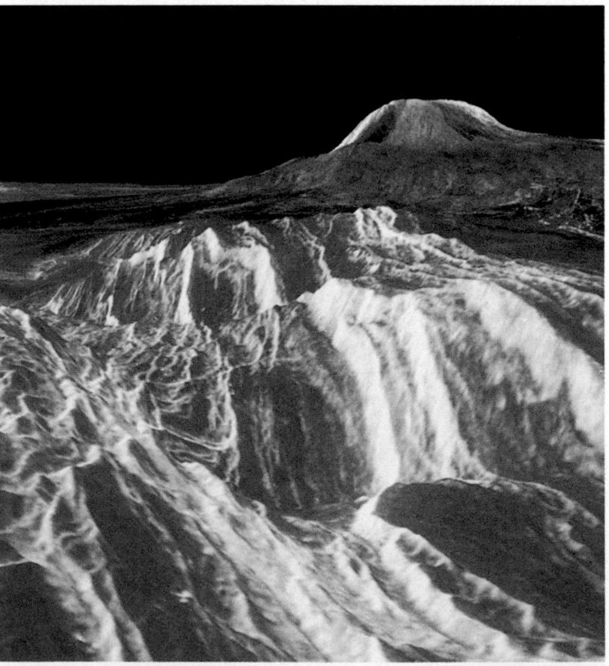

FIGURE 6-13 *Magellan* radar image of the surface of Venus. The Danu mountains, having individual peaks about 2 kilometers high, are depicted. (*Courtesy of NASA.*)

Spacecraft observations of Venus include NASA's *Pioneer* mission, the former Soviet Union's *Venera 15* and *16*, and NASA's *Magellan* mission. To penetrate the dense cloud of sulfuric acid droplets that surround the planet (Fig. 6-12), these explorations used radar imaging devices. The radar images revealed that 65 percent of the planet consists of rolling plains, about 27 percent is lowlands, and the remaining 8 percent is highlands. Among the highlands are mountains that tower 11 kilometers above the adjacent plains (Fig. 6-13). The images suggest that tectonics operate on Venus in a different way than on Earth. There is little evidence of any horizontal movements of crust. One does not see Earthlike tectonic features such as transform faults or deep trenches. Instead, features formed by vertical movements predominate. This suggests that areas of the crust were raised from below by rising plumes of viscous magma. In addition to this so-called "blob tectonics," volcanic eruptions and massive lava flows have been important in the evolution of the Venusian crust. Many of the volcanoes are enormous, measuring nearly 400 kilometers in diameter and rising 10 kilometers above surrounding lowlands. Craters are common. Most of these are larger than 1.5 kilometers in diameter, indicating only larger impacting bodies survived the passage through the atmosphere.

The *Venera 13* and *14* missions to Venus gave us a glimpse of a Venusian surface strewn with rust-colored slabby rocks and rubble. In addition to providing the photographs, one of the landers was able to drill out a few centimeters of the planet's surface, which bore a strong resemblance to oceanic basalts on Earth.

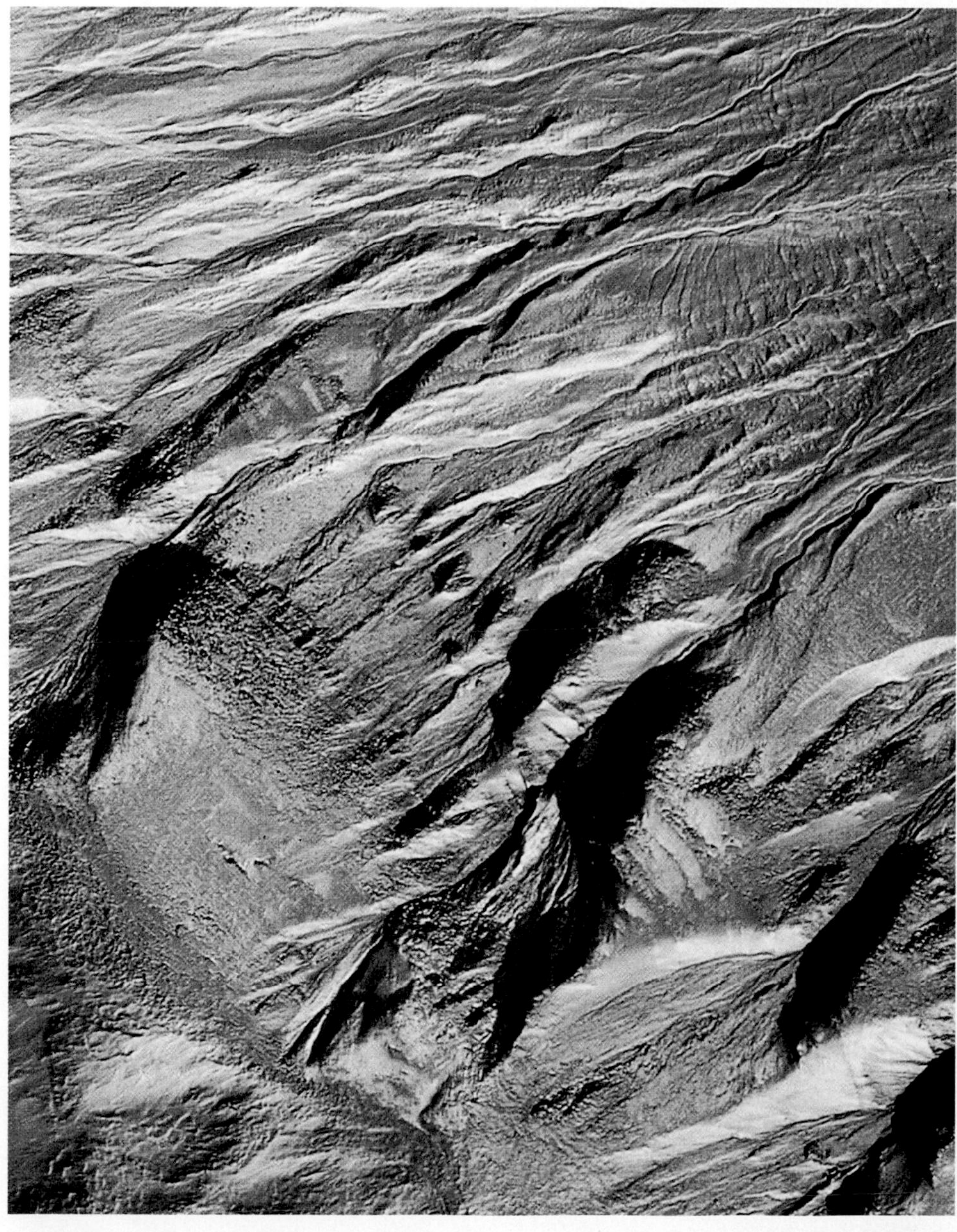

FIGURE 6-14 Heavily eroded canyonlands on Mars. This *Viking* photograph shows an area about 60 kilometers across. (*NASA/USGS, courtesy of Alfred McEwen.*)

Mars

Mars has always been of special interest to us because of speculation that some form of life, however humble and microscopic, may exist there. It is a planet only a little farther from the Sun than our own, it has an atmosphere that includes clouds and white polar caps composed of frozen carbon dioxide, it experiences seasonal changes, and it has a richly varied landscape (Fig. 6-14). Magnificent craters, colossal volcanic peaks, deep gorges, sinuous channels, dunes, and massive structures of layered rock like those in the Earth's Grand Canyon have been revealed to us by NASA's spacecraft and landers. There is ample evidence of erosion by running water, wind, and ice.

Mars' diameter is only about 50 percent that of the Earth, and its mass is only 10 percent of the Earth's. Carbon dioxide is the principal gas in its atmosphere, although small amounts of nitrogen, oxygen, and carbon monoxide are also present. Unlike Venus or the Earth, the planet has virtually no greenhouse effect. Temperatures at the equator range from about 21°C to −85°C. Mars has two small moons, named Phobos (meaning fear) and Deimos (meaning panic). A large asteroid named Gaspra also orbits the planet. The somewhat lower density of Mars, as compared with the other terrestrial planets, and its lack of a detectable magnetic field suggest that the planet has only a small iron-rich core. The rusty orange color of Martian surface materials (Fig. 6-15) is due to the presence of iron oxide. Devices on the *Viking* lander that analyzed the loose material on the Martian surface revealed that it was rich in iron, magnesium, and calcium. These elements were probably derived from parent rocks rich in ferromagnesian minerals.

The existence of water on Mars is of particular interest to those seeking evidence that life may have at one time existed on that planet. Indeed, the chemistry of life is water chemistry. It is the most abundant molecule in all organisms. When life was originating on Earth, water provided a medium for the essential movement and interaction of molecules. So little water exists today in the Martian atmosphere that if all of it were condensed, it would probably provide only enough water to fill a community swimming pool. Yet there is ample evidence of abundant liquid water on Mars early in its history. For example, images taken by the *Mars Global Surveyor* spacecraft in the year 2000 reveal massive structures of stratified rocks resembling layered sedimentary sequences deposited in seas and lakes here on Earth. Most of these stratified terrains occupy regions in or near craters, which could have served as basins for standing water. In addition to these apparent water-deposited sedimentary sequences, the surface of Mars is dissected with channels and canyons that have sinuous or braided patterns characteristically developed by streams on Earth (Fig. 6-16). Thus, in addition to bodies of standing water, flowing water was an important geologic agent early in Martian history. There may have been a large reserve of water frozen in the subsurface as a kind of permafrost. Some of the larger channels could have been eroded by water gushing out onto the surface as the permafrost melted. Many Martian channels dwarf our own Grand Canyon in size. Their formation would have required torrential floods of a magnitude difficult to visualize by Earth standards.

THE HADEAN HISTORY OF MARS Like the Earth, Mars was transformed from a protoplanet by accretion of a multitude of smaller bodies. This accretion

FIGURE 6-15 *Viking* lander photograph of the surface of Mars. The device on the right is part of the lander that was purposely dropped for scale. It is about 20 centimeters long. (*Courtesy of NASA.*)

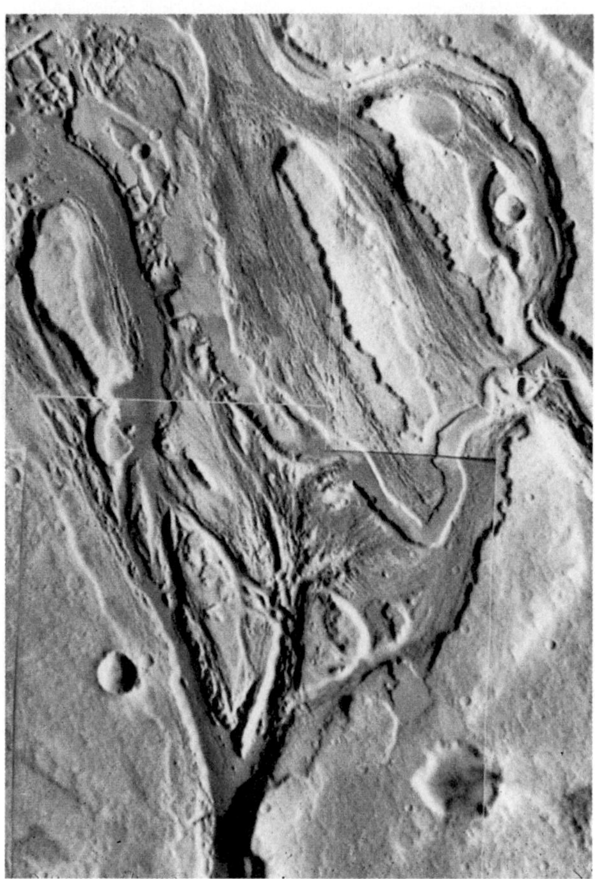

FIGURE 6-16 Mangala Vallis, a large channel system on Mars. The system resembles some dry river systems on Earth. (*Courtesy of NASA.*)

was followed by differentiation, during which the once homogeneous body became partitioned into masses that had different chemical compositions and physical properties. Very likely, the differentiation process involved melting, so that fluids could move from one region of the planet to another.

During the initial billion years or so of Martian history, the planet experienced heavy bombardment of meteorites and asteroids. During this episode, the exposed crust became densely cratered. The crust was disturbed not only by meteoric impact but by faulting and fracturing that resulted from volume changes in the mantle.

It is likely that a dense atmosphere formed on Mars as a result of volcanic outgassing (removal of embedded gases by heating). Perhaps as a result of a greenhouse effect, the planet was warmer at that time. Channels and stratified deposits indicate the presence of liquid water. Eventually the supply of atmospheric water and carbon dioxide became so depleted that the planet began to cool. Remaining carbon dioxide and water migrated to polar caps and subsurface reservoirs, where they have remained.

THE OUTER PLANETS

Beyond the orbit of Mars lies the ring of asteroids, and beyond the asteroids are the orbits of Jupiter, Saturn, Uranus, Neptune, and Pluto. The first four of these planets are similar to one another in their large sizes and relatively low densities (see Fig. 6-10). Pluto is too small to be considered a Jovian planet. Some believe it may be an escaped satellite of Neptune.

Jupiter

Jupiter, named for the leader of the ancient Roman gods, is the largest planet in our solar system. Its diameter is 11 times greater than that of the Earth, and its volume exceeds that of all other planets combined. Its mass is 318 times that of the Earth, yet it is only one fourth as dense. Jupiter rotates rapidly, making one full rotation in slightly less than 10 Earth hours. The rapid rotation has resulted in the planet's distinctive encircling atmospheric bands (Fig. 6-17). Hydrogen, helium, and lesser amounts of methane and ammonia are the predominant gases in Jupiter's atmosphere. The largest component of the planet's interior is in liquid form, probably composed of ultra-compressed hydrogen surrounding a rock core. With its 16 satellites, Jupiter resembles a miniature solar system.

Two *Voyager* spacecraft swept past Jupiter in 1979 and transmitted splendid images of the Planet's brightly colored atmosphere, the Great Red Spot, and five of Jupiter's large moons. The Great Red Spot is a huge vortex of a violent and protracted storm that extends deep into the cloud cover. In 1995, The *Galileo* spacecraft arrived at Jupiter, dropped a probe into the planet's atmosphere, and went into orbit around the Jupiter system The probe revealed that Jupiter's winds increase in velocity with depth, indicating that the energy that drives them comes from below. A ring of debris encircles the planet, making Jupiter the fourth planet in the solar system known to possess such a feature (Saturn, Uranus, and Neptune are the other planets having rings).

Among Jupiter's satellites are four large moons. Called the Galilean moons to commemorate their discoverer, they are Callisto, Ganymede, Europa, and Io (Fig. 6-18). **Callisto**, the outermost of the Galilean satellites, is riddled with craters and broad impact basins resulting from meteoric and asteroid impact. Callisto's nearest neighbor is **Ganymede**. Sinuous ridges and criss-crossing fractures scar Ganymede's cratered surface, suggesting fault movements. **Europa**, the brightest of the Galilean moons, is a predominantly rocky satellite about the size of our moon. In the year 2000, magnetic measurements of the Galilean satellites recorded by the Galileo

(A)

(B)

FIGURE 6-17 **(*A*) Jupiter, photographed by *Voyager 1* spacecraft from a distance of 33,000,000 kilometers (*B*) A close-up of Jupiter's Great Red Spot.** Note the turbulence to the west (left) of the Great Red Spot. (*Courtesy of NASA.*)

spacecraft indicated that Callisto, Europa, and Ganymede hold seas of liquid water beneath their frigid crusts. Because they have water, life may exist or may once have been present on these satellites. Whereas the surfaces of Callisto, Europa, and Ganymede are icy, **Io's** crust is pock-marked by volcanoes ejecting huge plumes of sulfur and sulfur dioxide. When the *Galileo* spacecraft approached within a few hundred kilometers of Io, it recorded about 100 actively erupting volcanoes.

FIGURE 6-18 **A composite of *Voyager* images of Jupiter and its four large Galilean satellites.** The image shows a spacecraft perspective on the way to Jupiter, so that the size of each satellite is not actual but reflects distance from the spacecraft. In order of increasing size we see Io, Europa, Ganymede, and Callisto. (*Courtesy of NASA and JPL.*)

Saturn

Saturn is the second largest body in the solar system. Its mass is equivalent to 75 Earth masses. Its density, however, is only about 70 percent that of water. Thus, if one could find a big enough swimming pool, Saturn would float in it. Measurements of Saturn's size and density suggest that the planet may have a central core of heavy elements (probably mostly iron), surrounded by an outer core of hot, compressed volatiles such as methane, ammonia, and water. These two core regions, however, constitute only a small part of the total volume of the planet, in which hydrogen and helium are the overwhelmingly predominant elements.

Voyagers 1 and *2* flew by Saturn in 1980 and 1981. Additional spacecraft are scheduled to approach Saturn in 2004. The *Voyager* missions provided truly spectacular views of Saturn's rings and moons (Fig. 6-19). Data transmitted back to Earth confirm that the planet has a magnetic field, radiation belts, and an internal source of heat.

There is perhaps no more beautiful sight in the solar system than the rings of Saturn. At one time Saturn was believed to have three rings, but spacecraft observations reveal that each of the three rings

FIGURE 6-19 Saturn and some of its moons are shown in this composite compiled from *Voyager 1* and *Voyager 2* photographs. Shown clockwise from the upper left, the satellites are Titan, Iapetus, Tethys, Mimas, Enceladus, Dione, and Rhea. (*Courtesy of NASA.*)

is really a complex of thousands of rings and rings within rings. They are composed of billions of particles of dust and ash, as well as larger fragments measuring tens of meters in diameter. All of these revolve around Saturn in the approximate plane of the equator. Saturn's 17 or more moons also revolve in this plane. Some suggest the rings are remnants of an exploded satellite. Others argue that they are a remnant of the particle cloud from which satellites form by accretion.

Uranus

Beyond the orbit of Saturn lies that of Uranus (Fig. 6-20). Although Uranus is about four times larger than the Earth, it is so distant that it appears only as a tiny greenish disk when viewed through large telescopes. The planet has a mean density of 1.3 g/cm^3. The most unusual characteristic of Uranus is the orientation of its axis of rotation, which lies nearly in the plane of its orbit. Thus, it rotates on its side, and its ring and satellite system follow the same orientation. Collision with another large body may have caused the unusual tilt of Uranus. Like Saturn and Jupiter, Uranus has rings. Five large moons and five recently detected smaller moons orbit the planet. Thick clouds of methane ice crystal lie beneath an atmosphere of hydrogen. Traces of methane mixed with the hydrogen are responsible for the planet's greenish color.

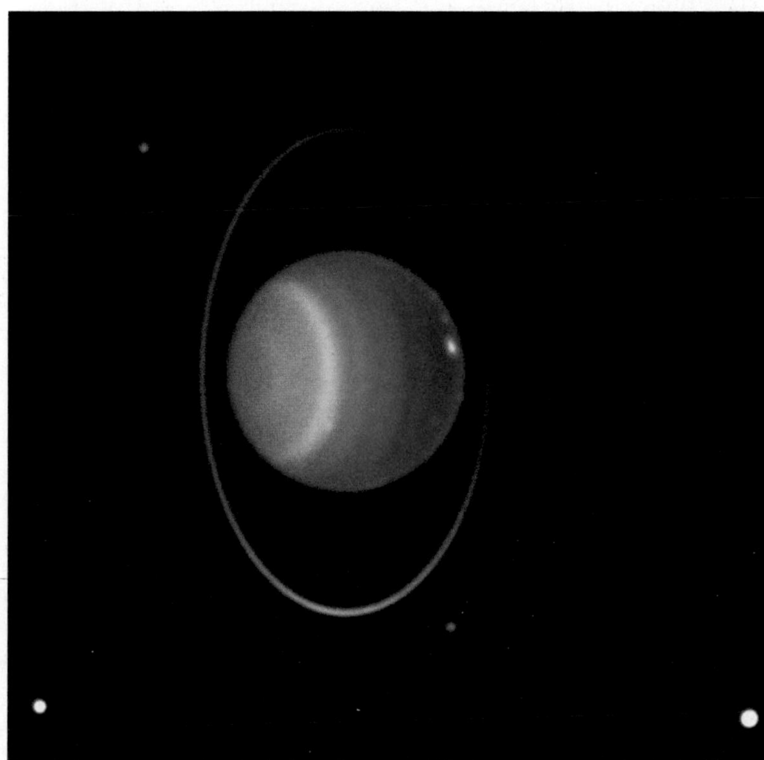

FIGURE 6-20 Uranus and its rings, imaged in the near-infrared with the Hubble Space Telescope. The image is shown in exaggerated false color and with a filter that lessens the intensity of the disk. We see the rings by reflected sunlight. One can also discern some of the moons of Uranus. (*Courtesy of NASA.*)

Neptune

Neptune and Uranus are called the twin planets. When viewed telescopically, both have a greenish color. Neptune's density of 1.7 g/cm^3 is only slightly greater than that of Uranus. Its atmosphere is also similar to that of Uranus. Hydrogen, helium, and methane have been detected spectroscopically. Two moons (Triton and Nereid) orbit the planet. Recently, evidence of a third moon and of a thin ring system was discovered. When *Voyager 2* approached Neptune in 1989, a huge counterclockwise rotating area of the atmosphere was found. It was named the Great Dark Spot (Fig. 6-21) and seemed to be analogous to Jupiter's Great Red Spot in size and location in the planet's southern hemisphere. The Great Dark Spot and other areas of apparent atmospheric circulation indicate Neptune has active weather systems.

Pluto

Far out, on the outer limits of our solar system, lies the orbit of the small planet Pluto. Its orbit is tilted an unusual 17° to the plane of the Earth's orbit (i.e., the ecliptic plane), and its orbit crosses that of Neptune. Because of the high inclination of Pluto's orbit, however, there is no danger of collision with Neptune. Its orbital peculiarity and small size suggest that Pluto may not be a true planet but rather a former satellite of Neptune that escaped the gravitational pull of that planet. Evidence for such a theory, however, is not conclusive. Pluto has a moon. It is named Charon after the mythological boatman who ferried souls across the river Styx to Hades.

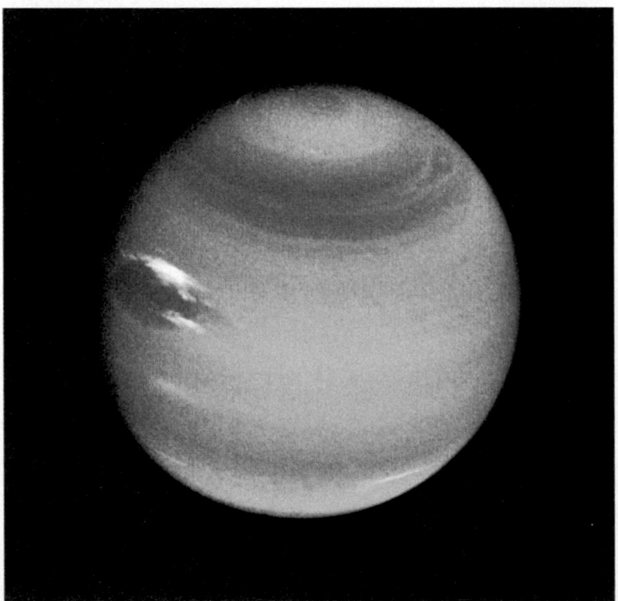

FIGURE 6-21 Neptune, as photographed in 1989 by the *Voyager* spacecraft. (*Courtesy of NASA.*)

PLANETS OUTSIDE THE SOLAR SYSTEM

With our improved space age technology, astronomers have found indications of over 90 planets that lie outside of the solar system, and 70 circling suns in our own Milky Way galaxy. They are all too far away to be observed directly, even with the Hubble Space Telescope. They are detected by observing their gravitational effect on the star they orbit. That gravitational effect causes a rhythmic wobble in the star. Analysis of the wobble permits scientists to deduce information about the mass and orbital path of the planet. Many have circular orbits resembling the orbital pattern of planets in our solar system.

THE HADEAN EARTH

Accretion and Differentiation

According to the solar nebular theory, the Earth and other planets were formed by accretion of dust and larger particles circulating within a turbulent dust cloud (Fig. 6-22). Thus, our planet would have acquired its materials while relatively cool and would have consisted initially of an unsorted conglomeration of metals and silicates from its center to its surface. The Earth, however, now is not homogeneous throughout. As noted in the previous section, neither are the other terrestrial planets of the solar system. They, along with the Earth, have experienced **differentiation**, a process whereby a planet becomes internally zoned or layered.

To accomplish differentiation of an initially homogeneous planet, the planet must have been heated and become at least partially melted. Under such conditions, much of the iron and nickel in the planet would percolate downward to form the dense core. The remaining iron and other metals would combine with silicon and oxygen and separate out to form the less-dense mantle that surrounds the core. Still lighter components would separate out of the mantle to form an uppermost crustal layer. Geologists do not insist on complete melting for this differentiation to have occurred. **Partial melting** on a vast scale would be a likely occurrence, as well as **solid diffusion** of ions mobilized by high temperatures and pressures. Indeed, it would be difficult to account for the present abundance of volatile elements on Earth if the planet had melted completely, for under such circumstances these materials would have been driven off. Even for partial melting and solid diffusion to have occurred, however, an enormous amount of heat would have been required.

The heat needed for differentiation may have been derived from several different processes. Some of it may have been accretionary heat derived from bombardment by particles and meteoric materials when the

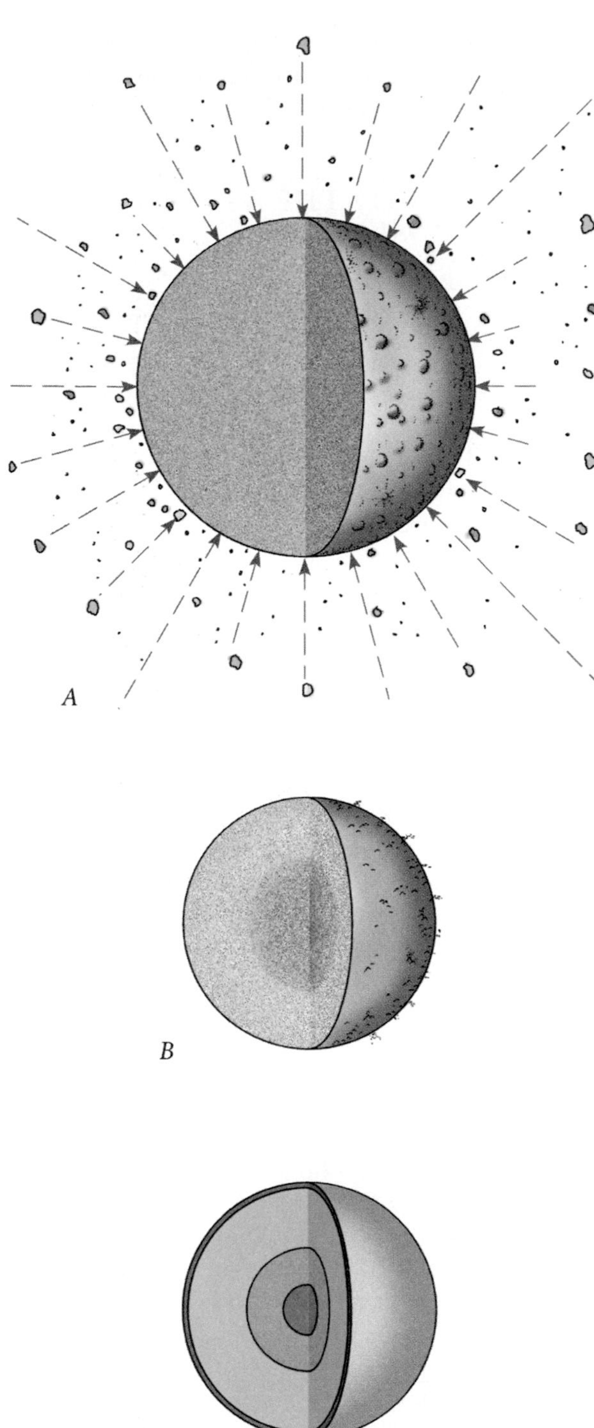

FIGURE 6-22 Conceptual diagrams of stages in the Earth's early history. (*A*) Growth of the planet by the aggregation of particles and meteorites that accreted and bombarded its surface. At this time, the Earth was composed of a homogeneous mixture of materials. (*B*) The Earth has shrunk because of gravitational compression. Temperatures in the interior have reached a level at which differentiation has begun. Iron (*red drops*) sinks toward the interior to form the core, whereas lighter silicates move upward. (*C*) The result of the differentiation of the planet is evident by the formation of core, mantle, and crust.

planet was still actively growing. Much of the heat generated by the great infall of impacting bodies would have been preserved because subsequent showers of material provided blankets of insulating debris. Also, as more and more material accumulated at the surface of the planet, the weight of overburden on underlying zones would have provided heat by gravitational compression. Compression also served to reduce the Earth's size.

Another extremely important source of heat for differentiation would result from the decay of radioactive isotopes incorporated within the materials of the planet during accretion. Much of this radioactively generated heat would have accumulated deep within the Earth, where the poor thermal conductivity of enclosing rocks provided insulation. There was more heat derived from radioactivity in the early days of the Earth's development than today because many of the isotopes having relatively short half-lives had not decayed and were generating heat.

Scientists estimate that about 100 million years after the solid Earth formed, there would have been sufficient heat from all these sources to melt iron at depths of 400 to 800 kilometers. One can envision myriad heavy, molten droplets of iron and nickel gravitating toward the planetary center, displacing lighter materials that had been there (see Fig. 6-22). As the mantle began to form, it too experienced differentiation.

An important consequence of the formation of the Earth's core was the development of our planet's magnetic field and magnetosphere. The field is generated by moving material in the liquid metallic core. As the liquid metal circulates it sets up an electric current, which in turn produces a magnetic field. The region around the Earth where the magnetic field dominates over the weak field generated by the sun is called the **magnetosphere**. The magnetosphere fends off much of the solar wind. When a magnetosphere is detected in other planets, it serves as evidence of the presence of a metallic core in those bodies.

The Hadean Crust

The Earth's original crust was dominated by dark iron and magnesium silicates. To understand the origin of the early crust, we must imagine the Earth during its Hadean infancy, when it seethed with heat generated from the decay of short-lived radioactive isotopes, from gravitational forces, and from the infall of meteorites. Even after the final stages of accretion, there was ample heat to partially or possibly totally melt the upper mantle. As a result of the melting, an extensive **magma ocean** may have covered the Earth's surface. (A somewhat similar magma ocean on the Hadean Moon produced the rocks of the lunar highlands.) When the surface patches of the Earth's magma ocean cooled, they solidified as rocks called **komatiites**. Komatiites form at temperatures even greater than the 1100°C required to produce basalt. Thus, they reflect the higher temperature gradients that prevailed during the late Hadean.

Komatiites are a form of **ultramafic** rock, a term used for rocks composed chiefly of olivine and pyroxene. They are therefore "ultra-rich" in iron and magnesium. Ultramafic rocks are less than 10 percent feldspar. Thus, they are even richer in iron and magnesium than so-called **mafic** rocks, which are as much as 50 percent calcium feldspar.

Even without total melting of the upper mantle to produce a magma ocean, it is probable that extensive partial melting in this zone produced sufficient quantities of magma to form an early crust. As this was occurring, high mantle temperatures and associated vigorous convection probably resulted in some form of plate movement and recycling of surface rocks. Indeed, because komatiites are somewhat denser than hot, partly molten upper mantle, there may have been a tendency for slabs of primordial komatiitic oceanic crust to sink as they were being conveyed laterally by underlying convection currents. With the passage of time, with continued cooling of the mantle, and with partial melting of ultramafic rocks in the upper mantle, komatiitic additions to the surface may have been replaced by mafic (basaltic) magmas. These may have been added to the expanding oceanic crust at spreading centers.

The Earth has two kinds of crust today, the mafic or oceanic crust discussed above and the less dense granitic continental crust. Could patches of continental crust have formed during the Hadean? In the opening paragraph of this chapter, mention was made of the discovery of 4.4-billion-year-old zircon grains in western Australia. The zircon grains were derived from some no-longer-existing body of igneous rock that experienced erosion. Grains loosened by weathering and erosion were transported and deposited in a layer of sandstone that was later metamorphosed to form a quartzite. The zircons were obtained from the quartzite. The zircons provide some startling information. For example, the heavy oxygen isotope values and the presence of microscopic inclusions of SiO_2 indicate the zircons crystallized from a granitic magma. The magma was probably produced over subduction zones by partial melting of descending plates. The resulting less dense granitic magma would then rise to become small islands or patches of continental crust. Evidence that these patches were above sea level is indicated by bedding features in the metamorphosed sandstone that resemble those found in modern stream deposits.

The oxygen isotope values also indicate that the rocks that were melted to form the zircon-producing magma must have interacted with liquid water at relatively low temperatures. The conclusion reached on the basis of this information is that the Earth had both patches of continental crust and a hydrosphere 4.4 billion years ago, during the early stages of the Hadean.

EVOLUTION OF THE ATMOSPHERE AND HYDROSPHERE

The Primitive Atmosphere

We have described how the Earth formed by aggregation of small and large bodies that were themselves once part of a nebula of gas and dust. Now we will review theories concerning the origin of the Earth's atmosphere. The particles and meteorites incorporated into the growing planet contained significant amounts of gases. Even greater amounts of gas may have been derived from the impact of comets, for these bodies are composed almost entirely of frozen gases, ice, and dust. The volatiles (substances easily driven off by heating) either were scattered uniformly throughout the accreting planet by the process of homogeneous accretion or were contained in the impacting bodies that arrived during the late stages of accretion. Among meteorites, volatiles are especially abundant in carbonaceous chondrites (Fig. 6-8). In addition to nitrogen, hydrogen, and carbon, some of the dark iron and magnesium silicates in the carbonaceous chondrites contain water. The water and gaseous elements would have been readily released by heat associated with accretion or by the melting and volcanism accompanying the Earth's differentiation.

The process by which water vapor and other gases are released from the rocks that held them and then vented to the surface is termed **outgassing**. Calculations indicate that most of the outgassing of water occurred within the first 1 billion years of Earth history. As evidence of this, the presence of an early ocean is clearly indicated by marine sedimentary rocks dating from as long ago as 3.8 billion years.

Before 3.8 billion years ago, the Earth's Hadean atmosphere was strongly reducing, meaning that it lacked free oxygen (i.e., O_2 gas). By Archean time, some neutralization may have occurred as limited amounts of oxygen became available from the breakdown of water molecules by sunlight (a process to be described in the next section). The Hadean atmosphere was, however, still anoxic, consisting mostly of carbon dioxide, water vapor, nitrogen, and lesser amounts of carbon monoxide, hydrogen sulfide, and hydrogen chloride. Actually, the volatiles released by modern volcanoes provide an approximation of the composition of the early atmosphere, with the exception that the primitive exhalations probably contained a higher percentage of hydrogen and possibly traces of ammonia and methane as well. Uncombined oxygen was not discharged into the early atmosphere, and if small amounts were released from the interior, these amounts would not have escaped immediate recombination with easily oxidized metals such as iron.

One might argue that the gases of our atmosphere were not degassed from the interior but were merely late accumulations of light volatiles left over from the original nebula. The Earth's relative scarcity of inert gases such as argon, neon, and krypton provides strong evidence for opposing this view. These chemically stable gases are too heavy to escape the Earth's gravity. If they were present in gases collected near the end of the accretionary period, they would be unable to escape and would still be here. The inert gases, however, are about a million times less abundant in our present atmosphere than in the Sun and other stars. This indicates that our atmosphere is not a residue of gases left over from the solar nebula but was derived from the planet's interior by degassing.

Once at the Earth's surface, the volatiles that had been vented were subjected to a variety of changes (Fig. 6-23). Water vapor was removed by condensation, and seas began to take form. The outgassing of carbon dioxide and other gases resulted in rainfall and ocean water that were considerably more acidic than today. When in contact with rocks, the acidic waters caused rapid chemical weathering and thereby brought calcium, magnesium, and other ions into solution. Much later, when the seas were less acidic and oxygen more prevalent in the atmosphere, these ions would be joined with carbon dioxide to form limestones, the calcareous structures of algae, and the shells of myriad marine organisms.

Formation of an Oxygen-Rich Atmosphere

We have seen that the early atmosphere of the Earth was reducing and anoxic (nonoxygenic). It was followed by an atmosphere rich in oxygen, an atmosphere that was, and still is, dependent on life for maintenance of its O_2 content. We are reminded of the significance of this atmosphere with every breath we take, yet the earlier nonoxygenic atmosphere was also of vital importance, for in it the earliest forms of life evolved.

The change from an oxygen-poor to an oxygen-rich atmosphere occurred by Proterozoic time as a result of two processes. The first was dissociation of water molecules into hydrogen and oxygen. Termed **photochemical dissociation**, this process occurs in the upper atmosphere when water molecules are split by high-energy beams of ultraviolet light from the sun. The reaction can be written as follows:

$$2\ H_2O + \text{uv radiation} \rightarrow 2\ H_2 + O_2$$

Before the Earth had an ozone layer to filter out ultraviolet radiation, some oxygen was probably produced by this process. Today, however, photochemical dissociation does not produce free oxygen at a rate sufficient to balance the loss of the gas by dissipation into space. Another far more important oxygen-generating mechanism came with the advent of life on Earth and, more specifically, with life that had evolved the remarkable capability of separating carbon dioxide into carbon and free oxygen. We now know this process as **photosynthesis**. In the section entitled "Life of the Archean" later in this chapter, we will examine more fully the

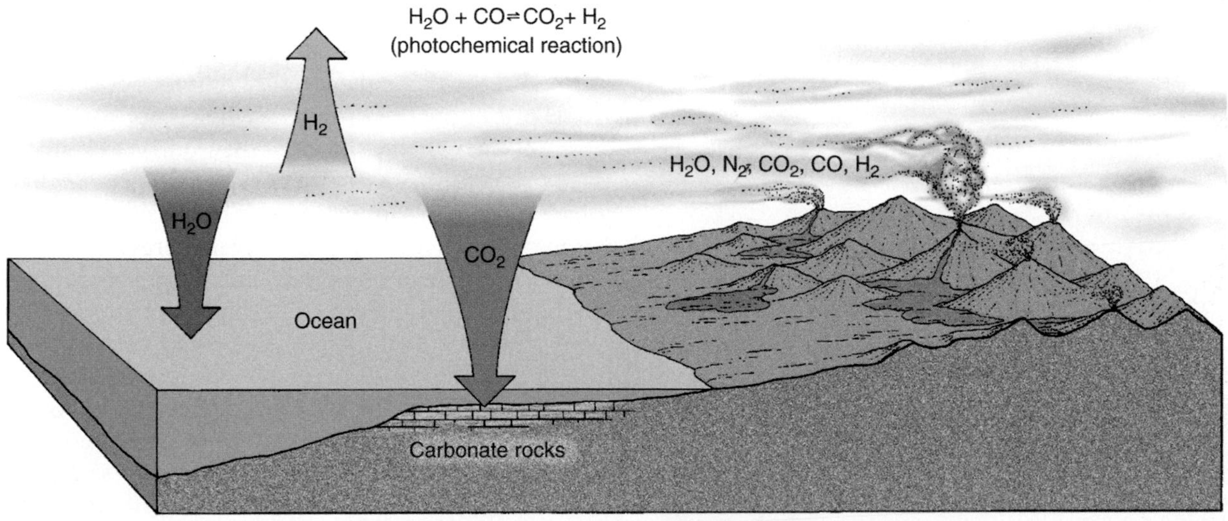

FIGURE 6-23 The relative amounts of gases in the primordial atmosphere were different from the abundances vented to the exterior of the Earth during differentiation. Nitrogen tended to be retained in the atmosphere, whereas much of the water vapor was lost to the oceans by condensation. Some CO_2 was combined with calcium and magnesium derived from weathering and extracted to form carbonate rocks. Hydrogen, being light, was lost to space.

role of photosynthesis in the evolution of early life forms. It is a process that has made the Earth a truly unique planet in our solar system.

Geologic Clues to the Nature of the Early Atmospheres

The most ancient rocks providing evidence of the atmospheric changes described previously are about 3.5 billion years old. Although they are metamorphic rocks, many of them still contain depositional features indicating that they were originally water-laid sediments. Mineral grains within these old rocks were formed or modified by weathering, and weathering cannot occur without an atmosphere. Further, it is highly probable that the atmosphere was not oxygenic, for there is an obvious absence of oxidized iron such as hematite. Minerals such as pyrite and uraninite are prevalent in some Archean sedimentary rocks, but in an oxygenic environment such as exists today, these minerals quickly decompose. In addition, Archean rocks are often dark in color because of the presence of carbon. Had oxygen been present, the carbon would have been oxidized.

The Archean atmosphere was rich in carbon dioxide. Carbon dioxide and water combine to form carbonic acid. In such an acidic environment, alkaline rocks such as dolomite and limestone do not develop, and this may account for the absence of carbonate rocks from this early stage of Earth history. Chert, a rock more tolerant of such conditions, is commonly found in ancient rock sequences.

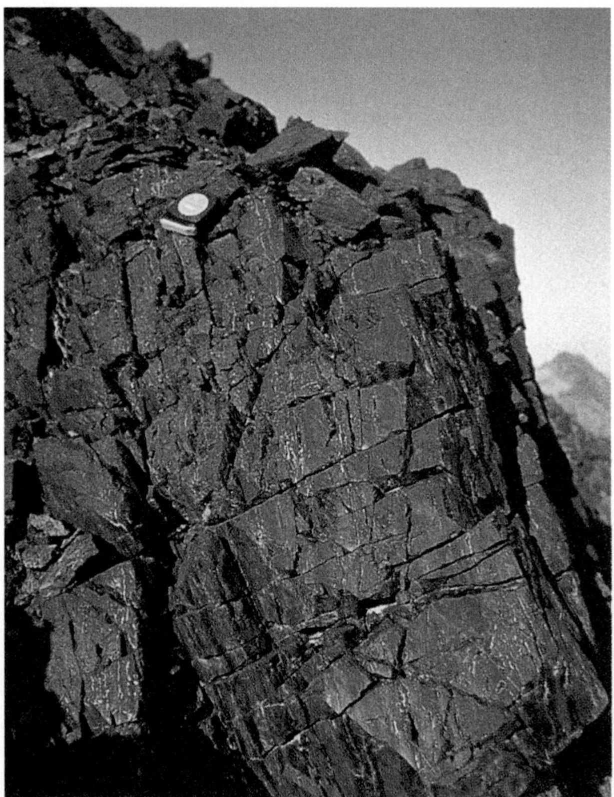

FIGURE 6-24 Banded iron formation. The red bands are hematite and are interbedded with chert. Wadi Kareim, Egypt. (*Courtesy of D. Bhattacharyya.*)

Gradual additions of oxygen in the Earth's early atmosphere have often been inferred from the presence of rocks called **banded iron formations**, or **BIFs** (Fig. 6-24). BIFs are cherts that exhibit an alteration of rust-red and gray bands. The former are colored by ferric iron oxide (Fe_2O_3), whereas the gray bands are not as rich in oxygen. Banded iron formations first appear in the geologic record about 3 billion years ago in the Archean and give way to unbanded iron deposits by about 1.8 billion years ago during the Proterozoic. They may have originated in more than one way. As we have noted, the early Earth was bathed in ultraviolet radiation. Under conditions of intense ultraviolet radiation, insoluble iron hydroxides can be precipitated from sea water that contains iron in solution. The iron hydroxides would then become a component of bottom sediment, later to be converted to ferric iron oxides.

Bacteria may also have played a role in the origin of BIFs. German microbiologists recently discovered that purple nonsulfate bacteria employ photosynthesis to break up dissolved iron carbonate in sea water and precipitate insoluble iron hydroxide. As suggested by the alternation of bands in BIFs, the process may have had a seasonal component. For example, seasonal upwelling of nutrient-rich bottom waters would enhance the bacterial action that produced the iron-rich bands. Such upwelling might also have brought bottom waters rich in dissolved iron to the surface of the sea, where the effects of ultraviolet radiation would be most intense.

Both of the above theories account for the accumulation of iron in Precambrian sedimentary rocks, but what was the ultimate source of the iron? Was it all derived from anoxic weathering of Archean rocks, or did it have a more dynamic origin? The discovery of submarine hydrothermal springs or vents prevalent along the great midoceanic ridges of the globe has provided another explanation for the production of such prodigious amounts of late Archean and early Proterozoic iron. It is well known that impressive quantities of iron and other elements are emitted at these vents (Fig. 6-25). The hot solutions originate when sea water seeps downward into fracture zones and encounters very hot rocks. Once heated, the scalding, chemically corrosive solutions dissolve iron and other elements from adjacent rocks. When the hot solutions rise to the surface of the sea floor and encounter cool ocean water, iron reacts with oxygen to form insoluble iron hydroxides that settle out and become components of deep-sea sediment.

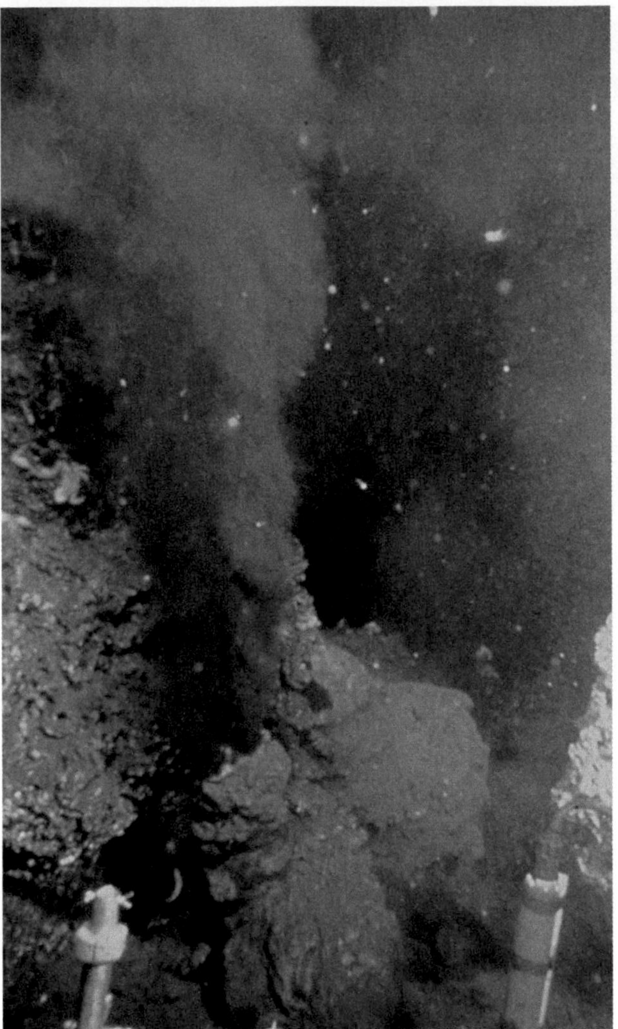

FIGURE 6-25 A submarine hydrothermal vent called a black smoker spouting from the East Pacific rise. Sea water becomes hot as it circulates through the hot basalt near a spreading center. The hot water dissolves sulfur, iron, zinc, copper, and other ions from the basalt. When the hot solution comes in contact with cold water, the ions precipitate as "smoke," consisting of tiny mineral grains. The hot, mineral-rich water sustains a diverse plant and animal community. *(Courtesy of D. Foster, Woods Hole Oceanographic Institution.)*

Although hydrothermal vents now in existence do not appear to provide enough iron to account for the enormous quantity in Precambrian banded iron formations, during the late Archean and early Proterozoic ample iron may have been provided by hydrothermal vents. At that time, the Earth's interior was hotter, convection in the mantle was more vigorous, and hydrothermal vents were more numerous. Geologist Ann Isley has calculated that these primordial submarine thermal springs could have provided 25 million tons of iron to the ocean each year. This would be about 10 times the amount needed to account for all the iron of the Precambrian banded iron formations.

Banded iron formations are not only important in the context of geologic history but are of enormous economic importance as well. These formations provide most of the iron that is mined worldwide, including the great iron ore deposits of the Ukraine, Brazil, Minnesota, Labrador, and western Australia. They are the material foundation of our industrial civilization.

In rocks younger than 1.8 billion years, evidence of free oxygen in the atmosphere is seen in Proterozoic red shales, siltstones, and sandstone derived from weathering of older rocks on the continents. One also finds carbonate rocks (dolomites and limestones) of about the same age. If carbon dioxide was still abundant at this time, these carbonate rocks could not have formed.

The Ocean

If the water in the atmosphere was outgassed from the interior, then the bodies of water that accumulated on the surface of the Earth were derived from the atmosphere after it had cooled sufficiently to condense and allow rain to fall. Liquid water probably began to collect on the Earth's surface as early as 4.4 billion years ago. With time, the amount of surface water increased, and depressions in the primordial crust began to fill. Oceans began to take form. Initially, the seas would have probably been acidic, for they would have taken up large amounts of carbon dioxide from the early atmosphere. Change to a more basic composition, however, may have occurred rather rapidly as emanations from submerged volcanoes neutralized the acidic waters with additions of sodium, calcium, and iron.

Could the enormous volume of oceanic water have come from the interior? Calculations provide an affirmative answer to this question. As noted earlier, vast amounts of water were locked within hydrous minerals that were ubiquitous within the primordial Earth. These waters, "sweated" from the Earth's interior and precipitated onto the uplands, immediately began to dissolve soluble minerals and carry the solutes to the sea. The initial high acidity of the oceans enhanced and speeded the process so that the oceans quickly acquired the saltiness that is their most obvious compositional characteristic. They have maintained a relatively consistent composition by precipitating surplus solutes at about the same rate as they are supplied. Of course, because of its high solubility, sodium remains in sea water longer than do other common elements. However, the fossil record of marine organisms suggests that even this element has not varied appreciably in sea water for at least the past 600 million years.

Having outgassed its present quantity of water early in its history, the Earth has been recycling that water ever since. Water is continuously recirculated by evaporation and precipitation—processes powered by the

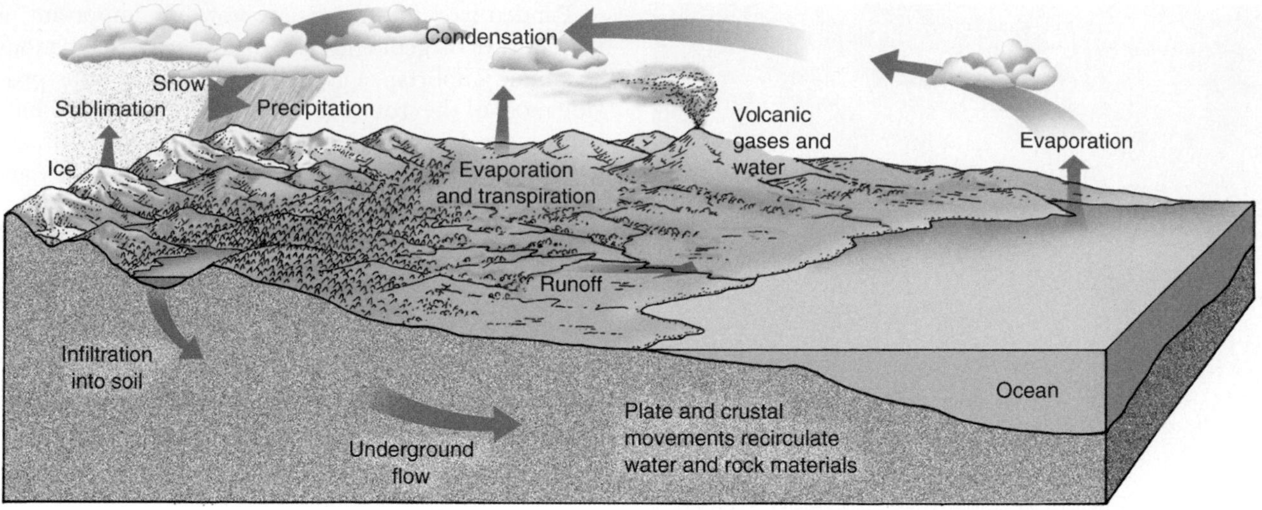

FIGURE 6-26 The constant circulation of water at and near the Earth's surface.

Sun and gravity. Some of the water in the oceans is temporarily lost by being incorporated into hydrous clay minerals that settle to the ocean floors. However, even this water has not permanently vanished, for the sediments may be moved to orogenic belts and melted into magmas that return the water to the surface in the course of volcanic eruptions (Fig. 6-26).

THE ARCHEAN

When 18th-century geologists first began to describe local sequences of strata, they sometimes encountered a primary or basement complex of metamorphic and igneous rocks that lay beneath fossil-bearing sedimentary strata. These older, mostly crystalline rocks came to be known as Precambrian, a derivative of the term *Cambrian* proposed for overlying younger rocks by Adam Sedgwick in 1835. To most of Sedgwick's contemporaries, correlating and even crudely estimating the age of the apparently unfossiliferous, altered, and deformed Precambrian rocks seemed an impossible task. Nevertheless, geologists recognized that Precambrian terranes were of great importance. They represented an enormous block of geologic time, formed the very cores of continents, and contained rich deposits of iron and other metals. For these reasons, it was inevitable that a few intrepid geologists would devote their lives to deciphering the complex relationships of Precambrian rocks.

One of these pioneers of Precambrian geology was Sir William Logan, of the Canadian Geological Survey. In the mid-1800s, Logan was able to associate groups of Precambrian rocks according to their superpositional and cross-cutting relationships. To substantiate these relationships, he inferred that those rocks that had experienced the most metamorphism were likely to be the oldest. Modern geologic investigations show that Logan erred in attempting to relate degree of metamorphism to age. Older rocks may escape metamorphism, and younger ones may be radically metamorphosed. Yet in southeastern Canada, where Logan mapped, one could find an older terrane of gneissic rocks as well as younger sequences of less altered metamorphic and sedimentary rocks. Because of this, it seemed reasonable to think of Precambrian time as divisible into an older **Archean Eon** and a younger **Proterozoic Eon**. All of the remainder of geologic time up to the present day was designated the Phanerozoic Eon. The term **Hadean** was used to refer to the interval between the birth of the planet and the beginning of the Archean.

Today, all geologists agree that the only reliable basis for correlation of unfossiliferous Archean and Proterozoic rocks is through isotopic dating techniques. Only after the absolute ages of rocks in a particular region are known can they be placed in correct chronologic order. Radiometric dating of Precambrian rocks has provided a basis for mapping Precambrian rocks and for deciphering their geologic history. The hundreds of dates already obtained permit a tentative calibration of the Precambrian geologic time table (Table 6-2). The beginning of the Proterozoic is placed at about 2.5 billion years ago. The Archean is generally considered to have begun about 3.96 billion years ago, a date that is close to the age of the oldest known crustal rocks.

Shields and Cratons

Although exposures of Precambrian rocks may occur in the deep canyons of some mountains and plateaus, the most extensive exposures are found in broadly un-

TABLE 6-2 Time Divisions for the Precambrian

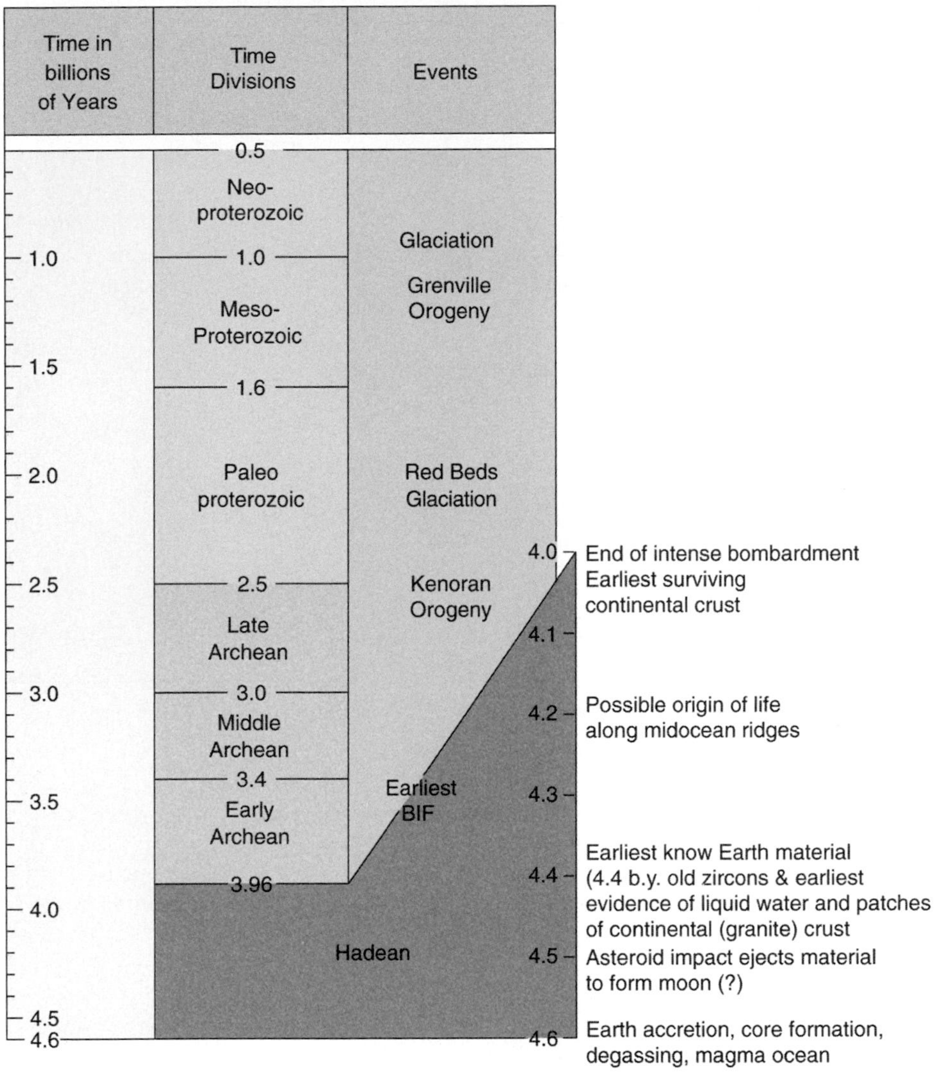

Time in billions of Years	Time Divisions	Events
	0.5	
	Neo-proterozoic	Glaciation
1.0	1.0	Grenville Orogeny
	Meso-Proterozoic	
1.5	1.6	
2.0	Paleo proterozoic	Red Beds Glaciation
2.5	2.5	Kenoran Orogeny
	Late Archean	
3.0	3.0	
	Middle Archean	
	3.4	
3.5	Early Archean	Earliest BIF
	3.96	
4.0	Hadean	
4.5		
4.6		

Hadean (b.y.)	Events
4.0	End of intense bombardment Earliest surviving continental crust
4.1	
4.2	Possible origin of life along midocean ridges
4.3	
4.4	Earliest know Earth material (4.4 b.y. old zircons & earliest evidence of liquid water and patches of continental (granite) crust
4.5	Asteroid impact ejects material to form moon (?)
4.6	Earth accretion, core formation, degassing, magma ocean

warped, geologically stable regions of continents called **shields**. Shields are the oldest parts of continents, and every continent has one or more shields (Fig. 6-27). Extending across 3 million square miles of northern North America is the great Canadian shield. Because continental glaciers of the last ice age have stripped away the cover of soil and debris, Precambrian rocks of the Canadian shield are extensively exposed at the surface.

To the south and west of the Canadian shield, Precambrian rocks are covered by sedimentary sequences of Phanerozoic age. These younger rocks are mostly flat-lying or only gently folded. Such stable regions where basement rocks are covered by relatively thin blankets of sedimentary strata are called **platforms**. The platform of a continent, together with its shield, constitutes that continent's **craton** (Fig. 6-28).

On the basis of differences in the trends of faults and folds, the style of folding, and the ages of component rocks, the Canadian shield has been divided into a number of **Precambrian provinces** (Fig. 6-29). The boundaries of these provinces are often marked by abrupt truncations in structural lineations, or they may be represented by bands of severely deformed rocks of former orogenic belts. In North America, seven Precambrian provinces are recognized: the Superior, Wyoming, Slave, Nain, Hearne, Rae, and Grenville. These provinces were once separate crustal segments that have been consolidated to form the larger North American craton. They are bound together by belts of deformed, metamorphosed, and intruded rocks that mark the location of collision of the various cratonic elements. For this reason, the belts of deformed rocks are called collisional orogens. It has been estimated that all of the Archean and Proterozoic elements of the Canadian shield were consolidated by about 1.8 to 1.9 billion years ago.

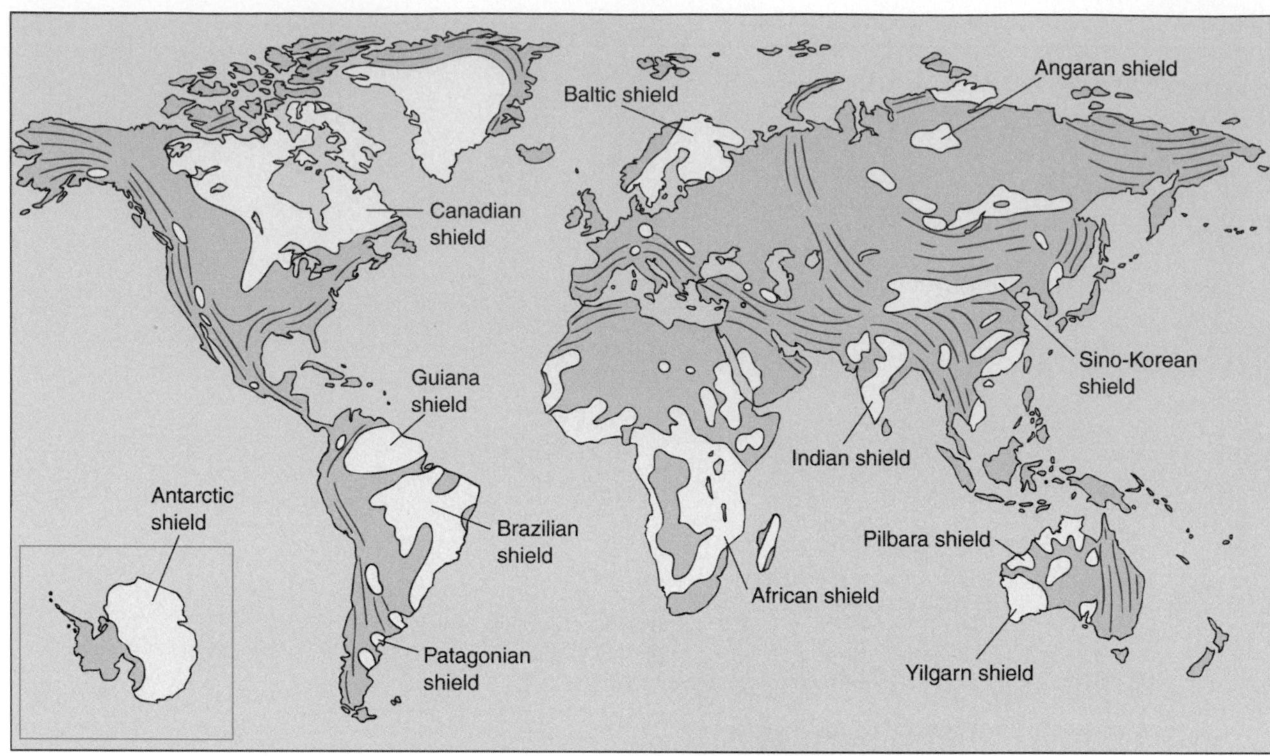

FIGURE 6-27 Major areas of exposed Precambrian rocks (shown in yellow). Trends of mobile belts are indicated by the red lines.

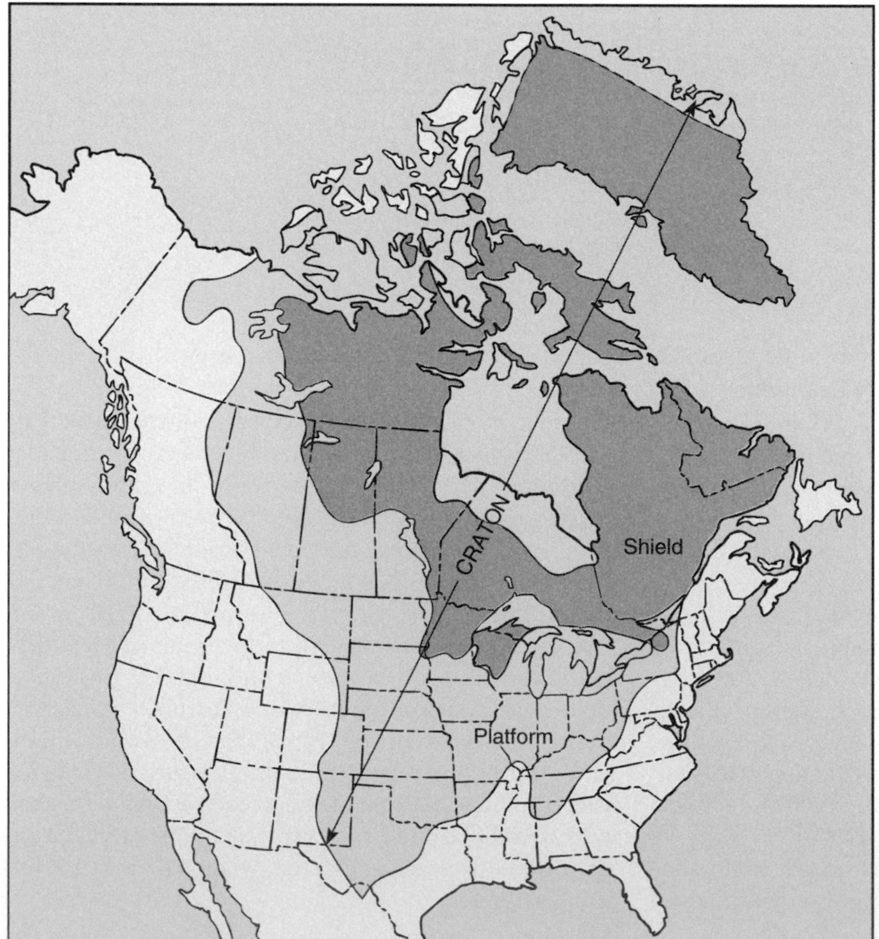

FIGURE 6-28 North American craton, shield, and platform.

? *What is the difference between shield and platform?*

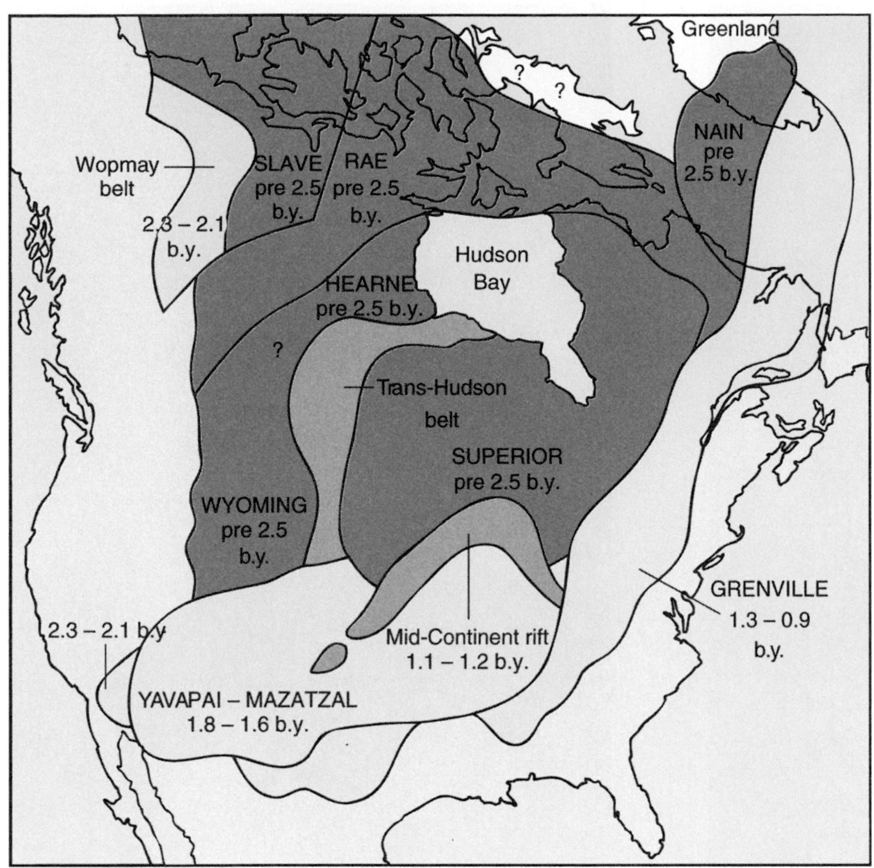

FIGURE 6-29 Precambrian provinces of North America. Archean provinces (those greater than 2.5 billion years [b.y.] old) are in red. Proterozoic provinces are in green. The Trans-Hudson belt is a zone along which rifting and then closure occurred to weld the Superior province to the Wyoming province. The Hearne and Wyoming provinces may actually be a single province. A major episode of rifting that occurred 1.2 to 1.0 billion years ago is represented by the midcontinent rift zone. The Wopmay belt was developed by rifting followed by closure of an ocean basin. (*Adapted from Hoffman, P. F. 1989. Precambrian geology and tectonic history of North America, in Geology of North America, Vol. A, Ch. 16. Boulder, CO: Geological Society of America.*)

Expansion of the Continental Crust

In contrast to the mafic nature of the oceanic crust, the continental crust is rich in feldspars, quartz, and muscovite. Whereas the first oceanic crust formed about 4.5 billion years ago by partial melting of ultramafic rocks in the upper mantle, the continental crust formed somewhat later (about 4.4 billion years ago) by partial melting of wet mafic rocks in the descending slabs of subduction zones (Table 6-3). As opposed to mafic, the continental crust is felsic (granitic) in composition. One of the oldest known patches of felsic crust is the Acasta Gneiss of Canada's Northwest Territories. Radiometric dates reported from zircon grains within these felsic rocks indicate that they are about 4.0 billion years old (Fig. 6-30). These old continental rocks are tonalite gneisses. They were formed by metamorphism of tonalite, a variety of the igneous rock diorite that is

TABLE 6-3 Summary of the Characteristics of the Earth's Early Oceanic and Continental Crust

	Oceanic Crust	Continental Crust
First appearance	About 4.5 billion years ago	About 4.4 billion years ago
Where formed	Ocean ridges (spreading centers)	Subduction zones
Composition	Komatiite–basalt	Tonalite–granodiorite
Lateral extent	Widespread	Local
How generated	Partial melting of ultramafic rocks in upper mantle	Partial melting of wet mafic rocks in descending slabs

Source: Condie, K. C. 1989. Origin of the Earth's crust. *Paleogeography, Paleoclimatology, Paleocology* 75:57–81, with permission.

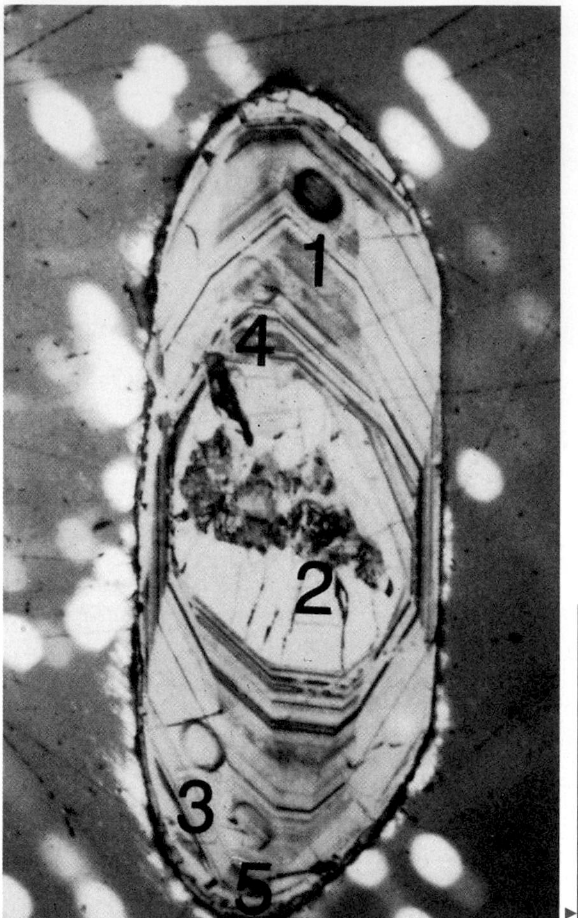

FIGURE 6-30 Photomicrograph of a zircon grain extracted from the Acasta Gneiss, Slave province, Northwest Territories of Canada. Such grains have yielded ages of about 4.0 billion years. The grain is 0.5 millimeters long. Its polished surface has been etched with acid to highlight crystal growth zones. Numbers refer to points selected for analysis. (*Courtesy of S. A. Bowring.*) ? *Why are zircon crystals particularly valuable in determining isotopic ages?*

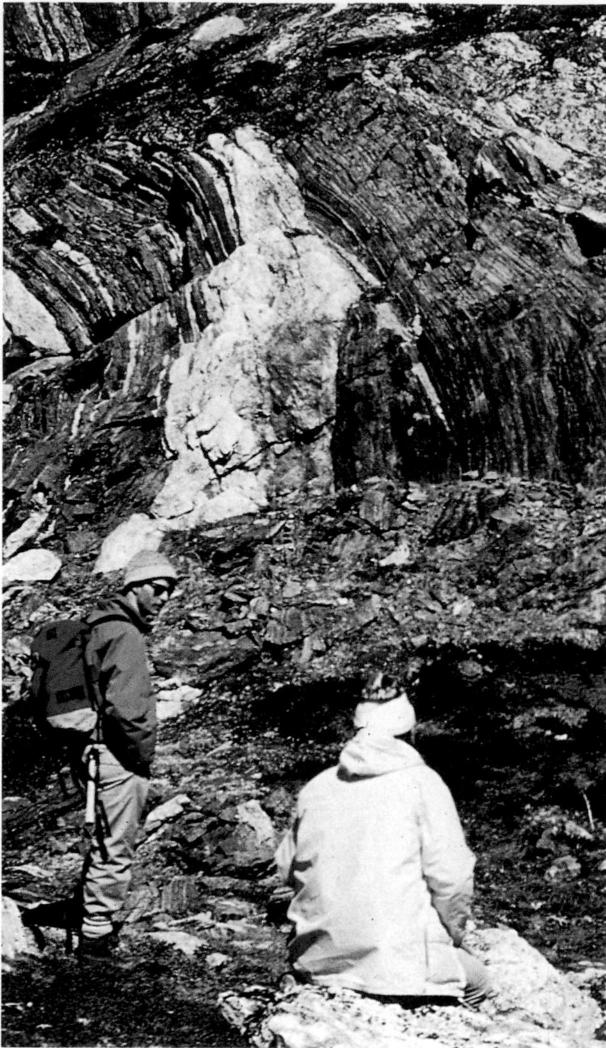

FIGURE 6-31 One of the oldest rocks on Earth, the Amitsoq Gneiss of the Isua Region of southwest Greenland. These Archean tonalite gneisses are about 3.8 billion years old. (*Courtesy of M.S. Smith.*) ? *Would tonalite be considered a felsic or mafic rock?*

at least 10 percent quartz. The Isua region of southwestern Greenland also contains very old tonalite gneisses (3.8 billion years old). The Greenland outcrops (Fig. 6-31) are more extensive in area and contain fragments of ultramafic rocks that may be remnants of a surface that was present before there was any felsic crust. Other patches of extremely old felsic crust have been found in Enderby Land, Antarctica (3.9 billion years old). In addition, in western Australia, the previously mentioned 4.4 billion-year-old detrital zircon grains have been found in Archean sedimentary rocks. Evidence of the presence of Archean continents that stood distinctly above sea level is also seen in a 3.46-billion-year-old erosional unconformity containing a fossil soil (paleosol) recently discovered in Australia's arid Pilbara region. The unconformity and its associated paleosol represent the oldest land surface known and give clear evidence that subaerial weathering and erosion were at work early in the Archean.

When we compare the ages obtained from the samples just described to the 4.6-billion-year age of our planet, it is apparent that the Earth had acquired patches of continental (tonalitic and granodioritic) crust a few hundred million years after its formation, or about 4.4 billion years ago. This is probably close to the maximum obtainable age for continental crust, for it is unlikely that any older rocks could have survived the intense bombardment by meteorites that occurred up to and even beyond that time. But how did these initial patches of continental crust form and grow? Their composition indicates an origin at subduction zones where slabs of sediment-covered, wet mafic crust plunged into the mantle, there to be partially melted to release felsic components. Because of their lower density, felsic melts ascended to form tonalites and granodiorites. The more mafic materials remained behind in the mantle. By about 3.0 billion years ago, granites formed by a similar mechanism involving partial melting of tonalitic, granodioritic, and sedimentary rock materials.

It is likely that the early Archean segments of continental crust were small, perhaps less than about 500 kilometers (310 miles) in diameter. These initial patches grew larger by several mechanisms that continue to operate today. One such mechanism may have involved the coming together of two tectonic plates bearing felsic crust at their encountering margins. Compressive deformation and suturing may have occurred to form a larger expanse of continental crust. Growth may also have occurred as island arcs and microcontinents were added to the host continent along subduction zones. A third method may have involved the welding of prisms of sediment onto continental margins. This process begins with the weathering and erosion of pre-existing rocks and deposition of the resulting sediment in the ocean adjacent to an upland area. Next, the thick accumulation of sediment is subjected to orogeny caused by an encounter with a converging tectonic plate. The sediment in the prism would have roughly the same bulk composition as granitic igneous rocks. Such material would be converted to felsic crystalline crustal rocks if subjected to intense metamorphism and melting. The repetition of such cycles of deposition and orogeny, along with suturing events and the incorporation of microcontinents, could account for continuing growth of continental crust. As indicated by the known distribution of 3.0- to 2.5-billion-year-old felsic rocks, very large continents had been assembled by late Archean time. Once large land masses were established, there was a possibility for more rigorous climates. Thus, near the end of the Archean, in 2.8-billion-year-old Witwatersrand sediments of South Africa, one finds evidence of the Earth's earliest episode of glaciation. Prior to this time the Earth may have been too warm for ice sheets to form.

Granulites and Greenstones

In general, Archean rocks of cratons can be grouped into two major rock associations. The first is the granulite association. It is composed largely of gneisses derived from strongly heated and deformed tonalites, granodiorites, and granites, as well as layered intrusive gabbroic rocks such as **anorthosites**. Few, if any, clues about the nature of parent materials can be found in these severely metamorphosed granulite associations.

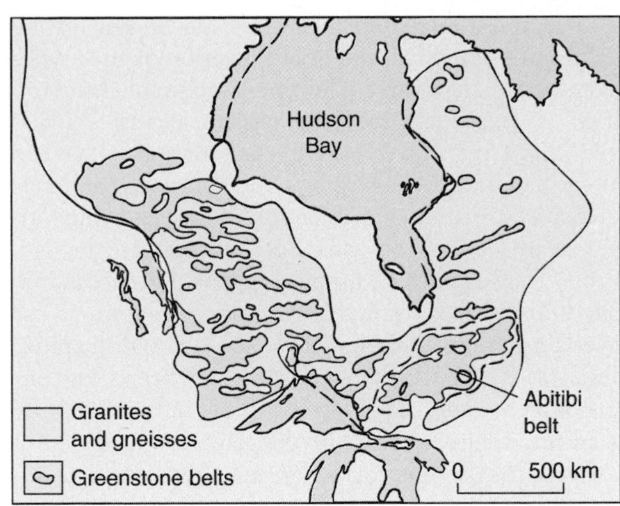

FIGURE 6-32 Greenstone belts of the Superior province. (*After Goodwin, A. M. 1968. Proc. Geol. Assoc. of Canada 19:1–14.*)

The second general group of Archean cratonic rocks are **greenstones**. Greenstone associations usually occur in roughly troughlike or synclinal belts. They are prominent features of Archean terranes of all continents. The Abitibi belt of the Superior province is the largest uninterrupted greenstone trend in the Canadian shield (Fig. 6-32). Like most other greenstone belts, it is composed of basaltic, andesitic, and rhyolitic volcanic rocks along with metamorphosed sediments derived by weathering and erosion of the volcanics. Lavas of greenstone belts exhibit pillow structures that indicate that they were extruded under water (Fig. 6-33).

FIGURE 6-33 Exposure of greenstone showing well-developed pillow structures, near Lake of the Woods, Ontario, Canada. The original basalts were extruded under water during Late Archean time. During the Pleistocene Ice Age, glaciers beveled the surface of the exposures and left behind the telltale scratches (glacial striations). (*Courtesy of K. Schultz.*)

The most intriguing characteristic of greenstone belts is the roughly sequential transition of rock types from ultramafic near the bottom to felsic near the top (Fig. 6-34). In many, the basal rock masses are komatiites (Fig. 6-35). Overlying basalts are somewhat less mafic and include low-grade metamorphic minerals such as chlorite and hornblende. These green minerals impart the color for which greenstones are named. Proceeding upward in a typical greenstone succession, one finds felsic volcanics above mafic rocks, which are overlain in turn by shales, graywackes, conglomerates, and, less frequently, banded iron formations. The entire greenstone sequence is often surrounded by granitic intrusive rocks. In the Superior province of Canada, the granitic rocks were emplaced during a deformational event that occurred about 2.6 to 2.7 billion years ago. That event, called the **Kenoran orogeny**, marks the close of the Archean Eon in Canada.

Archean Protocontinents

The upwardly changing sequence from ultramafic or mafic to felsic rocks in greenstone belts records the development of felsic crustal materials from older, more mafic source rocks. The sedimentary rocks that occur most commonly near the top of the greenstone sequence provide clues to the nature of the earliest patches of Archean continental crust. Among the sediments are elongate layers of coarse conglomerates containing pebbles of granite. Thus, felsic source rocks were already present at the Earth's surface. Graywackes and dark shales present in the greenstone sedimentary sequence appear to have been deposited in deep-water environments adjacent to protocontinental coasts. In the greenstone sedimentary association, we do not find blankets of well-sorted, ripple-marked, cross-bedded sandstones. The existence of broad lowlands that could be flooded by shallow interior seas is therefore unlikely. Nor is there any evidence of wide marginal shelf environments. This absence of blanket sandstones and the presence of graywackes and elongate bodies of conglomerates suggest that Archean protocontinents were steep-sided, relatively small, narrow landmasses with rugged shorelines.

Archean Tectonics

When did plate tectonics begin on Earth? Probably not during the first few million years, for the planet during that time was too hot for plates to form. Extreme heat would have caused the mantle to churn too fast for surface materials to cool and form lithospheric plates. However, by about 4 billion years ago, sufficient cooling had probably occurred to permit the initiation of plate formation. As evidence of this, rocks formed at about this time in Canada and Scandinavia include great parallel bands of continental crust that appear to have been moved against one another by plate tectonic processes. Similar patterns are recognized in geologically younger terranes where continents have increased in size by the additions of microcontinents along subduction zones. In addition, maps of the lithosphere prepared from seismic data reveal the locations of ancient subduction zones.

Once plate tectonics was in operation, it provided a mechanism capable of supplying emerging continents with mantle-derived materials that could be recycled and partially melted to form additions to continental crust. Archean tectonics would, however, have operated at a more dynamic level as compared to the Phanerozoic because of the greater amount of heat still present in the mantle. Rates of convection in the mantle would have been far greater, thermal plumes more numerous, midoceanic ridges longer and more numerous, and volcanism more extensive. These were the conditions under which greenstones and granulite associations developed. An early stage in the development of the greenschist–granulite association would

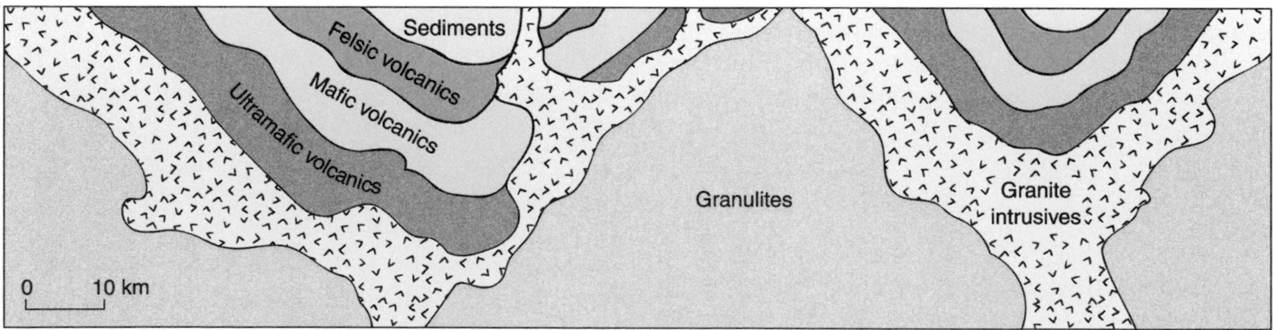

FIGURE 6-34 Generalized cross-section through two greenstone belts. Note their synclinal form and the sequence of rock types from ultrabasic near the bottom to felsic near the top. A late event in the history of the belt is the intrusion of granites. Ultramafic basal layers are particularly characteristic of greenstone belts in Australia and South Africa but do not occur in the Archean of Canada.

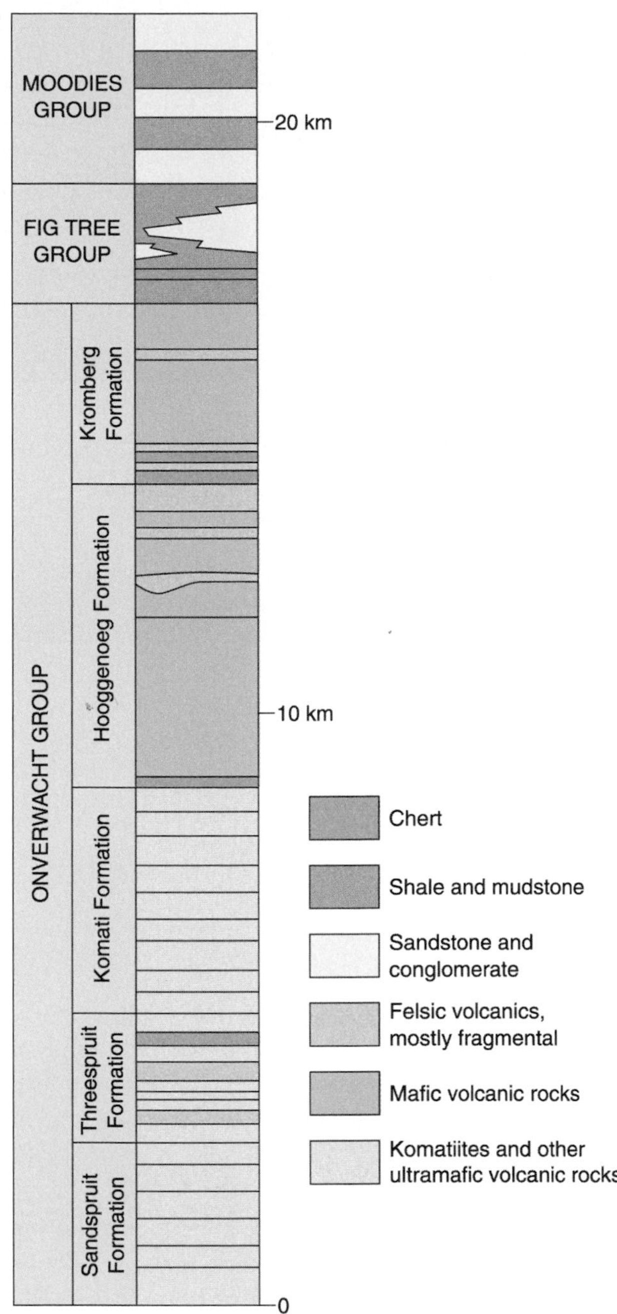

FIGURE 6-35 Stratigraphic sequence in a greenstone belt in Barberton Mountain Land, South Africa. (*After D. R. Lowe. 1980. Ann. Rev. Earth Planet. Sci. 8:145–167.*)

have been the growth of volcanic arcs adjacent to subducting plates (Fig. 6-36). Lavas and pyroclastics of the arcs would then have been weathered and eroded to provide sediment to the adjacent trench. In the subduction zones, this wet sediment would be carried downward along with oceanic crust and heated. Partial melting of the mix of materials may have produced increasingly more felsic magmas, and these would have worked their way upward, engulfing earlier volcanics. During or shortly after these events, compressional deformation and metamorphism altered the rocks to form the granulite terranes.

Whereas volcanic arcs were the sites for the birth of granulite associations, the back-arc basins were the probable locations for the development of greenstone belts. In the early stages of that development, ultramafic to mafic lavas ascended through rifts in the back-arcs to form the lower layers of the greenstone sequences. Later extrusions included increasingly more felsic lavas, pyroclastics, and sediments derived from the erosion of adjacent uplands. The final stage of development involved compression of the back-arcs to form the synclinal structures characteristic of greenstone belts. Compression may have occurred simultaneously with the magmatic activity that produced the granitic intrusions that surround the lower levels of greenstone belts.

If plate tectonics has been operating for so long a time on Earth, would there not have been supercontinents before Pangea? The answer is yes. A supercontinent named **Rodinia**, for example, existed about 600 million years ago during the late Proterozoic. Subsequently Rodinia fragmented, and the separated components assembled at the end of the Paleozoic Era to form Pangea.

Archean Rocks

Southern Africa's Barberton Mountain Land and a terrane in western Australia called the Pilbara block contain excellent records of Archean sedimentation. Here one finds large cratons rather than the small protocontinents discussed earlier. These cratons were fully formed and stable as early as 3 billion years ago. Because they have been stable since that early time, sedimentary rocks within them have often escaped severe metamorphism that might otherwise have obscured their environments of deposition. In southern Africa, some of these sedimentary rocks contain gold and have therefore been intensely investigated.

In the greenstone belts of Barberton Mountain Land, the geologic column begins in its lower part with ultramafic and mafic lava flows (Fig. 6-35), followed by pillowed oceanic basalts. These rocks give way upward to felsic lava flows and then graywackes and claystones, probably derived from the erosion of volcanic mountains and cindercones. Pyroclastic beds capped by chert layers also occur. The chert probably formed from the silica of volcanic ash. The general sequence from ultramafic or mafic volcanics to more felsic volcanics to sediments is sometimes repeated, and with each cycle the younger rocks of the sequence are more felsic. Sandstones containing quartz, orthoclase, and granitic detritus are found near the top of the geologic column. These detrital sediments indicate the presence of nearby granitic source areas exposed during episodes of

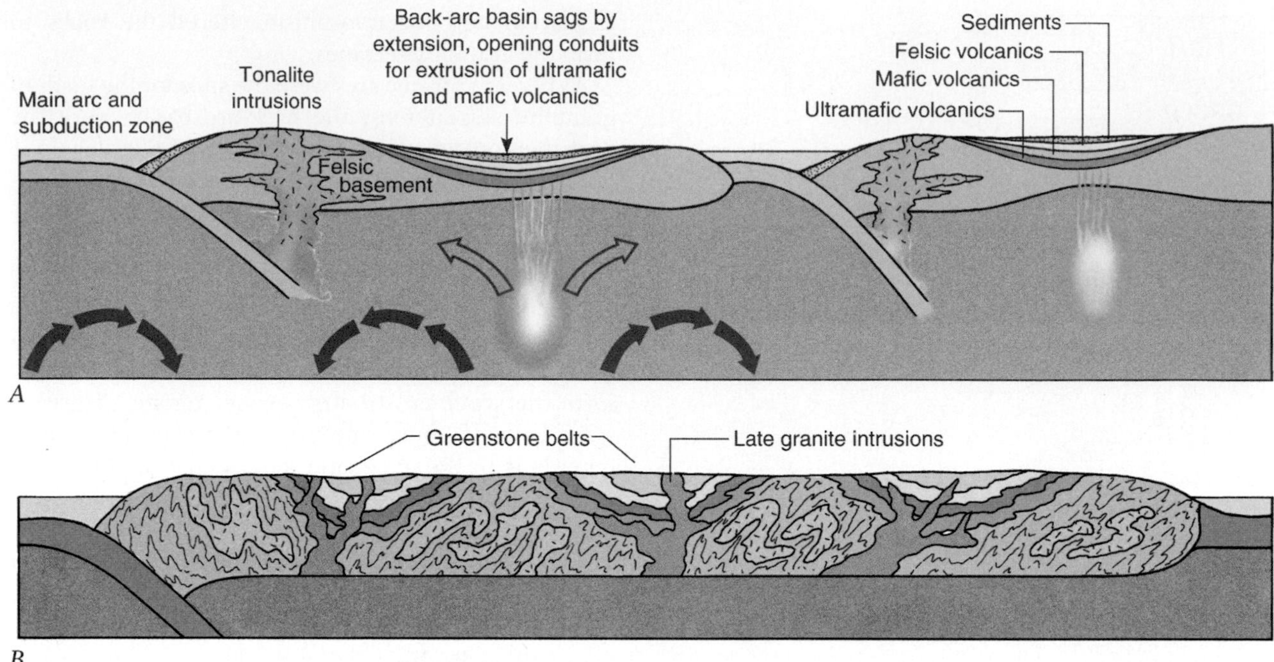

FIGURE 6-36 Plate tectonics model for the development of greenstone belts and growth of continental crust. (*A*) Plates are in motion, driven by convection cells in the upper mantle. Subduction provides for the emplacement of wedges of oceanic crust and for mixing and melting to provide tonalite intrusions. Behind the main arc, the back arc sags by extension, and the greenstone volcanic sequence is extruded. (*B*) Compression has occurred to create the greenstone belts with their synclinal form and to aggregate small continental patches into a larger continental mass. Later, granites are intruded in and around greenstone belts. (*Simplified from a model proposed by B. F. Windley, 1984. The Evolving Continents, 2nd ed. New York: John Wiley & Sons.*)

cratonic uplift. By the time these highly felsic clastic sediments were being deposited, a full range of depositional environments had developed. Deep-water continental slope and rise environments are recorded by graywacke exhibiting graded bedding and mudstones of the Fig Tree Group in Barberton Mountain Land. The overlying Moodies Group includes quartz sandstones and shales that were spread across shallow marine shelves, tidal flats, and deltas.

Intertidal clastic sedimentary rocks of the Pongola Group are above the Moodies rocks, and these are overlain by a remarkable sequence of river-borne conglomerates, sandstones, and shales known as the Witwatersrand Supergroup. The Witwatersrand is notable for several reasons. First, its areal extent (about 103,000 km^2, or 40,000 square miles), thickness (8 to 10 kilometers, or 4.9 to 6.2 miles), and nonmarine environment of deposition (fluvial, deltaic, and lacustrine) clearly denote that very large cratons were in existence by about 2.7 billion years ago. Second, rocks of the Witwatersrand contain gold. The gold occurs as placer gold. After being eroded from parent rock, gold particles were transported as stream deposits. Where stream velocity was sufficient, the lighter sedimentary particles were partly swept away, leaving heavier gold particles to sink to the bottom. Since the discovery of Witwatersrand gold in 1883, half the world's supply of the precious metal has come from fluvial and deltaic conglomerates of the Witwatersrand.

LIFE OF THE ARCHEAN

The splendid diversity of animals and plants on our planet is a consequence of billions of years of chemical and biologic evolution that began during the Precambrian. As described earlier, the basic materials from which microbial organisms could have been initially assembled may have arrived on Earth during the Hadean by way of carbonaceous chondrites. We would like to have fossil documentation of the earliest events in the progression from nonliving to living things, but that kind of direct evidence is lacking. Theories for the origin of life are based on our understanding of present-day biology, coupled with evidence obtained from the study of Archean rocks, meteorites, and the compositions of planets.

The Beginning of Life

Systematic progressions from simple to complex structures are a manifest characteristic of both living

Voyageurs National Park

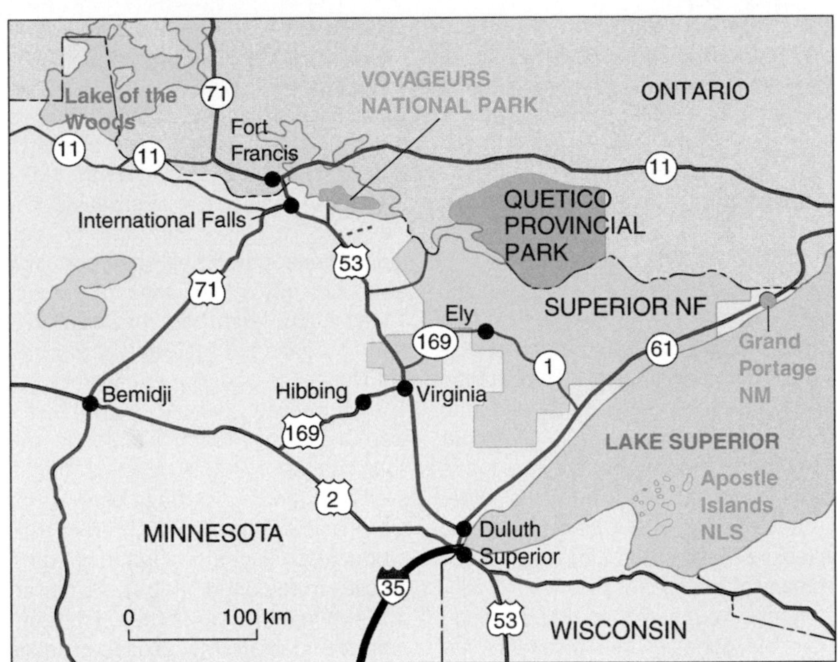

Location map of Voyageurs National Park. (*From Voyageurs Official Map and Guide, National Park Service. Washington, DC: GPO, 1993.*)

Voyageurs National Park lies along the Minnesota–Ontario border, near the southern margin of the Canadian Shield (see location map). The park contains more than 30 lakes bordered by rocky shorelines and containing bout 900 islands clothed in forests of fir, spruce, pine, aspen, and birch. Moose, deer, beaver, and wolves make their home on many of the islands. Cars are left at the park's entry stations, and visitors explore the park by boat. Voyageurs National Park takes its name from the colorful French-Canadian trappers who traveled the lakes in long birchbark canoes, carrying pelts to fur company facilities in Grand Rapids, Minnesota.

Rather like a history book in which all but the first and last chapters have been destroyed, the park provides a record of very early and very recent geologic events. Early events can be read from severely deformed and compressed schists and gneisses of Precambrian age. All of the exposed bedrock in the park originated during

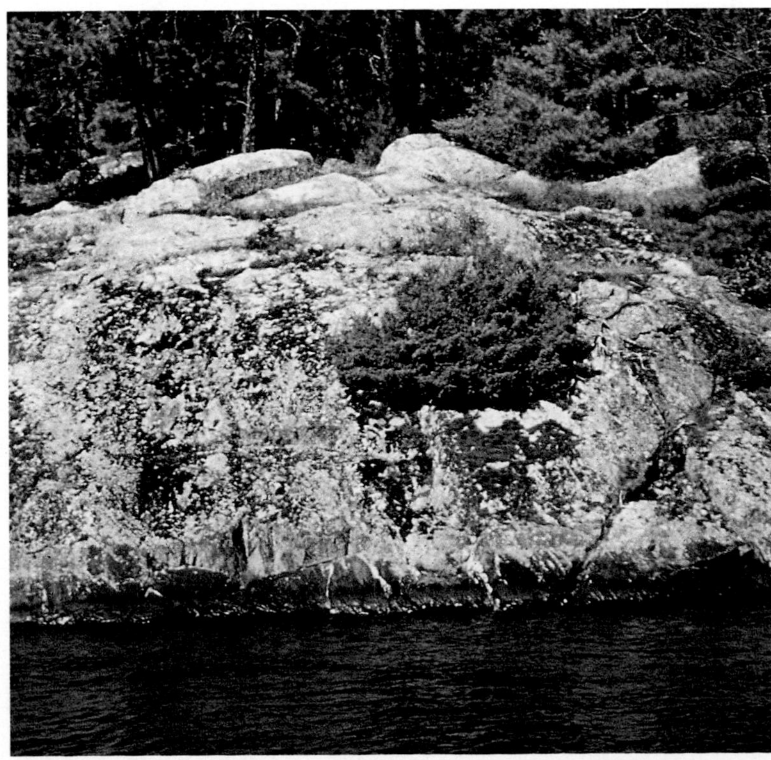

Exposure of Precambrian gneiss along the shore of Lake Kabetogama, Voyageurs National Park, Minnesota. (*Courtesy of R. F. Dymek.*)

the Precambrian (see accompanying figure), and most of the formations are Archean. Younger events are revealed in erosional and depositional features resulting from the work of glaciers.

The Archean rocks at Voyageurs National Park include both the granulite and greenstone associations described in this chapter. Mafic volcanics within the greenstone association have pillow structures, indicating that much of the Archean volcanic activity in this region occurred beneath the sea. Associated metamorphosed graywackes suggest rapid deposition in trenchlike basins.

Near the end of the Archean, the Precambrian Superior province in which Voyageurs National Park lies was compressed in a mountain-building event termed the Kenoran Orogeny. One consequence of the high temperatures produced during the orogeny was the melting of rocks at depth and the emplacement of a granitic batholith (the Vermillion batholith). Pink and gray granites of the batholith can be seen along the southern margin of Kabetogama Lake. Gold veins found in a fault zone at the western edge of the park probably were derived from hydrothermal solutions emanating from the batholith. The gold veins caused a short-lived gold rush in the 1890s.

The last stages of Archean geologic history in this region involved erosion of the mountain ranges produced during the Kenoran orogeny. The wearing away of upland areas continued into the Paleoproterozoic. Then, about 2100 million years ago, massive mafic dikes were intruded into the older Archean rocks. The dark rocks of the dikes are clearly seen along the shores of many of the islands.

Rocks younger than Precambrian may have once covered Voyageurs National Park. If so, it is likely that these rocks would have been scoured and scraped away by the great continental ice sheet of the Pleistocene Epoch. As will be described in Chapter 13, this gargantuan blanket of ice covered most of the northern half of North America. The ice was over a mile thick in places. As it advanced slowly southwestward, it was easily able to dislodge huge chunks of bedrock and transport billions of tons of debris. Today many large boulders from distant locations are scattered across the landscape of the Superior province. They have been dubbed glacial erratics. One huge erratic visible on the south shore of Cranberry Lake weighs over 200 tons. Rock fragments within the ice served like the teeth in a rasp, producing glacial scratches and grooves in the underlying Precambrian rocks. When the ice receded, it left behind the glaciated landscape of lakes and islands we know as Voyageurs National Park.

and nonliving systems. Awareness of such progressions has strongly influenced theories on the origin of life. Subatomic particles such as neutrons, protons, and electrons are built into atoms. Atoms in turn combine to form molecules, and molecules may be joined to form sheets, chains, and a variety of complex molecular structures. Perhaps in a somewhat similar fashion, the atoms of elements essential to organisms (carbon, oxygen, hydrogen, nitrogen, sulfur, and phosphorus) are built into organic molecules. At some stage in the progression, simpler molecules might have added components that would gradually transform them into the complex organic molecules essential to all living things. An analogy might be the manner in which small molecules of nitrogenous bases combine with sugar and phosphorous compounds to form the large deoxyribonucleic acid (DNA) molecule (see Fig. 4-14). In living organisms, DNA is found chiefly in the nucleus of the cell. It functions in the transfer of genetic characteristics and in protein synthesis. Some theorists suggest that such large and complex molecules might have organized themselves into bodies capable of performing specific functions. Such bodies in unicellular organisms are called **organelles**. Organelles might then have combined to form larger entities that grew, metabolized, reproduced, and mutated. These are the attributes of all living things. Could a similar progression have occurred in the early years of the Archean? We are uncertain about how some of the steps in the progression would have been accomplished, but the basic concept appears to be biochemically feasible.

Preliminary Considerations

The oldest direct evidence of life is found in rocks exposed in western Australia that are 3.5 billion years old. In these rocks, paleobiologist J. W. Schopf has identified the remains of several different types of microorganisms. Many are strikingly similar to cyanobacteria that are found all around the world today. These fossils strongly suggest that oxygen-producing organisms had already evolved by this early stage in Earth history. The microfossils were found in a chert bed that occurs within the Apex Basalt of Australia's Archean Warrawoona Group.

Less direct evidence of the presence of even older life has been found in 3.8-billion-year-old banded iron deposits in southern Greenland. These rocks contain ratios of Carbon 13 to Carbon 14 similar to those in present-day organisms.

As described earlier in this chapter, the Earth accreted from a dust cloud about 4.6 billion years ago, and an initial atmosphere was subsequently generated by volcanic outgassing. Presumably, the gases emitted were similar to those in volcanic eruptions today in that water and carbon dioxide were dominant, with lesser amounts of nitrogen, hydrogen, hydrogen sulfide, sulfur dioxide, carbon monoxide, methane, and ammonia. Note the absence of free oxygen in this list. The planet's earliest organisms must have been anaerobic—that is, they did not require oxygen for respiration. The environment in which they lived lacked a protective atmospheric shield of ozone, which is derived from oxygen and which absorbs ultraviolet radiation from the Sun. Perhaps early life escaped these lethal rays by living beneath protective ledges of rock or at greater water depths. Some biologists insist that in the early stages of the development of life, ultraviolet radiation may not have been a hindrance but rather served to impel molecules to interact and form complex associations.

The carbon, hydrogen, oxygen, nitrogen, phosphorus, and sulfur necessary to produce life are among the most abundant elements in the solar system. They were undoubtedly present on the primitive Earth. What had to be accomplished was the organization of those elements into the molecular components of an organism, and the development of a system for self-replication. Perhaps the first self-replicating molecule was RNA (ribonucleic acid). Like the DNA described earlier, RNA is a nucleic acid basic to life. However, unlike DNA, RNA is single-stranded and may have been a precursor to DNA.

There are four essential components of life. The first component is protein. **Proteins** are essentially strings of comparatively simple organic molecules called amino acids. Proteins act as building materials and as catalysts that assist in chemical reactions within the organism. The second of the basic components is nucleic acids, such as DNA and ribonucleic acid (RNA). Organic phosphorous compounds provide a third component of life. They serve to transform light or chemical fuel into the energy required for cell activities. An important fourth ingredient for life is some sort of container, such as a cell membrane. The enclosing membrane provides a relatively isolated chemical system within the cell and keeps the various components in close proximity so that they can interact.

From the previous description, it is apparent that amino acids have an important role in the development of the larger and more complex molecules. They are the building blocks of proteins. Two environmental circumstances in the early years of Earth history may have been important in the natural synthesis of amino acids. Prior to the accumulation of the ozone layer in the Earth's upper atmosphere, ultraviolet rays bathed the Earth's surface. As demonstrated in experiments, ultraviolet radiation is capable of separating the atoms in mixtures of water, ammonia, and hydrocarbons and of recombining those atoms into amino acids. A second form of energy capable of accomplishing this feat is electrical discharge in the form of lightning. Either together or separately, lightning and ultraviolet radiation may have stimulated the production of amino acids at shallow depths in lakes or ocean bodies.

In addition to amino acids naturally synthesized on Earth, these important building blocks of life may have arrived on Earth from outer space. Amino acids such as alanine and glycine, as well as nucleic acid bases, have been frequently identified in meteorites.

How did the amino acids present on Earth during the early Archean aggregate to form proteins? A credible hypothesis supported by laboratory experimentation suggests that the first proteins may have formed on the surfaces of clay particles. As described in Chapter 2, clay minerals have a layered structure like that of mica. The planes of atoms in the atomic structure of clays include metallic ions that, together with the other ions of the mineral, provide positive and negative charges on the surface of each clay layer. The ions attract and thus concentrate organic molecules in an orderly array and cause them to align and link themselves into proteinlike chain structures.

Test Tube Amino Acids and a Step Beyond

Scientists in the mid 19th century had succeeded in manufacturing some relatively simple organic compounds in the laboratory. However, it was not until 1953 that the laboratory synthesis of amino acids and other molecules of roughly similar complexity was announced. Stanley Miller, at the suggestion of Harold Urey, performed the now-famous experiment. He infused an atmosphere at that time thought to be like that of the Earth's earliest atmosphere into an apparatus similar to the one shown in Figure 6-37. It was a methane, ammonia, hydrogen, and water vapor atmosphere. As the mixture was circulated through the glass tubes, sparks of electricity (simulated lightning) were discharged into the mixture. At the end of only 8 days, the condensed water in the apparatus had become turbid and deep red. Analysis of the crimson liquid showed that it contained a bonanza of amino acids as well as somewhat more complicated organic compounds that enter into the composition of all living things. In additional experiments by other biochemists, it was shown that similar organic compounds could also be produced from gases (carbon dioxide, nitrogen, and water vapor) of the preoxygenic atmosphere. The main requirement for the success of the experiments seemed to be the lack, or near absence, of free oxygen. To the experimenters, it now seemed almost inevitable that amino acids would have developed in the Earth's prelife environment. Because amino acids are relatively stable, they probably increased gradually to levels of abundance that would enhance their abilities to join together into more complex molecules leading to proteins.

FIGURE 6-37 Stanley Miller and Harold Urey used an apparatus similar to this to replicate what they believed were the conditions of early Earth. An electric spark was produced in the upper right flask to simulate lightning. The gases present in the flask reacted together, forming a number of basic organic compounds. (*Courtesy of Dr. Stanley L. Miller, University of California, San Diego.*)

To come together and form proteinlike molecules, amino acids must lose water. This loss can be accomplished by heating concentrations of amino acids to temperatures of at least 140°C. Volcanic activity on the primitive crust would be capable of providing such temperatures. However, the biochemist S. W. Fox discovered that the reaction also occurred at temperatures as low as 70°C if phosphoric acid was present. Fox and his co-workers were able to produce proteinlike chains from a mixture of 18 common amino acids. They called these structures proteinoids and reasoned that billions of years ago they were the transitional structures leading to true proteins. This is not extravagant conjecture, for Fox was able to find proteinoids similar to those he created in his laboratory among the lavas and cinders adjacent to the vents of Hawaiian volcanoes. Apparently, amino acids formed in the volcanic vapors and were combined into proteinoids by the heat of escaping gases.

Hot, aqueous solutions of proteinoids will, on cooling, form into tiny spheres that show many characteristics common to living cells. These proteinoid microspheres (Fig. 6-38), as they are called, have a filmlike outer wall; are capable of osmotic swelling and shrinking; exhibit budding, as do yeast organisms; and can be observed to divide into "daughter" microspheres. They occasionally aggregate linearly to form filaments, as in some bacteria, and they exhibit a streaming movement of internal particles similar to that observed in living cells.

Although complete long-chain nucleic acids have not been experimentally produced under prelife conditions, short stretches of ordered sequences of nucleic acid components have now been produced in the laboratory. Some paleobiologists believe similar sequences could have been formed and accumulated on the surfaces of clay particles, as described earlier.

The Birthplace of Life

Although there is a possibility that life may have begun on land, perhaps in the warm waters of thermal springs, or even deep beneath the surface, there are several good reasons to believe that the ocean provided a particularly hospitable environment for early life. The sea contains the salts needed for health and growth. The waters of the oceans serve as universal solvents capable of dissolving a great variety of organic compounds, and currents in the oceans ceaselessly circulate and mix these compounds. Such constant motion would have favored frequent collisions of the vital molecules and would thereby have increased the probability of their combining into larger bodies. One can imagine the larger bodies adding components, increasing in complexity, and ascending in a thousand infinitesimally

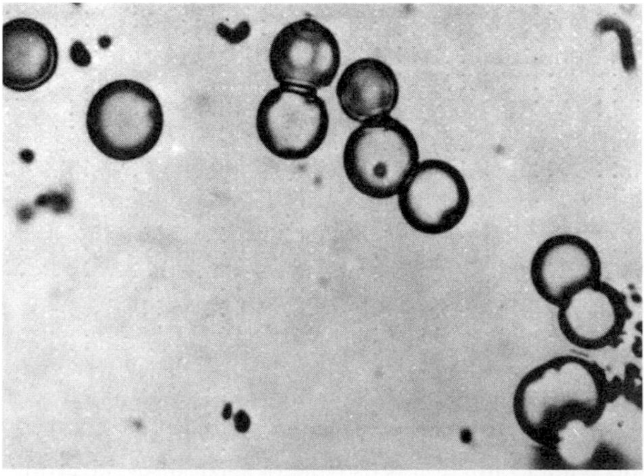

FIGURE 6-38 Proteinoid microspheres formed in hot aqueous solutions of proteinoids that have been permitted to cool slowly. The spheres are about 1 to 2 millimeters in diameter. (*Photo by S. W. Fox.*)

small steps from nonliving to transitional things of a biologic netherworld to the first living organisms. Such a sequence would be unlikely in most land environments today. Oxygen and present-day microbial predators would destroy the delicate structures. However, life originated before there were organisms to cause decay and before there was sufficient free oxygen to be troublesome.

The belief that the ocean was the birthplace of life has been with us for many decades. Since the time of Darwin, biologists have spoken of the primordial ocean as a rich organic broth containing all the necessary raw materials for life. Until recently, the more placid, illuminated upper parts of the ocean were considered likely sites for biogenesis. Indeed, the ocean-atmosphere interface is still the best candidate for the site of life's beginning. As another possibility, however, direct observations made from diving submersibles cruising near the surface of midoceanic ridges have provided evidence that life may have originated in total darkness, at abyssal depths, among jets of scalding water rising from submarine hydrothermal vents and volcanic chimneys. The evidence for such a stygian beginning of life consists of the presence of microbes called **hyperthermophiles** that thrive in sea water at temperatures that exceed 100°C. The organisms are apparently able to live deep within fissures below the actual vents, for they are often ejected in such great numbers that they cloud the surrounding waters. In the absence of light for photosynthesis, the organisms derive energy from chemosynthesis, a process within the cell that causes hydrogen sulfide to react with oxygen and to produce water and sulfur. The chemosynthetic microbes around vents of the midoceanic ridges form the base of a food chain that supports an astonishing variety of invertebrates, including shrimplike arthropods, crabs, clams, and tube worms as much as 10 feet long (Fig. 6-39).

Although once regarded as specialized forms of bacteria, hyperthermophilic microbes have DNA that is sufficiently different from that of bacteria to warrant their placement in a separate branch of the tree of life. In Chapter 4 we noted that all life can be organized into three domains, the Archaea, Bacteria, and Eukarya. The hyperthermophilic and related microbes that feed on hydrogen and produce methane belong in the domain Archaea. Other organisms, such as plants and animals, whose cell structure includes a nucleus and organelles, are placed in the domain Eukarya.

The tube worms of hydrothermal vents are truly remarkable creatures. They are able to grow more than 33 inches in a year, exceeding the growth rate of any known invertebrate. As they have no mouth or gut, they depend upon their symbiotic relationship with the thermophilic microbes for survival. Spongy tissue inside the worm is essentially a rich culture of archean organisms. Tube worm blood contains hemoglobin that binds hydrogen sulfide and transports it to these microbes in the spongy tissue. The organisms then oxidize the hydrogen sulfide into carbon compounds that nourish the tube worm.

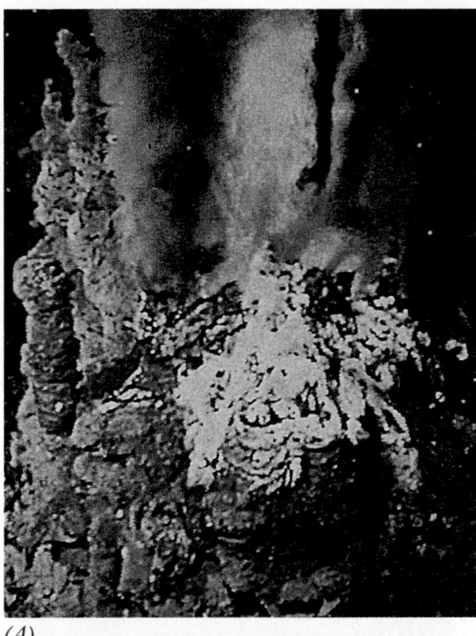

(*A*)

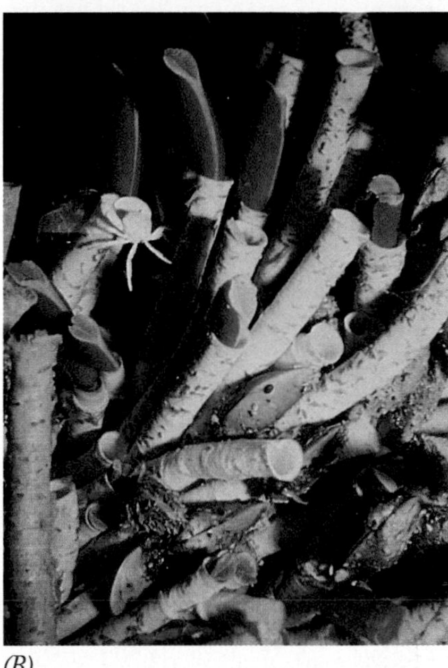

(*B*)

FIGURE 6-39 A hydrothermal vent (*A*) located on a midoceanic ridge in the eastern Pacific. As the escaping hot water encounters cold sea water, a variety of metallic sulfides are precipitated. In addition, the nutrient-rich vent water supports chemosynthetic bacteria that ultimately support a variety of invertebrates, including giant tube worms (*B*). (*A, courtesy of T. McCollum; B, copyright VU/WHOI/D. Foster/Visuals Unlimited.*)

One can envision the steps by which the Earth's first hyperthermophiles originated as beginning with the downward percolation of ocean water through fractures and fissures toward the zone of extremely hot rocks that supply the lavas of midoceanic ridge volcanoes (Fig. 6-40). There the water would be heated to about 1000°C, yet pressure would be so great that it would prevent boiling. The superheated water would react readily with surrounding rocks, extracting elements needed to construct organic molecules. Such molecules would begin to form as the hot waters gradually returned to the surface, cooling as they made their way upward. During that time, amino acids and other organic compounds might be synthesized and combined to form the first protocells. Growth, proliferation, and evolutionary processes operating on the protocells might then lead to primitive chemosynthetic microbes like those seen today in hydrothermal vent environments.

Life in Hostile Environments

Recent discoveries of organisms living in what we consider extremely harsh environments have provided a new view of how inevitable and ubiquitous life on Earth really is. The heat-loving hyperthermophiles are, of course, an example. In addition, microscopic life has been detected in polar locations where temperatures dip to −113°C, in hot barren deserts, in submicroscopic spaces between grains of solid rock lying more than 3 kilometers beneath the Earth's surface, and in rocks having temperatures in excess of 110°C. (One such microbe, found at a depth of 2.8 kilometers in the course of drilling for natural gas, was appropriately named *Bacillus infernus*, the "bacterium from hell.") These subterranean microbes have been dubbed **lithotrophs**. Most of them live off geochemical energy derived from hydrogen, iron, magnesium, and sulfur. Some thrive on rocks heated by radioactive decay. In the gold mines of South Africa prodigious numbers of microbes live prolifically at depths of over 3000 meters. Microbes have also been discovered in uncontaminated granites 3600 meters below the Earth's surface in Sweden. Did the ancestors of these microbes originate at the surface, and through eons of change work their way downward, or could life have originated at depth and moved upward as the Earth's primordial surface environment became gradually more hospitable? These questions are providing intriguing new questions, not only about the origin of life on Earth but also about the probability of life on other planets.

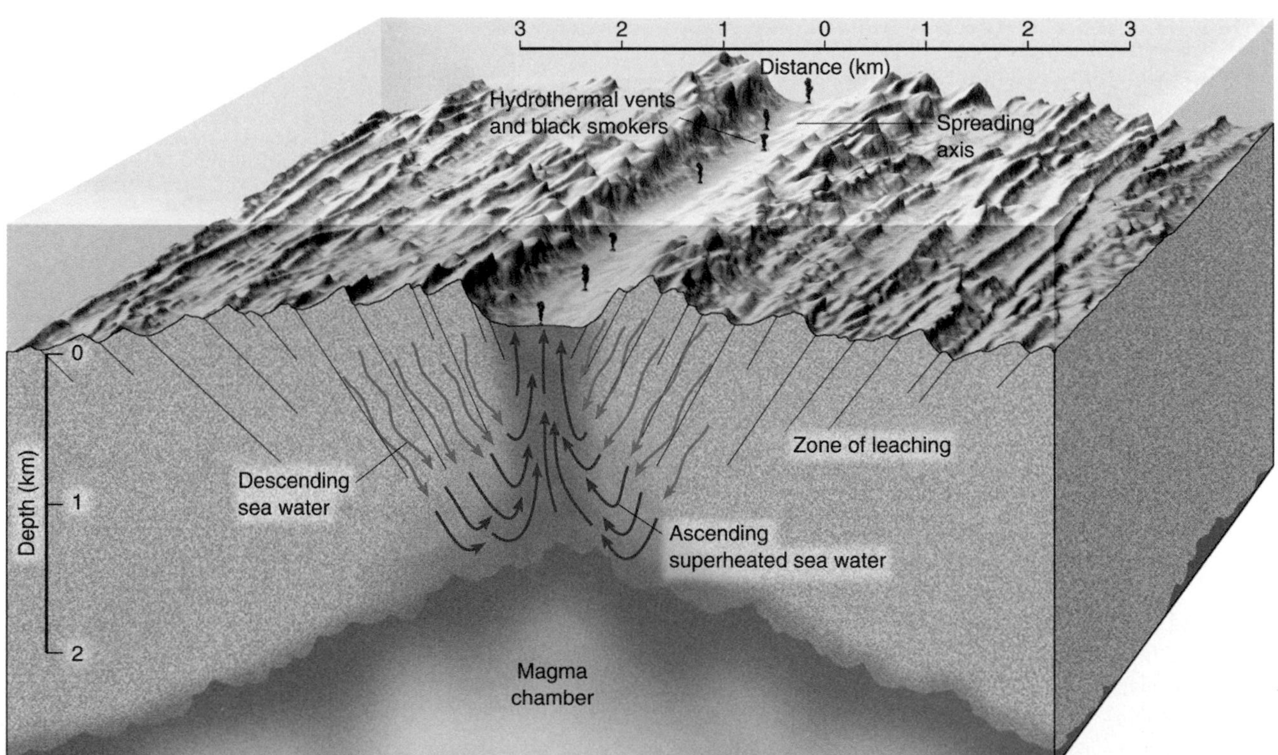

FIGURE 6-40 Section across the central part of a midoceanic ridge (spreading center) depicting the origin of hydrothermal vents on the ocean floor. Cool water (*blue arrows*) is gradually heated as it descends toward the hot magma chamber, leaching sulfur, iron, copper, zinc, and other metals from surrounding rocks. The hot water (*red arrows*) returning to the surface carries these elements upward, discharging them at the hydrothermal springs on the ocean floor, where they support a distinctive community of organisms.

Metabolic Pathways

It is unlikely that the first microorganisms would have evolved a means of manufacturing their own food but rather assimilated small aggregates of organic molecules that were present in the surrounding water. Some would probably have consumed even their developing contemporaries. Today organisms with this type of nutritional mechanism are termed **heterotrophs**. As suggested by the study of living microbial heterotrophs, the food gathered by the ancestral heterotrophs might have been externally digested by excreted enzymes before being converted to the energy required for vital functions. In the absence of free oxygen, there was only one way to accomplish this conversion—by **fermentation**. There are many variations of the fermentation process. The most familiar reaction involves the fermentation of sugar by yeast. It is a process by which organisms are able to disassemble organic molecules, rearrange their parts, and derive energy for life functions. A very simple fermentation reaction can be written as follows:

$$\underset{\text{Glucose}}{C_6H_{12}O_6} \rightarrow \underset{\text{Carbon dioxide}}{2\ CO_2} + \underset{\text{Alcohol}}{2\ C_2H_5OH} + \text{Energy}$$

Animal cells are also able to ferment sugar in a reaction that yields lactic acid rather than alcohol.

Eventually, the original fermentation organisms would have begun to experience difficult times. By consuming the organic compounds of their environment, they would eventually have created a food shortage; this scarcity, in turn, might have caused selective pressures for evolutionary change. At some point prior to the depletion of the food supply, organisms evolved the ability to synthesize their needs from simple inorganic substances. These were the first autotrophic organisms. Unlike the heterotrophs, **autotrophs** were able to manufacture their own food. The organisms that had developed this remarkable ability saved themselves from starvation. Their evolution proceeded in diverse directions. Some manufactured their food from carbon dioxide and hydrogen sulfide and were the probable ancestors of today's sulfur bacteria. Others employed the life scheme of modern nitrifying bacteria by using ammonia as a source of energy and matter.

However, more significant than either of these kinds of autotrophs were the photoautotrophs (Fig. 6-41), which were capable of carrying on photosynthesis. Their gift to life was the unique capability of dissociating carbon dioxide into carbon and free oxygen. The carbon was combined with other elements to permit growth, and the oxygen escaped to prepare the environment for the next important step in the evolution of primitive organisms. In simplified form, the reaction for photosynthesis can be written as follows:

$$6\ CO_2 + 6\ H_2O \xrightarrow{\text{Sunlight}} C_6H_{12}O_6 + 6\ O_2$$

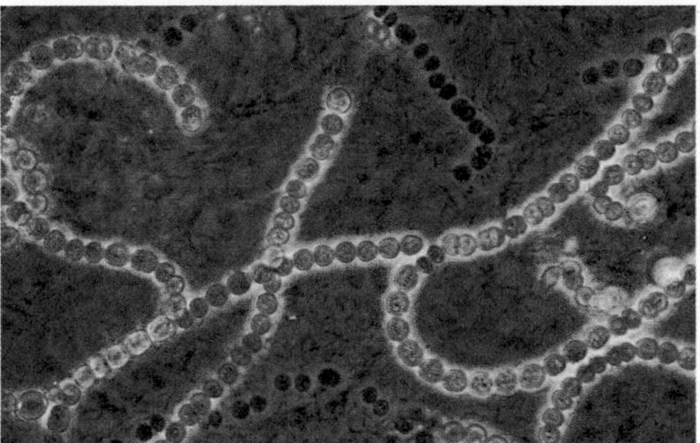

FIGURE 6-41 Photoautotrophs. These organisms, known as *Nostoc*, use light energy to power the synthesis of organic compounds from carbon dioxide and water. (*S. Stammers/Science Photo Library/Photo Researchers, Inc.*)

With the multiplication of the photoautotrophs, billions upon billions of tiny living oxygen generators began to change the primeval nonoxygenic atmosphere to an oxygenic one. Fortunately, the change was probably gradual, for if oxygen had accumulated too rapidly, it would have been lethal to developing early microorganisms. Some sort of oxygen acceptors were needed to act as safety valves and to prevent too rapid a buildup of the gas. Iron in rocks of the continental crust provided suitable oxygen acceptors. Eventually, organisms evolved oxygen-mediating enzymes that permitted them to cope with the new atmosphere.

After the superficial iron on the Earth's surface had combined with its capacity of oxygen, the gas began to accumulate in the atmosphere and hydrosphere. Solar radiation acted on atmospheric oxygen to convert part of it to ozone, and ozone in turn formed an effective shield against harmful ultraviolet radiation. Still-primitive and vulnerable life was thereby protected and could expand into environments that formerly had not been able to harbor life. The stage was set for the appearance of aerobic organisms.

Aerobic organisms use oxygen to convert their food into energy. The reaction, which can be considered a form of cold combustion, provides far more energy in relation to food consumed than does the fermentation reaction. This surplus of energy was an important factor in the evolution of more complex forms of life.

Prokaryotes, Eukaryotes, and Endosymbiosis

There is no unequivocal evidence to date the transition from chemical or prebiotic evolution to organic evolution. The most accurate statement that can be made in this regard is that the transition occurred at some time prior to 3.5 billion years ago. In rocks of that age, paleontologists have discovered the earliest fossil evidence of life. These oldest fossils belong to a category of microorganisms called **prokaryotes**

(Fig. 6-42*A*). Prokaryotes lack definite membrane-bounded organelles and also do not have a membrane-bounded nucleus in which genetic material is neatly arranged into discrete chromosomes (Table 6-4). Modern prokaryotes do possess cell walls, and most are able to move about. The cyanobacteria are prokaryotes that are capable of photosynthesis (Fig. 6-41).

Prokaryotes are asexual and thereby restricted in the level of variability they can attain. In sexual reproduction, as noted in Chapter 4, there is a union of gametes (egg and sperm) to form the nucleus of a single cell, the zygote. The formation of the zygote results in recombination of the parental chromosomes and leads to a multitude of gene combinations among the gametes that give rise to the next generation. In the asexually reproducing prokaryotes, in contrast, a cell divides from the parent and becomes an independent individual containing the identical chromosome complement of the parent cell. Unless mutation intervenes, the number and kinds of chromosomes in individuals produced by asexual reproduction are exactly the same as those in the parent. Thus, the possibilities for variation are based on mutations and therefore more limited than in sexually reproducing organisms. It is probably for this reason that prokaryotes have shown little evolutionary change through more than 2 billion years of Earth history. Nevertheless, such organisms represent an important early step in the history of primordial life.

Evolution proceeded from the prokaryotes to organisms with a definite nucleus, well-defined chromosomes, and the capacity for sexual reproduction. These more advanced forms were **eukaryotes** (see Fig. 6-30*B*). Unlike prokaryotes, eukaryotes contain organelles such as chloroplasts (which convert sunlight into energy) and mitochondria (which metabolize carbohydrates and fatty acids to carbon dioxide and water, releasing energy-rich phosphate compounds in the process). Evidence for the existence of eukaryotes now extends back to at least 2.2 billion years ago, and possibly 2.7 billion years ago.

Biologists believe that the organelles in eukaryotic cells were once independent microorganisms that entered other cells and then established symbiotic relationships with the primary cell. The process has been named

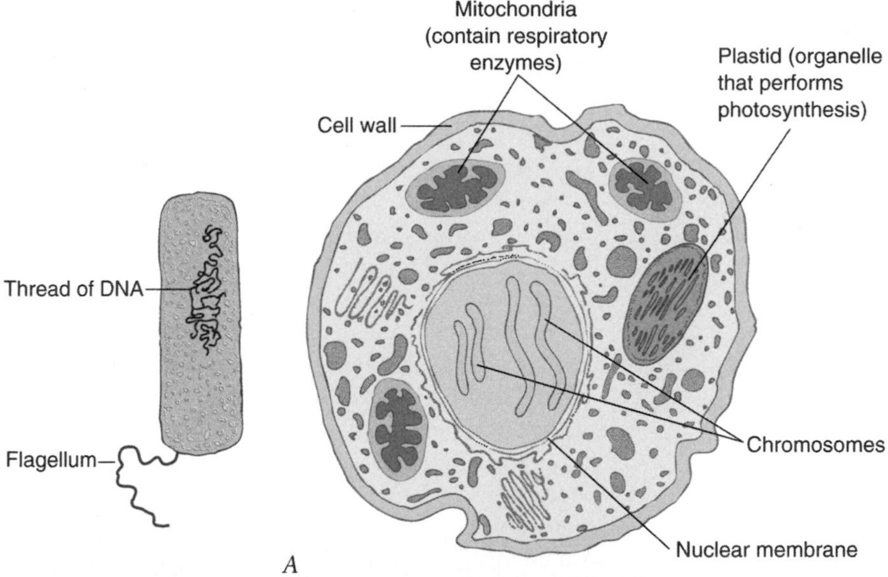

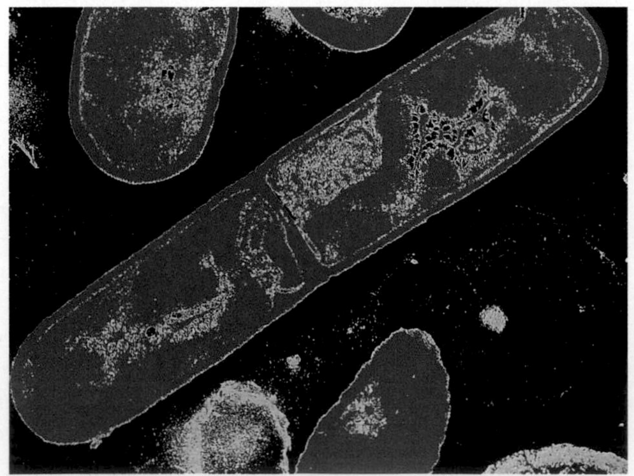

FIGURE 6-42 **Comparison of a prokaryotic cell and eukaryotic cell.** Note that the prokaryotic cell (*A*) lacks a true nucleus, whereas the eukaryotic cell contains a true nucleus and various organelles. Also, prokaryotic cells are smaller, ranging in size from 0.5 to 1.0 millimeters as compared with 10 to 100 millimeters for eukaryotic cells. (*B*) A false-color electron micrograph of the living prokaryote *Bacillus subtilus* that is about to complete cell division. Magnification × 335,000. (*B, Photo Researchers, Inc.*)

TABLE 6-4 Comparison of Some of the Characteristics of Prokaryotic and Eukaryotic Organisms

	Prokaryotes*	Eukaryotes
Organisms	Bacteria and cyanobacteria	Protists, fungi, plants, animals
Cell size	1 to 10 μm	10 to 100 μm
Genetic organization	Loop of DNA in cytoplasm	DNA in chromosomes in membrane-bounded nucleus
Organelles	No membrane-bounded organelles	Membrane-bounded organelles (chloroplasts and mitochondria)
Reproduction	Binary fission, dominantly asexual	Mitosis or meiosis, dominantly sexual

* Organisms of the Domain Archaea and Domain Bacteria.

endosymbiosis ("endo" means within and "symbiosis" means living together). During endosymbiosis, an anaerobic, heterotrophic prokaryote such as a fermentative bacterium might engulf, but not digest, aerobic bacterial cells. These *aerobic bacterial endosymbionts* survive and reproduce within the host cell. They persist in future generations of the host organism. They become organelles called chloroplasts (Fig. 6-43). The host cells would have the advantage of having an internal organelle capable of photosynthesis, whereas the endosymbiont would benefit from a protected environment within the host cell and the availability of vital raw materials and nutrients. Eventually, the endosymbionts would lose their ability to function outside of the primary cell. In a similar way, aerobic bacteria engulfed by host cells would ultimately become mitochondria, the organelles that produce most of the energy for the cell. Mitochondria depend on oxygen to generate energy. Therefore, they evolved only after the photosynthetic prokaryotes filled the world with oxygen. In yet another example of endosymbiosis, a nonmotile host may have acquired mobility by forming a symbiotic relationship with a whiplike organism such as a spirochete. The resulting organism would appear to have a flagellum for locomotion. As with the acquisition of chlorplasts and mitochondria, natural selection would favor such mutually beneficial symbiotic relationships.

Chloroplasts and mitochondria contain their own DNA, which is separate and different from the DNA of the cell nucleus. Also, their DNA is circular, forming a continuous loop like the DNA of bacteria and many viruses. On a limited level they can synthesize protein. They can also divide independently of the cell in which they reside. This evidence strongly supports the endosymbiotic theory that present-day mitochondria and chloroplasts in eukaryotic cells are descended from prokaryotes that came to live inside ancient host cells.

With the development of sexually reproducing eukaryotes, genetic variations could be passed from parent to offspring in a great variety of new combinations. This development was truly momentous, for it led to a dramatic increase in the rate of evolution and was ultimately responsible for the evolution of complex multicellular animals.

The Archean Fossil Record

Because of the prevalence of prokaryotic microorganisms in certain Archean sedimentary rocks, the eon might well be called the age of prokaryotes. Among the more widespread indications of these primitive organisms are fossils known as stromatolites. Stromatolites are laminar, organic sedimentary structures formed by the trapping of sedimentary particles and precipitation of calcium carbonate in response to the metabolic activities and growth of matlike colonies of cyanobacteria and fewer numbers of other prokaryotes. The study of present-day stromatolites indicates that they develop when fine particles of calcium carbonate settle on the sticky, filamentous surface of cyanobacterial mats to form thin layers (laminae). The cyanobacterial colonies then grow through the layer and form another surface for the collection of additional fine sediment. Repetition of this process along with precipitation of calcium carbonate produces a succession of laminae. The resulting structures can be cabbagelike, columnar, tabular, or branching in shape.

The relation between cyanobacterial activity and growth of stromatolites can be directly observed in modern stromatolites such as those at Shark Bay, Australia (Fig. 6-44). For ancient stromatolites, this is, of course, not possible. Thus, there is the possibility that some of these Archean stromatolitic buildups were

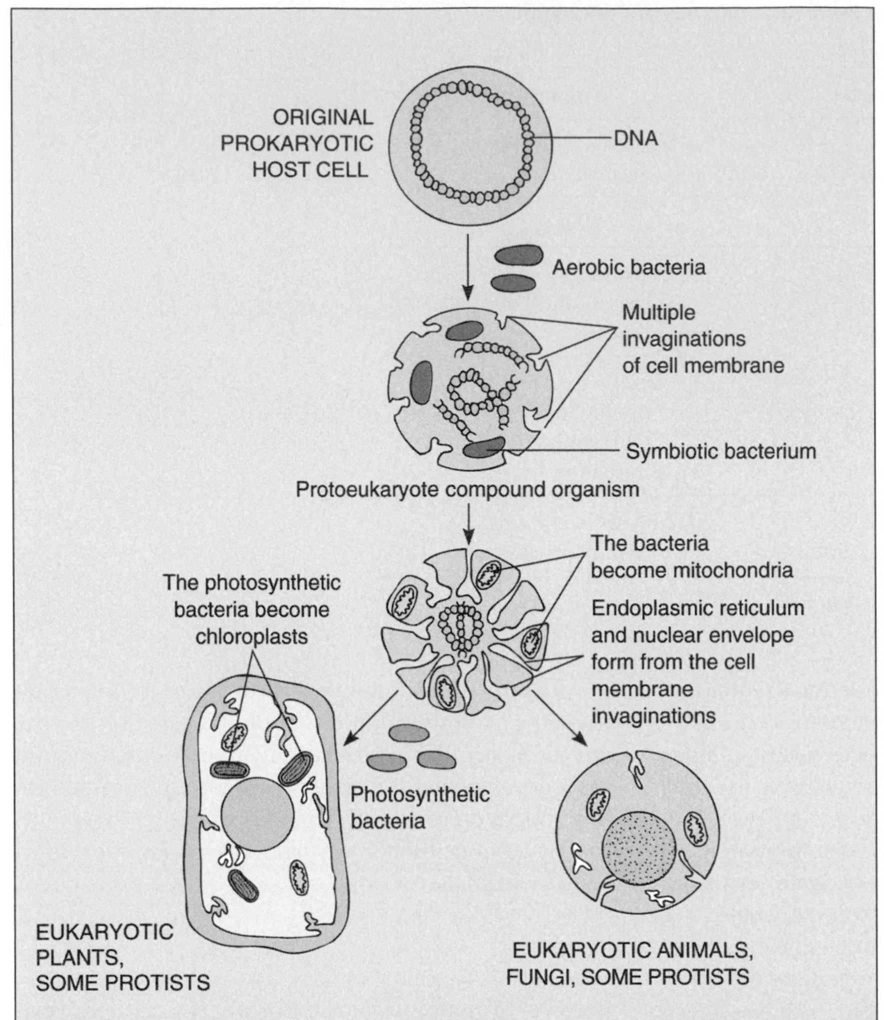

FIGURE 6-43 The endosymbiotic theory for the origin of eukaryotes.

(*A*)

(*B*)

FIGURE 6-44 Present-day and ancient stromatolites. (*A*) Present-day columnar stromatolites growing in the intertidal zone of Shark Bay, Australia. Metabolic activities of colonial marine cyanobacteria result in the formation of these structures. Fine particles of calcium carbonate settle between the tiny filaments of the matlike colonies and are bound with a mesh of organic matter. Successive additional layers result in the laminations that are the most distinctive characteristic of stromatolites. (*B*) Fossil stromatolites from Precambrian rocks exposed in southern Africa. (*A, courtesy of J. Ross; B, courtesy of J. W. Schopf, UCLA.*)

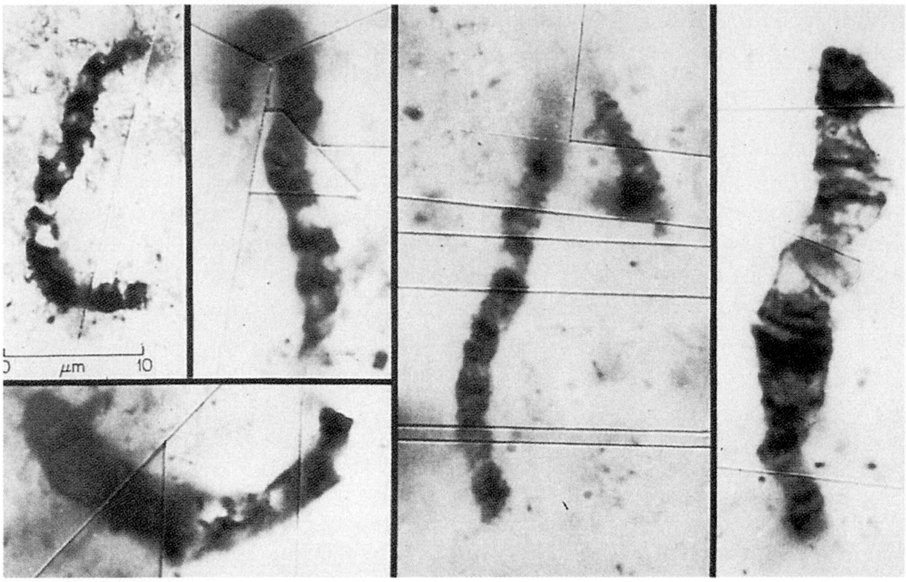

FIGURE 6-45 Filamentous prokaryotic microfossils from 3.5-billion-year-old black cherts of the Archean Warrawoona Group, Pilbara shield of western Australia. *(Courtesy of J. W. Schopf and B. M. Packer.)*

formed by chemical processes in ocean water saturated with calcium carbonate, and that any bacterial remains represent passive occupants of the areas in which stromatolites were growing.

Stromatolites are more common in Proterozoic rocks than Archean rocks. This may be a reflection of a rarity of the warm shelf environments conducive to the growth of marine stromatolites during the Archean. The oldest stromatolite-like fossils occur in 3.5-billion-year-old rocks of the Warrawoona Group of Australia's Pilbara shield. The black chert beds in the Warrawoona Group also contain a variety of filamentous fossil prokaryotes (Fig. 6-45). Somewhat younger Archean stromatolites are found in the 3.0-billion-year-old Pongola Group of southern Africa and the 2.8-billion-year-old Bulawayan Group of Australia. The Carbon 12 to Carbon 13 ratios in the Bulawayan rocks provide evidence of biologic fixation of carbon dioxide by photosynthetic organisms.

Some of the earliest discoveries of Archean microfossils were the result of careful study of thin sections and electron micrographs prepared from rocks of the southern African Fig Tree Group. These rocks are about 3.1 billion years old and consist mostly of dark cherts, their blackness providing a good indication of carbon derived from organic-walled microbes. Elso S. Barghoorn and J. William Schopf were among the first paleontologists to study the fossils of the Fig Tree Group. By examining thin sections of chert from the Fig Tree strata, they were able to find numerous spheroidal bodies very similar to present-day cyanobacteria. Because the samples were collected near the town of Barberton, Barghoorn and Schopf named the fossils *Archaeosphaeroides barbertonensis* (Fig. 6-46). The filaments of organic matter and hydrocarbons of organic origin found in Fig Tree rocks are considered evidence of the presence of microscopic life in the Archean. Many of the dark layers in the Fig Tree Formation are found to contain organic compounds regarded as breakdown products of chlorophyll, indicating the activities of organisms

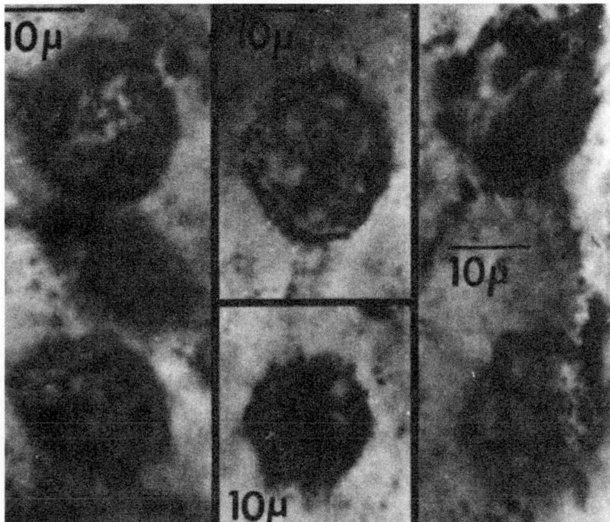

FIGURE 6-46 Spheroidal microfossils. (*Archaeosphaeroides barbertonensis*) in black chert of the early Precambrian Fig Tree Series. These fossils are 3.1 billion years old. The scale bars are in microns (1 μm is equal to 0.001 mm). *(Courtesy of J. W. Schopf.)*

capable of photosynthesis. Fossils similar to those discovered by Barghoorn and Schopf are now known from rocks 3.5 billion year old.

MOLECULAR FOSSILS OF EUKARYOTES In 1999, Australian researchers found evidence of eukaryotic life on Earth 2.7 billion years ago. Prior to this discovery, eukaryotes were thought to have appeared only about 1.6 billion years ago. The evidence does not consist of actual body fossils or trace fossils but of preserved organic molecules that only eukaryotes synthesize. They are called **molecular fossils**. The eukaryotic molecular fossils were carefully extracted from black shales of Archean age in the Pilbara Craton of northwestern Australia.

Although the molecular evidence indicates eukaryotes originated in the Archean, they did not begin to diversify until about 1.2 billion years ago, during the Proterozoic. Many believe the expansion of eukaryotic algae is related to the build-up of oxygen in upper layers of the ocean, as well as the production of nourishing quantities of nitrates.

SUMMARY

Because of the limited number of exposures of Hadean rocks, our understanding of Hadean events is limited. For this reason, geologists seek additional clues to the history of this primordial eon by studying the Moon, meteorites, and other planets. The planets in the solar system in order of increasing distance from the sun are Mercury, Venus, Earth, Mars, Jupiter, Saturn, Uranus, Neptune, and Pluto. These planets and other bodies in the solar system originated by condensation and accretion from a rotating nebula of gases and dust particles.

During the Hadean, the Earth was still very hot, convection was vigorous, and steaming vents and volcanoes blistered the planet's cratered surface. It was during this eon that the Earth's iron core separated from the silicate mantle and the Earth was enveloped in an atmosphere devoid of free oxygen but rich in carbon dioxide, water, and other volcanic gases. It was also a time during which the Earth, its Moon, and other planets in the solar system were being vigorously pelted by meteorites and asteroids. On Earth, craters formed by Hadean meteorite showers have been destroyed by later plate tectonic processes. However, proof of the bombardment is clearly evident on the crater-scarred surface of the Moon. Rocks from the lunar highlands are about 4 billion years old. At that time, there may have been sufficient heat generated by the vast infall of meteorites to cause melting and consequent resetting of radiometric clocks. Radiometric clocks on the Hadean Earth would have been reset in a similar way. Rocks of the Archean are most extensively exposed for study in the broadly unwarped, geologically stable regions of continents known as Precambrian shields. The shields can be divided into a number of Precambrian provinces that tend to be distinctive in age and structural characteristics. The margins of provinces may be marked by belts of deformed rocks called collisional orogens.

The original crust of the Earth was mafic in composition and was probably formed of extrusions originating from a partially or totally molten underlying mantle. From this original mafic crust, patches of continental crust of a more felsic composition were formed by partial melting of mafic rocks and by passing the weathered products of original mafic terranes through cycles of tectonic and orogenic activity involving melting and metamorphism.

Because of high heat flow from the Earth's interior during the early history of the Archean, it is likely that the crust was frequently disrupted and that continents were small. By late in the Archean, however, large continents had developed by accretion of microcontinents and consolidation of marginal depositional tracts during postdepositional orogenies.

Most Archean rocks can be grouped into two major rock associations. The first is the granulite association, which consists of gneisses derived from granodioritic and gabbroic rocks. The second is the greenstone association, which typically occurs in synclinal belts within the more extensive granulite terranes. Greenstones consist of mafic or ultramafic volcanics overlain by felsic volcanics and are capped by sedimentary rocks. In theory, both greenstone belts and granulite terranes may have developed in the course of Archean tectonics, in which granulite associations formed along subduction zones adjacent to volcanic arcs and greenstone belts developed in back-arc basins. Archean sedimentary rocks are found chiefly in greenstone belts and consist largely of graywackes, shales, claystones, conglomerates, and cherts.

The Archean was the time when life originated on Earth. The planet's first life arose in an environment deficient in free oxygen but containing carbon, hydrogen, nitrogen, and water, in which inorganic salts are dissolved. Complex organic molecules may be formed when such a mixture is activated by ultraviolet light or electrical discharges from lightning. These organic compounds and their derivatives are found in every living organism. They are the molecular basis of all organic structures, heritable information, and energy.

The discovery of bacteria thriving in total darkness at great depths in solid rock and along submarine hydrothermal vents has prompted new theories about the origin of life. Organisms living within the fractures beneath hydrothermal vents are called hyperthermophiles. Instead of using energy from the Sun as in photosynthesis, the hyperthermophiles use energy released from certain chemical bonds. The process is termed chemosynthesis. Water in the hydrothermal system percolates downward toward magma chambers, where it is superheated and able to dissolve compounds essential for life from surrounding rock. The solutions are then returned to the surface and, on cooling, may form organic molecules and combinations of molecules leading to protocells. In the Archean, the protocells may have evolved into the earliest microorganisms. Hydrothermal systems would have been far more prevalent in the Archean than today, providing ample opportunity for the evolution of chemosynthetic hyperthermophiles.

The first living cells might have subsisted on the chemical energy they derived by consuming other cells and molecules. They were heterotrophs and were followed by cellular organisms that could use the energy of the Sun through photosynthesis. Free oxygen accumulated as a byproduct of photosynthesis, making possible the evolution of organisms that use oxygen to process their food for energy.

Archean fossils have been found in rocks about 3.6 billion years old and sporadically in younger rocks of Archean age. The majority of Archean fossils are prokaryotes. They consist of cyanobacteria and other bacteria often associated with the distinctive laminated structures known as stromatolites. However, molecular fossils of eukaryotes have been detected in 2.7-billion-year-old Archean black shales.

QUESTIONS FOR REVIEW AND DISCUSSION

1. At successively greater distances from the Sun, name the planets of our solar system. In what galaxy are these planets located?

2. Compare Mercury, Venus, Earth, and Mars. What do they have in common and how do they differ?

3. Given that both the Earth and the Moon experienced heavy meteoric bombardment during the Hadean, why are there so few craters recording that event on Earth?

4. Why are we likely to learn more about the Hadean history of the Earth by studying the surface and rocks of the Moon than the Earth?

5. How did the internal layers of the Earth develop?

6. What are the principal kinds of meteorites? What is the particular significance of carbonaceous chondrites?

7. Distinguish between the terms *Precambrian shield*, *craton*, and *platform*.

8. Differentiate between the terms *mafic* and *felsic*. Give an example of a felsic extrusive rock and one of a mafic extrusive rock. Name two minerals common to each of these rock types.

9. How might patches of felsic continental crust have been derived from mafic rocks during the Archean?

10. Describe the structural configuration and general vertical sequence of rock types in a greenstone belt. In a plate tectonics scenario, where might greenstone belts form and how would they develop?

11. How do komatiites differ from mafic rocks such as basalt? Where do they occur in greenstone sequences?

12. What is the economic significance of the Witwatersrand rocks of Africa? What kind of sedimentary rocks occur in the Witwatersrand Supergroup, and what was their environment of deposition?

13. What geologic evidence suggests that free oxygen was beginning to accumulate in the Earth's atmosphere about 3 billion years ago?

14. Why are Archean rocks more difficult to correlate than rocks of the Phanerozoic? What is the method used to date and correlate most Archean rocks?

15. Discuss the role of symbiosis in the evolution of eukaryotes. What organelles may have originated by symbiosis?

16. What are hyperthermophiles? What does the presence of hyperthermophiles suggest regarding the possibility of finding organisms on other planets in the universe?

READINGS

Cone, J. 1991. *Fire Under the Sea*. New York: William Morrow & Co.

Goodwin, A. M. 1996. *Principles of Precambrian Geology*. London, England: Academic Press.

Holm, N. G. (ed.). 1992. *Origins of Life and Evolution of the Biosphere*. Boston: Kluwer Academic Publishers.

Kershaw, S. 1990. *Evolution of the Earth's atmosphere and its geological impact*. Geol. Today 6(2):55-60.

Margulis, L. 1992. *Symbiosis in Cell Evolution*. 3rd ed. San Francisco: W. H. Freeman & Co.

Margulis, L., Matthews, C., and Haselton, A. 2000. *Environmental Evolution: Effects on the Origin and Evolution of Life on Planet Earth*. 2d ed. Cambridge, Massachusetts: MIT Press.

Nisbet, E. G. 1987. *The Young Earth*. Winchester, MA: Allen and Unwin, Inc.

Pasachoff, J.M., and Filippenko, A. 2001. *The Cosmos: Astronomy in the New Millenium*. Philadelphia: Harcourt College Publishers.

Schopf, J. W. 2002. *Cradle of Life: Discovery of the Earth's Earliest Fossils*. Princeton, N.J.: Princeton University Press.

Shock, E. K. 1994. Hydrothermal systems and the emergence of life. *Geotimes* 39(3):13–14.

Skinner, C.H. and Bansfield, J.F. 1997 Microbes all around. *Geotimes* 42(8):16–19.

York, F. 1992. The earliest history of the Earth. *Sci. Am.* 268(1):90–96.

WEB SITES

The Earth Through Time Student Companion Web Site (www.wiley.com/college/levin) has online resources to help you expand your understanding of the topics in this chapter. Visit the Web Site to access the following:

1. Illustrated course notes covering key concepts in each chapter;
2. Online quizzes that provide immediate feedback;
3. Links to chapter-specific topics on the web;
4. Science news updates relating to recent developments in Historical Geology;
5. Web inquiry activities for further exploration;
6. A glossary of terms;
7. A Student Union with links to topics such as study skills, writing and grammar, and citing electronic information.

7

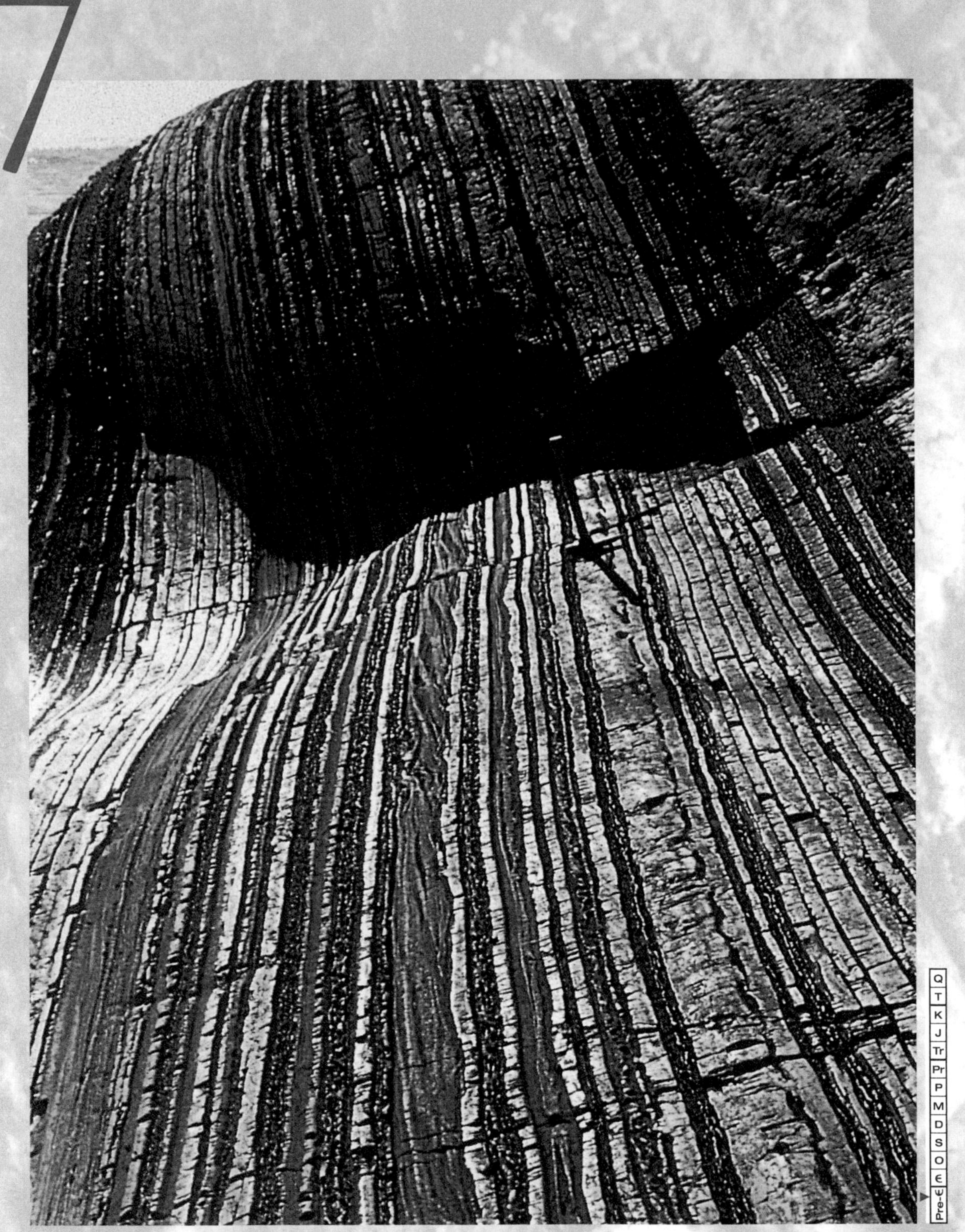

Early Proterozoic graywackes with graded bedding interstratified with shaly limestone, deposited in the aulacogen east of Great Slave Lake, Northwest Territories of Canada. (Courtesy of P. Hoffman, Harvard University.)

CHAPTER 7

The Proterozoic Eon

We crack the rocks and make them ring,
And many a heavy pack we sling;
We run our lines and tie them in,
We measure strata thick and thin,
And Sunday work is never sin,
By thought and dint of hammering.

From the poem *Mente et Malleo*, written in 1888 by geologist A. F. Lawson

OUTLINE

The Proterozoic Eon began 2.5 billion years ago and ended only 540 million years ago. This enormous period of time comprises 42 percent of Earth history. To facilitate the study of Proterozoic history, the eon can be divided into an early **Paleoproterozoic** Era (2.5 to 1.6 billion years ago), a Mid-Proterozoic unit termed the **Mesoproterozoic** (1.6 to 1.0 billion years ago), and the most recent **Neoproterozoic** (1.0 billion years ago to the beginning of the Paleozoic, 540 million years ago). The designation of the beginning of the Proterozoic at 2.5 billion years ago is somewhat arbitrary, but it is used because this is about the time when a more modern style of plate tectonics and sedimentation began to be prevalent. Because Proterozoic rocks are less altered, they are easier to study and interpret than rocks of the Archean. Difficulties persist, however, for they do not contain abundant fossils as do many Phanerozoic strata.

As described in the preceding chapter, a number of distinct cratonic elements, termed **Precambrian provinces**, had developed in North America during the Archean. The provinces were welded together to form a large continent called **Laurentia** during the Paleoproterozoic. The suturing occurred along belts of crustal compression, mountain building, and metamorphism called **orogens**. This process of aggregation and consolidation of the once-separate cratonic elements was completed by about 1.7 billion years ago. During the remainder of the Proterozoic, Laurentia continued to grow as a result of accretionary events. Accretion involves additions of crustal materials to continental margins. The additions may consist of sedimentary rocks deposited along the once-passive margins of continents that may later be compressed and sutured onto the continents, or it may involve the incorporation of microcontinents and island arcs that are carried to subduction zones adjacent to continents by sea-floor spreading.

The processes of accretion and aggregation of Archean crustal elements were in operation worldwide during the Proterozoic and resulted in the formation of several very large cratons. In addition, the Proterozoic was characterized by substantial amounts of lateral tectonic plate motion and accompanying rifting and sea-floor spreading. On and around the growing cratons, tectonic events occurred and environments of deposition developed that closely resemble those of the succeeding Phanerozoic Eon. By Proterozoic time, wide continental shelves and epeiric seas existed in which shallow-water clastic sediments and carbonate deposits accumulated. On the whole, such sediments are rare in the Archean. Late in the Proterozoic, continents were assembled into a supercontinent named **Rodinia**.

At various times during the Proterozoic, major parts of the Earth became frigid and covered with ice sheets. One of these icy episodes occurred during the Paleoproterozoic, between about 2.4 and 2.3 billion years ago. The second occurred in the Neoproterozoic, between 800 and 600 million years ago. However, not all of the Neoproterozoic glaciations were strictly contemporaneous, for some glacial deposits appear to have been deposited only 1 billion years ago.

PALEOPROTEROZOIC EVENTS IN NORTH AMERICA

Proterozoic Plate Tectonics

In the Canadian shield, accretionary Paleoproterozoic rocks occur primarily in areas that circumscribe the older Archean provinces. One such area lies along the western margin of the Slave province in Canada's Northwest Territories. Here, in a belt of deformation, called the Wopmay orogen (Fig. 7-1), there is evidence of (1) the opening of an ocean basin, (2) sedimentation along the resulting new continental margins, and (3) closure of the ocean basin through plate tectonic processes. This sequence of events, well documented in the Phanerozoic, is called a **Wilson cycle**, after J. Tuzo Wilson, one of the pioneers of the plate tectonics theory.

In the region of the Wopmay orogen, evidence of the initial opening of an ocean basin consists of numerous normal (tensional) faults (Fig. 7-2). Alluvial fan and fluvial deposits accumulated in the downfaulted blocks, and lavas flowed upward through the fault planes to become interlayered with the sediments. Continued opening produced an oceanic tract that slowly widened. As the ocean widened, the western edge of the Slave province became a passive continental margin on which two parallel zones of deposition developed. An eastern zone next to the Slave platform received coastal plain deposits that graded westward (seaward) to shallow marine quartz sandstones. These sandstones were subsequently metamorphosed and are now quartzites. The quartzites are overlain by massive stromatolitic dolomites of the Rocknest Formation (Fig. 7-3). To the west of these shallow continental shelf deposits, deeper-water turbidites of the continental rise were deposited.

After a period of relatively uneventful sedimentation along the Wopmay tract, the more dynamic episode of the Wilson cycle, involving ocean closure, began. As a result of westward subduction of the lead-

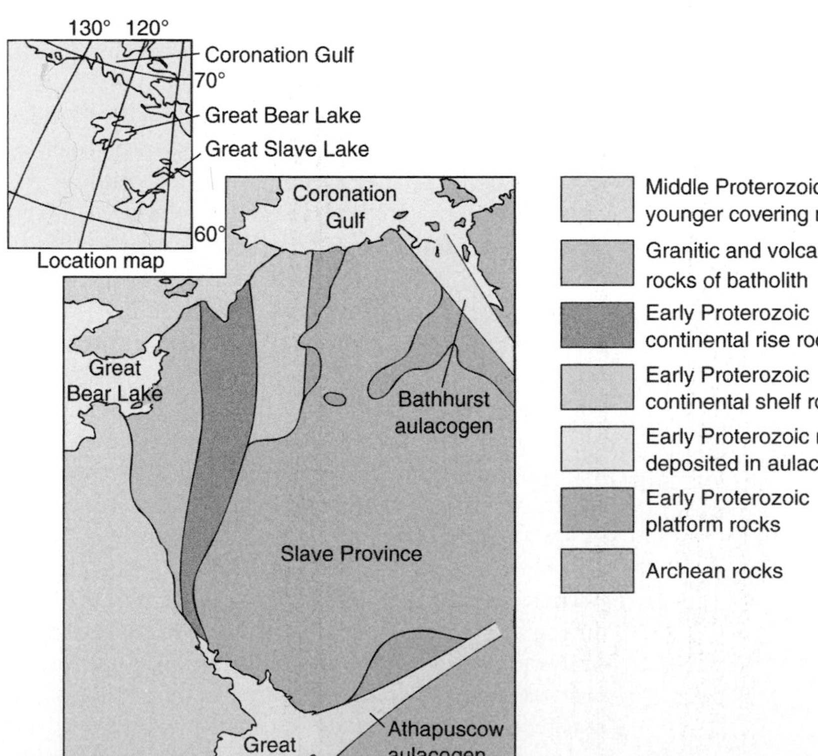

FIGURE 7-1 Tectonic elements in and around the Wopmay orogen, Northwest Territories, Canada. (*From Hoffman, P. 1973. Evolution of an early Proterozoic continental margin: The Coronation geosyncline and associated aulacogens of the northwestern Canadian shield.* Phil. Trans. Royal Soc. London A 273:547–581.)

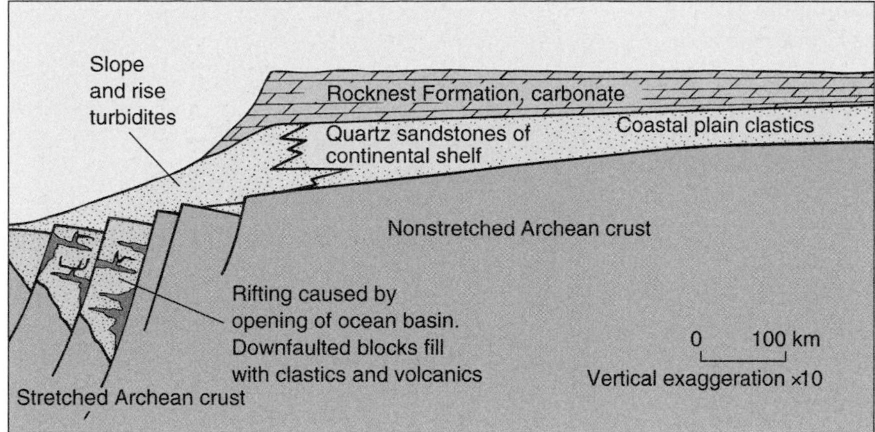

FIGURE 7-2 Relationship of rock units following Paleoproterozoic opening of an ocean basin along the western margin of the Slave craton. (*Simplified from Hoffman, P. 1989.* Geology of North America, *Vol. A, pp. 447–512. Boulder, CO: Geological Society of America.*)
? *What type of faults are beneath the turbidites? Are these the kind of faults associated with the leading edges of tectonic plates or the passive margins?*

ing edge of the Slave plate beneath a microplate approaching from the west, the continental shelf buckled downward, creating an elongate subsiding trough that ended carbonate deposition and served as a basin for the accumulation of deep-water clastics. When the basin had filled, deltaic and fluvial sands were deposited above the marine clastics. Subsequently, plate collisions caused additional folding and faulting.

The sequence of events in the Wopmay orogen closely parallels that of occurrences during the late Paleozoic evolution of the Appalachian Mountains. The history of both tracts began with deposition of alluvial deposits in fault-bound basins, followed by the opening of an oceanic tract; it progressed to the deposition of continental shelf and rise sediments along passive continental margins and ended with ocean closure, deformation, and deposition of nonmarine clastic sediments derived from the erosion of mountains produced during the closure episode.

Paleoproterozoic Aulacogens

Another similarity between Proterozoic tectonics and the tectonics of the Phanerozoic is the development of aulacogens. As described in Chapter 5, an **aulacogen** is an inactive rift of a radiating three-rift system that develops over an area of crust that is bulging upward, possibly because of an underlying mantle plume (see Fig. 5-48). As the bulge forms, the stretched crust breaks to form three rift zones that radiate from a common center. Two of these rifts may widen and become continental margins, whereas the third trends inland away from the continental margin, becomes the "failed arm" of the three-prong system, and fills with sediment.

(*A*)

(*B*)

FIGURE 7-3 Rocknest Formation. (*A*) Dolomite and interbedded shales of the Rocknest Formation, Northwest Territories, Canada. (*B*) This oblique aerial view impressively illustrates the folding of these dolomitic rocks that resulted from east–west crustal shortening and accretion that occurred during the orogenic phase of the development of the Wopmay orogen. (*Courtesy of P. Hoffman, Harvard University.*)

Two aulacogens have been detected along the western side of the Slave province (Fig. 7-1). The Athapuscow aulacogen extends northeastward from Great Slave Lake, and the Bathhurst aulacogen trends southeastward from Coronation Gulf.

The Trans-Hudson Orogen

The Wopmay is just one of many orogens that developed along the margins of Archean provinces during the Proterozoic. Another is the Trans-Hudson orogen, which separates the Superior province from provinces to the north and west (Fig. 7-4). Rocks of the Trans-Hudson belt record another Wilson cycle of initial rifting, opening of an oceanic tract, and subsequent closure along a subduction zone. This closure, with the accompanying severe folding, metamorphism, and intrusive igneous activity, welded the Superior plate to the Hearne and Wyoming plates that lay to the west.

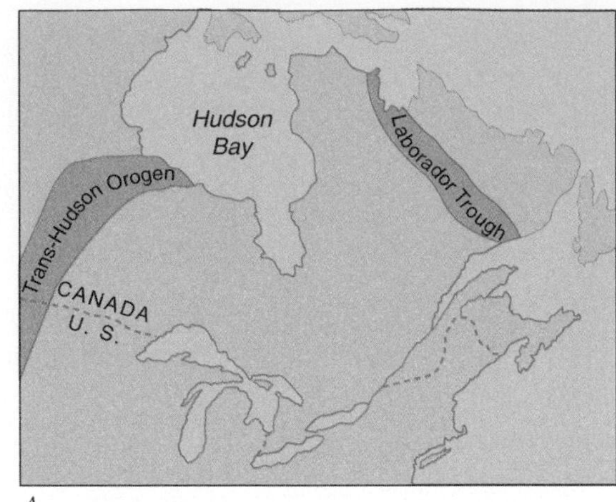

A

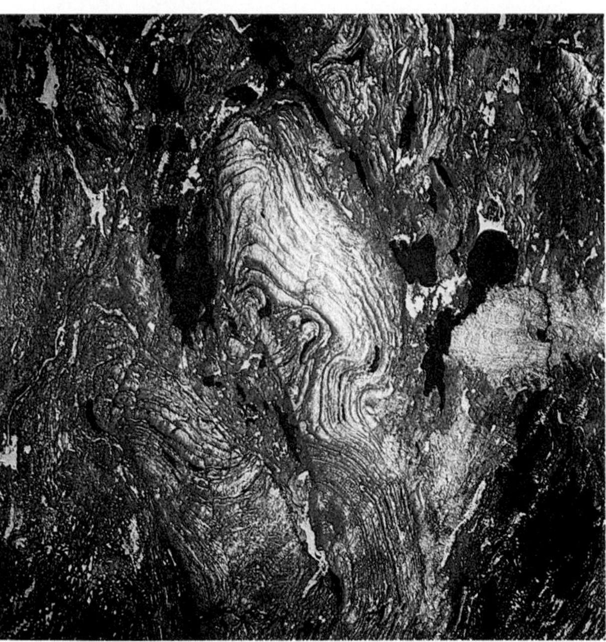

B

FIGURE 7-4 **(*A*) Location of the Labrador Trough and Trans-Hudson Orogen. (*B*) Fold belt of the Labrador Trough depicted in a Landsat image.** The belt was once a chain of mountains raised by the collision of two continents about 1.8 to 1.7 billion years ago. Since then, the mountains have eroded, exposing the deeper metamorphic and igneous rocks that were deformed into these folds by the collision. (*Courtesy of NASA.*)

A Paleoproterozoic Ice Age

In the region north of Lake Huron, rocks of the Paleoproterozoic are called **Huronian**. They consist mainly of coarse clastics that were eroded from older crystalline rocks. In addition, the Huronian Supergroup includes the **Gowganda Formation**, a rock unit that is notable because of conglomerates and laminated mudstones of glacial origin. Laminations in the mudstones represent regularly repeated summer and winter layers of sediment that are called **varves**. Varved mudstones typically form in meltwater lakes that are marginal to ice sheets. The varved sediments in the Gowganda Formation alternate with **tillites** (consolidated rocks composed of unsorted glacial debris), indicating periodic advances of ice into the marginal lakes. Cobbles and boulders in the tillite have been scratched and faceted, providing evidence of the abrasive action of the ice mass as it moved across the underlying bedrock. The Gowganda Formation is intruded by 2.1-billion-year-old igneous rocks. A basement of 2.6-billion-year-old crystalline rocks lies beneath the Gowganda Formation. Hence, in North America, glaciation occurred at some time between these two dates. Glacial conditions and accompanying cooler climates were widespread at this time, for rocks similar to those of the Gowganda are recognized in Finland, southern Africa, and India. This Paleoproterozoic episode of glaciation may have been the first our planet experienced, for during the Archean the planet may not have cooled sufficiently for the accumulation of continental ice sheets.

The Animikie Group

Paleoproterozoic rocks surrounding the western shores of Lake Superior include those of the **Animikie Group**. This group of formations is world famous for the bonanza iron ores it contains. The ore minerals are oxides of iron and thus record the presence of free oxygen in the Earth's atmosphere. By implication, the oxygen-generating process of photosynthesis was probably in vigorous operation by this time. Coarse sandstones and conglomerates deposited in shallow water lie near the base of the Animikie Group. These rocks are overlain by cyclic successions of chert, cherty limestone, shales, and banded iron formations (BIFs). BIFs are also known in Archean sequences, but these

ENRICHMENT

Banded Iron Formations: Civilization's Indispensable Treasure

Iron is the backbone of modern civiliztion, and much of the iron that we consume is derived from banded iron formations (BIF). These formations were deposited in a relatively brief geologic interval between 2,600 and 1,800 million years ago. They occur in huge bodies that exceed hundreds of meters in thickness and thousands of meters in lateral extent. The most common type of BIF consists of bands of hematite or magnetite alternating with lighter laminations of chert. Most researchers believe that these deposits were chemically or biochemically precipitated. Fossils of bacteria found in some of the deposits resemble present-day bacteria that precipitate ferric iron hydroxide in oxygen-deficient environments. There is little agreement, however, on the source of the iron. The exhalations of volcanoes could have been a source, or the iron may have been derived from deep weathering and erosion of nearby mafic crustal rocks.

Substantial parts of the BIF are useable even as a low-grade ore called taconite. Most of the North American high-grade BIF ores are now depleted, but taconite ores, in which the iron is concentrated into pellets, is now being shipped to smelters. The pellets are about 60 percent iron. The most important product made from the raw iron is steel, an alloy of iron with a small amount of carbon. At present, banded iron ores are being mined in the Labrador Trough, the Lake Superior region of North America, the Ukraine, and the Hammersley Range of western Australia

are not as extensive as those of the Paleoproterozoic (Fig. 7-5). Some of the Canadian deposits are over 1000 meters thick and extend over 100 kilometers. Within the banded iron sequence, there is a formation known as the Gunflint Chert, which contains an interesting assemblage of cyanobacteria and other prokaryotic organisms.

The Labrador Trough

East of the Superior province, one finds extensive outcrops of rocks that, like those of the Wopmay orogen, were deposited on the continental shelf, slope, and rise of a craton. These rocks occur along a curving, elongate structural depression known as the Labrador Trough. On the western side of the trough, quartz sandstones, dolomites, and iron formations lie above fluvial and alluvial deposits of fault-bound basins. Pillow lavas, basalts, mafic intrusives, and graywackes occupy an adjacent zone to the east. The eastern zone was subjected to intense folding (Fig. 7-4*B*), metamorphism, and westward thrust faulting during a mountain-building episode called the Hudsonian orogeny. The Hudsonian orogeny serves as the event that separates Paleoproterozoic from Mesoproterozoic geologic history.

A

B

FIGURE 7-5 Banded iron formations. (*A*) This banded iron formation is exposed at Jasper Knob in Michigan's Upper Peninsula. (*B*) Banded iron formations occur worldwide in Proterozoic rocks, as suggested by these at Wadi Kareim, Egypt. (*B, courtesy of D. Bhattacharyya.*) ? *What mineral imparts the red color?*

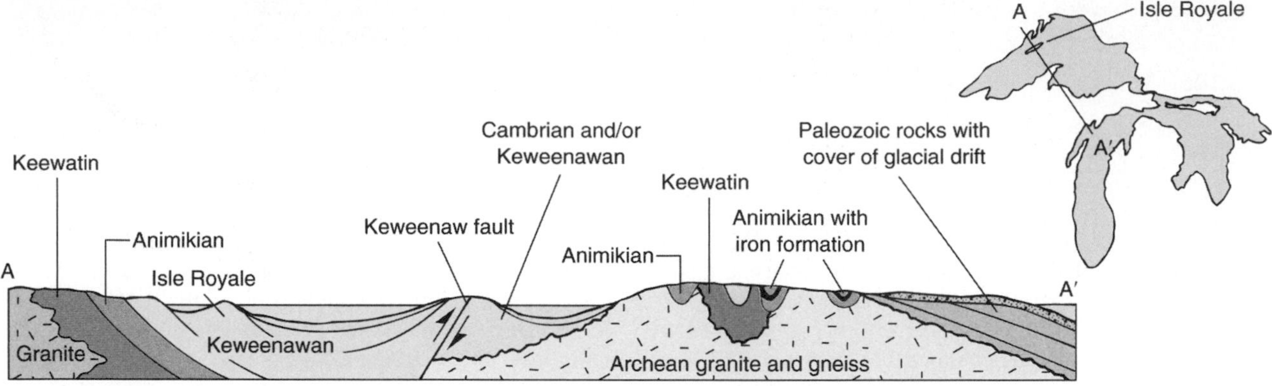

FIGURE 7-6 Geologic cross-section of Lake Superior. (*Adapted from Michigan Geologic Survey*, Bedrock of Michigan, 1968).

MESOPROTEROZOIC EVENTS

Keweenawan Sequence

Among the rocks deposited or emplaced during the Mesoproterozoic were those of the Keweenawan sequence (Fig. 7-6). These rocks rest on either crystalline basement or Animikian strata. They extend for hundreds of kilometers from the Lake Superior region southward beneath a cover of Phanerozoic rocks. Keweenawan rocks consist of quartz sandstones, arkoses, and conglomerates as well as basaltic volcanics. The lava flows are well known for their content of native copper. Holes originally formed as gas bubbles in the extruding lava provided voids in which the copper was deposited. In addition, copper fills small joints and pore spaces in associated conglomerates. Long before Europeans arrived, the Keweenawan copper deposits were worked by Native Americans.

The Keweenawan lavas accumulated to a thickness of several kilometers. Even with so great an outpouring, however, much of the supply of mafic magma was not tapped. It remained beneath the surface, where it crystallized to form the 12-kilometers-thick and 160-kilometers-long mass of the Duluth Gabbro. As noted earlier, basalts are rock types more characteristic of oceanic areas. When such a large quantity of mafic material comes to the surface within the central stable region of a continent, it often signals the presence of a rift zone along which a continent may break apart and admit an ocean. The tract where the break begins typically develops tensional faults along which mafic magma rises to the surface to form Keweenawan-like accumulations. Evidence from gravity and magnetic surveys, as well as samples from deep drill holes, indicates that the rift zones associated with Keweenawan volcanism developed about 1.2 to 1.0 billion years ago and extended from Lake Superior southward beneath covering sedimentary rocks into Kansas. If this system of rifts had been extended to the edge of the craton, an ocean tract would have formed within the rift system, and the eastern United States would have drifted away. Rifting ceased, however, before such a separation could occur.

Sandstones lie at the top of the Keweenawan sequence. These are overlain by nearly identical sandstones of the Upper Cambrian. Because of the similarity, it is difficult to recognize the upper boundary of the Keweenawan, even though Lower and Middle Cambrian rocks have been removed by erosion.

The Grenville Orogeny and the Assembly of Rodinia

The Grenville province of eastern Canada is the youngest Mesoproterozoic region of the Canadian shield and the last to experience a major orogeny. Exposures of Grenville rocks extend from the Atlantic coast of Labrador to Lake Huron. This, however, is only part of their true extent, for they continue beneath covering Phanerozoic rocks down the eastern side of the United States and westward into Texas (see Fig. 6-29). In the United States, the study of Grenville rocks is difficult, for they have been subjected to the destructive processes that accompanied the building of the Appalachian Mountains during the Paleozoic. Typical Grenville rocks consist of carbonates and sandstones (Fig. 7-7) that have been deformed, metamorphosed, and intruded by igneous bodies.

Deformation of Grenville sediments occurred 1.2 to 1.0 billion years ago during the **Grenville orogeny**. That orogeny was only a segment of a great episode of continental collisions involved in the formation of the supercontinent **Rodinia** (Fig. 7-8). At the time of the Grenville orogeny, the east coast of Laurentia (North America) lay adjacent to a block of western South America termed Amazonia, and the west coast lay next to Antarctica and Australia.

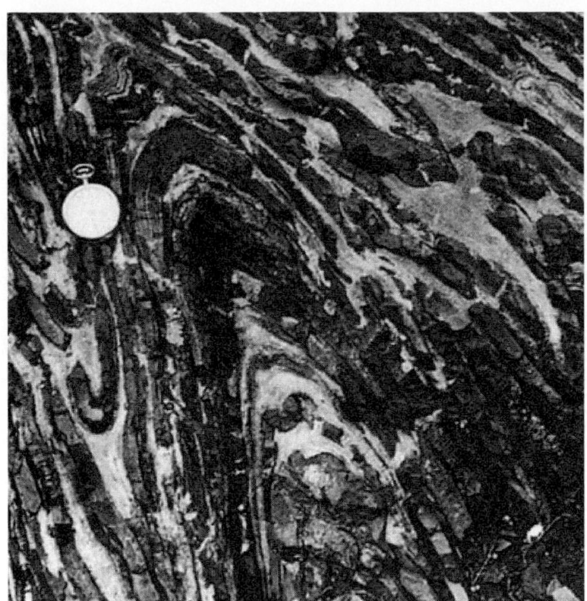

FIGURE 7-7 Folded sandstones (darker color) and carbonates of the late Proterozoic Grenville Group, Belmont Township, Ontario, Canada. (*Courtesy of the Geological Survey of Canada.*)

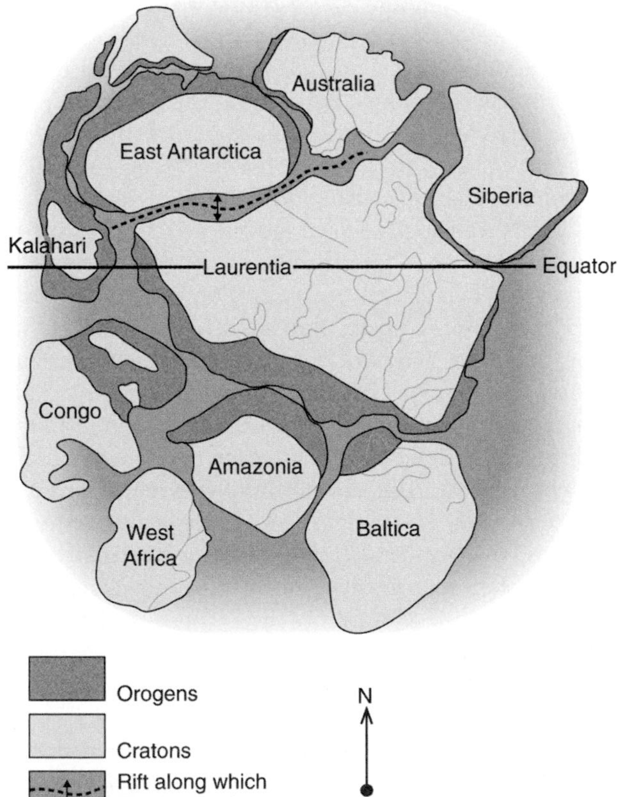

FIGURE 7-8 The Neoproterozoic supercontinent Rodinia as it began to break apart. (*After Hoffman, P. F. 1991. Science 252:1409–1412.*) ? *What continents bordered Laurentia during the Neoproterozoic?*

NEOPROTEROZOIC EVENTS

Neoproterozoic Plate Tectonics

Whereas Rodinia was assembled 1.2 to 1.0 billion years ago, it began to split apart about 750 million years ago. During this episode of supercontinent fragmentation, the Proto-Pacific Ocean (Panthalassa) was created. A great expanse of ocean now lay west of North America. The Neoproterozoic on the eastern margin of North America is represented by rocks that were deposited in basins and shelf areas surrounding the continent. Most of these sediments were later deformed during Paleozoic orogenic events. One such event involved rifting and the filling of rift valleys with lava flows and coarse clastics. The sea entered the rift valleys as the beginning of a widening proto-Atlantic ocean, and a new continental margin was established east of the Grenville orogen. The stage was then set for the deposition of the first Paleozoic shelf sediments.

While these events were occurring, Panthalassa continued to widen, and blocks of continental crust began to cluster around Africa. A new supercontinent (or at least a clustering of land masses) began to take form 800 to 700 million years ago. Deformation and metamorphism along convergence zones affected all Southern Hemisphere continents except India. The widespread episode of mountain building has been named the Pan-African orogeny.

The Neoproterozoic Global Ice House

Extensive continental glaciation occurred during the Neoproterozoic. Neoproterozoic rocks on all of the Earth's land masses except India and Siberia-Mongolia show the glacial striations produced by the movement of huge ice sheets. One also finds lithified, unsorted, boulder-bearing glacial debris called tillites, chunks of rock released from melting icebergs termed dropstones, and contrasting layers of sediment known as varves that represent seasonal changes in glacial lakes. The evidence is so far-ranging that geologists refer to our planet at the time of glaciation as "snowball Earth." More formally, the episode of frigid conditions has been named the **Varangian glaciation**, after Varangian Fiord in Norway, where Neoproterozoic tillites provided early evidence of this great ice age.

What conditions during the Neoproterozoic could have caused the Varangian glacial episode? One hypothesis proposes that plate tectonics may have had an important role in cooling the planet. Paleomagnetic data indicate that the continents were positioned at low and middle latitudes about 600 to 700 million years ago. There were no major land masses over the poles. In contrast to the wide equatorial region of heat-absorbing

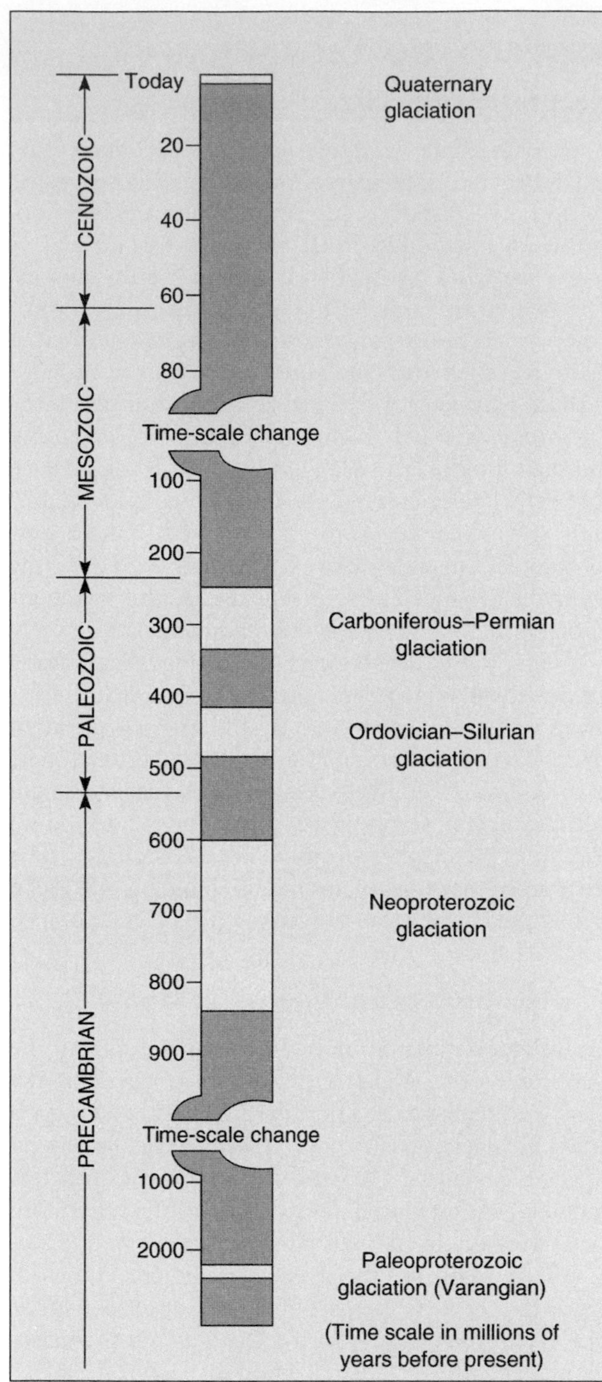

FIGURE 7-9 **In its long history, the Earth has experienced several major episodes of widespread continental glaciation.** Each of these major episodes contained shorter intervals of advance and retreat of the glaciers.

tropical ocean that exists today, the Neoproterozoic equator crossed extensive, highly reflective land surfaces. Heat lost by reflection alone may have been sufficient to cause cooling. As temperatures continued to fall, continental glaciers and ice caps that formed over the poles resulted in further reflection and loss of heat. Eventually even lands at the equator became ice-covered.

A second hypothesis relating to the origin of snowball Earth proposes that a decline in atmospheric carbon dioxide was responsible for lowering temperatures. The hypothesis is based on studies of the rates at which genes mutate. Once the rate has been determined, it can be used as a molecular clock to estimate the time in Earth history when particular species evolved. The results obtained from one such study suggest that fungi and lichens were present on Earth about a billion years ago (lichen is a symbiotic association of a fungus and a photosynthetic organism such as an alga or cyanobacterium). The molecular clock data also indicate vascular plants were already present on land near the time of cooling. The suggestion is made that these photosynthesizers caused the required decline in atmospheric carbon dioxide. Paleontologists, however, maintain a tentative attitude toward the molecular clock studies, for as yet, the fossil record has not provided confirmation.

An interesting aspect of the sediments deposited during the Varangian episode of glaciation is the existence of thick layers of limestone between Varangian tillites. Limestones normally originate in warm seas. How then could they be interspersed with glacial deposits associated with frigid conditions? One explanation of this seemingly incongruous association is that the wintry conditions would have inhibited photosynthetic uptake of carbon dioxide by stromatolites and other microbial plants. As a result, carbon dioxide my have accumulated in the atmosphere sufficiently to trigger episodes of greenhouse warming—thus the paradox of glaciers causing their own destruction.

The Neoproterozoic ice age was not the first for the Proterozoic. As described earlier, the Gowganda tillites and correlative formations on other continents reflect widespread glaciation slightly more than 2 billion years ago. Nor would Neoproterozoic glaciations be the last the Earth would witness, for glaciers advanced again over major portions of the continents during the Ordovician and Silurian, near the end of the Paleozoic, and again during the Quaternary period of the Cenozoic Era (Fig. 7-9). Each of these major episodes of glaciation included shorter intervals during which the ice sheets alternately advanced and regressed.

PROTEROZOIC ROCKS SOUTH OF THE CANADIAN SHIELD

Although outcrops are not as extensive as in Canada, Precambrian rocks also occur south and west of the Canadian shield. Particularly impressive successions are exposed in the Rocky Mountains and Colorado Plateau regions. These rocks have a complex history that began over 2.5 billion years ago with the development of an Archean terrane composed of strongly de-

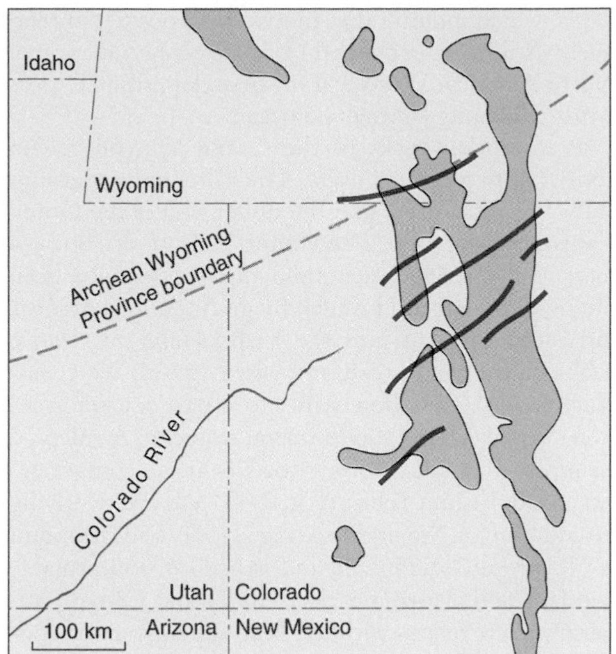

FIGURE 7-10 Shear zones (red) in Wyoming and Colorado developed when the Archean craton collided with island arc terranes during the Paleoproterozoic. Areas of lower Proterozoic outcrops are shown in green. (*Adapted from Karlstrom, K. E., and Bowing, S. A. 1988. J. Geol. 96:561–571.*)

formed and metamorphosed granitic rocks. Largely on the basis of the study of remnant patches of volcanic rocks and greenstones in Wyoming, it appears that this old Archean mass collided with one or more island arcs about 1.7 or 1.8 billion years ago. The line of collision is marked by shear zones (Fig. 7-10) in southern Wyoming and western Colorado and is characterized by severely crushed and brecciated rocks. Following this orogenic event, there was an extensive episode of magmatism during the Mesoproterozoic. Plutons were emplaced in a broad belt across North America from California to Labrador. The 1.4- to 1.5-billion-year-old plutons are composed primarily of anorthosites and granites. They are anorogenic, meaning that they are not associated with an orogenic cycle. This was the first time in the Earth's history that extensive anorogenic magmatism occurred. It is likely that the precondition for this widespread magmatic activity was the earlier development of a broad region of stable continental crust.

The next event south of the Canadian shield was a widespread episode of rifting that occurred about 1.4 to 0.85 billion years ago. Large fault-controlled depressions were formed, and in these depressions thick sequences of Neoproterozoic sediments accumulated. They include the Uinta Series of central Utah, the Pahrump Group of southeastern California, and the Belt Supergroup of Montana, Idaho, and

FIGURE 7-11 Undeformed, horizontal limestones of the Belt Supergroup comprise the upper third of Chief Mountain, located just outside of Glacier National Park, Montana. The lower two thirds of the mountain consists of deformed Cretaceous beds. A thrust fault separates the two and represents the surface over which the Proterozoic beds were pushed eastward over the weaker underlying Cretaceous strata. The fault is known as the Lewis thrust.

FIGURE 7-12 Neoproterozoic stromatolites in carbonates of the Belt Supergroup, Glacial National Park, Montana. (*Courtesy of E. Moldovanyi.*)

British Columbia. Because of the scenic features with which they are associated, the Belt rocks are of particular interest. In places, these rocks are over 12 kilometers thick. Especially impressive are the massive cliff-forming limestones that can be viewed at Waterton Lakes and Glacier National Parks (Fig. 7-11). Although exceptionally thick, the Belt rocks display features such as ripple marks and stromatolites (Fig. 7-12), which indicate that they were deposited in relatively shallow water along the passive western margin of North America. Westward of these depositional areas lay the widening Panthalassa ocean.

Precambrian rocks of the Grand Canyon region consist of two distinct units. The lower and older unit is the Vishnu Schist, and the upper unit is the Grand Canyon Supergroup. The Vishnu is a complex body of metamorphosed sediments and gneisses that have been intensely folded and invaded by granites. The granitic intrusions that cut into the Vishnu (and correlative rocks of the southwestern United States) were emplaced 1.4 to 1.3 billion years ago as part of a deformational event named the Mazatzal orogeny. A splendid example of an unconformity occurs at the contact between the Vishnu Schist (Fig. 7-13) and the overlying Grand Canyon Supergroup (Fig. 7-14). The latter unit is Neoproterozoic in age and is itself unconformably overlain by Paleozoic rocks. The Grand Canyon Supergroup correlates with the Belt Supergroup. It consists mostly of clastic rocks (sandstones, siltstones, and shales) that accumulated in a troughlike basin that extended into the craton. The Chuar Group of the Grand Canyon Supergroup contains small, circular carbonaceous structures believed to be the fossil algal spheres.

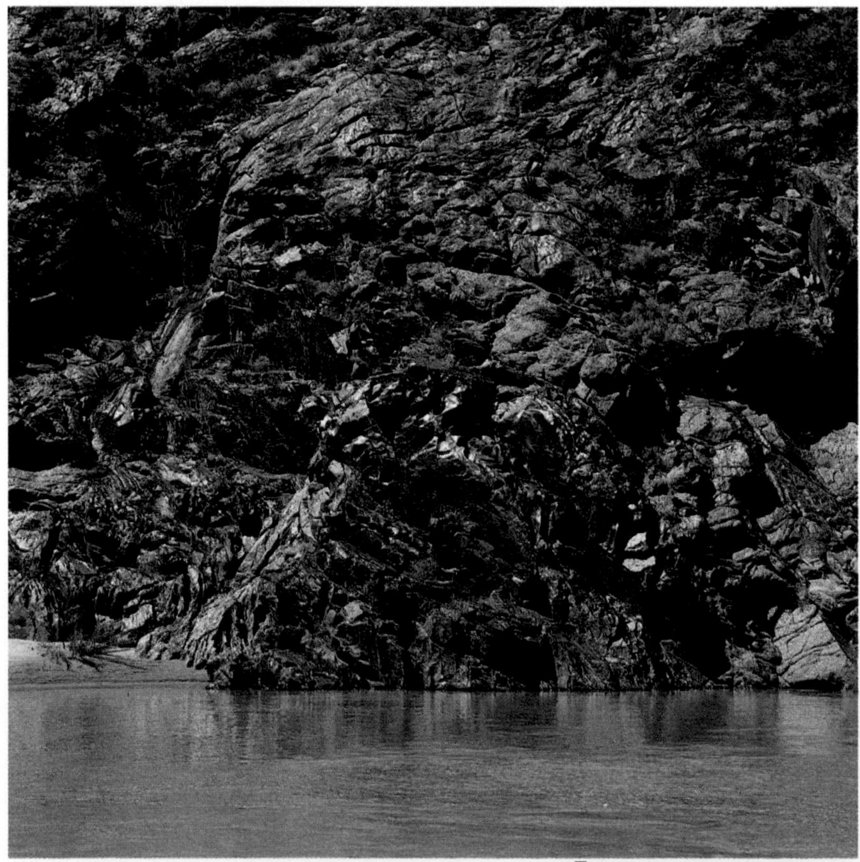

FIGURE 7-13 Vishnu Schist exposed along the walls of the Inner Gorge of the Grand Canyon of the Colorado River, Grand Canyon National Park. (*Photograph by J. Cowlin/Image Enterprises.*)

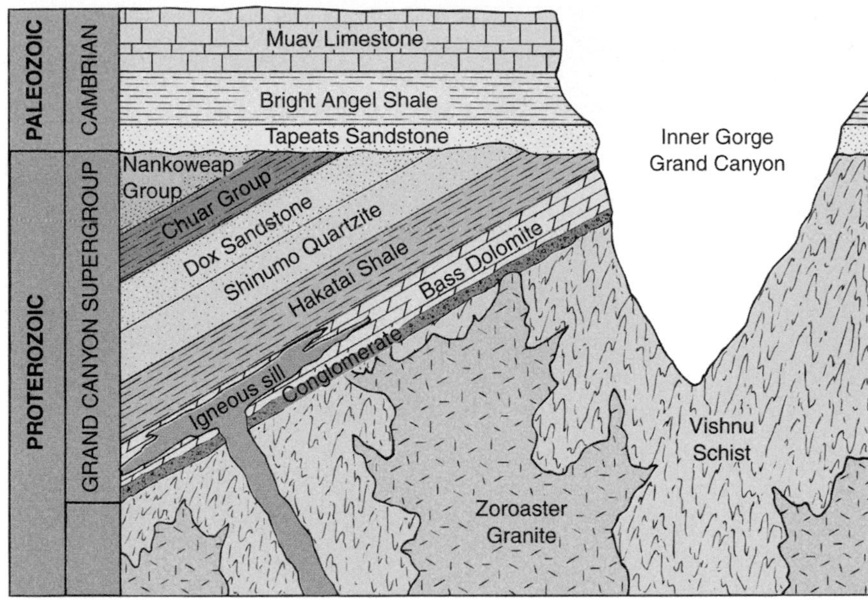

FIGURE 7-14 Vishnu Schist, Grand Canyon Supergroup, and other rocks in the Grand Canyon of the Colorado River. *Indicate by arrows and labels two kinds of unconformities in this cross-section. Is the conglomerate at the base of the Grand Canyon Group an expected lithology for the initial strata above an erosional unconformity?*

PRECAMBRIAN EVENTS OUTSIDE NORTH AMERICA

The Baltic Shield

Precambrian rocks are found in many of Europe's mountain ranges. However, the most extensive exposures form the surface of the Baltic shield (Fig. 7-15). The northwestern edge of the shield is bounded by an early Paleozoic orogenic belt. In general, there is an east–west trend in the age of rocks composing the Baltic shield. The oldest rocks are of Archean age and are exposed on the eastern side of the shield. These are followed by younger Proterozoic rocks in the central region of the shield. The youngest Precambrian rocks are found at the southwestern ends of Norway and Sweden. These rocks are approximately equivalent in age to the Grenville sequence in North America and may have been emplaced at about the time that western Europe began to drift away from Laurentia. A large downfaulted block (graben) near Oslo contains an interesting sequence of fossiliferous Paleozoic strata.

The extremely complex Precambrian history of the Baltic shield has been studied intensively by Swedish and Finnish geologists, who have recognized four great

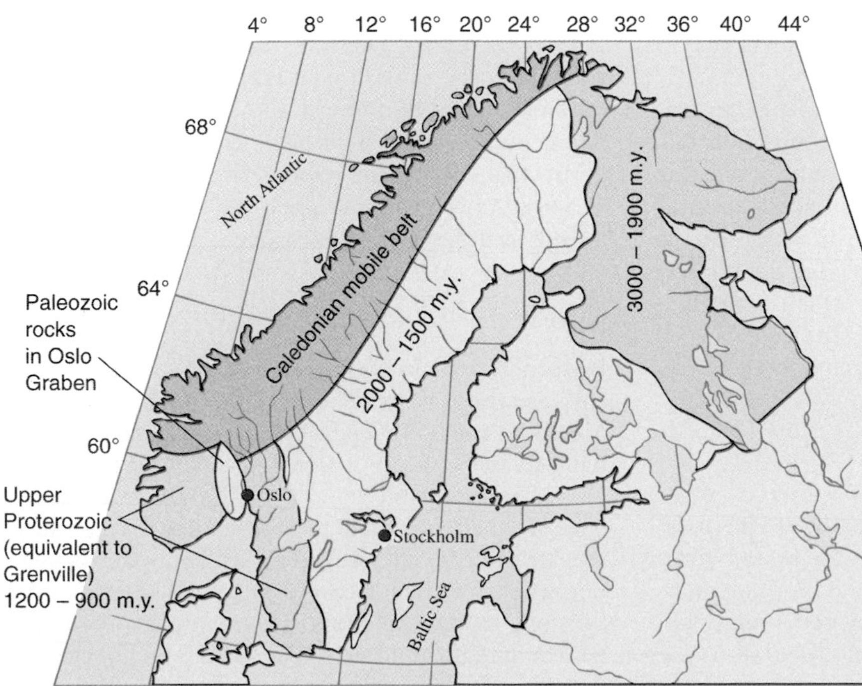

FIGURE 7-15 Region over which Precambrian rocks outcrop in the Baltic shield. Oldest (Archean) rocks are found in the far east, and youngest at the southwestern tips of Norway and Sweden. The Caledonian mobile belt is composed mostly of Paleozoic rocks.

sedimentary and volcanic cycles, separated by tectonic movements and batholithic intrusions. The rocks representing the two oldest cycles are mostly schists and gneisses that were once graywackes and volcanics. Conglomerates, ripple-marked quartz sandstones, dolomites, greenstones, slates, and iron ores characterize the third cycle. The most recent rocks escaped severe metamorphism and consist mostly of sandstones and lavas intruded by granites.

Siberia

The Siberian shield lies below the Arctic Ocean and includes the region between the Yenisei River on the west and the Lena River on the east. It is more difficult to study than either the Canadian or the Baltic shield because of a cover of younger rocks. Precambrian rocks, however, can be examined in exposed patches that penetrate that cover. To the east and south of the shield lie mobile belts that were later to be fashioned into the Ural and Himalayan mountains. Not unlike the Precambrian of other shields, the Siberian block has an older complex of gneisses, schists, and granites that provides evidence of a long and complicated Archean history. There is also a Proterozoic sequence of sedimentary rocks, volcanics, and granitic intrusives that dates from about 1.6 billion years ago. Russian geologists consider these rocks as belonging to a geologic system they have named the **Riphean**.

Sedimentary rocks of the Riphean consist of sandstones and siltstones as well as relatively extensive stromatolitic dolomites. The stromatolites, together with casts of salt and gypsum, indicate deposition occurred in shallow shelf areas near the equator. Distinctive varieties of stromatolites are used in correlating rocks of Riphean age across much of Eurasia.

For the very latest Proterozoic rocks, Russian geologists propose the term **Vendian System** (see Fig. 8-6). In the Ukraine, where it is best exposed, the Vendian incorporates strata from the lowest Neoproterozoic glacial deposits upward to the base of the Cambrian. Vendian rocks are notable for the impressions and burrows of soft-bodied metazoans (multicellular animals) they contain. Rarely, tiny tube-shaped fossils whose shells were originally composed of calcium carbonates are found.

South America

The number three seems to characterize the Precambrian geology of South America. There are three shields: the Guianan, the Brazilian, and the Patagonian (Fig. 7-16); also, the Precambrian rocks of South America are usually grouped into three divisions: Lower, Middle, and Upper Precambrian. As might be expected, gneisses and schists form the oldest (lower) sequence. The Middle Precambrian consists, in large part, of metamorphosed sediments, volcanics, and intrusions that appear to be the product of accretionary evolution. Metamorphic effects are less profound in the Late Precambrian. Quartz sandstones, conglomerates, volcanic flows, and ash beds comprise these Late Precambrian deposits. Because of a covering of younger sediments over parts of the basement as well as the dense forests that cover many regions, the Precambrian rocks of South America are still not as well known as those of the relatively barren Canadian and Baltic shields.

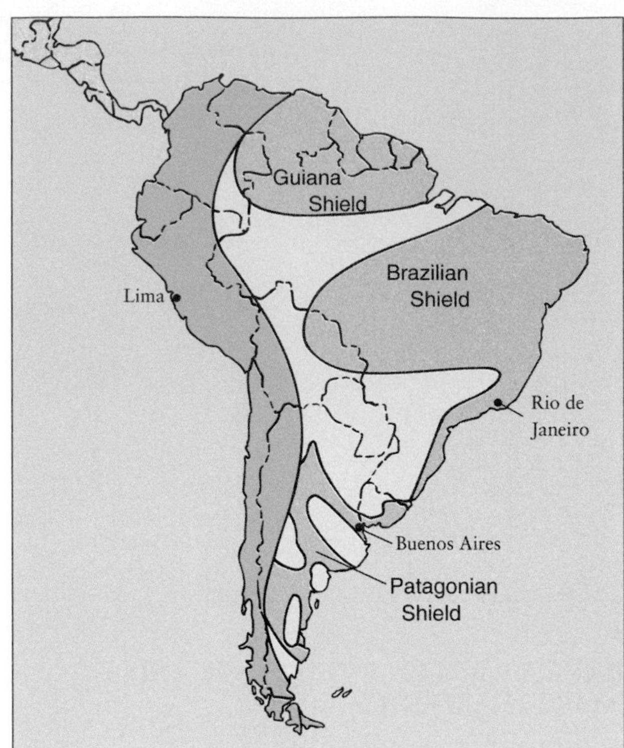

FIGURE 7-16 Regions of exposed Precambrian rocks in South America (shown in light red). The Andean mobile belt is colored green. Yellow areas consist of relatively undisturbed Phanerozoic strata. (*After Umgrove, J. H. F. 1947.* The Pulse of the Earth. *Nijholt: The Hague.*)

Africa

Precambrian rocks occur over nalf the surface of Africa and elsewhere lie beneath a veneer of Paleozoic and Mesozoic rocks. Indeed, Africa appears to be essentially a vast platform of abutting shield segments. It is mostly a very stable continent, but along its eastern edge are the remarkable fault structures that produced the African rift valleys. Africa's Precambrian rocks are renowned for their treasure of minerals; exploration for gold, diamonds, copper, uranium, and chromium has resulted in an improved understanding of the continent's complex geology. As described in Chapter 6, the Archean of Africa includes thick sections of komatiites and mafic lavas that are immersed in a sea of granitic intrusions. These igneous rocks constitute the Archean basement of South Africa. In most areas of South Africa, a nonconformity separates the lavas and batholithic granitic masses from the younger sequences of the Paleoproterozoic.

By about 1.9 to 1.1 billion years ago, four stable segments of the African craton had been formed (Fig. 7-17). Belts of metamorphosed granitic crust formed between these crustal blocks during the Neoroterozoic and welded them into a unified shield. The orogenic activity accompanying the welding of crustal elements continued until about 400 million years ago and resulted in the continent's taking on its present-day outlines.

India

From a geologic point of view, India can be separated into two distinctly different regions. To the north is the Himalayan orogenic belt that developed long after the Precambrian, whereas to the south lies the Indian shield. The shield was brought into juxtaposition with the Himalayan region in the late Mesozoic and early Cenozoic by northward drifting of what was once an Indian island continent.

As is the case in North America, geologists have been able to recognize a number of separate provinces within the shield that can be distinguished by their radiometrically determined ages and distinctive structural characteristics (Fig. 7-18). Five of these provinces are now recognized, and each is believed to be the craton of a separate continental fragment. Orogenic belts and zones of compressional faulting mark the suture zones where cratons were joined together. The joining of component cratons was apparently complete by about 1.5 billion years ago (Mesoproterozoic). Subsequently, the shield was remarkably stable, until Late Cretaceous time, when huge volumes of the Deccan basaltic lavas were extruded along numerous fissures. The outpouring of lava may have been a response to the separation of India from the Seychelles Islands.

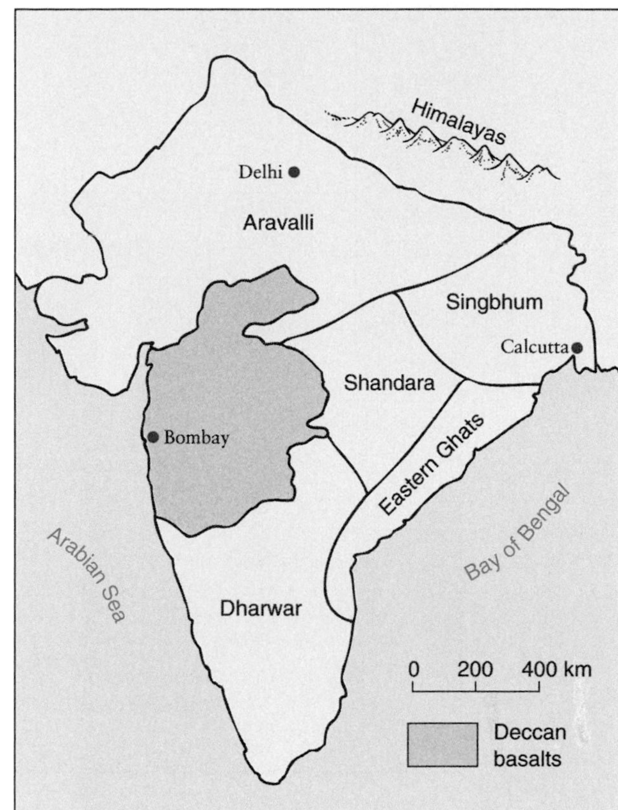

FIGURE 7-18 Precambrian provinces of India. (*After Nagvi, S. M., and Rogers, J. J. W. 1985.* Precambrian Geology of India. *Oxford, England: Oxford University Press.*)

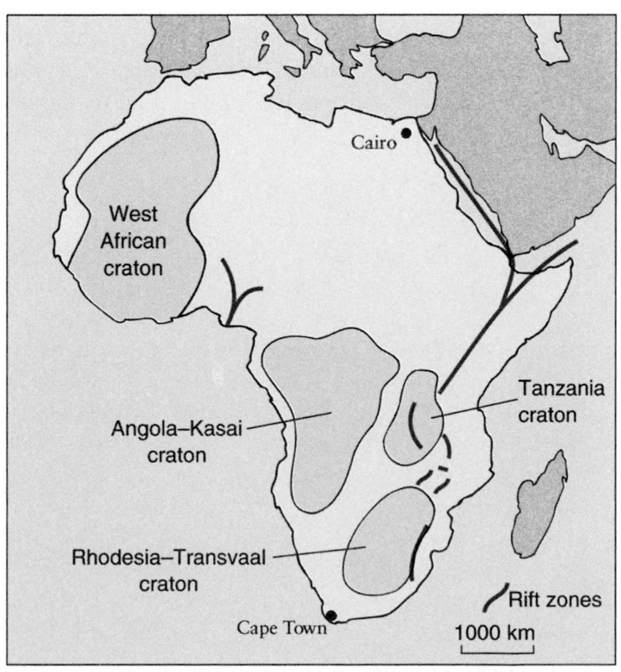

FIGURE 7-17 The four large cratonic segments of Africa. ? *Where on the map is there an aulocogen?*

Australia

Australia is composed of Phanerozoic orogenic belts along the eastern side of the continent and consists of shield and platform elements over most of the central and western parts of the continent. The largest area of exposed Archean rocks is in the Yilgarn shield in western Australia (Fig. 7-19). North of the Yilgarn shield lies the smaller Pilbara shield. Both of these Archean elements contain the greenstone belts and granulite expanses that typify Archean geology on other continents. At least two great episodes of accretion and mountain building are represented in these Archean rocks.

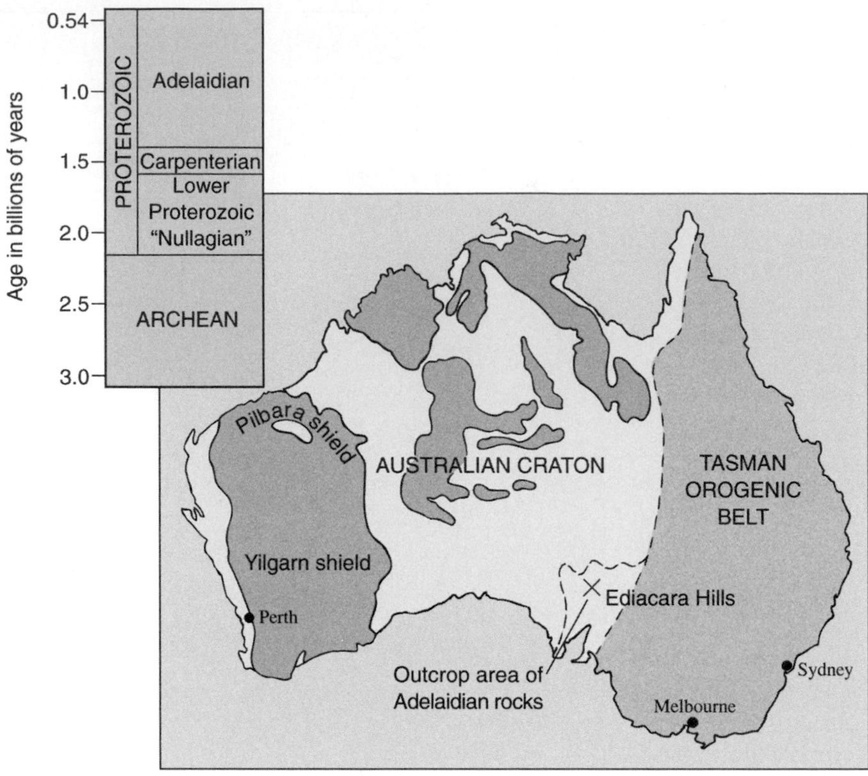

FIGURE 7-19 Central and western Australia consist of a vast cratonic region composed of Precambrian rocks (green) exposed as large patches wherever the cover of mostly horizontal Phanerozoic rocks is absent. The eastern third of the continent contains folded Paleozoic strata of the Tasman orogenic belt. Insert indicates Australian Precambrian time classifications. (*Generalized from geologic map of Australia, Department of National Development, Canberra, 1958.*)

The Proterozoic of Australia includes a thick, widespread sequence of relatively unaltered sedimentary rocks that span a time interval of over 1.5 billion years. Such rocks permit direct comparisons with modern sedimentation processes and provide many clues to paleoclimatic conditions in this part of the Proterozoic world. Included within the sedimentary sequence are extensive shallow marine deposits containing abundant stromatolites. Banded iron deposits of about the same age as those of the Canadian Animikian also occur. The Neoproterozoic sequence includes boulder beds deposited by glaciers, varved mudstones of marginal glacial lakes, and glacially striated pavements. Marine sediments slightly younger than the glacial deposits include the Rawnsley Quartzite (a metamorphosed sandstone), well known by geologists because of the impressions of soft-bodied metazoan animals it contains.

Antarctica

Beneath the frozen surface of Antarctica lies a Precambrian shield that was poorly understood until the advent of recent techniques for geophysical exploration. Over the past 4 decades, geologists have had an opportunity to examine the bedrock exposed around the margins of the huge Antarctic ice sheet and to visit mountain peaks (nunataks) that pierce the ice cover. Most of the rounded eastern half of Antarctica consists of Precambrian rocks. Archean rocks occupy the eastern margin of the continent, and Proterozoic rocks occur across much of the remainder of the shield. The western part of Antarctica consists of fold belts of late Cretaceous and younger age. These mountainous tracts show a marked similarity to those of the Andean orogenic belt of South America.

THE PROTEROZOIC FOSSIL RECORD

Life at the beginning of the Proterozoic was not significantly different from that of the preceding late Archean. Blue-green scums of photosynthetic cyanobacteria (Fig. 7-20) constructed filamentous algal mats

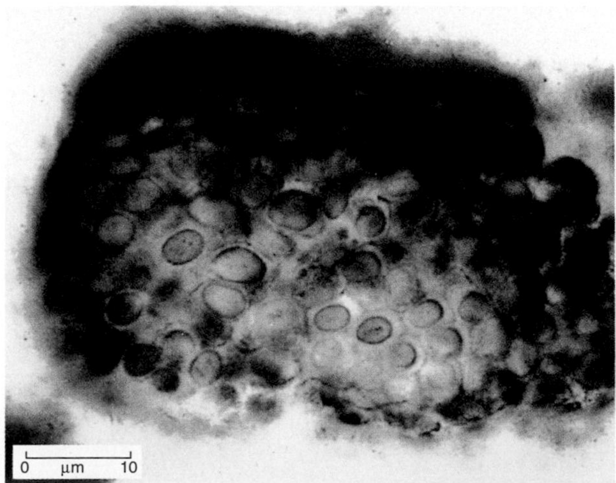

FIGURE 7-20 Paleoproterozoic colony of the cyanobacterium *Eoentophysalis*, photographed in a thin section of a Belcher Group stromatolitic chert, Belcher Islands, Canada. (*Courtesy of H. J. Hofmann.*)

ENRICHMENT

Heliotropic Stromatolites

Stromatolites are among the most abundant kinds of fossils found in Proterozoic rocks. Today, as during the Proterozoic, the photosynthetic cyanobacteria that construct stromatolites depend on sunlight for survival and growth. In a study of modern stromatolites, it has been observed that certain species form columnar laminated growths that are inclined toward the sun to gather the maximum amount of light on their upper surfaces. This tendency among organisms to grow toward the sun is called heliotropism. Stromatolites, however, may also exhibit growth responses to currents and rising sea level and thus must be examined carefully for evidence that a preferred growth orientation is related to sunlight. Stanley Awramic and James Vanyo (paleobiologist and astronomer, respectively) have studied 850-million-year-old stromatolites from Australia's Bitter Springs Formation for evidence of heliotropism. They believe they found such evidence in the species *Anabaria juvensis.* At the locality studied in central Australia, these stromatolites curve upward in a distinct sine-wave form that appears to record growth that followed the seasonal change in the position of the sun above the shallow sea in which the stromatolitic cyanobacteria lived. In their elongate sinuous form, each sine wave represented a year of growth. The stromatolites grew by the addition of a layer, or lamina, each day. Thus, by counting the laminae in the length of a single sine wave, one can obtain the number of days in the year. The results indicate that there were 435 days in a year 850 million years ago, when the stromatolites of the Bitter Springs Formation were actively responding to Proterozoic sunlight.

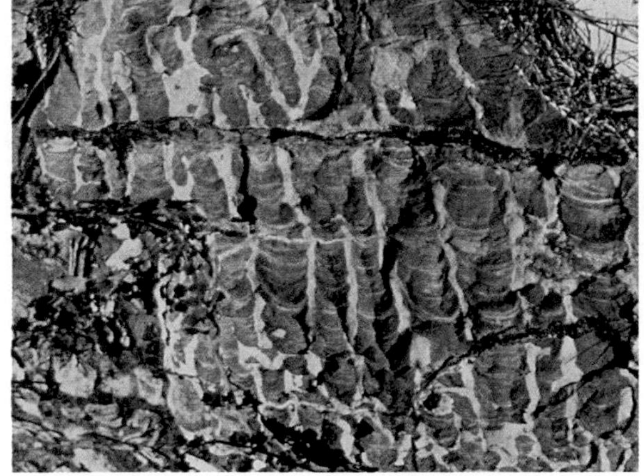

Some 850-million-year-old stromatolites from the Bitter Springs Formation of central Australia. Vertical dimension is 14 centimeters. (*Courtesy of J. W. Schopf.*)

around the margins of the ocean, and prokaryotes floated in the well-illuminated surface waters of seas and lakes as well. Anaerobic prokaryotes multiplied in environments deficient in oxygen, and these included the thermophiles of deep-sea hydrothermal springs. Stromatolites, which were relatively sparse during the Archean, proliferated during the Proterozoic (Fig. 7-21) but declined markedly by the end of the Neoproterozoic. Today stromatolites are rare, primarily because the microorganisms that build them are eaten by marine snails and other invertebrates. The decline of Proterozoic stromatolites may be similarly associated with overgrazing by newly evolving groups of small shelly fossils and other marine invertebrates. Even today, stromatolites survive only in environments unsuitable for most grazing invertebrates.

Molecular fossils of eukaryotes first appear near the end of the Archean about 2.7 billion years ago. Although prokaryotes still dominated during the Mesoproterozoic, eukaryotes became more abundant in the fossil record. The terminal stages of the Neoproterozoic witnessed the evolution of the Earth's first multicellular animals (metazoans).

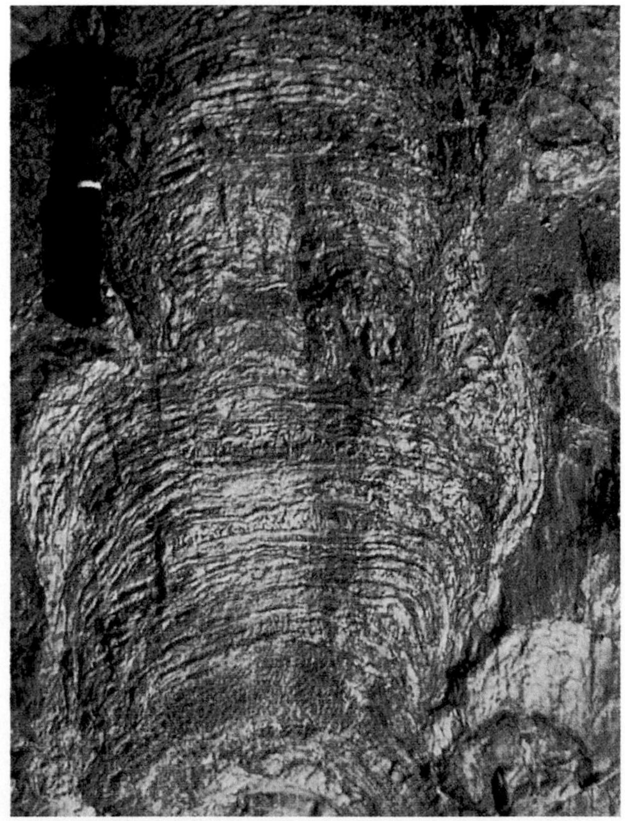

FIGURE 7-21 Two-billion-year-old columnar stromatolites from the Kona Dolomite near Marquette, Michigan. (*Courtesy of J. W. Schopf.*)

Microfossils of the Gunflint Chert

As mentioned in the previous chapter, the oldest currently known prokaryotes are those found by paleobiologist J. William Schopf in 3.5-billion-year-old

Archean rocks of Western Australia. Schopf reported his find in 1994. These were not the first Precambrian prokaryotes discovered. The initial discovery consisted of Proterozoic fossils detected in 1965 in a now-famous rock unit called the Gunflint Chert. This rock unit is exposed along the northwestern margin of Lake Superior. Radiometric dating methods provide an age of 1.9 billion years for the Gunflint Chert. The formation contains abundant and varied flora of so-called thread bacteria and cyanobacteria. All the fossils are preserved in chert. Unbranched filamentous forms, some of which are septate, have been given the name *Gunflintia* (Fig. 7-22). More finely septate forms, such as *Animikiea* (Fig. 7-23), are remarkably similar to such living algae as *Oscillatoria* and *Lyngbya*. Other Gunflint fossils, such as *Eoastrion* (literally, the "dawn star"), resemble living iron- and magnesium-reducing bacteria. *Kakabekia* and *Eosphaera* do not resemble any living microorganism, and their classification is uncertain. What does seem certain is that there was an abundance of photosynthetic organisms in the Paleoproterozoic, all actively producing oxygen and thereby altering the composition of the atmosphere. Not only do the Gunflint fossils resemble living photosynthetic organisms, but their host rock contains organic compounds regarded as the breakdown products of chlorophyll.

The Rise of Eukaryotes

The evolution of eukaryotes surely qualifies as one of the major events in the history of life. Eukaryotes possess the potential for sexual reproduction, and this provides enormously greater possibilities for evolutionary change. Unfortunately, the fossil record for the earliest eukaryotes is somewhat ambiguous. This is not surprising when one considers that the first eukaryotes were microscopic unicells whose identifying characteristics (enclosed nucleus, organelles, and so on) are rarely preserved. There is, however, one important clue to the identification of a fossil as a eukaryote. It is based on the size of cells. Living spherical prokaryotic cells rarely exceed 20 microns in diameter, whereas eukaryotic cells are nearly always larger than 60 microns. Such larger (and hence probably eukaryotic) cells begin to appear in the fossil record by about 2.7 to 2.2 billion years ago. This earliest evidence of eurkaryotic life is based on biochemical remnants of eukaryotes. These chemical clues are termed **molecular fossils**. They consist of compounds called steranes that are derived from molecules similar to cholesterol. Although eurkaryotes evolved very early, they did not begin to diversify until about 1.2 to 1.0 billion years ago. It may be that they were unable to expand until the oxygen

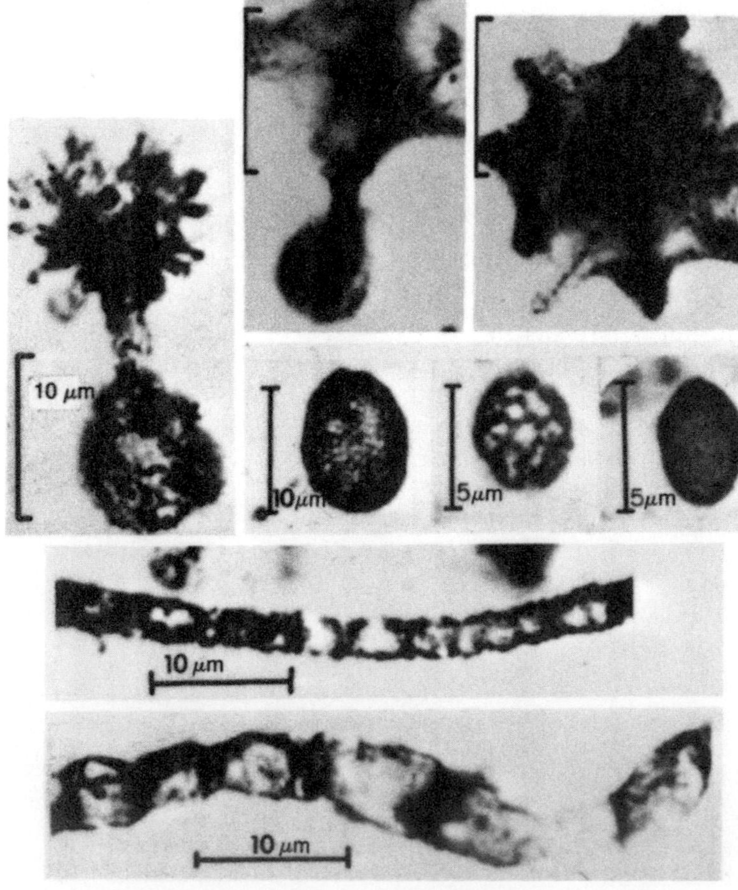

FIGURE 7-22 Fossil remains of microorganisms from the Gunflint Chert. The three specimens across the top with umbrella-like crowns are *Kakabekia umbellata*. The three subspherical fossils are species of *Huroniospora*. The filamentous microorganisms with cells separated by septa are species of *Gunflintia*. (*Courtesy of J. W. Schopf.*)

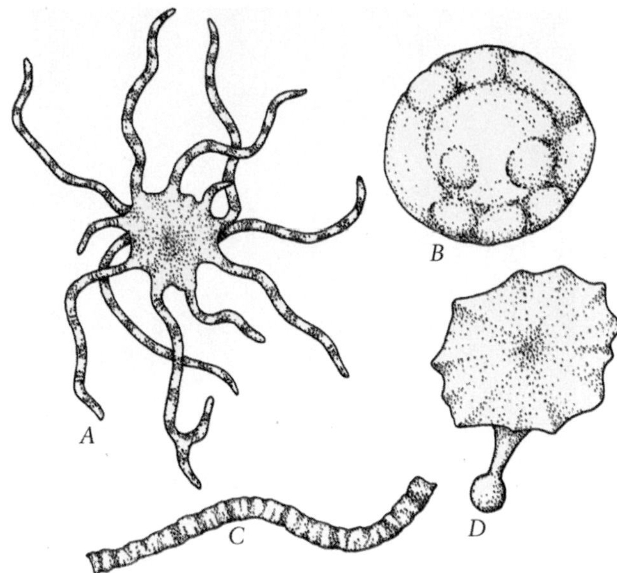

FIGURE 7-23 (*A*) *Eoastrion*, (*B*) *Eosphaera*, (*C*) *Animikiea*, and (*D*) *Kakabekia* from the Gunflint Chert. All specimens are drawn to the same scale. *Eosphaera* is about 30 millimeters in diameter.

content of the ocean reached a suitable level, or perhaps they did not diversify until the advent of sexual reproduction.

A group of organisms particularly useful in correlating Proterozoic strata are known as acritarchs. **Acritarchs** are unicellular, spherical microfossils with resistant single-layered walls. The walls may be smooth or variously ornamented with spines, ridges, or papillae (Fig. 7-24). Although their precise nature is uncertain, acritarchs appear to be phytoplankton that grew thick coverings during a resting stage in their life cycle. Some resemble the resting stage of unicellular, biflagellate, marine algae known as dinoflagellates. (Dinoflagellates today are one kind of organism causing the so-called red tide that periodically poisons fish and other marine animals.) Membrane-bounded nuclei can be detected within some acritarchs, and their size is comparable to that of living eukaryotes.

Acritarchs first appear in rocks about 1.6 billion years old. They reached their maximum diversity and abundance 850 million years ago and then suffered a steady decline. By about 675 million years ago, few remained. Their decline coincided with the Varangian glaciation near the end of the Proterozoic. A reduction of carbon dioxide and an increase in atmospheric oxygen accompanying glacial conditions may have been responsible for the extinction of all but a few species that managed to survive until Ordovician time.

In addition to acritarchs, protozoan eukaryotes were probably present in the Proterozoic. As is true today, these protozoans derived their energy by ingesting other cells. Amoebas and ciliates are living examples of

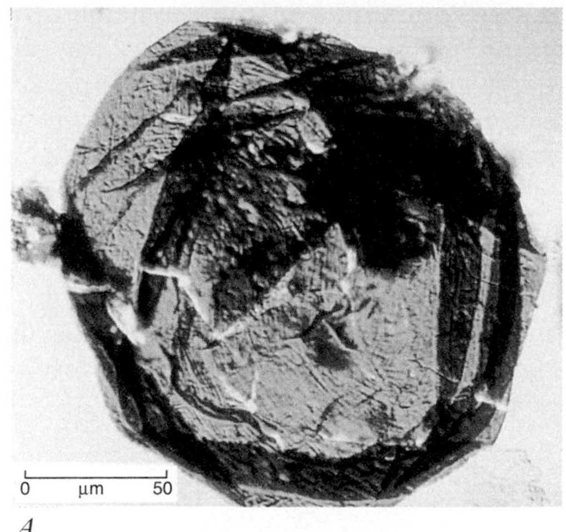

A

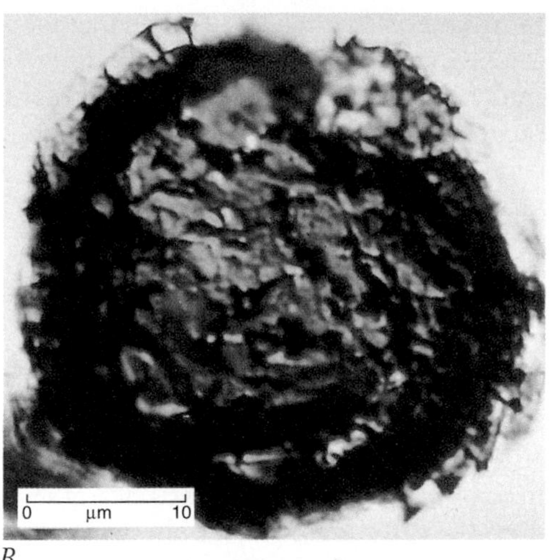

B

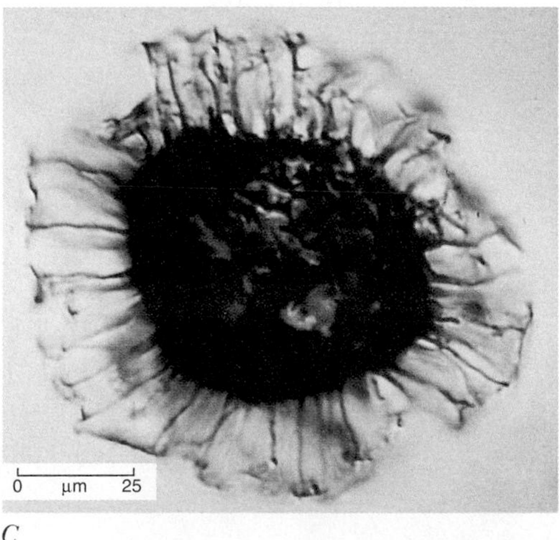

C

FIGURE 7-24 Acritarchs. (*A*) The striated acritarch *Kildinosphaera lophostriata* from the Chuar Group of Arizona (see Fig. 7-12). (*B*) A spiny acritarch named *Vandalosphaeridium walcotti*, also from the Chuar Group. (*C*) The spiny acritarch *Skiagia scottica* from the early Cambrian (about 560 million years old) of northern Greenland, depicted here for comparison with the late Proterozoic forms. (*Courtesy of G. Vidal, Micropaleontology Laboratory, Sweden.*)

protozoans. Protozoan fossils are common in many Phanerozoic rocks, largely because they had evolved readily preservable shells. Proterozoic protozoans were naked, shell-less creatures with little chance of preservation. The exception is tiny vase-shaped microfossils about 800 million years old found in both Arizona and Spitsbergen. The scarcity of Proterozoic protozoans, however, does not mean that they were not present in great abundance. It is likely that the algal mats and masses of decaying organic matter in late Proterozoic seas teemed with protozoans exploiting these rich sources of food.

Advent of the Metazoans

The Neoproterozoic fossil record for larger, multicellular animals is considerably better than that for protozoans. Indeed, since the 1960s, important discoveries of large, multicellular Neoproterozoic animals have been made on every continent. Most of the fossils consist of impressions in sedimentary rocks of animals that are clearly at the metazoan level of evolution (Fig. 7-25). **Metazoans** are multicellular animals that possess more than one kind of cell and have their cells organized into tissues and organs. Extraordinary evidence of the presence of metazoans about 570 ± 20 million years ago was recently discovered in the uppermost Neoproterozoic Doushantuo Formation of South China. The fossils do not consist of adult animals, but the eggs and embryos of metazoans. The preserved embryos clearly show precise patterns of cell cleavage at the blastula stage of embryonic development (Fig. 7-26). Phosphate impregnations and coatings provided the embryos with truly exquisite preservation.

FIGURE 7-25 Fossil metazoans from the Conception Group, Avalon Peninsula, Newfoundland. These early metazoans are similar to those of the Ediacaran fauna of Australia. The rocks in which they occur are thought to be Late Proterozoic in age. The elongate forms have been interpreted as a form of colonial soft coral, and the circular fossil has been called a jellyfish. There are, however, other interpretations. (*Courtesy of B. Stinchcomb.*)

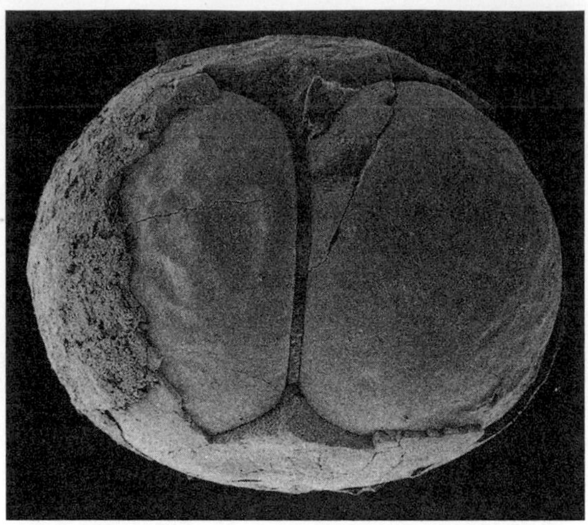

FIGURE 7-26 Metazoan embryo (570 ± 20 million years old) from the Doushantuo Formation of China at the two-cell stage. The specimen is about 1 millimeter in diameter. (Courtesy of Shuhai Xiao, Dept. of Geology, Tulane University.)

The first important discovery of large metazoans in Proterozoic rocks was made in Australia's Ediacara hills in the 1940s. The Australian fossils and remains of similar metazoans from other parts of the world have been named the **Ediacaran fauna**. The oldest members of this fauna have been discovered in China.

Although once thought to have disappeared by Cambrian time, several animals within the fauna survived into the early Cambrian. For example, Ediacaran fossils of definite Cambrian age have been reported in 510-million-year-old rocks in Ireland. A few more possible survivors of the Ediacaran fauna occur in the Burgess Shale of the middle Cambrian age. Thus, there was apparently no mass extinction of Ediacaran animals near the end of the Neoproterozoic as once proposed.

One can separate Ediacaran organisms into three groups based on general appearance. Some are discoidal, some frondlike, and some elongate. Discoidal forms such as *Cyclomedusa* (Fig. 7-27) were initially thought to be jellyfish. Another discoidal form, *Tribrachidium* (Fig. 7-28*B*), appears to have no modern counterpart and may be a member of an extinct phylum.

The second group of Ediacaran specimens are called frond fossils. They resemble living octocorals informally called sea pens (Fig. 7-29). Sea pens look rather like fronds of ferns, except that tiny coral polyps are aligned along the branchlets. The polyps capture and consume microscopic organisms that float by. Frond fossils similar to those from Australia are also known from Africa, Russia, and England. In those from England (*Charniodiscus*), the frond is attached to a basal, concentrically ringed disk that apparently served to hold the organisms to the sea floor. The disks are frequently found separated from the fronds, which suggests that many of the discoidal fossils common in

FIGURE 7-27 Impression of a soft-bodied discoidal fossil in the Ediacaran Rawnsley Quartzite of southern Australia. This organism has been interpreted as a jellyfish and named *Cyclomedusa*. Some paleontologists, however, believe it is unrelated to any living organism. (*Courtesy of B. N. Runnegar.*)

Ediacaran faunas around the world may not be jellyfish but anchoring structures of frond fossils.

The third group of Ediacaran fossils are ovate to elongate in form. They were originally regarded as impressions made by large flatworms and annelid worms. Typical of these fossils is *Dickinsonia* (Fig. 7-30), which attained lengths of up to a meter, and *Spriggina* (Fig. 7-31), a more slender animal with a distinctive crescent-shaped structure at its anterior end.

A particularly significant Ediacaran fossil bears the name *Kimberella* (Fig. 7-32). About four rather poorly preserved specimens of this creature were found at the Ediacaran site. They were tentatively judged to be the flattened remains of jellyfish. Then, in 1993, a fossil locality on the shores of the White Sea in northern Russia yielded over 30 specimens of this unusual organism. It was readily apparent from these fossils that *Kimberella* was no lowly jellyfish. It was, in fact, a complex invertebrate that ranked higher on the evolutionary scale than either jellyfish or flatworms. *Kimberella* exhibited evidence of a true coelum, or body cavity, in which the digestive tract and other internal organs were suspended. At 550 million years of age, it is the first animal known to possess this important characteristic found in all higher animals. In addition, it was bilaterally symmetric, had a limpetlike dorsal cover, and possessed a distinctive ruffled border that some interpret as the edge of a mantle (the organ that secretes the shell in mollusks). Evidently, *Kimberella* crept across the substrate, grazing on algae in much the same way as some living snails. Although *Kimberella* appears to be very mollusk-like, there is no evidence that it had a radula—the rasping tonguelike structure found in most mollusks. Whether or not it is a true mollusk, however, does not

A

B

C

FIGURE 7-28 Three members of the Ediacaran fauna from the Rawnsley Quartzite of Ediacara Hills in southern Australia. (*A*) *Pseudohizostomites*, a wormlike form of uncertain affinity. (*B*) *Tribrachidium*, an unusual discoidal form that appears to have no living relatives. (*C*) *Parvancorina*, possibly an arthropod. Specimen *A* is an imprint in the Rawnsley Quartzite; *B* and *C* are plaster molds made from the original fossils. (*All specimens courtesy of M. F. Glaessner.*)

FIGURE 7-29 Diorama of the sea floor in which lived Ediacaran metazoans. The large, frondlike organisms are interpreted here as soft corals, known today as sea pens. Silvery jellyfish are seen swimming about. On the floor of the sea, one can find *Parvancorina* and elongate, wormlike creatures. (*National Museum of Natural History, Smithsonian Institution.*) ? *What characteristics of the frond fossils places doubt on their identification as soft corals?*

detract from its importance as evidence that truly advanced and complex invertebrates existed on Earth about 10 million years before the great expansion of life called the "Cambrian explosion."

THE VENDOZA CONTROVERSY There is an interesting controversy with regard to Ediacaran fossils. Most have long been interpreted as Proterozoic members of such existing phyla as the Cnidaria (which includes jellyfish and corals) and Annelida. Indeed, many paleontologists still retain this opinion. However, that view of the affinities of Ediacaran organisms has been questioned by paleontologist Adolf Seilacher. Seilacher argues that the resemblance between living sea pens and frond fossils is only superficial. In support of his argument, he notes that the branchlets in the frond fossils are fused together and do not have passages through which water currents might pass. Living sea pens have such openings, and this permits the polyps on the branchlets to gather food particles efficiently in the passing flow of water. With regard to the discoidal fossils thought to be jellyfish, Seilacher notes that living jellyfish have radial structures at their centers and concentric structures around the periphery. This arrangement is opposite to that found in the discoidal Ediacaran fossils. Finally, the rather superficial resemblance of *Spriggina* and *Dickinsonia* to worms may have little significance, for there is no evidence in the fossils of organs that are essential to modern-day worms, such a mouth, gut, or anus.

FIGURE 7-30 Dickinsonia costata in the Ediacaran Rawnsley Quartzite of southern Australia. This fossil has been interpreted as a segmented worm. Divisions on the scale are in centimeters. (*Courtesy of B. N. Runnegar.*)

FIGURE 7-31 *Spriggina floundersi*, a bilaterally symmetric animal interpreted by some as a segmented worm from the Ediacaran Rawnsley Quartzite of southern Australia. The animal measures 3.5 centimeters in length. (*Copyright Ken Lucas / Visuals Unlimited*)

The Ediacaran animals may be fundamentally different from cnidarians and annelids in yet another way. Cnidarians and worms are rarely preserved in sandy deposits. Preservation of soft, delicate tissue is unlikely in such coarse sediment. Yet the Ediacaran animals have left distinct impressions in the enclosing rock. This suggests that they must have had tough outer coverings. The ribbed and grooved appearance of many Ediacaran impressions suggests to Seilacher that these animals had an exterior construction suggestive of an air mattress and that this construction provided the firmness needed to make an impression in sand.

Seilacher and his American colleague Mark McMenamin believe that Ediacaran animals lived with symbiotic algae in their tissues, as do modern corals. This relationship would have provided the animals with a way to derive nutrition from the photosynthetic activity of the algae. Perhaps also the broad, thin shapes of many of these animals gave them sufficient surface area to allow for the diffusion respiration essen-

FIGURE 7-32 Reconstruction of *Kimberella*. Specimens range up to 105 millimeters in length. (*After Fedonkin, M. A., and Waggoner, B. M. 1997. The Late Precambrian fossil Kimberella is a mollusk-like bilateral organism.* Nature 388(28):868–871). *Name an important characteristic of Kimberella that indicates it is a highly evolved multicellular animal.*

tial to animals that had not yet evolved complex circulatory, digestive, and respiratory systems.

The dissimilarities of Ediacaran creatures to animals that exist today support a view that they should not be placed in existing phyla. Seilacher suggests placing them in a separate taxonomic category, for which he proposed the name **Vendoza** (after Vendian, the final period of the Neoproterozoic in Russia). There are, however, many paleontologists who still believe that most of the Ediacaran animals are early members of existing phyla. The resolution of this controversy may have to await the discovery of fossils that provide more anatomic information.

The Ediacaran fauna survived for about 50 million years following their appearance about 630 million years ago. The fauna is important as a record of the Earth's first evolutionary radiation of multicellular animals. Although the record is still too patchy to trace phylogenetic relationships, certain Ediacaran animals were probably ancestral to Paleozoic invertebrates.

The Ediacaran fauna indicates that Vendian seas were populated largely by soft-bodied organisms. Some tiny shell-bearing fossils, however, have also been found. One genus, first discovered in Vendian rocks of Namibia, Africa, is named *Cloudina* after the American geologist Preston Cloud. *Cloudina* secreted a tubular, calcium carbonate shell only a few centimeters long (Fig. 7-33). It has been interpreted as a tube-dwelling annelid worm. Other small shelly fossils that occur worldwide in sediments of latest Proterozoic and earliest Cambrian include possible primitive mollusks, sponge spicules, and tiny tusk-shaped fossils called hyolithids.

Not all Precambrian fossils are body fossils (petrifactions, impressions, molds, or casts). Trails, burrows, and other trace fossils also provide important information about ancient life. Trace fossils of burrowing metazoans (Fig. 7-34) have been found in Vendian rocks of Australia, Russia, England, and North America. In every locality, they occur in rocks deposited after the late Proterozoic episode of glaciation (named

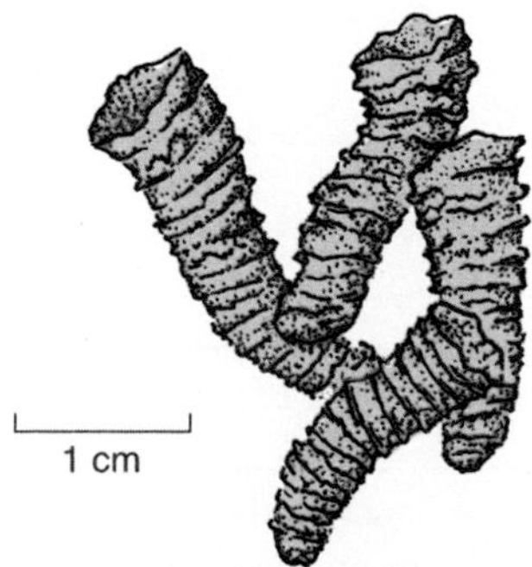

FIGURE 7-33 *Cloudina*, the earliest known calcium carbonate shell–bearing fossils. *Cloudina* was first described from the Proterozoic Nama Group of Namibia. It is believed to be a tube-dwelling annelid worm.

FIGURE 7-34 Trace fossils made by a possible mollusk as it crawled across soft sediment on the sea floor. The host rock occurs at the Proterozoic–Cambrian boundary in British Columbia, Canada. Originally, the traces had the form of elongate depressions. Sediment deposited on top of the original layer filled the depressions, so that the crawling traces are in convex relief. (*Courtesy of H. J. Hofmann.*)

ENRICHMENT

The 18-Hour Proterozoic Day

The Earth has not always had a 24-hour period of rotation on its axis. It has been slowing at the rate of about 2 seconds per 100,000 years. The slowdown means that days have been increasing in length through geologic time, and the number of days in the year has been decreasing. Tidal friction is the cause of the diminishing rate of spin. As the Earth rotates, there is a tendency for two tidal bulges to be carried around with it. However, the Moon's attraction prevents the Earth from dragging the bulges very far. Thus, the two tidal crests tend to act as rather inefficient brake bands clamped on either side of the rotating planet, dragging back on the planet as it rotates. The bulges experience friction along shallower areas of the ocean bottom and are also constrained by continents that stand in their way.

The slowdown in the Earth's rotation was evident in the calculations of astronomers decades ago. More tangible supporting evidence was provided in the early 1960s by paleontologist John Wells. Wells, an authority on fossil corals, recognized that the fine lines on the exoskeletons of coral organisms might represent daily growth increments. Thus, a coral might secrete one thin ridge of calcium carbonate each day. In addition, Wells could discern coarse monthly bands, presumably related to breeding cycles, during which time less calcium carbonate was secreted. There were also broader annual bands in corals that lived in regions of seasonal change. Wells counted the growth lines on several species of living corals and found the count "hovers around 360 lines in the space of a year's growth." Proceeding next to fossil corals successively older in age, he counted correspondingly larger numbers of growth lines in a yearly band. There were, for example, 398 growth lines in the annual band of Devonian corals known to be 370 million years old (see accompanying figures). Thus, during the Devonian, there would have been about 33 more days to the year than at present.

Intrigued by the Wells method, paleontologists attempted similar studies based on other calcium carbonate–secreting organisms. Although many of these investigations indicated the Earth once rotated faster than today, it was often difficult to achieve a high level of precision. Perhaps periodic changes in some sediments might indicate the shortened length of days long ago. In 1996, Charles Sonett employed sedimentary rocks known as tidal rhythmites to recognize changes in the Earth's rate of rotation. As seen along shorelines today, tidal rhythmites consist of alternating bands of dark and light silty deposits. The thickness of the bands reflects the neap and spring tides that mark the lunar month, and variations in the thickness of sets of bands can be related to seasonal change. Sonett's analysis indicates that during the Late Proterozoic, 900 million years ago, a day lasted only 18.2 hours. Without doubt, the shorter days would have resulted in rates of weathering, global temperatures, and weather patterns different from those we now experience. One can only speculate about the effect of shorter periods of heating and cooling, and light and darkness, on the life that was evolving 900 million years ago.

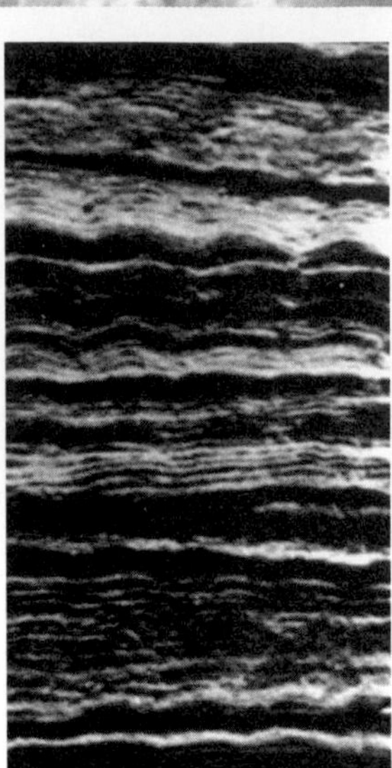

Growth banding displayed by a specimen of the extinct coral *Heliophyllum halli*. The finer lines may represent daily growth increments. There are approximately 200 growth lines per centimeter. Together with the annual bands, the growth increments can be used to estimate the number of days in a year at the time the animal lived.

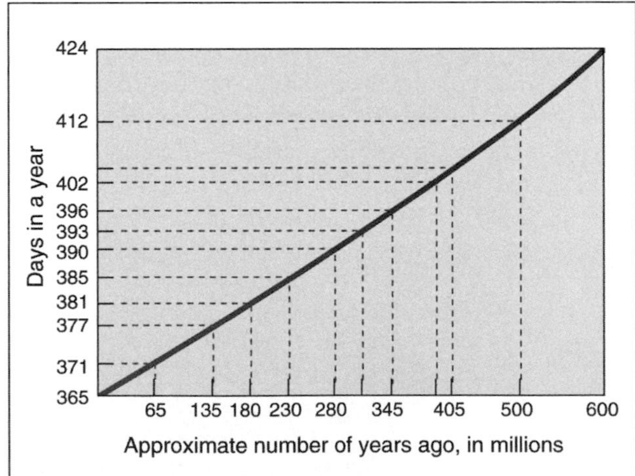

The changing length of the day through geologic time. (*From Wells, J. W. 1963. Nature 197:948–950.*)

the Varangian glaciation). It is interesting that the Proterozoic traces consist mostly of relatively simple, shallow burrows, whereas those of the overlying Cambrian are more complex, diverse, and numerous. One sees a similar increase in complexity, diversity, and abundance of metazoan body fossils as one passes from the Proterozoic into the Cambrian.

The sudden appearance of Ediacaran fossils near the end of the Proterozoic and the absence of such metazoans in older rocks may be related to the accumulation of sufficient free oxygen in the atmosphere to permit oxidative metabolism in organisms. On the other hand, the Ediacaran life may have evolved more gradually from earlier small and naked forms that were essentially incapable of leaving a fossil record. Perhaps the ancestral metazoans lived in "oxygen oases" in which marine plants were concentrated. After the atmospheric oxygen content had reached about 1 percent of the present level, these as yet undiscovered animals would have been free to leave their oases and spread widely in the seas. According to this view, the evolutionary development of metazoans may not have been abrupt, but their dispersal may have been rapid after suitable conditions became prevalent.

THE CHANGING PROTEROZOIC ENVIRONMENT

The Earth's early atmosphere and hydrosphere were largely devoid of free oxygen. Then, about 2 billion years ago, oxygen began to accumulate in the atmosphere. Before that time much of the oxygen being produced was incorporated into iron and other readily oxidized elements. These elements served as "oxygen sinks" by using up oxygen that might otherwise go into the atmosphere. Once the oxygen sinks had been largely filled, oxygen began to accumulate in the atmosphere. Very likely, increased plant photosynthetic activity contributed to the build-up (Fig. 7-35). As

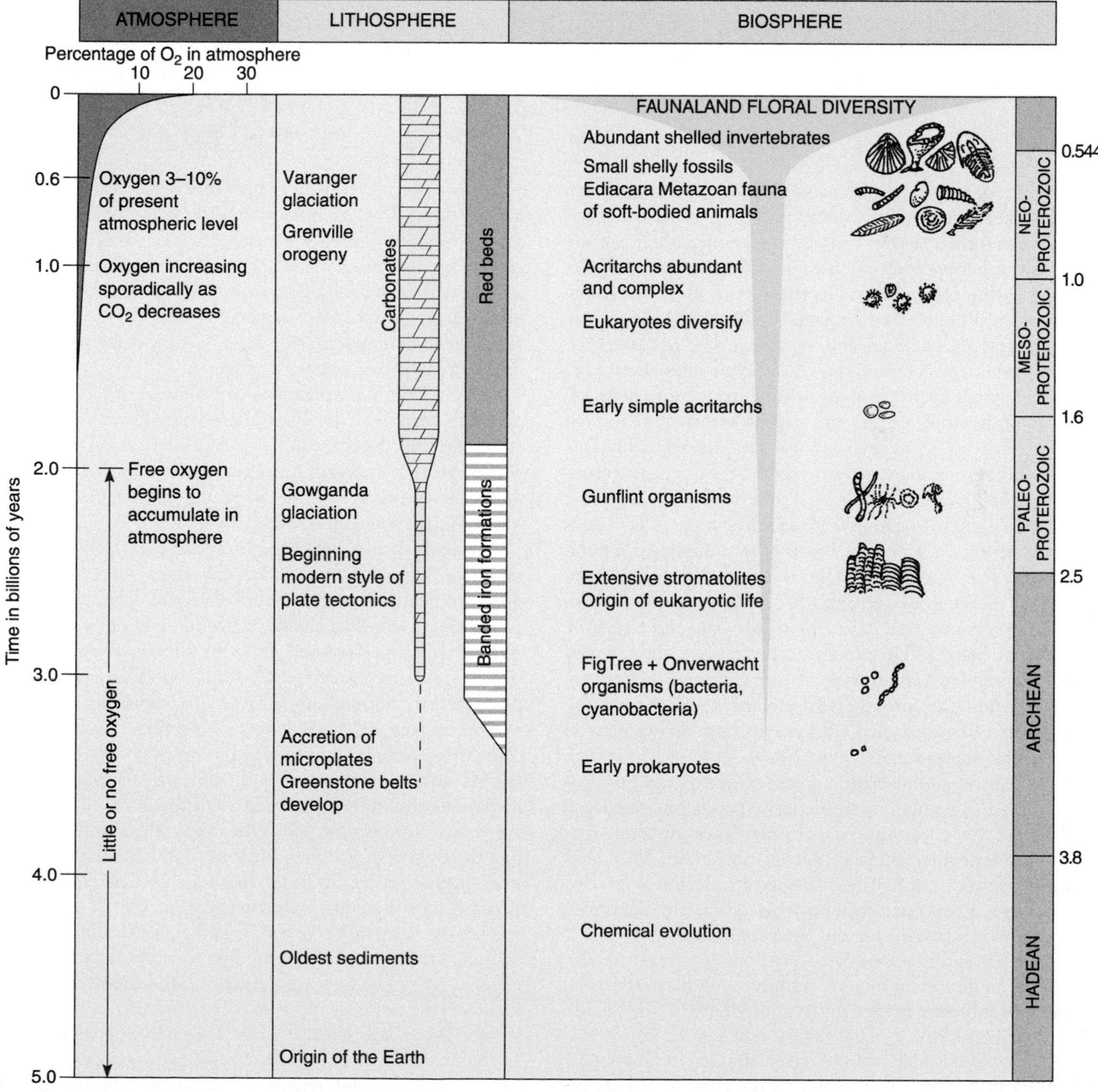

FIGURE 7-35 Correlation of major events in the history of the biosphere, lithosphere, and atmosphere.

levels of atmospheric oxygen increased, so also did the amount of oxygen in the sea. Nitrogen present in sea water could then be oxidized to form nitrate, an important nutrient for eurkaryotic algae. This may be one reason for the expansion of acritarchs and other eurkaryotes near the end of the Paleoproterozoic.

One result of the oxygen buildup was the accumulation on land of considerable amounts of ferric iron oxide, which stained terrestrial sediments a rust-red color. Such sedimentary rocks are known as **red beds** and are considered a valid indication of the advent of an oxygenic environment. Of course, the oxygen level probably rose slowly and very likely did not approach 10 percent of present atmospheric levels of free oxygen until the Cambrian Period.

In general, Proterozoic rocks provide evidence of a wide range of climatic conditions, but there is no indication that these climates were especially unique in comparison with those of the Phanerozoic eras. Thick limestones and dolomites with reeflike algal colonies were deposited along the equator, where warm, tropical conditions prevailed much as they do today. During the Early Proterozoic, the equator lay close to the northern border of North America. Precambrian evaporite deposits in eastern Canada and in Australia suggest that conditions were periodically rather arid during the Middle Proterozoic. In the lower latitudes, climates were more severe, as indicated by consolidated deposits of glacial debris (tillites) and glacially striated basement rocks. As previously discussed, the best known of these poorly sorted, thick, boulderlike deposits is the Paleoproterozoic Gowganda Formation. The ice returned dramatically during the Neoproterozoic when continental glaciers covered extensive areas of the globe, including some regions that lay within about 30 degrees of the equator.

SUMMARY

By the beginning of the Proterozoic about 2.5 billion years ago, many of the smaller cratonic elements of the Archean had come together to form a large craton. Lines of contact between the once separate cratonic elements were marked by orogenic belts formed during collisions. In North America, the large continent known as Laurentia was formed by this welding of smaller crustal segments. Laurentia and other large cratons of the Proterozoic experienced further growth largely as a result of accretion of sedimentary prisms along continental margins or by additions of microplates. In the Neoproterozoic, blocks of continental crust aggregated to form the supercontinent Rodinia. These events indicate that a modern style of plate tectonics was in full operation during the Proterozoic.

The Wopmay orogen of Canada's Northwest Territories is a well-studied example of Proterozoic sedimentation and tectonics. The Wopmay contains a sequence of sedimentary deposits reflecting the opening and closing of an ocean basin. Similar events occurred in the Trans-Hudson and Grenville orogens of Canada. The Neoproterozoic Grenville orogeny was a consequence of suturing of Laurentia to other crustal blocks during the formation of Rodinia. Aulacogens, the failed rifts of three-armed rift systems, are also evident in Proterozoic depositional basins.

Two episodes of glaciation occurred during the Proterozoic, each documented by ice-deposited conglomeratic beds (tillites), varved mudstones of meltwater lakes, and rocks that have been scratched and faceted by glacial action. The Gowganda Formation contains glacial deposits of a Paleoproterozoic ice age. The second episode of glaciation (the Varangian glaciation) occurred during the Neoproterozoic.

Banded iron formations (BIFs) are notable for two reasons. They are an important mineral resource, and they reflect the buildup of sufficient oxygen in the atmosphere to oxidize iron at the Earth's surface. The development of banded iron formations is followed by deposition of red beds beginning about 2.0 billion years ago. A major change in the Earth's atmosphere toward an increase in free oxygen is indicated by red bed deposition. The Keweenawan lavas of the Mesoproterozoic reflect an episode of severe rifting in the North American midcontinental region. The extensive systems of faults associated with rifting provided conduits for the extrusion of Keweenawan lavas. The rifts, however, did not extend to the edge of the continent and did not evolve into an opening oceanic tract. Mesoproterozoic sedimentary sequences that had accumulated along the eastern margin of Laurentia were deformed during the Neoproterozoic Grenville orogeny.

South of the Canadian shield, Proterozoic rocks are exposed in several of the Cordilleran ranges. Among these rocks, the Belt Supergroup was deposited along the rifted western margin of North America. To the west of these rifted depositional basins lay the widening Panthalussa Ocean, formed by the breakup of Rodinia.

Life at the beginning of the Proterozoic resembled that of the Late Archean. It consisted of stromatolites and other prokaryotes as well as simple acritarchs and primitive eukaryotes. Banded iron formations, reflecting additions of photosynthetically generated oxygen to the atmosphere, characterize the Paleoproterozoic. During the Mesoproterozoic, prokaryotes continued to dominate, stromatolites became more abundant, and the eukaryotes diversified. Acritarchs become more complex and diverse during the Neoproterozoic, but the most significant biological event was the appearance of metazoans (multicellular animals). The fossils of these animals are known from various localities around the world but were first discovered in Ediacara Hills in southern Australia. The fossils include large, discoidal, frondlike, or elongate impressions that have traditionally been considered to have been made by early members of such existing phyla as the Cnidaria or Annelida. Another view, based on their dissimilarity to existing phyla, is that they are a separate taxonomic designation, for which the name Vendoza has been proposed.

The Proterozoic seems to have had climates generally similar to those of the Phanerozoic. They ranged from warm tropical or subtropical conditions, suggested by the presence of thick, stromatolitic carbonate sequences, to the cooler conditions that accompanied the two episodes of continental glaciation.

QUESTIONS FOR REVIEW AND DISCUSSION

1. Describe the sequence of events recorded by the rocks of the Wopmay region of northwestern Canada.

2. What are aulacogens? How do they develop? Where do aulacogens exist today?

3. With regard to the history of the Earth's atmosphere, what is the significance of banded iron formations (BIFs)? What is the significance of the red beds that lie above sequences of BIFs?

4. When was the supercontinent Rodinia assembled? What orogenic event in eastern North America was the result of the assembly of Rodinia?

5. What kind of tectonic activity was the probable cause of the massive extrusions of Keweenawan lavas in the Lake Superior region? When did this occur?

6. How do eukaryotes differ from prokaryotes (see Chapter 6)? When do eukaryotes appear in the fossil record?

7. Paleontologist Andrew Knoll has stated that "cyanobacteria are the heroes of Earth history." Why do these lowly organisms deserve such praise?

8. Were acritarchs eukaryotic organisms? When did acritarchs reach their maximum diversity and when did they nearly become extinct? What climatic conditions may have contributed to their decline?

9. What are metazoans? What is the earliest known occurrence of abundant metazoans? With regard to their general appearance, what are the three major groups of Ediacaran metazoans?

10. What are the three major geochronologic divisions of the Proterozoic? During which of these divisions were Riphean rocks deposited? Vendian rocks?

11. When rafting through the Inner Gorge of the Grand Canyon of the Colorado River, what Proterozoic rock unit would you see exposed in the walls of the gorge?

12. When did continental glaciation occur during the Proterozoic? What is the evidence that such glaciation occurred? Why is it unlikely that continental glaciers would have formed during the earlier Archean?

13. What characteristics of the Ediacaran discoidal fossils suggest that they may not really be jellyfish, that the frond fossils may not be sea pens (soft corals), and forms such as *Dickinsonia* may not be worms?

14. Stromatolites were exceptionally widespread during the Proterozoic but became relatively sparse thereafter. What other organisms may have contributed to the post-Proterozoic decline of stromatolites?

15. Where was the Belt Supergroup deposited? What evidence indicates that these rocks were deposited in shallow coastal areas?

READINGS

Condie, K. C. 1992. *Plate Tectonics and Crustal Evolution*, 4th ed. Boston: Butterworth Heinemann.

Conway Morris, S. 1990. Late Precambrian and Cambrian soft-bodied faunas. *Ann. Rev. Earth Planetary Sci.* 18: 101–122.

Conway Morris, S. 1987. The search for the Precambrian–Cambrian boundary. *Sci. Am.* 75:157–167.

Dalziel, I. W. D. 1995. The Earth before Pangea. *Sci. Am.* 272(1):58–63.

Diver, W. L., and Peat, C. L. 1979. On the interpretation and classification of Precambrian organic-walled microfossils. *Geology* 7:401–404.

Glaessner, M. F. 1984. *The Dawn of Animal Life.* Cambridge, England: Cambridge University Press.

Goodwin, A. M. 1996. *Principles of Precambrian Geology*. London, England: Academic Press.

Hoffman, P. 1989. Precambrian geology and tectonic history of North America. In *Geology of North America*, Vol. A, pp. 447–511. Boulder, CO: Geological Society of America.

Margulis, L. 1982. *Early Life.* New York: Van Nostrand Reinhold.

McMenamin, M. A. S., and McMenamin, D. L. S. 1989. The *Emergence of Animals*. New York: Columbia University Press.

McMenamin, M. A. S, 1998. *The Garden of Ediacara*. New York: Columbia University Press.

Schopf, J. W. 1992. *Major Events in the History of Life*. Boston: Jones and Bartlett.

Seilacher, A. 1989. Vendoza: Organismic construction in the Proterozoic biosphere. *Lethaia* 22:229–239.

WEB SITES

The Earth Through Time Student Companion Web Site (www.wiley.com/college/levin) has online resources to help you expand your understanding of the topics in this chapter. Visit the Web Site to access the following:

1. Illustrated course notes covering key concepts in each chapter;
2. Online quizzes that provide immediate feedback;
3. Links to chapter-specific topics on the web;
4. Science news updates relating to recent developments in Historical Geology;
5. Web inquiry activities for further exploration;
6. A glossary of terms;
7. A Student Union with links to topics such as study skills, writing and grammar, and citing electronic information.

8

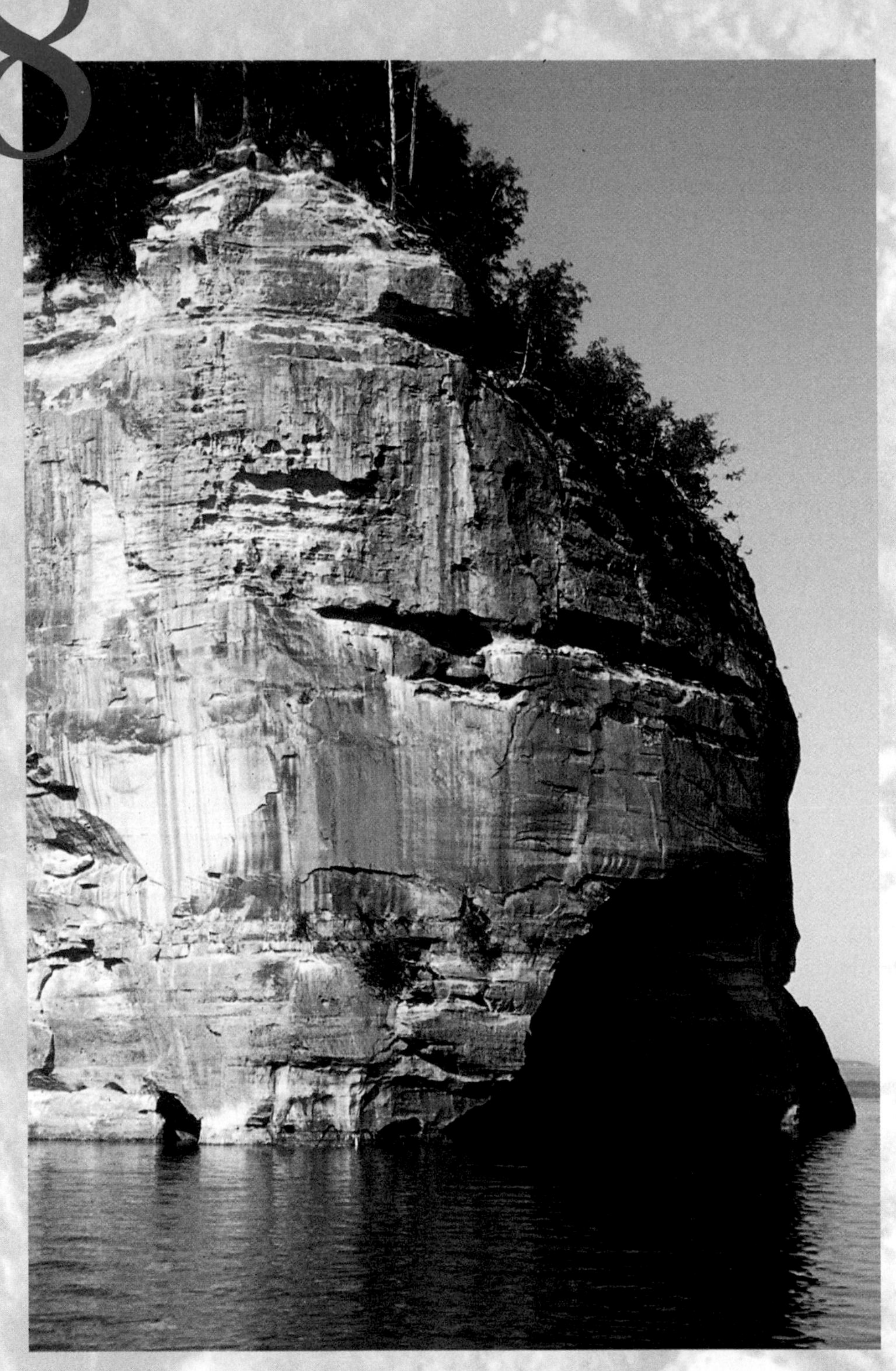

Cambrian sandstone in a vertical cliff that rises about 300 feet above the south shore of Lake Superior. Near Munising in the Upper Penninsula of Michigan. (Courtesy of R. F. Dymek.)

CHAPTER 8

Early Paleozoic Events

If in late Cambrian time you had followed the present route of Interstate 80, you would have crossed the equator at Kearney, Nebraska.
John McPhee, *In Suspect Terrain*, 1983

OUTLINE

Rocks of the Phanerozoic Eon have yielded their secrets more readily than those of the older Archean and Proterozoic. They are often more accessible, less altered, and more fossiliferous. In Chapter 1, we noted that the Phanerozoic includes three eras: the Paleozoic, Mesozoic, and Cenozoic. We now consider the geologic history of the first three geologic periods of the Paleozoic Era. The Cambrian, Ordovician, and Silurian periods lasted about 190 million years. In the subsequent 180 million years, the Devonian, Mississippian, Pennsylvanian, and Permian periods ran their course.

In general, the geologic history of the Paleozoic is characterized by long periods of sedimentation punctuated by intervals of mountain building. Just before the Paleozoic began, Rodinia had fragmented, and the return of the displaced segments to ultimately form Pangea was associated with tectonic plate convergences that affected most of the continents that now surround the Atlantic Ocean (Fig. 8-1). In North America, these events caused the Taconic, Acadian, and Allegheny orogenies, whereas in Europe, the names Caledonian orogeny and Hercynian orogeny are applied to these correlative Paleozoic mountain-building events.

LANDS, SEAS, AND OROGENIC BELTS

Clues to Paleogeography

Because of the many processes that have altered blocks of continental crust, reconstructing global geography for the Paleozoic becomes a formidable task involving synthesis of all paleomagnetic, paleoclimatic, geochronologic, and paleontologic data. Such a synthesis by geologist Ian Dalziel and colleagues indicates that during the late Neoproterozoic, Laurentia (North America) had begun to break away from the Precambrian supercontinent Rodinia. In fact, by about 570 million years ago, Laurentia had drifted so far to the south that it collided with what today is the Argentinian margin of South America. Fifty million years or so later during the Early Cambrian, Laurentia then severed its connection to South America and moved northward (Fig. 8-1*A*). By about the middle of the Ordovician (Fig. 8-1*B*), Gondwana had rotated counterclockwise, and Laurentia was again converging on South America. It was during this

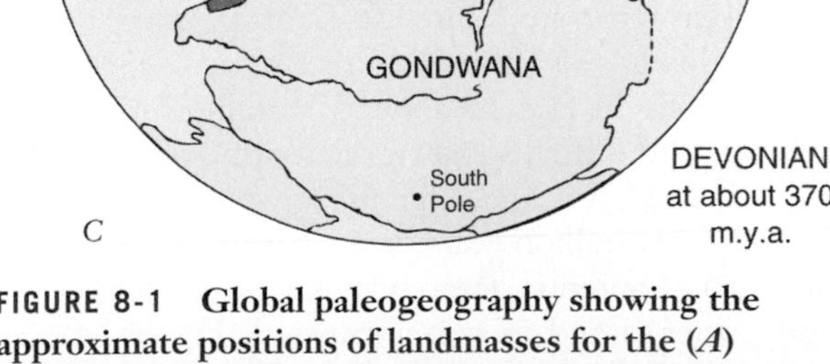

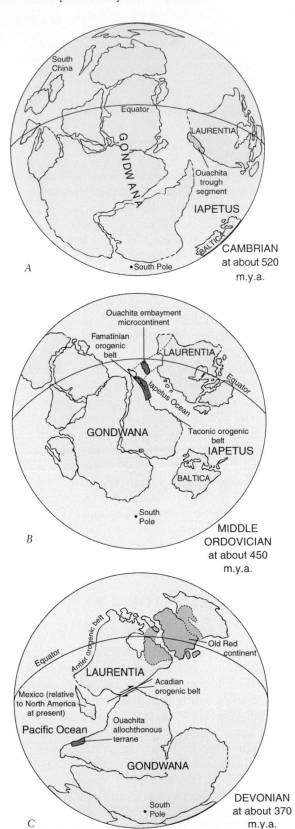

FIGURE 8-1 Global paleogeography showing the approximate positions of landmasses for the (*A*) Cambrian, (*B*) mid-Ordovician, and (*C*) Devonian. (m.y.a. = million years ago). *(After Dalziel, I. W. D., Salda, L. H., and Gahagan, L. M. 1994. Geol. Soc. Am. Bull. 106(2):243–252.)*

convergence that a crustal segment of the southern margin of Laurentia called the Ouachita Terrane was accreted to what is now termed the Precordillera of the Andes. Thus, according to Ian Dalziel's hypothesis, the Andean Precordillera is an allochthonous terrane that is distinctly North American in both lithology and the trilobite and brachiopod fossils it contains. The gap in the southern margin of North America where the Ouachita Terrane was lost to the Andes explains why the Appalachian Mountains appear to end in Alabama, only to show up again in the Ouachita Mountains of Arkansas. The allochthonous block separated along a well-known system of faults called the Ouachita Embayment fault system.

Another consequence of the closure of Laurentia and Gondwana was the Taconic orogeny of the Appalachians and the Famatinian orogeny in western South America. Both of these mountain-building events occurred between about 480 and 460 million years ago. By Devonian time, at about 370 million years ago, Laurentia had become separated from the Andean margin of South America. As it moved northward, it collided obliquely with the northwest corner of South America (Fig. 8-1*C*). The final Paleozoic event, the Allegheny orogeny, involved the collision of Laurentia with northwestern Africa. The gathering of continental masses to form Pangea was now largely completed.

During the Cambrian and Ordovician, the paleoequator crossed North America from northern Mexico to the Arctic (Fig. 8-2). Great thicknesses of carbonates occur within 30° of the paleoequator, and salt and gypsum accumulations are found in Ordovician strata of Arctic Canada at a paleolatitude of only 10°.

The Continental Framework

THE STABLE INTERIOR In Chapter 6, we noted that continents can be described in terms of cratons and mobile belts or orogens. A craton is the relatively stable part of the continent, consisting of a Precambrian shield and the extension of the shield that is covered by flat-lying or only gently deformed Phanerozoic strata. These cratonic strata were originally wave-washed sands, muds, and carbonates deposited in shallow seas that periodically flooded continental regions of low relief. The shallow, warm, and well-lighted seas were favorable habitats and allowed for major diversification of marine life. Such extensive inland seas do not exist on our planet today.

Here and there, the sedimentary layers deposited in Paleozoic epeiric seas are warped into broad synclines, basins, domes, and arches (Fig. 8-3). The resulting tilt to the strata is slight and is usually expressed in feet per mile rather than degrees. In the course of geologic history, the arches and domes stood as low islands in the epeiric seas or as barely awash submarine banks.

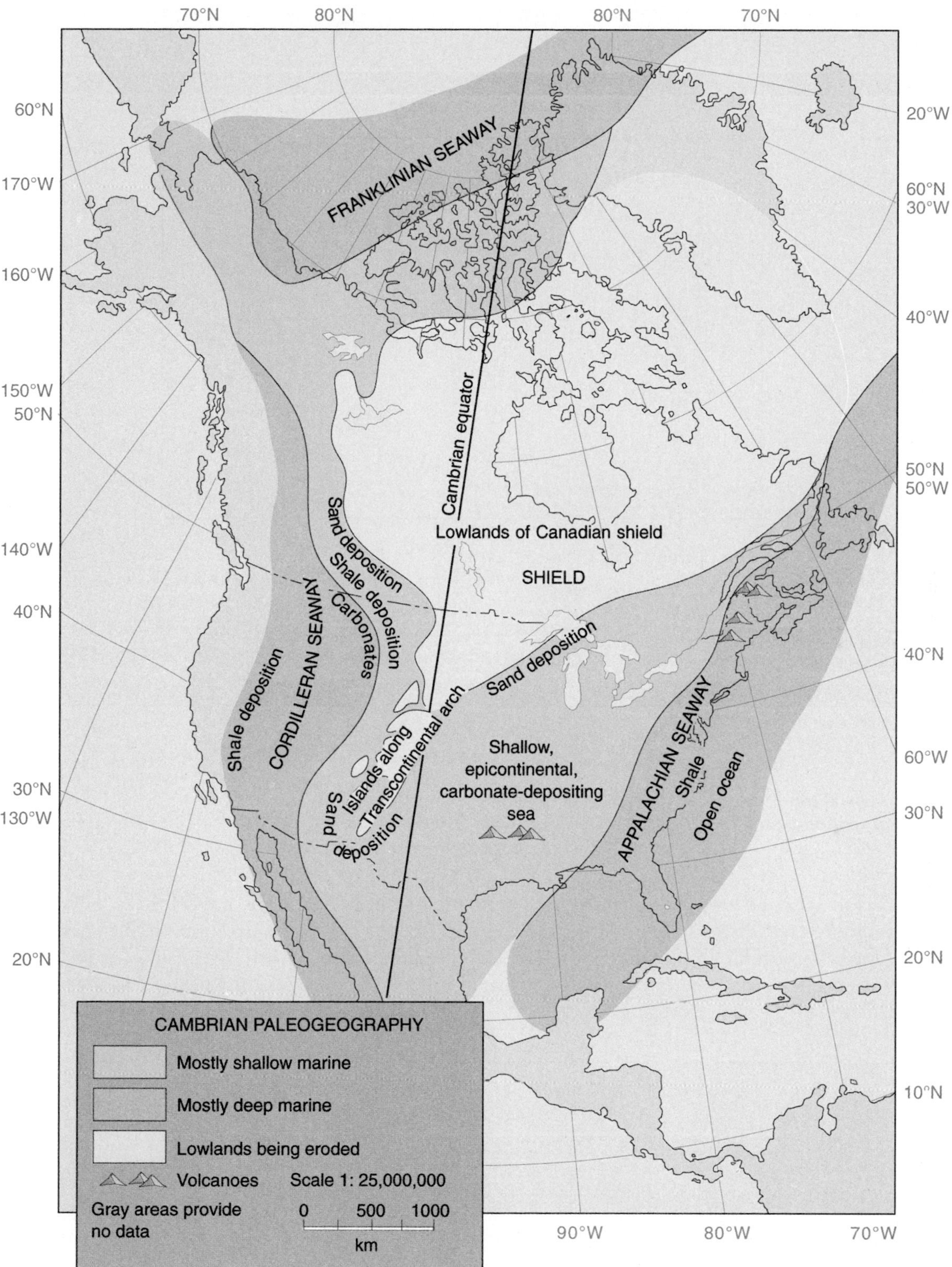

FIGURE 8-2 Paleogeographic and tectonic elements of North America during the Cambrian Period, showing position of the Cambrian paleoequator. *What were the conditions at the location of your home during the Cambrian period?*

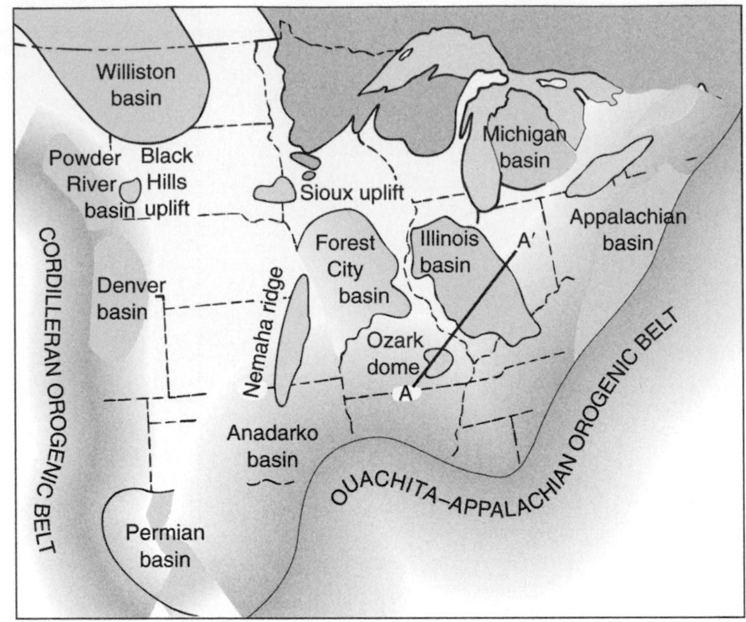

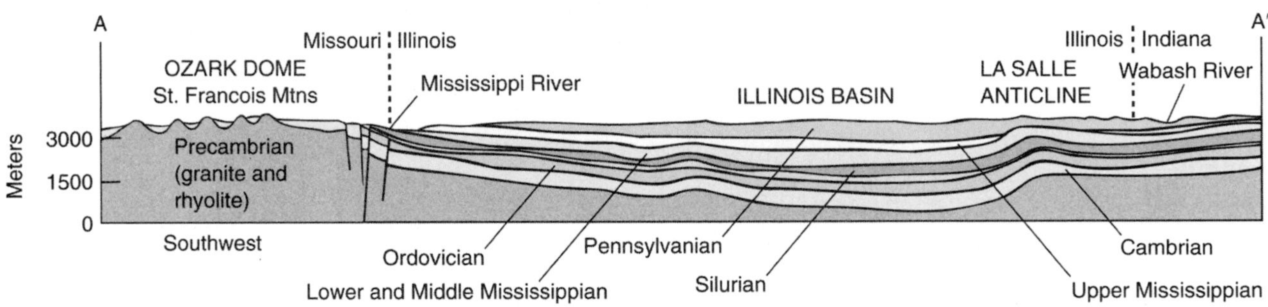

FIGURE 8-3 The central platform of the United States showing major basins and domes. Structural section below the map crosses part of the Ozark dome and the Illinois basin. The basins and domes developed at different times during the Phanerozoic.

Domes and arches seem to have developed in response to vertically directed forces unlike those that formed the compressional folds of mountain belts. Possibly, forces associated with plate convergence were transmitted to the craton, causing flexures (domes, arches) in cratonic crustal rocks. Where tensional forces were in operation, basins developed. Domes and basins may also have resulted from isostatic adjustments caused by density changes in the underlying lithosphere.

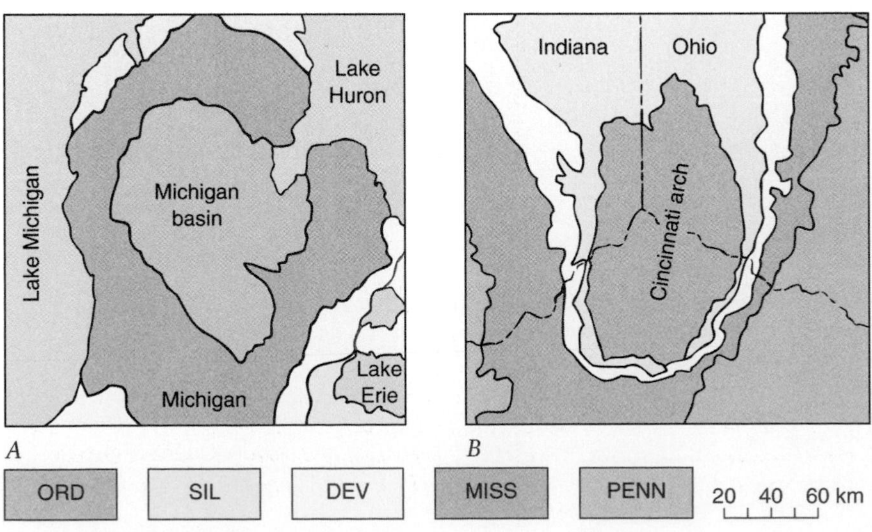

FIGURE 8-4 Michigan basin and Cincinnati arch. (*A*) In an erosionally truncated basin such as the Michigan basin, the youngest beds are centrally located. (*B*) In a domelike structure such as the Cincinnati arch, the oldest beds are located in the center.

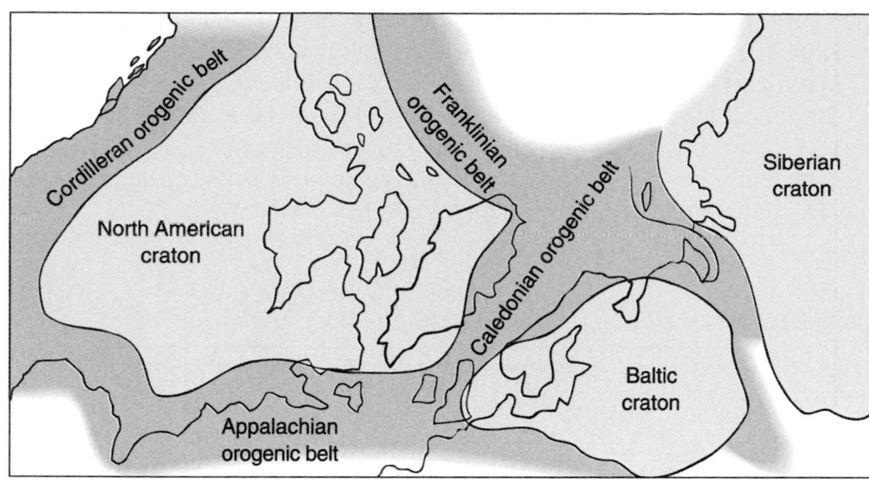

FIGURE 8-5 Cratons and mobile belts of North America and Europe.

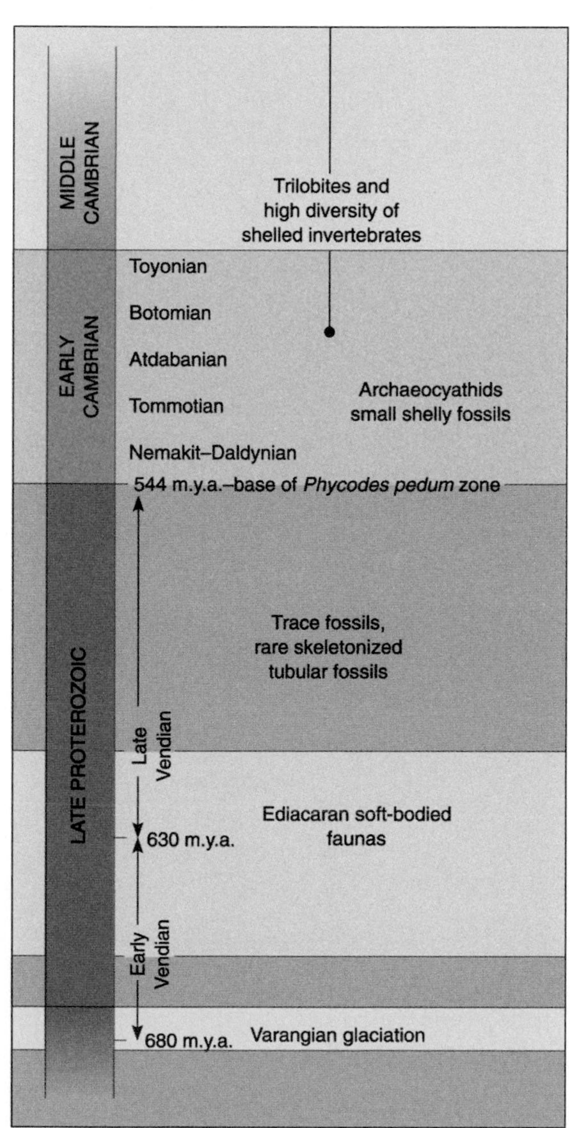

FIGURE 8-6 The Cambrian–Precambrian boundary is placed at the base of the Phycodes pedum zone, corresponding to a uranium–lead age determination of about 540 million years.

Domes and basins can often be recognized by their pattern of outcropping rocks. Erosional truncation of domes exposes older rocks near the centers and younger rocks around the peripheries. Sequences of strata over arches and domes tend to be thinner. Also, because these structures were periodically above sea level, they characteristically have many erosional unconformities. Basins, in contrast, were more persistently covered by inland seas, have fewer unconformities, and developed greater thickness of sedimentary rocks. In erosionally truncated basins, younger rocks are centrally located and older strata occur successively farther toward the peripheries (Fig. 8-4).

OROGENIC BELTS The North American craton is bounded on four sides by great elongate tracts that have been the sites of intense deformation, igneous

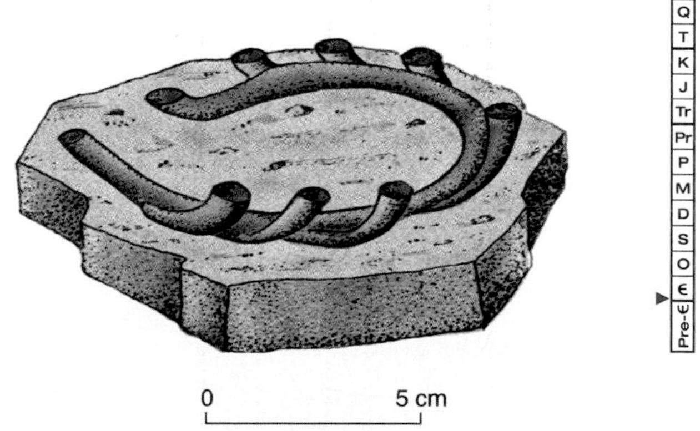

FIGURE 8-7 The trace fossil *Phycodes*. The fossil represents the feeding behavior of a bottom dwelling organism that produced horizontal cylindrical burrows with sequential upward shafts as it explored for food. The base of the *Phycodes pedum* biozone is a marker for the base of the Cambrian system. (*After Crimes, T. P. 1989. Trace Fossils, in The Cambrian–Precambrian Boundary. Oxford Monographs on Geology and Geophysics 12. Oxford, England: Clarendon Press.*)

activity, and earthquakes. One or more of such orogenic belts are present on all continents (Fig. 8-5). Most are located along present or past margins of continents, such as the North American Cordilleran belt. As described in Chapter 6, the passive margin of a continent may experience an early stage in which thick sequences of sediments accumulate along the continental shelf and rise. This stage may be followed by subduction of oceanic lithosphere along continental margins and may terminate in continent-to-continent collisions, as in the classic Wilson cycle. As we review the history of eastern North America later in this chapter, we will describe the geologic evidence of this sequence of events along the Appalachian orogen.

TABLE 8-1 Cratonic Sequences of North America*

Geologic Time		Cratonic Sequences (Center of craton → Margin of craton)	Orogenic Events	Biologic Events	Ice Ages
CENOZOIC		Tejas	Himalayan Alpine	Age of mammals	
		65 m.y.a.	Laramide	*Massive extinctions*	
MESOZOIC	Cretaceous	Zuni	Sevier	First flowering plants Climax dinosaurs and ammonites	
			Nevadan		
MESOZOIC	Jurassic			First birds Abundant dinosaurs and ammonites	
MESOZOIC	Triassic			First dinosaurs First mammals Abundant cycads	
		250 m.y.a.	Sonoma	*Massive extinctions* (including trilobites)	
LATE PALEOZOIC	Permian	Absaroka		Mammal-like reptiles	
LATE PALEOZOIC	Pennsylvanian		Alleghenian	Great coal forests Conifers First reptiles	
LATE PALEOZOIC	Mississippian	Kaskaskia		Abundant amphibians and sharks Scale trees Seed ferns	
			Antler		
LATE PALEOZOIC	Devonian			*Extinctions* First insects First amphibians First forests First sharks	
		410 m.y.a.	Acadian–Caledonian		
EARLY PALEOZOIC	Silurian	Tippecanoe		First jawed fishes First air-breathing arthropods	
EARLY PALEOZOIC	Ordovician		Taconic	*Extinctions* First land plants Expansion of marine shelled invertebrates	
EARLY PALEOZOIC	Cambrian	Sauk		First fishes Abundant shell-bearing marine invertebrates Trilobites	
		540 m.y.a.			
LATE PROTEROZOIC				Rise of the metazoans	

*The green areas represent sequences of strata. They are separated by major unconformities, indicated in yellow. Note that the rock record is most complete near cratonic margins, just as the time spans represented by unconformities are greatest near the center of the craton. Major biologic, orogenic, and glacial events are added for reference. (Cratonic sequence model after Sloss, L. L. 1965. *Bull Geol. Soc. Amer.* 74:93–114.)

THE BASE OF THE CAMBRIAN

At one time, recognition of the boundary between the Cambrian and the underlying Precambrian seemed relatively simple. The base of the Cambrian System could be identified by the first occurrence of shell-bearing multicellular animals. Among these, extinct marine arthropods known as trilobites were used to identify the Precambrian-Cambrian boundary. A problem became evident in the 1970s, however, when a distinctive group of shelly fossils was found beneath the lowest occurrence of strata containing the first trilobites. Although these animals had mineralized skeletons, they had not been discovered earlier, partly because of their very small size. They include early members of such existing phyla as Porifera, Brachiopoda, Mollusca, and possibly Annelida. Although such "small shelly fossils" were originally discovered in Siberia, similar fossils have also been found recently in other continents. Their discovery and the known occurrences of a variety of trace fossils have resulted in a new classification of Lower Cambrian chronostratigraphic units (Fig. 8-6). On the basis of uranium-lead isotope dates obtained from rocks in northeastern Siberia, it can be said that the Cambrian began 544 million years ago. The boundary is marked by the lowest (oldest) occurrence of feeding burrows of the trace fossil *Phycodes pedum* (Fig. 8-7). The initial stage of the Lower Cambrian has been designated the Nemakit-Daldynian. It is followed by the Tommotian, Atdabanian, Botomian, and Toyonian stages.

EARLY PALEOZOIC EVENTS

North America at the beginning of the Paleozoic Era was in a state of relative stability. It was as if the continent rested from the rigors of orogeny and severe climate that had occurred near the end of the preceding Neoproterozoic. We know little of the Paleozoic ocean basins, for those old sea floors have long since been consumed at subduction zones. However, oceans did exist and gradually spilled out of their basins onto low regions of the continents. Although a large part of the sedimentary record has been removed by erosion, enough remains to indicate that practically all of the Canadian shield was inundated at times. Initially, the dominant deposits were sands and clays derived from the weathering and erosion of igneous and metamorphic rocks of the shield. Later, as these quartz- and clay-producing terranes were reduced, limestones and dolomites became increasingly more prevalent. The limestones contain abundant remains of carbonate-secreting marine organisms. Their distribution indicates that shallow seas were common throughout much of the equatorial region of the Earth during the early Paleozoic. Indeed, the advances and retreats of these epeiric seas were the most apparent events in the early Paleozoic history of the continental interiors.

Cratonic Sequences

It is convenient to discuss the geologic history of the continental mass that now constitutes North America in terms of its cratonic and orogenic regions. It is also advantageous to subdivide the early Paleozoic of the craton on the basis of advances (transgressions) and retreats (regressions) of epeiric seas. Regressions exposed the old sea floors to erosion and produced extensive unconformities that are used as regional stratigraphic markers.

Chapter 1 included a discussion of the largely European basis for developing the geologic time scale and defining its geologic periods. Although the cratonic rocks of North America can be correlated with those of Europe on the basis of fossils, most of the stratigraphic breaks in deposition do not correspond to those that form European period boundaries. Indeed, if the relative geologic time scale had been worked out in North America instead of Europe, there might have been only two periods in the early Paleozoic rather than three. The American stratigrapher Laurence Sloss suggested a workable remedy for

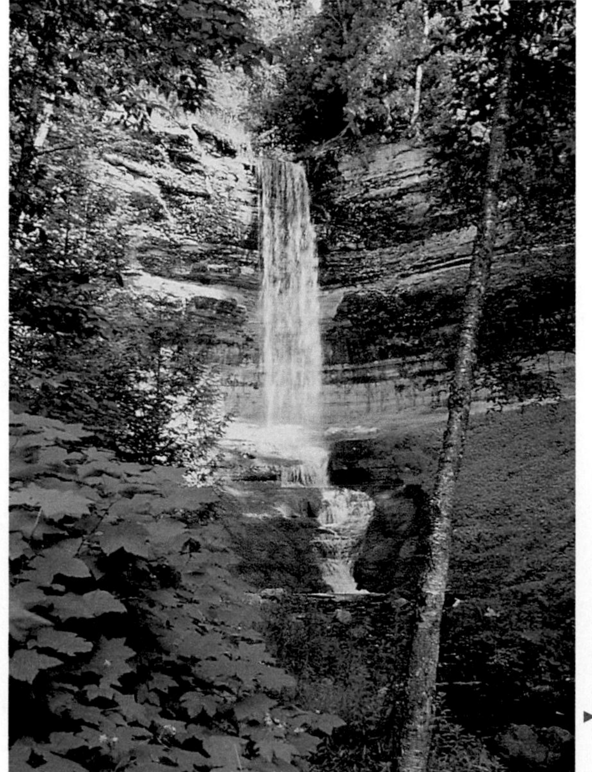

FIGURE 8-8 Cambrian sandstone of the Munising Formation provides the precipice for one of the many waterfalls in Pictured Rocks National Lakeshore in the Upper Peninsula of Michigan. The area was the inspiration for Henry Wadsworth Longfellow's narrative poem, The Song of Hiawatha. (*Photograph by R. F. Dymek.*)

this disparity. He proposed that the Paleozoic rocks of the North American craton be divided into sequences of sedimentary rocks that have unconformities at their upper and lower boundaries. Each sequence consisted of the sediments deposited as the sea transgressed an old erosional surface, reached its maximum inundation, and then retreated. For the Paleozoic, he named these the **Sauk**, **Tippecanoe**, **Kaskaskia**, and **Absaroka** sequences (Table 8-1).

Because the cratonic sequences seen in North America can also be found in other continents, it appears that worldwide changes in sea level were responsible for the repeated major transgressions and regressions. Widespread continental glaciation may also have been a cause of the eustatic changes. It is also reasonable that minor variations in the major cycle resulted from tectonic movements on the craton. Many geologists believe that the long-term changes in sea level were probably related to sea-floor spreading. For example, rapid spreading results in high midoceanic ridges. The ridges in turn displace water, causing a global rise in sea level.

There is an interesting correspondence between Sloss's cratonic sequences and the Vail sea level curves discussed in Chapter 3. The Vail curves are derived from high-precision seismic profiles and permit one to "see" deep beneath the surface. They indicate that unconformity-bounded sequences can be traced across the craton and into the thick, deep sections along the continental margins.

Sauk and Tippecanoe

During the earliest years of Sauk deposition, seas were largely confined to the continental margins (continental shelves and rises), so most of the craton was exposed and undergoing erosion. No doubt it was a bleak and barren scene, for vascular land plants had not yet evolved. Uninhibited by protective vegetative growth, erosional forces gullied and intricately dissected the surface of the land. Over a span of at least 50 million years, the Precambrian crystalline rocks were deeply weathered and must have formed a thick, sandy "soil." Eventually, marine waters spilled out of the marginal

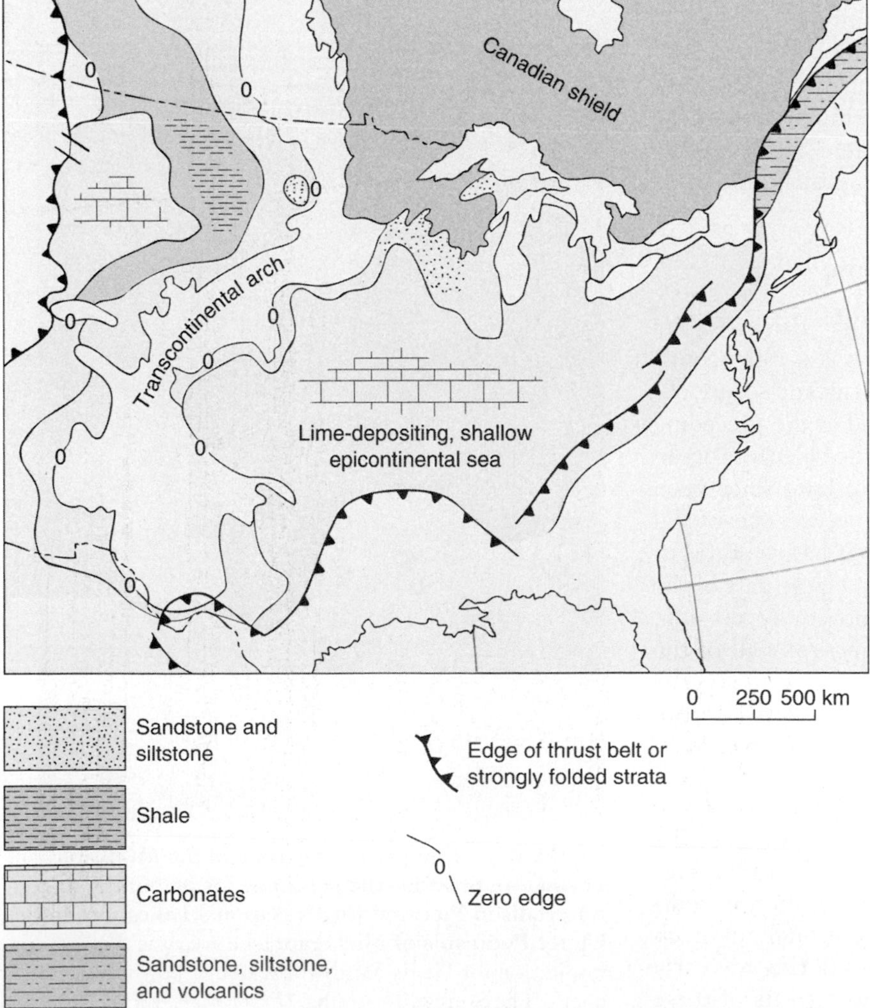

FIGURE 8-9 Upper Cambrian lithofacies map. (*Simplified and adapted from Stratigraphic Atlas of North America and Central America. Shell Oil Company, Exploration Department.*)

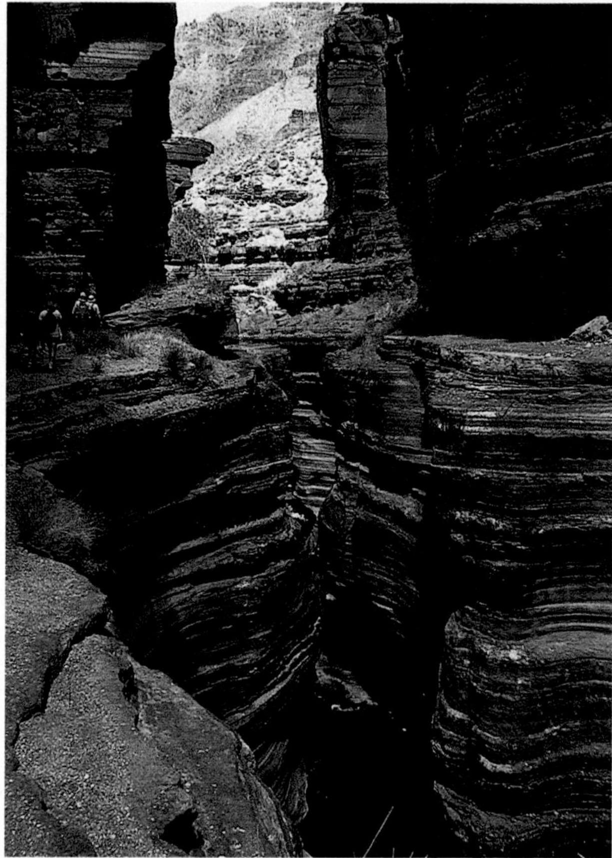

FIGURE 8-10 The Tapeats Sandstone exposed in the vertical walls of Deer Creek Canyon, Grand Canyon, Arizona. The Tapeats is composed mostly of sand grains deposited in a nearshore environment of the shallow sea that transgressed over a large area of the southwestern United States during the Cambrian Period. It rests on erosionally truncated Protozeroic rocks and grades upward and eastward into the Bright Angel Shale, which in turn grades into the Muav Limestone. (*Courtesy of N. Potter, Jr., and the American Geological Institute.*)

basins and began a slow encroachment over the eroded and weathered surface of the central craton.

The craton was not a level, monotonous plain during Sauk time but had distinct upland areas composed of Precambrian igneous and metamorphic rocks. During times of marine transgression, these uplands became islands in early Paleozoic seas and provided detrital sediments to surrounding areas. The absence of marine sediments over these upland tracts provides evidence of their existence and extent. One of the largest of the highland regions was the transcontinental arch (see Fig. 8-2), which extended southwestward across the craton from Ontario to Mexico.

By Late Cambrian time, seas extended across the southern half of the craton from Montana to New York. A vast apron of clean sand (Fig. 8-8) was spread across the sea floor for many miles behind the advancing shoreline. This sandy aspect, or facies, of Cambrian deposition was replaced toward the south by carbonates (Fig. 8-9). Here the waters were warm, clear, and largely uncontaminated by clays and silts from the distant shield. Marine algae flourished and contributed to the precipitation of calcium carbonate. Invertebrates, although present, did not contribute to the volume of sediment to the degree they did in later periods.

In the Cordilleran region, the earliest deposits were sands, which graded westward into finer clastics and carbonates. An excellent place to study this Sauk transgression is along the walls of the Grand Canyon of the Colorado River. Here one finds the Lower Cambrian Tapeats Sandstone (Fig. 8-10) as an initial strand line deposit above the old Precambrian surface. The Tapeats can be traced both laterally and upward into the Bright Angel Shale, which was deposited in a more offshore environment (Fig. 8-11). Next is the Muav Limestone, which originated in a still-more-seaward

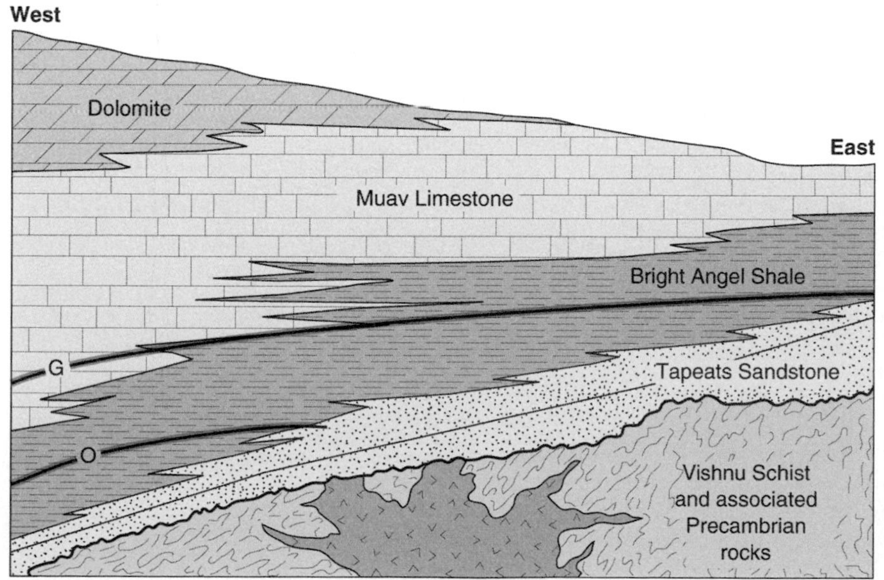

FIGURE 8-11 East–west section of Cambrian strata exposed along the Grand Canyon. The line labeled "O" indicates the top of the Lower Cambrian *Olenellus* trilobite zone. The line labeled "G" indicates the location of the Middle Cambrian *Glossopleura* trilobite zone. The section is approximately 200 kilometers long, and the section along the western margin is about 600 meters thick. (*Adapted from McKee, E. D. 1945. Cambrian Stratigraphy of the Grand Canyon Region. Washington, Carnegie Institute, Publication 563.*) ? *What does the vertical sequence of rock types indicate about changes in sea level in this region during the Cambrian?*

environment. As the sea continued its eastward transgression, the early deposits of the Tapeats were covered by clays of the Bright Angel Formation, and the Bright Angel was in turn covered by limy deposits of the Muav Limestone. Together these formations form a typical transgressive sequence, recognized by coarse deposits near the base of the section and increasingly finer (and more offshore) sediments near the top.

Cambrian rocks of the Grand Canyon region not only provide a glimpse of the areal variation in depositional environments as deduced from changing lithologic patterns, but they also illustrate that particular formations are not usually the same age everywhere they occur. Detailed mapping of the Tapeats, Bright Angel, and Muav units and correlation of trilobite occurrences indicate that deposition of the highest facies (Muav) had already begun in the west before deposition of the lowest facies (Tapeats) had stopped in the east. Correlations based on the guide fossils provided reliable evidence that the Bright Angel Formation was largely Early Cambrian in age in California and mostly Middle Cambrian in age in the Grand Canyon National Park area. This example illustrates the principle of temporal transgression, which stipulates that sediments deposited by advancing or retreating seas are not necessarily of the same geologic age throughout their areal extent.

The episode of carbonate deposition that was so characteristic of the southern craton continued into the early Ordovician almost without interruption. Then, near the end of the early Ordovician, the seas regressed, leaving behind a landscape underlain by limestones that was subjected to deep subaerial erosion. That erosion

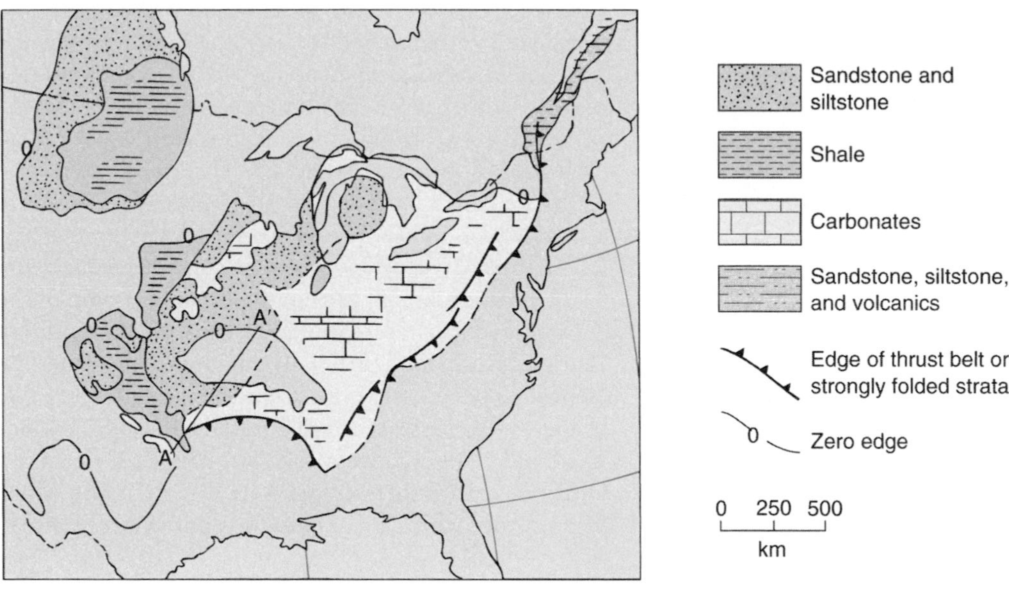

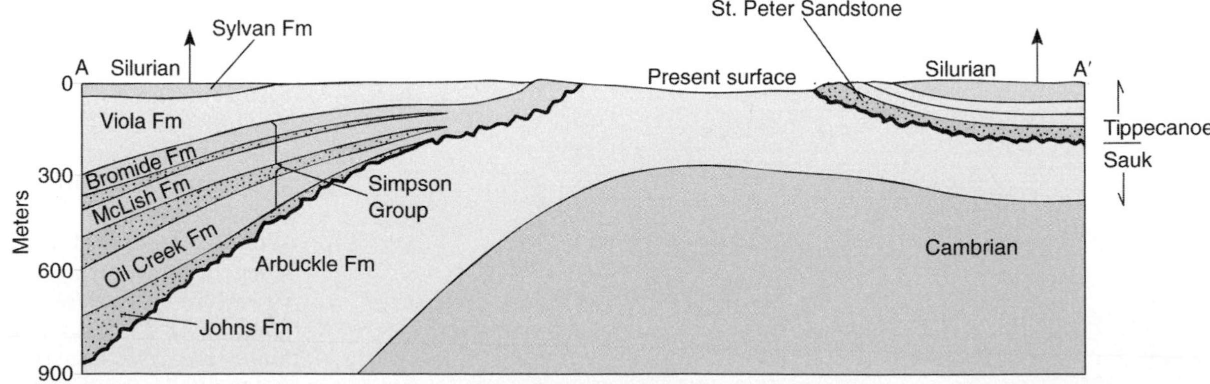

FIGURE 8-12 Lower Tippecanoe lithofacies map and cross-section along line A to A′. In the centrally located white area, Tippecanoe strata have been removed by erosion. Note the extensive unconformity (indicated by wavy line) between the Sauk and the Tippecanoe that is depicted on the cross-section.

Pre-Є Є O S D M P Pr Tr J K T Q

FIGURE 8-13 The St. Peter Sandstone is the massive bed directly beneath the overhanging ledge of horizontally bedded dolomite of the Joachim Formation. Both formations are Middle Ordovician in age. The St. Peter is the initial deposit of the transgressing Tippecanoe sea. The brown stains are iron oxide. This exposure is 30 miles southwest of St. Louis, Missouri.

Pre-Є Є O S D M P Pr Tr J K T Q

FIGURE 8-14 Restoration of some of the common invertebrates that populated the floors of shallow epeiric seas during the Ordovician. The large semispherical object on the left is the skeleton of a colonial coral. In front of the coral and elsewhere in view are "branching twig" colonial bryozoans, and in the left foreground are trilobites, one of which is being eaten by the cephalopod with a conical shell. The long-stemmed creatures with branching arms are crinoids. Many different species of the bivalved organisms known as brachiopods can be seen. (*Diorama in the National Museum of Natural History, Smithsonian Institution.*)

produced a widespread unconformity that geologists use as the boundary between the Sauk and the Tippecanoe sequences (Fig. 8-12).

The second major transgression to affect the platform occurred when the Tippecanoe sea flooded the region vacated as the Sauk sea regressed. Sediment deposited in the advancing Tippecanoe sea form the initial geologic formations of the Tippecanoe sequence. Again the deposits begin with great blankets of clean quartz sands. Perhaps the most famous of these Tippecanoe sandstones is the St. Peter Sandstone (Fig. 8-13), which is nearly pure quartz and is thus prized for use in glass manufacturing. Such exceptionally pure sandstones are usually not developed in a single cycle of erosion, transportation, and deposition. They are the products of chemical and mechanical processes acting on still older sandstones. In the St. Peter Formation, waves and currents of the transgressing Tippecanoe sea reworked Upper Cambrian and Lower Ordovician sandstones and spread the resulting blanket of clean

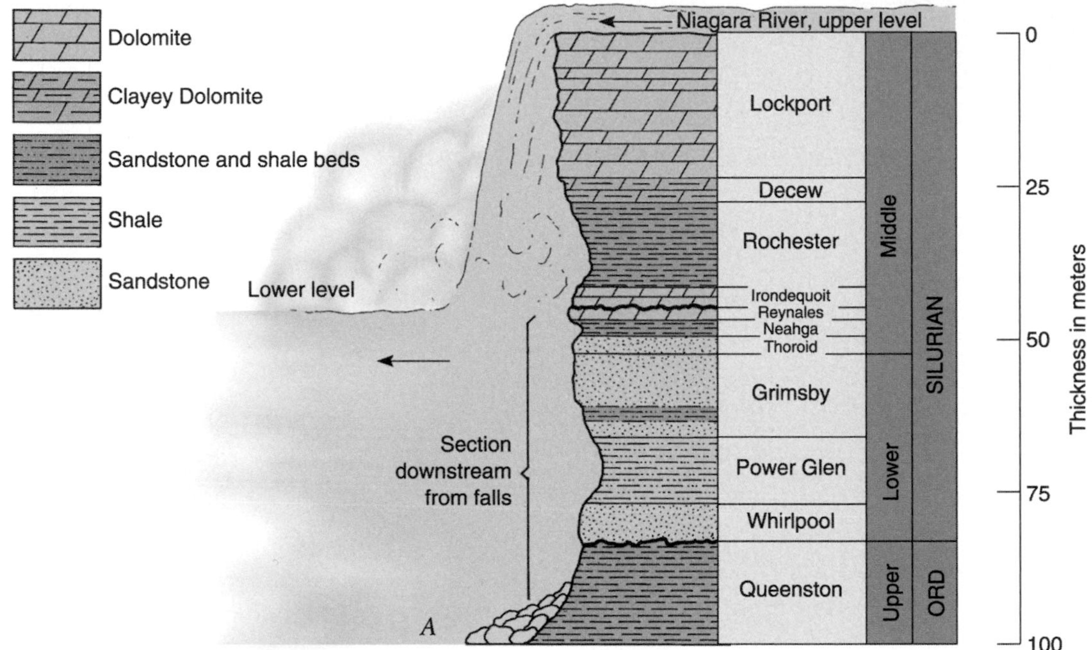

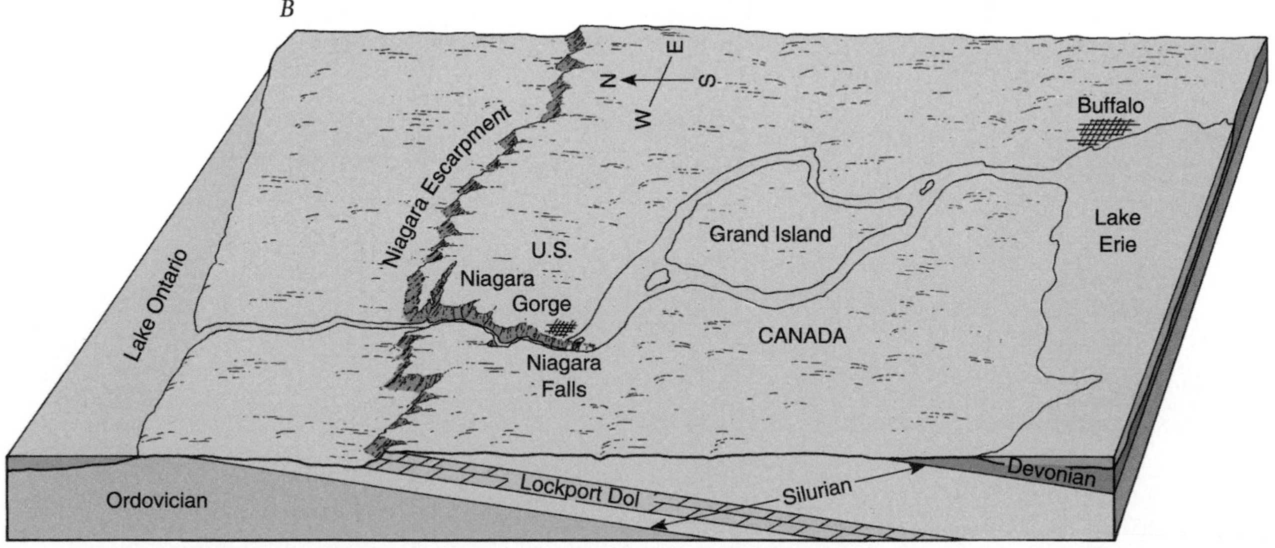

FIGURE 8-15 Stratigraphic section (*A*) and block diagram (*B*) of Niagara Falls. The Lockport Dolomite forms the resistant lip of the falls. The rocks dip gently to the south in this area, and where harder dolomite layers such as the Lockport Dolomite intersect the surface, they form a line of bluffs known as the Niagaran Escarpment. (*Map after E. T. Raisz.*)

FIGURE 8-16 Niagara Falls, formed where the Niagara River flows from Lake Erie into Lake Ontario. The classic section of the American Silurian System is exposed along the walls of the gorge below the falls. (*Guido Alberto Rossi/The Image Bank.*)

sand over an area of nearly 7500 square kilometers. As described in Chapter 3, sandstones can be described by their texture and composition as either mature or immature. These textural and compositional traits are, of course, related to the processes that produced the sandstone. Sandstones with a high proportion of chemically unstable minerals and angular, poorly sorted grains are considered immature, whereas aggregates of well-rounded, well-sorted, highly stable minerals (such as quartz) are considered mature. The St. Peter Sandstone is so pure that it is geologically unusual and perhaps should be termed an ultramature sandstone.

The sandstone depositional phase of the Tippecanoe was followed by widespread deposition of carbonates, often containing calcareous remains of brachiopods, bryozoans, crinoids, cephalopods, corals, and algae (Fig. 8-14). Some of these limey deposits were chemical precipitates, many were fossil fragment limestones (bioclastic limestones), and some were great organic reefs. Frequently, the carbonate sediment underwent chemical substitution of some of its calcium by magnesium and in the process was converted from limestone to dolomite.

In the region east of the Mississippi River, dolomites and limestones were gradually supplanted by shales. As we shall see, these clays are the peripheral sediments of the **Queenston clastic wedge**, which lay far to the east. The geologic section exposed at Niagara Falls is a classic locality in which to examine these rocks (Figs. 8-15 and 8-16).

Near the close of Tippecanoe time, landlocked, reef-fringed basins developed in the Great Lakes region (Fig. 8-17). During the Silurian, evaporation in the Michigan Basin caused the precipitation of salt and gypsum on an immense scale. The evaporites record arid conditions. In the salt-bearing Salina Group evaporites have an aggregate thickness of 750 meters. The precipitation of this amount of salt and gypsum, if theoretically confined to a single continuous episode of evaporation, would require evaporation of a column of sea water 1000 kilometers tall. Because such a column of water has never existed on Earth, it is likely that the evaporating basins had connections to the ocean through which periodic replenishment could occur. One setting in which this might have taken place would be in a basin that had its opening to the sea restricted by organic reefs, a raised sill, or a submerged bar. Evaporation within the basin would have produced heavy

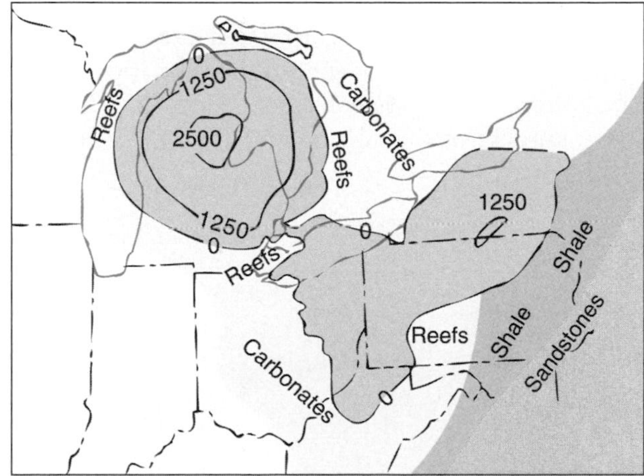

FIGURE 8-17 Isopach map showing thickness and area of evaporite basins during the late Silurian. Areas of evaporite precipitation were surrounded by carbonate banks and reefs. Gaps in the reefs and banks provided channelways for replenishment of the basins with normal sea water, thus replacing water lost by evaporation. (*Adapted from Alling, H. L., and Biggs, L. I. 1961. Bull. Am. Assoc. Petrol. Geol. 45:515–547.*)

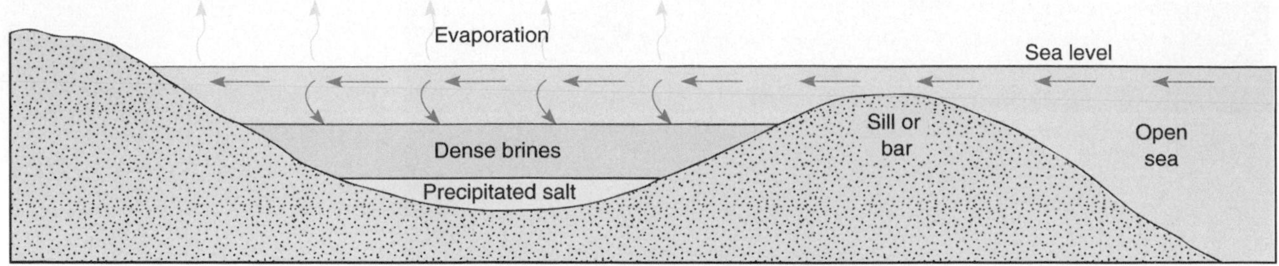

FIGURE 8-18 Cross-section illustrating a model for the deposition of evaporites in a barred basin. Sea water is able to flow more or less continuously into the basin over the partially submerged barrier. It is evaporated to form dense brines. Because of their density, the brines sink and are thus unable to return to the open sea because of the barrier. When the brine has become sufficiently concentrated, salts are precipitated.

brines. The brine would have sunk to the bottom and been prevented from escaping because of the sill or bar (Fig. 8-18). This type of feature is called a **barred basin**.

The Cordilleran Region

During much of early Paleozoic time, large areas of western North America lay beneath sea level. A coastal plain and shallow marine shelf extended westward from the Transcontinental arch. This relatively level expanse was bordered on the west by deep marine basins adjacent to the western continental margin. Many of the marine sedimentary basins of the Cordillera appear to have originated during the Neoproterozoic, when North America (Laurasia) separated from eastern Antarctica as part of the breakup of Rodinia. At that time, another continental block separated from Laurasia, drifted westward, rotated counterclockwise, and was ultimately incorporated into the Siberian platform (Fig. 8-19). Evidence of this relocation of part of California to Siberia consists of similarities in Proterozoic rocks of the two now widely separated segments as well as correspondence in the alignment of structural trends when the two blocks are placed in proper juxtaposition. In addition, there is geophysical evidence of the presence of a failed rift or aulacogen in Southern California. As is characteristic of aulacogens, this one is perpendicular to the possible line of separation between the two land masses.

Clearly, western North America was a passive margin during the early Paleozoic. The Belt Supergroup of Montana, Idaho, and British Columbia as well as the Uinta Series of Utah and the Pahrump Series of California were all deposited in fault-controlled basins formed during the late Neoproterozoic and early Paleozoic.

Some of the grandest sections of Cambrian rocks in the world are exposed in the Canadian Rockies of British Columbia and Alberta (Fig. 8-20). The formations have been erosionally sculpted into some of Canada's most magnificent scenery (see Jasper National Park box). Lower Cambrian rocks include ripple-marked quartz sandstones. These clastic sediments were derived primarily from the Canadian shield. By Middle Cambrian time, seas had transgressed farther

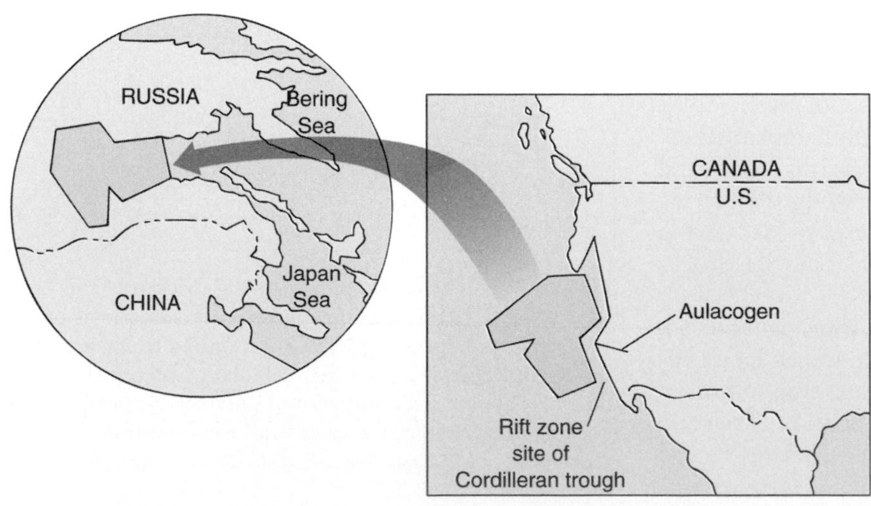

FIGURE 8-19 Development of a passive margin trough along the western edge of North America by detachment of a land mass that ultimately was incorporated into the Siberian Platform of Russia. (*Concept from Sears, J. W., and Price, R. A. 1978. Geology 6:227–270.*)

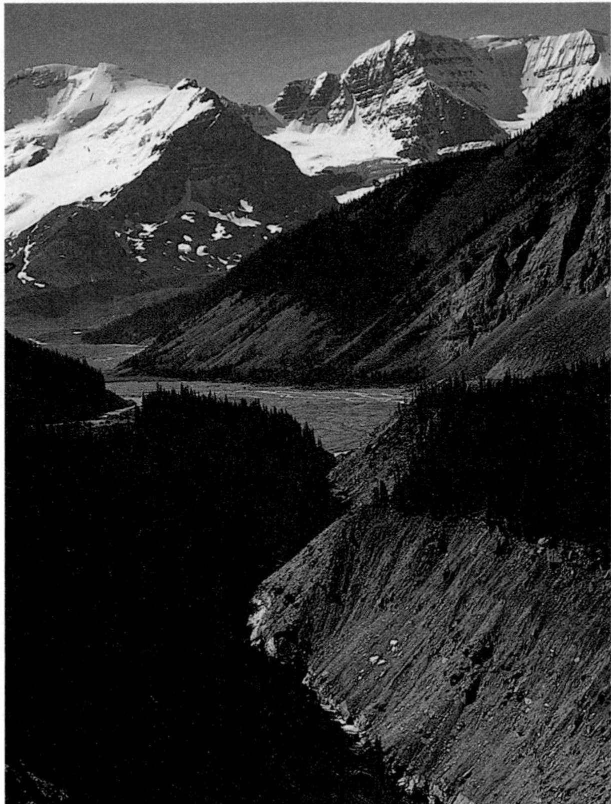

FIGURE 8-20 The Canadian Rocky Mountains in Jasper National Park. The V-shaped valley in the foreground has been cut into glacial deposits. In the middle distance one can see the braided Athabasca River. The mountains in the distance are part of a thrust sheet composed of lower Paleozoic, horizontal sedimentary rocks.

eastward, and shales and carbonates became more prevalent. An interesting section of Middle Cambrian rocks is exposed along the slopes of Kicking Horse Pass in British Columbia near the border with Alberta. One of the units in this section, the Burgess Shale, has excited the interest of paleontologists around the world because it contains abundant remains of Middle Cambrian soft-bodied animals. We will discuss these fossils in Chapter 10, along with other organisms of the early Paleozoic.

Patterns of sedimentation in the Cordilleran region during the Ordovician and Silurian indicate that the western edge of the continent was no longer a passive margin, as was the case in the late Proterozoic and earliest Cambrian. The Pacific plate was now moving against North America, and a subduction zone with an associated volcanic chain had formed (Fig. 8-21). A thick subduction zone complex, consisting of graywackes and volcanics, was laid down in the trench above the subduction zone, whereas east of the volcanoes, siliceous black shales and bedded cherts accumulated. The shales are noted for their rich content of fossil organisms called graptolites (Fig. 8-22). Graptolites are so prevalent in lower Paleozoic rocks that the sediments have been referred to as the "graptolite facies." The associated bedded cherts may have derived their silica by dissolution of particles of volcanic ash and the tiny shells of siliceous marine microorganisms.

Eastward of the subduction zone complex in the Cordillera, one finds a thick sequence of fossiliferous carbonates, shales, and sandstones (Fig. 8-23). These are the rocks of the continental shelf. Because many of the carbonates contain abundant fossils of marine invertebrates, they are sometimes dubbed the "shelly facies."

Silurian deposits of the Cordillera are generally similar to those of the Ordovician. Sediments of the subduction complex can be recognized in southeastern Alaska, California, and Nevada, where they occur in great thrust sheets. Nevada also contains exposures of Silurian carbonates deposited in shallower continental shelf areas.

Deposition in the Far North

The cold and sparsely inhabited northern part of North America is a difficult region in which to carry on geologic field studies. Thus, this region has received less attention than more hospitable parts of the continent. To add to the difficulty, lower Paleozoic outcrops are often

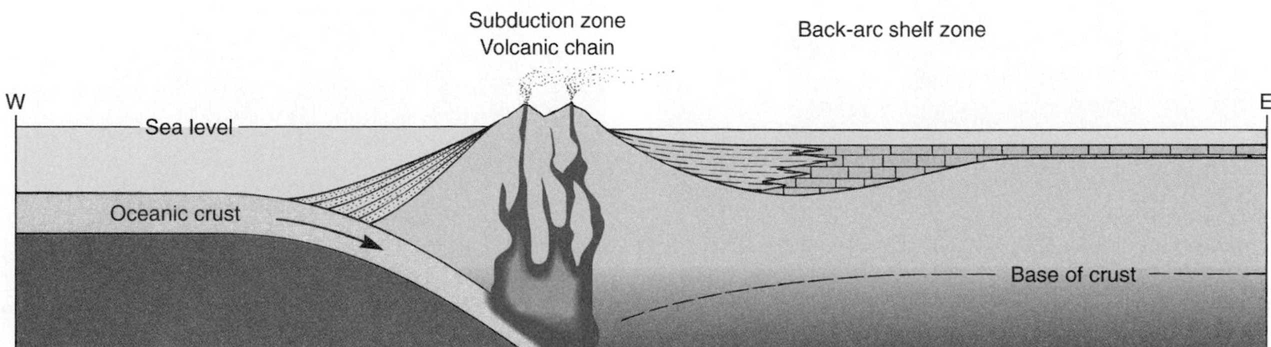

FIGURE 8-21 Interpretive cross-section of conditions across the Cordilleran region during early Paleozoic time.

Jasper National Park

For sheer scenic beauty, few areas in the world rival Jasper National Park. The park lies about 400 kilometers northeast of Vancouver, British Columbia. It extends for nearly 200 kilometers along the eastern slope of the Rocky Mountains. In this region, the Rocky Mountains are the result of erosional sculpting of multiple thrust sheets composed of strata ranging in age from Neoproterozoic to Jurassic. Rather like shingles on the roof of a house, each sheet has been thrust eastward over the one in front of it. As a result, the strata above each of the fault planes are older than those beneath. The bold east-facing cliffs are composed of erosion-resistant Cambrian quartzites and limestones. More easily eroded shaly beds form the more gentle slopes.

View toward the terminus of Athabasca Glacier, British Columbia. The glacier is one of several descending from the Columbia Ice Field.

Both in Jasper National Park and to the south in Banff National Park, the mountains support ice fields and glaciers. The Columbia Ice Field, for example, sends tongues of glacial ice into valleys on several sides. Notable among these is the Athabasca Glacier (Fig. *A*), which is actually on the northernmost tip of Banff National Park. Today's glaciers, however, do not compare at all to those of the Pleistocene, when ice filled all the intermontane basins and only the peaks of mountains protruded from the frigid landscape. As the ice moved slowly toward lower elevations, it carved a spectacular array of scenic features. Among these are hanging valleys, formed where small tributary glaciers, unable to erode their floors as deeply as those of the larger main glaciers, join the larger valleys high above their floors. After melting, the difference in elevation manifests itself as precipices whose majesty is enhanced by waterfalls and cascades. Bowl-shaped re-

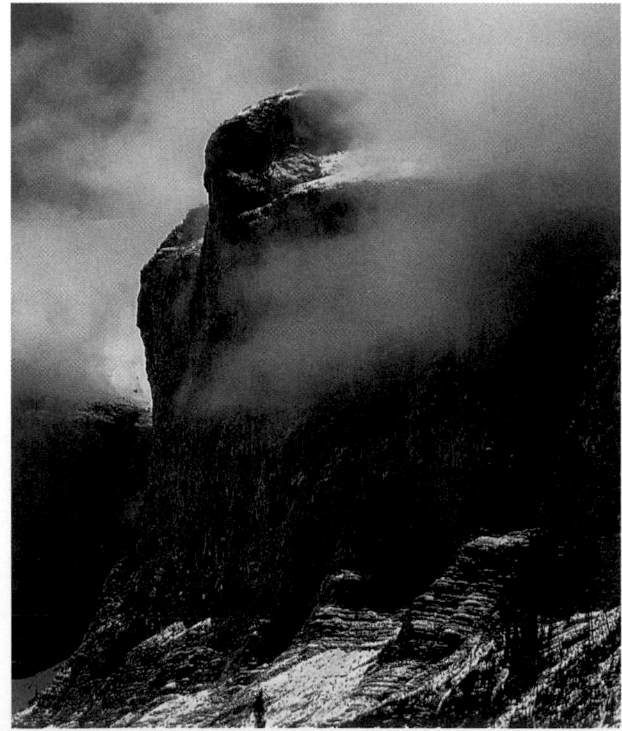

Towering wall of massive Cambrian limestone at the southeastern end of Maligne Lake, Jasper National Park.

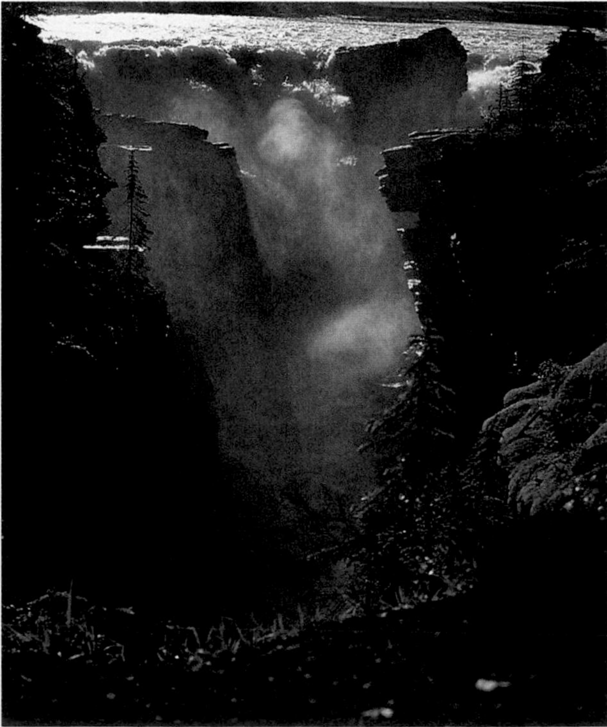

Gorge carved into quartzites of the Cambrian Gog Group by the Athabasca River.

cesses called cirques are seen near the heads of glaciated valleys, as are sharp-edged, often serrate ridges called arêtes. Where three or more cirques have been carved into the sides of a mountain, pointed peaks called horns have been formed.

The rocks sculpted by glacial ice reveal a fascinating geologic history. The first chapter of that history records the dismemberment of the supercontinent Rodinia about 750 million years ago. Following that event, the western margin of North America became a passive margin, receiving clastic sediments from rivers draining the continental interior. These Neoproterozoic sediments can be seen west of the town of Jasper where the Yellowhead Highway follows the Miette River. Appropriately, the rocks have been named the Miette Group. Above the Miette strata one finds shallow-water limestones (Fig. *B*), limy shales, and sandstones (now quartzites) of the Cambrian Gog Group (Fig. *C*). Some of these beds contain fossils of extinct marine arthropods called trilobites, as well as reef-building organisms called archaeocyathids. Brachiopods, bryozoans, and graptolites can be found in overlying Ordovician and Early Silurian strata. Late in the Silurian, however, the warm, shallow sea in which these animals lived receded, and the former sea floor was subjected to erosion. It was not until Late Devonian that the sea returned. The old sea floor now became the site of limestone deposition, and in some areas reefs were constructed by extinct corals, organisms called stromatoporoids, and calcareous algae.

The general pattern of quiet deposition continued in Jasper National Park through the remainder of the Paleozoic and into the Triassic. Subsequently, there were two great collisional events. The first occurred during the Jurassic, when a microplate that formed somewhere out in the Panthalassan Ocean docked against the western edge of Canada, crushing sediments along the continental margin and welding itself to North America. Then, during the Cretaceous, yet another allochthonous mass drifted toward Canada and smashed into the allochthonous terrane that preceded it. The ponderous compressional forces resulting from these collisions caused the sedimentary sequence to detach from underlying Precambrian granites and break into the thrust sheets that are the hallmark of the Rocky Mountains in Jasper National Park.

sparse, covered with younger rocks, and deformed or metamorphosed. Nevertheless, there is evidence that the northern edge of the continent formed in the late Proterozoic, when a continental mass split off from the North American craton. The rifting and displacement were part of the general dismemberment of Rodinia occurring at this time. By Ordovician time, conditions had stabilized enough to permit the development of a carbonate platform on the continental shelf. To the north, the shelf dropped off abruptly, providing a deep-water environment in which turbidities were trapped. The Silurian section in this region is one of the thickest in the world. It contains extensive

FIGURE 8-22 Branches (stipes) of the graptolite *Diplograptus*. *Diplograptus* is also common in dark shales of Ordovician age in both Europe and North America.

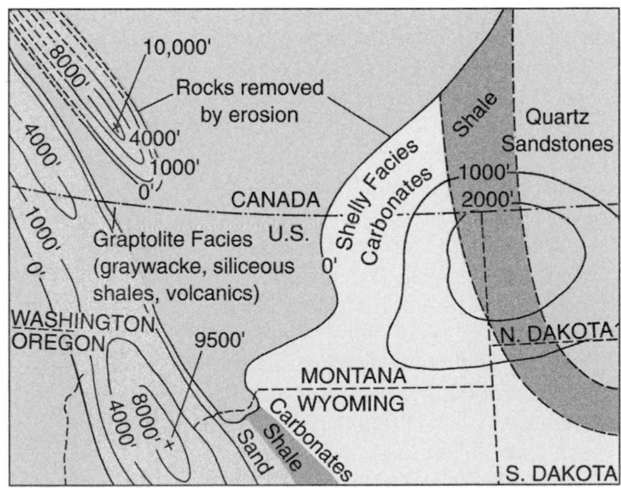

FIGURE 8-23 Ordovician sedimentation in the Montana Williston Basin region. Isopach lines indicate thickness of the Ordovician chronostratigraphic unit. Deep-water sediments reach thicknesses of 9500 feet (2900 meters) in Idaho and 10,000 feet (3080 meters) in British Columbia. (*After Sloss, L. L. 1950. Bull. Am. Assoc. Petrol. Geol. 34:423–445.*)

Shenandoah National Park

Shenandoah National Park in northwest Virginia is an excellent place to view the effects of the many orogenic events that have punctuated the geologic history of the Appalachian region. Much of the viewing can be done from overlooks along Skyline Drive (Fig. *B*), which runs the entire length of the park along the crest of the Blue Ridge Mountains. Seventy-one overlooks along the drive provide spectacular views of the Shenandoah Valley to the west and the Piedmont to the east. For those who prefer to explore on foot, there are 455 miles of trails, including a 95-mile segment of the Appalachian trail that extends from Maine to Georgia.

The oldest rocks in Shenandoah National Park originated during the Proterozoic Eon. These ancient metamorphics consist of gray–green gneisses of the Pedlar Formation. Exposures of the Pedlar rocks can be seen along Skyline Drive at the Tunnel Parking and Buck Hollow overlooks. About 1100 million years ago, an attractive, light-colored, coarsely crystalline granite intruded the Pedlar Formation (Fig. *C*). The intrusive body, named the Old Rag Granite after its exposure on Old Rag Mountain, can be seen from the Thorofare Mountain overlook and nearby trails. As is typical of many other granite bodies, the rock has developed a complex system of joints. Water flowing through the joints has eroded and chemically weathered the granite, slowly widening it and ultimately sculpting many of the separated blocks into huge, rounded boulders. The rounding is a consequence of corners having greater surface area for their volume than edges or faces, so they weather more rapidly (see Fig. *A*).

Erosion following the intrusion of the Old Rag Granite stripped away much of the cover of older, less resistant rocks. The relatively quiet episode of erosion ended near the end of the Proterozoic, when rifting occurred and basaltic lavas rose to the surface along tensional faults. The lavas accumulated to thicknesses in excess of 600 meters. Some of them exhibit pillow structure, indicating that they were extruded beneath the sea. In subsequent orogenic events, these basalts were metamorphosed to become the greenstones that cap many of the mountains in the park. Rifting and volcanic activity continued into the Cambrian Period, as a reflection of the breakup of Proto-Pangea that had begun near the end of the Proterozoic. Gravels, sands, and clays, carried to the rift zone by streams, were deposited along the western margin of the widening sea. Subsequently, when this passive margin became a convergent zone, these sediments were altered to metaconglomerates, phyllites, and quartzites. The rocks constitute formations of the Cambrian Chilhowee Group. Chilhowee quartzites, because of their resistance to erosion, cap many of the ridges in Shenandoah National Park. In addition, the earliest evidence of life in the park consists of tubes and burrows of wormlike organisms discovered in Chilhowee metasediments.

For the remainder of the Paleozoic, Shenandoah National Forest and the Appalachian orogen were subjected to periodic and often severe tectonism. As noted in Chapters 8 and 9, these mountain-making events began with the Ordovician Taconic orogeny and culminated in the great Allegheny orogeny at the end of the Paleozoic. There are a few patches of Mesozoic rocks in the park. These consist primarily of basaltic dikes. One can view these dikes, which have well-developed columnar jointing, from Skyline Drive at Franklin Cliffs and Crescent Rocks. Cenozoic geologic history in Shenandoah National Park has been dominated by erosion and periodic isostatic adjustments.

Spheroidally weathered Old Rag Granite, Shenandoah National Park. (*Photograph by R. F. Dymek.*)

Reference

Gathright, T. M. II. 1976. Geology of Shenandoah National Park, Virginia. Virginia Division of Mineral Resources, Bull. 86, Charlottesville, VA.

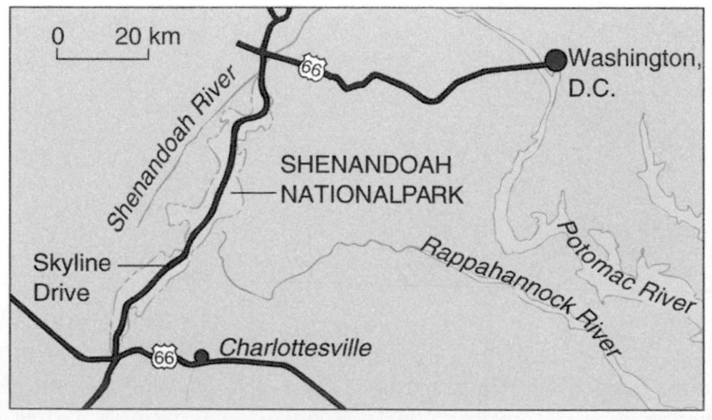

Location map, Shenandoah National Park.

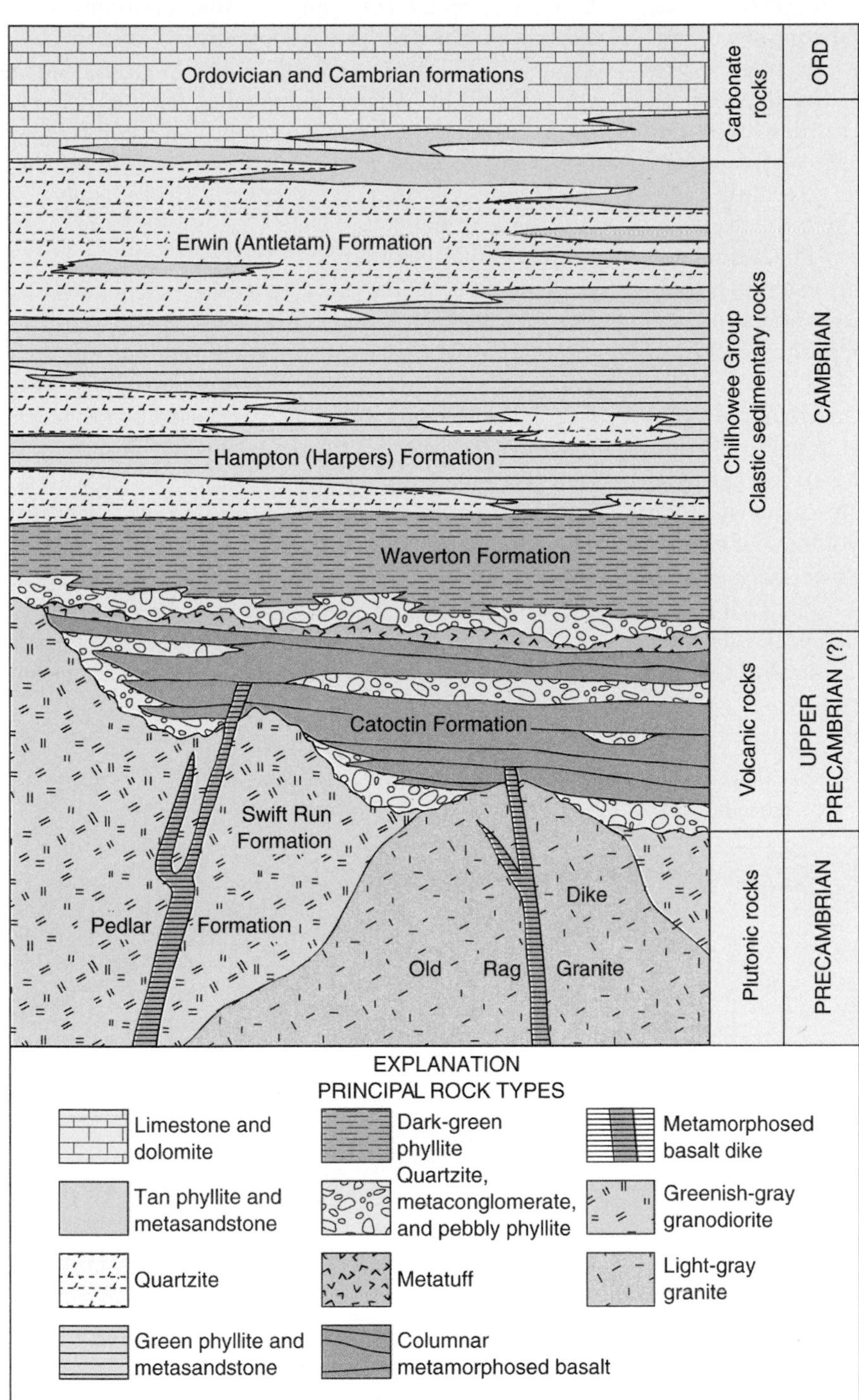

Generalized columnar section of rock formations, Shenandoah National Park. (*From Gathright, T. M. II. 1976. Geology of Shenandoah National Park, Virginia. Virginia Division of Mineral Resources Bulletin 86.*)

reef deposits, some of which stood more than 200 meters above the sea floor. There are thick deposits of evaporites as well. These rocks imply warm conditions, an interpretation strengthened by paleomagnetic data indicating that this presently frigid region lay within 15° of the early Paleozoic equator.

The Eastern Margin of North America

At the beginning of the Paleozoic, the eastern margin of North America was a passive margin, much like it is today. Relatively shallow-water deposits rich in the fossils of marine invertebrates were being spread across the continental shelf, while deeper-water sediments accumulated farther seaward along the continental rise. The initial deposits along the shelf were largely quartz sandstones and shales derived from the stable interior, which was still largely emergent in earliest Cambrian time. Gradually, however, the seas spread inland, and sandy strandlines migrated westward. The initial sandstone deposits of the shelf were soon to be blanketed by thick sequences of carbonate rocks. The carbonates formed a shallow bank or platform extending from Newfoundland to Alabama. Mudcracks and stromatolites indicate the carbonates were deposited at or near sea level.

The rather quiet depositional scenario of the Cambrian and Early Ordovician changed dramatically during the Middle Ordovician (Fig. 8-24). At that time, carbonate sedimentation ceased, the carbonate platform was downwarped, and huge volumes of graptolite-bearing black shales and immature marine sands spread westward over the dolomites. The Ordovician section in New England contains such coarse detritus, as well as volcanic pyroclastics and interbedded lava flows. The rock record indicates the development of a subduction zone with an accompanying volcanic chain along the eastern margin of North America. Apparently, the Proto-Atlantic (also called Iapetus) that had opened during the Neoproterozoic and Cambrian was beginning to close again. That closure resulted in a subduction–volcanic arc complex and ultimately carried the continents that were to compose Pangea into contact with each other.

The pulses of orogenic activity that had begun in the Early Ordovician were followed by several more intense deformational events in Middle and Late Ordovician that constitute the **Taconic orogeny** (Fig. 8-25). The Taconic orogeny was caused by the partial closure of the Iapetus Ocean. During that closure, an island arc, and possibly other terranes, collided with the formerly passive margin of North America. Later orogenic episodes resulted from further compression of opposing plates. In this scenario, Europe and Africa lay opposite and east of North America. An alternate hypothesis suggests that South America lay adjacent to the eastern margin of North America and that a southern

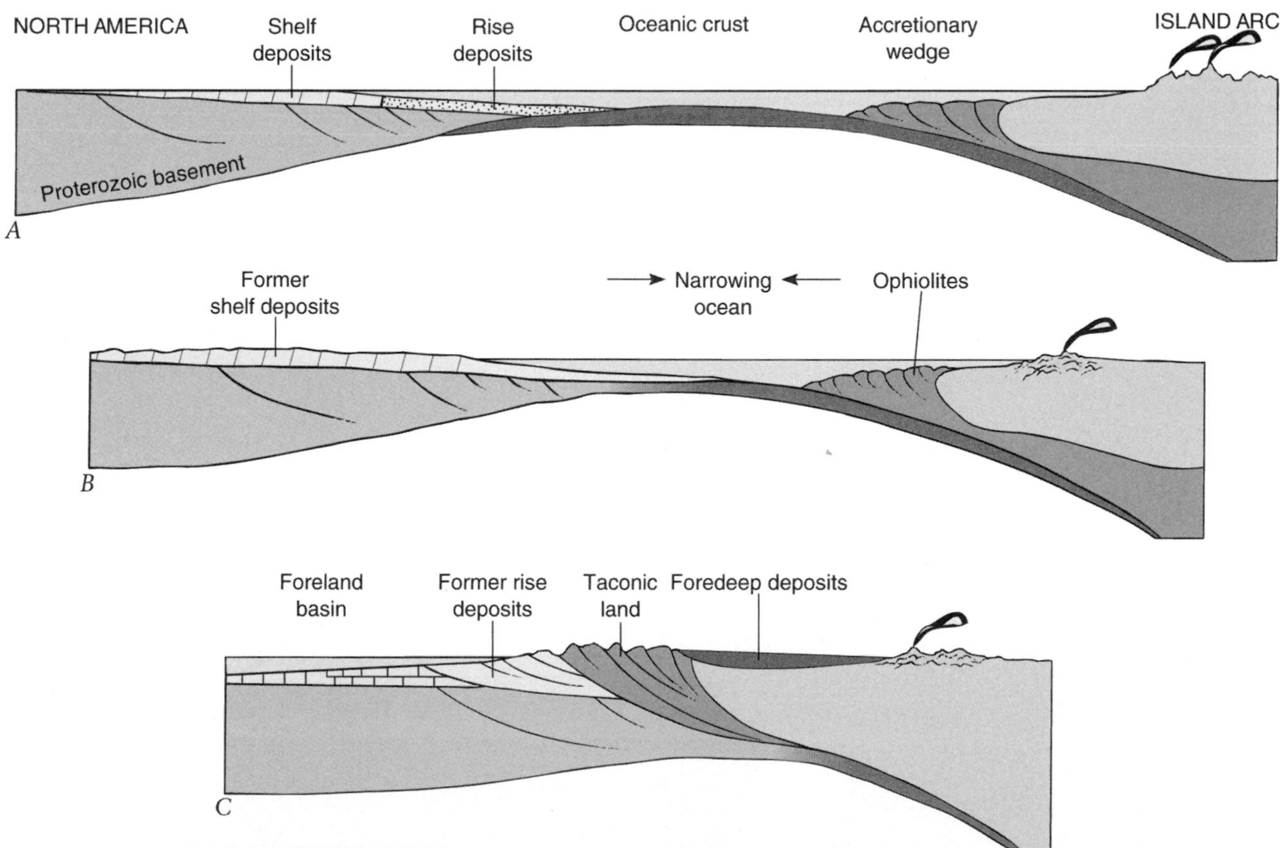

FIGURE 8-24 Plate tectonic forces that resulted in the Taconic orogeny. Following the Neoproterozoic break-up of Rodinia, a passive margin characterized the eastern border of North America (*A*). Subsequently, a large island arc converged on the passive margin and converted it to an orogenic belt with growing mountain ranges (*B* and *C*). (*Adapted from Rowley, D. B., and Kidd, S. F. 1981. J. Geol. 89:199–218.*)

ENRICHMENT

A Colossal Ordovician Ash Fall

The Ordovician succession of marine sedimentary rocks in eastern North America and northwestern Europe contains many widespread ash beds that have been altered to bentonites. Because these ash beds were deposited in what can be termed a geologic instant, they constitute excellent time planes for correlation and provide the basis for a variety of geologic studies that require comparison of sedimentary rocks that are of identical age.

The most extraordinary of the Ordovician ash falls, called the Millburg bed in North America and the Big Bentonite in Europe, is estimated to have thrown an amount of ash into the atmosphere that is equivalent to 1140 km^3 of dense rock if it were pulverized to ash-size particles. The bentonite layer, now compacted, is 1 to 2 meters thick over an area of several million square kilometers. The mineralogy of the ash indicates a single eruption, and its thickness distribution suggests that the parent volcano was centered somewhere between eastern North America and Sweden. So large a volume of ash would have required a sustained eruption over a period of 10 to 15 days. It was clearly the largest ash-producing eruption recorded in all of the past 590 million years, and yet it seems to have had little effect on Ordovician marine organisms. This observation casts doubt on theories that invoke atmospheric dust and ash as a primary cause of global extinctions.

Uranium-lead ages obtained from zircon crystals in the bentonite of the giant ash fall confirm that the beds now on opposite sides of the Atlantic are of the same age and that the eruption occurred 454 million years ago.

Reference

Huff, W. D., Bergstrom, S. M., and Kolata, D. R. 1992. Gigantic Ordovician volcanic ash fall in North America and Europe: Biological, tectonomagmatic, and event-stratigraphy significance. *Geology* 20:875–878.

segment of the Andes was contiguous with the Taconic mountains.

The effects of the Taconic orogeny are most apparent in the northern part of the Appalachians. Caught in the vise between closing lithospheric plates, sediments were crushed (Fig. 8-24), metamorphosed, and thrust northwestward along a great fault. In this fault, named **Logan's thrust**, continental rise sediments have been shoved over and across some 48 kilometers of shelf and shield rocks. Today we are able to view the remnants of this activity in the Taconic Mountains of New York.

FIGURE 8-25 Ordovician shale beds exposed at Black Point, Port au Port, Newfoundland are intensely folded as a result of the Taconic orogeny of Late Ordovician time. (*Courtesy of the Geological Society of Canada, Ottawa.*)

Ash beds, now weathered to a clay called **bentonite**, attest to the violence of volcanism. Masses of granite now exposed in the Piedmont help record the great pressures and heat to which the sediments of the subduction zone complex were subjected. But even if the igneous rocks had never been found, geologists would know mountains had formed, for the great apron of sandstones and shales that outcrop across Pennsylvania, Ohio, New York, and West Virginia must have had their source to the east in the rising Taconic ranges (Fig. 8-26).

From a feather edge in Ohio, this barren wedge of rust-red terrestrial clastics called the Queenston clastic wedge (Fig. 8-27) becomes increasingly coarser and thicker toward ancient source areas (Figs. 8-28 and 8-29). Streams flowing from those mountains were laden with sand, silt, and clay. As they emerged from the ranges, they spread their load of detritus in the form of huge deltaic systems that grew and prograded westward into the basin, often covering earlier shallow marine deposits. The deltaic aspect of many of these deposits accounts for the alternate name of Queenston delta for the Queenston clastic wedge. It has been estimated that over 600,000 km^3 of rock was eroded to produce the enormous volume of sediment in the Queenston clastic wedge. In order to produce this great volume of sediment, the Taconic ranges must have exceeded 4000 meters (13,100 feet) in elevation.

During the Silurian, the most intense orogenic activity seems to have shifted northeastward into the Caledonian belt. Meanwhile, erosion of the Taconic highlands continued. Early Silurian beds are often coarsely clastic, as evidenced by the Shawangunk

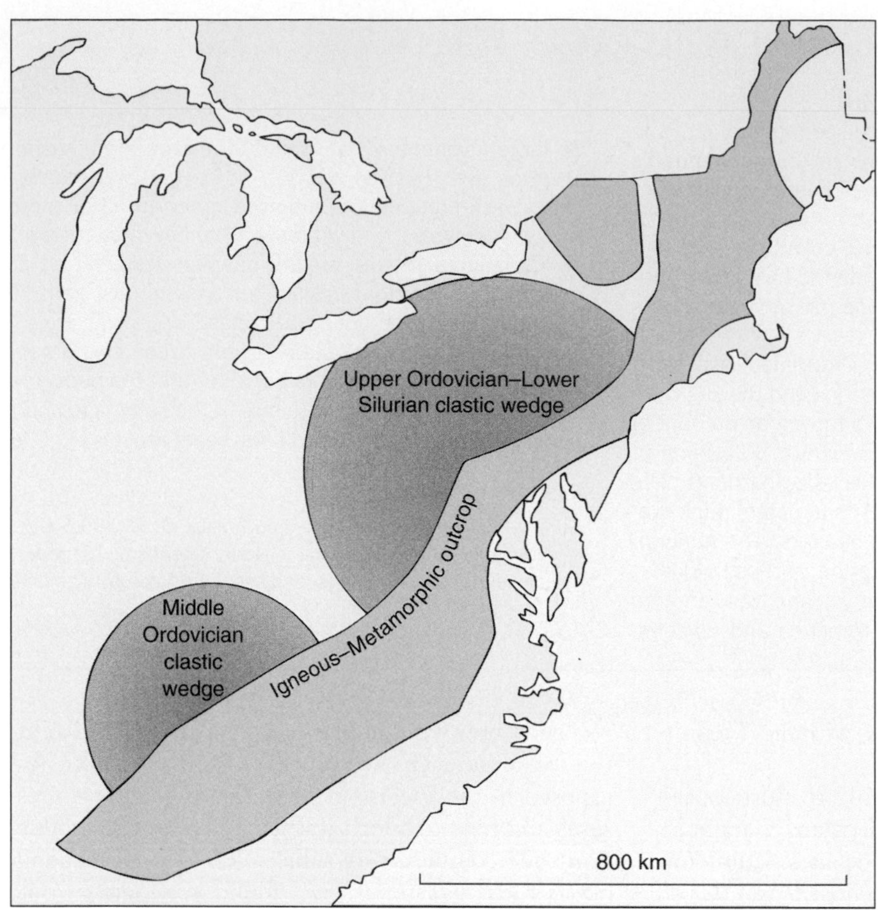

FIGURE 8-26 Great wedges of clastic sediments spread westward as a result of erosion of mountain belts developed during the early Paleozoic. *(After Hadley, J. B. 1964. In Tectonics of the Southern Appalachians. V. P. I., Dept. Geol. Sci. Memoir 1.)*

Conglomerate of New York (Fig. 8-30). These pebbly strata give way upward and laterally to sandstones such as those found in the Cataract Group of Niagara Falls. Silurian iron-bearing sedimentary deposits accumulated in the southern part of the Appalachian region. The greatest development of this sedimentary iron ore is in central Alabama, where there are also coal deposits of Pennsylvanian age. The coal is used to manufacture coke, which is needed in the process of smelting. Limestone, used as blast furnace flux, is also available nearby. The fortunate occurrence of iron ore, coal, and limestone in the same area accounts for the formerly important steel industry of Birmingham, Alabama (Fig. 8-31).

Extending across the southern margin of the North American craton is the Ouachita–Marathon depositional trough (see Fig. 8-30). Although over 1500 kilometers long, only about 300 kilometers of its folded strata are exposed. Additional information about the distribution of early Paleozoic rocks within the belt is derived mostly from oil wells and geophysical surveys.

Overall, nearly 10,000 meters of Paleozoic sediment fill the Ouachita–Marathon trough. However, of this thickness, only about 1600 meters were deposited in the Early Silurian. This indicates more rapid subsidence and deposition during the late Paleozoic. Unfossiliferous metamorphosed graywackes, quartzites, and cherty beds of probable Early Cambrian age are the oldest Paleozoic sediments known from the Ouachita belt. They indicate deposition along the continental rise. The next oldest rocks are Middle Cambrian igneous bodies that are 525 to 535 million years old. These are overlain unconformably by basal sandstones and fossiliferous limestones of platform facies. Ordovician and Silurian deposits in the Ouachita region can be differentiated into deeper water graptolitic facies and shallow water shelly facies of the continental shelf. The deeper water deposits lay to the south, whereas the shallower shelf sediments were deposited along the edge of the craton. Northward, the early Paleozoic section thins as it becomes part of the interior platform. Paleozoic rocks of the Ouachita belt are noted for their unusually siliceous and cherty character. Very likely, the abundant silica was derived from submarine weathering of volcanic ash ejected by volcanoes that lay to the south of the Ouachita trough.

The Caledonides

Scotland's ancient name, still used in poetry, is Caledonia. Eduard Suess used the name for the Scottish remnants of an early Paleozoic mountain range, the Caledonides. It is also the basis for the name of the Caledonian orogenic belt, which extends along the northwestern border of Europe (see Fig. 8-5).

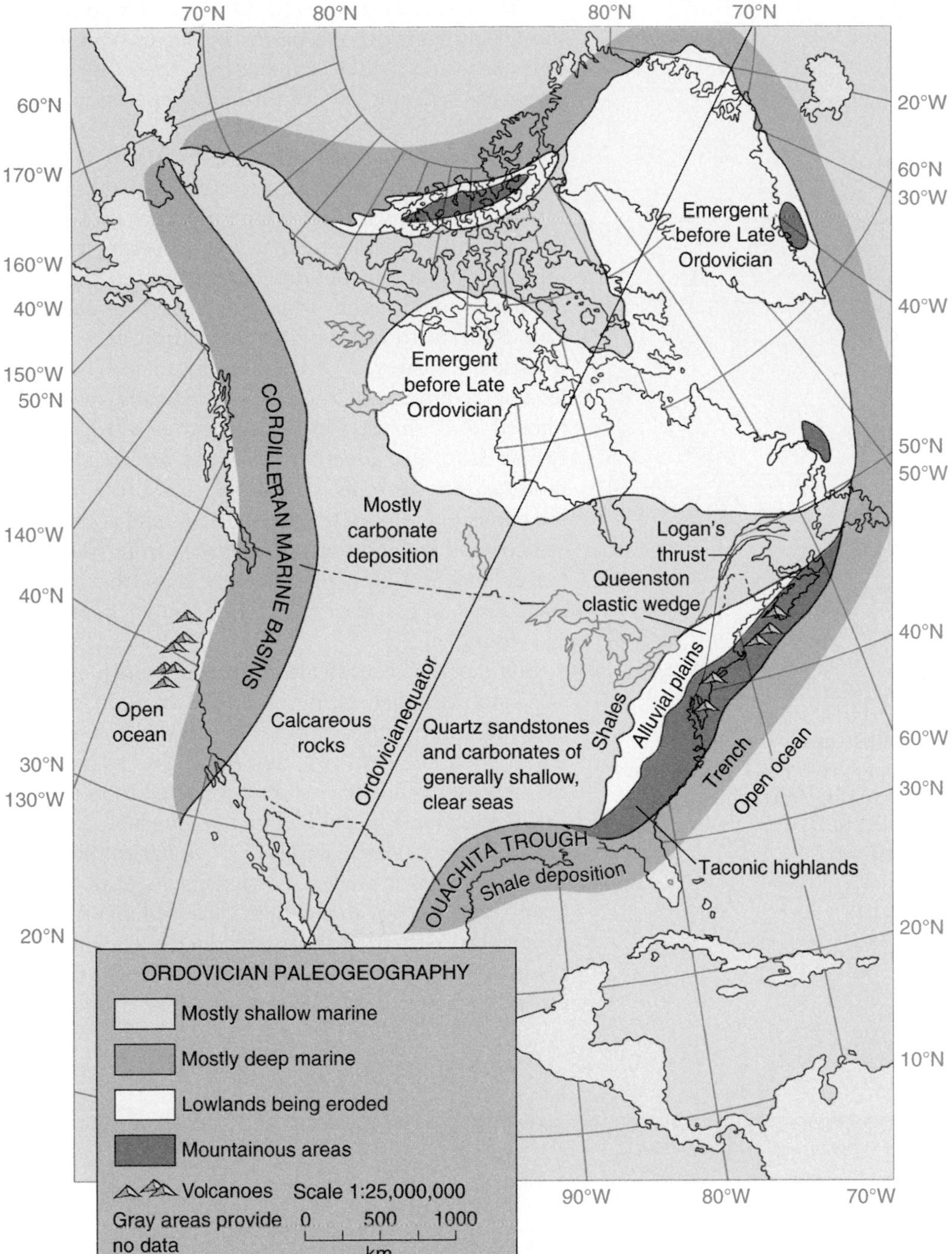

FIGURE 8-27 Paleogeographic map of Ordovician North America.

The Caledonian and Appalachian belts have had a generally similar history. This is not surprising, for both are really part of one greater Appalachian–Caledonian system. Both elongate depositional sites evolved as a result of a cycle of ocean expansion and contraction, such as is schematically illustrated in Figure 8-26. The cycle began with a Late Precambrian to Middle Ordovician episode of sea-floor spreading as the Iapetus Ocean widened to admit new oceanic crust along a spreading center. On the continental shelves and rises of the separating blocks, thick lenses of sediment accumulated. This depositional phase is marked by the development of two distinct facies. The graptolite facies consists of more than 6000 meters of volcanics, graywackes, and graptolitic shales. The graptolites prevalent in this facies are used for subdividing the Ordovician and Silurian into biostratigraphic zones. A shelly facies composed of clean sandstones and fossiliferous limestones identifies the shallower or continental shelf facies. Here and there along the margins of the Caledonian trough there are deposits of Upper Silurian freshwater shales that are noteworthy for their fossil remains of early fishes and strange arthropods known as eurypterids (Fig. 8-32).

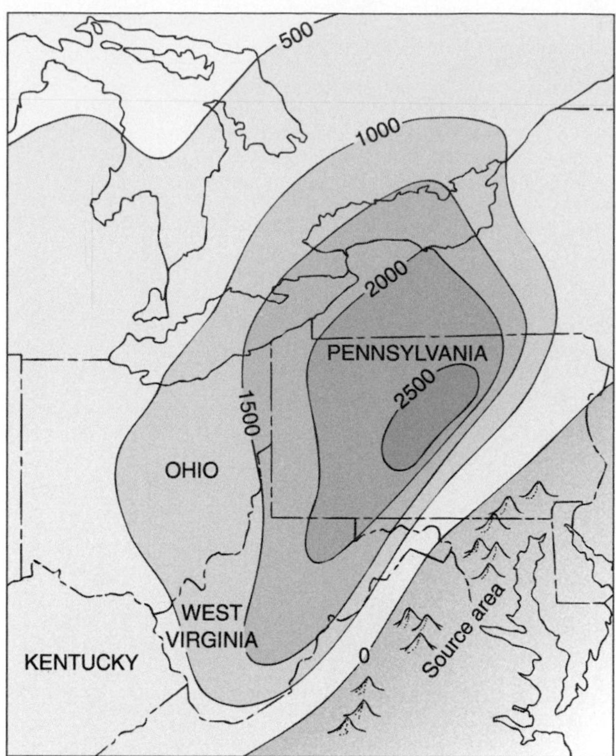

FIGURE 8-28 Isopach map illustrating regional variation in thickness of Upper Ordovician sedimentary rocks in Pennsylvania and adjoining states. *(After Kay, M. 1951. Geol. Soc. Am. Mem. 48.)* ? *Why are there no Upper Ordovician rocks in the area colored pink?*

The closure of the Iapetus Ocean and crumpling of the Caledonian marine basins began in Middle Ordovician time, when subduction zones developed along the margins of the formerly separating continents. This event is recognized from the volcanic rocks that occur in the Canadian Maritime Provinces, northwestern England, northeastern Greenland, and Norway. Little by little, the Iapetus Ocean closed until the opposing continental margins converged in a culminating mountain-building event termed the Caledonian orogeny. The orogeny reached its climax in Late Silurian to earliest Devonian time. It was most intense in Norway, where Precambrian and lower Paleozoic sedimentary rocks are drastically metamorphosed and thrust-faulted. Southwestward, in the British Isles, the effects were not as severe, although mountainous terrains also developed there. Today, the truncated folds of the Caledonians end at the western coast of Ireland but can be traced to northeastern Greenland and Spitzbergen and then the Acadian structures across the Atlantic in Nova Scotia and Newfoundland.

South of the unruly Caledonian belt there is evidence of a land mass that during Late Silurian and Devonian received sandy detritus from the growing Caledonides. This region, on which was deposited clays and sands of the Old Red Sandstone, is often referred to as the Old Red Continent (Fig. 8-33). The deposits here are in the form of a thick, clastic wedge very similar to the earlier Queenston clastic wedge. Fossil remains of plants and fresh-water fish are abundant in these sediments. Hugh Miller (1802–1856) was intrigued by these fossils while laboring in a quarry near Cromarty, Scotland. His interest led to a career as a literary geologist. Miller's *The*

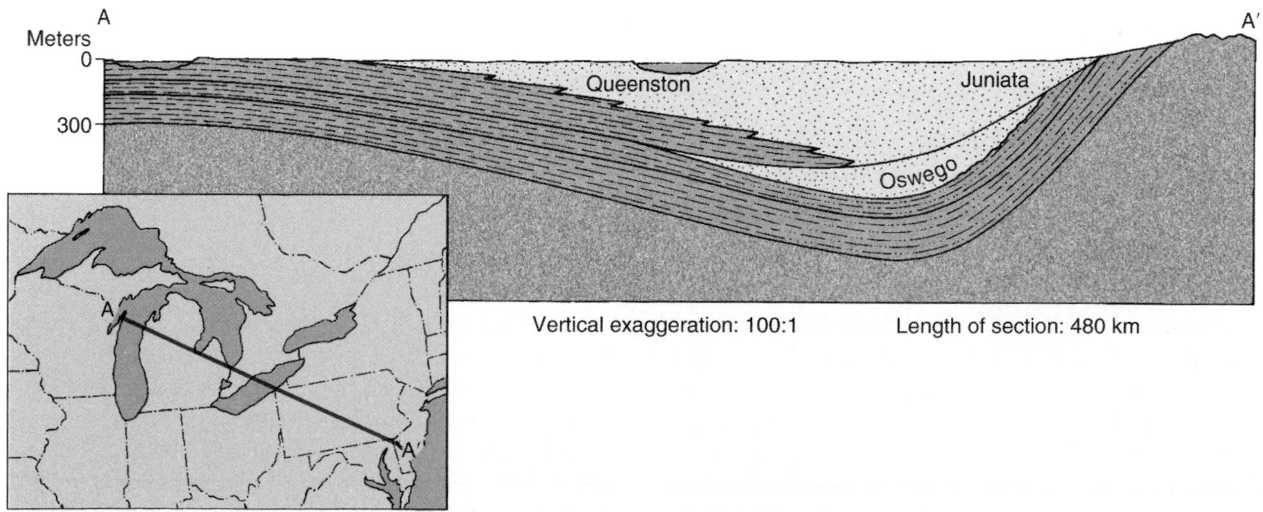

FIGURE 8-29 Restored section of Upper Ordovician rocks from the Atlantic to a point northeast of Lake Michigan. The interpretation given to the cross-section is that a rising highland area to the east supplied clastic sediments to the basin until it filled. Continued sedimentation forced the retreat of the sea westward and extended a clastic wedge toward the west.

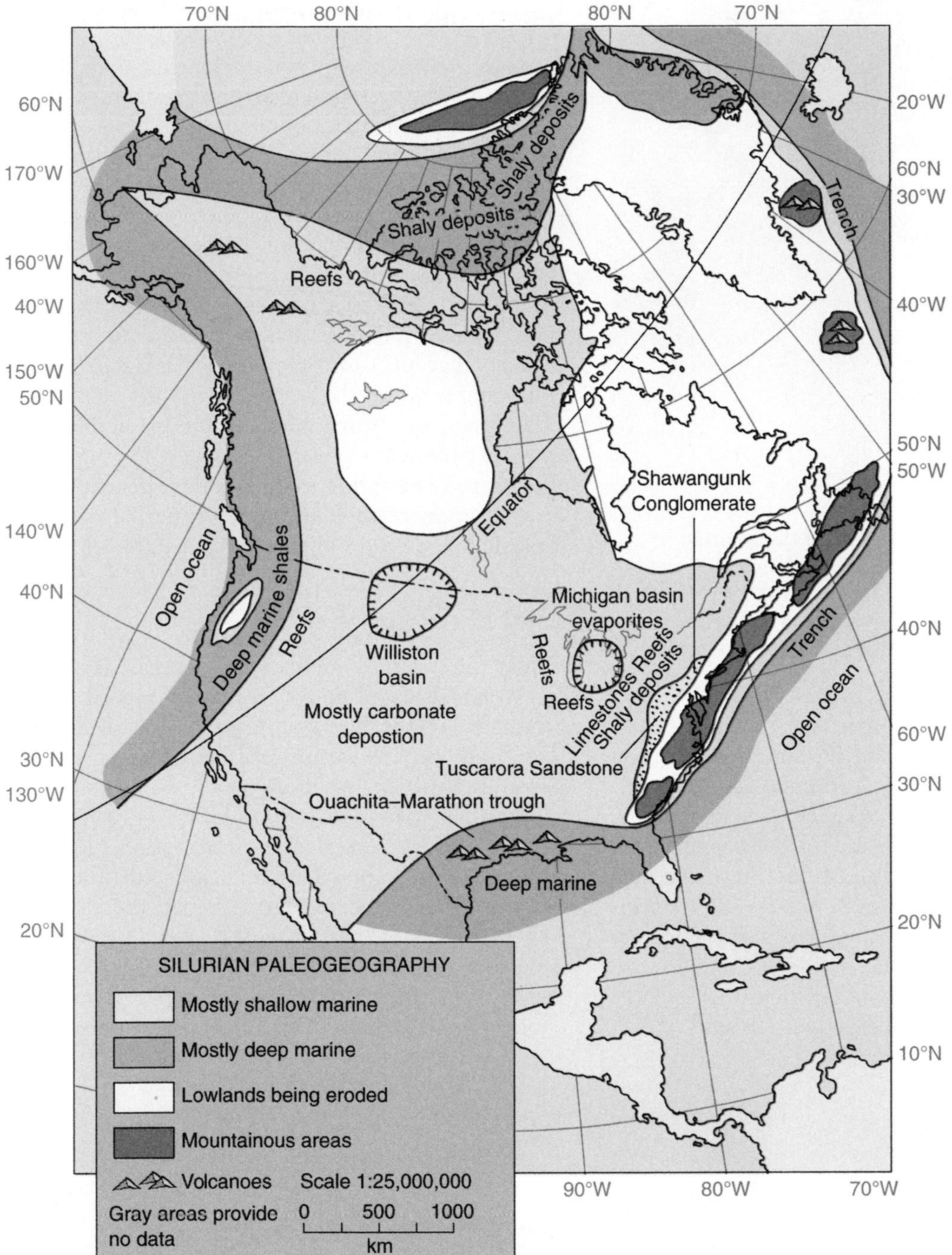

FIGURE 8-30 **Paleogeographic map of Silurian North America.**

Old Red Sandstone and *Footprints of the Creator* were "best sellers" in 19th-century Britain.

The Uralian Seaway

The elongate seaway in which Lower Paleozoic rocks of the Urals were deposited extended southward from the Russian Arctic along the course of the present Ural Mountains. The seaway separated the Russian platform on the west from the Siberian shield on the east. The western side of the Uralian belt contains early Paleozoic shelf carbonates, sandstones, and shales that thin toward the west and pinch out against the Russian platform (Fig. 8-34). Deep marine conditions characterized the eastern side of the Uralian tract. Here we find turbidites, shales, and volcanics.

Although the Uralian orogenic phase was not to come until the late Paleozoic, folded strata, unconformities, and mafic intrusions attest to plate convergence and subduction in the region. The history of the Urals seems to have followed an ocean expansion and contraction cycle (that is, a Wilson cycle) in some ways similar to that of the Caledonian belt. Paleomagnetic studies indicate that the Russian and Siberian platforms

FIGURE 8-31 Clinton iron ore from the Silurian Clinton Group near Birmingham, Alabama. The ore is an oölite of the iron oxide mineral hematite. The oöids and shells of marine fossils have been replaced by iron oxide. Nearby coal beds provide the fuel needed to produce steel from the iron ore.

were separated by a wide oceanic tract during the early Paleozoic. In the Silurian and Devonian, subduction zones developed between the two opposing continental masses. As oceanic crust was being subducted, the intervening ocean narrowed, and the continents began to converge on one another (see Middle Silurian in Fig. 8-35). By the end of the Paleozoic, and possibly continuing into the Triassic, the former depositional belt experienced its orogenic climax as the two continental blocks collided (see early Late Permian in Fig. 9-1). They have remained welded together ever since.

Asia Begins to Take Form

Asia east of the Urals is a very ancient and complex terrane formed by accretion of several crustal blocks. The sutures along which the continents collided are marked by ophiolite belts, narrow zones of high-pressure metamorphism, and volcanic rocks that characterize crustal tracts formed above subduction zones that descend at continental margins. Some geologists believe that as many as nine separate crustal blocks or microcontinents may have combined to form Asia. In the early Paleozoic, faunal and paleomagnetic evidence suggests that most of these blocks were separated by considerable widths of oceanic crust. By the Middle Ordovician, narrowing of these oceans and convergence of continental crustal blocks were in progress, although major episodes of collisional orogenesis did not occur in most regions until late in the Paleozoic and Triassic.

Southern Hemisphere Events

Unlike the stratigraphic record of the early Paleozoic for North America and Europe, the record for the Southern Hemisphere land masses is relatively poor. This may be the result of the continents being prevailingly above sea level, which would favor erosion rather than deposition. However, there is a fairly complete record of sedimentation along the north–south-trending Tasman belt of Australia (Figs. 7-18 and 8-36). Sedimentation began there just before the Paleozoic and provided a sequence of Neoproterozoic sandstones that grade into overlying Lower Cambrian formations without erosional unconformity. The relatively quiet period represented

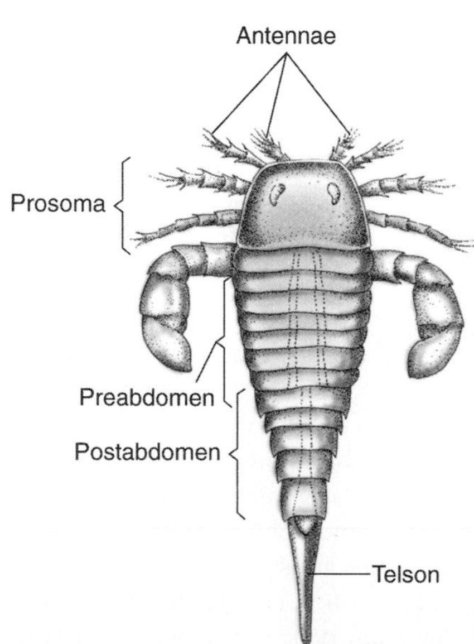

FIGURE 8-32 Two genera of eurypterids. *Eurypterus* (*A*) is noted for its broad, flipperlike paddles and blunt frontal margin. *Pterygotus* (*B*) is distinguished by a pair of formidable-looking frontal pincers. The animal swimming in the center background is a primitive jawless fish. (*Diorama courtesy of the National Natural History Museum, Smithsonian Institution.*)

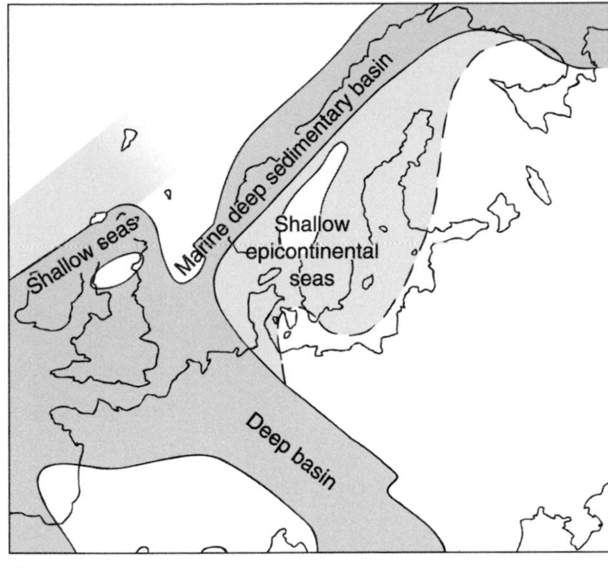

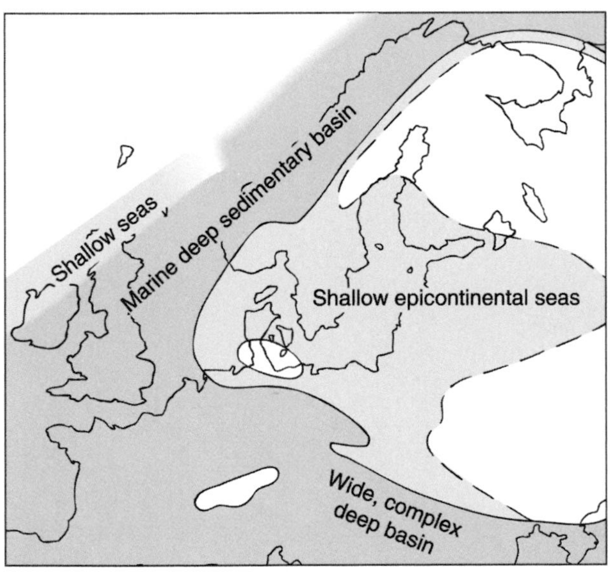

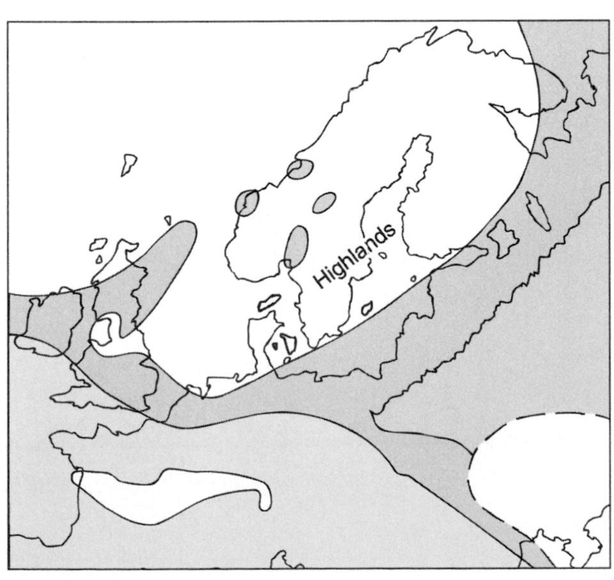

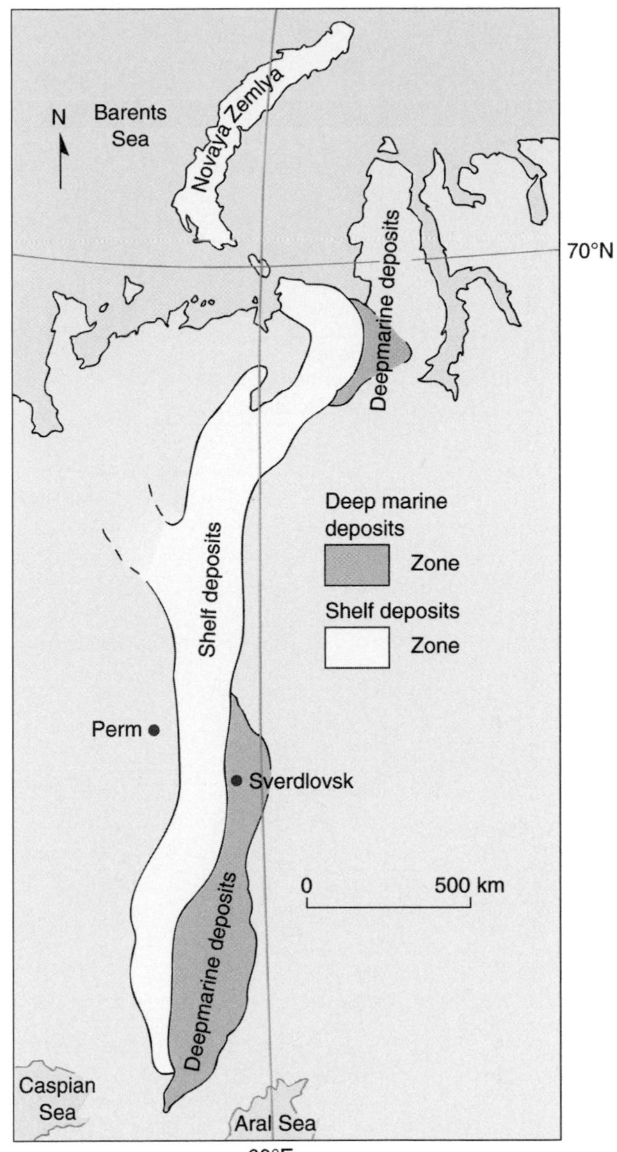

FIGURE 8-34 **The Uralian seaway.** (*Adapted from Read, H. H., and Watson, J. 1975. Introduction to Geology. New York: Halstead Press.*)

by these sediments was followed by a full-scale orogeny in the Late Cambrian. Thick sections of deformed graywackes and volcanics record the orogenic disturbances. During the Ordovician, a broad-shelf sea spread across central Australia and received epeiric sea deposits of well-sorted sandstones and limestones.

The African record of the early Paleozoic is far more obscure than that of Australia. The most complete

FIGURE 8-33 **Paleographic maps of Europe during the (*A*) Cambrian, (*B*) Ordovician–Silurian, and (*C*) Devonian.** (*After Brinkman, R. 1969. Geologic Evolution of Europe. Stuttgart: Enke; New York: Hafner Publishing Co.*)

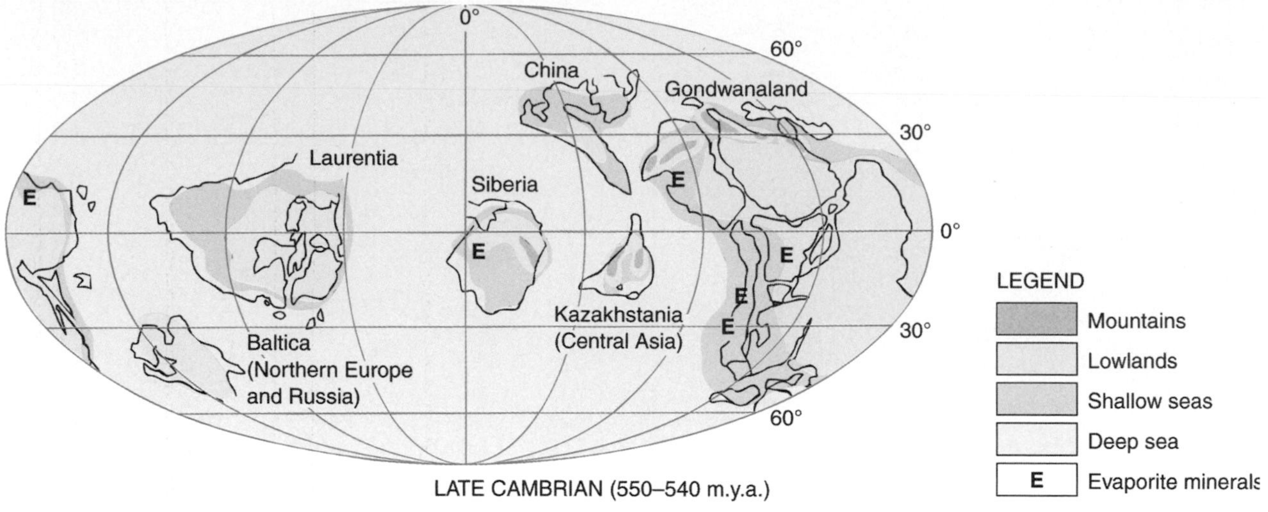

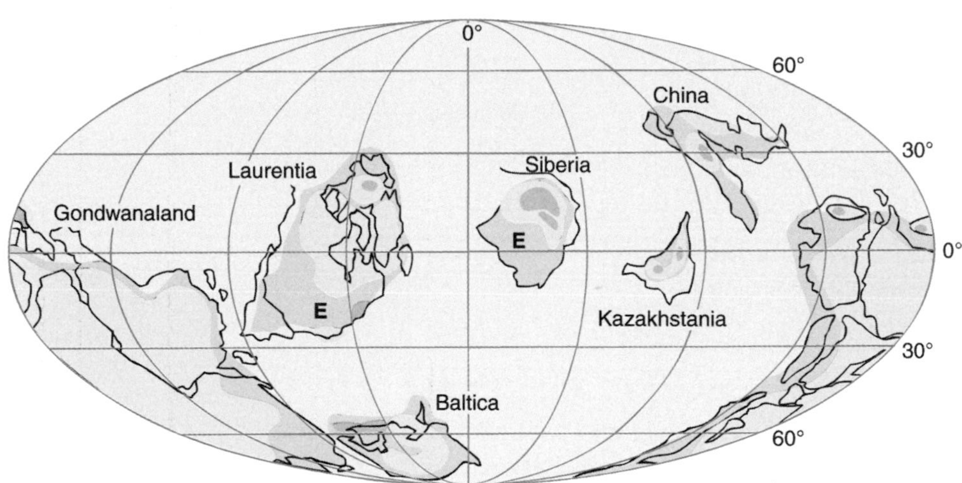

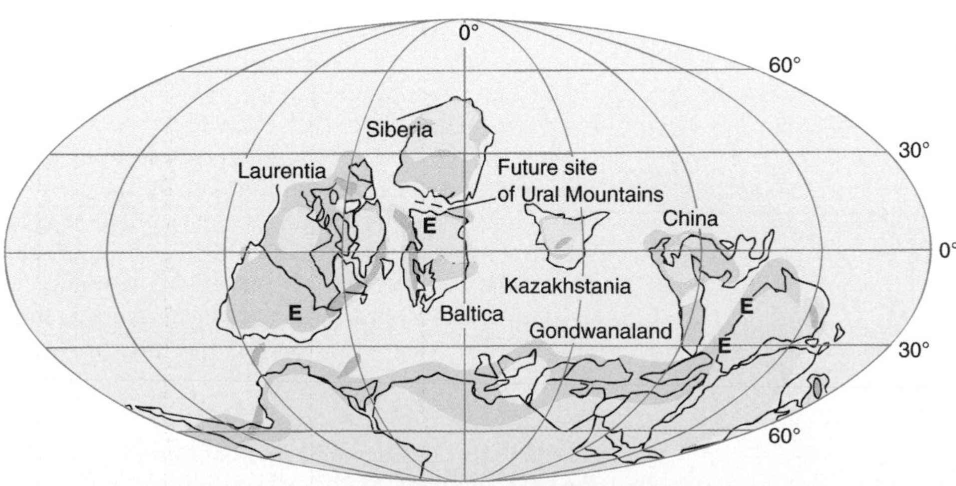

FIGURE 8-35 Global paleogeography of the early Paleozoic. *(From Scotese, C. R., Bambach, R. K., and Barton, C., et al. 1979. J. Geol. 87(3):217–277.)*

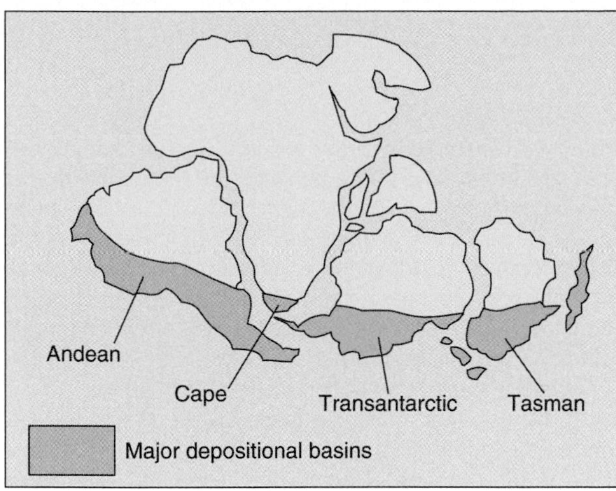

FIGURE 8-36 Principal depositional basins of Southern Hemisphere continents during early Paleozoic time.

sequences of strata are located north of the equator, where one finds Cambrian through Silurian rocks that were laid down in shallow embayments of an east-west trending seaway. Africa south of the equator was persistently above sea level and was being eroded during most of the early Paleozoic.

One feature in the geologic record of the northern part of Africa is rather astonishing. There is remarkable evidence of an Ordovician episode of glaciation in the area that is now the Sahara Desert. As indicated by paleomagnetic data, this part of Africa was close to the South Pole during the Ordovician.

Lower Paleozoic rocks in South America are also patchy in distribution and difficult to interpret. Strata deposited along the Andean tract have been greatly deformed and frequently obliterated by Mesozoic orogenic activity. Elsewhere, the Paleozoic sequence lies buried beneath an immense pile of sediment spread eastward from the eroding Andes Mountains. We do, however, have scattered bits of the record. We know that Cambrian and Ordovician seas occupied the western margin of South America and that graptolitic shales and shelly limestones were deposited there. Platform deposits were also laid down in shallow basins within the present Amazon region. In Ordovician rocks of Argentina, we find boulder beds and faceted cobbles that are products of continental glaciation.

ASPECTS OF EARLY PALEOZOIC CLIMATES

During the early Paleozoic, the Earth was characterized by latitudinal and topographic variations in climate, just as today. As we have seen, during episodes of great marine transgressions and low-lying terrains, climates were probably somewhat milder; during those periods when continents stood high and great ranges diverted atmospheric circulation patterns, climates were more diverse and extreme. However, there were also factors that would have made early Paleozoic climate rather unique. During the early Paleozoic, the Earth turned faster and the days were shorter. Tidal effects were stronger, and until the Late Ordovician, there were no vascular plants to absorb the sun's radiation. There may even have been changes in solar radiation or atmospheric composition that left no unambiguous cues. The total climatic effects of these possibilities can only be surmised.

At the very beginning of the Cambrian, global climate was probably somewhat cooler than the average for the entire early Paleozoic. The Earth was still recovering from the great Neoproterozoic ice age. Soon, however, climates became warmer. Thick fossiliferous limestones and extensive coral reefs, which today form only in the warmer regions of the ocean, reflect the warmer conditions. Paleogeographic reconstructions, such as those shown in Figure 8-34, suggest that North America, Europe, and even Antarctica were positioned astride or near the equator during the early Paleozoic. In addition, the study of cross-bedding in sandstones indicates that eastern North America lay in a zone of trade winds that blew from northeast to southwest with respect to the present directional grid. These winds thus moved in a direction opposite to those prevalent in eastern North America today.

There were severe as well as equable climates during the early Paleozoic. As noted above, extensive Late Ordovician glacial deposits in the region of the present Sahara Desert attest to frigid conditions. These deposits are sufficiently widespread that they must have been emplaced by continental glaciers and certainly must have been accompanied by a lowering of mean annual temperatures at middle and high latitudes. Study of the distribution of the glacial deposits, directions of glacial striations, and paleomagnetic data indicates that in Ordovician time the north African part of Gondwanaland was located over the South Pole.

During the Silurian, there were latitudinal variations in climate somewhat similar to those of today. Glacial deposits of this age are found only in higher latitudes, usually in excess of 65° from the paleoequator. Coral reefs, evaporites, and desert sandstones are found within 40° of the Silurian equator. Certainly there were regions of marked aridity during the Silurian. Epeiric seas became relatively less extensive as the Silurian drew to a close. We can speculate that the Iapetus Ocean formed by the division of Rodinia had already begun closing, and that the Caledonian crunch was about to take place. The curtain was ready to rise on the world of the late Paleozoic.

ENRICHMENT

The Big Freeze in North Africa

The news media inform us of critical environmental problems almost every day. We are ourselves to blame for many of these problems, including those caused by pollution, acid rain, deforestation, and modification of natural water systems. Yet environmental and ecological crises have occurred many times in the geologic past, long before there were humans to generate them. For example, consider the effects on life of the Ordovician Ice Age described in this chapter. We have evidence that the South Pole at this time was located in what is now a barren area between Algeria and Mauritania. Evidence of an enormous ice sheet in this region includes glacial grooves and striations, outwash plains, moraines, and meltwater channels that extend across thousands of square kilometers. There must have been a truly enormous buildup of continental ice, and that accumulation would surely have resulted in cooler temperatures in ocean regions far to the north. Eustatic and climatic changes would have had a pervasive effect on life. Indeed, paleontologists have long recognized that the Late Ordovician was marked by extinctions of large numbers of marine invertebrates, including entire families of bryozoans, corals, brachiopods, sponges, nautiloid cephalopods, trilobites, crinoids, and graptolites. Twenty-two percent of all families of known animals were wiped out. It was one of the greatest episodes of mass extinction in the geologic record.

It seems likely that the demise of these families of organisms resulted from general cooling of tropical seas, as well as the draining of epeiric and marginal seas as sea level was lowered. The loss of the shallow-water environments in which early Paleozoic invertebrates thrived and the crowding of species into remaining habitats would explain this Ordovician ecological disaster.

MINERAL RESOURCES OF THE LOWER PALEOZOIC

The lower Paleozoic systems contain several kinds of Earth materials that have economic importance. Among these are building stones, including slates once extensively quarried in New Jersey and Pennsylvania, and marbles derived from Ordovician limestones. Lower Paleozoic marble is still quarried in New England. The St. Peter Sandstone (see Fig. 8-12) has been and continues to be an important source of clean quartz for use in the manufacture of glass. As noted earlier, salt is also an important mineral resource of the lower Paleozoic (Silurian). In New York State, salt recovery and marketing are a major industry. The method used in recovery involves pumping water down wells into salt-bearing formations. The water dissolves the salt, which is then pumped to the surface to be reprecipitated in evaporating pans and refined.

The most notable ore deposits of the lower Paleozoic are sedimentary iron ores that occur in Newfoundland and in a band extending southward from New York into Alabama. At Birmingham, these iron ore beds were once the basis for a thriving steel industry. However, most of the mines are now closed. Nevertheless, a large amount of ore remains, and it may become economically feasible to reopen mines at some future date. Ores of lead and zinc have been mined in Missouri for many decades. Many of the early Paleozoic deposits, however, are now largely depleted, although exploration for new deposits continues. Petroleum has been recovered from Ordovician and Silurian rocks in Ohio, Oklahoma, and Texas, as well as in western Europe.

SUMMARY

The Paleozoic Era began about 540 million years ago. It began at a time when the supercontinent Rodinia had begun to pull apart, allowing oceanic tracts to open between separating land masses. The Iapetus Ocean was an early Paleozoic proto-Atlantic ocean that was widening at the beginning of the Cambrian but was to close during later Paleozoic as a result of convergence of tectonic plates bearing North America, western Europe, and Gondwanaland.

Most of the North America craton was above sea level and experiencing erosion at the beginning of the Paleozoic. The initial sites of marine deposition were along the rifted continental margins. Gradually, however, the seas spilled out of these marginal basins onto the continental platform. The sediments deposited on the floors of epeiric seas that covered the platform record cycles of transgression and regression. The thicker sections are preserved in basins. Thinner sequences with many unconformities developed over domes. The Sauk and Tippecanoe cratonic sequences record the transgressive–regressive cycles of the early Paleozoic. For much of the early Paleozoic, the transcontinental arch was a positive area, often above sea level, providing sediment to surrounding seas.

By Ordovician time, the relative calm that characterized the Cambrian was interrupted by a plate convergence and mountain-building episode called the Taconic orogeny. Periodic crustal disturbances associated with this orogeny raised mountains along the eastern margin of the North American craton. Huge volumes of detritus produced by rapid erosion of the mountains were spread westward by streams and deposited in a great apron of detrital sediment called the Queenston clastic wedge. Similar mountain-building activity was also underway in northwestern Europe, where the Caledonian depositional basin was compressed, intruded, and uplifted in the early stages of the Caledonian orogeny. The Taconic, Acadian, and Caledonian orogenies were all the result of plate convergence associated with the narrowing of the Iapetus Ocean.

The initial deposits of the early Paleozoic in western North America were continental shelf and rise deposits of the rifted continental margin. This early episode of relatively quiet sedimentation was interrupted during the Ordovician and Silurian by movement of the Pacific plate against the western edge of North America. The result of this movement was the development of a subduction zone and chain of volcanic islands along the western margin of the continent.

In general, the range of climatic conditions during the early Paleozoic was not significantly different from that of the recent geologic past. Cold climates are indicated by the presence of Ordovician glacial deposits in Southern Hemisphere continents. Warmer conditions prevailed in regions where reefs were extensive and richly fossiliferous limestones were deposited. Aridity is indicated by the immensely thick sections of evaporites in the region around the Great Lakes. Although the range of climates may not have been greatly different from that of today, there were many factors in the environment that do not exist today. In the early Paleozoic, the days were shorter and tides were stronger, and until the Late Ordovician, the lands had no cover of green plants.

QUESTIONS FOR REVIEW AND DISCUSSION

1. What are the names and locations of the major orogenic events of the early Paleozoic, and when did each occur?

2. What geologic evidence indicates that the transcontinental arch was above sea level during most of the early Paleozoic?

3. What is a barred basin? Why is a barred basin a particularly effective place for precipitation of evaporites? Where and when did such basins develop during the early Paleozoic?

4. What is a clastic wedge? Are clastic wedges primarily marine or nonmarine? Under what circumstances do they develop? What clastic wedge is associated with the Taconic orogeny?

5. Why are the unconformities that form the boundaries of cratonic sequences considered the result of eustatic lowering of sea level rather than local tectonic uplift of parts of the craton?

6. Why is the unconformity that separates the Sauk from the Tippecanoe not the same age at widely separated locations?

7. How does the history of the Cordilleran region differ from that of the Appalachian region during the early Paleozoic?

8. Domes and basins are characteristic structural features of the central stable regions of North America. Define each and explain how they differ in thickness of strata and unconformities. How might one distinguish between these two structures on a geologic map?

9. The guide fossil for the base of the Cambrian is a trace fossil (*Phycodes*). What are trace fossils?

10. What evidence might you seek in the field to confirm the following?

- **a.** the location of the paleoequator
- **b.** former conditions of extreme aridity
- **c.** former extensive episodes of glaciation
- **d.** mountain building associated with tectonic plate collision

11. Describe the motion of tectonic plates during the Taconic, Caledonian, and Acadian orogenies.

READINGS

Bambach, R. K., Scotese, C. R., and Ziegler, A. M. 1980. Before Pangaea: The geographies of the Paleozoic world. *Am. Scientist* 68(1):26–38.

Cowie, J. W., and Brasier, M. D. (eds.). 1989. *The Precambrian–Cambrian Boundary*. Oxford, England: Clarendon Press.

Dalziel, I. W. D., Salda, L. H. D., and Gahagan, L. M. 1994. Paleozoic Laurentia–Gondwana interaction and the origin of the Appalachian–Andean mountain system. *Geol. Soc. Am. Bull.* 106:243–252.

Dalziel, I. W. D. 1995. Earth before Pangea. *Sci. Am.* 272(1):58–63.

Frazier, W. J., and Schwimmer, D. R. 1987. *Regional Stratigraphy of North America*. New York: Plenum Press.

Landing, E. 1994. Precambrian–Cambrian boundary global stratotype ratified and a new perspective on Cambrian time. *Geology* 22(2):179–182.

Sloss, S. L. 1963. Sequences in the cratonic interior of North America. *Geol. Soc. Am. Bull.* 74:93–111.

Yorath, C., and Gadd, B. 1995. *Of Rocks, Mountains, and Jasper*. Toronto: Dundurn Press.

WEB SITES

The Earth Through Time Student Companion Web Site (www.wiley.com/college/levin) has online resources to help you expand your understanding of the topics in this chapter. Visit the Web Site to access the following:

1. Illustrated course notes covering key concepts in each chapter;
2. Online quizzes that provide immediate feedback;
3. Links to chapter-specific topics on the web;
4. Science news updates relating to recent developments in Historical Geology;
5. Web inquiry activities for further exploration;
6. A glossary of terms;
7. A Student Union with links to topics such as study skills, writing and grammar, and citing electronic information.

9

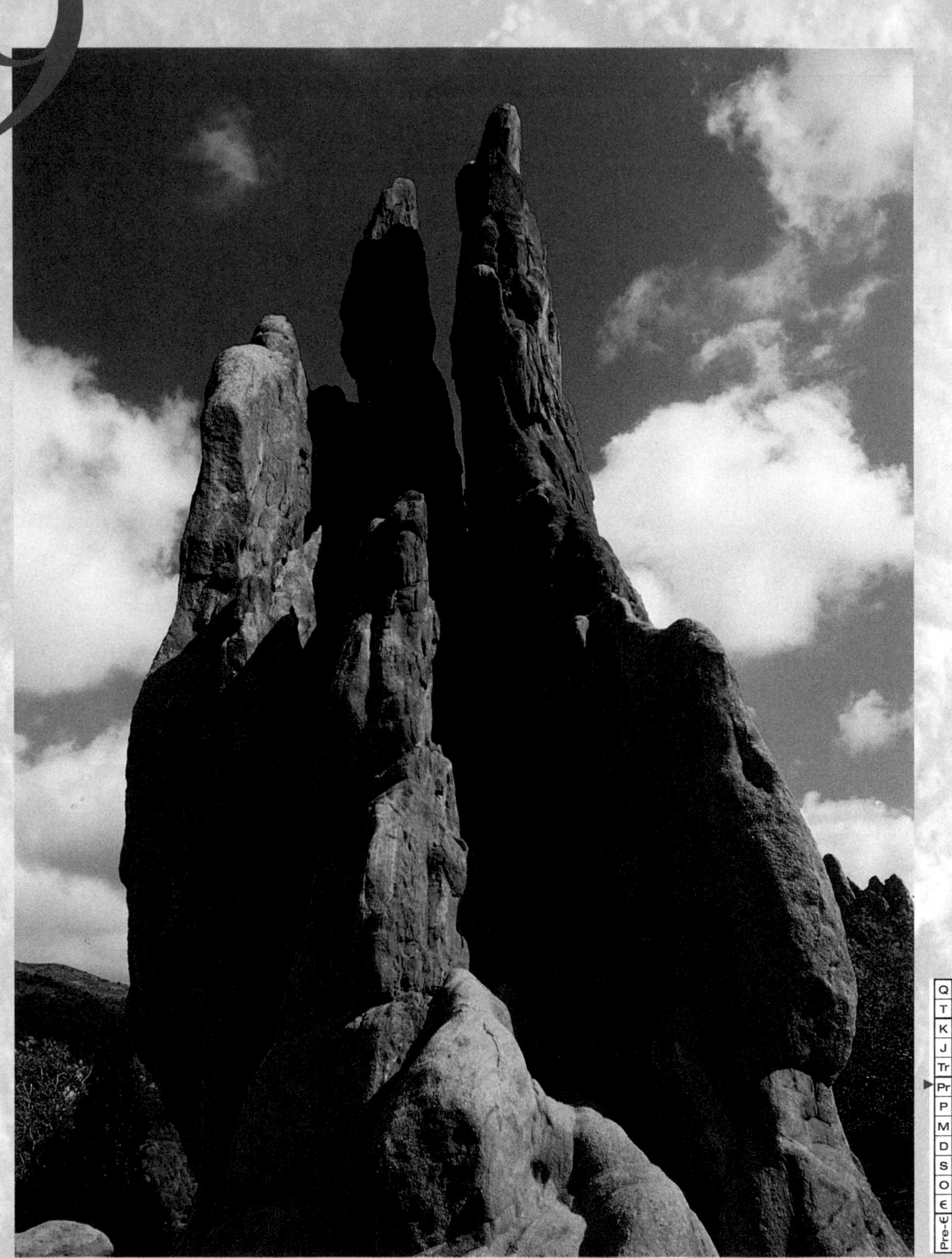

Red sandstone of the Permian Lyons Formation, Garden of the Gods, just west of Colorado Springs, Colorado. Erosion has sculpted these nearly vertical beds into dramatic pinnacles and spires, some of which stand 15 meters above their surroundings. Erosion of the Ancestral Rocky Mountains that lay to the west provided the sands that now compose the Lyons. The beds were tilted upward during the orogeny that produced the modern Rocky Mountains. (Photograph by D. Bewley.)

CHAPTER 9

Late Paleozoic Events

We need not be surprised if we learn from geology that the continents and oceans were not always placed where they now are, although the imagination may well be overpowered when it endeavors to contemplate the quantity of time required for such revolutions.
Charles Lyell, *The Student's Elements of Geology*, 1882

OUTLINE

Having examined events of the Cambrian, Ordovician, and Silurian Periods, we now proceed to events of the Devonian, Mississippian, Pennsylvanian, and Permian Periods of the Paleozoic Era. By Permian time, most of the separate landmasses of earlier periods were assembled into the great supercontinent of **Pangea**. Before the final joining of shorelines, the margins of larger continents, such as eastern North America, grew by accretion of island arcs and microcontinents. Ultimately, the larger continents collided. The **Caledonian** and **Acadian** orogenies brought North America and Europe together as a combined landmass named **Laurasia** (Fig. 9-1). Subsequently, the plate bearing Gondwanaland began to close on Laurasia. Contact of the two landmasses occurred late in the Carboniferous, generating the **Hercynian** orogeny of central Europe and the **Allegheny** orogeny of eastern North America. It may seem startling that mountains in the southern Appalachian region were the result of a collision with northwestern Africa. Similarly, the now deeply eroded and largely buried mountains of the Ouachita orogenic belt were produced by collision with the South American segment of Gondwanaland. The part of South America that is now Venezuela apparently formed the southern margin of that late Paleozoic Ouachita mountain system. The clustering of once-separate continents was nearly complete by the Late Permian.

In addition to continental collisions and recurrent orogenic activity, the late Paleozoic was a time of diverse sedimentation, progress in organic evolution, and changing climatic conditions. During this time, there was widespread colonization of the land by both large land plants and vertebrates. Amphibians and reptiles, along with spore-bearing trees and seed ferns, were especially evident in the landscapes of the late Paleozoic. Near the end of the era, conifers and reptiles able to tolerate drier, cooler climates became abundant. Marine invertebrates thrived in shallow epeiric seas that were especially extensive in North America during the Mississippian. In the central interior of the United States, limestones deposited in these inland seas are over 700 meters thick. They are among the Earth's most extensive sheets of limestone. Adjacent to the orogenic belts and in regions where the epicontinental seas were drained as a result of uplift of the land or lowering of sea level, sediment-laden streams deposited their loads of sand and mud.

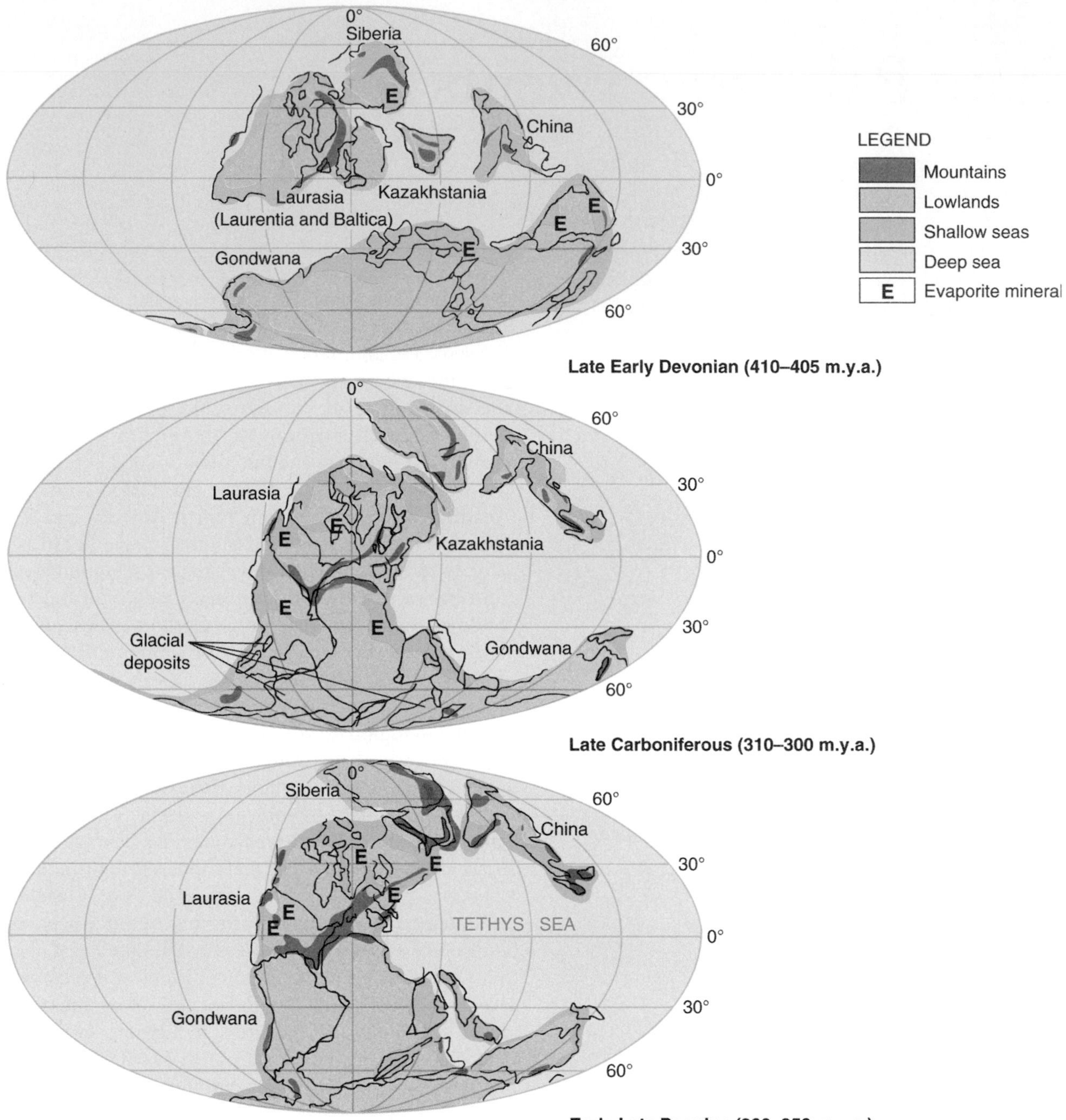

FIGURE 9-1 Global paleogeography of the late Paleozoic. *(From Scotese, C. R., Bambach, R. K., Barton, C., Van Der Voo, R., and Ziegler, A. M. 1979. J. Geol. 87(3):217–277.)* During the Pennsylvanian Period, what continent would you cross into if you traveled south from the present day location of Louisiana?

These nonmarine sediments contain the fossil plants and animals that aid us in inferring climatic conditions on the continents. The accumulation of abundant vegetation in swampy areas provided the raw materials from which the famous coal deposits of Pennsylvanian time were formed. Consolidated sands of desert dunes and layers of evaporites attest to arid conditions at many localities around the globe (Fig. 9-2). Tillites and glacial striations on bedrock provide evidence of glaciation. This is not unexpected, for the more southerly parts of Gondwanaland were located near the pole. Evidence of Late Carboniferous or Permian ice sheets can be found on the now separated landmasses of South America, Africa, Australia (Fig. 9-3), Antarctica, and India. Decades before geologists had an understanding of

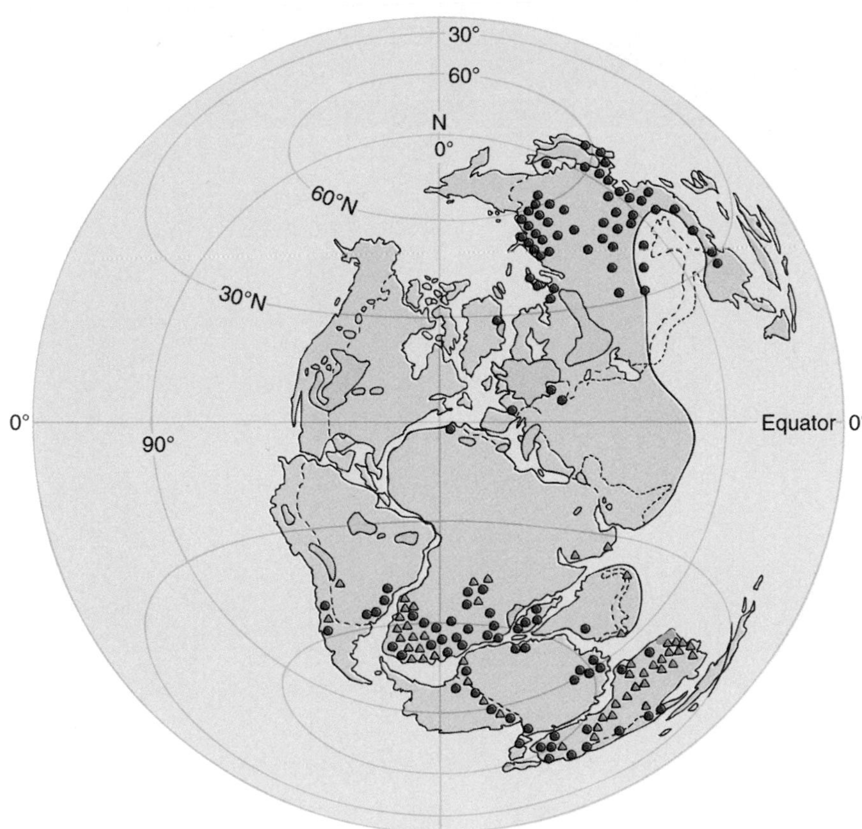

FIGURE 9-2 The distribution of glacial tillites (blue triangles), coal (red circles), and evaporites (irregular green areas) during the Permian, about 250 million years ago. (*From Dewey, G. E., Ramsey, T. S., and Smith, A. G. 1974. J. Geol. 82(5):539.*)

plate tectonics, ancient tillite deposits on these continents were cited as evidence of continental drift.

THE CRATON OF NORTH AMERICA

In Chapter 8, we noted that the last event of the early Paleozoic cratonic region had been the withdrawal of inland seas. Many early Paleozoic formations were subjected to erosion following the withdrawal. Erosion exposed older and deeper formations in those areas that had experienced uparching. Gradual flooding of that early Paleozoic erosional surface was the first event of the late Paleozoic. The new inland sea is termed the **Kaskaskia sea** (Fig. 9-4), and in it were deposited marine rocks of the Kaskaskia sequence (see Table 8-1). Except for minor regressions, the sea persisted over extensive areas of the craton until the end of the Mississippian Period. As the sea withdrew, erosion of the layers of sediment deposited on the Kaskaskia sea floor commenced but was soon interrupted by the advance of a second sea, the Absaroka.

FIGURE 9-3 Glacial striations. Rocks embedded in the base of a moving glacier gouged the striations into underlying bedrock. (*Courtesy of Gray Thompson.*)

The Kaskaskia Sequence

Along the eastern side of the North American craton, Kaskaskia sedimentation consists initially of clean quartz sands. The most famous of these sandy formations is the Oriskany Sandstone of New York, Pennsylvania, and Virginia (Fig. 9-5). Because of its purity, Oriskany sandstone is extensively used in making glass.

Although quartz is by far the dominant mineral in the Oriskany, the formation also contains a very small percentage of other silicate materials, which, like quartz, are resistant to weathering and erosion but are heavier than quartz. These so-called **heavy minerals** can be used to determine the kinds of source rocks and source areas from which clastic sediments were derived. Two distinctly different assemblages of heavy minerals occur in the Oriskany Sandstone. In the area

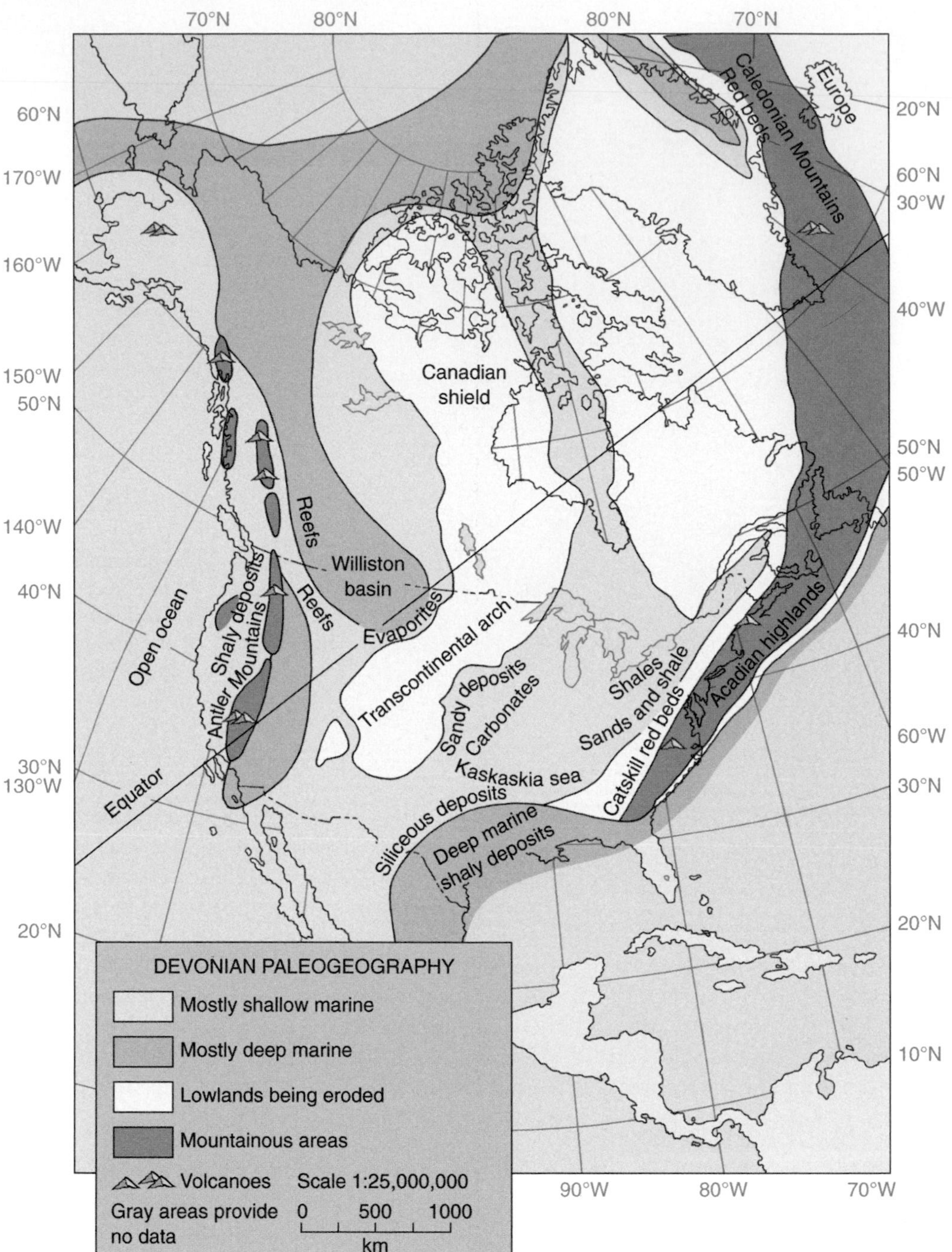

FIGURE 9-4 Paleography of North America during the Devonian Period.

south of New York, the formation contains exceedingly small amounts of only the most stable heavy minerals: tourmaline, zircon, and rutile. The grains in these sandstones are exceptionally well-rounded and worn, and less stable heavy minerals are lacking. This part of the Oriskany is considered to have derived from older clastic sedimentary units to the east and north. In contrast, the Oriskany in New York State contains a heavy mineral assemblage of relatively unstable pyroxene, amphibole, biotite, and garnet. Garnet is ordinarily derived from metamorphic source rocks, whereas the other members of the assemblage are usual components of a variety of igneous and metamorphic rocks. The grains are unaltered. One may infer, therefore, that the sands composing the Oriskany of the New York area may have had an igneous or metamorphic source area and were not recycled from older sandstones. It is another example of the kind of geologic detective work that permits geologists to reconstruct ancient conditions.

The Oriskany Sandstone is the initial deposit of a transgressing sea, for the sands are spread across an

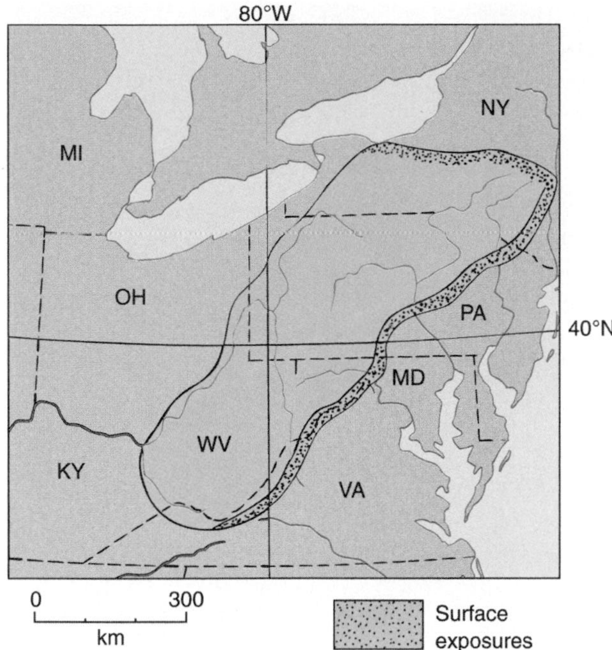

FIGURE 9-5 Areal extent of the Oriskany Sandstone. Surface exposures occur along the northern and eastern borders of the area shown in green.

erosional unconformity. As this marine advance continued, limy sediments were deposited over the Oriskany, and corals began to build reef structures. In areas where water circulation was restricted, salt and gypsum were deposited. During the Middle and Late Devonian of the region east of the Mississippi Valley, carbonate sedimentation gave way to shales. The change to clastic deposition was a consequence of mountain building associated with the Acadian orogeny in the Appalachians. As will be described shortly, highlands formed during this orogeny were rapidly eroded, and clastics were transported westward to form an extensive apron of sediments that are coarser and thicker near their eastern source area and give way to thin but widespread black shales in the east-central region of the United States.

In the far western part of the craton, Middle and Upper Devonian rocks are largely limestones, although there are shales as well. In a depressed area known as the Williston basin (from South Dakota and Montana northward into Canada), extensive reefs developed. Arid conditions and restricted circulation in reef-enclosed basins resulted in the deposition of impressive thicknesses of gypsum and salt. The reefs provided permeable structures into which petroleum migrated, resulting in some of Canada's richest oil fields.

During the Late Devonian, mountains existed along the eastern margin of North America. A clastic wedge called the Catskill clastic wedge was built along the western side of the mountains. Conglomerates and sandstones were the characteristic deposits of the wedge, but farther to the west, suspended muds were carried into the shallow sea that covered the platform. These muds were deposited as a thin but extraordinarily widespread sheet of dark shale. There are many local names for the shale, but the formal name most commonly used is **Chattanooga Shale**. Because the shale is both widespread and easy to recognize, it has served as a marker in regional correlation. It also contains uranium for age determination and has been found to be 350 million years old.

In addition to its uranium content, the Chattanooga is rich in finely disseminated pyrite and organic matter. It is black or dark gray and contains few, if any, fossils of bottom-dwelling invertebrates. These characteristics suggest that the Chattanooga accumulated in stagnant, oxygen-deficient waters. Today, such environments are much more limited in size. It is difficult to conceive how a sea occupying an area as vast as the depositional

A

B

FIGURE 9-6 Crinoidal limestone. (*A*) Specimen of limestone (Burlington Formation) in which crinoid fragments (mostly stem plates) stand out in relief because of weathering. (*B*) Reconstruction of crinoids growing on the floor of the Kaskaskia (Mississippian) sea. (*B, courtesy of the National Museum of Natural History, Smithsonian Institution.*)

area of the Chattanooga could sustain such oxygen-deficient conditions. One interesting suggestion is that phytoplankton in the Chattanooga sea proliferated catastrophically, consuming and thereby depleting the oxygen supply. Without the oxygen required for decay of organic matter, carbon would accumulate and contribute to the black color characteristic of the Chattanooga Shale.

During the passage from Devonian to Mississippian time, highland source areas that provided the Chattanooga Shale were reduced, and the quantity of muddy sediment decreased. Carbonates then became the most abundant and widespread kind of sediment in the epeiric seas of the platform. Cherty limestones (see Fig. 2-31), shelly limestones, limestones composed of the remains of billions of crinoids (Fig. 9-6) and other invertebrates, and limestones containing myriads of oöids (Fig. 9-7) formed extensive beds across the central and western parts of the craton. You may recall from Chapter 2 that oöids are spherical grains formed by the precipitation of calcium carbonate around a nucleus such as a tiny particle of calcium carbonate. They form in clear, shallow sea water where currents exist to move particles about on the sea floor. Salinity somewhat greater than normal appears to enhance the formation of oöids.

The sea in which these crinoidal, oölitic, and other types of limestones were deposited was the most extensive that North America had experienced since Ordovician time (Fig. 9-8). The widespread blanket of carbonates deposited in this sea is often referred to as the great Mississippian lime bank. They record the last great Paleozoic flooding of the North American craton.

In Late Mississippian time, as the Kaskaskia seas began their final withdrawal, large quantities of detrital sediments and thin, scattered limestones were deposited. Today, these rocks are reservoirs for petroleum in Illinois and therefore have been studied extensively. Detailed maps of thicknesses of some of the sandstone units show that they are thickest along branching, sinuous trends that suggest old stream valleys developed on the former sea floor. Studies of grain size, cross-bedding (Fig. 9-9), and current-produced sedimentary structures indicate that the detritus was derived from the northern Appalachians and was transported southwestward across the central interior.

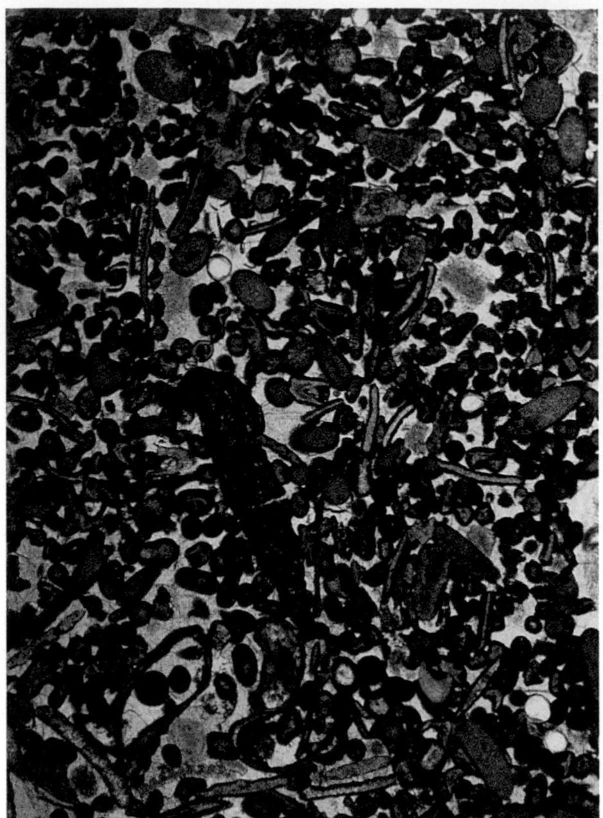

A

B

FIGURE 9-7 Oölitic limestone. (*A*) Photomicrograph of Salem Limestone (Mississippian). The Salem consists primarily of fossil debris and oöids cemented with clear calcite. It is extensively quarried in southern Indiana and used as trim and building stone in public buildings throughout the United States. Width of section is 2.5 millimeters. (*B*) *Uintatherium*, an Eocene mammal, carved in Salem Limestone. The sculpture is one of many carvings of fossils that decorate the geology building at Washington University in St. Louis.

The Absaroka Sequence

When the Kaskaskia seas had finally left the craton at the end of Mississippian time, the exposed terrain was subjected to erosion that resulted in one of the most widespread regional unconformities in the world. Erosion removed entire systems of older rocks over

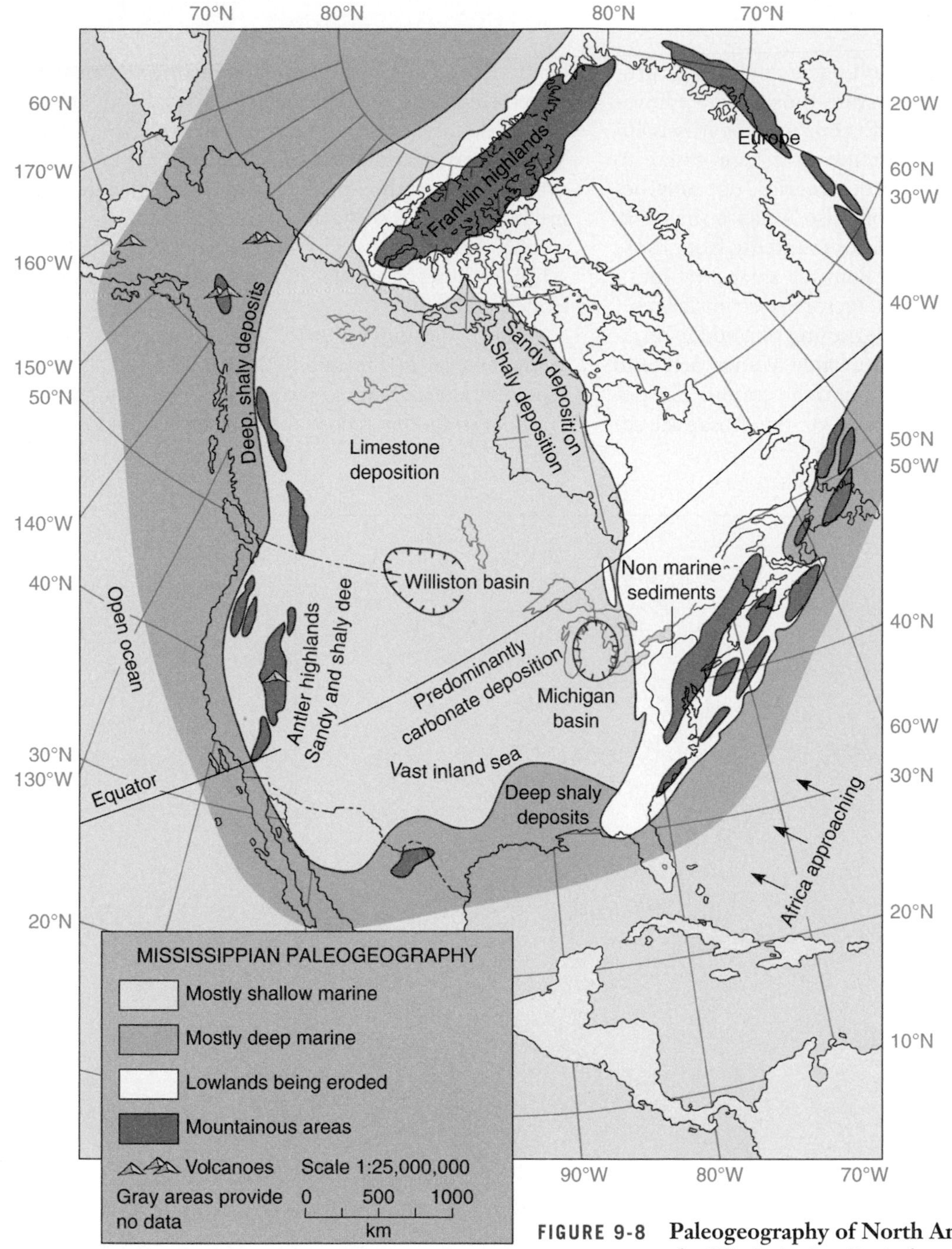

FIGURE 9-8 **Paleogeography of North America during the Mississippian Period.**

FIGURE 9-9 **Cross-bedding in sandstones of the Upper Mississippian Tar Springs Formation, southern Illinois.** *(Photography courtesy of J. C. Brice.)*

arches and domes. The resulting unconformity provides the criterion for separating those strata equivalent to the Carboniferous of Europe into the Mississippian and Pennsylvanian systems. It is appropriate to use these two names in North America, not only because of the unconformity but also because the rocks above the erosional hiatus differ markedly from those below. Indeed, the overlying Pennsylvanian strata are a consequence of very different tectonic circumstances.

It was not until near the beginning of Middle Pennsylvanian time that the seas were able to encroach onto the long-exposed central region of the craton. The deposits of this seaway are those of the Absaroka sequence. In general, the Pennsylvanian rocks near the eastern highlands are thicker, and virtually all are terrestrial sandstones, shales, and coal beds (Fig. 9-10). This eastern section of Pennsylvanian rocks gradually thins away from the Appalachian belt and changes from predominantly terrestrial to about half marine rocks and half nonmarine rocks. Still farther west, the Pennsylvanian outcrops are largely marine limestones, sandstones, and shales.

One of the most notable aspects of Pennsylvanian sedimentation in the middle and eastern states is the repetitive alternation of marine and nonmarine strata. A group of strata that record the variations in depositional

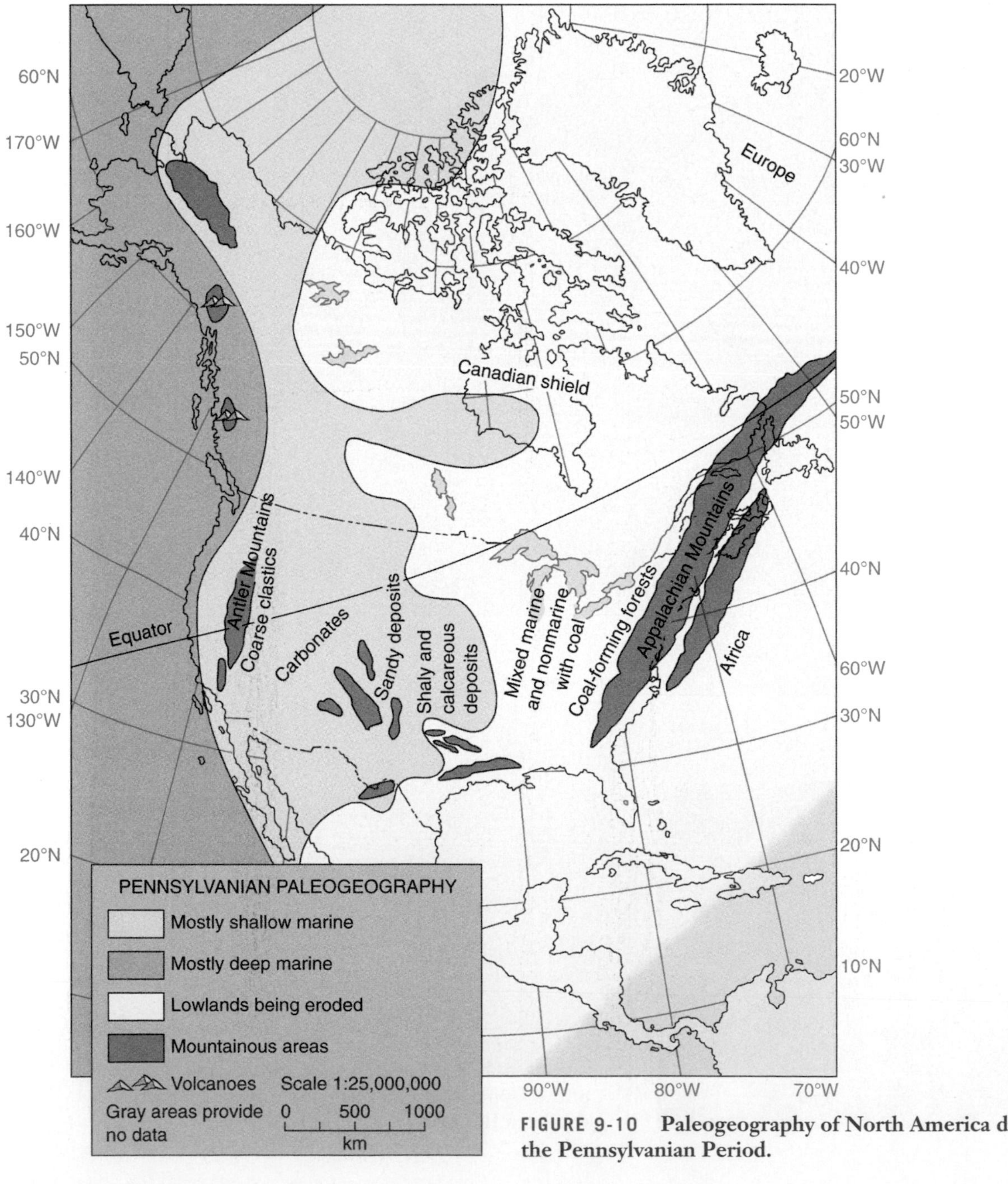

FIGURE 9-10 **Paleogeography of North America during the Pennsylvanian Period.**

environments caused by a single advance and retreat of the sea is called a **cyclothem.** A typical cyclothem in the Pennsylvanian of Illinois contains 10 units (Fig. 9-11). Units 1 through 5 are continental deposits, the uppermost of which is a coal bed (Fig. 9-12). The strata deposited above the coal bed represent an advance of the sea over an old vegetated area.

Cyclothems are notable not only for their cyclicity but also because they can be correlated over great distances. Those of the Appalachian region, for example, can be correlated to cyclothems in western Kansas. In Missouri and Kansas, at least 50 cyclothems are recognized within a section only 750 meters thick, and some of these extend across thousands of square kilometers.

It is apparent that cyclothems are the result of repetitive and widespread advance and retreat of seas. But what was the cause of such oscillations? One explanation involves periodic regional subsidence of the land to a level slightly below sea level, so that marginal seas could spill onto swampy lowlands. A relatively short time later, subsidence might cease, and the shallow sea might be filled with sediment, forcing the sea out and reestablishing nonmarine conditions. Alternatively, temporary regional uplifts could have caused withdrawal of the sea. Finally, worldwide (eustatic) changes in sea level related to continental glaciation might have produced repetitive marine invasions and regressions. We have ample evidence that large regions of Gondwanaland were covered with glacial ice from Late Mississippian through Permian time. When the ice sheet grew in size, sea level would be lowered because of water removed from the ocean and precipitated as snow onto the continent. During warmer episodes, meltwater returning to the ocean would raise sea level. These sea-level oscillations could result in the shifting back and forth of shorelines that produced cyclothems. Additional support for this idea comes from recognition of cyclothems of the same age on continents other than North America, indicating that the cause of the cyclicity was global rather than local.

Ordinarily, cratonic areas of continents are characterized by stability. The southwestern part of the

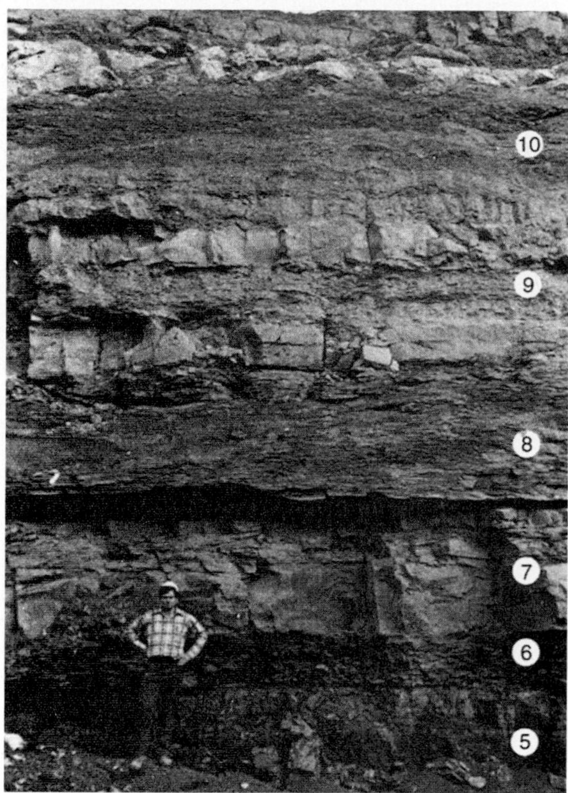

FIGURE 9-12 Part of an Illinois cyclothem. The lowermost layer is the coal seam (cyclothem bed 5), followed upward by shale (bed 6) near the geologist's hand, limestone (bed 7), shale (bed 8), another limestone (bed 9), and the upper shale (bed 10). Part of another sequence caps the exposure. This cyclothem is part of the Carbondale Formation. *(Photograph courtesy of D. L. Reinertsen and the Illinois Geological Survey.)* ? *Would rocks deposited above bed 10 be predominantly marine or nonmarine?*

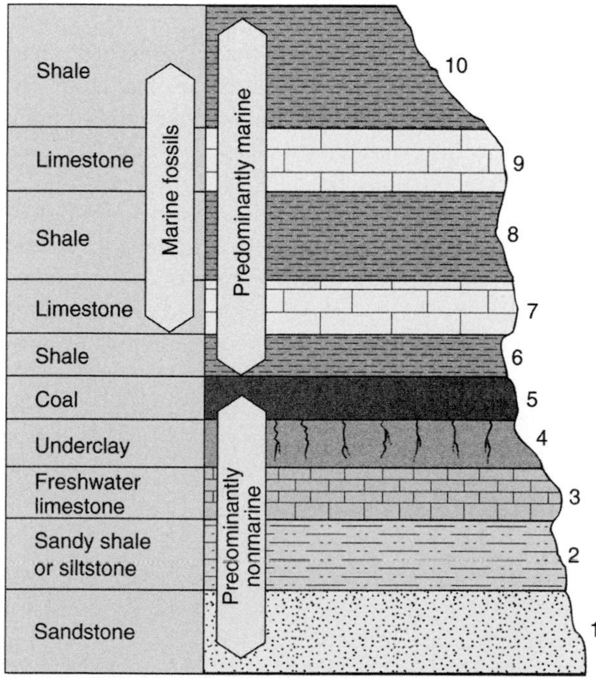

FIGURE 9-11 An ideal coal-bearing cyclothem, showing the typical sequence of layers. Many cyclothems do not contain all 10 units, as in this illustration of an idealized sequence. Some units may not have been deposited because changes from marine to nonmarine conditions may have been abrupt and/or units may have been removed by erosion following marine regressions. The number 8 bed usually represents maximum inundation and, correlated with the same bed elsewhere, provides an important correlative stratigraphic horizon. ? *If you came across a limestone that was part of a cyclothem, how might you ascertain that it was a marine rather than a freshwater limestone?*

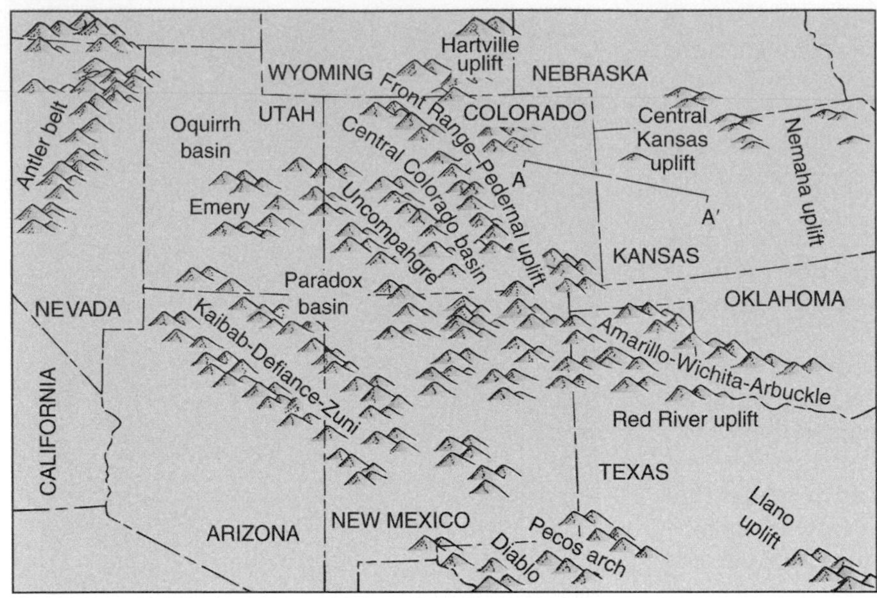

FIGURE 9-13 Location of the principal highland areas of the southwestern part of the craton during Pennsylvanian time.

North American craton provides an exception to this general rule, for during the Pennsylvanian this was a region of mountain building. The resulting highlands formed the ancient Uncompahgre mountains in southwest Colorado and the Oklahoma mountains of western Oklahoma. These mountains and related uplifts appear to have resulted from movements of crustal blocks along large, nearly vertical reverse faults. The Colorado mountains included a range that extended north–south across central Colorado (the Front Range–Pedernal uplifts) and a segment curving from Colorado into eastern Utah (the Uncompahgre uplift). The central Colorado basin lay between these highland areas (Fig. 9-13). A separate range, the Zuni–Fort Defiance uplift, extended across northeastern Arizona on the southwestern perimeter of the Paradox basin. To the east of the Colorado mountains lay the southeastward-trending Oklahoma mountains. Eroded stumps of this once-rugged range form today's greatly reduced Arbuckle and Wichita mountains. Remains of the Amarillo mountains are now buried beneath younger rocks and are known from rock samples recovered during exploratory drilling for oil.

On the basis of the tremendous volume of sediment eroded from the Uncompahgre mountains, it is likely that they attained heights in excess of 1000 meters. They also probably were subjected to repeated episodes of uplift. Erosion of these highlands eventu-

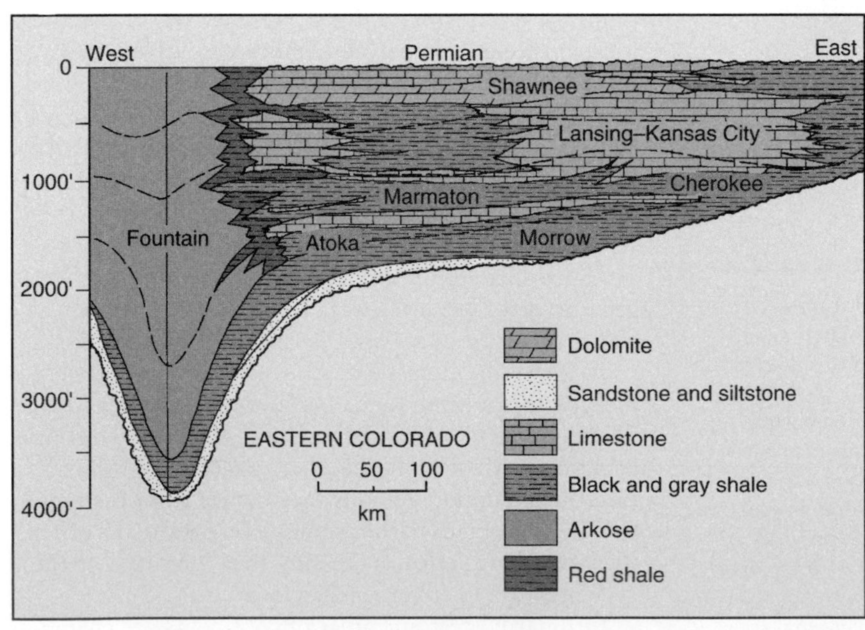

FIGURE 9-14 Pennsylvanian cross-section across eastern Colorado and western Kansas, showing great accumulation of coarse arkosic sandstones east of the Uncompahgre mountains. Line of section is along A—A′ in Figure 9-13. (*From Stratigraphic Atlas of North and Central America. Shell Oil Company, Exploration Department.*)

ENRICHMENT

Reefs

Reefs are of great interest to geologists for many reasons. They comprise immense masses of carbonate material generated entirely by marine invertebrates and algae. Reefs are often well-preserved and therefore provide intact representations of complex ancient and recent biologic communities, from which a wealth of ecologic information can be obtained. Petroleum geologists are aware that ancient reefs, because of their porosity, serve as excellent traps for petroleum, and therefore they carefully map their locations and trends.

Reefs are shallow-water, wave-resistant buildups of calcium carbonate formed around a rigid framework of skeletal material. For most of the Phanerozoic, that framework has been built by corals, although the families of corals that construct reefs have varied through time as various groups evolved or became extinct. Before there were abundant reef-building corals, calcareous algae and bryozoa were the principal frame builders. The calcareous skeletons of other reef-dwelling organisms such as crinoids, brachiopods, mollusks, sponges, and algae add mass to the reefs and help to trap interstitial accumulations of sand and lime mud.

In most reefs, such as the spectacular Permian Capitan Reef of western Texas (see Figs. 9-17 and 9-19), one can recognize three major reef components or facies. The first is the **reef core** itself, which is formed of organisms that have built the reef upward from the shallow sea floor to sea level. Waves constantly crash along the front or windward side of the reef core. In response to this kind of vigorous wave action, organisms living along the reef front form low-growing or encrusting skeletal structures. Waves continuously break off pieces of this part of the reef, and the resulting debris forms a steeply dipping apron of rubble that resembles a submarine talus deposit. This part of the reef is referred to as the **fore-reef facies**. A more sheltered area known as the **back-reef facies** is located landward of the reef core. Here carbonate sands, muds, and oöids accumulate among more erect growing frame builders. The back-reef is also known for the extraordinary diversity of its biologic community.

ally exposed their Precambrian igneous and metamorphic cores. As erosion and weathering continued, great wedge-shaped deposits of red arkosic sediments spread onto adjacent and intervening basins (Fig. 9-14). A small part of this massive accumulation of clastic sediment can be visited in the Red Rocks Amphitheatre just a few miles west of Denver and at the "flatirons" near Boulder, Colorado (Fig. 9-15).

Geologists have long puzzled over this episode of deformation of the craton, but plate tectonics has recently provided a reason for the events. It seems likely that the collision of Gondwana with North America along the site of the Ouachita orogenic belt generated stress in the bordering area of the craton to the north. Crustal adjustments to relieve these stresses resulted in the deformation that produced the highlands and associated basins.

The basins that lay adjacent to the ancient Uncompahgre mountains are important in working out the geologic history of the southwestern craton because of the excellent stratigraphic record they contain. The Paradox basin, located immediately southwest of the Uncompahgre Mountains, is especially interesting. It

Pre-Є | Є | O | S | D | M | P | Pr | Tr | J | K | T | Q

FIGURE 9-15 Steeply dipping beds of the Fountain Formation form the "flatirons" just west of Boulder, Colorado. The red arkosic sandstones, conglomerates, and mudstones of the Fountain Formation were deposited during the late Pennsylvanian and early Permian. The source of Fountain Formation sediment was the ancestral Rocky Mountains that lay to the west. The formation was tilted upward to form the flatirons, as well as dramatic ridges and hogbacks of the Colorado Front range, during the orogeny that produced the modern Rocky Mountains. (*Photograph courtesy of S. D. Levin.*)

was inundated by the Absaroka sea in Early Pennsylvanian time. The initial Pennsylvanian deposits, largely shales, were deposited over karst topography developed on Mississippian limestones. By Middle Pennsylvanian time, the western access of the Absaroka sea to the Paradox basin had become restricted, and thick beds of salt, gypsum, and anhydrite were deposited (Fig. 9-16). Fossiliferous and oölitic limestones developed around the periphery of the basin, and patch reefs grew abundantly along the western side. The association of porous reefs and lagoonal deposits resulted in strata suitable for the entrapment of petroleum.

Near the end of the Pennsylvanian, the Paradox basin was filled to above sea level by arkosic sediments shed from the recently uplifted Uncompahgre highlands. Also at this time, the Absaroka sea, which had begun its transgression at the beginning of the Pennsylvanian, began a slow and irregular regression near the end of the same period. The withdrawal was still incomplete in Early Permian time, so marine sediments continued to be deposited in a narrow zone from Nebraska to western Texas. Fossiliferous limestones characterized these inland seas, although near highlands in Colorado, Texas, and Oklahoma, coarse clastics accumulated to sufficient thicknesses to bury surrounding uplands.

The deposits along what was the eastern edge of the seaway have been eroded away, but at several places along the western side one can observe the change from richly fossiliferous beds below to barren shales, red beds, and evaporites above. The thick and extensive salt beds of Kansas provide testimony to the gradual restriction and evaporation of Permian seas in the central United States. The last Permian marine

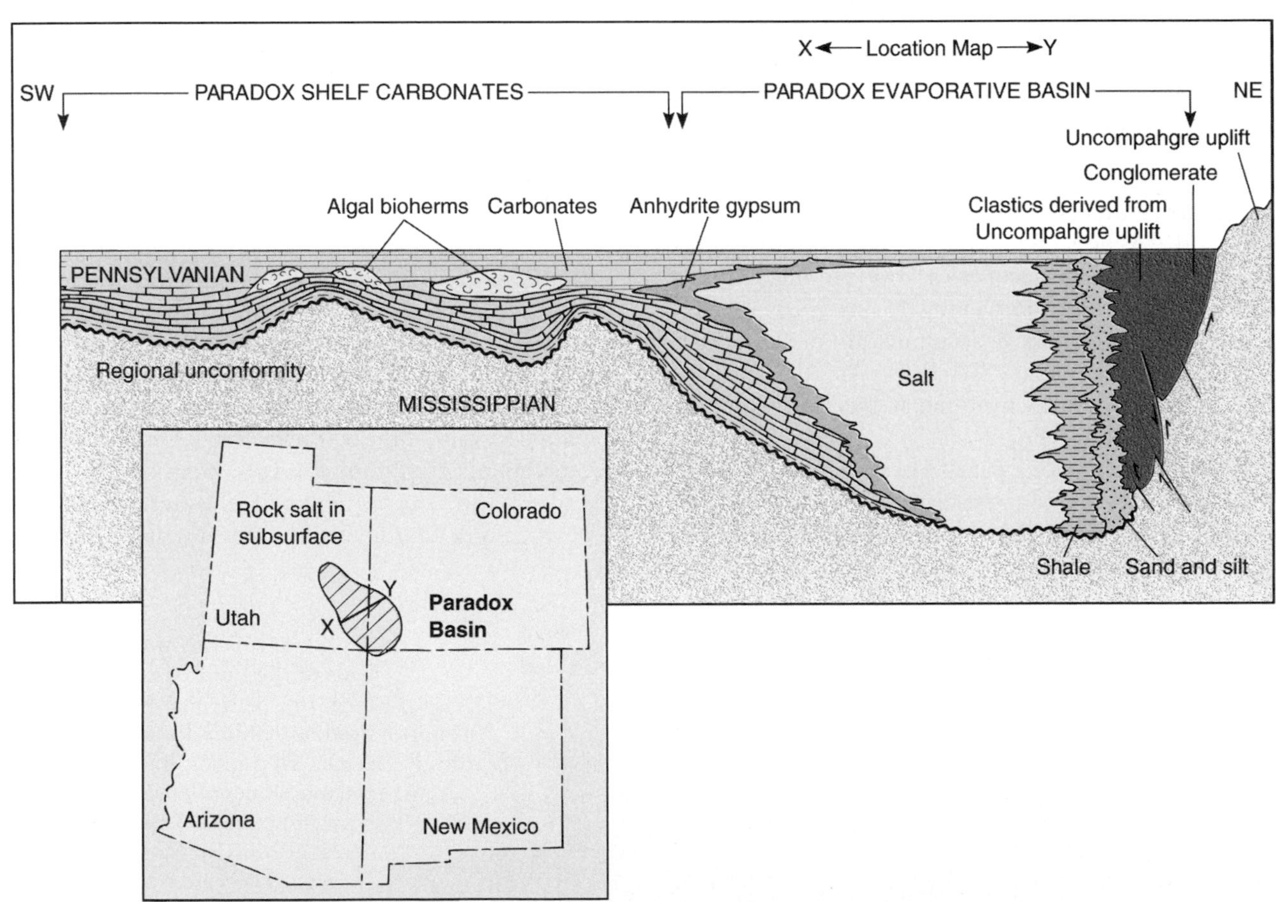

FIGURE 9-16 Generalized cross-section through the Paradox basin, an evaporative basin of Pennsylvanian age in southeastern Utah and southwestern Colorado. Erosion of uplands resulting from the Uncompahgre Uplift produced the coarse, arkosic clastics at the northeastern side of the cross-section. These clastics merge with evaporites of the Paradox basin and then carbonates of the Paradox shelf. Reeflike masses of calcareous algae called algal bioherms occur within the carbonate section. Because of their porosity, petroleum and natural gas have accumulated in many of the bioherms. *(Simplified from Baars, D. L. et. al. 1988. Basins of the Rocky Mountain region. Sedimentary Cover–North American Craton: U.S. DNAG (Decade of North American Geology) D-2:198–220.)*

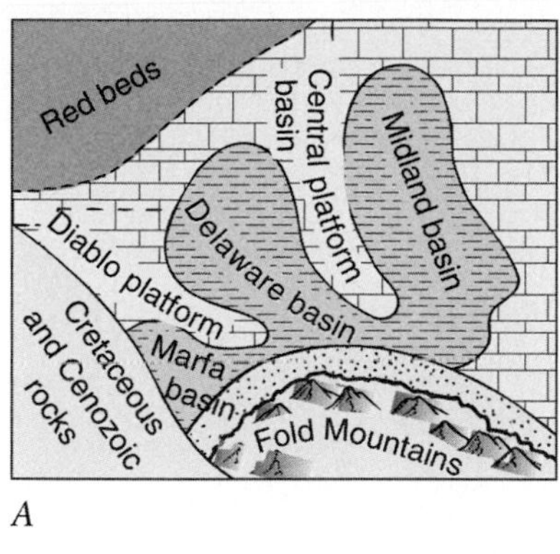

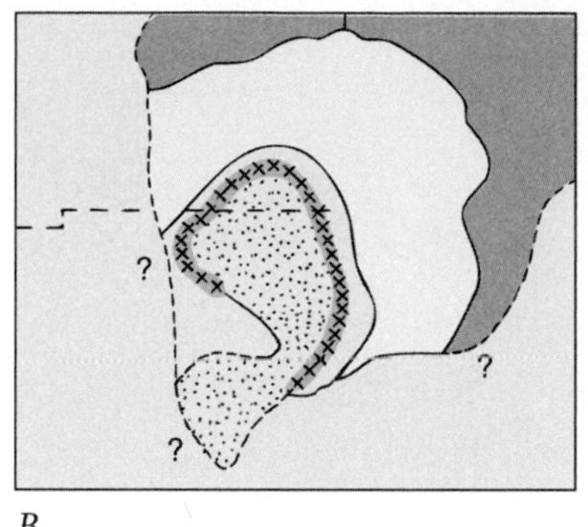

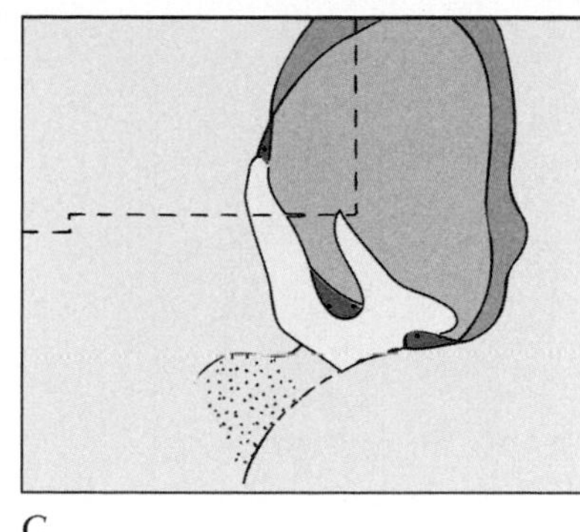

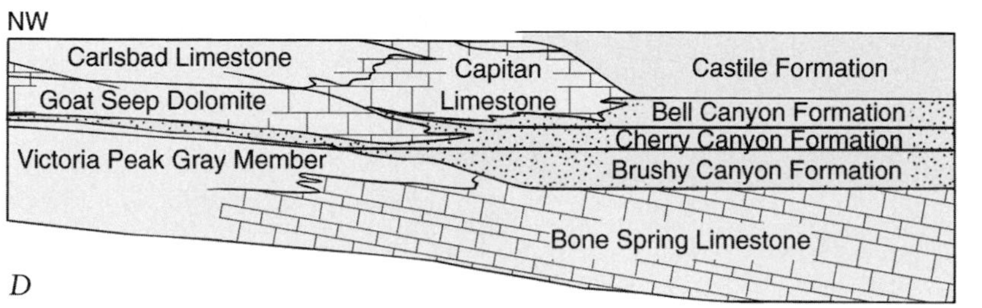

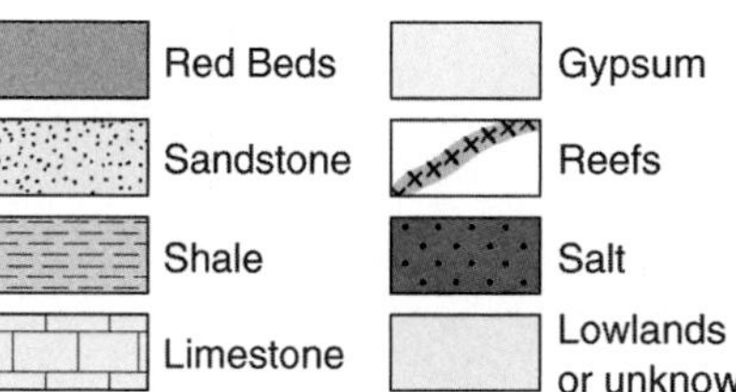

FIGURE 9-17 The Permian of North America can be divided into four stages. The oldest is the Wolfcampian, which is followed by Leonardian, Guadalupian, and Ochoan. *A*, *B*, and *C* show the paleogeography and nature of sedimentation in West Texas during the (*A*) Wolfcampian, (*B*) Guadalupian, and (*C*) Ochoan. (*D*) A simplified cross-section of Leonardian and Guadalupian sediments of the Guadalupe Mountains indicates the relationship of the reef to the other facies. (*After many sources, but primarily King, P. B. 1948. U.S. Geological Survey Professional Paper 215.*)

deposition occurred in the western part of Texas and southeastern New Mexico, where a remarkable sequence of interrelated lagoon, reef, and open-basin sediments was deposited (Figs. 9-17 and 9-18). In this region, several irregularly subsiding basins developed between shallowly submerged platforms. Dark-colored limestones, shales, and sandstones were deposited in the deep basins, whereas massive reefs formed along the basin edges. Behind the reefs, in what must have been lagoons, the deposits are thin limestones, evaporites, and red beds. Late in the Permian, the connections of these basins to the south became so severely restricted that the waters gradually evaporated, leaving behind great thicknesses of gypsum and salt.

Much paleoenvironmental information has been obtained from study of the West Texas Permian rocks. From the lack of medium- and coarse-grained clastics, one may assume that surrounding regions were low-lying. The gypsum and salt suggest a warm, dry climate in which basins were periodically replenished with sea water and experienced relatively rapid evaporation. Mapping of the rock units has helped to establish an estimated depth of about 500 meters for the deeper basins and only a few meters for the intrabasinal platforms. The basin deposits were dark in color and rich in organic carbon as a likely consequence of accumulation under stagnant, oxygen-poor conditions. Perhaps upwelling of these deeper waters may have provided an influx of nutrients on which phytoplankton and reef-forming algae thrived. Along with the algae, the reefs contain the skeletal remains of over 250 species of marine invertebrates. Today, because of their relatively greater resistance to erosion, these ancient reefs form the steep El Capitan promontory in the Guadalupe Mountains of West Texas and New Mexico (Fig. 9-19). Here one can examine the forereef composed of broken reef debris that formed a submarine talus deposit caused by the pounding of waves along the southeastern side of the reef.

THE EASTERN MARGIN OF NORTH AMERICA

During the latter half of the Paleozoic Era, the Appalachian and Ouachita belts experienced their culminating and most intense episodes of orogenesis. The crumpling of these former depositional tracts was a consequence of the reassembly of formerly separated continents into Pangea. By Devonian time,

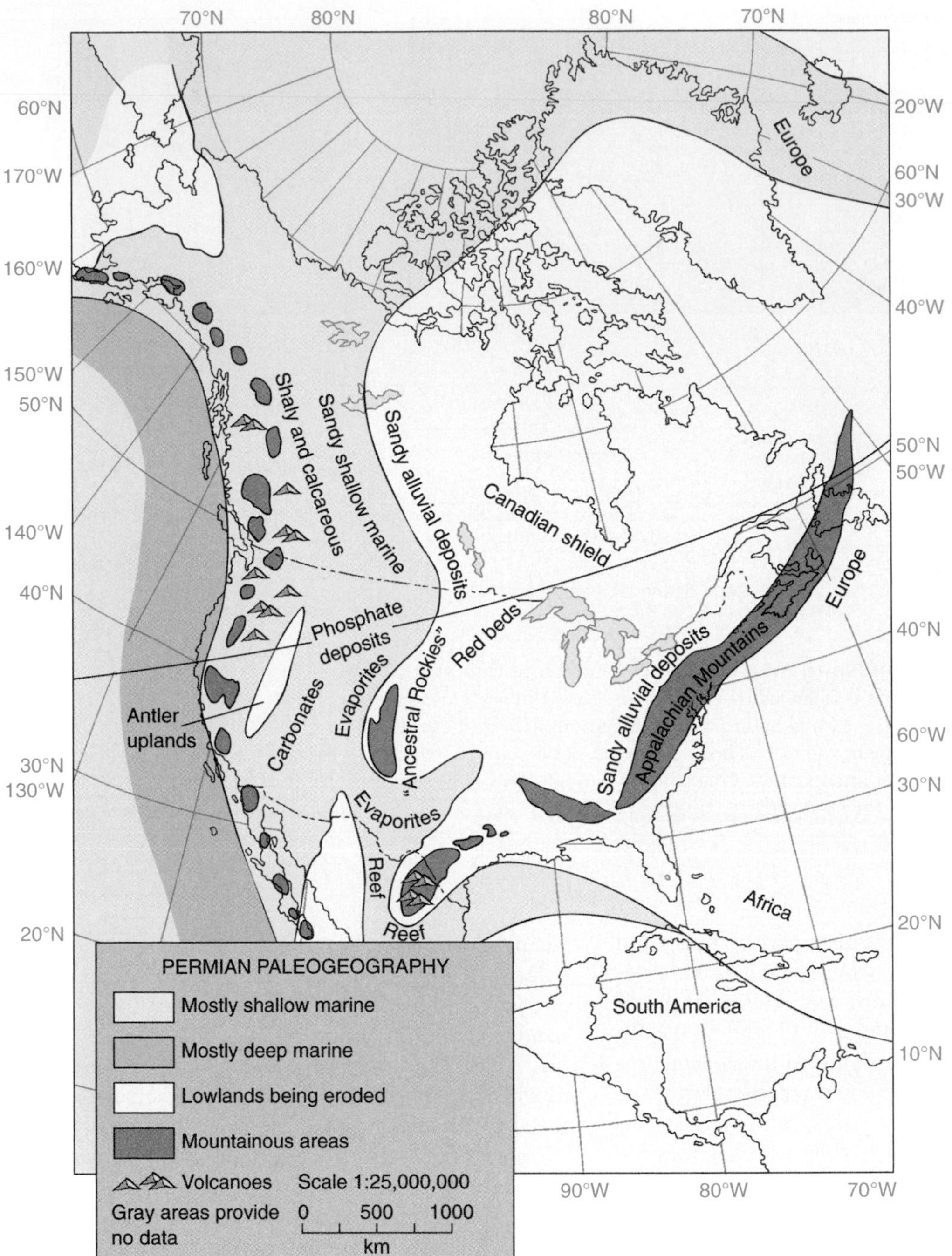

FIGURE 9-18 **Generalized paleogeographic map for the Permian Period.**

the Appalachians had taken the shock of a collision with a displaced continental fragment named the **Avalon terrane** (Fig. 9-20). Yet an additional collision occurred in late Carboniferous when northwest Africa moved against the southern part of the Appalachian belt. That encounter was the cause of the **Allegheny orogeny**. This great smash-up not only raised old basins of sedimentation but also transmitted compressional forces into the interior, causing deep-seated deformations such as those that raised the Colorado and Oklahoma mountains.

The effects of the Devonian deformation, called the Acadian orogeny, are clearly seen in a belt extending from Newfoundland to West Virginia. Here one finds thick, folded sequences of turbidites interspersed with rhyolitic volcanic rocks and granitic intrusions. The intensity of the compression that affected these rocks is reflected in their metamorphic minerals, which indicate that mineralization occurred at temperatures exceeding 500°C and at pressures equivalent to burial under 15,000 meters of rock. The overall result of the Acadian orogeny was to demolish forever the marine

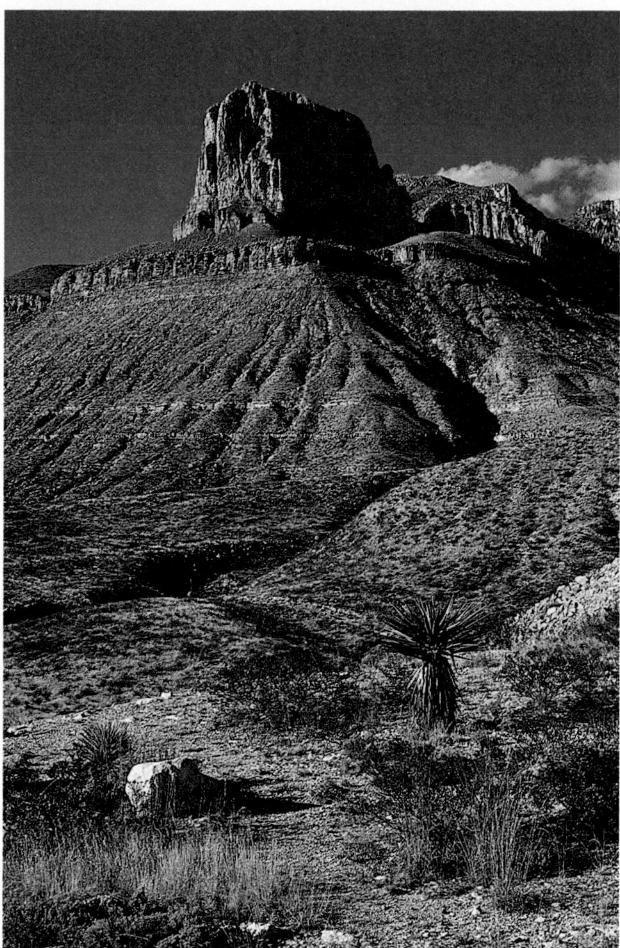

FIGURE 9-19 Example of a facies change in rocks at the southern end of the Guadalupe Mountains, about 100 miles east of El Paso, Texas. The prominent cliff is called El Capitan. It is composed of a Permian limestone reef deposit, rich in sponges and associated marine invertebrate fossils. Behind the reef there once existed lagoons with abnormally high salinity, as suggested by evaporites, dolomites, and virtually unfossiliferous limestones. These lagoonal beds are exposed along the ridge on the right side of the photograph. Thus, a massive reef limestone facies lies adjacent to a lagoonal facies. To the south of the reef lay a marine basin in which normal marine sediments were deposited. The slope beneath the limestone cliff is composed of shales of the Cherry Canyon Formation. (*Copyright Mickey Gibson /Earth Scenes.*) ? *What is the probable kind of sedimentary rock underlying the slope of the Cherry Canyon Formation?*

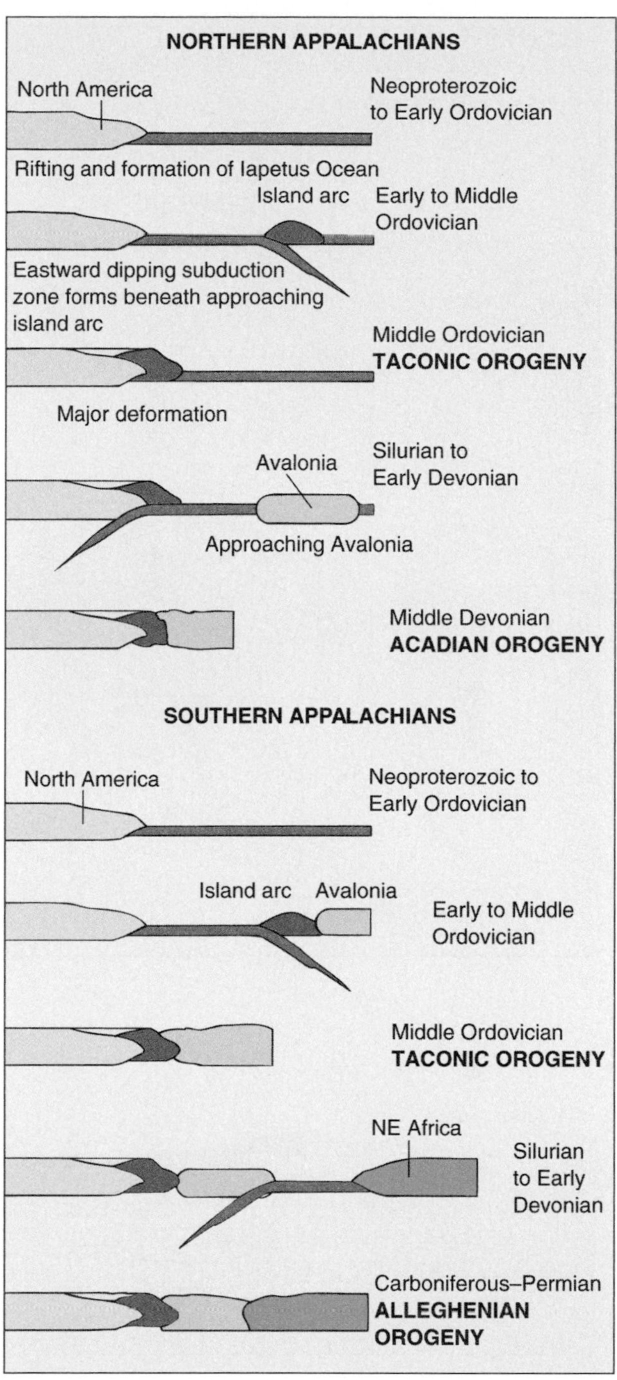

FIGURE 9-20 Simplified diagrammatic plate tectonic sequence involved in the evolution of the northern and southern Appalachians. (*Adapted from Taylor, S. R. 1989, GSA Memoir 172.*)

depositional basin and establish in its place mountainous areas in which erosion was the prevalent geologic process. Here and there in isolated basins among the mountains, Devonian nonmarine sediments were deposited. However, the greatest volume of erosional detritus was spread outward from the highlands as a great wedge of terrigenous sediment called the **Catskill clastic wedge** (Figs. 9-21 and 9-22).

The Catskill Clastic Wedge

For many reasons, the Devonian rocks of the Catskill clastic wedge (also called the Catskill delta) have excited the interest of geologists for well over a century. This apron of sediments provides an ideal area for the examination of facies from a varied assortment of both marine and nonmarine depositional environments

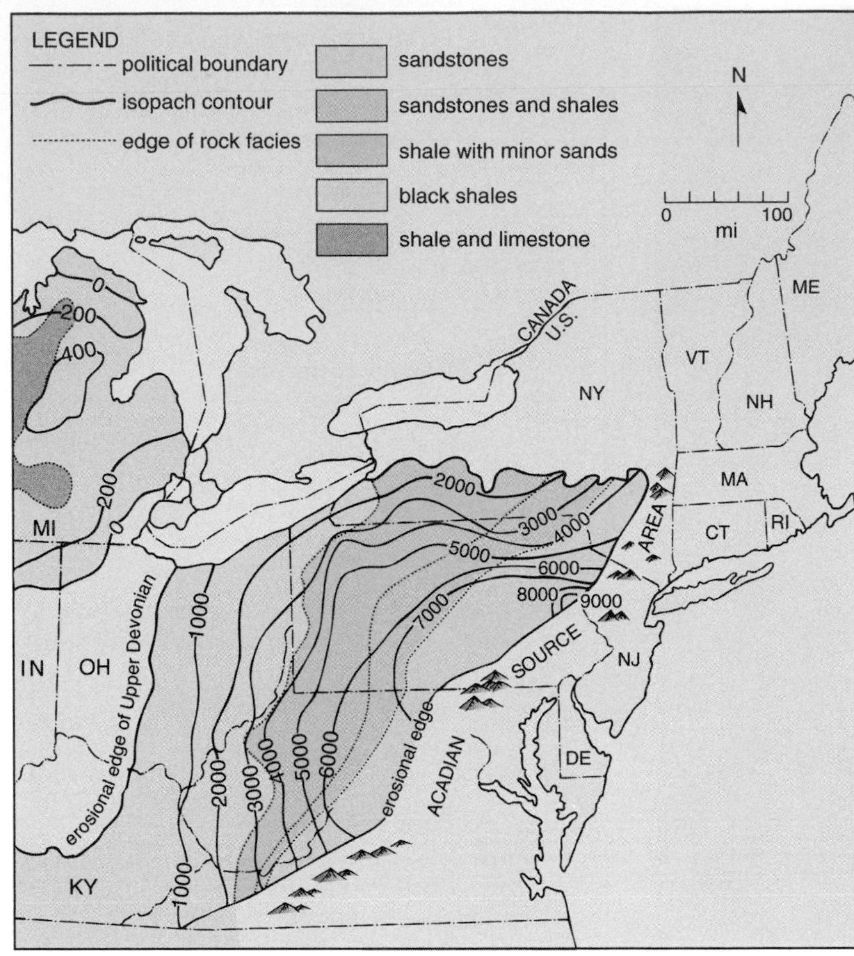

FIGURE 9-21 Isopach and lithofacies map of Upper Devonian sedimentary rocks in the northeastern United States. Thicknesses are given in feet. *(Modified from Sevon, W. D., 1985. Geol. Soc. Am. Special Paper 201:79–90; and Ayrton, W. G. 1963. Pa. Geol. Survey Rpt. 39(4):3–6.)*

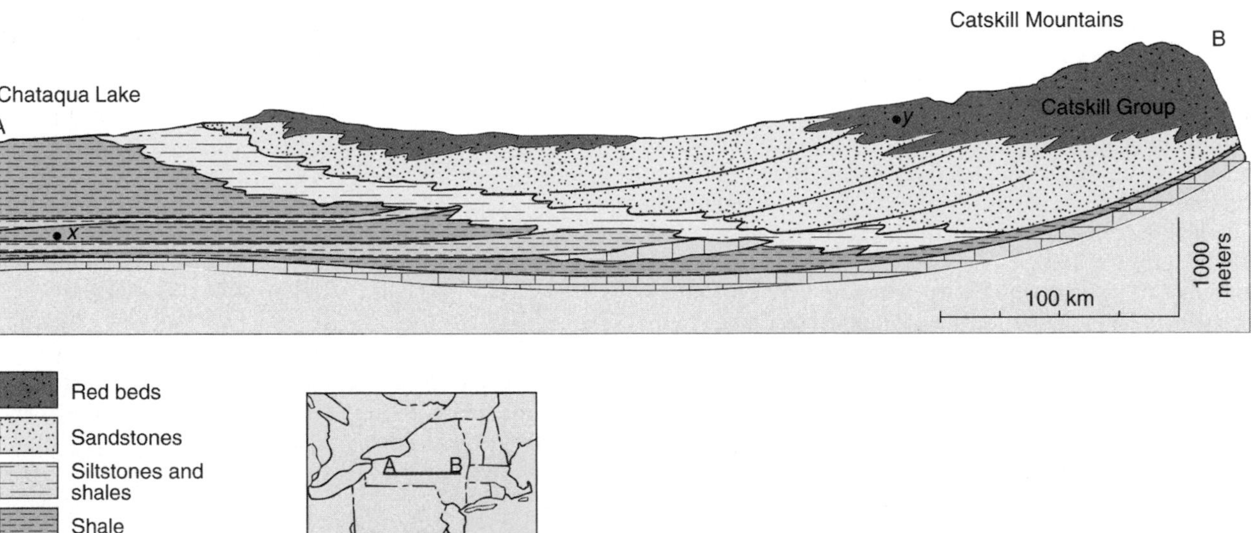

FIGURE 9-22 East–west section across the Devonian Catskill clastic wedge, New York. Note that continental red beds interfinger with nearshore marine sandstones, and these in turn grade toward the west into offshore siltstones and shales. Continental deposits prograded upon the sea, pushing the shoreline progressively westward. (Based on several classic studies by Chadwick, G. H., and Cooper, G. A., completed between 1924 and 1942.) ? *Are the shale beds at point* ***x*** *older, younger, or about the same age as the sandstone beds at point* ***y****?*

A *B*

FIGURE 9-23 Sedimentary rocks of the Catskill clastic wedge. (*A*) Catskill Formation sandstone filling an abandoned stream channel (scale divisions are 10 cm). (*B*) Tidal flat deposits of the Catskill clastic wedge near Altoona, Pennsylvania. (*Courtesy of W. D. Sevon.*) ? *What is their approximate angle of dip?*

(Fig. 9-23). Catskill facies characteristically exhibit rapid lateral changes from sandstones to shales. Such relationships form traps for petroleum and are the reason for the thousands of wells drilled into Catskill strata. The flat slabs of red sandstone have also provided vast tonnages of flagstone for buildings in eastern cities. But for historical geologists, the most pervasive reason for studying Catskill rocks is the clues they provide to the locations and time of occurrence of phases of the Acadian orogeny. Most of these clues come from the study of subsidiary clastic wedges within the larger Catskill complex. The location of these wedges indicates that the Acadian orogeny was caused by the convergence of the Avalon terrane, and possibly other terranes, against the irregular eastern margin of the North American craton. Additional pulses of orogenic activity followed as the westward-moving Avalon terrane encountered parts of the cratonic margin that projected eastward as promontories. The first orogenic pulse occurred during the early Middle Devonian, when the Avalon terrane collided with a promontory in the present vicinity of the St. Lawrence valley. Orogenic pulses followed in succession throughout the remainder of the Devonian as the Avalon terrane encountered the more southerly promontories of regions now occupied by the central and southern Appalachians.

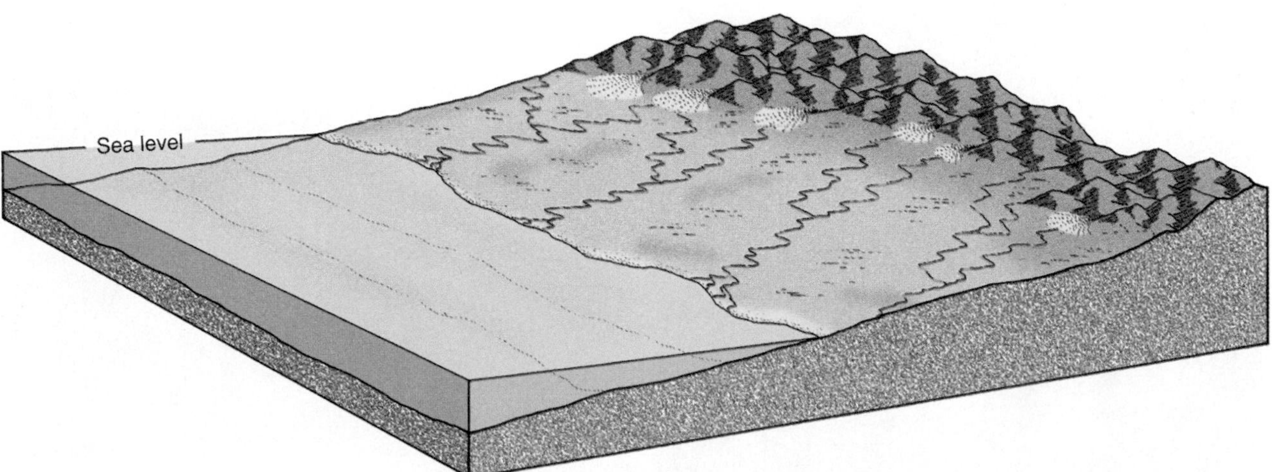

FIGURE 9-24 A panoramic view of the Catskill clastic wedge as seen from a location above south-central Pennsylvania. Shoreline trends for about 300 kilometers in a northeastern direction. (*Modified from Woodrow, D. L. 1985. Geol. Soc. Am. Special Paper 201:51–63.*)

In regard to Catskill facies, the older term, Catskill delta, is not appropriate, for these sediments were not deposited as large deltas of one or two major streams. Study of the directional properties of the fluvial sandstones indicates deposition from many small streams, all flowing westward out of the Acadian highlands (Fig. 9-24). In many areas, marine processes reworked the sediment supplied by streams so rapidly that even small deltas did not develop.

Unlike the earlier Queenston rocks described in Chapter 8, the Catskill sediments were laid down at a time when land plants were abundant (Fig. 9-25) and provided a green mantle for the alluvial plains and hills. Fossil remains of these plants are most frequently encountered in the sediments deposited along former stream valleys. They indicate a tropical climate in this part of the Devonian world.

Nonmarine Catskill sedimentary rocks are dominated by sandstones and shales in which the contained iron is deeply oxidized. The oxide takes the form of the mineral hematite and causes the reddish-brown hues so characteristic of Catskill nonmarine rocks. The majority of these so-called red beds represent deposits of braided or meandering streams.

The marine components of Catskill sediments consist mostly of fine sands and clays carried westward to an adjoining basin called the Catskill sea. Waves and currents shifted these clastics along the shoreline, forming lagoonal areas, bars, and beaches. More basinward, deeper areas were characterized by deposition of dark clays interspersed with occasional sandy turbidite deposits.

Geologists in both North America and western Europe recognize the many similarities between the nonmarine rocks of the Catskill clastic wedge and those that were spread across Europe south of the Caledonian orogeny. As noted in the previous chapter, this region of largely alluvial deposition is named the Old Red continent after its most famous formation, the Old Red Sandstone (Fig. 9-26).

After the Devonian

Mississippian strata crop out in the Appalachian region from Pennsylvania to Alabama. Nonmarine shales and sandstones predominate, indicating that erosion of the mountainous tracts developed during the Acadian orogeny was still in progress. Some of the finer clastics were spread westward onto the craton to form the widespread blanket of Lower Mississippian black shales mentioned earlier. Particularly coarse sediments, including conglomerates, were deposited as part of the **Pocono Group**. Pocono sandstones (Fig. 9-27) form some of the resistant ridges of the Appalachian Mountains in Pennsylvania. Westward, the Pocono section thins and changes imperceptibly into marine siltstones and shales. Thus, there developed another great clastic wedge of alluvial deposits that sloped westward and merged into deltas along the coast of the epeiric sea. The plains and deltas, standing only slightly above sea level, were backed by the rising mountains of the Appalachian fold belt.

Pennsylvanian rocks of the Appalachians are characterized by cross-bedded sandstones and gray shales that were deposited by rivers or within lakes and swamps. Coal seams are, of course, prevalent in the Pennsylvanian System of the eastern United States. They reflect luxuriant growths of mangrove-like

FIGURE 9-25 Fossil plant root traces in mudstones of the Catskill clastic wedge (Oneonta Formation), near Unadilla, New York. *(Courtesy of W. D. Sevon.)*

FIGURE 9-26 Tilted (dipping) beds of Old Red Sandstone pierce the famed emerald green Irish landscape in the highlands called the Macgillycuddy's Reefs near the southwestern coast of Ireland. The Old Red Sandstone is over 6500 meters thick in this area.

FIGURE 9-27 Exposure of the Lower Mississippian Purslane and Rockwell formations (Pocono Group) in a highway cut through Sideling Hill, western Pennsylvania. The section includes thick, cross-bedded fluvial sandstones as well as a few dark, coaly shales. (*Courtesy of J. D. Glaser, Maryland Geological Survey.*) ? *What kind of geologic structure is seen here? During what orogenic event was it produced?*

forests. This was an ideal environment for coal formation. Vegetation that accumulated in the poorly drained swampy areas was frequently inundated and killed off. Immersed in water or covered with muck, the dead plant material was not destroyed by oxidation to carbon dioxide and water. It was, however, attacked by anaerobic bacteria. These organisms broke down the plant tissues, extracted the oxygen, and released hydrogen. What remained was a fibrous sludge with a high content of carbon. Later, such peatlike layers were covered with additional sediment—usually siltstones and shales—and then compressed and slowly converted to coal.

The Allegheny Orogeny

Bordering the interior platform of North America on the east from Newfoundland to Georgia is the Appalachian orogenic belt. (Its southwestward extension is called the Ouachita orogenic belt.) The Appalachian belt contains the record of three major orogenic events and many minor disturbances. Study of the rocks deformed during these events has been the basis not only for classic theories of mountain building but also for new ideas that view the Appalachian belt as a collage of microcontinents and other suspect terranes accreted to the eastern edge of North America by plate convergence.

The Appalachian depositional basin (see Fig. 8-2) can be neatly divided into a western belt composed of shales, limestones, and sandstones and an eastern belt composed largely of graywackes, volcanics, and siliceous shales. The latter are often highly metamorphosed and intruded by granite masses of batholithic proportions. Today these rocks underlie the Blue Ridge and Piedmont Appalachian provinces, whereas rocks of the western belt occur beneath the Valley and Ridge province and Appalachian plateaus (Fig. 9-28).

All three of the orogenic episodes that molded the Appalachian provinces occurred during the Paleozoic. As described in a previous chapter, the first of these was the Taconic orogeny that occurred during the Ordovician Period. The second was the Acadian orogeny of the Devonian Period. The culminating third event occurred as Gondwana (Africa and South America) converged on North America and Europe (Fig. 9-29). In North America, the orogenic activity resulting from these collisions is termed the Allegheny orogeny. This great episode of mountain building began during the Mississippian Period and continued throughout the remainder of the Paleozoic. It affected a belt that extended across 1600 kilometers, from southern New York to central Alabama. Like the earlier Acadian orogeny, the Allegheny deformations were the result of ocean closure and the consolidation of continents to form Pangea.

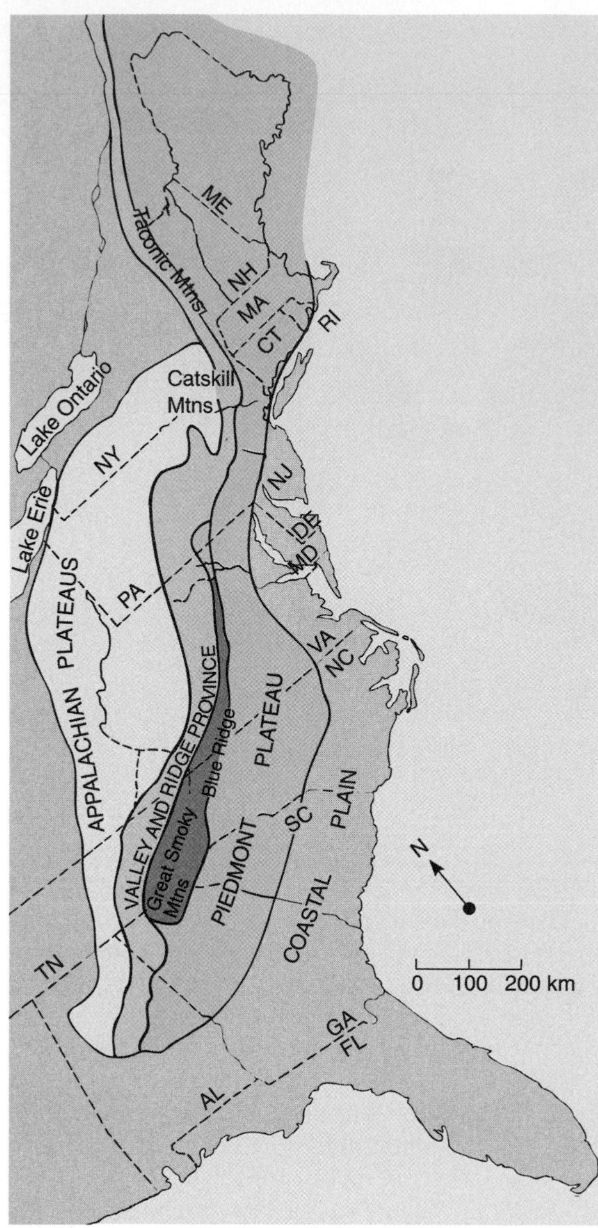

FIGURE 9-28 Physiographic provinces of the eastern United States.

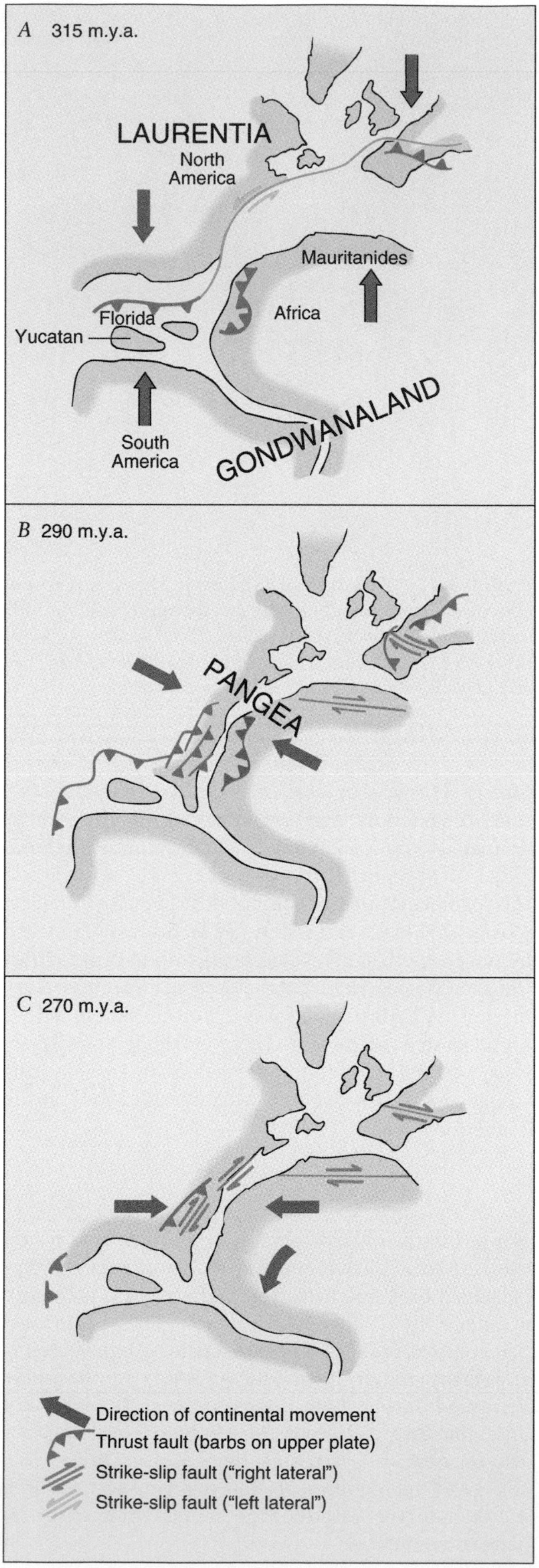

The effects of the Allegheny orogeny were profound and included Permian compression of early continental shelf and rise sediments as well as strata deposited along the bordering tract of the craton. The great folds now visible in the Ridge and Valley Province were developed during this orogeny. Less visible at the surface but

FIGURE 9-29 Plate tectonic model for late Paleozoic continental collisions, proposed by P. E. Sacks and D. T. Secor, Jr. (*A*) Early Pennsylvanian, (*B*) Late Pennsylvanian, (*C*) Permian. (*Adapted from Sacks, P. E., and Secor, D. T., Jr. 1990. Science 250:1702–1705.*)

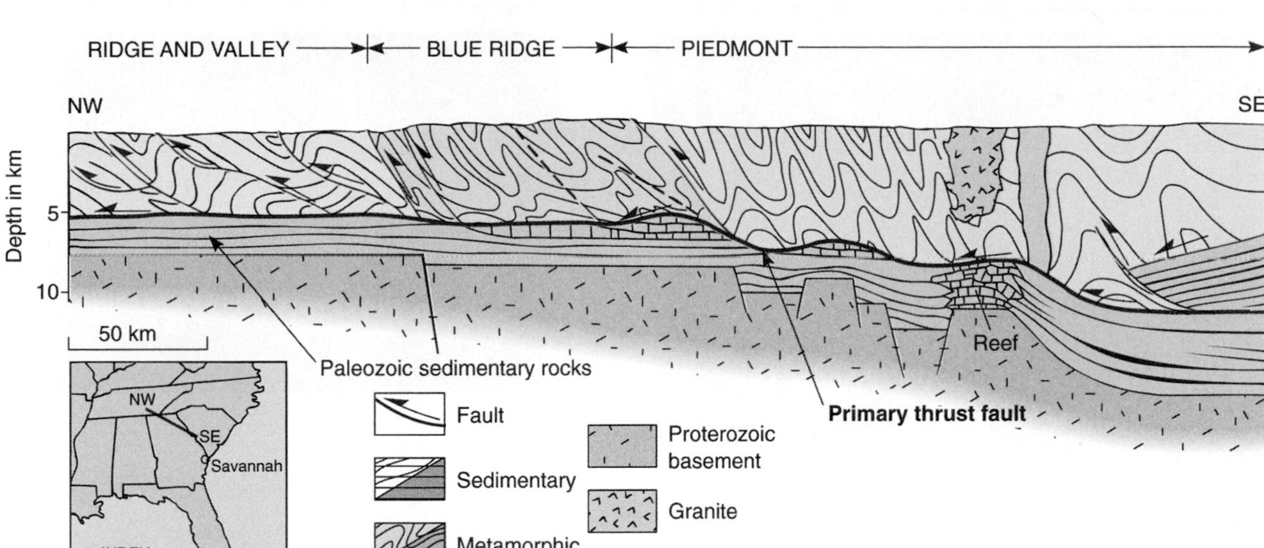

FIGURE 9-30 Cross-section through the southern Appalachians, based on deep seismic-reflection profiles. (*Adapted from Cook, F. A. et al. 1979. Geology 7:563–567.*)

no less impressive are enormous thrust faults formed along the east side of the southern Appalachians. Many of the folds are asymmetrically overturned toward the northwest, and the surfaces of the thrust faults are inclined southeastward, suggesting the entire region was moved forcibly against the central craton (Fig. 9-30). This kind of deformation, in which basement rocks are largely unaffected and the overlying "skin" of weaker sedimentary rocks breaks into multiple thrust faults, is known as **thin-skin tectonics**.

Erosion of the mountains produced during the Allegheny event produced another great blanket of nonmarine sediments. These mostly red sandstones and gritty shales compose the **Dunkard Group** (Fig. 9-31) and **Monongahela Group** of Permian–Pennsylvanian age.

The Ouachita Deformation

Deformation of the Ouachita orogenic belt (Fig. 9-32) was caused by the collision of the northern margin of Gondwanaland (northern South America and perhaps part of northwestern Africa) with the southeastern margin of the North American craton (see Fig. 9-29). That deformation began rather late in the Paleozoic, for the rock record indicates that from Early Devonian to Late Mississippian time the region experienced slow deposition interrupted by only minor disturbances. Carbonates predominated in the more northerly shelf zone, whereas cherty rocks known as novaculites accumulated in the deeper marine areas (Fig. 9-33).

Novaculites are hard, even-textured, siliceous rocks composed mostly of microcrystalline quartz. They are formed from bedded cherts that have been subjected to heat and pressure. Arkansas novaculite is used as a whetstone in sharpening steel tools.

Rates of deposition increased immensely near the end of the Mississippian, when over 8000 meters of graywackes and shales were spread into the depositional

FIGURE 9-31 The Waynesburg Sandstone of the Dunkard Series exposed along the Ohio River in Meigs County, Ohio. (*Photograph courtesy of the Ohio Division of Geological Survey.*)

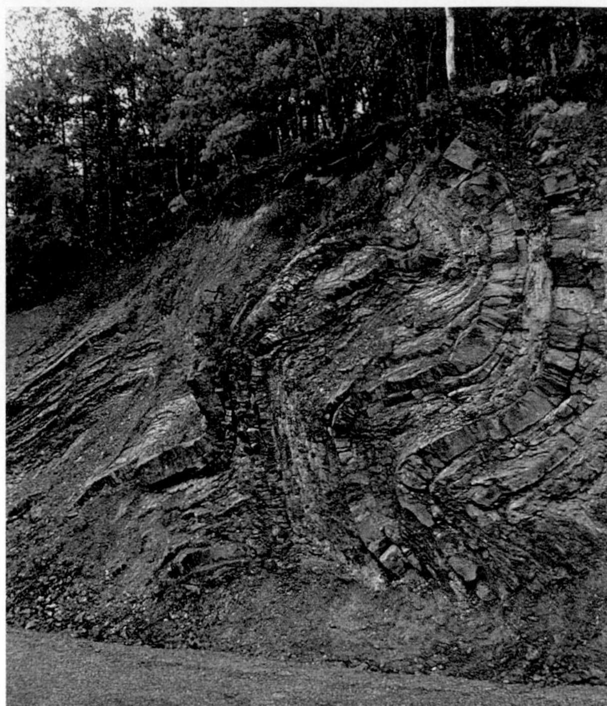

FIGURE 9-32 Late Paleozoic strata in the Ouachita Mountains of Arkansas, showing the effects of compressional deformation associated with the Ouachita orogenic episode. *(Courtesy of E. L. Shock.)*

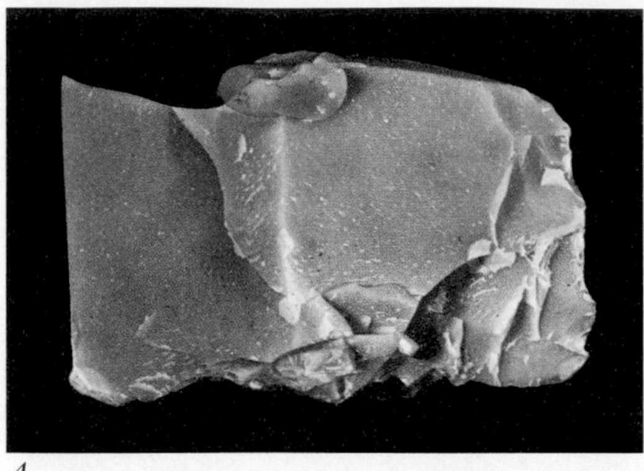

A

B

FIGURE 9-33 A hand specimen of Arkansas novaculite (*A*), and a close-up view of deformed Arkansas novaculite (*B*). The prominent bed in *B* is 8 centimeters thick. *(Image* B *copyright E. R. Degginger / Earth Scenes.)*
? *What is novaculite used for?*

basin. The flood of clastic sediment continued into the Pennsylvanian, forming a great wedge of debris that thickened and became coarser toward the south, where growing mountain ranges were undergoing rapid erosion (Fig. 9-34). Radiometric dating of now deeply buried basement rocks from the Gulf Coastal states indicates that these rocks were metamorphosed during the late Paleozoic and were the likely source for the Pennsylvanian clastics. The coarse sediments document several pulses of orogenesis that ultimately produced mountains along the entire southern border of North America. In the faulting that accompanied the intense folding, strata of the continental rise were thrust northward onto the rocks of the shelf. By Permian time, stability returned, and strata of this final Paleozoic period are relatively undisturbed.

Since the time of active mountain building, erosion has leveled most of the highlands, leaving only the Ouachita Mountains of Arkansas and Oklahoma and the Marathon Mountains of southwest Texas as remnants of once-lofty ranges. Although the Ouachita belt has been traced over a distance of nearly 2000 kilometers, only about 400 kilometers are exposed. Thus, the actual configuration of the belt has been determined by the examination of millions of well samples and other data obtained during drilling activities associated with petroleum exploration.

SEDIMENTATION AND OROGENY IN THE WEST

The late Paleozoic history of the western or Cordilleran belt was almost as lively as that of the Appalachian. You may recall from Chapter 8 that during the early Paleozoic, a passive margin existed along the western side of the North American craton. The familiar belts of continental shelf, slope, and rise sediments were deposited along that relatively quiet margin. More dynamic conditions began in the Devonian, when subduction of oceanic lithosphere beneath the western margin of the continent was initiated. This was

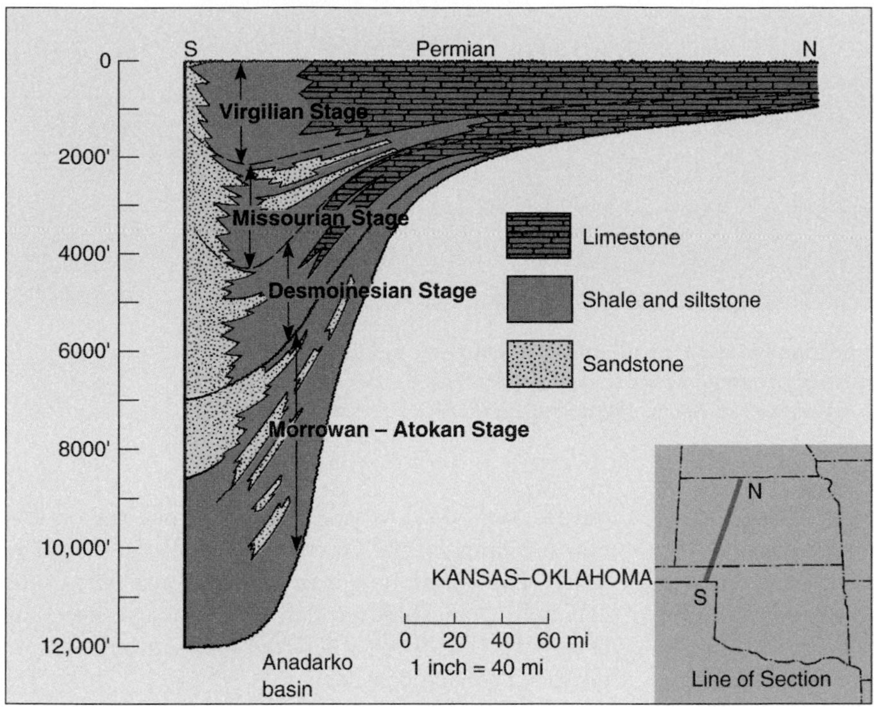

FIGURE 9-34 Geologic section of Pennsylvanian rocks across Oklahoma and Texas, showing the thick wedge of sediment shed from mountain ranges to the south. (*Adapted from Stratigraphic Atlas of North and Central America. Shell Oil Company, Exploration Department.*) *What is the predominant Pennsylvanian rock in the panhandle of Oklahoma? What type of sedimentary rock predominates in northern Kansas? Why?*

the beginning of a disturbance known as the **Antler orogeny**. During the Antler orogeny, a volcanic island arc converged on the western margin of North America, crushing sediment in the intervening basin. The convergence was accompanied by thrust faulting on a massive scale. The effects are clearly seen in the Roberts Mountains Thrust Fault of Nevada and at other locations northward into Idaho (Fig. 9-35). One can recognize continental rise and slope deposits that have been thrust as much as 80 kilometers over shallow-water sediments of the former continental shelf.

The Antler orogeny, which had begun in the Late Devonian, continued actively into the Mississippian and Pennsylvanian. Highlands created by the orogeny provided detrital sediment that was transported into adjacent basins (Fig. 9-36). Especially thick sequences of Pennsylvanian and Permian shelf sediments accumulated in the area now occupied by the Wasatch and Oquirrh mountains (Fig. 9-37). In the latter area, the Oquirrh Group is over 9000 meters thick.

Mississippian and Pennsylvanian deposits west of the Antler highlands include a great volume of coarse detritus and volcanic rocks. These materials were swept eastward from a volcanic arc that lay along the western side of North America (Fig. 9-38). Over 2000 meters of sandstones, shales, lavas, and ash beds are found in the Klamath Mountains of northern California. Volcanic rocks in western Idaho and British Columbia attest to a continuation of vigorous volcanism from the Mississippian through the Permian. Crustal deformation along the west side of the Cordilleran belt is indicated by an extensive angular

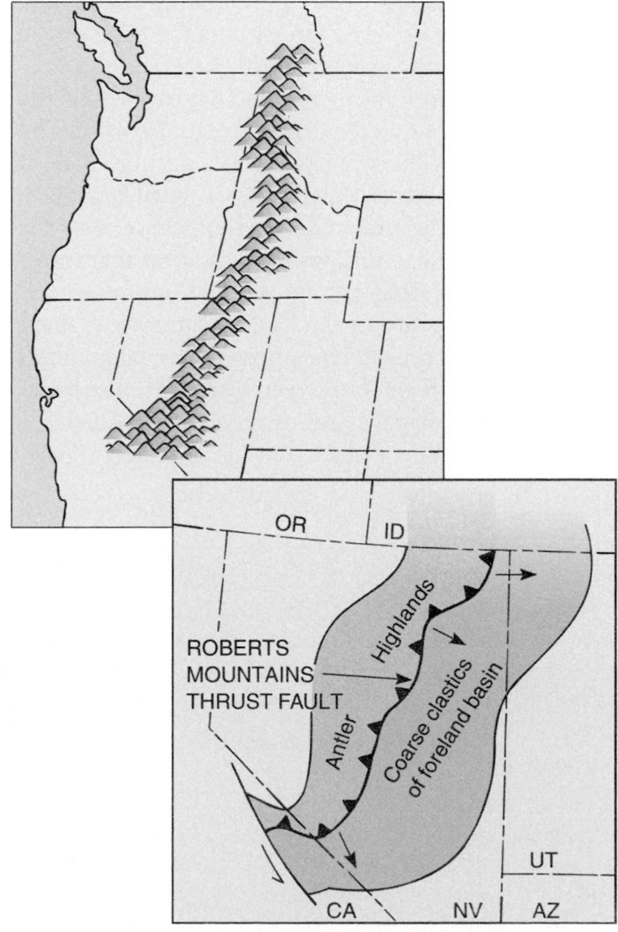

FIGURE 9-35 Extent of highland areas associated with the Antler orogeny, and location of the Roberts Mountains Thrust Fault.

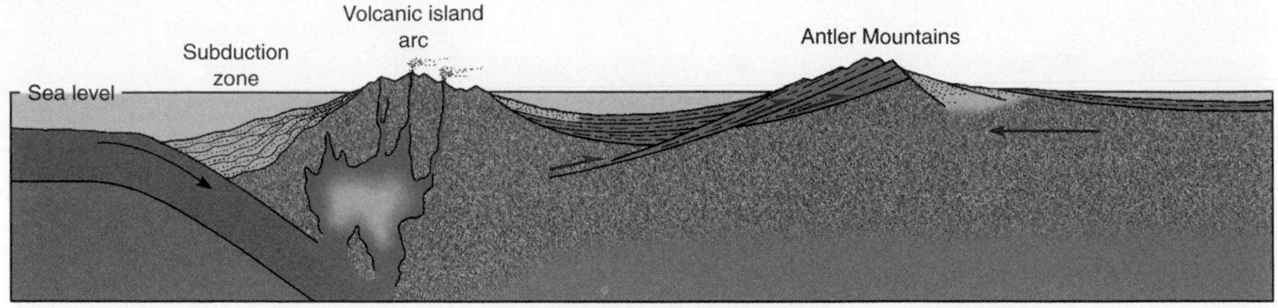

FIGURE 9-36 An interpretation of conditions in the Cordilleran orogenic belt in Early Mississippian time, shortly after the Antler orogeny. (*Based on diagrams by Poole, G. F. 1974. Society of Economic Paleontologists and Mineralogists Special Publication 22:58–82.*)

unconformity between Permian and Triassic sequences. These Permian–Triassic disturbances of the Cordilleran region have been named the **Cassiar Orogeny** in British Columbia and the **Sonoma Orogeny** in the southwestern United States. Like the earlier Antler Orogeny, the Sonoma event was probably caused by the collision of an eastward-moving island arc against the North American continental margin in west-central Nevada. Oceanic rocks and remnants of the arc were forced into the edge of the continent and became part of North America.

Permian conditions in the shelf area east of the Antler uplift were quieter than those to the west. The region was occupied by a shallow sea during much of the Permian Period. Within that sea, several large platform deposits accumulated. One of these was the Kaibab Limestone, an imposing formation that forms the vertical cliffs along the rim of the Grand Canyon. Beneath and eastward of the Kaibab Limestone, there are red beds that reflect deposition on coastal mudflats and floodplains. Sand dunes were prevalent nearby, as indicated by the massive, extensively cross-bedded Coconino Sandstone. Driven by northeasterly paleowinds, sand dunes also spread across the region that now lies north of the Grand Canyon of the Colorado River. The dunes, now frozen in time, are part of the DeChelly Sandstone, which displays the sweeping cross-beds (Fig. 9-39) and pitted quartz grains characteristic of eolian deposits.

While the Kaibab was being deposited in the southwest, a relatively deep marine basin was developing to the north in the general area now occupied by Wyoming, Montana, and Idaho. In this basin, sediments of the Phosphoria Formation were deposited. Although the Phosphoria includes beds of cherts, sandstones, and mudstones, it takes its name from its many layers of dark phosphatic shales, phosphatic limestones, and phosphorites. Phosphorite, a dark gray concretionary variety of calcium phosphate, is mined from the formation and used in the manufacture of fertilizers and other chemical products. The unusual concentration of phosphates may have been produced by upwelling of phosphorus-rich sea water from deep parts of the basin. Metabolic activities of microorganisms may have assisted in the precipitation of phosphate salts.

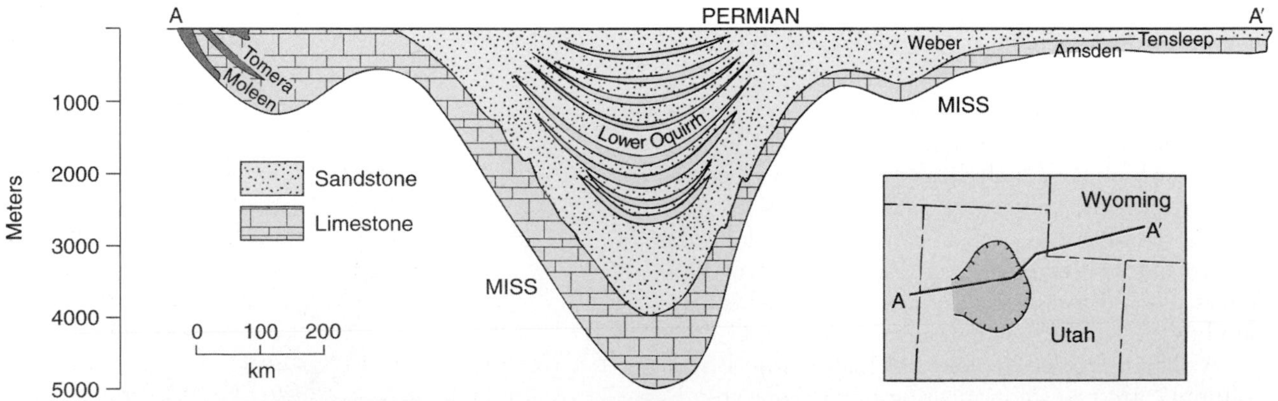

FIGURE 9-37 Section across the Oquirrh basin. (*From Stratigraphic Atlas of North and Central America. Shell Oil Company, Exploration Department.*) ? *What mountain-building episode provided a highland source area for this great accumulation of sandstone?*

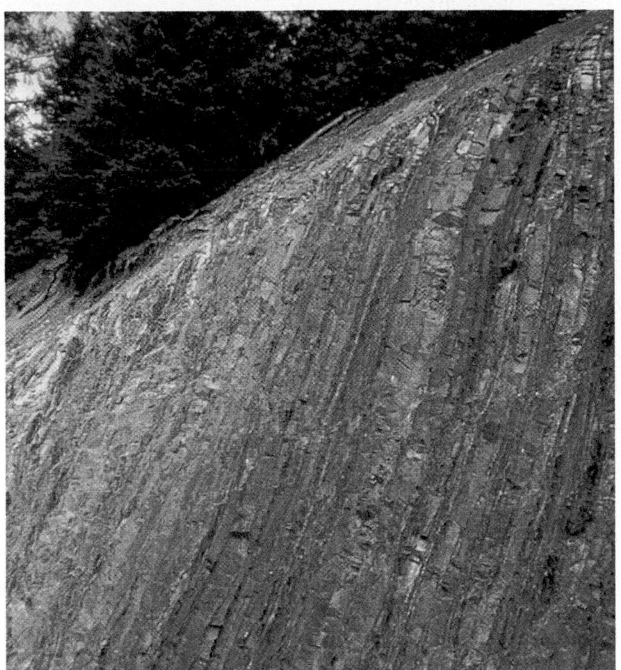

FIGURE 9-38 The Mississippian Bragdon Formation of the eastern Klamath Mountains of California. The Bragdon is composed of siltstones, sandstones, and felsic ash shed from a volcanic island arc that formed next to North America during late Paleozoic time. (*Courtesy of M. Miller.* ? *What is the approximate angle of dip in these beds?*

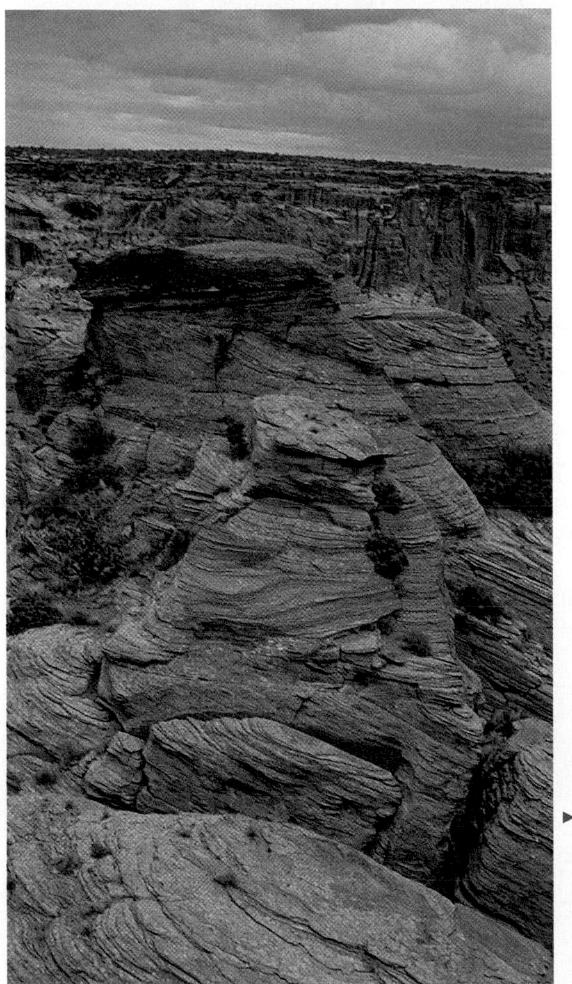

FIGURE 9-39 Cross-bedded DeChelly Sandstone of Permian age, Canyon de Chelly National Monument, northeastern Arizona. (*Photograph by R. F. Dymek.*)

EUROPE DURING THE LATE PALEOZOIC

During the late Paleozoic, Europe was bordered by the Uralian seaway on the east and the Hercynian on the south. Along the northern margin of Europe, the Caledonian orogeny had produced the Old Red Continent. Uplands provided a source for clastic sediments. These were swept into numerous basins to accumulate to thicknesses exceeding 10,000 meters. Judging from the many interlayers of ash and lava, volcanic activity was frequent on the Old Red continent. Fossil fishes and plant remains indicate that Europe's climate was tropical and possibly semiarid at the time.

South of the Caledonian mountains, the continental deposits gradually thin and grade into marine shales and limestones (see Fig. 8-34*C*). For a time, quiet prevailed in western Europe. However, in the Late Devonian and Early Mississippian, the depositional sites were intensely folded, metamorphosed, and intruded by granites as continents continued the collisions that were to produce Pangea. The late Paleozoic orogenic event has been named the **Hercynian orogeny**. It produced a great range of mountains across southern Europe. The Hercynian orogeny occurred at approximately the same time as the Allegheny orogeny in North America. For the most part, the eroded stumps of the Hercynian mountains are now buried, but here and there patches of the younger covering rocks have been eroded away, revealing the intensely folded, faulted, and intruded older rocks that lie beneath.

From the Hercynian uplands, gravel, sand, and mud were carried down into basins and coastal environments.

The clastic deposits were quickly clothed in dense tropical forests. Burial and slow alteration of vegetative debris from these forests provided the material for coal formation in the great European coal basins. Plant fossils found in these coal seams are of the tropical kind and differ from the more temperate Gondwana flora of the Southern Hemisphere.

Although the greatest amount of Hercynian deformation occurred toward the end of the Early Carboniferous (Mississippian) in Europe, spasms of unrest continued in both the Late Carboniferous and the Permian. These later episodes of folding correlate with similar deformations in the southern Appalachian and Ouachita orogenic belts of North America.

No less important than the Hercynian basins in Europe was the lengthy Uralian tract, which extended along a belt now occupied by the Ural Mountains. The Uralian basins were already in existence at the beginning of the Paleozoic Era. Paleomagnetic evidence indicates that the Russian and Siberian platforms were widely separated during early Paleozoic time and that they began to converge in the middle part of the era and ultimately collided by the end of the Paleozoic (see Fig. 9-1). Indeed, the western Urals consist largely of oceanic material scraped off against the edges of the converging plates. The collision resulted in the formation of a great mountain system along the entire Uralian orogenic belt. It was the event that unified ancestral Siberia with eastern Europe.

In central Europe and parts of Russia, there were temporary Late Permian marine incursions that precipitated evaporites. The famous German potassium salts are a product of one of these inland seas, which has been named the **Zechstein sea**. Thus, aridity was as characteristic of western Europe during the Permian as it was of Texas and New Mexico.

GONDWANA DURING THE LATE PALEOZOIC

During the late Paleozoic, the great landmass of Gondwana remained fairly intact. It moved across the South Pole and more fully entered the side of the Earth on which Laurasia was located. In its northward migration, Gondwana closed the ocean that separated it from Laurasia, causing the Hercynian orogeny of Europe and the Allegheny orogeny of North America. Orogenic activity associated with subduction zones was also evident in the late Paleozoic history of Gondwana, particularly along the Andean belt of South America and the Tasman belt of eastern Australia.

The most dramatic paleoclimatologic event of Gondwana's late Paleozoic history was the growth of vast continental glaciers (see Fig. 9-2). Extensive layers of tillite (Fig. 9-40) and the scour marks of glaciers (see Fig. 9-3) have been found at hundreds of locations in South America, South Africa, Antarctica, and India. There are indications of at least four and possibly more glacial advances, suggesting a pattern of cyclic glaciation not unlike that experienced by North America and Europe less than 100,000 years ago. The orientation of striations chiseled into bedrock by the moving ice suggests that the glaciers moved northward from centers of accumulation in southwestern Africa and eastern Antarctica. During the warmer interglacial stages and in outlying less frigid areas, *Glossopteris* and other plants tolerant of the cool, damp climates grew in profusion and provided the materials for thick seams of coal.

FIGURE 9-40 Upper Carboniferous tillite at base of slope (beneath hammer) composed of Precambrian basalt, Kimberley, South Africa. (*Courtesy of W. Hamilton, U.S. Geological Survey.*) ? *How does tillite differ from consolidated stratified drift?*

In time the ice receded, and Permian nonmarine red beds and shales were deposited on the Gondwana craton. As we shall see in the next chapter, some of these sediments contain the fossil remains of the ancestors of the Earth's first mammals.

CLIMATES OF THE LATE PALEOZOIC

The main climatic zones of the late Paleozoic paralleled latitudinal lines, just as they do today. Of course, the continents were located differently. The South Pole was in South Africa and the North Pole was in open ocean. The equator extended northeastward across Canada and southward across Europe (Fig. 9-41). Coal beds containing tropical plants, as well as evaporite deposits, coral reefs, and desert red beds developed within 40° of the paleoequator. Warmth-loving amphibians and reptiles lived in forests containing dense growths of ferns and scale trees.

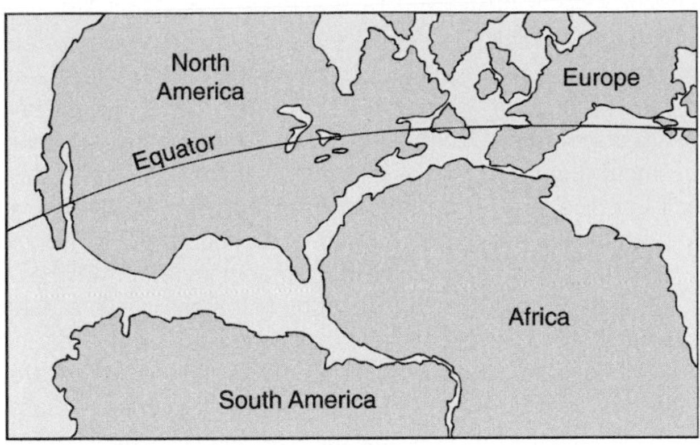

FIGURE 9-41 Approximate relationship of continents to the equator during the Carboniferous (Mississippian–Pennsylvanian).

Such balmy conditions were not characteristic of the Gondwana continents. A pole location over South Africa certainly contributed to the cold that gripped southern continents near the end of the Paleozoic. Atmospheric carbon dioxide may also have been a contributing cause. Carbon dioxide keeps our planet warm by trapping heat radiated from the Earth's surface. The phenomenon is known as the greenhouse effect. The more carbon dioxide, the warmer the atmosphere. Reduce the amount of atmospheric carbon dioxide, and the atmosphere becomes cooler. If cooling of the atmosphere during the Carboniferous was caused by a decrease in carbon dioxide, what could have caused that decrease? The Carboniferous is notable for prodigious amounts of plant matter produced and buried to form huge volumes of coal. We are aware that the conversion of dead vegetation to coal requires *burial*, for if organic matter is left exposed to air, carbon in that material will be combined with atmospheric oxygen and removed as carbon dioxide. Burial of organic matter thus prevents formation and addition of carbon dioxide to the atmosphere. With less carbon dioxide in the atmosphere, perhaps the greenhouse effect was reduced sufficiently to contribute to the late Paleozoic Gondwana ice age.

MINERAL PRODUCTS OF THE LATE PALEOZOIC

Although a great variety of mineral deposits were formed during the late Paleozoic, the fossil fuels (coal, oil, and gas) are particularly significant. Coal occurs in all post-Devonian systems. In Northern Hemisphere continents, it is particularly characteristic of the Late Carboniferous (Pennsylvanian). Thick deposits of Pennsylvanian coal occur in the Appalachians, the Illinois basin (Fig. 9-42), and the industrial heartland of Europe. Some sequences of Pennsylvanian strata may include many coal beds, as in West Virginia, where 117 different layers have been named. In the anthracite district of western Pennsylvania, orogenic compression has partially metamorphosed coal into an exceptionally high-carbon, low-volatile variety that is prized for its industrial uses. Permian coal seams are found in China, Russia, India, South Africa, Antarctica, and Australia.

Commercial quantities of oil and gas are frequently found in Upper Paleozoic strata. Devonian reefs within the Williston basin of Alberta and Montana have been exceptionally productive reservoir rocks for petroleum. Devonian petroleum has also been produced in the Appalachians. Indeed, in 1859, the first U.S. oil well was drilled into a Devonian sandstone. (Oil was struck at a depth of only 20 meters.) Carboniferous formations of the Rocky Mountains, midcontinent

A

B

FIGURE 9-42 Mining coal underground and at the surface. (*A*) A remote-controlled continuous mining machine with a flooded bed scrubber for dust control is removing coal in a West Virginia underground mine. (*B*) Large stripping shovel in the pit of an Illinois surface mine. (*Courtesy of Peabody Coal Co.*)

Acadia National Park

Acadia National Park, in Maine, is a delightful place in which to observe geologic features and processes. Thanks to the work of Pleistocene glaciers in stripping away soil and debris, one can walk directly on bedrock in many interior parts of the park. Along the shoreline, granite, diorite, and schist are revealed in wave-eroded cliffs and platforms. At Acadia one can witness the effects of igneous activity, explore glacial moraines, examine exotic rocks on cobble beaches, and observe the powerful action of waves in sculpting a rugged coast. All of this is easily reached by traveling southeast out of Bangor, through the town of Ellsworth to Bar Harbor (Fig. *A*).

Ice sheets covered Acadia and the rest of New England repeatedly during the Pleistocene Epoch. Each advance of the ice destroyed features formed by the previous advance, leaving features of the final glaciation clearly in evidence. The most apparent of these features are glacial scratches, glacial polish, and scars made in bedrock as rock debris carried along in the base of the glacier chipped out pieces of the underlying surface. There are also numerous erratics (Fig. *B*), including one called "Mammoth Rock" on the summit of South Bubble Mountain. Valleys in Acadia have the parabolic cross-section characteristic of valleys that have been glaciated. This distinctive shape results from the ability of glacial ice to erode, not only at the base of the valley as streams do but also well up onto the valley sides. As the glaciers moved down valleys, some areas were excavated deeper than others. Later, the deeper excavations filled with water to form so-called trough lakes (Sea Cove Pond, Eagle Lake, Jordan Pond). When the great Ice Age came to a close, meltwater that returned to the ocean caused sea level to rise. The ocean flooded into valleys and low areas, producing the remarkably irregular New England coast. Appropriately, it is called a drowned coastline.

Along the drowned coastline of Acadia one finds the most dramatic scenic attractions in the park. Here one achieves a vivid impression of the enormous energy released by waves as they batter sea cliffs and erupt in spray (Fig. *C*). Thunder Hole, south of Bar Harbor, provides a spectacular view of wave erosion in action. Thunder Hole is a deep gorge eroded along joints by pounding waves. When seas are heavy, water surges into the narrow chasm, compressing the trapped air, which, on release, resounds like the clap of thunder. Surging waves along the coast lift and carry gravel, which is hurled against cliffs, undercutting them and causing collapse. These processes ultimately produce sea caves, sea arches, and wave-cut platforms. In quieter areas, the waves deposit their load, forming cobble and shingle beaches.

A particularly intriguing aspect of Acadia's geology relates to the derivation of its most ancient rocks. Acadia is a part of an exotic or suspect terrane, other components of which have been traced from Nova Scotia down into the southern Appalachians. As noted in Chapter 6, suspect terranes are pieces of continental crust or island arc moved by sea-floor spreading from a distant place of origin to the margin of a continent and then sutured to the host continent at its subduction zone. They can be recognized as suspect terranes because they differ from their host continent in rock type, age, fossils, and paleomagnetic orientation.

The suspect terrane that includes Acadia National Park has been named **Avalonia**. It contains early Paleozoic

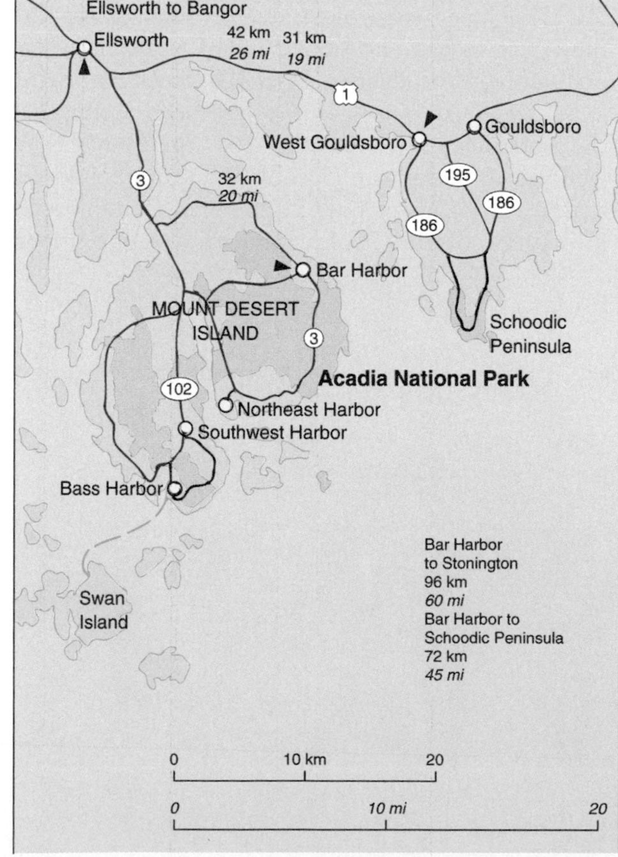

FIGURE A Location map for Acadia National Park. (*Courtesy of U.S. National Park Service.*)

FIGURE B Glacial erratic boulder on Cadillac Mountain, Acadia National Park. An erratic is a rock fragment that has been carried by glacial ice and deposited at some distance from the outcrop from which it was derived. Usually, erratics rest on rock of different lithology and age. (*Photograph by C. Toops.*)

FIGURE C Granite exposed along the rugged coast of Mount Desert Island, Acadia National Park. The impact of waves exerts pressures of as much as 6000 pounds per square foot, readily fragmenting and eroding huge masses of rock.

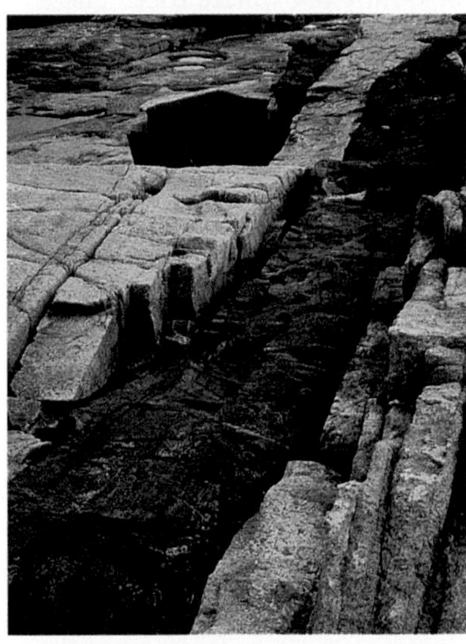

FIGURE D Basalt dike cutting across lighter colored Cadillac Granite on Acadia's Schoodic Peninsula, the only part of the park on the mainland.

rocks with trilobite and graptolite faunas distinctly different from those of adjacent areas of North America. In fact, these faunas more closely resemble those of Europe (Baltica).

When a suspect terrane arrives at a continent, it is said to have docked (rather like ships that arrive at a port one by one). Suspect terranes older than Avalonia had already docked against North America during the Ordovician Taconic orogeny. Avalonia came along somewhat later, during the Devonian. Rocks of the terrane indicate it is a vestige of an island arc that arrived at the subduction zone during the Acadian orogeny. After Avalonia had docked, at least three additional terranes were incorporated into the Appalachian orogen.

The Ordovician and Silurian rocks of Acadia are mostly metamorphosed sediments and volcanic ash formed long before Avalonia had moved to North America. From oldest to youngest, these rocks consist of the Ellsworth Schist, Cranberry Series, and Bar Harbor Series. An erosional unconformity marks the upper boundary of these three units. Intruded into these units (and hence younger) are rocks of Devonian age. The earliest intrusives are diorites that form dikes and massive sills. Some of the dioritic magma penetrated to the surface and erupted as fiery lava flows. After most of the dioritic magma had been emplaced and had crystallized, granitic magmas penetrated the diorites and metasediments. The initial fine-grained granites gave way to coarser-grained and finally medium-grained granites, reflecting variations in the cooling history of the intruded bodies. After these granitic rocks had been emplaced, basaltic lavas rose into fracture systems. Their dark color and sharp contacts make them clearly visible against the lighter granites. The best area in which to view the basalt dikes is in the part of the park called the Schoodic Peninsula (Fig. *D*). All of this igneous activity was accompanied by compressional deformation of the eastern margin of North America. Ultimately, however, compression gave way to tensional forces as the break-up of Pangea began following the Permian Period. Erosion and intermittent uplifts were the principal post-Paleozoic events until the coming of the glaciers during the Pleistocene.

(Fig. 9-43), and Appalachians also contain oil reservoirs. Wells drilled into reefs and sandstones of the Permian basin of West Texas have yielded huge amounts of oil. Oil trapped in Upper Paleozoic strata beneath the North Sea is produced for use in Europe.

Arid, warm climatic conditions, particularly in northern continents during the Late Paleozoic, provided a suitable environment for deposition of sodium and potassium salts. In Late Permian time, enormous amounts of phosphates (Fig. 9-44) were also deposited in Montana, Idaho, Utah, and Wyoming. Shales of the Phosphoria Formation provide phosphate needed as plant food.

Mountain building, with its attendant igneous activity, is nearly always accompanied by the emplacement of metallic ores. The Hercynian and Allegheny orogenies of the late Paleozoic generated ores of tin, copper, silver, gold, zinc, lead, and platinum. Deposits of all the precious metals, as well as copper, zinc, and lead, are found in the Urals of Russia and in China, Japan,

FIGURE 9-43 Drilling for oil in upper Paleozoic rocks of eastern Kansas. ? *What purpose does the derrick serve when drilling for oil?*

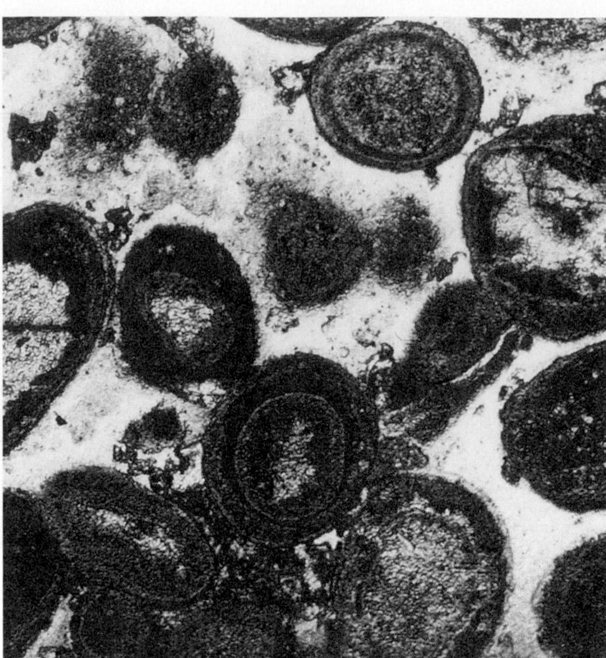

FIGURE 9-44 Pelletal, oölitic phosphate rock. The phosphatic pellets and oöids are cemented by microcrystalline quartz. This sample was obtained from an outcrop of phosphate rock in Madison County, Montana (average diameter of oöids, 0.8 millimeter). ? *As world populations increase, why do phosphate deposits become increasingly important?*

Burma, and Malaya. Tin, tungsten, bismuth, and gold are mined in Australia and New Zealand. It is self-evident that late Paleozoic rock sequences, with their stores of both metallic and nonmetallic economic minerals and their content of fossil fuels, are vitally important to the welfare of modern civilizations.

SUMMARY

The assembling of continents that had begun in the early Paleozoic continued throughout the remainder of the era. As plates converged, they carried island arcs and microcontinents to the margins of Europe and eastern North America, and these land areas were welded to their larger host continents. The collisions produced the Taconic and Acadian orogenies. As oceanic closure continued into the late Paleozoic, North America and Baltica collided, resulting in the Acadian orogeny. The subsequent Allegheny orogeny occurred near the end of the era as northwestern Africa moved against North America. To the south, the plate bearing South America converged on the underside of North America and resulted in the Ouachita orogenic activity.

Whereas the process of bringing continents into contact resulted in mountain building along their converging margins, the interiors were affected primarily by broad epeirogenic movements and associated advances and withdrawals of epicontinental seas. In North America, these shifts in shorelines are reflected in the marine and nonmarine sediments that compose the Kaskaskia and Absaroka cratonic sequences. Kaskaskia rocks are predominantly fossiliferous limestones, with sandstones and shales increasing in volume toward the eastern and southern depositional areas. As the Kaskaskia seas withdrew, deltaic and fluvial deposits spread across the old sea floor. The regression resulted in a great regional unconformity that marks the boundary between Mississippian and Pennsylvanian systems in North America. The most distinctive feature of Pennsylvanian sediments is their cyclic nature. In the midcontinent, they consist of alternate marine and nonmarine groupings. To the east, the Pennsylvanian sediments were largely terrestrial, whereas marine deposition prevailed in the western part of the craton. Near the end of the Paleozoic, epeiric seas gradually withdrew. Evaporite and red bed sequences were deposited in the Permian basins of New Mexico and West Texas.

Mountain building was not confined to eastern North America during the late Paleozoic. In the west, crustal disturbances raised a range of mountains that extended from Arizona to Idaho. Thick deposits of sand and gravel attest to the former existence of these ranges. There was further activity during the Pennsylvanian, when uplands were formed in Texas, Arkansas, and Oklahoma. A chain of mountains (the Colorado mountains) also developed across Colorado. These Colorado mountains were the source of sandstones that today form the famous red rocks of the Garden of the Gods along the Front Range of the Rockies.

The Old Red continent, which extended across Ireland, Wales, England, Scotland, and Scandinavia, was the dominant feature of Europe at the beginning of the Devonian. For a time, relative stability prevailed in Europe, and then much of the continent was thrown into an episode of folding, volcanism, and intrusion as it collided with Gondwana. These disturbances were part of the Hercynian orogeny. The collision of Gondwana with Laurussia was not only the source of the Hercynian orogeny in central Europe but also the cause of the Allegheny orogeny along the eastern and southern margin of North America.

Gondwana was ringed by orogenic belts during the late Paleozoic, and the Andean and Tasman segments experienced late Paleozoic orogenesis. The central regions were occasionally covered with inland seas in which were deposited fossiliferous limestones and shales. During the Permian, when the northern landmasses were relatively warm and sometimes arid, Gondwana was cooler. The Permian South Pole was located in what is now South Africa, and all of the now-separated Gondwana continents exhibit the scars of widespread Permian glaciations.

QUESTIONS FOR REVIEW AND DISCUSSION

1. Describe the relation between the orogenies that produced the Appalachian and Ouachita mountains and the movement and position of tectonic plates.

2. What is the plate tectonic reason for the approximate time equivalence between the Allegheny and Hercynian orogenies?

3. In the study of an ancient mountain range, how might a geologist recognize displaced terrane? Having recognized a terrane as displaced (alien or exotic), how might he or she determine that the terrane was from an island arc? From a microcontinent?

4. Why is it logical to divide the late Paleozoic into two cratonic sequences? What are the names and durations of those sequences?

5. When did the Old Red continent develop and where was it located? What kind of sediment was being deposited across the Old Red continent?

6. How do the deposits of the Old Red continent compare with those of the Catskill clastic wedge?

7. What is the geologic evidence of the occurrence of a geographically extensive episode of continental glaciation in the Southern Hemisphere during the late Paleozoic?

8. What is the paleoenvironmental significance of an association of red sandstone, salt, and gypsum deposits? Where and when during the late Paleozoic did deposition of this association of rocks form?

9. What is a cyclothem? Prepare a list and discuss each of the eustatic and tectonic conditions that might account for the cyclicity for which cyclothems are named.

10. Describe the environmental conditions under which the Chattanooga Shale was deposited. What problem is associated with hypotheses for the origin of this far-ranging blanket of dark shales?

11. Approximately where was the South Pole located during the Permian Period? How have geologists ascertained that position? What confirmation for the pole position can be obtained from the global distribution of corals?

12. Name three metallic and three nonmetallic mineral resources extracted from Upper Paleozoic rocks.

13. Explain the occurrence of oil and gas in the calcareous algal reeflike structures found in the shelf carbonates of the Paradox basin.

READINGS

Bally, W., and Palmer, A. R. (eds.). 1989. The geology of North America; an overview. *Decade of North American Geology*, vol. A. Boulder, CO: Geological Society of America.

Bambach, R. K., Scotese, C. R., and Ziegler, A. M. 1980. Before Pangaea: The geographies of the Paleozoic world. *Am. Scientist* 68(1):26–28.

Bird, J. M., and Dewey, J. F. 1970. Lithosphere plate–continental margin tectonics and the evolution of the Appalachian orogeny. *Bull. Geol. Soc. Am.* 81:1031–1060.

Burchfiel, B. C., Cowan, D. S., and Davis, G. A. (eds.). 1992. Tectonic overview of the Cordilleran orogen in the western U.S. *Decade of North American Geology*, vol. G-3. Boulder, CO: Geological Society of America.

Frazier, W. J., and Schwimmer, D. R. 1987. *Regional Stratigraphy of North America*. New York: Plenum Press.

Hatcher, R. D., Thomas, W. A., and Viele, G. W. (eds.). 1989. The Appalachian–Ouachita orogen in the United States. *Decade of North American Geology*, vol. F-2, Boulder, CO: Geological Society of America.

Moores, E. M., and Twiss, R. J. 1995. *Tectonics*. New York: W. H. Freeman & Co.

Rodgers, J. J. W. 1993. *A History of the Earth*. Cambridge, England: Cambridge University Press.

Woodrow, D. L., and Sevon, W. D. (eds.). 1985. The Catskill Delta. *Geological Society of America Special Paper 201*. Boulder, CO: Geological Society of America.

WEB SITES

The Earth Through Time Student Companion Web Site (www.wiley.com/college/levin) has online resources to help you expand your understanding of the topics in this chapter. Visit the Web Site to access the following:

1. Illustrated course notes covering key concepts in each chapter;
2. Online quizzes that provide immediate feedback;
3. Links to chapter-specific topics on the web;
4. Science news updates relating to recent developments in Historical Geology;
5. Web inquiry activities for further exploration;
6. A glossary of terms;
7. A Student Union with links to topics such as study skills, writing and grammar, and citing electronic information.

10

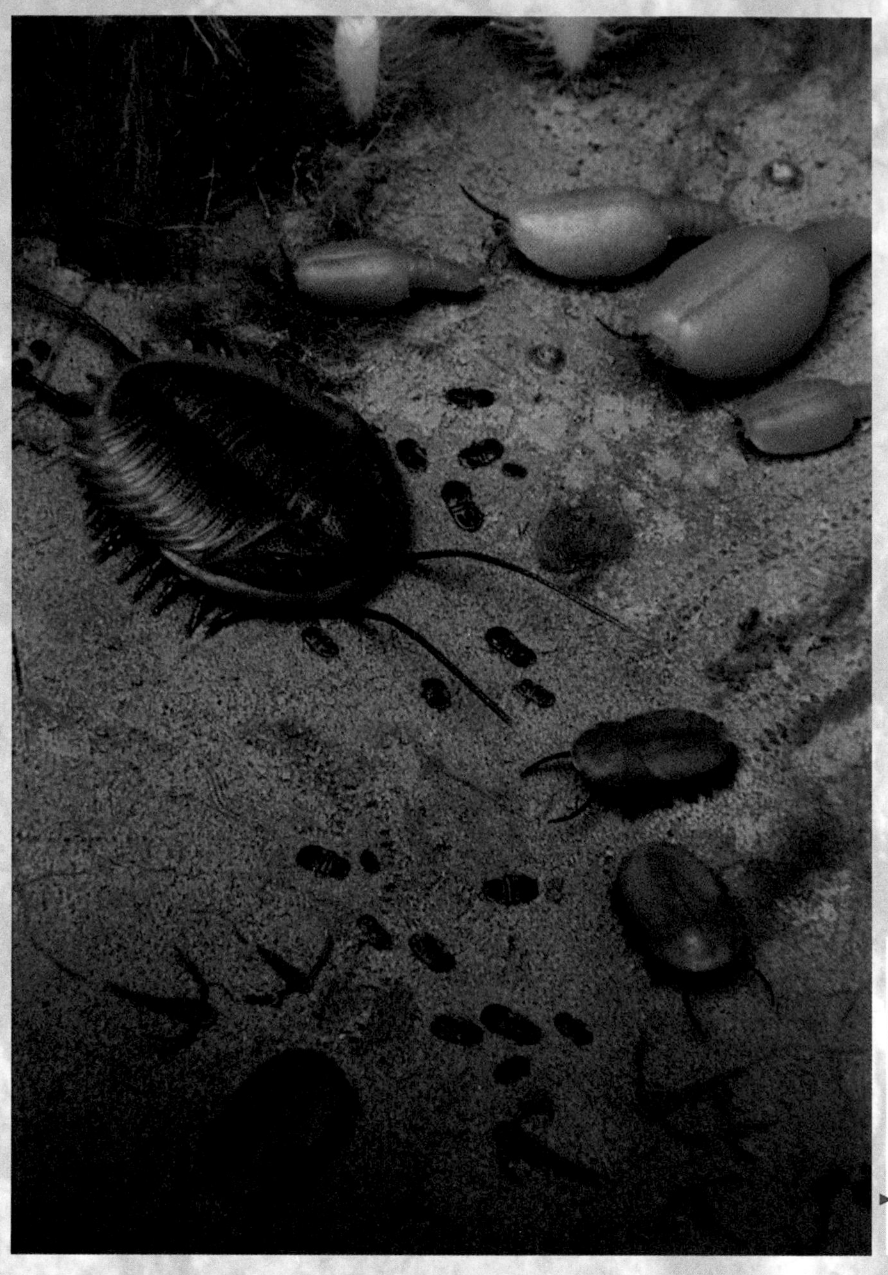

Reconstruction of some of the arthropods preserved in the Cambrian Burgess Shale of British Columbia. The organisms provide a view of marine life as it existed 530 million years ago.

CHAPTER 10

Life of the Paleozoic

When you were a tadpole and I was a fish
In the Paleozoic Time,
And side by side on the ebbing tide
We sprawled through the ooze and slime,
Or skittered with many a caudal flip
Through the depths of the Cambrian fen,
My heart was rife with the joy of life,
For I loved you even then.
Langdon Smith, *Evolution*

OUTLINE

The pace of evolution appears to have gone into double-quick time at the start of the Phanerozoic Eon. The rich fossil record stands in startling contrast to that of the Precambrian. We have noted that most of the Precambrian was a world of microbes. Metazoans appear only in uppermost Proterozoic strata, such as those at Ediacara Hills in Australia. Except for a few "small shelly" creatures, most Precambrian animals had not acquired the ability to secrete shells.

Certainly the improvement in the fossil record so evident in Cambrian strata can be attributed to the spread of shell-building abilities. During the Cambrian, shell-bearing trilobites and brachiopods were particularly abundant, but an even greater expansion of shelly and coralline animals followed in the early Ordovician. Epeiric seas that spread across the cratons of the world provided a multitude of habitats and opportunities for diversification and expansion of invertebrates. Almost every phylum of shell-bearing invertebrates living today had originated by Ordovician time. The proliferation of animals with shells was probably related to the protection and support for soft tissues that shells provide.

Later in the Paleozoic, as invertebrates continued their remarkable success in populating the seas, plant evolution progressed steadily to the point at which the transition from water to land was accomplished. Vascular land plants spread across the continents, ultimately producing densely forested regions. From streams and swamps of forests, vertebrates with legs instead of fins came ashore.

Animals with backbones—the vertebrates—also made their appearance during the Paleozoic. The earliest vertebrates were jawless fishes now known from Cambrian rocks in China. Subsequently, fishes with articulated lower jaws developed and triggered a Devonian burst of evolution that gave rise to all the major groups of fishes. Some of these groups still exist. Also in the Devonian, an advanced lineage of fishes with stout fins and primitive lungs provided the stock from which four-legged animals, the **tetrapods**, evolved. Although capable of walking on land, these tetrapods were linked to swamps, lakes, and streams by eggs that survive only in water. Cladistic taxonomists prefer the term **Tetrapoda** for land vertebrates having this style of reproduction, although in informal use they are referred to as amphibians.

An evolutionary innovation known as the **amniotic egg** liberated vertebrate land dwellers from their reproductive dependency on water bodies. With an egg that would survive on land, the nonamniotic tetrapods provided the lineage from which early **amniotes** evolved (see Figure 10–73). By Permian time, amniotes called therapsids were present. Therapsids possessed certain morphologic features of mammals. The arrival of true mammals, however, was a Mesozoic event.

Life during the Paleozoic was not free of calamity. The era is marked by intervals when the global environment became inhospitable or when catastrophic events occurred. At such times, many entire families of animals suffered extermination. Such biological disasters are termed mass extinctions. During the Paleozoic Era, mass extinctions ocurrred late in the Ordovician, Devonian, and Permian periods.

THE ARRIVAL OF ANIMALS WITH SHELLS

As indicated by the intriguing Ediacaran fauna described in Chapter 7, multicellular animals were already present before the Paleozoic Era. Fossil discoveries in Namibia, Africa, have shown that the Ediacaran animals, although characteristic of the late Precambrian, ranged across the Cambrian boundary. The Ediacaran creatures, however, were soft-bodied and therefore infrequently preserved. When animals began to secrete hard parts, the probability of their leaving preserved remains improved immensely. As noted in the previous chapter, the first animals to achieve this milestone in the history of life are collectively called **small shelly fossils** (Fig. 10–1). They are also referred to as the **Tommotian fauna** because their shells are found in rocks of the Cambrian Tommotian Stage. Many of the small shelly fossils had phosphatic shells.

Although they are usually found in strata at the base of the Cambrian System, small shelly fossils have been shown to date back to the Vendian Stage of the late Proterozoic (Fig. 10–2). They were widely distributed around the globe. The tiny fossils are rarely more than a few millimeters in size and include shells and skeletal elements of mollusks, sponges (known from spicules), and a variety of small shelly invertebrates of uncertain classification that secreted tubular or cap-shaped shells. *Cloudina* (See Fig. 7-32), an organism that lived from Neoproterozoic time into the Cambrian, secreted an annulated calcareous tube, whereas the phosphatic shell of *Anabarites* (Fig. 10–3) resembles three "tubes" open to one another along the length of the shell. Each tube is surmounted by a narrow keel that may have provided stability for the animal on the soft mud of the sea floor. The coiled shell of *Aldanella*

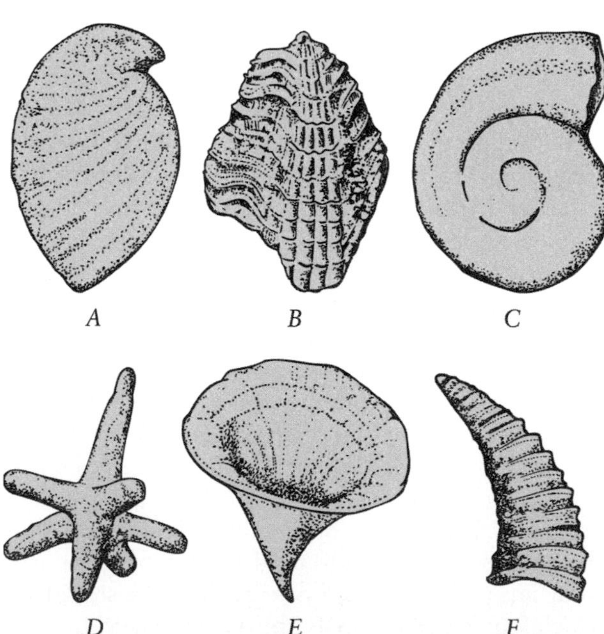

FIGURE 10-1 Late Precambrian and Early Cambrian shell-bearing fossils from Siberia. (*A*) *Anabarella*, × 20, a gastropod; (*B*) *Camenella*, × 18, affinity uncertain; (*C*) *Aldanella*, × 20, a gastropod; (*D*) sponge spicule, × 30; (*E*) *Fomitchella*, × 45, affinity uncertain; and (*F*) *Lapworthella*, × 20. (*After Matthews, S. J., and Missarzhevsky, V. V. J. 1975. Geol. Soc. London 131:289–304.*)

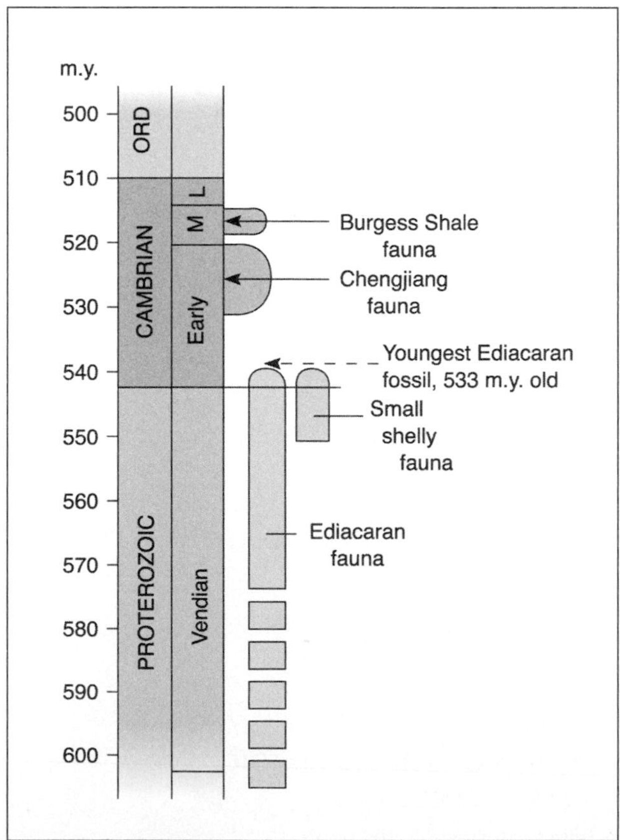

FIGURE 10-2 Geologic time scale across the Proterozoic–Cambrian boundary showing position of the Ediacaran, Chengjiang, and Burgess Shale faunas.

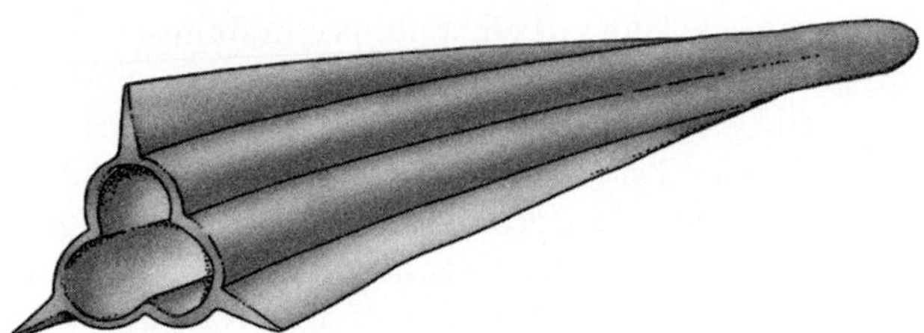

FIGURE 10-3 ***Anabarites*, an early Cambrian shelled fossil.** The shell is only 5 millimeters in length. *(After a drawing by Dianna L. Schulte McMenamin in McMenamin, M. A. S., and McMenamin, D. L. S. 1990.* The Emergence of Animals. *New York: Columbia University Press.)*

was secreted by a Tommotian gastropod (snail). Another form, *Lapworthella* (see Figure 10-1), has the shape of a curved cone and is ornamented with grooves and ridges. It was probably not the single shell of an animal, but rather one of many similar skeletal elements that covered all or part of the body. Such an interpretation is favored by the discovery of *Lapworthella* elements fused together in side-by-side arrangement.

Most of the small shelly fauna had disappeared by the end of the first stage of the Lower Cambrian. Immediately thereafter, large-shelled invertebrates became abundant and diverse. Familiar fossils such as trilobites and brachiopods dominate the fossil record. But there were many soft-bodied creatures as well. Some of these are known from extraordinary preservation at only a few localities, such as the Chengjiang fossil site in China and the Burgess Shale site in British Columbia. Fossil sites containing such extraordinary fossils in abundance are termed **lagerstätten**, from a German word for "mother lode."

Windows into the Past: The Burgess Shale and Chengjiang Lagerstätten

High on a ridge near Mount Wapta, British Columbia, there is an exposure of Middle Cambrian Burgess Shale that contains one of the most important faunas in the fossil record (Fig. 10–4). The fossils in the Burgess Shale are reduced to shiny, black impressions on the bedding planes of the black shale (Fig. 10–5). Most are the remains of animals that lacked hard parts. Altogether, they form an extraordinary assemblage (Fig. 10–6) that includes four major groups of arthropods (trilobites, crustaceans, and members of the taxonomic groups that include scorpions and insects) as well as sponges, onycophorans (see Fig. 10–4C), crinoids, mollusks, three phyla of worms, and possibly corals, chordates, and many species that defy placement in any known phylum.

Although trilobites had been discovered in the so-called "Mount Stephens Trilobite Beds" on nearby Mount Stephens as early as 1885, the richly diverse

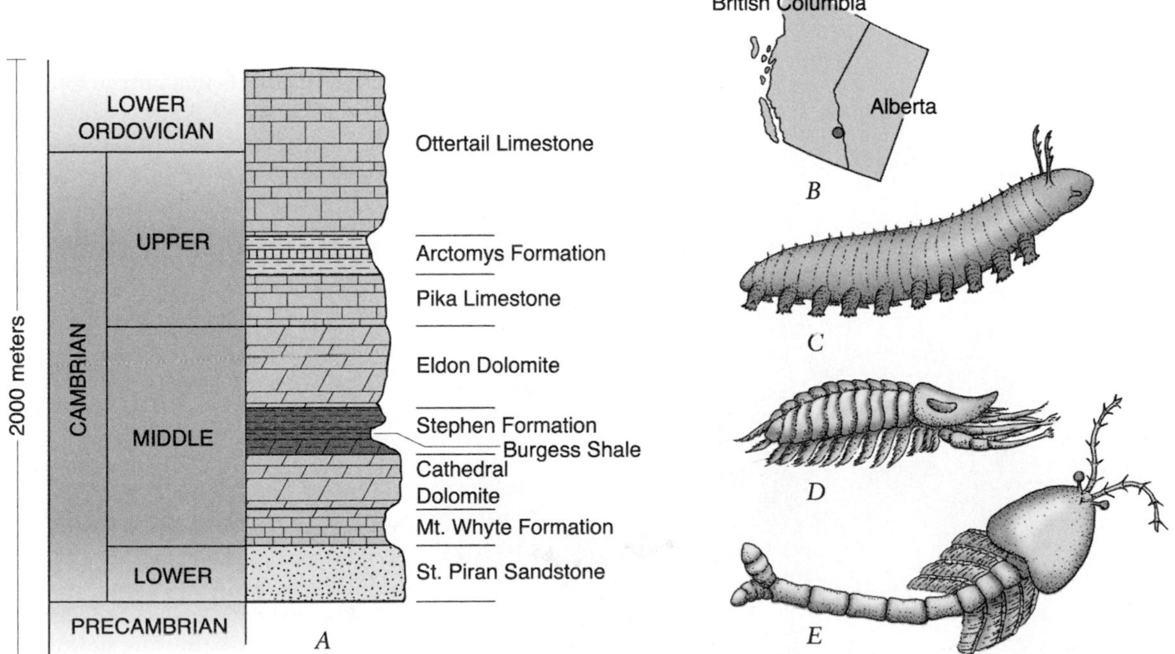

FIGURE 10-4 The Cambrian geologic column (*A*) at Kicking Horse Pass, British Columbia (*B*) where Walcott discovered the Burgess Shale Fauna. *Aysheaia* (*C*) is an invertebrate called an onycophoran or velvet worm. They are of particular interest because they appear to be intermediate in evolution between segmented worms and arthropods. *Leanchoila* (*D*) and *Waptia* (*E*) are among the many kinds of arthropods found at this locality.

FIGURE 10-5 ***Haplophrentis.*** This photograph illustrates the nature of preservation of Burgess Shale fossils. *Haplophrentis* had a tapering shell surmounted by a lid or operculum. The shell could be closed for protection. The lateral blades on either side may have served as props. The length of the shell is 2 centimeters. (*Courtesy of the U. S. National Museum of Natural History, Smithsonian Institution; photograph by Chip Clark.*)

Burgess Shale fauna was first noticed in August of 1909 by Charles D. Walcott. The initial discovery consisted of fossil-bearing slabs of rock that had been carried downslope by a snow slide. The discovery was made at the end of the field season, and rock exposures higher up the slope were already blanketed with snow. As a result, Walcott and his crew of fossil collectors (including Mrs. Walcott and two sons) were forced to leave the site and return the following year. At that time, they traced the fossiliferous slabs upward to their source in the Burgess Shale. Quarrying was begun immediately and continued during the 1912, 1913, and 1917 field seasons.

Altogether, about 60,000 specimens were collected and stored in the U.S. National Museum of Natural History, where Walcott served as secretary of the Smithsonian Institution. In the 1960s, these fossils were re-examined by Harry B. Whittington, who then joined paleontologists of the Geological Survey of Canada in reopening the quarry and assembling a second collection. Whittington devoted the next 15 years to an incisive and critical scrutiny of the Burgess Shale fauna. He and his colleagues in Britain were able to describe and depict the anatomy of the Burgess animals with great clarity. During the study, it became apparent that earlier interpretations of many Burgess animals as ancestors to existing taxonomic groups

FIGURE 10-6 The Burgess Shale diorama at the U.S. National Museum of Natural History. The reconstruction is based on actual fossil remains of organisms. It depicts a benthic marine community of the Middle Cambrian. A steep submarine escarpment forms the background. Slumping along this wall contributed to the preservation of the Burgess fauna by burying the organisms. Along the wall, one can see green and pink vertical growths of two types of algae. The large purplish creatures that resemble stacks of automobile tires are sponges (*Vauxia*). The blue animals are trilobites (*Olenoides*). The brown arthropods with distinct lateral eyes are named *Sidneyia*. Climbing out of the hollow on the sea floor are crustaceans called *Canadaspis*. The yellow animals swimming toward the right above the sea floor are *Waptia*. *Opabinia* has crawled out of the left side of the hollow. Burrowing worms are visible in the vertical cut at the bottom. (*Courtesy of the U. S. National Museum of Natural History, Smithsonian Institution; photograph by Chip Clark.*)

were incorrect and that Burgess fauna included creatures so different as to warrant placement in new taxonomic categories.

Among the sometimes bizarre and unique creatures of the Burgess fauna were small, elongate animals that are currently believed to be chordates. Chordates are animals that, at some stage in their development, have a notochord (an internal supportive rod) and a nerve cord that extends along the dorsal side (upper side) of the notochord. Chordates like ourselves, in which the notochord is supplanted by a series of vertebrae, are called vertebrates. The early chordate found in the Burgess Shale has been named *Pikaia* (Fig. 10–7) after Mount Pika, near the Burgess Shale discovery site. Fossils of *Pikaia* exhibit two features upon which their designation as chordates is based. They possess a notochord and a series of V-shaped muscles along the sides of the body, as is characteristic of the musculature in fish. These muscles, working against the rigid but flexible notochord, provide a sinuous, fishlike body motion required for swimming.

For a time *Pikaia* had the distinction of being the oldest known chordate. However, probable close relatives of *Pikaia*, but about 10 million years older, were found at the Chengjiang fossil site in Yunnan Province, China. These early chordates have been named *Cathaymyrus* and *Yunnanozoon*. They are 535 million years old. The Chengjiang fossils also include the world's oldest known fish. In 1999, paleontologists Shu

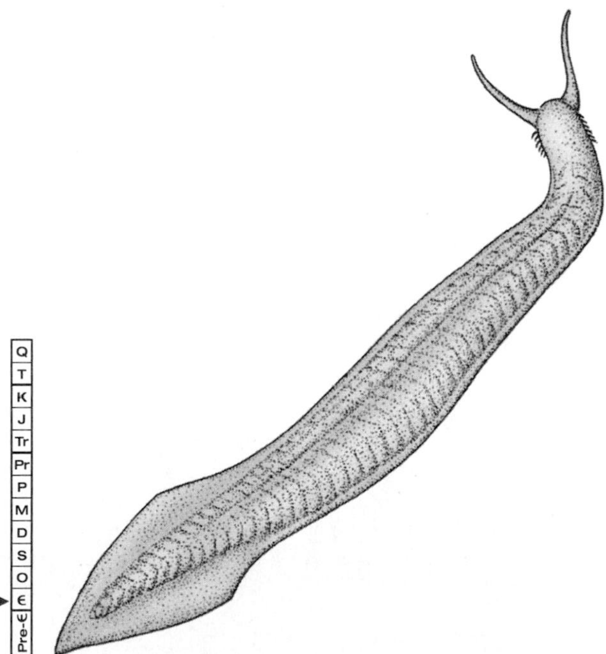

FIGURE 10-7 Reconstruction of *Pikaia*, an early member of our own phylum, the Chordata. Note the rod along the animal's back that appears to be a notochord (length is about 4 centimeters). **?** *Name another chordate feature seen in Pikaia fossils.*

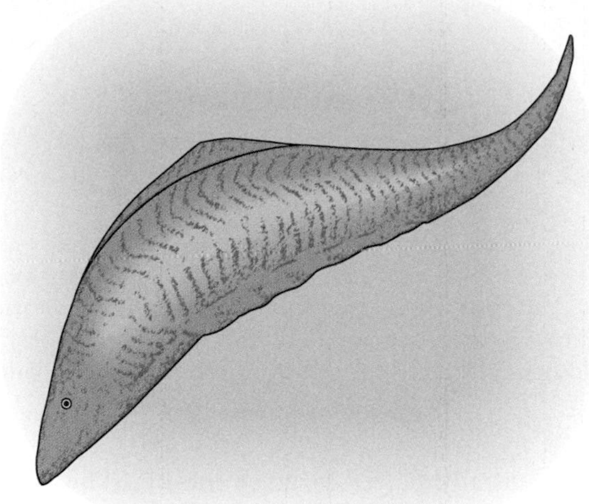

FIGURE 10-8 The Early Cambrian fish *Myllokunmingia*. This fossil fish, only about 3 centimeters in length, was discovered in the richly fossiliferous beds near the town of Chengjiang in China.

Degan, of Northwest University in Xi'an, and Simon Conway Morris, of Cambridge University, published descriptions of two tiny (3–4 centimeters long) Cambrian fishes named *Haikouichthys* and *Mylokunmingia* (Fig. 10–8).

Even as each day brings new discoveries from China, the Burgess Shale fauna continues to be important because of the perspective it gives us on the diversity of Cambrian life. That diversity includes such incomparable creatures as *Anomalocaris*, *Opabinia*, *Marrella*, and *Hallucigenia*. More than a half-meter long, *Anomalocaris* (Fig. 10–9) was a huge predator. On its head were two stalked eyes and a pair of feeding appendages used to capture and carry prey

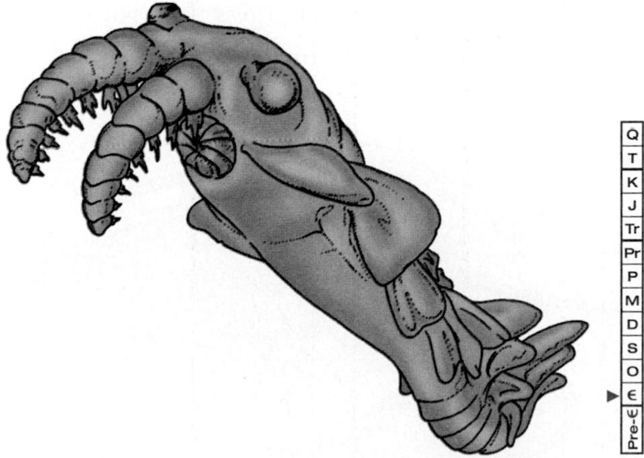

FIGURE 10-9 Anomalocaris. This giant predator (often 60 centimeters in length) captured its prey with its huge frontal appendages and passed the victims back to the circular mouth with its outer and inner circles of teeth. The side flaps were used in swimming, like underwater wings.

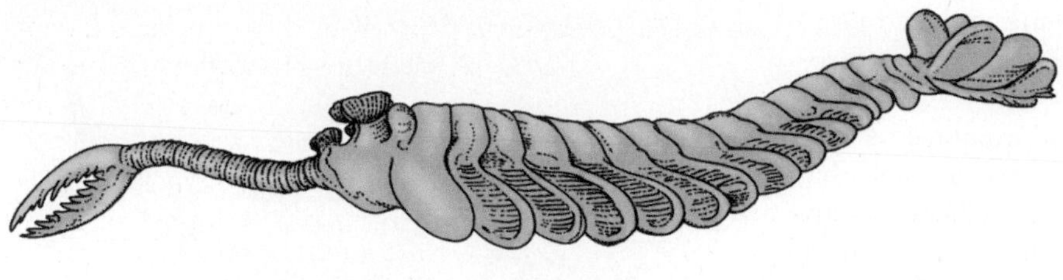

FIGURE 10-10 ***Opabinia.*** This strange Burgess Shale animal had a frontal "nozzle" with a jaw-like structure at its end, used for gathering food. There were five eyes on its head. Each side was covered with overlapping lobes bearing narrow stripes interpreted as gills. This specimen was 7 centimeters long. (*Adapted from Whittington, H. B. 1975.* Philosophical Transactions of the Royal Society of London, *Vol. B.*)

to its ventrally placed circular mouth. The mouth was ringed with sharp teeth that were likely responsible for wounds observed in the fossils of other members of the community. *Anomalocaris*, along with a similar arthropod named *Lagginia*, are considered members of a new taxonomic class of arthropods, the Dinocarida. Paleontologist Desmond Collins regards these fierce creatures as "the invertebrate equivalent of the dinosaurs."

If *Anomalocaris* was the largest animal of its time, *Opabinia* (Fig. 10–10) might well qualify as the strangest. The creature had five eyes and a flexible nozzle equipped with nippers used to capture victims. Conceivably predators such as *Opabinia* and *Anomalocaris* caused selective pressures in prey that favored the spread of protective shells.

Perhaps the most elegant of the Burgess Shale arthropods was *Marrella* (Fig. 10–11), named after Walcott's friend, the British geologist John E. Marr. *Marrella* is readily recognized by the four spines that extend backward from its head, or cephalon. Over 15,000 specimens of this tiny fossil, dubbed the "lace crab" by Walcott, have been recovered, making it the most abundant animal in the Burgess Shale fauna.

Hallucigenia (Fig. 10–12) is a perplexing Burgess Shale fossil. Its original interpretation suffered from the limited amount of information that could be gleaned from its compressed and carbonized remains. In early reconstructions, *Hallucigenia* was depicted as walking on pairs of stiltlike legs, rather like some sea urchins are able to walk on movable spines. A single row of tentacles was thought to extend along the back of the animal. Subsequently, the Swedish paleontologist Lars Ramskold was able to flake away some of the covering shale from a specimen and reveal a claw at the tip of one of the supposed tentacles. Claws normally are found on feet, so reconstructions were flipped over to show the former legs as paired dorsal spines and the former tentacles as paired legs. Although only one row of legs had been detected, the second row was assumed to lie in the shale matrix beneath the body. That interpretation was validated in 1991 when Ramskold reexamined the best specimen at the Smithsonian Institution and discovered parts of the second leg row. *Halucigenia* is considered an onycophoran. They are an interesting group represented today by the velvet worm. Because onycophorans share many characters of both annelids and arthropods, they are considered a sort of "missing link" between the two groups.

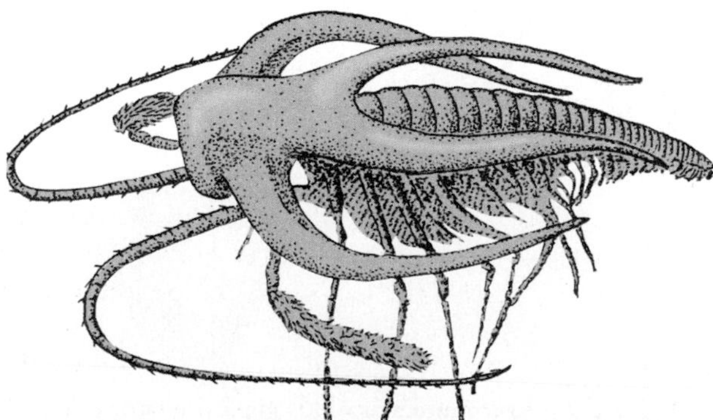

FIGURE 10-11 ***Marrella*, the most common arthropod in the Burgess Shale fauna.** The animal was about 2 centimeters in length.

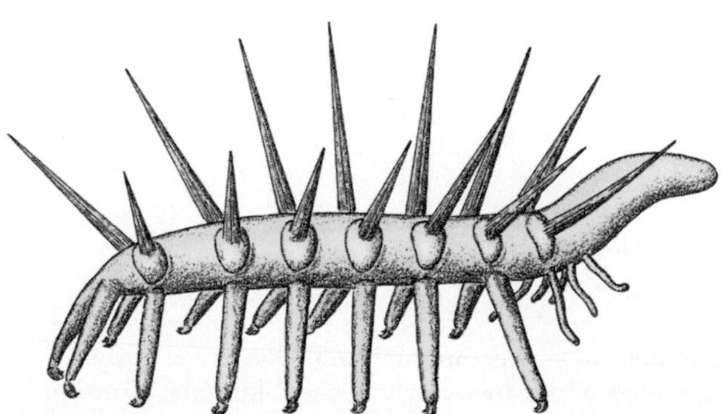

FIGURE 10-12 The early Cambrian Burgess Shale fossil *Hallucigenia*.

The Cambrian Explosion

Prior to 1984, the Burgess Shale offered the only extensive assemblage of abundant and diverse animals having bodies too delicate for easy preservation. In that year the Lower Cambrian Chengjiang fauna was discovered. Over 80 species of invertebrates have been recovered from the Chengjiang site. The fossils exhibit extraordinary preservation. Jellyfish show the detailed structure of tentacles, radial canals, and muscles. Even on the soft bodies of worms, eyes, annulations, digestive organs, and patterns on the "skin," or ectoderm, are readily recognized. As with the Burgess fauna, many of the Chengjiang organisms belong to such existing phyla as the Cnidaria, Porifera, Annelida, Brachiopoda, and Arthropoda. Species identical or closely similar to those from the Burgess Shale are common. Like the fauna of the Burgess Shale, there are also creatures that cannot be placed with certainty in any established taxonomic category.

Clearly, Chengjiang and Burgess Shale fossils provide a marvelous panoramic view of life at the beginning of the Paleozoic. They reveal that the Cambrian seas teemed with a splendid diversity of animals that had already attained an advanced stage of evolution. The startlingly abrubt appearance of so many varied animals, and the rapidity with which they radiated, is the basis for the expression "the Cambrian explosion." That explosion, also dubbed "evolution's big bang" in the popular press, lasted only 10 million years. From the fossil evidence at least, it would seem that all the principal animal phyla except the Bryozoa appeared during the geologically short interval from about 535 to 520 million years ago and that evolution since that time has progressed primarily by modifications of groups that originated near the beginning of the Paleozoic. Some molecular biologists, however, are currently challenging this fossil-based hypothesis. Aware that genes mutate at a relatively constant rate, these scientists analyze DNA sequences in animals to determine how long ago their ancestors originated. The analysis suggests that the branching of several phyla occurred over a billion years ago. However, there are no fossils to confirm this hypothesis.

Continuing Diversification

The Cambrian biologic radiation was followed in Ordovician time by a tripling of global biodiversity, particularly among trilobites, brachiopods, bivalve mollusks, gastropods, and coralline animals. Overall, the number of families of marine organisms increased from about 160 to 530 during the Ordovician Period. The new groups spread widely to become the dominant marine invertebrates throughout the remainder of the Paleozoic.

The increase in Ordovician biodiversity coincided with increased levels of mountain building and volcanism at many places around the world. Possibly, such geologic activity may have increased the influx of terrestrial materials into the marine environment beneficial for biologic productivity. It also may have resulted in an increase in habitats so as to enhance speciation. In any case, the Ordovician faunas had evolved a variety of lifestyles and adapted to a multitude of habitats.

There were **epifaunal** animals that lived on the surface of the sea floor, as well as **infaunal** creatures that burrowed beneath the surface. Infaunal animals evolved rapidly during the Cambrian. One can recognize their presence in the way Cambrian sediments were disturbed by bioturbation. In the preceding Neoproterozoic, fine layers of sediment were usually undisturbed. A covering of cyanobacterial mats probably protected large tracts of the ocean floor. Extensive bioturbation began in the Cambrian and marked the evolution of infaunal burrowers. Some refer to the dramatic change in the fabric of sea floor sediments as the "Cambrian substrate revolution."

In the early Paleozoic, there were borers as well as burrowers, attached forms and mobile crawlers, swimmers, and floaters. Some were **filter-feeders** that strained tiny bits of organic matter or microorganisms from the water; others were **sediment-feeders** that passed the mud of the sea floor through their digestive tracts to extract the nutrients within. There were animals that grazed on algae that covered parts of the ocean bottom and carnivores that consumed these grazers. A host of scavengers processed organic debris and thereby aided in keeping the seas clean and suitable for life. Some classes of invertebrates maintained a single mode of life, whereas groups within other classes adapted to several different lifestyles. Snails, for example, included scavengers, herbivores, and carnivores. With the passage of geologic time, many evolutionary changes and extinctions were to occur among the invertebrates, but these changes tended to occur within the taxonomic groups that had been established during the early Paleozoic.

UNICELLULAR ANIMALS (PROTISTANS)

The principal groups of Paleozoic unicellular animals with a significant fossil record are the **foraminifera** and **radiolarians**. Foraminifera first appear in the Cambrian and survive to the present. Foraminifera build their tiny shells (called tests) by adding chambers singly, in rows, in coils, or in spirals. Chambers open to one another by a series of ports, or *foramina*, from which the animals derive their name. Some species construct the test of tiny particles of silt, whereas others secrete tests made of calcium carbonate. Early Paleozoic foraminifera were mostly simple, saclike, tubular, or loosely coiled species with tests composed of sedimentary particles. They became more numerous and varied by Carboniferous time. The increase is strikingly evident among the **fusulinids**, a group whose shells superficially resemble grains of wheat (Fig. 10–13). Although many genera of fusulinids are similar in external appearance, they have complex internal features that are distinctive for groups living at particular stages of geologic time. Fusulinids had global distribution during the Pennsylvanian and Permian and were sometimes so abundant that they constituted a high percentage of the bulk volume of entire strata. Because they evolved rapidly and because they are so widespread and abundant, fusulinids are among the most important invertebrate guide fossils of the late Paleozoic.

In modern oceans, the tests of planktonic calcareous foraminifera rain down continuously on the sea floor, much like an unending limy snow. The accumulated debris forms a deep-sea sediment called globigerina ooze because of the prevalence of the remains of species of *Globigerina* and related genera.

Radiolarians are single-celled planktonic organisms that have been present on Earth at least since the beginning of the Paleozoic Era. Their threadlike pseudopodia project from an ornate lattice-like skeleton of opaline silica. In some regions of the oceans today, radiolarian skeletons accumulate to form deposits of radiolarian or siliceous ooze. Although radiolarians do occur in lower Paleozoic rocks (Fig. 10–14), they are rare and not yet useful for stratigraphic correlation. They are more abundant in Mesozoic and Cenozoic rocks and have been shown to be particularly useful in correlating Pleistocene deep-sea deposits. Of special interest is the observation that Pleistocene radiolarian extinctions appear to be associated with dates of reversal in the Earth's magnetic field.

FIGURE 10-13 Fusulinid limestone of Permian age from the Sierra Madre of West Texas. (*Courtesy of W. D. Hamilton, U. S. Geological Survey.*) ? To what phylum of invertebrates do these organisms belong? During what geologic period did fusulinids become extinct?

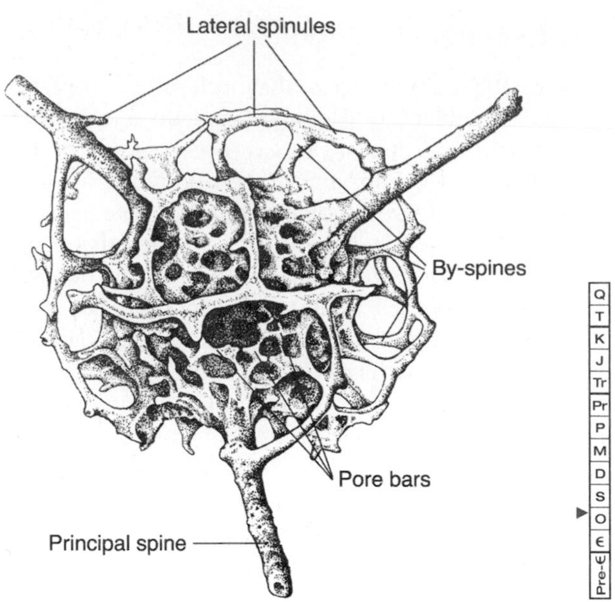

Q T K J Tr Pr P M D S O Є Pre-Є

FIGURE 10-14 A late Ordovician radiolarian (× 150) from the Hanson Creek Formation of Nevada. Because of the fragile and delicate structure of the radiolarian skeleton, it is unusual to find well-preserved radiolarians in rocks of the early Paleozoic. (*From Dunham, J. B., and Murphy, M. A. 1976. J. Paleo. 50(5):883.*)

PALEOZOIC METAZOAN INVERTEBRATES

Cup Animals: Archaeocyathids

Archaeocyatha means "ancient cups." It is an appropriate name for somewhat enigmatic Cambrian animals that constructed conical or vase-shaped skeletons out of calcium carbonate (Fig. 10–15). **Archaeocyathids** possessed features suggestive of both sponges and corals, yet these features are not distinctive enough to warrant placement in either group. They grew on the shallow sea floor, often in association with stromatolites. The growths resemble reefs, although they were not nearly as massive and extensive as later coral reefs. Most of the archaeocyathid build-ups were no more than about 30 meters in diameter and about 3 meters in height. However, Australian archaeocyathid reefs have been found that are over 60 meters thick and extend horizontally in narrow bands for over 200 kilometer. Although abundant during the Early Cambrian, archaeocyathids declined thereafter and became totally extinct by the end of the Middle Cambrian. They are

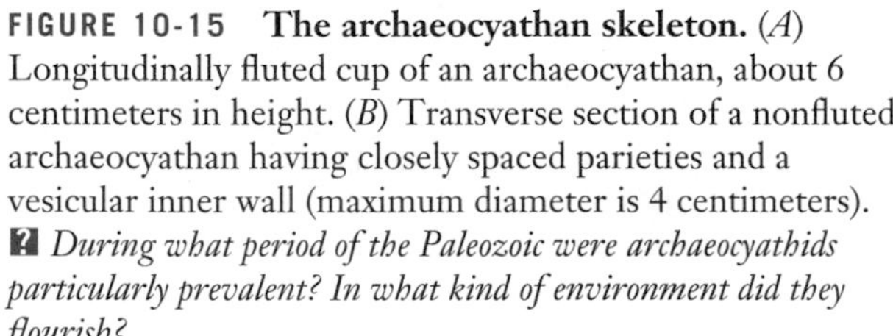

FIGURE 10-15 The archaeocyathan skeleton. (*A*) Longitudinally fluted cup of an archaeocyathan, about 6 centimeters in height. (*B*) Transverse section of a nonfluted archaeocyathan having closely spaced parieties and a vesicular inner wall (maximum diameter is 4 centimeters). ? *During what period of the Paleozoic were archaeocyathids particularly prevalent? In what kind of environment did they flourish?*

exceptionally useful in the stratigraphic correlation of Lower and Middle Cambrian strata.

Sponges

Among the many stationary animals to colonize the early Paleozoic sea floor were the sponges. Sponges are members of the phylum **Porifera**. They appear to have evolved from colonial flagellated, unicellular creatures and thus provide insight into how the transition from unicellular to multicellular animals may have occurred.

Sponges are a relatively conservative branch of invertebrates that have a long history. Cambrian representatives of all but one modern class of Porifera are known as fossils. Sponges (Fig. 10–16), such as *Protospongia* (Cambrian), *Nevadocoelia* (Ordovician), and the Silurian sponges *Astylospongia*, *Caryospongea*, and *Palaeomanon*, are widely used stratigraphic markers. *Astraeospongia* (Silurian) is notable for its prominent star-shaped spicules (Figure 10–17). *Hydnoceras* (Fig. 10–18) and *Prismodictya* are among the glass sponges found in Devonian rocks of eastern North Amerrica. Their siliceous spicules form quadrate patterns on the surfaces of fossil specimens.

Sponges have always been predominantly marine, although a few modern species live in fresh water. *Spicules* formed in the walls and cell layers of sponges are distinguishing characteristics, and provide both protection and support. The skeleton of a sponge may be composed of spicules made of silica or of calcium carbonate, or the sponge may be supported by a proteinaceous material termed *spongin*. As would be expected, the mineralized spicules are more commonly preserved. They are frequently found by geologists examining rocks that have been disaggregated for study. Spicules are also important in the classification of sponges.

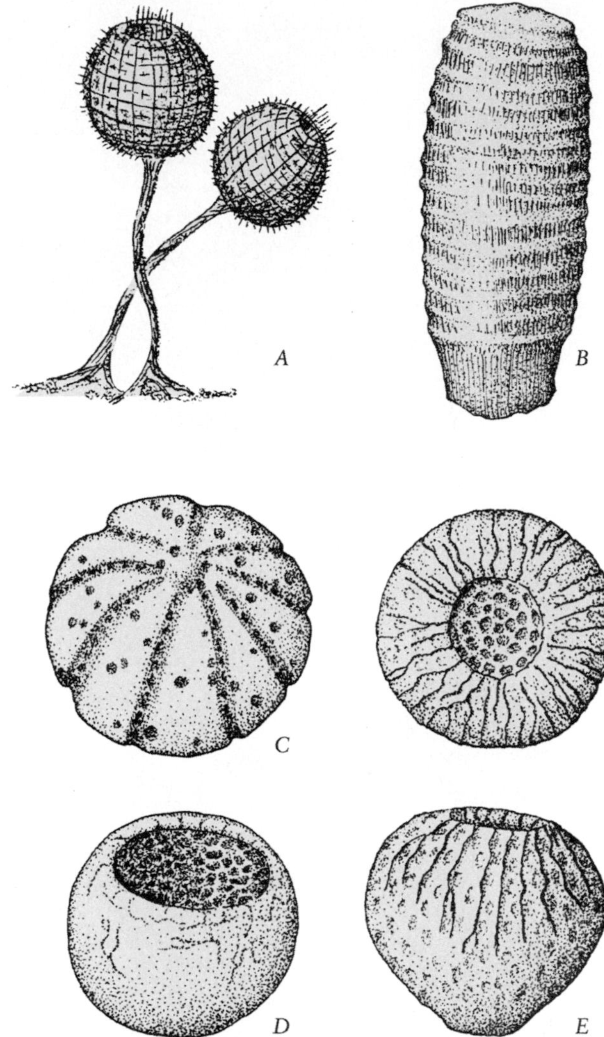

FIGURE 10-16 Early Paleozoic (Silurian) sponges. (*A*) *Protospongea* (Cambrian), (*B*) *Nevadocoelia* (Ord.), (*C*) *Caryospongea* (Sil.), (*D*) *Palaeomanon* (Sil.), (*E*) top and side views of *Astylospongea* (Sil.).

Although sponges vary greatly in size and shape, their basic structure (Fig. 10–19) is that of a highly perforated vase modified by folds and canals. The body is attached to the sea floor at the base, and there is an excurrent opening, or *osculum*, at the top. In the simplest sponges, the wall consists of two layers of cells. Facing the internal space is a layer of collar cells (choanocytes), and on the outside is a protective wall of flat cells that resemble the bricks of a worn masonry pavement. Between these two layers one finds a gelatinous substance called *mesenchyme*. Here amoeboid cells go about the

FIGURE 10-17 The Silurian sponge *Astraeospongea*. This fossil sponge takes its name from its star-shaped, six-rayed spicules, some of which can be discerned in the photograph.

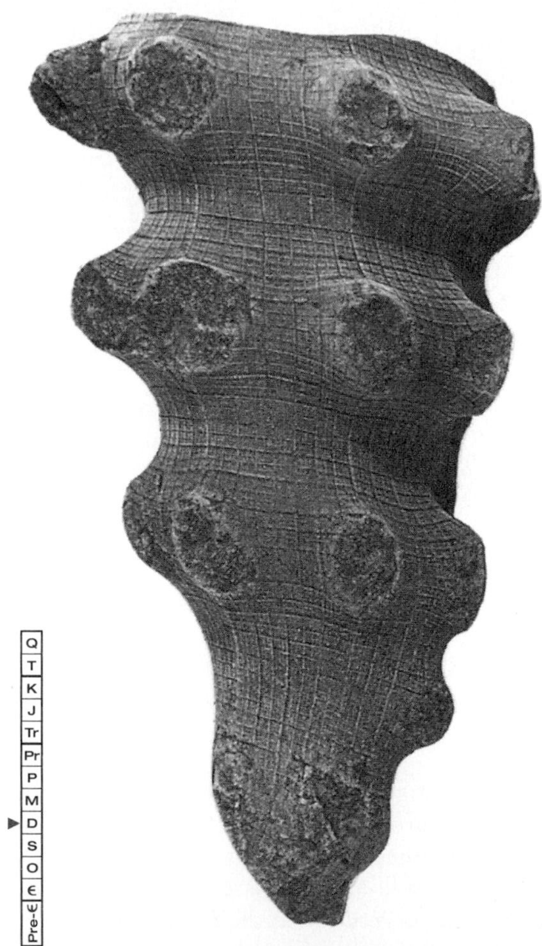

FIGURE 10-18 The Devonian siliceous sponge *Hydnoceras*. The specimen is 15 centimeters in height. *(Photograph courtesy of J. Keith Rigby, Brigham Young University.)* ? *What cells in the sponge produce the spicules of which this sponge skeleton is composed? Describe the pattern made by the spicules on the surface of the fossil.*

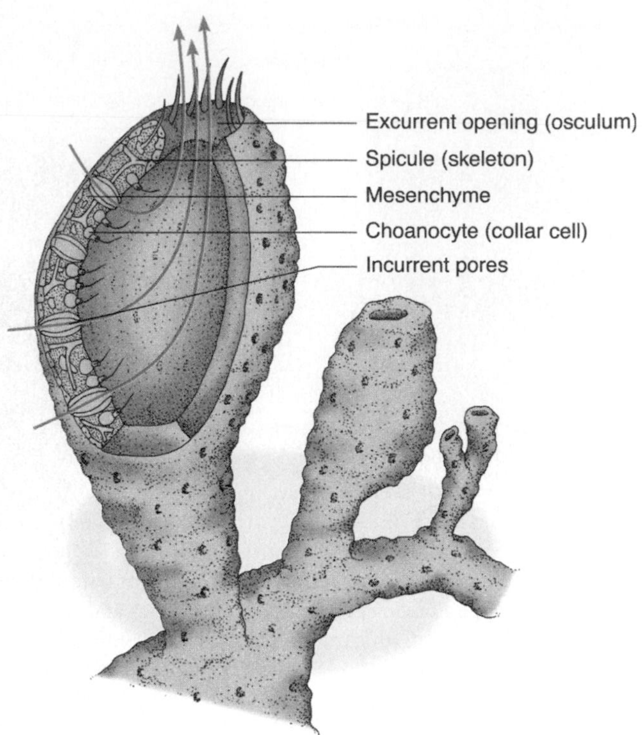

FIGURE 10-19 Schematic diagram of a sponge having the simplest type of canal system. The path of water currents is indicated by arrows.

work of secreting the spicules. Sponges lack true organs. Water currents moving through the sponge are created by the beat of *flagella* attached to collar cells. These currents bring in suspended food particles, which are ingested by the collar cells. In a simple sponge, water enters through the pores, flows across sheets of choanocytes in the central cavity, and passes out through the osculum.

A group of early Paleozoic sponges of particular interest because of their reef-building capabilities are the **stromatoporoids** (Fig. 10–20). These organisms,

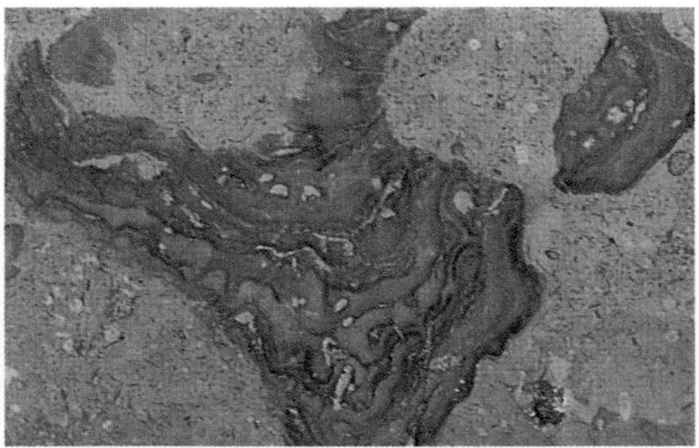

FIGURE 10-20 Polished slab of limestone containing the stromatoporoid named *Stromatoporella*, from Devonian rocks of Ohio.

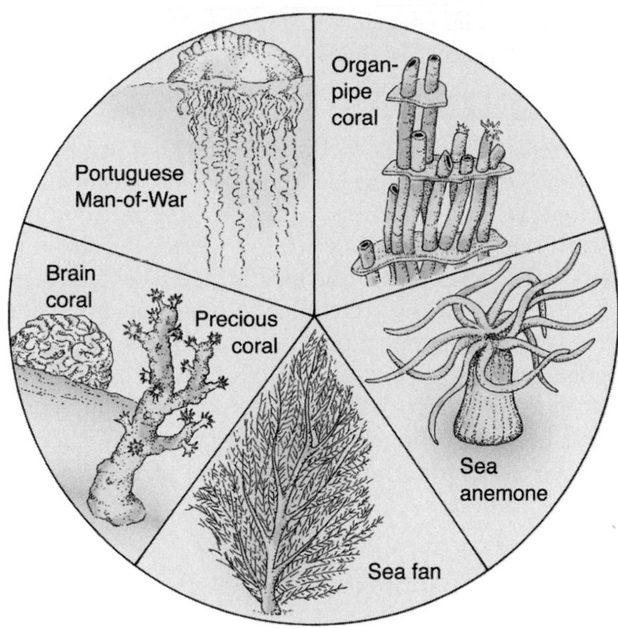

FIGURE 10-21 Diversity among some common cnidarians. (*From Levin, H. L., 1975.* Life Through Time. *Dubuque, IA: William C. Brown Co.*)

which bear a superficial resemblance to stromatolites, constructed fibrous, calcareous colonies consisting of pillars that support thin laminae. Stromatoporoids are common in reef-associated carbonate rocks of the Silurian and Devonian. They often grew profusely and in close association with corals, brachiopods, and other invertebrate reef-dwellers. Silurian coral-stromatoporoid reefs are well known to geologists primarily because of extensive study of reef exposures on Gotland Island, off the Baltic coast of Sweden, and in the region around and southwest of the Michigan basin. Stromatoporoid reef development during the Devonian was similar to that in the Silurian, with blanket-like, massive, and cylindric stromatoporoid colonies forming a considerable part of the total mass of the reef structures.

Corals and Other Cnidaria

Sea anemones, sea fans, jellyfish, the tiny *Hydra*, and reef-forming corals are all representatives of the phylum **Cnidaria** (formerly Coelenterata). It is a phylum known for the great diversity and beauty of its members (Fig. 10–21). The cnidarian body wall is composed of an outer layer of cells, the *ectoderm*; an inner layer, the *endoderm*; and a thin, intermediate layer, the *mesoglea*. Primitive sensory cells, gland cells that secrete digestive enzymes, flagellated cells, and nutritive cells to absorb nourishment are located in the endoderm. A distinctive feature of cnidarians is the presence of stinging cells, which, when activated, can inject a paralyzing poison. These specialized cells, called *cnidocytes*, are a unique characteristic of all Cnidaria.

Body form in Cnidaria may be either *polyp* or *medusa* (Fig. 10–22). The medusoid form is seen in the jellyfish, which resembles a bell or umbrella in shape. Jellyfish have a concave undersurface that contains a centrally located mouth. For this reason, it is designated the oral surface. All jellyfish have cnidocyte-laden tentacles, most commonly located around the margin of the umbrella. Jellyfish swim by rhythmic contractions of the umbrella.

In the polyp form, as exemplified by *Hydra*, there is a circle of tentacles around the mouth. Corals and sea anemones are among the frequently encountered cnidaria with this polyp form. In stony corals, the polyp secretes a calcareous cup, in which it lives. The animal may live alone or may combine with other individuals to form large colonies. The cup, or theca, may be divided by vertical plates called *septa*, which serve to separate layers of tissue and provide support. As the animal grows, it also secretes horizontal plates termed

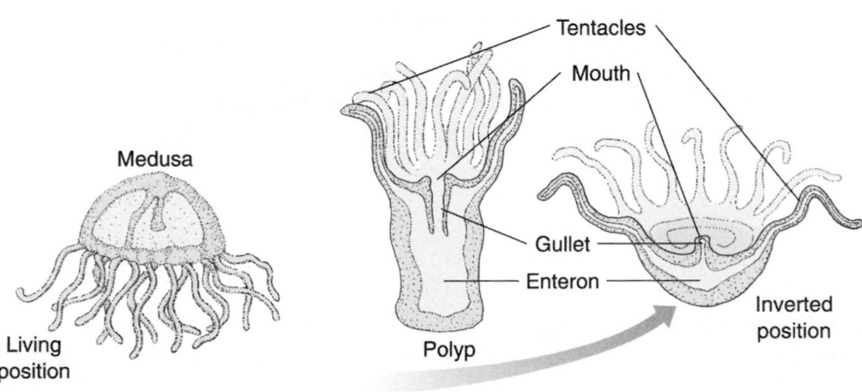

FIGURE 10-22 Comparison of polyp and medusa forms in cnidarians.

tabulae. Corals are identified according to the nature of their septa, tabulae, and other skeletal features. For example, after the development on an initial embryonic set of six protosepta, the Paleozoic rugose corals, or Rugosa, insert new septa at only four locations of the six spaces between protosepta. In other Paleozoic corals, septa are absent or poorly developed, and tabulae are the most important features. These are the Tabulata. Tabulate corals include many interesting colonial forms, such as the Paleozoic honeycomb and chain corals. The rugose corals and tabulate corals became extinct at the end of the Paleozoic and were followed in the Mesozoic by corals of the order Scleractinia. In scleractinian corals, septa are inserted in all six of the spaces between protosepta.

The fossil record for Cnidaria begins with fossil jellyfish impressions discovered in upper Precambrian rocks. The phylum is poorly represented in Cambrian rocks. In the succeeding Ordovician, the record improves dramatically, for the lime-secreting anthozoans begin to expand and diversify. The first stony corals were the tabulates, recognized by their simple, often clustered or aligned tubes divided horizontally by transverse tabulae. These include the honeycomb coral *Favosites* (Fig. 10–23). Tabulate corals were the principal reef formers during the Silurian. They declined after the Silurian, but their role in reef building was assumed by the rugose corals. Both solitary and compound colonies of rugosids (Fig. 10–24) are abundant in the fossil record of the Devonian and Carboniferous, but this group declined and became extinct during the Late Permian.

Reefs built on the North American platform by Paleozoic corals were extensive and often exceeded 250 meters in total thickness. Most of these reefs lie beneath younger rocks and were discovered while drilling for oil. Because of their porosity, buried Paleozoic reefs became sites for the accumulation of petroleum. Aside from their economic importance, coral reefs aid the paleogeographer by providing a way to approximately locate former equatorial regions. Nearly all living coral reefs lie within about 30A7 of the equator. It is a reasonable assumption that ancient coral reefs formed at similar latitudes.

Moss Animals: Bryozoa

Bryozoans are minute, bilaterally symmetric animals that grow in colonies that frequently appear as crusts or twiglike branches. The individuals, called zooids, are housed in a capsule, or zooecium, which is often preserved as a result of its being calcified. The zooecia appear as pinpoint depressions on the outside of the colony, or zoarium. The zooid has a complete, U-shaped digestive tract with mouth surrounded by a tentacled feeding organ called the lophophore.

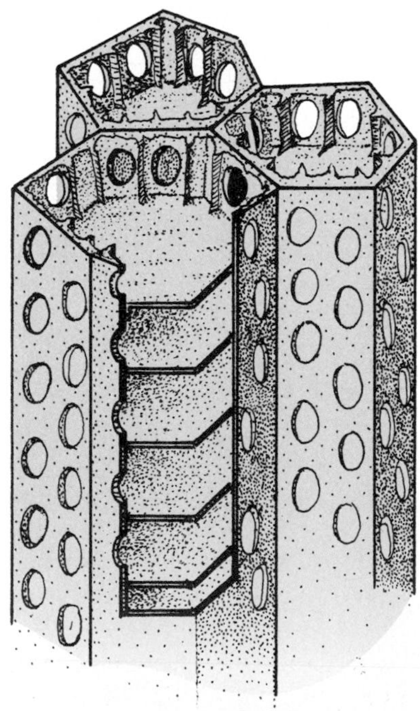

FIGURE 10-23 Tabulate colonial coral. (*A*) Skeletal structure of *Favosites*.Thecae are about 3 millimeters wide. (*B*) A colony of *Favosites*. Note closely spaced tubulae.

FIGURE 10-24 Devonian rugose corals. (*A*) The solitary horn coral *Zaphrenthis* with clearly visible radiating septa in the hornlike theca. (*B*) The compound (colonial) rugose coral *Lithostrotionella*. (*C*) A polished slab of the compound coral *Hexagonaria*. Water-worn fragments of this coral are found along the shore of Lake Michigan at Petoskey, Michigan, and this accounts for its being called Petoskey stone. Although not a rock, Petoskey stone is the designated state rock of Michigan. (*D*) Reconstruction of compound and solitary rugose corals on the floor of a Devonian epeiric sea. (*Diorama photograph courtesy of the U. S. National Museum of Natural History, Smithsonian Institution.*) *What was the purpose or function of the septa in rugose corals?*

Today, there are more than 4000 living species of bryozoans, but nearly four times that number are known as fossils. Their earliest unquestioned occurrence is from Lower Ordovician strata of the Baltic, but they did not become abundant until Middle Ordovician and Silurian time. In rocks of these ages, their remains sometimes make up much of the bulk of entire formations. Like the corals, bryozoa contributed to the framework of reefs. One of the commonest of all the early Paleozoic bryozoans was *Fistulipora*, different species of which develop massive, incrusting, or branching colonies. The genus *Hallopora* (Fig. 10–25*A*) is one of many branching twig bryozoans. Star-shaped patterns on the surface of *Constellaria* provided the inspiration for this bryozoan's generic name. Late Paleozoic bryozoans include varieties that constructed lacy, delicate, fanlike colonies (Fig. 10–25*B*). Associated with these so-called fenestellid colonies were the bizarre Mississippian corkscrew bryozoans known by the generic name *Archimedes* (Fig. 10–25*C* and *D*).

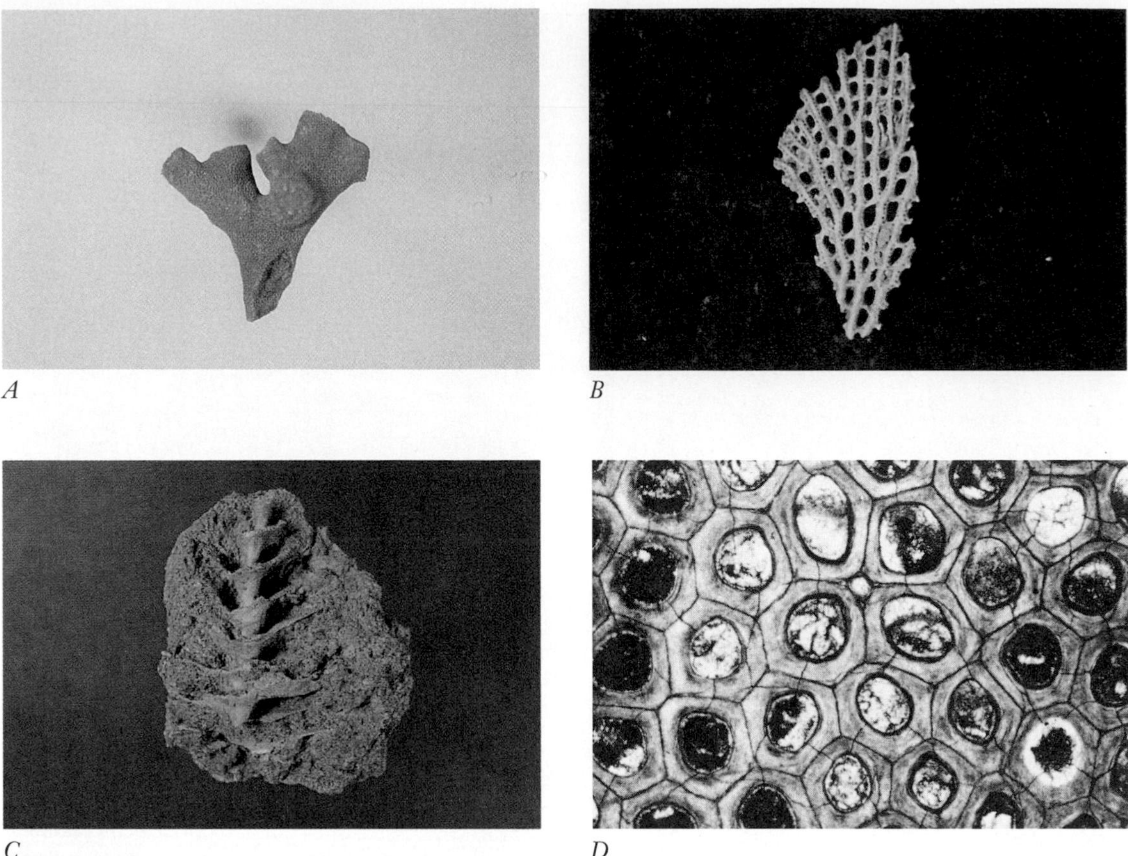

A B C D

FIGURE 10-25 Paleozoic bryozoans. (*A*) The branching twig bryozoan *Hallopora* from the Ordovician of Kentucky. (*B*) *Fenestella*, a lacy bryozoan from Devonian limestones at the Falls of the Ohio River. (*C*) *Archimedes*, with part of the spirally encircling frond of lacy bryozoan colony attached and visible. (*D*) Section of *Batostoma* cut in a plane at right angle to axis of zooecia. ? *Where were the zooecia located in* Fenestella*?*

Brachiopods

Brachiopods are probably the most abundant, diverse, and useful fossils readily found in Paleozoic rocks. They are entirely marine animals characterized by a pair of enclosing *valves*, which together constitute the shell of the animal (Fig. 10–26*A*). They resemble clams in this regard, but in symmetry of the valves and soft-part anatomy, brachiopods are very different from bivalves. Brachiopod valves are almost always symmetric on either side of the midline, and the two valves differ from each other in size and shape. The valves of a clam are right and left, whereas those of a brachiopod are dorsal and ventral. Brachiopod valves may be variously ornamented with radial ridges or grooves, spines, nodes, and growth lines. Calcium carbonate is the usual hard tissue of brachiopods, although the valves of some families are composed of mixtures of chitin and calcium phosphate. This is particularly true of the group called **inarticulates** (Fig. 10–26*C* and 10–27).

In the **articulate** brachiopods, the valves are hinged along the posterior margin and are prevented from slipping sideways by teeth and sockets. The inarticulates lack this definite hinge, and their valves are held together by muscles. Inarticulates characteristically have simple spoon-shaped or circular valves. Although brachiopod larvae swim about freely, the adults are frequently anchored or cemented to objects on the sea floor by a fleshy stalk (pedicle) or by spines. Some simply rest on the sea floor. One of the more conspicuous soft organs of brachiopods is the *lophophore*, a structure consisting of two ciliated, coiled tentacles whose function is to circulate water between the valves, distribute oxygen, and remove carbon dioxide (see Fig. 10–26). Water currents generated by cilia on the lophophore move food particles toward the mouth and short digestive tract.

Brachiopods still live in the seas today, although in far fewer numbers than during the Paleozoic. Earliest brachiopods were almost entirely the chitinous inarticulate types. These increased in diversity during the Ordovician but then declined. Very few species remain today.

The articulates also first appeared in the Cambrian but became truly abundant during the succeeding

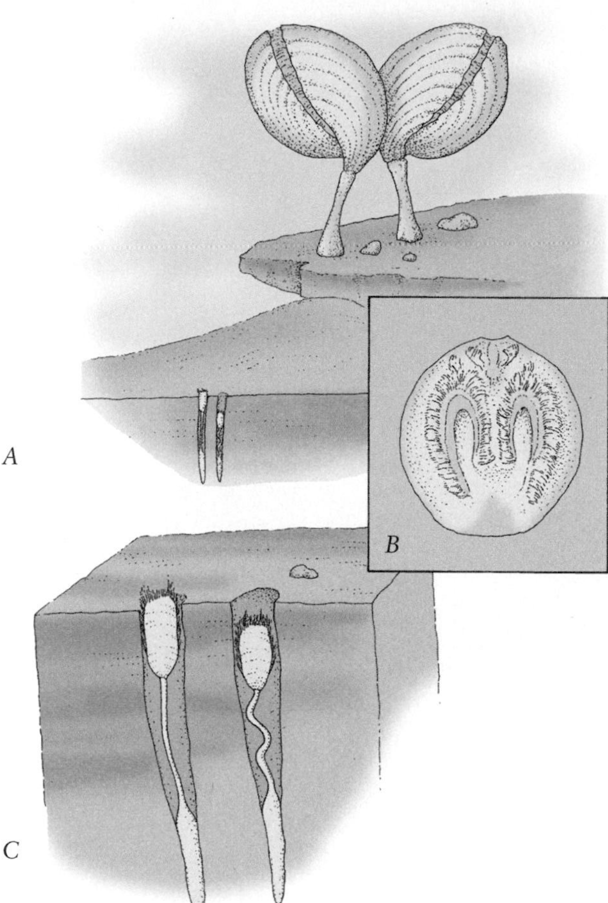

FIGURE 10-26 Living positions of articulate and inarticulate brachiopods. (*A*) The articulate living brachiopod *Magellania* attached to the sea floor by pedicle and with lophophore barely visible through the gape in the valves. (*B*) Interior of brachial valve showing ciliated lophophore. (*C*) The inarticulate brachiopod *Lingula*, which excavates a tube in bottom sediment and lives within it. The pedicle secretes a mucus that glues the animal to the tube.

Pre-Є | Є | O | S | D | M | P | Pr | Tr | J | K | T | Q

FIGURE 10-27 The Cambrian inarticulate brachiopod *Dicellomus* (maximum diameter 2.2 centimeters).

Ordovician Period, when there was a great expansion of all sorts of shelled invertebrates. Across the floors of early Paleozoic epeiric seas, large and small aggregates of these filter-feeders could be found. Today, their valves compose much of the volume of thick formations and provide the stratigrapher with essential markers for correlation. Among the articulate brachiopods that were particularly abundant during the early Paleozoic were the strophomenids, orthids, pentamerids, and rhynchonellids (Fig. 10–28). During the Devonian Period of the late Paleozoic, spiriferid brachiopods became particularly abundant. Spiriferids take their name from the calcified internal spirals that supported the lophophore (Fig. 10–29). The most characteristic brachiopods of the Carboniferous and Permian were the productids. Productids were distinctively spinose (Fig. 10–30), with large, inflated ventral valves. They were so numerous during the Carboniferous that the period might well be dubbed the age of productids.

Mollusks

A stroll along almost any seashore will provide evidence that mollusks are today's most familiar marine invertebrates. Such well-known animals as snails, clams, chitons, tooth shells, octopods, and squid are included within the phylum Mollusca (Fig. 10–31). Although most mollusks possess shells, some, such as the slugs and octopods, do not. The various classes of Mollusca differ considerably in external appearance, yet they have fundamental similarities in their internal structure. There is a muscular portion of the body called the *foot* that functions primarily in locomotion. In cephalopods, the foot is modified into tentacles. A fleshy fold called the *mantle* secretes the shell. In aquatic mollusks, respiration is accomplished by means of gills. Well-developed organs for digestion, sensation, and circulation attest to the advanced stage of evolution that mollusks have attained.

Included within the phylum Mollusca are a group of marine animals that may be collectively referred to as **placophorans**. Placophorans are relatively primitive mollusks that have multiple paired gills and, in shelled forms, a creeping foot much like that seen in snails. The most familiar placophorans are the polyplacophorans, represented by chitons (Fig. 10–32), which possess eight overlapping calcareous plates covering an ovoid, flattened body. Chitons are highly adapted for adhering to and grazing on algae that covers rocks and shells. Their fossil record is scant but lengthy. It begins in the Cambrian and extends to the present.

As indicated by their name, **monoplacophorans** are placophorans that have a single shell, resembling a flattened or short cone. Their fossil record extends back to the Cambrian, but they disappeared from the fossil

FIGURE 10-28 Some early Paleozoic brachiopods. The first three are Ordovician strophomenid brachiopods: (*A*) *Rafinesquina*, (*B*) *Strophomena*, and (*C*) *Leptaena*. (*D*) *Hebertella* is an Ordovician orthid brachiopod; (*E*) *Lepidocyclus*, is an Ordovician rhynchonellid; and (*F*) is an internal mold of *Pentamerus*, a Silurian pentamerid.

record in the Devonian. In 1952, however, 10 living specimens of a monoplacophoran were dredged from a deep ocean trench off the Pacific coast of Costa Rica. The specimens, later named *Neopilina* because of their similarity to the Silurian monoplacophoran *Pilina* (Fig. 10–33), displayed a segmental arrangement of gills, muscles, and other organs. This indicates mollusks branched off the invertebrate family tree at some point above the branch represented by segmented annelid worms, and that the unsegmented condition seen in

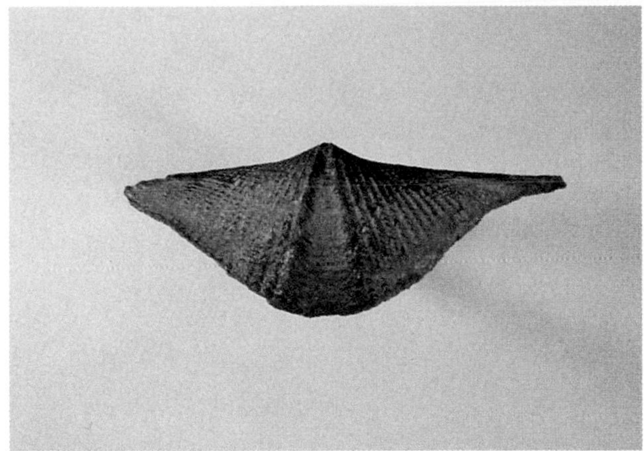

A

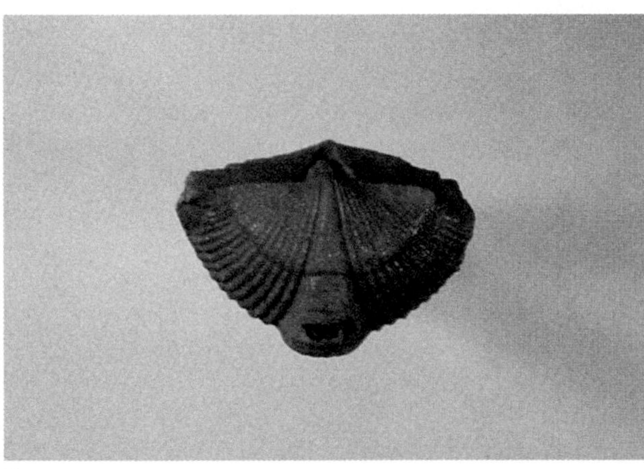

B

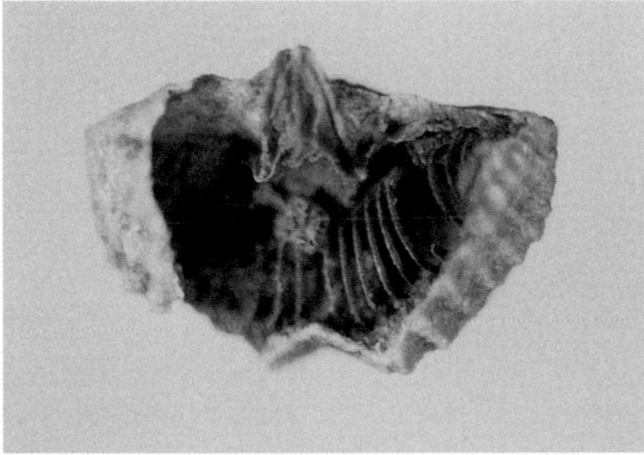

C

FIGURE 10-29 Devonian spiriferid brachiopods. (*A*) *Mucrospirifer* (*B*) *Platyrachella* (*C*) spiriferid brachiopod with shell broken to reveal the internal spiral supports for the lophophore (natural size).

FIGURE 10-30 Ventral (upper photograph) and side (lower photograph) views of the Permian spinose productid brachiopod *Marginifera ornata* from the Salt Range of West Pakistan. Valves (not including spines) are about 2 centimeters wide. (*Courtesy of R. E. Grant, U.S. Geological Survey.*) ? *What was the probable purpose or function of the spines?*

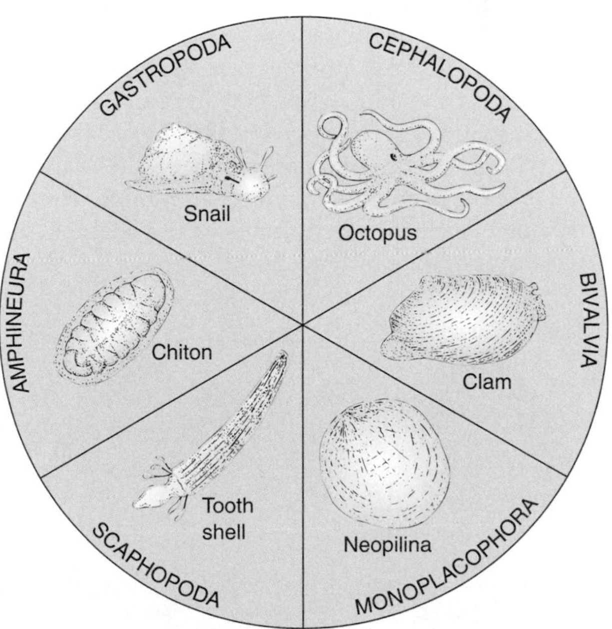

FIGURE 10-31 Some common members of the phylum Mollusca. (*From Levin, H. L. 1975. Life Through Time. Dubuque, IA: William C. Brown Co.*)

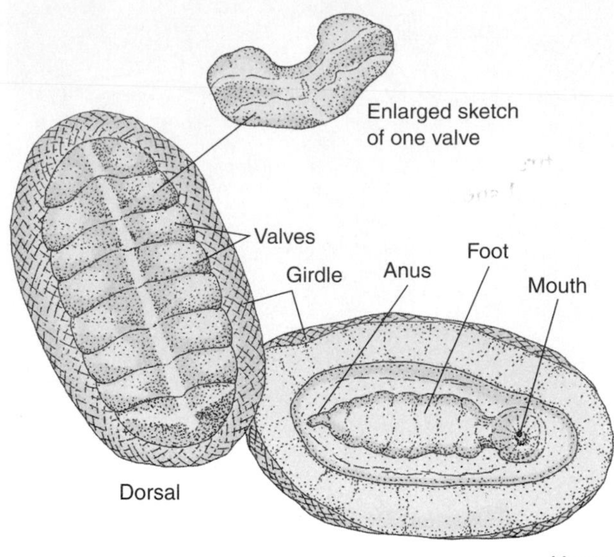

FIGURE 10-32 The common Atlantic Coast chiton.

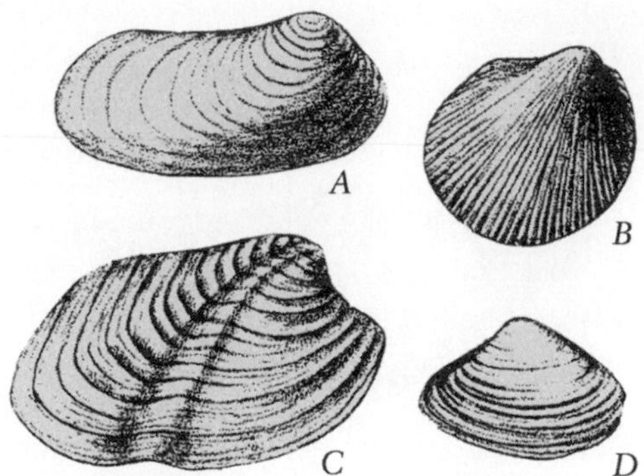

FIGURE 10-34 Paleozoic bivalves. (*A*) *Allorisma* (Miss.–Penn.), (*B*) *Cardiopsis* (Miss.–Penn.), (*C*) *Grammysia* (Sil.–Miss.), and (*D*) *Ctenodonta* (Ord.–Sil.), all × 1. ? *How does the symmetry of these bivalves differ from that typically seen in articulate brachiopods?*

most mollusks is not primitive but rather a secondary development. Many paleontologists regard the monoplacophorans as ancestral to more familiar mollusks such as bivalves, gastropods, and cephalopods.

The **Bivalvia** or **Pelecypoda** (Fig. 10–34) are a class of mollusks that includes clams, mussels, cockles, and oysters. Bivalves differ from other mollusks in having two calcareous valves joined on the dorsal side by a tough elastic ligament. Unlike brachiopods, which utilize muscles to open their valves, the valves of these mollusks are opened by the hinge ligament. When the adductor muscles close the valves, the elastic ligament is stretched. With relaxation of the adductors, the ligament causes the valves to open. Most bivalves (oysters are an exception) have valves that are mirror images of each other. Bivalves originated in the Cambrian but did not become notably abundant until the Carboniferous and Permian, when they spread widely in the shallow seas of the time.

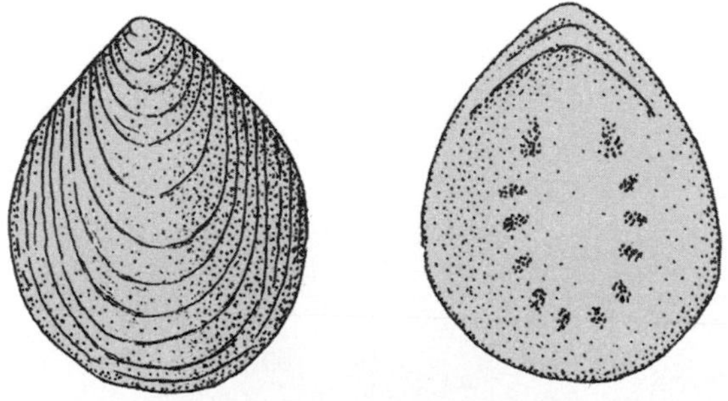

FIGURE 10-33 The monoplacophoran *Pilina*. (*A*) External view of the shell. (*B*) Internal view showing muscle scars. Maximum diameter is 3.5 centimeters.

Gastropods first occur abundantly in Lower Cambrian strata. Earliest forms constructed small, conical shells. During later Cambrian and Ordovician time, gastropods with the more familiar coiled conchs became commonplace (Fig. 10–35). Many early gastropods coiled in a plane, but in later forms spiral coiling was more common. Gastropods of the Ordovician and Silurian had shapes similar to those of living species. By Pennsylvanian time, gastropods had become abundant and diverse. During this period, the oldest known air-breathing gastropods of the class

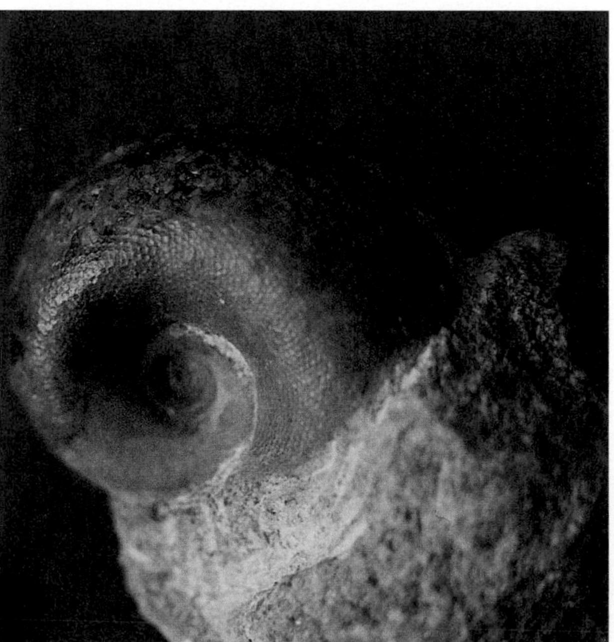

FIGURE 10-35 Ordovician gastropod with reticulate pattern on the surface of the shell. Average diameter 4 centimeters. ? *How would the interior of this gastropod conch (shell) differ from that of a similarly coiled cephalopod conch?*

Pulmonata appeared. The succeeding Permian Period ended with widespread extinctions among marine invertebrates, and many families of gastropods were decimated.

The **cephalopods** may be the most complex of all invertebrates. Today, this marine group is represented by the squid, cuttlefish, octopods, and the attractive chambered nautilus. The nautilus in particular provides us with important information about the soft anatomy and habits of a vast array of cephalopods known only by their preserved shells. In the genus *Nautilus* (Fig. 10–36), one finds a bilaterally symmetric body, a prominent head with paired image-forming eyes, and tentacles developed on the forward portion of the foot. Water is forcefully ejected through the tubular "funnel" to provide swift, jet-propelled movement.

A

B

FIGURE 10-36 ***Nautilus*, a modern nautiloid cephalopod.** (*A*) Conch of a nautiloid sawed in half to show large living chamber, septa, and septal necks through which the siphuncle passed. (*B*) Living animal photographed at a depth of about 300 meters. (*Photographs courtesy of W. Bruce Saunders.*)

At first glance, the shells of cephalopods resemble gastropod shells, but the resemblance is only superficial. Although there are exceptions, most gastropod shells coil in a spiral, whereas the cephalopod shell characteristically coils in a plane. More importantly, the planispirally coiled cephalopod shell is divided internally into a series of chambers by transverse partitions, or *septa*. Gastropods lack septa. The bulk of a cephalopod's soft organs reside in the final chamber. Where the septa join the inner wall of the conch, suture *lines* (or simply *sutures*) are formed. These lines are enormously useful in the identification and classification of cephalopods. For example, cephalopods placed in the subclass **Nautiloidea** have straight or gently undulating sutures, whereas the **Ammonoidea** have more complex sutures. Specific configurations of sutures characterize each ammonoid species and can be used in precise stratigraphic correlation of cephalopod-bearing strata.

The oldest fossils classified as cephalopods have small, conical conchs. They occur in Lower and Middle Cambrian rocks of Europe. The class gradually increased in number and diversity and became ubiquitous inhabitants of Ordovician and Silurian seas. Indeed, by Silurian time, a great variety of conch forms—from straight to tightly coiled nautiloids (Fig. 10–37)—had developed. Some of the elongate forms were giants that exceeded 4 meters in length. The first signs of a decline in nautiloid populations can be detected in Silurian strata. After Silurian time, the group continued to dwindle until today only a single genus, *Nautilus*, survives.

During the Devonian, the first ammonoid cephalopods appeared. These were the **goniatites**, characterized by angular and generally zigzag sutures without any additional crenulations (Fig. 10–38). The goniatites persisted throughout the late Paleozoic and gave rise to the ceratites and **ammonites** of the Mesozoic.

Arthropods

The **arthropod** phylum is enormous. It includes such living animals as lobsters, spiders, insects, and a host of other animals that possess chitinous exterior skeletons, obviously segmented bodies, paired and jointed appendages, and highly developed nervous systems and sensory organs. Members of Arthropoda that have left a particularly significant fossil record are the trilobites, ostracods, and eurypterids.

Trilobites (Fig. 10–39) were swimming or crawling arthropods that take their name from a division of the

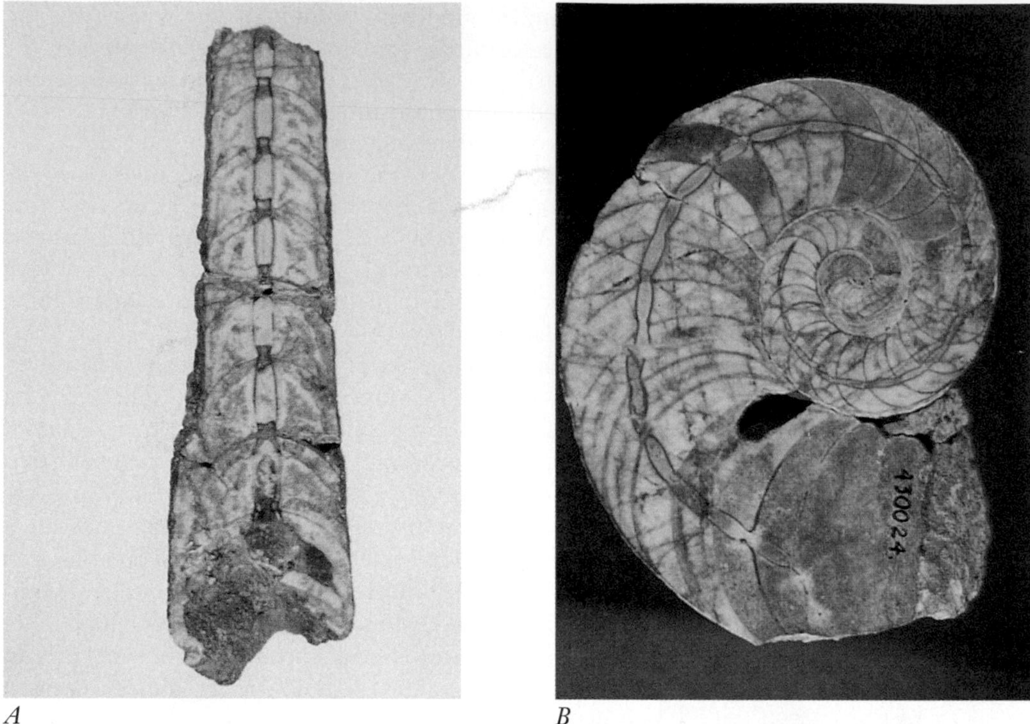

FIGURE 10-37 Variation in conch shape among early Paleozoic nautiloid cephalopods. Both of these specimens are from the Silurian of Bohemia. (*A*) A sawed and polished section of the straight conch of *Orthoceras potens* showing septa and siphuncle. (*B*) Sawed and polished section of *Barrandeoceras*, exhibiting a coiled form. Specimen *A* is 22.5 centimeters in length; *B* has a diameter of 18 centimeters.

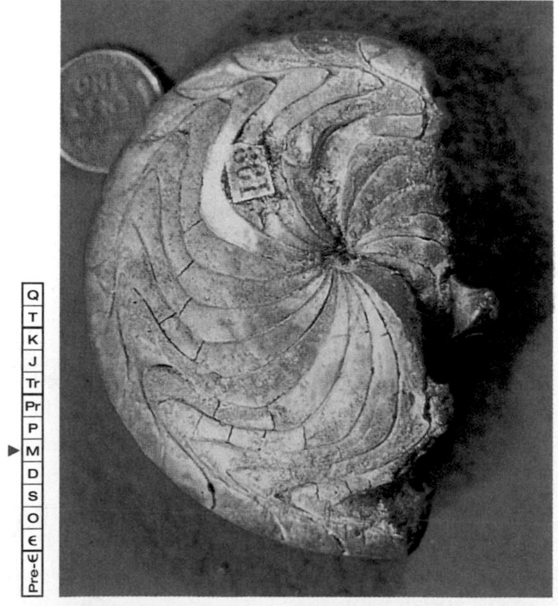

FIGURE 10-38 Goniatite ammonoid cephalopod exhibiting zigzag sutures. Diameter is 4.6 centimeters.

dorsal surface into three longitudinal segments, or *lobes*. There is, for example, a central axial lobe and two lateral (pleural) lobes (Fig. 10–40). There is also a transverse differentiation of the shield into an anterior *cephalon*, a segmented *thorax*, and a posterior *pygidium*. The skeleton was composed of chitin strengthened by calcium carbonate in parts not requiring flexibility. In some forms, the skeleton was sufficiently flexible to permit the animal to roll into a tight ball in order to protect soft parts from predators (Fig. 10–41). As in many other arthropods, growth was accomplished by molting. Although some trilobites were sightless, the majority had either single-lens eyes or compound eyes composed of a large number of discrete visual bodies.

Although trilobites were not the first animals to develop mineralized skeletons (small shelly fossils preceded them in this regard), they are nevertheless prominent among the Early Cambrian invertebrates that achieved this condition. The Early Cambrian has been divided into five stages, designated the Nemakit-Daldynian (the oldest), Tommotian, Atdabanian, Botomian, and Toyonian (the youngest). Trilobites make their appearance in rocks of the early Atdabanian stage of northern Europe, Siberia, and Morocco. Body fossils of trilobites have not been discovered in

FIGURE 10-39 A gallery of trilobites. (*A*) The Middle Cambrian guide fossil *Ogygopsis klotzi* from Mount Stephens, British Columbia, near the site of the famous Burgess Shale quarry (length 6.0 centimeters). (*B*) *Ellipsocephalus* from the Middle Cambrian of Bohemia (average length 2.8 centimeters). (*C*) *Isotelus* from Ordovician limestones in New York State (length 3.8 centimeters). (*D*) *Dalmanites* from Silurian beds in Indiana (length 3.2 centimeters). ? *In which of the photographs is there reason to believe that pygidia evolve by fusion of post-thoracic segments?*

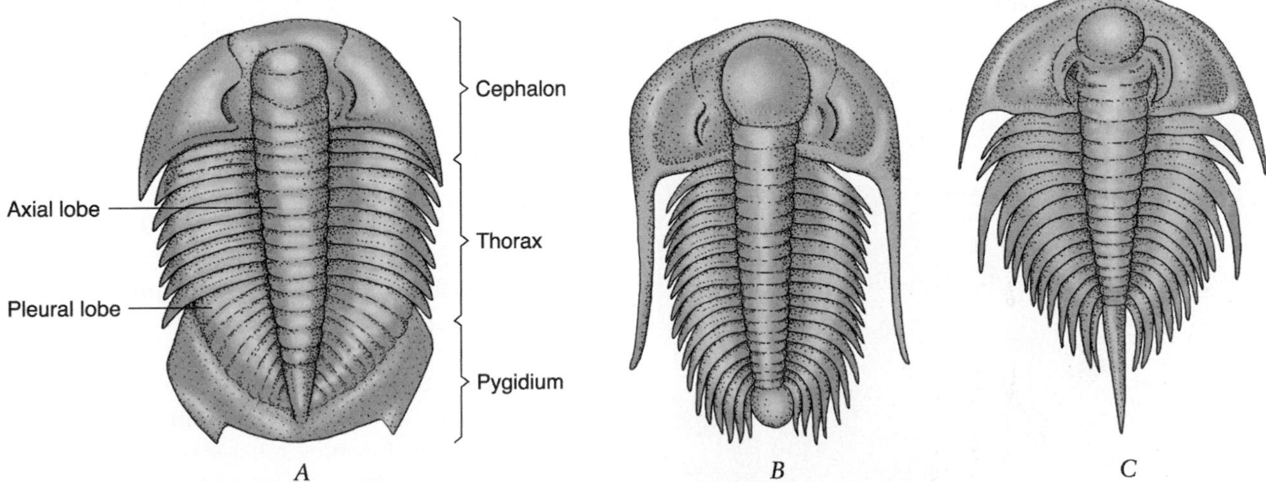

FIGURE 10-40 Three well-known Cambrian trilobites. (*A*) *Dikelocephalus minnesotensis* (Upper Cambrian), (*B*) *Paradoxides harlani* (Middle Cambrian), and (*C*) *Olenellus thompsoni* (Lower Cambrian).

ENRICHMENT

The Eyes of Trilobites

"The eyes of trilobites eternal be in stone, and seem to stare about in mild surprise at changes greater than they have yet known," wrote T. A. Conrad. Indeed, trilobites may have been the first animals to look out upon the world, for they possessed the most ancient visual systems known. The eyes of trilobites include both simple and compound types. A simple eye appears as a single, tiny lens, resembling a small node. A few receptor cells probably lie beneath each simple lens. Simple eyes are rare among trilobites. Far more abundant are compound eyes composed of a large number of individual visual bodies, each with its own biconvex lens. Each lens is composed of a single crystal of calcite. One of the properties of a clear crystal of calcite is that in certain orientations, objects viewed through the crystal produce a double image. Trilobites, however, were not troubled by double vision because the calcite lenses were oriented with the principal optic axis (the light path along which the double image does not occur) normal to the surface of the eye.

In some compound trilobite eyes, termed holochroal, the many tiny lenses are covered by a continuous, thin, transparent cornea. Beneath this smooth cover there may be as many as 15,000 lenses. In other trilobites, such as *Phacops* and *Calliops* (see figure in this box), there are discrete individual lenses, each covered by its own separate cornea and separated from its neighbors by a cribwork of exoskeletal tissue. Such compound eyes are termed schizochroal.

FIGURE A Side view of the trilobite *Phacops rana*, showing the discrete visual bodies of the large compound eye. Length is 3.2 centimeters.

Trilobite eyes can sometimes provide clues useful in determining the habits of certain species. Eyes along the anterior margin of the cephalon, for example, may indicate active swimmers. Eyes located on the ventral side of the cephalon may indicate that the trilobite was a surface-dweller. In the majority of trilobites, the eyes are located about midway on the cephalon in a position appropriate for an animal that crawled on the sea bottom or occasionally swam above it. Trilobites that either lacked eyes or were secondarily blind appear to have been adapted for burrowing in the soft sediment on the ocean floor.

Tommotian rocks, but probable trilobite tracks indicate they may have been present. Perhaps trilobites existed before the Atdabanian but were not preserved because they lacked mineralized exoskeletons.

FIGURE 10-41 An enrolled specimen of *Flexicalyme* from Ordovician limestones in Ohio.

Following their appearance in the Atdabanian, trilobites expanded rapdily. The exoskeletons of many of the early species are characterized by a large number of thoracic segments and pygidia that were either very small or completely absent. An example is *Olenellus* (Fig. 10–40*C*), a genus widely used in correlation of Early Cambrian strata of northeastern North America, Scotland, and Greenland. *Paradoxides* (Fig. 10–40*B*), *Ellipsocephalus* (Fig. 10–39*B*), and a host of other genera became abundant during the Middle Cambrian. They were followed in the Late Cambrian by the "scoop-tailed" trilobite *Dikelocephalus* (Fig. 10–40*A*). Near the end of the Cambrian, trilobites declined in both numbers and diversity. Factors contributing to the decline may have included predation (by anomalocarids and cephalopods) or possible adverse environmental conditions associated with the restriction of epeiric seas.

The survivors of the late Cambrian episode of extinction provided the stock for a second radiation of trilobites that began in the Ordovician and continued until Middle and Late Devonian when they experienced a second episode of hard times. Only a few groups were able to carry on into the Carboniferous.

The final blow for trilobites came during the Middle Permian, when all of the remaining families disappeared. Although they were unable to survive beyond the Middle Permian, trilobites cannot be regarded as biologic failures, for they were important animals on our planet for over 300 million years.

In Chapter 4 we noted that the best fossils for stratigraphic correlation have short geologic ranges and wide geographic distribution. The former criterion was amply met by the trilobites, but the latter was not always achieved. Most trilobites were bottom dwellers (benthonic), living in shallow shelf areas. Thus, they tended to be provincial (confined to a particular area). For this reason, trilobite biozones are often combined with the biozones of other groups living at the same time. Although the provinciality of trilobites may have diminished their usefulness in far-ranging correlations, it enhanced their value in making paleogeographic reconstructions. For example, early in the Paleozoic, Europe and North America were separated by the Iapetus Ocean. Trilobite fauna living in shelf areas on either side of the ocean were distinctly different. They were provincial faunas. The Iapetus Ocean, however, subsequently began to close. Eventually, the marginal seas of both continents converged. Formerly separated trilobite faunas intermingled and lost their distinctive differences. The resulting cosmopolitan fauna thus provided evidence of tectonic plate convergence.

FIGURE 10-42 Ostracods (*Leperditia fabulites*) in the Ordovician Plattin Limestone, Jefferson County, Missouri. Ostracods are the smooth, bean-shaped fossils. These average about 1 centimeter in length and are thus large ostracods. Two strophomenid brachiopods are also present on the surface of the limestone.

Arthropod companions to the trilobites in Paleozoic seas were the small, often bean-shaped **ostracods** (Fig. 10–42). At first glance, ostracods appear so different from trilobites as to cause one to question their classification within the same phylum. The ostracods have a bivalved shell vaguely suggestive of a tiny clam. However, this bivalved carapace encloses a segmented body from which extend seven pairs of jointed appendages. Adult animals are about 0.5 to 4 millimeters in length. The valves are composed of both chitin and calcium carbonate and are hinged along the dorsal margin.

Ostracods first appeared early in the Cambrian and many species still exist today. They occur in both marine and freshwater sediments. Because of their small size, they are brought to the surface in wells drilled for oil, and along with other microfossils, such as foraminifera and radiolaria, are used by geologists in correlating the strata of oil fields.

FIGURE 10-43 The fossil eurypterid *Eurypterus lacustris* from the Bertie Waterlime of Silurian age, Erie County, Pennsylvania. This well-preserved specimen is about 28 centimeters long. (*Courtesy of Ward's Natural Science Establishment, Inc., Rochester, NY.*)

Eurypterids (Fig. 10–43) are a group of arthropods that, because of their rarity, are less useful in stratigraphic studies than are either trilobites or ostracods. Nevertheless, they are impressive early Paleozoic aquatic invertebrates. Although many were of modest size, some were nearly 3 meters long. Had these giants survived, they would be suitable subjects for a Hollywood monster film. Five pairs of appendages and a fearful-looking pair of pincers (chelicerae) extended from the body. Some were also equipped with a terminal spine. Eurypterids were predators that ranged across portions of the sea floor and brackish estuaries from Ordovician until Permian time but were especially abundant during the Silurian and Devonian.

Spiny-Skinned Animals: Echinoderms

Just as modern seas abound with starfish, sea urchins, and sea lilies, so were the oceans of the early Paleozoic populated with members of the phylum Echinodermata (Fig. 10–44). **Echinoderms** are animals with mostly five-way symmetry that masks an underlying primitive bilateral symmetry. They have an endoskeleton consisting of calcium carbonate plates that may bear spines. A unique characteristic of the phylum is the presence of a system of soft-tissue tubes—the **water vascular system**—which functions in respiration and locomotion (Fig. 10–45). Members of the phylum are exclusively marine, typically bottom-dwelling, and either attached to the sea floor or able to move about slowly.

Among the many classes of the phylum Echinodermata, the **Asteroidea** (starfish), **Ophiuroidea** (brittle stars), **Echinoidea** (sea urchins), **Edrioasteroidea**, **Crinoidea** (crinoids), **Blastoidea** (blastoids), and **Cystoidea** (cystoids) are the most abundant and useful in geologic studies. The mostly attached and sessile crinoids, blastoids, and cystoids were particularly characteristic of the Paleozoic Era, whereas the more mobile asteroids and echinoids were more common in later eras.

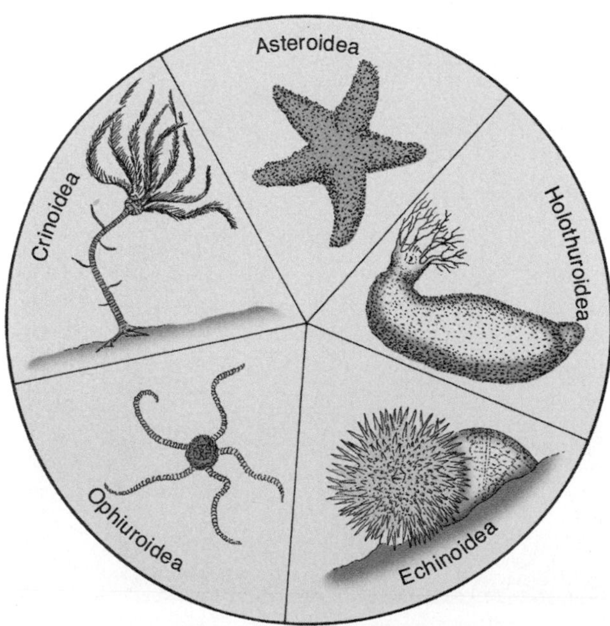

FIGURE 10-44 Representative living echinoderms. (*From Levin, H. L. 1975.* Life Through Time. *Dubuque, IA: William C. Brown Co.*)

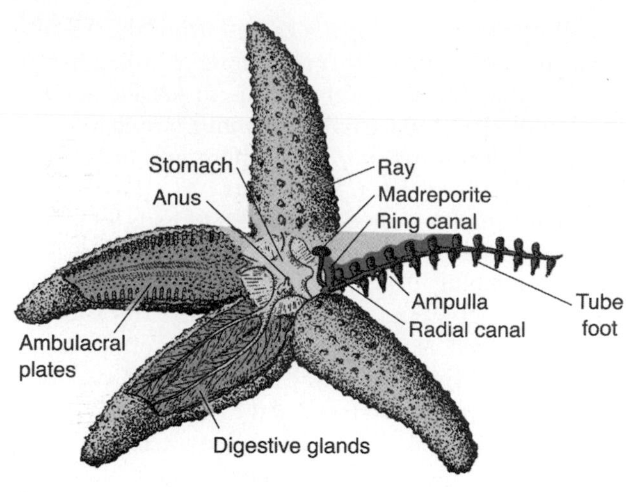

A

B

FIGURE 10-45 (*A*) **Partially dissected starfish showing elements of the water vascular system and other organs.** (*B*) Underside of starfish showing tube feet bordering ambulacral grooves.

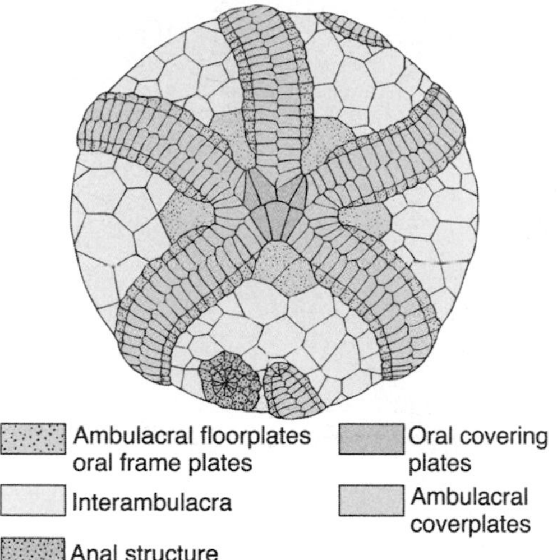

FIGURE 10-46 ***Edrioaster bigsbyi*, a Middle Ordovician edrioasteroid.** Specimen is 45 millimeters in diameter. (*From Bell, B. M. 1977.* J. Paleo. 51(3):620.)

FIGURE 10-47 **The spiraled, spindle-shaped Early Cambrian echinoderm *Helicoplacus*.** The specimen is about 2.6 centimeters long. (*After Durham, J. W., and Caster, K. E. 1963.* Science 140:820–822.)

Echinoderms appear to have evolved late in the Proterozoic, although fossil remains of Proterozoic forms are few and often enigmatic. The Ediacaran fauna, for example, includes a globular fossil named *Arkarua* that has five rays on its surface and may be related to edrioasteroids. Edrioasteroids (Fig. 10–46) are considered by many paleontologists to be ancestral to starfish and sea urchins. They construct globular, discoidal, or cylindrical tests (shells), many of which have concave lower surfaces to facilitate attachment to hard substrates. Edrioasteroids appear early in the Cambrian. By the end of the Carboniferous, the group had become extinct.

By Early and Middle Cambrian time, several classes of rather peculiar echinoderms appeared, but this initial radiation was apparently not successful. None of these became abundant, and three became extinct before the end of the Cambrian. One of the most bizarre of these early echinoderms was *Helicoplacus* (Fig. 10–47), a form that has plates and food grooves arranged in a spiral around its spindle-shaped body.

The stemmed or stalked echinoderms first occur in Middle Cambrian strata but do not become abundant until Ordovician and Silurian time. Stalked forms called cystoids (Fig. 10–48) are the most primitive among this group. The striking pentamerous (five-way) symmetry that is evident in most echinoderms is often less well developed in cystoids. Beginning students of paleontology often recognize cystoids by the characteristic pattern of pores on the plates of the calyx. Although cystoids range from Cambrian to Late Devonian, they are chiefly found in Ordovician and Silurian rocks.

FIGURE 10-48 **A well-preserved specimen of the Silurian cystoid *Caryocrinites ornatus* from the Lockport Shale of New York.** (*From Sprinkle, J. 1975.* J. Paleo. 49(6):1062–1073.)

Unlike some of the cystoids, stalked echinoderms known as blastoids (Fig. 10–49) have a orderly arrangement of plates. The five radial areas (ambulacral areas) are prominent and bear slender branches or brachioles along their margins. Unfortunately, the delicate brachioles are rarely preserved in fossil specimens. Blastoids also have a well-developed and unique water vascular system. There are no blastoids living today. They first appeared in Silurian time, expanded in the Mississippian, and declined to extinction in the Permian.

Crinoids, like most cystoids and blastoids, are composed of three main parts: the cup-shaped calyx (which contains the vital organs), the arms, and the stem, with its anchoring holdfast (Fig. 10–50). The arms bear ciliated food grooves and, like the brachioles of blastoids, serve to move food particles toward the mouth. Crinoids are found in rocks that range in age from the Ordovician Period to the Holocene Epoch of the Cenozoic Era. In some areas, Ordovician, Silurian, and Carboniferous rocks contain such great quantities of disaggregated plates of crinoids that they are named crinoidal limestones. So abundant are crinoid remains in Mississippian rocks that the period has been dubbed the age of crinoids (Fig. 10–51).

Although one group of crinoids survived the hard times at the end of the Paleozoic and gave rise to those still living today, most died out before the end of the Permian. The blastoids also died out before the end of the Paleozoic. Free-living echinoderms such as echinoids (Fig. 10–52) and starfish were locally abundant during the late Paleozoic but not as numerous or diverse as they would become in post-Paleozoic time.

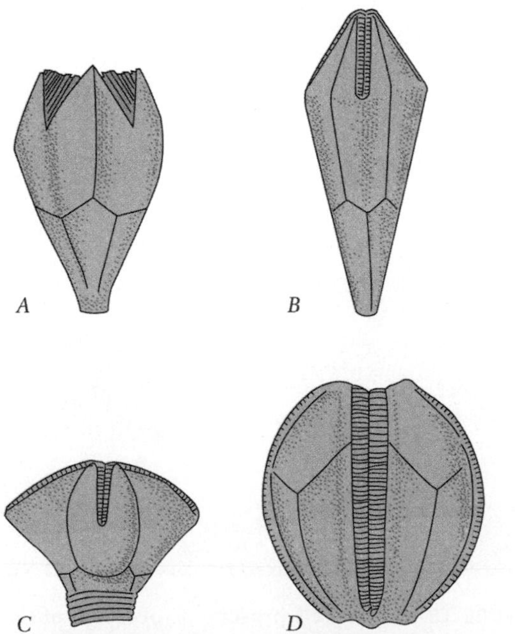

FIGURE 10-49 Some common Paleozoic blastoids. (*A*) *Codaster* (Dev.), (*B*) *Troosticrinus* (Sil.), (*C*) *Orophocrinus* (Miss.), and (*D*) *Cryptoblastus* (× 2).

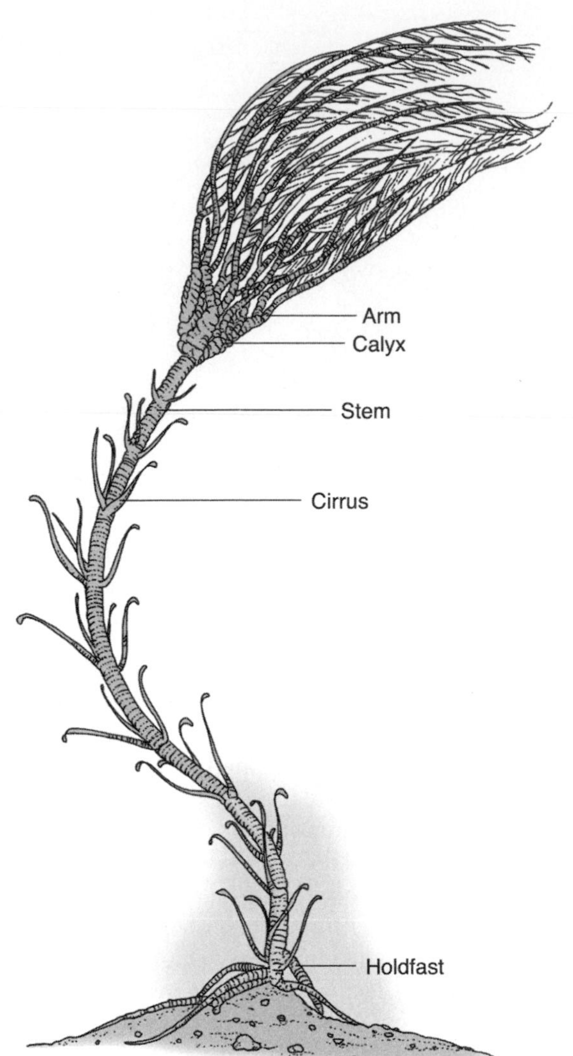

FIGURE 10-50 Crinoid in living position on sea floor.

THE ECHINODERM–CHORDATE CONNECTION Because we ourselves are chordates, we have a particular interest in the close evolutionary relationship between chordates and echinoderms. Many lines of evidence support this relationship. In the early development of the chordate and echinoderm embryo, a small sphere of cells is formed, called the **blastula**. A group of cells move inward on the blastula to form an opening called the **blastopore**. In echinoderms and chordates the blastopore develops into the *anus*. (The mouth develops later from another opening.) Animals like the echinoderms and chordates that form the anus in this way are termed **deuterostomes**. Along with starfish and sea urchins, you and I are deuterostomes. The companion term **protostomes** refers to animals like

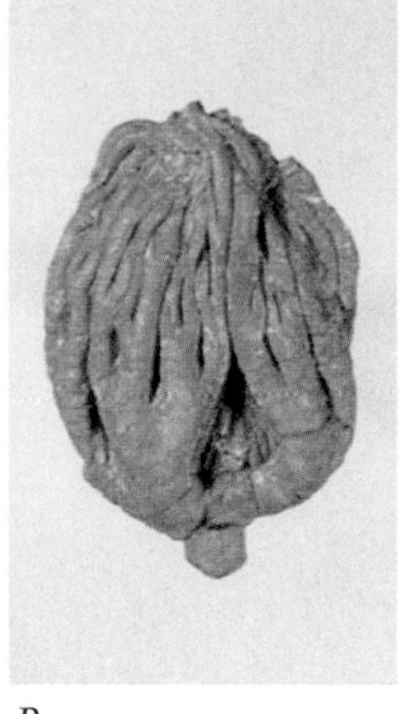
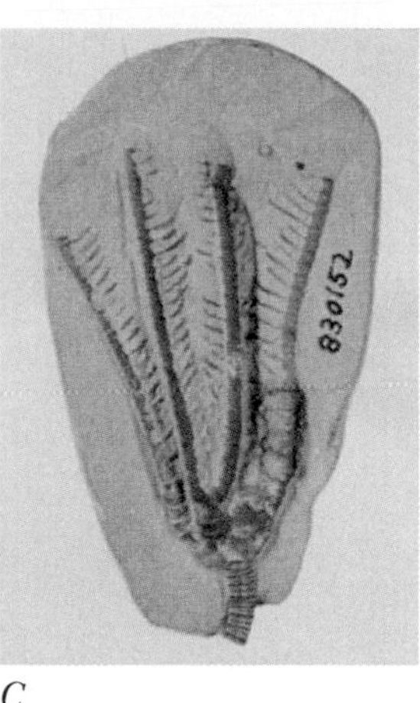

A *B* *C*

FIGURE 10-51 **Crinoids.** (*A*) Diorama depicting crinoids living on the floor of a Mississippian epeiric sea. (*B*) Calyx and arms of the Mississippian crinoid *Taxocrinus* (height 5.1 centimeters). (*C*) *Scytalocrinus* (height 4.5 centimeters). Both fossil crinoids are from the Keokuk Formation of Mississippian age, near Crawfordsville, Indiana. (*Diorama photograph courtesy of the U. S. National Museum of Natural History, Smithsonian Institution.*)

arthropods, mollusks, and annelid worms in which the blastopore develops into a *mouth*. Protostomes and deuterostomes also differ in their patterns of early cell division. In chordates and echinoderms the early cells cleave one above the other in an arrangement termed *radial* (Fig. 10–53). In protostomes cell divisions are *spiral* so that successive rows are nestled neatly within the depressions between earlier formed cells. Also in embryologic development, the mesodermal layer of cells as well as certain other elements of the body arise in the same way in both echinoderms and chordates. Even the larvae of echinoderms resemble those of primitive chordates. Biochemistry has provided a final line of evidence of a relationship between echinoderms and chordates by revealing chemical similarities associated with muscle activity and the chemistry of oxygen-carrying pigments in the blood.

FIGURE 10-52 **The large Mississippian echinoid *Melonechinus* from the St. Louis Limestone, St. Louis, Missouri.** Average diameter of specimens is 12 centimeters. (*Photograph by J. Simon.*)

Continental Invertebrates

Because of the greater hazards of postmortem destruction, the fossil record of land-dwelling invertebrates is not as complete as that for marine invertebrates. Nevertheless, as plants invaded the continents, animals were able to follow. Arthropods, with their sturdy legs and protective exoskeletons, were probably the marine animals best preadapted for coming ashore. Early invaders may have lived within the wet, nutritious plant debris washed up along the beaches of early Paleozoic seas. Their fossil record begins with possible millipede trace fossils in Ordovician rocks. More direct evidence consists of arthropod cuticles and bristles recovered along with spores from early Silurian terrestrial rocks. The same samples also contain fecal pellets believed to have been left by arthropods feeding on fungi and algae. Body fossils of centipedes and millipedes occur in late Silurian deposits of Britain. Near Gilboa, New York, Devonian rocks have yielded the fossil remains of mites, primitive wingless arthropods called archaeognaths, centipedes, the earliest known spiders,

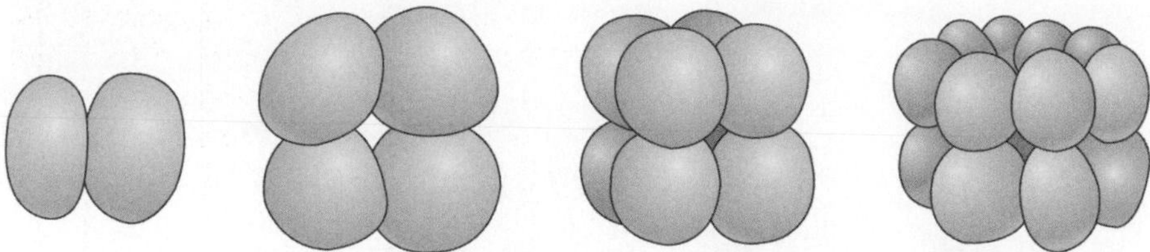

Radical cleavage, as seen in the development of the deuterostome embryo

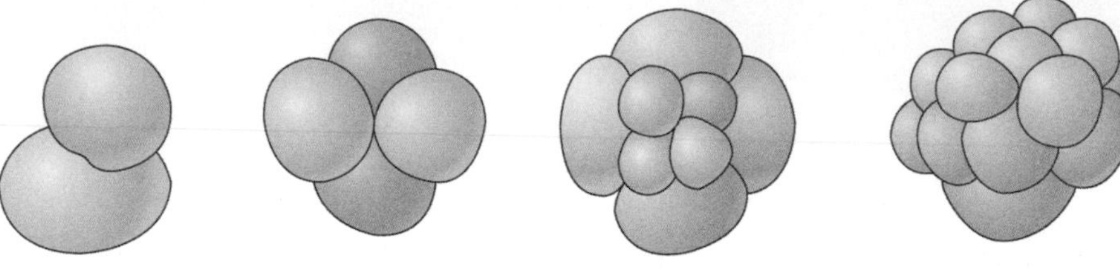

Spiral cleavage, as seen in the development of the protostome embryo

FIGURE 10-53 Comparison of patterns of embryologic cleavage in deuterostomes and protostomes.

and spiderlike predators known as trigonotarbids. Except for the primitive bristletails, insects do not become common until about 315 million years ago (late Mississippian). The subsequent Pennsylvanian Period witnessed the arrival of flying insects, like the dragonfly *Mischoptera* (Fig. 10–54). The coal swamp insect fauna of the Pennsylvanian period also included the largest insect known, a monstrous dragonfly with a wingspread of over 2 feet. Other insect giants included 4-inch-long cockroaches that foraged for their food in forest litter and rotting vegetation. Although never abundant, stream- and lake-dwelling eurypterids persisted throughout the late Paleozoic. In Nova Scotia, an interesting collection of land snails has been collected from hollows within the stumps of fossil trees. Scorpions and centipedes, as well as fish, amphibians, plants, insects, other arthropods, and mollusks, have been found in abundance in concretionary Pennsylvania sedimentary rocks exposed along Mazon Creek in northern Illinois. One can even discern impressions of soft parts in many of these splendidly preserved organisms.

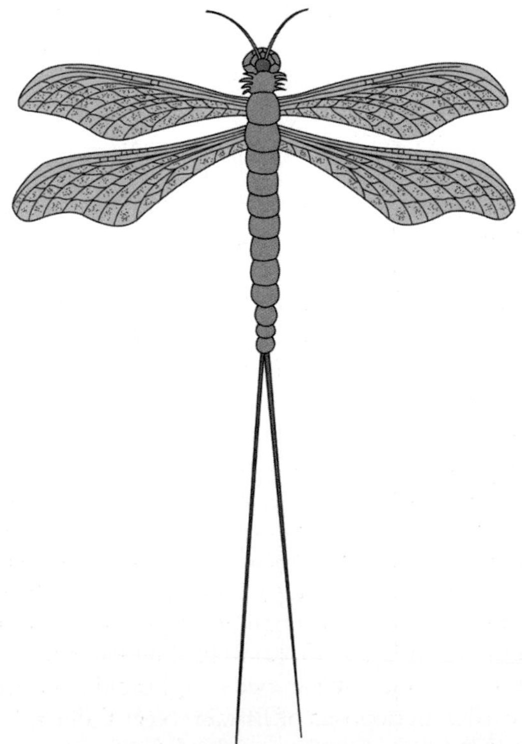

FIGURE 10-54 ***Mischoptera*, a Pennsylvanian age dragonfly** (length 8.4 centimeters).

Of all the land invertebrates, it is apparent that the arthropods and gastropods have been persistently successful. Many paleontologists believe this may be because of attributes already evolved by their aquatic ancestors. As protection against desiccation, arthropods had evolved relatively impervious exoskeletons. Snails derived similar benefits from their shells. Both groups included very active animals with sufficient mobility to seek out food aggressively.

GRAPTOLITES

While myriads of benthic invertebrates populated the sea floors of the early Paleozoic, there were also great numbers living in the water column above the ocean floor. Many of these mostly planktonic creatures lacked skeletons and may never be revealed to us. One group, the graptolites (Fig. 10–55), had preservable chitinous skeletons that housed colonies of tiny individual organisms.

The structure of the graptolite skeleton is distinctive. The graptolite animals live in tiny cups, or *thecae*, that are aligned along a stem, or *stipe*. The stipes may be solitary or formed into a system of two or more branches. In general, evolution progressed from

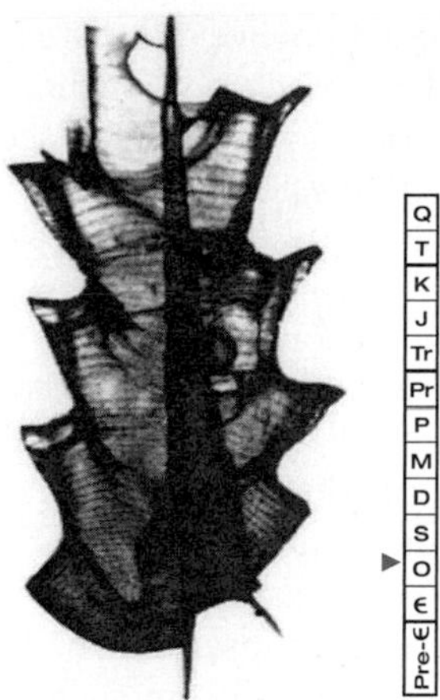

FIGURE 10-55 Part of a stipe of the Ordovician graptolite *Orthograptus quadrimucronatus* (× 15). The specimen has been bleached for better visualization of internal features. The cuplike thecae are clearly visible on either side. One can also see the inverted cone of the sicula. The nema would rise from the pointed end of the sicula. (*From Herr, S. R. 1971.* J. Paleo. 45(4):628–632.)

multibranched forms to those having only a single stipe. The entire colony is referred to as a rhabdosome. Usually positioned at the lower end of the rhabdosome is a single thin filament—the *nema*. Some graptolites were attached to floating objects by the threadlike nema. Where the lower end of the nema reaches the base of a stipe, there is a conical *sicula*, which may have served as the theca for the first individual. From the sicula, subsequent individuals and their thecae were added by budding.

Fossil graptolites characteristically occur as flattened and carbonized impressions in dark shales. Rarely, uncompressed specimens are found. These reveal an unexpected relationship to primitive living chordates called **pterobranchs**. Both groups secrete tiny enclosed tubes, and in both the unique structure of the thecae is so similar that a close relationship is virtually certain.

Graptolites made their appearance at the end of the Cambrian but did not become abundant until the Ordovician. They are no longer present in the fossil record by the end of the Mississippian. In 1989, however, a specimen was recovered from South Pacific waters that *may* be a surviving species. Marine biologist Noel Dilly named this apparent survivor *Cephalodiscus graptoloides*. Subsequently, additional specimens were found in the waters surrounding Bermuda. The protein in the skeletal material and the general morphology of these organisms closely resemble those of early Paleozoic graptolites.

During the Ordovician and Silurian, graptolites were so abundant that they may have formed large floating masses. They were apparently carried about by ocean currents and thus achieved worldwide distribution. Because of their wide dispersal and rapid rate of evolution, they are among the most important guide fossils for Ordovician and Silurian sedimentary rocks.

VERTEBRATE ANIMALS OF THE PALEOZOIC

Vertebrate Attributes

Vertebrates are animals having a segmented dorsal **vertebral column**. The vertebral column is composed of individual units termed **vertebrae**. Arches of the vertebrae encircle and protect a hollow nerve cord referred to as the **spinal cord**. In addition to the vertebral column, vertebrates have a cranium or skull that houses a brain.

Vertebrates are a diverse group that includes both water-dwelling and land-dwelling **tetrapods**. The term tetrapod comes from two Greek words meaning "four feet." Tetrapods include not only vertebrates that walk on four legs (quadrupedal) but also those that walk on only their hind legs (bipedal), those whose forelimbs have been modified into wings, and those whose limbs have become flippers for swimming.

An important reproductive criterion serves to separate vertebrates into two major groups. The first group consists of fishes and amphibians, both of which must be in water, or at least wet, in order to reproduce. Their eggs are naked in that they lack a covering. Such eggs must be fertilized externally. The remaining group of vertebrates evolved a method of reproduction that requires internal fertilization and an enclosed egg called an **amniotic egg**. Thus, modern amphibians like frogs and salamanders can be termed **nonamniotic vertebrates**, whereas all higher vertebrates are **amniotic vertebrates** ("amniotes").

The embryo in an amniotic egg is enveloped in an **amniotic membrane**, which allows oxygen to enter but which retains water (Fig. 10–56). The membrane encloses the embryo in a cushioning watery environment. There is also a yolk sac containing nutrients to nourish the embryo. Another membrane, the **allantois**, holds the embryo's waste products. Finally, the **chorion** regulates the exchange of oxygen and carbon dioxide. Most amniotes have a shell or leathery covering to protect the embryo, but in mammals, which are also amniotes, the embryo grows within an amnion that develops within the mother's womb. For all amniotes, the remarkable amniotic egg provided freedom from dependency on water bodies. It facilitated the exploitation of diverse terrestrial environments and has been hailed as an extraordinary milestone in the evolution of vertebrates.

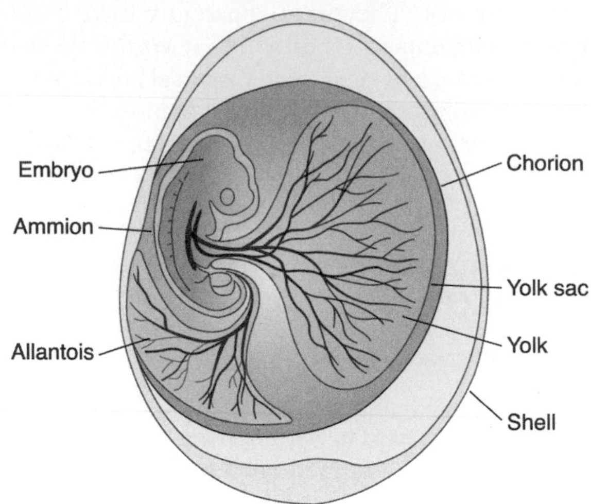

FIGURE 10-56 The amniotic egg. The amnion, from which the egg takes its name, encloses the embryo in water. The allantois serves as a reservoir for waste, the chorion regulates the exchange of gases, and the yolk (also present in fish and amphibian eggs) serves as a storage area for fats, proteins, and other nutrients needed by the developing embryo.

The Rise of Fishes

A momentous biologic event occurred in early Paleozoic time. It was the birth of the chordate line. In our earlier discussion of the Burgess Shale fauna, we briefly described chordates as animals that have (at least at some stage in their life history) a stiff, elongate supporting structure. In addition, all chordates have a dorsal, central nerve cord, gill slits, and blood that circulates forward in a main ventral vessel and backward in the dorsal. The supportive structure in primitive chordates is called a notochord and has been studied by generations of biology students in such animals as the lancelet *Branchiostoma* (Fig. 10–57). In taxonomic hierarchy, vertebrates are simply those chordates in which the notochord is supplemented or replaced by a series of cartilaginous or bony vertebrae.

Because of similarities in embryologic development, some biologists believe that the ancestors of the vertebrate lineage may lie somewhere among the echinoderms. The theoretic evolutionary progression that was to lead to vertebrates may have begun with sedentary, filter-feeding animals that had exposed cilia located along their arms. From such a beginning there may have evolved filter-feeders with cilia brought inside the body in the form of gills. In a subsequent stage, the organisms may have become free-swimming, gilled creatures, in appearance not unlike *Cathaymyrus* from the Lower Cambrian of China and *Pikaia* from the Burgess Shale. Such animals were the precursors of forms more clearly resembling fish.

The vertebrates that we informally call fishes include five taxonomic classes (Fig. 10–58). They are the jawless fishes, or **Agnatha**; two groups of archaic jawed fishes, the **acanthodians** and **placoderms**; the cartilaginous fishes, or **Chondrichthyes**; and the familiar **Osteichthyes** with their highly developed bony skeleton. It is the first three of these categories—the Agnatha, Acanthodii, and Placodermi—that are most frequently found in rocks of the early Paleozoic.

The oldest known agnathids were discovered in the Early Cambrian fossil beds near Chengjiang, China. The fossils, named *Myllokunmingia* (see Fig. 10–8) and *Haikouichthys*, have several features that favor their

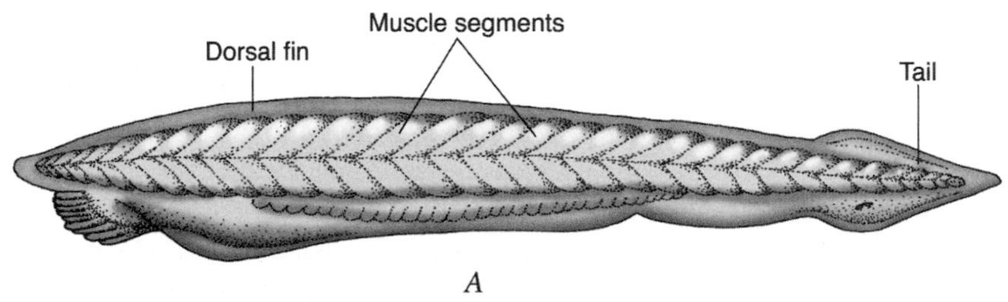

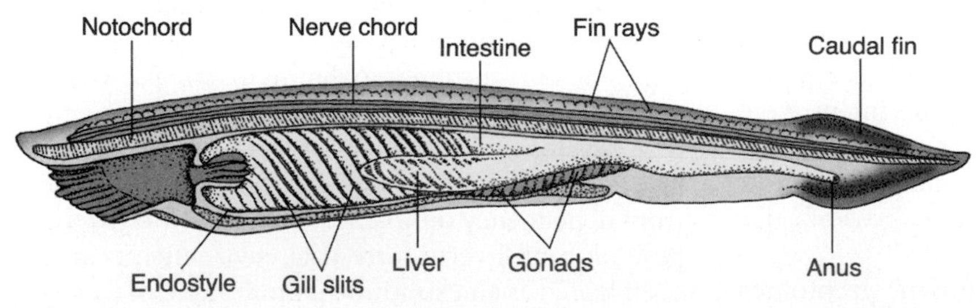

FIGURE 10-57 *Branchiostoma*, a member of the subphylum Cephalochordata. (*A*) External view; (*B*) longitudinal section (length 3 to 5 centimeters).

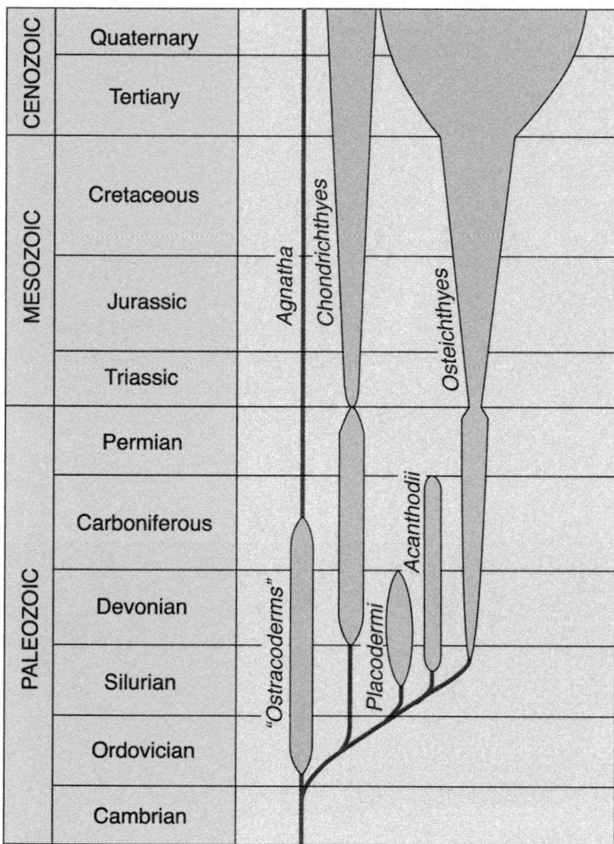

FIGURE 10-58 Evolution of the five major categories of fishes. The width of the vertical red areas indicates the approximate relative abundance of each group.

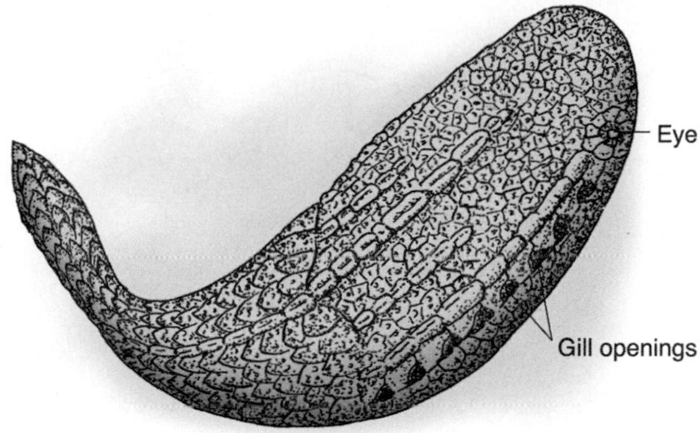

FIGURE 10-59 The Ordovician agnathan *Astraspis* from the Harding Sandstone of Colorado (× 0.8). *(After Elliot, D. K. 1987. A reassessment of* Astraspis desiderata, the oldest North American vertebrate. *Science 237:190–192.)*

placement in the Vertebrata. These include the prevously mentioned V-shaped musculature, a relatively complex skull, gill supports, and fin supports. The tiny fish rather resemble the larvae of that living agnathid, the lamprey. Both *Myllokunmingia* and *Haikouichthys* were covered by soft tissue. Armored agnathids were also present during the early Paleozoic. The rather clumsy-looking *Astrapis* (Fig. 10–59), with its jawless slit-like mouth and row of gill openings, is one of these earliest agnathans. Early Paleozoic agnathids are collectively termed **ostracoderms** ("shell skins"), even though some forms lacked heavy bony armor. The ostracoderms comprise a rather diverse group that included the unarmored forms *Theolodus* and *Jamoytius* as well as such armored creatures as *Hemicylaspis* and *Pteraspis* (Fig. 10–60).

The purpose of the bony armor that is the hallmark of many of the ostracoderms is still being debated. A widely held view is that the armor provided protection against predators. Another theory stipulates that the dermal armor was not primarily for protection but rather was a device for storing seasonally available phosphorus. According to this idea, phosphates could be accumulated as calcium phosphate in the armor during times of greater availability and then used during periods when supply was deficient. This cache of phosphorus may have been vital for early vertebrates to maintain a suitable level of muscular activity.

Hemicyclaspis (Fig. 10–60*D*) is one of the most widely known of the ostracoderms. It is recognized by its large, semicircular head shield, on the top of which can be found four openings: two for the upward-looking eyes, a small pineal opening, and a single nostril. The pineal opening may have housed a light-sensitive third eye. The mouth was located ventrally in a position suitable for taking in food from the surface layer of soft sediment. Depressed areas along the margin of the head shield covered a system of nerves and may have had a sensory function.

The ostracoderms continued into the Devonian but did not survive beyond that period. For the most part they were small, sluggish animals restricted to mud-straining or filter-feeding modes of life. They were to be replaced gradually by fishes that had developed bone-supported, movable jaws. The evolution of the jaw was no small accomplishment, for it enormously expanded the adaptive range of the vertebrates. Fishes with jaws were able to bite and grasp. These new abilities led to more varied and active ways of life and to new sources of food not available to the agnathids.

There are currently two theories that attempt to account for the origin of the jaw. The older is that jaws formed by modification of an anterior pair of gill arches, the thin strips of bone or cartilage that support the soft tissue of the gills. More recently it has been proposed that jaws arose from a structure called the velum, which functions in respiration and feeding in larval lampreys. Both theories are based on the study of the anatomy and embryology of living fishes, for the first jaws in the fossil record give little evidence of their origin.

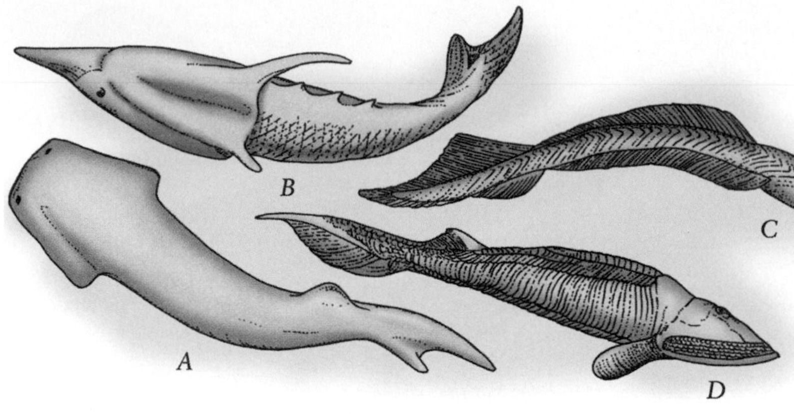

FIGURE 10-60 Early Paleozoic ostracoderms. (*A*) *Thelodus*, (*B*) *Pteraspis*, (*C*) *Jamoytius*, and (*D*) *Hemicyclaspis*, drawn to the same scale.

The oldest fossil remains of jawed fishes are found in nonmarine rocks of the Late Silurian. These fishes, called **acanthodians** (Fig. 10–61), became most numerous during the Devonian and then declined to extinction in the Permian. Acanthodians were archaic jawed fishes and were distinct from the great orders of modern fishes. Another group of archaic fishes includes the placoderms, or "plate-skinned" fishes. They too arose in the late Silurian, expanded rapidly during the Devonian, and then in the latter part of that period began to decline and be replaced by the ascending sharks and bony fishes.

There was considerable variety among the placoderms. The most formidable of these plate-skinned fishes was a carnivorous group called **arthrodires**. *Dunkleosteus* (Fig. 10–62) was a Devonian arthrodire whose length exceeded 9 meters and whose huge jaws could be opened exceptionally wide to engulf even the largest of available prey. Other placoderms, called **antiarchs** (Fig. 10–63), had the heavily armored form and mud-grubbing habits of their ostracoderm predecessors.

Among the early Paleozoic acanthodians and placoderms were the ancestors to the cartilaginous and bony fishes. In rocks of Late Silurian and Devonian age, fossils of entire **chondrichthyans** (cartilaginous fishes) and **osteichthyans** (bony fishes) are known from many

FIGURE 10-62 The gigantic armored skull and thoracic shield of the formidable late Devonian placoderm fish known as *Dunkleosteus*. *Dunkleosteus* was over 10 meters (about 30 feet) long. The skull shown here is about 1 meter tall. It is equipped with large bony cutting plates that functioned as teeth. Each eye socket was protected by a ring of four plates, and a special joint at the rear of the skull permitted the head to be raised, thereby making an extra large bite possible. *Dunkleosteus* ruled the seas 350 million years ago. (*Courtesy of the U.S. National Museum of Natural History, Smithsonian Institution; photograph by Chip Clark.*)

FIGURE 10-61 The Early Devonian acanthodian fish *Climatius*. (*After Romer, A. S. 1945.* Vertebrate Paleontology. *Chicago: University of Chicago Press.*)

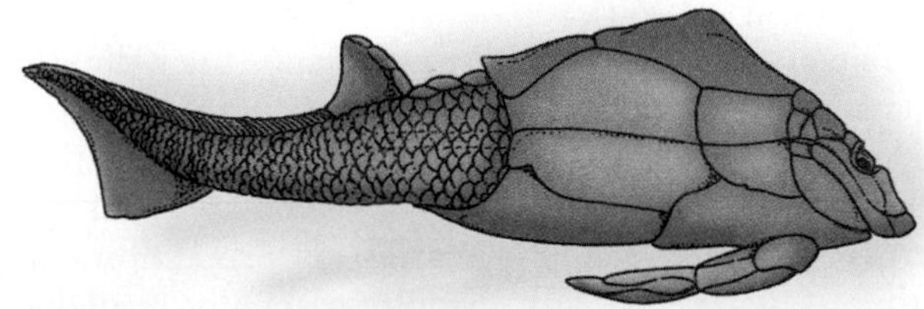

FIGURE 10-63 The Devonian antiarch fish *Pterichthyodes*. (*From Romer, A. S. 1945.* Vertebrate Paleontology. *Chicago: University of Chicago Press, p. 54, fig. 38.*)

localities around the world. Sharks, rays, and skates are among the familiar chondrichthyans populating modern seas. Among the better known of the late Paleozoic sharks were species of *Cladoselache* (Fig. 10–64*A*). Remains of this shark are frequently encountered in the Devonian shales that crop out on the southern shore of Lake Erie. During the Late Carboniferous, a group represented by *Xenacanthus* (Fig. 10–64*B*) managed to penetrate the freshwater environment. A third group of Paleozoic cartilaginous fishes were the bradyodonts. These had flattened bodies like modern rays and blunt, rounded teeth for crushing shellfish. Apparently, modern sharks arose from cladoselachian ancestors but retained their archaic traits until the Jurassic.

Because of the role of bony fishes in the evolution of tetrapods (four-legged animals) and because they are the most numerous, varied, and successful of all aquatic vertebrates, their evolution is of particular importance. Bony fishes may be divided into two categories: the familiar ray-fin, or **actinopterygians**, and the lobe-fin, or **sarcopterygians**.

As implied by their name, ray-fin fishes lack a muscular base to their paired fins, which are thin structures supported by radiating bony rays. Unlike the Sarcopterygii, they do not possess paired nasal passages that open into the throat. The ray-fins began their evolution in Devonian lakes and streams and quickly expanded into the marine realm. They became the dominant fishes of the modern world. The more primitive Devonian ray-finned bony fishes are well represented by the genus *Cheirolepis* (Fig. 10–65). From such fishes as these evolved the more advanced bony fishes during the Mesozoic and Cenozoic.

The second category of bony fishes, the sarcopterygians, is characterized by fishes with sturdy, fleshy lobe-fins and a pair of openings in the roof of the mouth that led to clearly visible external nostrils. Such fish were able to rise to the surface and take in air, which was passed on to functional lungs. Lungs and gills do not seem to occur together in modern fishes, but in late Paleozoic fishes the combination was not uncommon. Studies of living examples of sarcopterygians indicate

FIGURE 10-64 **Models of (*A*) the Devonian marine shark *Cladoselache* and (*B*) the Pennsylvanian freshwater shark *Xenacanthus*.**

FIGURE 10-65 ***Cheirolepis*, the ancestral bony fish that lived during the Devonian Period.**

that lungs probably began their evolution as sac-like bodies developed on the ventral side of the esophagus and then became enlarged and improved for the extraction of oxygen.

Two major groups of sarcopterygians lived during the Devonian. They are designated the **dipnoans** and **Crossopterygians**. The dipnoans, represented in the Devonian by *Dipterus* (Fig. 10–66), were not on the evolutionary track that was to lead to tetrapods. They are, nevertheless, an interesting group that includes living freshwater lungfishes of Australia, Africa, and South America. Their restricted presence south of the equator suggests that Gondwanaland was the probable center of dispersal for dipnoans. The taxonomic term Dipnoi means "double breather." The name was suggested by the observation that living species are able to breathe by means of lungs during dry seasons. At such difficult times, they burrow into the mud before the water is gone. When the lake or stream is dry, they survive by using their accessory lungs, and when the waters return, they switch to gill respiration. The dipnoans are an enduring group. One species living today in Queensland, Australia, has a fossil record that extends back 100 million years.

Because of the arrangement of bones in their muscular fins (Fig. 10–67), the pattern of skull elements (Fig. 10–68), and the structure of their teeth (Fig. 10–69), fossil crossopterygians are considered the ancestors of those earliest invaders of land habitats, the nonamniotic tetrapods. A rather advanced fish that exemplifies Devonian crossopterygians is *Eusthenopteron* (Fig. 10–70). In this genus, the paired fins were short and muscular. Internally, a single basal limb bone, the humerus (equivalent to the femur in the hindlimbs) articulated with the girdle bones and was followed by two bones, the ulna and radius, for the pectoral fins and two others, the tibia and fibula, for the posterior fins.

Of course, the robust skeleton and sturdy limbs of the crossopterygians did not evolve because fishes had miraculously taken on a desire for life on land. Very likely, these were adaptations to permit the animal to leave a body of water that was drying out or stagnating and move to another body of water that might offer a better chance of survival. Another theory suggests that legs evolved to assist movement in shallow water.

During the Devonian, two distinct branches of crossopterygians had evolved. One of these, the rhipidistians, led ultimately to nonamniotic tetrapods, and the other led to fishes called coelacanths. Coelacanths were thought to have undergone extinction during the Late Cretaceous, but their survival to the present was documented beginning in 1938 by several catches of the coelacanth *Latimeria* (Fig. 10–71) off the coast of the Comoro Islands (the Indian Ocean west of Madagascar). Several other discoveries of the rare fish were made within the past decade. *Latimeria* is a large fish, reaching nearly 2 meters in length. It swims slowly, using its paired fins in a manner resembling the way four-legged animals walk on land. Sharp movement of the tail fin provides the burst of speed needed to catch its prey.

Conodonts

For more than a century, tiny fossils known as **conodont elements** have been known to geologists. Over this span of time, however, the nature of the animal that bore the elements and the function they served have been a persistent puzzle. Even without this knowledge, however, conodont elements have served admirably as guide fossils. They are found in a variety

FIGURE 10-66 ***Dipterus*, a Devonian lungfish.**

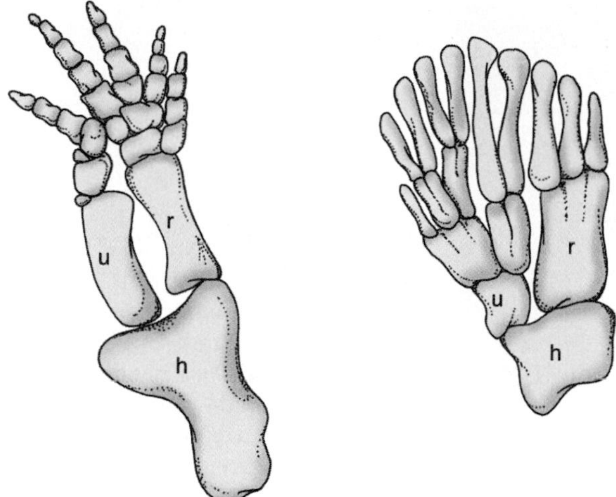

FIGURE 10-67 Comparison of the limb bones of a crossopterygian fish (upper right) and an early amphibian. Some early amphibians may have had more than five digits. (*From Levin, H. L. 1975.* Life Through Time. *Dubuque, Iowa: William C. Brown Co.*)

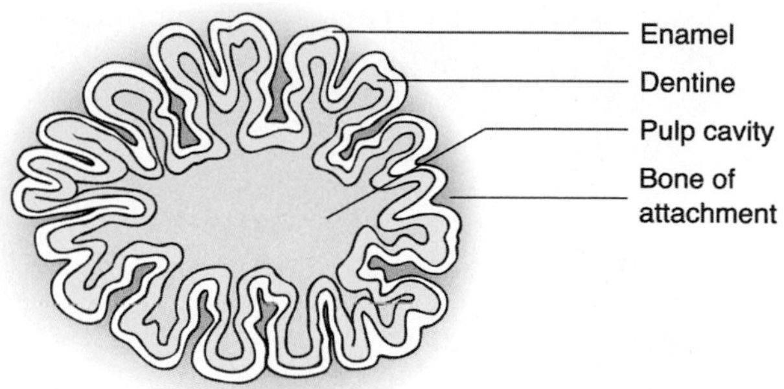

FIGURE 10-69 Cross-section of a crossopterygian tooth, clearly exhibiting the distinctive pattern of infolded enamel. This is also a characteristic of the teeth of the amphibian descendants of crossopterygian fishes. (*From Levin, H. L. 1975.* Life Through Time. *Dubuque, Iowa: William C. Brown Co.*)

of marine sedimentary rocks that range from latest Proterozoic to Triassic in age.

Conodont elements consist of cones, bars, and blades bearing tiny denticles or cusps (Fig. 10–72), or variously shaped ridged structures. The elements are usually less than a millimeter in size and composed of a durable calcium phosphate variety of the mineral apatite. Fossil occurrences consist mostly of disassociated elements. In order to determine which elements occurred together in the parent organism, one must find conodont elements in natural associations of elements that have not been scattered after death. When, with good fortune, one finds such an assemblage, it is often possible to reconstruct the way differently shaped elements were arranged in the complete apparatus.

But what was the nature of the conodont animal? Since the discovery of conodont elements by Christian Pander in 1856, proposals have been advanced that they were grasping, filtration, or support structures for such diverse animals as worms, fish, snails, cephalopods, and arthropods. Recently uncovered evidence, however, indicates that the conodont animal was a chordate. Several fossils of conodont animals have been reported over the past 3 decades, but the most informative fossil was found in 1995 in the Ordovician Soom Shale of South Africa. It consists of the remains of a creature with an elongate eel-like body that in life was about 40 millemeters long (see Fig. 10–72). The creature had distinctive eye musculature, similar to that in primitive jawless fish. Like the living cephalochordate *Branchiostoma* (the amphioxus or lancelet), muscles along each side of the conodont animal were V-shaped units lying sidewise, with the point forward. Scanning electron microscopy of the fossilized fibrous muscle tissue revealed similarities to the muscles in fish. The Soom Shale specimen, named *Promissum*, had conodont elements at its anterior end, and it had a longitudinal band that can be interpreted as a notochord.

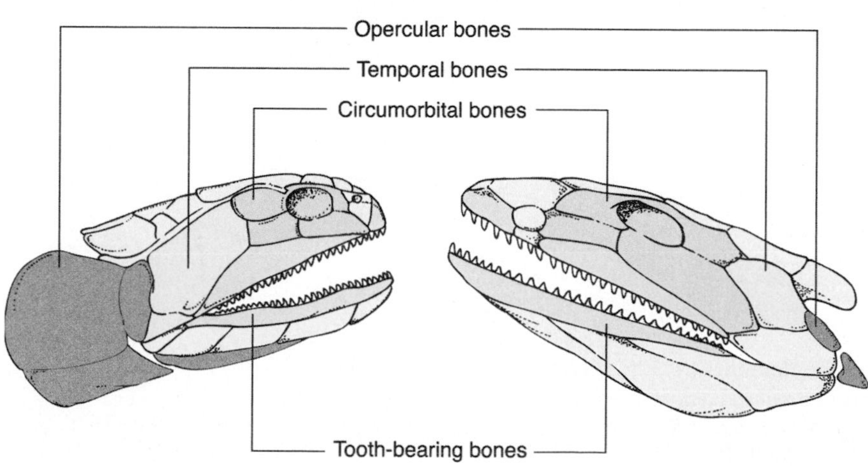

FIGURE 10-68 Comparison of skulls and lower jaws of a crossopterygian (*left*) and the Devonian amphibian *Ichthyostega*. (*From Levin, H. L. 1975.* Life Through Time. *Dubuque Iowa: William C. Brown Co.*) ? *Which of the senses (sight, hearing, smell, taste, and touch) would have been improved by the longer distance from nostril to eye orbit developed in Ichthyostega?*

FIGURE 10-70 The Devonian crossopterygian lungfish *Eusthenopteron* had sturdy fins whose structure foreshadowed that of four-footed land animals that were to be their descendants. (*St. Louis Science Center diorama.*)

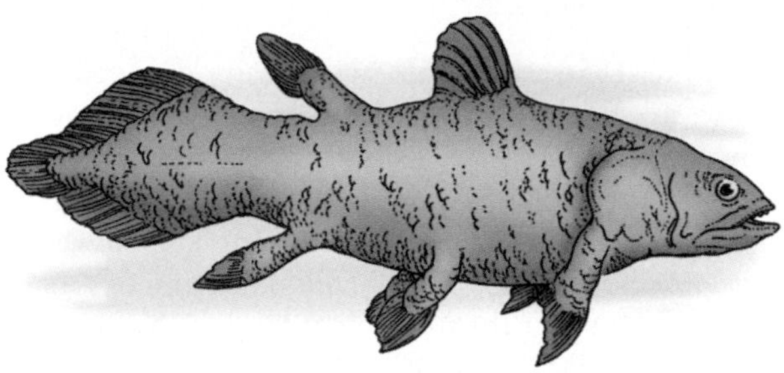

FIGURE 10-71 *Latimeria*, a surviving coelacanth living in the ocean near Madagascar. *Latimeria* is a large fish, nearly 2 meters in length. (*From Levin, H. L. 1975.* Life Through Time. *Dubuque, Iowa: William C. Brown Co.*)

The above evidence strongly favors chordate status for the conodont animal. But what function was served by the conodont elements? Were they supports for food-gathering organs? Did they act as sieves for filtering out fine food particles from water? Or were they used for chewing food? If conodont elements did function as teeth, they should show the effects of wear associated with grasping food. Scanning electron microscope images of some conodont elements reveal fine scratches, pitting, and other surface textures, suggesting the elements were used in processing food.

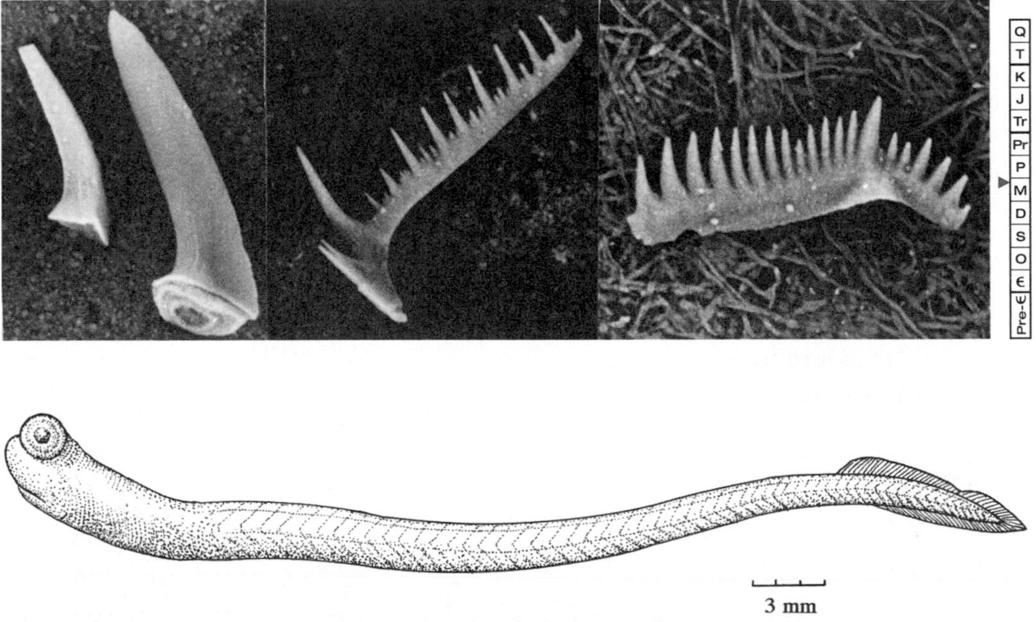

FIGURE 10-72 Shapes of three conodont elements (× 50); (*above*) and a restoration of the conodont animal (*below*) as interpreted from fossil remains in the Lower Carboniferous Granton Shrimp Bed of Scotland. (*Photographs courtesy of K. Chauff; restoration after Aldridge, R. J. et al. 1993.* Phil. Trans. R. Soc. London B 340:405–421.)

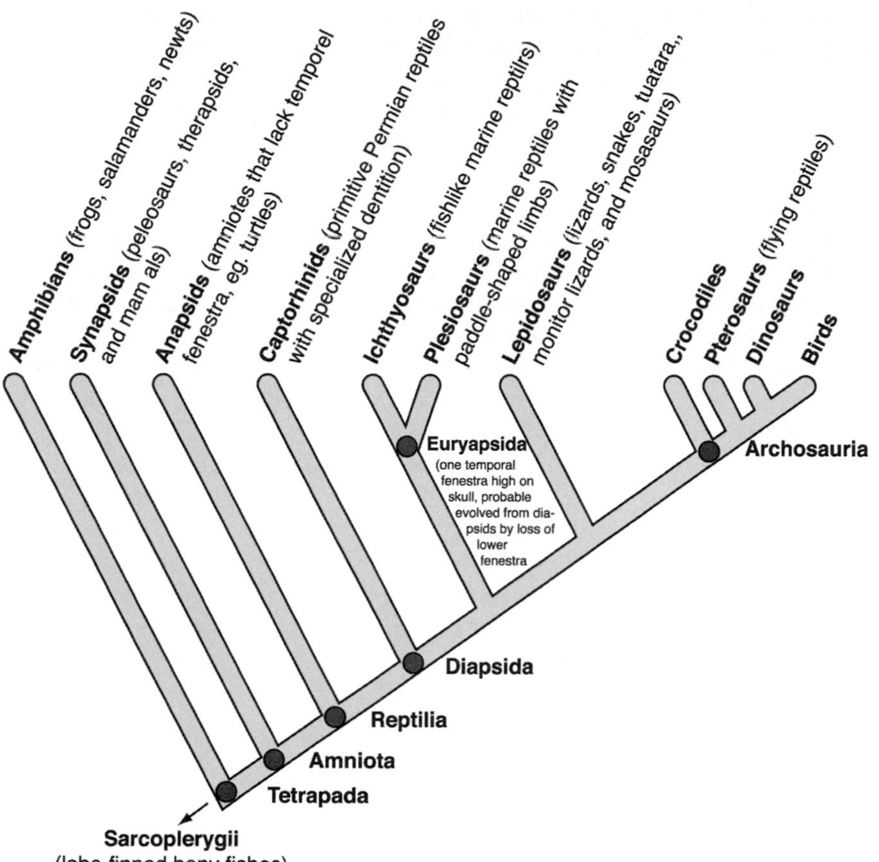

FIGURE 10-73 Cladogram showing relationships among premammalian land vertebrates.

Others, however, have a laminar structure, indicating new material was laid down in layers from an enveloping layer of soft tissue.

The Advent of Tetrapods

A CLADISTIC CLASSIFICATION OF TETRAPODS **Tetrapods** are four-footed (Gr. *Tetra* = four + *pous* = foot) vertebrates. The Linnaean system places tetrapods into the familiar classes Amphibia, Reptilia, Mammalia, and Aves. This traditional scheme, however, does not trace evolutionary pathways as objectively as does the cladistic method described in Chapter 4. Cladistic analyses identify groups of species that all have a trait or traits that evolved as new features or "evolutionary novelties." Descendents of the first animal to develop a new feature usually retain it and therefore can be recognized as a descendant of the anestral species in which the new feature appeared. Thus, any given animal can be assigned to a branch, or **clade**, that truly reflects its ancestry.

Figure 10–73 is a simple cladogram showing the names and relationships of major groups of tetrapods. It serves not only as a guide to evolutionary pathways but also as a reference for some of the unfamiliar names you will encounter as you read about life of the geologic past. The terms **Anapsida, Synapsida, Diapsida,**

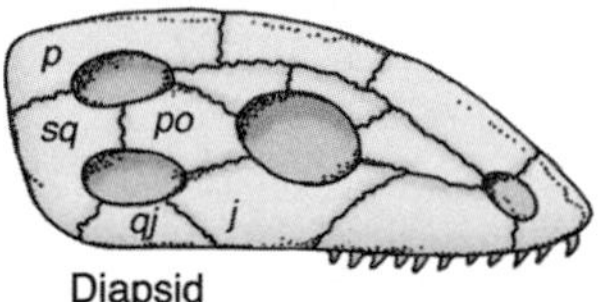

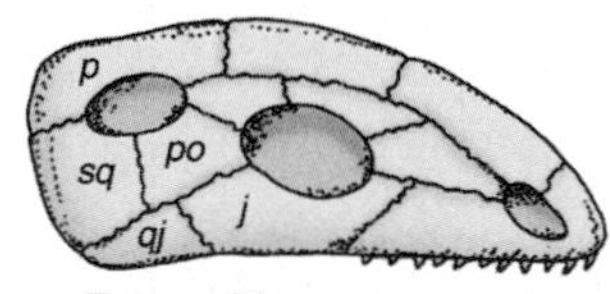

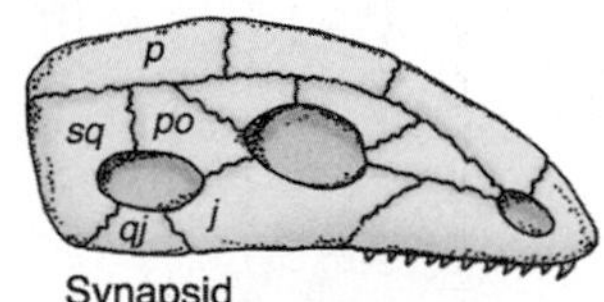

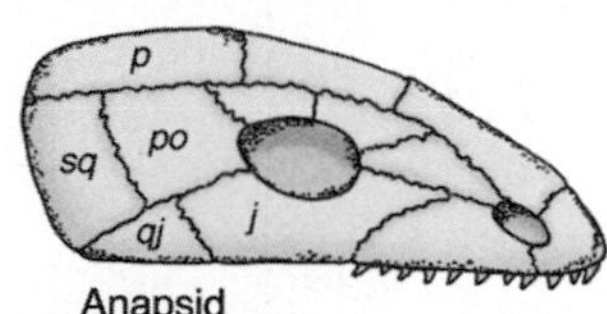

FIGURE 10-74 Amniote skull types (*p*, parietal; *sq*, squamosal; *po*, postorbital; *j*, jugal; *qj*, quadratojugal).
? *Which of the above skull types identifies dinosaurs?*

and **Euryapsida** that appear on the chart are derived from traditional Linnaean nomenclature. They refer to the position and number of openings on the sides of the skull (Fig. 10-74). The openings are termed **temporal fenestrae**. Some groups such as the early reptiles and the turtles lack these "holes in the head." They are designated Anapsida (informally, "anapsids"). Among anapsids, the muscles that move the jaws extend from each side of the skull through holes in the palate (bones forming the roof of the mouth) to the lower jaw. In the remaining groups temporal fenestrae provided space for bulging muscles. Diapsids (including lizards, snakes, crocodiles, flying reptiles, and dinosaurs) have two temporal fenestrae on each side of the skull. Only a single temporal fenestra bounded above by the squamosal and postorbital bones identifies the synapsids. If a single temporal fenestra is developed higher on the skull so that the squamosal and postorbital bone lies beneath the opening, the group is termed euryapsids.

COMING ASHORE: BASAL TETRAPODS It required tens of millions of generations to convert the crossopterygian fishes into animals that could live comfortably on solid ground (Fig. 10–75). Even so, the conversion was not complete, for the first amphibian tetrapods continued to return to water to lay their fishlike, naked eggs. From these eggs came fishlike larvae, which, like fish, used gills for respiration.

A number of changes accompanied the shift to land dwelling. A three-chambered heart developed to route the blood more efficiently to and from the lungs. The limb and girdle bones were modified to overcome the constant tug of gravity and to better hold the body above the ground. The spinal column, a simple structure in fishes, was transformed into a sturdy but flexible bridge of interlocking elements. To improve hearing in air rather than under water, the old hyomandibular bone, used in fishes to prop the braincase and upper jaw together, was transformed into an ear ossicle—the stapes. The fish spiracle (a vestigial gill slit) became the eustachian tube and middle ear. To complete the auditory apparatus, a tympanic membrane (eardrum) was developed across a prominent notch (otic notch) in the rear part of the skull.

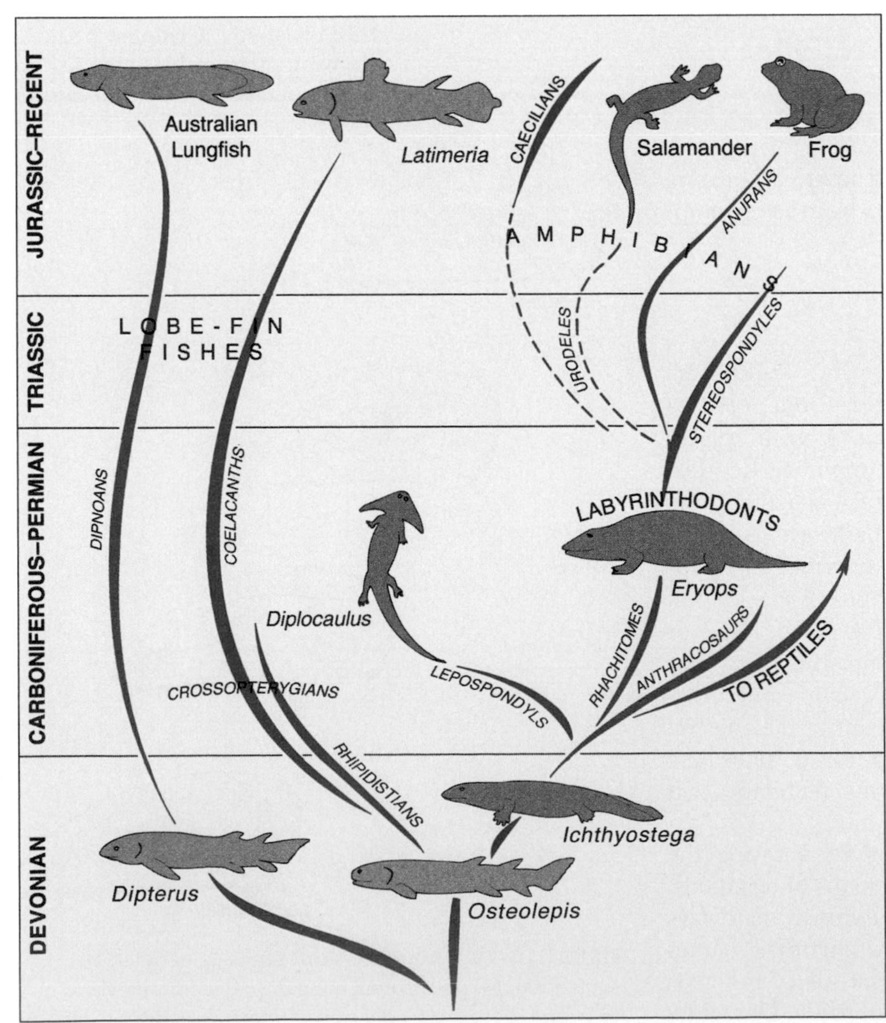

FIGURE 10-75 The evolution of amphibians and lobe-fin fishes. (*From Colbert, E. H., and Morales, M. 1991.* Evolution of the Vertebrates, 4th ed. *New York: John Wiley. With permission of the author, artist Lois Darling, and the publisher.*)

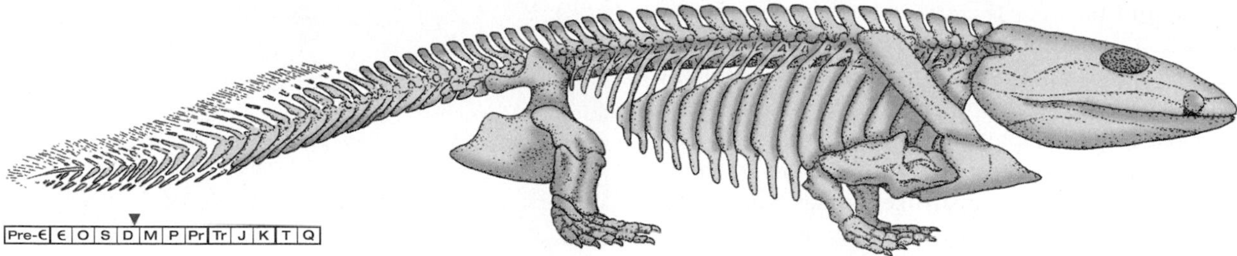

FIGURE 10-76 The skeleton of *Ichthyostega* still retains the fishlike form of its crossopterygian ancestors. (*From Levin, H. L. 1975.* Life Through Time. *Dubuque, Iowa: William C. Brown Co.*)

The fossil record for basal tetrapods begins in the Late Devonian with a group called **ichthyostegids** (Fig. 10–76). As suggested by their name, these creatures retained many features of their fish ancestors, including the tail fin, bony gill covers, fishlike vertebrae, and skull bones that closely resemble the bones of crossopterygians. The teeth, like those of crossopterygians, were characterized by labyrinthic folding of enamel. For this reason, the tetrapods that followed the ichthyosaurs are informally called **labyrinthodonts** (see Fig. 10–69). More precisely, they are called **temnospondyls**. During the Carboniferous, large numbers of labyrinthodonts wallowed in swamps and streams, eating insects, fish, and very likely, one another.

Except for the need to return to water in order to reproduce, many Carboniferous and Permian labyrinthodonts lived out most of their lives on solid ground. *Cacops* (Fig. 10–77) was one such land-dweller. Like many labyrinthodonts, *Cacops* had a heavy skull pierced by five openings, namely, two eye orbits, two nostril openings, and a pineal opening that contained a median light receptor. As noted above, a prominent notch (otic notch) at either side of the back of the skull accommodated a large eardrum. *Eryops* was a bulky, heavy-limbed Permian temnospondyl that weighed in at about 300 pounds. In addition to the temnospondyls, a group termed anthracosaurs inhabited Carboniferous and Permian forests and fresh water bodies. Anthracosaurs appear to be close relatives of early reptiles. Most anthracosaurs were terrestrial and had deeper skulls and longer legs than typical temnospondyls, but others became secondarily adapted for life in the water as crocodile-like fish-eaters.

Amniotes

The evolutionary advance from fish to basal tetrapods was no trivial biologic achievement, yet the Paleozoic was the time of another equally significant event. Among the evolving land vertebrates were some that had developed a way to reproduce without returning to the water. This new manner of reproduction came with the development of the amniotic egg.

The oldest fossils of amniotes were recovered from 300-million-year-old swamp deposits of Nova Scotia. Named *Hylonomus*, these slender reptiles were only about 24 centimeter long, including their tails. *Paleothyris*, found in beds 300 million years old, was a similar reptile. Both *Hylonomus* and *Paleothyris* resembled modern terrestrial insectivorous lizards. Their skeletal

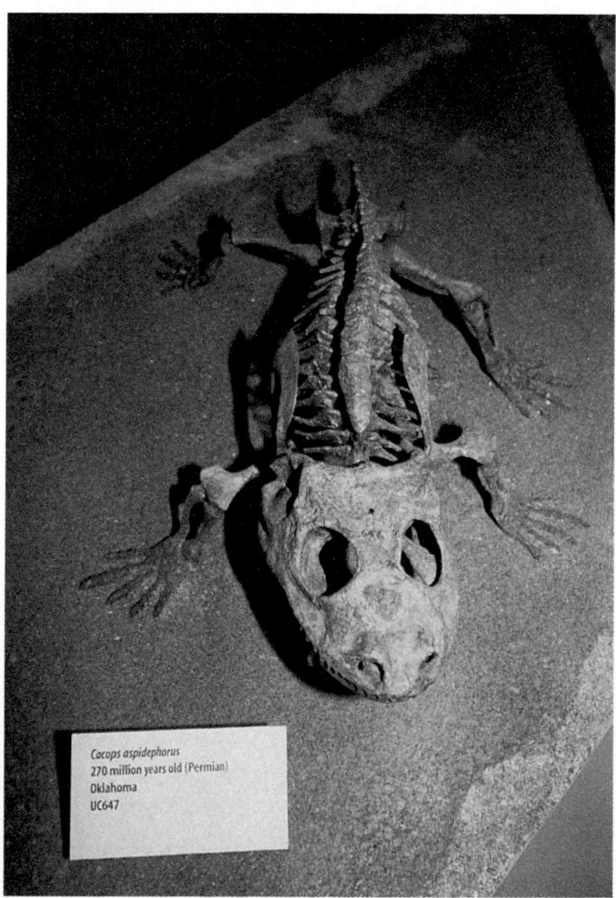

Cacops aspidephorus
270 million years old (Permian)
Oklahoma
UC647

FIGURE 10-77 *Cacops*, a small labyrinthodontic amphibian from the Lower Permian. (*Photograph of a specimen on exhibit at the Field Museum in Chicago.*)

Reptilia and includes the anapsids, diapsids, and archosaurs. The second branch consists of the **Synapsida**, an important group of amniotes traditionally termed "mammal-like reptiles." The traditional view, still favored by some paleontologists, is that synapsids are a subclass of the reptilia. Cladistic analyses, however, show that the "mammal-like reptiles" should not be designated "reptiles" because they diverged from ancestors completely different from *Hylonomus* and other true reptiles. Both the first synapsids and the first reptiles make their appearance at the same time in the Early Pennsylvanian.

The small and relatively primitive groups of the Pennsylvanian provided the stock from which a succession of more specialized Permian amniotes evolved. The most spectacular of these were synapsids called **pelycosaurs**, several species of which sported erect "sails" supported by rodlike extensions of the vertebrae. The pelycosaurs were a varied group. Some, such as *Edaphosaurus*, were plant-eaters, whereas others, as indicated by their great jaws and sharp teeth, ate flesh. *Dimetrodon* (Figs. 10–78 and 10–79) was one such predator. There have been many attempts to explain the function of the pelycosaurian sail. The most reasonable explanation is that it acted in temperature regulation by sometimes serving as a collector of solar heat and at other times as a radiator giving off surplus heat. Because pelycosaurs had several mammalian skeletal characteristics, they are thought to be early representatives of the reptilian group from which an advanced group of synapsids known as **therapsids** arose. In this regard, it is interesting that they were attempting a kind of body temperature control.

For paleontologists, therapsids are among the most fascinating of fossil vertebrates. They were widely dispersed during Permian and Triassic. There

FIGURE 10-78 Permian reptiles. The prominent sailback reptile in the left foreground, with a larger skull and daggerlike teeth, is the carnivore *Dimetrodon*. The sailbacks with smaller heads and blunt cheek teeth, in the foreground at right and in the distance, are plant-eaters of the genus *Edaphosaurus*. (*Copyright J. Sibbick.*) ? *Is it likely that the two sailbacks in the foreground will have a peaceful encounter?*

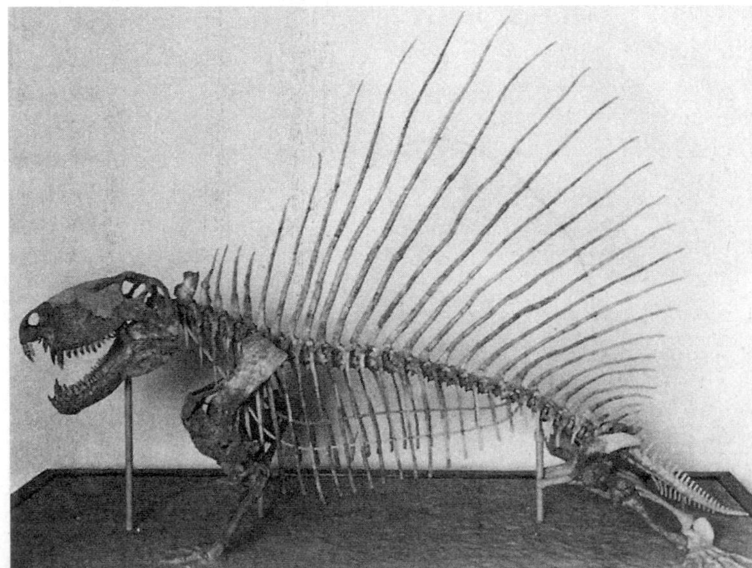

FIGURE 10-79 Mounted skeleton of the Permian "sail-reptile" *Dimetrodon gigas*. (*Courtesy of the U. S. National Museum of Natural History, Smithsonian Institution.*)

Q T K J Tr ▶Pr P M D S O Є Pre-Є

FIGURE 10-80 Mammal-like reptiles. The scene depicts three carnivorous forms (*Cynognathus*) about to attack a plant-eating therapsid reptile (*Kannemeyeria*). (*Courtesy of The Field Museum of Natural History, Chicago; painting by C. R. Knight.*)

is an excellent record of these reptiles and their contemporaries in the Karoo basin of South Africa and in the younger mid- to late Triassic of Madagascar. Discovered as recently as 1999, the Madagascar fossils are exceptionally well preserved and include many previously unknown species These new discoveries occur in terrestrial sediments deposited in lowland areas along a rift zone formed as Madagascar separated from Africa. Additional discoveries of therapsids have been made in Russia, South America, Australia, India, and Antarctica. Therapsids were predominantly small to moderate-sized animals that displayed at least the beginnings of several mammalian skeletal traits. There were fewer bones in the skull than generally found in reptile skulls, and there was a mammal-like enlargement of the lower jaw bone (dentary) at the expense of more posterior elements of the jaw. A double ball-and-socket articulation had evolved between the skull and neck. Teeth show a crude but recognizable differentiation into incisors, canines, and cheek teeth. The limbs were in more direct vertical alignment beneath the body, and the ribs were reduced in the neck and lumbar region for greater overall flexibility. These mammal-like features are well developed in the therapsid *Cynognathus* (Fig. 10–80).

One group of therapsids became particularly common during the Triassic. These were the cynodonts (meaning "dog-toothed"). Mammal-like traits such as those described above for therapsids were further refined in this group. Cheek teeth, for example, were modified for efficient slicing of food so that it could be more easily swallowed and more efficiently digested. Efficient food processing allows an animal to obtain energy from food more rapidly. Cynodonts had also developed a bony palate, permitting them to breathe while simultaneously chewing their food. Such an adaptation would have great importance to an animal approaching the mammalian condition of endothermy. Finally, examination of bone on the snout portion of the skull revealed probable whisker pits, suggesting a covering of hair to slow the rate of heat loss.

PLANTS OF THE PALEOZOIC

The history of plants begins among microbial groups of the Precambrian. That algal growths were abundant is indicated by stromatolites that were already present in the late Archean, expanded during the Proterozoic, and continued to occur in limestones of all younger periods. Cambrian, stromatolitic reefs were widespread (Fig. 10–81), but they became more restricted during and after the Ordovician. Then as now, they may have been able to form only in areas lacking in marine invertebrates that graze on stromatolite organisms. Many

FIGURE 10-81 Hummocky stromatolites of Cambrian age exposed in limestones along the banks of the Black River in Missouri.

such grazing invertebrates had appeared by late Cambrian time.

The evolution of cyanobacteria, such as those responsible for the stromatolitic growth, was an important first step in the colonization of continental environments. Although most Precambrian stromatolites were formed in shallow marine environments, some grew in freshwater bodies as well. Thus, stromatolitic cyanobacteria were hardy organisms that were able to migrate into brackish bays and estuaries and eventually into streams and lakes.

The appearance of green algae, or **chlorophytes**, was the probable next step in the evolutionary journey toward land plants. The close relationship between chlorophytes and land plants is suggested by the adaptation of some species to freshwater bodies and even moist soil. In addition, both green algae and land plants possess the same kind of green pigment and produce the same kind of carbohydrate during photosynthesis.

Another relatively common group of marine algal fossils found in lower Paleozoic rocks are the **receptaculids** (Fig. 10–82). Receptaculids are fossils produced by organisms whose classification is uncertain. Many investigators believe they were constructed by lime-secreting algae. In general appearance, they remind one of the seed-bearing central area of a sunflower, causing amateur fossil collectors to call them "sunflower corals." However, they are neither sunflowers nor corals. Nor are they sponges, a designation once considered because of their spicule-like rods and circulatory passages. Although most frequently found in Ordovician rocks, members of this group also occur sparsely in Silurian and Devonian strata.

Today's land plants include bryophytes (mosses, liverworts, and hornworts) and tracheophytes, or vascular plants. Tracheophytes, such as trees, ferns, and flowering plants, have vascular tissues that provide for transport of water and nutrients. Tubes and vessels convey fluid from one part of the plant to another. The importance of the vascular system is apparent when we remember that above-ground moisture is undependable in most land areas. There is nearly always a greater persistence of moisture and plant nutrients within the pore spaces of soil. Beneath the surface, however, there is no light for photosynthesis. The vascular system of tracheophytes permits part of the plant to function underground, where there is water but no light, and another part to grow where water supplies are uncertain but sunlight for photosynthesis is ensured.

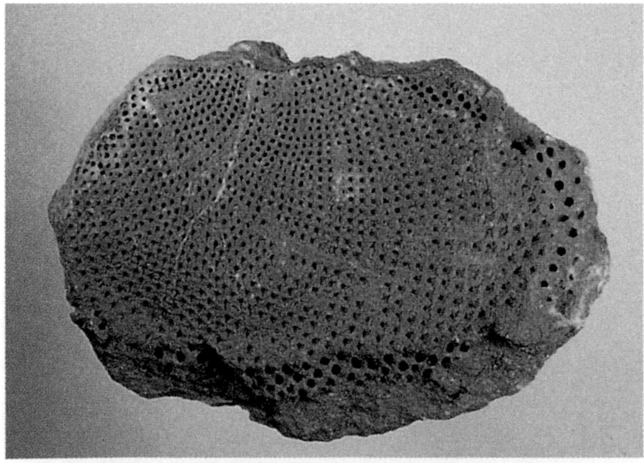

FIGURE 10-82 ***Fisherites*, a receptaculid from the Ordovician Kimmswick Formation of Missouri.** Sometimes called sunflower corals, receptaculids are the fossil remains of a type of green algae and not related to corals. About half of a complete specimen is shown here. Note the pattern of intersecting spiral lines that define small rhombic openings (about 10 × 16 centimeters).

The earliest fossil evidence for the invasion of continents by plants consists of groupings of four spore bodies in tetrahedral arrangement. These are called tetrads and have been obtained from rocks as old as Ordovician. They are thought to be the reproductive and dispersal elements of the progenitors of both bryophytes and tracheophytes. By Silurian time (about 425 million years ago), spores with a distinctive three-rayed scar appeared. They are called trilete spores and are the characteristic reproductive bodies of primitive tracheophytes.

When one surveys the entire history of vascular plants, three major advances become apparent. Each involved the development of increasingly more effective reproductive systems. The first advance led to seedless, spore-bearing plants, such as those that were ubiquitous in the great coal-forming swamps of the Carboniferous. The second saw the evolution of seed-producing, pollinating, but nonflowering plants (gymnosperms). This was a late Paleozoic event. The evolution of plants with both seeds and flowers (angiosperms) came late in the Mesozoic Era.

The spread of early Paleozoic land plants profoundly altered the environment. Their roots slowed erosion of the land. Decaying vegetation promoted the formation of soils. But of even greater importance, plants provided food and shelter for emerging land animals.

The transition from aquatic plants to land plants was apparently difficult, for it was late in coming. The first unquestioned remains of vascular plants occur in rocks of Silurian age. As seen in fossils of *Cooksonia* (Fig. 10–83), they were small, leafless plants with thin, evenly branching stems surmounted by the spore-bearing bodies called sporangia. By Devonian time, species of *Aglaophyton* (Fig. 10–84) and *Rhynia* (Fig. 10–85) had made their appearance. True woody vascular tissue can be discerned in the fossils of *Rhynia*.

Although small, the early plants paved the way for the evolution of large trees. Because of the evolution of wood, plants were able to stand against the pull of gravity and the force of strong winds. By Late Devonian, lofty, well-rooted, leafy trees (Fig. 10–86) had made their appearance. One such forest containing thick stands of

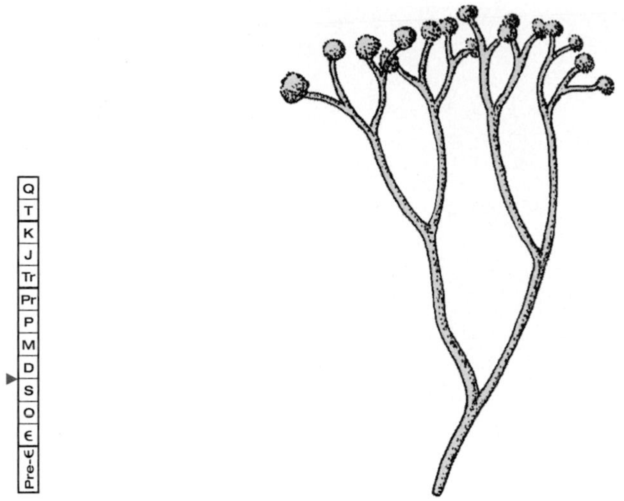

FIGURE 10-83 The Late Silurian and Early Devonian vascular plant *Cooksonia*. Height is approximately 4 centimeters. (*After Edwards, D. et al. 1992. A vascular conducting strand in the early land plant* Cooksonia. Nature 357:683–685.)

FIGURE 10-84 Restoration of *Aglaophyton*, an early land plant from the Devonian of Scotland. *Aglaophyton* grew to heights of about 20 to 25 centimeters and had a simple conducting strand for moving water up the stem. (*Adapted from Kidston, R., and Lang, W. H. 1917–21. On Old Red Sandstone plants showing structure, from Rhynie Chert Bed, Aberdeenshire*. Transactions Royal Society of Edinburgh, *vols. 51–52.*)

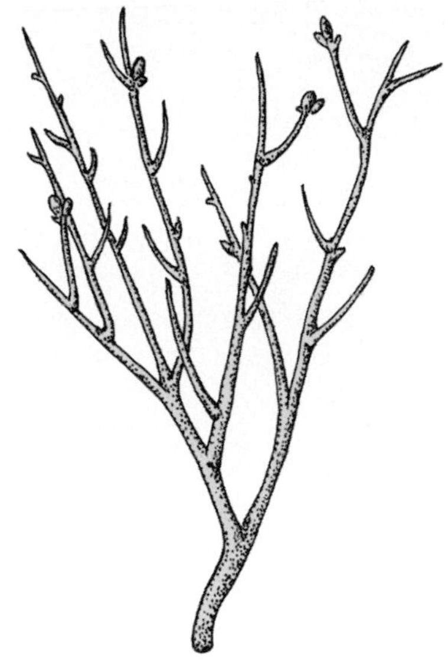

FIGURE 10-85 The Middle Devonian vascular land plant *Rhynia*. Xylem, the vascular tissue that conducts water and dissolved minerals, has been identified in fossils of *Rhynia*. Because of its thick-walled cells, xylem also gives strength to a plant. A major group of early land plants, the rhynophytes, takes its name from *Rhynia*. (*Adapted from Kingston, R., and Lang, W. H. 1917–21. On Old Red Sandstone plants showing structure, from the Rhynie Chert Bed, Aberdeenshire*. Transactions Royal Society of Edinburgh, *vols. 51–52.*)

spore-bearing trees covered the present area of the Sahara Desert 370 million years ago. Another Devonian forest stood near the present site of Gilboa, New York. Some of the Gilboa trees were over 7 meters tall, but they were to be dwarfed by their Carboniferous descendents.

Among the moisture-loving plants of the Carboniferous were the so-called scale trees, or **lycopsids**. Today, they are represented by their smaller survivors, the club mosses, of which *Lycopodium* is a member. Small size was not particularly characteristic of the late Paleozoic lycopsids. The forked branches of *Lepidodendron* reached 30 meters into the sky. The elongate leaves of the scale trees emerged directly from the trunks and branches. After being released, they left a regular pattern of leaf scars. In *Lepidodendron*, the scars are arranged in diagonal spirals (Fig. 10–87), whereas species of *Sigillaria* have leaf scars in vertical rows.

Another dominant group of plants that grew side by side with the Carboniferous lycopsids were the sphenopsids. Living sphenopsids include the scouring rushes, or horsetails. Fossil sphenopsids, such as *Calamites* (Fig. 10–88) and *Annularia* (Fig. 10–89), possessed slender, unbranching, longitudinally ribbed stems with a thick core of pith and whorls of leaves at each transverse joint. At the top, a cone bore the spores that would be scattered in the wind.

True ferns (Fig. 10–90) were also present in the coal forests. Many were tall enough to be classified as trees.

FIGURE 10-86 Restoration of a Middle Devonian forest in the eastern area of the United States. (*A*) An early lycopod, *Protolepidodendron*. (*B*) *Calamophyton*, an early form of the horsetail rush. (*C*) Early tree fern, *Esopermatopteris*. (*Courtesy of The Field Museum of Natural History, Chicago; painting by C. R. Knight, photo by R. Testa.*)

Like the lycopsids and sphenopsids, they reproduced by means of spores carried in regular patterns on the undersides of the leaves.

Seed plants made their debut during the late Paleozoic. They probably arose from Devonian fernlike plants. These plants, appropriately dubbed "seed ferns," had fern-like leaves, but unlike true ferns they reproduced by means of seeds. They were the dominant plant group of the Late Paleozoic.

One of the most widely known Carboniferous and Permian plants of Southern Hemisphere continents is *Glossopteris* (see Fig. 5–27). Most species of *Glossopteris* were plants with thick, tongue-shaped leaves. Because of certain anatomic traits and because of their association with glacial deposits, *Glossopteris* and associated plants are thought to have been adapted to cool climates. Glossopteris has traditionally been considered a seed fern, but recent investigations suggest that some of the plant remains attributed to *Glossopteris* may have been derived from other kinds of plants.

Fossils of seed plants are also present in the Northern Hemisphere. Cordaites (primitive members of the conifer lineage) were abundant, some towering to 50 meters. Their branching limbs were crowned with clusters of large straplike leaves (Fig. 10–91). These and somewhat more modern cone-bearing conifer plants spread widely during the Permian, perhaps as a consequence of drier climatic conditions. The first ginkgoes made their appearance during the Permian. They became abundant during the Jurassic. Today, only a single species, *Ginkgo biloba*, remains as a survivor of this once-flourishing group. The tree is readily identified by its fanlike leaves.

FIGURE 10-87 Fossilized bark of the Carboniferous tree *Lepidodendron*. Leaf scars are about 1 centimeter wide. (*Photograph courtesy of Wards Natural Science Establishment, Inc., Rochester, NY.*)

FIGURE 10-88 ***Calamites*, a sphenopsid.** Plants shown are about 3 to 5 meters tall.

Extinction overtook many plant groups near the end of the Permian Period. Many species of lycopsids, seed ferns, and conifers disappeared. Small ferns that grow in damp areas, however, were not profoundly affected by the crisis.

FIGURE 10-89 ***Annularia*, an abundant sphenopsid of Pennsylvania age.**

FIGURE 10-90 ***Pecopteris*, a true fern from the Pennsylvanian of Illinois** (the penny is for scale).

FIGURE 10-91 **End of a branch of *Cordaites*, showing the straplike leaves of these trees.** Not uncommonly, the leaves attained lengths of 1 meter. The clustered bodies produced the plant's male gametes. (*Adapted from Grand'Eury, C. 1877. Flore Carbonifère de Départment de la Loire et du centre de la France*. Mem. Acad. Sci. Institut France. *24:624 pp.*)

MASS EXTINCTIONS

For most of the Paleozoic, the Earth was populated by a rich diversity of life. There were, however, times when the planet was less hospitable, and large groups of organisms suffered extinction (Fig. 10–92). Early geologists saw evidence of these mass extinctions in the fossil record and used the abrupt termination of fossil ranges to define the boundaries between geologic

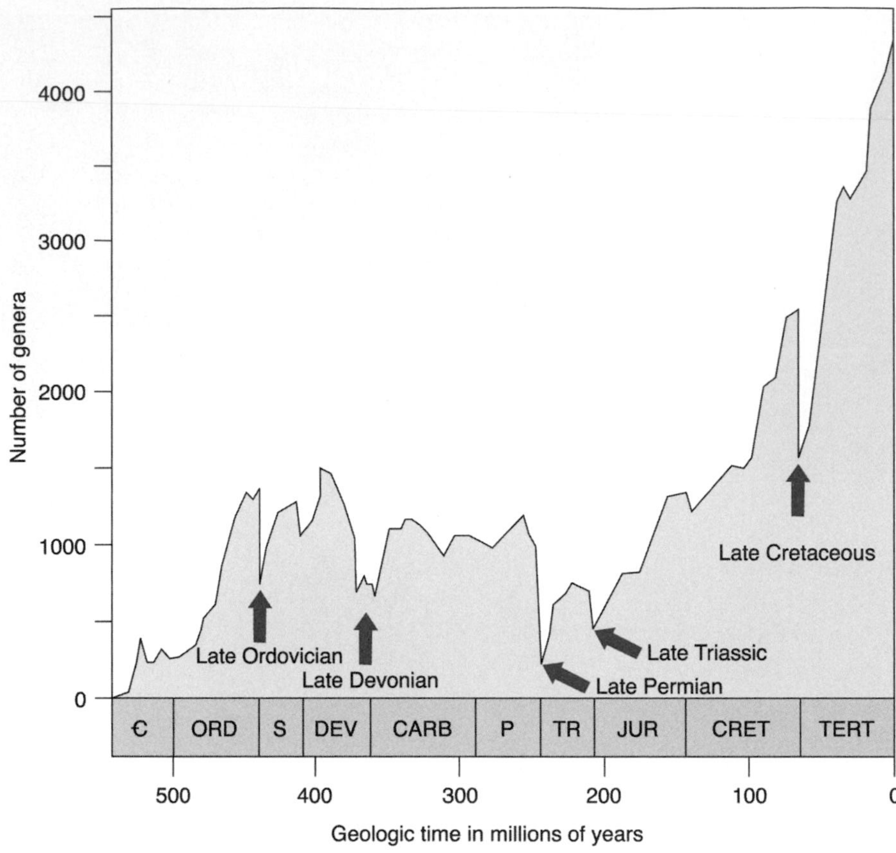

FIGURE 10-92 Diversity of marine animals, compiled from a database recording first and last occurrences of more than 34,000 genera. The graph depicts five major episodes of mass extinction (global extinctions over a short span of geologic time). (*Adapted from Sepkoski, J. J., Jr. 1994.* Geotimes *39(3):15–17.*)

systems. Although there have been many less severe episodes of mass extinction, five were particularly catastrophic. Of these, one occurred at the end of the Ordovician, one late in the Devonian, one near the end of the Permian, one late in the Triassic, and one that marks the end of the Cretaceous. Here we will examine only the Paleozoic extinction events.

Late Ordovician Extinctions

Extinctions near the end of the Ordovician Period occurred in two phases. During the first phase, planktonic (floating) and nektonic (swimming) organisms such as graptolites, acritarchs, many nautiloids, and conodonts, as well as such benthic creatures as trilobites, bryozoa, corals, and brachiopods, were the principal victims. In the second phase, several trilobite groups that had survived the first wave of extinction perished. At the same time, corals, conodonts, and bryozoans were severely reduced in numbers and diversity. Both phases of extinction appear to be related to global cooling associated with the growth of Gondwana ice caps. As described in Chapter 8 (Enrichment box, "The Big Freeze in North Africa"), an extensive ice sheet covered part of what is now northern Africa during the late Ordovician. The fossil record reveals that habitable zones of many plant and animal groups were shifted and compressed toward the equator in response to cooling at higher latitudes. With cooler conditions, tropical organisms were hit the hardest. As ice accumulated in the continental glaciers, sea level was lowered. This caused a loss of epeiric seas and shallow marginal shelf environments. These seas had been optimum areas for the proliferation of many groups of marine invertebrates. Paleobiologist Steven Stanley has shown that when such a loss of shallow shelf environments is accompanied by cooling, mass extinctions are an expected consequence. The second phase of Ordovician mass extinction accompanied the global rise in sea level caused by melting of ice sheets and warmer conditions. Organisms tolerant of cooler conditions were now under stress and many became extinct.

Late Devonian Extinctions

Following the late Ordovician crisis, life expanded again, building slowly in the early Silurian but attaining rich levels of diversity during the Devonian. By the end of the Devonian, however, marine invertebrates were again confronted with adverse environmental change. That there was environmental stress is reflected in the decimation of once-extensive Devonian reef communities. Reef-building tabulate corals and stromatoporoids are rarely seen in rocks deposited during the remainder of the Paleozoic. Of scores of rugose corals, only a few species survived. In addition, brachiopods, goniatites, trilobites, conodonts, and placoderms were severely reduced in numbers and variety. Altogether, some 70 percent of the ocean's families of invertebrates disappeared.

Because the Late Devonian extinctions occurred over a span of 20 million years, it is unlikely that a sudden event, such as the impact of an asteroid, is likely to have been the cause. Nor is there evidence of heavy metal concentrations that would result from a shattered asteroid. More likely, the cause was some sort of ecological crisis. One notices, for example, the extensive forests of huge trees that spread across the continents for the first time during the Devonian. The abundance of vegetation would have produced organic acids that may have accelerated rates of weathering. Rapidly decomposing bedrock may have released huge volumes of phosphorus and other minerals that, on reaching the ocean, may have produced explosive growths of algae. Bacteria consuming the remains of dead algae would have robbed the ocean of oxygen. The process, termed eutrophication, causes fish kills today in lakes subjected to similarly excessive algal growth. Evidence of such anoxic conditions is seen in the extensive tracts of Devonian black shales, a rock type that characteristically develops in oxygen-deficient waters. Also, many late Devonian limestones are rich in the isotope carbon-13 and relatively poor in carbon-12. An explosive growth of algae would explain this, for algae use more carbon-12 during photosynthesis than carbon-13. Thus, the percentage of the heavier isotope is increased in ocean water and incorporated into limestones and the shells of marine invertebrates. Finally, the eutrophication hypothesis may explain why corals were so severely decimated. Corals live in warm regions adjacent to land areas densely covered by tropical forests.

Some geologists prefer a hypothesis that puts the blame for the Late Devonian crisis on continental glaciation. Glacial drift, erratics, and striations found in Brazil provide evidence that by Late Devonian time, South America had drifted to a position over the South Pole. Global cooling and reduction of shallow-water environments may once again have been the basic reason for the die-off.

Late Permian Extinctions

POSSIBLE TERRESTRIAL CAUSES The decimation of life at the end of the Permian Period has been described by Smithsonian paleontologist Douglas Erwin as "the mother of mass extinctions." The loss of biologic diversity at this time exceeded that of all other extinctions, including the one that involved the demise of the dinosaurs 66 million years ago. By some estimates, more than 90 percent of all pre-existing marine species disappeared or were severely reduced. On land, spore-bearing ferns and other plants gave way to conifers, cycads, ginkgoes, and other gymnosperms. The newer groups probably evolved in response to cooler, drier climates. Vegetative changes would have disrupted the food web for land animals, causing a detrimental ecologic chain reaction. Many families of basal tetrapods, primitive reptiles, and synapsids disappeared. Of all groups, however, tropical marine invertebrates experienced the most extensive losses. Fusulinids that had once blanketed large areas of the sea floor disappeared. Rugose corals, many families of crinoids, productid brachiopods, lacy bryozoa, and many groups of ammonoids became extinct. Trilobites were not involved in this episode of extinction, for they had disappeared in an earlier stage of the Permian.

Many factors may have contributed to the late Permian mass extinction. At the time of the extinctions, the supercontinent Pangea had completed its development. More rigorous climatic conditions existed across the great continent, as is characteristic of the interiors of large continents today. Epeiric seas had been drained or were very limited, and this reduced the number of favorable sites for the proliferation of shallow marine invertebrates. Frigid polar regions existed at both the north and south ends of Pangea. Organisms accustomed to warm waters shifted toward the lower latitudes, but many did not survive. Large numbers of tropical organisms could not surive the cooling, which also prevented construction of organic reefs and thereby impeded the formation of limestones. Pangea also blocked equatorial circulation in the ocean, disrupting the life zones of many organisms. The late Permian was also a time of extraordinary volcanic activity. One of the greatest episodes of flood basalt volcanism in the history of the Earth occurred in Siberia at the end of the Permian. Carbon dioxide and dust released into the atmosphere during that volcanic episode may have triggered a biologically destructive greenhouse effect.

A POSSIBLE EXTRATERRESTRIAL CAUSE Perhaps the mass extinction at the end of the Permian was not the result of events on Earth but was caused by an extraterrestrial body striking the Earth. Recently obtained evidence derived from the analysis of sedimentary rocks at the P-T (Permian-Triassic) boundary indicate just such a cataclysmic event. The sediments contain a high concentration of soccer ball-shaped carbon molecules called *fullerenes* (also dubbed buckyballs because they resemble the geodesic dome invented by Buckminster Fuller). Atoms of helium and argon are trapped within the fullerenes, and the ratio of the two is distinctly unlike any that occur in the Earth's crust. Also, the isotope of helium found in the fullerenes is extraterrestrial helium-3, whereas the terrestrial isotope is helium-4. Significantly, similar fullerenes have been found in two well-known carbonaceous chondrites, the Allende and

Murchison chondritic meteorites. The evidence suggests an impact by an extraterrestrial body.

In Chapter 12, the widely publicized impact event that caused the extinction of the dinosaurs is described. If you read ahead in the section entitled "The Terminal Cretaceous Crisis," you will learn that the evidence of that impact includes not only the discovery of the impact crater but also the presence of shocked quartz grains and the element iridium, of probable extraterrestrial origin. At this time, shocked quartz and iridium have not yet been found at the P-T boundary, but their absence can be explained. You will recall that the immense Panthalassa Ocean existed at the end of the Permian. If the impacted body had landed in the ocean, the crater on the ocean floor would have been conveyed to a subduction zone and destroyed. As the ocean floor is made of basalt, there would be no quartz to be "shocked." Finally, if the impacting body was not an asteroid but a comet composed almost entirely of ice (containing few if any rock fragments), one might not detect iridium. On the positive side of the question, fullerenes are now being recovered from rocks at the boundary between the Cretaceous and Tertiary.

If an extraterrestrial body did impact the Earth at the end of the Permian, the impact itself would be only the beginning of a series of disastrous events. Tsunamis might have devastated low-lying continental margins. The impact would throw so much water vapor and dust into the atmosphere as to block the sun's radiation. Disturbance of the crust would likely trigger rampant volcanism, which in turn would change the composition of the atmosphere, harmfully affect climate, and cause ocean anoxia. All living things would be in danger of extermination.

SUMMARY

The fossil record of the Paleozoic is immensely better than that of the preceding Proterozoic. Shell building is one important reason for the improvement. Archaeocyathids, brachiopods, and trilobites were abundant during the Cambrian, and by Ordovician time, all major invertebrate phyla were well established. The best fossils of soft-bodied invertebrates are those of the Early Cambrian Chengjiang fossil site and the Middle Cambrian Burgess Shale fauna, which include many previously unknown arthropods, echinoderms, sponges, and cnidarians as well as animals believed to be the earliest known chordates and jawless fishes. Among the other important invertebrate groups of the Paleozoic were foraminifera (fusulinids expanded during the Pennsylvanian and Permian), sponges, corals (tabulate and rugose), bryozoa, mollusks, echinoderms (especially blastoids and crinoids), and graptolites.

The first fossils of vertebrate animals are found in rocks of Cambrian age. They are remains of a group of jawless fishes known as agnathids. Ostracoderm agnathids expanded during the Ordovician and Silurian. Archaic jawed fishes appear late in the Silurian but do not begin to dominate the marine realm until Devonian time. This was also the time when cartilaginous (chondrichthyan) and bony (osteichthyan) fishes began their rise to dominance. A special group of bony fishes called crossopterygians arose from osteichthyan stock. These fishes could breathe air by means of accessory lungs and possessed muscular fins that provided short-distance overland locomotion. The bones within the fins of crossopterygians resembled those of primitive amphibians. Other skeletal traits clearly suggest that these Devonian fishes were the ancestors of the first tetrapods—the ichthyostegids. From a start provided by ichthyostegids, tetrapods called labyrinthodonts underwent a successful adaptive radiation that lasted until the close of the Triassic. Long before their demise, however, they provided a lineage from which synapsids and reptiles evolved.

The transition from nonamniotic basal tetrapods to amniotic reptiles involved the development of the amniotic egg, an evolutionary advancement that liberated land vertebrates from the need to return to water bodies to reproduce. The first known amniotic reptiles are Pennsylvanian in age and include such forms as *Hylonomus* and *Petrolacosaurus*. During very late Pennsylvanian and Permian, amniotes underwent an elaborate evolutionary radiation that produced the finback reptiles and therapsids as well as several other major reptilian groups.

The earliest Paleozoic precursors to plants were mainly unicellular and microbial, with the most obvious larger fossils being stromatolites. Fossil spores indicate that by late Ordovician, vascular land plants were present. It is probable that they evolved from the green algae (chlorophytes). Psilophytes were followed in the Devonian and Carboniferous by lycopsids, sphenopsids, and seed ferns. In temperate regions of Gondwana, a flora dominated by the *Glossopteris* developed. Permian forests included plants more tolerant of drier and cooler conditions, such as conifers and ginkgoes.

Life did not expand steadily throughout the Paleozoic Era. There were difficult times when major groups of organisms became extinct. These mass extinctions occurred at the end of the Ordovician, Devonian, and Permian periods. During the extinction at the end of the Permian, nearly half of the known families of animals disappeared. The decimation of marine animals was particularly dramatic and included the loss of fusulinids, spiny productid brachiopods, rugose corals, two orders of bryozoans, and many taxa of stemmed echinoderms, including blastoids. Among the vertebrates, over 70 percent of vertebrate families perished. Fortunately, a few groups survived the Permian crisis and were able to continue their evolution during the Triassic. The cause of the Permian biologic crisis is not definitely known. Many believe it was the result of terrestrial causes such as the final consolidation of the great Pangean supercontinent, the high elevation of many regions on Pangea, reduction of warm epeiric seas, global cooling, development of ice caps, and other sharp climatic changes stemming from all of these factors. Other scientists put the blame on the catastrophic impact of a large meterorite or asteroid.

QUESTIONS FOR REVIEW AND DISCUSSION

1. Which of the Chengjiang and Burgess Shale fossils are considered to be chordates? What is the basis for this assignment?

2. To what geologic system(s) of the Paleozoic would rocks containing the following fossils be assigned?
   a. fusulinids
   b. archaeocyathids
   c. *Archimedes*

3. The phylum Cnidaria was formerly termed Coelenterata. Why is Cnidaria an appropriate name for these animals? What classes of Cnidaria lived only during the Paleozoic?

4. How do the sutures of nautiloids differ from those of ammonoid goniatites?

5. What are conodont elements? What other animal group has hard tissue (tooth or bone) of the same composition as conodont elements?

6. What groups of Paleozoic invertebrates had become extinct by the end of the Paleozoic?

7. How are shallow shelf and epeiric sea environments affected by episodes of continental glaciation?

8. Define and discuss the principal characteristics of the Echinodermata, Trilobita, Mollusca, Brachiopoda, Bryozoa, and Porifera.

9. Discuss the advantages that accrued to invertebrates that evolved shells.

10. Distinguish between the following:
    a. ostracoderms and placoderms
    b. osteichthyes and chondrichthyes
    c. crossopterygians and labyrinthodonts
    d. infaunal and epifaunal invertebrates
    e. synapsids and diapsids

11. Discuss those characteristics of therapsid reptiles that indicate that they were on the main line of evolution toward mammals.

12. What group of algae is considered the probable ancestors of the first land plants (tracheophytes)? For what reasons is this group considered ancestral to land plants?

13. Why was the evolution of a vascular system critical to the invasion of the lands by plants? What were the first plants to make the transition, and what were their characteristics?

14. Describe in general terms the appearance of synapsid reptiles.

READINGS

Benton, M. J. 1997. *Vertebrate Paleontology*. 2nd. ed. London: Chapman & Hall.

Bowring, S. A., et al. 1993. Calibrating rates of Early Cambrian evolution. *Science* 261:1293–1298.

Briggs, E. G., Erwin, D. H., and Collier, F. J. 1994. *The Fossils of the Burgess Shale*. Washington, DC: Smithsonian Institution Press.

Cleal, C.J., and Thomas, B.A. 1999. *Plant Fossils: The History of Land Vegetation*. Suffolk, U.K.: Roydell Press.

Conway, M. S. 1987. The search for the Precambrian–Cambrian boundary. *Am. Scientist* 75:157–167.

Cowen, R. 1995. *History of Life*, 3rd ed. Boston: Blackwell Scientific Publications.

Erwin, D. H. 1993. *The Great Paleozoic Crisis: Life and Death in the Permian*. New York: Columbia University Press.

Gould, S. J. 1989. *Wonderful Life*. New York: W. W. Norton Co.

Janvier, P. 1996. *Early Vertebrates*. Oxford, England: Clarendon Press.

King, N.R. 1998 (Revised Printing, 2000). *Paleontology of Higher Vertebates*, Dubuque, Iowa: Kendall/Hunt.

Long, J. A. 1995. *The Rise of Fishes*. Baltimore: Johns Hopkins University Press.

Lipps, J.H., and Signor, P.W. 1992. *Origin and Early Evolution of the Metazoa*. New York: Plenum Press.

Stanley, S. M. 1987. *Extinction*. New York: W. H. Freeman & Co.

Stewart, W. N., and Rothwell, G. W. 1993. *Paleobotany and the Evolution of Plants*, 2nd ed. Cambridge, England: Cambridge University Press.

Stokstad, E. 2001. Exquisite Chinese Fossils Add New Pages to Book of Life. *Science* 291:232–236, Jan. 12.

Ward, P.D. 2000. *Rivers in Time: Search for the Clues to the Earth's Mass Extinctions*. New York: Columbia University Press.

Whittington, H. B. 1985. *The Burgess Shale*. New Haven: Yale University Press.

WEB SITES

The Earth Through Time Student Companion Web Site (www.wiley.com/college/levin) has online resources to help you expand your understanding of the topics in this chapter. Visit the Web Site to access the following:

1. Illustrated course notes covering key concepts in each chapter;
2. Online quizzes that provide immediate feedback;
3. Links to chapter-specific topics on the web;
4. Science news updates relating to recent developments in Historical Geology;
5. Web inquiry activities for further exploration;
6. A glossary of terms;
7. A Student Union with links to topics such as study skills, writing and grammar, and citing electronic information.

11

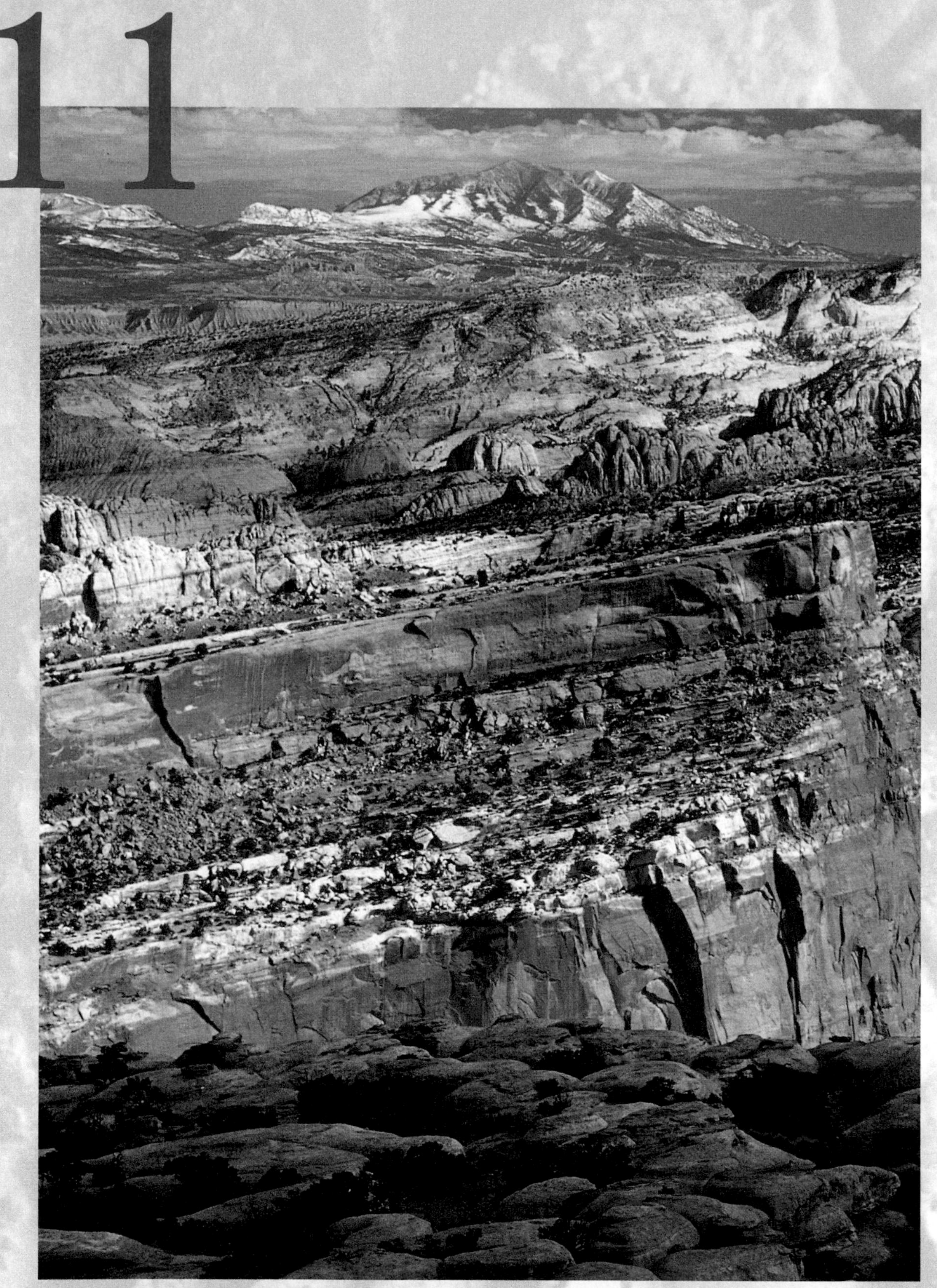

View of dipping Jurassic and Cretaceous sandstones along Waterpocket Monocline, Capitol Reef National Park, Utah. (Photograph by S. Smith.)

CHAPTER 11

The Mesozoic Era

The Earth, from the time of the chalk to the present day, has been the theater of a series of changes as vast in their amount as they were slow in their progress. The area on which we stand has been first sea and then land for at least four alternations and has remained in each of these conditions for a period of great length.
Thomas Huxley, 1868, *On a Piece of Chalk*

OUTLINE

The extinction of Paleozoic animals and plants that occurred at the close of the Permian Period did not go unnoticed by the founders of geology. The fossil evidence of this time of crisis provided a natural boundary that was used to mark the end of the Paleozoic. The overlying younger strata contained distinctly different populations of animals and plants that form a middle chapter in the history of life. That middle chapter was named the Mesozoic Era by the English geologist John Phillips in 1840. Like the end of the Paleozoic, the Mesozoic ended in a biologic crisis that marks its separation from the youngest era of all—the Cenozoic.

The Mesozoic Era lasted an estimated 185 million years, ending approximately 66 million years ago. The three periods of the Mesozoic are unequal in duration. The Triassic lasted about 47 million years, the Jurassic 68 million years, and the Cretaceous about 70 million years.

During the lengthy span of the Mesozoic, many new groups of plants and animals evolved and experienced often spectacular radiations. It was the era in which two new vertebrate classes, the birds and the mammals, appeared. It was also a time in which the supercontinent Pangea, which had formed in the previous era by the joining of ancestral continents, was gradually dismembered. As the continents moved slowly apart, the oceanic rifts between them broadened. A process of fragmentation and drift had begun that would ultimately lead to the present physical geography of the planet.

THE BREAK-UP OF PANGEA

Any discussion of the history of the Mesozoic Era must have Pangea as its starting point. The same forces that had drawn the plate segments together to form this great supercontinent were still in operation at the beginning of the Mesozoic; however, they now moved in different directions or had shifted their locations. In general, the dismemberment of Pangea seems to have occurred in four stages. The **first stage** began in the Triassic, with rifting and volcanism along normal fault systems. Normal faults resulting from crustal extension often flatten with depth as they extend into the lower crust. Such faults are called **listric normal faults**. The term is derived from the Greek word *listron*, meaning *shovel*, in reference to their curved snow-shovel shape.

The development of normal faults accompanied by fissure and vent eruptions is the expected consequence of the breaking apart of tectonic plates. In this instance, the forces were associated with the separation of North America and Gondwana. As the rifting progressed, Mexico was decoupled from South America, and the eastern border of North America parted from the Moroccan bulge of Africa (Fig. 11–1). Oceanic basalts were added to the sea floor of the newly formed and gradually widening Atlantic Ocean. The largely tensional geologic structures as well as the volcanic and clastic sedimentary rocks in the Triassic System of both the eastern United States and Morocco exhibit many striking similarities. These now widely separated regions were on opposite sides of the same axis of spreading and were affected by similar forces and events.

Recently, geologists used seismic techniques to locate the 300-million-year-old zone of convergence, or suture zone, formed by the convergence of Africa and North America. The zone, which trends roughly east-southeast across southern Georgia, was left behind when the two continents subsequently parted company about 190 million years ago. The discovery was particularly interesting to paleontologists, who had postulated the convergence and later separation of the two continents from fossil evidence.

Throughout Triassic time, Laurasia north of the Maritime Provinces remained intact. Sea-floor spreading was apparently more rapid in the northern region of the newly opening Atlantic. South America maintained its hold on Africa throughout most of the Triassic.

The **second stage** in the break-up of Pangea involved the rifting and opening of narrow oceanic tracts between southern Africa and Antarctica. This rift extended northeastward between Africa and India and developed a branch that separated northward-bound India from the yet unseparated Antarctica–Australia landmass. The rifting was accompanied by outpouring of great volumes of basaltic lavas.

In **stage three** of the break-up of Pangea, the Atlantic rift began to extend itself northward, and clockwise rotation of Eurasia closed the eastern end of the Tethys Sea, a forerunner of the Mediterranean. By the end of Jurassic time, an incipient breach began to split South America away from Africa. The cleft apparently worked its way up from the south, creating a long seaway reminiscent of the present Red Sea. Australia and Antarctica remained intact, but India had moved well along on its long voyage to Laurasia. By Late Cretaceous time, about 70 million years ago, South America had completely separated from Africa, and Greenland began to pull away from Europe. However, northeasternmost North America remained attached to Greenland and northern Europe.

The **fourth stage** in the dismemberment of Pangea was not a Mesozoic event. It occurred early in the Cenozoic, when the North Atlantic rift slowly penetrated northward until the North American and Eurasian continents were completely separated. Also during this final era of Earth history, Antarctica and Australia separated. This final separation of continents occurred only about 45 million years ago. The total span of time involved in the fragmentation of Pangea was about 150 million years.

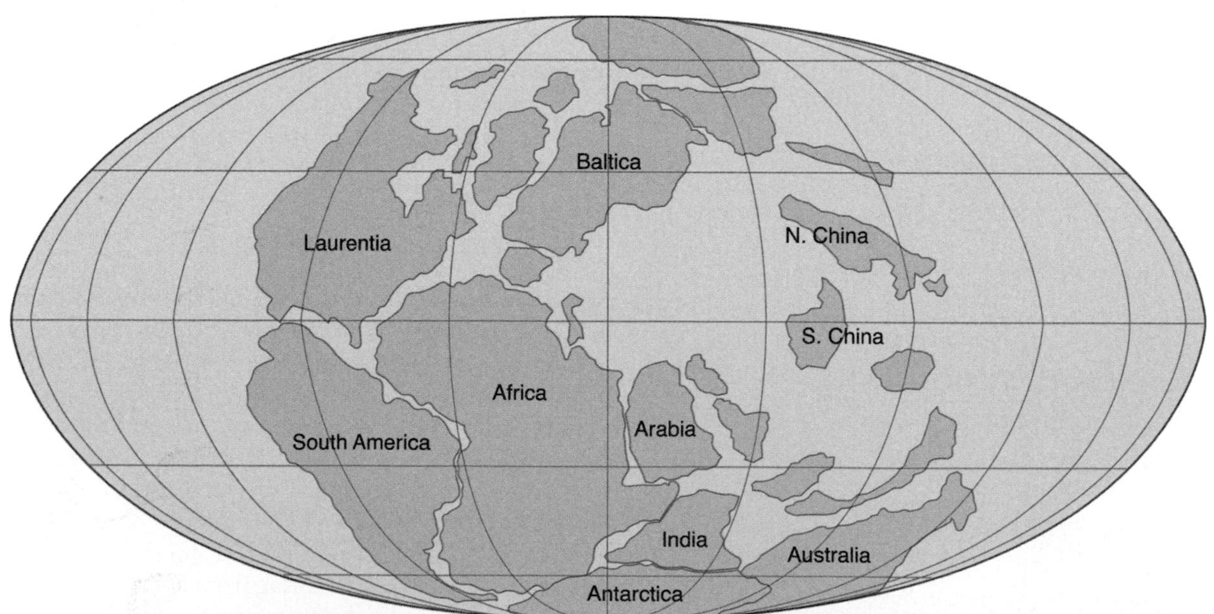

FIGURE 11–1 Paleogeographic reconstruction of the world about 180 million years ago, when the break-up of Pangea was beginning. (After Scotese, C. R. and McKerrow, W. S. 1990. Paleogeography and Biogeography, *Geol. Soc. London Mem. 12:1–21.*)

THE MESOZOIC HISTORY OF NORTH AMERICA

To the East and South

TRIASSIC AND JURASSIC At the beginning of the Mesozoic, conditions in the eastern part of North America were much the same as they had been near the end of the Paleozoic. The rugged Appalachian ranges that had been raised during the Allegheny orogeny were undergoing vigorous erosion. Coarse clastics derived from the uplands filled intermontane basins and other low areas during the Early and Middle Triassic. Then, during the Late Triassic and Early Jurassic, North America began to experience the tensional forces that precede actual separation of continents. As a result, a series of fault-bounded troughs developed along the eastern coast from Nova Scotia to North Carolina (Figs. 11–2 and 11–3). Many of these faults represent reactivation of lateral strike-slip, and reverse faults formed during late Paleozoic orogenic activity. The older faults provided zones of weakness that facilitated rifting during the subsequent Triassic. Downfaulted blocks provided traps for the accumulation of great thicknesses of poorly sorted arkosic red sandstones and shales. These sediments are nonmarine and indicate that insufficient separation had occurred during the Late Triassic to permit ocean water to enter the riftzone. Rocks deposited in the downfaulted basins constitute the Newark Supergroup of Late Triassic and Early Jurassic age.

The poor sorting of the coarser Newark clastics and their high content of relatively unweathered feldspar grains indicate transportation and deposition by streams flowing swiftly down from granitic highlands that bordered the fault basins (Fig. 11–4). In some places, drainage in the basins became impounded and lakes formed. Remains of freshwater crustaceans, fish, and even the footprints of early dinosaurs (Fig. 11–5) are frequently found on the lithified surfaces of Newark sediments. Mudcracks and raindrop impressions provide further evidence of terrestrial conditions. As the Newark sediments were being deposited, tremendous volumes of lava flowed from fissures, while at the same time volcanoes spewed ash onto the surrounding hills and plains. Volcanic activity was not confined to eastern North America, for contemporaneous eruptions were occurring in Africa, Europe, and South America as well. The widespread igneous activity was the expected consequence of the fragmentation of Pangea. Globally, the outpourings of black basaltic lavas covered over 7,000,000 square kilometers of terrain and may have released enough carbon dioxide to contribute to Late Triassic mass extinctions.

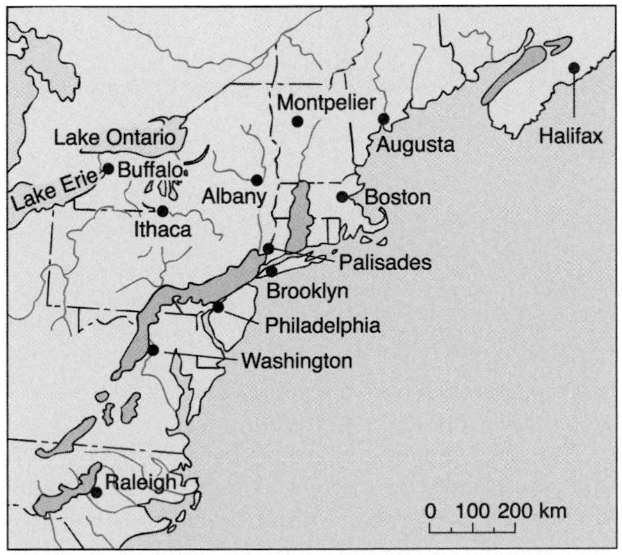

FIGURE 11–2 Outcrop areas of Triassic rocks in eastern North America. Green areas show troughlike deposits of Late Triassic age.

Three particularly extensive lava flows and an imposing sill are included within the Newark Supergroup in the New Jersey–New York area. The exposed edge of a vertically jointed sill forms the well-known Palisades of the Hudson River (Fig. 11–6). Radiometric dates obtained for the Palisades basalt indicate that it solidified about 190 million years ago. Although most of the Newark beds are Triassic, recent spore and pollen studies suggest that in some areas Newark sediments were still accumulating at the beginning of the Jurassic.

By Late Triassic time, the fault-block mountains that had been produced as a result of rifting had been severely reduced by erosion. The highlands were further worn away during the Jurassic and Early Cretaceous until only a broad, low-lying erosional landscape remained. That old erosional surface is called the "Fall Line surface." It can be seen along the Fall Line, which developed at the contact between resistant rocks of the Appalachian Piedmont and more easily eroded rocks of the coastal plain. The name is derived from the presence of waterfalls along the contact between the two rock masses with differing resistance to erosion.

South of the old Appalachian–Ouachita tracts, a new depositional province, the Gulf of Mexico, began to take form. The initial deposits were mostly evaporites and do not lend themselves to radiometric dating. However, analyses of pollen from salt taken from this sequence suggest an age of Late Triassic. Evaporites continued to be deposited well into the Jurassic, indicating aridity in the Gulf region (Fig. 11–7). Apparently, the Gulf of Mexico was rather like a great evaporating basin, concentrating the waters of the Atlantic Ocean and precipitating salt and gypsum to thicknesses

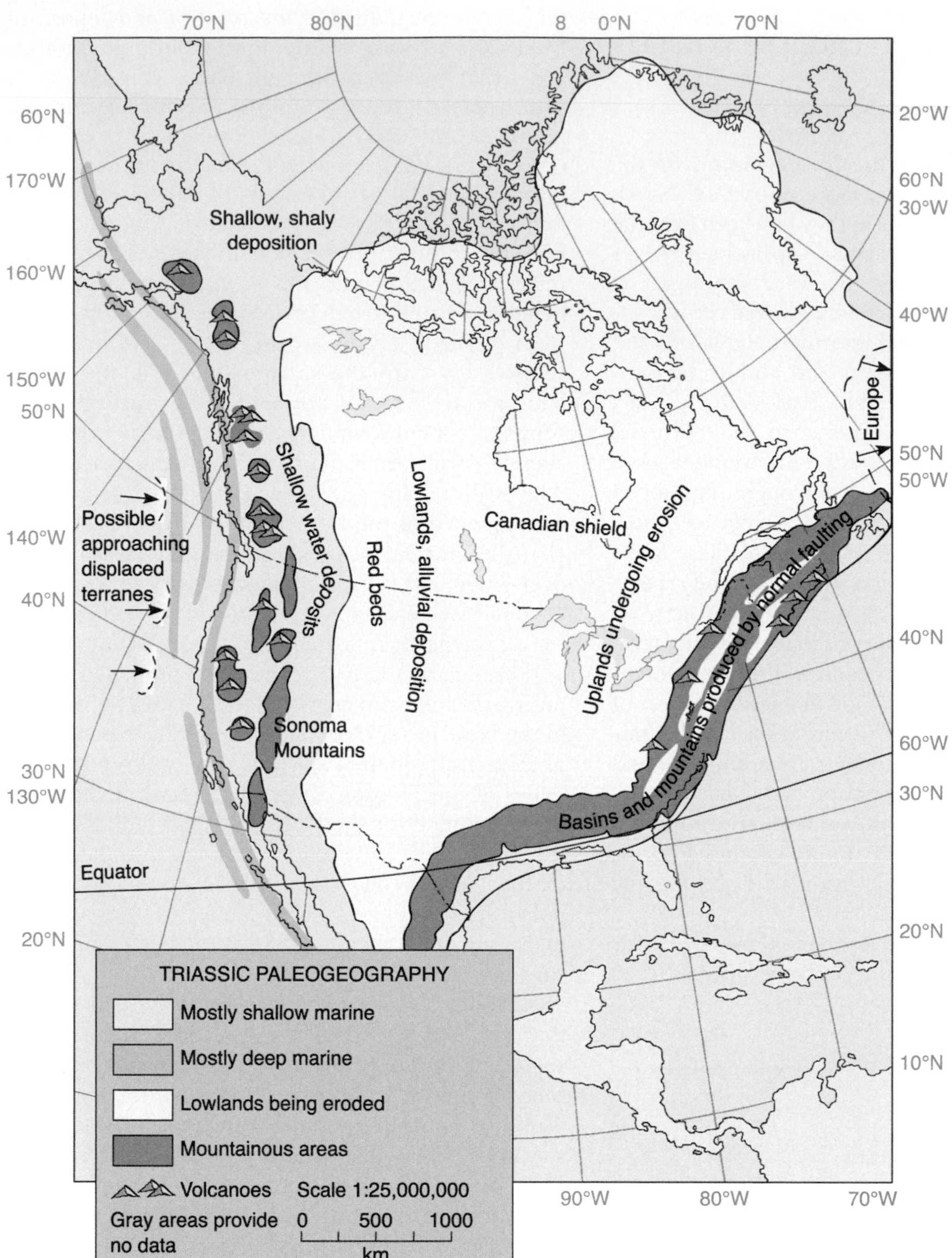

FIGURE 11–3 Generalized paleogeographic map for the Triassic of North America.
What was the cause of the faulting along the eastern margin of the continent?

exceeding 1000 meters (3280 feet) (Fig. 11–8). The Jurassic evaporite beds are the source of the **salt domes** of the Gulf Coast. Salt domes are economically important structures associated with the entrapment of petroleum. When compressed by a heavy load of overlying strata, salt tends to flow plastically. As salt moves upward in the direction of lesser pressures, it produces folds and faults in overlying strata. The permeable, often faulted beds that slope downward away from the salt or are arched over the dome result in traps for petroleum accumulation (Fig. 11–9).

Evaporite conditions abated later in the Jurassic, and several hundred meters of normal marine limestones, limy muds, and sandstones accumulated in the alternately transgressing and regressing seas of the Gulf embayment. Although there are surface exposures of these rocks where they extend into northeastern Mexico, in the United States they are deeply buried

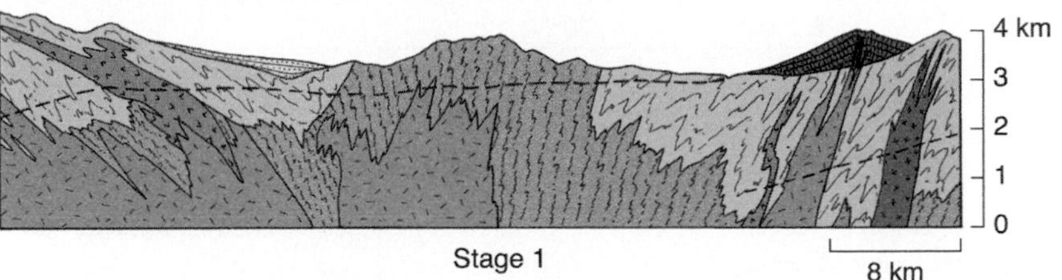

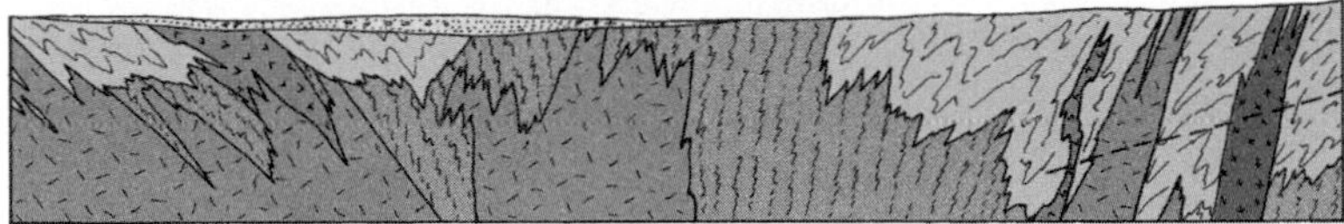

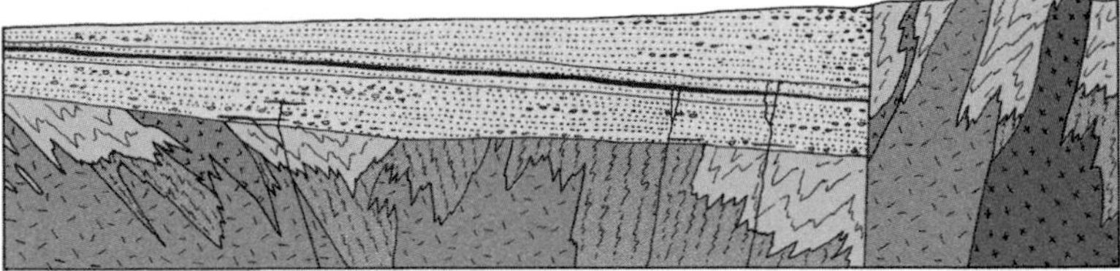

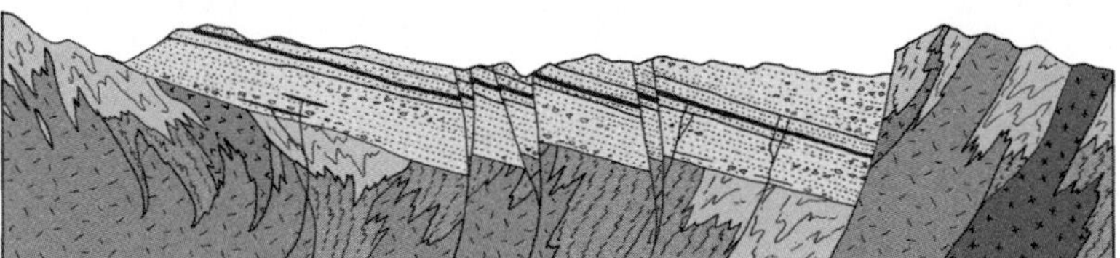

FIGURE 11–4 Four stages in the Triassic history of the Connecticut Valley. (Stage 1) Erosion of the complex structures developed during the Allegheny orogeny. (Stage 2) Mountains have been eroded to a low plain, and Triassic sedimentation has begun. (Stage 3) Newark sediments and basaltic sills, flows, and dikes accumulate in troughs resulting from faulting. (Stage 4) In Late Triassic time, the area is broken into a complex of normal faults as part of the Palisades orogeny. (*Modified from the classic 1915 paper by J. Barrell, Central Connecticut in the Geologic Past*, Conn. Geol. Nat. Hist. Bull. No. 23.)

FIGURE 11–5 Slab of Newark Group sandstone containing the cast of a footprint made by a three-toed Triassic reptile. The footprint is about 20 centimeters long. The linear ridges in the slab are casts of mud cracks.

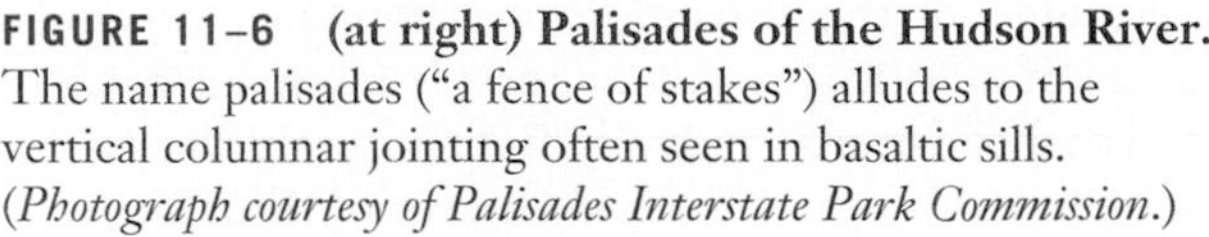

FIGURE 11–6 (at right) Palisades of the Hudson River. The name palisades ("a fence of stakes") alludes to the vertical columnar jointing often seen in basaltic sills. (*Photograph courtesy of Palisades Interstate Park Commission.*)

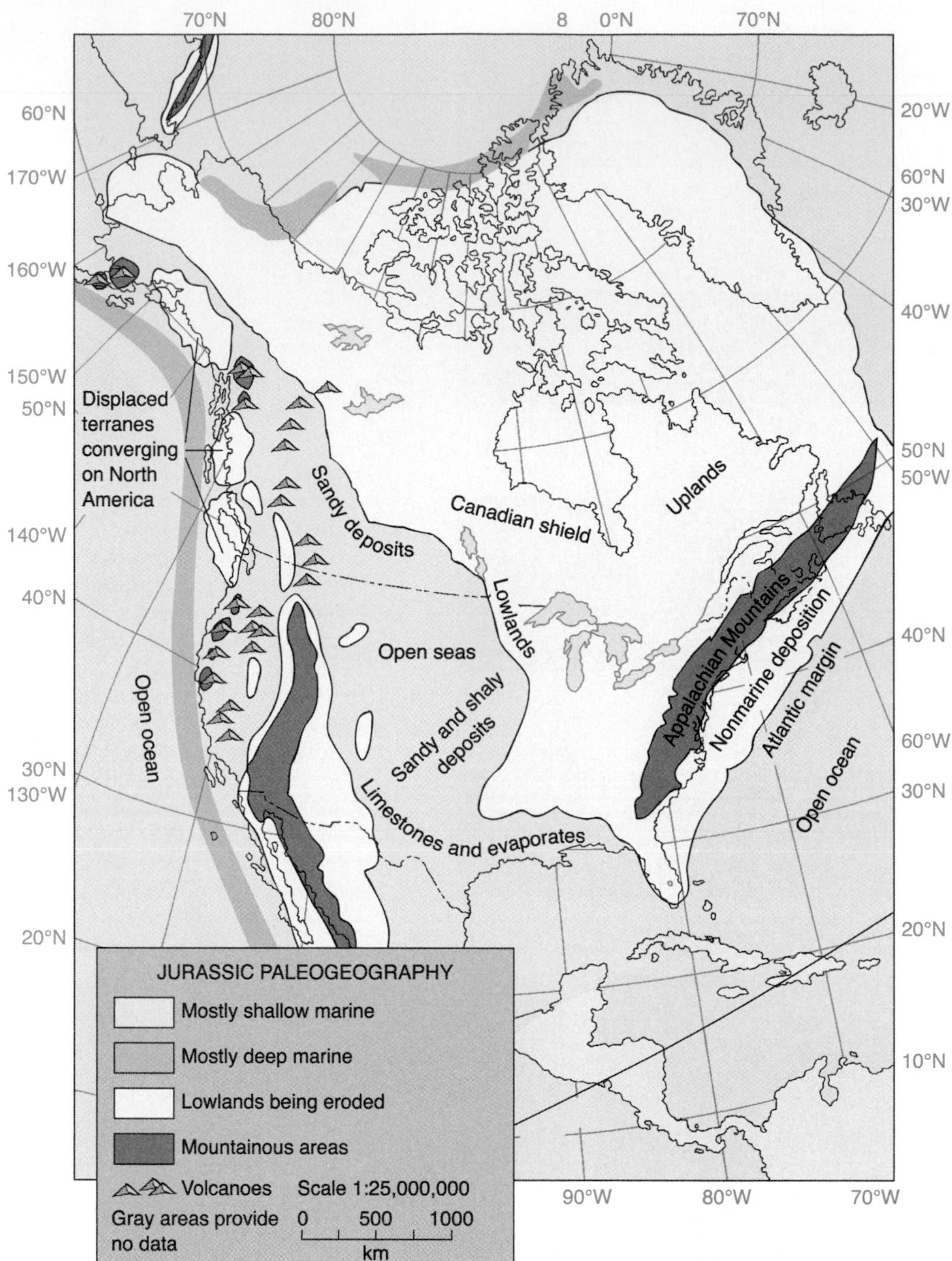

FIGURE 11–7 Generalized paleogeographic map for the Jurassic of North America.
? *Describe the conditions at the site of your school during the Jurassic Period.*

beneath a cover of Cretaceous and Cenozoic sediments. Were it not for the drilling activities of oil companies, the nature of these rocks could only be inferred from geophysical data.

CRETACEOUS The Cretaceous was a time of great marine inundations of landmasses, and the Atlantic and Gulf coastal regions received their full share of flooding. Early in Cretaceous time, the Atlantic coastal plain, which had been experiencing erosion since the beginning of the Mesozoic, began to subside. At about the same time, the Appalachian belt was elevated. On the subsiding coastal plains, alternate layers of marine and deltaic terrestrial deposits accumulated and were gradually built into a great wedge of sediments that thickened seaward (Fig. 11–10).

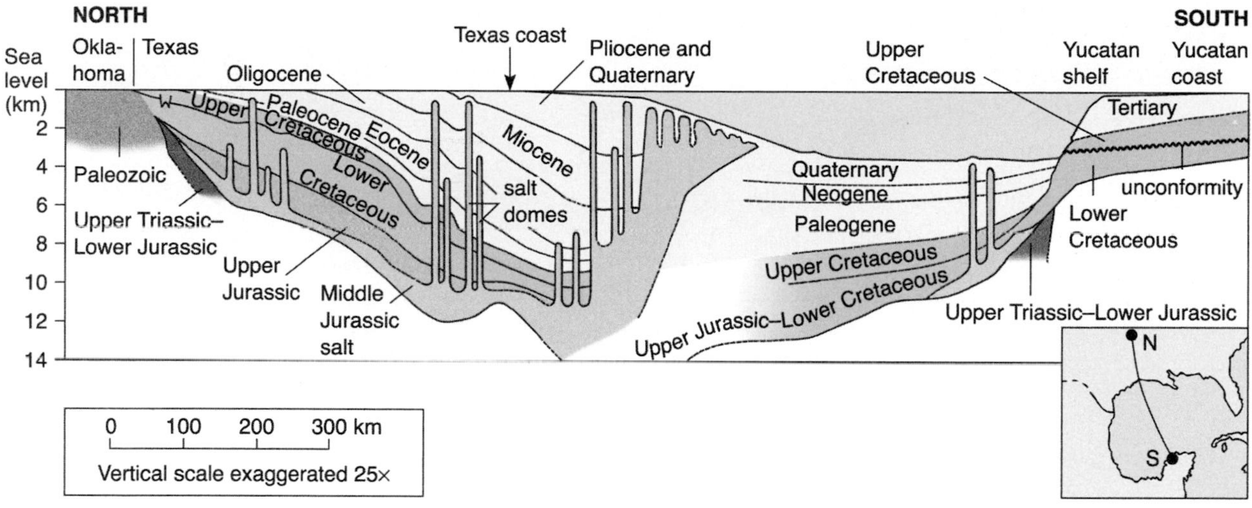

FIGURE 11-8 North–South cross-section of the Gulf of Mexico basin. (*Adapted from Salvadore, A.* 1991. Geology of North America, *pp. 1–12. Boulder, CO: Geological Society of America.*)

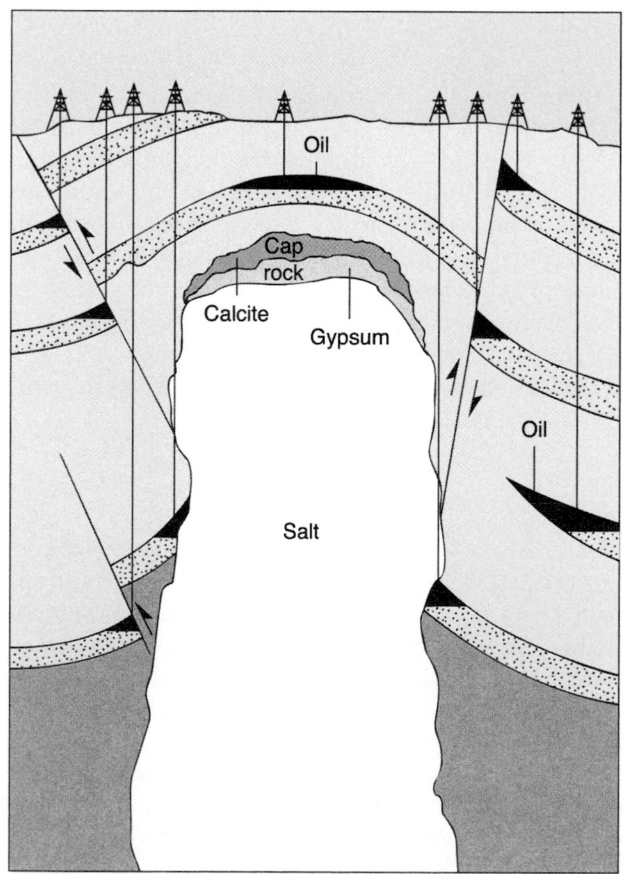

FIGURE 11-9 Salt dome, illustrating possibilities for oil entrapment in domelike structures (top center) by faults and by pinchout of oil-bearing strata.

Today, the thinner eastern border of the wedge is exposed in New Jersey, Maryland, Virginia, and the Carolinas. But the greatest volume of Cretaceous sediments was deposited farther east along the present continental shelf.

To the south, the Florida region was a shallow submarine bank during the Cretaceous. The oldest strata consist of limestones, but later in the period, terrigenous clastics were brought into the area by streams flowing from the southern Appalachians. These source areas were gradually worn down, and carbonates were deposited above the older sandstones and shales.

The Cretaceous is noteworthy as a period of the Mesozoic in which carbonate reefs were extensive in the warmer regions of both the Eastern and Western hemispheres. Among the invertebrates that contributed their skeletal substance to these reefs, a group of bivalves called **rudistids** (Fig. 11–11) were important. In many Cretaceous reefs, the shells of these creatures, some of which mimic the appearance of corals, form the bulk of the reef framework. Because of their high porosity and permeability, rudistid reefs are reservoir rocks for petroleum.

At the end of the Jurassic, the region northwest of Florida consisted of low-lying coastal plains. Early in the Cretaceous, however, these relatively level lands were invaded by marine waters moving northward from the ancestral Gulf of Mexico. Nearshore deposits such as sandstones are overlain by finer sedimentary rocks that are characteristic of deeper water and provide clear evidence of the northward migration of the shoreline. The advance of the Cretaceous sea was not uniform. An extensive regression occurred near the end of the first half of the period; however, flooding

A
Cape Hatteras Lighthouse
B
0
5
10
15
km
Continental basement
Salt dome
Oceanic basement
50 100 150 200 250 300 350 400
km
Tertiary
Cretaceous
Jurassic

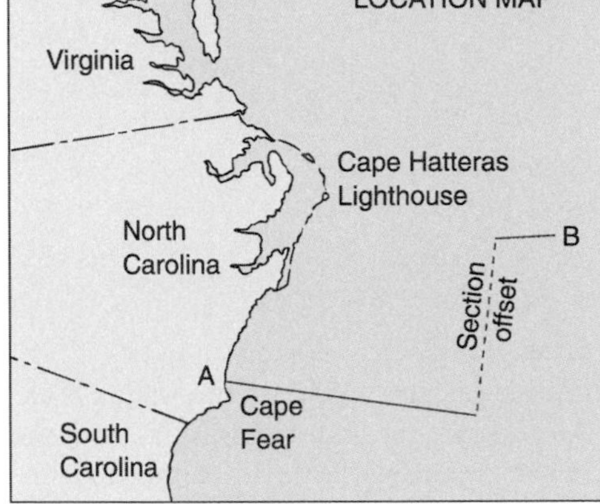

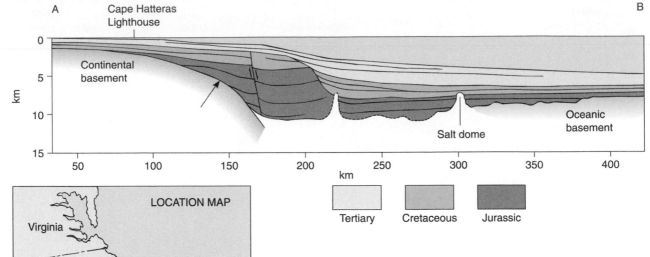

FIGURE 11-10 Section across the Atlantic continental margin east of North Carolina, compiled from seismic reflection and drillhole data. This part of the Atlantic continental margin is called the Carolina Trough. It was formed by pulling apart of North America and Africa parallel to the trend of older Paleozoic faults. (*Modified from Klitgord, K. D., Hutchinson, D. R., and Schouten, H. 1988. U.S. Atlantic Continental Margin: Structural and Tectonic Framework.* Geol Soc. Amer. Decade of North American Geology, *1–2:19–55. Boulder, CO: Geological Society of America.*) ? *Is the fault the kind to be expected on the passive margin of a tectonic plate? Explain.*

FIGURE 11-11 ***Eoradiolites davidsoni*, a Cretaceous rudistid bivalve from Texas.** Rudistids appeared in the Jurassic and proliferated during the Cretaceous, when they were important reef-making organisms. They bear a close resemblance to some Paleozoic corals. This specimen is 8.6 centimeters tall.

resumed in Late Cretaceous time, when a wide seaway occupied a tract from the Gulf of Mexico to the Arctic Ocean (Fig. 11–12). Among the Upper Cretaceous sediments deposited in this inland sea, **chalk** was particularly prevalent. Chalk is a white, fine-grained, soft variety of limestone that is composed mainly of the microscopic calcareous platelets (called coccoliths) of golden-brown algae. Chalk is common among beds of Cretaceous age in many other parts of the world (Fig. 11–13). Indeed, the Cretaceous System takes its name from *creta*, the Latin word for chalk.

To the West

As the eastern border of North America was experiencing crustal extension caused by separation from Europe and Africa, compressional forces prevailed in the west. While the newly formed Atlantic Ocean was widening, North America moved westward, overriding the Pacific plate. Thus, deformation of western North America was caused by events in the east. It has been shown, for example, that the pace of tectonic activity in the North American Cordillera was most intense when sea-floor spreading was most rapid in the Atlantic.

ACCRETIONARY TECTONICS During the Mesozoic, a subduction zone existed along the western margin of North America. This was not a simple Andean-type

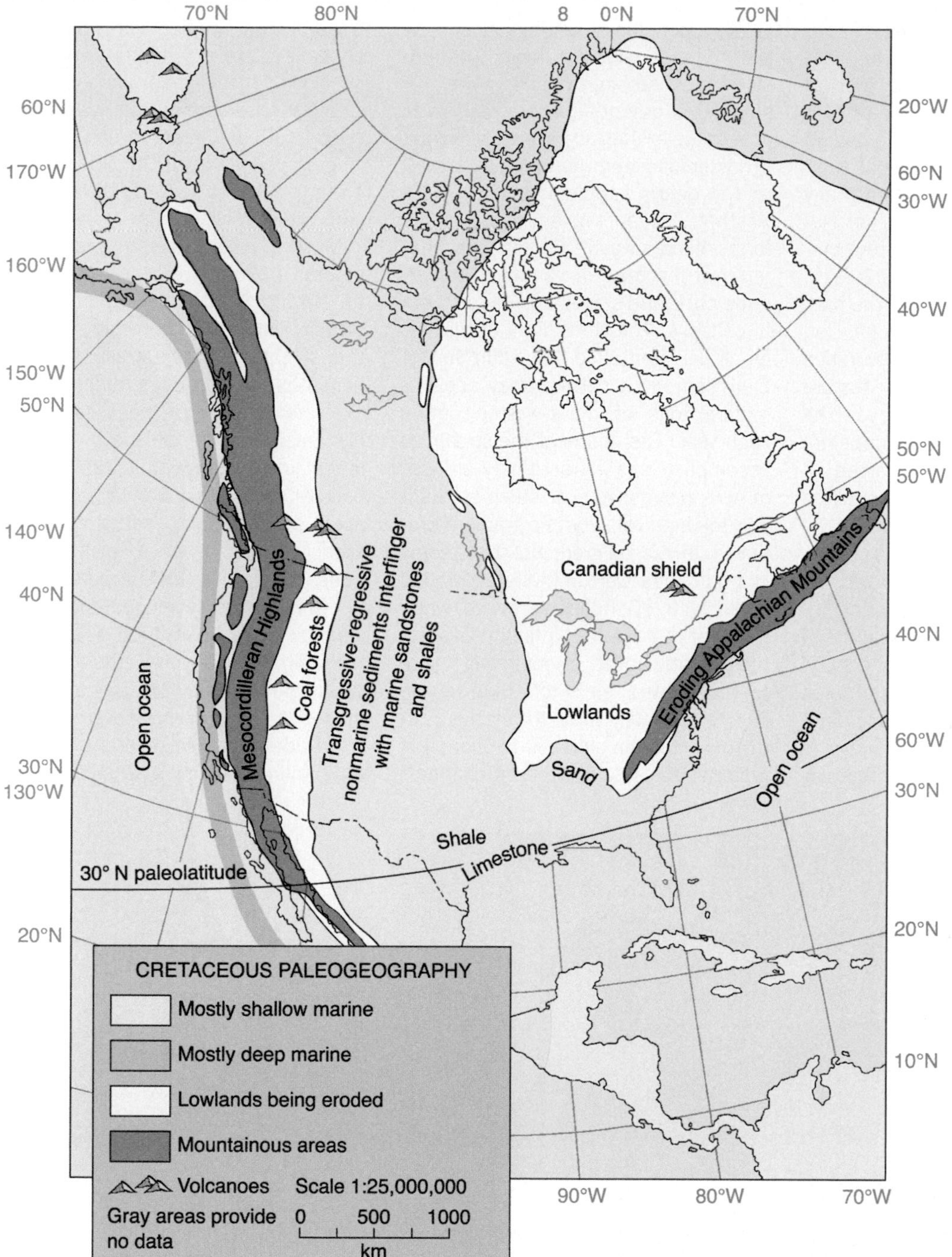

FIGURE 11-12 Generalized paleogeographic map for the Cretaceous of North America.

subduction zone, however, for the advancing Pacific plate carried considerably more than ocean basalts and sea-floor sediments. Entire sections of volcanic arcs, fragments of distant continents, and pieces of oceanic plateaus were carried to North America's western margin as well. These fragments, now incorporated into the Cordilleran belt, constitute displaced or **allochthonous terranes**, described in Chapter 5. The better documented of these terranes can be identified by their distinctive age, rock assemblages, and mineral resources, which indicate the part of the world from which the displaced fragments came. Many of the fragments have major faults along their boundaries.

Altogether, more than 50 allochthonous terranes have been recognized in the Cordillera. They may constitute as much as 70 percent of the total Cordilleran region. Thus, the growth of North America has occurred not only by accretion of materials

Zion National Park

Towering "temples" of Navajo Sandstone, colorful cliffs and arches, slot canyons, and the tumbling roar of streams greet the visitor to Zion National Park. For those who have had an introductory course in historical geology, the park is a place to see many rock units of the Mesozoic. Zion National Park is in the southwestern corner of Utah adjacent to Interstate 15. There is a scenic drive within the park that extends through spectacular Zion Canyon from the park's south entrance northward to an erosional feature called the Temple of Sinawava. The drive includes a mile-long tunnel in which windows have been carved so travelers can stop and view interesting features.

Zion Canyon has been (and continues to be) fashioned by mass wasting and running water. Mass wasting involves the movement of rock and soil downhill under the influence of gravity. In Zion it includes the sliding of large blocks of rock downslope, as resistant caprock is undermined by erosion and weathering. Rockslides and rockfalls are common. Mass wasting alone, however, cannot carve a canyon. The erosive and transportational power of streams is essential. The stream acts as a conveyor belt, carrying away what it has eroded from its channel as well as the debris supplied to it by mass wasting. Streams even cause some forms of mass wasting by removing material at the base of valleys to undermine cliffs and slopes and cause them to collapse. The North Fork of the Virgin River controlled the sculpting of Zion Canyon. It carries away about 3 million tons of mass-wasted debris each year. The ability of the stream to accomplish this extraordinary amount of work results from its steep gradient and the high velocity this gradient produces. Increased gradients for streams in this region were initially developed about 12 to 10 million years ago, when much of the Colorado Plateau was uplifted. With uplift, streams were able to flow faster, entrench themselves, and, along with their tributaries, cut deeply into the plateau. In Zion, the major streams have cut their channels so rapidly that the floors of some tributary valleys are at a higher level than the floor of the main valley. Such hanging valleys are normally a characteristic of glaciated areas. Where hanging valleys occur in Zion, waters of the tributaries cascade down to the main valley in waterfalls.

Most of the interesting erosional features in Zion have been carved from the Navajo Sandstone, which is part Triassic and part Jurassic in age. Among these features are towering monoliths referred to as temples. Zion takes its name from these cathedral-like features, reminiscent of the citadel in Palestine that was the nucleus of Jerusalem. Beneath resistant layers of sandstone, softer rock has often spalled away, leaving inset arches (blind arches). Free-standing arches have also developed, the largest of which is the Kolob Arch. The beauty of the arches, canyon walls, and temples is enhanced by an array of colors: shades of red, brown, and green from iron oxides; gray-green and bright orange from lichens; and shiny black

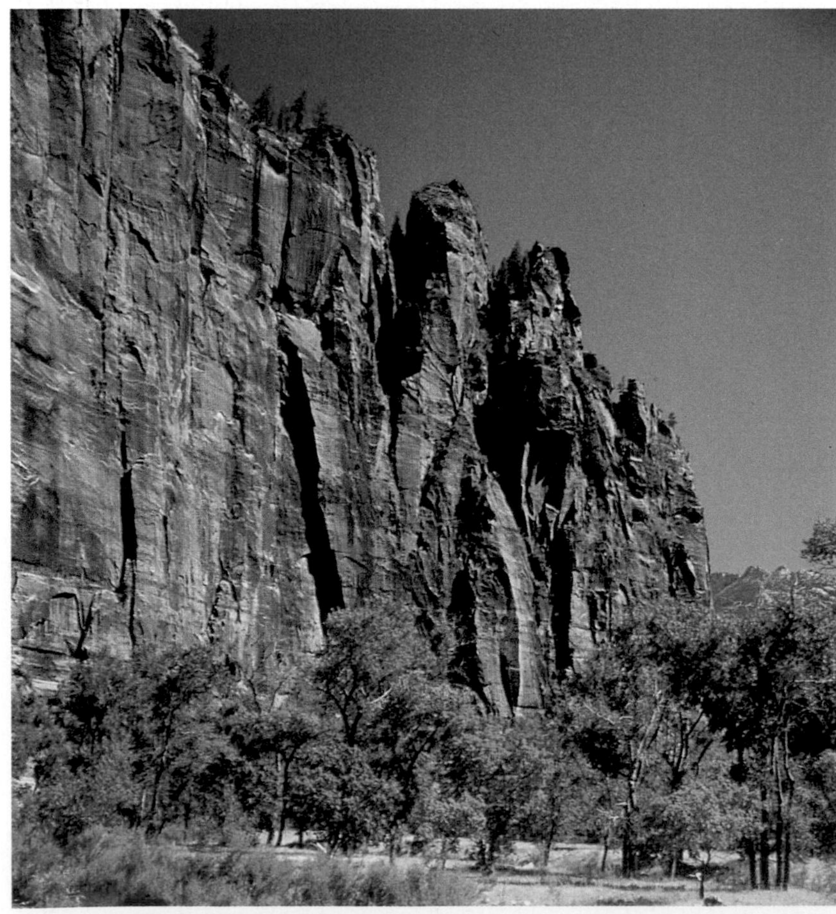

Navajo Sandstone in the North Fork of the Virgin River, Zion Canyon of Zion National Park, Utah. (*Photograph by R.F. Dymek.*)

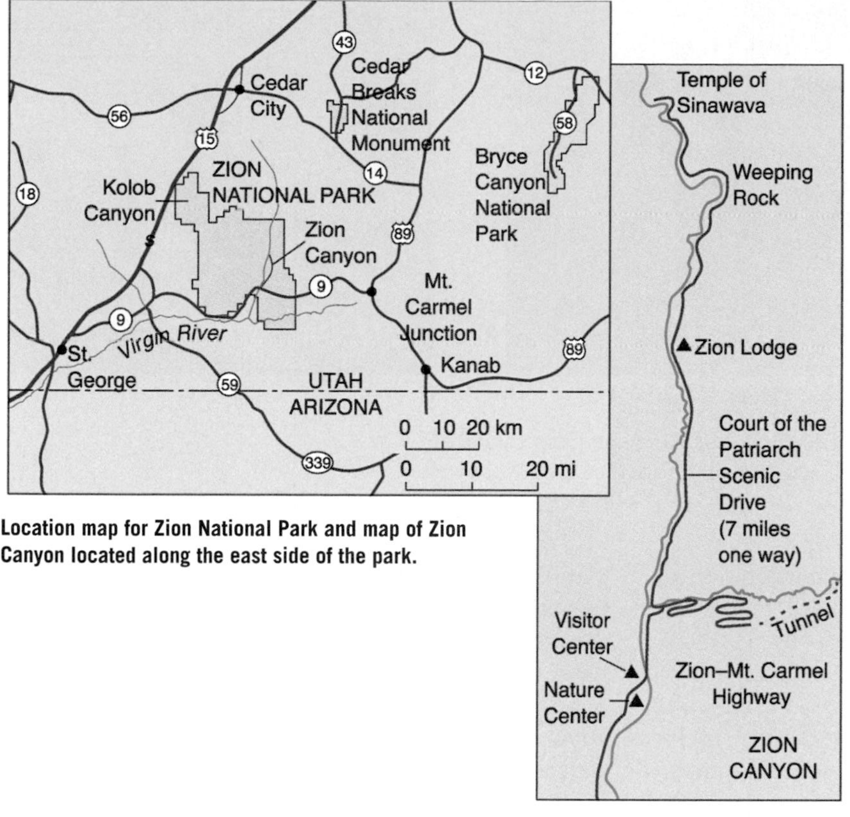

Location map for Zion National Park and map of Zion Canyon located along the east side of the park.

from the iron and manganese oxides of desert varnish.

With clearly exposed rock units on all sides, Zion is a nearly ideal place to interpret geologic history. The lowermost (hence oldest) rock unit in the park is Permian in age. It is the Kaibab Limestone, a yellow-gray rock unit exposed in Hurricane Cliffs north of the Kolob Visitor Center. The Kaibab sediments were deposited in a shallow sea. That sea had withdrawn by the end of the Permian, and above the old sea bed shales and sandstone of the Triassic Moenkopi Formation were deposited. During this time, there were minor incursions of the sea, and gypsum was precipitated in the coastal lagoons associated with these incursions. After deposition of the Moenkopi, the region experienced uplift. Rejuvenated streams with steep gradients were able to carry gravels into the region, and these comprise the Shinarump Conglomerate at the base of the Chinle Formation. Most of the Chinle Formation consists of sands, shales, and beds of volcanic ash. Tree branches, logs, and stumps occur in the Chinle. These have been converted to petrified wood identical to that in the logs of Petrified Forest National Park. Apparently, much of the silica required for silicification of the wood was derived from dissolution of volcanic ash in the Chinle.

Above the Chinle beds in Zion National Park, one finds ripple-marked and cross-bedded siltstones of the Moenave Formation. The Moenave in turn is overlain by siltstones of the Kayenta Formation. The footprints of dinosaurs have been discovered along the bedding surfaces of the Kayenta.

The Navajo Sandstone is next in the sequence of formations in Zion National Park. It is over 220 meters (670 feet) thick in places and is the principal cliff former in the park. Lower beds of the Navajo show bedding features indicating that they were deposited in shallow water, but the greater part of the unit in Zion displays the kind of curved and wedge-shaped cross-bedding characteristic of wind-transported sediment. This interpretation is strengthened by the prevalence of frosted quartz grains in the Navajo. The frosted appearance is the result of grain-to-grain impact during wind transport. Cross-bedding in the Navajo sandstones is magnificently displayed in Zion at Checkerboard Mesa (see Fig. 11–25). One can envision the depositional area as a great coastal desert somewhat like the present Sahara, where dunes migrated along the coastal plain of a shallow sea.

Streams laden with red mud provided the material for the fluvial shales of the Temple Gap Formation, which rests on the sandstones of the Navajo Formation. These beds are covered by shallow-water limestones of the Jurassic Carmel Formation. The culminating events in Zion's history include extrusion of basaltic lavas from about 1.4 to 0.25 million years ago and intermittent episodes of further uplift.

A

B

FIGURE 11-13 Chalk. (*A*) Sea cliffs composed of chalk along the Dorset coast of England. (*B*) Close-up view of the chalk. The dark nodules are flint, a gray or black variety of chert.

along subduction zones but also by incorporation of huge crustal fragments formed elsewhere and subsequently conveyed to North America by sea-floor spreading. Some of these fragments were evidently so large that they caused changes in subduction zone configuration. Others, composed of low-density rocks, were too buoyant for subduction. These were scraped off the subducting plate, splintered, and thrust onto the continental margin. In contrast to subduction, the process whereby one rock mass rides up and over another is called **obduction**. Stated differently, in subduction, one plate stays at a constant level and the other plate is forced downward under it; in obduction, one plate remains at a constant level, while the other plate is forced upward over it.

The growth of a continent by the progressive incorporation of crustal fragments in the aforementioned manner is termed **accretionary tectonics**. In an oversimplified analogy to accretionary tectonics, one can imagine icebergs in an ocean being driven onshore by powerful winds. Each crashes into the coast and is then struck from behind by subsequent arrivals until a jumbled region of displaced icebergs is constructed. In geology, an analogous region of displaced terranes is called a **tectonic collage**.

TRIASSIC AND SEDIMENTATION The Cordilleran region as it existed during the Mesozoic can be divided into a western belt, containing thick volcanic and siliceous deposits, and a wide eastern tract, adjacent to the more stable interior of the continent. Nearly 800 meters of clastic sediments and volcanics exposed in southwestern Nevada and southeastern California attest to the instability of the western zone. Geologists speculate that the belt may well have resembled the Indonesian island arc of today.

The initial orogenic event of the Cordilleran region is difficult to date precisely, although evidence of deformation at or near the Permian–Triassic boundary can be found from Alaska to Nevada. In the United States, the event has been named the **Sonoma orogeny**. Its effects can be studied in west-central Nevada. The Sonoma orogeny occurred when an eastward-moving volcanic arc collided with the Pacific margin of North America. At that time, North America was bordered by a westward-dipping subduction zone. As a result of the collision, the island arc and its adjacent accretionary wedge of sediment was welded onto the western edge of North America, adding as much as 300 kilometers to the continent's east-west dimension. Oceanic rocks were thrust eastward onto the eroded structures of the old Antler orogen (Fig. 11–14). The larger of the thrust sheets moved scores of kilometers eastward and were themselves broken into many smaller internal faults, with each fault slice bringing a repeated stratigraphic sequence to the surface.

The Triassic rocks of the far western part of the Cordillera include great thicknesses of volcanics and graywackes derived from the island arc. It is uncertain, however, whether these rocks were deposited where they are now found or whether they were part of a displaced terrane, called the **Sonoma terrane**, that originated in some part of the Pacific.

Early Triassic rocks of the eastern Cordillera consist of shallow marine sandstones and limestones. The thickest section of these marine strata occurs in southeastern Idaho, where nearly 1000 meters of Lower Triassic sediments accumulated. Eastward from the Cordillera, these marine beds interfinger with continental red beds. Although seas remained in Canada during the Middle Triassic, in the United States they regressed westward, leaving vast areas of the former sea floor subject to erosion.

Upper Triassic formations rest on the unconformity resulting from this erosion. The Upper Triassic series consists of continental deposits transported by rivers

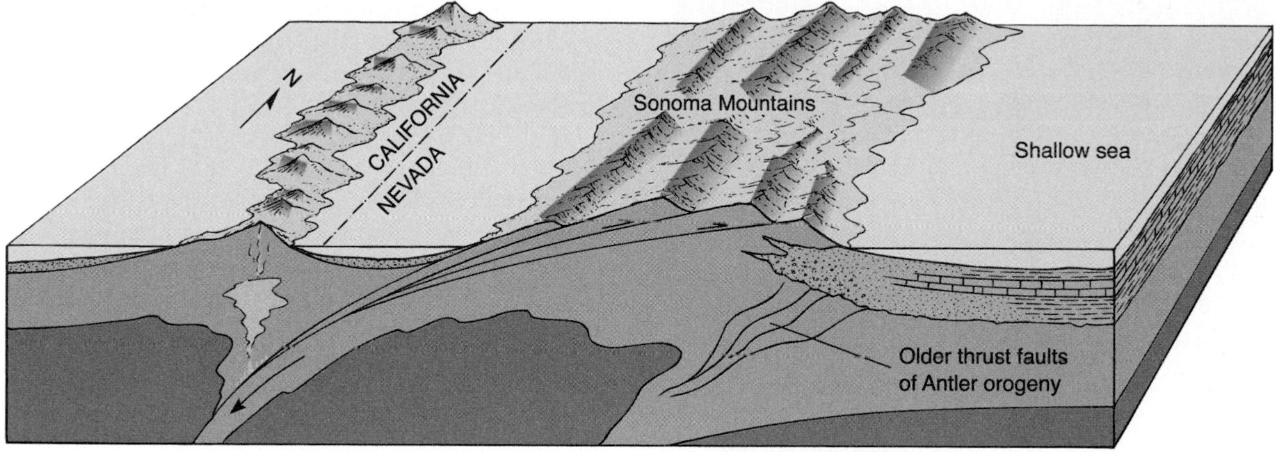

FIGURE 11-14 Tectonic conditions and paleogeography along the Cordilleran region during the Late Permian and Early Triassic.

flowing westward across an immense alluvial plain. The lowermost strata consist of the sandy **Moenkopi Formation**, followed by the pebbly **Shinarump Formation** (Fig. 11–15) derived from uplifted neighboring areas in Arizona, western Colorado, and Idaho. Above the Shinarump lie the vividly colored shales and sandstones of the **Chinle** and **Kayenta** formations. The sediments of these two formations were deposited in stream valleys and lakes. They alternate with sand dune accumulations of the Upper Triassic to Lower Jurassic **Navajo Sandstone** (Fig. 11–16) and the Lower Jurassic **Wingate Sandstone** (Fig. 11–17). These formations are beautifully exposed in the walls of Zion Canyon in southern Utah. They display sweeping cross-bedding, such as is formed in

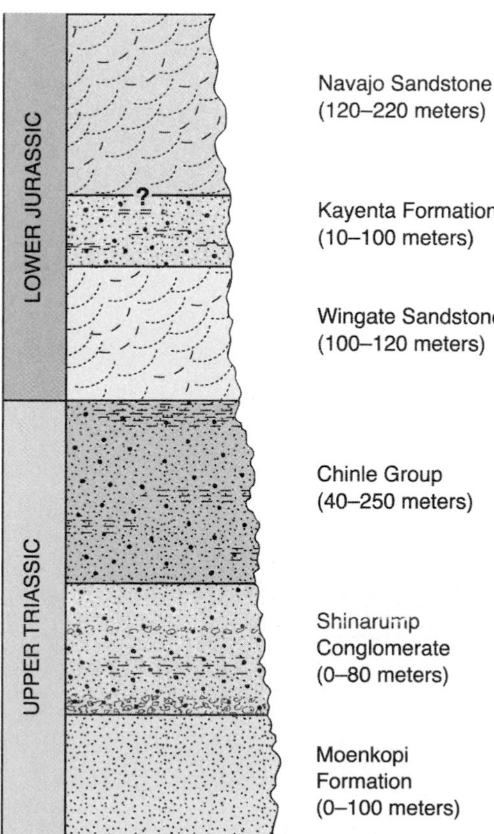

FIGURE 11-15 Generalized geologic section of Upper Triassic and Lower Jurassic sedimentary rocks of central Utah.

FIGURE 11-16 Panoramic view of Triassic and Jurassic formations in Zion National Park, Utah. The lowermost slope is formed by shales of the Moenkopi Formation. Above that slope is a prominent ledge of Shinarump sandstones. The next slope above the Shinarump is the one formed in Chinle Formation. The cliffs at the top are Navajo and Carmel formations. *(Photograph courtesy of the U.S. National Park Service.)* **?** *Which of the units depicted in Figure 11–15 are not present in this locality? What are some possible reasons for their absence?*

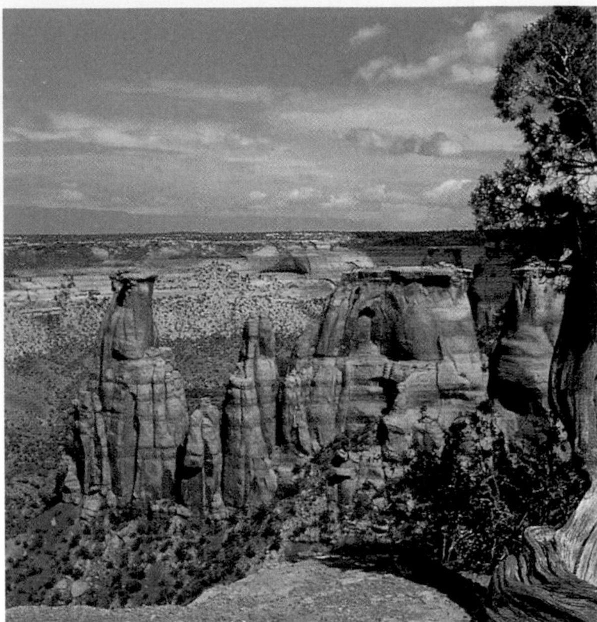

FIGURE 11-17 The erosional features of the foreground are sculpted from the Lower Jurassic Wingate Sandstone, beneath which lies the red Chinle Formation. The cliffs in the far background are also Wingate Sandstone. Colorado National Monument near Grand Junction, Colorado. (*Courtesy of R. J. Weimer, Colorado School of Mines.*)

sandy deposits transported by winds. (However, such so-called festoon cross-bedding is not always caused by winds, for submarine dunes may have similar form.)

The Painted Desert of Arizona is developed mostly in Chinle rocks (Fig. 11–18). The formation is known throughout the world for the petrified logs of conifers it contains. Each year, thousands of tourists examine these logs, now turned to colorful agate, in Petrified Forest National Park (Fig. 11–19). Apparently, during times of Triassic floods, the trees were left on sandbars or trapped in log jams and covered by sediment. Percolating solutions of underground water subsequently replaced the wood with silica.

JURASSIC TO EARLY TERTIARY TECTONICS Most of the orogenic activity in the Cordillera during the Mesozoic was a result of the continuing eastward subduction of oceanic lithosphere beneath the continental crust of the North American plate. That subduction varied in rate, in inclination, and, to a small degree, in direction. It resulted in eastward-shifting phases of deformation, which initially affected the far western part of the Cordillera and then proceeded eastward to reach the margin of the craton.

The deformational and magmatic activity associated with the western tract is termed the **Nevadan orogeny**. During the Triassic, and increasingly during the Jurassic and Cretaceous, graywackes, mudstones,

FIGURE 11-18 Typical exposure of the Triassic Chinle Formation, Painted Desert, Arizona. Some of the red bands represent ancient soils (paleosols). (*Photograph by J. Cowlin.*)

FIGURE 11-19 Petrified logs, Petrified Forest National Park, Arizona. The petrified logs and wood fragments are *Araucarioxylon*. They have been weathered from the Chinle Formation of Triassic age. (*Courtesy of L. F. Hintze.*)

FIGURE 11-20 Strongly deformed bedded cherts of the Franciscan Formation. The exposure is on the flank of Mt. Davidson near the center of San Francisco, between Interstate 280 and Portola Drive. Horizontal distance in photograph is 12 meters. (*Courtesy of B. C. Park-Li.*)

cherts, and volcanics that had been swept into the subduction zone were severely folded, faulted, and metamorphosed. The crumpled and altered rock sequences trapped between the converging plates are appropriately termed a **mélange**, which means "a jumble." The Franciscan fold belt of California (Fig. 11–20) provides a good example of a mélange. In addition to the deformation of the sedimentary rock sequences, an enormous volume of granitic rock was generated above the subduction zone and intruded overlying rocks repeatedly during the Jurassic and Cretaceous. The Sierra Nevada (Fig. 11–21), Idaho, and Coast Range batholiths are impressive evidence of the vast scale of this magmatic activity during and after the Nevadan orogeny (Fig. 11–22).

Somewhat before the Sierra Nevada batholith was emplaced during the Cretaceous, another series of deformations had begun east of the present-day Sierra Nevada. This second phase of the tectonic development of the Cordillera primarily affected shallow-water carbonates and terrigenous clastics deposited over a time span from Middle Jurassic to earliest Cenozoic. This new orogeny has been named the **Sevier**. During the Sevier orogeny, strata were sheared from underlying Precambrian rocks and broken along parallel planes of weakness to form multiple, imbricated, low-angle thrust faults (Fig. 11–23). The French word *décollement* ("unsticking") has been used to describe this kind of structure, in which older rocks are thrust on younger ones in multiple, nearly parallel

FIGURE 11-21 Granodiorite of the Sierra Nevada batholith exposed in Yosemite Valley, California.

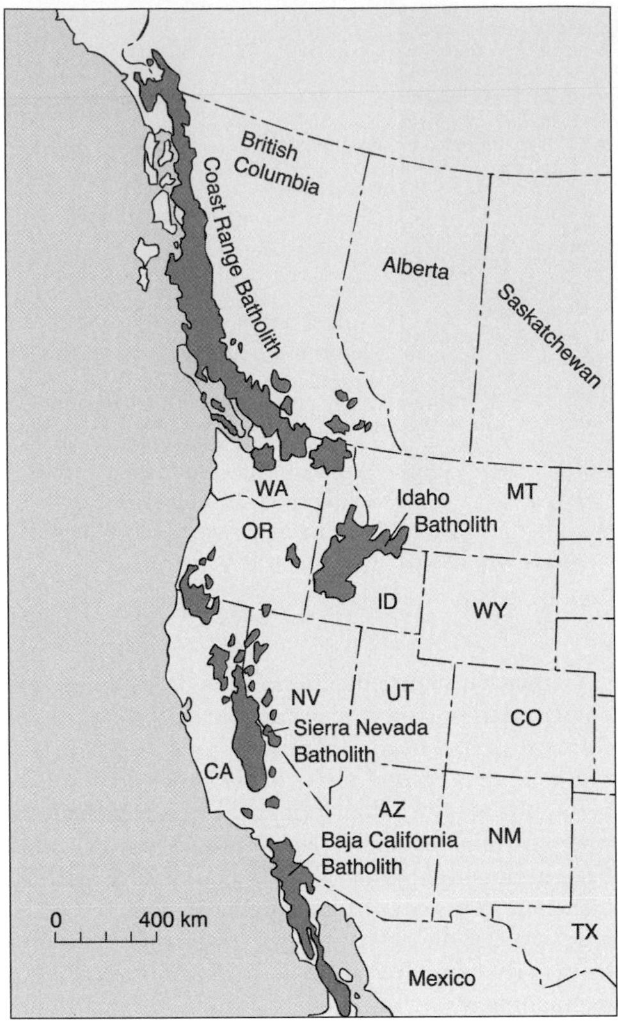

FIGURE 11-22 Mesozoic batholiths in west-central North America.

slabs of crust. It has been estimated that the compressional structures produced during the Sevier orogeny resulted in over 100 kilometers of crustal shortening in the Nevada-Utah region. In addition to the major thrust faults, several large folds are known, as well as intricately folded strata that were deformed within the major thrust units.

Although the term Sevier is sometimes reserved for deformational episodes in the Nevada-Utah region, a similar style of deformation occurred to the north in Montana, British Columbia, and Alberta. In each of the major ranges in this region is a fault block composed of Paleozoic strata of the shelf that have been thrust toward the east along westwardly dipping fault surfaces (Fig. 11–24).

Magmatic activity along the far western edge of the North American plate had diminished near the end of the Cretaceous, and by early Tertiary time, much of the major thrusting of the marginal shelf region had subsided. Once again, deformational events shifted eastward to the cratonic region, where the Rocky Mountains of New Mexico, Colorado, and Wyoming are now located. These more eastward disturbances are called the **Laramide orogeny**. High-angle reverse faults (which become thrust faults with depth) developed during the Laramide orogeny, but the more characteristic kind of structures are broadly arched domes, basins, monoclines, and anticlines. Strata composing the domes and anticlines are draped over central masses of Precambrian igneous and metamorphic rocks. In several instances, erosion has stripped away the cover of strata, exposing the central crystalline mass. Resistant layers of inclined beds that surround the central cores stand as mountains today. Many of these uparched areas appear to have been produced by movements along underlying faults in the basement rocks. It has been suggested that these controlling structures were in turn developed by drag when the eastward-moving subducted oceanic lithosphere scraped the underside of the cratonic margin of North America.

Most of the structures of the present Rocky Mountains are the result of Laramide phases of orogeny. The landscape we see today, however, is the result of repeated episodes of Cenozoic erosion and uplift. Erosion, acting on the geologic structures already present, sculpted the final scenic design.

JURASSIC SEDIMENTATION During the Jurassic, when the mélange was being formed in the subduction zone complex, depositional sites inland experienced a less dramatic kind of sedimentation. Early Jurassic deposits consist of clean sandstones, such as the Navajo Sandstone, which is Late Triassic and Early Jurassic in age. Large-scale cross-bedding is well developed in the Navajo (Fig. 11–25). However, thin beds of fossiliferous limestone and evaporites occur locally and indicate that at least some parts of the formation were deposited in water. The Navajo and associated sand bodies were probably deposited in a nearshore environment, and it is likely that some of the deposits were part of a coastal dune environment (Fig. 11–26). They do not seem to have been laid down on the floors of a vast interior desert, as has often been postulated. Judging from studies of cross-bedding orientations, the source area for these clean sandstones was probably the Montana–Alberta region of the craton. Somewhat older but similar quartz sandstones were recycled and spread southward into Wyoming and Utah and then westward across Nevada.

Marine conditions became more widespread in the Middle Jurassic At that time the entire west-central part of the continent was flooded by a wide seaway

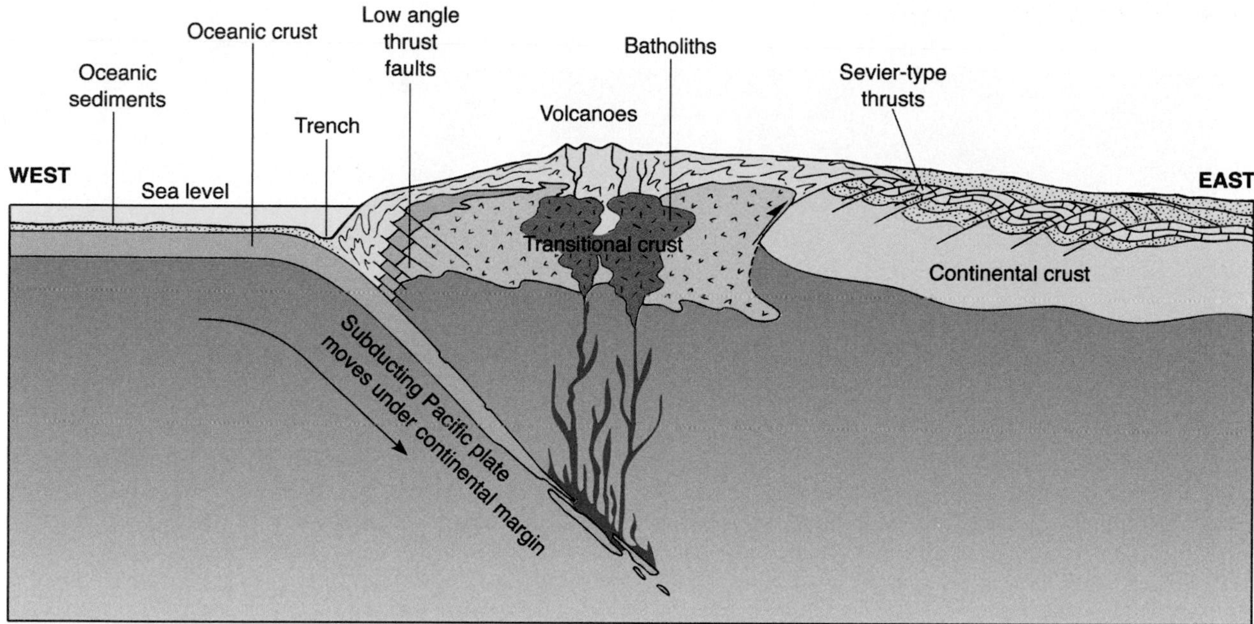

FIGURE 11-23 An advanced stage in the evolution of the North American Cordillera, with structures developing as a consequence of underthrusting of the continent by the Pacific oceanic plate. Note the multiple, imbricated, low-angle thrusts on the east side of the section. (*The diagram is simplified from Dewey, J. F., and Bird, J. M. 1970.* J. Geophys. Res. 75(14):2638.) ? *Where along the cross-section would one find an ophiolite suite of rocks?*

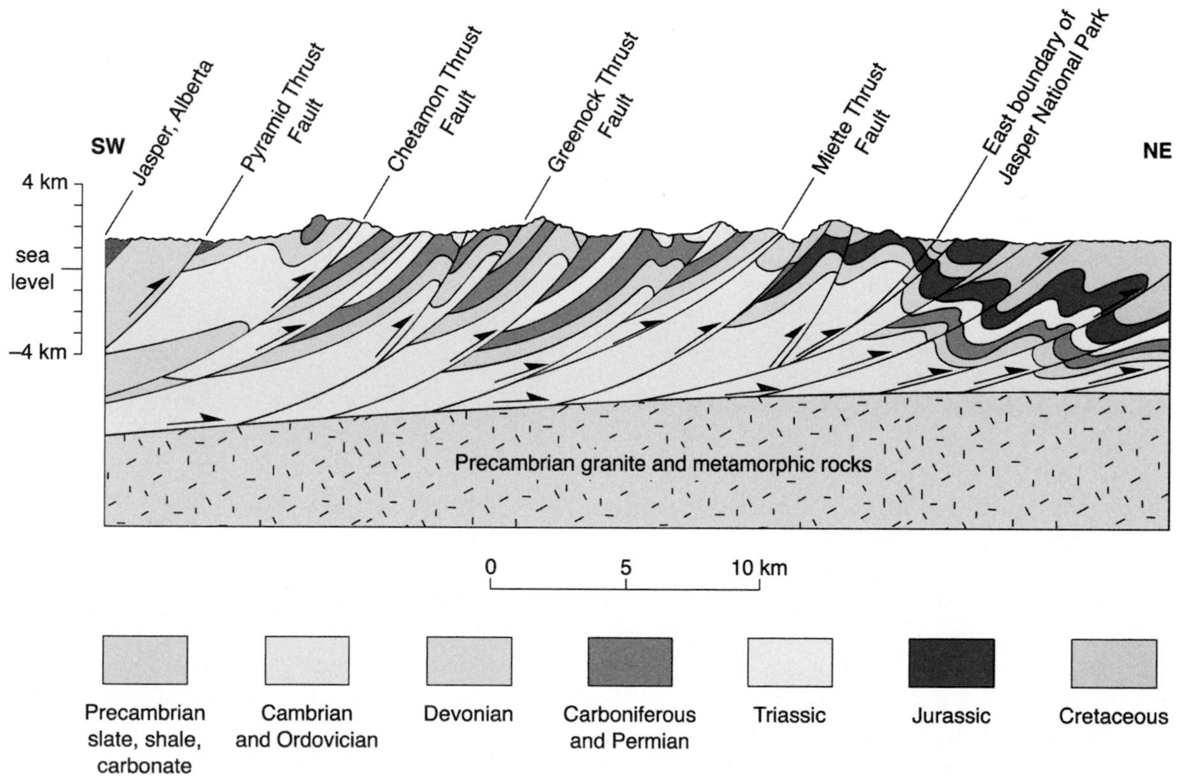

FIGURE 11-24 Structural cross-section extending 40 kilometers northeastward from the town of Jasper, Alberta, Canada. The section shows Sevier type deformation consisting of multiple, imbricated thrust faults. The faults flatten at depth and merge into a single plane of detachment along the surface of the Precambrian igneous and metamorphic basement. (*Adapted from structural section accompanying map entitled* Geology of Jasper National Park and Surrounding Regions, *compiled by C. J. Yorath for the Geological Survey of Canada.*)

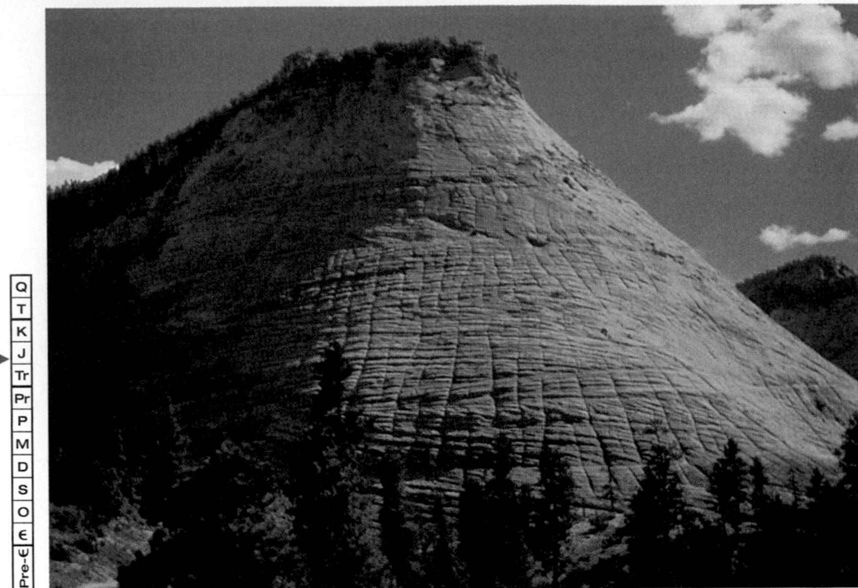

FIGURE 11-25 Cross-stratification in the Navajo Sandstone, Checkerboard Mesa, Zion National Park, Utah. *(Courtesy of L. F. Hintze.)*

that extended well into central Utah (Fig. 11–27). This great embayment has been dubbed the **Sundance Sea**. Within the Sundance Sea were deposited the sands and silts of the **Sundance Formation**, famous for its fossil content of Jurassic marine reptiles. Sediment for the Sundance Formation and overlying younger Jurassic rock units was derived from the Cordilleran highlands that lay to the west. These highlands continued to grow throughout the Jurassic, ultimately extending from Mexico to Alaska. Eventually the Sundance Sea regressed, leaving behind a vast, swampy plain across which meandering rivers built floodplains. These deposits compose the **Morrison Formation** (Fig. 11–28), which extends across millions of square kilometers of the American West.

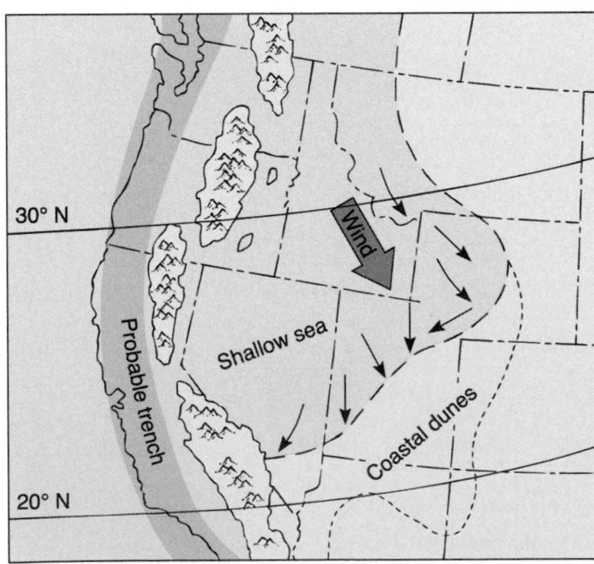

FIGURE 11-26 Paleogeographic map for the early Jurassic of the western United States, showing general extent of sea and land as well as paleolatitudes. (*From Stanley, K. O., Jordan, W. M., and Dott, R. H. 1971.* Bull. Am. Assoc. Petrol. Geol. 55(1):13.)

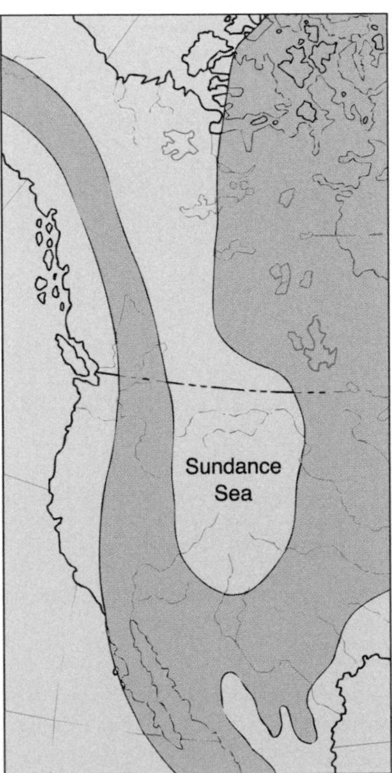

FIGURE 11-27 Region in the western North America inundated by the Middle Jurassic Sundance Sea. (Land areas are shown in tan, marine areas in blue.)

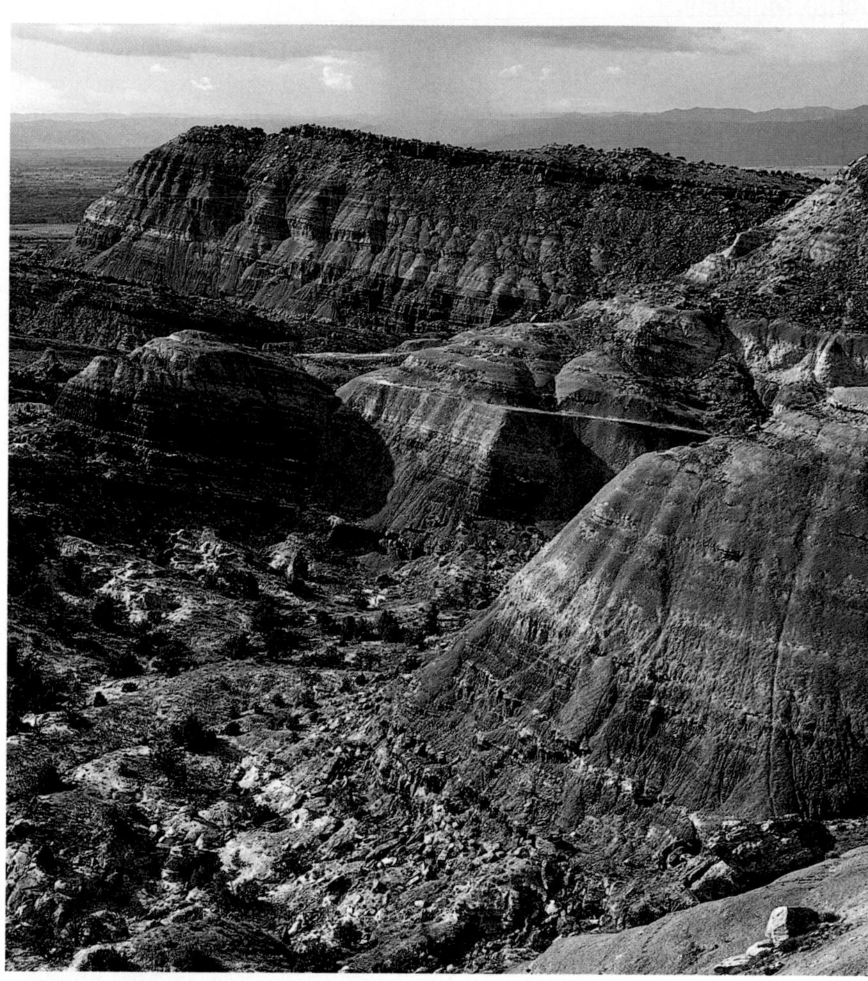

FIGURE 11-28 The Jurassic Morrison Formation near Grand Junction, Colorado. *(Copyright Francois Gohier/Photo Reswerchers, Inc.)*

Enclosed within the floodplain deposits of the Morrison are the bones of more than 70 species of dinosaurs, including some of the largest land animals ever to have existed.

CRETACEOUS SEDIMENTATION During the Cretaceous, the Pacific border region of North America was a land of lofty mountains formed during the orogenesis that had begun in the Jurassic. Erosion of these ranges brought sediments into adjoining rapidly subsiding basins, many of which were open to the Pacific Ocean. In some places, over 15,000 meters of volcanics and clastics accumulated. Cycles of folding and intrusion occurred repeatedly, with exceptional deformations taking place during Middle and Late Cretaceous time.

In the more eastward regions of the Cordillera, the Cretaceous began with the advance of marine waters both northward from the ancestral Gulf of Mexico (Fig. 11–29) and southward from arctic Canada. These

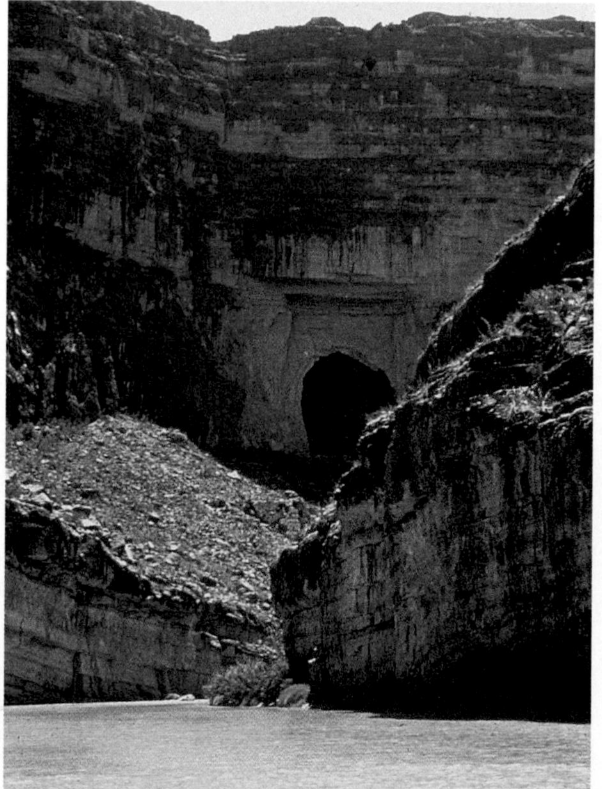

FIGURE 11-29 Imposing cliffs of Early Cretaceous marine limestones along the canyon of the Rio Grande River, Big Bend National Park, southwestern Texas. *(Photograph by R. F. Dymek.)*

Grand Staircase–Escalante National Monument

Grand Staircase–Escalante National Monument was established by presidential proclamation in 1996. The monument encompasses 1.9 million acres of breathtaking buttes and mesas, labyrinthic canyons, and multihued rock formations piled one upon the other like layers of a cake. The Grand Staircase section of the monument takes its name from these layers, named "a grand staircase" by Clarence Dutton in his 1880 *Report of the Geology of the High Plateaus*.

Grand Staircase–Escalante National Monument is located in southern Utah about 290 miles south of Salt Lake City. Access to the monument is provided by Highway 89 on the south and Highway 12 on the North (Fig. A). The monument can be divided into three distinct sections: the *Grand Staircase*, the *Kaiparowits Plateau*, and the *Canyons of the Escalante*. The Grand Staircase, which comprises the western third of the monument, consists of a series of level areas ("steps" or benches) interrupted by cliffs (termed risers). As one proceeds northward from Highway 89, the first riser encountered form the Chocolate Cliffs, which consist of members of the Triassic Moenkopi Formation (Fig. B). The Moenkopi is capped unconformably by the hard, conglomeratic Shinarump Sandstone Member of the Chinle Formation. This resistent layer forms the surface of the bench leading to the next riser, the Vermillion Cliffs. Stream and lake sediments of the Jurassic Moenave and Kayenta formations provide the gray, beige, and pink colors of the Vermillion Cliffs. The imposing White Cliffs form the next riser. Bedrock here consists of the Navajo Sandstone capped by a thin limestone member of the Carmel Formation. An excellent view of the White Cliffs can be obtained by driving for a short distance north of Highway 89 on Kitchen Corral Wash road. The Gray Cliffs, composed of resistant beds of Cretaceous sandstones, are the next risers to be encountered. The Pink Cliffs of the staircase are found mostly north and west of the monument. The limestones and marls of the Pink Cliffs have been sculpted into the erosional features of nearby Bryce Canyon. Above the Pink Cliffs lies the Paunsaugunt Plateau. the uppermost bench of the staircase.

To the north of the Grand Staircase lie buttes, mesas, and steeply incised gorges of the Kaiparowits Plateau. Kaiparowits is a Paiute name meaning "Big Mountain's Little Brother." Although this section is topographically a dissected plateau, its underlying architecture is that of a broad structural basin. As one climbs upward from the floor of one of the Kaiparowits canyons, the Cretaceous Dakota Sandstone is seen, followed by the Tropic Shale, Straight Cliffs Formation, and Waheap Formation. Along the western margin of the Plateau, steeply dipping beds of a monocline are sculpted into a dramatic line of hogbacks named The Cockscomb.

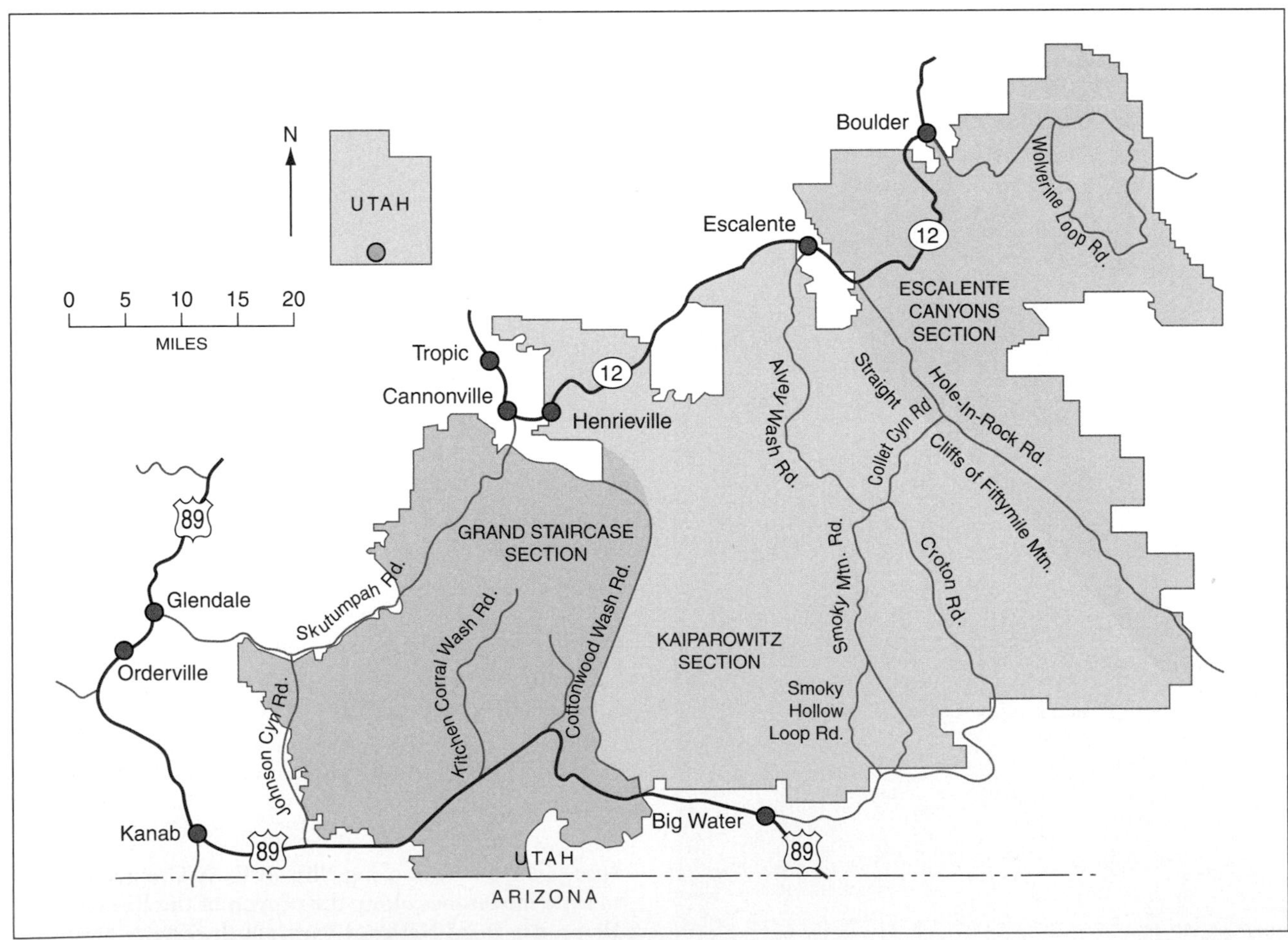

FIGURE A **Location map for the Grand Staircase–Escalante National Monument.**

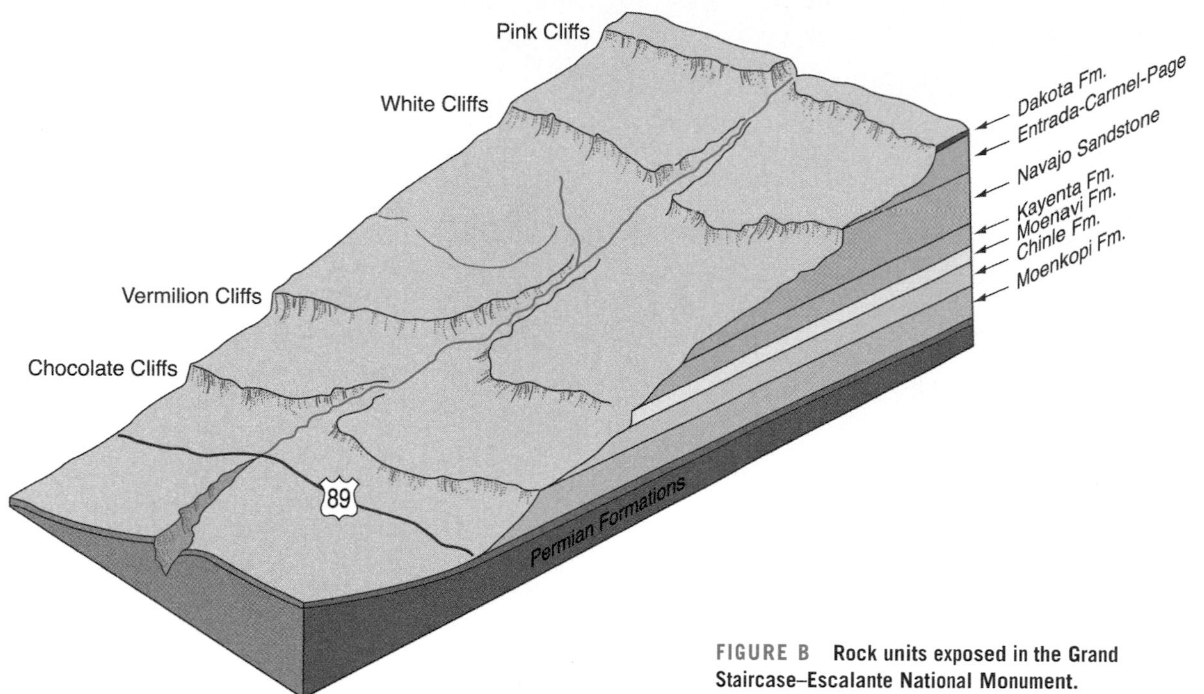

FIGURE B Rock units exposed in the Grand Staircase–Escalante National Monument.

The Canyons of the Escalante lie northward of The Kaiparowits Plateau. It is a wondrous area of narrow canyons and colorful formations eroded by the Escalante River and its tributaries. Scenic features abound among the Canyons of the Escalante. A hike along Calf Creek from a campsite off Highway 12 will permit visits to high waterfalls and superb exposures of the Navajo and Kayenta Formations. Hole-in-the-Rock Road leading from Highway 12 will take intrepid hikers to Devil's Garden, where erosion of the Jurassic Entrada Formation has produced goblins, rock babies, and hoodos. Dinosaur tracks can also be seen on the bedding surfaces of the Entrada. Ten miles west of the town of Escalante along Highway 12, visitors can view the sweeping cross-bedding of the Navajo Sandstone.

Nearly 270 million years of the Earth's history is recorded by the rocks and fossils of the Grand Staircase–Escalante National Monument. During the Permian, the region was a low coastal area occasionally inundated by a shallow sea. Sediments deposited during the Permian are exposed above along Kaibab Gulch southwest of The Cockscombs. The Permian units include the Hermit Shale, Coconino Sandstone, Toroweap Formation, White Rim Sandstone, and Kaibab Limestone. Above the Permian strata, one encounters the Triassic Moenkopi Formation. Both marine and terrestrial fossils can be found in members of the Moenkopi. The terrestrial fossils of the Moenkopi include footprints and bones of reptiles as well as skeletal parts of labyrinthodont amphibians. The Chinle Formation, widely known for the petrified wood it contains, rests on the Moenkopi. In addition to petrified logs, the Chinle contains fossils of labyrinthodonts and crocodile-like

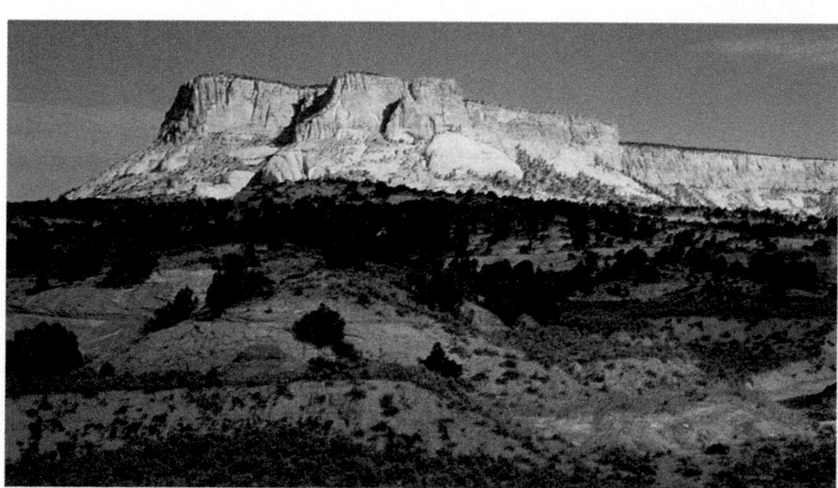

FIGURE C Navajo Sandstone exposed in the White Cliffs of the Grand Staircase section. The Navajo is capped by a thin limestone member of the Carmel Formation (*From Doelling, H. H., Blackett, R. E., Hamblin, A. H., Powell, J. D., and Pollock, G. L.,* 2000. Geology of the Grand Staircase–Escalante National Monument, Utah, *in Sprinkel, D. A., Chidsey, T. C. Jr., and Anderson, P. B., eds.* Geology of Utah's Parks and Monuments. *Utah Geological Association Publication 28, pp. 189–231, with permission.)*

FIGURE D **Erosionally etched cross beds in the Navajo Sandstone exposed in the Escalante Canyons section of the monument.** (*From Doelling, H. H., Blacket, R. E., Hamblin, A. H., Powell, J. D., and Pollock, G. L., 2000.* Geology of the Grand Staircase–Escalante National Monument, Utah, *in Sprinkel, D. A., Chidsey, T. C. Jr., and Anderson, P. B., eds.* Geology of Utah's Parks and Monuments. *Utah Geological Association Publication 28, pp. 189–231, with permission.*)

reptiles called phytosaurs. Chinle sedimentary rocks are of stream and lake origin. The formation is surmounted by stream channel deposits of the Shinarump.

In ascending order, the Lower Jurassic rocks in the Monument include the Wingate Sandstone–Moenavi Formation, the Kayenta Formation, and the Navajo Sandstone. Together, these formations comprise the Glen Canyon Group. The San Rafael Group of Middle Jurassic age overlies the Glen Canyon formation. In ascending order, the San Rafael Group includes the Page Sandstone, Carmel Formation, Entrada Sandstone, Henrieville Sandstone, and Romana Mesa Sandstone. Cretaceous rocks in the Monument area are exposed in the Kaiparowits Plateau. Perhaps the most famous of the Cretaceous formations is the Dakota Sandstone. Although the lower beds of the Dakota appear to be of fluvial origin, the uppermost beds are marine and contain oyster coquinas of *Ostrea* and *Exogyra*.

From the above brief sketch of the Grand Staircase–Escalante, it is apparent that the Monument contains a virtual cornucopia of fascinating geologic features. Many of the best sites can be reached only by unpaved roads. To prepare for a true adventure in the Monument, you can obtain maps and information on weather, roads, and necessary supplies from the web site, **http://www.ut.blm.gov/monument/visitor**.

Reference

Doelling, H.H., Blackett, R.E., Hamlin, A.H., Powell, J.D., and Pollock, G.L. *Geology of Grand Staircase–Escalante National Monument, Utah,* in Sprinkle, D.A., Chidsey, T. C. Jr., and Anderson, P.B., eds. 2000. *Geology of Utah's Parks and Monuments.* Utah Geologic Association Publication 28. Salt Lake City: Publishers Press.

Early Cretaceous invasions did not meet, so an area of dry land existed in Utah and Colorado. Their advance was reversed by a general regression that resulted in the unconformity used to separate the Cretaceous into an Early and a Late division. Flooding that followed in the Late Cretaceous was the greatest of the entire Mesozoic Era. This time, the embayment from the north joined with the southern seaway and effectively separated North America into two large islands.

Sedimentation in Cretaceous epicontinental seas was largely controlled by local conditions. Along the Gulf Coast, limestones and marls (clayey limestones) accumulated (Fig. 11–30). North of Texas, however, the great bulk of sediments consisted of terrigenous clastics supplied by streams flowing from the Cordilleran highlands.

In terms of plate tectonics, the depositional basin that received these sediments (as well as smaller amounts from the craton) can be termed a **foreland basin** (Fig. 11–31). The foreland basin of the western interior of North America was immense. It extended from the Arctic to the Gulf of California. In places it was over 1600 kilometers wide. It was bounded on the west mostly by fold and thrust belts, such as the Sevier, and on the east by the craton. Many of the rock formations in the basin yield commercial quantities of oil, natural gas, and coal. Exploration for these resources has provided hundreds of well logs and detailed surface studies used to recognize and map Cretaceous rocks of the Piedmont, coastal plains, shorelines, marine shelf areas, and deep-water tracts of the central basin. The picture that has emerged shows coarse, terrigenous

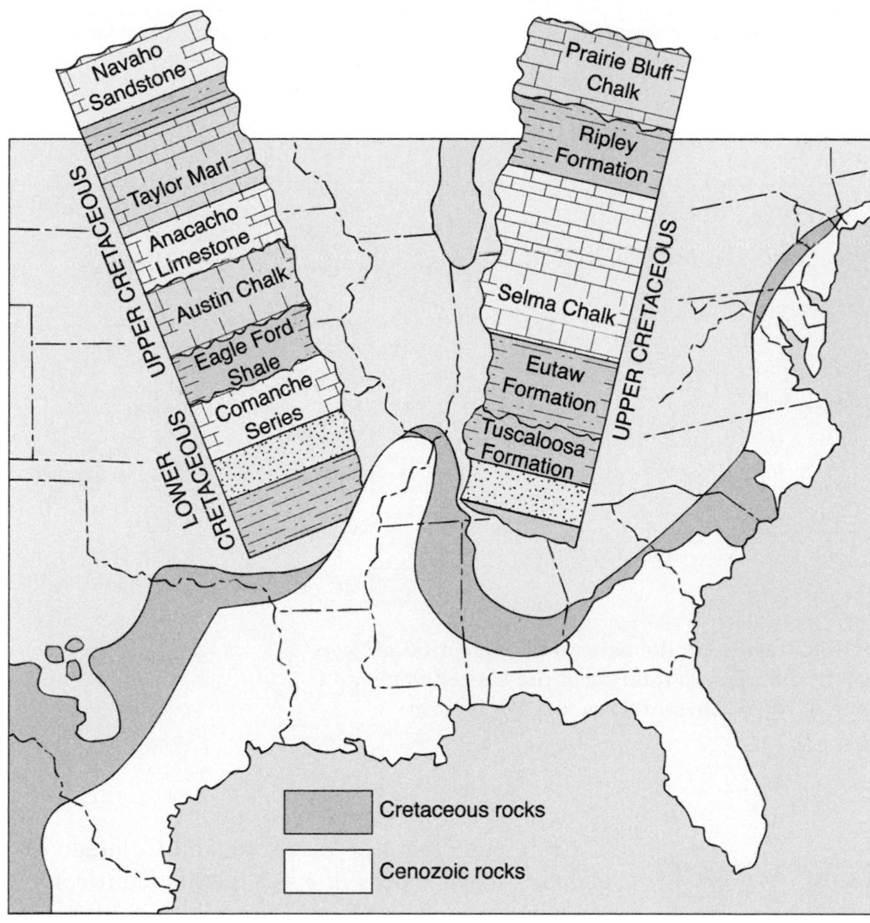

FIGURE 11-30 Area of outcrop of Cretaceous limestone and marl rocks in the Atlantic and Gulf coastal plains, and generalized columnar sections of the Texas (*left*) and Alabama (*right*) Cretaceous.

clastics immediately adjacent to the Cordilleran highlands. These interfinger with coal-bearing continental deposits that similarly interfinger with marine rocks farther to the east (Fig. 11–32). The marine and nonmarine facies shift repeatedly with the alternate advances and regressions of the Cretaceous sea. As is normally true, the transgressive phase is marked by sandstone beds, such as those of the Cretaceous **Dakota Group** (Fig. 11–33). The red, yellow, and brown layers of the Dakota are exposed in many places along the eastern front of the Rocky Mountains, where the inclined beds form prominent ridges called *hogbacks* (Fig. 11–34). In parts of the Great Plains, the Dakota sandstones are an important source of underground water.

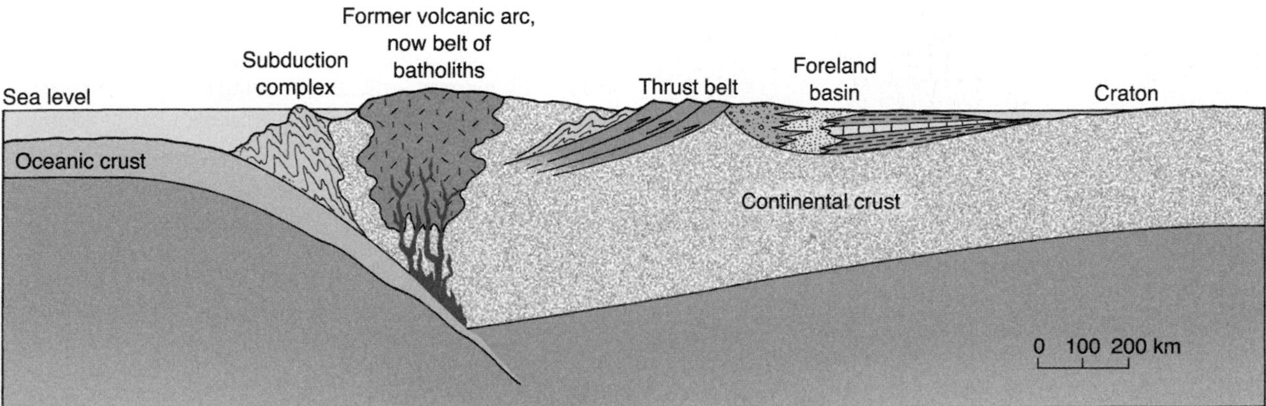

FIGURE 11-31 Conceptual cross-section indicating major tectonic features present during the Cretaceous across the western United States.

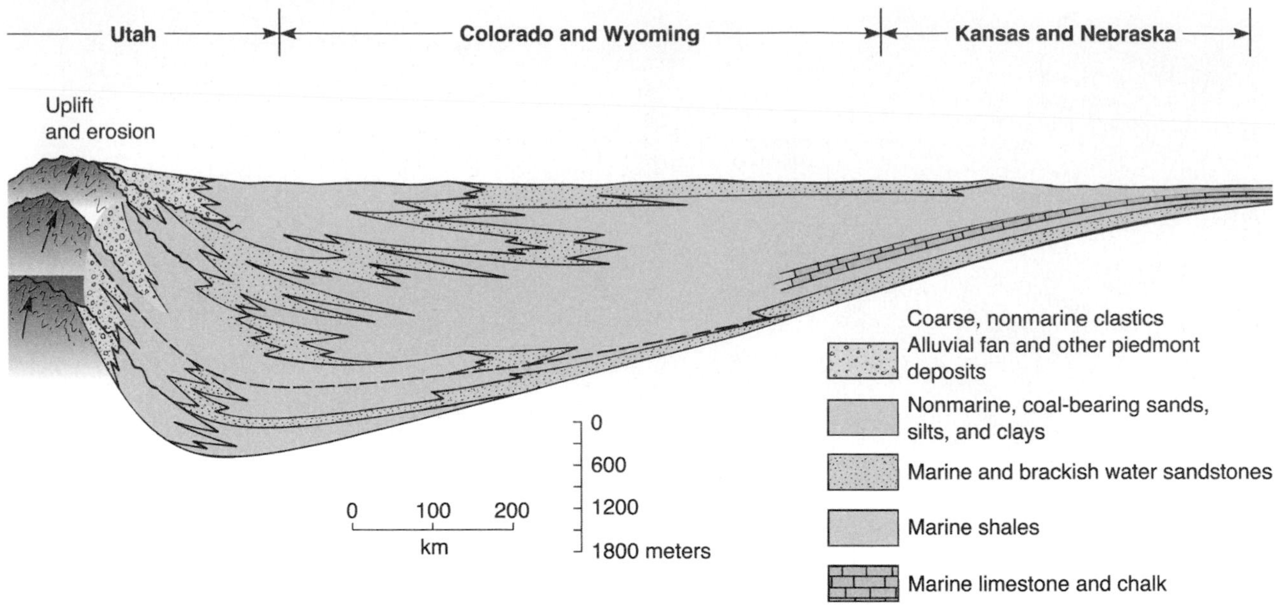

FIGURE 11-32 Generalized cross-section illustrating the pattern of deposition of Cretaceous rocks in the back-arc basin of the western interior of the United States. (*Simplified from King, P. B. 1977.* Evolution of North America. *Princeton, NJ: Princeton University Press.*)

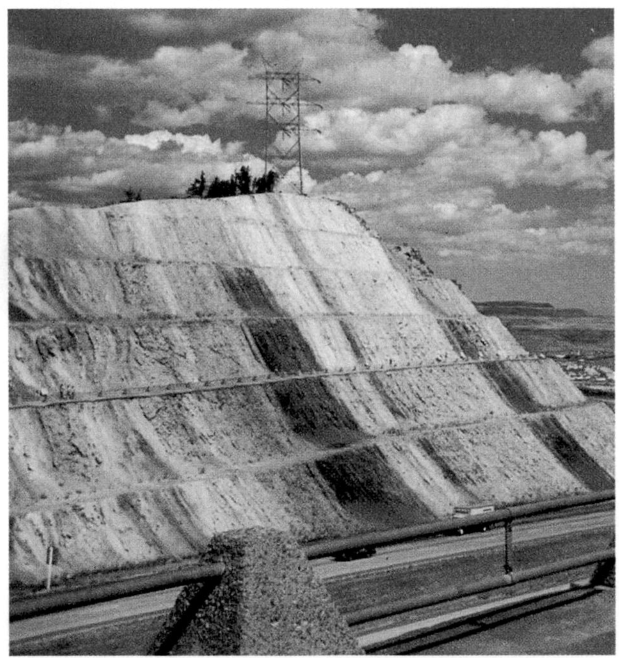

FIGURE 11-33 The north side of the impressive Interstate 70 road cut west of Denver, Colorado. The rocks visible in this view are part of the Dakota Group. They are dipping steeply toward the east (right). Sandstones of the Dakota Group serve as an important source of artesian groundwater beneath the Great Plains. They have also yielded oil and gas in eastern Colorado, Nebraska, and Wyoming.

Cretaceous rocks of Wyoming and Colorado also include extensive beds of a soft, plastic, light-colored clayey rock called bentonite. **Bentonite** is composed of clay minerals that formed by the alteration of volcanic ash. Volcanoes in the Idaho region during the Late Jurassic and Cretaceous explosively ejected tremendous volumes of ash that were carried into adjacent states by westerly winds. The resulting ash beds, subsequently converted to bentonites, represent single geologic events of short duration and can be traced for great distances, even across changing facies. Hence, bentonite beds are of great value in chronostratigraphic correlation. Volcanic ash also provided a source for the silica in Cretaceous siliceous shales and bedded cherts.

Cretaceous carbonate formations include soft, clayey limestones and chalky shales of the **Niobrara Formation** (Fig. 11–35). The Niobrara has yielded the remains of a variety of marine creatures, including enormous numbers of oysters, a large Cretaceous diving bird (Fig. 11–36), marine reptiles, and the large flying reptile *Pteranodon*. Toward the end of the Cretaceous, the seas that supported these creatures began a slow withdrawal. This regression of the Cretaceous epeiric sea from central North America was contemporaneous with the Laramide deformation that produced the Rocky Mountains. As the seas withdrew, coal-bearing deltaic and other continental sediments were spread across the old sea floor.

FIGURE 11-34 Prominent north–south trending hogbacks formed by the eastward-dipping Dakota Sandstone of Cretaceous age, a few miles south of Golden, Colorado. The Dakota Sandstone at this location is a beach deposit of the Cretaceous sea. Well-developed ripple marks and occasional dinosaur footprints are preserved in the sandstone. (*Courtesy of U.S. Geological Survey.*)

EURASIA AND THE TETHYS SEAWAY

Extending eastward from Gibraltar and across southern Europe and Asia to the Pacific is the great Alpine–Himalayan mountain belt. Most of the rocks composing the ranges within the belt were laid down in an important depositional trough known as the **Tethys seaway**. Long after their deposition, these rocks were deformed into mountain ranges as northward-moving segments of Gondwanaland collided with Eurasia.

The geologic history of the Tethys seaway can be traced back to the Paleozoic. But unlike the Appalachians, for which most deposition and tectonic activity occurred during the Paleozoic, the Tethys experienced its most eventful geologic history in the Mesozoic and Cenozoic.

During the Triassic Period, limestone deposition predominated within the Tethys (Fig. 11–37). Within the lime-depositing sea, bivalves, cephalopods, crinoids, reef corals, and calcareous algae thrived and left a rich fossil record. Of particular importance for

FIGURE 11-35 An exposure of Cretacous Niobrara Formation in Gove County, Kansas. (*Copyright Larry Miller/Photo Researchers, Inc.*)

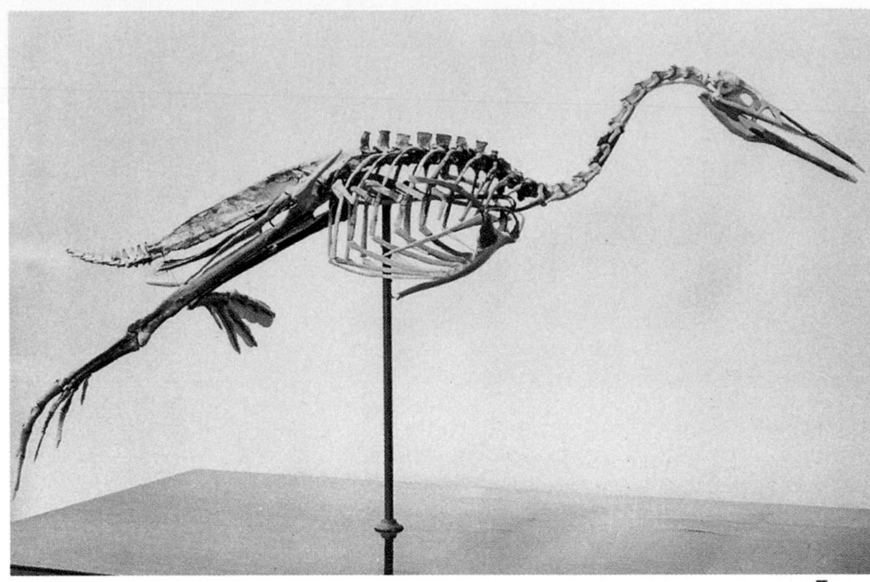

FIGURE 11-36 Skeleton of the Cretaceous diving bird *Hesperornis*. This bird was over a meter tall when standing. Propulsion was provided by its large, webbed feet. A notable primitive trait of this bird was the retention of teeth. (*Courtesy of the U. S. National Museum of Natural History, Smithsonian Institution.*)

biostratigraphic correlation of Triassic strata are the abundant ammonoid cephalopods (see Chapter 12).

North of the Tethys, the development of a highland tract known as the Vindelician arch separated marine sedimentation to the south from very different sedimentation in north-central Europe. In that northern region, the Triassic record begins with red and brown nonmarine clastic rocks that resemble those deposited under arid conditions during the preceding Permian Period. These continental deposits are overlain by shoreline sandstones, marls, evaporites, and limestones deposited during a temporary marine invasion of central Europe. The sea did not remain for long. Nor was it able to reach as far north as Great Britain, where a nonmarine sequence known as the New Red Sandstone was being deposited. As the sea withdrew, continental sedimentation much like that of the earliest Triassic resumed.

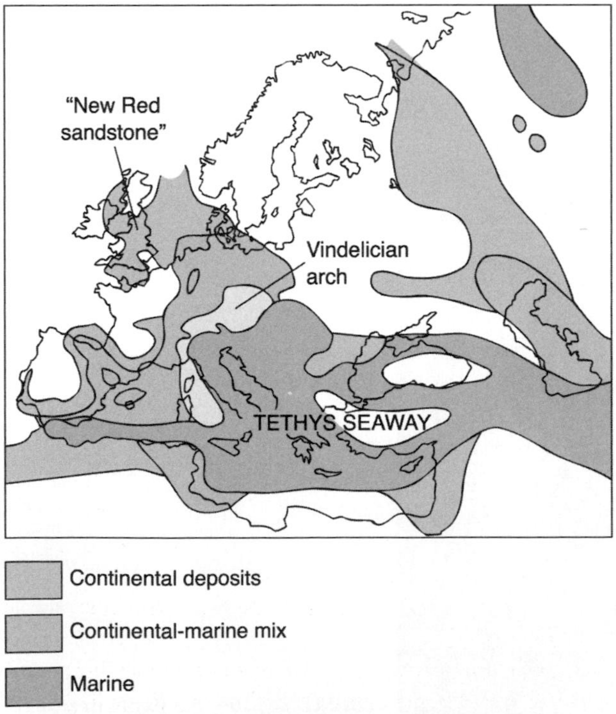

Marine conditions were far more widespread during the Jurassic than they had been during the Triassic. Shallow seas advanced from both the Tethys and the Atlantic and spread across Europe, leaving a rich sedimentary record in the basins that lay between the old Hercynian uplands. Although the predominant Jurassic sediment is limestone, there is often a gradation to fine clastics adjacent to the highlands. Eventually, marine conditions extended from the Tethys across Russia and into the Arctic Ocean. The invasion was short-lived, however, and shallow seas had drained from the continent by the end of the Jurassic. Marine conditions persisted in the Tethys, and a complicated pattern of facies changes developed in response to block-faulting on the sea floor.

The Cretaceous is the most eventful geologic period of European Mesozoic history. During this time, the African plate was moving northward, narrowing the basin of the Tethys seaway. Along the southern margin of Europe, formerly quiet depositional tracts of the

"New Red sandstone"
Vindelician arch
TETHYS SEAWAY
Continental deposits
Continental-marine mix
Marine

FIGURE 11-37 Paleogeographic map of the Triassic in Europe. (*Modified from Brinkmann, R. 1969.* Geologic Evolution of Europe. *New York: Hafner Publishing Co.*)

ENRICHMENT

Cretaceous Epeiric Seas Linked to Sea-Floor Spreading

During the Cretaceous, continents were extensively inundated by inland seas. Marine sedimentary rocks in North America, Europe, Australia, Africa, and South America indicate that about a third of the land area now emergent was under water. Because all the continents experienced marine transgressions at about the same time, the cause must have been a eustatic rise in sea level and not the subsidence of the land. However, there is little evidence of glaciation during the time that submergence was at its peak. It is therefore unlikely that meltwater from ice caps or continental glaciers was the cause of the eustatic rise in sea level. An alternative would be for crustal materials to take up space in the ocean and thereby cause a rise in sea level. This might occur if there were an increase in the rate at which oceanic crust is formed at midocean ridges.

Geologists have been able to estimate the volume of new ocean crust produced during 10-million-year intervals as far back as the beginning of the late Cretaceous. These studies indicate that the rate at which ocean crust was being produced along midocean ridges was exceptionally rapid. Newly formed ocean crust is hotter than older crust and therefore occupies more volume. This is evident on the present ocean floor, where the surface of newly formed crust stands 2.8 kilometers below sea level, as contrasted with an average of 5 kilometers for older, cooler oceanic crust. Also, when spreading rates are high (producing high-volume crust), the corresponding subduction of cooler, low-volume crust is also greater. Taken together, these sea-floor spreading processes decrease the space for water in the ocean basins, causing sea level to rise and resulting in marine transgressions across formerly emergent regions of the continents. In North America, the rise in sea level was responsible for epeiric seas that advanced southward from Alberta, Canada, and joined an embayment that spread northward from the Gulf of Mexico. At its peak, that seaway was over 1600 kilometers wide and divided our continent into two separate landmasses.

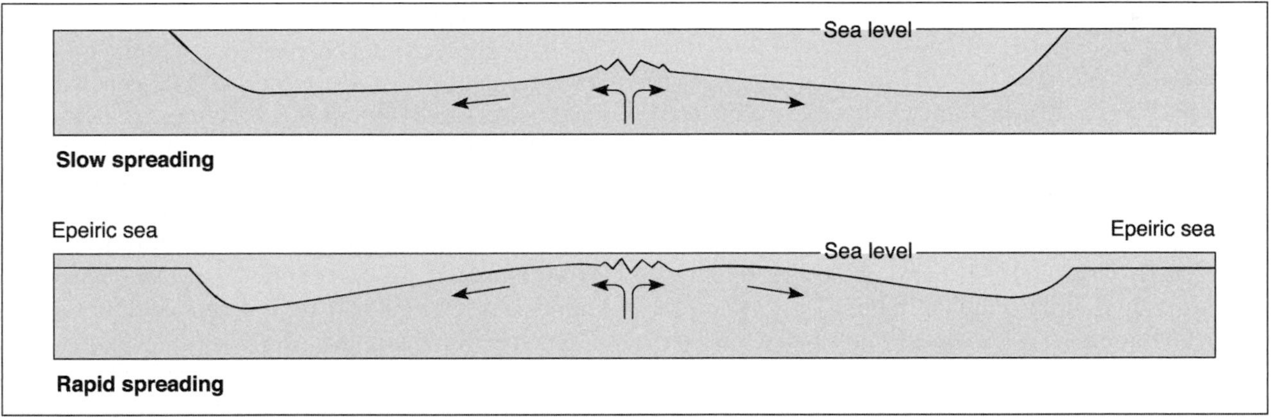

Diagram indicating how rapid sea-floor spreading can cause displacement of sea water onto continental margins.

Tethys were subjected to powerful compressional forces. The record of this event is seen in complex overturned folds and thrust faults and in a zone of ophiolites that trend east-west across southern Europe. As described in Chapter 5, ophiolites represent deep-sea sediments and rocks of oceanic crust that have been strongly compressed, metamorphosed, and intruded as they are carried into subduction zones. Subduction of Tethyan oceanic crust beneath Europe as the African plate approached created the Alpine ophiolite zones as well as the deformation and volcanic activity that characterized the southern margin of Europe during the Cretaceous.

East of the Alpine region, the Tethys was being squeezed out of existence as the Indian plate moved toward Asia. The first major collision of the two continents occurred after the Cretaceous, about 50 million years ago. Within another 5 million years, the Tethys tract was totally subducted. During the closure, at least two microcontinents were incorporated into the Himalayan Mountains. This great mountain system continues to be tectonically active today, with evidence of northward movement and episodes of uplift.

The Late Cretaceous is noteworthy not only for its mountain-building events but also as a time of extensive marine transgressions. More of Europe was inundated during the Late Cretaceous than at any other time since the Cambrian (Fig. 11–38). A great embayment from the Tethys worked its way northward until it ultimately joined with a similar southward encroachment

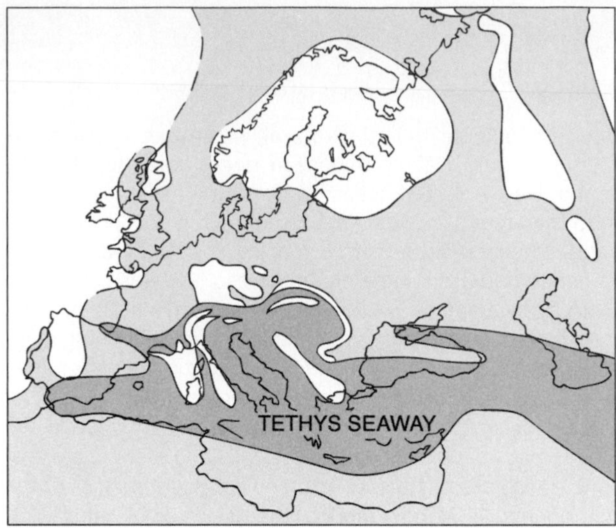

FIGURE 11-38 Areas in western Europe covered by seas during the Cretaceous. (*Simplified from Brinkmann, R. 1969.* Geologic Evolution of Europe. *New York: Hafner Publishing Co.*)

from the North Atlantic. One result of the linking of the two seaways was a rapid interchange of marine organisms from the two formerly separated regions. In the Tethys–Alpine region, block-faulting that had begun during Late Jurassic time continued until about the Middle Cretaceous At that time further compression from the south resulted in folding of the thick sequence of marine Tethys sedimentary rocks. The axis of these folds parallels the east-west trend of the Tethys. Most of the great anticlines and synclines remained submerged, but the tops of some of the folds rose above the level of the waves, forming elongate islands along the north side of the narrowing Tethys. Erosion of the anticlinal islands produced clastic sediments that were periodically swept downward by turbidity currents into the adjoining basins.

An event that determined the geographic configuration of the northern coastlines of Spain and France occurred in Europe between Late Jurassic and Late Cretaceous time. As indicated by paleomagnetic data, there was no Bay of Biscay separating western France from Spain prior to the Late Jurassic. The Bay of Biscay opened as the Iberian Peninsula (Spain and Portugal) rotated about 35° from its former line of contact with France. The upheaval of the Pyrenees was another consequence of this rotation. The dating of the rotational event is confirmed by faults of Late Jurassic age on the northern border of the Bay of Biscay and by the fact that the oldest rocks in the bay are Late Cretaceous in age.

East of the Alpine region, the Tethys extended as a great loop past Burma and southeastward into Indonesia. Triassic and Jurassic deposits of the eastern Tethys consist of shales, fossiliferous limestones, and dolomites. During the Cretaceous, volcanic activity occurred throughout what is now the Himalayan region. Some areas seem to have subsided, and terrigenous rocks resembling those moved by submarine landslides or turbidity currents merged with marine volcanic tuffs, breccias, and radiolarian cherts.

THE GONDWANA CONTINENTS

Africa

Near the beginning of the Mesozoic, separations had already developed between the eastern coast of America and northwestern Africa. However, as yet no rift existed between Africa and South America (see Fig. 11–1). Thus, there was no South Atlantic Ocean. Most of Africa was relatively stable throughout the Mesozoic, with only relatively minor transgressions along its northern and eastern borders.

Although the Mesozoic rock record in the more interior parts of Africa is not exceptional, some regions yield particularly interesting geologic information. One such region is the Karoo basin at the southern end of Africa (Fig. 11–39). The Karoo was a continental basin that was formed late in the Carboniferous and received swamp, lake, and river deposits until late in the Triassic Period. Rocks of the Karoo sequence are known to paleontologists around the world for the fossil therapsids they contain.

As the Triassic drew to a close, terrestrial siltstones and sandstones of the Karoo sequence were covered by layer after layer of low-viscosity lava, which flowed onto the surface from fissures and, less commonly, from volcanoes to the southeast. The old Karoo landscape was buried beneath more than 1000 meters of basalt. Geologists speculate that these great outpourings of lava were associated with the pulling away of the Gondwana segments that once adjoined South Africa. Such fragmentation would have caused severe fracturing of the continental crust, thereby providing multiple avenues for upwelling molten rock. The extrusions continued well into the Jurassic Period. Contemporaneous lava floods and volcanism also occurred on the separating landmasses of South America, Australia, and, somewhat later, Antarctica.

Africa experienced little tectonic activity throughout the remainder of the Jurassic and Cretaceous. Marginal seas extended along the eastern edges of the continent. Periodic advance and retreat of these seas brought an alternation of marine and nonmarine beds.

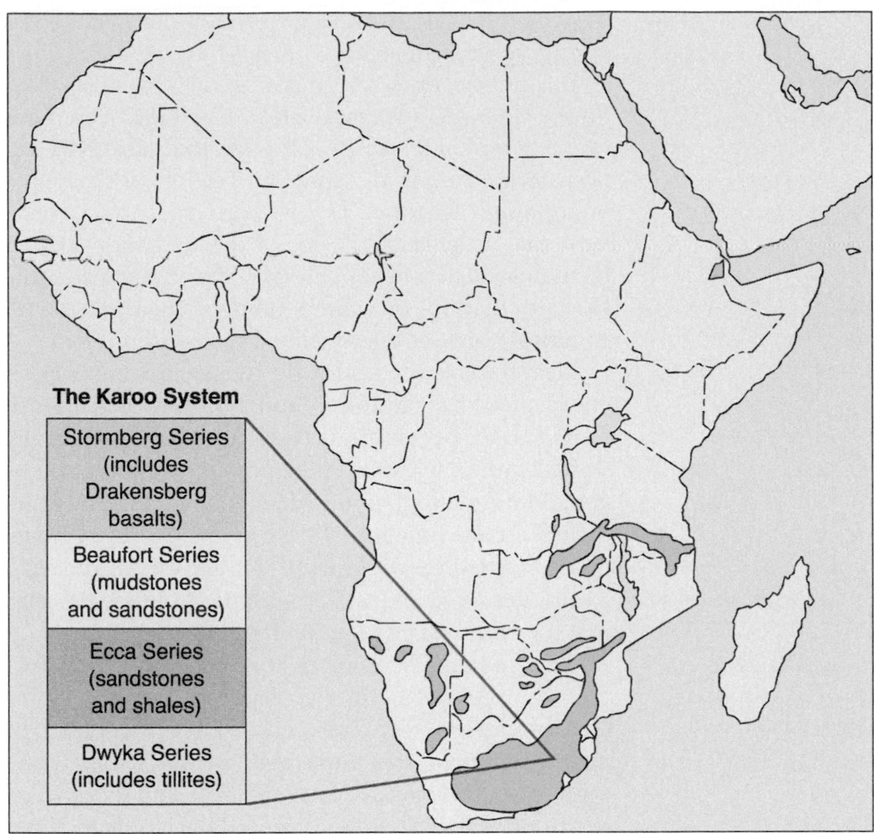

FIGURE 11-39 Outcrops of the Karoo System (green) of South Africa and the four principal units of the Karoo section.

Australia and New Zealand

Australia was still attached to Antarctica during the Triassic, and both continents were predominantly emergent. Terrestrial sandstones and shales were deposited. Except for brief marginal embayments, deposition of lake and stream sediments continued to dominate until the middle of the Early Cretaceous. At that time, a marine transgression brought sandstones and chalk beds into the interior.

New Zealand during the Mesozoic was geologically more restless than Australia. From Mesozoic time to the present, this large island has been positioned along the boundary between the southwestern part of the Pacific plate and the eastern edge of the Indian plate. The Mesozoic rock sequence includes turbidites, siliceous rocks, and abundant volcanics. Near the end of the Jurassic, a series of orogenic events occurred as a result of the collision and subduction of the Pacific plate beneath the island. In the course of the orogeny, mountains were erected, and deep, unstable trenches developed. The activity continued into the Cenozoic. It continues intermittently, even to the present day.

India

During the Mesozoic, India moved steadily northward on its remarkable voyage to Laurasia. The Indian continent itself remained tectonically stable, even though it was moving at a rapid rate toward its ultimate collision with Laurasia in Cenozoic time. The landmass experienced only relatively minor marginal incursions of the sea. Across the interior, erosion of upland areas spread terrigenous clastic sediments into the lowlands and plains. Dinosaur bones and superb fossils of plants are dug from these sandstones and shales. By Cretaceous time, India was nearing the Tethyan region. Here and there, the floor of the Tethyan trough was buckled into elongate ridges that were indications of the coming tectonic collision. While these ridges were developing, the northwestern half of India was flooded with immense quantities of low-viscosity basaltic lava (Fig. 11–40). These now solidified lavas are flood basalts of the Deccan Traps. Radiometric dates obtained from the basalts indicate that the outpourings continued from Cretaceous time well into the Cenozoic. It is likely that these basalt floods record the passage of India across a fixed hot spot in the mantle.

The Deccan Traps cover about 500,000 square kilometers of western and central India with an aggregate volume in excess of 1,000,000 cubic kilometers. As such, they comprise the greatest volume of continental basalt on the Earth's surface. Erosion has removed much of the Deccan volcanics, and their original extent and volume were once far greater. The extensive fiery

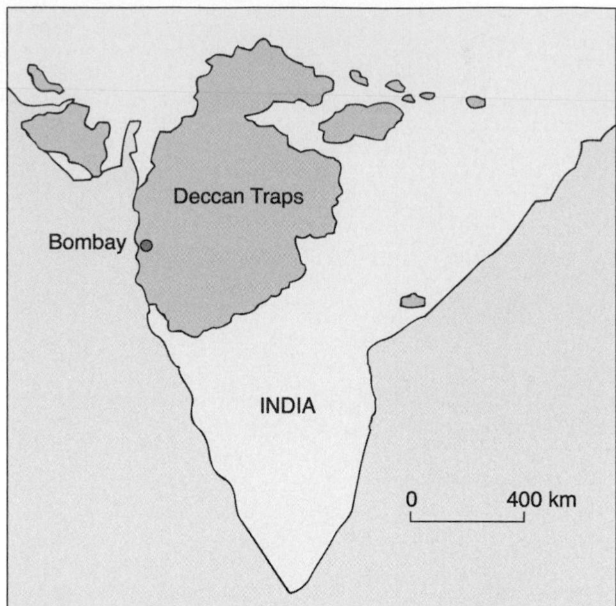

FIGURE 11-40 Present-day distribution of Deccan Trap basalts. Much of the original cover of lava has been lost to erosion, as indicated by the isolated remnants to the east of the large outcrop area.

extrusions and huge volumes of carbon dioxide released along the web of fissures may have triggered climatic warming and chemical changes deleterious to life both on land and in the oceans. As will be noted in the next chapter, Deccan volcanism, accompanied by the eruption of huge volumes of lava in Brazil and on the ocean floor, may have contributed to the terminal Cretaceous episode of mass extinction.

South America

Much like Africa, South America stood well above sea level at the beginning of the Mesozoic. Along the western margin of the continent, two parallel zones of deposition were already established in Triassic time. The more western tract was characterized by turbidites, conglomerates, and siliceous sediments laid down in the deeper waters of the continental rise and slope. Eastward of these deep-water facies, carbonates and shales were deposited across shelf zones. The approximate boundary between the two suites of rocks can now be traced along the central ranges of the Andes. In addition to these marine tracts, broad basins in the interior accumulated continental deposits. Particularly remarkable are the eolian and fluvial sandstones and siltstones that spread across the southeastern region of the continent during the Triassic. The beds contain a rich fauna of Triassic vertebrates. Lava flows of the same age as the Karoo rocks of Africa are also prevalent in the upper part of this stratigraphic sequence.

By Jurassic time, the initial narrow split between South America and Africa had widened into a configuration resembling the present-day Red Sea. New sea floor formed along the spreading center of the developing South Atlantic. The western side of the continent was at the leading edge of the westward-moving tectonic plate. It was opposed by the Pacific plate, the forward margin of which plunged beneath South America as an eastward-inclined subduction zone.

Near the end of the Jurassic, the Andean belt experienced deformation and volcanism. This activity appears to have begun in the south and then sporadically shifted with time toward the north. The climax of orogenic deformation occurred in Late Cretaceous and early Tertiary time and was accompanied by regional uplift that led to widespread withdrawal of the seas. Evaporites were deposited in scattered isolated basins that for a time retained marine waters.

The frequency and intensity of deformational and volcanic activity along the Andean mobile belt increased during the Cretaceous. In the subduction zone created by the movement of the Pacific plate against South America, oceanic crust and its cover of sea-floor sediment were carried down to deep zones and melted. The magmas in turn worked their way back toward the surface as great intrusions and outpourings of andesite. Deformation and igneous activity have continued even to the present and formed many of the geologic structures now seen in the towering peaks of the Andes.

Antarctica

Antarctica was emergent during the Mesozoic. Eastern areas of the now frigid landmass were sites of continental deposition. Beds of volcanic ash and lava flows occur between layers of lake and stream deposits. In contrast, the rocks of western Antarctica are similar to those of equivalent age in the Andes. The strata of the Antarctic Peninsula most likely represent part of an orogenic belt that was continuous with the Andes to the north.

ECONOMIC RESOURCES

Uranium Ores

Rocks of the Mesozoic System, like those of earlier eras, contain a wealth of important mineral resources. Notable among such resources are the nuclear fuels. In the United States, uranium ores are derived chiefly from continental Triassic and Jurassic rocks of New Mexico, Colorado, Utah, Wyoming, and Texas. The chief ore mineral is carnotite (Fig. 11–41), a yellow, earthy oxide of uranium that usually occurs within

FIGURE 11-41 The bright yellow uranium mineral carnotite, disseminated through and encrusting Shinarump sandstone. Specimen is 9 centimeters long.

FIGURE 11-42 The dark layers in this road cut are coal beds in the Mesaverde Formation, Upper Cretaceous, near Castlegate, Utah.

the pore spaces of sandstones, as encrustations, or as replacements of fossil wood. In some instances, petrified logs have provided amazing concentrations of not only uranium but also vanadium and radium. These are striking exceptions, however, and most of the ores are of very low concentration.

The uranium in Triassic and Jurassic beds of the western United States was originally dissolved from consolidated volcanic ash (tuff) and transported in groundwater. Precipitation of the ore occurred when the uranium-bearing solutions encountered rocks rich in organic matter in which chemically reducing (rather than oxidizing) conditions existed.

Fossil Fuels

Coal, petroleum, and natural gas are derived from the partially decomposed remains of ancient organisms included within sedimentary rock. For this reason, they are called **fossil fuels**. Fossil fuels are the primary sources of energy for our industrial economy. Eighty-seven percent of the energy consumed in the United States is derived from the fossil fuels. As we have seen, the Carboniferous contains tremendous reserves of coal, but rocks of Mesozoic age also have significant amounts of coal, as well as oil and natural gas. The Jurassic System, for example, contains workable deposits of coal in Siberia, China, Australia, Tasmania, Spitzberger, and North America. In North America, thick seams of Jurassic coal occur in British Columbia and Alberta. Cretaceous coal underlies more than 300,000 square kilometers of the Rocky Mountain region (Fig. 11–42). Much of this coal is now being mined, particularly because of its relatively low content of environmentally offensive sulfur.

In addition to coal, Mesozoic rocks supply large quantities of oil and gas to energy-hungry industrial nations. The oil provinces of the Middle East and North Africa contain more oil than the combined reserves of all other countries. Middle East petroleum comes primarily from thick sections of Jurassic and Cretaceous sedimentary rocks that accumulated in shelf and reef environments bordering the Tethys seaway. Other areas of petroleum production from Mesozoic rocks occur in the western United States (Fig. 11–43), Alaska, Arctic Canada, the Gulf coastal states, western Venezuela, and Southeast Asia and beneath the North Sea and the eastern offshore area of

FIGURE 11-43 Oil being pumped from Jurassic and Cretaceous formations in the Bighorn basin of Wyoming. (*Photograph courtesy of L. E. Davis.*)

ENRICHMENT

Chunneling in the Cretaceous: The Channel Tunnel

For nearly 3 centuries, visionaries had dreamed of constructing a tunnel between England and France. That dream became reality in October 1990, when English and French tunnelers who had been working toward one another met deep beneath the floor of the English Channel. They had excavated and installed the lining of the first of a three-part tunnel system that was completed by the summer of 1993. The Channel Tunnel is the longest undersea tunnel on Earth. It has two "running tunnels" flanking a central "service tunnel." High-speed trains carry passengers through the running tunnels from Folkestone, England (southwest of Dover), to France in 30 minutes.

About 85 percent of the "chunnel," as it has been dubbed, was excavated in a Lower Cretaceous formation known as the Chalk Marl. The term marl indicates that the chalk contains clay, and the mixture of chalk and clay results in a rock that is relatively strong, stable, and impermeable. It is an excellent natural material in which to excavate a tunnel. In addition, throughout most of the tunnel route, the Chalk Marl is over 25 meters thick and thus easily accommodates the 8-meter diameter of the running tunnels. The Chalk Marl is overlain by the Upper Cretaceous Gray and White Chalk formations. Beneath the Chalk Marl is the Gault Clay, a unit unsuitable for tunneling because it is plastic and inherently weak.

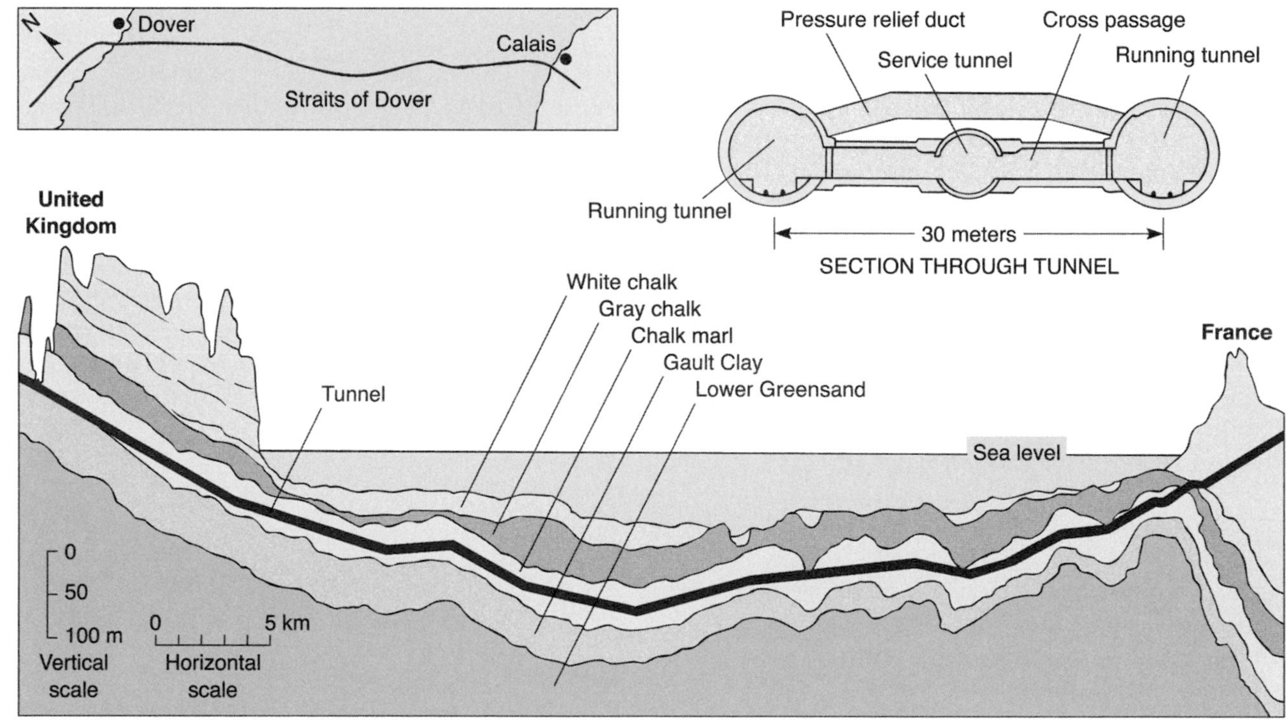

Cross-section along the route of the Channel Tunnel and cross-section of the tunnel. The cross-section showing the channel route has large vertical exaggeration. (*Adapted from* Tunnelling '88, *Institute of Mining & Metallurgy.*)

Australia. Notwithstanding the size of some recent discoveries in remote parts of the world, it appears likely that the present rates of oil and gas consumption will cause exhaustion of the world's resources in less than a century. Therefore, oil and gas must be replaced by other energy sources in the future, especially if we are to conserve these valuable materials for the chemical industry. The time may come when petroleum will be too valuable to burn.

Metallic Mineral Deposits

Metalliferous deposits were formed widely throughout the active orogenic belts of the Mesozoic world. A variety of metals now found in the Rocky Mountains, the Pacific coastal states, and British Columbia were emplaced during batholithic intrusions that accompanied Jurassic and Cretaceous orogenies. The California gold rush of 1849 was a consequence of the discovery

FIGURE 11-44 The Morenci open-pit copper mine in Arizona. The pit occupies the former location of the town of Morenci, Arizona. The Metcalf copper mine is in the background. Copper minerals at this location are disseminated throughout porphyritic igneous rock. (*Courtesy of E. Melchiorre.*)

of gold-bearing gravels eroded from gold-bearing quartz veins emplaced during the final stages of the Sierra Nevada batholithic intrusion.

The copper, silver, and zinc veins of the Butte, Montana, and Coeur d'Alene, Idaho, mining districts were emplaced as a result of Cretaceous igneous activity. One belt of copper-bearing rocks extends from Denver into southern Arizona. These deposits and similar ones in Utah and Nevada occur in igneous rock that has a porphyritic texture (large crystals of feldspar and quartz in a finer matrix) and hence are called *porphyry copper deposits* (Fig. 11–44). The copper minerals, principally native copper and copper-iron sulfides, are disseminated through rock that has been pervasively fractured during episodes of explosive volcanic activity. Fractures produced by the explosions provided passages for hot, mineralizing solutions from which the ore was precipitated. Porphyry copper deposits occur not only in western North America but also in South America, the Philippines, Iran, and eastern Europe.

Important nonmetallic deposits of the Mesozoic include sulfur and salt. Both are produced from the salt domes mentioned earlier (see Fig. 11–9). Diamonds are also obtained from Mesozoic igneous rocks. Siberia's diamond-bearing intrusions are believed to have penetrated the upper crust during the Triassic and Jurassic. Similar diamond pipes in Africa are probably of Cretaceous age.

SUMMARY

Within the approximate 160 million years of the Mesozoic Era, the pattern of lands and seas was extensively altered as large segments of Pangea were separated. The break-up appears to have begun during the Triassic, with the splitting off of North America from Gondwana. By Late Jurassic time, rifts had developed between all of the Gondwana segments except Australia and Antarctica. These two continents, like northwestern North America, Greenland, and northwestern Europe, retained a hold on one another until early in the Cenozoic.

At the beginning of the Mesozoic, most of North America was emergent. In the east, crustal tension was expressed by the development of large-scale block-faulting and volcanism. In downfaulted valleys, arkosic sandstones, conglomerates, and lacustrine shales accumulated. While this was occurring, the ancestral Gulf of Mexico began to take form. During the Jurassic and Cretaceous, thick sequences of limestones, evaporites, sandstones, and calcareous shales accumulated along what are now the Gulf and Atlantic coastal states.

Continental deposits, particularly red beds, characterize the Triassic of the Rocky Mountain regions of North America. Marine conditions persisted along the Pacific margin, where volcanic rocks and graywackes accumulated to great thicknesses. During the Jurassic, this western belt was strongly deformed as North America began to override the subduction zone at the leading edge of the eastward-moving Pacific plate. The advancing Pacific plate carried large fragments of volcanic arcs, oceanic plateaus, and microcontinents to the western margin of North America. These crustal fragments became incorporated into the Cordillera as a tectonic collage of displaced terranes. Growth of a continental margin in this way is referred to as accretionary tectonics.

In general, deformation of the North American Cordillera during the Mesozoic began along the Pacific coast and moved progressively eastward, until by Late Cretaceous time, inland seas that had occupied the western interior of North America were displaced. Majestic mountain ranges stood in their place. Crustal unrest continued to affect the western states even during the early epochs of the Cenozoic Era.

The focus of Eurasian Mesozoic events was the great east–west trending Tethys mobile belt. The initial Mesozoic deposits of the western Tethys were predominantly carbonates. Block-faulting developed along parts of the belt during the Jurassic. This middle part of the Mesozoic witnessed marine invasions that completely spanned the continent from the Tethys to the Arctic. During the Cretaceous, the Tethys was affected by powerful compressional forces as the northward-moving oceanic plate bearing peninsular India moved toward and under the Eurasian plate. One result of these movements was the development of a series of anticlinal welts along the northern side of the Tethyan seaway. Rapid erosion of these elongated, uplifted tracts produced thick sequences of coarse clastics in the intervening troughs. Along the southern margins of the Tethys, conditions were more stable, and relatively undisturbed sequences of carbonates and sandstones were deposited.

The Mesozoic history of the interior regions of Africa is read primarily from continental deposits. Among these, none is more famous than the beds of the Karoo Basin of South Africa. The Karoo sequence includes both Permian and Triassic strata, some of which have yielded splendid collections of therapsids that once inhabited southern Africa.

During the Mesozoic, a largely emergent peninsular India was progressing steadily toward its Eurasian destination. By Cretaceous time, India had approached sufficiently to cause initial crustal buckling in the eastern Tethys.

The Jurassic was an important geologic period for South America, which during this time began to separate from Africa. The separation continued during the remainder of the Mesozoic, and a new ocean, the South Atlantic, was born. The western margin of the continent formed the leading edge of the South American plate and was characterized by volcanic activity, folding, and intrusions.

The rocks of the Mesozoic have supplied the world with a variety of important economic resources. These include such fossil fuels as coal, petroleum, and natural gas as well as nuclear fuels. Orogenic activity during the Mesozoic resulted in the emplacement of such critical metals as copper, zinc, chromite, gold, silver, lead, and mercury.

QUESTIONS FOR REVIEW AND DISCUSSION

1. What geologic event precedes the separation of a continent into two or more parts? Cite examples from Mesozoic geologic history.

2. Describe the general structural and environmental conditions associated with the deposition of the Newark Group. Why did normal faulting predominate rather than reverse faulting? Did older structures influence the location of Triassic faults? Explain.

3. Discuss the characteristics of structures produced by the Sevier orogeny.

4. What environmental conditions account for the presence of Jurassic evaporites in the Gulf Coast region? How are these evaporites related to petroleum traps in overlying Cretaceous and Cenozoic strata?

5. Explain what is meant by the term accretionary tectonics. Provide examples of the occurrence of accretionary tectonics from the Mesozoic history of North America. How are allochthonous terranes related to accretionary tectonics?

6. Discuss the possible effects of Deccan and extensive ocean floor volcanism on global climate near the end of the Mesozoic Era.

7. Describe and contrast obduction with subduction and cite examples of the occurrence of each.

8. What is chalk? In what Mesozoic geologic period was it particularly abundant? How is the origin of chalk related to marine plankton?

9. What is the total duration in years of the Mesozoic Era? Which period of the Mesozoic was the longest?

10. What was the probable source area for the sediments of the Morrison Formation of the Rocky Mountain region? What evidence indicates that the Morrison Formation was a continental rather than a marine formation?

11. At what time during the Mesozoic were epicontinental (epeiric) seas most extensive? During which period were such marine incursions most limited?

12. What was the Tethys seaway? What water bodies today are remnants of this seaway? Where might one go to examine the rocks deposited within the Tethys?

READINGS

Ager, D. V. 1980. *The Geology of Europe*. New York: John Wiley.

Bally, A.W., and Palmer, A.R. (eds.) 1989. *Geology of North America: An Overview*. Decade of North American Geology (DNAG), vol. A. Boulder, Colorado: Geological Society of America.

Greenwood, P. H. (ed.). 1981. *The Evolving Earth*. Cambridge, England: Cambridge Press.

Jones, D. L., Cox, A., Coney, P., and Beck, M. 1982. The growth of western North America. *Sci. Am.* 247(5):70–128.

Kearey, P., and Vine, F. J. 1996. *Global Tectonics*, 2nd ed. Oxford, England: Blackwell Science Ltd.

Saleeby, J. B. 1983. Accretionary tectonics of the North American Cordillera. *Ann. Rev. Earth Planet Sci.* 15:45–73.

Salvadore, A. 1991. *Geology of North America*, pp. 1–12. Boulder, CO: Geological Society of America.

WEB SITES

The Earth Through Time Student Companion Web Site (www.wiley.com/college/levin) has online resources to help you expand your understanding of the topics in this chapter. Visit the Web Site to access the following:

1. Illustrated course notes covering key concepts in each chapter;
2. Online quizzes that provide immediate feedback;
3. Links to chapter-specific topics on the web;
4. Science news updates relating to recent developments in Historical Geology;
5. Web inquiry activities for further exploration;
6. A glossary of terms;
7. A Student Union with links to topics such as study skills, writing and grammar, and citing electronic information.

12

The Cretaceous meat-eater Tyrannosaurus, *shown charging its prey. Tyrannosaurus was about 12 meters in length, could run at an estimated 40 km per hour, and used its 15-cm-long teeth to bring down and eat its prey (Copyright J. Sibbick.)*

CHAPTER 12

Life of the Mesozoic

Tyrannosaurs, enormous bipedal caricatures of men, would stalk mindlessly across the sites of future cities and go their way down into the dark of geologic time.
Loren Eiseley, The Immense Journey, *1959*

OUTLINE

In this chapter, we direct our attention to the Earth's inhabitants during the Mesozoic Era. Among those inhabitants, none have captured the interest of the public more than the dinosaurs. Yet with all our fascination for dinosaurs, we should not forget that it was during this era that birds and mammals also began their evolution. Many new families of both marine invertebrates and marine reptiles and fishes made their debut. Planktonic foraminifera appear in the Cretaceous and expand dramatically. Their calcareous skeletons (called tests) along with the calcium carbonate plates of coccolithophores (coccoliths) accumulated by the billions in limy sediments of the ocean. In the world of plants, angiosperms arrive on the scene at the boundary between the Jurassic and Cretaceous. With their arrival, modern trees bearing seeds, nuts, and fruits expanded across the continents. As today, the distribution, evolution, and abundance of all these many forms of life were dependent on climate. Climate, in turn, was affected by the changing locations of continents, major marine transgressions and regressions, and the formation of great mountain systems.

MESOZOIC CLIMATES

Climate is influenced by a large number of factors, but the primary global control of climate is the balance between incoming and outgoing radiation from the sun. That balance is affected by a variety of factors, including the configuration and dimensions of the oceans and continents; the development and location of mountain systems and land bridges between continents; changes in snow, cloud, or vegetative cover (which affect the amount of solar radiation reflected back into space from the Earth); the carbon dioxide content of the atmosphere (which can trigger greenhouse warming); the location of the poles; the amount of radiation-blocking aerosols thrown into the atmosphere by volcanoes; and such astronomic factors as changes in the Earth's axis and orbit. Cool climates seem to have characterized many continental areas during the final days of the Paleozoic Era. The vastness of the Pangea supercontinent, the upheaval of mountains, overall increases in elevation, and withdrawal of inland seas in many regions were contributing causes of the generally cooler conditions. Gradually, however, the climate warmed. Glaciers in Africa, Australia, Argentina, and India began to melt as these continents drifted away from the

South Pole. During most of the Jurassic and Cretaceous, climates for most continental regions were warm and equable. Also during these periods, extensive regions were covered by epeiric seas, and this contributed to warmer conditions. Water bodies retain heat far better than do land areas. Furthermore, ocean waters are in constant circulation, distributing that warmth around the globe. Hence, when the proportion of ocean to land increases, warmer climates may result. The reverse holds true as well, and this was probably an important factor causing cooler climates at the end of the Paleozoic.

During the Triassic, the continents were still tightly clustered. The paleoequator extended from central Mexico across the northern bulge of Africa (Fig. 12–1). As noted in Chapter 11, the Triassic was a time of general emergence of the continents. Mountains, thrust upward at the end of the Paleozoic, inhibited the flow of moist air into the more centrally located regions, causing widespread aridity. Evaporites, dune sandstones, and red beds accumulated at both high and low latitudes and attest to relatively dry and warm conditions.

Reconstructions based in part on paleomagnetic studies suggest that during the Jurassic, the continents were at the approximate latitudinal positions they occupy today. Marine waters extended northward into a great trough formed by the opening of the Atlantic Ocean, and in many places shallow inland seas spilled out of the deep basins onto the continents. An arm of the Protopacific extended westward as the great Tethys seaway. It is not unreasonable to assume that warm,

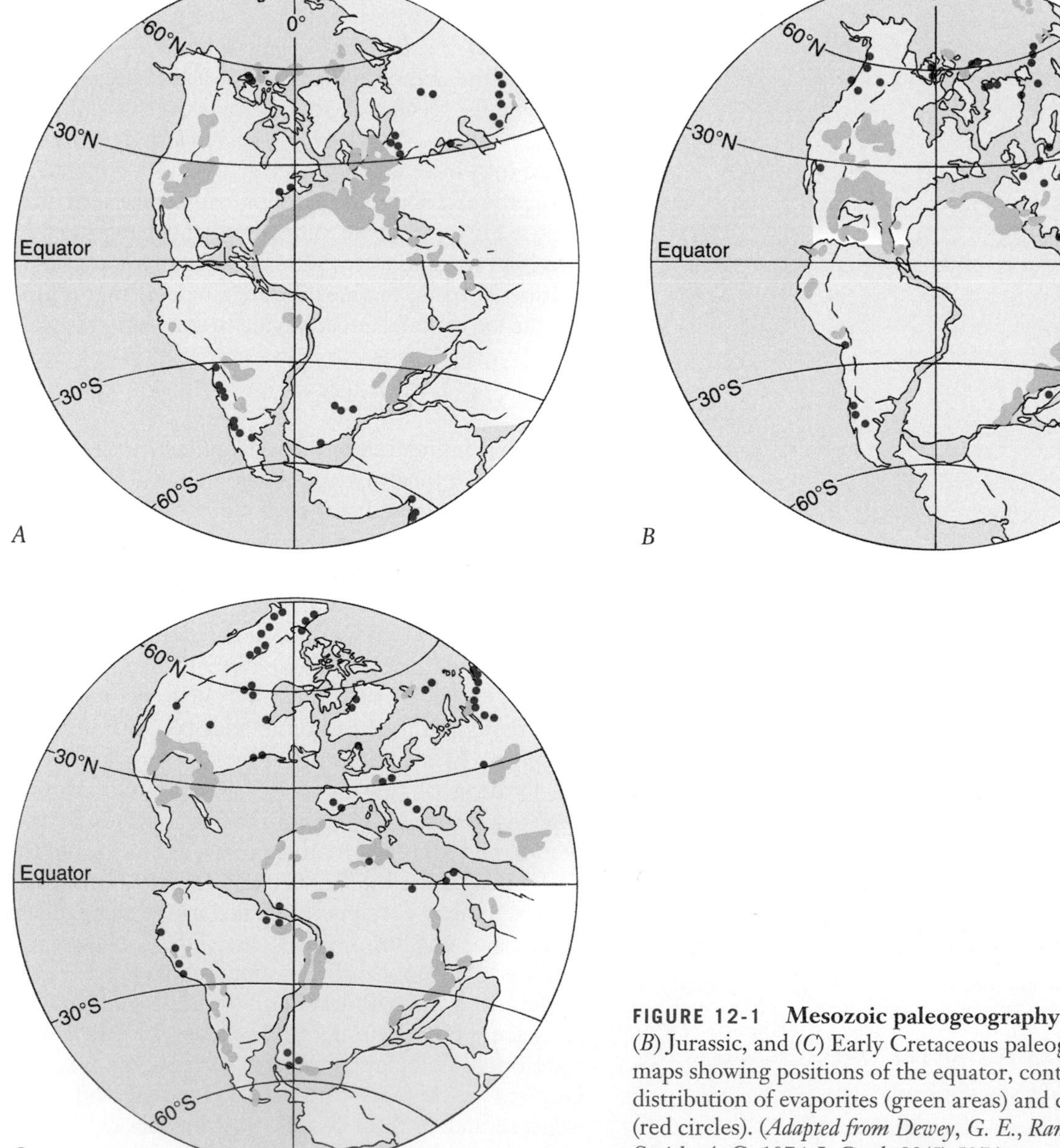

FIGURE 12-1 Mesozoic paleogeography. (*A*) Triassic, (*B*) Jurassic, and (*C*) Early Cretaceous paleogeographic maps showing positions of the equator, continents, and distribution of evaporites (green areas) and coal deposits (red circles). (*Adapted from Dewey, G. E., Ramsey, T. S., and Smith, A. G. 1974.* J. Geol. 82(5):537.)

westward-flowing equatorial currents may have penetrated far into the Tethys. These same equatorial currents, deflected by the east coast of Pangea, were also shunted to the north along coastal Asia and to the south along northeastern Africa and India. To complete the cycle, the cooled currents may have returned to the equator along the west side of the Americas. The presumed ocean and wind currents and the rather extensive coverage of seas, both upon and adjacent to continents, brought mild climates to many regions during the Jurassic. There is also evidence, however, of aridity and climatic patterns characterized by monsoons. Glacial deposits of this age are not known, and coal beds occur in Antarctica, India, China, and Canada. Paleobotanical evidence suggests that tropical conditions prevailed over regions that are now temperate.

During most of the Cretaceous, climatic conditions were generally warm and stable as they had been during the Jurassic. A remarkably homogeneous flora spread around the world, with subtropical plants thriving in latitudes 70° from the equator. Coal beds formed on nearly every continent and even at very high latitudes. However, such pleasant conditions did not persist, and toward the end of the Cretaceous, the climatic pendulum began a slow backward swing toward more rigorous conditions. The Late Cretaceous South Pole was centrally located in Antarctica, and the North Pole was located at the north edge of Ellesmere Island. Europe and North America had moved somewhat farther north, and the widespread inland seas had begun to recede. As noted in the previous chapter, the worldwide regressions were accompanied in some regions by major orogenic disturbances.

There are several lines of evidence that favor the interpretation of a terminal Cretaceous cooling. Terrestrial plants provide some clues. The tropical cycads underwent a sharp reduction, and ferns declined in both North America and Eurasia. Hardier plants such as conifers and angiosperms extended their realms. Evidence comes also from paleotemperature studies based on the oxygen isotope method described in Chapter 4. The oxygen isotope ratios obtained from open ocean planktonic calcareous organisms indicate a decline in ocean temperatures beginning about 80 million years ago. If there was indeed a dip in worldwide mean annual temperatures, it might have had a deleterious effect on plant and animal life. For this reason, some paleontologists speculate about its relationship to the extinctions that occurred at the end of the Mesozoic.

MESOZOIC INVERTEBRATES

Marine Invertebrates

At the end of the Paleozoic, about 95 percent of all species of marine animals became extinct. Many more species came perilously close to extinction. Over 100 years ago, the English geologist John Phillips noted the abrupt disappearance of marine invertebrates and used it to define the boundary separating rocks of Paleozoic and Mesozoic Eons. The Mesozoic resurgence of marine organisms was somewhat tardy, as indicated by the limited nature of Early Triassic faunas. However, after a few groups had become established, marine life expanded quickly. The bivalves became increasingly prolific from Middle Triassic on and eventually surpassed the brachiopods in colonization of the sea floor. Among the most successful bivalves were the oysters, represented by such genera as *Gryphaea* (Fig. 12–2) and *Exogyra* (Fig. 12–3). Some members of the oyster

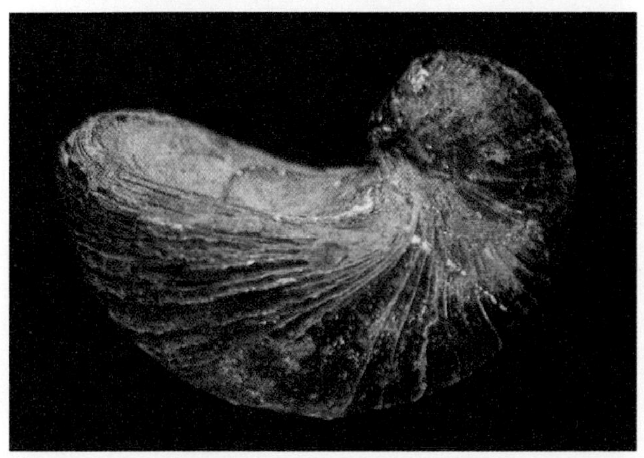

A

B

FIGURE 12-2 Mesozoic oysters. (*A*) The bivalve *Gryphaea* is found in shallow marine Jurassic and Cretaceous strata around the world. It is a member of the oyster family and is characterized by a large left valve that is arched up and over the smaller right valve. ? *How does the shape of this bivalve differ from that of clams and mussels?* (*B*) Fossils of *Gryphaea* exposed by weathering of the Tununk Shale Member of the Mancos Shale of Cretaceous age. (*B, courtesy of L. Hintze.*)

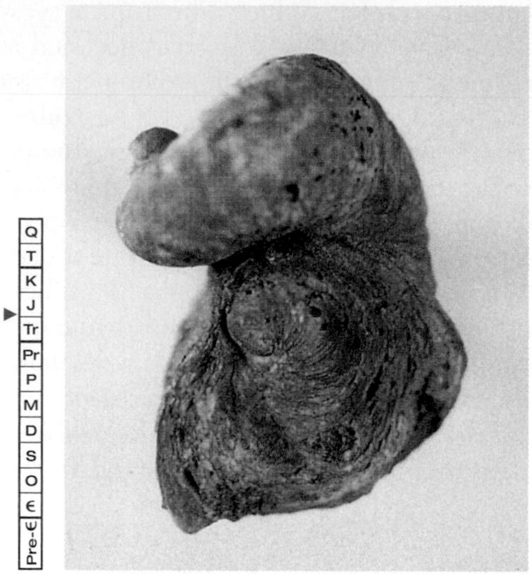

FIGURE 12-3 ***Exogyra arietina*** **from the Cretaceous Del Rio Clay of Texas.** *Exogyra* differs from *Gryphaea* in having the beak of the larger left valve twisted to the side so as not to overhang the right valve. Both genera occur in enormous numbers in shell banks of Cretaceous age along the Gulf and Atlantic coastal plains (actual size).

group became giants of their kind. Other bivalves grew conical shells that strikingly resembled some rugose corals and gastropods of the Paleozoic. In these forms, called **rudistids** (Fig. 12–4), the left valve formed a small lid that closed the open end of the cone. Rudistids formed reefs during the Jurassic and Cretaceous.

In those parts of the Mesozoic marine environment where the water was warm and clear, shallow corals proliferated. During the Late Jurassic, for example, the Tethyan seaway was the site of major coralline evolution and reef building. Corals of the Mesozoic are called **scleractinians.** Scleractinians are divided into two important groups: the **hermatypic scleractinians** that build reefs and the **ahermatypic** or nonreef scleractinians. As is true today, hermatypic scleractinians had rather restrictive environmental requirements. Most could exist only in clear water of normal salinity no deeper than about 50 meters, with temperatures no lower than about 20°C. One reason hermatypic corals require shallow water is that they have a symbiotic relationship with zooxanthellae algae that live within the coral polyp and are dependent on sunlight.

The great reefs of stony corals offered food and shelter to a host of other kinds of oceanic life. Surviving groups of brachiopods clung to the reef structures, as did bivalves, algae, bryozoans, and other sedentary creatures. Gastropods grazed ceaselessly along the reef structures, while crabs and shrimp scuttled about, seeking food in the recesses and cavernous hollows of the reefs. In the quieter lagoonal areas behind the reefs and on the floors of the epicontinental seas, starfish, sea urchins, and crinoids such as *Pentacrinus* (Fig. 12–5)

FIGURE 12-4 **The Late Cretaceous rudist pelecypod *Coralliochama orcutti* from Baja California (× 0.5).** (*A*) Cluster of specimens from the reef deposit. (*B* and *C*) Two single specimens. (*From Marincovich, L., Jr. 1975.* J. Paleontol. 49(1):212–223. *Used with author's permission.*)

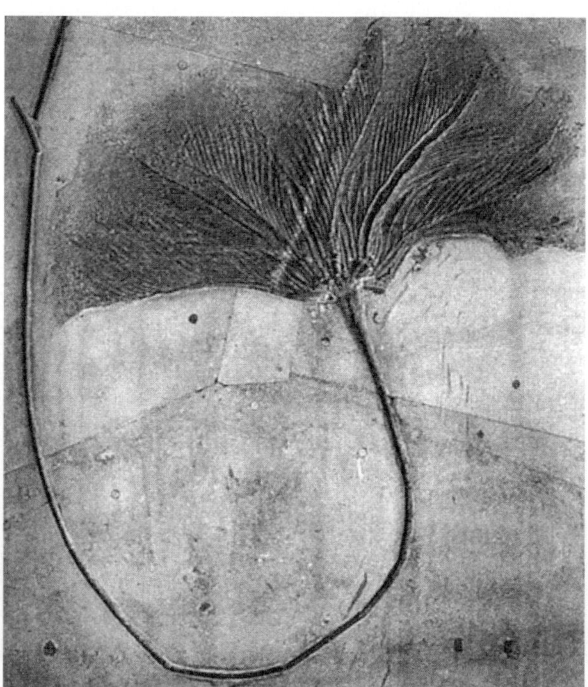

FIGURE 12-5 **Large, well-preserved specimen of the crinoid *Pentacrinus subangularis* from Lower Jurassic strata near Holzmaden, Germany.** The extended arms of this giant crinoid span about 1.2 meters. (*Washington University collection, photograph by J. Simon.*)

thrived. Not all crinoids were attached to the sea floor by stems. Bottom-dwelling stemless varieties such as *Uintacrinus* (Fig. 12–6) became abundant during the Cretaceous. Richly fossiliferous layers of *Uintacrinus* occur in the Cretaceous chalk formations of Kansas and England.

Following the extinction of the rugose corals at the end of the Permian, there was an Early Triassic interval barren of stony corals. As a result, one cannot determine the precise ancestor or ancestors of the scleractinians by means of fossils. One hypothesis is that scleractinians arose from rugose corals, as suggested by such features as septa, thecae, and tabulae that both groups possess. Scleractinians, however, use a different calcium carbonate mineral in their skeletons (aragonite, rather than calcite as in rugosids), and have a unique mode of septal insertion. These differences have promoted the view that scleractinians evolved from a soft polyp group resembling sea anemones.

Those cousins of the crinoids, the echinoids, became far more diverse and abundant in the Mesozoic than they had ever been in the preceding era. Some Lower Cretaceous formations in the Gulf Coast region contain prodigious remains of these spiny creatures. Collectors in Europe prize the silicified echinoids obtained from Cretaceous chalk beds. The regular sea urchins were especially numerous. In these forms, the symmetry is fivefold, and the shell, or test, is nearly spherical (Fig. 12–7*A*). The regular forms were overtaken by the irregular echinoids during the Cretaceous (Fig. 12–7*B*). These are mostly flattened bilateral echinoids that live as burrowers in the sediment of the sea floor. Starfish and ophiuroids (Fig. 12–8), although not as common as echinoids, were also abundant in Mesozoic seas.

For a paleontologist specializing in marine invertebrates, the Mesozoic might appropriately be designated the Age of Ammonoids. Not only were these mollusks abundant, but they were so varied that they are exceptionally useful in worldwide correlation of Mesozoic rocks. As pelagic swimming animals, many

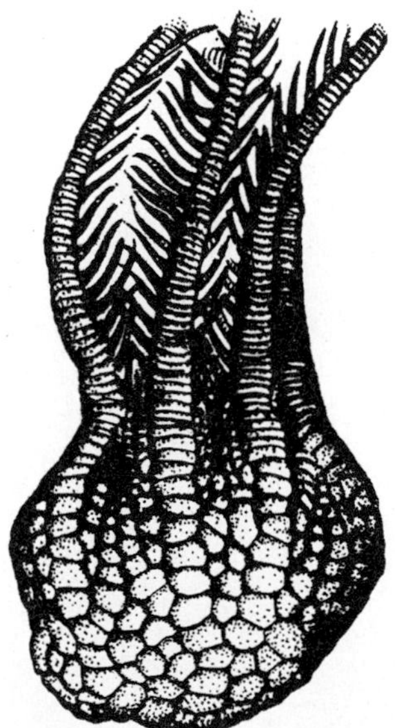

FIGURE 12-6 The Late Cretaceous crinoid *Uintacrinus*. This unusual crinoid possessed a large calyx surmounted by 10 long arms. Because it lacked an attachment stem, it was formerly thought to be a free swimmer or floater. Recent studies, however, indicate *Uintacrinus* lived on the sea floor with the massive calyx embedded in sediment. Some of its arms extended upward to intercept food particles, and others extended along the sea floor so as to provide support.

A

B

FIGURE 12-7 Mesozoic echinoids. (*A*) The regular fossil echinoid *Cidaris* from marine Jurassic strata. (*B*) *Hemiaster*, a Cretaceous irregular echinoid.

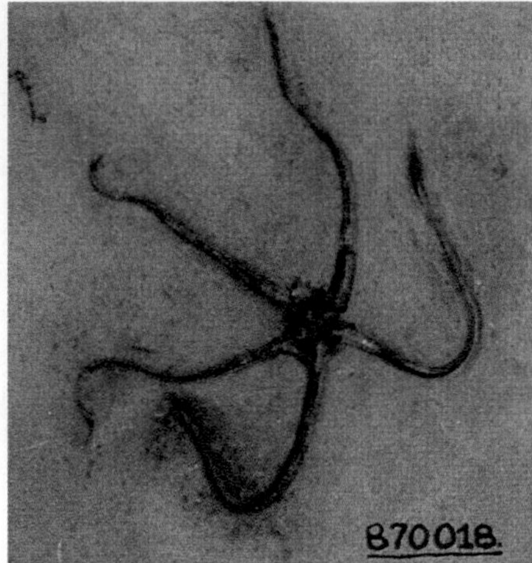

A

B

FIGURE 12-8 Two members of the echinoderm class Stelleroidea. (*A*) An ophiuroid (brittle star or serpent star) from the Triassic of England. Slab is about 10 centimeters wide. (*B*) An imprint of an asteroid (starfish) in a California Cretaceous sandstone. The starfish is about 15 centimeters in diameter.

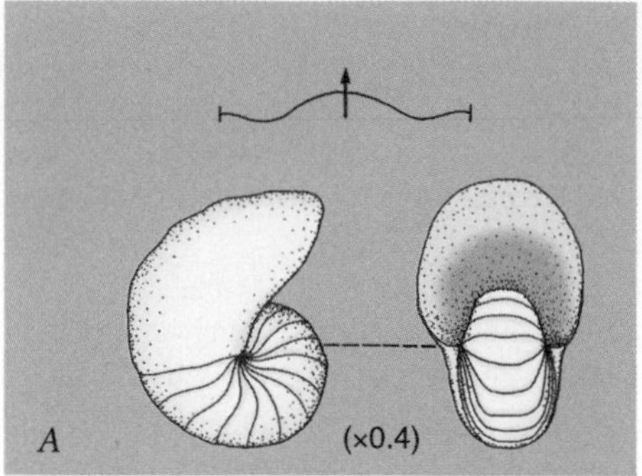

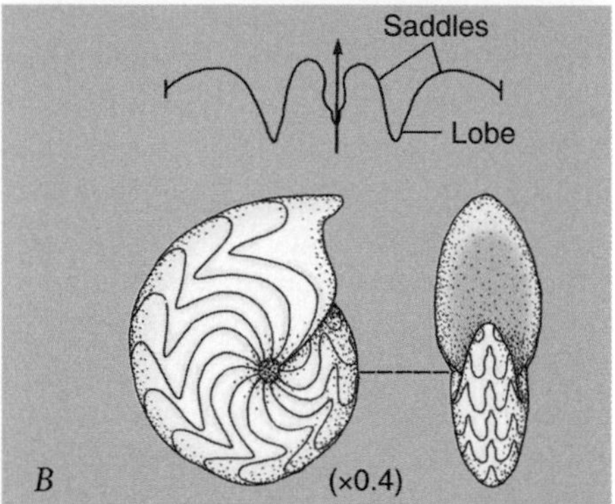

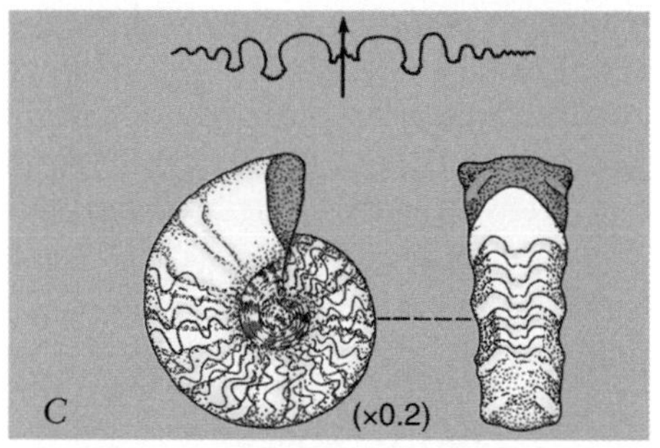

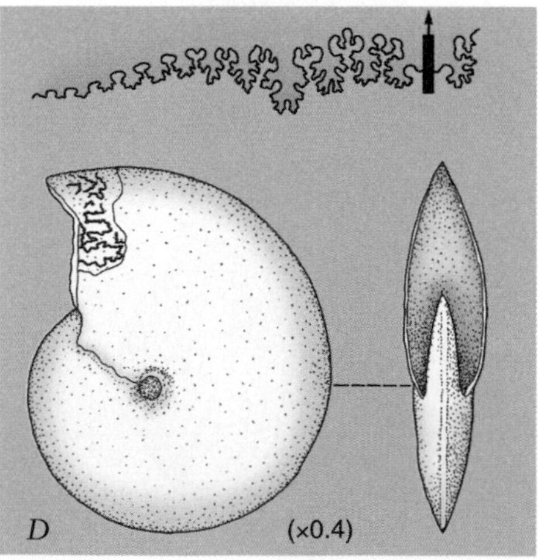

FIGURE 12-9 Sutures of cephalopods. (*A*) Nautiloid cephalopod (arrow at midventral line pointing toward conch opening). (*B*) Ammonoid cephalopod with goniatitic sutures. (*C*) Ammonoid with ceratitic sutures. (*D*) Ammonoid cephalopod with ammonitic sutures. (*From Twenhofel, W. J., and Shrock, R. R. Invertebrate Paleontology. New York: McGraw-Hill Book Co., 1935.*) ? *The wrinkled patterns of sutures trace the similarly wrinkled edges of septa. What may have been the purpose of the wrinkled septal margins?*

ammonoids attained global distribution, and even after death, their floating gas-filled shells were widely dispersed by currents. Zones developed on the basis of ammonoid guide fossils permit correlation of Mesozoic time—rock units with a level of precision often surpassing that of isotopic techniques. You may recall that two orders of cephalopods arose during the Paleozoic Era: the **Nautiloidea**, having relatively straight sutures, and the **Ammonoidea**, having wrinkled sutures. *Sutures* of cephalopods are lines formed on the inside of the shell where the edge of each chamber's partition, or *septum*, meets the inner wall. Wrinkled sutures are a reflection of *septa* that, like the edges of a pie crust, are fluted. On the basis of the complexity of the suture patterns (Fig. 12–9), **ammonoids** can be subdivided into three groups: **goniatites** (see Fig. 10–38), which lived from Devonian to Permian time; **ceratites**, which were abundant in Permian and Triassic marine areas (Fig. 12–10); and **ammonites** (Fig. 12–11). Ammonites, although represented in all three Mesozoic periods, were most prolific during the Jurassic and Cretaceous.

Knowledge of the exact suture pattern of ammonoid cephalopods is necessary for their identification and hence their use in correlation. The function of the septal fluting that produced the suture patterns provides an interesting subject for speculation. One theory is that, like the corrugated steel panels used in buildings, septa that were fluted at their margins provided greater strength. Comparison studies based on the living cephalopod *Nautilus* have revealed that the gas-filled chambers exert only a slight outward pressure, whereas the water pressure on the outside of the conch wall is considerable. The septal fluting may have helped the animal to withstand the differences in pressure. Proponents of this concept call attention to the fact that ammonoid conchs (unlike nautiloid conchs) tend to thin toward the adult chambers. In theory, to compensate for that thinner and presumably weaker conch wall, the septa in many species became more closely spaced and more intricately fluted.

Not all investigators agree that the fluted margins of ammonoid septa functioned to resist hydrostatic pressure on the conch. Another proposal is that the septal folds served as places that could be filled with soft tissue so as to provide multiple, circumferential anchorage sites. In this way, the light, buoyant conch could be held firmly to the heavy body of the cephalopods and prevent it from being torn away.

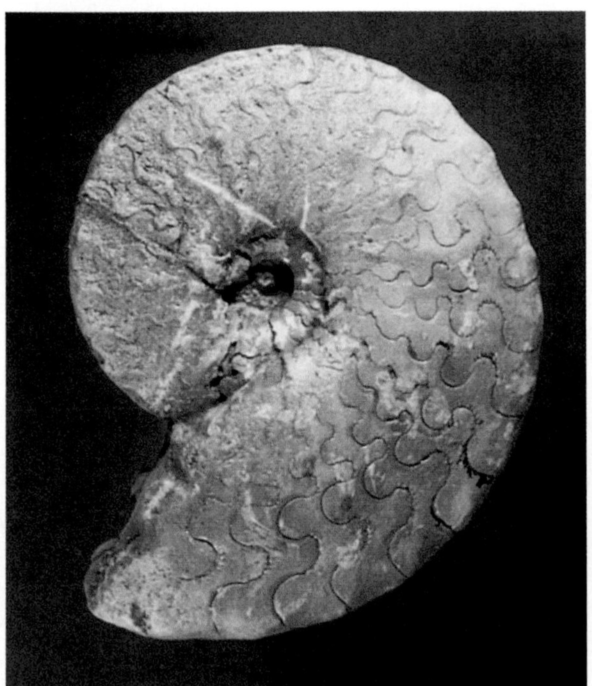

FIGURE 12-10 An ammonoid cephalopod displaying ceratitic sutures (note the tiny serrations on the lobes of the sutures). This Triassic specimen was recovered from the Upper Muschelkalk strata of Germany. (*Photograph by E. Holdener.*)

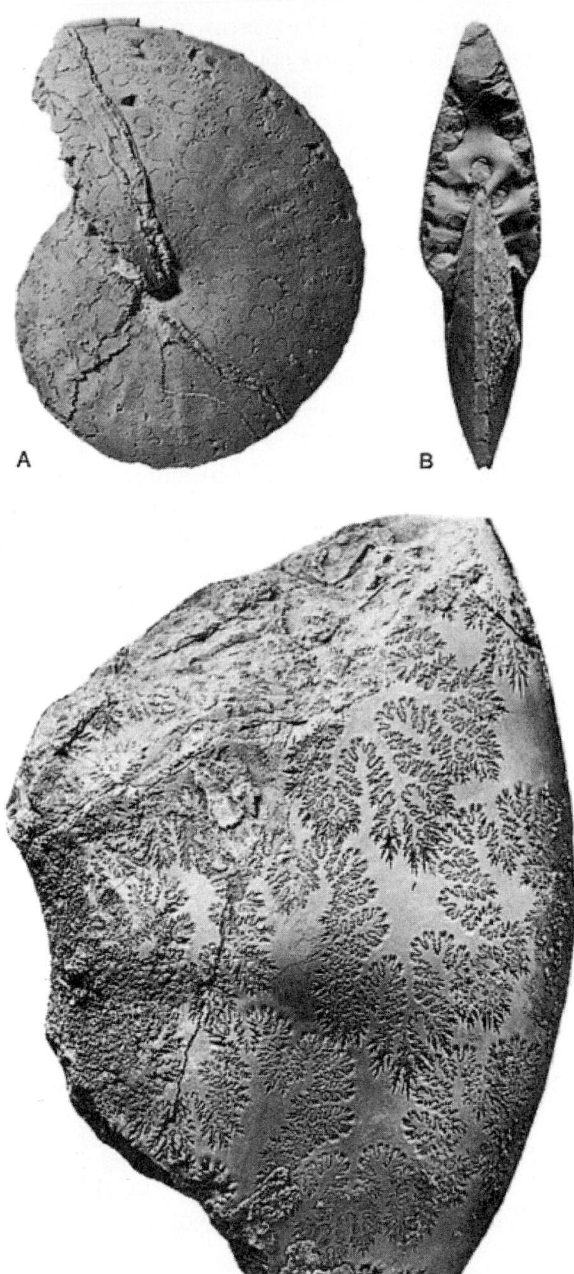

FIGURE 12-11 Cretaceous ammonoid cephalopods from the Mancos Shale of New Mexico. (*A*) *Hoplitoides sandovalensis* (vertical diameter of 9.3 centimeters). (*B*) Apertural view of another individual of same species as in *A*. (*C*) Part of outer whorl of *Tragodesmoceras socorroense* (maximum height of 16 centimeters). (*Photograph courtesy of W. A. Cobban, U.S. Geological Survey.*)

The great variety of Mesozoic ammonoids is an indication of their success in adapting to a variety of marine environments. They seem to have expanded not only within the shallow epicontinental seas but also in the open oceans. Most ammonites coil in a plane. Such coiling is termed *planispiral.* During the Cretaceous, however, many became aberrant in shape, so that the normally planispiral forms were joined by species with open spirals, straightened conchs, and even some that coiled in a helicoid fashion, like that of a snail. However, near the end of the Cretaceous, the entire diverse assemblage began to decline and, rather mysteriously, became extinct by the end of the era. Only their close relatives, the nautiloids, survived.

Another group of cephalopods that became particularly common during the Mesozoic were the squid-like **belemnites** (see Fig. 4–47). The belemnite conch was inside the animal. Its pointed end was at the rear, and the forward part was chambered. A few remarkable specimens from Germany are preserved as thin films of carbon and clearly show the 10 tentacles and body form. X-ray study of these fossils has revealed details of internal anatomy as well. Much like the modern-day squid, the belemnites were probably able to make rapid reverse dashes by jetting water out of a funnel located at the anterior end.

The belemnites were highly successful during the Jurassic and Cretaceous. Triassic belemnites may well have been the ancestors to the squids, which were also numerous during the Jurassic and Cretaceous. Octopods, because they lack a shell, have a less adequate fossil record. However, their presence is affirmed by an imprint of an octopus found in strata of Late Cretaceous age from Lebanon.

Marine **gastropods** were also abundant during the Mesozoic. Many are found in sediments that represent old beach deposits. Then, as now, cap-shaped limpets grazed across wave-washed boulders, while a variety of snails with tall, helicoid conchs crawled about on the surfaces of reefs. For the most part, the gastropod fauna had a modern appearance and included many colorful and often beautiful forms that have present-day relatives.

Modern types of marine **crustaceans**, such as crayfish, lobsters, crabs, shrimps, and ostracods, were abundant by Jurassic time. At some localities, barnacles grew in profusion on reefs and wave-washed rocks.

Among the single-celled protozoans that crawled or floated in Mesozoic seas were the **radiolarians** (Fig. 12–12) and **foraminifera** (Figs. 12–13 and 12–14). As noted in Chapter 10, both of these groups appeared in the Paleozoic. Radiolarians make their lattice-like skeletons from opaline silica. In some regions today, radiolarian and diatom skeletal remains accumulate on the sea floor to form extensive deposits of siliceous ooze and in the past have contributed to the formation of chert beds.

The tests of foraminifers often were more readily preserved than those of radiolarians. The forams, as they are often called, left an imposing Mesozoic fossil record and one of great importance in stratigraphic correlation. They are especially important in petroleum

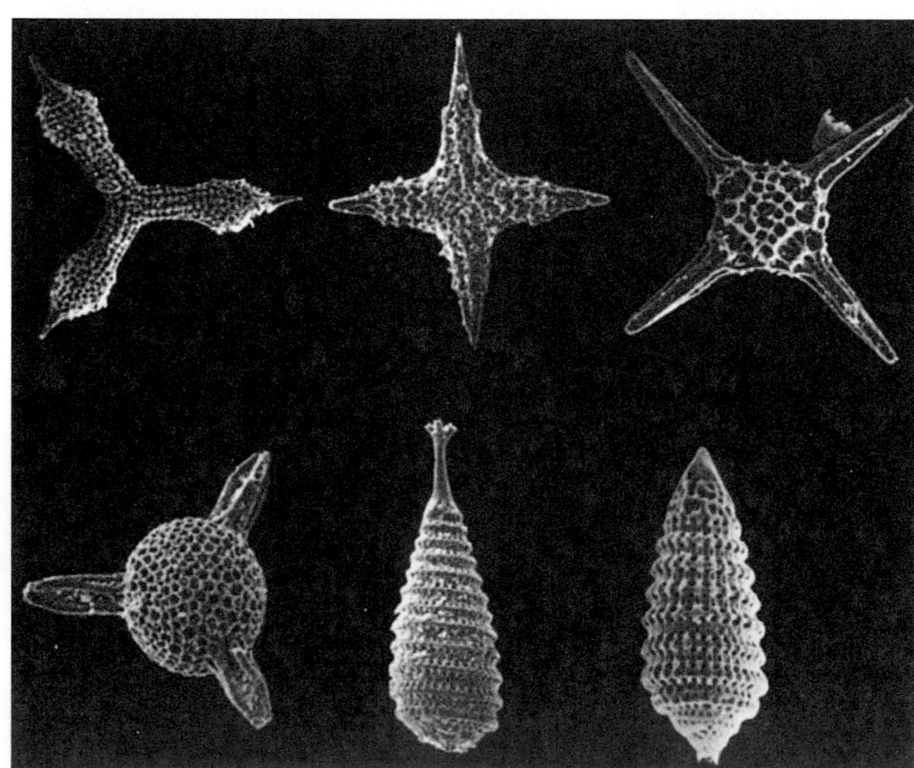

FIGURE 12-12 **Scanning electron photomicrographs of Jurassic radiolarians from the Coast Ranges of California.** Top row, left to right: *Paronaella elegans*, *Crucella sanfilippoae*, and *Emiluvia antiqua*. Bottom row, left to right: *Tripocyclia blakei*, *Parvicingula santabarbarensis*, and *Parvicingula hsui*. (*From Pessagno, E. A., Jr. 1977.* Micropaleontology 23(1):56–113. ? *In what taxonomic kingdom are radiolarians placed?*

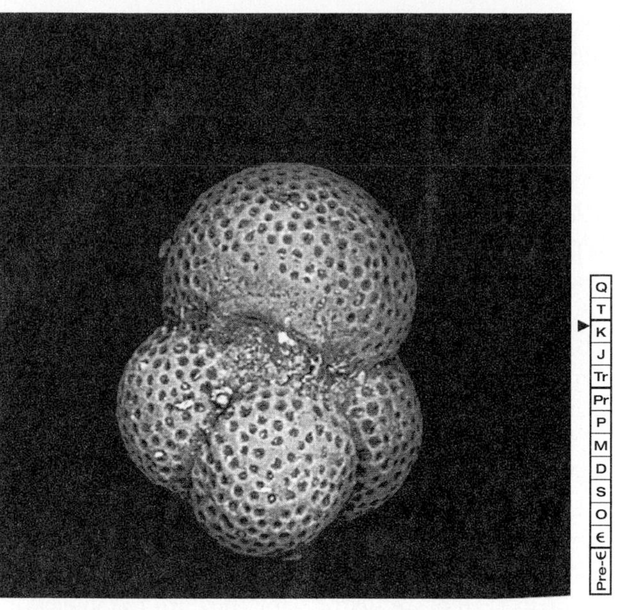

FIGURE 12-13 Electron micrograph of the planktonic foraminifer *Globigerinoides*. The height of the specimen is about 0.09 millimeters.

exploration. Because of their small size and strong tests, large numbers of foraminifers can be obtained unbroken from the small pieces of rock recovered while drilling for oil. They are then used in tracing stratigraphic units from well to well. Forams are also sensitive indicators of water temperature and salinity. They therefore provide data useful in reconstructing ancient environmental conditions.

Foraminifers were only meagerly represented in the Triassic but began to proliferate thereafter. By Cretaceous time they attained their greatest number of species. Foraminifera continued to be prolific members of the marine biota well into the Cenozoic. Nearly all were bottom-dwelling species until Cretaceous time, when planktonic foraminifers colonized the upper levels of the ocean in prodigious numbers. Their empty tests, accompanied by myriad coccoliths, blanketed the sea floor and became part of thick beds of chalk and marl that characterize marine Cretaceous sections in many parts of the world.

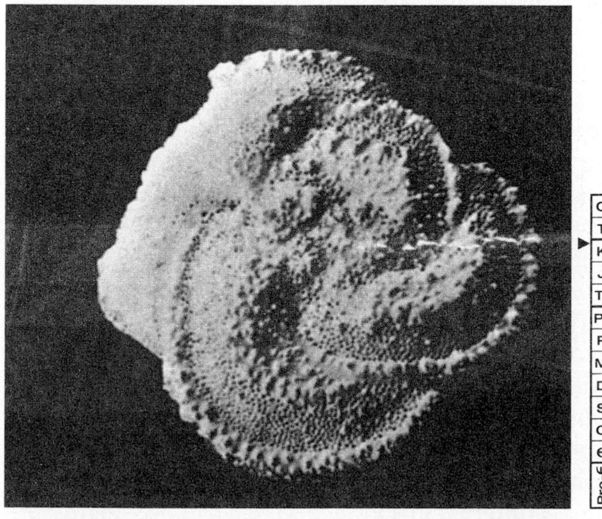

FIGURE 12-14 The Cretaceous planktic foraminifer *Globotruncana* (× 100).

Terrestrial Invertebrates

Although paleontologists have assembled an enormous body of information about the marine invertebrates of the Mesozoic, relatively little is known about continental groups. Fossils of air-breathing snails are rarely found. Freshwater clams and snails are more commonly fossilized. Freshwater **crustaceans**, including nonmarine species of ostracodes, are frequently found in Mesozoic lake bed deposits. It is reasonable to assume that many varieties of worms existed, but they left few traces. The spiders, millipedes, scorpions, and centipedes that had been abundant in Carboniferous forests were undoubtedly also present in the Mesozoic, although their remains are rare. Among the insects, butterflies and moths (**Lepidoptera**); ants, bees, and wasps (**Hymenoptera**); termites (**Isopetra**); mantises (**Mantodea**); and earwigs (**Dermopera**) had appeared before the end of the Jurassic. Many of the best insect fossils known are collected from the Solnhofen Limestone, an unusual Jurassic formation in Bavaria. Even the Solnhofen collection is not truly representative, however, because most of the fossils are of larger species. Smaller insects are more difficult to find and are not as readily preserved. The fossil record of Cretaceous insects is also sparse. Insects preserved in amber (fossilized tree sap) of Cretaceous age are known from New Jersey, Arkansas, Canada, and Alaska. Specimens include bees, wasps, ants, beetles, flies, and mosquitoes. The list of Mesozoic insects does not include fleas. They appear in the early Cenozoic at the same time that their hosts, the mammals, were ascending.

MESOZOIC VERTEBRATES

The Rise of Modern Amphibians

Among the tetrapods of the late Paleozoic, the group known as **temnospondyls** were able to survive the wave of extinctions at the end of the Permian. About 17 families of temnospondyls are known from the Triassic. Thereafter, they declined, with only two lineages surviving into the Jurassic and one into the Early Cretaceous. Their successors were the modern amphibians: frogs, toads, newts, salamanders, and caecilians.

The Lissamphibia consist of three groups: the Anuria, which includes frogs and toads; the Urodela,

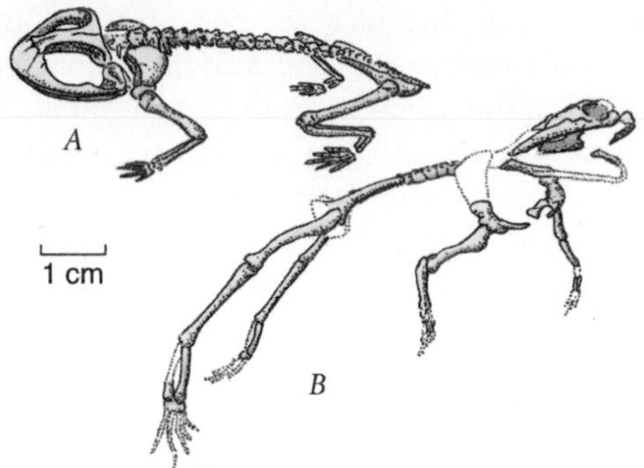

FIGURE 12-15 Restorations of the skeletons of (*A*) the ancestral Triassic frog *Triadobatrachus*, and (*B*) *Prosalirus*, an early modern frog from the Lower Jurassic Kayenta Formation of Arizona. (Triadobatrachus *after Rage, J. C., and Rocek, Z. 1989. Redescription of* Triadobatrachus massinoth, *an anurian amphibian from the Early Triassic.* Palaeontographica A206:1–16; Prosalirus *from Shubin, N. H., and Jenkins, F. A. 1995. An Early Jurassic jumping frog.* Nature 377:49.)

represented by salamanders and newts; and strange limbless caecilians. *Triadobatrachus* (Fig. 12–15*A*), from the Early Triassic of Madagascar, is the oldest known frog. Although its skull was decidedly froglike, the postcranial skeleton was not as highly modified for jumping as that of its Jurassic descendants (Fig. 12–15*B*). Salamanders and newts are less specialized than frogs. They retained four walking legs and evolved a flattened tail for swimming. The oldest known salamander is *Karaurus*, from Late Jurassic beds in Kazakhstan.

The caecilians should take the prize for the oddest of modern amphibians. Externally they resemble earthworms, whereas internally they are supported by as many as 200 vertebrae. They live in tropical regions, where they find their food in ponds or by burrowing through leaf litter. The oldest caecilian thus far discovered is *Eocaecilia*, from the Early Jurassic of Arizona. This creature still retained very reduced limbs.

The Triassic Transition

The general unrest, broad uplifts, and upheavals that occurred during the Carboniferous and Permian periods caused regressions of epicontinental seas, resulted in a variety of continental environments, and generally provided the environmental stimulus needed to maintain the spread and diversification of land vertebrates. Although marine faunas changed abruptly in passing from the Paleozoic to the Mesozoic, there was considerable continuity among land animals. The main Carboniferous amphibians continued into the Early Cretaceous before becoming extinct. The mammal-like therapsids were also able to cross the era boundary. The most progressive therapsids succeeded their Permian precursors to become contemporaries of primitive Triassic mammals.

Many new reptile groups appeared in the Triassic. Among these were the ancestors of the first turtles. Well-preserved turtles have been recovered from Upper Triassic rocks in Germany. Although they appear similar to modern turtles, many of these early forms retained teeth in their jaws. The Triassic was also the time during which many lineages of marine reptiles appeared. The **rhynchocephalians**, represented today by the tautara of New Zealand, also were abundant. Most interesting of all, however, were reptiles known as **archosaurs**. The Archosauria are a large and important group of diapsids that include crocodiles, the extinct flying reptiles, dinosaurs, and thecodonts. The thecodonts have a distinguished place in vertebrate history, for they are the ancestors of the dinosaurs.

The Basal Archosaurs

In chapter 10, we noted that the classification of reptiles included a group called diapsids, recognized by the presence of two temporal fenestra in each side of the skull. The diapsids consist of two groups, the **lepidosaurs** (lizards, snakes, and their ancestors) and the **archosaurs.** Archosaurs include ornithischians and saurischians (informally, the dinosaurs) as well as pterosaurs (flying reptiles) and crocodilians. Cladistic analyses indicate birds are archosaurs as well.

Several groups of archosaurs were present during the Triassic. They are termed "basal archosaurs" because they are at the starting point of archosaurian evolution. As exemplified by *Hesperosuchus* (Fig. 12–16), the basal archosaurs were typically small, agile, lightly constructed animals with long tails and short forelimbs. They had already developed the unique habit of walking on their hind legs. This bipedal mode was an important innovation. Bipedalism permitted basal archosaurs to move about more speedily than their sprawling ancestors. Because their forelimbs were not used for support, they could be employed for catching prey; even more important, they could be modified for flight.

Although many of the basal archosaurs were bipedal sprinters, some reverted to a four-footed stance and evolved into either armored land carnivores or large crocodile-like aquatic reptiles called **phytosaurs** (Fig. 12–17). Occasionally in the history of life, organisms of separate lineages that initially were unlike gradually become more and more similar in form. The once distinctly different groups, in fact, change over many generations so that they are better adapted to a particular environment. The evolutionary process responsible for

FIGURE 12-16 ***Hesperosuchus* from the Triassic of the southwestern United States.** Adult *Hesperosuchus* was about 4 feet long. (*Illustration by Carlyn Iverson.*)

the trend toward similar form in unrelated organisms is called **convergence**. Phytosaurs and crocodiles are good examples of evolutionary convergence. Indeed, the most visible distinction between the two groups is the position of the nostrils, which are at the end of the snout in crocodiles but were just in front of the eyes in phytosaurs. Phytosaurs were among the largest land animals of the Triassic. Some attained lengths of 11 meters (about 35 feet).

The Dinosaurs

Of all the vertebrates that have ever lived on this planet, few are more fascinating than the dinosaurs (Fig. 12–18). Dinosaurs are the most awesome and familiar of prehistoric beasts. As mentioned above, these headliners of the Mesozoic Era include two groups: the **Saurischia** (lizard-hipped) and the **Ornithischia** (bird-hipped). As suggested by these names, the

FIGURE 12-17 ***Rutiodon*, a Triassic phytosaur.** Like many other phytosaurs, *Rutiodon* grew to lengths of 10 or more feet. (*Illustration by Carlyn Iverson.*) ? *What living reptile is an example of convergent evolution with Rutiodon?*

CERATOPSIANS
Horned dinosaurs
IGUANODONTS
Large Jurassic and Cretaceous ornithopods
HADROSAURS
Cretaceous duck-billed ornithopods
PACHYCEPHALOSAURS
Dome-headed dinosaurs
COELUROSAURS
Small to medium theropods
SAUROPODS
Large to immense sauropodomorphs
CARNOSAURS
Large to immense theropods
PROSAUROPODS
Triassic to Jurassic sauropodomorphs
CERATOSAURS
Theropods with nasal horn
STEGOSAURS
Plated dinosaurs
ANKYLOSAURS
Armored dinosaurs
SAUROPODOMORPHS
Herbivorous saurischian dinosaurs
THEROPODS
Carnivorous saurischian dinosaurs
ORNITHISCHIANS
Bird-hipped dinosaurs
SAURISCHIANS
Lizard-hipped dinosaurs
DINOSAURIA

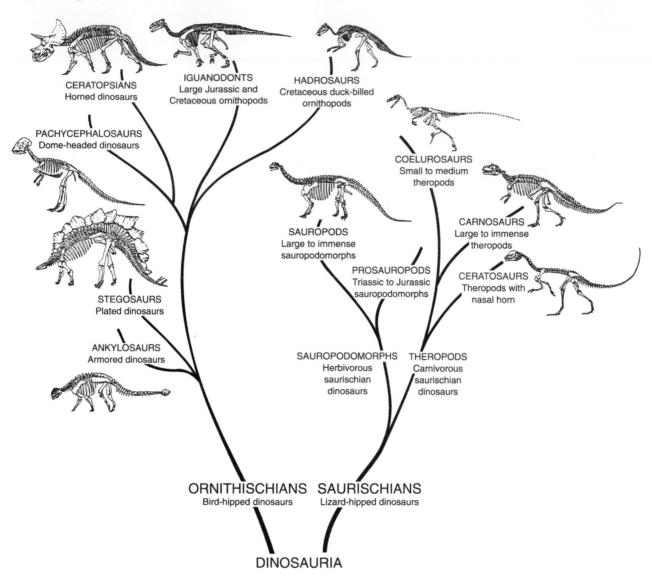

FIGURE 12-18 Major groups of dinosaurs. ? *Which of the above groups were predators?*

arrangement of bones in the hip region provides the criterion for the twofold classification. The reptile pelvis is composed of three bones on each side. The uppermost bone is the **ilium**, which is firmly clamped to the spinal column. The bone extending downward and slightly backward is the **ischium**. Forward of the ischium is the **pubis**. In the saurischians, the arrangement of the three pelvic bones is triradiate, as it was in their basal archosaurian ancestors (Fig. 12–19, and see Fig. 12–22). However, in the ornithischians the pubis is

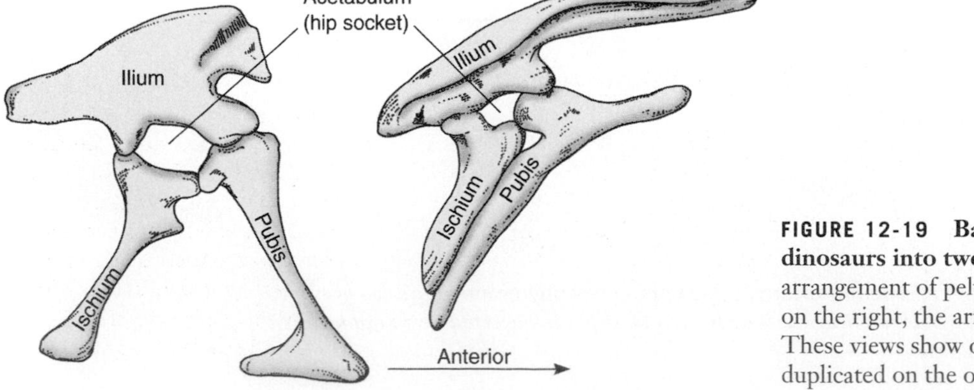

FIGURE 12-19 Basis for the division of dinosaurs into two groups. On the left is the arrangement of pelvic bones in the Saurischia; on the right, the arrangement in Ornithischia. These views show one side only; the bones are duplicated on the other side.

FIGURE 12-20 ***Herrerasaurus*, from the Triassic of Argentina, is one of the oldest known true dinosaurs.** It was a bipedal flesh-eater about 3 meters long. (*Copyright J. Sibbick.*)

swung downward and backward so that it is parallel to the ischium, as in birds. Often there is an additional forward process on the pubis.

Another difference between ornithischians and saurischians is evident in the placement of teeth. In saurischians, teeth either extended around the entire margins of the jaws or were limited to the frontal area. In contrast, ornithischians lack teeth in the front of both upper and lower jaws. In addition, the forward part of the jaws assume a beak-like shape. To complete the beak-like aspect, a predentary bone is an addition to the forward part of the lower jaw.

SAURISCHIA The Saurischia can be divided into theropods and sauropodomorpha. Theropods are typically bipedal meat eaters, and sauropodomorpha are typically large quadrupedal herbivores. The earliest dinosaurs were theropods, represented by *Eoraptor* and *Herrerasaurus* (Fig. 12–20), discovered in Triassic beds of Argentina. As indicated by potassium-argon dating of volcanic ash beds in the rock sequence containing the fossils, these predators roamed the Earth about 225 million years ago during the late Triassic. *Eoraptor* was only about a meter in length. Its carnivorous lifestyle is obvious when one views its clawed three-fingered hands and jaws armed with curved, serrated teeth.

Coelophysis (Fig. 12–21) is a theropod whose bones have been recovered from the Triassic Chinle Formation. One locality at Ghost Ranch, New Mexico, yielded the skeletons of over 100 skeletons of *Coelophysis* apparently killed during some local catastrophe such as a flash flood. In this large collection, males could be distinguished from females on the basis of skeletal differences, providing evidence that dinosaurs showed sexual dimorphism.

During the Cretaceous period, a distinctive group of theropods called ornithomimosaurs appeared. *Ornithomimus* (Fig. 12–22), with its long neck, slender hind limbs, and toothless beaklike jaws, is a good representative of this group. *Ornithomimus* (the word means "bird mimic") must have resembled a featherless ostrich and probably lived in much the same way.

In part because of such films as *Jurassic Park* and *The Lost World*, the large meat-eaters of the Jurassic and Cretaceous have become more familiar to some of us than many animals seen in zoos. Among the favorites are *Allosaurus* (Fig. 12–23) of the Jurassic and the ferocious Cretaceous beasts *Deinonychus* (Fig. 12–24), *Tyrannosaurus* (Fig. 12–25), and *Velociraptor*.

Tyrannosaurus has always been the most famous of the theropods. It attained lengths of over 13 meters and weighed in excess of 4 metric tons. Its teeth were

FIGURE 12-21 **The small, agile theopod *Coelophysis* lived about 220 million years ago, during the Late Triassic.** *Coelophysis* was about 3 meters in length. These fast, agile, bipedal predators may have pursued their prey in packs, and there is evidence that they occasionally even ate juveniles of their own species. (*Copyright J. Sibbick.*)

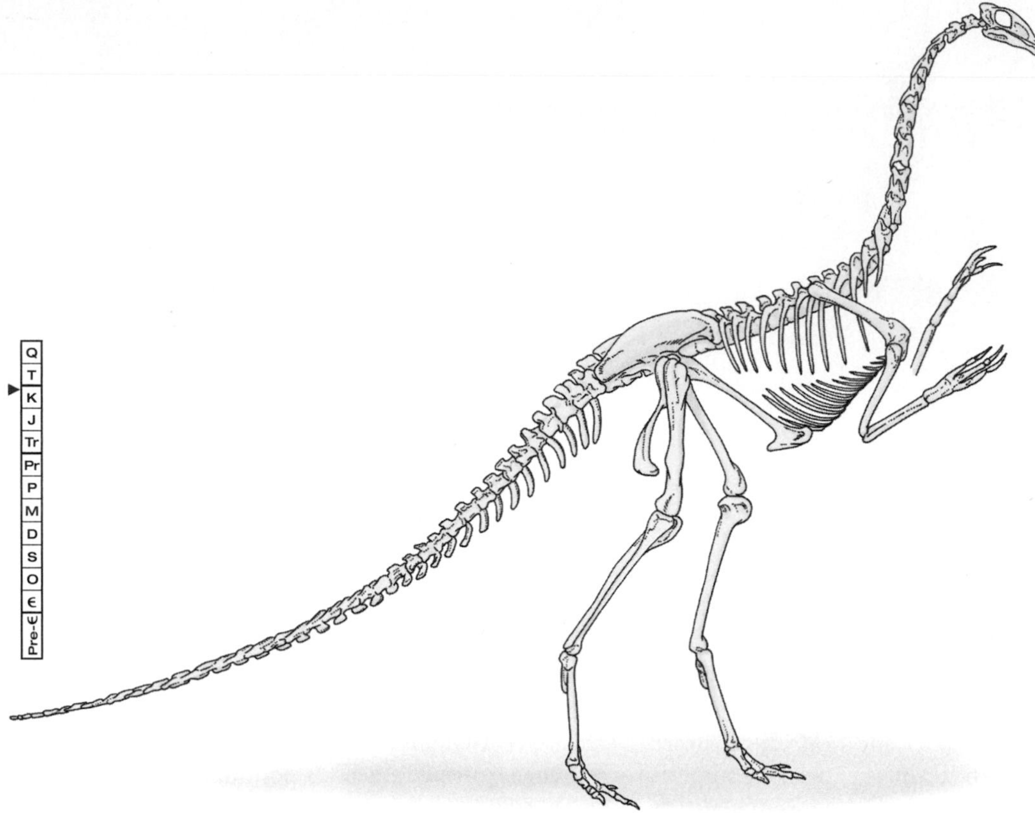

FIGURE 12-22 ***Ornithomimus*, a Cretaceous dinosaur that resembled an ostrich in size and form.** (*From Osborn, H. F. 1917*. Bull. Am. Mus. Nat. Hist. 35:733–777.) *Label the pubis, ischium, and ilium on the drawing.*

large, curved, laterally compressed, and serrated for efficiently slicing through meat. The powerful jaws could exert over 3000 pounds of biting force. (By comparison, a lion is capable of only 937 pounds of biting force.) Accurate and rapid movement of the head was accomplished because of a mobile skull-to-neck attachment. The eyes of *Tyrannosaurus* were large and stereoscopic, suggesting that the animal located its prey visually. (There is no evidence for the *Jurassic Park* premise that they could not detect prey unless it

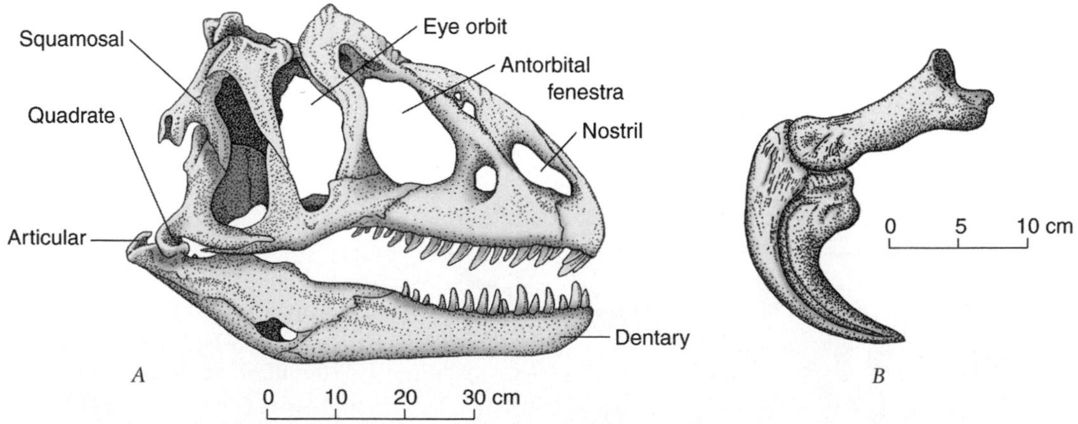

FIGURE 12-23 **Teeth and claw of *Allosaurus* verify its predaceous habit.** (*A*) Skull of the carnivorous Late Jurassic theropod *Allosaurus*. Note the long jaws and daggerlike teeth, which Allosaurus used effectively in dealing with prey. The skull consisted largely of a framework of sturdy arches to which powerful muscles were attached. (*B*) One of the three powerful curved claws on the manus (hand) of *Allosaurus*. Such claws were clearly effective for tearing at the flesh of prey. (*From Osborn, H. F.* 1917. Bull Am. Mus. Nat. Hist. 35:733–771.)

FIGURE 12-24 ***Deinonychus*** **was a fierce predator with large, serrated teeth and a greatly enlarged claw on the second digit of its hind feet.** The size of the claw compelled the animal to run on only two toes. Shown here are three of these "terrible clawed" predators attacking the large ornithischian *Tenontosaurus*. (*Copyright J. Sibbick.*)

FIGURE 12-25 **Three giant Cretaceous predatory dinosaurs on the prowl.** From left to right, they are *Daspletosaurus*, *Tyrannosaurus*, and *Tarbosaurus*. A recent study of living relatives of dinosaurs provides evidence that the fleshy nostril of theropods and many other dinosaurs had a more forward location than is traditionally illustrated. The up-front location may have provided a better olfactory sense. (*Copyright J. Sibbick.*)

moved.) As the great beast hunted (or possibly searched for a carcass to scavenge), it held its body nearly horizontal, pivoting its weight on the pelvic region. The tail was also held horizontally so as to serve as an effective counterbalance to the forward part of the body.

Until 1995, *Tyrannosaurus* was regarded as the largest known terrestrial carnivore. In that year, however, the more massive *Gigantosaurus* was discovered in Argentina. Giant theropods also roamed Africa, as indicated by the bones of *Carcharodontosaurus saharicus* (the shark-toothed reptile from the Sahara).

The herbivorous sauropodomorpha are subdivided into **prosauropods** and **sauropods**. It is likely that the smaller prosauropods were the ancestors of the sauropods. Prosauropods lived from Late Triassic to Early Jurassic. Their remains have been found on all continents except Antarctica. As seen in *Plateosaurus* (Fig. 12–26), the forelimbs in Prosauropods were shorter than the hindlimbs. Nevertheless, the animal was basically quadrupedal. Very likely it could rear up on its hind legs in order to reach food on higher branches of trees. When in this upright posture, sauropods could also survey the horizon for the approach of menacing predators.

During the Early Jurassic, the prosauropods were replaced by the more famous sauropods. These giants of the Jurassic and Cretaceous are the beasts most people associate with the word dinosaur. The most impressive were gargantuan, long-necked, long-tailed beasts that required a sturdy four-legged stance to support their tremendous bulk. *Apatosaurus* (formerly known as *Brontosaurus*) is a 30-ton sauropod favorite of schoolchildren. Another favorite is *Brachiosaurus* (Fig. 12–27). Although awesome in size, *Apatosaurus* and *Brachiosaurus* were relative lightweights when compared to the 80- to 100-ton *Supersaurus* (Fig. 12–28).

Most sauropods were "high browsers" whose long necks permitted browsing on the high foliage of trees. The longer forelimbs of *Brachiosaurus* allowed an even higher reach into the treetops. To avoid a burden of weight on the neck, the heads of sauropods were relatively small.

For many years, paleontologists speculated that, even with their pillar-like legs, these largest of land animals could not have supported their own weight continuously. It was therefore surmised that they dwelt in the buoyant waters of lakes and streams. However, there is little evidence to support this theory. Sauropod footprints and foot structure indicate that they walked

FIGURE 12-26 ***Plateosaurus*, a Late Triassic prosauropod.** (*Copyright J. Sibbick.*)

FIGURE 12-27 The mounted skeleton of the huge Jurassic sauropod *Brachiosaurus* on display at the Field Museum in Chicago.

on the tips of the toes on the front feet, with the heels of their hind feet resting on large pads, as seen in living elephants. They were clearly land-dwellers whose massive limbs provided adequate support on dry land.

Large size afforded certain advantages to the sauropods. Predators often avoid encounters with huge animals. In addition, great size in reptiles also serves to slow changes in body temperature. The ratio of surface area to mass for an animal decreases as size increases. Consequently, a large animal has a proportionately smaller surface for heat loss, and just as a large pot of water loses its heat more slowly than a small pot, so does a large animal lose its heat more slowly than a small animal. Physiologists call animals that retain their body heat because of their immense size gigantotherms.

Sauropods had begun their major expansion during the Early Jurassic, and by the end of that period they were at the peak of their diversity, abundance, and size. In North America, they declined somewhat after the Jurassic but did manage to "hold their heads high" in diminished numbers until the close of the Cretaceous. A reduction in sauropods during the Cretaceous did not occur in Southern Hemisphere continents, where they continued as dominant land herbivores.

ORNITHISCHIA The other major dinosaur line, the Ornithischia, evolved near the end of the Triassic and thrived throughout the remaining Mesozoic. Ornithischians were plant-eaters. The teeth in the forward part of the jaws were replaced by a beak suitable

FIGURE 12-28 The largest land animals ever to evolve are the sauropods. The sauropod at the left is *Seismosaurus*. At the right is *Supersaurus*. The sauropod at the center was formerly named *Ultrasaurus* but subsequently was determined to be *Supersaurus* as well. (*Copyright J. Sibbick.*)

FIGURE 12-29 *Iguanodon* (in rear) was a herbivorous ornithischian dinosaur that lived during the Early Cretaceous. Approaching from the right is the carnivorous dinosaur *Baryonx*. The boy in the painting is for scale. (*From* A Gallery of Dinosaurs and Other Early Reptiles, *by David Peters, Alfred A. Knopf, Inc. Copyright 1989, with permission.*)

for cropping vegetation. The group included both quadrupedal and bipedal varieties, with the bipedal condition considered more primitive. Even the most advanced quadruped ornithischians had shorter forelimbs, indicating their descent from bipedal forms.

The bipedal group of ornithischians is known as **ornithopods**. Their evolutionary history began in the Triassic, with lightly built, small species that lived primarily on dry land. A representative large Jurassic ornithopod is *Camptosaurus*, a bipedal dinosaur of medium size with a heavy tail, short forelimbs, and long hind legs. The articulation of the jaw was arranged to bring the teeth together at the same time, a design frequently seen in herbivores down through the ages. Leaves and stems were cropped by the forward, beaklike part of the jaws and passed backward to the cheek teeth for chopping and chewing. From camptosaurid-like ancestors, the larger Cretaceous ornithopods developed. Among these was *Iguanodon* (Fig. 12–29), one of the first dinosaurs to be scientifically described. *Iguanodon*, sometimes called the "thumbs-up dinosaur" because of the horny spike that substituted for a thumb, is thought to have been a gregarious animal that may have moved about in herds. Evidence of this comes from a Belgian coal mine, where 29 of these dinosaurs were found together as a result of their having fallen into an ancient fissure.

By Cretaceous time, ornithopods had moved into a variety of terrestrial environments. A particularly successful group was the trachodonts or **hadrosaurids**. Hadrosaurids were large ornithischians in which the forward part of the skull was broad, flat, and toothless. It therefore resembled the bill of a duck (Fig. 12–30), hence the nickname "duck-billed dinosaurs" for members of this group. Behind the toothless forward part of the jaw were rows of lozenge-shaped teeth well adapted for chewing coarse vegetation. The teeth were cemented together to form rasplike grinding surfaces.

An interesting peculiarity of some hadrosaurs was the development of bony skull crests, often containing tubular extensions of the nasal passages (Figs. 12–31

FIGURE 12-30 Mounted skeletons of two Cretaceous duck-billed dinosaurs (hadrosaurs) on display at the American Museum of Natural History in New York.

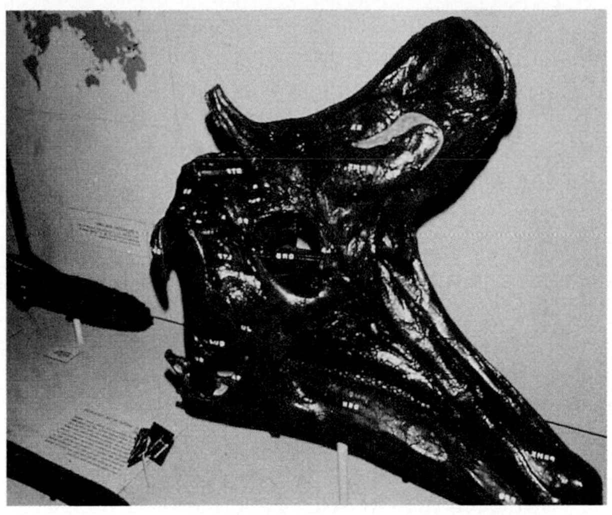

A

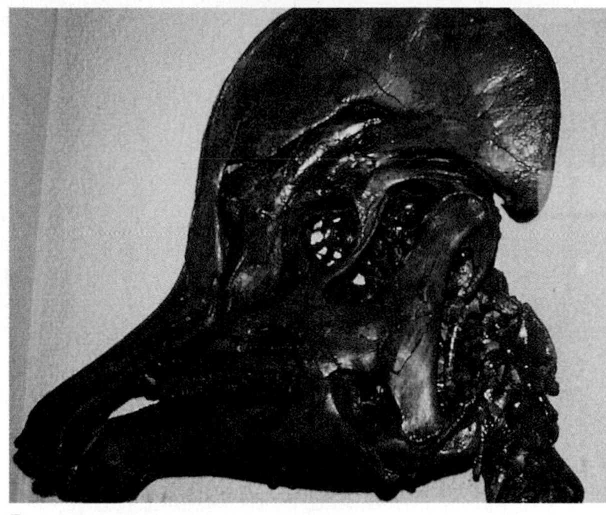

B

FIGURE 12-31 Skulls of two crested hadrosaurs, or "duck-billed" dinosaurs, of Cretaceous age. (*A*) Skull of the hadrosaur *Lambeosaurus* with its peculiar hatchet-shaped crest. (*B*) *Corythosaurus* had a helmet-shaped crest. These crests may have functioned as vocal resonators for bellowing. The skulls are approximately 0.75 meter (30 inches) long. (*Courtesy of the U.S. National Museum of Natural History, Smithsonian Institution.*)

and 12–32). It has been suggested that the cranial crests were used to catch the eye of sexual partners of the same species and could be further employed as vocal resonators used to call a potential breeding partner.

The bipedal ornithischians include a group called **pachycephalosaurs** exemplified by the "bone-head" dinosaur *Pachycephalosaurus* (Fig. 12–33). The skull in *Pachycephalosaurus* and other pachycephalosaurs consisted mostly of solid bone with only a small space to accommodate a brain of unimpressive size. One cannot help but wonder about the purpose of the thick skulls of the pachycephalosaurians. One idea is that

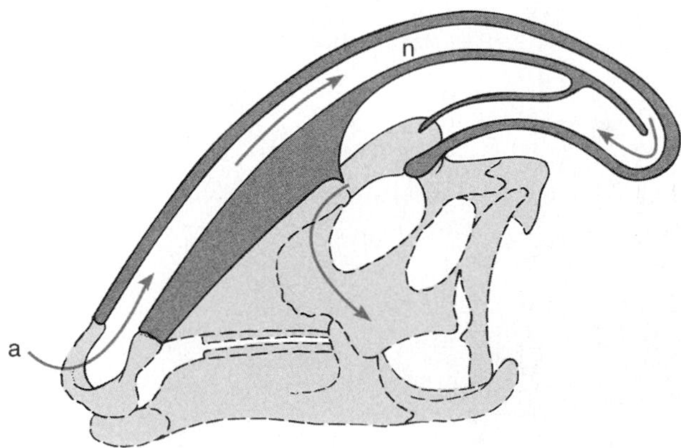

FIGURE 12-32 Internal structure of the skull crest of *Parasaurolophus cyrtocristatus*. The superficial bone of the left side has been removed to expose the left nasal passage (n). Air enters nostrils at a, moves up and around partition in crest, and from there moves down and back to internal openings in the palate. (*From Hopson, J. A. 1975. Paleobiology 1:24.*)

FIGURE 12-33 Reconstruction of the head of *Pachycephalosaurus*, the "bone-head" dinosaur. (*Courtesy of the U.S. National Museum of Natural History, Smithsonian Institution.*)

these animals used their skulls as battering rams to butt heads against one another during competition for territory or females. Abrasions found on the skulls and tendons that braced the neck vertebrae support this interpretation. Similar head-butting behavior is seen today in bighorn sheep.

The best known of the quadrupedal ornithischians are the stegosaurs (Fig. 12–34). *Stegosaurus*, the classic stegosaur, had two pairs of heavy spikes mounted on the tail that were used for defense. However, the more identifying feature of stegosaurs were the plates that stood upright along the back. The plates were not attached to the spinal column but were held in place by ligaments and muscles. Scientists have debated the purpose of these plates for many years. One theory is that the plates functioned in the regulation of body temperature by serving as body heat dissipaters. Indeed, the arrangement, size, and shape of the plates and the presence of branching surface grooves that may have contained blood vessels would favor the thermoregulatory suggestion. The plates may have served as sexual display, camouflage, or protection as well. There is disagreement about whether the plates were arranged in a single or double row. This is because they were not attached to the skeleton. When the animal died, the plates came loose and were scattered about.

Stegosaurs were the dominant quadrupedal ornithischians of the Jurassic. Their contemporaries in the Jurassic were the **nodosaurs** and **ankylosaurs**, both of which surpassed the stegosaurs during the Cretaceous. Ankylosaurs were bulky, squat ornithischians bearing closely fitted bony plates that protected the entire length of their 6-meter backsides. The head was small, and in some forms it was also covered with armor. As seen in *Euoplocephalus*, in Fig. 12–35, a bony tail club may have been used in defense, perhaps to damage the legs of an attacking theropod. **Nodosaurids** were somewhat less ponderous beasts that could therefore move about more quickly. They also lacked a tail club. *Edmontonia* (Fig. 12–35) is a representative nodosaurid.

FIGURE 12-34 Stegosaurs, the "plated dinosaurs" of the Jurassic. The best known of this group is *Stegosaurus* (upper left), which weighed from 1 to 2 tons. Its relatively small head terminated in a toothless, narrow beak suitable for cropping plants. Chewing plant food was the task of leaf-shaped teeth in the cheek regions. Although it was a large animal, the brain of *Stegosaurus* weighed only about 2.6 ounces. The distinctive diamond-shaped plates were covered with grooves and canals that mark the location of blood vessels. Thus, the plates may have functioned in temperature regulation, somewhat like solar panels and radiators. Other stegosaurs in this image are *Tuojiangosaurus* (upper right), *Dacentrurus* (lower right), *Lexovisaurus* (lower center), and *Kentrosaurus* (lower left). (*Copyright John Sibbick.*)

FIGURE 12-35 The Cretaceous ankylosaurs were the most heavily armored of the dinosaurs. Among the ankylosaurs were *Euoplocephalus* (rear, with tail club raised) and *Edmontonia* (foreground, with large spikes). *Edmontonia* is representative of the nodosaurid family of ankylosaurs. (*Copyright J. Sibbick.*)

The fourth group of quadrupedal ornithischians are the **ceratopsians.** These beasts take their name from horns that grew on the face of all but the earliest forms. Typically, ceratopsians possessed a median horn just above the nostrils, and in some species an additional pair projected from the forehead. The head was quite large in proportion to the body and displayed a shieldlike bony frill at the back of the skull roof. One can readily understand a display and defensive function for the horns, but what purpose was served by the shield? Might it have protected the animal from a frontal attack by a large theropod? Although beneath its covering of skin the frill was made of bone, the bone did not form a continuous plate but rather contained a pattern of large and small holes. The larger holes probably served to reduce the weight of the shield and hence pressure on the neck. It has been suggested that the smaller holes accommodated an array of blood vessels whose function was to help the animal radiate excess body heat, or, if needed, absorb heat from the environment. In addition, the frill provided attachment sites for the large ceratopsian jaw muscles. All ceratopsians possessed jaws that had the shape of a parrot's beak (Fig. 12–36). Judging from the scars that mark the shield bones of some ceratopsians, they were often attacked by predatory theropods. No doubt they frequently emerged the victor.

FIGURE 12-36 Horns, huge heads, parrotlike beaks, and frills characterize the certatopsians. At left is the well known *Triceratops*. *Pachyrhinosaurus* is shown in the center, and multihorned *Styracosaurus* is at the right. (*Copyright J. Sibbick.*)

ENRICHMENT

Can We Bring Back the Dinosaurs?

It would certainly be an awesome experience if we could visit a zoo or theme park and safely view a living, breathing *Tyrannosaurus* or *Triceratops*. Michael Crichton described such an imaginary place in his 1990 science fiction novel *Jurassic Park*, which appeared 3 years later as a Hollywood movie. The novel describes a dinosaur-populated theme park constructed on an isolated tropical island. Dinosaurs dwelling in the park were copies of real animals, produced by cloning of their Mesozoic counterparts.

Cloning has been accomplished in several kinds of living animals (the cloned sheep named Dolly is a famous example). The process generally involves transplanting DNA from a somatic cell (a *body cell*, not involved in reproduction) to an egg that has been stripped of its nucleus. With DNA from the animal to be cloned, the egg completes development without fertilization by a spermatozoan. The result is a precise copy of the animal that contributed the DNA.

In *Jurassic Park*, the prerequisite DNA was obtained from dinosaur blood. Preserved, unequivocal dinosaur blood has never been found. This is no problem for Crichton, for in his story dinosaur blood cells are extracted from the belly of Mesozoic mosquitoes that had dined on dinosaur blood before being encased and preserved in amber (see Fig. 4–6). Missing segments of the vital molecule were added from the DNA of frogs.

The premise of *Jurassic Park* is certainly intriguing, but in the real world it faces formidable obstacles. First of all, we have no way of knowing whether or not Mesozoic mosquitoes feasted on dinosaur blood. Perhaps they preferred the blood of the small furry mammals that were scurrying about. Assuming, however, that they did have a preference for dinosaurs, would blood extracted from the belly of an amber-encased mosquito come from one species of dinosaur, or from two or more? Some critical DNA detective work would be required to make the necessary identifications.

Another problem relates to the extreme improbability that DNA would survive intact for the tens of millions of years that have elapsed since dinosaurs roamed the Earth. Lengthy sections would have decomposed or been destroyed. It would be necessary to reconstruct missing parts, filling the gaps by specifically and repetitively copying segments beween defined nucleotide sequences. If we make the extravagant assumption that we would be able to produce a dinosaur gene segment, or even a few hundred segments, these would still represent only a tiny fraction of the billions of segments that were once present in complete dinosaur DNA. To add to the complexity, there are specific proteins that coat chromosomes and bind to key locations. These proteins govern the way genes are expressed. Without them, the chromosomes are ineffective.

But let us assume that all of the above problems are solved and we have well preserved DNA with correctly linked proteins. When the DNA is implanted into the egg of another species, will it develop into a dinosaur embryo? Eggs are not just passive containers waiting to receive DNA from any provider. They contain specific directions about how cell division is to proceed and how the embryo is to be positioned within the confines of a shell of particular size and shape. It is improbable that the egg of a living reptile or bird would so closely resemble that of a *Tyrannosaurus rex* that it would foster complete development of that ancient beast. We have witnessed some truly astonishing advances in molecular biology over the past several decades. It may be that someday some of the seemingly insurmountable obstacles to dinosaur cloning may find solutions. For the foreseeable future, however, don't expect to see dinosaurs on your next visit to the zoo.

WERE DINOSAURS WARM-BLOODED? Since the late 18th century, when dinosaurs were first studied scientifically, they have been regarded as reptiles and hence **ectothermic**, or cold-blooded. Ectothermic animals have little or no ability to maintain a uniform body temperature by physiologic processes. Some ectotherms, however, may regulate their body temperature to a certain degree by seeking either sun or shade in response to temperature needs. In living reptiles, the pineal gland may play a role in directing this type of behavior. In extinct reptiles, certain anatomic features, such as the sail in *Dimetrodon*, the plates on the back of *Stegosaurus*, or the head frill on *Triceratops*, may have served to catch the sun's rays or dissipate body heat.

In contrast to largely ectothermic animals, **endothermic** animals, such as mammals and birds, maintain a constant body temperature by physiologic production of heat internally and the radiation of excess heat away from the body. Like other animals, mammals produce heat by oxidizing food. However, when body temperature rises, special physiologic mechanisms regulated by the hypothalamus (part of the brain) help to dissipate the heat. These mechanisms include expansion of the blood vessels in the skin, perspiration, and (in furry animals) panting. When temperatures fall, other mechanisms—such as restriction of blood vessels in the skin and shivering—minimize heat loss.

In the 1960s, the idea that dinosaurs might have been warm-blooded or endothermic was suggested by John Ostrom. (Thomas Huxley made a similar suggestion in the 1860s.) Ostrom's former student, Robert Bakker, became an enthusiastic supporter of dinosaur endothermy. Bakker knew that birds and dinosaurs are similar in general anatomic design. If, like birds, dinosaurs were truly endothermic animals, then it would appear logical to restructure the classification of vertebrates by removing dinosaurs from the Linnaean class Reptilia and erecting a new taxonomy that would include dinosaurs and birds in the Archosauria. In such a reclassification of vertebrates, dinosaurs still "live" today. They are birds.

Bakker supported his theory of dinosaurian endothermy with several other lines of evidence. One relates to the way dinosaurs stood and walked. Today's lizards and salamanders are ectotherms and most have a sprawling stance, with their limbs directed more or less to the side. In contrast, the limbs of mammals and birds are held directly beneath the body. Dinosaur stance resembles that of mammals and birds, and hence, by correlation, dinosaurs were endothermic. But is this an entirely valid correlation? Dinosaur posture may be only an evolutionary solution for supporting their enormous weight.

Another observation considered by Bakker and others to favor dinosaur endothermy relates to the microscopic structure of bones. The bones of some dinosaur species are richly vascular, like the bones of mammals. Most reptiles have less vascular bones, indicating a poorer supply of blood to the bone tissue. The correlation, however, is not absolute. Some of the bones in living crocodiles have considerable vascularity, and crocodiles are primarily ectothermic. Perhaps the high amount of vascular bone in dinosaurs evolved in response to requirements related to growth rates or size and was not related to endothermy.

Isotope analysis of bone also provides a test of dinosaur endothermy. In ectothermic animals, there are large differences in the oxygen isotope content of bones of the extremities as compared to bones of the body core. Endothermic vertebrates do not exhibit this disparity. Analyses of bone from Cretaceous theropods, ceratopsians, and hadrosaurs show isotope variability similar to that in warm-blooded vertebrates.

Among the additional arguments for dinosaur endothermy is one that makes a correlation between the proportions of predators to prey among mammals (endotherms) as opposed to living reptiles (ectotherms). Today's endothermic communities consist of about 3 percent predators and 97 percent plant-eating prey. It clearly takes a lot of food to fuel the energy requirements of an endothermic predator such as a lion or wolf. In the ectothermic community, 33 percent of the animals are predators and 66 percent are prey. Determining the proportion of predators to prey among dinosaurs is rather tricky because the fossil record cannot be as precise as data obtained from living communities. At present, the evidence is contradictory. Early studies indicated that the proportion of predators to prey among dinosaurs was rather like that seen in endothermic communities, but more recent studies provide ratios more like that of ectothermic communities.

DINOSAURS AND BIRDS An obvious characteristic of birds are *feathers*. If dinosaurs and birds are closely related, would not one expect to find evidence of feathers at least on some dinosaurs? Unlike bones, feathers do not have a high probability of being preserved as fossils. Nevertheless, in 1996, Chinese paleontologists discovered the remains of a small carnivorous theropod with a covering of hollow fibers that resemble simple feathers. They named their fossil *Sinosauropteryx*. In their report, the Chinese scientists suggested the fibers were protofeathers that served as insulation in trapping body heat. Subsequent to the discovery of *Sinosauropteryx*, many other feathered theropods have been found, including *Caudipteryx and Protarchaeopteryx* in China.

We can expect the debate about dinosaur warm-bloodedness and their relation to birds to continue into the coming decade. However, even if it is demonstrated that dinosaurs were not warm-blooded and were truly reptiles, this does not mean that some did not achieve a measure of body temperature regulation. Their large size provided the means for achieving this feat. As noted earlier, size determines the rate at which an animal loses heat to the outside environment. Large animals have less surface area per unit volume, so they lose heat very slowly and can maintain a more or less consistent internal temperature for a long time, especially in an environment where cold weather is not a factor.

DINOSAUR PARENTING Dinosaurs reproduced by laying eggs. Clutches of dinosaur eggs have been found at dozens of localities around the world. But did dinosaurs care for the eggs after they were laid, and did they subsequently nurture the hatchlings? Recent evidence that at least some dinosaurs cared for their young has come principally from localities in Montana and in Mongolia.

Dinosaur eggs from Montana occur in the Cretaceous Two Medicine Formation in the western part of the state. Dubbed "Egg Mountain" by its discoverer John (Jack) R. Horner, the fossil site includes an entire hatchery of hadrosaurian dinosaurs complete with nests, clutches of eggs, hadrosaur embryos, and nestlings. There is evidence that hadrosaur babies were nurtured by their parents; lived within the social structure of large herds; were warm-blooded; and, in general, behaved more like birds than like living reptiles.

Excavations within the Two Medicine Formation revealed that hadrosaurs had hollowed out bowl-shaped nests in soft fluvial sediments and that within each nest they had laid about 20 eggs in neatly arranged circles. Plant impressions in the sediment suggest that the eggs were covered with decaying vegetation so that fermentation would provide warmth. Nests were about 7.5 meters (24.6 feet) from

Dinosaur National Monument

The story of Dinosaur National Monument begins with the famous industrialist Andrew Carnegie, who decided he wanted something really big for the new exhibit hall of his Carnegie Museum in Pittsburgh. A huge dinosaur would fit the "big" requirement very well, so Carnegie dispatched a paleontologist named Earl Douglass to the Uinta Mountains to find an impressive specimen. Douglass had been searching for fossils of ancient mammals in the Uintas and was therefore somewhat familiar with the region. In the summer of 1909, as he hiked along stream valleys, scrutinizing the banks and canyon walls for exposed bone, he sighted eight intact vertebrae of the huge sauropod *Apatosaurus*. The bones were embedded in a steeply dipping bed of the Jurassic Morrison Formation. As Douglass examined the exposure further, it became apparent that he had come upon a bonanza of dinosaur bones. The fossil site was dubbed "dinosaur ledge." Douglass supervised the excavation of bones at dinosaur ledge from 1909 to 1924. In that period, over 350 tons of specimens were shipped east by rail. Skeletons from the quarry are on display today, not only at the Carnegie Museum but also at the National Natural History Museum in Washington, D.C., the American Museum of Natural History in New York, the Denver Museum of Natural History, and the University of Utah. More than 1500 fossil bones can still be viewed in the rock of dinosaur ledge, which forms a wall of the Dinosaur Quarry Visitor Center (see accompanying figure).

Such a paleontological treasure needed to be preserved. Therefore, in 1915, President Woodrow Wilson designated the quarry "Dinosaur National Monument." In 1938, the site was enlarged to include the spectacular canyons of the Green and Yampa rivers (see accompanying map). The most scenic is Colorado's Ladore Canyon and Utah's Split Mountain Gorge. At places the walls of these canyons tower 1000 to 3000 feet above the valley floors. One can drive to Dinosaur National Monument either from Boulder, Colorado, or from Vernal, Utah, on I-40. Carefully watch for the UT 149 sign in Jensen, Utah. Then proceed about 7 miles north on UT 149 to Dinosaur Quarry Visitor Center.

The fossils at Dinosaur Quarry are in the Brushy Basin Member of the Morrison Formation. The rocks in which the bones are embedded are lithified muds and sands of floodplain and sandbars. Accompanying the dinosaur bones are remains of Jurassic ferns, cycads, and conifers, as well as aquatic insects, crocodiles, crustaceans, clams, and fish. The plants and abundance of large herbivorous dinosaurs indicate a warm, moderately humid climate. Ordinarily, such deposits do not provide such a high yield of vertebrate remains. Apparently, bloated carcasses of dinosaurs as well as bones and dismembered body parts of dinosaurs killed during seasonal floods or other catastrophic events were rafted downstream and trapped in sandbars or

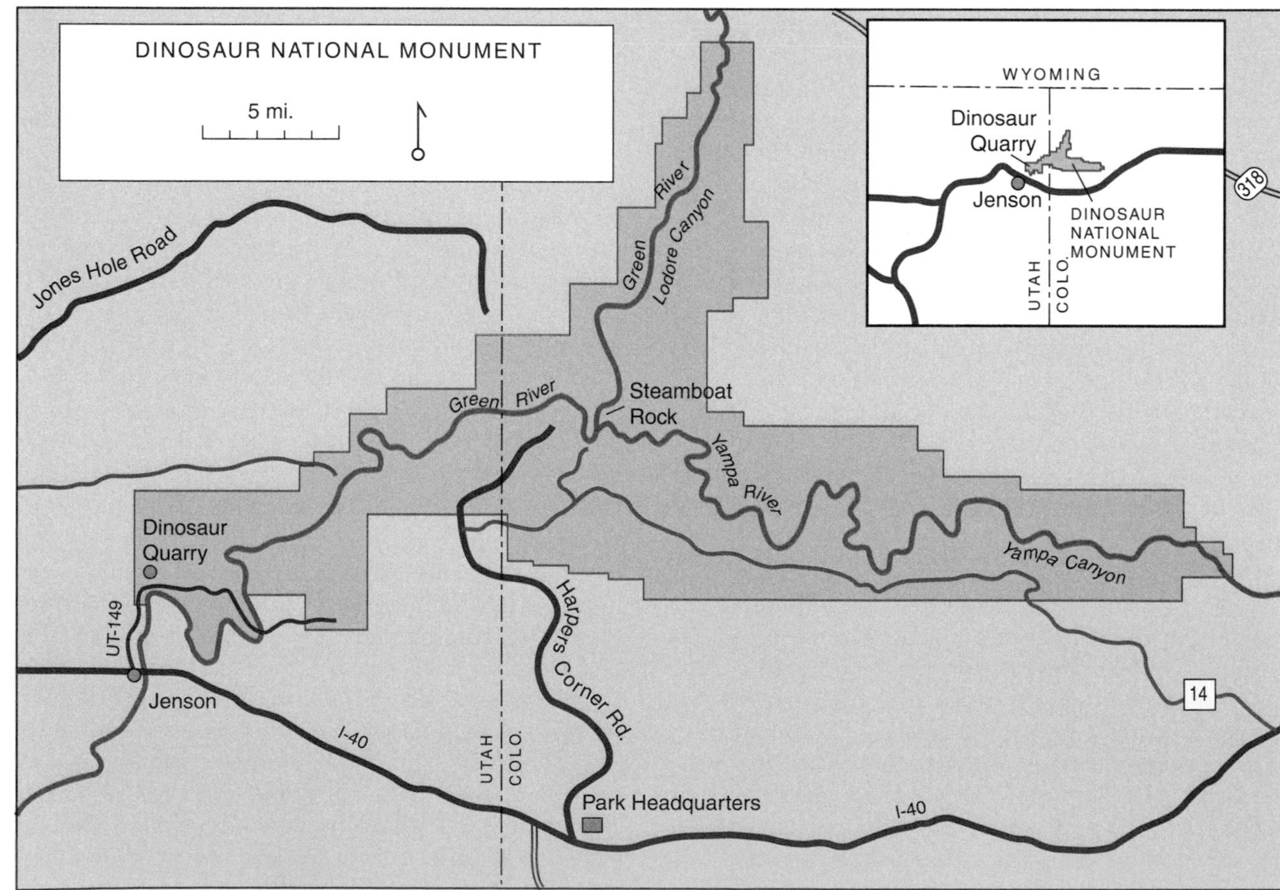

Location maps for Dinosaur National Monument and Dinosaur Quarry.

A

B

The building at the west entrance to Dinosaur National Monument that encloses the quarry face (*A*) and a view of a museum technician examining exposed dinosaur bones (*B*). (*Courtesy of National Park Service, Dinosaur National Monument.*)

deltaic areas. Indeed, the area appears to have been a burial ground for animals from a drainage area of hundreds of square miles.

The steep dip of the Morrison Formation at Dinosaur National Monument is a consequence of folding and faulting during the Laramide Orogeny, when the Uinta Mountains were formed. Erosionally carved from a great east-west-trending anticlinal arch, the range extends for 150 miles. It is only about 35 miles wide, however. Several large faults parallel the range. The Green and Yampa rivers owe their spectacular canyons to the fact that they eroded downward from strata that lay above the Uinta anticlinal arch. As a result, their courses were established before reaching the steeply dipping rocks of the arch. They were able to cut across these underlying older structures. Such streams are termed *superposed.*

one another, a distance approximating the length of the adult parent. Horner found two lines of evidence suggesting that the parent hadrosaurs, named *Maiasaura* (from the Greek word for "good mother lizard"), nurtured their young. He observed that many of the nests contained the bones of juveniles that were about a meter long. Thus, babies, which were only about 30 centimeters long at the time they hatched, stayed in their nests, where food was brought to them until they had grown sufficiently to fend for themselves. In addition, the teeth of these juveniles exhibited distinct signs of wear, suggesting that they had hatched earlier and had been feeding for some time while still in the nest. The interpretation that newly hatched babies stayed in the nest until their size tripled also supported the interpretation that *Maiasaura* was warm-blooded. It would take a baby crocodile 3 years to triple in size. It is inconceivable that *Maiasaura* babies would have stayed in their nests 3 years. Like today's ostriches, they probably attained their meter-long size in just a few months. Such rapid growth occurs only in endothermic vertebrates.

In 1984, the hunt for dinosaur remains in western Montana provided Horner and his field party with their best evidence that *Maiasaura* lived and traveled together in enormous herds. The evidence consisted of a bed of rock containing the bones of about 10,000 individual *Maiasaura*. The bone bed is about $1\frac{1}{4}$ miles by $\frac{1}{4}$ mile wide. A layer of volcanic ash occurs above the bone bed. Its presence suggests that the herd was killed by suffocating ash, lethal gases, or possibly mud flows associated with a nearby volcanic eruption. Such volcanic events were commonplace in the Rockies during the Cretaceous.

The relatively gentle plant-eating hadrosaurs were not the only dinosaurs to hover, like birds, over their nests. Predators appear to have had the parenting instinct as well. Evidence for this interpretation was discovered in 1993 by Mark A. Norell and his team while searching for dinosaur remains in Cretaceous rocks of the Gobi Desert of Mongolia. The evidence consisted of a nest of dinosaur eggs arranged in a semicircle and including the nearly complete skeleton of an embryonic theropod. The tiny dinosaur was about to hatch and thus resembled a miniature adult. It was readily identified as the embryo of the predatory dinosaur *Oviraptor*.

The fossil eggs found by Norell are identical to those found in 1922 in the Gobi Desert by the famous fossil hunter Roy Chapman Andrews (Fig. 12–37). At

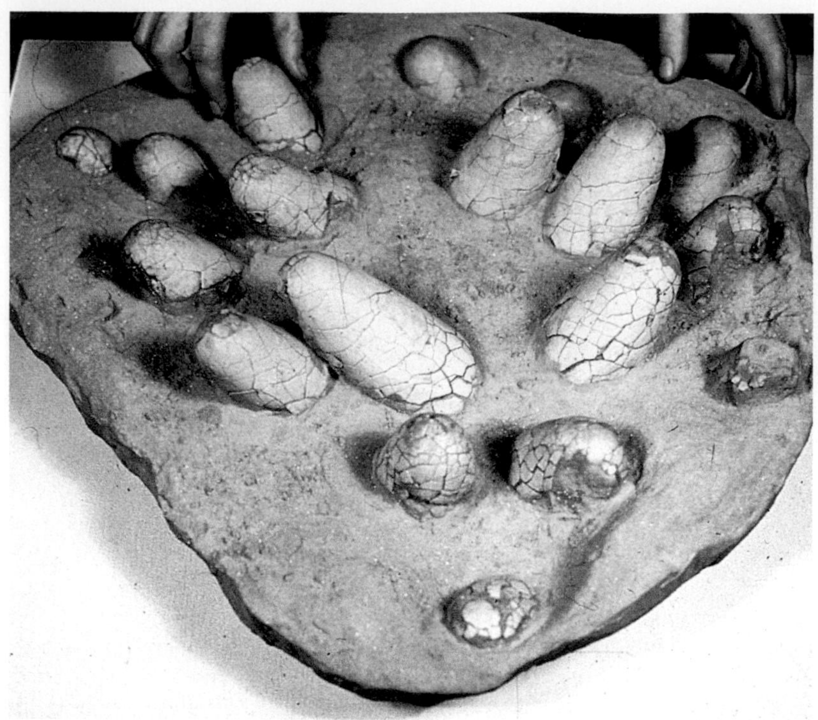

FIGURE 12-37 Fossil dinosaur eggs from the Upper Cretaceous of Mongolia. Recent discovery of eggs containing fully formed embryo skeletons indicates that the eggs belonged to the theropod dinosaur Oviraptor. (*Courtesy of the U.S. National Museum of Natural History*, Smithsonian Institution.)

the time, the eggs were thought to belong to the ceratopsian dinosaur *Protoceratops*. On top of the nest were the bones of *Oviraptor*, so named because that theropod was assumed to have died while attempting to steal or eat the eggs. Actually, *Oviraptor* was falsely accused, for the eggs were its own. It died while protecting or incubating them. As evidence of this, several *Oviraptor* skeletons were recently found squatting over a nest in precisely the manner of birds. The discovery provides evidence of birdlike nest brooding among dinosaurs.

Aerial Archosaurs of the Mesozoic

Again and again in the history of life, the descendants of a small group of animals that were initially adapted to a narrow range of ecologic conditions have, by means of evolutionary processes, radiated into peripheral environments. As the new groups diverged from their ancestral lineage, they changed in ways that made them better suited to their new surroundings. This process, known as **adaptive radiation**, is well demonstrated by the Mesozoic vertebrates. The adaptive radiation began in the Late Carboniferous and ultimately produced the enormous variety of large and small, herbivorous and carnivorous, dry land and aquatic animals of the Mesozoic. The radiation did not end with terrestrial vertebrates, however, for during the Mesozoic, vertebrates invaded the marine environment and even managed to fly.

The first reptiles to attempt to conquer the air probably either were gliders similar to present-day flying lizards or flew as bats do. For such animals flight provided an easy way to move from treetops to the ground or from branch to branch. The earliest of these aerial reptiles was *Coelurosauravus*. The skin membranes that served as wings for *Coelurosauravus* were supported along the sides of the body by 50 centimeters long, slender bones extending outward from each side of the chest. At one time these elongate bones were thought to be ribs, but a fossil discovered in 1996 reveals that the bones did not attach to the rest of the skeleton. When not used in gliding, the wings of *Coelurosauravus* could be closed like a Japanese fan. The Triassic lizard *Icarosaurus* (named after Icarus, the ill-fated son of Daedalus) was a somewhat similar reptile.

One of the more unusual aerial animals of the early Mesozoic was found in Kyrgyzstan in 1969. It was given the name *Longisquama*. Only 10 inches long, the creature had long, flattened ribs that appear to have been covered by skin so as to serve as a wing for gliding flight. The lightness of the ribs and the curvature of their tips further supports such an interpretation.

The Triassic beds that contained the remains of *Longisquama* also yielded a reptile that was not a passive glider but an active flyer. Its name is *Sharovipteryx* (Fig. 12–38). In *Sharovipteryx*, the skin membranes served as a wing (or parachute). Presumably, the animal could have manuevered while gliding by changing the position of its hind limbs, extended between the elbows and the knees and from the rear legs to the tail.

A discovery from Jurassic beds in the former Soviet Union confirmed the generally accepted theory that active, wing-flapping reptilian flyers must have been able to maintain a constant high internal temperature. Without that kind of metabolism, cold air would reduce their power of exertion and they would have difficulty staying

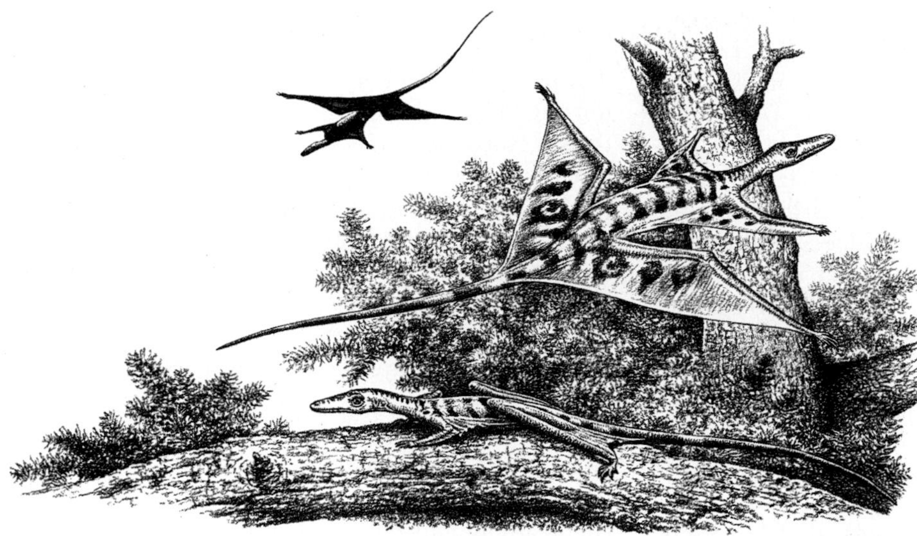

FIGURE 12-38
Sharovipteryx, a primitive, gliding, diapsid reptile from the Triassic of central Asia. From the snout to the tip of the tail, this early glider was only about 24 centimeters long. (*Copyright J. Sibbick.*)

aloft. The evidence was found in a well-preserved pterosaur that had a covering of soft hair. Appropriately, it was named *Sordes pilosus*, meaning the "hairy devil." The fur appears to have been longest on the animal's underside, prompting speculation that the hair also served to incubate eggs or insulate hatchlings.

The **pterosaurs** may appear to us as ugly, graceless creatures, but their existence from Late Triassic until Late Cretaceous attests to their adaptive success. The Jurassic and Cretaceous pterosaurs that are most familiar typically had rather large heads and eyes and long jaws. In most forms the jaws were lined with thin, slanted teeth. The bones of the fourth finger were lengthened to help support the wing, whereas the next three fingers were of ordinary length and terminated in claws. The wing was a sail made of skin stretched between the elongate digit, the sides of the body, and the hind limbs. There were two general groups of pterosaurs. The earlier groups were the rhamphorhynchoids, which had long tails. In some rhamphorhynchoids the tail terminated in a diamond-shaped vane. *Eudimorphodon* (Fig. 12–39) typifies this group. The

FIGURE 12-39 **Pterosaurs are the most famous of the flying reptiles.** Shown here are *Peteinosaurus* (left, with rudderlike membrane at the end of its tail) and *Eudimorphodon* (right foreground). The long, sharp teeth of *Eudimorphodon* were used effectively in catching and holding slippery fish. *Eudimorphodon* was about 60 centimeters in length. (*Copyright J. Sibbick/Salamander Picture Library.*)

more advanced were tail-less pterodactyloids. The latter group is exemplified by *Pteranodon* (Fig. 12–40), species of which had an astonishing wingspan of over 7 meters, yet its body was about the size of that of a goose. The skeleton was lightly constructed, as is fitting for an aerial vertebrate. The animals probably soared along much like oversized sea birds, snapping up various sea creatures in their toothless jaws. Relative to body size, pterosaurs had somewhat larger brains than some of their land-dwelling relatives. Perhaps this was a result of the higher level of nervous system control and coordination needed for flight.

A prize for large size among pterosaurs must certainly go to *Quetzalcoatlus northropi* (Fig. 12–41) from Upper Cretaceous beds of western Texas. This giant, named after an Aztec god that took the form of a feathered serpent, had an estimated wingspan of 12 meters (nearly 40 feet). Like other very large pterosaurs, *Quetzalcoatlus* probably soared about much like a modern condor, using thermal air currents and winds to help keep it aloft.

Whereas *Quetzalcoatlus* was the largest pterodactyloid, perhaps the most peculiar was *Pterodaustro*, from Cretaceous lake sediments of Argentina. *Pterodaustro* had long, curved jaws. The upper jaw held rounded teeth for crushing the shells of invertebrate prey. The lower jaw was set with hundreds of long, wire-thin teeth (Fig. 12–42) that could be used for straining food from water. While feeding, it is probable that *Pterodaustro* would fold back its wings, dip its curved snout into the water, and sweep the lake bottom for mollusks and crustaceans. On lifting its head, water would drain through the wire-thin teeth, leaving the food behind to be swallowed.

A Return to the Sea

The marine habitat is one in which the archosaurs were not notably successful. Only one archosaurian group—the sea crocodiles—was able to invade the oceanic environment. Other groups, such as the ichthyosaurs, plesiosaurs, mosasaurs, and sea turtles, however, were very successful in adapting to the oceanic environments (Fig. 12–43). Many fed on ammonoids, sharks, and modern bony fishes (teleosts) that had preceded them in populating the seas.

Not unexpectedly, the invasion of the marine environment required many modifications in form and function. Paddle-shaped limbs and streamlined bodies evolved to allow efficient movement through the water. Because these reptiles were unable to abandon air breathing and reconvert to gills, their lungs were modified for greater efficiency. In those sea-going reptiles that were unable to lay their eggs ashore, reproductive adaptations provided for birth at sea.

Marine reptiles that had paddle-shaped limbs for locomotion were already present during the Triassic Period. One group, the **nothosaurs**, were just beginning to take on adaptations that would be perfected in their descendants, the plesiosaurs. The nothosaurs were joined in the Triassic by a group of mollusk-eating, flippered reptiles known as **placodonts** (Fig. 12–44). These bulky animals had distinctive pavement-type teeth in the jaws and palate that they used for crushing the shells of the marine invertebrates upon which they fed.

By far the best known of the paddle swimmers were the **plesiosaurs** (Fig. 12–45). Their earliest remains are found in Jurassic strata. Plesiosaurs had short, broad bodies and large, many-boned flippers. In some

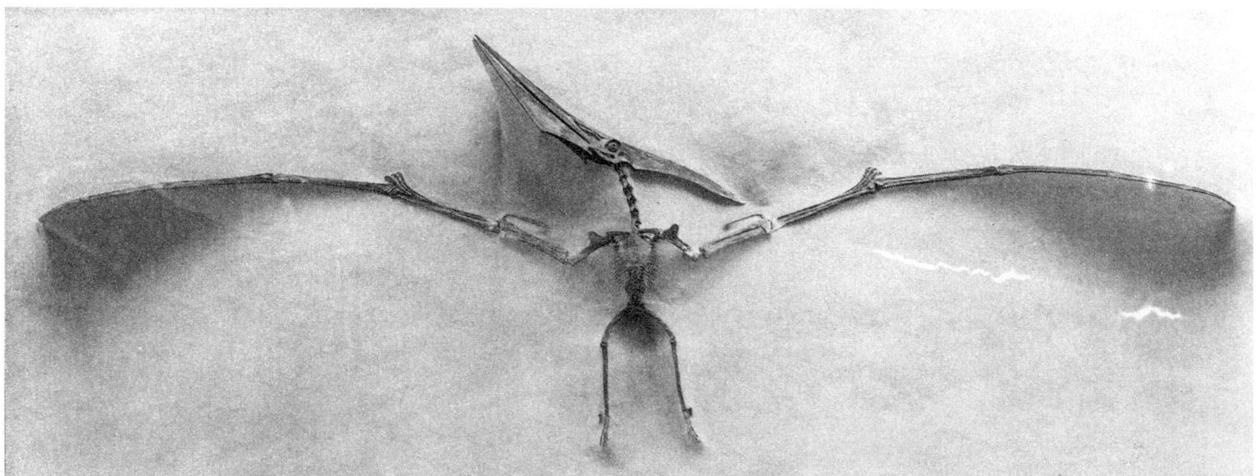

FIGURE 12-40 The skeleton of the pterodactyl *Pteranodon ingens* from the Cretaceous Niobrara Formation of Kansas. The crested point at the back of the skull may have functioned as a weathervane in keeping the head facing forward during flight. As the crest differs somewhat between males and females, it may have had a role in sexual identification and display. (*Courtesy of the U.S. National Museum of Natural History, Smithsonian Institution.*)

FIGURE 12-41 The gigantic Cretaceous pterosaur *Quetzalcoatlus* had a wingspan of about 12 meters. (Copyright J. Sibbick/Salamander Books.)

FIGURE 12-42 The distinctive pterosaur *Pterodaustro* used its over 400 slender flexible teeth as a sieve to trap the tiny organisms on which it fed. (*Copyright J. Sibbick/Salamander Books.*)

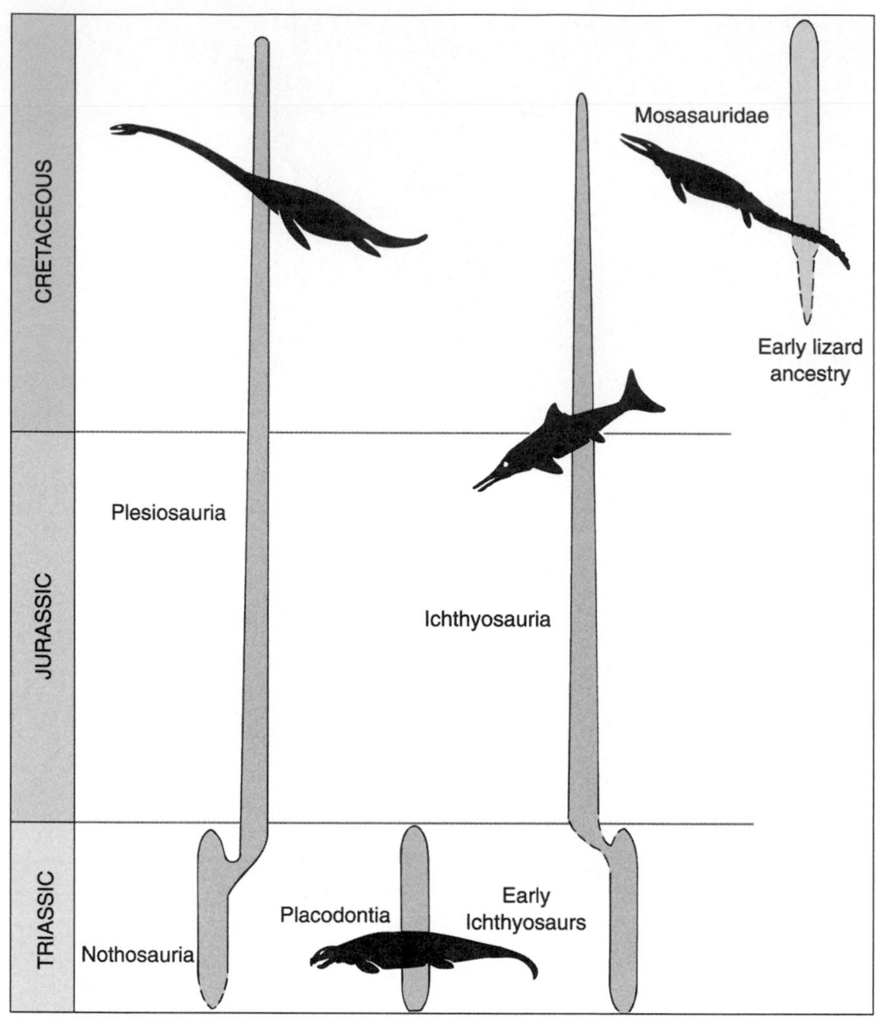

FIGURE 12-43 Stratigraphic ranges of major groups of Mesozoic marine reptiles.

species, the neck was extraordinarily long and was terminated by a smallish head. Slender, curved teeth, suitable for ensnaring fish, lined the jaws. *Elasmosaurus*, a well-known long-necked Cretaceous plesiosaur, attained an overall length in excess of 12 meters. On viewing the skeleton of a long-necked plesiosaur, Thomas Huxley remarked that the animal reminded him of "a snake threaded through a turtle."

In addition to the long-necked types, there were many plesiosaurs characterized by short necks and large heads. It is likely that the short-necked plesiosaurs were aggressive divers. *Kronosaurus*, a giant,

FIGURE 12-44 Triassic nothosaurs and placodonts. Nothosaurs such as *Ceresiosaurus* (*top left*) and *Nothosaurus* (*right foreground*) were medium-sized marine reptiles with long necks and jaws set with sharp teeth for snaring fish. Placodonts, such as *Placodus* (*bottom left*) were massively constructed reptiles specialized for feeding on mollusks that they plucked from the sea floor. *Placodus* was about 3.5 meters long. (*Copyright J. Sibbick.*)

FIGURE 12-45 Cretaceous long-necked plesiosaurs. *(From a painting by David Peters, with permission.)*

short-necked form from the Lower Cretaceous of Australia, had a skull size that probably exceeds that of any known reptile: it was 3 meters long.

The most fishlike in form and habit of all the marine reptiles were the Triassic to Early Cretaceous **ichthyosaurs** (Fig. 12–46). In many ways, they were the reptilian counterparts of the toothed whales of present-day oceans. Ichthyosaurs had fishlike tails in which vertebrae extended downward into the lower lobe, boneless dorsal fins to help prevent sideslip and roll, and paddle limbs for steering and braking. The head was a pointed entering wedge suitable for cutting rapidly through the water. A ring of bony plates surrounded the large eyes and may have helped protect them against water pressure. Clearly, ichthyosaurs were active predators with good vision and the ability to move swiftly through the water.

The mosasaurs, a group of giant marine lizards related to monitor lizards, were a highly successful group of Cretaceous sea-dwellers. A typical mosasaur had large, sharp teeth, an elongate body, and porpoise-like flippers. The creatures propelled themselves through the water by the sculling action of their long, vertical, flattened tails and the rhythmic undulations of their long bodies. The lower jaw had an extra hinge at mid-length (Fig. 12–47), which greatly increased its flexibility and gape. Mosasaurs were primarily fish-eaters, but some frequently dined on large mollusks. Shells of cephalopods have been found with puncture marks that precisely match the dental pattern of their mosasaurian foes.

Perhaps less spectacular than the mosasaurs, but far more persevering, were the sea turtles. In this group we find a trend toward increased size. The Cretaceous turtle *Archelon*, for example, reached a length of nearly 4 meters. As an adaptation to its aquatic habitat, the carapace of the marine turtle was greatly reduced, and the limbs were modified into broad paddles.

As briefly indicated earlier, the marine crocodiles (Fig. 12–48) were the only archosaurian group that entered the sea. They became relatively common during the Jurassic, but only a few remained by Early Cretaceous. It is likely that they did not fare well in competition with the mosasaurs.

The Birds

From the time of Darwin, naturalists have been aware of the structural similarities between birds and reptiles. These similarities prompted the statement by Thomas Huxley that birds are only "glorified reptiles" that have gained wings and feathers and lost their teeth. However, such comparisons depreciate the marvelous attainments of birds, not the least of which are their superior powers of flight and high level of endothermy. Both of these attributes are related to the evolution of feathers. Developmentally, feathers are homologous with reptilian scales. In their earliest development, they may not have functioned in flight, but instead for insulation, camouflage, or display.

FIGURE 12-46 **Readily recognized by their dolphin-like bodies, the ichthyosaurs were the supreme marine reptiles of the Mesozoic.** Shown here is the large Jurassic ichthyosaur *Grendelius*, which could boast an overall length of 4 meters. (*Copyright J. Sibbick.*)

FIGURE 12-47 **Mounted skeleton of a Cretaceous mosasaur.** These giant marine lizards attained lengths of over 9 meters. (*Courtesy of the U.S. National Museum of Natural History, Smithsonian Institution.*)

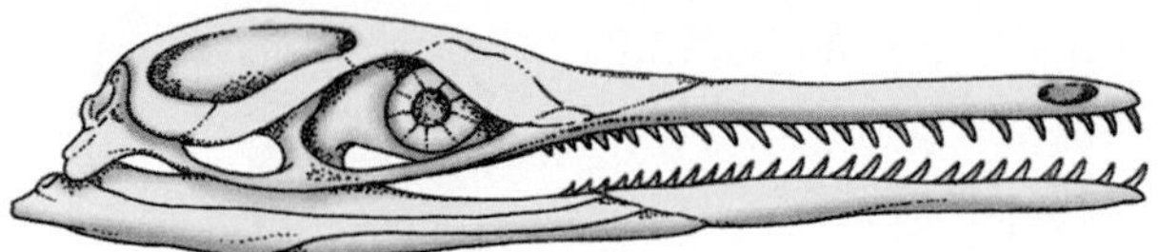

FIGURE 12-48 Toothy skull of the Jurassic marine crocodile *Geosaurus*. Length of skull is about 45 centimeters.

Although the question as to whether birds evolved from basal archosaurs or small theropods has not yet been fully resolved, there is considerable evidence of theropod ancestry. Both groups were already bird-like in their bipedal stance and in the structure of their forelimbs and hindlimbs, shoulder girdle, and skull. Several theropods are known to have had feathers, hollow bones, and keeled breastbones. Proponents of the view that birds are of theropod ancestry support this with a compilation of over 150 characteristics shared by birds and dinosaurs.

The first undisputed fossil bird to be discovered, *Archaeopteryx*, is a close evolutionary link between small, bipedal theropods and modern birds. *Archaeopteryx* (Fig. 12–49) was about the size of a crow. With the exception of its distinctly fossilized feathers (arranged on the forelimbs as in modern flying birds), its skeletal features were largely dinosaurian. The jaws bore teeth, and the creature had a long, feathered, but otherwise lizardlike tail. Unlike the wings of modern birds, in which the bones of the digits coalesce for greater strength, the primitive wings of *Archaeopteryx* retained claw-bearing, free fingers for climbing and grasping. The light-weight sternum lacked a keel, indicating that the heavy muscles needed for sustained flight were lacking. Because of this, some paleontologists believe *Archaeopteryx* was a structurally primitive evolutionary side branch and that it did not give rise to more advanced fliers.

Archaeopteryx lived about 147 million years ago. A rather similar bird, also having a long tail and claws on its wings, was recently discovered in China. The fossil was recovered from rocks several millions of years younger than the strata that contained the remains of *Archaeopteryx*. Named *Confuciusornis*, this is the earliest known bird to possess a toothless beak. Beaks are an advantage to a flying vertebrate because they are lighter than jaws bearing teeth.

Small, delicate, hollow-boned animals are not readily preserved, and thus the fossil record for early birds is not good. Although sparse, fossils of Cretaceous birds suggest an abundance of small, tree-dwelling, perching species; flightless, swift-running ground-dwellers; and aquatic swimmers and divers, such as *Hesperornis*, that relied on their webbed feet for swimming (see Fig. 11–36).

A

B

FIGURE 12-49 The Jurassic bird *Archaeopteryx*. (*A*) The skeleton of *Archaeopteryx* in the Solnhofen Limestone of Germany. The Solnhofen beds formed from lime mud deposited on the floor of a tropical lagoon. The specimen resides in the Berlin Museum of Natural History. (*John D. Cunningham/Visual Unlimited*). (*B*) Restoration of *Archaeopteryx*. (*From a painting by Rudolph Freund; Courtesy of the Carnegie Museum of Natural History.*)

The Mammalian Vanguard

While the Mesozoic reptiles were having their heyday, small, furry animals were scurrying about in trees and undergrowth awaiting their day of supremacy. These shrewlike creatures were the primitive mammals. On the basis of rare and often minuscule remains, they are known from all three systems of the Mesozoic. Among the earliest of the mammals were *Megazostrodon*, *Eozostrodon*, and the more widely known *Morganucodon* (Fig. 12–50), from Upper Triassic rocks of southern Wales. As in mammals, these tiny creatures had a battery of teeth that clearly included incisors, canines, premolars, and molars. Further, the teeth grew from milk teeth, suggesting that the young were suckled. Whereas reptiles have a single ear bone (the stapes), more efficient hearing was achieved in these early mammals by two additional ear bones, the malleus and incus. As in the cynodonts, evidence of whiskers indicates a covering of hair. The articulation of the jaw to the skull was mammalian, and the lower jaw was functionally a single bone, as in mammals.

Tooth morphology is of particular importance in the identification of early mammals. The group called **docodonts**, for example, had multicusped molar teeth, which suggests that they may have been the stock from which present-day **monotremes** evolved. Very likely, they fed on insects, as did many of these primitive mammals. The **triconodonts** can be recognized by cheek teeth in which three cusps are aligned in a row (Fig. 12–51). Some triconodonts were as large as cats and may have preyed on smaller vertebrates. **Symmetrodonts** had molars constructed on a more or less triangular plan. As suggested by their name, **multituberculates** had teeth with many tubercles or cusps on their molars. They may have been the first entirely herbivorous mammals. Their chisel-like incisors and the gap between the incisors and the cheek teeth gave them a decidedly rodentlike appearance (Fig. 12–52) and indicate that they probably had rodentlike habits as well.

The multituberculates were a persevering lineage. They appeared during the Late Jurassic and thrived for 100 million years. They even survived the biologic catastrophe that killed off the dinosaurs and did not disappear until the Eocene. In Mesozoic rocks of the Northern Hemisphere, their remains comprise as much as 75 percent of all mammal fossils. Most were ground-dwellers, but some lived in trees, having evolved reversible rear feet so that they could climb down headfirst like squirrels.

Taxonomists divide mammals into two groups. The first are called **prototherians**, and include the triconodonts, multituberculates, and the monotremes. Monotremes are represented today by the platypus and spiny anteater of Australia. The second group consists of **therians**. Marsupials and placental mammals are therians. The earliest known placental was recently discovered in the Lower Cretaceous Yixian Formation of China. Named *Eomaia*, the tiny placental mammal lived 125 million years ago. As indicated by the structure of its feet and limbs, *Eomaia* was well adapted for grasping and climbing among the branches of trees.

For the mammals, the Mesozoic was a time of evolutionary experimentation. During the Mesozoic,

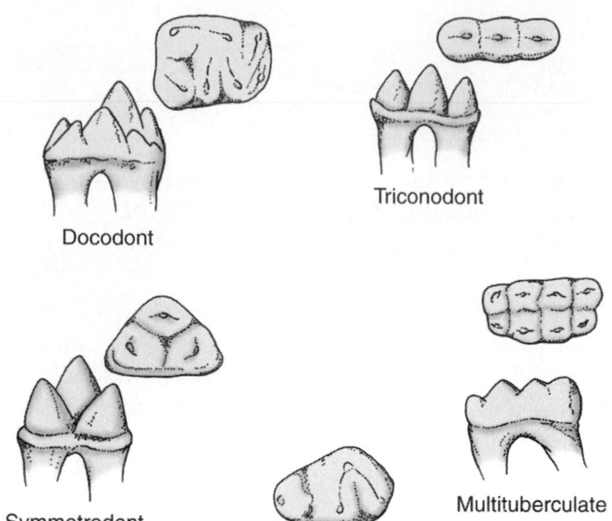

FIGURE 12-51 Molar teeth of Mesozoic mammals. Side views of lower molars (as viewed from inside the mouth) and top views of the oral surfaces.

FIGURE 12-50 Restoration of *Morganucodon*, an early mammal from the Late Triassic of Wales.

FIGURE 12-52 The rodentlike multituberculate *Taeniolabis*.

they effectively and unobtrusively lived among the great archosaurs while simultaneously their nervous, circulatory, and reproductive anatomy improved. Equipped with an exceptionally reliable system for control of body temperature, they were able to thrive in cold as well as warm climates. As the reptile population declined near the end of the era, mammals quickly expanded into the many habitats vacated by the dinosaurs. As will be described in Chapter 14, hoofed placental plant-eaters become most prevalent during the Cenozoic Era. This group, however, first appears during the Late Cretaceous. The discovery of primitive hoofed placentals in the 85-million-year-old Bissekty Formation of Uzbekistan has pushed the date for the origin of this important group back about 20 million years.

MESOZOIC PLANT LIFE

The existence of animal life on Earth is ultimately dependent on plant life. This generalization was as valid during the Mesozoic as it is today. Then, as now, plants made up the broad base of the food pyramid. Their nutritious starches, oils, and sugars made possible the evolution and continuing existence of animals. Plants are a fundamental part of the Earth's essentially self-sustaining ecologic system. The operation of the system is dependent on oxygen and carbon dioxide. Animal respiration provides the carbon dioxide needed for plant photosynthesis, whereas plants supply—as a by-product of photosynthesis—the oxygen needed by animals. In the geologic past, variations in plant productivity may have caused corresponding changes in the amount of carbon dioxide and oxygen in the atmosphere. Such variations may have favored the evolution of some animals over others and may have been responsible for the demise of particular groups.

Marine Phytoplankton

Because photosynthetic organisms that live in the ocean are suspended in water, they do not require the vascular and supportive systems that characterize land plants. Most of these organisms are unicellular, although they may grow in impressive colonies and aggregates. They are part of that vast realm of floating organisms termed **plankton**. Those having internal organelles called **chloroplasts**, and hence the ability to photosynthesize their food, are generally called **phytoplankton**. The more familiar of these organisms are members of the Protoctista.

The geologic record of the most important fossil groups of phytoplankton is shown in Figure 12–53. As indicated on the chart, phytoplankton that did not secrete mineralized coverings predominated in the pre-Mesozoic eras. These include both cyanobacteria and

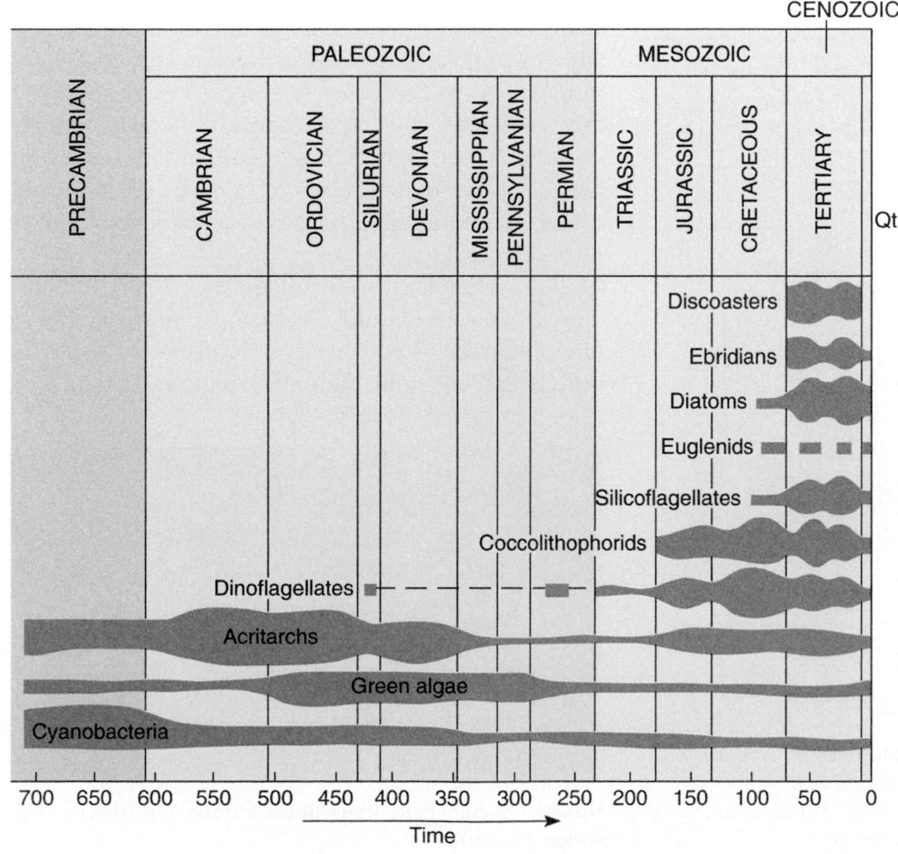

FIGURE 12-53 Geologic distribution and abundances of phytoplankton. (From Tappan, H., and *Leoblich*, *A. R.*, *Jr.* 1970. Geol. Soc. Am. Special Paper 127:257).

green algae, as well as a group of cellulose-covered unicellular algae of uncertain affinity that are loosely called **acritarchs**. The most abundant phytoplankton groups of the Mesozoic were the coccolithophorids, dinoflagellates, silicoflagellates, and diatoms.

Dinoflagellates are frequently encountered as fossils and are important aids in Mesozoic and Cenozoic stratigraphy. From the Jurassic on, they were among the primary producers in the marine food chain. Dinoflagellates are unicellular organisms that have a cell wall composed of a substance called sporopollenin, like that in pollen. For propulsion, the organism is equipped with two flagella: one is longitudinal and whiplike, and the other is transverse and ribbon-like. During their life cycle, dinoflagellates develop a motile planktonic form and a cyst phase that is formed within the motile organism. Only dinoflagellate cysts (Fig. 12–54), which have a covering extremely resistant to decay, are known as fossils.

The coccolithophorids also began their expansion during the Early Jurassic. Unlike the dinoflagellates, these calcium carbonate-secreting organisms have a splendid fossil record. Their abundant remains have formed many of the extensive coccolith limestones of the Mesozoic and early Cenozoic. Today, they are frequently present in the deep-sea sediment known as calcareous ooze. The coccolithophorid organism is one of several varieties of unicellular golden-brown algae. These algae deposit calcium carbonate internally on an organic matrix and construct tiny, shieldlike structures called **coccoliths**. Once formed, the coccoliths move to the surface of the cell and form a calcareous armor (Fig. 12–55). Coccoliths measure only from 0.002 to 0.01 millimeter in diameter. However, with the aid of the electron microscope, it is possible to discern their intricate construction. With such magnification, they are seen to consist of one, or sometimes two, superimposed elliptical or round plates that are concave on one side to fit snugly around the surface of a spherical cell. Each disc or plate is itself composed of smaller elements that may be triangular, rhombic, or variously shaped. The crystalloids are uniformly arranged, usually in a circular, radial, or spiral plan that is often astonishing in its precision (see Fig. 4–39). Because they are commonly fossilized, have experienced frequent evolutionary changes through time, and are readily dispersed by oceanic currents, coccoliths are extremely useful in stratigraphic correlation of Cretaceous to Holocene rocks.

The earliest known silicoflagellates (Fig. 12–56) and diatoms appeared in the Cretaceous. Along with other phytoplankton, they experienced a decline at the end of the Cretaceous, and then all groups expanded again

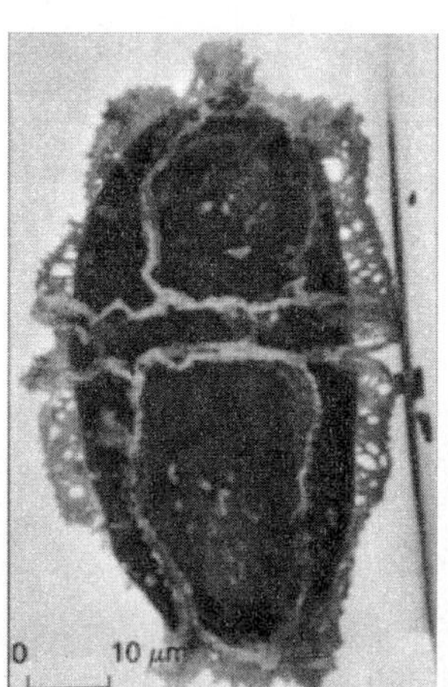

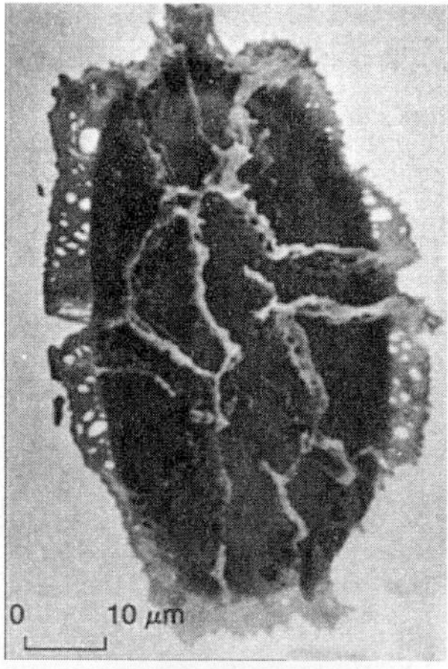

Pre-€ | € | O | S | D | M | P | Pr | Tr | J | K | T | Q

FIGURE 12-54 Fossil dinoflagellate cyst, *Prionodinium alveolatum*, from the Cretaceous of Alaska. (*From Leffingwell, H. A., and Morgan, R. P. 1977.* J. Paleontol. 51(2):292.)

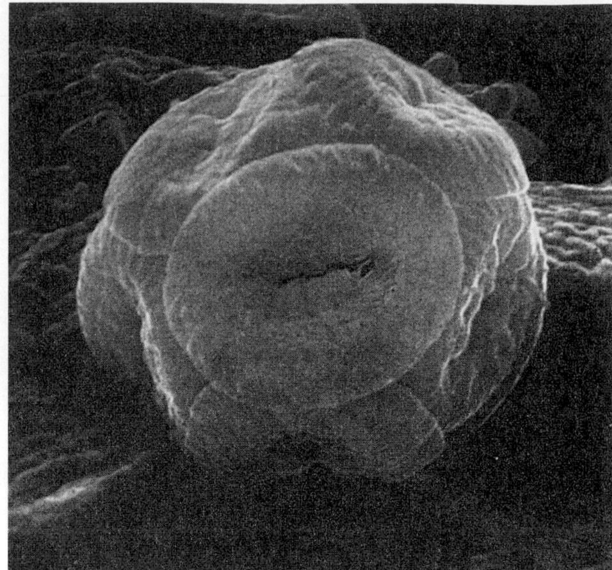

Pre-€ | € | O | S | D | M | P | Pr | Tr | J | K | T | Q

FIGURE 12-55 Scanning electron micrograph of a Late Cretaceous coccosphere (× 6000). ? *What are the individual plates that compose the coccosphere called?*

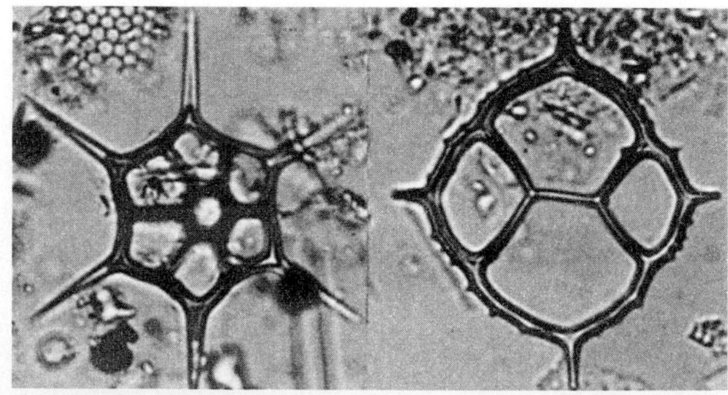

FIGURE 12-56 Silicoflagellates from the mid-Atlantic Ridge (× 3800).

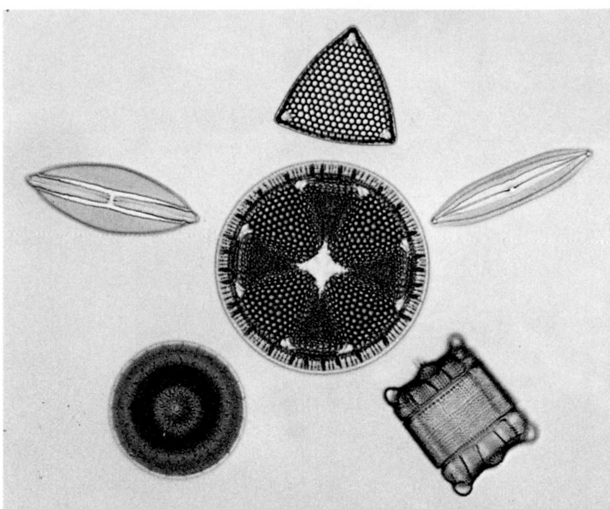

FIGURE 12-57 Modern marine diatoms. The shells (tests) of diatoms are composed of silica and consist of two perforated structures that overlap like the two parts of a pill box. (*Copyright M. Abbey/Photo Researchers, Inc.*) ? *In what taxonomic kingdom are diatoms placed?*

into the early epochs of the Cenozoic. The silicoflagellates and diatoms, along with the coccolithophorids, are members of the phylum Chrysophyta.

As suggested by their name, silicoflagellates are flagella-bearing organisms that secrete delicate siliceous skeletons in the form of simple, latticelike frameworks. Radiating spines characterize most genera, whereas others have stellate form. They range in size from 0.02 to 0.1 millimeters.

Like the silicoflagellates, **diatoms** secrete siliceous coverings (Fig. 12–57). The covering is called a **frustule**. It is usually composed of an upper part (the epitheca) and a lower part (the hypotheca) that fit together like a lid on a box. The frustule may be circular, cylindric, triangular, or a variety of other, often beautiful shapes. Today, marine diatoms are prevalent in the cooler regions of the oceans. In the past, a proliferation of diatoms was often associated with volcanic activity. Volcanic ash dissolved in sea water released silica which was then available to diatoms for constructing their skeletons.

Terrestrial Plants

Although a vertebrate paleontologist might refer to the Mesozoic as the Age of Reptiles, paleobotanists would argue that the term *Age of Cycads* would be equally appropriate. The cycads (Fig. 12–58) are seed plants in which true flowers have not been developed. Jurassic cycads included tall trees with rough branches marked by the leaf bases of earlier growths and by crowns of leathery pinnate leaves. Cycads experienced a marked decline in the Late Cretaceous, and only a few have survived to the present time. One such survivor is the sago palm, often used as a house plant.

FIGURE 12-58 A cycad growing in a South African forest. The pineapple-like structures at the top are pollen cones. (*W. H. Hodge/Peter Arnold, Inc.*)

In Chapter 4, it was noted that there were three important episodes in the evolution of land plants. The first stage led to the development of the spore-bearing leafy, treelike plants. The second involved the evolution of such nonflowering, pollinating seed plants as cycads, ginkgoes (Fig. 12–59), seed ferns, and conifers. All but the seed ferns have living representatives. Six groups of conifers were present during the Jurassic and Cretaceous, including large numbers of pines. In 1994, the remains of over 39 huge pines were discovered in Wollemi National Park, Australia. Dubbed the "Wollemi pines," many of the trees are species known only from this locality.

The third great episode of plant history is marked by the appearance of species having enclosed seeds and flowers. Such plants are known as **angiosperms**. Pollen grains resembling those produced by angiosperms provide the earliest evidence of these flowering plants. The record for angiosperms was improved in 2002 when fossil remains of a 140 million year old fruit-bearing plant was discovered in lake sediments in China. The fossil was named **Archaefructus sinensis**, meaning "ancient fruit from China." By Middle Cretaceous, angiosperms had become widespread. Forested areas included stands of birch, sycamore, magnolia, holly,

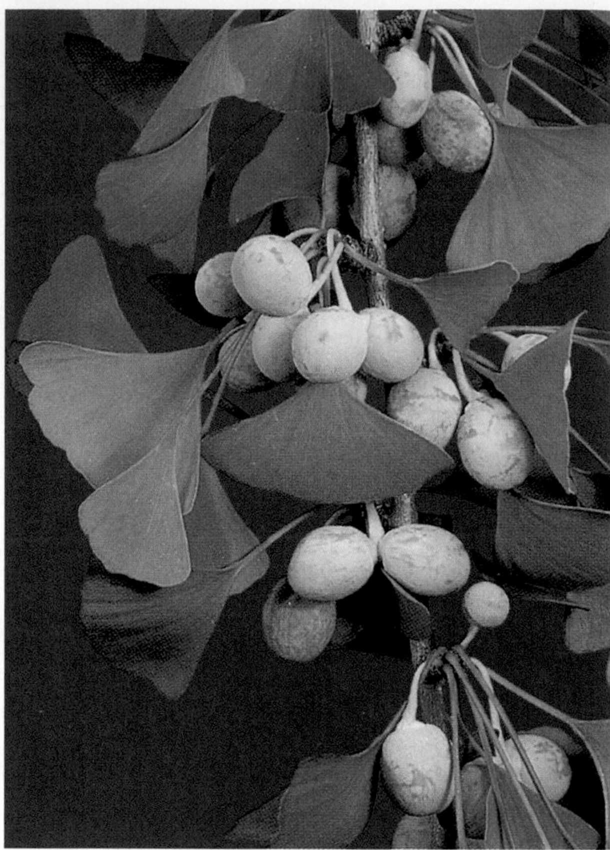

FIGURE 12-59 ***Ginkgo biloba*, the ginkgo or maidenhair tree (so called because the leaves resemble those of the maidenhair fern).** Note the naked, fleshy seeds that grow on female trees. Fossils of these plants that are over 200 million years old are nearly identical to living forms. (*W. H. Hodge/Peter Arnold, Inc.*)

FIGURE 12-60 **Fossil leaf of a Cretaceous sassafras tree from the Dakota Sandstone, Elsworth, Kansas.** The slab is about 12.5 centimeters (5 inches) wide.

palm, maple, walnut, beech, sassafras (Fig. 12–60), poplar, and willow trees. Before the period came to a close, angiosperms had surpassed the nonflowering plants in both abundance and diversity. Flowering trees, shrubs, and vines expanded across the lands and, except for the absence of grasses, gave the landscape a modern appearance.

Angiosperms provide many examples of **coevolution** with Mesozoic insects, dinosaurs, mammals, and birds. Coevolution occurs when two or more different organisms develop a close and reciprocal relationship in which the evolution of one organism is partially dependent on the evolution of the other. The coevolution of insects and flowering plants is a well-known example. Angiosperms, by encouraging insect visits, are able to use insects as delivery agents for pollen. This provides a far more efficient means of pollen dispersal than the more random wind pollinations, which require each plant to disperse tens of millions of pollen grains rather than a few dozen. The insect pollinators are encouraged to visit particular flowers to obtain the nectar and pollen needed for food. Having had success with a flower of a certain form, color, and scent, they move to find others of the same kind to repeat the favorable experience. In this way, pollen is transported on the hair, legs, and bodies of insects from plant to plant of the same species. The selective competition for efficient pollinators has induced a constantly changing range in variations among both plants and insects. In the angiosperms, the need for each plant to be recognizably different results in a spectacular floral variety that has persisted from the Cretaceous to the present.

In addition to their interdependence with insects, it is likely that early angiosperms developed coevolutionary relationships with birds, mammals, and even dinosaurs. Many of the great sauropods subsisted largely on a diet of ferns, horsetails, tree-sized club mosses, and conifers. Such plants are slow to grow and slow to regenerate. At times, regeneration may not have kept pace with dinosaur consumption, leaving over-browsed areas that could be quickly invaded by the evolving lineages of shrublike angiosperms. Unlike gymnospermal plants, the angiosperms reproduce, disperse, and grow rapidly. Thus, they are highly resistant to over-browsing. The presence of this expanding angiospermal food supply undoubtedly influenced the evolution of the ornithischians that were so numerous during the Cretaceous. The duckbills took on habits not unlike today's antelope and bison, whereas ceratopsians may have lived rather like rhinoceroses. The new plants provided nutritious fruits and nuts for these dinosaurs, and the dinosaurs in turn aided in the dispersal of angiosperm seeds by passing them unharmed through their digestive tracts.

THE TERMINAL CRETACEOUS CRISIS

Just as the end of the Paleozoic was a time of crisis for animal life, so also was the conclusion of the Mesozoic. Primarily on land but also at sea, extinction overtook many seemingly secure groups of vertebrates and invertebrates. In the seas, the plesiosaurs and mosasaurs perished. The ammonoid cephalopods and their close relatives, the belemnites, as well as the rudistid bivalves, disappeared. Entire families of echinoids, bryozoans, planktonic foraminifers, and calcareous phytoplankton became extinct. On land, the most noticeable losses were among the great clans of archosaurs. Gone forever were the magnificent dinosaurs and soaring pterosaurs. Turtles, snakes, lizards, crocodiles, and the New Zealand reptile *Sphenodon* (the tuatara) are the only reptiles that survived the great biologic crash. Altogether, the Late Cretaceous catastrophe eliminated about one fourth of all known families of animals.

The question of what caused the decimation in animal life at the end of the Mesozoic continues to intrigue paleontologists. Scores of theories, some scientific and many preposterous, have been offered. Those that have the most credibility attempt to explain the *simultaneous* extinctions of both marine and terrestrial animals and seek a single or related sequence of events as a cause. In general, the theories tend to fall into two broad categories. The first of these relies on some sort of extraterrestrial interference, such as an encounter with an asteroid or comet. Such events are considered catastrophic in that their effects are concentrated within a relatively short span of time. The second group of hypotheses proposes that the extinctions were the result of events that originated here on Earth.

Extraterrestrial Causes of Extinctions

ASTEROID IMPACT Since the time that geologists first became aware of the extinctions that mark the Cretaceous-Tertiary boundary, there have been attempts to place the blame on meteorites, comets, or lethal cosmic radiation. Tangible evidence to support those ideas, however, was lacking. This situation changed in 1977, when geologist Walter Alvarez discovered a thin layer of clay at the Cretaceous-Tertiary boundary outside of the town of Gubbio, Italy (Fig. 12–61). Alvarez sent samples of the clay to his father Luis, a physicist, who had the clay analyzed. The result of the analysis was startling. The samples contained approximately 30 times more of the metallic element iridium than is normal for the Earth's crustal rocks. Where could this high concentration of iridium have come from? Iridium is probably present in the Earth's core and perhaps the mantle, but how could the metal from so deep a source find its way into a clay layer at the Cretaceous-Tertiary boundary? While volcanism is a possibility, iridium also occurs in extraterrestrial objects such as asteroids and meteorites. The father and son Alvarez team favored this theory of extraterrestrial origin for the iridium in the clay layer and proposed that an iridium-bearing asteroid (Fig. 12–62) had crashed into the Earth at the end of the Cretaceous. The explosive, shattering blow from the huge body (presumed to be over 10 kilometers in diameter) would have thrown dense clouds of iridium-bearing dust and other impact ejecta into the atmosphere. Mixed and transported by atmospheric circulation, the dust might have formed a lethal shroud around the planet, blocking the Sun's rays and thereby causing the demise of marine and land plants on which all other forms of life ultimately depend. As the dust settled, it would have formed the iridium-rich clay

FIGURE 12-61 The coin marks the location of the iridium-rich layer of clay that separates Cretaceous and Tertiary rocks near Gubbio, Italy. The gray Cretaceous limestone below the coin contains abundant fossil coccolithophores, but few of these phytoplankton remain in the Tertiary beds above the iridium-rich clay layer. (*Lawrence Berkeley Laboratory, University of California.*)

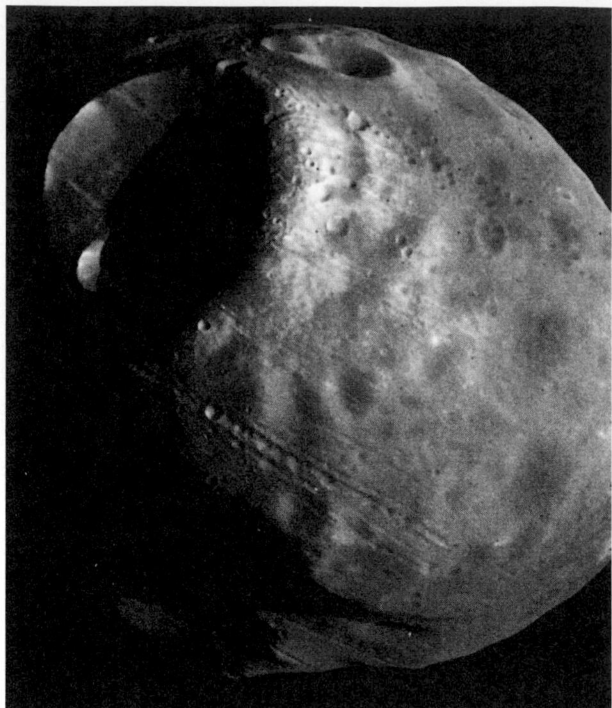

FIGURE 12-62 A montage of three separate images of the Martian moon Phobos, taken by Viking I from a distance of about 620 kilometers. The moon has an average diameter of about 20 kilometers and is approximately twice the size of the asteroid that is proposed to have struck the Earth at the end of the Cretaceous. (*Courtesy of the National Space Science Data Center.*)

layers found at Gubbio and subsequently in Denmark, Spain, New Zealand, North America, Austria, Haiti, and the former Soviet Union and in sediment layers beneath the Atlantic and Pacific oceans (Fig. 12–63).

In addition to the iridium-rich clay layer found at various sites around the world, there are other kinds of evidence of the impact of a large extraterrestrial body at the end of the Cretaceous. One of these is the widespread occurrence of **shocked quartz** (Fig. 12–64) at the Cretaceous–Tertiary boundary. These mineral grains are recognized by distinctive parallel sets of microscopic planes (called shock lamellae) that were produced when high-pressure shock waves, such as those emanating from the impact of a large meteorite, travel through quartz-bearing rocks. Often, in the same stratum containing grains of shocked quartz, one also finds tiny glassy spherules thought to represent droplets of molten rock thrown into the atmosphere during the impact event. They are termed **tektites**. A rare, dense, high-pressure silicate mineral known as **stishovite**, found at Meteor Crater and other known impact structures, is also found in the boundary clay. It is taken as evidence of sudden extremely high pressures, such as those that would be associated with the impact of an asteroid. Finally, sediments at the Cretaceous–Tertiary boundary often include a layer of soot that may be the residue of vegetation burned during widespread fires caused by extraterrestrial impact.

Any large extraterrestrial object that explodes on striking the Earth is called a **bolide**. Bolides have collided with the Earth many times in the geologic past, but their scars left on continents have mostly been

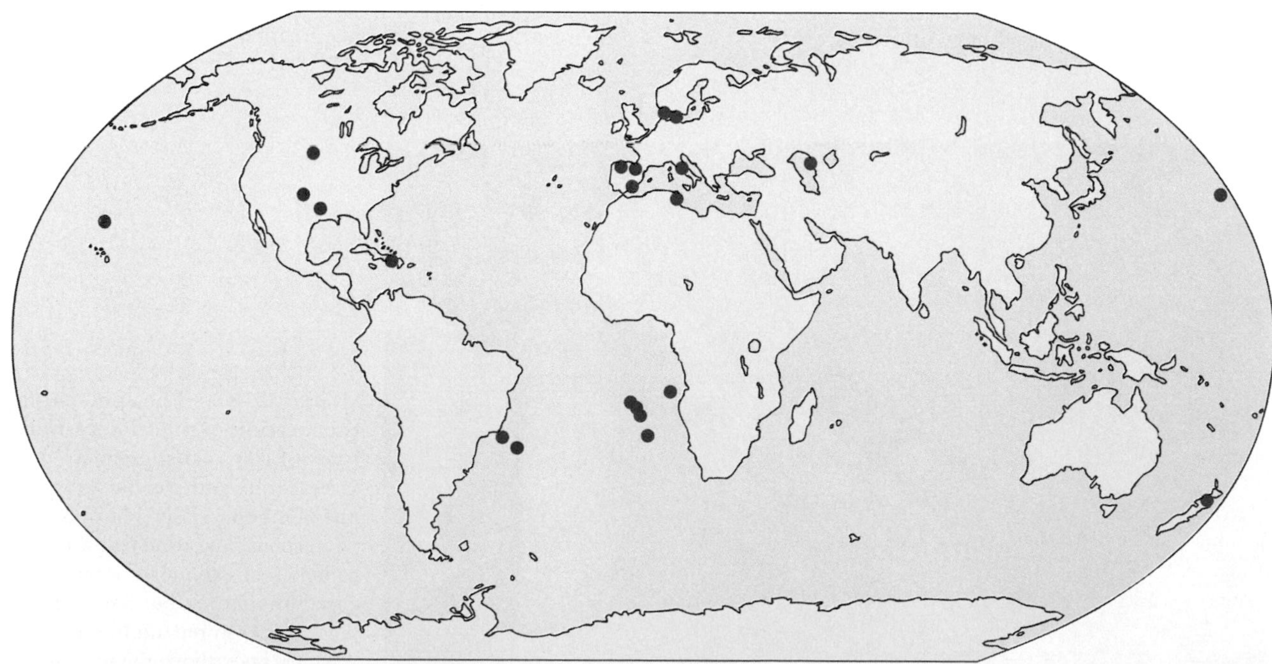

FIGURE 12-63 Occurrences of the iridium-rich sediment layer at the Cretaceous–Tertiary boundary. (*From Alvarez, W., et al. 1990.* Geol. Soc. Am. Special Paper 190:305–315.)

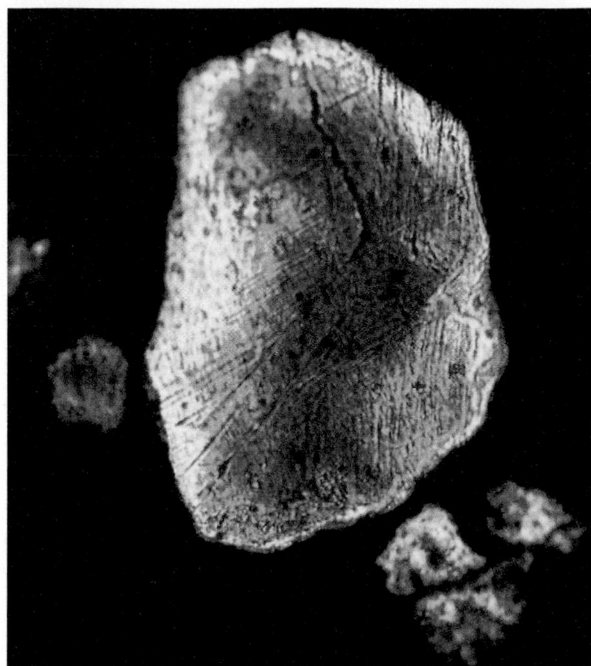

FIGURE 12-64 A shocked quartz grain from the Cretaceous–Tertiary boundary layer. Note the parallel intersecting sets of microscopic planes, called shock lamellae. These intersecting lamellae are similar to those observed in quartz from rock fragments clearly known to have been subjected to meteorite impact. Maximum diameter of grain is 0.3 millimeters. (*Courtesy of J. S. Alexopoulos.*)

obliterated by the relentless action of weathering and erosion. Nevertheless, a few can be discerned on photographs taken from spacecraft or during geologic investigations. For example, geologists have discovered a large crater of Jurassic age left by bolide impact on the floor of the ocean north of Norway. It has been named the Mjølnir Crater and is about 40 kilometers in diameter. Sediment within and around the crater contains shocked quartz and high concentrations of iridium.

THE CHICXULUB STRUCTURE The Mjølnir Crater is of Jurassic age. Where, then, is the crater produced by the bolide that allegedly caused the mass extinctions at the end of the Cretaceous? Currently, the best candidate has been located in the Gulf of Mexico, just offshore from the Yucatan Penninsula. It is named for the nearby town of Chicxulub (Fig. 12–65). At this location, magnetic and gravity surveys, as well as cores and logs of oil wells, reveal a buried, circular, craterlike structure about 180 kilometers in diameter. Andesitic rock exists in the central core of the structure. The andesite has an isotopic and chemical composition similar to that of tektites that are abundant in the Cretaceous– Tertiary (K/T) boundary layer at many locations in the Caribbean region.

Further evidence of bolide impact is found in core samples of rocks penetrated during the drilling for oil in and around the Chicxulub structure. Prominent in the core samples are coarse breccias that occur both

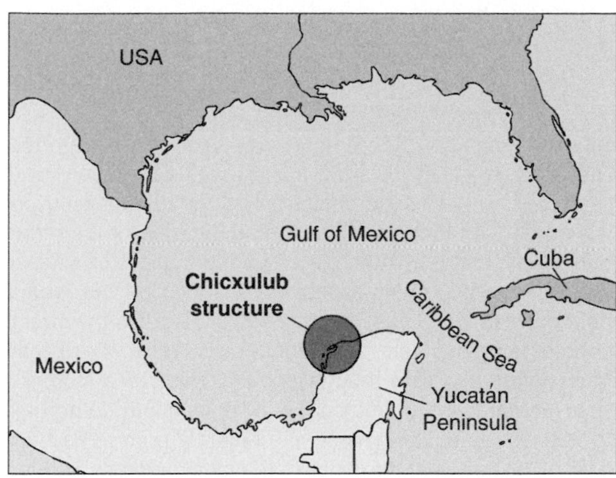

FIGURE 12-65 Location map for the Chicxulub structure.

above and interbedded with the andesite. The breccias contain shocked quartz and appear to be part of a blanket of severely fragmented rock such as would be produced by a massive impact.

If a large asteroid impacted our planet at the end of the Mesozoic, one would expect to find layers of sedimentary rock containing the fossil remains of some of the asteroid's victims. In 1996, paleontologist William Zinsmeister reported making such a discovery near the tip of the Antarctic Peninsula on Seymour Island. There, in strata immediately above the iridium-rich Cretaceous-Tertiary boundary clay, he found the remains of a huge fish kill. Their doom would appear to be directly related to the event that produced the iridium-rich boundary clay.

Not only fish but also plankton in the ocean would have been affected by the proposed impact. In 1997, a core sample was taken from beneath the ocean floor about 320 kilometers east of Jacksonville, Florida. The core represents the entire time span of the bolide event. At its base, there is a layer of white sediment containing the shells of hundreds of microfossils (mostly foraminifers). They represent life just before the impact. Above the fossiliferous white layer is a green clay thought to represent the dust and ash from the impact explosion. The green layer is capped by a red clay that contains no fossil at all, suggesting the plankton had been exterminated. Finally, at the top of the core, fossils again appear, indicating that several thousand years later recovery was beginning to take place.

Does the timing of the proposed impact correlate with extinctions that were widespread at the end of the Cretaceous? Rocks that had been melted by the event were subjected to $^{40}Ar/^{39}Ar$ analysis and provided an age of 65.2 ± 0.4 million years. In addition, these same so-called melt rocks acquired remanent magnetism that indicates they solidified during the episode of reverse geomagnetic polarity known to exist at the time of deposition of the K/T boundary layer.

ENRICHMENT

Is There a Bolide Impact in Our Future?

After reading about the possibility that a large extraterrestrial body smashed into the Earth about 65 million years ago, causing a biologic catastrophe, one cannot help but wonder about the chance of a similar disaster for the future human population. Although often the subject of science fiction, a bolide impact capable of destroying civilization is not an unrealistic cause for concern. Meteorites the size of cobbles and boulders are frequent visitors to the Earth. Most do little harm, although globally about 16 buildings a year are damaged by meteorites. (Police could do nothing to help the French motorist who recently reported a huge hole in his parked automobile made by a meteorite.) Only a small chunk of extraterrestrial rock ruined the Frenchman's car. What might be the result of an impact like the one that produced Meteor Crater of Arizona (see figure) only about 50,000 years ago? Thought to have had a diameter of about 30 meters, that meteor excavated a crater 1.2 kilometers wide while releasing energy equivalent to a 20-megaton atomic bomb. Meteor Crater is only one of about 130 such impact structures now recognized on Earth.

A meteorite in the size range from 50 to several hundred meters is believed to strike the Earth every 200 to 300 years. In 1908, one such object thought to have been about 60 meters in diameter entered the Earth's atmosphere and exploded above the Tunguska Valley in Siberia. The explosion, which was heard as far away as London, flattened and burned vegetation across 50 square kilometers of forested area. That event, however, would be trivial compared to the catas-trophe produced by a 1000-meter-wide impacting body. Such a smashup, releasing the energy equivalent of a million megatons of TNT, would eject a gigantic cloud of fiery rock and dust into the atmosphere. Great firestorms would sweep across the continents. Dust and smoke would blot out the Sun and plunge the world into total darkness. The loss of sunlight for months would prevent photosynthesis and cause mass extinctions of plants and animals both on land and in the ocean. At the very least, the event would have the potential to wipe out over a quarter of the planet's human population.

As an indication that the possibility of future bolide impacts should be taken seriously, space scientists are currently examining ways to detect, track, and intercept bodies headed our way. The program has been dubbed Spacewatch and includes a global network of telescopes for early detection of approaching large meteorites and asteroids. If the incoming body is spotted sufficiently early, it may be possible to push it into a nonthreatening trajectory by exploding a nuclear device off to its side. On the side nearest the explosion, heat from radiation would vaporize the surface of the meteorite, and the resulting jet of vapor might act as a rocket in nudging it off course. (It would be important that the bomb not strike the body directly, for this might cause it to shatter and result in a lethal rain of rock fragments.) Other plans are still being formulated as scientists grapple with possible ways to safeguard civilization against a future cosmic catastrophe.

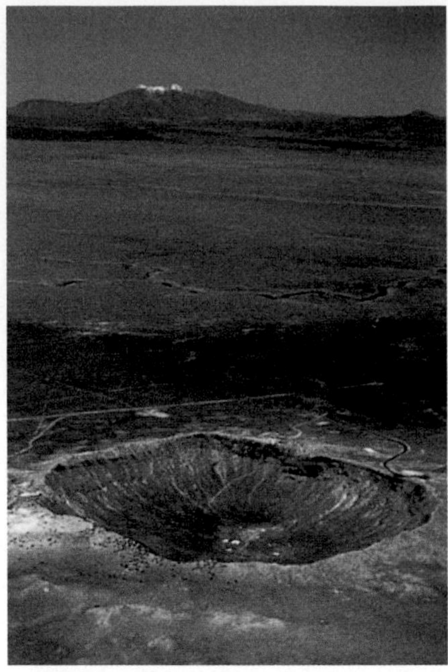

Aerial view of Meteor Crater, formed by the impact of a large meteorite about 50,000 years ago. Coconino County, Arizona. (*Courtesy of W. B. Hamilton.*)

If the Chicxulub structure is indeed an impact crater, it is the largest known ever to have been blasted out of the Earth. The bolide that produced it would have had an estimated diameter of about 10 kilometers. As it splashed down, it would have produced a wave over 34 meters high that would flood land areas in and around the Caribbean and Gulf of Mexico. Rocks would be vaporized, and the atmosphere would fill with dust, water vapor, and carbon dioxide from melted limestones. Could life on Earth sustain such an event?

AN OPPOSING VIEW On first examination, the bolide-impact hypothesis seems a tidy way to account for the extinction of dinosaurs and many of their animal and plant contemporaries. Like all hypotheses, however, it must stand the test of scrutiny by the scientific community. Were the Alvarezes correct in assuming that the iridium was derived from an impacting asteroid, or could there be another explanation? Geologists have found evidence that iridium in clays such as that at Gubbio may have its source in the Earth's mantle, from which it can move to the surface by way of conduits and blast into the atmosphere as iridium-rich volcanic ash and dust. Volcanism was indeed prevalent during the late stages of the Cretaceous. Especially significant were the tremendous outpourings of lava in India.

Known as the **Deccan traps**, these lava flows cover a large part of the Indian Peninsula. They were extruded about 65 million years ago and are associated with the separation of tectonic plates. Intensive volcanic activity was also vigorous in many other parts of the world, including western North America, Greenland, Great Britain, Hawaii, and the western Pacific. At the same time a conspicuous magnetic reversal occurred. Such reversals are thought to reflect some major event at the core-mantle boundary. Whether that event deep within the Earth caused global volcanism and whether a bolide triggered the reversal are unanswered questions.

Could the widespread volcanism have caused lethal global warming by pumping huge amounts of carbon dioxide into the Late Cretaceous atmosphere? A clue to the amount of atmospheric carbon dioxide can be found in the stomata of fossil leaves. (Stomata are small pores in leaves that are flanked by cells which take in carbon dioxide and expell oxygen.) Paleobotanist David J. Beerling discovered that fossil leaves of ferns living immediately after the terminal Cretaceous extinction event exhibited a marked increase in the number of stomata indicating a sudden increase in atmospheric carbon dioxide. He suggested that the increase was caused by a bolide impact that vaporized calcium carbonate sediment so as to release large amounts of carbon dioxide. The carbon dioxide concentration persisted for only a short time. If Deccan volcanic activity were responsible for carbon dioxide induced lethal global warming, there would be indications of persistent carbon dioxide over a span of about 2 million years. The evidence from counts of stomata on fossil leaves indicates a short burst of carbon dioxide concentration. Thus, as a cause of mass extinctions, bolide impact is favored over the effects of volcanism.

COSMIC RAYS If there were no actual impacts of comets, asteroids, or large meteorites to cause extinctions at the end of the Cretaceous, might there still have been other, more subtle but equally effective extraterrestrial causes? One culprit might have been an influx of abnormally high amounts of cosmic radiation. Such radiation might have destroyed organisms directly or severely impaired their reproductive capabilities. Most proponents of this idea suggest that reversals of the Earth's magnetic field (a phenomenon known to occur) may have temporarily eliminated the planet's protection from harmful rays. The hypothesis, however, has some aspects that are difficult to defend. It does not explain why marine organisms, which could descend a few meters and escape radiation damage, were decimated at levels far in excess of those among unprotected land plants. Also, several known periods of polar reversals do not correlate to episodes of mass extinction. In addition, the Mesozoic extinctions spanned a time interval far greater than that encompassed by documented magnetic reversals.

Another possibility for radiation damage unrelated to magnetic reversals might have been the arrival of a blast of lethal rays from a nearby supernova. Supernovas are colossal stellar explosions that radiate about as much energy as 10 billion suns and can affect the surface of planets located as far as 100 light-years away (1 light-year is approximately 8898 billion kilometers). Astronomers estimate that, on the average, a supernova may occur within 50 light-years of the Earth every 70 million years. If such an event did take place, there seems little doubt that plant and animal communities would have been severely affected. For the present, however, we must await firm evidence that such a stellar explosion did occur about 65 million years ago.

Terrestrial Causes of Extinctions

Several hypotheses that seek to explain the demise of plants and animals near the end of the Cretaceous propose that events on Earth upset the ecologic balance between organisms and the environment to which they had become adapted. Recall from Chapter 11 that continents during the Cretaceous were extensively covered by shallow, warm, epeiric seas. Many geologists believe these inundations were caused by displacement of ocean water when midoceanic ridges were raised as a consequence of accelerated rates of sea-floor spreading. Whatever the cause, the warm inland seas that resulted were very favorable for marine life and helped to moderate and stabilize climates on the continents as well. Under such conditions, plants and animals experienced remarkable increases in variety and abundance. However, these favorable conditions were soon to change. Studies of the stratigraphic sequence across the Cretaceous–Tertiary boundary indicate a global lowering of sea level at the end of the Mesozoic. Perhaps this change in sea level was caused by a slowing of rates of sea-floor spreading. Whatever its cause, the change in sea level spelled disaster to animals and plants of formerly extensive shallow coastal areas and especially to the phytoplankton adapted to the shallow sea environment. Without the moderating effects of vast epeiric seas, landmasses would also have experienced harsher climatic conditions and more extreme seasonality. Many families of organisms that had become adjusted to the previous environment might not have been able to adapt to the changes and, one by one, would have met their end. As is characteristic within the complex ecologic web, their demise would affect other organisms, resulting in waves of extinctions among directly or indirectly dependent species.

There are many other variations on the general theme of environmental factors that can cause extinctions. Some geologists do not believe a loss of epeiric seas could have caused the mass extinction at the K/T boundary. They invoke other events that put stress on the global ecosystem. For example, there is evidence that the Arctic Ocean was landlocked during the Cretaceous, with the result that it became a freshwater body.

Near the end of the Cretaceous, tectonic plate movements broke the land barrier containing the Arctic water, allowing low density, nonsaline water to flow southward over denser, salty Atlantic waters. Marine plankton living in the vital upper zones of the ocean would probably be unable to tolerate the change in water chemistry and marine circulation. A chain of ecologic disasters would have been a possible result.

As described earlier, the major hypothesis for a terrestrial cause of Cretaceous extinctions proposes that the die-off resulted from widespread volcanism. Volcanoes produce dust and aerosols, such as sulfuric acid, that block solar radiation and cause a decline in temperatures. The sulfuric acid would result in acid rain, and such precipitation might have changed the alkalinity of the oceans, placing lethal stress on plankton, other invertebrates, and life higher in the food chain. Thus, volcanism can be considered harmful to animals both on land and in the sea.

There are still other hypotheses to account for the Late Cretaceous extinctions, but the examples provided here illustrate the complexity of the problem. The debate between those favoring sudden extraterrestrial causes and those supporting purely terrestrial hypotheses will surely continue for decades. Whatever the final outcome, it is a fact that hard times near the end of the Mesozoic doomed the dinosaurs, pterosaurs, ammonites, and over three fourths of the known species of marine plankton. The extinctions, however, did not happen at the same time for all groups. Indeed, many groups died out sporadically over an interval of 0.5 to 5 million years. This fact is considered an argument against a sudden extraterrestrial catastrophe. Perhaps the real cause of extinctions will be found in a combination of the two categories of hypotheses. The paleontologic evidence indicates that extinctions were widespread in the Late Cretaceous. The debilitated and diminished survivors of environmental hard times may have been dealt a *coup de grace* by an extraterrestrial event. Fortunately, small mammals, birds, lizards, crocodiles, turtles, many fish groups, many deciduous plants, and certain mollusks survived the hard times and proliferated during the following Cenozoic Era.

SUMMARY

Climates of the Mesozoic were in general mild and equable, except for occasional intervals of aridity and an episode of cooler conditions near the end of the era. In the widespread Mesozoic seas, coccoliths and diatoms flourished, as did such invertebrate groups as ammonoids, belemnites, oysters, and other bivalves, echinoderms, corals, and foraminifers. On land, cycads and conifers were common in Triassic and Jurassic forests. In the Cretaceous Period, the flowering plants expanded and with them a multitude of modern-looking insects. The changing composition of plant populations was matched by innovations among terrestrial vertebrates. The most dramatic of these changes involved the evolution of dinosaurs, pterosaurs, and birds from the small bipedal archosaurs of the Triassic.

Dinosaurs were the ruling vertebrates of the Mesozoic. Both carnivorous and herbivorous varieties occupied a variety of habitats. Based on differences in pelvic structure, two groups of dinosaurs, the Saurischia and Ornithischia, are recognized. Certain species of dinosaurs as well as the pterosaurs may have been endothermic, as indicated by bone structure, posture, predatory-prey relationships, and the presence of feathers on some theropods. The archosaurian and reptilian dynasty extended to the oceans as well, where ichthyosaurs, plesiosaurs, mosasaurs, and large turtles competed successfully with sharks and bony fishes. The Mesozoic is also noteworthy as the era during which mammals and birds appeared. The birds, amply feathered for insulation and flight, may have evolved from small carnivorous dinosaurs (theropods). The oldest remains that are unquestionably those of a true bird are of *Archaeopteryx*, from the Jurassic.

Mammals made their debut during the Triassic. In general, most of these primitive mammals remained small and inconspicuous. One group of primitive prototherian mammals gave rise to marsupial and placental mammals during the Cretaceous.

Like the Paleozoic, the Mesozoic Era closed with an episode of extinctions. Dinosaurs, pterosaurs, plesiosaurs, mosasaurs, ammonoid cephalopods, belemnites, many groups of bivalves, and large numbers of foraminiferal groups became extinct. Geologists are not agreed on the cause of terminal Cretaceous extinctions. There is, however, an impressive body of evidence that a large meteorite or asteroid struck the Earth about 66 million years ago, and the effects of the impact may have caused or contributed to the mass extinction. Others argue that the extinctions can be attributed to extensive volcanic activity, withdrawal of epeiric seas, changes in oceanic circulation, or climatic changes.

QUESTIONS FOR REVIEW AND DISCUSSION

1. During what geologic period of the Mesozoic is chalk particularly abundant? What group of organisms provides skeletal remains for chalk deposition? Why are chalk formations rare in Paleozoic rocks?

2. What are diatoms? How do they differ in composition and morphology from coccolithophorids? How might the distribution of diatoms and coccolithophorids be related to the acidity or alkalinity of ocean water?

3. How did the terrestrial plant flora of the Jurassic differ from that of the Cretaceous? What environmental conditions might have driven the change in flora?

4. What are ammonoid cephalopods? What attributes of ammonoids result in their having special value as guide or index fossils?

5. What are foraminifers? Why are they of particular value to petroleum geologists involved in the correlation of subsurface strata?

6. Discuss the differences between the marine invertebrate faunas of the Mesozoic and those of the Paleozoic. What Paleozoic groups are not seen in the Mesozoic?

7. Discuss several lines of evidence that indicate that certain Mesozoic reptile groups were endothermic. Why would endothermy be less important for the giant saurischians as compared to small dinosaurs?

8. What attributes of Cretaceous mammals may have contributed to their survival during the biologic crisis at the end of the Mesozoic?

9. Discuss the differences between world geography at the end of the Permian as compared to the end of the Cretaceous.

10. What two classes of vertebrates appear for the first time during the Mesozoic Era?

11. Cite two examples of evolutionary convergence among animals living during the Mesozoic Era.

12. Describe the particular evolutionary importance of the following: (a) basal archosaurs (b) *Archaeopteryx*, and (c) *Eomaia*.

13. Prepare a list of the reptilian groups that survived the mass extinction at the end of the Cretaceous Period.

14. Oceans cover about 71 percent of the Earth's surface, yet evidence of impact craters on the ocean floor is rarely seen. Why?

15. What evidence at the boundary between the Cretaceous and Tertiary systems at many localities favors the bolide impact hypothesis for the extinction of the dinosaurs and many other animal groups? What arguments can be advanced against this popular hypothesis?

16. When were the Deccan traps extruded? Discuss the possible role of this and other synchronous volcanism as a cause of mass extinctions.

17. Formulate a simple, single-sentence definition for each of the animals or animal groups listed below:

anapsid	ornithischian	phytosaur
ammonoid	ankylosaur	theropod
rudist	ornithopod	plesiosaur
teleost	stegosaur	ceratopsian
diapsid	saurischian	pterosaur
mosasaur	belemnite	thecodont

READINGS

Alvarez, W., and Asaro, F. 1990. What caused the mass extinction? An extraterrestrial impact. *Sci Am*. 246:78–84.

Archibald, J. D. 1996. *Dinosaur Extinction and the End of an Era*. New York: Columbia University Press.

Bakker, R. T. 1986. *The Dinosaur Heresies*. New York: Wm. Morrow & Co.

Carpenter, K., Hirsch, K. F., and Horner, J. R. eds. 1994. *Dinosaur Eggs and Babies*. Cambridge, England: Cambridge University Press.

Chatterjee, S. 1997. *The Rise of Birds*. Baltimore: Johns Hopkins University Press.

Colbert, E. H., and Morales, M. 1991. *Evolution of the Vertebrates*. New York: John Wiley & Sons.

Courtillot, V. E. 1990. What caused the mass extinction? A volcanic eruption. *Sci. Am*. 246:85–92.

Desmon, A. J. 1976. *The Hot-Blooded Dinosaurs*. New York: Dial Press.

Dodson, P. 1996. *The Horned Dinosaurs*. Princeton: Princeton University Press.

Feduccia, A. 1996. *The Origin and Evolution of Birds*. New Haven, CT: Yale University Press.

Gillette, D. D. 1994. *Seismosaurus the Earth Shaker*. New York: Columbia University Press.

Horner, J. R. 1984. The nesting behavior of dinosaurs. *Sci. Am*. 250:130–137.

Lucas, S. G. 2000. *Dinosaurs, The Textbook*, 3rd ed. Dubuque, IA: Wm. C. Brown.

Norell, M. A., Gaffney, E. S., and Dingus, L. 1995. *Discovering Dinosaurs*. New York: Alfred A. Knopf, Inc.

Paul, G. S. 1988. *Predatory Dinosaurs of the World*. New York: Simon & Schuster.

Prothero, D. P., and Schoch, R. M., eds. 1994. *Major Features of Vertebrate Evolution*. Knoxville: University of Tennessee Press.

Weishample, D. B., Dodson, P., and Osmolska, H. 1990. *The Dinosauria*. Berkeley, CA: University of California Press.

WEB SITES

The Earth Through Time Student Companion Web Site (www.wiley.com/college/levin) has online resources to help you expand your understanding of the topics in this chapter. Visit the Web Site to access the following:

1. Illustrated course notes covering key concepts in each chapter;
2. Online quizzes that provide immediate feedback;
3. Links to chapter-specific topics on the web;
4. Science news updates relating to recent developments in Historical Geology;
5. Web inquiry activities for further exploration;
6. A glossary of terms;
7. A Student Union with links to topics such as study skills, writing and grammar, and citing electronic information.

13

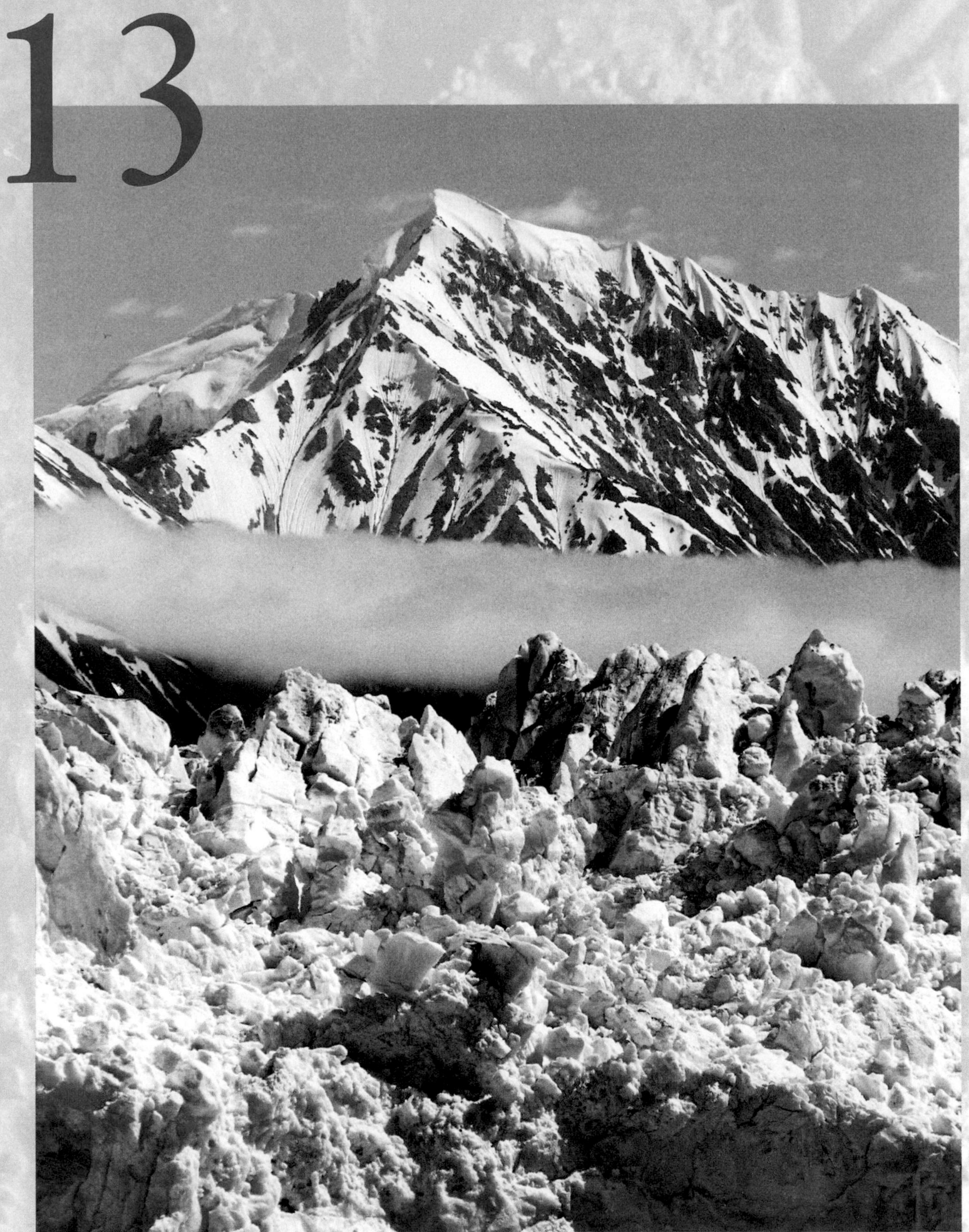

Terminus of the Hubbard Glacier at the head of Yakutat Bay, Alaska. The Hubbard Glacier is one of the largest in Alaska, extending nearly 90 miles. (Photograph by Dean Mitchell, with permission.)

CHAPTER 13

The Cenozoic Era

Many an aeon moulded earth before
her highest, man, was born,
Many an aeon too may pass when
earth is manless and forlorn.
Earth so huge and yet so bounded—
pools of salt and plots of land—
Shallow skin of green and azure—
chains of mountains, grains of sand.
A. Tennyson, *Locksley Hall: Sixty Years After*, 1866

OUTLINE

The Cenozoic is the final era of geologic history. It is the era during which the continents and their landscapes acquired their present form. The evolution of animals and plants of the present world was shaped and modified by Cenozoic events.

There are two sets of subdivisions used for the Cenozoic (Table 13–1). In a scheme long used by geologists, the era is separated into the **Tertiary** and **Quaternary** periods. Many geologists, however, prefer a timetable that divides the era into **Paleogene**, **Neogene**, and **Quaternary** periods, arguing that this represents a more equal division of the epochs and also a more natural way of dividing Cenozoic rocks of Europe. The Paris basin is the type area for most of the Cenozoic epochs. In that area, a major unconformity representing a marine regression does indeed serve as a boundary between the Paleogene and Neogene systems.

The Cenozoic was and is a time of considerable tectonic plate motion and sea-floor spreading. It has been estimated that approximately 50 percent of the present ocean floor has been renewed along midocean ridges during the past 65 million years. Much of this new ocean floor was emplaced in the expanding Atlantic and Indian oceans. As this widening progressed, the Americas moved westward. The area that is now California came into contact with the northward-moving Pacific plate and thereby produced the San Andreas fault system. South America moved firmly against the Andean trench and actually bent and displaced it. Orogenic and volcanic activity was vigorous along the western backbone of the Americas and resulted in the formation of the Isthmus of Panama, which today links North and South America. The Panamanian land bridge blocked the westward progression of the North Atlantic Current, causing it to swing northeastward as the Gulf Stream. The warm Gulf Stream had the effect of bringing warmer climates to northwestern Europe. The mild climate of the British Isles and the fact that harbors in Norway are ice-free throughout the year are a consequence of the influence of the Gulf Stream.

Also during the Cenozoic, the North Atlantic rift extended toward the North Pole, separating Greenland from Scandinavia and destroying the land connection between Europe and North America. During the Late Eocene, Australia separated from Antarctica

TABLE 13–1 Geochronologic Terminology Used for Divisions of the Cenozoic Era

Era	Period		Epoch	
Cenozoic	Quaternary		Holocene	
			Pleistocene	1.75 m.y.
	Tertiary	Neogene	Pliocene	5.3 m.y.
			Miocene	23.5 m.y.
		Paleogene	Oligocene	38.7 m.y.
			Eocene	53 m.y.
			Paleocene	65 m.y.

Source: After the 1989 Global Stratigraphic Chart of the International Union of Geological Sciences.

(Fig. 13–1) and began its journey to its present location. This was the only major continental break-up during the Cenozoic, and it appears to have affected climates around the globe. Prior to the separation of the two continents, Antarctica was warmed by ocean currents flowing toward it from warmer regions to the north. By Oligocene time, however, a frigid northward flowing current had developed in the widening rift between the two continents. The cold current deflected the warmer waters that had formerly given Antarctica a less severe climate. Around the now isolated continent of Antarctica, cold circumpolar currents were set in motion, and soon the continent began to assume the level of frigidity that characterizes it today. Water made dense because of its extreme cold sank to the bottom of the ocean surrounding Antarctica and began to drift northward along the ocean floor, exterminating benthonic invertebrates adapted to warmer conditions as it did so. It is likely that the transfer of frigid waters northward contributed not only to cool conditions during the Late Eocene and Oligocene but also to the eventual development of the Pleistocene Ice Age.

During the Cenozoic, a branch of the Indian Ocean opened between Arabia and Africa, creating the Gulf of Aden and the Red Sea (Fig. 13–2). But the most dramatic Cenozoic tectonic event was the collision of Africa and India with Eurasia. This titanic smashup transformed a large part of the Tethys seaway into lofty mountain ranges, among which are the present Alps and Himalayas.

The interior regions of continents stood relatively high during the Cenozoic, and as a result, marine transgressions were limited. As indicated by the distribution of tropical and subtropical plants, warm climates were characteristic of the early Cenozoic. By mid-Cenozoic, however, temperate floras began to spread across the continents, and plants requiring warmer conditions retreated to lower latitudes. A reflection of the change was the development of extensive grasslands. The cooling trend continued throughout the Cenozoic, culminating in Pleistocene glaciation.

BEFORE THE ICE AGE

Eastern United States

The eastern margin of North America experienced little orogenic activity during the Cenozoic. Erosion continued in the Appalachians. Periodically, as the uplands were beveled by erosion, broad, isostatic uplifts occurred. Streams, revitalized by the uplifts, sculpted a new generation of ridges and valleys. The most recent of the uplifts in the Appalachian belt was accompanied by gentle tilting of the Atlantic coastal plain and adjacent parts of the continental shelf. Terrigenous clastic sediments were carried out of the highlands by streams and deposited on the plains and, on occasion, were reworked by waves and currents of marine transgressions. The Cenozoic section of marine rocks is thinner near the Appalachian source areas and becomes thicker and less clastic toward the region that lies offshore (Fig. 13–3). Southward, in the vicinity of Florida, terrigenous clastics were less available. Carbonate sediments accumulated to thicknesses of more than 2500 meters along a series of subsiding, elongate coralline platforms that were probably much like the Bahama Banks (Fig. 13–4). In the final ages of the Tertiary, uplift along the northern end of this tract raised the land area of Florida above the waves.

Gulf Coast

The best record of Cenozoic strata in North America is found in the Gulf of Mexico coastal plain (Fig. 13–5). Altogether, eight major transgressions and regressions are recorded in this region. The Paleocene transgression brought marine waters as far north as southern Illinois. Frequently, during marine regressions, nearshore deltaic sands were deposited above offshore shales. The resulting interfingering of permeable sands and impermeable clays provided ideal conditions for the eventual entrapment of oil and gas. in this regard, Cenozoic rocks of the Gulf Coast are famous for the petroleum they once yielded in quantities far greater than today. Much of the oil was trapped in structures associated with salt domes (see Fig. 11–9). Oil wells drilled in the Gulf Coast have penetrated more than 200 of these salt domes, and many more exist beneath the continental shelf.

"A wedge of sediments that thickens seaward" is a particularly suitable description for the Cenozoic formations of the Gulf Coast. Geophysical measurements

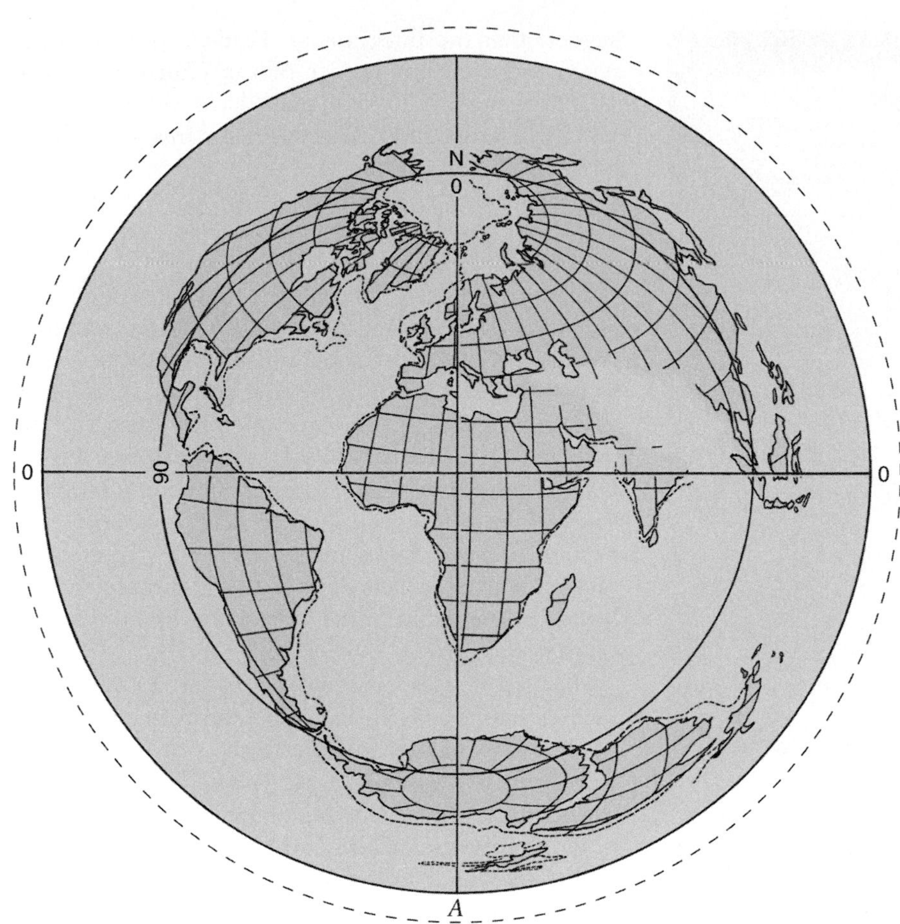

FIGURE 13-1 Eocene Global Paleography. (*A*) Location of major landmasses during the Eocene about 50 ± 5 million years ago. Note that Antarctica and Australia are still connected. *Although Antarctica lay astride the South Pole, it did not experience extreme cold. How do we know this, and what would have caused its mild Eocene climate?* (*B*) The world as it is today plotted on the same (Lambert Equal Area) projection. (*From Briden, J. C., Drewey, G. E., and Smith, A. G. 1974.* J. Geol. *82:556–558.*)

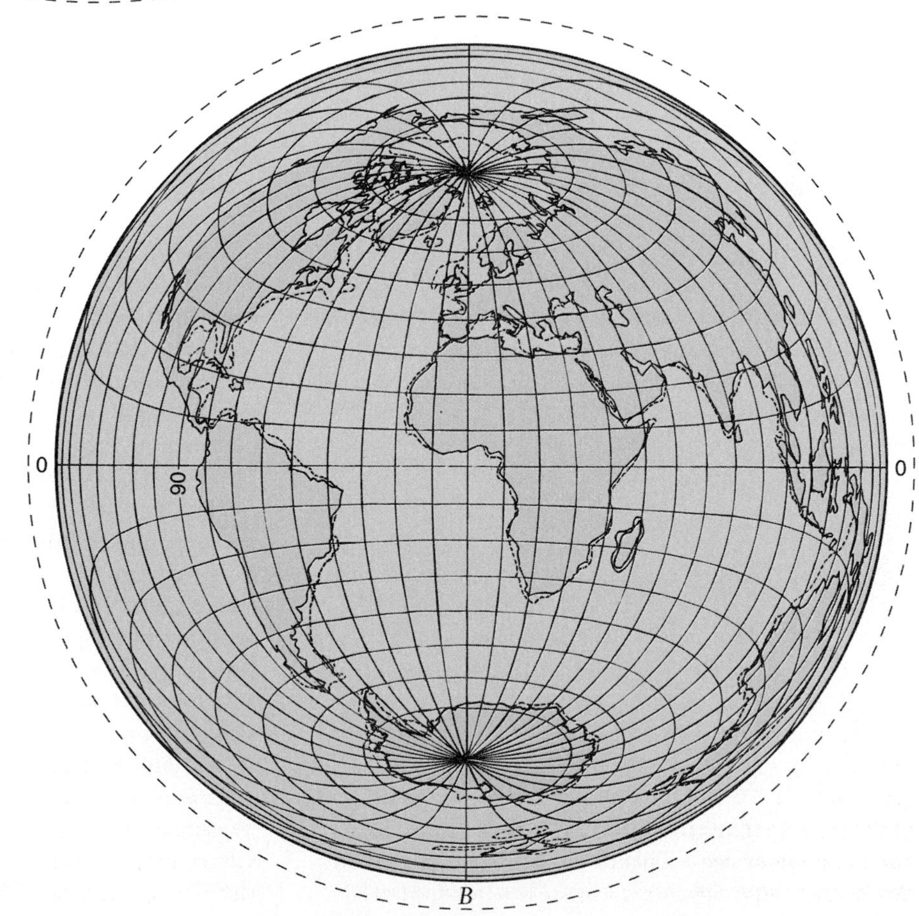

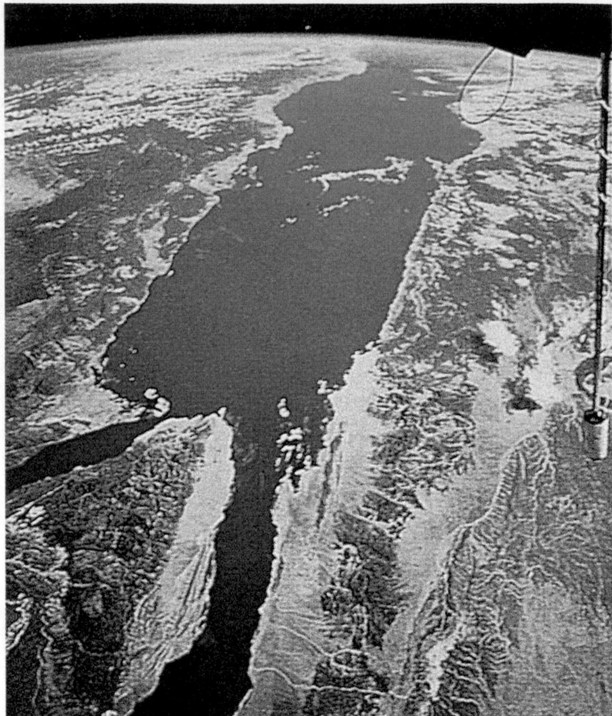

FIGURE 13-2 The Red Sea, shown here in a photograph taken from the Gemini spacecraft, formed about 30 million years ago when the Arabian Peninsula broke away from Africa. The seaway continues to widen today. The view is toward the south. The wedge of land in the lower left is the Sinai Peninsula, bordered on the right by the Gulf of Suez and Egypt, and on the left by the Gulf of Aqaba and Saudi Arabia. (*Courtesy of NASA.*)

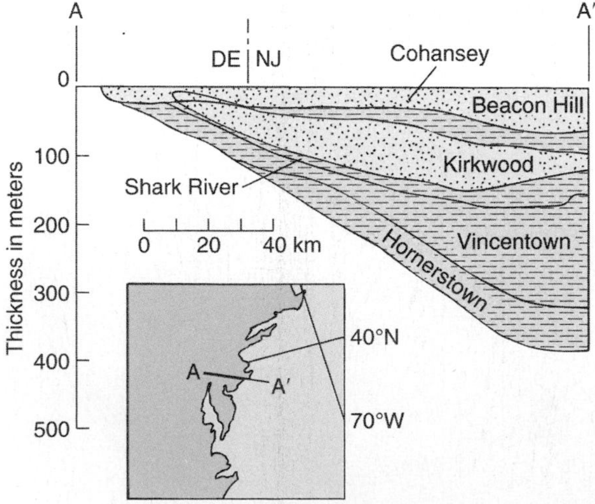

FIGURE 13-3 Cross-section of Tertiary strata across the New Jersey coastal plain. (*From Stratigraphic Atlas of North and Central America, courtesy of Shell Oil Company.*)

suggest that the thickness of Tertiary sediments beneath the northern border of the Gulf may exceed 10,000 meters. The area must have been subsiding rapidly in order to provide space for this great thickness of sediment.

Rocky Mountains and High Plains

While marine sedimentation was under way along the eastern coastal regions, terrestrial deposition prevailed in the Rocky Mountain region. The exception was a single marine incursion (the Cannonball Sea) during the Paleocene (Fig. 13–6). The strata recording the presence of this seaway are found in western North Dakota and consist of dark shales containing over 150 species of invertebrates. There is no evidence of a connection between the Cannonball Sea and either the Gulf of Mexico or the Arctic, and it is therefore believed to be a vestige of the more extensive Late Cretaceous epeiric sea.

The Late Cretaceous and Early Tertiary mountain building described in Chapter 11 were largely responsible for the major structural features of the Western Cordillera. However, the present topography of this region is due primarily to erosion following uplifts that began during the Miocene. As always, erosion acting on the tilted and folded hard and soft layers was the final factor in shaping the landscape. Following Late Cretaceous orogenesis, erosional debris was trapped largely in basins between the mountain ranges. These intermontane terrestrial sediments provide the record of the earlier Tertiary epochs. Later uplifts and erosion resulted in the spectacular relief of the Rockies and caused the detritus of the older basins and newer uplands to be spread over the plains that lay to the east. The result was the creation of a vast apron of nonmarine Oligocene-through-Pliocene sands, shales, and lignites that underlies the western high plains. Beds of ash interspersed among the sediments attest to periodic episodes of volcanic pyrotechnics.

The lower Tertiary sedimentary rocks formed from continental sediments deposited in intermontane basins consist of gray siltstones and sandstones, as well as carbonaceous shales, lignite, and coal. These rocks are well represented in the Fort Union Formation, a unit that can be recognized in several intermontane basins in the Rocky Mountain region. The Fort Union is approximately 1800 meters thick (5900 feet). In its lower levels, it contains immense tonnages of low-sulfur coal. The coal beds are indicators of widespread swampy conditions during Paleocene time.

Eocene deposits of the Rocky Mountains include such well-known formations as the Claron (formerly Wasatch) and Green River. The sediments of the Claron Formation were deposited by streams, whereas the Green River is a lake deposit. The lowermost beds

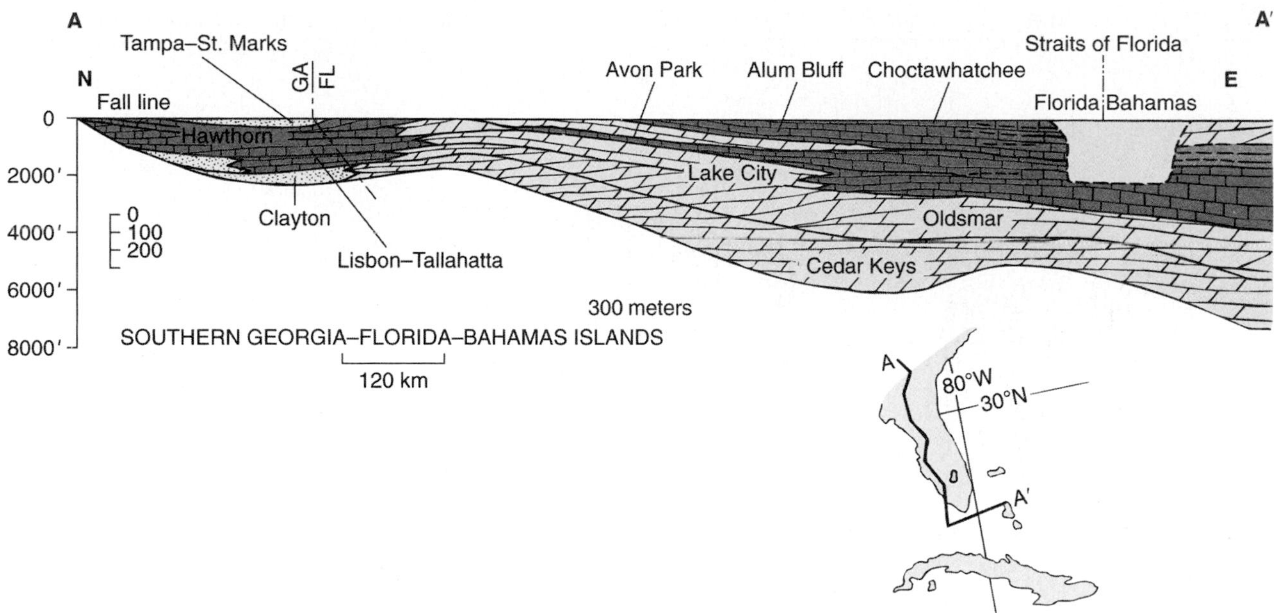

FIGURE 13-4 **Cross-section of Tertiary strata from southern Georgia to the Bahamas.** (*From Stratigraphic Atlas of North and Central America, courtesy of Shell Oil Company.*)

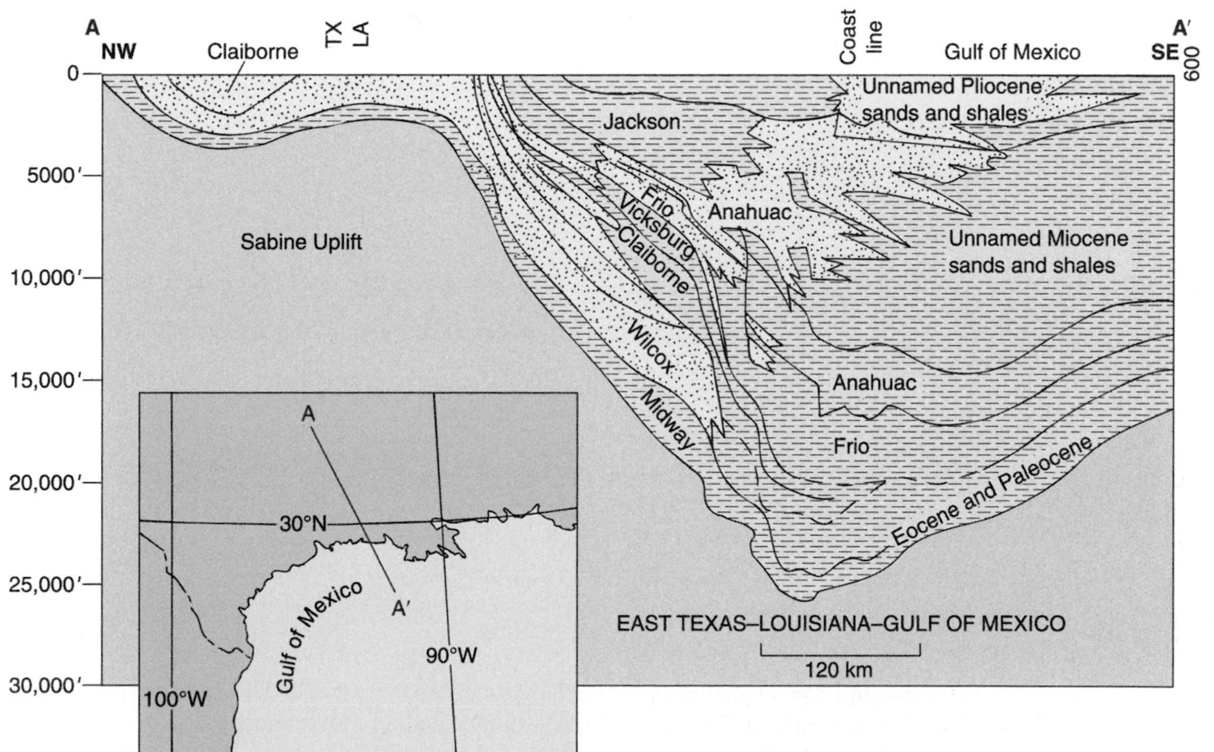

FIGURE 13-5 **Cross-section of Tertiary strata across the Gulf coastal plain and Gulf of Mexico.** (*From Stratigraphic Atlas of North and Central America, courtesy of Shell Oil Company.*)
? *Fossils of marine invertebrates indicate that most of the formations shown here accumulated at shallow to moderate depths of water. How, then, could 27,000 feet of sediment accumulate in water depths that rarely exceeded 1000 feet?*

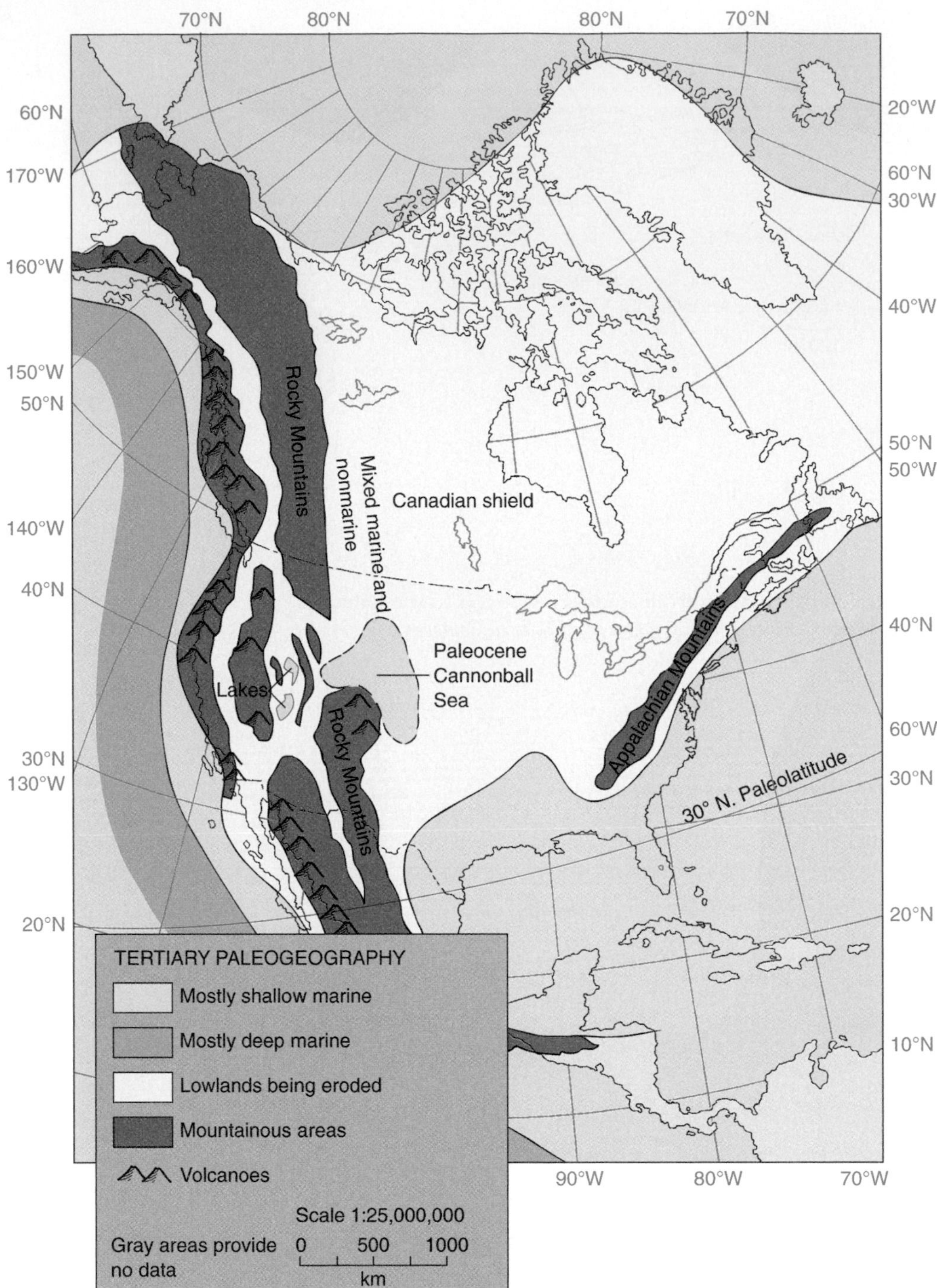

FIGURE 13-6 Paleogeographic map of North America during the early Tertiary.

of the Wasatch are often coarse and gravelly, but at higher levels finer clastics are most common. These include red mudstones interbedded with mudstones and lenticular sandstones. The bright colors contrast markedly with the underlying drab, gray shales of the Fort Union Formation. At Bryce Canyon (Fig. 13–7) and Cedar Breaks in Utah, the attractively colored Cedar Breaks Formation is eroded into the pinnacles of a magnificent Badlands topography.

Many of the streams flowing from the Wind River and Owl Creek mountains during the early Cenozoic had no outlet. As a result, water filled many of these basins, forming large lake basins. Some of these lakes, represented by its lacustrine deposits, formed in the

FIGURE 13-7 Cedar Breaks National Monument in Utah. Here the Eocene Cedar Breaks Formation has been eroded into steep ravines, pinnacles, and razor-sharp divides. Early explorers in the region used the term "breaks" to describe the change in topography where an elevated level area "breaks down" by eroding to a lower elevation. (*Copyright Charles Ott/Photo Researchers, Inc.*)

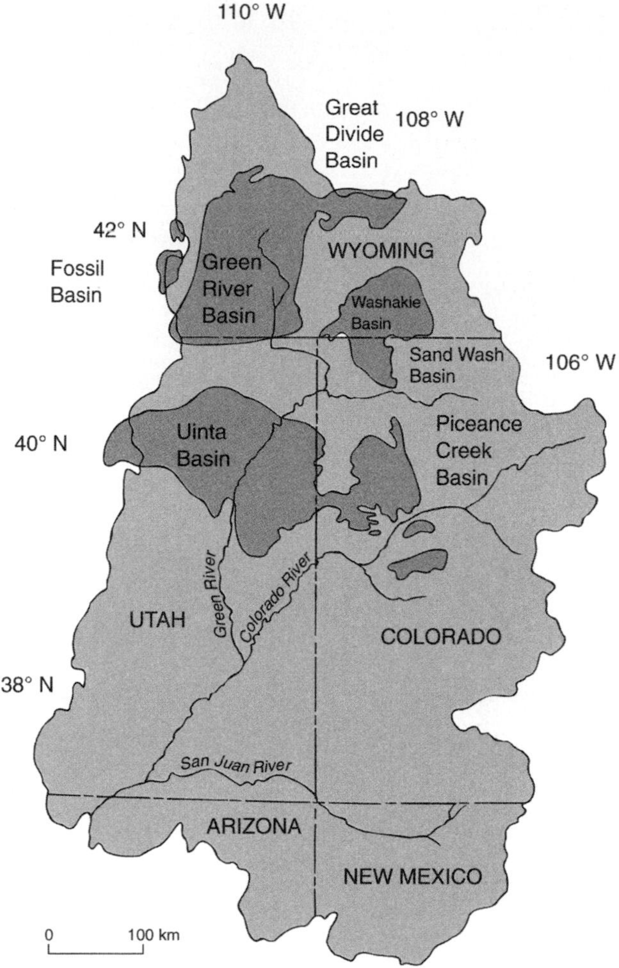

FIGURE 13-8 Cenozoic basins of Colorado, Utah, and Wyoming containing important oil shale deposits. Map boundary is the Upper Colorado River drainage basin. (*Simplified from Rickert, D. A., Ulman, W. J., and Hampton, E. R. [eds.] 1979.* Synthetic Fuels Development, *U.S. Geologic Survey publication.*)

Green River Basin of southwestern Wyoming (Fig. 13–8). Sediments deposited in the basin comprise the Green River Formation, noted earlier (Fig. 13–9). Over 600 meters of freshwater limestones and fine, evenly laminated shales were deposited in the Green River basin. The laminations are varves, each of which consists of a thin, dark, winter layer and a lighter-colored summer layer. By counting of the varves, it has been determined that the Green River sediments were deposited over a period of more than 6.5 million years. Within the fine shales are well-preserved fishes (Fig. 13–10), insects, and plant fragments. In addition, the shales are rich in waxy hydrocarbons. For this reason, they are called **oil shales** and can be processed to yield petroleum. The Claron Formation, also of Eocene age, underlies and interfingers laterally with the Green River beds. Several oil fields in Wyoming pump oil from the Claron Formation. The source rock for the hydrocarbons was probably the Green River Shale.

The Uinta basin south of the Green River basin is notable because it is structurally the deepest of those in the Colorado Plateau. It received a particularly thick succession of Tertiary sediments (Fig. 13–11). These formations can still be observed in exposures in the interior of the basin, where they have been only slightly affected by post-Eocene erosion. Strata along the northern edge of the basin are sharply upturned, forming hogbacks.

During the late Eocene and Oligocene, explosive volcanic activity blanketed the region that today includes

Pre-Є | Є | O | S | D | M | P | Pr | Tr | J | K | T | Q

FIGURE 13-9 The Green River Formation exposed in a road cut about 10 miles west of Soldier Summit, Utah. The exposure here consists of dark shales and thin, lighter-colored limestones.

FIGURE 13-10 Fossil of an Eocene freshwater fish (*Diplomystis*) from the Green River Formation, Wyoming. The fish is 8 centimeters long.

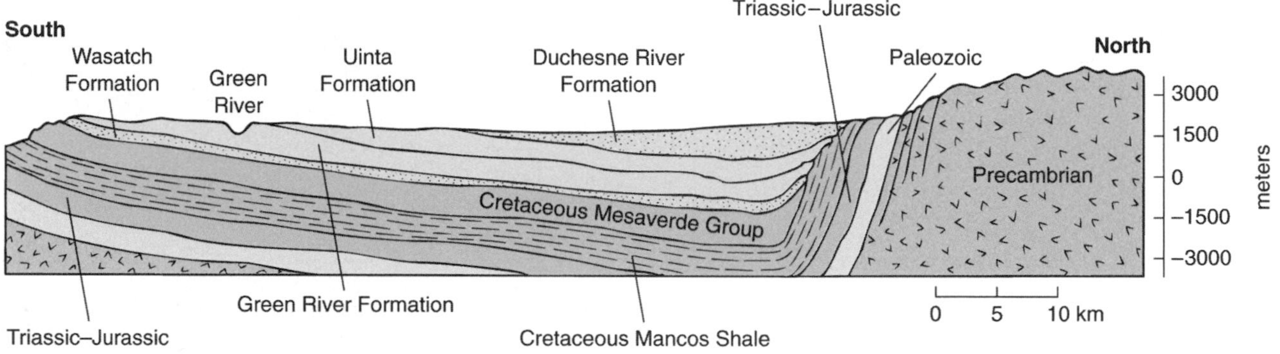

FIGURE 13-11 North–south cross-section through the Uinta basin. Tertiary units are above the Mesaverde Group.

Yellowstone National Park (Fig. 13–12) and the San Juan Mountains with layers of volcanic ash. For paleontologists, however, the more interesting rocks of this epoch are floodplain deposits of the White River Formation. This famous formation contains entire skeletons of Tertiary mammals in extraordinary number, variety, and preservation. If one recalls newsreels of recent floods, it is not difficult to understand how Oligocene mammals became buried in the sediment of overflowing streams. The clays, silts, and ash beds of the White River Formation are the sediments from which the Badlands of South Dakota have been sculpted (see Fig. A, Badlands National Park, South Dakota [Enrichment box]).

Another particularly interesting occurrence of Oligocene beds is exposed near Florissant, Colorado. Explosive volcanic activity in this area produced a great

FIGURE 13-12 Yellowstone Falls and canyon in Yellowstone National Park, Wyoming. Rocks exposed in the canyon walls consist of Lower Tertiary lava flows and volcanic ash, often altered to bright colors by hydrothermal activity. (*Courtesy of L. E. Davis.*)

Badlands National Park, South Dakota

Travelers driving west on Interstate 90 toward Rapid City, South Dakota, will miss an extraordinary vista if they fail to take Exit 131 and travel only 8.5 kilometers southward on Highway 240 to Badlands National Park. Immediately on entering the park, visitors are in a strange and alien land of steep ravines and ethereal spires. They are now in the world's best example of badlands topography (Fig. A).

Badlands are areas nearly devoid of vegetation, where erosion dissects the land into a labyrinth of chasms, razor-sharp ridges, and pinnacles. Bedrock of impermeable clays and shales, the absence of a protective cover of plants, and infrequent heavy rains are the primary conditions under which badlands form. All three of these conditions exist in Badlands National Park.

FIGURE A The Badlands of South Dakota. This rugged terrain has been carved from flat-lying Eocene and Oligocene shales, mudstones, and ash beds. (*Courtesy of R. F. Dymek.*) *What bedrock and climatic conditions favor the formation of badlands?*

The rocks in which Badlands National Park have been fashioned consist of horizontal layers of clay, shale, and volcanic ash. Much of the ash has been weathered to a clayey sedimentary rock called bentonite. In addition, there are beds of river sandstones. The sandstone layers are more difficult to erode. They often form a caprock at the top of buttes and provide an overhanging ledge for erosional features called mushroom rocks.

Running water has been the most important agent in the formation of the badlands. Rain falling suddenly during cloudbursts cannot infiltrate the impermeable clays and shales. It runs off and becomes channeled into numerous small rivulets that erode a dense system of ever-enlarging, coalescing gullies and ravines. The impact of raindrops as they pelt the soft surfaces dislodges particles of rock, speeding the erosional processes. Bentonite in the beds also facilitates erosion. The clay in bentonite swells enormously when it becomes wet. On drying, bentonite crumbles and disintegrates, forming masses of easily eroded sediment. The White River has taken its name from the whitish color imparted to the water by submicroscopic particles of bentonitic clays.

The actual formation of the topography seen in Badlands National Park began with an episode of regional uplift late in the Pliocene Epoch. As a result of the uplift, major streams were rejuvenated and entrenched themselves into the underlying poorly resistant beds. White River became one of these entrenched streams. The process provided steeper gradients to tributary streams that flowed southward into White River. The southern perimeter of the uplifted area that paralleled the White River Valley to the north became intricately gullied and dissected, forming an early badlands topography that was subsequently eroded away. As tributary streams continued to extend

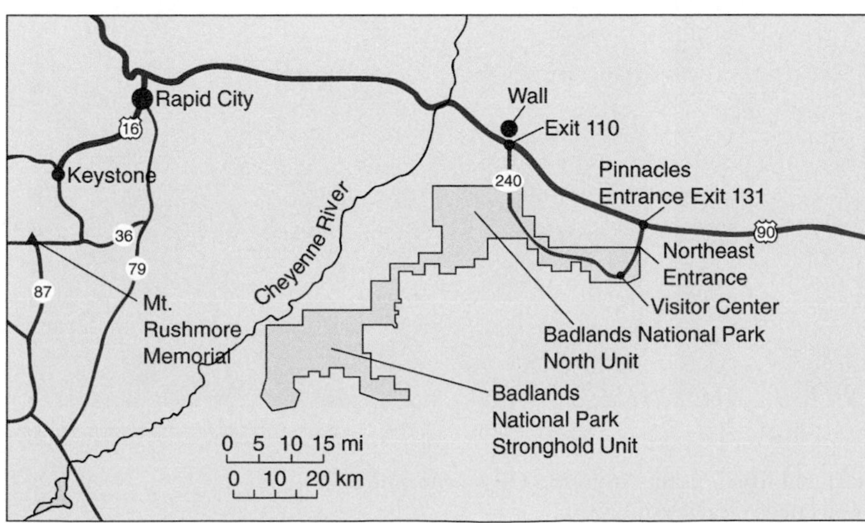

FIGURE B Location Map for Badlands National Park, South Dakota.

their channels northward into the upland area, the badlands we view today began their development.

Badlands National Park has an interesting geologic history. The oldest formation in the park is the Pierre Shale. The beds of the Pierre Shale were deposited in the shallow but extensive sea that covered much of western North America during the Late Cretaceous. The Cretaceous ended with regional uplift, and as the sea receded, erosion produced the unconformity on which the principal beds of the badlands were deposited. These are sediments of the Oligocene White River Group. They consist primarily of the deposits of slow-moving streams that flowed across a landscape characterized by wide floodplains, marshlands, lakes, and ponds. At this time, there was abundant plant food for a rich fauna that included turtles, lizards, alligators, huge titanotheres, aquatic rhinoceroses, three-toed horses, early camels, entelodonts, oreodonts, and tapirs. Many of these herbivores were prey for predatory members of the dog and cat families. Figure 14–44 provides a panoramic view of the environment and its inhabitants. The rich fossil discoveries in Badlands National Park have prompted some geologists to refer to the Oligocene as "The Golden Age of Mammals."

deal of ash, which settled to the bottom of a neighboring lake, burying thousands of insects (Fig. 13–13), leaves, fish, and even a few birds. The Florissant flora, known not only from tree trunks in their original position and leaves but also from spores and pollen, indicate subhumid conditions for the region and elevations of between 300 and 900 meters.

The Miocene was a time of continued fluvial and lacustrine sedimentation in intermontane basins as well as on the plains east of the mountains. By Miocene time, climates had become somewhat cooler. There were expanding areas of grasslands populated by Miocene camels, horses, rhinoceroses, deer, and other grazing mammals. In the central and southern Rockies, Miocene formations include beds of volcanic ash as well as interspersed lava flows attesting to vigorous volcanic activity. The well-known gold deposits at Cripple Creek, Colorado, are mined from veins associated with

FIGURE 13-13 Fossil spider (upper left) and fossil insects from the Oligocene tuff beds near Florissant, Colorado. (Tuff is consolidated volcanic ash.)

FIGURE 13-14 The jagged Teton Range has been cut by water and ice erosion from a crustal block uplifted along a nearly vertical fault. Movement along the fault began in the Pliocene and continued sporadically through the Pleistocene. (*Courtesy of E. Moldovanyi.*)

a Miocene volcano. Regional uplift of the Rockies also began in the Miocene and was accompanied by increased rates of erosion. Great volumes of terrigenous detritus eroded from uplifted areas, filled intermontane basins, and spread eastward, where it contributed to the construction of the Great Plains. Some of these crustal movements continued throughout the remaining epochs of the Cenozoic and brought some of the highest peaks of the Rockies to spectacular elevations.

As indicated earlier, the present Rocky Mountain topography is largely the result of uplift and erosion that began in the Miocene. Much of the erosional detritus was spread on top of the White River deposits and equivalent beds over extensive areas of South Dakota and Nebraska. The general character of the sedimentary layers does not change in the overlying Pliocene. Remains of plants and animals in some of these sediments indicate that the Pliocene was even cooler and drier than the Miocene had been. The cooling was an early indication that an ice age was about to begin.

At particular localities west of the Great Plains, Tertiary geologic events produced some of the most striking scenic attractions in the world. Normal faulting and volcanism accompanying late Tertiary uplifts were responsible for some of these features. In northwestern Wyoming, the lofty Teton Range (Fig. 13–14) was elevated along a normal fault. Displacement along the fault reached nearly 6000 meters. The magnificent eastern face of the Tetons is a fault scarp that rises 2.5 kilometers to an elevation of over 4000 meters.

Basin and Range Province

The Basin and Range Province is a large-scale physiographic feature that is the result of late Tertiary crustal extension (stretching). The Province occupies a broad zone from Nevada and western Utah southward into central Mexico (Fig. 13–15). This region had been

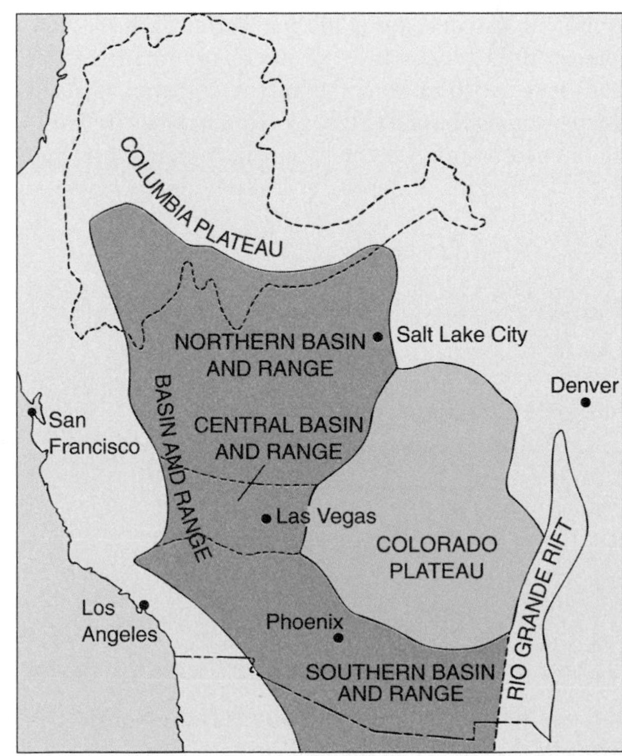

FIGURE 13-15 Location of the U.S. part of the Basin and Range Province, the Colorado Plateau, and the Columbia Plateau. (*Modified from Wernicke, B. 1992.* Geology of North America *G–3:553–581. Boulder, CO: Geological Society of America.*)

ENRICHMENT

Hellish Conditions in the Basin and Range Province

If the Poet Dante (1265–1321) had been strolling in the Basin and Range Province during the Oligocene and Early Miocene, his vision of the lower world might have included turbulent clouds of white-hot ash, jets of searing gases, and fiery explosions of incandescent molten rock. During the Mid-Cenozoic, the region was indeed the scene of such volcanic activity. As one hikes the trails of mountain ranges in the province, evidence is nearly always underfoot in the form of hardened ash falls and solidified lava beds. Volcanic rocks also lie between the mountain ranges but are blanketed by layers of sediment.

The volcanic rocks of the Basin and Range Province are derived from high-silica and hence highly viscous magmas. They have the same general composition as the granodiorites of the Sierra Nevada. Such magmas produce volcanism of the explosive type, in which hot gases, steam, ash, and pyroclastics of all sizes are discharged with enormous force. Nuée ardentes characteristically accompany such eruptions. As these fiery turbulent clouds rushed swiftly down slopes, they incinerated all life and ultimately deposited their load of ash on the surrounding lowlands. Glassy particles in these fiery clouds were still so hot and plastic when they came to rest that they fused together to form hard, welded rocks called ignimbrites. The volume of ash expelled was truly incredible. A single ash fall covered over 10,000 kilometers to a depth of over 180 meters.

Commonly, high-silica magmas cool slowly at depth and form great batholiths composed of granite and granodiorite. The magma of the Basin and Range Province, however, breached the crust and produced the intense episode of volcanism. What may have been the cause? The answer may lie in a hypothesis that relates events in the region to plate tectonics. During the Mid-Cenozoic, the North American plate was moving westward at the rapid rate of about 15 centimeters per year. It overrode the Pacific plate so quickly that the plate was unable to descend into the mantle. Rather, it slid nearly horizontally under North America. The sea-floor plate was close to its origin. It had only recently moved from its spreading center, and consequently it was still very hot. As the hot sea-floor plate moved along beneath what is now the Basin and Range Province, it might have provided much of the thermal energy responsible for the wave of volcanism that engulfed this region of North America during the Mid-Cenozoic.

folded and overthrust during the Mesozoic and existed as a structural arch during the early Tertiary. Beginning in the Miocene, the arch subsided between great normal faults that developed on both the west and the east sides. Similar faults with general north–south alignment also developed in the interior of the region. The uplifted blocks formed linear mountain ranges that became sources of sediment for the adjacent down-dropped basins (Figs. 13–16 and 13–17). Faulting opened fissures for the escape of molten rock from bodies of magma at depth. Volcanoes along the western side of the province produced extensive lava flows, whereas in the eastern region, volcanoes erupted explosively, covering the landscape with ash and pumice.

Subsequent geologic history in the region is recorded in coarse terrigenous clastics washed out of the mountains. These Miocene to Holocene sediments

FIGURE 13-16 The California and Nevada Nopah Range of the Basin and Range Province.

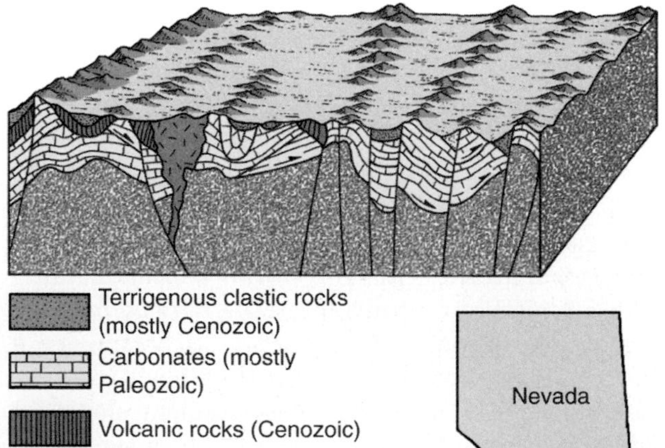

FIGURE 13-17 Geologic section across the Basin and Range Province along line A–A′ in southern Nevada.

blocked the passes and caused lakes to develop. Miocene sediments often include gypsum and salt layers formed when these lakes evaporated.

The cause of the tensional faulting that produced the Basin and Range Province is still being debated. On the basis of paleomagnetic evidence, some geologists believe that uparching, crustal extension, and faulting in this region occurred when westward-moving North America overrode part of the Pacific oceanic plate and a spreading center (the East Pacific rise) that was being subducted along the coast of California. When the subducted spreading center reached the region beneath eastern Nevada and western Utah, it caused uplift and stretching of the crust in east-west directions. Recently, however, this idea has fallen into disfavor because of a lack of evidence of progressive eastward deformation that should accompany the passage of a spreading center beneath a continent.

As an alternative theory, many geologists now believe that the normal faulting in the Basin and Range is simply the way the crust adjusted to the change along the California coast when oblique shearing of the edge of the continent during the early Miocene replaced the earlier subduction zone. Yet another model proposes extension and uplift of the crust from the remnants of an oceanic plate that had been carried beneath the Basin and Range region by an earlier episode of subduction. When subduction ceased, the oceanic slab may have formed a partially molten buoyant mass that pressed upward against the overlying crust and caused tensional faulting and escape of lava along fault and fracture zones. Finally, there is the possibility that the crustal extension and tensional faulting are related to convectional movements beneath the continental plate, similar to those that cause the breaking apart of continents.

Colorado Plateau

One of the most magnificent regions of uplift in the American West is the Colorado Plateau (Figs. 13–18 and 13–19). Somehow this block of crust had remained undeformed during the Mesozoic orogenies, for its Paleozoic and Mesozoic rocks are relatively flat-lying. The region formed a buttress. However, folds and faults developed ground the margins of the plateau producing encircling highlands. The plateau was repeatedly raised during early to middle Pliocene time (about 5 to 10 million years ago). Steep faults developed on the plateau during its rise and provided avenues for the escape of volcanic materials. The San Francisco Peaks near Flagstaff, Arizona, are a group of impressive recent volcanoes and cinder cones built above the level of the plateau (Fig. 13–20).

The best-known feature resulting from the linked processes of uplift and erosion on the Colorado Plateau is the Grand Canyon of the Colorado River (Fig. 13–21). This awesome monument to the forces of erosion is more than 2600 meters (8500 feet) deep in some places. The river has penetrated through Phanerozoic strata into crystalline Precambrian basement rocks.

Columbia Plateau and Cascades

Unlike the Colorado Plateau, which is constructed of layered sedimentary rocks, the Columbia Plateau in the northwestern corner of the United States has been built by volcanic activity (Fig. 13–22). During the late Tertiary and Quaternary, basaltic lavas erupted along deep fissures in the region. The liquid rock spread out and buried more than 500,000 km^2 of existing topography under layer after layer of lava. In some places, these low-viscosity lavas flowed 170 kilometers from

FIGURE 13-18 Canyon of the Dolores River, Colorado Plateau, south of Grand Junction, Colorado.

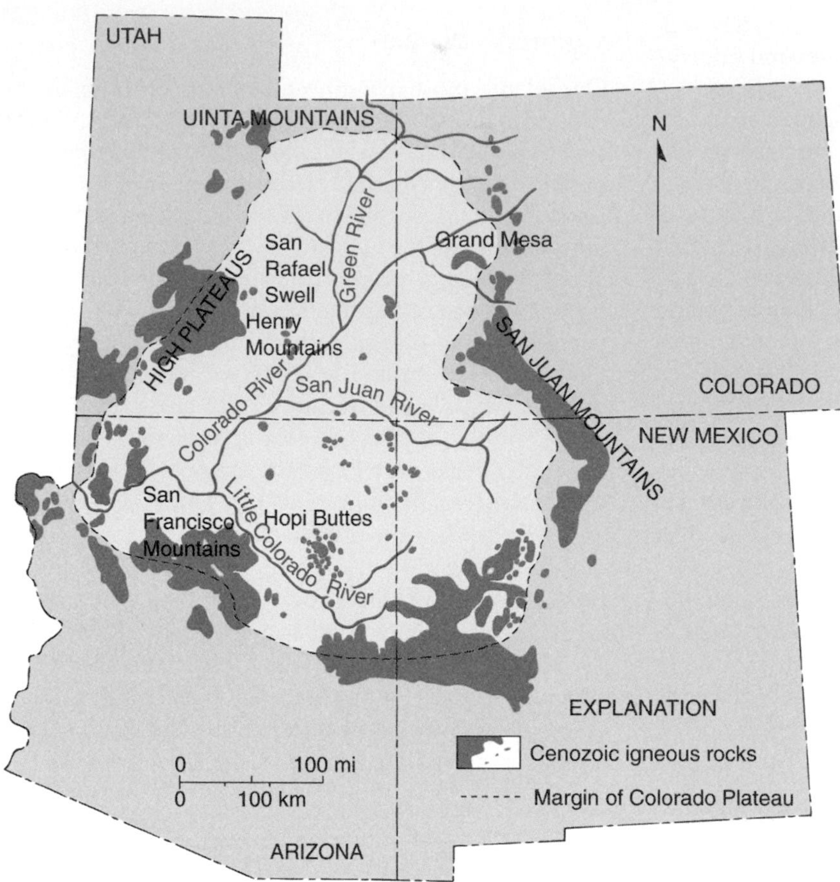

FIGURE 13-19 Location of volcanic and other centers of igneous activity of Cenozoic age in the Colorado Plateau and its periphery. (*From Hunt, C. B. 1956* Cenozoic Geology of the Colorado Plateau, *U.S. Geological Survey Professional Paper 279.*)

FIGURE 13-20 Vertical aerial photograph of a large cinder cone in the San Francisco volcanic field of northern Arizona. The solidified flow issuing from the cone is 7 kilometers long and more than 30 meters thick. (*Courtesy of U.S. Geological Survey.*) ? *Did the lava flow from the volcano before or after the extrusion of the pyroclastics that built the cinder cone?*

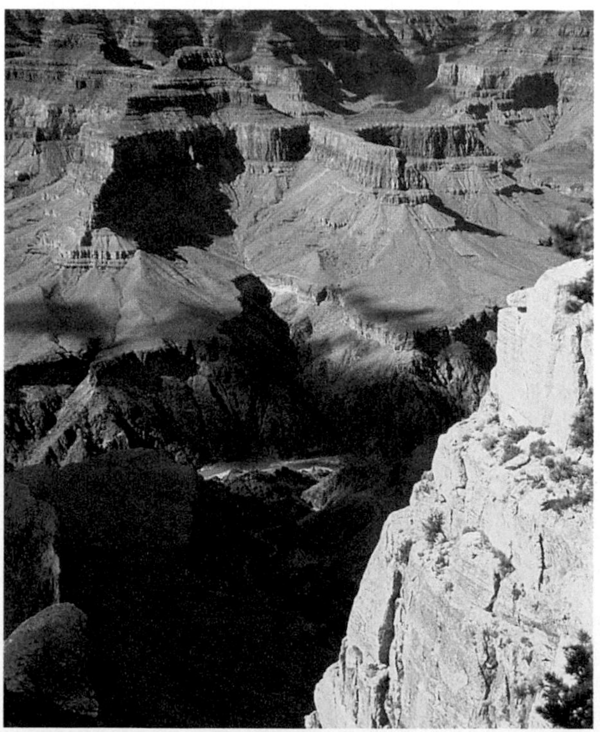

FIGURE 13-21 Grand Canyon of the Colorado River. This great gorge is largely the result of uplift accompanied by stream erosion during the Cenozoic Era. (*Photograph by R. F. Dymek.*)

FIGURE 13-22 Basalt lava flows of the Columbia Plateau. The Snake River is in the foreground. ? *What is the origin of the closely spaced vertical cracks seen on the cliff faces?* (William E. Ferguson.)

their source. Their combined thickness exceeds 2800 meters.

West of the Columbia Plateau lies an uplifted belt that was also the site of extensive volcanic activity. Here, however, the outpourings of more viscous lavas resulted in the mountains of the Cascade Range. Volcanism in this region began about 4 million years ago. Volcanic eruptions have continued in this region into the present century, as was made dramatically obvious during the 1980 eruption of Mount St. Helens (Fig. 13–23).

The recent activity of Mount St. Helens and the older eruptions that gave us the volcanic peaks of the Cascades are surface manifestations of an ongoing collision between the American plate and the small Juan de Fuca plate of the eastern Pacific, which moves eastward toward the coasts of Oregon and Washington. As the Juan de Fuca plate plunges beneath the coastline, molten rock generated as the plate moves downward rises to supply lava for the volcanoes (Fig. 13–24).

Mount St. Helens is not the only famous volcanic mountain in the Cascades, nor is it the only peak having periodic eruptions. As recently as 1914, Mount Lassen extruded lava and ash for a year and then exploded violently on May 19, 1915, producing a nuée ardente ("fiery cloud") that roared down the mountainside, destroying everything in its path. Other well-known Cascade peaks include Mount Baker, Mount Hood, and Mount Rainier. The last major eruption of Mount Rainier was about 2000 years ago, although minor disturbances occurred frequently in the late 1800s. Crater Lake in Oregon (Fig. 13–25) was formed from the volcanic cone of Mount Mazama after eruptions about 6000 years ago had drained a large portion of molten rock from the magma chamber beneath the volcano. Unsupported by underlying molten rock, the roof of the magma chamber collapsed

A

B

FIGURE 13-23 (*A*) Mount St. Helens prior to the May 18, 1980 eruption and (*B*) during the eruption. During the initial eruption, clouds of steam and ash were blown toward the northeast from the prominent plume, which reached about 20,000 meters in altitude. The eruption produced an amount of ash roughly equivalent to that ejected during the 79 A.D. eruption of Mount Vesuvius that buried Pompeii. (*Courtesy of U.S. Geological Survey.*)

(Fig. 13–26). The circular depression with a relatively flat floor and steep surrounding walls produced by the collapse is called a **caldera**, from the Spanish word for kettle. Other calderas form as a result of violent explosive activity.

Sierra Nevada and California

The ranges of the Sierra Nevada lie south of the Cascades. The rocks of these mountains were folded and intruded during the Nevadan orogeny. For most of the Tertiary, the peaks were steadily reduced by erosion until their granitic cores lay exposed at the surface. Then, in Pliocene time and continuing into the Pleistocene, the entire Sierra Nevada block was raised along normal faults (Fig. 13–27). Over time, its high eastern front was lifted an astonishing 4000 meters. The de-

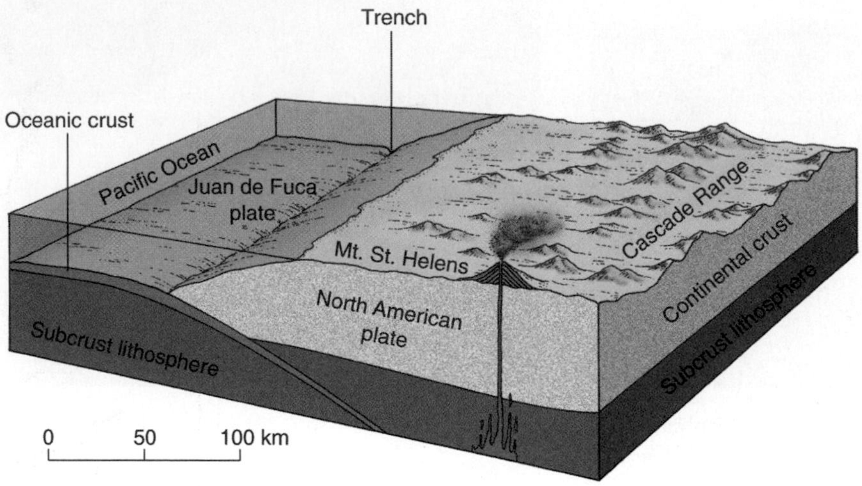

FIGURE 13-24 **As the small Juan de Fuca plate plunges beneath Oregon and Washington, molten rock is generated and rises to supply the volcanoes of the Cascade Range.**

pressed western side formed the California trough. As these movements were under way, rejuvenated streams and powerful valley glaciers eroded the block to form the magnificent landscapes of the present-day Sierra.

The southern Sierra Nevada ranges differ from many other great ranges, in that they are not supported by an exceptionally thick crust. Unlike the traditional model, in which mountain ranges behave isostatically like icebergs with a thick root below, there was probably not a sufficient thickness of low-density crust to provide for the observed amount of uplift. It has been proposed that molten rock beneath the ranges provided at least part of the uplifting force.

In the early epochs of the Cenozoic, the region west of the Sierra Nevada was most directly affected by subduction tectonics. As described in the next section, however, this region underwent a change from subduction tectonics to strike-slip tectonics at about the beginning of Miocene time. Complex fault movements resulting from this shift in tectonic style created islands and intervening sedimentary basins in which fine-grained marine clastics, diatomites, and bedded cherts accumulated. Also during the Miocene, folding and uplift resulted in marine regressions, and by the end of the Tertiary, seas were restricted to a narrow tract along the western edge of California. Ultimately, Quaternary uplifts caused the removal of this seaway as well.

The Tertiary stratigraphic succession of California is impressive. No less imposing are sections of Tertiary rocks far to the north in Alaska and the arctic islands of Canada. Coal is found associated with these strata. Along the coastal mountains and Aleutian chain, Tertiary sands and shales are interspersed with layers of pyroclastics and lava flows.

FIGURE 13-25 **Crater Lake, Oregon.** The small cinder cone projecting above the lake surface was built during the last eruption and is called Wizard Island. Crater Lake is an example of a caldera, a feature recognized by its roughly circular outline, relatively flat floor, and steep surrounding walls.

West Coast Tectonics

The western edge of North America during most of the Cenozoic was the site of an eastward-dipping subduction zone. This subduction was in one way or another responsible for the batholiths, compressional structures, volcanism, and metamorphism that accompanied Mesozoic and early Tertiary orogenies. The oceanic plate that was being fed into the subduction zone has been named the Farallon plate. During the Cenozoic, the Farallon plate was being consumed at the subduction zone faster than it was receiving additions at its spreading center. As a result, most of the Farallon plate and part of the East Pacific rise that generated it were gobbled up at the subduction zone along the western edge of California (Fig. 13–28). Today, the small Juan de Fuca plate near Oregon and Washington and the Cocos plate off the coast of Mexico are the remnants of the once more extensive Farallon plate.

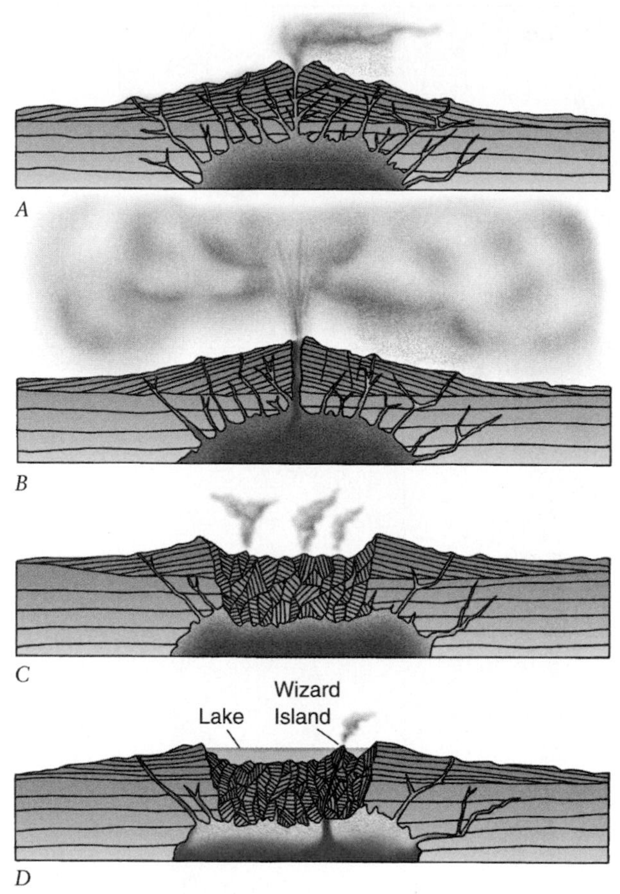

The building of ancient Mount Mazama. Magma chamber is filled and supplying magma to the volcano.
A
Often explosive eruptions begin to exhaust the reservoir of magma, leaving the upper part of the magma chamber empty.
B
Collapse of the summit into the vacated chamber below.
C
Lake
Wizard Island
Minor eruptions build small volcanoes on caldera floor, and caldera gradually fills with water to form lake.
D

FIGURE 13-26 The origin of the caldera at Crater Lake, Oregon. (*After Williams, H.,* Crater Lake: The Story of Its Origin. *Bulletin, Dept. of Geological Sciences, vol. 25. University of California Publications, Berkeley, 1941.*)

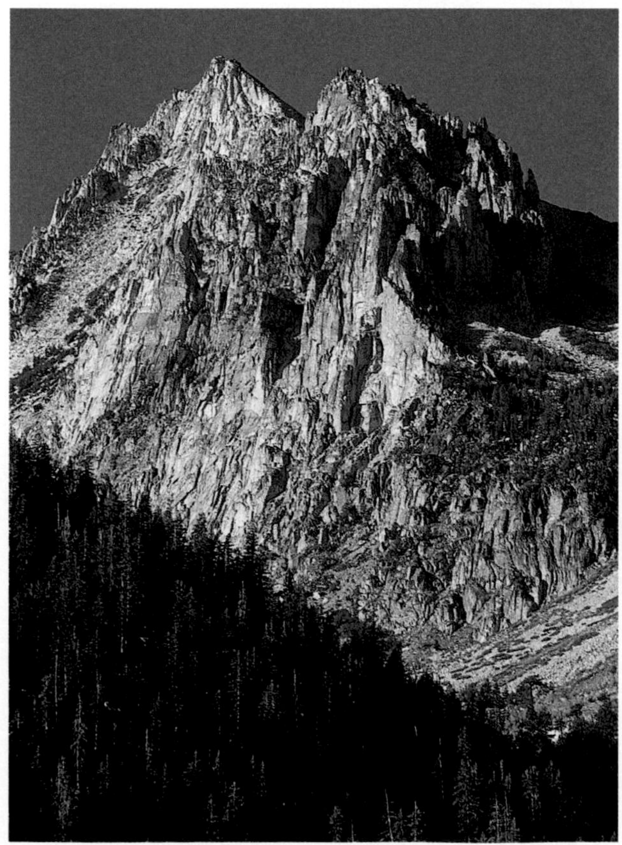

FIGURE 13-27 Uplifted and exposed portion of the Sierra Nevada batholith, Yosemite National Park.

With the loss of the Farallon plate near California, the North American plate was brought into direct contact with the Pacific plate, and a new set of plate motions came into operation. Before making contact with the West Coast, the Pacific plate had been moving toward the northwest. As a result, when contact was made, the Pacific plate did not plunge under the continental margin but rather slipped along laterally, giving rise to the San Andreas fault. No longer did the California sector of the West Coast have an Andean type of subduction zone (Fig. 13–29). Strike-slip movements now characterized the southwestern margin of the United States.

An important result of the shearing and wrenching of the West Coast was the tearing away of Baja California from the mainland of Mexico about 5 million years ago. Because it was located west of the San Andreas fault system, Baja California was sheared from the American plate and now accompanies the northward movement of the Pacific plate.

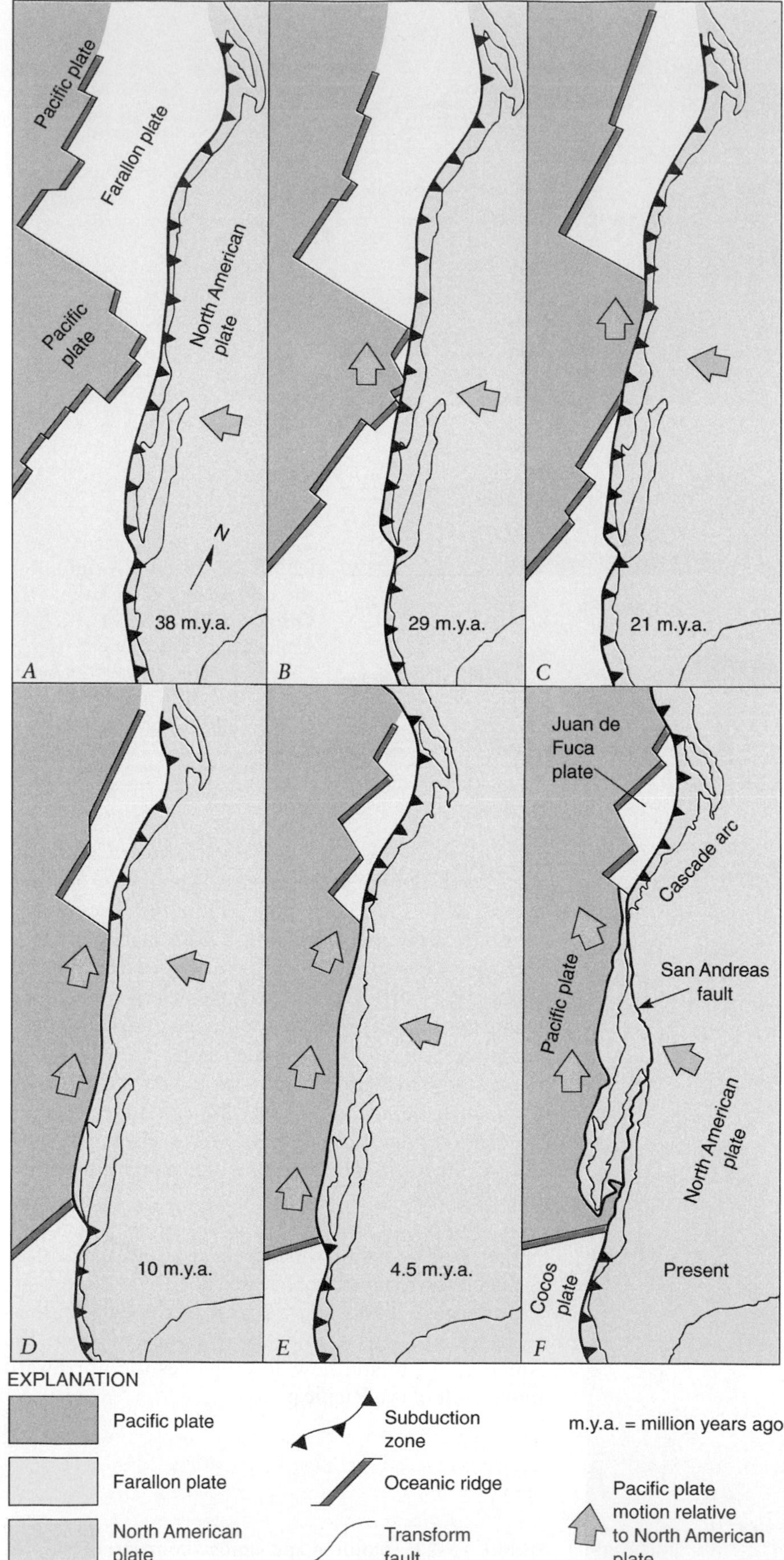

FIGURE 13-28 Schematic model of the interaction of the Pacific, Farallon, and North American plates for six time intervals during the Cenozoic. Note how the Farallon plate was largely subducted by Late Cenozoic time, leaving only remnants to the north (Juan de Fuca plate) and to the south (Cocos plate). The San Andreas and associated faults were caused by right-lateral movements beginning about 29 million years ago. (*Adapted from Nilsen, T. H.* San Joaquin Geological Society Short Course No. 3, 1977).

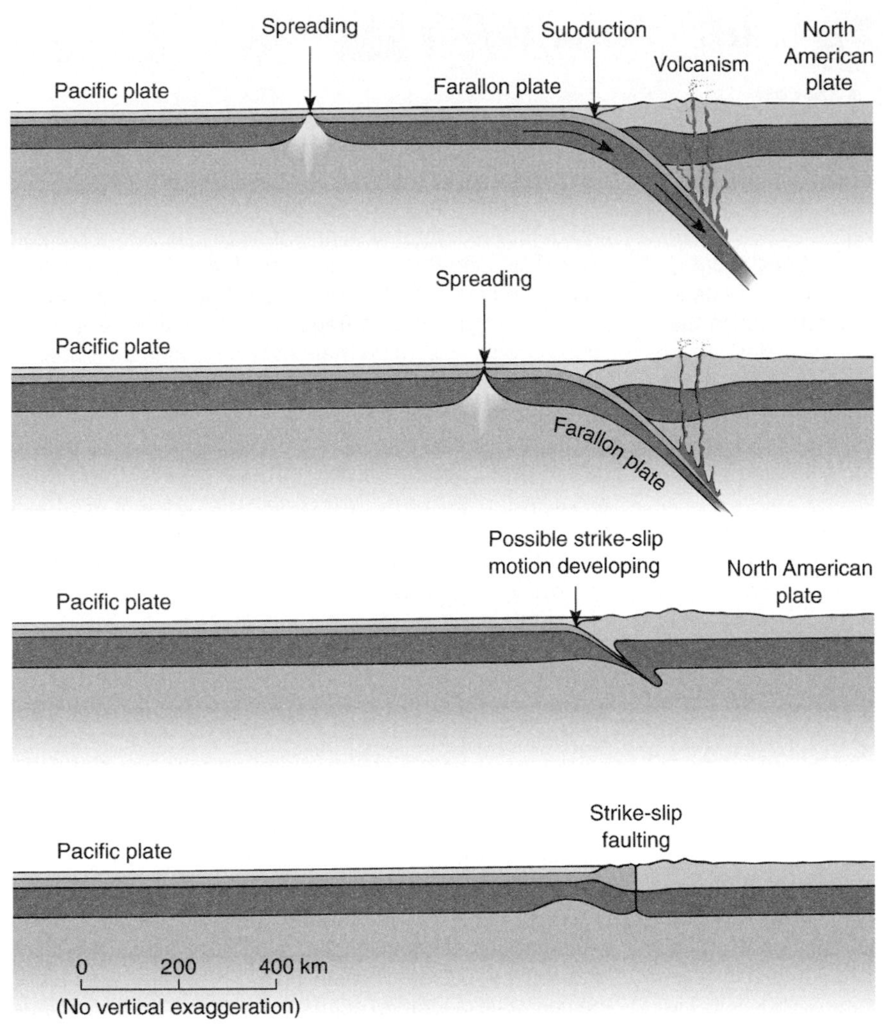

FIGURE 13-29 Sequence of cross-sections of California and its offshore area, illustrating the subduction of the Farallon plate. In this model, the Pacific plate is considered fixed as the spreading center (East Pacific Rise) encounters the continental margin. (*After Atwater, T.* 1970 Bull. Geol. Soc. Am. *81*[*12*]:*3513–3536*.)

South America

Extensive deformation, metamorphism, and emplacement of granitic masses had characterized the Andean belt during the Cretaceous. Folding and volcanism continued to be widespread during the Cenozoic as well and was especially intense during the Miocene. Subsequently, the Andean highlands were eroded to low relief and then were uplifted in late Pliocene. Erosion of the uplifted surface is responsible for the present relief and topography of the Andes. Cenozoic instability in the Andes has been related to the continued movement of the floor of the eastern Pacific Ocean beneath South America. An oceanic trench off the Pacific coast and an eastwardly inclined seismic zone characterized by deep-focus earthquakes (a Benioff zone) attest to this interpretation.

As a consequence of the frequent upheavals that beset the western side of South America, epicontinental seas were either marginal or limited. Sedimentation was primarily terrestrial, as silt, sand, and gravel that eroded from the Andes was washed down into the Amazon and Orinoco basins and across the Pampas. Rift-like basins created by block faulting served as collecting sites for prodigious thicknesses of clastic sediments.

The Tethyan Realm

The conversion of a major seaway separating Eurasia and Gondwanaland to a spectacular array of mountains and plateaus must be considered one of the greatest events of the Cenozoic Era. During most of the era, the Tethys seaway lay to the south of Europe. Its waters spilled onto the northern margin of Africa as an epicontinental sea. Disturbances prophetic of impending tectonic upheavals began along the Tethys during the Eocene, with large-scale folding and thrust-faulting. It was at approximately this time that the northward-moving African block first encountered the western underside of Europe and crumpled the strata that now compose the Pyrenees and Atlas mountains. Then, with a sort of scissor-like movement, the Alpine region began to be squeezed. At first, marine conditions persisted

ENRICHMENT

When the Mediterranean Was a Desert

In 1970, geologists Kenneth J. Hsu and Brian B. F. Ryan were collecting research data while aboard the oceanographic research vessel Glomar Challenger. An objective of this particular cruise was to investigate the floor of the Mediterranean and to resolve several questions about its geologic history. The first question was related to evidence that the invertebrate fauna of the Mediterranean had changed abruptly about 6 million years ago. Most of the older organisms were nearly wiped out, although a few hardy species survived. A few managed to migrate into the Atlantic. Somewhat later, the migrants returned, bringing new species with them. Why did the near extinction and migrations occur?

A second question concerned the origin of an enormous buried gorge that lay beneath—and nearly parallel to—the present course of the Rhone River, which flows from Switzerland to France. The buried canyon is nearly 1000 meters deep. What had happened in the geologic past to give a stream sufficient erosive powers to cut so deeply?

The third task for the Glomar Challenger's scientists was to try to determine the origin of the dome-like masses buried deep beneath the Mediterranean sea floor. These structures had been detected years earlier by echo-sounding instruments, but they had never been penetrated in the course of drilling. Were they salt domes such as are common along the United States Gulf Coast, and if so, why should there have been so much solid crystalline salt beneath the floor of the Mediterranean?

Finally, there was the presence—again detected on seismic profiles—of a hard layer of sedimentary rock 100 meters or so below the present sea floor. This hard layer was a strong acoustic reflector (layer of sediment or rock capable of reflecting sound waves) and was easily recognized nearly everywhere in the Mediterranean. Most deep-sea deposits are soft, poorly reflective oozes. What was the composition and origin of the hard layer?

With such questions as these clearly before them, the scientists aboard the Glomar Challenger proceeded to the Mediterranean to search for the answers. On August 23, 1970, they recovered a sample from the surface of the hard layer. The sample consisted of pebbles of hardened sediment that had once been soft, deep-sea mud, as well as granules of gypsum and fragments of volcanic rock. Not a single pebble was found that might have indicated that the pebbles came from the nearby continent. In the days following, samples of solid gypsum were repeatedly brought on deck, as drilling operations penetrated the hard layer at its predicted depth. Clearly, the strong acoustic reflector was a layer of gypsum. Furthermore, the gypsum was found to possess peculiarities of composition and structure that suggested it had formed on desert flats. Sediment above and below the gypsum layer contained tiny marine fossils, indicating open-ocean conditions. As they drilled into the central and deepest part of the Mediterranean basin, the scientists took solid, shiny, crystalline salt from the core barrel. Interbedded with the salt were thin layers of what appeared to be windblown silt.

The time had come to formulate a hypothesis. The investigators theorized that about 20 million years ago, the Mediterranean was a broad seaway linked to the Atlantic by two narrow straits. Crustal movements closed the straits, and the landlocked Mediterranean began to evaporate. Increasing salinity caused by the evaporation resulted in the extermination of scores of invertebrate species. Only a few organisms especially tolerant of very salty conditions remained. As evaporation continued, the remaining brine became so dense that the calcium sulfate of the hard layer was precipitated. In the central deeper part of the basin, the last of the brine evaporated to precipitate more soluble sodium chloride. Later, under the weight of overlying sediments, this salt flowed plastically upward to form salt domes. Before this happened, however, the Mediterranean was a vast

Workers on the research vessel *Glomar Challenger* add a length of drill pipe as they drill into the ocean floor.

"Death Valley" 3000 meters deep. Then about 5.5 million years ago came the deluge. As a result of crustal adjustments and faulting, the Strait of Gibraltar opened, and water cascaded spectacularly back into the Mediterranean. Turbulent waters tore into the hardened salt flats, broke them up, and ground them into the pebbles observed in the first sample taken by the Challenger. As the basin was refilled, normal marine organisms returned. Soon layers of oceanic ooze began to accumulate above the old hard layer.

The salt and gypsum, the faunal changes, and the unusual gravel provided abundant evidence that the Mediterranean was once a desert. But how did this knowledge relate to the buried gorge beneath the Rhone Valley? The answer was now clear. If the Mediterranean had been emptied, the surrounding lands would have stood high above the floor of the basin. The gradients of streams would have markedly increased, and their swiftly flowing waters would have incised rapidly, eroding deep canyons. Such canyons should also have been cut by other streams that once flowed into the Mediterranean. As news of the Glomar Challenger's findings spread, geologists who had worked in the region alerted the investigators to the presence of gorges in North Africa. A Russian geologist recalled a 240-meter chasm that had been encountered at the time the foundations had been prepared for the Aswan Dam in Egypt. Oil company geologists in Libya sent word of many similar gorges that they had detected from seismograph records. A final piece of the puzzle seemed to fall into place.

between the emerging folds. Dark siliceous shales, poorly sorted sandstones, and cherts accumulated between elongate submarine banks. In the Alps, these marine sediments are referred to as **flysch**.

By Oligocene time, compression from the south caused enormous recumbent folds to rise as mountain arcs out of the old seaway and to slide along thrust faults over lands that lay to the north of the Tethys. The great folds with faults along their undersides were pushed one on top another as spectacular monuments to the forces involved in plate collisions. North of these folded thrust structures lay a topographic depression that received piedmont deposits eroded from the mountains. These terrestrial clastics, termed **molasse** by European geologists, resemble similar deposits swept eastward from the Rocky Mountains during the Cenozoic.

Even after the Oligocene, Alpine compressions continued. During the Pliocene, further thrusts from the south carried the older folded belts northward over the molasse deposits and crumpled strata to form the Jura folds. Today these folded rocks form the northern front of the Alps. The thrusting was followed by spasmodic regional uplifts that continue to the present day.

The part of the Tethys that extended far to the east of the Alps experienced its own episode of mountain-making during the Cenozoic. Volcanism, folding, thrusting, and emplacement of massive granitic intrusions began early in the era and increased markedly during the Miocene. Great elongate tracts of the former sea floor were squeezed into folds and thrust southward. Much of the early deformation occurred at or near sea level, but ultimately the sea-floor section was forced up above sea level. A broad, subsiding trough formed along the northern edge of the newly arrived peninsular block of India. On this lowland, more than 5000 meters of continental sediments were deposited. The strata contain an important fossil record of Cenozoic mammals and plants. In the two final epochs of the era, regional (epeirogenic) uplifts brought the plateaus and ranges to lofty elevations and caused the retreat of those marginal seas that still remained. The uplift may very well have been caused by the continued movement of the Indian block northward, beneath the southern edge of the Asian plate.

The history of the Tethys provides another example of a region of the Earth's crust that has undergone extensive geographic and physiographic change. The once quiet seaway that bordered Eurasia on the south was transformed into a structurally complex region of rugged highlands that includes the Alps, Appennines, Carpathians, Caucasus, and Himalayas (Figs. 13–30 and 13–31).

FIGURE 13-30 Shivling Peak in the Himalayas. The mountain has been eroded by glaciers into a glacial horn. (*Courtesy of D. Bhattacharyya.*)

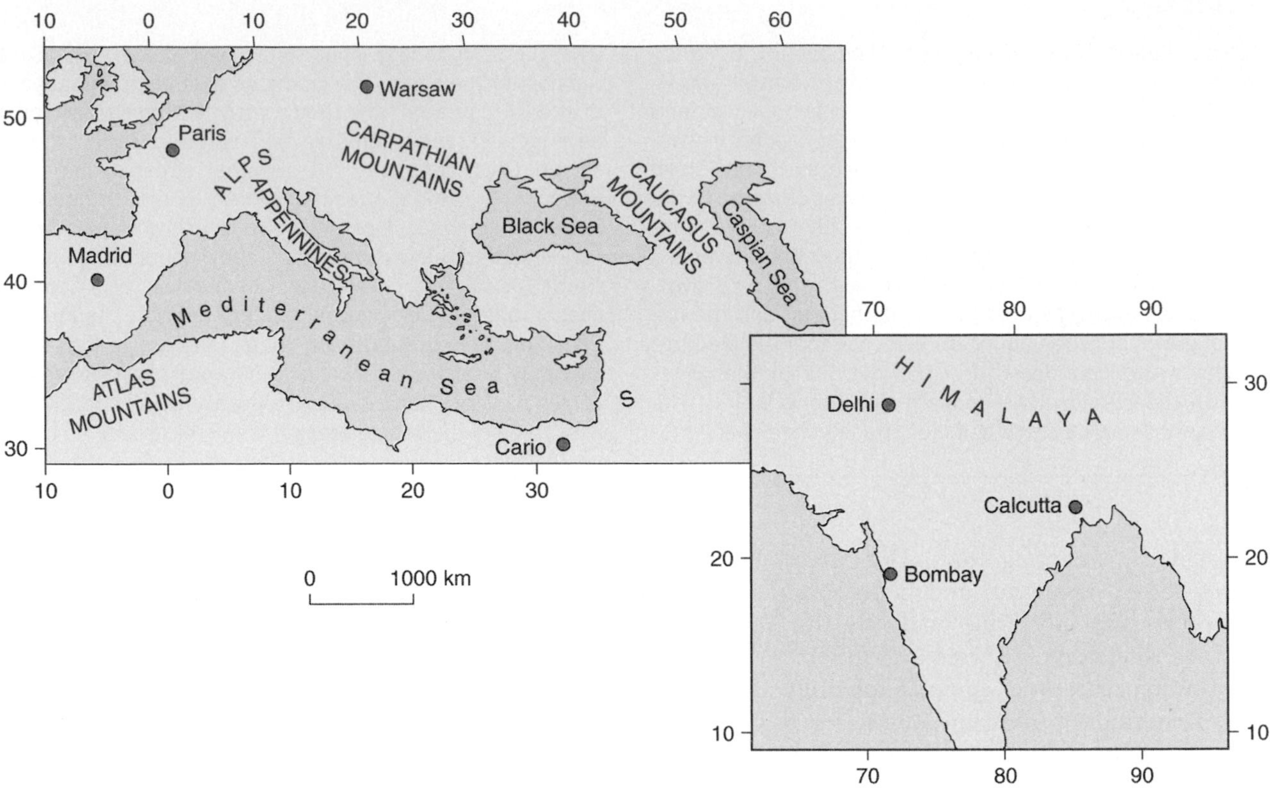

FIGURE 13-31 Location map for the Alps, Apennines, Atlas, Caucasus, and Himalaya mountains.

One event in the history of the Tethys was not directly related to orogeny in the region. As described in the Enrichment box entitled "When the Mediterranean Was a Desert," there was an interval about 5.5 million years ago when the Mediterranean dried up and may actually have resembled the present Death Valley region of California. This extraordinary change occurred near the end of the Miocene and has been named the **Messinian event** in that the Messinian is the name given to the final stage of the Miocene in Italy. The Messinian event was an indirect result of a sudden but temporary expansion of the Antarctic ice sheet and closure of the Straits of Gibraltar by the counterclockwise rotation of the Iberian Peninsula. Water needed to form the ice came from the ocean, and the loss in volume of ocean water lowered mean global sea level by as much as 50 meters. This decline left the Mediterranean isolated from other oceans. Without a way to replenish water lost by evaporation, the Mediterranean dried up. Evidence of this event comes primarily from an extensive evaporite bed encountered in cores that penetrated deep into the Mediterranean sea floor.

North of the Tethys

Important geologic events were also occurring in regions north of the Tethys during the Cenozoic. In the early part of the era, lavas were extruded in Scotland, Ireland, Svalbard, Greenland, and Baffin Island. In Ireland, these lavas are vertically jointed and form the famous Giant's Causeway (Fig. 13–32). The separation of Greenland from Europe provided the rift zone and associated fractures along which lavas rose to the surface.

Across the English Channel in northeastern Europe there is a record of repeated early Cenozoic marine incursions. The most extensive occurred during the Oligocene. Like the earlier transgressions, waters flooded toward the southeast from the North Sea. Rocks deposited during the alternate marine transgressions and regressions have been thoroughly studied in the Paris basin. The Paris basin is not only a sedimentary basin but a structural basin as well. This is quickly apparent on a geologic map of the area (Fig. 13–33), which shows younger Tertiary deposits surrounded by older Mesozoic rocks. After the Oligocene, general uplift prevented further marine

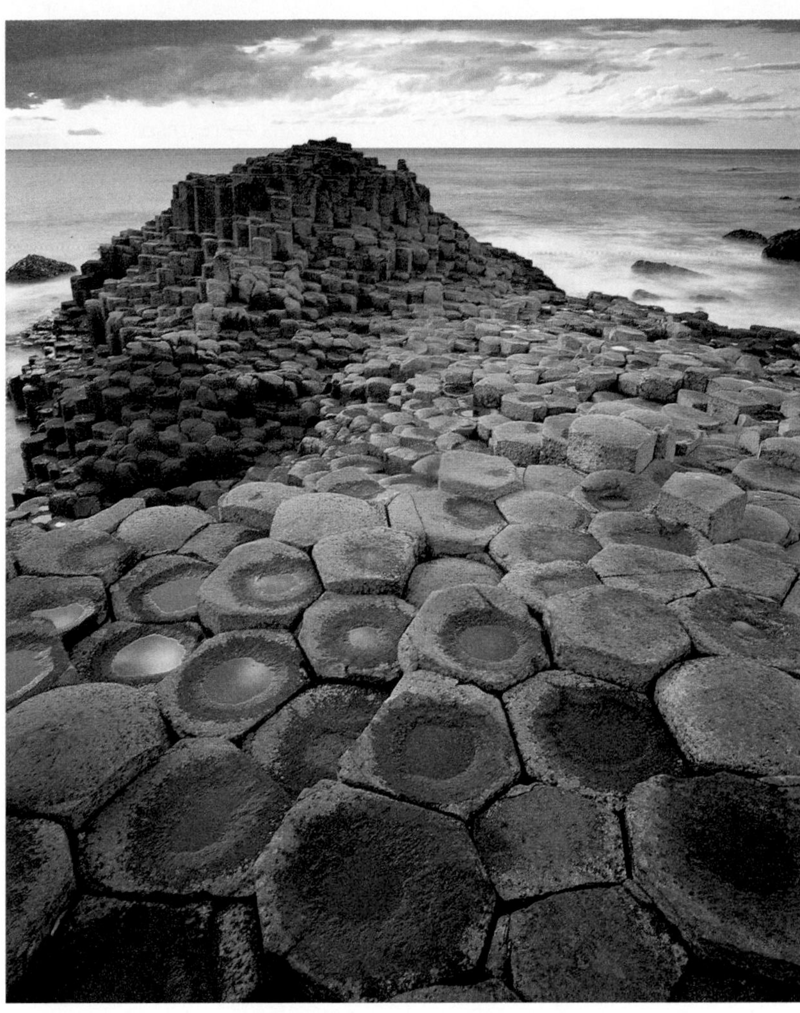

FIGURE 13-32 The Giant's Causeway, near Portrush, Antrim, Northern Ireland. This is one of the best-known locations in the world for viewing columnar jointing in basaltic rock. (*Photograph by T. Till, with permission.*)

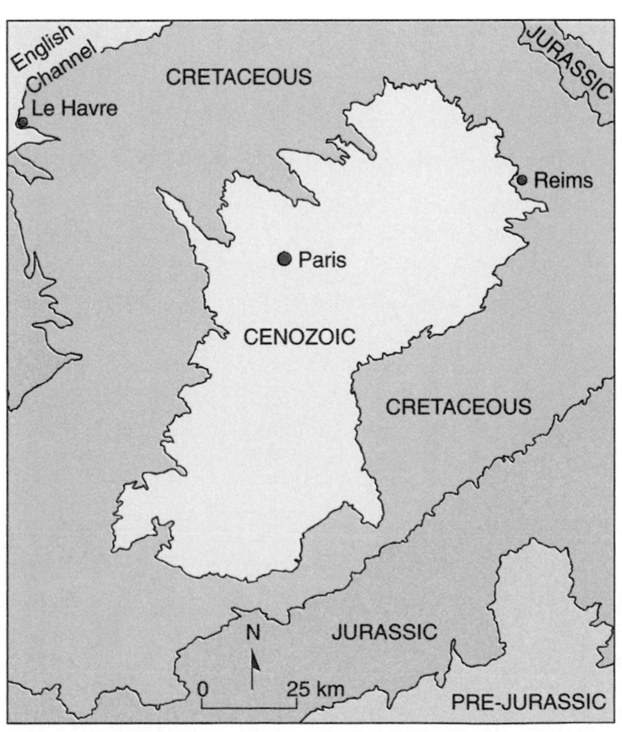

FIGURE 13-33 Geologic sketch map of the Paris Basin. ? *How does the outcrop pattern confirm that the structure is a basin rather than a dome?*

incursions. The subsequent late Cenozoic is read largely from lake and stream deposits.

Africa

The southern margin of the Tethys seaway formed the upper boundary of Africa during most of the Cenozoic. The region had a much quieter tectonic history than did the northern Tethys. In Libya and Egypt, the formations are flat or only moderately folded carbonates; the only severely folded mountain ranges are in western North Africa, which experienced pulses of orogeny throughout the Cenozoic.

The larger part of Africa that lay south of the Tethys was generally emergent during the Cenozoic. It was a time in which erosion prevailed. The most conspicuous changes that occurred were those associated with the rift valleys and uplands along the eastern side of the continent. During the Late Cenozoic, much of eastern Africa was arched upward nearly 3000 meters above sea level. Fracturing and faulting occurred across the crest of the arch. The famous rift valleys of eastern Africa formed as narrow slivers of crust that had slipped downward between great fault blocks. As the faulting continued, volcanoes developed along the fault trends. Mount Kenya and Mount Kilimanjaro are two of the most notable of these volcanoes, but there are many others as well. In addition to the volcanoes, a series of elongate lakes (Fig. 13–34) formed within the downfaulted blocks. Today, Lake Malawi and Lake Tanganyika are splendid examples of these fault-controlled inland water bodies. This region of lakes, volcanoes, and fault-controlled ranges is well known for its exceptional scenic grandeur. As will be described in Chapter 15, the strata contain volcanic ash beds that permit isotope dating of the bones of early hominids and their vertebrate contemporaries.

The Western Pacific

Most of Australia was comparatively stable during the Cenozoic Era. Marine deposits were restricted to peripheral basins. Near the center of the continent, coal-bearing terrestrial clastics are found. However, the general lack of tectonic activity that prevailed over most of the continent was not characteristic of the extreme southeastern border, where Paleocene and Eocene volcanism was intense. This activity was apparently associated with the separation of Antarctica from Australia an estimated 50 or 60 million years ago (see Fig. 13–1).

The Cenozoic history of New Zealand and the region north of Australia was also marked by exceptional instability. Throughout the islands that extend from the Aleutians down through Japan, the Philippines, Indonesia, and into New Zealand, folding, volcanism, and spectacular upheavals were common occurrences. Tropical rains and storms caused rapid erosion and

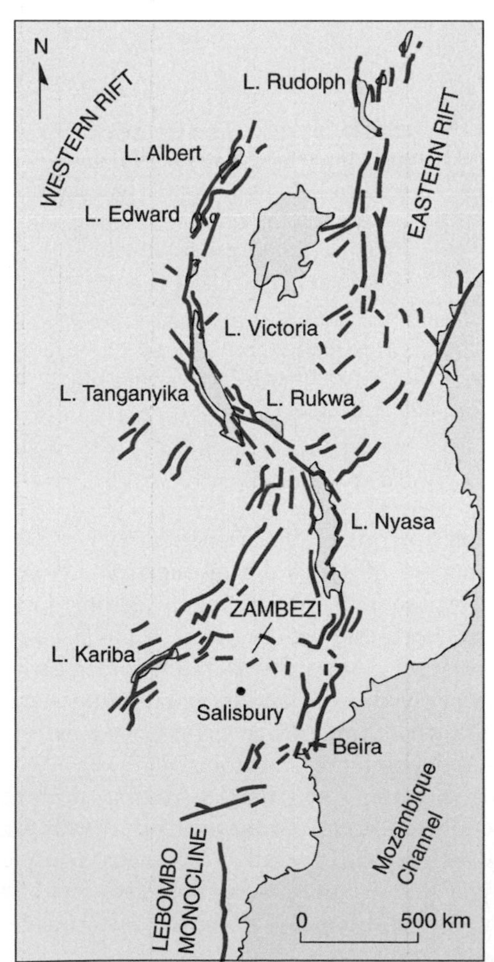

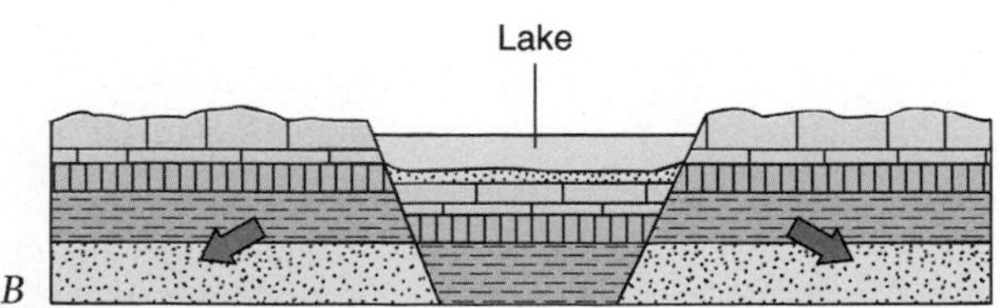

FIGURE 13-34 Location and origin of African rift valleys. (*A*) Eastern African rift valleys and associated lakes. (*B*) Schematic illustration of the formation of rift valleys from the action of tensional forces in the crust. (A, *from Read, H. H., and Watson. J. 1975* Introduction to Geology. *London, England: Macmillan & Co., p. 237, Figure 8.3.*)

FIGURE 13-35 Lightning strikes during the eruption of Galunggung volcano, West Java, Indonesia, December 3, 1982. *(Photo by Ruska Hadian, volcanology survey of Indonesia. Courtesy of U.S. Geological Survey.)*

deposition of coarse sediments. Deep sedimentary basins received the erosional detritus and frequently spread it seaward as vast deltas.

The enormous volumes of Cenozoic andesites and basalts found in New Zealand clearly indicate its participation in the instability. Indeed, a belt of active volcanoes persists today on the northern island. Late in the Cenozoic, vertical movements raised a great fault segment an incredible 1.8 kilometers along the eastern side of the northern island.

The crustal activity that characterized the western Pacific during the Cenozoic has continued to the present. If one were to speculate about the most active region of the globe during the near geologic future, the sinuous arcs from the Kamachatka Peninsula through Japan and down into Indonesia (Fig. 13–35) would be likely candidates. (The Caribbean region might qualify as a second choice.)

Antarctica

During much of the early Cenozoic, the now frigid continent of Antarctica had a more pleasant semitropical climate. As noted in the introduction to this chapter, during the Paleocene and Eocene, Antarctica was warmed by ocean currents that originated in temperate zones. During those early epochs, Australia had not yet separated from Antarctica. By Miocene time, however, cold currents flowed through the wide oceanic tract that opened between Antarctica and Australia. Warm waters were forced northward, and snow began to accumulate on a landmass now separated from its Gondwana neighbors. As Antarctica became ice-bound, deep-marine clastics and volcanic rocks reflecting tectonic instability accumulated along its Pacific side. The crustal unrest and volcanism was the result of the eastern movement of an oceanic plate beneath the western margin of Antarctica.

THE GREAT PLEISTOCENE ICE AGE

The final two epochs of the Cenozoic Era are the Pleistocene and Holocene (or Recent). They represent only about 2 million years of geologic time, but for humans they are an exceedingly significant interval. It was during the Pleistocene that primates of our own species evolved and rose to a dominant position. Second, during the Pleistocene, more than 40 million km^3 of snow and ice were dumped on about one third of the land surface of the globe (Figs. 13–36 and 13–37). Such an extensive cover of ice and snow had profound effects not only on the glaciated terrains themselves (Fig. 13–38) but also on regions at great distances from the ice fronts. Climatic zones in the Northern Hemisphere were shifted southward, and arctic conditions prevailed across northern Europe and the United States. Mountains and highlands in the Cordilleran and Eurasian ranges were sculpted by spectacular mountain glaciers, rather like those depicted in Figure 13–39. While the snow and ice accumulated and spread in higher latitudes, rainfall increased in lower latitudes, with generally beneficial effects on plant and animal life. Even as late as the beginning of the Holocene, presently arid regions in northern and eastern Africa were well watered, fertile, and populated by nomadic tribes. Peoples of middle and late Pleistocene time hunted along the fringes of

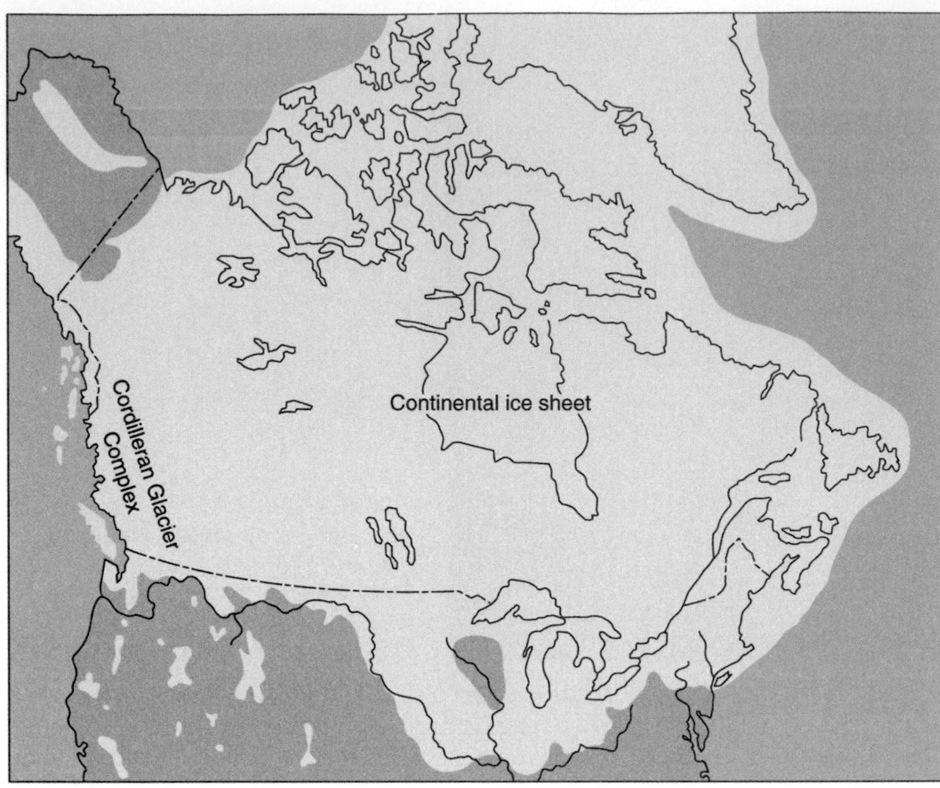

FIGURE 13-36 Areal coverage of continental glaciers in North America during the latest glacial advance, about 18,000 years ago. (*Courtesy of Thompson, G.R. and Turkl, J. 1997,* Modern Physical Geology, *Philadelphia: Saunders College Publishing.*)

the continental glaciers, where game was abundant and the meat kept longer with less danger of spoilage. Animal furs provided warm clothing, and following the discovery of fire, the caves could be warmed against the arctic winds.

In addition to the glaciations, the Pleistocene was a time of recurring crustal unrest. Volcanoes were active in New Mexico, Arizona, Idaho, Mexico, Iceland, Svalbard, and the Pacific borders of both North and South America.

Pleistocene crustal uplifts occurred in the Tetons, the Sierra Nevada, and the ranges of the central and northern Rocky Mountains. Pulses of uplift during the Pleistocene also characterized the Alps, the Himalayas,

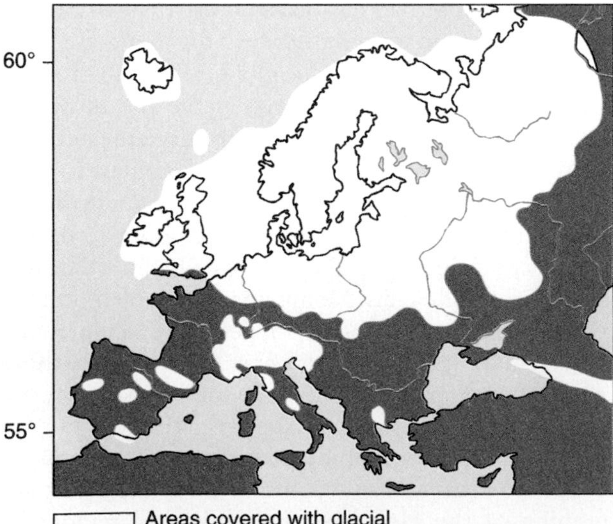

FIGURE 13-37 Areal extent of glaciers in Europe during the Pleistocene Epoch.

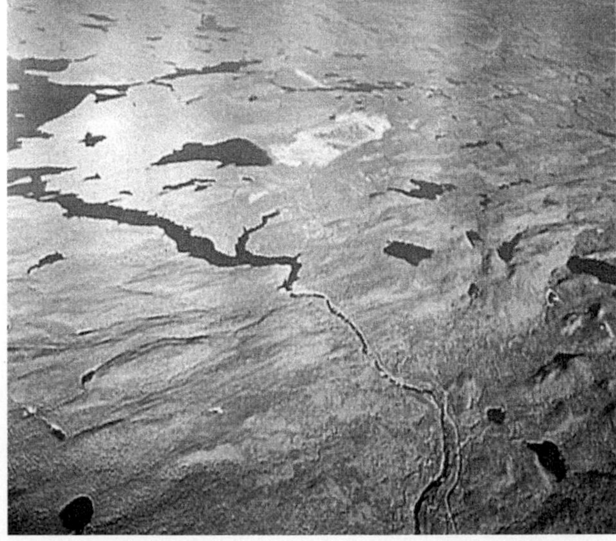

FIGURE 13-38 Glaciated Canadian Shield north of Montreal, Canada. (*Courtesy of the Royal Canadian Air Force.*) ? *Draw a line on the photograph along which the glacier moved.*

FIGURE 13-39 Oblique aerial view of glaciers in the Fairweather Mountains of Alaska. Note the debris flowlines. *(Courtesy of M. E. Dalechek, U.S. Geological Survey.)*

and the ranges that lay between. All of this crustal and climatic activity may suggest that the Pleistocene was a unique time in geologic history. However, this is not the case, for widespread glaciations also occurred in the Precambrian, late in the Ordovician, in the Permo-Carboniferous, and possibly during the Oligocene and Pliocene as well. Indeed, study of relatively accessible Pleistocene and Holocene glacial deposits has greatly improved our ability to interpret the vestiges of more ancient glaciations. Carbon and oxygen isotope analyses of foraminifers from deep-sea cores have provided evidence of several episodes of ocean water cooling that may not necessarily have been associated with glaciation on land.

Pleistocene and Holocene Chronology

It is a popular but biostratigraphically untenable conception that the Pleistocene Epoch began with a sudden worldwide onslaught of frigid conditions and that the lower boundary of the Pleistocene Series is easily recognized by sedimentologic and paleontologic clues to that frigidity. As with some other epochs of the Cenozoic, the Pleistocene was defined by Charles Lyell in 1839 according to the proportion of extinct to living species of mollusk shells in layers of sediment. At the time, Lyell referred to rocks subsequently named Pleistocene as "Newer Pliocene."

Lyell's scheme for naming the epochs seems straightforward, but it is no simple matter at all to find suitably fossiliferous sediments in various parts of the globe that can be confidently correlated to Lyell's type section in eastern Sicily. The most widely accepted time for the beginning of the Pleistocene is 1.8 million years ago. This date, however, does not coincide everywhere with the beginning of glaciation (as indicated by glacial sediments). The most extensive glaciation appears to have begun about 1 million years ago. Also, the low temperatures and abundant precipitation needed for extensive continental glaciation occurred at different times in different places.

For Pleistocene marine beds, fossils are indispensable for correlation. In some cases, a horizon correlative to the standard section can be recognized by the earliest appearance of certain cooler water mollusks and foraminifers. When studying cores of deep-sea sediments, one can frequently recognize the basal Pleistocene oozes by the extinction point of fossils called **discoasters**. Discoasters (Fig. 13–40) are calcareous, often star-shaped fossils believed to have been

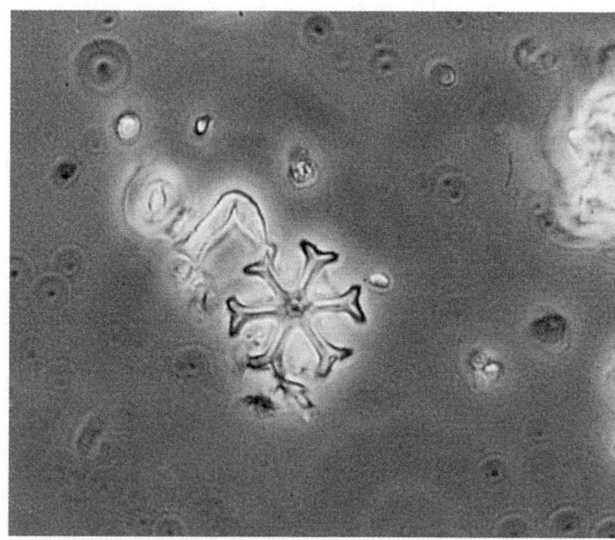

FIGURE 13-40 Discoaster (*Discoaster challengeri*), seen through a light microscope at a magnification of × 1200.

TABLE 13–2 Classic Nomenclature for Glacial and Interglacial Stages of the Pleistocene Epoch

North America	Alpine Region	Years before Present
		— 10,000
WISCONSINIAN	Würm	— 75,000
Sangamon	Riss-Würm	— 125,000
ILLINOIAN	Riss	— 265,000
Yarmouth	Mindel-Riss	— 300,000
KANSAN	Mindel	— 435,000
Aftonian	Günz-Mindel	— 500,000
NEBRASKAN	Günz	— 1800,000
Pre-Nebraskan	Pre-Günz	

In North America, the glacial stages are Nebraskan, Kansan, Illinoian, and Wisconsinian. These terms correspond approximately to the Günz, Mindel, Riss, and Würm in Europe.

produced by golden-brown algae related to coccoliths. In continental deposits, the fossil remains of the modern horse (*Equus*), the first true elephants, and particular species of other vertebrates are used to recognize the deposits of the lowermost Pleistocene.

Before the mid-1970s, geologists customarily thought of the Pleistocene as divisible into four distinct glacial stages, with intervening interglacial stages (Table 13–2). Recent investigations indicate, however, that there may have been as many as 30 periods of severely frigid conditions over the past 3 million years. Widespread cold climates thus existed well before the beginning of the Pleistocene, and indeed the first subtle evidence of the coming cold began to appear 40 million years ago. Antarctica has been frozen into an ice age for at least the past 15 million years.

The end of the Pleistocene (and beginning of the Holocene) is considered to be the time of melting of the ice sheets to approximately their present extent and the concomitant rise in sea level. This would mean that the Pleistocene ended about 8000 years ago. However, there is some argument about the date of this boundary. Some geologists prefer to set the Pleistocene-Holocene boundary at the midpoint in warming of the oceans, in which case the Ice Age would have ended between 11,000 and 12,000 years ago.

Whatever date is confirmed for the end of the Pleistocene Epoch, it is evident from historical records and carbon-14 dating of old terminal moraines that cold spells have recurred periodically into the Holocene. Climates on our planet are in delicate balance with atmospheric, geographic, and astronomic variables. A change in any one of these variables is likely to affect climate. Meteorologists, for example, have found a very good correlation between periods of minimum sunspot activity and episodes of colder conditions on Earth. One well-documented period of cooler and drier conditions in the Northern Hemisphere occurred between A.D. 1540 and 1890, when temperatures were often 2° to 4°F cooler than today. These four centuries encompass the so-called **little ice age**. During the little ice age, frigid conditions periodically extended across most of Europe and in the United states as far south as the Carolinas. Arctic sea ice intruded far to the south of its present margin. The cold caused loss of harvests, with resulting famine, food riots, and warfare in Europe. In the middle of the 14th century, harsh conditions of the little ice age caused the demise of a once flourishing Norse colony in Greenland.

Stratigraphy of Terrestrial Pleistocene Deposits

Glacial deposits are difficult to correlate. They often consist of chaotic mixtures of coarse detritus completely lacking in fossil remains. Moving glaciers may carry away even these materials, leaving only a bare surface of scoured and polished bedrock. However, if the ice does not advance for a time, a terminal moraine may be deposited in front of the glacier (Fig. 13–41). As the ice melts, a widespread ground moraine is left behind. Moraines may consist of **till**, an unsorted mixture of particles from clay to boulder size dropped directly by the ice (Fig. 13–42). **Stratified drift**, on the other hand, has been washed by meltwater, is better sorted, and has many of the characteristics of stream deposits.

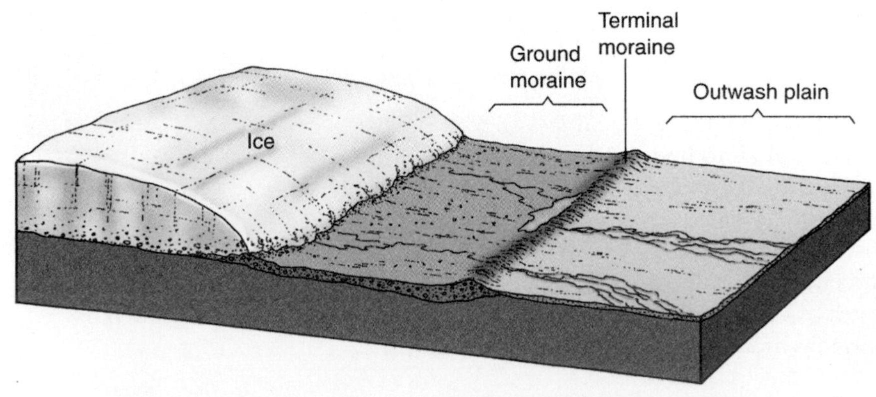

FIGURE 13-41 Cross-section of a terminal moraine, ground moraine, and outwash plain. ? *How might sediment of the outwash plain differ from sediment of the ground moraine?*

FIGURE 13-42 Glacial till in the moraine of a valley glacier, Kenai Peninsula, Alaska. (*Courtesy of U.S. Geological Survey.*)

In either case, however, the stratigrapher is faced with similar-looking masses of debris that are woefully deficient in distinguishing characteristics. As a result, several mutually substantiating criteria must be used in making correlations. The degree to which one sheet of glacial debris has been dissected by streams may indicate that it is older or younger than another sheet. The depth of oxidation and amount of chemical weathering may also provide some estimate of the relative age of interglacial soils. Fossil pollen grains in thick exposures of bog or lake deposits often clearly reflect fluctuations of climate and can be used to mark times of glacial advance and retreat. Varved clays deposited in lakes near the glaciers can sometimes be correlated to similar sediments of other lakes and, in addition, may provide an estimate of the time required for deposition of the entire thickness of lake deposits. However, the most accurate means of dating and correlating Pleistocene sediments is to extract pieces of wood, bone, or peat and determine their age by radiocarbon dating techniques. Unfortunately, even this method has its limitations, the most significant of which is the relatively short half-life (5570 years) of carbon-14. This limits the method to materials less than about 100,000 years old and thus restricts the use of the carbon-14 technique to deposits of the last glacial stage.

Pleistocene Deep-Sea Sediments

Sediments deposited on the ocean floor during the Pleistocene are easier to date and correlate than terrestrial deposits. The ocean basins are sites of a relatively continuous sedimentary record, the column of sediments contains abundant fossil remains, and the deposits can be dated by relating them to paleomagnetic data and to radiometric isotopes having short half-lives.

Continuous sections of deep-sea sediments can be obtained by means of piston coring devices. These tools provide cores more than 15 meters long. The cores can then be analyzed in various ways. One approach is to use the oxygen isotope ratios obtained from the shells of calcareous foraminifers within the cored sediment to determine the volume of ocean water that was stored in glacial ice. Because of its greater mass, less oxygen-18 is evaporated and precipitated as snow. Hence, during an ice age when ice is accumulating globally, there will be relatively greater percentages of oxygen-18 relative to oxygen-16 in sea water and in the shells of marine invertebrates. Plotted against depth in a deep-sea core, these isotope ratios indicate variations in ice volume and temperature with time (Fig. 13–43). Indications of cooler conditions can then be correlated with declines in worldwide sea level.

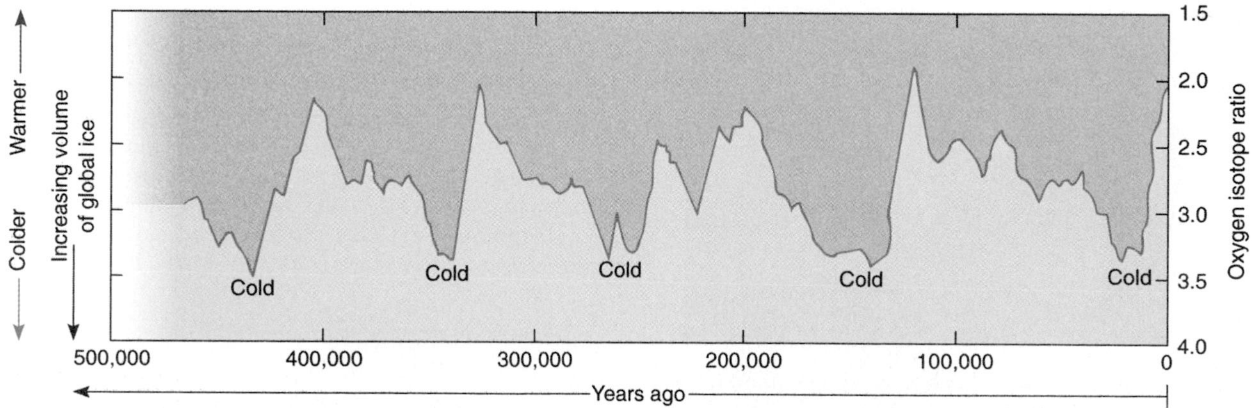

FIGURE 13-43 Curve reflecting variations in the global volume of ice (and, indirectly, paleotemperatures) during the past 500,000 years. Data are from radiometric dating and isotope measurements of cores from the Indian Ocean. (*Data from Hays, J. D., and Shackleton,* N. J. 1976. Science *194:1121–1132.*)

Another method used in the correlation of Pleistocene marine sediments is to plot the relative abundance of foraminifers known to be especially sensitive to temperature. For example, the tropical species *Globorotalia menardii* (Fig. 13–44) is alternately present or absent within Pleistocene cores from the equatorial Atlantic. The absence of the species in part of a core is taken to indicate an episode of cooler climates and glaciation. Carbon-14 dates in the upper part of such cores permit one to determine the rates of sedimentation.

Another foraminifer, *Globorotalia truncatulinoides*, also permits one to recognize alternate cooling and warming of the oceans (Fig. 13–45). This species is trochospirally coiled, which means that it coils in a spiral with all the whorls visible on one side but only the last whorl visible on the other side. Individuals that coil to the right dominate in warmer water, whereas left-coiled individuals prefer colder water. During glacial advances, when ocean temperatures declined, the right-coiled populations of *G. truncatulinoides* in middle and low latitudes were replaced by populations in which left coiling predominated. The record of such changes is clearly apparent in the deep-sea cores.

Remanent magnetism in some deep-sea sediments is sometimes sufficiently strong and stable that a particular section of the core can be correlated to a magnetic

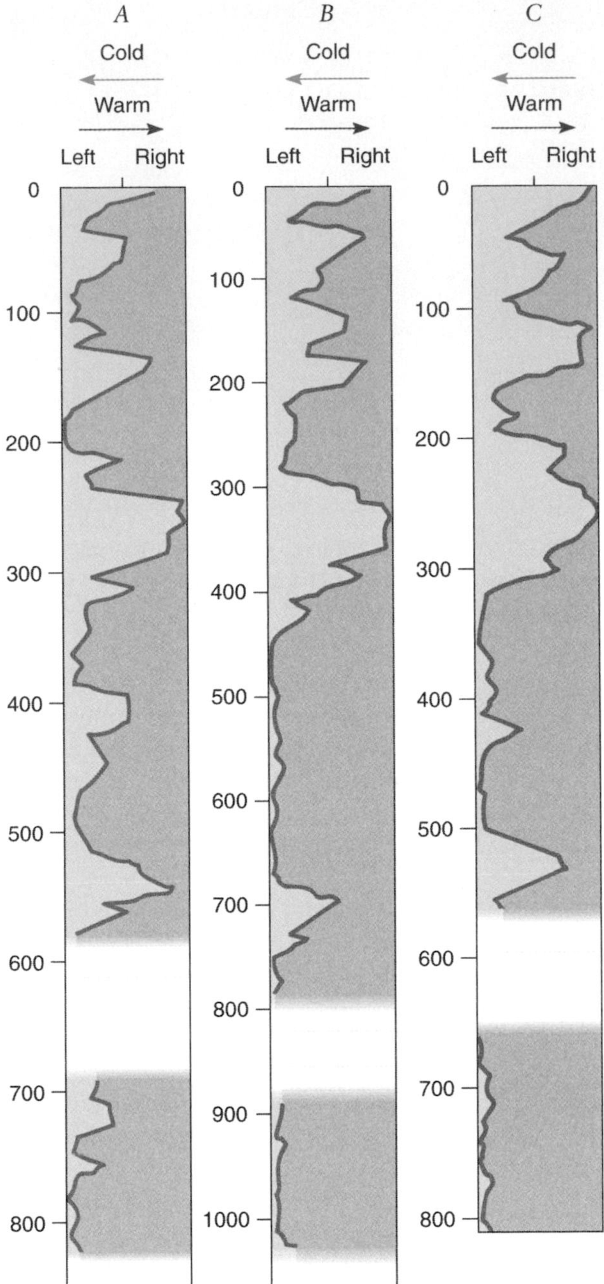

FIGURE 13-45 Curves depicting the percentages of right- and left-coiled *Globorotalia truncatulinoides*. The scales run from 100 percent right-coiling forms at the right margins of each column to 100 percent left-coiling forms at the left margins. Left-coiling forms prefer colder waters. Numbers to the left of the columns are deep-sea core depths in centimeters. Cores were not recovered in white areas. (*Adapted from Erickson, D. D., and Wollin, E.* Science *162[3859]:233.*)

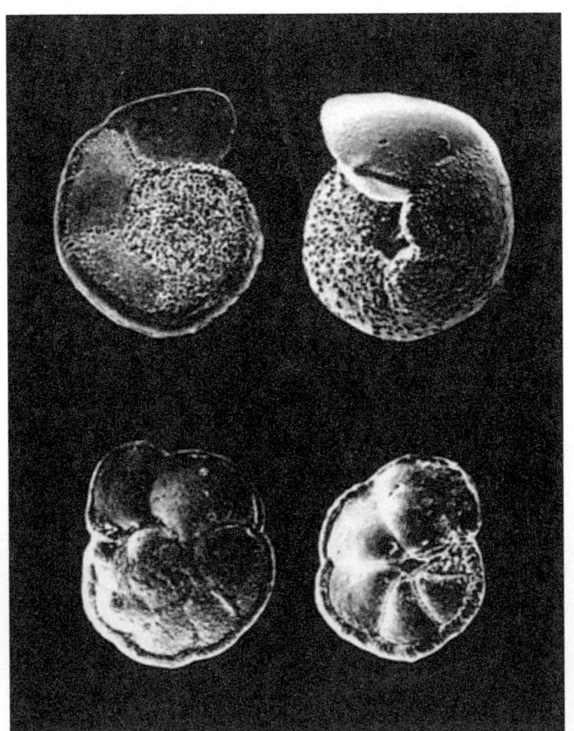

FIGURE 13-44 Two well-known species of planktonic foraminifers used in correlating deep-sea sediments. At the top are views of both sides of *Globorotalia truncatulinoides*. Below are views of opposite sides of *Globorotalia menardii* (all × 50).

reversal documented on land in volcanic rocks of known age. However, the analysis is likely to be in error if the original orientation of detrital magnetic grains has been disturbed by burrowing organisms (a process termed bioturbation).

Effects of Pleistocene Glaciations

Earlier it was noted that during maximum glacial coverage, over 40 million km^3 of ice and snow lay on the continents. Removal of an amount of water from the oceans equivalent to that much ice had a multitude of effects on the Pleistocene environment. It has been estimated that sea level may have dropped at least 75 meters during maximum ice coverage. Extensive tracts of the present continental shelves became dry land and were covered with forests and grasslands. The British Isles were joined to Europe, and a land bridge stretched from Siberia to Alaska. During interglacial stages, marine waters returned to the low coastal areas, drowning the flora and forcing terrestrial animals inland.

The glaciers had a direct impact on the erosion of lands and the creation of glacial land forms. The great weight of the ice depressed the crust of the Earth over large parts of the glaciated area, in some places to a level of 200 to 300 meters below the preglacial position. With the removal of the last ice sheet, the downwarped areas of the crust gradually began to return to their former positions. The rebound is dramatically apparent in parts of the Baltic, the Arctic, and the Great Lakes region of North America, where former coastal features are now elevated high above sea level (Fig. 13–46).

As the great continental glaciers advanced, they obliterated old drainage channels and caused streams to erode new channels. These dislocations are especially evident in the north–central United States. Prior to the Ice Age, the northern segment of the Missouri River drained northward into the Hudson Bay, and the northern part of the Ohio River flowed northeastward into the Gulf of St. Lawrence. The lower Ohio drained into a preglacial stream named the Teays River. Today, geologists recognize the former location of the Teays by a thick linear trend of sands and gravels. Those parts of the Missouri and Ohio rivers that once flowed toward the north were turned aside by the ice sheet and forced to flow along the fringe of the glacier until they found a southward outlet. The present trend of the Missouri and Ohio rivers approximates the margin of the most southerly advance of the ice. The equilibrium of streams was also affected, for with glacial advances there was a lowering of sea level and thus an increase in stream gradients near coastlines. A reverse effect occurred with wasting away of the ice sheets.

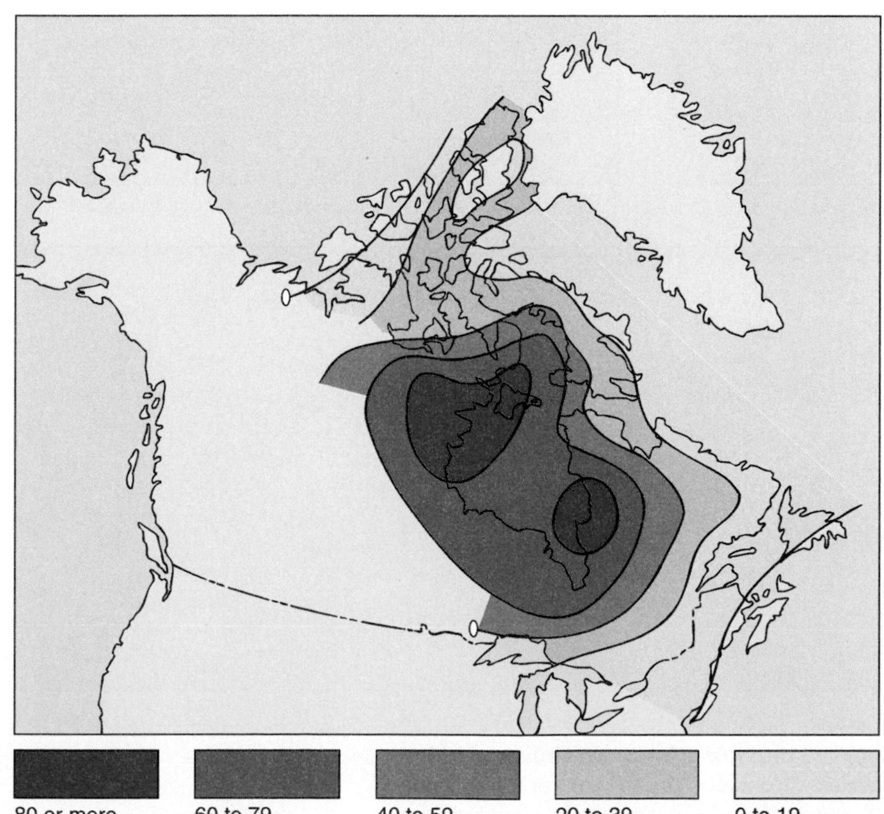

FIGURE 13-46 Postglacial uplift of North America, determined by measuring elevation of marine sediments 6000 years old. (*Simplified and adapted from Andrews, J. T. 1969.* The pattern and interpretation of restrained, postglacial and residual rebound in the area of Hudson Bay, *in* Earth Science Symposium on Hudson Bay, Ottawa, *1968, Canadian Geological Survey Paper 68–53, p. 53.*)

Prior to the Pleistocene, there were no Great Lakes in North America. The present floors of these water bodies were lowlands. Glaciers moved into these lowlands and scoured them deeper. As the glaciers retreated, their meltwaters collected in the vacated depressions (Fig. 13–47). South of the lake basins, huge blocks of ice containing a huge load of glacial clastics melted and in doing so produced a hummocky topography composed of irregular hills and numerous small lakes called **kettles**. Somewhat later, sands deposited along the southern shores of Lake Michigan were carried eastward by gusty winds and deposited as dunes around the southeastern bend of the lake. Although many of these Pleistocene dunes have been lost to developers, some remain and can be seen at Indiana Dunes National Lakeshore (Fig. 13–48).

Niagara Falls, between Lake Erie and Lake Ontario, was formed when the retreating ice of the last glacial stage uncovered an escarpment formed by southwardly tilted resistant strata. Water from the Niagara River tumbled over the edge of the escarpment, which is supported by the Lockport Limestone. Weak shales that lie beneath the Lockport are continuously undermined, causing southward retreat of the falls (see Fig. 8–15).

Another large system of ice-dammed lakes covered a vast area of North Dakota, Minnesota, Manitoba, and Saskatchewan. The largest of these lakes has been named Lake Agassiz in honor of Jean Louis Rodolphe Agassiz, the great French naturalist who settled in North America and who had initially insisted on the existence of the Ice Age. Today, rich wheatlands extend across what was once the floor of the lake.

Other lakes developed during the Pleistocene, occupying basins that were not near ice sheets. These were formed as a consequence of the greatly increased precipitation and runoff that characterized regions south of the glaciers. Such water bodies are called **pluvial**

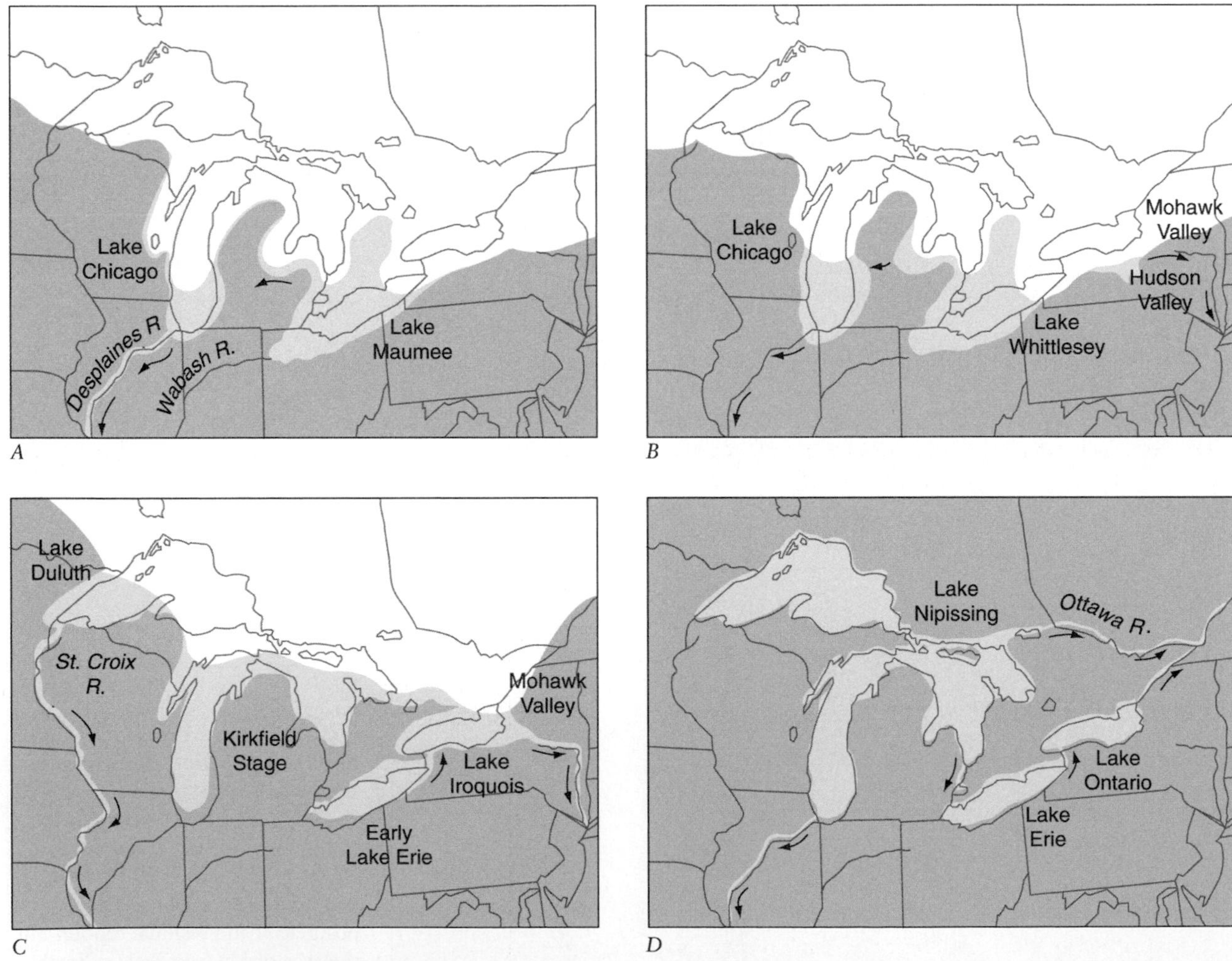

FIGURE 13-47 Continental glaciers moved into low-lying areas and scoured them deeper. As they retreated northward, meltwater filled the depressions to form the Great Lakes. (*Courtesy of Thompson, G.R. and Turk, J. 1977,* Modern Physical Geology, *Philadelphia: Saunders College Publishing.*)

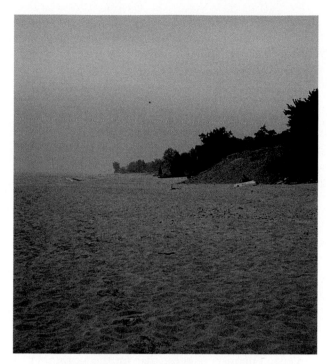

FIGURE 13-48 Sand carried to the southeastern shores of Lake Michigan during the Pleistocene formed this beach and sand dunes behind the beach at Indiana Dunes National Lakeshore.

FIGURE 13-49 Conglomeratic scabland deposits exposed near Lyons Ferry, Washington. (*Courtesy of U.S. Geological Survey.*)

lakes (which comes from the Latin *pluvia*, meaning "rain"). Pluvial lakes were particularly numerous in the northern part of the Basin and Range Province of North America, where faulting produced more than 140 closed basins. So-called pluvial intervals, when lakes were most extensive, were generally synchronous with glacial stages, whereas during interglacial stages, many lakes shrank to small saline remnants or even dried out completely. Lake Bonneville in Utah was one such lake. It once covered more than 50,000 km^2 and was as deep as 300 meters in some places. Parts of Lake Bonneville persist today as Great Salt Lake, Utah Lake, and Sevier Lake.

A particularly spectacular event associated with the formation of Pleistocene lakes occurred in the northwestern corner of the United States. Lobes of the southwardly advancing ice sheet repeatedly blocked the Clark Fork River. The impounded water formed a long, narrow lake extending diagonally across part of western Montana. Called Lake Missoula, the lake contained an estimated 2000 km^3 of water. With the recession of the glacier, the ice dam broke, and tremendous floods of water rushed out catastrophically across eastern Washington, causing severe erosion and depositing huge volumes of gravel, boulders, and cobbles (Fig. 13–49). The dissected region is appropriately termed the channeled scablands (Fig. 13–50).

The glacial conditions of the Pleistocene also had an effect on soils. In many northern areas, fertile topsoil was stripped off the bedrock and transported to more southerly regions, which are now among the world's most productive farmlands. Because of the flow of dense, cold air coming off the glaciers, winds were strong and persistent. Fine-grained glacial sediments that had been spread across outwash plains and floodplains were picked up and transported by the wind and then deposited as thick layers of windblown silt called loess. Such deposits blanket large areas of the Missouri River valley, central Europe, and northern China.

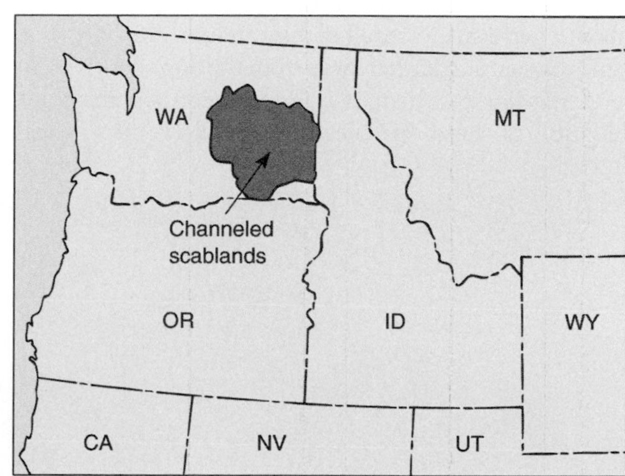

FIGURE 13-50 Location of the channeled scablands.

Cause of Pleistocene Climatic Conditions

The results of oxygen isotope research indicate that world climates grew progressively cooler from the Middle Cenozoic to the Pleistocene. The culmination of this trend was not a single sudden plunge into frigidity but rather an oscillation of glacial and interglacial stages. Any theory that adequately explains glaciation must consider not only the long-term decline in worldwide temperatures but the oscillations as well. Further, the theory must include reasons for the ideal combination of temperature and precipitation required for the buildup of continental glaciers. Although geologists, physicists, and meteorologists have speculated about the cause of the Ice Age for more than a hundred years, no undisputed cause has been found. Indeed, it seems likely that Pleistocene climatic conditions came upon us as a result of several simultaneous factors.

A widely accepted theory for the temperature fluctuations that may have caused the Ice Age was developed by the Yugoslavian mathematician Milutin Milankovitch (1879–1958). After 30 years of careful study, Milankovitch convincingly proposed that irregularities in the Earth's movements and their influence on the amount of solar radiation received by the Earth could account for glacial and interglacial stages of the Pleistocene. His calculations were based on three variables: the Earth's axial tilt, precession, and orbital eccentricity. With regard to the first of these variables, Milankovitch recognized that the tilt of the Earth's axis varies between about 22° and 24° over a period of 41,000 years. This results in a corresponding variation in the seasonal length of days and in the amount of solar radiation received at higher latitudes. **Precession**, the second variable, refers to the way the axis of rotation moves slowly in a circle that is completed about every 26,000 years. The effect is equivalent to tilting a rapidly spinning top, whose axis responds by describing a cone in space (Fig. 13–51). The third variable is the eccentricity of the Earth's orbital path around the Sun, which over an interval of about 100,000 years varies about 2 percent. As a result of this variation, the Earth is at times closer to or farther away from the Sun.

According to Milankovitch's calculations, the combination of these variables periodically results in less solar radiation received at the top of the Earth's atmosphere, and this might suffice to cause cooling and recurrent glaciations. The Milankovitch cycles correspond rather well to the timing of episodes of glaciation over the past 100,000 years. That the Milankovitch cycles do have a role in causing the alternation of glacial and interglacial stages is supported by oxygen isotope analyses of the calcium carbonate shells of foraminifera recovered from deep-sea cores. These analyses confirm an interval of about 100,000 years between the times of coldest temperatures over the past 600,000 years.

If, however, the Milankovitch effect has been in operation through most of the Phanerozoic eon, why haven't there been Pleistocene-like glaciations continuously throughout geologic time? Apparently, other factors must be involved. One such factor might be a variation in the amount of solar energy reflected from the Earth back into space rather than being absorbed. The fraction of solar energy reflected back into space is termed the Earth's **albedo**. At present, it is about 33 percent. Theorists suggest that by the beginning of Pleistocene time, when continents were fully emergent and hence highly reflective, temperatures may have been lowered enough for the Milankovitch effect to begin to operate. This is not an unreasonable suggestion, for only a 1 percent lessening of retained solar energy could lead to as much as an 8°C drop in average surface temperatures. This would be sufficient to trigger a glacial advance if ample precipitation were available over continental areas. Still other geologists speculate that absorption of solar energy was hindered by cloud cover, volcanic ash, and dust in the atmosphere or fluctuations in carbon dioxide. A decrease in carbon dioxide content would cause a corresponding decrease in the warmth-gathering greenhouse effect, for example. On the other hand, it is also possible that a buildup of carbon dioxide might trigger glaciation, for as warming occurred, there might be more rapid evaporation and an increase in highly reflective cloud cover.

Other hypotheses for Ice Age origin stress the need for both ample snowfall and the preservation of that precipitation on suitably located continental areas. Ample snowfall may have been produced by the

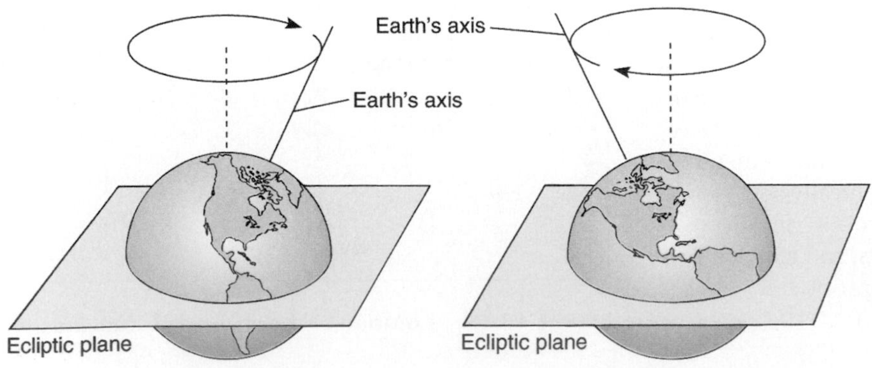

FIGURE 13-51 Two positions in the precession of the Earth's axis. The period of a complete precession is 26,000 years. Thus, the two positions shown are separated by 13,000 years. (*From Pasachoff, J. 1985.* Contemporary Astronomy, *3rd ed. Philadelphia: Saunders College Publishing.*)

ENRICHMENT

Oil Shale

The Eocene oil shale (see Fig. 13–53) that occurs in parts of Wyoming, Utah, and Colorado is rich in an oil-yielding organic compound known as kerogen. When heated to 480°C, the kerogen in oil shale vaporizes. The vapor can then be condensed and will form a thick oil, which, when enriched with hydrogen, can be refined into gasoline and other products in much the same way as ordinary crude oil. The retort in which the shale is heated resembles a giant pressure cooker that fuels itself with the very gases generated during heating.

Many oil shales yield from half to three quarters of a barrel (42 U.S. gallons) of oil per ton of rock. If one were to mine only the shale layers thicker than 10 meters, they would yield an impressive 540 billion barrels of oil. If this oil were to be produced during the next decade, it would appreciably reduce the United States' dependency on foreign crude oil. Production of oil from oil shale, however, requires the use of vast amounts of water in a region of North America where water is in short supply.

Two additional problems have prevented full-scale mining and retorting of oil shales. The first is the waste disposal problem. In the process of being crushed and retorted, the shale expands to occupy about 30 percent more volume than was present in the original rock. Geologists call this the popcorn effect. Where can this great volume of light, dusty material be placed? The second problem relates to air quality, for the processing of huge tonnages of oil shale is likely to release large amounts of dust into the atmosphere. Unfortunately, the shales occur in arid regions, and there is little water available for use in processing the rock to contain the dust and provide for revegetation of the land. Until these problems are solved, production of oil from oil shale will be slow. Perhaps this will work to our advantage, for we may need this oil for processed goods a century or two from now, when the internal combustion engine may be as rare as the horse and buggy are today.

northward deflection of the Gulf Stream after its westward movement was blocked by the formation of the Isthmus of Panama about 3.5 million years ago. Precipitation resulted when the moist air was shunted northward, fell on land areas as snow, and accumulated to form continental glaciers.

Although scientists are not yet ready to formulate a complete and unified theory of the cause of the Pleistocene's multiple glaciations, it appears that they may be very near the answer. Few deny that the Milankovitch effect, albedo, and pole positions have played a role in creating Pleistocene climatic conditions. The problem is to determine the relative importance of these factors and how they relate to one another. If Pleistocene-like climatic cycles are fundamentally the result of changes in the Earth's orbital geometry, one can make some interesting speculations about future climate. Calculations indicate, for example, that the long-term climatic trend over the next 20,000 years is toward extensive Northern Hemisphere glaciation and, of course, cooler conditions. What effect human activities, such as the burning of coal, oil, and gas, might have on this trend is uncertain.

CENOZOIC CLIMATES

The worldwide cooling that culminated in the great Ice Age actually had begun during the Early Cenozoic. However, the decrease in temperatures was not entirely uniform. There were warming trends during the Late Paleocene and Eocene, as indicated by fossils of palm trees and crocodiles found in Minnesota, in Germany, and near London. Trees that are today characteristic of moist temperature zones thrived in Alaska, Spitzbergen, and Greenland. Coral reefs grew in regions 10° to 20° north of their present optimal habitat. Cooling resumed during the Oligocene, and coral reefs began a slow retreat toward the equator. On land, temperate and tropical forests were also displaced toward lower latitudes. Another pulse of warmer conditions occurred during the Miocene, after which climates grew steadily cooler, as if in anticipation of the approaching Pleistocene glaciations.

Although the most extensive glaciations of the Cenozoic were those of the Pleistocene, glaciation occurred on a more limited scale at other times during the Cenozoic. Deep-sea cores taken near Antarctica indicate that an ice sheet had already begun to form during the Eocene. In addition, Miocene glacial conditions are indicated both by terrestrial glacial deposits in Antarctica and ice-rafted glacial deposits in Miocene deep-sea cores from the Bering Sea. The accumulation of Miocene glacial ice and the resulting lowering of sea level contributed to the drying up of the Mediterranean Sea during the Messinian event. Evidence of Pliocene glaciation has been recognized in the Sierra Nevada, Iceland, South America, and the former Soviet Union.

MINERAL RESOURCES OF THE CENOZOIC

Most of the known accumulation of oil has been found in Tertiary reservoir rocks. Paleocene petroleum reservoirs occur in Libya and beneath the North Sea.

FIGURE 13-52 Drilling for offshore oil. Shown here is a huge semisubmersible drilling rig. (*Courtesy of Reading and Bates Drilling Company.*)

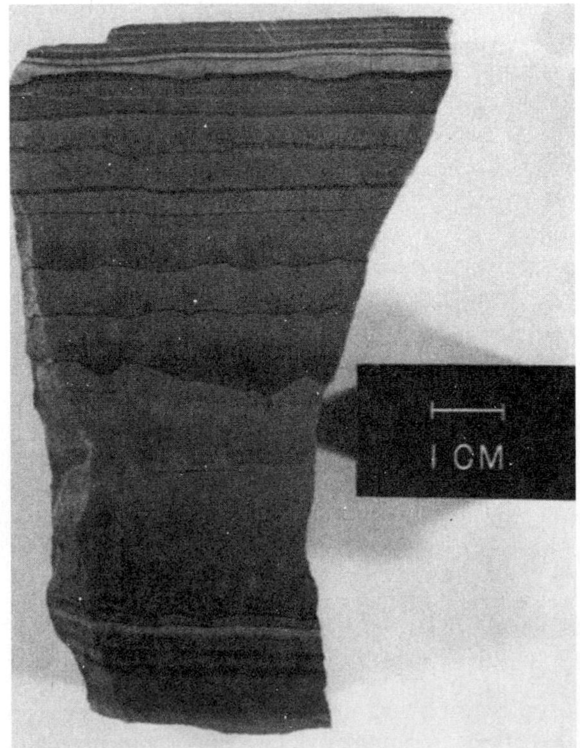

FIGURE 13-53 Eocene oil shale from northeastern Colorado. (*Courtesy of U.S. Geological Survey.*)

Petroleum trapped in Eocene permeable limestones and sandstones is being pumped in Texas, Louisiana, Iraq, the former Soviet Union, Pakistan, and Australia. Strata of Oligocene age yield oil in western Europe, Burma, California, and the Gulf Coast states of the United States. Miocene oil-bearing sandstones occur on every continent except Australia. The Miocene deposits are extraordinarily productive in Saudi Arabia, Kuwait, California, Texas, and Louisiana. Offshore oil fields in the Gulf Coast (Fig. 13–52) and California tap oil held in rocks of Pliocene age. Finally, the largest reserves of oil shale in the world occur in the Eocene Green River Formation of the western United States (Fig. 13–53).

Coal also occurs in Cenozoic rocks. Most Cenozoic coal is of the lignitic or sub-bituminous variety, but because of its low content of sulfur, it is extensively used. The coal beds of the Paleocene Fort Union Formation are mined in the Dakotas, Wyoming, and Montana.

They represent the largest recoverable fossil fuel deposits in the United States. Scattered coal beds, mostly of Eocene age, are exploited along the Pacific coast of the United States.

Metals mined from Cenozoic rocks include placer gold gathered by dredging and hydraulic mining of gravel deposits in California. The gold in these deposits has been eroded from older Jurassic gold-bearing quartz veins in the Sierra Nevada. About 60 percent of the world's supply of tin is also supplied by Cenozoic stream deposits. Manganese, an essential additive in the manufacture of steel, is mined extensively from lower Tertiary rocks of the former Soviet Union. Another metal used in making certain types of steel is molybdenum. One deposit in Climax, Colorado, at one time supplied 40 percent of the world's molybdenum. Other metals, such as copper, silver, lead, mercury, and zinc, are frequently found in association with Tertiary intrusive rocks of western North America, the Andes, and Tertiary orogenic belts along the western margins of the Pacific Ocean. Similar Cenozoic rocks were formed in southern Europe and Asia as a consequence of the deformation of the Tethys belt.

Nonmetallic materials from Cenozoic rocks are also important resources. Diatomite (Fig. 13–54), the white, porous rock formed from the silica coverings (frustules) of diatoms, is mined in great quantities in California. It is used as a filter, in abrasives, and in certain building materials. Cenozoic building stone, clay, phosphates, sulfur, salt, and gypsum are quarried in North America, Europe, the Middle East, and Asia.

FIGURE 13-54 Massive, jointed diatomite of the Bruneau Formation of middle Pleistocene age, Elmore County, Idaho. *(Courtesy of U.S. Geological Survey.)*
? *What properties of diatomite account for its use as a filter?*

SUMMARY

A compilation of the great historical events of the Cenozoic would certainly include the formation of the Alpine–Himalayan mountain system, the separation of Australia from Antarctica, the drift of continents to their present locations, the uplift of mountains around the perimeter of the Pacific, the development of an ice age, and the evolution of *Homo sapiens*. For most of the duration of the Cenozoic, the continents stood at relatively high elevations; epicontinental and marginal seas were of limited extent.

Along the eastern side of North America, the Appalachian chain was subjected to cycles of erosion and uplift that served to sculpt the surface to its present topography. Along the Atlantic and Gulf coastal plains, repeated transgressions and regressions reworked clastic sediment transported there from the interior. Florida was a shallow, limestone-forming, submarine bank until it was uplifted late in the era. In the Gulf of Mexico, Cenozoic deposits accumulated in thicknesses in excess of 9000 meters as the depositional basin was continuously downwarped.

In western North America, compressional forces that had produced many of the structures of the Cordillera ceased, and in their place vertical crustal adjustments occurred. The erosional debris worn from the highland areas first filled the basins between ranges and then spread eastward to form the broad clastic wedge of the Great Plains. Lakes formed in intermontane lowlands. Many of these served as collecting sites for oil shales. Crustal movements later in the Tertiary caused the elevation of the Sierra Nevada along a great fault and produced the Basin and Range Province. Intense volcanic activity accompanied these crustal movements. Volcanism continues today along those sectors of the West Coast of North America that lie adjacent to subduction zones. The Cascades, which are fed by lavas from the subducting Juan de Fuca plate, provide a present day example. Also in the far west, the Farallon plate and its spreading center were overridden by the westward-moving continent. As a result, the formerly convergent plate boundary was changed to a shear boundary characterized by strike-slip, rather than subduction. The development of the San Andreas fault system was a direct result of this change in tectonics style.

In the Tethys region, there was intense deformation during the Cenozoic as major tectonic plates moved against one another. Orogenies that accompanied the compressions and uplifts along the southern margin of the Tethys region resulted in the Atlas Mountains, whereas along the northern border, the Alps, Carpathians, Apennines, Pyrenees, and Himalayas were constructed from the sediments that had accumulated in the former Tethys seaway. Orogenic activity was also prevalent over much of the Middle East and Far East and along the western borders of the Pacific. Africa witnessed Cenozoic marine deposition along its northern border. However, the most dramatic occurrences in Africa were the continuing development of the great rift valleys that extend along the eastern side of the continent.

It is apparent from fossil evidence and oxygen isotope studies that the Earth's climate was on a fluctuating cooling trend during the Cenozoic. The culmination of that trend was the beginning of extensive glaciation about 1 million years ago. Variations in climate during the Pleistocene caused the vast ice sheets that had formed on the northern continents to alternately advance and recede. There were at least four major advances, with many lesser fluctuations before and within each. The ice, at times attaining thicknesses of 3000 meters, depressed the Earth's crust, smoothed and rounded the landscape, disrupted drainage patterns, altered climates of adjacent regions, and was responsible for the development of numerous lakes, not the least of which are the Great Lakes. As the ice alternately accumulated and melted, sea level was caused to rise and fall, with profound effects on low-lying coastal areas. It is a possibility that today we live in an interglacial stage and that some thousands of years from now the ice will again accumulate in Canada and northern Europe.

Although a wealth of gold, tin, copper, silver, and other metals is found in deposits or associated with intrusions of Cenozoic rock, the greatest resource of the era has been petroleum. Indeed, most of the world's petroleum has been found in Cenozoic strata.

QUESTIONS FOR REVIEW AND DISCUSSION

1. Describe the manner in which each of the following major physiographic features developed:
   a. Mountains of the Basin and Range
   b. Great Plains
   c. Columbia and Snake River Plateau
   d. Red Sea
   e. Teton Range
   f. Cascade Range
   g. Channeled Scablands
2. What Eurasian mountain ranges resulted from the compression and upheaval of large areas of the Tethys seaway?
3. What epochs of the Cenozoic Era are included in the Tertiary Period, the Quaternary Period, the Paleogene Period, and the Neogene Period?
4. What is the economic importance of the Green River Formation? The Fort Union Formation?
5. What is the origin of Lake Nyasa and Lake Tanganyika in eastern Africa?
6. What is the origin of Great Salt Lake in Utah? Why is it so salty?
7. In a 30-meter-long deep-sea core that penetrated most of the Pleistocene section, how might you differentiate sediment that had been deposited during a glacial stage from that of an interglacial stage?
8. What land bridge, important in the migration of Ice Age mammals, resulted from the lowering of sea level during glacial stages?
9. What is the explanation for the gradual rise in land elevations that has occurred within historic time around Hudson Bay, the Great Lakes, and the Baltic Sea?
10. What changes in movement of tectonic plates along the western border of the United States were responsible for the development of the San Andreas Fault?
11. What physical evidence indicates that the Mediterranean had dried up during the Neogene?
12. What is the evidence that the Gulf of Mexico experienced profound subsidence during the Cenozoic?
13. Discuss the advantages and disadvantages associated with the exploitation of Green River oil shales for their content of hydrocarbons.
14. Discuss conditions on Earth that might result in increased albedo. What are the possibilities that such conditions existed at the beginning of the Pleistocene?
15. What conditions during the Pleistocene favored the formation of extensive loess deposits? Where did this sediment come from? What accounts for the observation that the grains in loess are narrowly restricted in size, being mostly $\frac{1}{16}$ or $\frac{1}{32}$ millimeters in diameter?
16. The percentage of the isotope oxygen-18 relative to the percentage of oxygen-16 in the calcium carbonate shells of foraminifers increased during the colder glacial stages of the Pleistocene. Why?

READINGS

Ager, D. V. 1980. *The Geology of Europe*. New York: Halstead Press.

Atwater, T. 1970. Implications of plate tectonics for the Cenozoic evolution of western North America. *Bull. Geol. Soc. Am.* 81:3513–3536.

Covey, C. 1984. The earth's orbit and the ice ages. *Sci. Am.* 250(2):58–77.

Curtis, B. F., ed. 1975. Cenozoic history of the southern Rocky Mountains. *Geol. Soc. Am. Memoir* 144:1–279.

Ehlers, J. 1996. *Quaternary and Glacial Geology*. New York: John Wiley and Sons.

Fiero, B. 1986. *Geology of the Great Basin*. Reno: University of Nevada Press.

Goudie, A. 1992. *Environmental Change*, 3rd ed. Oxford: Oxford University Press.

Hsu, K. J. 1983. *The Mediterranean was a Desert*. Princeton, NJ: Princeton University Press.

Levenson, T. 1989. *Ice Time: Climate Science and Life on Earth*. New York: Harper and Row.

Match, C. L. 1976. *North America and the Great Ice Age*. New York: McGraw-Hill.

Muller, R.A., and MacDonald, G.J. 2000. *Ice Ages and Astronomic Causes*. New York: Springer-Praxis.

WEB SITES

The Earth Through Time Student Companion Web Site (www.wiley.com/college/levin) has online resources to help you expand your understanding of the topics in this chapter. Visit the Web Site to access the following:

1. Illustrated course notes covering key concepts in each chapter;
2. Online quizzes that provide immediate feedback;
3. Links to chapter-specific topics on the web;
4. Science news updates relating to recent developments in Historical Geology;
5. Web inquiry activities for further exploration;
6. A glossary of terms;
7. A Student Union with links to topics such as study skills, writing and grammar, and citing electronic information.

14

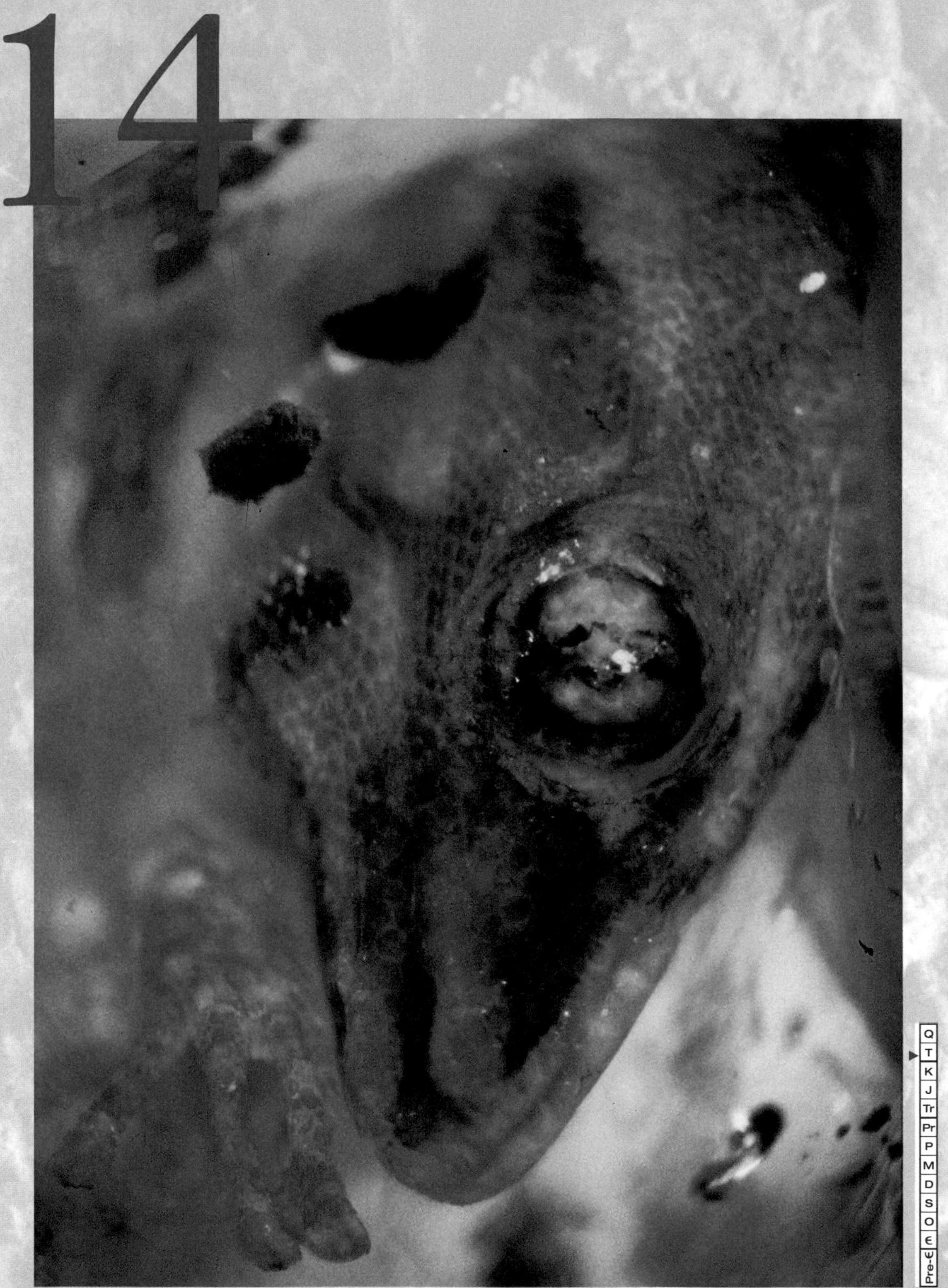

Cenozoic lizard from the Dominican Republic, La Toca Mine, preserved in amber. The specimen is 30 to 40 million years old. (Photograph by G. O. Poinar, Jr.)

CHAPTER 14

Life of the Cenozoic

OUTLINE

Because biologic developments of the Cenozoic have occurred so recently, and because Cenozoic fossils are stratigraphically topmost, better preserved, and more accessible, we have more facts about life of this era than about the far lengthier ones that preceded it. Armed with this more complete data, paleontologists are better able to see the effects of environmental and geographic changes on the evolution of animals and plants. Continental fragmentation clearly stimulated biologic diversity and resulted in distinctive faunal radiations on separated landmasses, as well as in isolated marine basins. In the seas, bivalves, gastropods, crustaceans, echinoids, scleractinian corals, and bony fishes flourished. The ammonites and such Mesozoic marine reptiles as ichthyosaurs, plesiosaurs, and mosasaurs were no longer present in Cenozoic seas. On land, major groups of mammals that have surviving species were present, including rodents, bats, carnivores, elephants, bison, camels, horses, and rabbits. South America, Australia, and Antarctica were separated from North America and Eurasia during most of the Cenozoic. As a result, a distinctive assemblage of mammals showing convergent evolution with Northern Hemisphere species evolved on these southern continents. When the Panamanian land bridge developed, many species of South American marsupials were driven to extinction by migrants from the north. Some South American mammals that were able to reach North America also caused extinctions but to a lesser extent. Among the many interesting evolutionary developments of the Cenozoic, however, none seem more intriguing than the changes experienced by primates, which by late Tertiary had produced species believed to be the ancestors of humans. In the Pleistocene Epoch, our own species, *Homo sapiens*, appeared.

PLANTS AND THE MAMMALIAN RESPONSE TO THE SPREAD OF PRAIRIES

Although they are the most recent plant group to evolve, the flowering plants are now the most widespread of all vascular land plants. Angiosperm floras did not explode upon the lands until mid-Cretaceous (Fig. 14–1). There were few spectacular floral innovations during the Cenozoic. Rather, this was a time of steady progress toward the development of today's

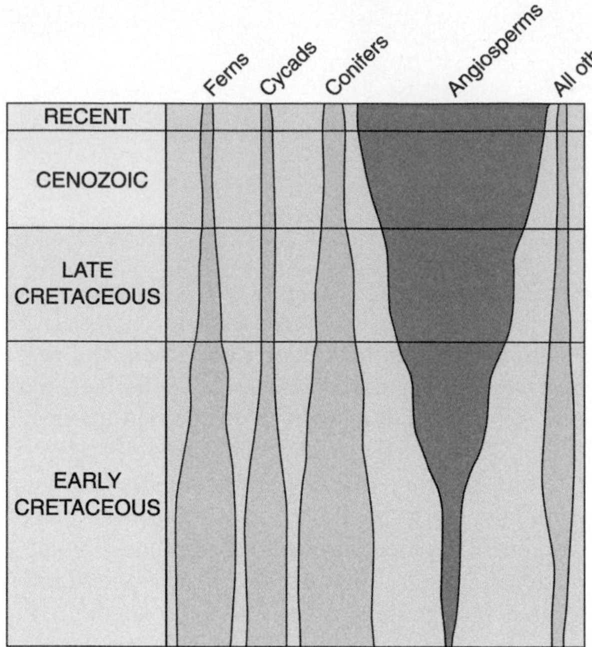

FIGURE 14-1 Relative proportions of genera in Cretaceous to recent floras.

complex plant populations. The Miocene is particularly noteworthy as the epoch during which grasses expanded and grassy plains and prairies spread widely over the lands. In response to the proliferation of this particular kind of forage, grazing mammals began their remarkable evolution.

Numerous evolutionary modifications among herbivorous mammals can be correlated to the development of extensive grasslands. Especially evident were changes in the dentition of grazing mammals. Grasses are abrasive materials. Many contain siliceous secretions, and because they grow close to the ground, they are often coated with fine particles of soil. To compensate for wear that results from chewing grasses, the major groups of herbivores evolved high-crowned cheek teeth that continue to grow at the roots during part of the animals' lives. To provide space for these high-crowned teeth, the overall length of the face in front of the eyes increased. Enamel, the most resistant tooth material, became folded, so that when the tooth was worn, a complex system of enamel ridges formed on the grinding surface. In general, incisors gradually aligned into a curved arc for nipping and chopping grasses.

In the open plains environment, it was difficult to hide from predators. As a result, many herbivores evolved modifications that permitted speedy flight from their enemies. Limb and foot bones were lengthened, strengthened, and redesigned by forces of selection to prevent strain-producing rotation and permit rapid fore-and-aft motion. To achieve greater speed, the ankle was elevated, and, like sprinters, animals ran on their toes. In many forms, side toes were gradually lost. Hoofs developed as a unique adaptation for protecting the toe bones while the animals ran across hard prairie sod. Some Herbivores evolved the four-chambered stomach in response to selection pressures favoring improved digestive mechanisms for breaking down the tough grassy materials. The response of mammals to the spread of grasses provides a fine example of how the environment influences the course of evolution.

MARINE PHYTOPLANKTON

As noted in Chapter 12, entire families of phytoplankton experienced extinction at the end of the Cretaceous. Only a few species in each major group survived and continued into the Tertiary. However, the survivors were able to take advantage of decreased competitive pressures and rapidly diversified. In general, peaks in diversity of species were reached in the Eocene and Miocene. A decrease in diversity has been recorded in the intervening Oligocene. Diatoms (Fig. 14–2), dinoflagellates (Fig. 14–3), and coccolithophorids (Fig. 14–4) provided the most abundant populations of Cenozoic marine phytoplankton.

INVERTEBRATES

The invertebrate animals of the Cenozoic seas included dense populations of foraminifers, radiolarians (Fig. 14–5), mollusks of all classes, scleractinid corals, bryozoans, and echinoids. The combined fauna had a decidedly modern aspect. Once successful groups such as ammonites and rudistid bivalves were no longer present. No new major groups of invertebrates appeared during the Cenozoic.

Zooplankton

Foraminifers were enormously prolific and diverse during the Cenozoic (Fig. 14–6) and included large numbers of benthonic as well as planktonic forms (Fig. 14–7). Among the larger genera, nummulitic foraminifers that were the size and shape of coins thrived in the clear warm waters of parts of the U.S. Gulf Coast, the western Atlantic, and Tethys seaway. The tests of these organisms have accumulated to form thick beds of nummulitic limestone. The ancient Egyptians quarried this rock and used it to construct the Pyramids of Gizeh. Even the famous Sphinx was carved from a large residual block of nummulitic limestone. The incredible numbers and variety of foraminifers account for their extensive use in

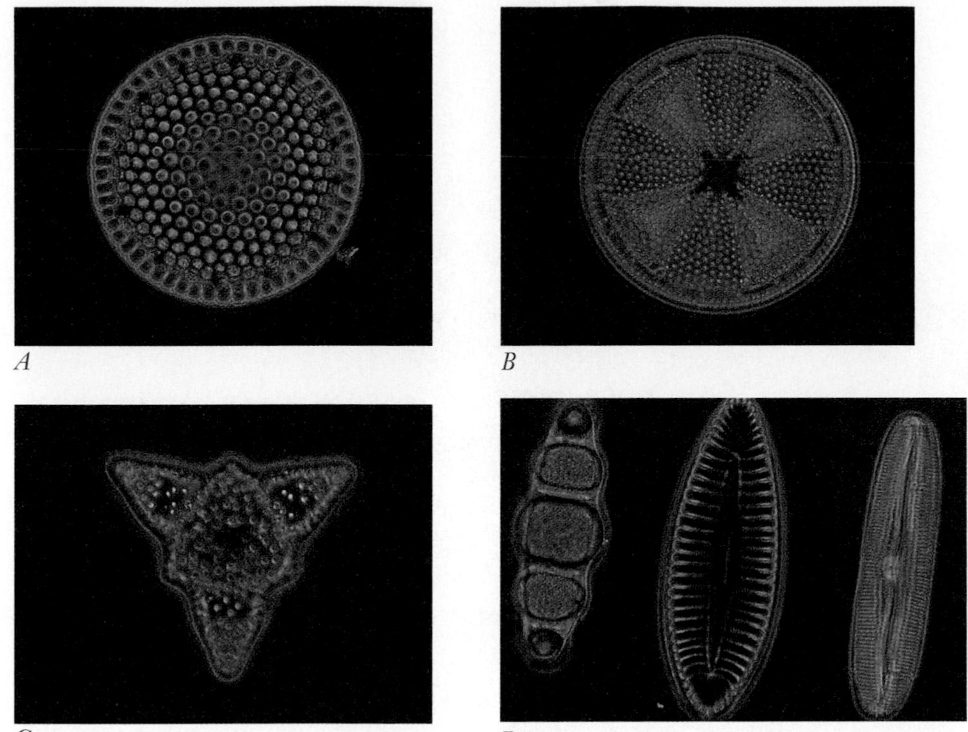

FIGURE 14-2 Cenozoic diatoms exhibiting various frustule shapes, including discoidal, triangular, and spindle. (*A*) *Arachnoidiscus*. (*B*) *Actinoptychus*. (*C*) *Triceratum*. (*D*) From left to right, *Biddulphia*, *Surirella*, and *Pinnulria*. Magnification × 50. ? *What is the composition of frustules of diatoms?*

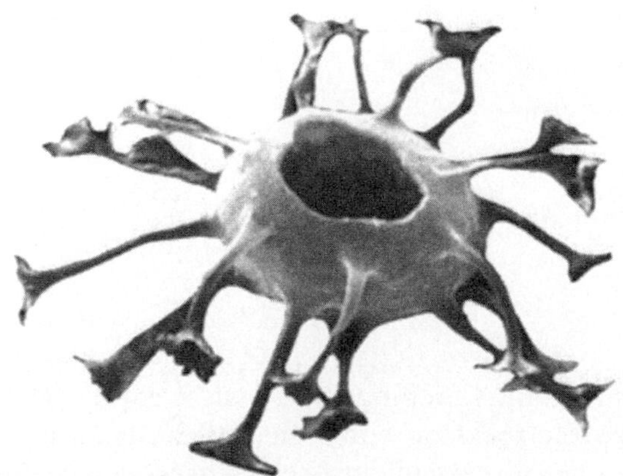

FIGURE 14-3 Dinoflagellate as seen with the aid of a scanning electron microscope. From Paleocene sedimentary rocks of Alabama. Magnification × 450. (*Courtesy of Standard Oil Company of California; photograph by W. Steinkraus.*)

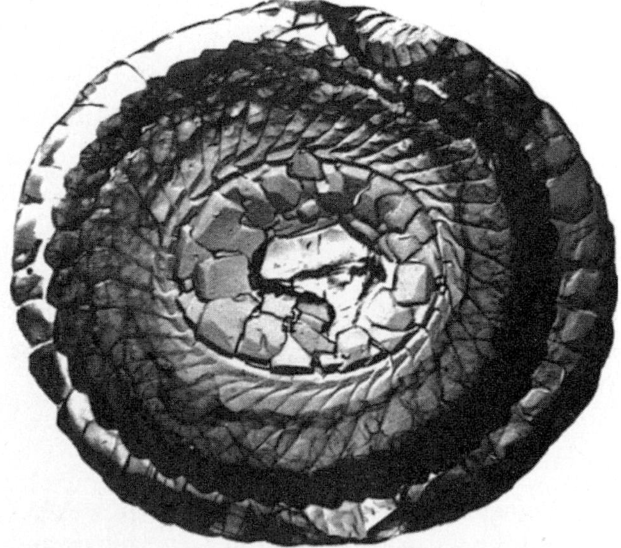

FIGURE 14-4 A coccolith of the coccolithophoroid *Coccolithus*, as seen with the aid of the transmission electron microscope. Maximum diameter is 16 millimeters.

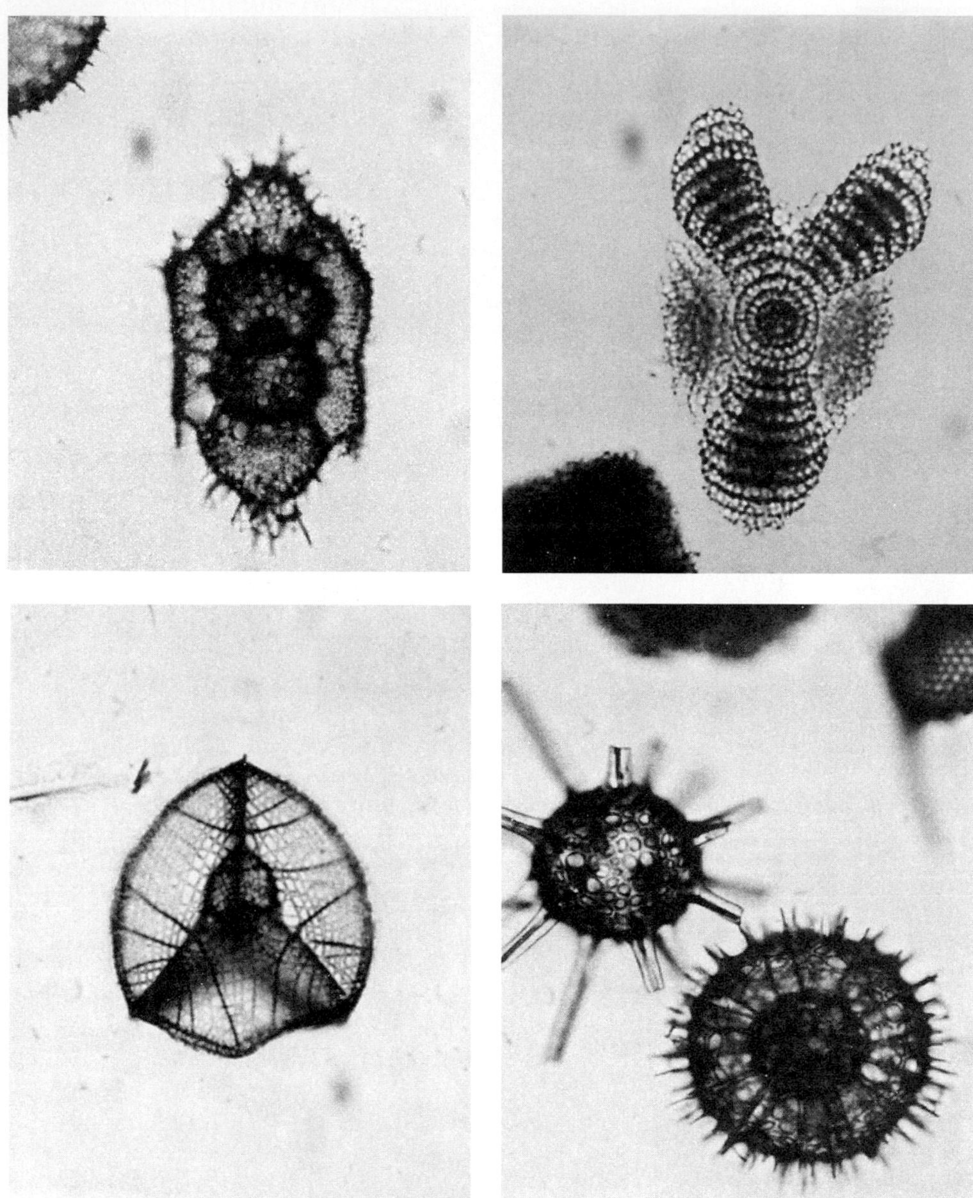

FIGURE 14-5 Quaternary radiolarians from the tropical Pacific Ocean. Magnification × 100. (*Courtesy of A. Sanfilippo, Scripps Institute of Oceanography.*)

correlating Cenozoic strata in oil fields of the Gulf Coast, California, Venezuela, the East Indies, and the Near East.

Corals

Scleractinian corals (Figs. 14–8 and 14–9) grew extensively in the warmer waters of the Cenozoic oceans. Fossil solitary corals are commonly found in shallow-water deposits of Early Tertiary age in Europe and the United States Gulf coastal region. Reef corals were most extensively developed in parts of the Tethyan belt, West Indies, Caribbean, and Indo-Pacific regions. Careful comparison of coral species in Cenozoic rocks on either side of the Isthmus of Panama has given geologists clues to when the Atlantic and Pacific oceans were connected across this present-day barrier. Continuous deposition of reef limestones occurred throughout the Cenozoic in atolls of the open Pacific. **Atolls** are ring-like coral reefs that grow around a central lagoon. In 1842, Charles Darwin brilliantly discerned the relationship between atolls and volcanic islands. In his subsidence theory, he proposed that, because of their great weight, volcanic islands subside slowly. Furthermore, such subsidence is usually slow enough to permit the

FIGURE 14-6 Diversity of form in Cenozoic foraminifers. Obtained from the Monterey Formation of Miocene age in California. Magnification × 50.

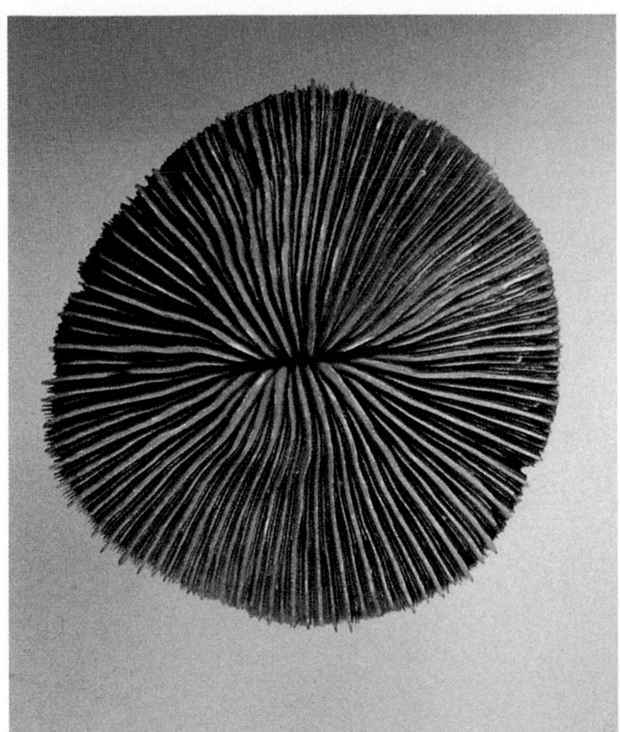

FIGURE 14-8 *Fungia*, a solitary coral that takes its name from its resemblance to the underside of the cap of a mushroom. It has a geologic range from the Miocene to the present and is common today in the seas of the Indo-Pacific region. This specimen is about 9.0 centimeters in diameter.

corals that grow around the fringe of the sinking island to grow upward and maintain their optimum habitat near the ocean's surface. Eventually, the central island would be submerged, but the encircling, living reef would persist (Fig. 14–10). Several deep borings into atolls have validated Darwin's theory. A boring drilled at Eniwetok Island, an atoll in the Marshall Islands of the western Pacific, encountered the basaltic summit of the volcano at depths of about 1200 meters. Eocene reefs lie directly above these igneous rocks.

It is likely that the subsidence associated with the development of some atolls is related to sea-floor spreading rather than to the weight of the volcanic mass, as Darwin thought. Spreading begins along the crest of a midoceanic ridge, and as the plates move away from the ridge, they subside very slowly due to thermal contraction and isostatic adjustment. The subsidence is estimated at about 9 centimeters per year for

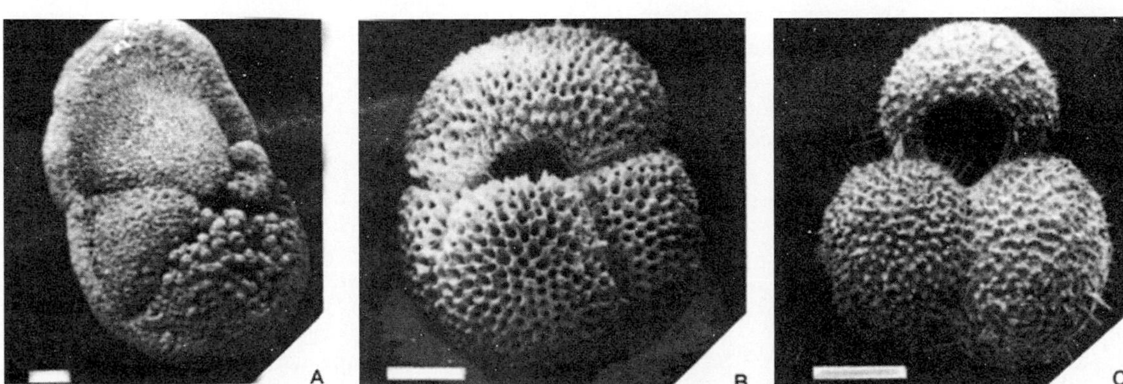

FIGURE 14-7 Present-day planktic foraminifers. (*A*) *Globorotalia tumida*. (*B*) *Globigerinoides conglobatus*. (*C*) *Globigerinoides rubra*. (*Courtesy of C. G. Adelseck, Jr., and W. H. Berger, Scripps Institute of Oceanography.*)

FIGURE 14-9 The living coral *Montastrea cavernosa*, with polyps extended for feeding. (*Copyright Jeff Rotman Photography.*)

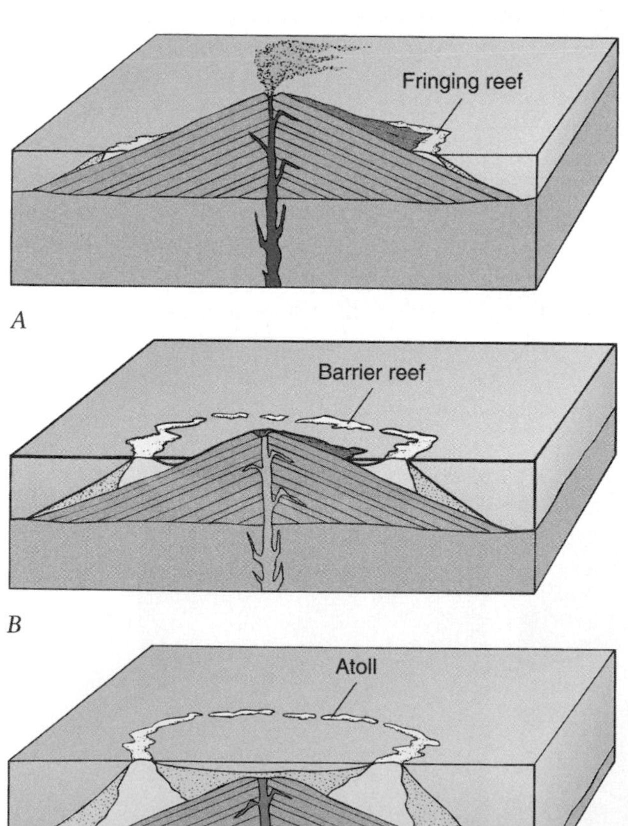

the first million years. Thus, a volcano formed at the crest of the midoceanic ridge is carried down the slight incline of the ocean plate until it is completely submerged. Coral growth that is able to keep pace with the subsidence will result in the formation of barrier reefs and atolls (Fig. 14–11).

There is an alternate method for atoll formation. It is possible that scleractinian corals might build fringing reefs around the perimeters of volcanic islands that had been erosionally truncated during a glacial period of low sea level. If sea level subsequently rose (as it did when the great Pleistocene ice sheets began to melt), then the corals would build the reef upward in order to stay at their optimum shallow living depth. Thus, some atolls may be primarily the result of subsidence of the island, whereas others may have been triggered by a rise in sea level. Still others resulted from a combination of eustatic rise in sea level and subsidence of the island.

FIGURE 14-10 Three stages in the development of an atoll. (*A*) In initial stage, a fringing reef develops around the shoreline of a volcanic island. (*B*) The island begins to subside, and corals build upward to stay in their optimal shallow-water life zone. The result is the development of a barrier reef backed by lagoons. (*C*) As subsidence continues, the original land area has become inundated, and a circle of reefs and coralline islands remains.

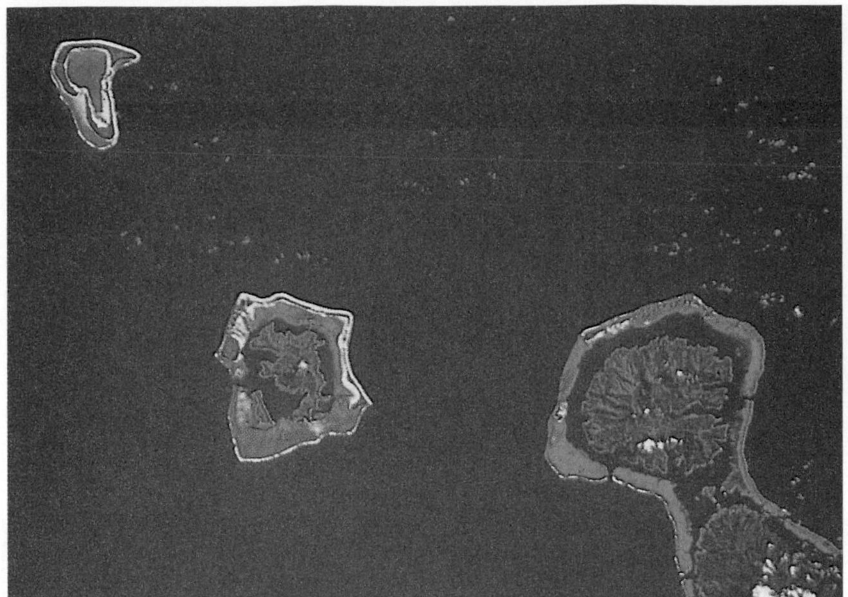

FIGURE 14-11 The Leeward Islands of the French Society Islands provide an illustration of the origin of coral reefs in deep oceans. At the bottom right, corals have constructed a fringing reef around the volcanic islands of Tahaa and Raiatea. Moving northward, the volcanoes are older and surrounded by barrier reefs. Bora Bora, near the center of the photograph, illustrates this phase of reef development. Tupai, visible in the top left of the photograph, is an atoll, and all that remains of the volcano around which the reef originally grew lies below the shallow floor of the central lagoon. (*Courtesy of L. E. Davis.*)

Mollusks

Cenozoic shells of mollusks look very much like those likely to be found along coastlines today. Bivalves and gastropods are the dominant groups of Cenozoic mollusks (Fig. 14–12). Their range of adaptation is truly outstanding. Arcoids, mytiloids, pectinoids, cardioids, veneroids, and oysters were particularly abundant bivalves. Although the climax of the evolution of chambered cephalopods had passed with the demise of the ammonites, nautiloids similar to the modern chambered nautilus lived in Tertiary seas, as they do today. Shell-less cephalopods, such as squid, octopi, and cuttlefish, were also well represented in the marine environment, although their fossil record is understandably sparse.

Echinoderms, Bryozoans, Arthropods, and Brachiopods

Many other invertebrates have continued successfully through the Cenozoic. Echinoderms, mainly free-moving types like echinoids, were particularly prolific. Among the echinoids, sand dollars (clypeasteroids) appear in the Eocene and spread rapidly. Their attractive flat shells (tests) are a favorite of shell collectors. In life, sand dollars lie just beneath a thin blanket of sediment. Cilia located along food grooves bring food to the mouth, located at the center of the underside of the animal.

Bryozoans are common in Tertiary rocks and are still abundant today in many parts of the ocean. A group known as cheilostomes (Jurassic to Holocene) are well represented in many Cenozoic marine formations. Also during the Cenozoic, modern crustaceans became firmly established both in freshwater bodies (Fig. 14–13) and in the oceans. Other noteworthy arthropods of the Cenozoic are the insects. Even though insects live mostly on land and have delicate exoskeletons, over 5000 fossil species have been described from Cenozoic strata. One of the world's best locations for unearthing fossil insects is at Florissant Fossil Beds National Monument, located about 56 miles west of Colorado Springs, Colorado. Over 100,000 insect specimens have been collected at this site, representing 1100 species. The rock in which the fossils occur is a finely textured volcanic ash. Particles of ash settled onto the floor of Oligocene Lake Florissant, burying insects (and plant debris as well). Even delicate features like the veinlets of insect wings are preserved. The Eocene Green River shales also contain well-preserved insect fossils. Finally, splendidly preserved insects are recovered from Cenozoic amber deposits (see Fig. 4–6).

Unlike the insects, brachiopods declined in abundance and diversity during the Cenozoic. Fewer than 60 genera survive today. They consist mostly of terebratulids (Fig. 14–14*A*), rhynchonellids, and such inarticulate brachiopods as *Lingula* (Fig. 14–14*B*).

VERTEBRATES

Fishes and Amphibians

Bony fishes that evolved to the highest level of ossification and skeletal perfection are the teleosts. Teleosts have achieved an enormous range of adaptive radiation during the Cenozoic. That radiation has produced such varied forms as perch, bass, snappers, seahorses, sailfishes, barracudas, swordfishes, flounders,

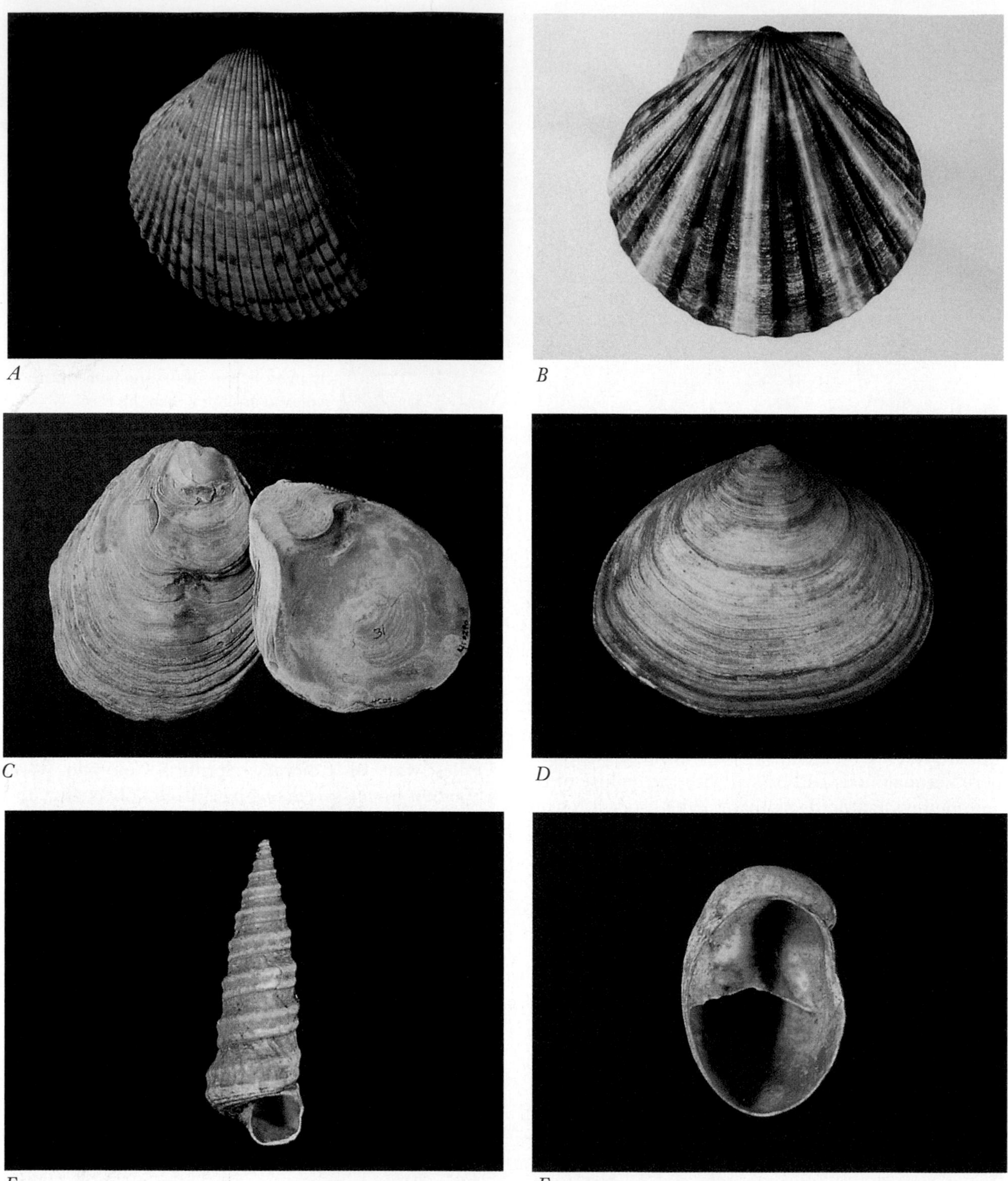

FIGURE 14-12 Common Cenozoic bivalves (*A*, *B*, *C*, *D*) and gastropods (*E*, *F*). (*A*) *Cardium*. (*B*) *Pecten*. (*C*) *Ostrea*. (*D*) *Mya*. (*E*) *Turritella*. (*F*) *Crepidula*, a small marine snail having a horizontal shelf that extends across the posterior part of the shell.

FIGURE 14-13 The freshwater shrimp *Bechleja rostrata* from the Eocene Green River Formation. *(Photograph courtesy of Rodney M. Feldmann, Kent State University. From Feldmann, M. et al. 1981. J. Paleontol.* 55[4]:788–799.)

flyingfish, and others too numerous to mention. The Green River strata of Wyoming are well known for their content of beautifully preserved Eocene teleosts (see Fig. 13–10). Eocene strata at Monte Bolca in Italy also yield well-preserved teleost fossils (Fig. 14–15).

In addition to the bony fishes, the cartilaginous sharks were at least as common in the Tertiary as they are today. Some of these sharks were more than 12 meters long and had teeth as big as a man's hand.

Amphibians throughout the Cenozoic have resembled modern forms. All have been small-bodied, smooth-skinned creatures not at all like their Paleozoic ancestors. Frogs, toads, and salamanders were relatively abundant. The first frogs appeared during the Triassic, and by Jurassic time they were already completely modern in appearance. Thus, they have continued almost unchanged for more than 200 million years.

Reptiles

By the beginning of the Cenozoic, the dinosaurs and most of their marine and aerial contemporaries had disappeared. The reptiles that have managed to survive and continue to the present include the tuatara, which inhabits islands off the coast of New Zealand, as well as

A

B

FIGURE 14-14 Articulate and inarticulate brachiopods. (*A*) Cluster of present-day articulate terebratulid brachiopods, *Terebratulina septentrionalis*. (*B*) *Lingula*, a persistent primitive inarticulate brachiopod with a thin shell of proteinaceous material and a long, fleshy, muscular pedicle.

FIGURE 14-15 The Eocene flat-bodied fish *Gasteronomus* from the Mt. Bolla fossil locality near Verona, Italy. The specimen is 42 centimeters long. *(Photograph by J. Simon, with permission.)*

turtles, lizards, and snakes (Fig. 14–16). The tuatara, formally known as *Sphenadon*, resembles a large lizard. It is the sole survivor of a group of ancient reptiles known as rhynchocephalians that evolved and diversified during the Triassic.

Cenozoic turtles are the descendants of a lineage that can be traced back into the Late Permian. Turtles are readily recognized by everyone because of their distinctive adaptations. The most apparent of these adaptations is the shell. In turtles, the ribs have expanded to form a broad carapace. On the underside, a bony growth called the plastron provides a similar covering. Both carapace and plastron are covered with a horny sheath. The jaws in turtles are toothless and covered by a beak that is used effectively in slicing through plants or flesh.

Both lizards and snakes belong to group of reptiles known as the squamates. Squamates are by far the most varied and numerous of living reptiles. Lizards are the ancestors of snakes. In fact, snakes are essentially modified lizards in which the limbs are lost, the skull bones modified into a highly flexible and mobile structure for engulfing prey, and the vertebrae and ribs greatly multiplied. As evidence of their tetrapod ancestry, some snakes retain vestiges of rear limb and pelvic bones. The evolution of limbless reptiles may have been driven by the habitat in which these animals lived. For example, forward movement of a small tetrapod living in an area of dense vegetation might have been impeded by having legs entangled by plants. Also, for vertebrates seeking refuge or food in small crevaces, there is little need nor space for limbs.

Fossil snakes have been found in rocks as old as Early Cretaceous. They began to diversify in the Early Miocene. It was at that time that poisonous snakes evolved. Specialized teeth or fangs are used by poisonous snakes to inject their prey. One type of poison (neurotoxin) affects parts of the nervous system that

A

B

C

D

FIGURE 14-16 Surviving reptiles. (*A*) Monitor lizard. (*B*) Three-toed box turtle. (*C*) *Alligator mississippiensis*. (*D*) Hognose snakes emerging from their shells. (B, *Copyright Z. Leszczynski/Animals Animals*; C, *E. Reschke*; D, *L. Stone/The Image Bank*.)

control breathing and heart action, whereas a second type (hemotoxin) destroys red blood cells and causes disintegration of small blood vessels. It has been suggested that the evolution of poisonous snakes and the feeding behavior and skull structure of all snakes may be tied to the diversification of the Cenozoic mammals that served as prey.

Crocodiles

Crocodilians are archosaurians that also began their evolution during the Triassic and were very successful contemporaries of the dinosaurs throughout the Mesozoic. Modern crocodilians include the broad-snouted alligators, the narrow-snouted crocodiles, and the very narrow-snouted gavials. The nostrils of crocodilians are at the ends of their long snouts. Thus, they are able to breath by simply keeping the end of the snout above water.

Birds

The Cenozoic record for birds is generally poor because they are rarely preserved. The rather fragmentary record indicates, however, that birds have been essentially modern in basic skeletal structure since the beginning of the Cenozoic. Distinctive skeletal features of birds include the fusion of bones of the "hand" to help support the wing, development of a vertical plate or keel on the sternum for attachment of the large muscles leading from the breast to the wings, and fusion of the pelvic girdle and vertebrae to provide rigidity during flight. Other characteristics include a body covering of feathers, light and porous bones, jaws in the form of a toothless horny beak, a four-chambered heart, and constant body temperature. The adaptive radiation of birds has produced a rich variety of groups, including songbirds such as robins, upland birds (pheasants), forest birds (owls), oceanic birds (albatrosses), wading birds (plovers), flightless aquatic birds (penguins), and flightless land birds (ostriches).

The fossil record for large flightless terrestrial birds is somewhat better than for small flying varieties. *Diatryma* (Fig. 14–17), from the Eocene of North America, was 2 meters tall, weighed nearly 300 pounds, and had massive legs, vicious claws, and a formidable beak. Although many paleontologists believe *Diatryma* was a predator, a more controversial view is that the huge bird was a scavenger or a browsing herbivore.

South America produced another group of predatory flightless birds, including *Andalgalornis* (Fig. 14–18), whose head was comparable in size to the head of a horse. Some flightless birds were vegetarians. Among these, huge moas lived until relatively recent time in New Zealand. Some moas were over 3 meters tall and laid eggs that had a 2-gallon capacity. Prior to settlement of New Zealand by Europeans, the moas were exterminated by Maoris hunters.

Perhaps the most famous of Cenozoic flightless birds was the dodo, which lived on the island of Mauritius east of Madagascar until about 1700 A.D. At about that time, the dodo was either exterminated by sailors

FIGURE 14-17 ***Diatryma*, one of the large, ground-dwelling, predatory birds that evolved during the early Cenozoic.** *Diatryma* stood 2 meters tall and had a huge, powerful beak. The artist has depicted *Diatryma* as a predator. (*Copyright J. Sibbick.*)

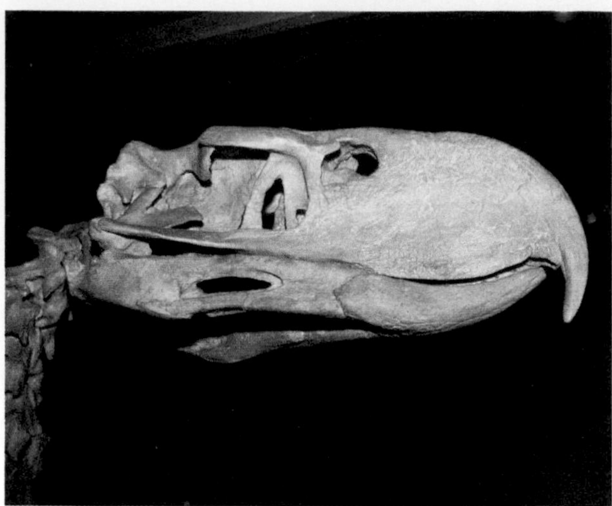

FIGURE 14-18 Skull of *Andalgalornis*. The skull of the giant flightless predatory bird *Andalgalornis ferox* from the Pliocene of Argentina. (*Photograph of specimen on display in The Field Museum of Natural History, Chicago.*)

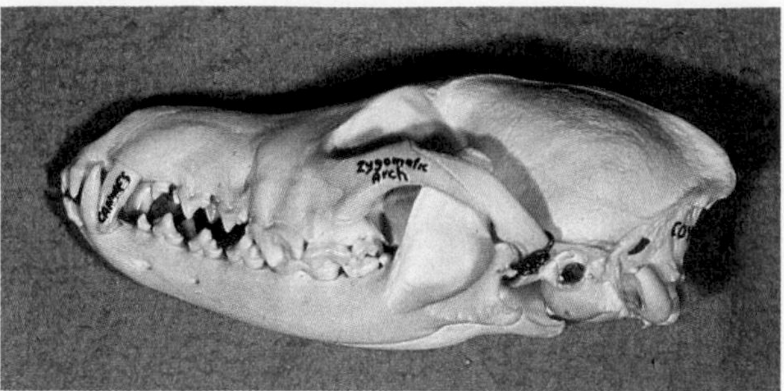

FIGURE 14-19 Skull of a coyote. One bone on either side, the dentary, forms the lower jaw. As is characteristic of most mammals, teeth are not identical in shape. Instead, there are incisors at the front of the jaw, followed by daggerlike canines for stabbing, carnassial teeth that shear past one another like scissors for cutting meat into smaller pieces, and more robust back teeth for crushing tougher objects and bone. Note also the ample size of the braincase.

searching for provisions or was unable to survive the influence of animals brought to Mauritius by Europeans. The African ostrich, the South American rhea, and the emus and cassowaries of Australia are surviving flightless land birds.

Mammals

During the Cenozoic, mammals came to dominate the Earth in much the same way as dinosaurs had done during the preceding Mesozoic era. As we noted in Chapter 12, however, the evolution of mammalian traits had already begun among the therapsids of the Permo-Triassic. The Karoo beds of Africa, for example, contain bones of near-mammals that had almost made the transition from reptile to mammal.

Just as birds are readily recognized by the possession of feathers, so are mammals by the possession of hair. Hair, however, is not an exclusive characteristic of mammals, for it was apparently also present in some therapsids and early flying reptiles such as *Sordes pilosus*, the so-called "hairy devil" of the Jurassic. Like feathers, hair functions as an insulating body cover. Mammals are also recognized by the presence of mammary glands. However, neither body hair nor mammary glands are particularly helpful to the paleontologist, who must recognize mammalian remains on the basis of skeletal characteristics. Among these, the lower jaw is particularly useful. It consists of a single bone, the dentary, on either side (Fig. 14–19). In reptiles and birds, there are always several bones in the lower jaw. Another mammalian trait is the bony mechanism that conveys sound across the middle ear. In mammals, there is a chain of three little bones (ossicles) rather than a single one, as in other tetrapods. Two of these ossicles, the incus and malleus, were derived from bones of the reptilian jaw. Unfortunately, these auditory ossicles are so small that they are rarely preserved.

Typically, mammals have seven cervical, or neck, vertebrae, regardless of the length of the neck. The skull is usually recognizable by the expanded braincase, and teeth are nearly always of different kinds, serving different functions in eating. As an aid to their endothermic metabolism, mammals have a well-developed secondary palate that separates the oral cavity from the nasal passages. This makes simultaneous breathing and feeding possible. Without the secondary palate, an infant mammal could not suckle.

Paleontologists speculate that more rigorous climatic conditions during the Permo-Triassic favored selection of the mammalian traits of warm-bloodedness and postnatal care. All evidence indicates that the first mammals were diminutive creatures, and small animals lose heat rapidly. A high rate of heat loss could have been partly offset by insulating fur and an ample supply of food. If the earliest mammals laid eggs, as do primitive mammalian monotremes today, then the newly hatched young might also have had to cope with loss of body heat. They were very likely kept warm by snuggling next to soft fur that may have formed an incubation patch on the female. Perhaps at the same time, they were nourished by secretions from glands that preceded the development of true mammary glands. Unfortunately, fossils provide few clues to the reproductive characteristics of early mammals, and theories will continue to rely heavily on inference.

EARLIEST MAMMALS The mammalian fossil record begins with rare and difficult-to-find small jaw fragments, tiny teeth, and scraps of skull bone. The fossils, however, are sufficient to indicate that the earliest members of our own taxonomic class were very small; that they evolved from mammal-like reptiles such as those previously described from the late Paleozoic; and that, as indicated by tooth patterns, they were insect-eaters. Evidence from brain casts made of their inner cranial walls also suggests that the parts of their brains dealing with smell and hearing were particularly well developed, as in nocturnal animals. Perhaps the most significant and ironic fact about these tiny creatures is that they came on the scene at about the same time as the dinosaurs that they were destined to succeed. For 140 million years of dinosaur dominance, they quietly thrived as if waiting for just the right conditions for their own biologic triumph.

Cenozoic mammals benefited from their efficient nervous and reproductive systems, reliable systems of body temperature control, larger brains, and relatively high levels of animal intelligence. Armed with these attributes, mammals quickly expanded into the habitats vacated by the dinosaurs and found new pathways of adaptation as well (Table 14–1). By Paleocene time, 18 major groups of mammals had appeared, and throughout the early Tertiary, there was an intricate interplay of appearances of new groups and extinctions of older groups that ultimately produced more than twice that number. A brief review of some of the more interesting forms suggests the spectacular variety of the Cenozoic mammalian fauna.

TABLE 14-1 Evolutionary Modifications of Mammals for a Variety of Habitats*

Habitat	Limbs	Teeth	Other Features
Primitive walking (ambulatory)	Limbs generalized; feet flat on ground	Incisors for grasping; canines for piercing; premolars for cutting; molars for grinding	Tongue, stomach, and intestine generalized; cecum small
Running herbivores (cursorial) Horse, cow	Limbs elongate; toes elongate and reduced in number; hoof of horse is one toenail; in cow, two toenails	Incisors for nipping grass; cheek teeth heavy and ridged to withstand wear of grass	Stomach (cow) or intestine (horse) large and complex; cecum long
Running carivores (cursorial) Wolf, cheetah	Limbs elongate; often walk on toes; claws often long and sharp	Canines well developed; premolars sharp for cutting flesh	Stomach large; intestine short
Heavy body Elephant	Bones in limbs massive; flat jointed; toes in circle around pad	Cheek teeth massive and ridged for grinding plant materials	Nose elongated into trunk to reach food on ground or in trees
Tree-Climbing (arborial) Tree squirrel, monkey	Limbs elongate; wide range of motion; toes long and flexible	Cheek teeth usually smooth surfaced and flattened for crushing	Tail sometimes prehensile; diet usually fruit or vegetation
Diet and small, colonial insects Anteaters, spiny echidna	Claws enlarged for digging after insects	Reduced or absent	Tongue long, sticky, and extensile
Burrowing (fossorial) Moles, marsupial moles Various rodents	Limbs short and stout; forefeet often enlarged, shovel-like; claws long and sharp	Teeth various, depending on diet (vegetation or insects and worms)	Pinna of ears usually short or absent; fur short; tail short
Aquatic Whales, seals, sea cows	Limps shortened; feet enlarged and paddle shaped or absent	Often reduced to simple, coneline structures, sometimes absent	Tail often modified into swimming organ
Volant Flying phalangers, flying squirrels	Limps extendible to sides to hold gliding membrane outstretched	Various but usually show herbivorous adaptations	Tail often flattened for use in balancing
Flying Bats	Forelimbs and especially fingers of forelimbs elongate to support wing	Various for diets of insects, fruits and nectar, or other specific foods	Tongue, stomach, and intestine various, correlated with diet

*From Cockrum, E. L., and McCauley, W. J. 1965. *Zoology*. Philadelphia: W. B. Saunders Co.

A

B

FIGURE 14-20 Two present-day monotremes. (*A*) The duckbill platypus of Australia and Tasmania. (*B*) The spiny anteater of Australia. (*Copyright T. McHugh/Photo Researchers, Inc.*)
? *If monotremes lay eggs, why are they considered mammals?*

MONOTREMES The most primitive of all living mammals are called monotremes. These relics of an older time still lay eggs in the reptilian manner. However, unlike reptiles, monotremes have primitive mammary glands and provide their newly hatched offspring with nourishment, if only for a brief period. The oldest fossil remains of monotremes are from beds of the lower Cretaceous. The platypus of Australia and Tasmania and two species of spiny anteaters of New Guinea and Australia (Fig. 14–20) are the only monotremes still in existence.

MARSUPIALS Marsupials are mammals, most of which nurture their young in a special pouch, or **marsupium** (Fig. 14–21). The most familiar to us today are the kangaroos, wallabies, wombats, phalangers, bandicoots, and koalas of Australia (Fig. 14–22) and the opossums (Fig. 14–23) of North and South America. Compared to the placentals, they are a dwindling group.

Marsupials had their greatest success in Australia and South America. Both continents were more or less isolated during most of the Tertiary, and marsupials were able to evolve without excessive competition from placentals. Although a few middle Tertiary Australian mammal sites have been found, the fossil record of Australian marsupials is not good until Pleistocene time. An older and more complete record exists in South America, where marsupials evolved from opossumlike ancestors and produced an array of carnivorous and herbivorous

A

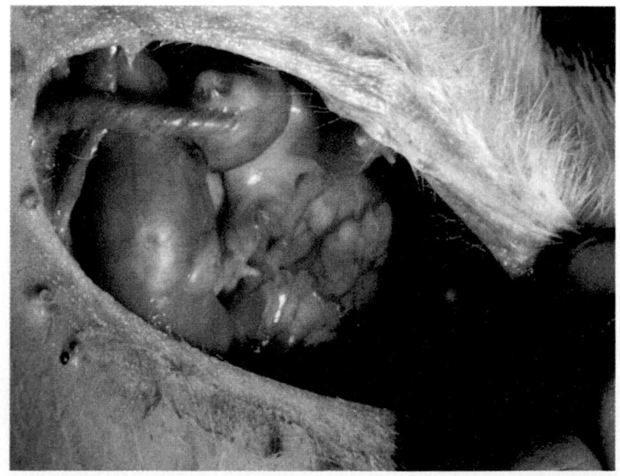

B

FIGURE 14-21 The short-nosed bandicoot, a common marsupial of Australia. (*A*) Advanced offspring that have left the pouch return for nursing. (*B*) A fetal bandicoot attached to a teat in the mother's pouch.

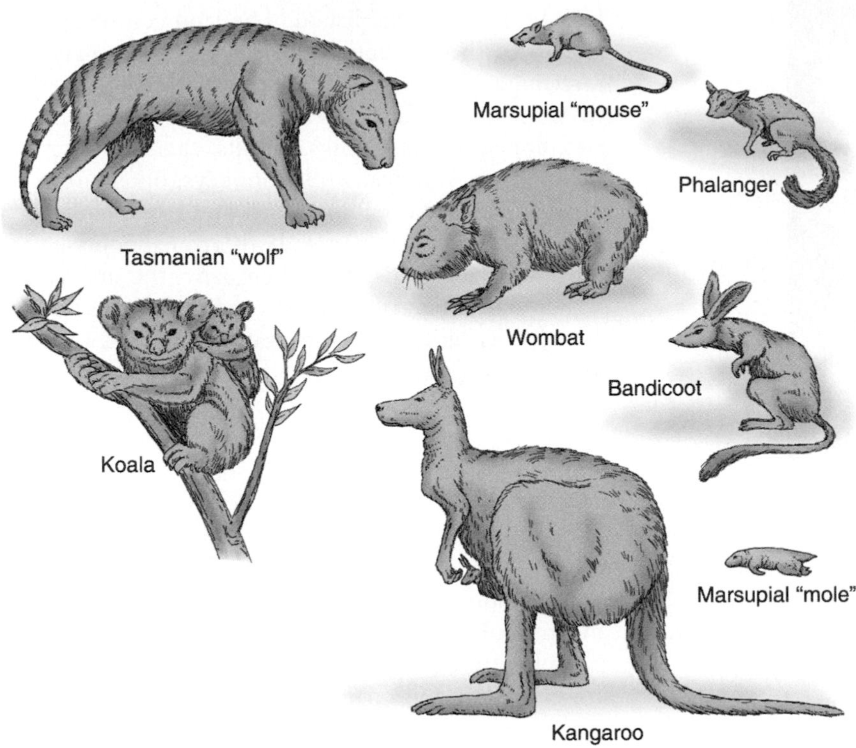

FIGURE 14-22 Adaptations of Australian marsupials for various habitats is evident in their diversity. (The Tasmanian wolf is now thought to be extinct.)

types. Many of these animals had an uncanny resemblance to placental plant-eaters and carnivores in North America. One example is the similarity between the South American marsupial sabertooth *Thylacosmilus* (Fig. 14–24) and the North American placental sabertooth *Smilodon*. There were also marsupial moles and dog-like, bear-like, and hippo-like marsupials. These animals serve as excellent examples of **convergent evolution**, in which there is similar development of unrelated animals for a particular mode of life.

FIGURE 14-23 The American opossum *Didelphis*. This interesting marsupial still retains many of the dental and skeletal characteristics present in its Cretaceous ancestors. (*The Field Museum of Natural History, Chicago*.)

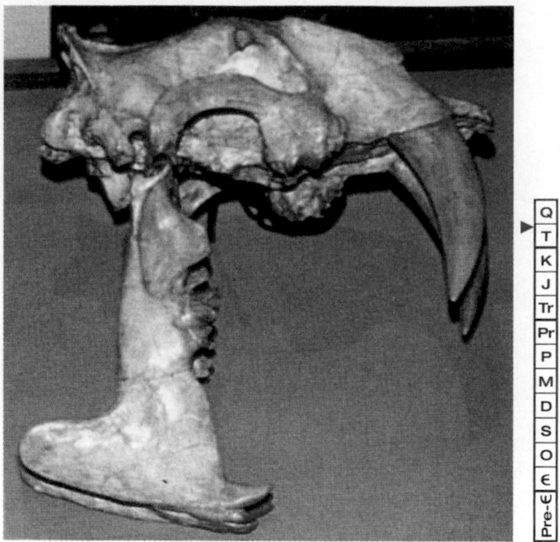

FIGURE 14-24 ***Thylacosmilus*, a Pliocene South American carnivorous marsupial comparable to the placental saber-toothed cats.** The bladelike upper canine teeth were about 18 cm long and were protected when the jaws were closed by a deep flange of bone in the lower jaw.

About 3 million years ago, the Central American land bridge was established. The land connection between North and South America permitted a major interchange of marsupials from the south and placentals from the north. Camels, elephants, deer, bears, peccaries, horses, tapirs, skunks, rabbits, cats, and dogs entered South America, while monkeys, opossums, rats, shrews, sloths, porcupines, and anteaters made their way northward. For a time, both groups flourished in their new locations, but eventually the marsupials began to decline. All of the hoofed marsupials became extinct, as did the placental ground sloths and glyptodonts (Fig. 14–25).

PLACENTAL MAMMALS

Insectivores Placental mammals appear during the Cretaceous as small, unspecialized insectivores. Modern moles are members of the order Insectivora, but the tiny shrew (Fig. 14–26) is more representative of the kind of animal from which other orders of Cenozoic placentals evolved. The descendants of the insectivores include the edentates, bats, primates, rodents, flesh-eating mammals, a host of herbivores, and various marine mammals.

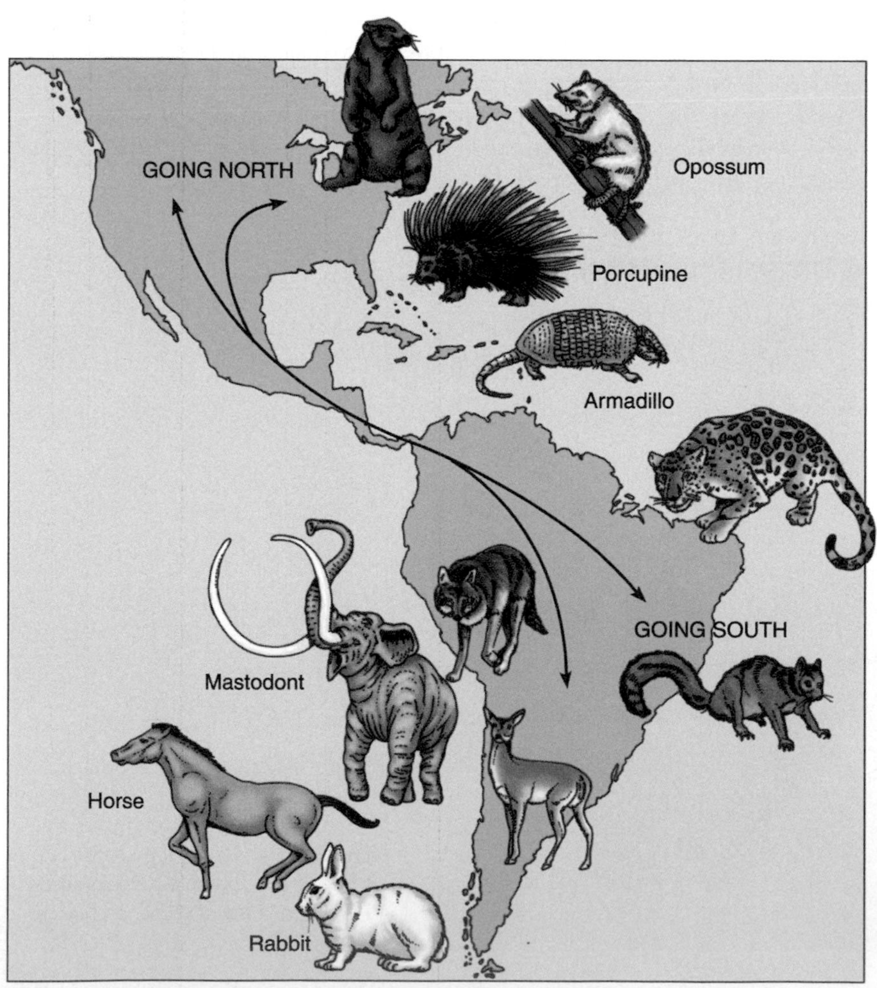

FIGURE 14-25 When North and South America were connected by a land bridge during the Late Pliocene, sloths, anteaters, caviomorph rodents, armadillos, porcupines, opossums, ground sloths, and glyptodonts migrated north. At the same time jaguars, squirrels, rabbits, saber-tooths, elephants, deer, wolves, and horses headed south. The event has been named "the great American interchange."

FIGURE 14-26 The common tree shrew most resembles the ancient insectivores that gave rise to the primates. (*W. Garst/Tom Stack & Associates.*)

Edentates Armadillos, tree sloths, and South American anteaters are edentates that have not become extinct. Fossil species of armadillos have been found in rocks as old as Paleocene. Among the extinct edentates are the **glyptodonts** (Fig. 14–27), which survived in South America until about 7000 years ago, and ground sloths. *Glyptodon* was about the size of a Volkswagen "Beetle." The stocky mammal was a walking fortress with a spike-covered knob on its tail for bludgeoning predators.

Quite unlike the glyptodonts were the ungainly Cenozoic ground sloths. This group of edentates includes truly colossal Pleistocene beasts such as *Megatherium* (Fig. 14–28). From head to tail *Megatherium* was 6 meters long. They were tall enough to browse on tree branches while resting on their hind legs. Dung and mummified body tissues of ground sloths have been found in association with human artifacts in caves in arid regions of North America.

An unusual adaptation among sloths is seen in *Thalassanus natans*. This creature lived in the sea and fed on aquatic plants in much the same manner as living manatees. Fossils of *Thalassanus* from Peru are found side by side with the skeletal remains of fish, whales, and sea lions.

Rodents and Rabbits The rodents have been exceptionally successful Cenozoic animals. Today, they probably outnumber all other mammals and have invaded nearly every habitat of the Earth. Their success has been possible because of a remarkable range of adaptations that permit survival in diverse habitats from the arctic to the tropics. A rapid rate of breeding and small size also engendered success. Their diversification has produced burrowers such as the marmot, partially aquatic animals such as the muskrat and beaver, desert-dwelling jerboas and kangaroo rats, and arboreal squirrels.

FIGURE 14-27 Restoration of a scene during the Middle Pleistocene in Argentina. The heavily armored animals are glyptodonts. On the left is a giant ground sloth. (*The Field Museum of Natural History, Chicago. Painting by C. R. Knight.*)

FIGURE 14-28 The great ground sloth *Megatherium*, which lived during the Pleistocene and was nearly as large as a present-day elephant. This specimen is on display at The Field Museum of Natural History in Chicago.

Rodents are the dominant placentals in the gnawing-nibbling category. They have distinctive adaptations of teeth and jaws to reflect their specialization. They lack canines and have only two prominent and opposing pairs of continuously growing incisors, one in the lower jaw and one in the upper. The outer surface of each incisor is made of enamel and is thus harder than the inner surface. When the rodent is chewing, the incisors are sharpened as the harder enamel layer persists to form a chisel edge. A gap, or diastema, lies between the chisel incisors and the cheek teeth. The latter are long, ridged, and efficient tools for grinding coarse food. Rodent jaw muscles are large, and the jaws have a flat articulation so that teeth can grind fore and aft rather than up and down as in humans.

The taxonomic divisions of the Rodentia are based primarily on the disposition of the muscles associated with movement of the jaws. A group called **protorogomorphs**, which originated in the Paleocene, were ancestors of the later true squirrels and old world porcupines. They are represented in the fossil record by *Paramys* (Fig. 14–29), which resembled a large squirrel, and by the unusual *Ceratogaulus*, the only rodent to have developed horns. Surviving protorogomorphs include the small and thick-furred sewellels.

Squirrels are the characteristic **sciuromorph** rodents, but this group also includes chipmunks and marmots. Beavers and their relatives are classified as **castorimorphs**. *Paleocastor* of the Pleistocene was a member of the beaver family that left casts of peculiar corkscrew burrows in Nebraskan fossil beds. The largest castorimorph was *Casteroides*, of the Pleistocene, which was about the size of a bear.

FIGURE 14-29 The Paleocene rodent *Paramys*, which resembled a large squirrel in size and form.

The old-world porcupines belong to the **hystricomorph** group, whereas new-world porcupines, as well as capybaras, agoutis, chinchillas, and spiny rats, are called **caviomorphs**. However, a group better known than these are the **myomorphs**, which evolved from either sciuromorphs or hystricomorphs during the Oligocene. Those amiable pets, the hamsters (*Mesocricetus*), fall into this category, along with rats, mice, dormice, pocket gophers, jerboas, and kangaroo rats. Among this inventory of rodents, rats and mice have formed a one-sided partnership with humans. They invade our buildings, eat our food, convey some of our diseases, and voyage with us to the far corners of

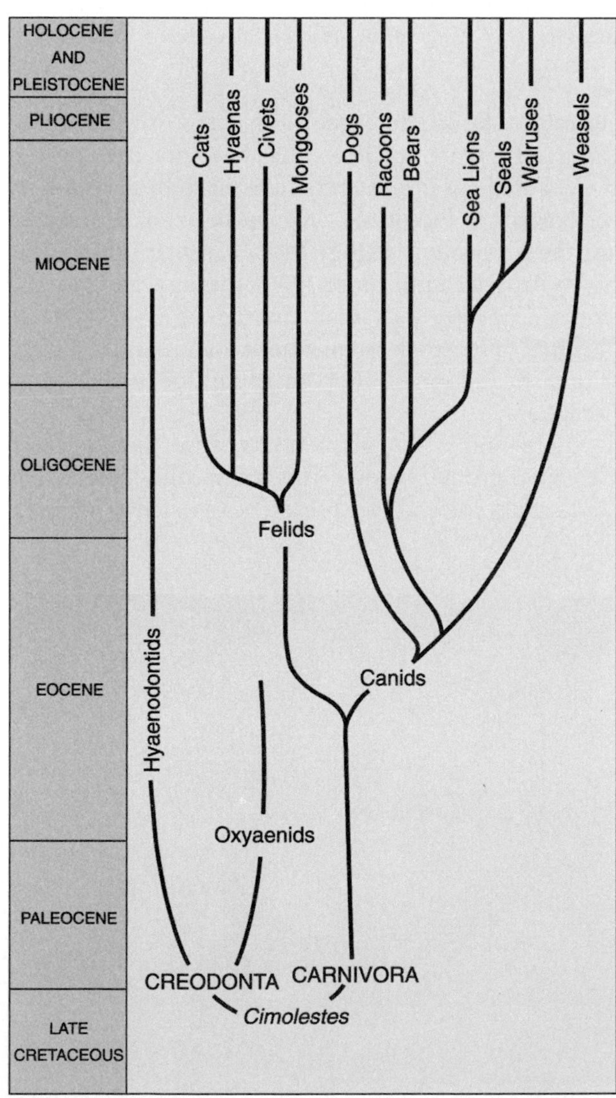

FIGURE 14-30 General relationships of flesh-eating mammals. (*Based on Wyss, A. R., and Flynn, J. J. 1983. A phylogenetic analysis and definition of the Carnivora, in Szalay, E. S., Novacek, M. J., and McKenna, M. C. [eds.].* Mammal Phylogeny. *Vol. II. New York: Springer-Verlag.*)

the world. Rodent persistence, tolerance, and tenacity are legendary.

Rabbits, hares, and pikas are called **lagomorphs**. Many people are surprised to learn that rabbits are not rodents, even though they have enlarged rodentlike incisors. Rodents, however, have only a single pair of chisel-like incisors above in opposition to a single pair below. Lagomorphs have two upper pairs and one lower pair. In addition, the cheek teeth are simpler and more numerous. They are elongate prisms with transverse ridges for cutting rather than the rodent manner of crushing. In the postcranial skeleton, the tail is reduced to a mere vestige, and hindlegs are strengthened and elongated for hopping. The lagomorphs are divided into two separate groups. On the one hand are the compact, short-legged pikas, or coneys. These animals lack the long ears of rabbits. Sometimes called mouse rabbits, the pikas are only about 8 inches long. They are mountain dwellers and have been recorded on Mount Everest at elevations of 17,000 feet. In contrast to the pikas, hares and rabbits have developed into swift runners with long ears for acute reception of sound. Along with that of the rodents, the fossil record for rabbits begins in Paleocene strata of Asia.

Bats Bats were also evolving during the Cenozoic. Their teeth have been discovered in Paleocene strata of France. A well-preserved skeleton was recovered from Lower Eocene strata in Wyoming and indicates that by that time, bats were already very similar to their living relatives. Bats are the only mammals to have achieved true flight. They have greatly elongated finger bones for the support of the membrane that forms the wing. Insect-eating bats fly mostly at night, when insect prey are abundant in the air. Although many bats live in temperate zones, they are most abundant in the tropics. Vampire bats, which live in South America, feed on the blood of other mammals. Their front teeth are adapted to pierce the skin of their prey. When blood oozes from the wound, it is lapped up by the bat. In addition to the insectivorous and vampire bats, there are bats that live on the nectar of flowers and those that eat fruit, like to so-called "flying foxes" of the Old World tropics.

Flesh Eaters The earliest known flesh-eating placental mammals are Late Cretaceous in age. They are represented by a small weasel-like animal given the name *Cimolestes*. Although small in size, *Cimolestes* was clearly a predator. Certain of the cheek teeth, the **carnassials**, were developed into blades able to shear past one another and slice meat into small pieces. Enlarged canines were the principal killing weapon.

By the end of the Cretaceous, flesh-eating mammals had split into two independent orders (Fig. 14–30). The first, named **Creodonta**, was composed of relatively small-brained animals with short limbs, clawed toes, and long nails. *Patriofelis* (Fig. 14–31) is an exam-

FIGURE 14-31 **The Eocene creodont, *Patriofelis*.** (*Courtesy of the U.S. National Museum of Natural History, Smithsonian Institution.*)

ple of an Eocene creodont of a group called **oxyaenids**. *Hyaenodon* is a creodont whose name was suggested by the animal's superficial resemblance to a hyena.

The second order of flesh eaters, the **Carnivora**, had evolved larger brains than the Creodonta. Also, their carnassial teeth were positioned farther to the front of the jaws. By Miocene time, these more progressive predators had replaced the Creodonta. Their expansion accelerated during the remainder of the Cenozoic, producing a host of now familiar cats, civets, mongooses, and true hyenas. Famous among the Pleistocene Carnivora was the large stabbing cat *Smilodon* (Fig. 14–32). This robust flesh-eater appears to have preyed on large herbivores, whereas the swift and agile biting cats probably sought their food among herds of grazing animals. Dogs, raccoons, bears, and weasels were contemporaries of the cats. Indeed, wild dogs were doing very well on Earth long before humans came along to make companions of them. The modern dog *Canis* in the form of the dire wolf roamed widely during the Ice Age.

As is true of carnivores today, Cenozoic carnivores were an essential element in the evolutionary process. To survive, they had to equal or better the speed and cunning of the herbivores on which they fed. The herbivores, in turn, responded to the carnivore threat by evolving adaptations for greater speed and defense. Then, as now, predators were not villains but necessary constituents of the total biologic scheme. They were able to cull out the weak, deformed, or sickly animals and thereby help to counteract the effects of degenerative mutation and overpopulation.

Not all the flesh-eaters of the Cenozoic are land dwellers. Seals, sea lions, and walruses obtain their food in, or at the edge of, the sea. As is evident from their sharp, pointed teeth, early Cenozoic seals and sea lions consume fish. Walruses, however, feed upon mollusks, the shells of which they crush with their powerful jaws and broad, flat cheek teeth.

Ungulates Most animals with hoofs (or descendents of animals with hoofs) are rather loosely called **ungulates**. A partial list of ungulates would include horses, tapirs, rhinoceroses, cattle, sheep, goats, hippopotamuses, pigs, giraffes, antelopes, camels, and deer. In addition to these familiar animals, ungulates include some surprisingly different groups (Fig. 14–33). Whales, for example, can be loosely considered ungulates in that they are descended from the hoof-bearing mammals (Fig. 14–34). **Sirenians**, popularly known as manatees and dugongs, are also ungulates, as are the hyraxes (coneys or dassies) that bear a resemblance to woodchucks but have toes that end in tiny hoofs. **Proboscideans**, such as mastodonts and elephants, are ungulates. The group also includes clumsy-looking, semiaquatic extinct beasts called **desmostylids** and huge mammals called **arsinotheres** that sported an enormous pair of side-by-side horns over their skull and snout (Fig. 14–35).

Primitive ungulates were already widespread by late Paleocene time. Their general characteristics are well represented by the sheep-sized herbivore named *Phenacodus* (Fig. 14–36). *Phenacodus* had cheek teeth with broad grinding surfaces, rather like the teeth of primitive members of the horse family. The outermost of its five toes were shorter than the middle three, but *Phenacodus* did not have the modifications needed for running on its toes for greater speed. It was a flat-footed or plantigrade ungulate. The plantigrade stance was also characteristic of the enormous six-horned *Uintatherium*

A

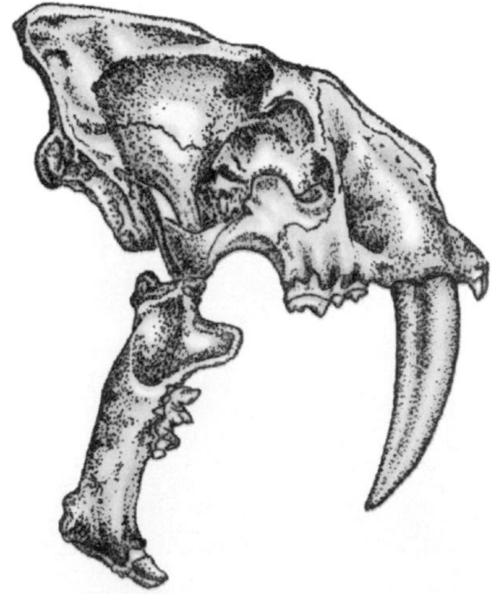

B

FIGURE 14-32 Restoration (*A*) and skull (*B*) of *Smilodon*.

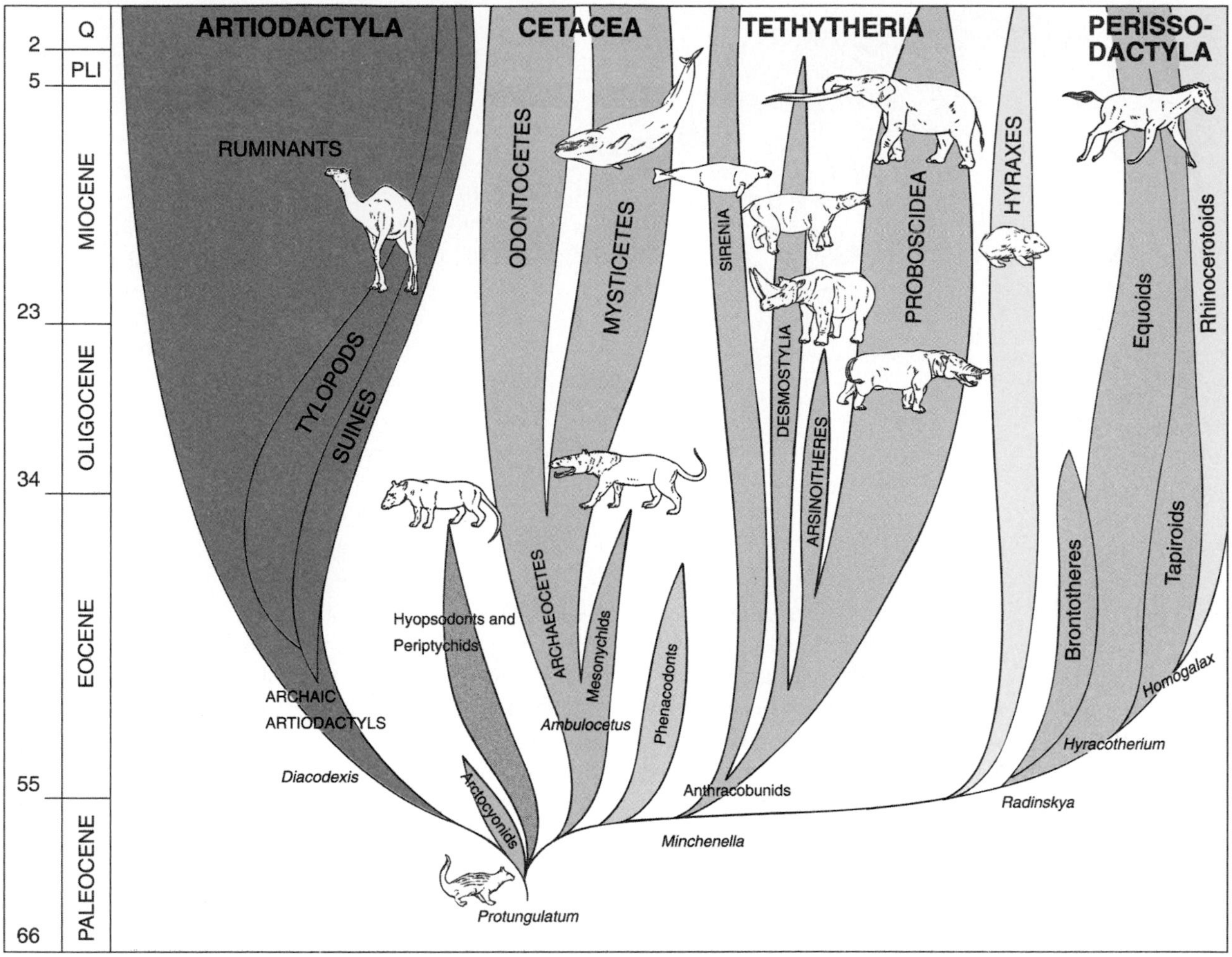

FIGURE 14-33 Phylogenetic tree of the major groups of hoofed animals. (*From Prothero, D. R. 1994. Mammalian evolution, in Prothero, D. R., and Schoch, R. M. (eds.). Major Features of Vertebrate Evolution. Short Courses in Paleontology No. 7, University of Tennessee Press and the Paleontological Society, with permission.*)

FIGURE 14-34 The Eocene whale *Basilosaurus*. The tendency toward increase in size was already evident in this whale, which was over 20 meters long. (*Illustration by Caryln Iverson.*)

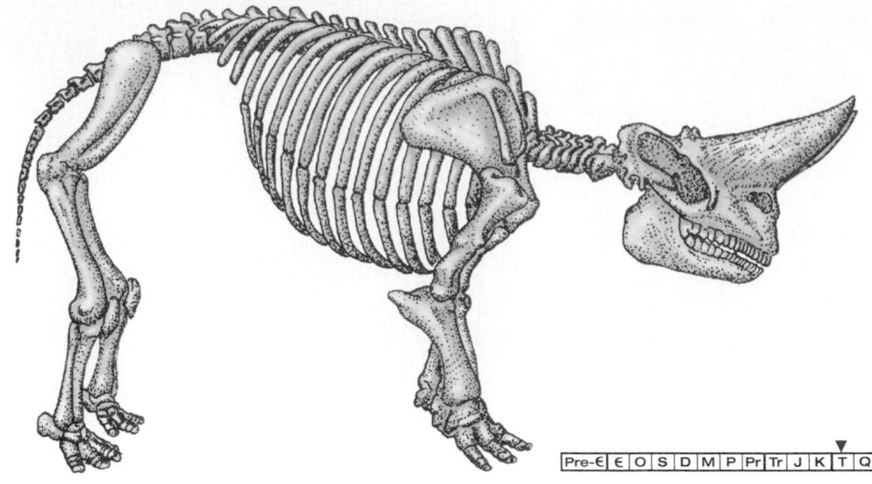

FIGURE 14-35 ***Arsinoitherium*, a huge browsing placental mammal of the Oligocene, whose most striking feature was a pair of massive bony horns.** (*After Andrews, C. W. 1906.* A descriptive catalogue of Tertiary Vertebrata of the Fayum, *Egypt. London.*)

(Fig. 14–37*A*). This Eocene herbivore is readily recognized by the bony protruberances on its head and the 15-centimeters-long canines possessed by males.

The two largest categories of today's hoofed animals are the **perissodactyls** and the **artiodactyls**. Perissodactyls include modern horses, rhinoceroses, and tapirs, as well as extinct forms known as chalicotheres, brontotheres, and such less familiar groups as lophiodonts, amynodonts, and hyracodonts (Fig. 14–38). Because many perissodactyls have either one or three toes on each foot, they are informally called odd-toed ungulates. Tapirs, however, and some extinct perissodactyls had four toes on their forefeet. Regardless of this exception, the axis of the foot, along which the weight of the body was primarily supported, lay through the third or middle toe. Reduction of lateral toes is characteristic of perissodactyls, and in the modern horse, only the single, central toe remains (Fig. 14–39). Perissodactyls, and most artiodactyls as well, are digitigrade, meaning they run or walk on their toes in order to attain a longer stride and greater speed.

The family tree of horses (see Fig. 4–19) presently appears to have its roots in a small Paleocene herbivore from China named *Radinskya*. It is uncertain if this primitive creature is a true perissodactyl or whether it is ancestral to Eocene browsing horses such as *Protorohippus* and *Orohippus* (see Fig. 14–37*F*). The latter were small horses about 40 centimeters tall with four toes on the forefeet and three on the hind feet. The molar teeth were bluntly cusped for browsing and had not yet developed the high crowns and ridged patterns that characterize modern grazers.

The rich fossil record of the horse family provides ample evidence of evolutionary change. One can readily follow the change from horses of small body size, short skulls, small cranial capacity, and low-crowned

FIGURE 14-36 The Early Tertiary archaic herbivorous mammal *Phenacodus*. Like most of the Early Tertiary herbivores, *Phenacodus* walked on all five toes in a method termed plantigrade. Speed was not essential to this forest-dwelling browser. (*Courtesy of the U.S. National Museum of Natural History, Smithsonian Institution.*)

FIGURE 14-37 **Eocene mammals.** (*A*) *Uintatherium* is the large six-horned and tusked animal in the upper right. Other animals are (*B*) the small, fleet rhinoceros *Hyrachus*; (*C*) *Trogosus*, a gnawing-toothed mammal; (*D*) *Mesonyx*, a hyenalike flesh eater; (*E*) *Stylinodon*, a gnawing-toothed mammal; (*F*) three early members of the horse lineage (*Orohippus*); (*G*) a saber-toothed mammal, *Machaeroides*; (*H*) *Patriofelis*, an early carnivore; and (*I*) *Palaeosyops*, an early titanothere. Restorations are based on skeletal remains from the Middle Eocene Bridger Formation of Wyoming. (*Courtesy of the U.S. National Museum of Natural History, Smithsonian Institution, J. H. Matternes mural.*)

teeth to larger animals with fewer toes, longer skulls, larger brains, and complexly ridged high-crowned teeth for rendering grasses more digestible. As new fossil species were discovered, it became clear that the phylogenetic tree of horses was actually more bushlike than treelike. There were, for example, over a dozen distinct branches of diverse kinds of horses in the Miocene alone.

Mesohippus (Fig. 14–40) exhibits many of the general characteristics of the horses that lived in North America during the Oligocene. About the size of a greyhound, this horse had lost the fourth toe on its forefeet. Its teeth show an early pattern of ridges on their occlusional surfaces but were still low-crowned. *Mesohippus* preferred a less abrasive diet of succulent leaves rather than grasses. In Mesohippus and in subsequent horses, the front portion of the skull and lower jaws became deeper in order to make space for the long roots of the cheek teeth. In addition, the battery of molars was shifted toward the front of the muzzle out of the way of the eye orbits and jaw articulation.

In most Miocene horses, three toes were retained, but the larger central one bore most of the animal's weight. *Merychippus* (Fig. 14–41) was a horse about a meter tall at the shoulder and was well adapted for life on the spreading grasslands. *Merychippus* lived on into the Pliocene alongside another group of three-toed horses termed hipparions. Having evolved in North America, hipparions migrated into Eurasia and Africa. *Pliohippus* is one of several genera of horses present during the Late Miocene and Pliocene. The lateral toes in *Pliohippus* were reduced to useless vestiges. By about 3.5 million years ago, even these vestigial toes were absent in the branch of the horse family that gave rise to the modern horse *Equus*. Several different species of Equus thrived during the Pleistocene. Some, such as *Equus occidentalis*, were the size of small ponies, but others, such as *Equus giganteus*, were larger than the great draft horses of today. They spread across the length of North America, Eurasia, and Africa. Only a few thousand years ago, horses suffered extinction in North America. (The continent was later restocked

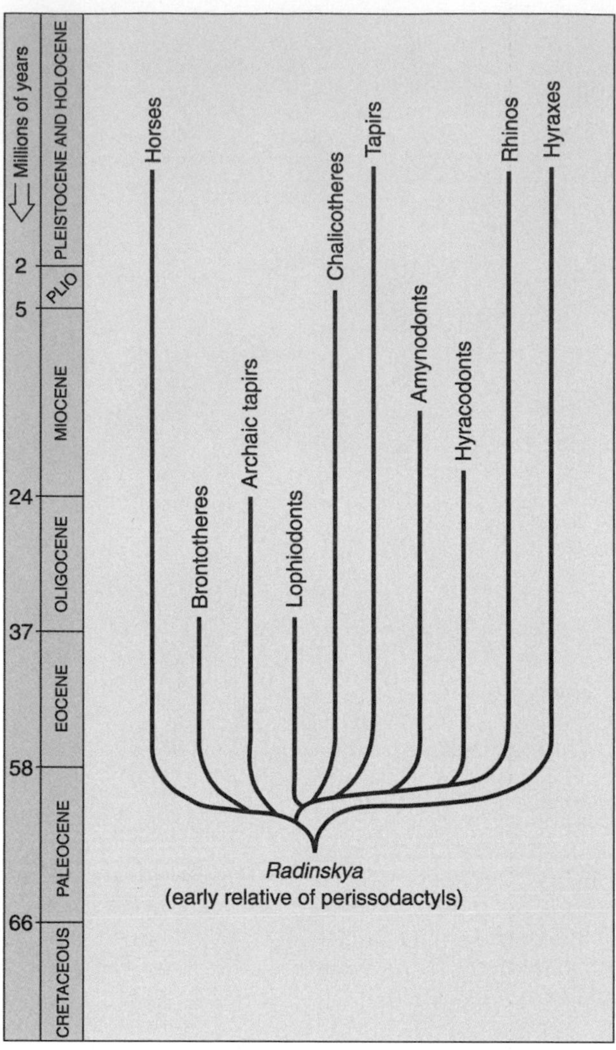

FIGURE 14-38 Simplified evolutionary tree of perissodactyls. (*Modified from Prothero, D. R. 1994. Mammalian evolution, in Prothero, D. R., and Schoch, R. M. (eds.).* Major Features of Vertebrate Evolution. *Short Course in Paleontology No. 7, University of Tennessee Press and the Paleontological Society.*)

with the progeny of domestic horses that had escaped from the Spanish explorers in the 16th century.) The cause of the extinction of horses in North America is somewhat of a mystery. Some believe it was brought on by contagious disease, whereas others speculate that the cause was overkill by prehistoric human hunters.

Among the other surviving perissodactyls are the tapirs and rhinoceroses. Tapirs retain the primitive condition of four toes on the forefeet and three on the rear, as well as low-crowned teeth. They are forest-dwelling, leaf-eating animals whose fossil record begins in the Oligocene.

Because of the great size and power of rhinoceroses, they are more impressive to humans than are tapirs. The early rhinoceroses, however, were not great ponderous beasts, but small, slender browsers about the size of a wolf. Like early horses and modern tapirs, they had four toes on the front feet and three on the hind feet. The skull was also relatively small, and there was no horn on the snout. By late in the Middle Eocene, three distinctive groups of rhinoceroses had diverged from the parent stock. One branch included small,

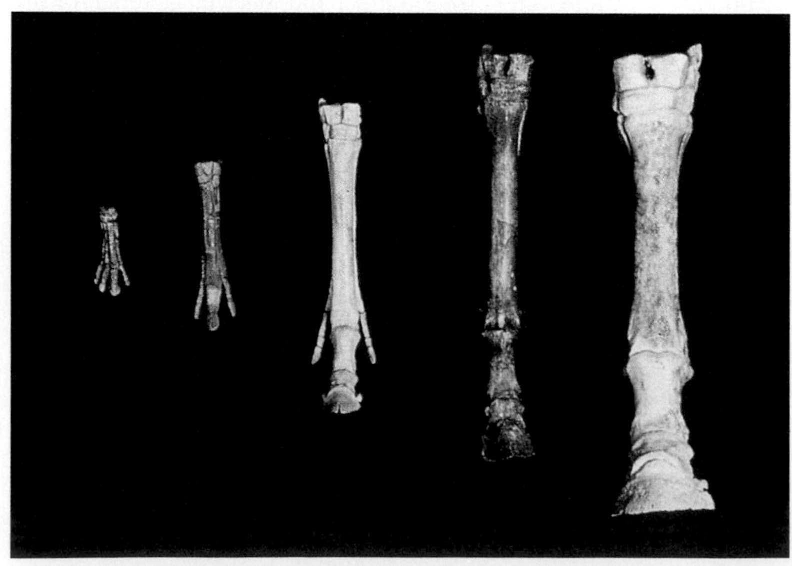

FIGURE 14-39 Evolution of the lower foreleg in horses, beginning at the far left with an Eocene horse and ending with the modern horse at the far right. (*Courtesy of the U.S. National Museum of Natural History, Smithsonian Institution.*)

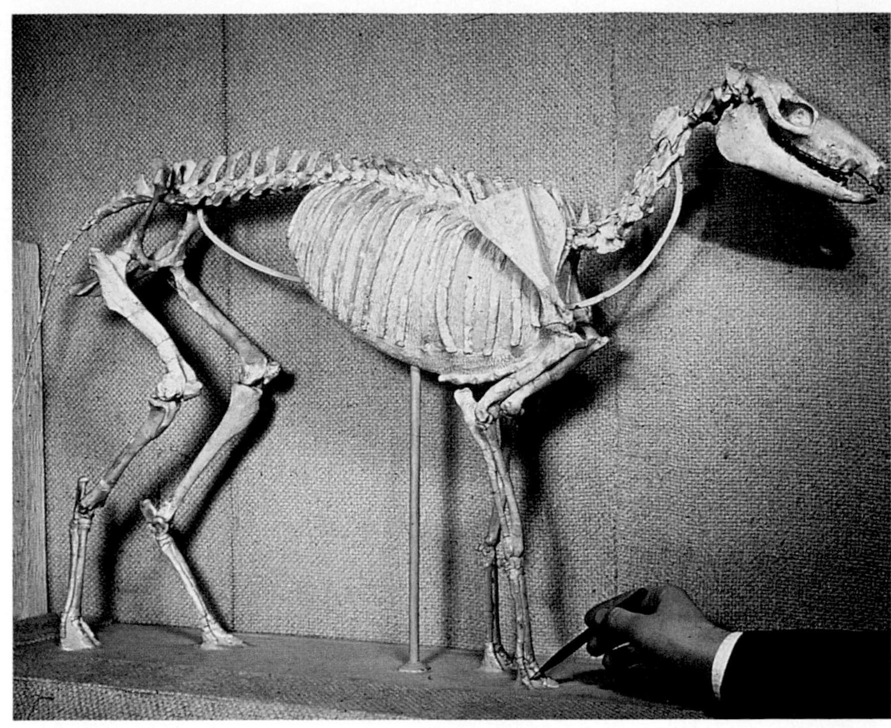

FIGURE 14-40 Skeleton of *Mesohippus*. This Oligocene horse was about the size of a collie. There were three toes on its feet, with the center toe larger than those on either side. (*Formerly on exhibit at the Museum of Comparative Zoology, Harvard University.*)

long-legged rhinoceroses called hyracodonts but also produced *Paraceratherium*, a colossal rhinoceros that stood 5 meters tall at the shoulders (Fig. 14–42). Another branch contained water-dwelling rhinoceroses called amynodonts that lived much like hippopotamuses do today. Neither the hyracodonts nor amynodonts survived beyond the Oligocene, but a third lineage, the rhinoceratids, produced an array of species that thrived well into the Pleistocene Epoch. By the end of the Ice Age, however, all but the five modern species of rhinoceroses had become extinct.

The horns of rhinoceratids are of interest because, unlike the horns of most other horned animals, they are formed of tightly matted hair and are therefore not preserved in prehistoric species. Paleontologists recognize the location of horns on fossil rhinoceratids by the presence of a roughened area on the skull where the horn was attached.

Two particularly interesting groups of perissodactyls known only from fossils are the **brontotheres** (Fig. 14–43*G*) and **chalicotheres** (Fig. 14–44*A*). Because of their large size and huge, blunt, forked horns

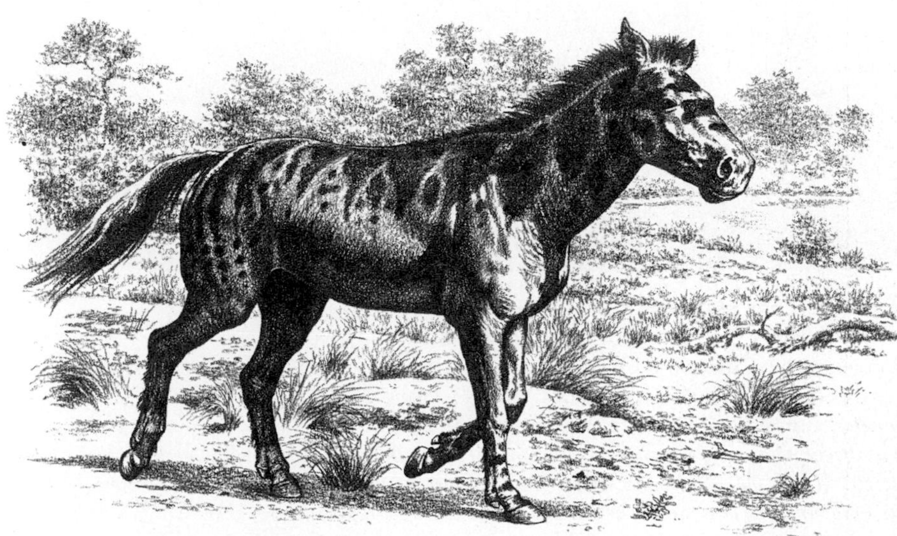

FIGURE 14-41 *Merychippus*, a Miocene member of the horse family, was about the size of a small pony. It still had three toes, but the side toes were reduced and of little use. (*Copyright J. Sibbick.*)

FIGURE 14-42 The colossal, hornless rhinoceros *Paraceratherium* was over 4 meters tall at the shoulder and weighed about 15 tons. This huge perissodactyl probably browsed on the leaves of trees. (*Copyright J. Sibbick.*)

FIGURE 14-43 Restoration of some Late Eocene mammals based primarily on fossils from the Chadron Formation of the White River Group of South Dakota and Nebraska. (The White River Formation above the Chadron is Oligocene in age.) The flora is based on the somewhat younger plant fossils from the Florissant beds of Colorado. (*A*) *Trigonia*, an early rhinoceros. (*B*) *Mesohippus*, a three-toed horse. (*C*) *Aepinocodon*, a remote relative of the hippopotamus. (*D*) *Archaeotherium*, an entelodont. (*E*) *Protoceras*, a horned ruminant. (*F*) *Hyracodon*, a small rhinoceros. (*G*) *Brontops*, a brontothere. (*H*) Oreodonts named *Merycoidodon*. (*I*) *Hyaenodon*, a carnivore. (*J*) *Poëbrotherium*, an ancestral camel. (*Courtesy of the U.S. National Museum of Natural History, Smithsonian Institution, J. H. Matternes mural.*)

FIGURE 14-44 Mural depicting an assemblage of early Miocene mammals. (*A*) The chalicothere *Moropus*. (*B*) The small artiodactyl *Merychyus*. (*C*) *Daphaenodon*, a large wolflike dog. (*D*) *Parahippus*, a three-toed horse. (*E*) *Syndyoceras*, an antelope-like animal. (*F*) *Dinohyus*, a giant, piglike entelodont. (*G*) *Oxydactylus*, a long-legged camel. (*H*) *Stenomylus*, a small camel. (*Courtesy of the U.S. National Museum of Natural History, Smithsonian Institution, J. H. Matternes mural.*)
? *Compare Figures 14–38, 14–44, 14–45, and 14–56. What trend in Cenozoic climate is indicated?*

over their snouts, museum reconstructions attract considerable attention at natural history museums. It seems likely that brontotheres used their horns for sexual display and possibly also for jousting with other males for access to receptive females. When needed, they may also have been used in defense.

Chalicotheres differed from all other Cenozoic perissodactyls in having three claws rather than hoofs on their feet. These odd creatures were rather similar to a horse in the appearance of the head and torso. The forelegs, however, were considerably longer than the hind legs, and the back sloped rearward. They lived from Eocene into the Pleistocene.

Even-toed ungulates, or **artiodactyls**, have been more successful than the perissodactyls in terms of survival, variety, and abundance. Those still living include pigs, peccaries, deer, hippopotamuses, goats, sheep, camels, llamas, giraffes, and scores of different kinds of cattle and antelope. Artiodactyls are clearly important to humans, for they supply meat and milk for our dinner table and wool for clothing. As even-toed mammals, artiodactyls have either four or two toes. The ankle and leg bones form a double pulley, rather than the single-pulley characteristic of perissodactyls. Unlike advanced perissodactyls, molar and premolar teeth are dissimilar.

Fossils of the earliest artiodactyls have been recovered from Lower Eocene strata in Pakistan, indicating that, like the perissodactyls, this great group of ungulates originated in the Old World before rapidly spreading around the globe. By late Tertiary, they had clearly achieved numerical and varietal superiority over other herbivores.

Among the Tertiary artiodactyls that became extinct, the **oreodonts** and the **entelodonts** are particularly interesting. Oreodonts were short grazing animals that roamed the grassy plains of North America in great numbers (see Fig. 14–43*H*). It is surprising that despite their outward appearance, they are members of the taxonomic group that includes camels.

Entelodonts, some of which were as large as an American bison, were repulsive-looking, hoglike beasts. Their trademarks were bony protruberances that grew along the sides of the skull and jaws (Fig. 14–43*D*, 14–44*F*, and 14–45).

The radiation of artiodactyls that began in the Eocene produced such surviving groups as swines, camels, and ruminants. Of these three, the swine family has remained the most primitive. They have kept all four toes, even though most of the weight is supported by the two middle toes. The group includes not only pigs but also the more lightly constructed and primarily South American peccaries. Hippopotamuses are the only modern amphibious artiodactyls. Although their name means "river horse," they are related to the pig family.

People are sometimes surprised to learn that the camel lineage originated and experienced most of its evolution in North America. Early members of the

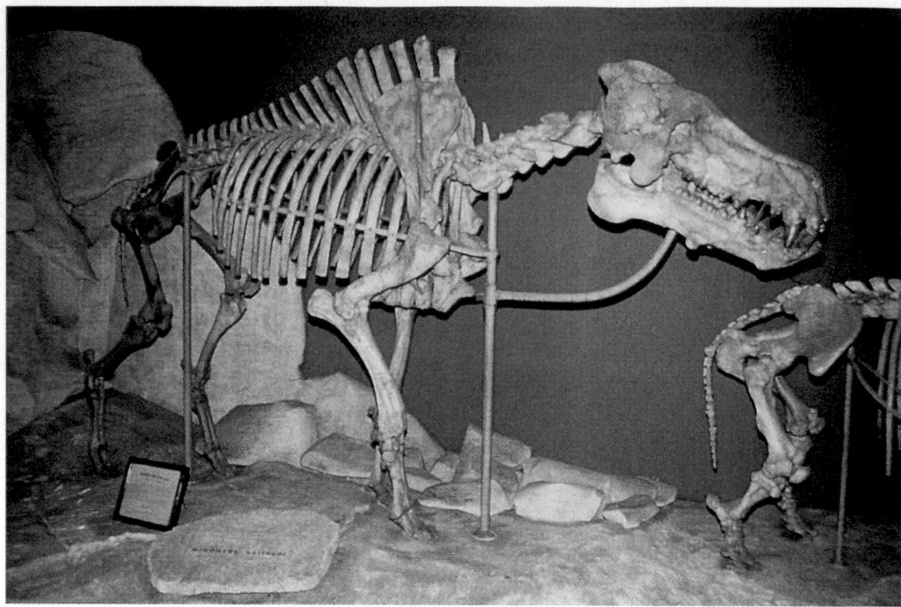

FIGURE 14-45 ***Dinohyus*, a giant piglike entelodont from the Miocene of America.** The animal was about 3 meters long. (*Courtesy of the Denver Museum of Natural History.*)

group were present by Middle Eocene. They were relatively small and, like the early horses, possessed low-crowned teeth. As they evolved through the Tertiary, they lost their side toes and developed high-crowned teeth and longer legs for rapid running. We see these changes occurring in a rich diversity of Tertiary species, with some camels assuming a gazelle-like appearance and others, like *Oxydactylus* (Fig. 14–46) and *Aepycamelus*, characterized by long necks and legs, giving them a superficial resemblance to modern giraffes. A particularly unusual member of the camel lineage is *Synthetoceras* (see Fig. 14–47*B*). Once thought to be a cud-chewing ruminant, Synthetoceras had a pair of horns above the eyes and a large Y-shaped horn on the snout. By late in the Pliocene Epoch, ancestors of modern llamas, alpacas, guanacos, and vicuñas were able to make their way across the Panamanian land bridge into South America. Further migrations occurred during the Pleistocene, when the ancestors of the dromedaries and Bactrian camels crossed the Bering land bridge into the old world. Early in the Holocene, extinction overtook the camels of North America.

FIGURE 14-46 **Skeleton of the Miocene camel *Oxydactylus*.** This small, graceful camel was only one of a varied group of camels that populated the grasslands of North America. (*Courtesy of the Denver Museum of Natural History.*)

Ruminants are the most varied and abundant of modern-day artiodactyls. This group takes its name from the *rumen*, the first of four compartments in their multichambered stomach. For the most part, they are cud chewers. The earliest ruminants were small, delicate, four-toed animals called tragulids. They are represented today by the skinny-legged and timid mouse deer of Africa and Asia. *Hypertragulus* (Fig. 14–48), whose remains are found in Lower Oligocene strata, was such an animal.

Ruminants such as sheep, cattle, giraffes, and deer are called pecorans. Deer are primarily browsers that have made the forests their principal habitat ever since their first appearance in the Oligocene. The early stages of deer evolution are exemplified by *Blastomeryx*, a small, hornless creature that lived in North America during the Miocene. Subsequent evolution involved

FIGURE 14-47 A variety of early Pliocene mammals. (*A*) *Amebelodon*, the shovel-tusked mastodon. (*B*) *Synthetoceras*. (*C*) *Cranioceras*. (*D*) *Merycodus*, an extinct pronghorn antelope. (*E*) *Epigaulis*, a burrowing, horned rodent. (*F*) *Neohipparion*, a Pliocene horse. (*G*) The giant camel *Megatylopus* and smaller *Procamelus*. (*H*) *Prosthennops*, an extinct peccary. (*I*) The short-faced canid *Osteoborus*. (*Courtesy of the U.S. National Museum of Natural History, Smithsonian Institution, J. H. Matternes mural.*)

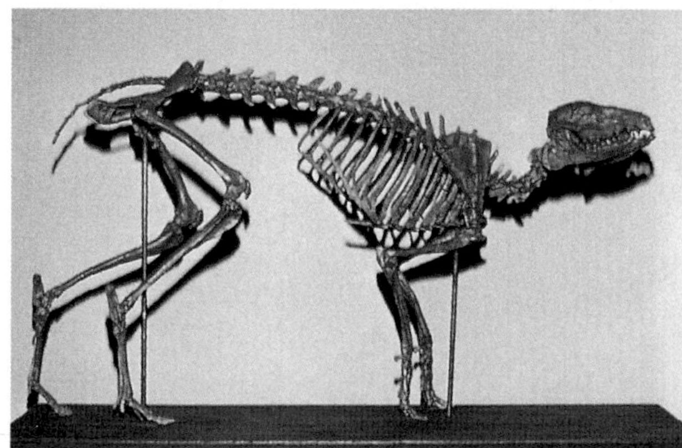

FIGURE 14-48 The primitive ruminant *Hypertragulus* from Lower Oligocene beds of North America. The tiny animal stood only about 30 centimeters high at the shoulder. (*Courtesy of the Denver Museum of Natural History.*)

increases in size and the development of antlers. The culmination of this trend is represented by the Pleistocene *Megaloceros*, whose antlers measured more than 3 meters from tip to tip (Fig. 14–49).

The giraffe family probably branched off from the deer lineage sometime during the Miocene and became specialized in browsing on leaves of trees that grew rather sparsely in areas where most other herbivores were grazers. *Palaeotragus*, the ancestral giraffe, looked much like the modern okapi of Africa. The exceptionally long neck by which we recognize modern giraffes was a late modification and is not typical of extinct giraffes, most of which had necks no longer than seen on a deer.

Bovids, a group that includes cattle, sheep, goats, muskoxen, and antelopes, are presently the most numerous of ruminants. Miocene strata provide the earliest fossil record of bovids. Bison are bovids that hold a particular interest for Americans because of the vast numbers that once roamed our western prairies. Seven species of bison lived in North America during the Pleistocene. Some attained great size and had horns that measured 2 meters from end to end.

FIGURE 14-49 ***Megaloceros*, the giant Irish "elk," was actually a deer whose remains are frequently found in the peat bogs of Pleistocene age in Ireland.** From tip to tip, the antlers were about 3.6 meters in breadth. (*Photograph taken in the Sedgwick Museum, Cambridge University, Cambridge, England.*)

A discussion of Cenozoic mammals would not be complete without mention of a major group of ungulates termed the tethytherians. The tethytherian group (see Fig. 14–33) includes proboscideans (elephants, mammoths, and mastodonts), the extinct heavy-horned arsinotheres (see Fig. 14–35), a group of peculiar aquatic animals called desmostylids, and the manatees and dugongs. The manatee and elephant would seem unlikely relatives on the basis of their outward appearance, but these two groups share a number of skull and dental characteristics that confirm their close relationship.

Although the manatees are familiar to most of us, people appear to be more fascinated with elephants. The trunk, of course, is the distinguishing feature of elephants, although earliest proboscideans lacked a trunk. The trunk's most important function is to bring food to the elephant's mouth. Other animals as tall as elephants reach food on the ground easily because of their long necks. Proboscideans, however, with their short, muscular necks needed to support their massive heads, have evolved their own unique anatomic solution to food gathering. Paleontologists are able to follow the development of the trunk in early proboscideans by noting the position of the external nasal opening at the front of the skull. Those openings recede toward the rear of the skull in sequential stages of trunk development. The trunk itself was formed by elongation of the upper lip and nose. The nasal passages that extend through the trunk are encased by over 500 muscles, and these provide ample power and dexterity.

A second proboscidean trademark is the tusks, which evolved by elongation of the second pair of incisors. Proboscideans use their tusks in both defense and offense. They are also an aid to browsing, for they are employed to hold down branches pulled into reach by the trunk.

The earliest proboscidean remains were found in Eocene rocks of North Africa. The trunk had not yet appeared in these early forms. *Barytherium* and *Moeritherium* (from fossil beds along Lake Moeris south of Cairo, Egypt) are representatives of the initial stages in the history of proboscideans. The skeletal remains of *Moeritherium* reveal an animal about the size of a pigmy hippopotamus. Upper and lower incisors were already formed into mini-tusks (Fig. 14–50).

From relatively primitive animals such as the moeritheres, proboscideans with trunks evolved and quickly divided into several branches. One branch led toward a group of Miocene and Pliocene animals called **dinotheres** (Fig. 14–51). The tusks of dinotheres were distinctive in that they were present only in the lower jaws and curved downward and

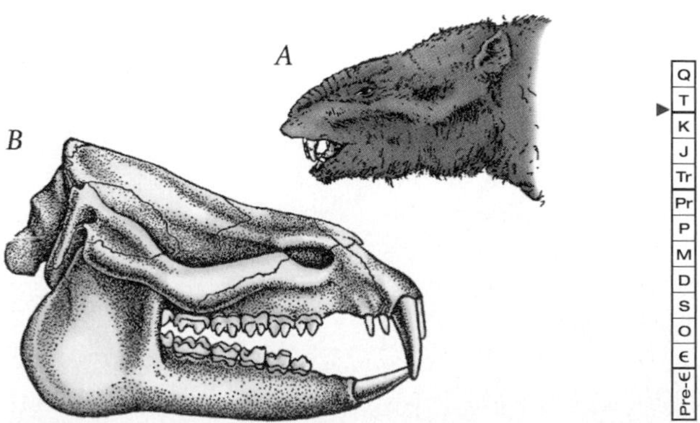

FIGURE 14-50 Restoration of the head (*A*) and the skull (*B*) of *Moeritherium*. (A, *from Levin, H. L. 1975.* Life Through Time. *Dubuque: Wm. C. Brown; B, from Andrews, C. W. 1906. A descriptive catalogue of the Tertiary Vertebrata of the Fayum, Egypt. London.*)

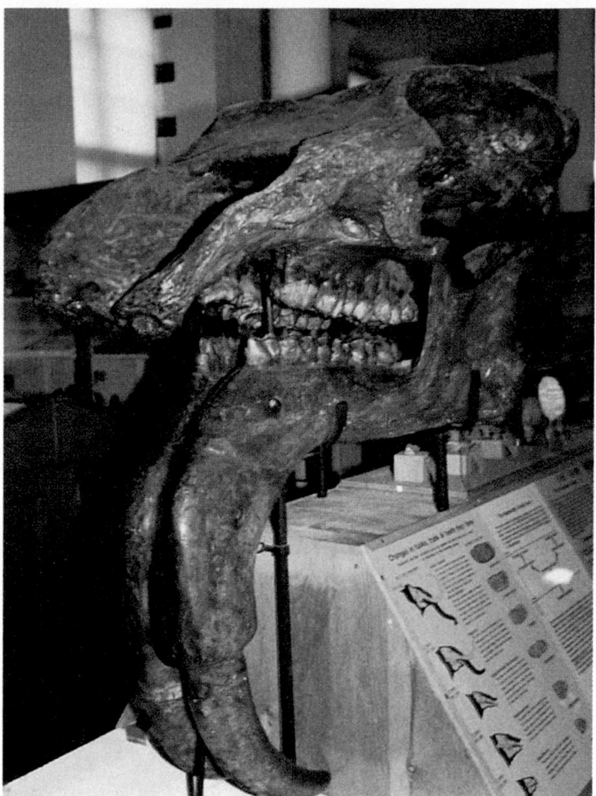

FIGURE 14-51 Skull of *Dinotherium*, a Miocene proboscidean. Length of skull is about 1.2 meters. (*Photograph taken in the Sedgwick Museum, Cambridge University, Cambridge, England.*)

backward, an orientation presumably useful for uprooting plants and digging for roots and tubers.

Another branch of the proboscidean family tree produced mastodonts and elephants. *Palaeomastodon*, in the Oligocene of North Africa, was probably representative of the group from which true mastodonts evolved. The skull was long, the trunk was short, and the second incisors in both the upper and the lower jaws had developed into tusks. By Miocene time, a larger proboscidean named *Gomphotherium* had found its way into North America by way of the Bering isthmus. Subsequent proboscidean evolution produced a variety of long-jawed mastodonts. *Trilophodon*, from the Pliocene, had lower jaws almost 2 meters long. The most bizarre proboscidean was the "shovel-tusked" *Ambelodon* (see Fig. 14–47*A*) of the Pliocene. The tusks of the lower jaws in these beasts were flattened and formed two sides of a broad, scooplike, ivory shovel. Mastodonts with shorter jaws and longer trunks are best represented by species of *Mammut* (Fig. 14–52), which were common throughout North America during the great Ice Age and survived until comparatively recent times. Mammut was heavier but not as tall as the

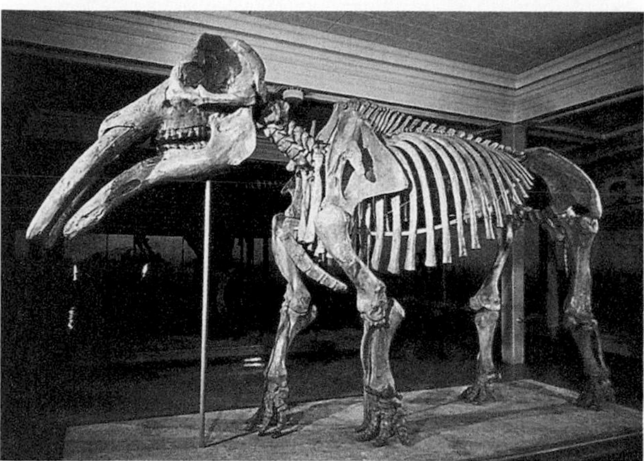

FIGURE 14-52 A four-tusked mastodont of the Late Miocene and Early Pliocene. The mounted skeleton is in the collection of the Denver Museum of Natural History.

African elephants. It had the sharply crested molars typical of mastodonts, a stocky build, and great curving tusks nearly 2 meters long.

True elephants and mammoths evolved from gomphotheres in the Old World. Their line of evolution can be traced backward to a Miocene animal named *Stegolophodon*. Next on the evolutionary scene was Stegodon, a Pliocene form with a short, tuskless lower jaw and molar teeth already containing the numerous cross-ridges that characterize true elephants. Such teeth are more suitable for a diet of abrasive grasses (Fig 14–53). Further along in the evolution of the elephant, the trunk grew longer, the height of the skull increased enormously, and the jaws became so short that they could accommodate no more than eight grinding teeth at one time. As the molars of elephants gradually wear down, new teeth form in the posterior region of the jaws and push forward until the older, worn teeth break out. The last molars appear when the animals are 40 to 50 years old.

Mammoth is a term loosely applied to the Ice Age elephants of North America, Europe, and Africa. They were a magnificent group of animals that included the famous woolly mammoths (Fig. 14–54) drawn by our own ancestors on the walls of their caves. The great imperial mammoth reached heights of 4.5 meters and ranged widely across California, Mexico, and Texas. The Columbian mammoth had immense spiral tusks that in older individuals overlapped at the tips, thus becoming useless for digging purposes. Many of the late Pleistocene proboscideans were hunted by humans. Then, about 8000 years ago, all but two genera of elephants became extinct. Their continued survival, as is the case with many wild animals, will depend on our good judgment and that of the governments we support.

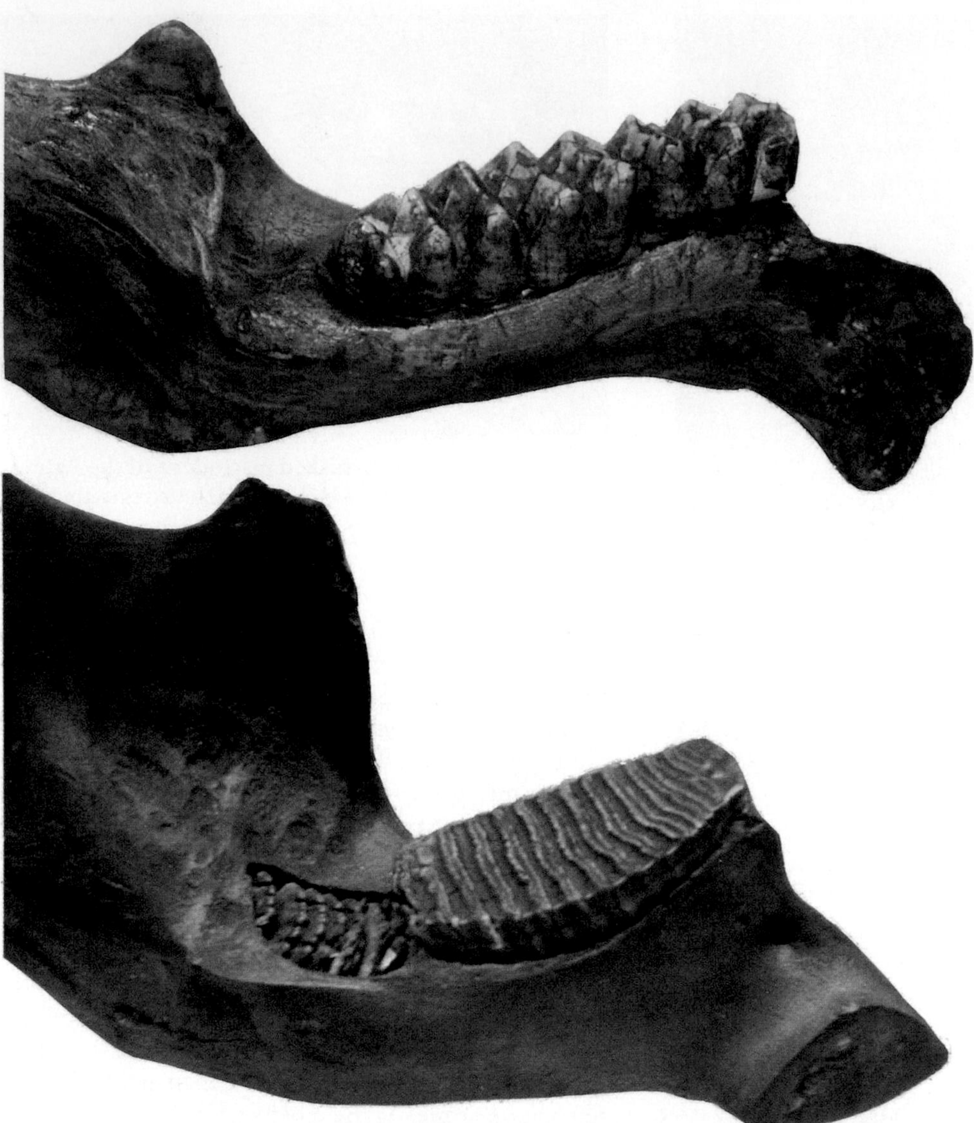

FIGURE 14-53 Cheek teeth in jaws of mastodont (*A*) and mammoth (*B*). The paired cusps of the mastodont teeth were used for crushing vegetation, as is typically seen in browsing mammals. The mammoth tooth, with its parallel ridges of infolded enamel, was well adapted for grazing on tough grasses and cereals. (*The Field Museum of Natural History, Chicago.*)

FIGURE 14-54 The woolly mammoth. In the Late Pleistocene, these magnificent animals lived along the borders of the continental glaciers. Their remains have been found frozen in the tundra of northern Siberia. (*Courtesy and copyright of The Field Museum of Natural History, Chicago. Painting by C. R. Knight, photograph of painting by R. Testa, with permission.*)

ENRICHMENT

How the Elephant Got Its Trunk

Among the many mammals that evolved during the Cenozoic, few inspire more interest than the proboscideans—mastodonts, mammoths, and elephants. Our interest seems to relate not just to their great size but also to their most distinctive feature, the trunk. The elephant's trunk is a highly muscularized, much elongated organ containing nostrils that extend downward through its entire length to the tip. The Indian elephant uses a one-finger process to pick up small objects with the tip of the trunk, while the African elephant uses a two-finger process.

The question, "How did the elephant get its trunk?" might be readily answered by a child on reading Rudyard Kipling's *The Elephant's Child.* According to this story, an overly inquisitive baby elephant tried to find out what the crocodile had for dinner. When the baby elephant lowered his head along the bank of the great, gray-green, greasy Limpopo River, the jaws of the crocodile closed on his nose. There ensued a fierce struggle in which the "bicolored-python-rock-snake" came to the aid of the poor baby elephant, but in the tug-of-war, the baby elephant's nose was stretched into the trunk we know today. Actually, the elephant achieved its wonderfully long and prehensile trunk by means of evolutionary processes that operated over many generations and resulted in an enormous elongation of the nose and upper lip. Scientists refer to such extraordinary growth of a body part as hypertrophy, and the elephant got its trunk by hypertrophy of the nose and upper lip. But why?

As one views the bones of proboscideans, beginning with the relatively small and primitive groups of the Eocene through the giant Pleistocene mastodonts and mammoths, several trends are apparent. One obvious trend was the increase in size, not only of the body but also of the huge proboscidean head. Supporting the weight of the ponderous head required a short, powerful neck. As the proboscideans grew taller, reaching food at ground level when equipped with a short neck would have been difficult. Selection paralleling that of the large head and short neck resulted in the development of the trunk. The elephant's trunk is a structural adaptation for feeding that could be used not only for raising food from the ground but also for stripping leaves from branches and stuffing them into the mouth. Accompanying the evolution of the trunk in proboscideans, the lower jaw became shorter, allowing the trunk to hang downward. Tusks in the lower jaw that were present in early proboscideans became smaller and gradually disappeared.

Cetaceans Among all mammals, none have so completely adapted to life in the sea as the cetaceans (whales and porpoises). When we see their sleek bodies moving perfectly through the water, it is difficult to grasp the fact that these amazing mammals may be descendants of hoof-bearing land dwellers known as artiodactyls. Living artiodactyls include ruminants (such as cows, goats, sheep, deer, and giraffes), as well as camels, pigs, and hippopotamuses. Those supporting the hypothesis that whales are derived from early artiodactyls cite recent discoveries from Pakistan of 55-million-year-old aquatic cetacean precursors that have intact ankle bones of a form characteristic only of artiodactyls. An earlier discovery named *Pakicetus* (Fig. 14–55) also had the double pulley artiodactyl ankle structure. *Pakicetus* lived in Eocene lakes, streams, and estuaries about 50 million years ago. The nonmarine shales in which its bones were found indicate that the animal had not yet ventured into the open ocean. That transition was apparently made by the discovery in India of *Himalayacetus*. The remains of this cetacean were found within sediments recognized as marine because of their content of oysters and foraminifera.

The evidence of the link between artiodactyls and whales is supported not only by fossil evidence but also by DNA sequencing studies. Molecular biologists further believe that, among artiodactyls, the hippopotamuses are closer to whales than any other group of artiodactyls (Fig. 14–56). Some paleontologists, however, are not yet ready to confirm that specific relationship.

In rocks only 10 million years younger than those yielding the remains of *Pakicetus*, yet another highly interesting ancestral whale has been found. It has been given the name *Ambulocetus*. *Ambulocetus* was about the size of a large sea lion. Hind limbs that terminated in webbed feet provided the principal propulsion, with an assist for turning and diving from well-developed front flippers. Certain characteristics of the back vertebrae suggest it may have moved about on land with the up-and-down motion seen in seals. This motion may have been a sort of preadaptation for the up-and-down tail propulsion seen in modern whales. In *Basilosaurus* (Fig. 14–34) from the Late Eocene of Egypt, we get a glimpse of the trend toward giganticism. This whale attained a length of 60 feet. Tiny hind limbs complete with toes were still present in *Basilosaurus*, but these rear limbs were too small and weak to have been of use in swimming. Nor would they have been able to support the animal's impressive weight on land. Some have suggested this ancient whale used its hind limbs to grasp and guide its partner during copulation. Others suggest the hind limbs were useless vestigial structures.

Modern whales arose from the group that included *Basilosaurus* and divided into two lineages: the toothed

FIGURE 14-55 The ancestral whale *Pakicetus* from the Eocene of Pakistan. Because the remains of Pakicetus were found in stream-deposited sediments, it is doubtful that this primitive whale ever ventured into the open ocean. It may, however, have lived in coastal estuaries. (Pakicetus was about 3 meters long.) (*Copyright J. Sibbick.*)

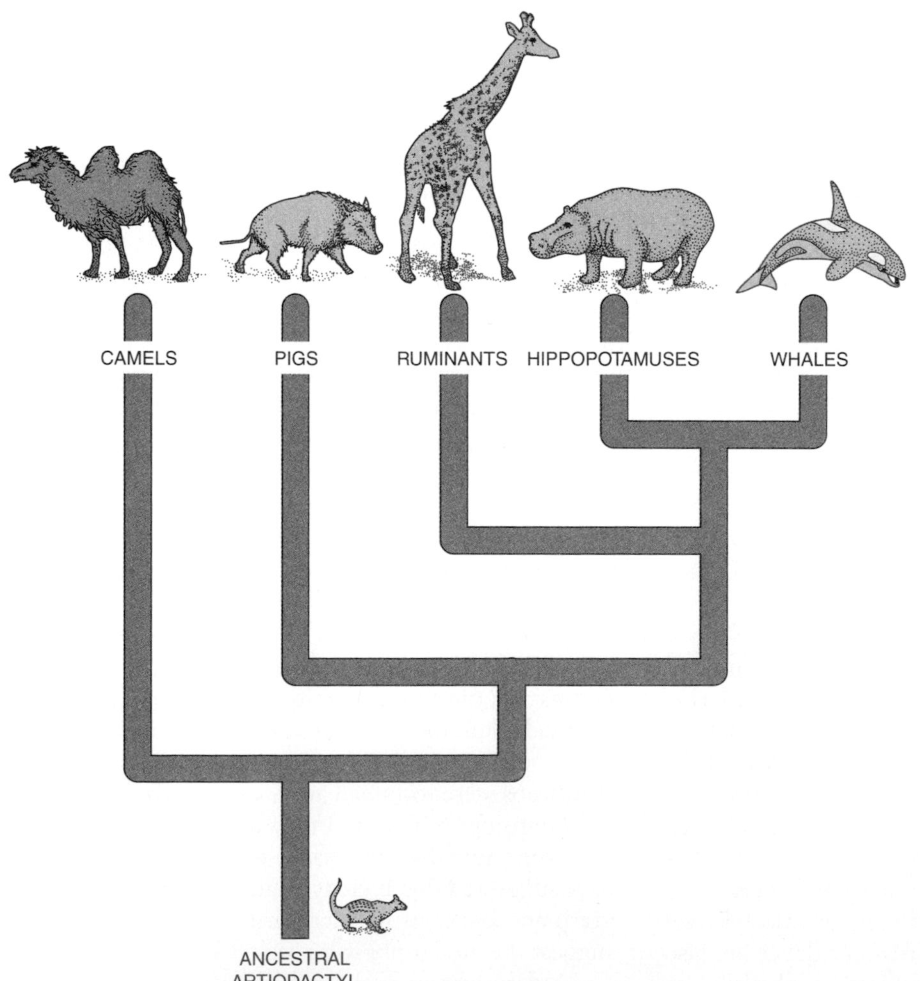

FIGURE 14-56 The close relationship between hippopotamuses and whales is indicated on this evolutionary tree of major artiodactyl groups.

whales and the whalebone (baleen) whales. Toothed whales, which made their debut during the Oligocene, today include porpoises, killer whales, and sperm whales like Melville's famous Moby Dick. The titanic blue whale, right whale, humpback, gray, and Greenland whale are all representatives of the second cetacean group. These plankton-feeding giants first appeared in the Miocene. Instead of teeth, they possess ridges of hardened skin that extend downward in rows from the roof of the mouth. The ridges are fringed with hair, which serves to entangle the tiny invertebrates on which these cetaceans feed—thus, the paradox of the largest of all animals feeding on some of the smallest. The great blue whales far exceed in mass even the largest dinosaurs, for some have attained weights of over 135 metric tons.

DEMISE OF THE PLEISTOCENE GIANTS

At the time of maximum continental glaciation (approximately 17,000 years ago), the Northern Hemisphere supported an abundant and varied fauna of large mammals, comparable to that which existed in Africa south of the Sahara several decades ago (Fig. 14–57). The fauna included giant beavers, mammoths, mastodonts, elk, most species of perissodactyls, many even-toed forms, and ground sloths. From all available evidence, most of these great beasts maintained their numbers quite well during the most severe episodes of glaciation, but they experienced rapid decline and extinction in the period around 8000 years ago, when the climate became milder. What might have been the cause of the extermination of these large Pleistocene mammals? Explanations tend to fall into two categories. In one, theories attempt to explain the extinctions as being related to human overkill; a second interpretation relates the losses to climatically controlled environmental changes.

The overkill concept is based on the inference that early humans had developed the ability to hunt in highly organized social groups and skillfully bring down large numbers of big animals on open prairies and tundras. Human predators may have killed then, as they do now, in excess of their needs. Unlike such predators as wolves, early human hunters probably did not seek out the weak or sick to kill but brought down large numbers at one time, including the best animals in the herds. In support of this idea, paleontologists note that most Pleistocene extinctions involved large terrestrial animals. Marine genera, protected by the sea from human predation, continued to thrive. Small mammals also survived; although frequently hunted, they were difficult to exterminate because of their more rapid breeding rate, their greater numbers, and the probability that they were killed only one at a time. The big, gregarious herbivores were easily accessible and reproduced more slowly. Entire herds could be driven off cliffs or into ravines, thus providing opportunities for slaughter of hundreds at the same time.

Another line of evidence favoring overkill stresses the observation that an early wave of extinction began in Africa and southern Asia, where predatory humans first became prevalent. The better known extinctions of

FIGURE 14-57 An assemblage of animals that lived in Central Alaska about 12,000 years ago, during the Late Pleistocene. Fossil remains of this period are abundant and indicate a fauna in which grazing animals predominated, with the remaining fauna composed of browsers and predators. *(Courtesy of the U.S. National Museum of Natural History, Smithsonian Institution, J. H. Matternes mural.)*

the colder north did not occur until much later, when humans had moved into this region after having first devised the means to better clothe and shelter themselves.

The overkill concept has not, however, gone unchallenged. Many paleontologists feel that human populations were too small and that nomadic tribes shifted too frequently to account for the extermination of entire species of mammals. Some animals, such as the cave lion, were probably never hunted as food, and yet they too became extinct along with the mammoths and mastodonts. Moreover, primitive tribes that today still live at the Stone Age level of development do not endanger the species survival of the animals they hunt.

These arguments have led some paleontologists to favor theories that relate extinctions to climatic warming. The great beasts of the Pleistocene flourished during the cold 100,000-year glacial stages, even though those cooler episodes were interrupted by interglacial stages lasting only about 10,000 years. They were primarily adapted to cooler climates. With the onset of warmer conditions following the last glacial advance, their reproductive capacity may have been curtailed. Geologist D. M. McLean notes that this would be particularly true for the females of large mammals because of their relatively smaller surface area-to-volume ratio. (They have less surface area to give off stored body heat than do smaller animals.) McLean proposes that pregnant females may have experienced increased blood flow to peripheral tissues simply in order to cool down. This would lessen blood flow to the uterus, thereby reducing the supply of oxygen, nutrients, and water to the developing embryo. Heat generated by embryonic metabolic processes would also be difficult to dissipate. The final result would be elimination of many huge Pleistocene mammals, as well as evolutionary selection for smaller animals.

SUMMARY

The Cenozoic was a time of gradual approach to conditions of the present. The warm, humid climates common during much of the Mesozoic were replaced by cooler and dryer conditions. As a result, there were pervasive changes in land vegetation as dense jungles gave way to deciduous forests, parklike woodlands, scrublands, and grasslands.

Among marine invertebrates, large discoidal foraminifera such as *Nummulites* were characteristic of areas in the Tethys seaway and the tropical western Atlantic. Smaller planktonic foraminifers proliferated and underwent a marked diversification during the Cenozoic. Gastropods and bivalves were the dominant higher marine invertebrates. Reef corals grew extensively across tracts of shallow warm water in the Tethys belt, the Caribbean, the West Indies, and around the many volcanic islands of the Pacific.

Dinosaurs did not survive into the Cenozoic. Among reptiles, only turtles, crocodiles, lizards, snakes, and rhynchocephalians survived to modern times. Although the fossil record for birds is less than adequate, it is likely that most modern groups were present by Eocene time. Nearly all the continents seem to have had one or more species of large, flightless, flesh-eating birds during the Cenozoic.

With considerable justification, the Cenozoic has been called the "Age of Mammals." During the Cenozoic, mammals expanded rapidly into the multitude of vacated habitats. From ancestral shrewlike creatures, either directly or indirectly, the higher groups of placental mammals evolved. In the vanguard were small-brained archaic herbivores and carnivores.

Following this early Cenozoic radiation, a second wave of placentals began to evolve in the middle part of the era, and most of the modern mammalian orders became established. This adaptive radiation was as remarkable as had been that of the earlier Mesozoic reptiles. Mammals in the form of bats conquered the air, whereas others, such as whales, seals, and walruses, returned to the sea, where the evolution of vertebrates had begun long ago.

The Carnivora diversified and became dependent on the multitude of advanced ungulates that populated forests and grasslands. Ungulates included two great categories: the perissodactyls and artiodactyls. The former include horses, rhinoceroses, and tapirs, as well as the extinct titanotheres and chalicotheres. Among artiodactyls, or even-toed ungulates, the large, piglike entelodonts and smaller oreodonts failed to survive beyond the Oligocene. However, living artiodactyls include swine, camels, hippopotamuses, and a diverse host of ruminants. As all of these ungulates evolved during the Cenozoic, they underwent specialization of teeth for grazing and limbs for running across relatively open plains. Grass tends to wear down teeth rapidly. To compensate, grass-eaters evolved teeth with complicated ridges of hard enamel and exceptionally high crowns.

Similar changes also occurred in members of the proboscidean line. The earliest proboscideans are recorded as fossils in Eocene beds in Egypt. Their radiation produced an impressive array of variously tusked and specialized mastodonts and mammoths. The Pleistocene remains of those great beasts are frequently found in bog deposits in the eastern United States, Alaska, and Siberia.

The Pleistocene was a time of splendid diversity among terrestrial mammals. The fauna was probably more varied and impressive than that of Africa a century ago. The Pleistocene was an age of giants, in which colossal ground sloths; beavers over 2 meters tall; giant dire wolves; bison; tall, rangy mammoths; and woolly rhinoceroses ranged far and wide across the continents. One might think an assemblage of beasts that had proved their stamina against the difficulties of an ice age would persist for at least as long as the dinosaurs. Yet widespread extinction was the dominant biologic theme at the end of the Pleistocene. Overkill by human hunters may have contributed to the demise of many Pleistocene mammals, or the extinction of these animals may be related to an inability to adjust to climatic warming.

QUESTIONS FOR REVIEW AND DISCUSSION

1. What is the depth of water preferred by reef corals? How does water depth control coral growth? Explain the development of atolls.

2. Which groups of marine phytoplankton proliferated during the Cenozoic Era?

3. How did the cephalopod faunas of the Cenozoic differ from those of the preceding era?

4. Which group of bony fishes is particularly characteristic of the Cenozoic?

5. What evidence suggests whales are descendents of ungulates?

6. How might a paleontologist determine that a fossil lower jaw containing teeth belonged to a mammal rather than a reptile?

7. What are ruminants? What advantages may be inherent in the ruminant type of digestion?

8. How is it possible to trace the development of the trunk or proboscis in the skeletal remains of proboscideans when the trunk itself is not preserved?

9. If evolutionary success can be measured in terms of diversity of a particular group of animals, how would you rate perissodactyls relative to artiodactyls?

10. Prepare a list of the changes that have occurred in horses during their Cenozoic evolutionary history.

11. Prepare arguments for and against the premise that the extinction of large Pleistocene mammals was a consequence of overkill by early human hunters.

12. Describe the adaptations seen in Cenozoic herbivores that are related to the spread of prairies.

13. Cenozoic whales demonstrate a remarkable example of evolutionary convergence with the marine reptiles known as ichthyosaurs of the Mesozoic. What characteristics of whales, however, are distinctly not reptilian?

READINGS

Benton, M. J. 1997. *Vertebrate Paleontology*, 2nd ed. London: Chapman and Hall.

Carroll, R. L. 1988. *Vertebrate Paleontology and Evolution*. New York: W. H. Freeman & Co.

Colbert, E. H., and Morales, M. 1991. *Evolution of the Vertebrates*, 4th ed. New York: John Wiley & Sons.

MacFadden, B. J. 1992. *Fossil Horses*. Cambridge, England: Cambridge University Press.

Martin, P. S., and Wright, H. E. 1967. *Pleistocene Extinctions*. New Haven, CT: Yale University Press.

Match, C. L. 1976. *North America and the Great Ice Age*. New York: McGraw-Hill.

Prothero, D. R., and Schoch, R. M. 1994. *Major Features of Vertebrate Evolution*. Short Courses in Paleontology No. 7, University of Tennessee Press and the Paleontological Society.

Savage, R. J. G., and Long, M. R. 1987. *Mammalian Evolution: An Illustrated Guide*. London, England: British Museum of Natural History.

Szalay, F. S., Novacek, M. J., and McKenna, M. C., eds. 1993. *Mammal Phylogeny*. New York: Springer-Verlag.

WEB SITES

The Earth Through Time Student Companion Web Site (www.wiley.com/college/levin) has online resources to help you expand your understanding of the topics in this chapter. Visit the Web Site to access the following:

1. Illustrated course notes covering key concepts in each chapter;
2. Online quizzes that provide immediate feedback;
3. Links to chapter-specific topics on the web;
4. Science news updates relating to recent developments in Historical Geology;
5. Web inquiry activities for further exploration;
6. A glossary of terms;
7. A Student Union with links to topics such as study skills, writing and grammar, and citing electronic information.

15

Right hand of the prosimian Europolemus, *which lived in Europe during the Middle Eocene. The opposable thumb and broad fingertips provide evidence for placement of this small creature in the Order Primates. Thus, the primate lineage extends backward in time to about 50 million years ago. The specimen is the collection of the Senckenberg Research Institute. (With permission of National Geographic.)*

CHAPTER 15

Human Origins

It walked on its hind feet, like something out of the vanished Age of Reptiles.
The mark of the trees was in its body and hands.
It was venturing late into a world dominated by fleet runners and swift killers.
By all the biological laws this gangling, ill-armed beast should have perished,
but you who read these lines are its descendant.
Loren Eiseley, 1971, *The Night Country*

OUTLINE

Evolution during the late epochs of the Cenozoic produced a large-brained mammal capable of shaping and controlling its own environment. That mammal, of course, was our own species, *Homo sapiens*. Unlike previous vertebrates, this remarkable creature profoundly changed the surface of the planet, modified the environment in ways both beneficial and destructive, and had such pervasive effects on populations of other creatures as to alter the entire biosphere. For these reasons, it is fitting that the final chapter in our history of the Earth examines the evolution of humankind.

In this chapter we will review the general characteristics of the primates, with whom we share many characteristics. For example, it is apparent that we resemble such primates as the great apes in basic body structure. On the other hand, we have become quite distinct from apes in many ways. We have a larger, more complex brain. We stand and walk erect and have structural modifications of our vertebral column, legs, and pelvic bones to make such erect posture possible. Our face is flatter, our teeth are less robust, and we are capable of extraordinary manual dexterity. This attribute has contributed to our ability to manufacture and use sophisticated tools. Moreover, *Homo sapiens* exceeds all other primates in a level of intelligence that has led to language, culture, and aesthetic sensibility.

PRIMATES

Primate Characteristics

What traits should an animal have to qualify as a primate? The question cannot be answered easily, for primates have remained structurally generalized compared with most other groups of placental mammals. They retain the primitive number of five digits, have teeth that are not specialized for dealing with either grain or flesh, and have never developed hoofs, horns, trunks, or antlers. In the course of their evolution from shrewlike insectivores, their principal changes have involved progressive enlargement of the brain and certain modifications of the hand, foot, and thorax that relate to their early life in the trees and their manner of obtaining food. These adaptations were not insignificant, however, for they enabled the human primate to shape a mode of life qualitatively different from that of any other animal.

The fictional pig Snowball in George Orwell's book *Animal Farm* remarked that "the distinguishing mark of man is the hand, the instrument with which he does all his mischief." Although it expresses a biased point of view, the phrase is correct in its assessment of the importance of the primate grasping hand with its opposable thumb (Fig. 15–1). This characteristic not only permitted primates a firm grip on their perches but also allowed them to grasp, release, and manipulate food and other objects. The forearms retained their primitive mobility, in which rotation of the ulna and radius upon one another permitted the hands to be turned at various angles and even reversed in position.

The development of the grasping, mobile hand was accompanied by improvement in visual attributes. The eyes of primates became positioned toward the front of the face so that there was considerable overlap of both fields of vision. The result was an improved ability to judge distances. It would seem that grasping hands and feet and good binocular vision are obvious adaptations for an animal that leaps from branch to branch and seeks its food from precarious boughs. However, many tree-dwelling animals such as squirrels, civets, and opossums lack the short face, close-set eyes, and opposable digits. Yet they get along very well. For this reason, anthropologists have recently suggested that the visual attributes of primates originated as predatory adaptations that allowed early insectivorous primates to gauge accurately the distance to insect prey without movement of the head. By being able to grasp narrow supports securely with its feet, the animal was able to use both of its mobile hands to catch the prey as it flew or scampered away. Claws, which are used by animals such as squirrels in moving about on relatively wide branches, would not be as advantageous for climbing about on thin boughs and vines and might not provide a sufficiently secure hold for the quick catch of a moving insect. Thus, it seems that binocular vision and the grasping hand may not have been originally related to arboreal locomotion alone, but to precision and safety in the capture of visually located prey in the insect-rich canopy of tropical forests.

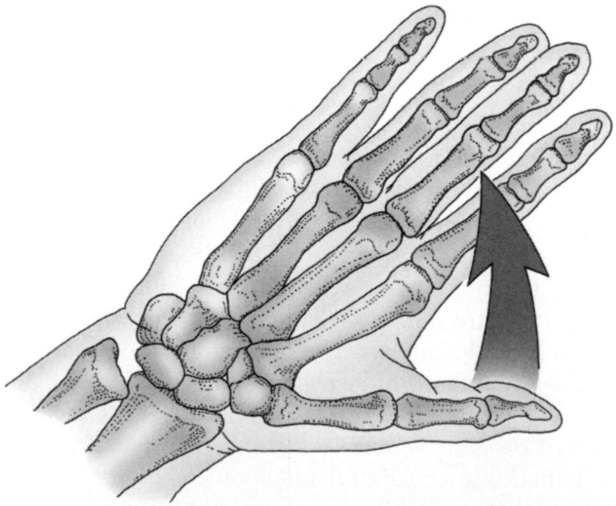

FIGURE 15-1 The right hand of a human (palm up). The human hand is not used in locomotion and can be used to manipulate small objects between the fingers and thumb.

Other evolutionary modifications of primates were related primarily to changes in the eyes and limbs. On the outer margin of each eye orbit, a vertical ridge of bone developed (postorbital bar) that protected the eyes from bulging jaw muscles and accidental impact. As the eyes became positioned more closely together, the snout was reduced so that the face became flatter. In response to a brachiating habit (swinging from branches), forelimbs and hindlimbs diverged in form and function, and an inadvertent predisposition toward upright posture developed.

Modern Primates

The two major groups of primates are the **Prosimii** (which include tree shrews, lorises, and tarsiers) and the **Anthropoidea** (which includes monkeys, apes, and humans) (Table 15–1). The Prosimii retain a greater number of primitive characters relative to the Anthropoidea. Tree shrews (see Fig. 14–26), for example, possess clawed feet and a long muzzle with eyes in the lateral positions. The lemurs (Fig. 15–2), which are confined largely to Madagascar, also have long snouts and lateral eyes, although these are positioned more to the front than those in the tree shrew. Some of the digits are clawed, whereas others are equipped with flattened nails. The East Indian tarsier (Fig. 15–3) has a relatively flat face. As befits a nocturnal animal, the eyes are large and positioned toward the front to provide stereoscopic vision. The fingers and toes terminate in nails rather than claws.

In the more progressive Anthropoidea, the trends initiated by prosimians are further developed. Monkeys are the more primitive members of the Anthropoidea. They are grouped into New World and Old World forms (Fig. 15–4). The **Ceboidea**, or New World monkeys, are an early branch not involved in the eventual evolution of humans. Included among the ceboids are the spider monkey and marmoset, as well as the familiar little capuchin, or organ-grinder monkey. New World monkeys can be identified by their flattish faces, widely separated nostrils, and prehensile tails. Most are small in comparison to their Old World cousins.

<table>
<tr><th>Order</th><th>Suborder</th><th>Superfamily</th><th colspan="2">Common Names of Representative Forms</th></tr>
<tr><td rowspan="10">Primates</td><td rowspan="4">Prosimii</td><td>TUPAIOIDEA</td><td colspan="2">Tree shrew</td></tr>
<tr><td>LEMUROIDEA</td><td colspan="2">Lemur</td></tr>
<tr><td>LORISOIDEA</td><td colspan="2">Bush baby, Slender loris</td></tr>
<tr><td>TARSIOIDEA</td><td colspan="2">Tarsier</td></tr>
<tr><td rowspan="6">Anthropoidea</td><td>CEBOIDEA</td><td colspan="2">Howler monkey, Spider monkey, Capuchin, Common marmoset, Pinche monkey</td></tr>
<tr><td rowspan="2">CERCOPITHECOIDEA</td><td colspan="2">Macaque, Baboon, Wanderloo, Common langur, Proboscis monkey</td></tr>
<tr><td>Family</td><td></td></tr>
<tr><td rowspan="3">HOMINOIDEA</td><td>HYLOBATIDAE</td><td>Gibbon, Siamang</td></tr>
<tr><td>PONGIDAE</td><td>Orangutan, Chimpanzee, Gorilla</td></tr>
<tr><td>HOMINIDAE</td><td>Humans</td></tr>
</table>

The oldest fossils of New World monkeys occur in Oligocene beds of South America. Possibly Eocene precursors of this group managed to travel from Africa to South America on rafts of vegetation. Although the two continents were closer together in the Eocene, it would nevertheless have been a hazardous journey. Considering the probability of success, the late George G. Simpson coined the term "sweepstakes route" for this kind of migration.

The more advanced Old World monkeys, or **Cercopithecoidea**, are widely distributed in tropical regions of Africa and Asia. They include the familiar macaque, or rhesus monkey, of laboratories and zoos; the Barbary ape of Gibraltar; langurs; baboons; and

FIGURE 15-2 The ring-tailed lemur of Madagascar. (*Frans Lanting/Minden Pictures.*)

FIGURE 15-3 ***Tarsius syrichta*, a Phillipine arboreal prosimian with large, forwardly directed eyes.** (*Copyright Ron Austing/Photo Researchers, Inc.*)

mandrills. In this group of monkeys, the nostrils are close together and directed downward (as in humans), and the tail is not prehensile.

The anthropoid apes (Fig. 15–5) are tail-less primates. Modern species probably evolved from the same ancestral stock that produced humans. DNA evidence indicates the divergence probably took place between about 7 and 5 million years ago. In this regard, comparative analysis of the DNA of humans, chimpanzees, gorillas, and other apes, as well as similarities in such proteins as hemoglobin and myoglobin, indicate that the chimpanzee is our closest relative. There is a 98 percent correlation in DNA sequences between chimpanzees and humans.

The more primitive branch of the tail-less apes is composed of the **Hylobatidae**, best represented by those long-armed acrobats, the gibbons. Orangutans, chimpanzees, and gorillas are grouped within the **Pongidae**. The red-haired orangutan, or "wild man," of Borneo and Sumatra, is considerably larger than the

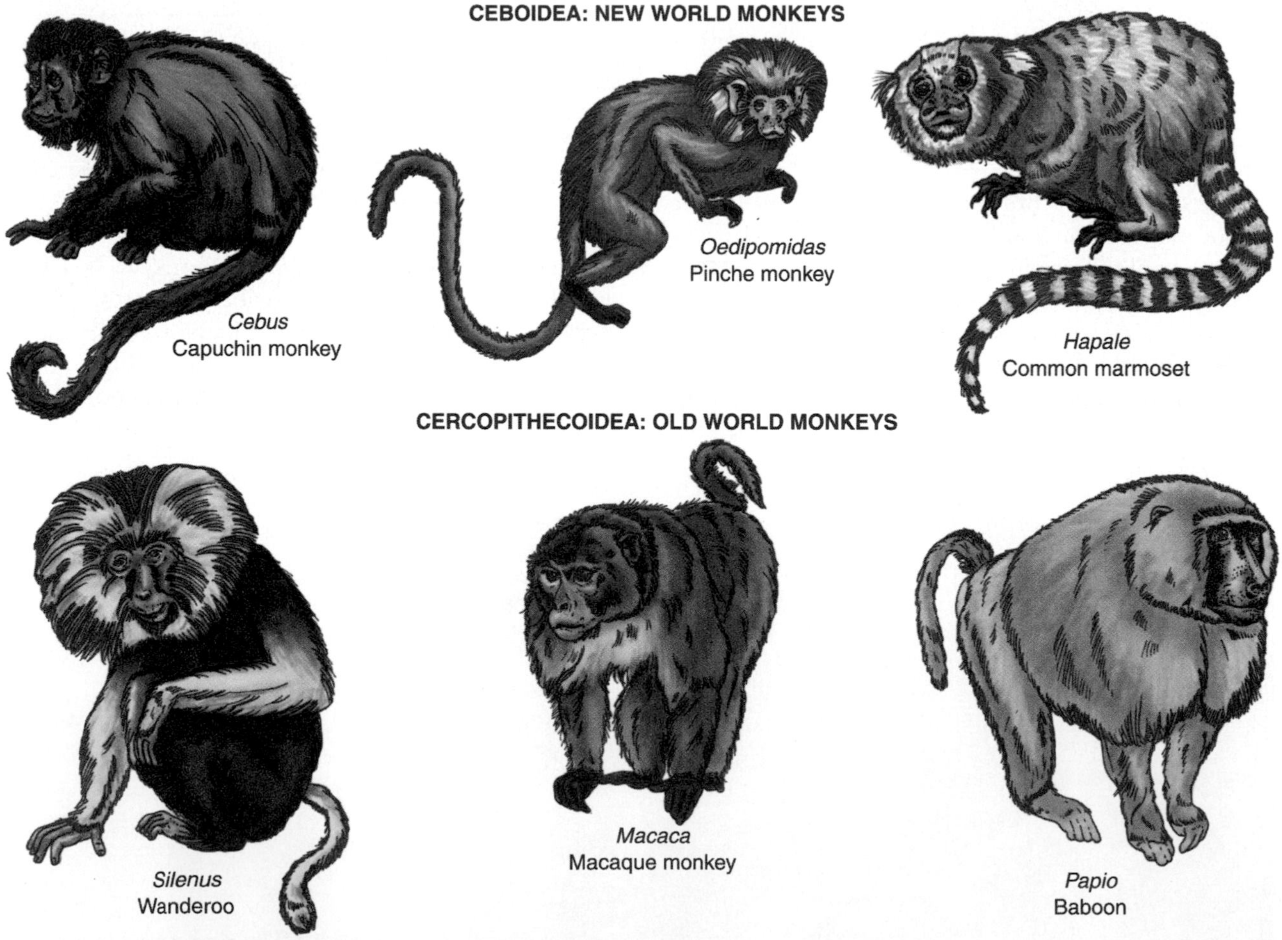

FIGURE 15-4 **Representative New World and Old World monkeys.** Most New World monkeys have prehensile tails that can function almost as effectively as another limb. (*From Cockrum, E. L., et al. 1966. Zoology. Philadelphia: W. B. Saunders Co.*)

A

B

C

D

FIGURE 15-5 The apes. (*A*) Gibbons are considered by many scientists as the most perfectly arboreal of the anthropoid apes. (*B*) Orangutans are solitary apes that seldom leave the protection of the trees. (*C*) A gorilla knuckle-walking. (*D*) Chimpanzees live in groups and have a complex social behavior. (A, *Copyright 1991 A. G. Nelson/Dembinsky Photo Associates*; B, *Tom McHugh/Photo Researchers, Inc.*; C, *John D. Cunningham/Visuals Unlimited*; D, *Copyright 1988, Z. Leszczynski/Animals Animals.*)

gibbon and has a larger brain as well. Chimpanzees and gorillas are similar in many ways and tend to be less pronounced in their arboreal adaptations. Gorillas are essentially ground dwellers and spend only a small fraction of their time in the trees.

THE PROSIMIAN VANGUARD

The early fossil record for primates includes a creature named *Purgatorius*, known only from a few teeth discovered in the Hell Creek Formation at Purgatory Hill in Montana. These finds indicate that the earliest primates were contemporaries of the last of the dinosaurs, at least in the tropical latest Cretaceous environments of North America. The fossil record for primates improves somewhat in the early epochs of the Cenozoic. *Plesiadapis* (Fig. 15–6), found in Paleocene beds of both the United States and France, has the distinction of being the only genus of primate (other than that of *Homo*) that has inhabited both the Old and New World. The presence of this prosimian on the now widely separated continents is one of many clues indi-

FIGURE 15-6 The Paleocene prosimian *Plesiadapis*.
? *What features of the skull of this prosimian are rodentlike?*

cating that the continents were not yet completely separated by the widening Atlantic in the Paleocene. *Plesiadapis* was a rather distinctive and specialized primate and represents a sterile offshoot of the primate family tree. The incisors were rodent-like and were separated from the cheek teeth by a toothless gap, or diastema. Fingers and toes terminated in claws rather than nails. The rodentlike characteristics and habits of the plesiadapiformes may have contributed to their extinction by late Eocene time. About that time, rodents were having their own radiation. Rodents are highly successful as gnawing, seed-eating animals and are able to reproduce at a rate which no prosimian can achieve.

General trends in prosimian evolution during the Eocene Epoch (55 to 37 million years ago) involved reduction in the length of the muzzle, increase in brain size, shifting of the eye orbits to a more forward position, and development of a grasping big toe. These trends are evident in the fossil remains of *Cantius* from the Wind River basin of Wyoming. In the Wyoming stratigraphic sequence, one can trace the evolution of successive species of *Cantius*. Near the top of the section, a new genus, *Notharctus* (Fig. 15–7), occurs as the apparently direct descendant of *Cantius*. *Notharctus* has been known from other localities for over a century. In general form, it appears to be just the kind of lemurlike animal from which monkeys and apes are derived. Prosimian populations during the Late Paleocene and Eocene also included tarsiers. *Tetonius* was an Eocene form with tarsioid traits of closely spaced large eyes and a shortened muzzle.

Both tarsiers and lemurs were abundant and widely dispersed on Northern Hemisphere continents during the Eocene. However, with the advent of cooler Oligocene climates, they virtually deserted North America. In the Eastern Hemisphere they were forced southward into the warmer latitudes of Asia, Africa, and the East Indies. Surviving prosimians are much reduced in variety and number. The prevailing opinion is that many have been replaced by the monkeys.

THE EARLY ANTHROPOIDS

The next step in the evolutionary advance of primates was the appearance of the first anthropoids. Discoveries in the Fayum region of Egypt have provided a wealth of information about the early anthropoids. The fossils there occur in several Oligocene horizons and include well over 100 specimens. Although many of the skull fragments and teeth retain subtle vestiges of prosimian ancestry, none of the remains are from prosimians. They are fossils of primates that have reached what may be called the monkey stage of organization. One primate discovered at Fayum is *Aegyptopithecus* (Fig. 15–8), a relatively robust arboreal primate with monkey-like limbs and tail, a brain larger than that of *Notharctus*, and eye orbits rotated to thc front of the skull. Bone fragments of *Aegyptopithecus zeuxis* have

FIGURE 15-7 The Eocene prosimian *Notharctus*. (*Illustration by Carlyn Iverson.*)

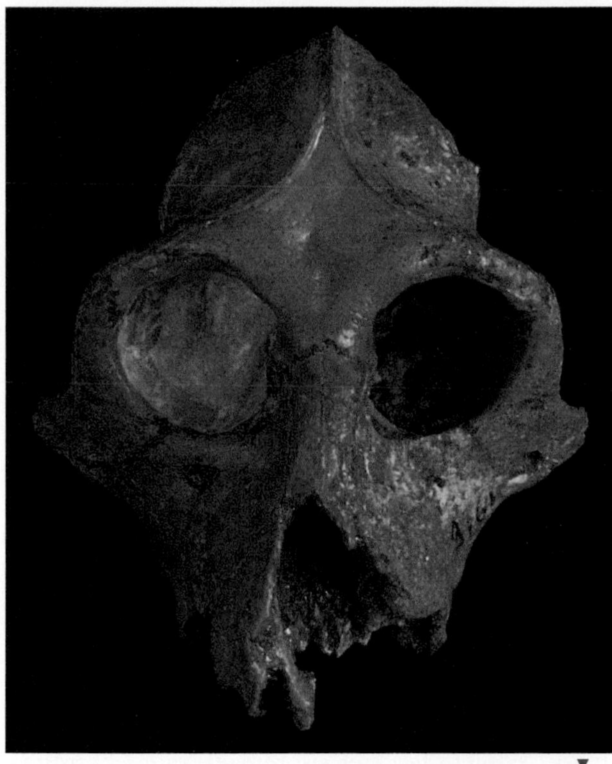

FIGURE 15-8 ***Aegyptopithecus zeuxis*** **from Oligocene fossil beds in the Fayum district of Egypt, near Cairo.** *Aegyptopithecus* is considered an early ape that preceded the Miocene apes exemplified by *Dryopithecus*. (*Courtesy of D. T. Rasmussen.*)

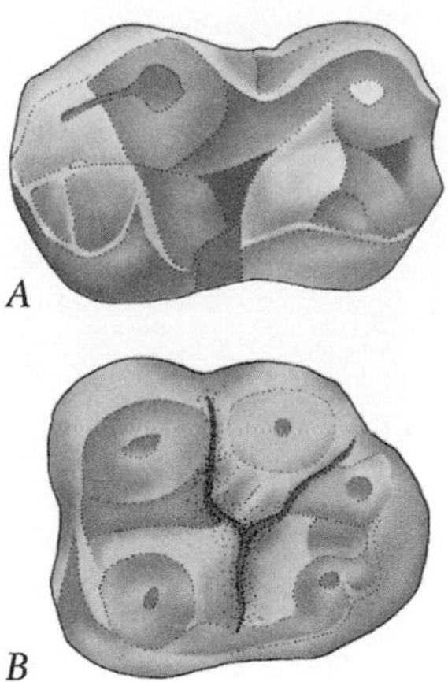

FIGURE 15-9 **General pattern of cusps on the molars of Old World monkeys.** (*A*) The lower molar of a baboon, showing a cusp at each corner. (*B*) The lower molar of a chimpanzee, showing the "lazy Y-5" pattern characterized by a Y-shaped depression that separates five cusps.

been dated as 33 to 34 million years old, indicating that the prosimian—anthropoid transition had taken place by Oligocene time. Egypt was not as arid during the Oligocene as it is now. It was well watered, and winding streams traversed lush tropical forests.

Because evolution is a continuing process in which each animal is a transitional link between older and younger species, it is a difficult and often arbitrary task to assign the fragment of a fossil jaw or tooth to either the monkey or the ape category. One of many clues used in attempting to make this distinction is the pattern of cusps on certain molar teeth. In Old World monkeys, specific molars have four cusps. Among apes (and humans), these same molars have five cusps, with an intervening Y-shaped trough. Molars belonging to *Aegyptopithecus* display the Y-5 pattern (Fig. 15–9), which is one reason anthropologists consider the genus to be the possible very early ancestor of Miocene apes. It seems safe to state that the Oligocene Epoch may have witnessed not only the transition from prosimians to anthropoids but also the differentiation of apes from monkeys.

During the Miocene, which lasted from 25 to 5 million years ago, plate tectonics had a significant influence on the evolution of primates. The unfragmented block containing Africa and Arabia was drifting northward and ultimately collided with Eurasia. (The Arabian portion of the continent then rotated counterclockwise to eventually produce the Red Sea.) As a result of these movements, east-west circulation of tropical currents across the former Tethys seaway was prevented, and East Africa became cooler and dryer. Extensive grass-covered savannas replaced former areas of dense forests. These environmental changes provided selective pressures that stimulated the adaptive radiation of Old World primates. During the Late Miocene, primates appear that gave rise to the pongids on the one hand and the hominids on the other.

Among the new players that came on the scene during the Miocene was a group called **dryomorphs** (formerly called dryopithecines). The dryomorphs take their name from *Dryopithecus fontani*, a species first discovered in France in 1856. The dryomorphs varied in size and appearance and include forms bearing different generic names from France, Spain, Greece, Hungary, Turkey, India, Pakistan, and Africa. Many anthropologists consider the important species *Proconsul africanus* to be either a very early dryomorph or the immediate dryomorph ancestor. *Proconsul africanus* was discovered by Mary and Louis Leakey in 1948 on an island in Lake Victoria, Kenya. The skull, jaws, and teeth of this small primate are ape-like (Fig. 15–10). Its rather long trunk, arms, and finger bones, however, are monkey-like. Much in the manner of modern monkeys, *Proconsul* probably scampered about on all four legs and lived on a diet of relatively soft fruit. The somewhat more advanced Middle Miocene dryomorph descendants of *Proconsul* are probably ancestral to both modern African apes

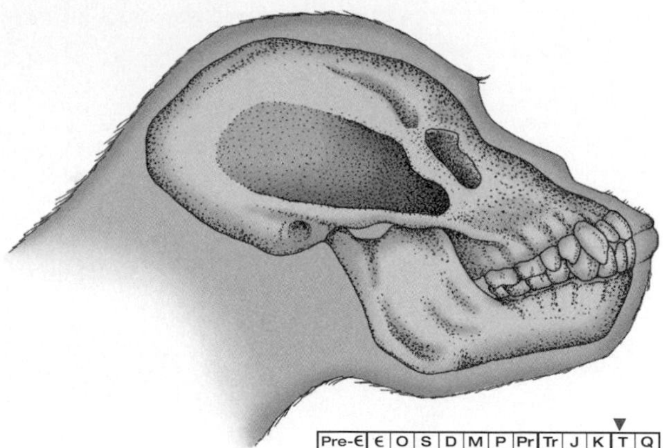

FIGURE 15-10 Skull of *Proconsul* (a dryomorph) from Lake Victoria, Kenya.

and the first hominids, the australopithecines. Among these more advanced, but still pre-hominid creatures, is the newly discovered genus *Equatorius*.

By about 18 million years ago, Africa had converged on Eurasia. Monkeys and apes migrated into Eurasia and diversified. Miocene apes, called ramamorphs, bearing such names as *Ramapithecus*, *Sivapithecus*, and *Gigantipithecus*, established themselves. These animals were more apelike than Proconsul.

The study of Miocene primates clearly indicates an absence of orderly sequential change along a single trend. The family tree of primates is a complex of parallel and diverging branches. It is more bush-like than tree-like. For this reason, the attempt to trace the ascent of humans up through the many bifurcations and dead ends is an exciting but difficult task. Paleontologists must be ready to reinterpret the story each time new fossil material is uncovered.

THE AUSTRALOPITHECINE STAGE AND THE EMERGENCE OF HOMINIDS

The story of the emergence of humans (hominids) begins with Raymond Dart's 1924 discovery of fossil remains of an immature primate in a South African limestone quarry. Dart named the fossil *Australopithecus africanus* (Fig. 15–11). In succeeding years, many additional skeletal fragments of species of *Australopithecus* have been found. In particular, fossil sites in eastern Africa have become increasingly important because of the veritable bonanza of hominid bones they have yielded. The new collections of fossils have given paleontologists an unsurpassed record of human evolution over the past 4 million years. Some of the eastern African sites, such as Olduvai Gorge (Fig. 15–12), are now famous as a result of the lifelong research programs of the Leakeys. Their efforts, and those of their son Richard, have provided fossils of relatively heavy-bodied, robust australopithecines as well as lighter types. Among the latter are remains of creatures that may well belong within our own genus.

Eastern Africa during the Cenozoic experienced a lively history of volcanic activity. As a result, many of the fossil sites have interspersed layers of volcanic ash and lava flows. This fact has been of great benefit to the paleoanthropologists, for samples of ash can often be dated by the potassium-argon method. Dates from two succeeding ash beds would then provide a close estimate of the age of fossils found in the intervening layers of sediment.

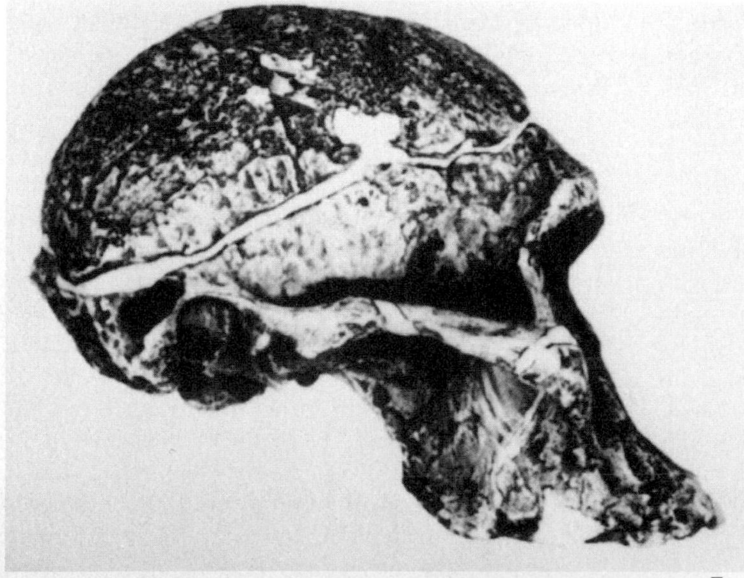

FIGURE 15-11 *Australopithecus africanus* from the Transvaal of South Africa. (*Photograph of Wenner-Gren Foundation replica by David G. Gantt.*)

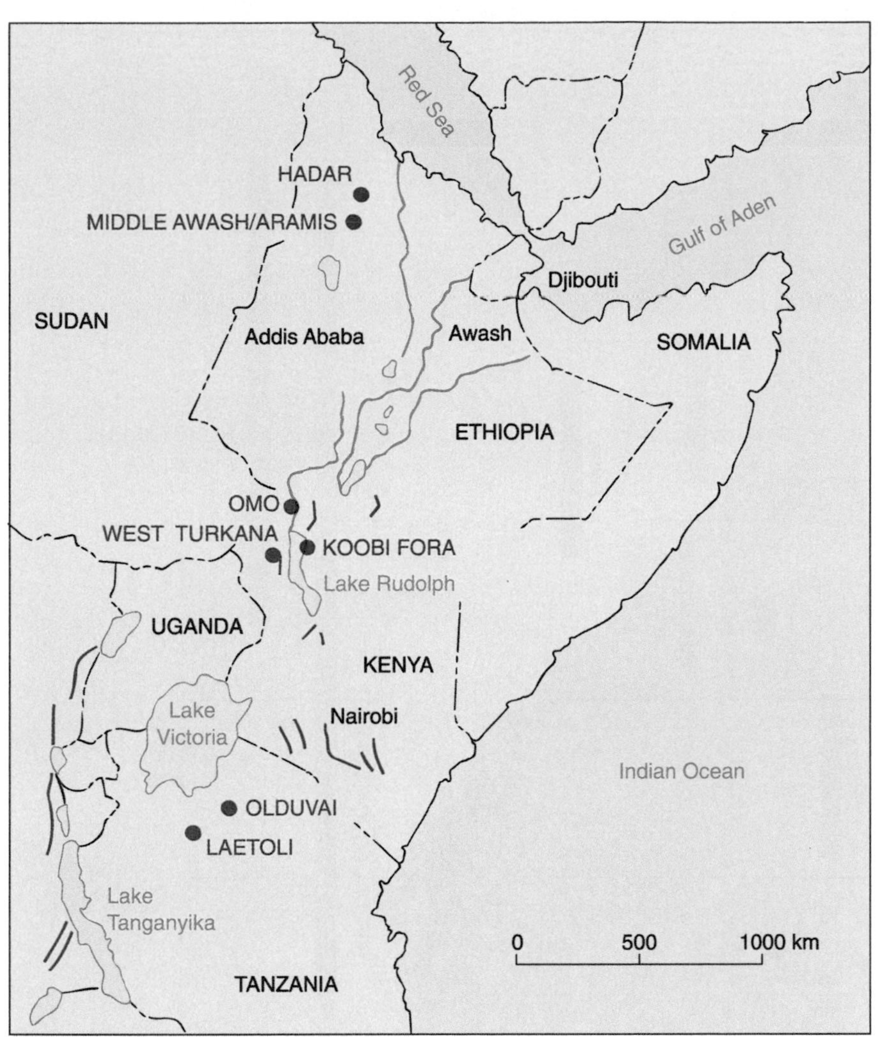

FIGURE 15-12 Location of some particularly rich hominid fossil sites in East Africa. *What is there about the geology and climate of this area that facilitates the discovery and dating of fossils?*

The oldest fossils of a hominid were recently discovered in the central African nation of Chad. Named *Sahelanthropus tchadensis*, this 4-foot tall primate lived 6 to 7 million years ago. *Sahelanthropus* was an unexpected discovery, for previously known early hominids were inhabitants of eastern and southern Africa. In addition, the Chad fossils are considerably older than the east African hominids. Prior to the discovery of *Sahelanthropus*, the earliest known hominid was the 5.8 to 5.2 million year old Ethiopian hominid named *Ardipithecus ramidus* (Fig. 15–13).

Specimens younger than *Ardipithecus ramidus* are being found with each new field season, and these discoveries have provided much new information about the complex, often controversial, and bushy phylogenetic tree of early hominids. Among the recent discoveries are over 20 specimens of *Australopithecus anamensis* from Kenya's great alkaline Lake Turkana (formerly Lake Rudolf). *Australopithecus anamensis* lived after *Ardipithecus ramidus*. Specimens range in age from 3.9 to 4.2 million years old. The species appears to be a good intermediate between *Australopithecus ramidus* and the famous fossil *Australopithecus afarensis* (Fig. 15–14) discovered by Donald Johanson and nicknamed "Lucy" after the popular Beatles song, "Lucy in the Sky with Diamonds." Since its original discovery in 1974, *Australopithecus afarensis* has been found at several eastern African sites. The additional specimens include both large and small individuals, prompting a debate as to whether they are different species or merely represent a single species in which the males were much larger than the females. Assuming it is a case of sexual dimorphism, the females were about 3½ feet tall and weighed about 60 pounds. Males were about a foot taller and weighed about 100 pounds. Skull shape in these hominids was more apelike than human, with the jaw thrust forward and no chin. Brain size was roughly comparable to that of a modern chimpanzee.

It is evident from the shape of the pelvic and leg bones in *Australopithecus afarensis* that this species walked erect. Further evidence of bipedalism came with the discovery of footprints in layers of volcanic ash

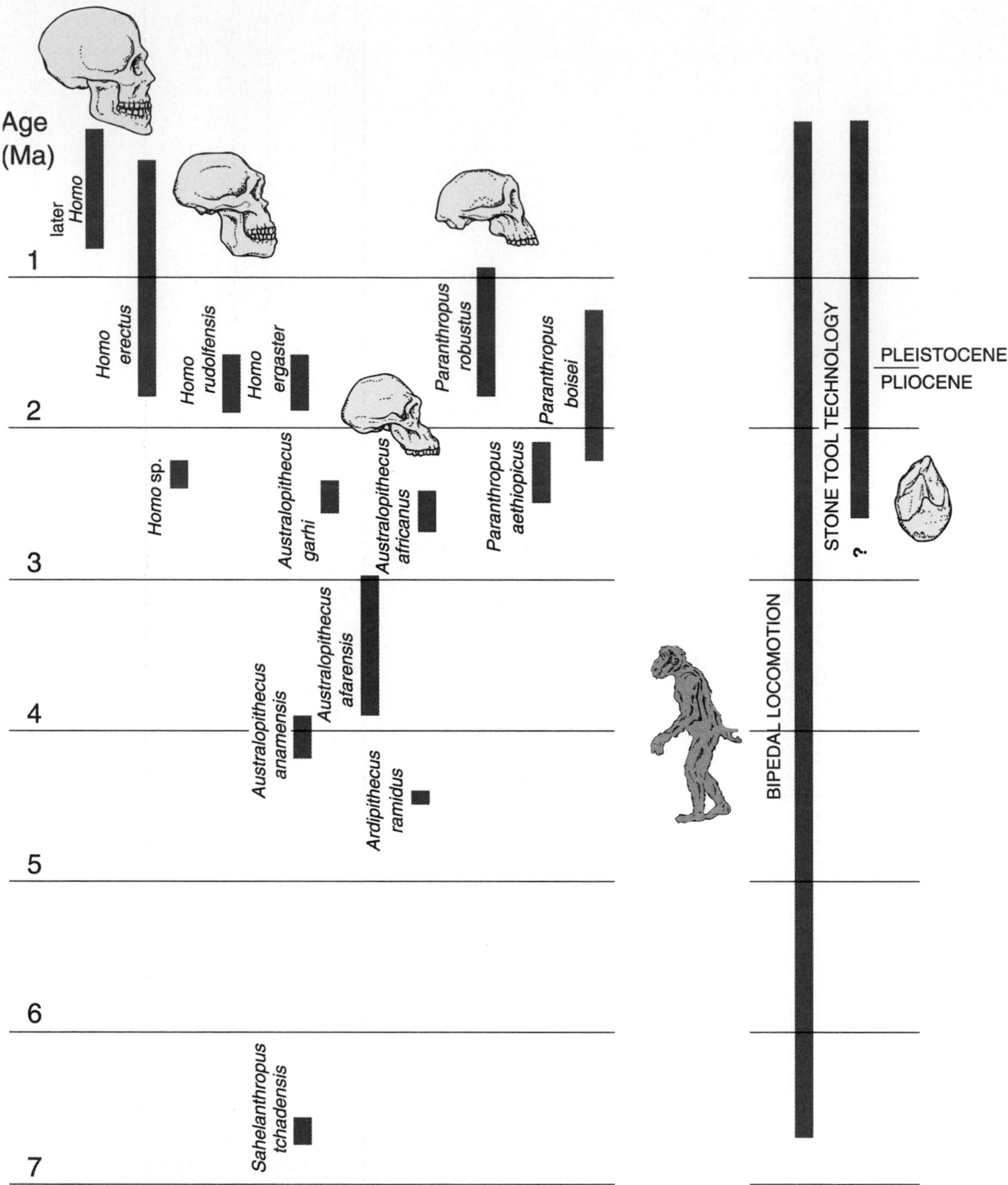

FIGURE 15-13 Chart showing the times during which hominids lived. *Sahelanthropus tchadensis*, discovered in 2002, is the earliest hominid presently known. *Sahelanthropus* lived in a region of central Africa which is now the nation of Chad. (*After Wood, B. 1996.* Human Evolution, Bioessays *18:945–954.)*

(tuff) at Laetoli (Fig. 15–15). The footprints can be traced over a distance of 9 meters and were made by two contemporaries of Lucy as they walked side by side over a layer of soft, moist, volcanic ash. (Subsequently, another ashfall formed a protective seal over the footprints.)

Eastern African fossil sites such as Koobi Fora and Omo have provided an extraordinary number of australopithecine specimens, and a few teeth and bones of *Homo* as well. The Omo digs are located along the Omo river in remote southwestern Ethiopia. Here one finds a nearly continuous section of sediments and

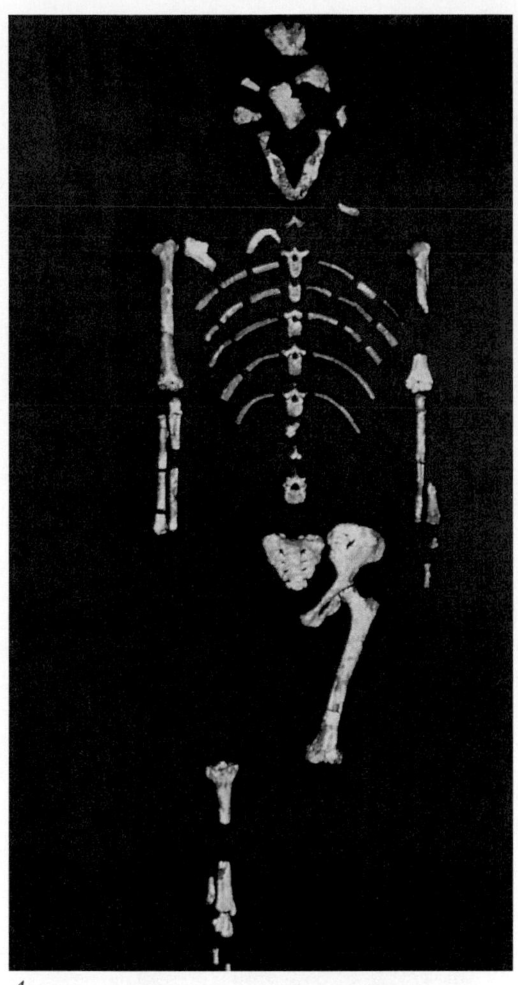
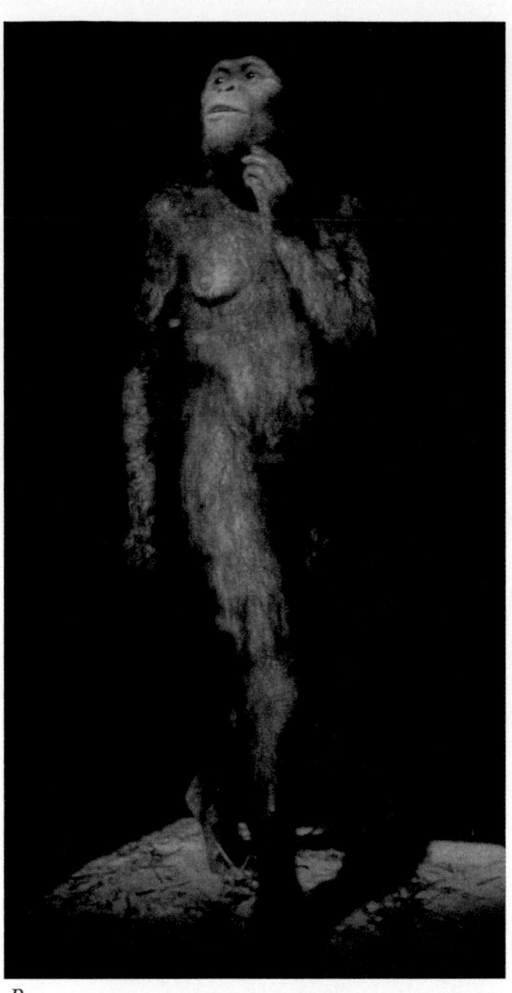

A *B*

FIGURE 15-14 ***Australopithecus afarensis*, informally known as "Lucy."** Lucy was a young, erect-walking hominid that lived in East Africa about 3.5 million years ago. The name Lucy for the original discovery may be inappropriate, for anthropologists at the University of Zurich have recently proposed that Lucy may have been a male. (*A*) The specimen shown here is the oldest known, most complete skeleton of an erect-walking human ancestor. (*B*) A reconstruction of Lucy on display at the St. Louis Zoo. (A, *Courtesy of the Cleveland Museum of Natural History*; B, *St. Louis Zoo.*)

volcanics that span 2.2 million years. The Koobi Fora locality is on the eastern side of Lake Turkana in Kenya (Fig. 15–16). It has been investigated by Richard Leakey and Glynn Isaac. Altogether, the Omo, Koobi Fora, Olduvai, Hadar, and Laetoli localities have given us sufficient skulls, jaws, teeth, and other bones to indicate that the australopithecines were not an entirely homogeneous group. However, certain shared characteristics appear in nearly all of the specimens unearthed. From the structure of their pelvic girdles, it is known that they stood upright in a fashion more human-like than apelike. Australopithecine dentition was essentially human, although the teeth were more robust than in modern humans. In contrast to these human-like characteristics, the australopithecine cranial capacity of about 600 cm^3 was far less than the 1400 cm^3 to 1600 cm^3 measured in modern humans.

In the past, paleoanthropologists have distinguished two kinds of australopithecines. Those with lighter, smaller teeth and somewhat smaller body size, such as *Australopithecus afarensis*, were termed gracile, as opposed to heavier, larger-toothed, robust types. The robust australopithecines include *Australopithecus robustus* (Fig. 15–17) and *Australopithecus boisei*. Both of these species appear to represent sterile side branches of hominid evolution that did not give rise to known descendants. Cheek teeth in these robust types had enormous grinding surfaces, suggesting a diet that included coarse plant food.

A *B*

FIGURE 15-15 Evidence of bipedalism. (*A*) Footprints probably made by *Australopithecus afarensis* about 3.2 million years ago, as a pair of these hominids walked across wet volcanic ash. The footprints confirm skeletal evidence that the species had a fully erect posture. (*B*) Lucy and a male companion leave telltale footprints in wet volcanic ash as they stroll across the Late Pliocene landscape of eastern Africa. (A, *Copyright John Reader/Science Photo Library/Photo Researchers, Inc.*; B, *American Museum of Natural History, with permission.*)

FIGURE 15-16 The Koobi Fora site in eastern Africa, which has provided an exceptional fossil record of some of the steps in hominid evolution. Not only have many australopithecine remains been found here but also the skull of an early member of the genus *Homo*. In addition to the hominid remains unearthed at Koobi Fora, paleontologists found pebble choppers and flake tools associated in deltaic deposits with the bones of a hippopotamus. The position of both the tools and hippo bones suggests that this was a butchering site. Apparently one day about 1.8 million years ago, some australopithecines came across the body of a hippopotamus that had recently died. They feasted on its carcass, using the sharp edges of flaked chert to get at the meat. (*Courtesy of K. Behrensmeyer.*)

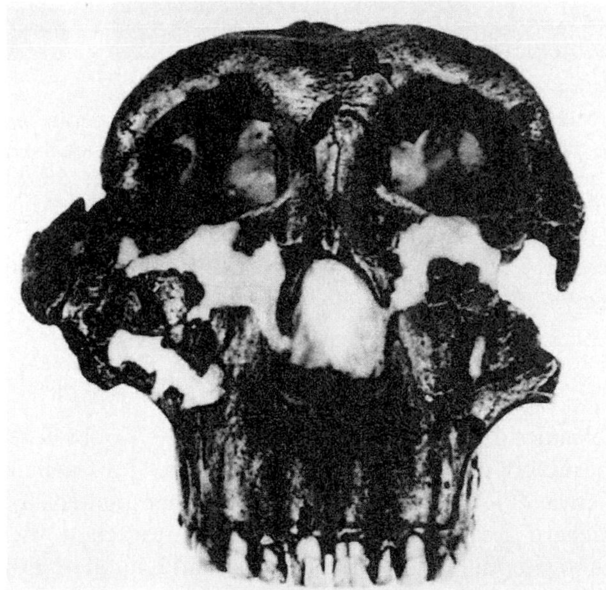

FIGURE 15-17 *Ausralopithecus robustus.* Because of the observation that these hominids had larger molars and premolars than did *Australopithecus africanus*, some paleoanthropologists believe they were primarily plant-eaters and were taxonomically distinct from *A. africanus*. Recent interpretations favor inclusion of these creatures under the generic name *Australopithecus*. This specimen was derived from Bed I on the south side of the main ravine at Olduvai Gorge. (*Photograph of Wenner–Gren Foundation replica by David G. Gantt.*)

The transition from gracile australopithecines to hominids of the genus *Homo* is not marked by striking anatomic differences. As compared to australopithecines, early species of *Homo* exhibited increased height of the cranial vault and a somewhat larger cranial capacity. The opening at the base of the cranium through which the spinal cord joins the brain (the foramen magnum) had a slightly more anterior position. Premolars were narrower, and the length of the row of cheek teeth was reduced. Of particular significance, crude stone tools are found in association with fossils of *Homo*.

The oldest remains of *Homo* are nearly 2.5 million years old. Jaw fragments and teeth occur near stone tools of the same age. By about 2 million years ago, *Homo rudolfensis* (Fig. 15–18) and *Homo ergaster* (also called *Homo habilis*) were roaming the savannas of Africa. What environmental conditions may have triggered the evolution of *Homo* from *Australopithecus*? There is now ample evidence that about 2.7 million years ago, Africa began to grow cooler and drier. As a result, rainforests gave way to broad expanses of grasslands. In the more open and difficult environment, hominids would have experienced selective pressures that eventuated in improved bipedalism, better cerebral abilities, and the development of a stone tool technology.

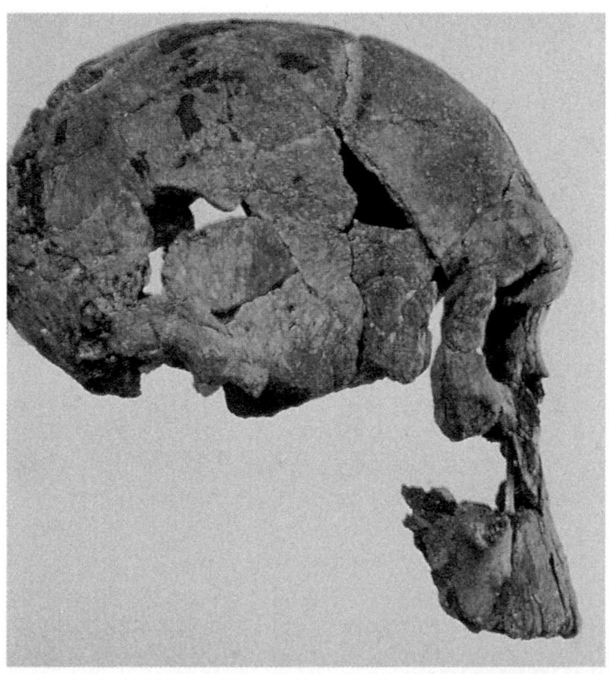

FIGURE 15-18 Lateral view of a nearly complete skull of the Koobi Fora specimen KNM-ER 1470, which has been assigned the name *Homo rudolfensis*. From front to rear, the braincase measures about 15 centimeters. (*Courtesy of the National Museums of Kenya.*)

THE *HOMO ERECTUS* STAGE

The next stage in hominid evolution is represented by *Homo erectus*, formerly known as *Pithecanthropus erectus* from Java and *Sinanthropus pekinensis* (Fig. 15–19) from China. *Homo erectus* is the first hominid known to have moved out of Africa into Eurasia. Noteworthy discoveries include a 1.8-million-year-old lower jaw with teeth from the Caucasus Mountains of eastern Georgia, a largely complete 1.5-million-year-old skeleton from the western Turkana fossil site, and a 750,000-year-old skull cap from Bed II of Olduvai Gorge (Fig. 15–20). If the dating of the Georgian fossil is correct,

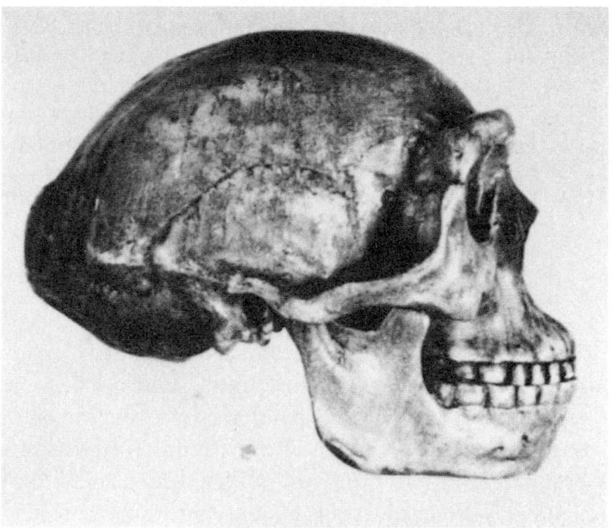

FIGURE 15-19 Replica of *Homo erectus*, previously known as *Sinanthropus pekinensis*. (*Photograph of Wenner–Gren Foundation replica by David G. Gantt.*)

FIGURE 15-20 Upper portion of skull of a specimen of *Homo erectus* from Olduvai Gorge. This specimen was found associated with stone implements of the type known as Chellan. (*Photograph of Wenner–Gren Foundation replica by David G. Gantt.*)

ENRICHMENT

Good and Bad News About Being Upright

Humans are unique among primates by having a body that is structured for standing fully erect and for walking on two legs. Our nonhuman ancestors, however, walked on all fours, and the evolutionary transition from those ancestors to hominid bipedalism has not been without problems. The human vertebral column, for example, is somewhat imperfect as a vertical support. That imperfection is one reason why so many people are plagued by lower back pain. Under the weight of the upper body, and provoked by lifting and unusual movements, intervertebral discs may become herniated and protrude, causing pressure against nerve structures. Severe pain and disability may follow, the treatment for which often requires delicate neurosurgery. In addition, evolutionary changes associated with our upright stance have increased the distance between the lowest ribs and the top of the pelvis. This has given us our distinctive waist, but it has also resulted in a weakening of the abdominal wall. As a result, humans are particularly prone to hernias (ruptures).

Although these conditions affect many of us at various times during our lives, the advantages of erect posture nevertheless far outweigh the disadvantages. An upright stance and bipedalism in human evolution freed the hands for the kind of precise manipulation that led to the manufacture of tools and the advent of technology. For humans, hands are not merely motor organs. They are also sensory organs that can investigate by touch, "seeing" into places the eye may not be able to reach. Hands explore and react to what has been discovered. They can gesture, give instruction, indicate directions, and even express emotion. If we merely walked on our hands, what chance would we have had to evolve the complex human brain that provides for abstract thought, symbolic communication, and the development of culture?

then *Homo erectus* must have found his way into Europe far earlier than previously thought. The remarkable discovery of *Homo erectus* from western Turkana has been dubbed the "Turkana Boy." Apparently this unfortunate 11- to 13-year-old perished in a marsh and was quickly covered with mud, preventing dismemberment of the body by scavengers. The skeleton reveals a long-limbed, tall, and slender individual. Adults would be about the same size or larger than many natives of eastern Africa today. Below the neck, the skeleton was remarkably modern.

Although the axial skeleton and limbs of *Homo erectus* were quite similar to those of modern humans, the skull was not. Cranial capacity ranged from about 775 cm^3 to nearly 1300 cm^3. In comparison, brain capacity in modern *Homo sapiens* ranges from 1200 cm^3 to 1500 cm^3, and thus the brain size of *Homo erectus* overlaps at least the lowermost range of modern peoples. Most would agree that *Homo erectus* represents a stage in the evolution of hominids during which relatively rapid increases in brain size had begun. No doubt the expansion of the brain involved the reshaping not only of the cranium but also of the birth canal. Selection would very likely have favored large pelvic inlets to accommodate fetuses with larger heads and brains.

Aside from its larger cranial capacity, the skull of *Homo erectus* was massive and rather flat. Heavy, bony, supraorbital ridges existed over the eyes (see Fig. 15–19). The forehead sloped, and the jaw jutted forward at the tooth line in a condition termed **prognathous**. A jutting chin was lacking, and the nose was broad and flat. These were rather primitive traits. However, except for their being somewhat more robust, the teeth and dental arcade in *Homo erectus* were essentially modern.

There is ample evidence that *Homo erectus* made good use of his larger brain. From the bones of other animals found at living sites, it is clear that these hominids were excellent hunters. They were also skilled at making simple implements of flint and chert (Fig. 15–21), such as axes and scrapers. Some also appear to have engaged in cannibalism, either for food or as a part of a ritual system. Paleoanthropologists do not know if *Homo erectus* spoke a distinctive language, wore clothes, or built dwellings. There is vague evidence that they hunted together in bands. What appear to be slaughter sites have been found in Europe. Unfortunately, these sites did not yield any bones of *Homo erectus*. A few fossil localities in Europe and China contain traces of carbon. However, this cannot be taken as indisputable evidence that *Homo erectus* had learned to use fire, for the carbon may have been the product of naturally occurring brushfires.

Fossils of *Homo erectus* of differing age and location show a degree of variability. Some paleoanthropologists consider this reason enough to divide *Homo erectus* into two species, *Homo erectus* and *Homo ergaster*. Others, however consider the variability no greater than that which exists today among members of our own species.

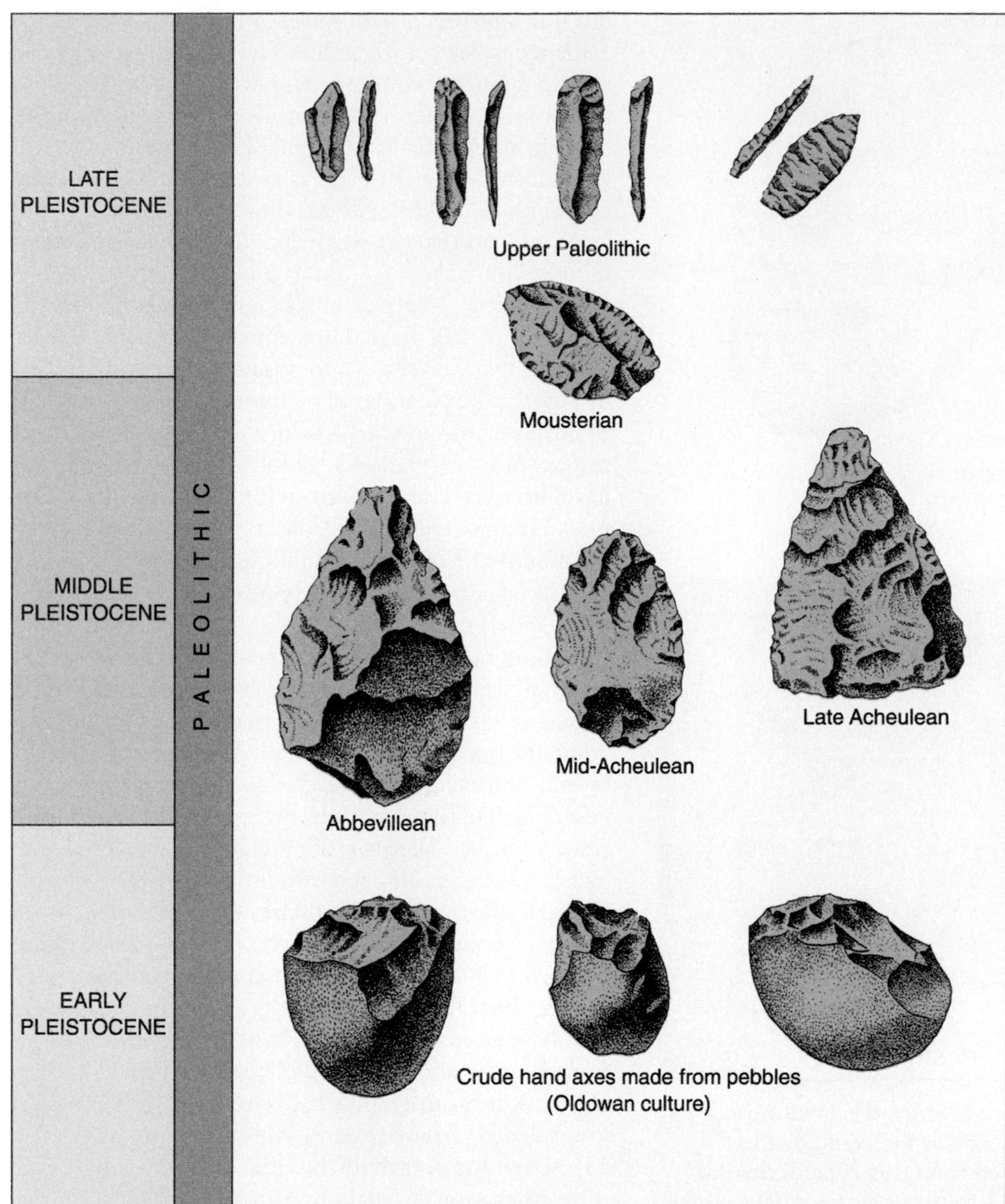

FIGURE 15-21 Progressive improvement in making tools from stone during the Pleistocene. The crude stone tools of the Early Pleistocene were produced by australopithecines. *Homo erectus* produced the better-shaped tools of the Middle Pleistocene. The Upper Paleolithic tools included carefully chipped blades and points. The next stage, not shown here, would be the Neolithic, characterized by refined and polished tools of many kinds.

FINAL STAGES OF HUMAN EVOLUTION

The Neandertals

From the *Homo erectus* stage of the Middle Pleistocene it is only a short step to Late Pleistocene hominids termed Neandertals (Fig. 15–22). Some paleoanthropologists have regarded the Neandertals as a subspecies of *Homo sapiens*, designated *Homo sapiens neanderthalensis*. Recent DNA analysis of Neandertal bones more than 30,000 years old, however, indicate that neandertals have quite different DNA than *Homo sapiens*. This DNA evidence supports the view that Neandertals were a separate species, different enough that they

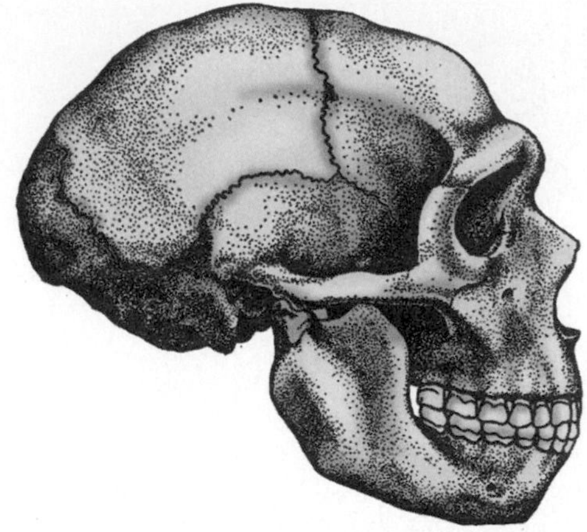

FIGURE 15-22 Right lateral view of the skull of a classic Neandertal.

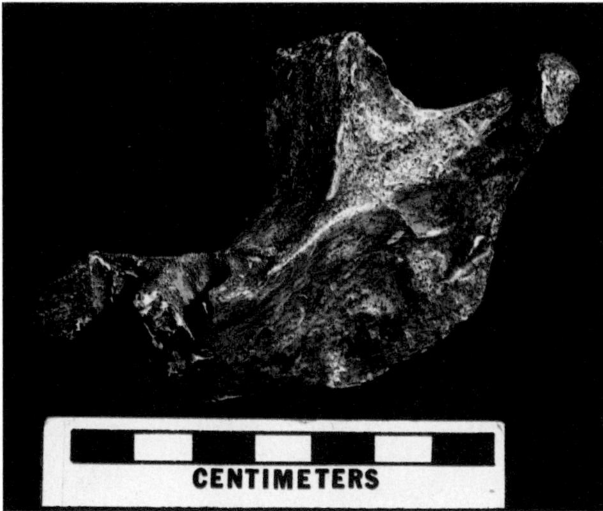

FIGURE 15-23 Neandertals roamed central Europe as recently as 28,000 years ago, as indicated by the radiocarbon date obtained from this Neandertal jaw bone from a cave site in Croatia. (*Courtesy of the Croatian Academy of Sciences.*)

did not interbreed with modern humans. If the DNA evidence is correct, then Neandertals represent a dead end in human evolution, and were not ancestral to anatomically modern humans. Radiocarbon dating of remains in Croatia indicate that Neandertals were still present in central Europe as recently as 28,000 years ago (Figure 15–23). Thus, there were thousands of years of co-existence beween Neandertals and early modern humans.

The initial Neandertal specimen was found in the Neander Valley near Dusseldorf, Germany. Subsequently, many additional fossils have been found to indicate that the Neandertal people ranged across the entire expanse of the Old World. With their heavy brow ridges and prognathous, chinless jaws, Neandertals have become the very personification of the "cave man." Indeed, the face seems a brutish carryover from the Middle Pleistocene. In addition to the protruding mid-facial region, the pubic bone of the pelvis is longer in Neandertals.

In most other features of the postcranial skeleton, Neandertal skeletons were similar to our own, and their brain size equaled or exceeded that of present humans. Thus, the once popular depiction of Neandertals as bent-kneed, flatfooted, bullnecked brutes with curved backs is false. Much of that early interpretation was derived from restoration of a skeleton of an elderly Neandertal who had been severely afflicted with osteoarthritis.

Most of the so-called classic Neandertals were sturdy people of small stature. They appear to have adapted well to life in the cold climates near the edge of the ice sheet. Relatively short limbs and a bulky torso may have been an advantage in helping them conserve body heat. In this regard, their enlarged nasal cavities may have been an adaptation for warming frigid air as it was inhaled. Many Neandertals lived in caves (Fig. 15–24) and successfully hunted contemporary cold-tolerant mammals, including cave bears, mammoths,

FIGURE 15-24 Reconstruction of a Neandertal family group. (*Courtesy of the U.S. National Museum of Natural History, Smithsonian Institution.*)

ENRICHMENT

Neandertal Ritual

The Neandertals lived from about 125,000 to about 40,000 years ago. Their culture included chipped flint tools, crude carvings, the use of fire, and deliberate burial of their dead in carefully prepared graves. Some of the burial sites contain indications that these ancient people may have had religious beliefs. The La Chapelle fossil site in southwestern France, for example, contains the body of a Neandertal male placed in ritual position within a shallow grave with the leg of a bison on his chest. Flint tools were also placed in the grave, possibly in the belief that they could be used in an afterlife.

At the Ferrassie Neanderthal site in France, an apparent family cemetery was discovered within a rock shelter. Two adults, presumed to be a father and mother, are buried head to head. Nearby, three small children and a newborn infant are buried. Once again, flint flakes and bone splinters were placed in the grave of the adult male, and a heavy, flat stone was placed over his head and shoulders. Perhaps the stone was to protect the body, or possibly to restrain the deceased from returning to life.

That Neandertals had a special attitude about death is also evident at a fossil site known as Teshik-Tash in Uzbekistan. Anthropologists working at Teshik-Tash uncovered an array of goat horns surrounding the buried body of a 9-year-old child.

Yet another particularly interesting Neandertal fossil site is located in Shanidar, Iraq. The remains of nine Neandertals have been uncovered at Shanidar, and four of these were deliberately buried. One of the burials contains an assortment of pollen grains, suggesting that flowers may have been placed in the grave. Although not related to funerary ritual, the Shanidar site includes the remains of a 30-year-old male who had been born with a crippled arm. The arm had been amputated below the elbow, and this hardy Neanderthal survived the surgery. The excessive wear on his teeth suggests that he used them to compensate for the missing limb.

wooly rhinoceroses, reindeer, bison, and the fierce ancestors of modern cattle known as aurochs. They were not devoid of culture and manufactured a variety of stone spear points, scrapers, borers, knives, and saw-edged tools. In addition, Neandertals made ample use of fire and apparently could ignite one at will in the hearths they excavated in the floors of caves.

The use and control of fire had many advantages for Pleistocene hominids. It provided light in caves that otherwise would have been avoided because of the near helplessness of humans in deep darkness. It gave warmth to creatures that had evolved in the tropics but had moved northward into colder realms, and it gave protection at night from predators. During the frigid winters of the Ice Age, fire provided the means to thaw meat that had become quickly frozen after a kill. Conceivably, cooking may have been a simultaneous result of the thawing process. Cooked foods provided the added advantages of promoting easier digestion and destroying harmful microorganisms.

In addition to their use of fire, Neandertals constructed shelters of skins, sticks, and bones in areas where caves may not have been available. That they also cared for the sick, pondered the nature of death, and believed in an afterlife is suggested by their custom of burying artifacts along with the dead. This is not to say that they were a kind and gentle folk. A cave in southwestern France contained the bones of at least six Neandertals than had been deliberately butchered. Could cannibalism have been widespread among Neandertals, and if so, was it a regular habit or did it occur only during episodes of famine?

Cro-Magnon

About 34,000 years ago, during the fourth glacial stage, humans closely resembling modern Europeans moved into regions inhabited by the Neandertals and in a

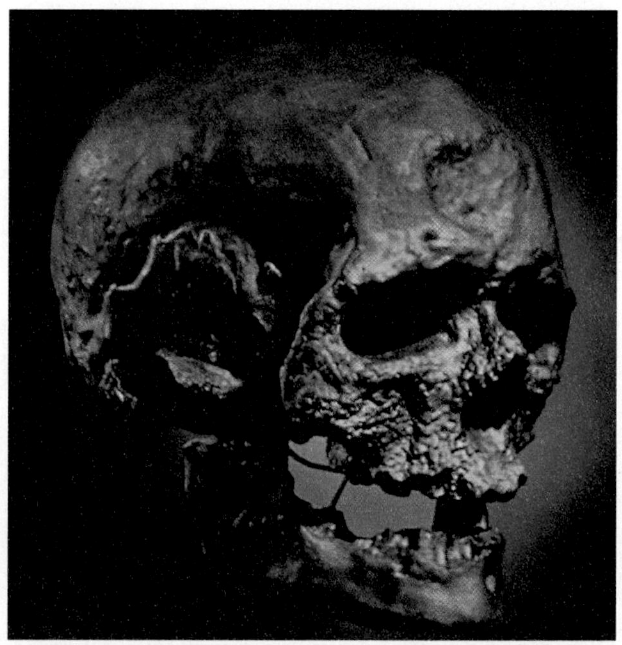

FIGURE 15-25 Skull of a Cro-Magnon human dated at about 30,000 years old. As in present-day humans, the face of Cro-Magnon was vertical rather than projecting. The heavy supraorbital ridges of the European Neandertals are not present, and there is a prominent chin. (*Photograph of Wenner–Gren Foundation replica by David G. Gantt.*)

FIGURE 15-26 A Cro-Magnon hunters. *(Field Museum of Natural History, Chicago.)*

short span of time completely replaced them, probably by tribal warfare and by competition for hunting grounds. These early members of our own species are informally dubbed **Cro-Magnon** (Figs. 15–25 and 15–26). They were mostly taller than Neandertals, had a more vertical brow, and had a decided projection to the chin (Fig. 15–27*C*). In short, Cro-Magnon's bones were modern, and anthropologists have recognized definite Cro-Magnon skull types among today's western and northern Europeans, as well as North Africans and native Canary Islanders.

The cultural traditions of the Neandertals appear to have been continued and further developed by Cro-Magnon. The variety and perfection of stone and bone tools were increased. Finely crafted spear points, awls, needles, scrapers, and other tools are found in caves once inhabited by these people. The walls and ceilings of these caves also retain splendid paintings and drawings. Carvings and sculptures of obese women (Fig. 15–28), probably used in fertility rites, were produced from fragments of bone or ivory, as were small, elegant engravings and statues of mammoths and horses. Cro-Magnon people enjoyed wearing body ornaments and frequently fashioned necklaces from pieces of ivory, shell, and teeth. Burial of the dead became a truly elaborate affair. Hunters were buried with their weapons and children with their ornaments. This apparent concern for an afterlife and the sense of self-awareness that resulted in art and complex ritual suggest that the beginning of the age of the philosopher had arrived.

Through most of his early history, *Homo sapiens* was a wandering hunter and gatherer of wild edible plants. However, about 15,000 to 10,000 years ago, near the beginning of the Holocene Epoch, tribes began to do-

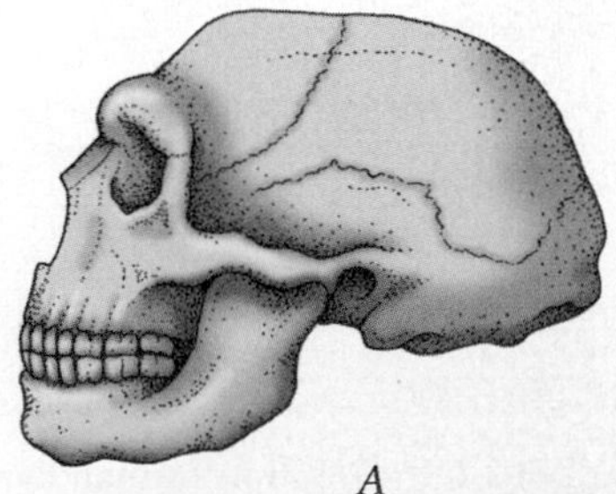

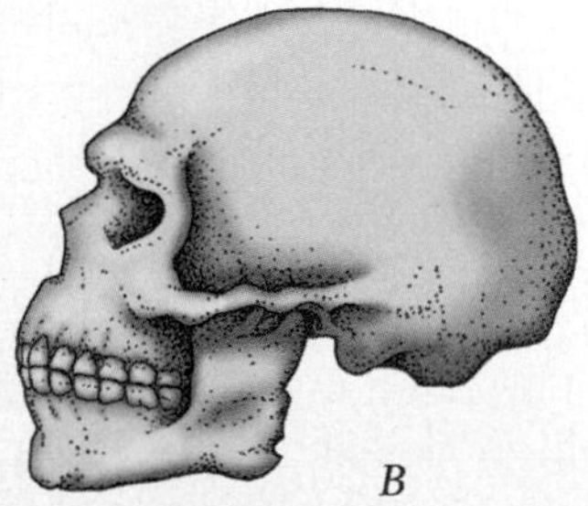

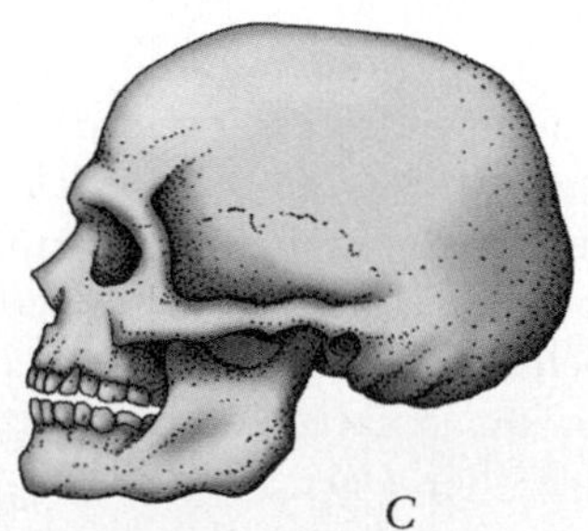

FIGURE 15-27 Comparison of skulls. (*A*) Neandertal; (*B*) skull from a rock shelter on the slope of Mount Carmel that appears to show both Neandertal and Cro-Magnon features; and (*C*) Cro-Magnon. The Mount Carmel skull is intermediate in both form and age between Neandertal and Cro-Magnon.

A

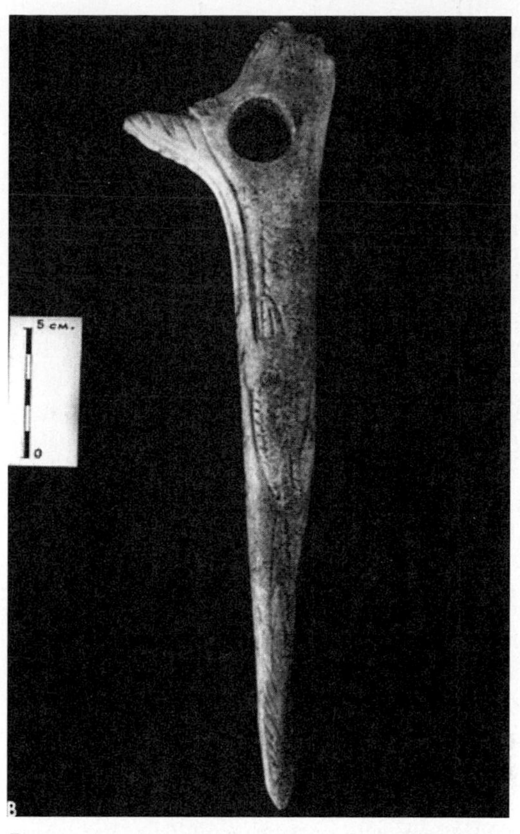

B

FIGURE 15-28 Prehistoric art of Late Pleistocene *Homo sapiens*. (*A*) Venus figure originally found in Austria ("The Venus of Willendorf"). (*B*) Thong-stropper used either to work hide thongs or to straighten arrow shafts. The tool is made from an antler and is intricately carved. (*Courtesy of Musée de l'Homme, Paris.*)

mesticate animals and cultivate plants. They learned to grind their tools to unprecedented perfection and to make utensils of fired clay. With more reliable sources of food, permanent settlements were developed, and individuals, spared the continuous demands of searching for food, were able to build and improve their cultures. Languages improved, and symbols developed into forms of writing. The era of recorded history had begun.

HUMANS ARRIVE IN THE AMERICAS

The fossil record for hominids in the New World is discontinuous and often ambiguous. More often than is desirable, anthropologists must trace human presence by discarded fragments of tools or weapons rather than actual skeletal remains. These clues, however, suggest that a few bands of humans were able to cross the Bering Straits and enter into Alaska during the late phase of the Wisconsin glacial stage. Earlier, mammoth, reindeer, dire wolves, bears, and other mammals had used the same route. Some of these animals provided food for bands of migrating humans. As glaciated areas continued to melt, a large ice-free corridor (the Canadian Plains Corridor) was opened between the still glaciated Rocky Mountains and the retreating continental glacier that had covered much of Canada. Tribal groups were able to migrate southward along the corridor. Precisely when these migrations occurred is difficult to determine and often hotly debated. Very crude flints, tentatively reported to be artifacts, have been found in alluvial fan sediments in California and may be around 30,000 years old. Charred bones of a dwarf mammoth believed to have been deliberately roasted in a campfire have been found on Santa Rosa Island off the coast of California. Carbon-14 dating indicates that the bones are about 28,000 years old. At Tule Springs, Nevada, fire pits determined to be 24,000 years old have been found in questionable association with flaked flints. In Mexico, an obsidian blade was discovered under a 23,000-year-old log.

There are a number of more recent sites that yield indirect evidence of **paleoindians**. Projectile points are found in definite association with extinct Late Pleistocene elephants and bison in the western United States. Sites near Clovis and Sandia, New Mexico, have yielded the distinctive spear points and tools of two separate cultures that may have lasted from about 13,000 to 11,000 years ago. A somewhat younger group of paleoindians lived from about 11,000 to 9000 years ago. This group has been named the Folsom Culture. Folsom people manufactured short, finely flaked points that were mounted on shafts. These flints have been found in kill sites in association with extinct species of bison.

The search for skeletal parts of early humans in America has provided fewer discoveries than has the search for their artifacts. However, there have been a few significant finds. A skull unearthed near Laguna Beach, California, has been dated by radiocarbon techniques as between 15,680 and 18,620 years old. Charred human bones from a cave on Marmes Ranch in southeastern Washington State have been dated at 11,500 to 13,000 years old. A projectile point, a bone needle, and the roasted bones of rabbits, deer, and elk were discovered in association with the remains of "Marmes Man."

Evidence that humans had made their way all the way down into southern South America includes 12,600-year-old jaw bones and teeth from Guitarrero Cave in northern Peru, as well as pieces of wood and stone from Monte Verde in southern Chile that were clearly shaped by human hands.

POPULATION GROWTH

One of the many lessons of historical geology is that the advent of *Homo sapiens* represents only one recent and momentary event along the sinuous, branching, 500-million-year evolution of vertebrates. We humans are linked by similarities of structure and body chemistry to lungfish struggling out of stagnating Devonian pools, to small, shrewlike precursors of the primate order, and to axe-carrying hunters that wandered along the margins of great ice sheets. The descendants of those and other Pleistocene hunters have emerged as the dominant species of higher life presently on this planet. Like other animals, *Homo sapiens* has been shaped by the combined powers of genetic change and environment. However, our species has quickly become a pervasive force in modifying the very physical and biologic environment from which it was shaped. Humanity has come to rely heavily on science to improve its lot, but has had great difficulty in finding ways to manage the resulting technology and to control the burgeoning problems arising from too few resources for too many humans.

About 10,000 years ago (a mere moment to a geologist), there were more than 6 million humans on the Earth. By the year A.D. 1, the number had jumped 50-fold to 300 million. In 1970, the figure stood at 3.6 bil-

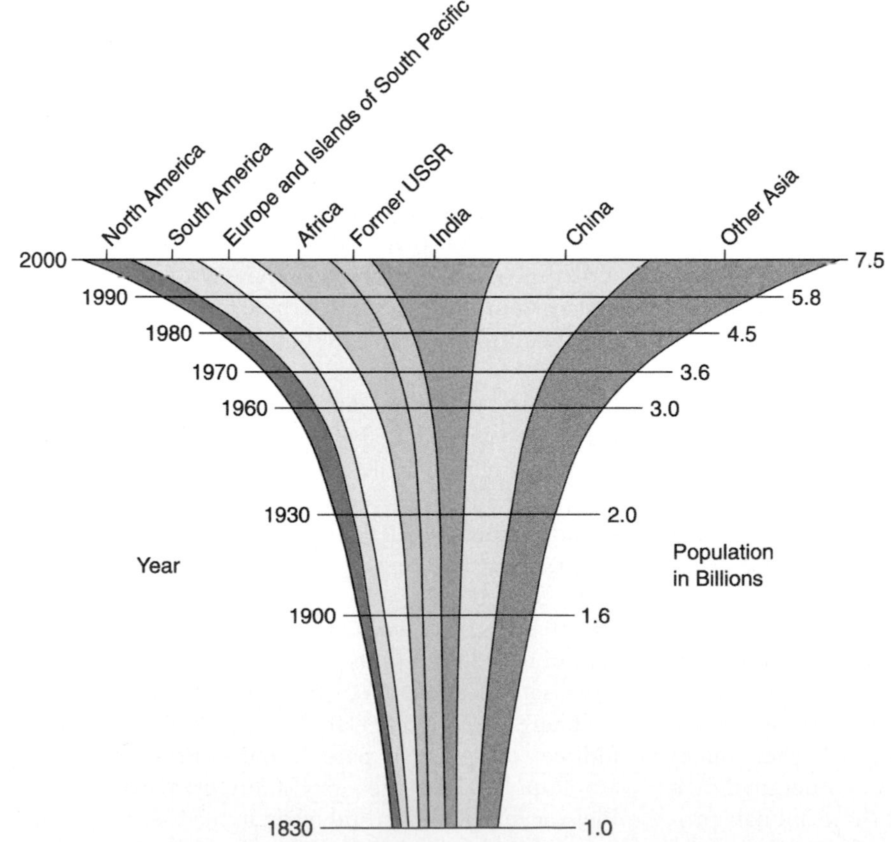

FIGURE 15-29 World population growth to the year 2000. *(From U.S. Dept. of State information circular.)*

lion. By 1995 there were 5.7 billion people requiring food, shelter, clean air to breath, and energy for their survival (Fig. 15–29). That number is expected to double by the year 2050. Can the Earth sustain such an enormous population? The resources of our planetary home are limited, and environments are readily damaged. Already we have squandered about a fifth of the topsoil needed to grow food. Over the past 5 decades we have lost a third of the globe's forested areas, and about an eighth of all agricultural land has been altered to barren desert, partly as a result of overgrazing associated with the demand for meat. There has been loss of soil productivity stemming from accumulation of mineral salts in soils caused by the extensive practice of irrigation. Water shortages have become critical in many regions around the world. Climate altering greenhouse gases have increased by more than a third, and there has been a 5 percent reduction of the ozone layer that protects all life from lethal ultraviolet radiation. Our mineral resources are also finite. As they are consumed and become rare or require more expensive extraction and refining, costs rise and standards of living decline in even the richest of nations.

With concerns such as these, few would doubt that the survival of *Homo sapiens* will depend on our ability not only to stabilize world population but also to protect agricultural lands, reforest the Earth, conserve and recycle mineral resources, and halt pollution of our environment. The task is vast in scale and complex. Yet if we fail in the endeavor, other animals may replace our species and add their own distinctive chapters to the history of life on Earth.

SUMMARY

During the early epochs of the Cenozoic, the main groups of placental mammals were differentiated. One group, the probable descendants of primitive insectivores, took to the trees of early Cenozoic forests, where food was plentiful and natural enemies few. The earliest of these primates have been named prosimians, literally "pre-monkeys." They were widespread during the Paleocene and Eocene but decreased during the Oligocene Epoch. One reason for the decline among prosimians was the appearance and differentiation of monkeys and apes. Looking back over early Cenozoic primate evolution, we can see the beginnings of the transformation of long-snouted, squirrel-like prosimians into progressive anthropoids. The head was reshaped and rounded, the muzzle was lost, and the face was flattened. Eyes were moved forward to provide good binocular vision, not only for quickly estimating the distance of nearby insect prey but also for accurately gauging the distance to the next bough. Manual dexterity became important for hunting, feeding, and gripping branches. Forelimbs with clawed digits became arms and hands equipped with an opposable thumb and an improved sense of touch. Swinging from branch to branch by the arms paved the way toward the erect posture that would be exploited at a later stage in primate history.

During the Miocene, a group of apes appeared that were the probable ancestors of both modern apes and the hominid line. They were the dryomorphs (the dryopithecines of older nomenclature). Proconsul africanus was a Miocene primate whose skull, face, and teeth indicate it was an ancestor of the subsequent apes, but whose limbs were not yet apelike. Other dryomorphs of the Miocene were more completely apelike and were widespread throughout Europe and Asia. In many ways, the Miocene dryomorphs were the evolutionary products of changing late Cenozoic environments. Dense, junglelike forests that had cloaked entire regions of Africa were being replaced by grassy, parklike habitats called savannas. The semi-erect dryomorphs and later the australopithecines were quick to respond with a restructuring of the axial skeleton, pelvic girdle, and hind feet. With this improved, erect stability, the hands were completely free for manipulative purposes. These creatures probably took advantage of the protection inherent in unified group action and may have had the ability to communicate, at least to the same degree as do modern apes. All of these functions led ultimately to greater development of the cerebral cortex, the part of the brain associated with dexterity, memory, learning, and thought. Great reproductive advantage must have accrued to those species that made the best use of their brains. Although growth in brain size was slow until the end of the Pliocene, it accelerated phenomenally during the Pleistocene.

The oldest known hominid is Sahelanthropus tchadensis from central Africa. Radiometric dates are not available for this early hominid, but correlation of other bones from the Sahelanthropus site to radiometrically dated fossils of eastern Africa, indicate an age of 6 to 7 million years. The australopithecines appeared on Earth in the Pliocene, about 4.4 million years ago. Their cranial capacity was about 600 cm^3, only about a third as large as that of humans. However, their dentition closely resembled that of modern people. Australopithecines were the ancestors to the first members of the genus Homo, to which we belong. Early members of the genus include such species as Homo ergaster and Homo rudolfensis.

The next stage of human evolution witnessed the appearance of Homo erectus. These fully erect-walking, primitive people had a brain capacity that ranged from about 775 cm^3 to 1300 cm^3. By Middle Pleistocene time Homo erectus had spread widely throughout Europe, Asia, and Africa and had progressively improved their tool-making abilities.

The Late Pleistocene was marked by the appearance of the very distinctive and famous Neandertal people.

Remains of the Neandertals have been discovered in dozens of European sites, as well as in the Middle East and Africa. The Neandertal skull was more massive than our own, had a less prominent forehead and chin, and had more prominent ridges above the eyes. The dozens of skulls already collected indicate that the Neandertal peoples varied as widely in characteristics as do modern human populations. Neandertals had brains as large as or larger than modern humans. Their tools were more finely constructed than those of Homo erectus, they fashioned objects for aesthetic purposes, they apparently had a religious concern about an afterlife, and they knew the use of fire.

Fire was of immense importance to Pleistocene humans. Because of the light that fire provides, the length of time during which they could effectively accomplish tasks was increased. Fire provided protection during the night and permitted Homo erectus to occupy cave shelters with less danger from great bears and other predators. (This alone has been a boon to paleoanthropologists, in that it narrows down the number of places that are likely to have good fossils.) Fire also gave early humans the means to thaw and cook food. Its most important value, however, was in the warmth it provided. Fire enabled humans to extend their range into frigid latitudes.

Anatomically modern Homo sapiens were present during the late Pleistocene. They continue down to the present day in the form of you and me. An early group, known as Cro-Magnon, replaced the Neanderthals in Europe and continued their cultural traditions. Even finer tools and weapons in bone and flint were manufactured by these peoples. They exhibited remarkable artistic skills, painting pictures of the animals vital to their survival on the walls of their caves and carving voluptuous Venus-type statues to encourage fertility. Humans came to the Americas at a rather late date. Artifacts and sparse skeletal remains indicate their presence somewhat over 20,000 years ago. Their spearheads are sometimes found among the remains of the extinct large mammals that they hunted.

Unlike our Ice Age ancestors, 20th-century humans face difficult problems caused by the population explosion. Overpopulation is important because it is a major contributor to all other environmental and resource problems. More people use more oil and more metals; cut down more trees; deplete more farmland; and add more sewage to rivers, more garbage to landfills, and more pollutants to the air. For these reasons, finding ways to achieve population stability should have high priority among the nations of the world.

QUESTIONS FOR REVIEW AND DISCUSSION

1. Distinguish between the following:
   a. Prosimii and Anthropoidea
   b. Hylobatidae and Pongidae
   c. Cercopithecoidea and Ceboidea
2. Early arboreal primates such as Notharctus had close-set eyes and grasping hands. Other than facilitating movement among tree branches, what function might these attributes have provided?
3. During what geologic period do we find the earliest known remains of primates? When does the first member of the genus Homo appear?
4. What physical properties of chert and flint render these rocks useful in the manufacture of spear points, axes, and scrapers by early humans?
5. Why is it easier to find fossils and to date them in the eastern African fossil sites than in many other parts of the world?
6. What distinctly humanlike traits were possessed by Australopithecus? What apelike characteristics were retained?
7. What features of Cro-Magnon people differentiate them from Neanderthal people?
8. Discuss the general changes in the skulls of fossil primates as one progresses from an animal such as Plesiadapis to modern humans.
9. Discuss the route by which Homo sapiens entered the New World. Would the migrations have been easier during a glacial or interglacial epoch? Why?
10. Prepare text for a debate in which you first argue for and then against the premise that life will be more difficult for humans during the year 2050 than it is today.

READINGS

Brace, C. L. 1991. *The Stages of Human Evolution*, 4th ed. Englewood Cliffs, NJ: Prentice Hall.

Conroy, G. C. 1997. *Reconstructing Human Origins: A Modern Synthesis*. New York: W. W. Norton.

Fleagle, J. G. 1988. *Primate Adaptation and Evolution*. San Diego: Academic Press.

Keyfitz, N. 1989. The growing population. *Sci. Am.* 261(3).

Lewin, R. 1984. *Human Evolution*. New York: W. H. Freeman & Co.

Lewin, R. 1993. *The Origin of Modern Humans*. New York: Scientific American Library.

Tattersall, I. 1993. *The Human Odyssey: Four Million Years of Human Evolution*. Englewood Cliffs, NJ: Prentice Hall.

WEB SITES

The Earth Through Time Student Companion Web Site (www.wiley.com/college/levin) has online resources to help you expand your understanding of the topics in this chapter. Visit the Web Site to access the following:

1. Illustrated course notes covering key concepts in each chapter;
2. Online quizzes that provide immediate feedback;
3. Links to chapter-specific topics on the web;
4. Science news updates relating to recent developments in Historical Geology;
5. Web inquiry activities for further exploration;
6. A glossary of terms;
7. A Student Union with links to topics such as study skills, writing and grammar, and citing electronic information.

APPENDIX A

Classification of Living Things

The classification presented below recognizes six kingdoms of organisms. Two of these, the Archaeobacteria and Eubacteria, are components of the superkingdom Prokaryota. The remaining Protista, Fungi, Plantae, and Animalia are grouped under the superkingdom Eukaryota. The classification has been simplified in that some groups lacking a fossil record or not mentioned in the text are omitted.

SUPERKINGDOM PROKARYOTA

Kingdom Archaeobacteria Prokaryotes with primitive genetic codes, including methane-producing bacteria (methanophiles), bacteria tolerant of extremely salty water (extreme halophiles), and bacteria tolerant of high temperatures (extreme thermophiles). In the three-domain classification, archaeobacteria are equivalent to the archaea.

Kingdom Eubacteria Advanced prokaryotes that include digestive and disease-causing bacteria as well as photoautotrophs like the cyanobacteria, formerly called blue-green algae. Cyanobacteria are represented in the fossil record by **stromatolites**. (Archean to Holocene).

SUPERKINGDOM EUKARYOTA

Kingdom Protista Solitary or colonial unicellular eukaryotes that do not form tissues.

Phylum Euglenophyta Euglenoid organisms. Members of this phylum illustrate the impossibility of distinguishing "plants" from "animals" among the Protista.

Phylum Chrysophyta Golden-brown algae. This phylum includes the **diatoms** and **coccolithophorids**, both of which are abundant and useful microfossils. (Triassic to Recent)

Phylum Chlorophyta Green algae. Includes the stoneworts and charophytes, as well as *Halimeda*, which grows in modern carbonate-depositing environments. (Cambrian to Holocene)

Phylum Pyrrophyta Dinoflagellates and cryptomonads.

Phylum Sarcodina Protozoans with pseudopodia, including the living *Amoeba* as well as important microfossil groups such as **Foraminifera** and **Radiolaria**. (Cambrian to Holocene)

Kingdom Fungi Multicellular eukaryotes that exist mostly as decomposers, such as molds and mushrooms. Some are parasites that infest plants. Definite fungal remains occur in rocks as old as Devonian, although some questionable remains have been reported from the Proterozoic. (Proterozoic? to Holocene)

Kingdom Plantae

Phylum Rhodophyta Red algae, usually marine, multicellular.

Phylum Phaeophyta Brown algae, often with large bodies, as in seaweed and kelp.

Phylum Bryophyta Liverworts, mosses, and hornworts. (Devonian to Holocene)

Phylum Psilophyta Primitive, spore-bearing, leafless, rootless, vascular land plants. (Mid-Paleozoic to Holocene)

Phylum Lycopodophyta (= Lycopsida) Spore-bearing vascular plants with true roots and leaves, including the scale trees (lycopsids like *Lepidodendron* and *Sigillaria*) of the Carboniferous. (Silurian to Holocene)

Phylum Equisetophyta (= Sphenopsida) Horsetails, scouring rushes, including the Carboniferous **sphenopsids** such as *Calamites* and *Annularia*. (Devonian to Recent)

Phylum Polypodiophyta The true ferns, or **pteropsids**. (Devonian to Recent)

Phylum Pinophyta (= Gymnospermophyta) The "**gymnosperms**," including conifers, cycads, and many evergreen plants that do not develop true flowers. (Middle Paleozoic to Holocene)

- *Class Pteridospermophyta* Seed ferns, known from such Carboniferous fossil plants as *Neuropteris* and *Glossopteris*. (Devonian to Holocene)
- *Class Bennettitopsida* The extinct **cycadeoids**. (Triassic to Early Cretaceous)
- *Class Cycadopsida* **Cycads**. (Triassic to Holocene)
- *Class Ginkgopsida* **Ginkgoes**. (Mesozoic to Holocene)
- *Class Pinopsida* **Conifers**, as well as the extinct **Cordaites**. (Late Paleozoic to Holocene)

Phylum Magnoliophyta The flowering plants, or "**angiosperms**." Seeds enclosed in an ovary; fertilization involving flowers. (Cretaceous to Holocene)

Kingdom Animalia

Phylum Porifera **Sponges**, multicellular aquatic invertebrates that lack tissue level of organization, gather their food from the flow of water through numerous pores and canals, and possess special flagellated collared cells called choanocytes. (Proterozoic to Holocene)

Class Demospongiae Skeleton of siliceous spicules, an organic substance called spongin, or a combination of these materials. (Proterozoic to Holocene)

Class Hyalospongea The "glass sponges," so named because their skeleton is composed of siliceous (opaline) spicules. (Cambrian to Holocene)

Class Sclerospongea Skeleton of siliceous spicules and spongin underlain by an encasement of aragonite. The class includes the **stromatoporoids**, Paleozoic reef-building poriferans that build structures composed of calcareous lamellae supported by connecting pillars and partitions. (Cambrian to Holocene)

Class Calcispongea (= Calcarea) Sponges that secrete calcareous spicules. (Devonian to Holocene)

Phylum Archaeocyatha Cup- or vase-shaped fossils of extinct sponge-like organisms. Important as Cambrian reef-builders and widely used in correlation of Cambrian strata. (Cambrian)

Phylum Cnidaria (= Coelenterata) Jellyfishes, anemones, and coals; radially symmetric, aquatic invertebrates with specialized cells called cnidocysts that produce capsules used in capturing prey. True tissues are developed. (Latest Proterozoic to Holocene)

Class Hydrozoa The *Hydra* and hydra-like cnidarians.

Class Scyphozoa True jellyfishes. The Scyphozoa include **conularids**, jellyfish that build pyramidal shells. (Proterozoic to Holocene)

Class Anthozoa Corals and sea anemones. Included are the extinct **tabulate** and **rugose** corals and the extant **scleractinid** corals (also called hexacorals).

Phylum Bryozoa Tiny, colonial, filter-feeding, mostly marine invertebrates with lophophore and digestive tract with mouth and anus. (Ordovician to Holocene)

Phylum Brachiopoda Lophophore-bearing marine invertebrates with shell composed of two unequal valves. The larger valve is called the pedicle or ventral valve, and the smaller the brachial or dorsal valve. Often attached to substrate by a flexible stalk called a pedicle. (Cambrian to Holocene)

Class Inarticulata Shell lacking teeth and sockets and without hinge line. Most composed of chitinophosphatic material. (Cambrian to Holocene)

Class Articulata Shell with definite hinge, usually bearing teeth and sockets. (Cambrian to Holocene)

Order Orthida Unequally biconvex shells with straight hinge line. (Cambrian to Permian)

Order Strophomenida Wide, straight hinge line, one valve concave. (Ordovician to Jurassic)

Order Pentamerida Large, biconvex shells, overhanging beak, short hinge line. (Cambrian to Devonian)

Order Rhynchonellida Strongly ribbed (plicate), strongly biconvex shells with prominent beak. (Ordovician to Holocene)

Order Spiriferida Interior of brachial valve with spiral lophophore supports. (Ordovician to Jurassic)

Order Productida Often large concavo-convex shells with elaborate development of spines over shell surface. (Devonian to Permian)

Order Terebratulida Interior of brachial valve with loop-shaped lophophore supports. (Devonian to Holocene)

Phylum Phoronida Lophophore-bearing, slender, worm-like marine animals that secrete and live within a chitious, leathery tube. (Devonian to Holocene)

Phylum Annelida Segmented worms. The phylum includes polychaetes (marine sandworms and tubeworms), earthworms, and leeches. (Proterozoic to Holocene)

Phylum Arthropoda Animals with paired, jointed appendages, an armor-like exoskeleton, segmented body, and respiratory systems consisting of gills for many aquatic forms and tracheae for terrestrial forms. (Cambrian to Holocene)

Subphylum Trilobita Invertebrates with body divided into three parts (cephalon, thorax, and pygidium) and divided longitudinally by a central lobe and two lateral or pleural lobes. (Cambrian to Permian)

Subphylum Chelicerata Arthropods having claw-like feeding appendages (chelicerae) and no antennae. Includes horseshoe crabs and arachnids (spiders, scorpions, ticks, and mites). (Cambrian to Holocene)

Subphylum Crustacea Arthropods, which like trilobites, have biramous appendages. Lobsters, crabs, barnacles, and ostracodes are all included in the Crustacea. (Cambrian to Holocene)

Subphylum Uniramia Arthropods with unbranched (uniramous) appendages and a single pair of antennae. (Cambrian to Holocene)

Class Onycophora Living and ancient velvet worms, including ***Aysheia*** from the Burgess Shale fossil locality.

Class Myriapoda Uniramids (centipedes and millipedes) having head and elongate trunk bearing either one pair of legs on each segment (most centipedes), or two pair (most millipedes). (Carboniferous to Holocene)

Class Hexapoda Insects, all characterized by a body consisting of head, thorax, and abdomen, six legs, and tracheal tubes for respiration. (Devonian to Holocene)

Phylum Mollusca Like the annelids and arthropods, mollusks are protostome animals (animals in which the blastopore develops into the mouth and the anus forms secondarily). The shell characteristic of most mollusks is secreted by the mantle, which also covers the visceral mass. (Cambrian to Holocene)

Class Monoplacophora Primitive mollusks with cap- or spoon-shaped shells, multiple pairs of gills. Monoplacophorans are thought to have evolved from annelid worms and may be ancestral to all other classes of mollusks. The living form ***Neopilina*** and the Silurian ***Pilina*** are examples. (Cambrian to Holocene)

Class Polyplacophora Chitons, marine mollusks with shells composed of eight articulating calcarous plates. (Cambrian to Holocene)

Class Scaphopoda Tusk shells; mollusks with curved tubular shells open at both ends. (Ordovician to Holocene)

Class Gastropoda Mollusks having a twisted body and a shell that, when present, is coiled. Includes snails, slugs, and abalones. (Cambrian to Holocene)

Class Bivalvia (= Pelecypoda) Shells composed of two valves (right and left) that are hinged dorsally. Includes clams, mussels, oysters, and scallops. (Cambrian to Holocene)

Class Cephalopoda Mollusks with foot divided into tentacles that surround the central mouth of the large head. Large, well-developed, image-forming eyes. Includes modern *Nautilus*, squid, cuttlefishes, and octopods. (Cambrian to Holocene)

Subclass Nautiloidea Cephalopods with external shell divided into chambers by septa and having simple suture lines. (Cambrian to Holocene)

Subclass Ammonoidea Cephalopods with external shell divided into chambers by septa with fluted margins, resulting in complexly folded suture lines. (Devonian to Cretaceous)

Subclass Coleoidea Cephalopods that lack a shell or have an internal shell. Includes belemnites (Mississippian to Tertiary) as well as living and fossil squid, cuttlefish, and octopods (Mississippian to Holocene).

Order Belemnoidea Skeleton consisting of small chambered structure surrounded by a thick rostrum. Shell developed internally like the shell of a cuttlefish. The **belemnites**. (Mississippian to Early Tertiary)

Order Sepioidea Cuttlefishes. (Jurassic to Holocene)

Order Teuthoidea Squids. (Jurassic to Holocene)

Order Octopoda Octopods. (Cretaceous to Holocene)

Phylum Hyolitha Marine, bilaterally symmetrical, solitary metazoans, typically with conical shells closed by a lid. (Cambrian to Devonian)

Phylum Echinodermata Spiny-skinned marine invertebrates that are radially symmetrical as adults (bilateral in larval stage) and have calcareous, spine-bearing skeletal plates and unique water vascular systems. Echinoderms are **deuterostomes**, in that the anus develops from the blastopore. (See Phylum Mollusca above for comparison to *protostome*.) (Cambrian to Holocene)

Class Asteroidea Starfishes. Animals with a central disk from which radiate five or more arms. (Ordovician to Holocene)

Class Ophiuroidea Brittle stars and serpent stars having a central disk from which abruptly radiate five or more slender arms that lack ambulacral grooves. (Ordovician to Holocene)

Class Echinoidea Sea urchins and sand dollars. Have solid shell without arms; often covered with spines. (Ordovician to Holocene)

Class Holothuroidea Sea cucumbers. Echinoderms with elongate, flexible bodies and a mouth surrounded by a circle of modified tube feet serving as tentacles. (Cambrian to Holocene)

Class Crinoidea Sea lillies and feather stars. Many sessile and attached to sea floor by stem and "root." (Cambrian to Holocene)

Class Blastoidea Blastoids commonly have bud-shaped compact thecae constructed of 13 or 14 plates. (Silurian to Permian)

Classes Rhombifera and *Diploporita* Component groups of the former Class Cystoidea. These are stalked echinoderms with thecae characterized by distinctive pore patterns.

Phylum Hemichordata As suggested by their name, which means "half chordates," hemichordates share certain features with chordates that may include a pharynx perforated by gill slits or pores and a hollow dorsal nerve cord. The phylum includes the extinct **Class Graptolithina**, as well as living acorn worms that have larval forms similar to those of echinoderms.

Phylum Chordata Bilaterally symmetric animals with notochord, dorsal hollow neural tube, and gill clefts in the pharynx. Include sea squirts, lancelets, and possibly conodont-bearing animals.

Subphylum Urochordata Sea squirts or tunicates. Larval forms have notochord in tail region.

Subphylum Cephalochordata *Branchiostoma* ("amphioxus"); small marine animals with fish-like bodies and musculature, notochord, and dorsal neural tube.

Subphylum Vertebrata Animals with a backbone composed of vertebrae, well-defined head, ventrally located heart, and well-developed sense organs. Includes fishes, amphibians, reptiles, synapsids, birds, and mammals.

Class Agnatha Jawless fishes, including living lampreys and hagfishes, as well as extinct armored jawless **ostracoderms**. (Cambrian to Holocene)

Class Acanthodii Earliest group of fishes with jaws. Often called "spiny sharks" because of their many spine-supported fins. (Middle to Late Paleozoic)

Class Placodermi Extinct jawed fishes with heavy bony armor and a specialized joint at rear of head. (Middle Paleozoic)

Class Chondrichthyes Fishes with cartilaginous skeletons, including sharks, rays, chimaeras, and skates. (Middle Paleozoic to Holocene)

Class Osteichthyes Fishes in which bone becomes the most abundant skeletal material. (Devonian to Holocene)

Subclass Actinopterygii Bony fishes in which the fins are supported by bony rays, hence "ray-finned" fishes. Actinopterygians are by far the dominant fishes in the modern world. (Devonian to Holocene)

Subclass Sarcopterygii Lobe-finned fishes. Stout, muscular fins are supported by bony skeletal elements. Many with lungs and the ability to breath air as well as features of the skull and teeth that indicate their relationship to early tetrapods. Include modern **lungfishes** and **coelacanths**. (Devonian to Holocene)

Order Crossopterygii Lobe-finned fishes most closely related to tetrapods.

Order Dipnoi Lungfishes, including the Devonian fish *Dipterus* and living lungfish of South America, Africa, and Australia. (Devonian to Holocene)

Class Amphibia Amphibians; the Earth's earliest tetrapods (vertebrates with four limbs.) Include living frogs, toads, salamanders, and newts as well as prominent Carboniferous labyrinthodontic amphibians known as **temnospondyls**. *Eryops*, pictured in the text, has the broad flat skull typical of temnospondyls.

Amniota Reptiles, birds, and mammals are all amniotes in that they develop an egg with membranes around the embryo (allantois, chorion, and amnion) that serve to retain water and provide for gas exchange, support, and protection. Thus amniotes do not require water bodies for reproduction. Amniotes evolved during the Carboniferous.

Tetrapoda Vertebrates with two pairs of walking limbs. The term tetrapods is derived from two Greek words meaning "four-footed." (Carboniferous to Holocene)

Class Reptilia Amniotes that include turtles, crocodiles, and many ancient vertebrates with distinctive elements of the skull. Excluded from the Reptilia are the so-called mammal-like reptiles (synapsids), mammals, and birds.

Subclass Anapsida The Anapsida are reptiles that lack special openings in the side of the skull to accommodate jaw muscles. This group includes many extinct reptiles of the Carboniferous to Triassic, as well as *Mesosaurus*, the geographic distribution of which provides evidence of drifting continents.

Subclass Testudinata Turtles, which also have anapsid skull structure, but in addition have shells and unique postcranial skeletal features. (Triassic to Holocene)

Subclass Diapsida Reptiles with two temporal openings in the skull. (Carboniferous to Holocene)

Infraclass Lepidosauromorpha Lizards, snakes, the tuatara (*Sphenodon*), and their extinct relatives. Relatives of early lepidosauromorphs may include such Mesozoic marine reptiles as the **ichthyosaurs** and **plesiosaurs**.

Infraclass Archosauria The archosaurs include the dinosaurs, the pterosaurs, and the crocodilians. They differ from lepidosauromorphs in having an opening or fenestra in front of the eye orbit, a fenestra in the lower jaw, laterally compressed and serrated teeth (in flesh eaters), no teeth on the palate, and limbs which supported the body in semi-upright or upright posture. Included within the Archosauria are archaic ancestral archosaurs, the more advanced **Ornithischia** (ankylosaurs, stegosaurs, ceratopsians, ornithopods) and **Saurischia** (sauropods, theropods), and according to cladistic analyses, birds (**Aves**).

Class Synapsida The synapsids include Late Paleozoic amniotes such as the finbacks and other Late Permian and Early Triassic vertebrates erroneously termed "mammal-like reptiles." (They are not reptiles). Mammals, the warm-blooded synapsids with body hair and mammary glands in females for nourishing offspring, range from Triassic to Holocene.

Subclass Pelycosauria Archaic synapsids of the Carboniferous and Permian, the most familiar of which are the carnivorous and herbivorous finbacks of the Permian Period.

Subclass Therapsida Progressive synapsids approaching the mammalian stage of evolution. Mammals, for example, have one dentary bone on either side of the lower jaw, and therapsids show enlargement of the dentary at the expense of other jaw bones. They also show development of a secondary palate and differentiation of teeth. (Permian to Triassic)

Subclass Mammalia Synapsids in which each half of the lower jaw consists of only a single bone, the dentary; also possess body hair and mammary glands. (Triassic to Holocene).

Superorder Monotremata (Eutheria) Monotremes are egg-laying mammals. Representatives are the platypus and echidna (spiny anteater).

Superorder Marsupialia (Metatheria) Marsupials are pouched mammals, including kangaroos, koalas, and opposums. (Cretaceous to Holocene)

Superorder Eutheria (= placental mammals) Mammals that retain their young for a long period within the body of the female, where they are nourished by means of a complex structure called the placenta. (Cretaceous to Holocene)

Order Insectivora Primitive insect-eating placental mammals, including moles and shrews. (Cretaceous to Holocene)

Order Chiroptera Bats. (Early Tertiary to Holocene)

Order Primates Prosimians (e.g., lemurs, tarsiers), New World and Old World monkeys, apes, and humans. (Early Tertiary to Holocene)

Order Edentata Living armadillos, anteaters, and tree sloths, as well as extinct **glyptodonts** and **ground sloths**. (Early Tertiary to Holocene)

Order Rodentia Squirrels, mice, rats, beavers, and porcupines. (Early Tertiary to Holocene)

Order Lagomorpha Rabbits, hares, and pikas. (Early Tertiary to Holocene)

Order Creodonta Extinct, archaic, carnivorous placental mammals. (Paleocene to Pliocene)

Order Carnivora Progressive carnivorous placentals, including living and ancient dogs, cats, bears, hyenas, seals, sea lions, and walruses. (Early Tertiary to Holocene)

Order Perissodactyla Odd-toed ungulates (hoofed mammals), including horses, rhinoceroses, and tapirs, as well as the extinct **titanotheres** and **chalicotheres**. (Early Tertiary to Holocene)

Order Artiodactyla Even-toed ungulates, including antelopes, cattle, deer, giraffes, camels, llamas, hippopotamuses, and pigs, as well as extinct **oreodonts** and **entelodonts**. (Early Tertiary to Holocene)

Order Proboscidea Elephants and extinct **mastodons** and **mammoths** (Early Tertiary to Holocene).

Order Cetacea Whales and porpoises, both of which are descendents of ungulates. (Paleocene to Holocene)

APPENDIX B

Formation Correlation Charts for Representative Sections

The following charts may be used to ascertain the approximate equivalence of some of the better known rock formations deposited during the Phanerozoic Eon. The vertical scale does not indicate thickness but rather is an approximation of time. Chronostratigraphic terms are indicated on either side of the chart. The diagonal lines indicate formations missing by nondeposition or erosion. Fm = formation (generally including two or more lithologic types); Ss = sandstone; Sh = shales; Ls = limestone; Dol = dolomite; Qtzite = quartzite; Congl = conglomerate; Grp = group (two or more formations); Mem = member. Solid horizontal lines between the formations indicate well-established boundaries, whereas broken lines indicate boundaries in question.

CAMBRIAN

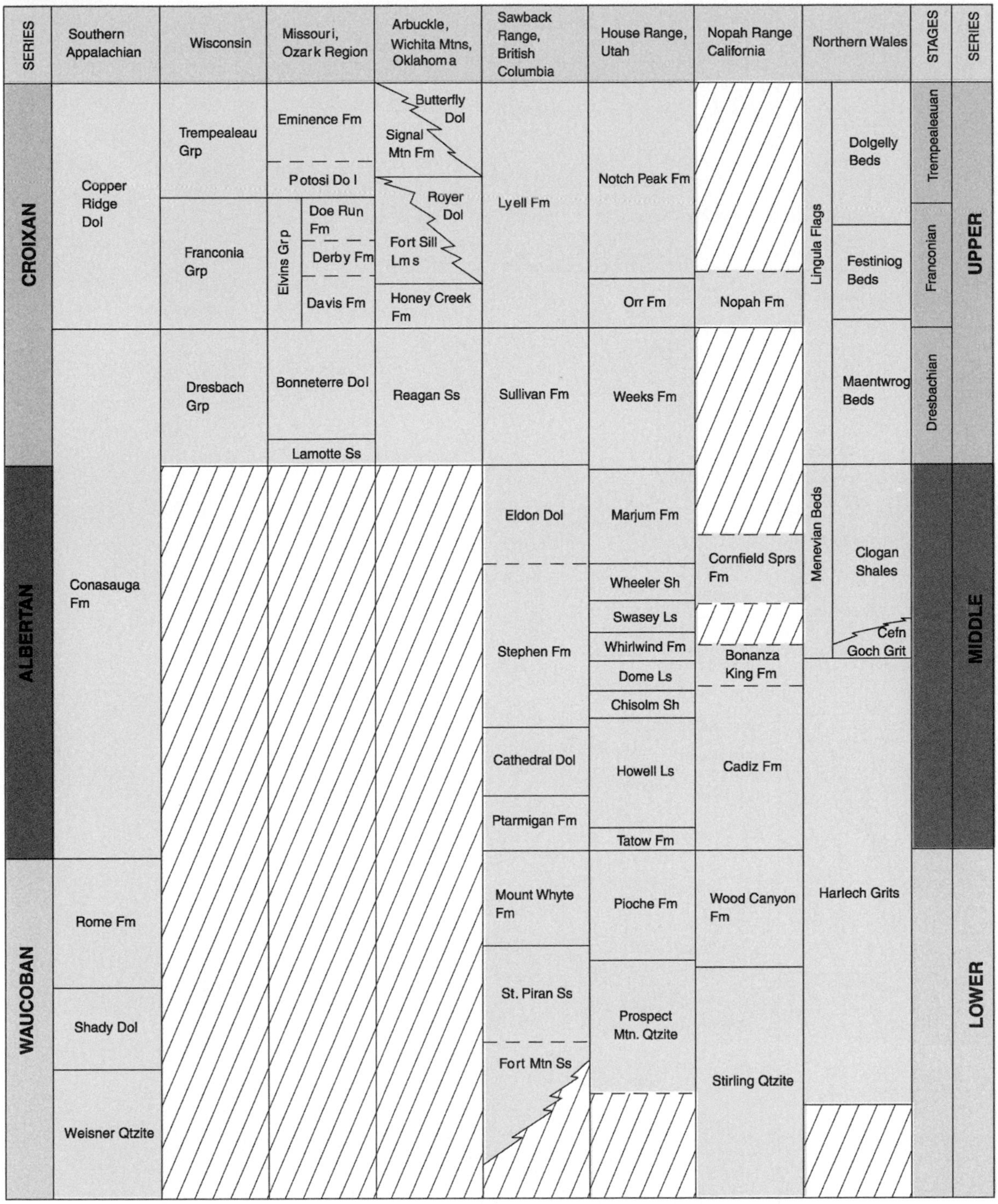

SERIES	Southern Appalachian	Wisconsin	Missouri, Ozark Region	Arbuckle, Wichita Mtns, Oklahoma	Sawback Range, British Columbia	House Range, Utah	Nopah Range California	Northern Wales	STAGES	SERIES
CROIXAN	Copper Ridge Dol	Trempealeau Grp	Eminence Fm Potosi Dol	Butterfly Dol Signal Mtn Fm Royer Dol	Lyell Fm	Notch Peak Fm		Lingula Flags Dolgelly Beds	Trempealeauan	UPPER
CROIXAN	Copper Ridge Dol	Franconia Grp	Elvins Grp: Doe Run Fm Derby Fm Davis Fm	Royer Dol Fort Sill Lms Honey Creek Fm	Lyell Fm	Notch Peak Fm Orr Fm	Nopah Fm	Lingula Flags Festiniog Beds	Franconian	UPPER
CROIXAN	Conasauga Fm	Dresbach Grp	Bonneterre Dol Lamotte Ss	Reagan Ss	Sullivan Fm	Weeks Fm		Lingula Flags Maentwrog Beds	Dresbachian	UPPER
ALBERTAN	Conasauga Fm				Eldon Dol Stephen Fm Cathedral Dol Ptarmigan Fm	Marjum Fm Wheeler Sh Swasey Ls Whirlwind Fm Dome Ls Chisolm Sh Howell Ls Tatow Fm	Cornfield Sprs Fm Bonanza King Fm Cadiz Fm	Menevian Beds Clogan Shales Cefn Goch Grit Harlech Grits		MIDDLE
WAUCOBAN	Rome Fm Shady Dol Weisner Qtzite				Mount Whyte Fm St. Piran Ss Fort Mtn Ss	Pioche Fm Prospect Mtn. Qtzite	Wood Canyon Fm Stirling Qtzite	Harlech Grits		LOWER

ORDOVICIAN

SERIES	STAGES	Taconic Area of Western Vermont	Northwestern New York	Virginia	Missouri	Oklahoma	Colorado	Utah	GREAT BRITAIN SERIES
CINCINNATIAN	Gamachian								
	Richmondian		Queenston Sh (Red beds)	Juniata Fm	Girardeau Ls Orchard Ck Sh Thebes Ss Maquoketa Sh Fernvale Ls	Polk Ck Sh	Fremont Ls	Fish Haven Dol	ASHGILL
	Maysvillian		Oswego Ss						
	Edenian		Lorraine Grp	Martinsburg Fm					
CHAMPLAINIAN	Mohawkian: Trentonian	Snake Hill Sh Rysedorf Congl Normanskill Sh	Utica Sh Trenton Ls	Bays Fm	Kimmswick Ls Decorah Grp	Big Fork Chert	Harding Ss		CARADOC
	Mohawkian: Black River		Black River Ls	Edinburg Fm	Plattin Grp				
	Chazian			Lincolnshire Ls Whistle Ck Sh	Joachim Ls	Womble Sh		Swan Peak Qtzite	LLANDEILO
	White-rockian			New Market Ls	St. Peter Ss Everton Fm				LLANVIRN
CANADIAN	Stages not yet established	Deepkill Sh			Smithville Fm Powell Fm Cotter Dol	Blakely Ss			
				Knox Dol (upper beds)	Jefferson City Grp Roubidoux Fm	Mazarn Sh	Manitou Ls	Garden City Ls	ARENIG
		Shaghticoke Sh			Gasconade Dol	Crystal Mtn Ss			TREMADOC

SILURIAN

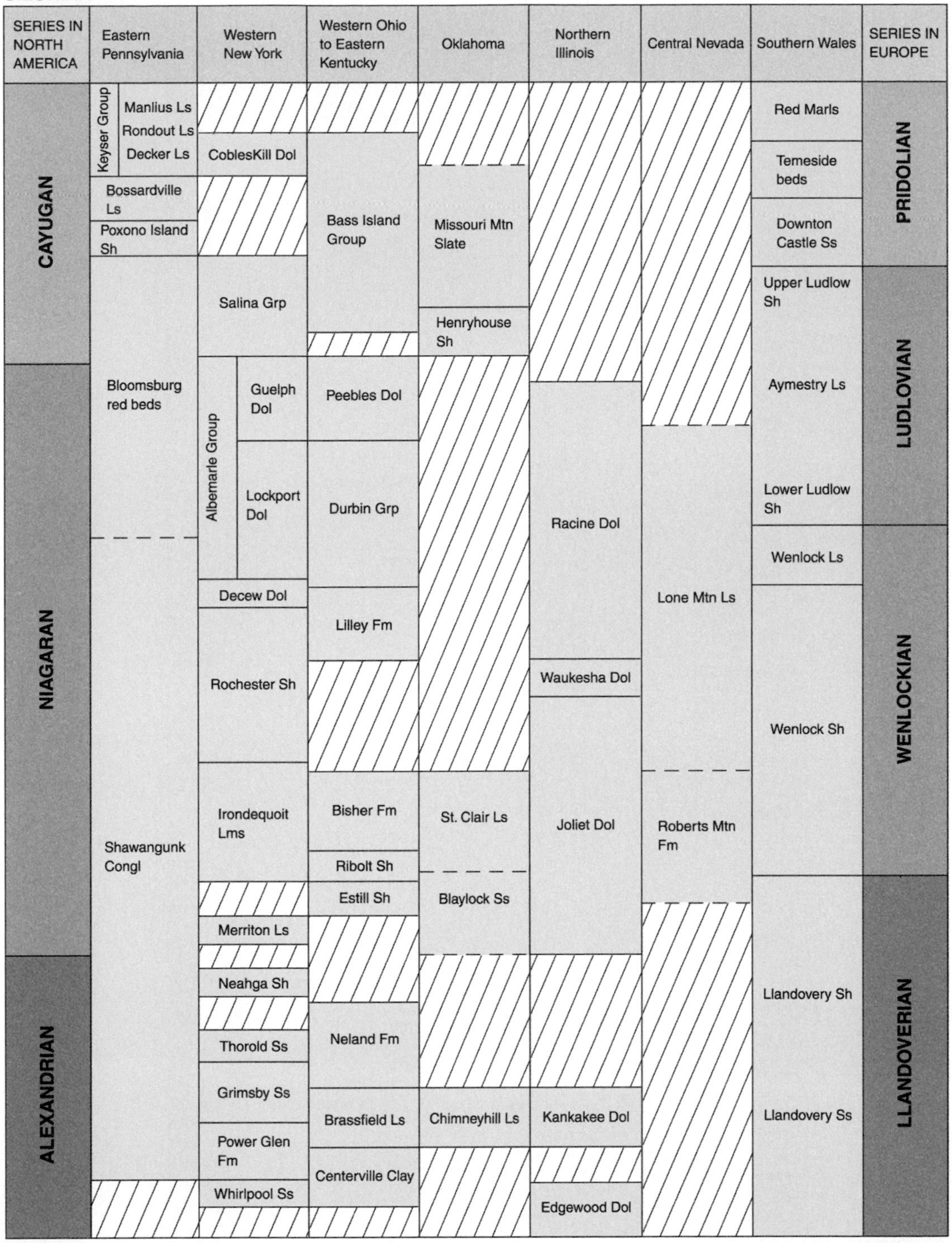

SERIES IN NORTH AMERICA	Eastern Pennsylvania	Western New York	Western Ohio to Eastern Kentucky	Oklahoma	Northern Illinois	Central Nevada
CAYUGAN	Keyser Group: Manlius Ls, Rondout Ls, Decker Ls Bossardville Ls Poxono Island Sh	CoblesKill Dol Salina Grp	Bass Island Group	Missouri Mtn Slate Henryhouse Sh		
NIAGARAN	Bloomsburg red beds Shawangunk Congl	Albemarle Group: Guelph Dol, Lockport Dol Decew Dol Rochester Sh Irondequoit Lms Merriton Ls	Peebles Dol Durbin Grp Lilley Fm Bisher Fm Ribolt Sh Estill Sh	St. Clair Ls Blaylock Ss	Racine Dol Waukesha Dol Joliet Dol	Lone Mtn Ls Roberts Mtn Fm
ALEXANDRIAN		Neahga Sh Thorold Ss Grimsby Ss Power Glen Fm Whirlpool Ss	Neland Fm Brassfield Ls Centerville Clay	Chimneyhill Ls	Kankakee Dol Edgewood Dol	

Southern Wales	SERIES IN EUROPE
Red Marls Temeside beds Downton Castle Ss	PRIDOLIAN
Upper Ludlow Sh Aymestry Ls Lower Ludlow Sh	LUDLOVIAN
Wenlock Ls Wenlock Sh	WENLOCKIAN
Llandovery Sh Llandovery Ss	LLANDOVERIAN

DEVONIAN

SERIES NA	STAGES NA	New York Catskill Mtns	New York Western	Virginia, Maryland, Pennsylvania	Western Tennessee	Nevada (Eureka Area)	Devon, England	Scotland	SERIES IN EUROPE
UPPER	BRADFORDIAN		Oswayo Grp Cattaraugus Congl	Catskill Red beds	Chattanooga Shale	Pilot Shale	Pilton Beds Baggy Beds Upcott Beds Pickwell Downs Ss	Upper Old Red Ss	FAMENNIAN
	CASSADAGIAN		Conneaut Grp Canadaway Grp Java Fm				Morte Slates		
	COHOCTONIAN	Wittenburg Congl Walton Sh	West Falls Fm Sonyea Fm	Chemung Ss Brallier sh		Devils Gate Ls	Ilfracombe Beds		FRASNIAN
	FINGER-LAKESIAN	Oneonta Fm	Genesee Fm	Harrell Fm					
MIDDLE	TAGHANICAN	Hamilton Grp		Burket Sh Tully Fm		Nevada Fm			GIVETIAN
	TIOUGHNIOGAN		Hamilton Grp	Mahantango Fm					
	CAZENOVIAN			Marcellus Sh			Hangman Grits		
	ONESQUETHAWAN	Onondaga Ls	Onondaga Ls	Onondago Ls			Lynton Beds		EIFELIAN
LOWER	ESPUSIAN	Schohari Fm Carlisle Fm Esopus Sh	Bois Blanc Fm	Needmore Sh	Camden Chert				EMSIAN
	DEERPARKIAN	Oriskany Ss		Oriskany Ss Shriver Chert	Harriman Fm Flat Gap Ls			Lower Old Red Ss	SIEGENIAN
	HELDERBERGIAN	Helderberg Grp Rondout Ls		Licking Ck Ls New Scotland Ls Elbow Ridge Ss Keyser Ls	Ross Fm				GEDINNIAN

MISSISSIPPIAN

STAGES IN NORTH AMERICA	West Virginia and Pennsylvania	Alabama	Mississippi Valley (Composite sequence)	Ohio	Arkansas Ouachita Mtns	Montana (Central)	Lower Carb. England (Bristol)	SERIES IN EUROPE
CHESTERIAN	Mauch Chunk Sh	Pennington Sh Bangor Ls Gasper Ls	Elvira Group: Vienna Ls, Tar Sprs Ss Glen Dean Ls Hardinsburg Ss Golconda Fm Cypress Ss Paint Ck Fm Bethel Ss Renault Ls Aux Vases Ss		Jackfork Ss Stanley Sh Hot Springs Ss	Lower Amsden Fm Big Snowy Group: Heath Fm, Otter Fm	Upper Cromwell Ss Hotwells Ls	VISEAN
MERAMECIAN	Mauch Chunk Sh	Ste. Genevieve Ls Tuscumbia Ls	Ste. Genevieve Ls St. Louis Ls Salem Ls Warsaw Ls	Maxville Ss		Big Snowy Group: Kibby Fm Charles Ls	Clifton Down Ls	VISEAN
OSAGEAN	Pocono Grp	Fort Payne Chert	Keokuk Ls Burlington Ls Fern Glen Fm	Logan Fm Cuyahoga Fm		Madison Group: Mission Canyon Ls, Woodhurst Ls	Goblin Combe Oolite Clifton Down Muds Gully Oolite	VISEAN
KINDERHOOKIAN	Pocono Grp	Chattanooga Sh (upper)	Gilmore City Ls Sedalia Ls Chouteau Ls Maple Hill Ls Dev-Miss Unassigned: Louisiana Ls, Saverton Sh, Grassy Creek Sh	Cuyahoga Fm Sunbury Sh Berea Ss Bedford Sh	Arkansas Novaculite (Upper & Middle Members)	Madison Group: Woodhurst Ls, Paine Ls	Black Rock Fm Lower Limestone Sh Shire Hampton Beds	VISEAN TOURNAISIAN

PENNSYLVANIAN

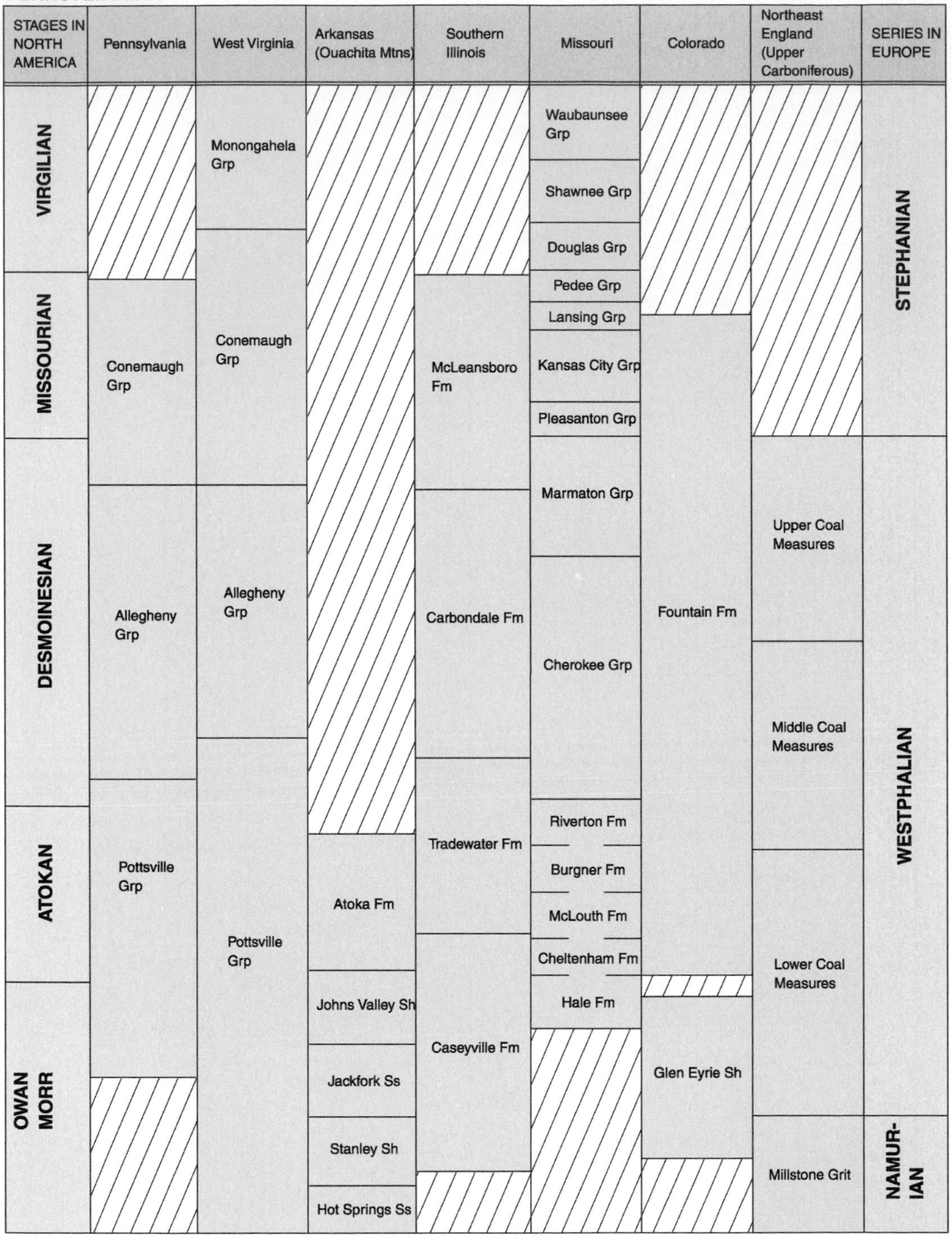

STAGES IN NORTH AMERICA	Pennsylvania	West Virginia	Arkansas (Ouachita Mtns)	Southern Illinois	Missouri	Colorado	Northeast England (Upper Carboniferous)	SERIES IN EUROPE
VIRGILIAN		Monongahela Grp Conemaugh Grp			Waubaunsee Grp Shawnee Grp Douglas Grp			STEPHANIAN
MISSOURIAN	Conemaugh Grp	Conemaugh Grp		McLeansboro Fm	Pedee Grp Lansing Grp Kansas City Grp Pleasanton Grp	Fountain Fm		STEPHANIAN
DESMOINESIAN	Allegheny Grp	Allegheny Grp Pottsville Grp	Atoka Fm	Carbondale Fm Tradewater Fm	Marmaton Grp Cherokee Grp	Fountain Fm	Upper Coal Measures Middle Coal Measures	WESTPHALIAN
ATOKAN	Pottsville Grp	Pottsville Grp	Atoka Fm Johns Valley Sh	Tradewater Fm Caseyville Fm	Riverton Fm Burgner Fm McLouth Fm Cheltenham Fm	Fountain Fm	Lower Coal Measures	WESTPHALIAN
OWAN MORR	Pottsville Grp	Pottsville Grp	Johns Valley Sh Jackfork Ss Stanley Sh Hot Springs Ss	Caseyville Fm	Hale Fm	Glen Eyrie Sh	Lower Coal Measures Millstone Grit	WESTPHALIAN NAMUR- IAN

PERMIAN

STAGES IN NORTH AMERICA	Ohio and West Virginia	Western Texas	Arizona	Wyoming	South Africa	Germany	Russia (Ural Mtns)	STAGES IN EUROPE
OCHOAN		Dewey Lake Fm Rustler Fm Salado Fm Castile Fm			Beaufort Series	Zechstein Fm	Tartarian	TARTARIAN (in part)
GUADALUPIAN		Bell Canyon Fm Cherry Canyon Fm Brushy Canyon Fm		Phosphoria Fm		Upper Rotliegend	Kazanian Kungurian	KAZANIAN KUNGURIAN
LEONARDIAN		Bone Spring Ls	Kaibab Ls Toroweap Fm Coconino Ss Hermit Sh		Ecca Series	Lower Rotliegend	Artinskian	ARTINSKIAN
WOLFCAMPIAN	Dunkard Grp	Hueco Ls	Supai Fm		Dwyka Series		Sakmarian	SAKMARIAN ASSELIAN

TRIASSIC

NA SERIES	STAGES	Connecticut Valley	Pennsylvania (York Co.)	Wyoming	Utah–Idaho	Colorado–South Dakota	Arizona	Nevada	Germany (Type Section)
UPPER	Rhaetian	Newark Grp: Portland Arkose		Nugget Ss	Ankareh Fm	Dolores Fm (Redbeds)	Chinle Fm	Gabbs Fm	Keuper: Rhät
	Norian	Newark Grp: Meriden Fm	Newark Grp: Gettysburg Fm	Chugwater Fm: Popo Agie Mem	Deadman Ls				Keuper: Gyps-keuper
	Carnian	Newark Grp: New Haven Arkose	Newark Grp: New Oxford Fm	Chugwater Fm: Crow Mtn Ss; Alcova Ls	Higman Grit (Congl)		Shinarump Congl	Luning Fm	Keuper: Lettenkohle
MIDDLE	Ladinian			Chugwater Fm				Grantsville Fm	Muschelkalk
	Anisian			Chugwater Fm				Excelsior Fm	
LOWER	Sythian			Chugwater Fm: Red Peak Mem; Dinwoody Fm	Thaynes Fm: Woodside Ss Mem; Portneuf Ls Mem; Woodside Red beds; Dinwoody Fm		Moenkopi Fm	Candelaria Fm	Bunter

JURASSIC

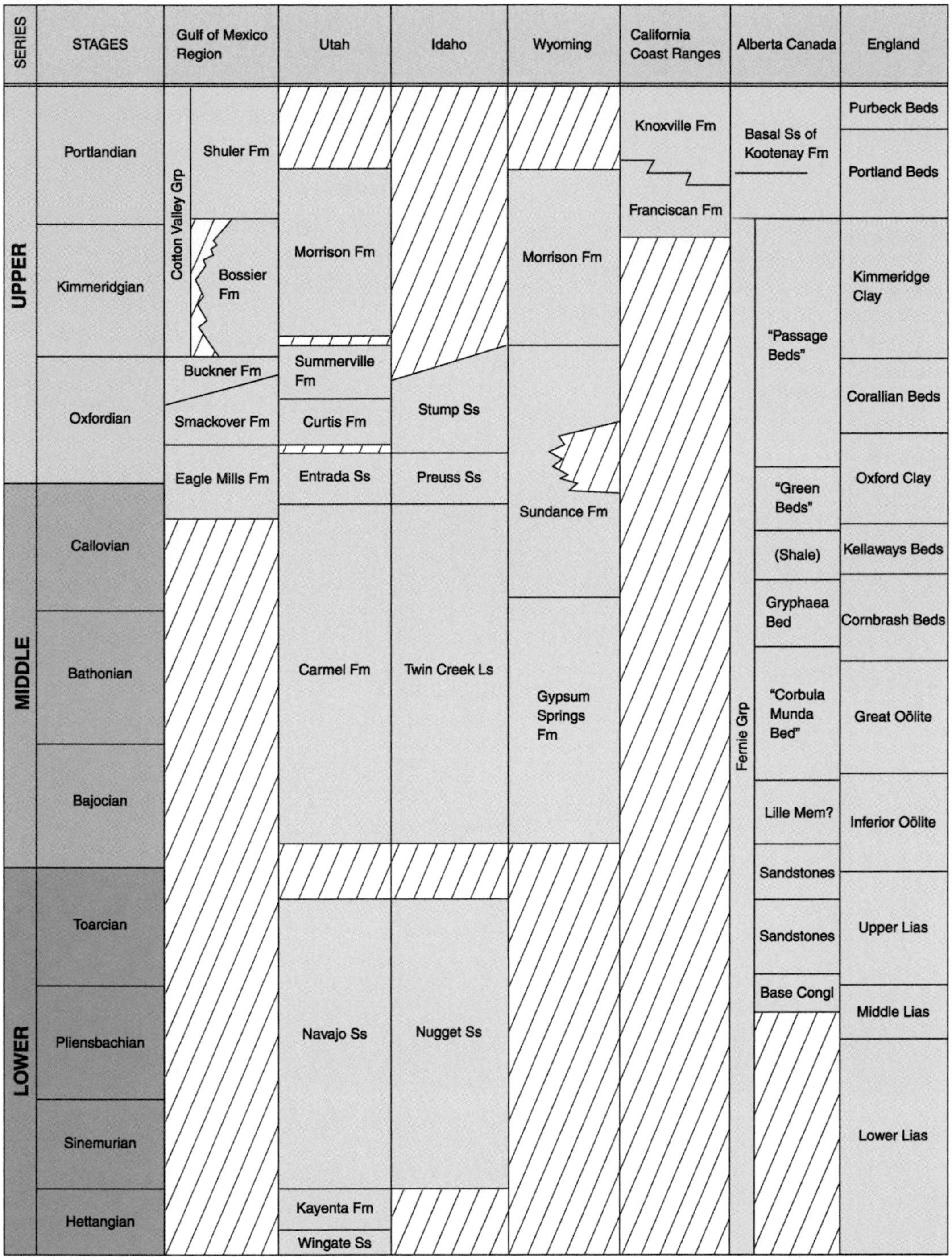

SERIES	STAGES	Gulf of Mexico Region	Utah	Idaho	Wyoming	California Coast Ranges	Alberta Canada	England
UPPER	Portlandian	Cotton Valley Grp: Shuler Fm				Knoxville Fm; Franciscan Fm	Basal Ss of Kootenay Fm	Purbeck Beds; Portland Beds
UPPER	Kimmeridgian	Cotton Valley Grp: Bossier Fm	Morrison Fm		Morrison Fm		Fernie Grp: "Passage Beds"	Kimmeridge Clay
UPPER	Oxfordian	Buckner Fm; Smackover Fm; Eagle Mills Fm	Summerville Fm; Curtis Fm; Entrada Ss	Stump Ss; Preuss Ss	Sundance Fm		Fernie Grp: "Passage Beds"; "Green Beds"	Corallian Beds; Oxford Clay
MIDDLE	Callovian	Eagle Mills Fm	Entrada Ss; Carmel Fm	Preuss Ss; Twin Creek Ls	Sundance Fm		Fernie Grp: "Green Beds"; (Shale)	Kellaways Beds
MIDDLE	Bathonian		Carmel Fm	Twin Creek Ls	Gypsum Springs Fm		Fernie Grp: Gryphaea Bed; "Corbula Munda Bed"	Cornbrash Beds; Great Oölite
MIDDLE	Bajocian		Carmel Fm	Twin Creek Ls	Gypsum Springs Fm		Fernie Grp: Lille Mem?	Inferior Oölite
LOWER	Toarcian		Navajo Ss	Nugget Ss			Fernie Grp: Sandstones; Sandstones	Upper Lias
LOWER	Pliensbachian		Navajo Ss	Nugget Ss			Fernie Grp: Base Congl	Middle Lias
LOWER	Sinemurian		Navajo Ss	Nugget Ss				Lower Lias
LOWER	Hettangian		Kayenta Fm; Wingate Ss					Lower Lias

CRETACEOUS

SERIES	STAGES	New Jersey	Alabama (Tuscaloosa)	Texas	Kansas (Wallace Co.)	Colorado	Montana	England
UPPER CRETACEOUS	Maastrichtian	Monmouth Grp	Prairie Bluff Ripley Fm	Navarro Grp: Escondido Fm		Animas Fm McDermott Fm	Hell Ck Fm Fox Hills Ss Bear Paw Sh	
UPPER CRETACEOUS	Campanian	Matawan Grp: Wenonah Sand Marshall-town Fm Englishtown Sand Woodbury Clay	Selma Chalk	Navarro Grp: Corsicana Marl Taylor Marl	Pierre Sh	Kirkland Fm Fruitland Fm Pictured Cliffs Ss Lewis Sh Mesaverde Grp	Judith River Fm Claggett Sh	Upper Chalk
UPPER CRETACEOUS	Santonian	Merchant-ville Clay	Eutaw Fm				Eagle Ss Telegraph Ck Fm	
UPPER CRETACEOUS	Coniacian	Magothy Fm		Austin Chalk	Niobrara Fm	Mancos Sh		
UPPER CRETACEOUS	Turonian			Eagle Ford Shale	Carlile Sh Greenhorn Ls			Middle Chalk
UPPER CRETACEOUS	Cenomanian	Raritan Fm	Tuscaloosa Fm	Woodbine Ss	Graneros Sh Dakota Ss	Dakota Ss	Colorado Sh	Lower Chalk
LOWER CRETACEOUS	Albian			Washita Grp Fredericksburg Grp	Kiowa Sh Cheyenne Ss			Upper Greensand
LOWER CRETACEOUS	Aptian			Trinity Grp		Burro Canyon Fm	Kootenai Fm	Lower Greensand
LOWER CRETACEOUS (Neocomian)	Barremian							Wealden Beds
LOWER CRETACEOUS (Neocomian)	Hauterivian							
LOWER CRETACEOUS (Neocomian)	Valanginian							
LOWER CRETACEOUS (Neocomian)	Berriasian							

CENOZOIC

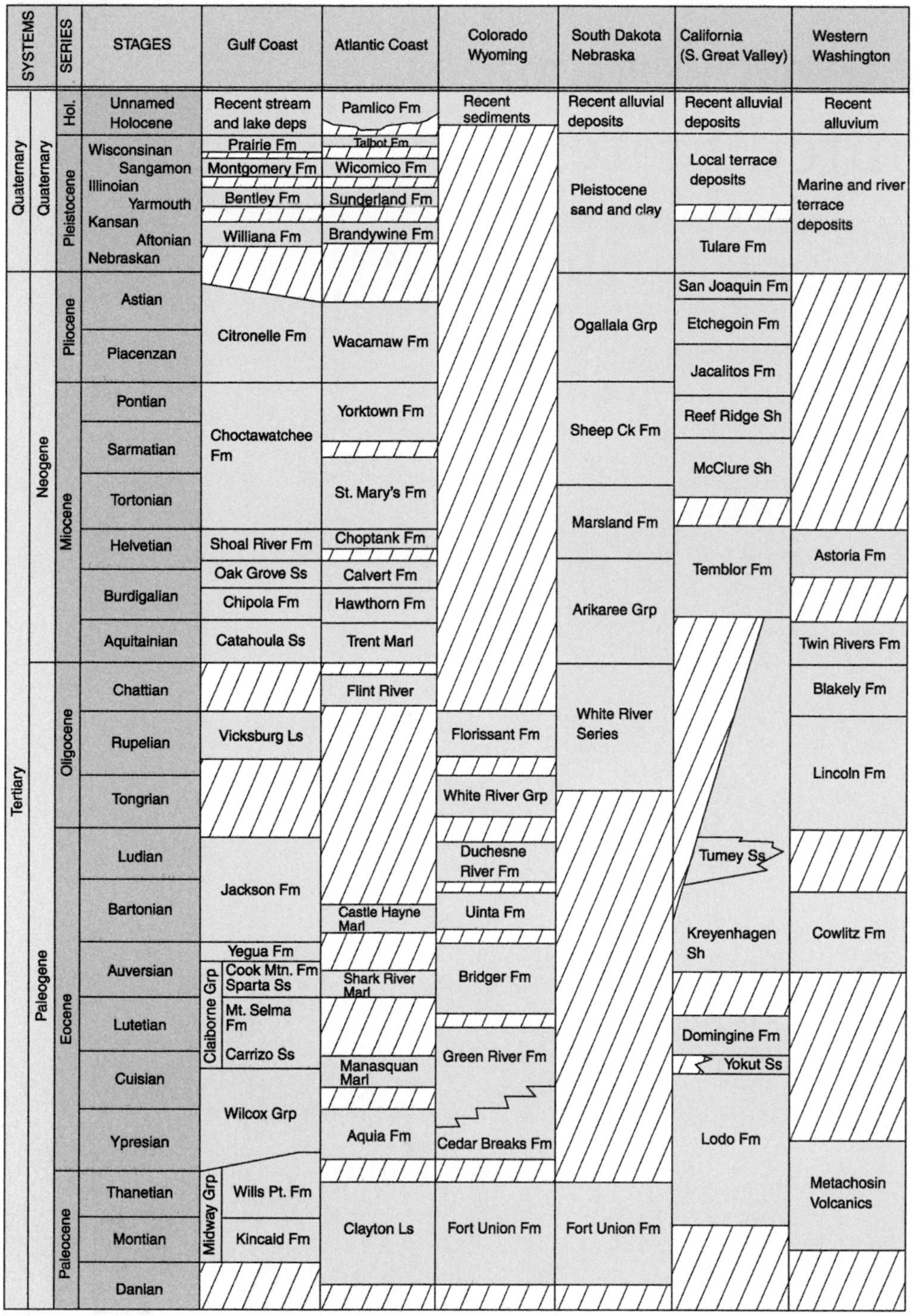

SYSTEMS		SERIES	STAGES	Gulf Coast	Atlantic Coast	Colorado Wyoming	South Dakota Nebraska	California (S. Great Valley)	Western Washington
Quaternary	Quaternary	Hol.	Unnamed Holocene	Recent stream and lake deps	Pamlico Fm	Recent sediments	Recent alluvial deposits	Recent alluvial deposits	Recent alluvium
		Pleistocene	Wisconsinan	Prairie Fm	Talbot Fm		Pleistocene sand and clay	Local terrace deposits	Marine and river terrace deposits
			Sangamon	Montgomery Fm	Wicomico Fm				
			Illinoian						
			Yarmouth	Bentley Fm	Sunderland Fm				
			Kansan						
			Aftonian	Williana Fm	Brandywine Fm			Tulare Fm	
			Nebraskan						
Tertiary	Neogene	Pliocene	Astian	Citronelle Fm			Ogallala Grp	San Joaquin Fm; Etchegoin Fm	
			Piacenzan		Wacamaw Fm			Jacalitos Fm	
		Miocene	Pontian	Choctawatchee Fm	Yorktown Fm		Sheep Ck Fm	Reef Ridge Sh	
			Sarmatian					McClure Sh	
			Tortonian		St. Mary's Fm		Marsland Fm		
			Helvetian	Shoal River Fm	Choptank Fm			Temblor Fm	Astoria Fm
			Burdigalian	Oak Grove Ss; Chipola Fm	Calvert Fm; Hawthorn Fm		Arikaree Grp		
			Aquitainian	Catahoula Ss	Trent Marl				Twin Rivers Fm
	Paleogene	Oligocene	Chattian		Flint River		White River Series		Blakely Fm
			Rupelian	Vicksburg Ls		Florissant Fm			Lincoln Fm
			Tongrian			White River Grp			
		Eocene	Ludian	Jackson Fm		Duchesne River Fm		Tumey Ss	
			Bartonian		Castle Hayne Marl	Uinta Fm		Kreyenhagen Sh	Cowlitz Fm
			Auversian	Yegua Fm; Claiborne Grp: Cook Mtn. Fm, Sparta Ss	Shark River Marl	Bridger Fm			
			Lutetian	Claiborne Grp: Mt. Selma Fm		Green River Fm		Domingine Fm	
			Cuisian	Claiborne Grp: Carrizo Ss	Manasquan Marl			Yokut Ss	
			Ypresian	Wilcox Grp	Aquia Fm	Cedar Breaks Fm		Lodo Fm	
		Paleocene	Thanetian	Midway Grp: Wills Pt. Fm	Clayton Ls	Fort Union Fm	Fort Union Fm		Metachosin Volcanics
			Montian	Midway Grp: Kincaid Fm					
			Danian						

Physiographic Provinces of the United States

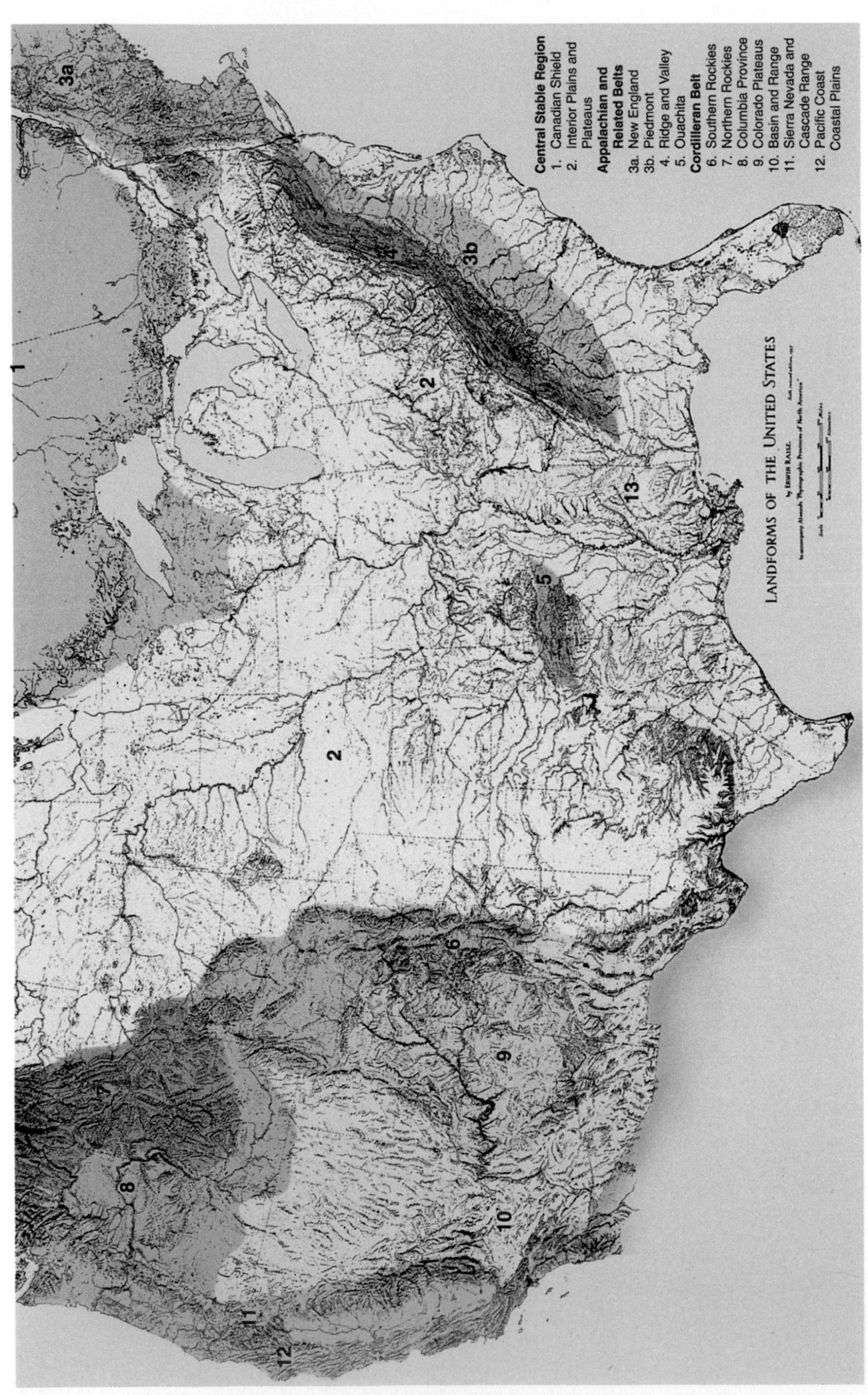

World Political Map

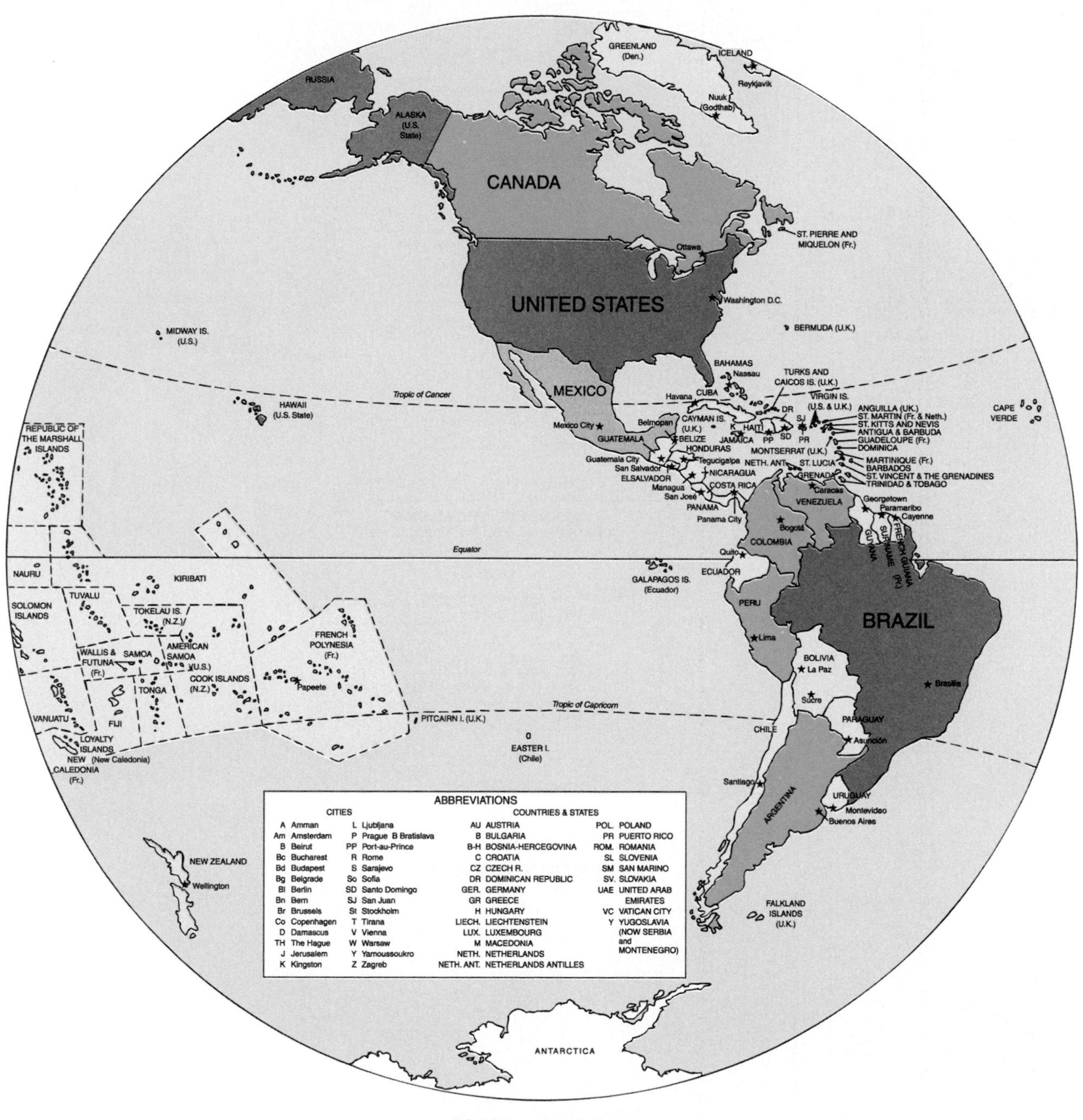

WESTERN HEMISPHERE

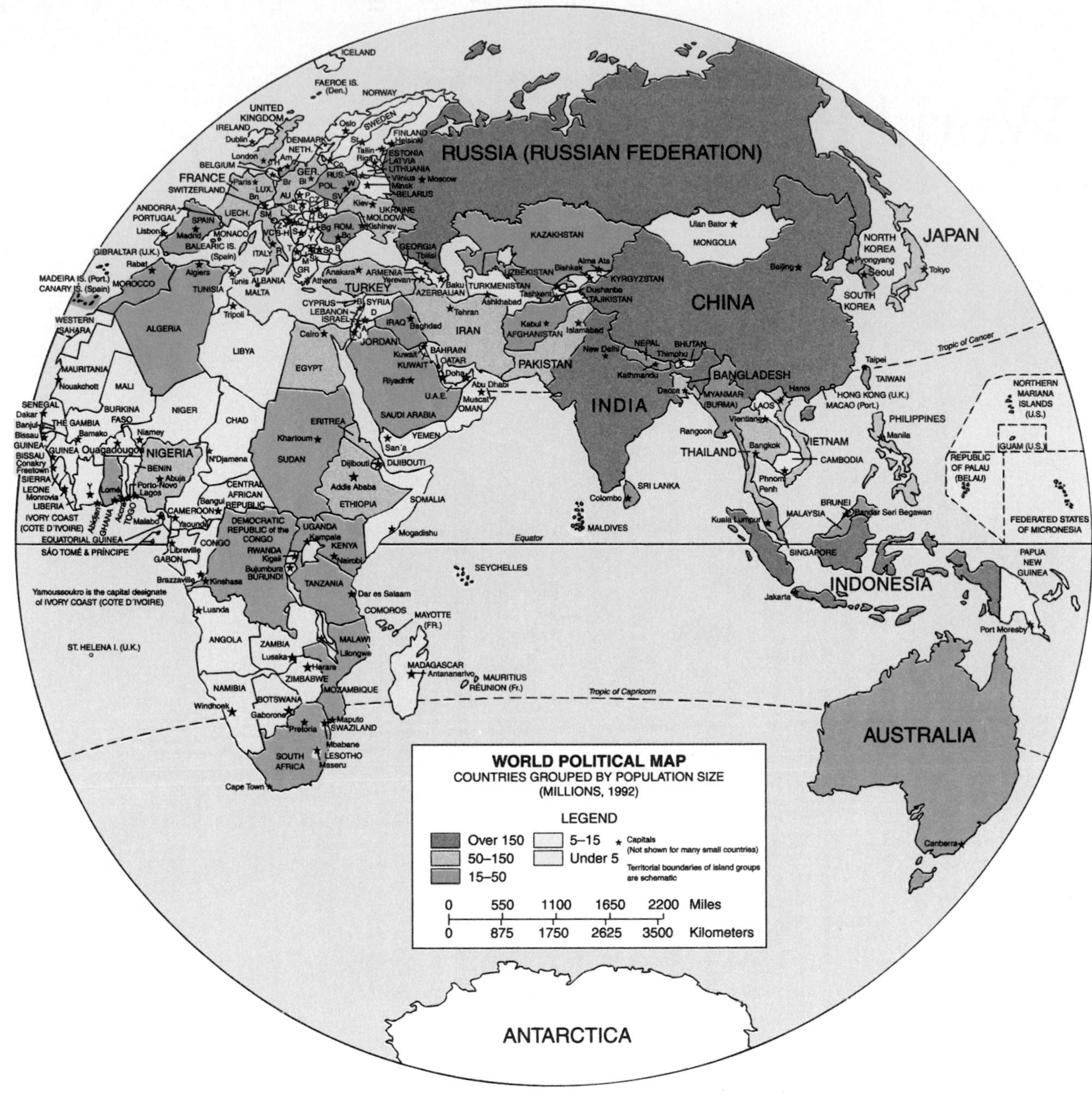

EASTERN HEMISPHERE

APPENDIX D

Periodic Table and Symbols for Chemical Elements

Scientists generally group or organize things that are similar to better undestand them and determine how they relate to one another. An important step in the organization of elements was made in 1869 by Dimitri Mendeleev. Mendeleev showed that when elements are arranged in order of their increasing atomic weights, their physical and chemical properties tend to be repeated in cycles. The arrangement of elements was depicted on a chart called the periodic table of elements, a modern version of which is shown in Table D–1. In the periodic table, each box contains the symbol of the element and its atomic number. Except for two long sequences set apart at the bottom, the elements appear in the increasing order of their atomic numbers. The vertical columns are called *groups* and contain elements with similar properties. For example, in column VIIA, one finds fluorine, chlorine, bromine, and iodine. All of these elements are colored and highly reactive and share other similarities. Fluorine, however, is chemically the most active, chlorine somewhat less active, bromine still less active, and iodine the least active of the four. Except for hydrogen, all of the elements in group IA are soft, shiny metals and very reactive chemically. The horizontal rows on the table are called *periods* and contain sequences of elements having electron configurations that vary in characteristic patterns. It is thus apparent that the periodic table shows relationships among elements rather well and certainly better than an arbitrary listing.

Table D–2 provides the chemical names for the symbols of the elements used in the periodic table.

TABLE D-1 Periodic Table of the Elements

Metals | Semimetals | Nonmetals

Key: 26 — Atomic number (Z); Fe — Element symbol; 55.845 — Atomic mass of naturally occurring isotopic mixture; for radioactive elements, numbers in parentheses are mass numbers of most stable isotopes

Transition elements

I	II											III	IV	V	VI	VII	VIII
1 H 1.0079																	2 He 4.0026
3 Li 6.941	4 Be 9.0122											5 B 10.811	6 C 12.011	7 N 14.007	8 O 15.999	9 F 18.998	10 Ne 20.180
11 Na 22.990	12 Mg 24.305											13 Al 26.982	14 Si 28.086	15 P 30.974	16 S 32.066	17 Cl 35.453	18 Ar 39.948
19 K 39.098	20 Ca 40.078	21 Sc 44.956	22 Ti 47.87	23 V 50.942	24 Cr 51.996	25 Mn 54.938	26 Fe 55.845	27 Co 58.933	28 Ni 58.693	29 Cu 63.546	30 Zn 65.39	31 Ga 69.723	32 Ge 72.61	33 As 74.922	34 Se 78.96	35 Br 79.904	36 Kr 83.80
37 Rb 85.468	38 Sr 87.62	39 Y 88.906	40 Zr 91.224	41 Nb 92.906	42 Mo 95.94	43 Tc (98)	44 Ru 101.07	45 Rh 102.91	46 Pd 106.42	47 Ag 107.87	48 Cd 112.41	49 In 114.82	50 Sn 118.71	51 Sb 121.76	52 Te 127.60	53 I 126.90	54 Xe 131.29
55 Cs 132.91	56 Ba 137.33	71 Lu 174.97	72 Hf 178.49	73 Ta 180.95	74 W 183.84	75 Re 186.21	76 Os 190.23	77 Ir 192.22	78 Pt 195.08	79 Au 196.97	80 Hg 200.59	81 Tl 204.38	82 Pb 207.2	83 Bi 208.98	84 Po (209)	85 At (210)	86 Rn (222)
87 Fr (223)	88 Ra (226)	103 Lr (262)	104 Rf (261)	105 Db (262)	106 Sg (263)	107 Bh (262)	108 Hs (265)	109 Mt (268)	110 Uun (269)	111 Uuu (272)							

Lanthanide series

57 La 138.91	58 Ce 140.12	59 Pr 140.91	60 Nd 144.24	61 Pm (145)	62 Sm 150.36	63 Eu 151.96	64 Gd 157.25	65 Tb 158.93	66 Dy 162.50	67 Ho 164.93	68 Er 167.26	69 Tm 168.93	70 Yb 173.04

Actinide series

89 Ac (227)	90 Th 232.04	91 Pa 231.04	92 U 238.03	93 Np (237)	94 Pu (244)	95 Am (241)	96 Cm (247)	97 Bk (247)	98 Cf (251)	99 Es (252)	100 Fm (257)	101 Md (258)	102 No (259)

TABLE D-2 Elements and Their Chemical Symbols

Actinium	Ac	Erbium	Er	Mercury	Hg	Scandium	Sc
Aluminum	Al	Europium	Eu	Molybdenum	Mo	Seaborgium	Sg
Americium	Am	Fermium	Fm	Neodymium	Nd	Selenium	Se
Antimony	Sb	Fluorine	F	Neon	Ne	Silicon	Si
Argon	Ar	Francium	Fe	Neptunium	Np	Silver	Ag
Arsenic	As	Gadolinium	Gd	Nickel	Ni	Sodium	Na
Astatine	At	Gallium	Ga	Niobium	Nb	Strontium	Sr
Barium	Ba	Germanium	Ge	Nitrogen	N	Sulfur	S
Berkelium	Bk	Gold	Au	Nobelium	No	Tantalum	Ta
Beryllium	Be	Hafnium	Hf	Osmium	Os	Technetium	Tc
Bismuth	Bi	Hassium	Hs	Oxygen	O	Tellurium	Te
Bohrium	Bh	Helium	He	Palladium	Pd	Terbium	Tb
Boron	B	Holmium	Ho	Phosphorus	P	Thallium	Tl
Bromine	Br	Hydrogen	H	Platinum	Pt	Thorium	Th
Cadmium	Cd	Indium	In	Plutonium	Pu	Thulium	Tm
Calcium	Ca	Iodine	I	Polonium	Po	Tin	Sn
Californium	Cf	Iridium	Ir	Potassium	K	Titanium	Ti
Carbon	C	Iron	Fe	Praseodymium	Pr	Tungsten	W
Cerium	Ce	Krypton	Kr	Promethium	Pm	Ununnilium	Uun
Cesium	Cs	Lanthanum	La	Protactinium	Pa	Unununium	Uuu
Chlorine	Cl	Lawrencium	Lr	Radium	Ra	Uranium	U
Chromium	Cr	Lead	Pb	Radon	Rn	Vanadium	V
Cobalt	Co	Lithium	Li	Rhenium	Re	Xenon	Xe
Copper	Cu	Lutetium	Lu	Rhodium	Rh	Ytterbium	Yb
Curium	Cm	Magnesium	Mg	Rubidium	Rb	Yttrium	Y
Dubnium	Db	Manganese	Mn	Ruthenium	Ru	Zinc	Zn
Dysprosium	Dy	Meitnerium	Mt	Rutherfordium	Rf	Ziconium	Zr
Einsteinium	Es	Mendelevium	Md	Samarium	Sm		

APPENDIX E

Convenient Conversion Factors

To Convert from	To	Multiply by*
Centimeters	Feet	0.0328 ft/cm
	Inches	0.394 in/cm
	Meters	0.01 m/cm
	Microns (micrometers)	10,000 μm/cm
	Miles (statute)	6.214×10^{-6} mi/cm
	Millimeters	10 mm/cm
Feet	Centimeters	30.48 cm/ft
	Inches	12 in/ft
	Meters	0.3048 m/ft
	Microns (micrometers)	304,800 μm/ft
	Miles (statute)	0.000189 mi/ft
	Yards	0.3333 yd/ft
Kilometers	Miles	0.6214 mi/km
Gallons (U.S., liquid)	Cubic centimeters	3785 cm^3/gal
	Cubic feet	0.133 ft^3/gal
	Liters	3.785 L/gal
	Quarts (U.S., liquid)	4 qt/gal
Grams	Kilograms	0.001 kg/g
	Micrograms	1×10^6 μg/g
	Ounces	0.03527 oz/g
	Pounds	0.002205 lb/g
Inches	Centimeters	2.54 cm/in
	Feet	0.0833 ft/in
	Meters	0.0254 m/in
	Yards	0.0278 yd/in
Kilograms	Ounces	35.27 oz/kg
	Pounds	2.205 lb/kg
Meters	Centimeters	100 cm/m
	Feet	3.2808 ft/m
	Inches	39.37 in/m
	Kilometers	0.001 km/m
	Miles (statute)	0.0006214 mi/m
	Millimeters	1000 mm/m
	Yards	1.0936 yd/m
Miles (statute)	Centimeters	160,934 cm/mi
	Feet	5280 ft/mi
	Inches	63,360 in/mi
	Kilometers	1.609 km/mi
	Meters	1609 m/mi
	Yards	1760 yd/mi
Ounces	Grams	28.3 g/oz
	Pounds	0.0625 lb/oz
Pounds	Grams	453.6 g/lb
	Kilograms	0.454 kg/lb
	Ounces	16 oz/lb

*Values are generally given to three or four significant figures.
Source: Turk, J., and Turk, A. 1977. *Physical Science*. Philadelphia: W. B. Saunders Company.

APPENDIX F

Exponential or Scientific Notation

Exponential or scientific notation is used by scientists all over the world. This system is based on exponents of 10, which are shorthand notations for repeated multiplications or divisions.

A positive exponent is a symbol for a number that is to be multiplied by itself a given number of times. Thus, the number 10^2 (read "ten squared" or "ten to the second power") is exponential notation for $10 \cdot 10 = 100$. Similarly, $3^4 = 3 \cdot 3 \cdot 3 \cdot = 81$. The reciprocals of these numbers are expressed by negative exponents. Thus $10^{-2} = 1/10^2 = 1/(10 \cdot 10) = 1/100 = 0.01$.

To write 10^4 in longhand form you simply start with the number 1 and move the decimal four places to the right: 10000. Similarly, to write 10^{-4} you start with the number 1 and move the decimal four places to the left: 0.0001.

$$\underbrace{1\ 000\ 000}_{6\text{ places}} = 10^6.$$

Similarly, the decimal place of the number 0.000001 is six places to the left of 1:

$$\underbrace{0.000001}_{6\text{ places}} = 10^{-6}.$$

APPENDIX G-1

Rock Symbols

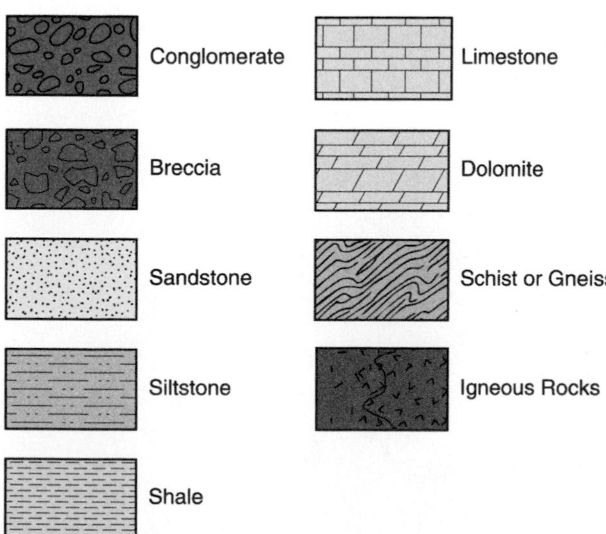

The standard geologic symbols used by geologists and in diagrams in this book are indicated above.

Bedrock Geology of North America

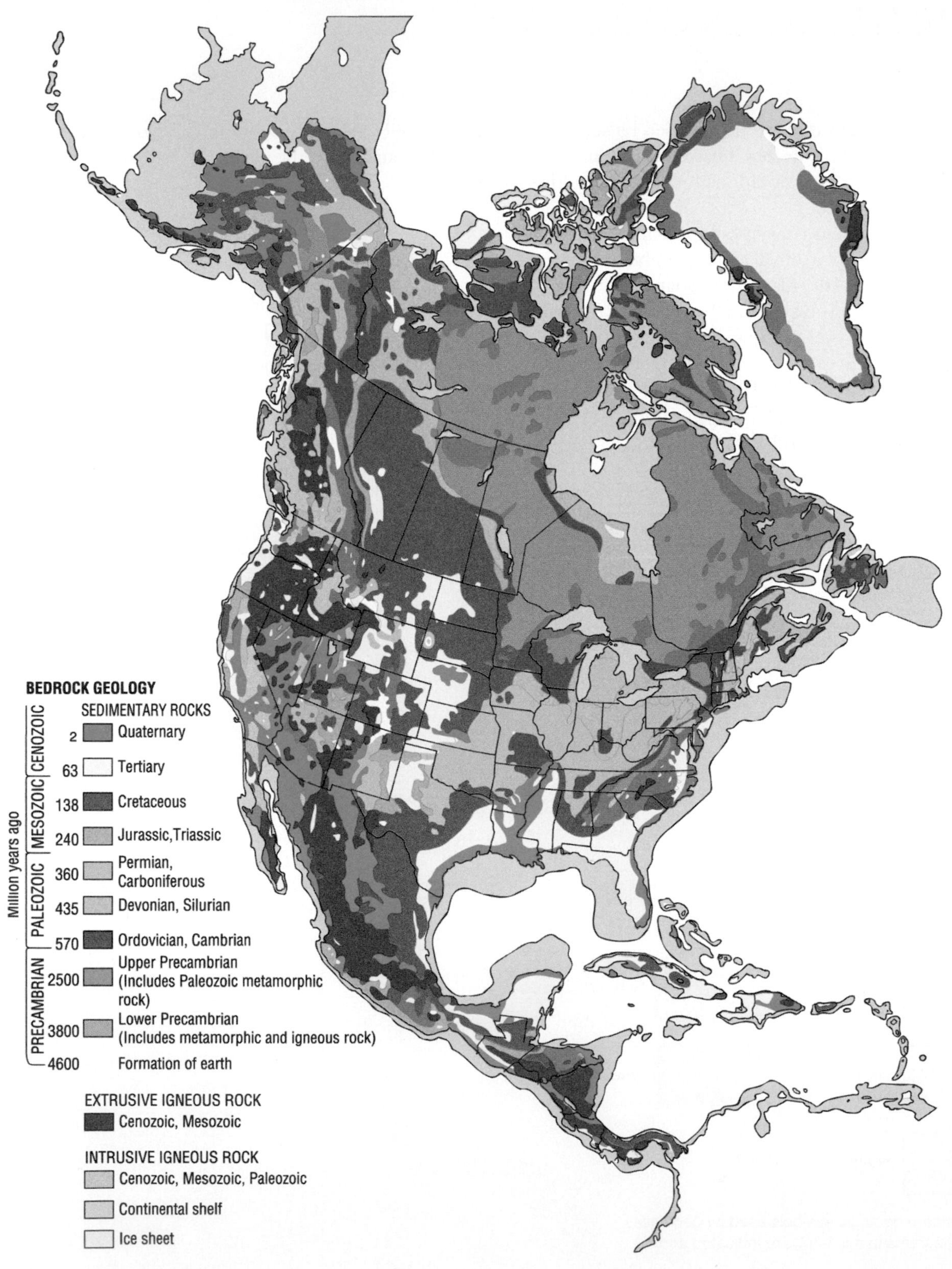
BEDROCK GEOLOGY
Million years ago
SEDIMENTARY ROCKS
CENOZOIC
2 Quaternary
63 Tertiary
MESOZOIC
138 Cretaceous
240 Jurassic,Triassic
PALEOZOIC
360 Permian, Carboniferous
435 Devonian, Silurian
570 Ordovician, Cambrian
PRECAMBRIAN
2500 Upper Precambrian (Includes Paleozoic metamorphic rock)
3800 Lower Precambrian (Includes metamorphic and igneous rock)
4600 Formation of earth
EXTRUSIVE IGNEOUS ROCK
Cenozoic, Mesozoic
INTRUSIVE IGNEOUS ROCK
Cenozoic, Mesozoic, Paleozoic
Continental shelf
Ice sheet

Common Rock-Forming Silicate Minerals

Silicate Mineral		Composition	Physical Properties
Quartz		Silicon dioxide (silica, SiO_2)	Hardness of 7 (on scale of 1 to 10)*; will not cleave (fractures unevenly); specific gravity: 2.65
Potassium feldspar group		Alluminosilicates of potassium	Harness of 6.0–6.5; cleaves well in two directions; pink or white; specific gravity: 2.5–2.6
Plagioclase feldspar group		Aluminosilicates of sodium and calcium	Hardness of 6.0–6.5; cleaves well in two directions; white or gray; may show striations on cleavage planes; specific gravity: 2.6–2.7
Muscovite mica	Ferromagnesian minerals	Aluminosilicates of potassium with water	Hardness of 2–3; cleaves perfectly in one direction, yielding flexible, thin plates; colorless; transparent in thin sheets; specific gravity: 2.8–3.0
Biotite mica	Ferromagnesian minerals	Aluminosilicates of magnesium, iron, potassium, with water	Hardness of 2.5–3.0; cleaves perfectly in one direction, yielding flexible, thin plates; black to dark brown; specific gravity: 2.7–3.2
Pyroxene group	Ferromagnesian minerals	Silicates of aluminum, calcium, magnesium, and iron	Hardness of 5–6; cleaves in two directions at 87° and 93°; black to dark green; specific gravity: 3.1–3.5
Amphibole group	Ferromagnesian minerals	Silicates of aluminum, calcium, magnesium, and iron	Hardness of 5–6; cleaves in two directions at 56° and 124°; black to dark green; specific gravity: 3.0–3.3
Olivine	Ferromagnesian minerals	Silicates of magnesium and iron	Hardness of 6.5–7.0; light green; transparent to translucent; specific gravity: 3.2–3.6
Garnet group		Aluminosilicates of iron, calcium, magnesium, and manganese	Hardness of 6.5–7.5; uneven fracture; red, brown, or yellow; specific gravity: 3.5–4.3

*The scale of hardness used by geologists was formulated in 1822 by Frederich Mohs. Beginning with diamond as the hardest mineral, he arranged the following table:

10 Diamond	8 Topaz	6 Feldspar	4 Fluorite	2 Gypsum
9 Corundum	7 Quartz	5 Apatite	3 Calcite	1 Talc

GLOSSARY

For terms not included in this glossary, students may wish to consult the *Glossary of Geology* (1997, edited by J. A. Jackson, 4th ed., Falls Church, Virginia, American Geological Institute).

Absaroka sequence A sequence of Permian–Pennsylvanian sediments bounded both above and below by a regional unconformity and recording an episode of marine transgression over an eroded surface, full flood level of inundation, and regression from the craton.

Absolute geologic age The actual age, expressed in years, of a geologic material or event.

Acadian orogeny An episode of mountain building in the northern Appalachians during the Devonian Period.

Acanthodians The earliest known vertebrates (fishes) with a movable, well-developed lower jaw, or mandible; hence, the first jawed fishes.

Accretionary terrane A block of continental crust having fault boundaries that is geologically distinct from surrounding terranes.

Adaptation A modification of an organism that better fits it for existence in its present environment or enables it to live in a somewhat different environment.

Adaptive radiation The diversity that develops among species as each adapts to a different set of environmental conditions.

Adenosine diphosphate (ADP) A product formed in the hydrolysis of adenosine triphosphate that is accompanied by release of energy and organic phosphate.

Adenosine triphosphate (ATP) A compound that occurs in all cells and that serves as a source of energy for physiologic reactions such as muscle contraction.

Aerobic organism An organism that uses oxygen in carrying out respiratory processes.

Age The time represented by the time-stratigraphic unit called a stage. (Informally, may indicate any time span in geologic history, as "Age of Cycads.")

Agnatha The jawless vertebrates, including extinct ostracoderms and living lampreys and hagfishes.

Algae Any of a large group of simple plants (thallophyta) that contain chlorophyll and are capable of photosynthesis.

Allegheny orogeny The late Paleozoic episodes of mountain building along the present trend of the Appalachian Mountains. (Also termed the Appalachian orogeny.)

Alluvium Unconsolidated, poorly sorted detrital sediments ranging from clay to gravel sizes and characteristically fluvial in origin.

Alpha particle A particle equivalent to the nucleus of a helium atom, emitted from an atomic nucleus during radioactive decay.

Alpine orogeny In general, the sequence of crustal disturbances beginning in the middle Mesozoic and continuing into the Miocene that resulted in the geologic structures of the Alps.

Amino acids Nitrogenous hydrocarbons that serve as the building blocks of proteins and are thus essential to all living things.

Ammonites Ammonoid cephalopods having more complex sutural patterns than either ceratites or goniatites.

Ammonoids An extinct group of cephalopods with coiled, chambered conch(s) and having septa with crenulated margins.

Amniotic egg An egg produced by reptiles, birds, and monotremes in which the developing embryo is maintained and protected by an elaborate arrangement of shell membranes, yolk, sac, amnion, and allantois.

Amphibians "Cold-blooded" vertebrates that utilize gills for respiration in the early life stages but that have air-breathing lungs as adults.

Amphibole A ferromagnesium silicate mineral that occurs commonly in igneous and metamorphic rocks.

Anaerobic organism An organism that does not require oxygen for respiration but rather makes use of processes such as fermentation to obtain its energy.

Andesite A volcanic rock that in chemical composition is intermediate between basalt and granite.

Angiosperms An advanced group of plants having floral reproductive structures and seeds in a closed ovary. The "flowering plants."

Anthropoidea The suborder of primates that includes monkeys, apes, and humans.

Anticline A geologic structure in which strata are bent into an upfold or arch.

Antler orogeny A Late Devonian and Mississippian episode of mountain building involving folding and thrusting along a belt across Nevada to southwestern Alberta.

Archaea The domain that includes thermophylic (heat-dependent), halophylic (salt-dependent), methane-producing, and sulfur-producing bacteria. (The remaining two domains are the Bacteria and Eurkarya.)

Archaeocyatha A group of Cambrian marine organisms having double, perforated, calcareous walls and conical-to-cylindric skeletal form. **Archaeocyathids** lived during the Cambrian.

Archean Pertaining to the division of Precambrian time beginning 3.8 billion years ago and ending 2.5 billion years ago.

Archosaurs Advanced reptiles of a group called diapsids, which includes thecodonts, "dinosaurs," pterosaurs, and crocodiles.

Arcoids A group of bivalves (pelecypods) exemplified by species of Arca.

Artiodactyl Hoofed mammals that typically have two or four toes on each foot.

Asteroid One of numerous relatively small planetary bodies (less than 800 kilometers in diameter) revolving around the Sun in orbits lying between those of Mars and Jupiter.

Asthenosphere The zone between 50 and 250 kilometers below the surface of the Earth, where shock waves of earthquakes travel at much reduced speeds, perhaps because of less rigidity. The asthenosphere may be a zone where convective flow of material occurs.

Atoll A ring-like island or a series of islands formed by corals and calcareous algae around a central lagoon.

Atom The smallest divisible unit retaining the characteristics of a specific element.

Atomic fission A nuclear process that occurs when a heavy nucleus splits into two or more lighter nuclei, simultaneously liberating a considerable amount of energy.

Atomic fusion A nuclear process that occurs when two light nuclei unite to form a heavier one. In the process, a large amount of energy is released.

Atomic mass A quantity essentially equivalent to the number of neutrons plus the number of protons in an atomic nucleus.

Atomic number The number of protons in the nuclei of atoms of a particular element. (An element is thus a substance in which all of the atoms have the same atomic number.)

Aulocogen The failed arm of a triple junction rift.

Australopithecines A general term applied loosely to Pliocene and early Pleistocene primates whose skeletal characteristics place them between typically ape-like individuals and those more obviously human.

Autotroph An organism that uses an external source of energy to produce organic nutrients from simple inorganic chemicals.

Bacteria Unicellular, prokaryotic microorganisms belonging to the Kingdom Monera. One of the three great domains of organisms that include also the Archaea and the Eukarya.

Basin A depressed area that serves as a catchment area for sediments (basin of deposition). A structural basin is an area in which strata slope inward toward a central location. Structural basins tend to experience periodic downsinking and thus receive a thicker and more complete sequence of sediments than do adjacent areas.

Belemnites Members of the molluscan Class Cephalopoda, having straight internal shells.

Benioff seismic zone An inclined zone along which frequent earthquake activity occurs and that marks the location of the plunging, forward edge of a lithospheric plate during subduction.

Benthic Pertaining to a bottom-dwelling organism.

Bentonite A layer of clay, presumably formed by the alteration of volcanic ash.

Beta particle A charged particle, essentially equivalent to an electron, emitted from an atomic nucleus during radioactive disintegration.

Bioturbation The disturbance of sediment by burrowing, boring, and sediment-ingesting organisms.

Bivalvia A class of the Phylum Mollusca (also known as the Class Pelecypoda).

Blastoids Sessile (attached) Paleozoic echinoderms having a stem and an attached cup or calyx composed of relatively few plates.

Bolide A meteorite, asteroid, or comet that explodes on striking the Earth.

Brachiating Swinging from branch to branch and tree to tree by using the limbs, as among monkeys.

Brachiopod A bivalved (doubled-shelled) marine invertebrate. Brachiopods were particularly common and widespread during the Paleozoic and persist in fewer numbers today.

Breccia A clastic sedimentary rock composed largely of angular fragments of granule size or larger.

Bryozoa A phylum of attached and incrusting colonial marine invertebrates.

Burgess Shale fauna A beautifully preserved fossil fauna of soft-bodied Cambrian animals discovered in 1910 by Charles Walcott in Kicking Horse Pass, Alberta, Canada.

Caledonian orogeny A major early Paleozoic episode of mountain building affecting Europe that created an orogenic belt, the Caledonides, extending from Ireland and Scotland northwestward through Scandinavia.

Carbon-14 A radioactive isotope of carbon with an atomic mass of 14. Carbon-14 is frequently used in determining the age of materials less than about 50,000 years old.

Carbonate A general term for a chemical compound formed when carbon dioxide dissolved in water combines with oxides of calcium, magnesium, potassium, sodium, and iron. The most common carbonate minerals are calcite (which forms the carbonate rock limestone) and dolomite.

Carbonization The concentration of carbon during fossilization.

Cast (natural) A replica of an organic subject, such as a fossil shell, formed when sediment fills a mold of that object.

Catskill delta A build-up of Middle and Upper Devonian clastic sediments as a broad, complex clastic wedge derived from the erosion of highland areas formed largely during the Acadian orogeny.

Ceboidea The New World monkeys, characterized by prehensile tails, and including the capuchin, marmoset, and howler monkeys.

Centrifugal force The apparent outward force experienced by an object moving in a circular path. Centrifugal force is a manifestation of inertia, the tendency of moving things to travel in straight lines.

Ceratites One of the three larger groups of ammonoid cephalopods having sutural complexity intermediate between goniatites and ammonites.

Ceratopsians The quadrupedal ornithischian dinosaurs characterized by the development of prominent horns on the head.

Cercopithecoidea The Old World monkeys (of Asia, southern Europe, and Africa), including macaques, guenons, langurs, baboons, and mandrills.

Cetaceans The group of marine mammals that includes whales and porpoises.

Chalicotheres Extinct perissodactyls having robust claws rather than hoofs.

Chalk A soft, white, fine-grained variety of limestone composed largely of the calcium carbonate skeletal remains of marine microplankton.

Chert A dense, hard sedimentary rock or mineral composed of submicrocrystalline quartz. Unless colored by impurities, chert is white, as opposed to flint, which is dark or black.

Chlorophyll The catalyst that makes possible the reaction of water and carbon dioxide in green plants to produce carbohydrates. Photosynthesis is the reaction.

Choanichthyes That group of fishes that includes both the dipnoans (lungfishes with weak pelvic and pectoral fins) and crossopterygians (lungfishes with stout lobe-fins).

Chondrichthyes The broad category of fishes with cartilaginous skeletons that is exemplified by sharks, skates, and rays.

Chondrites Stony meteorites that contain rounded silicate grains or chondrules. Chondrules are believed to have formed by crystallization of liquid silicate droplets.

Chromosome A thread-like, microscopic body composed of chromatin. Chromosomes appear in the nucleus of the cell at the time of cell division. They contain the genes. The number of chromosomes is normally constant for a particular species.

Chromosphere One of the concentric shells of the Sun, lying above the photosphere and telescopically visible as a thin, brilliant red rim around the edge of the sun for a second or so at the beginning and end of a solar eclipse.

Cladistic phylogeny An approach to biologic classification in which organisms are grouped according to similarities that are derived from a common ancestor.

Clastic texture Texture that characterizes a rock made up of fragmental grains such as sand, silt, or parts of fossils. Conglomerates, sandstones, and siltstones are clastic rocks; the individual clastic grains are termed clasts.

Clastic wedge An extensive accumulation of largely clastic sediments deposited adjacent to uplifted areas. Sediments in the wedge become finer and the section becomes thinner in a direction away from the upland source areas. The Queenston and Catskill "deltas" are examples of clastic wedges.

Coccolithophorids Marine, planktonic, biflagellate, golden-brown algae that typically secrete coverings of discoidal calcareous platelets called coccoliths.

Colorado mountains Highlands uplifted in Pennsylvanian time in Colorado. Sometimes inappropriately termed the "ancestral Rockies."

Conodonts Small, tooth-like, stratigraphically useful fossils composed of calcium phosphate and found in Cambrian to Triassic rocks. Conodonts are elements of an apparatus used by primitive chordates in capturing and chewing food.

Contact metamorphism The compositional and textural changes in a rock that result from heat and pressure emanating from an adjacent igneous intrusion.

Convergence (in evolution) The process by which similarity of form or structure arises among different organisms as a result of their becoming adapted to similar habitats.

Cordaites A primitive order of tree-like plants with long, blade-like leaves and clusters of naked seeds. Cordaites were in some ways intermediate in evolutionary stage between seed ferns and conifers.

Core The central part of the Earth that lies beneath the mantle.

Coriolis effect The deflection of winds and water currents to the right in the Northern Hemisphere and to the left in the Southern Hemisphere as a consequence of the Earth's rotation.

Cosmic rays Extremely high-energy particles, mostly protons, that move through the galaxy and frequently strike the Earth's atmosphere.

Craton The long-stable region of a continent, commonly with Precambrian rocks either at the surface or only thinly covered with younger sedimentary rocks.

Creodonta Primitive, early, flesh-eating placental mammals.

Crinoids Stalked echinoderms with a calyx composed of regularly arranged plates from which radiate arms for gathering food.

Cross-bedding (cross-stratification) An arrangement of laminae or thin beds transverse to the planes of stratification. The inclined laminae are usually at inclinations of less than 30A7 and may be either straight or concave.

Crossopterygii That group of choanichthyan fishes ancestral to earliest amphibians and characterized by stout pectoral and pelvic fins as well as lungs.

Crust (Earth) The outer part of the lithosphere; it averages about 32 kilometers in thickness.

Crustacea A subphylum of the phylum Arthropoda that includes such well-known living animals as lobsters and crayfishes.

Curie temperature The temperature at which a cooling mineral acquires permanent magnetic properties that record the surrounding magnetic field orientation and strength at the time of cooling. Magnetic properties of a mineral above the Curie temperature will change as the surrounding field changes. Below the Curie temperature, magnetic characteristics of the mineral will not alter.

Cyanobacteria Prokaryotic, photosynthetic microorganisms that possess chlorophyll and produce oxygen. Formerly termed blue-green algae.

Cycadales A group of seed plants that were especially common during the Mesozoic and were characterized by palm-like leaves and coarsely textured trunks marked by numerous leaf scars.

Cyclothem A vertical succession of sedimentary units reflecting environmental events that occurred in a constant order. Cyclothems are particularly characteristic of the Pennsylvania System.

Cystoids Attached echinoderms with generally irregular arrangement and number of plates in the calyx and perforated by pores or slits.

Deccan traps A thick sequence (3200 meters) of Upper Cretaceous basaltic lava flows that cover about 500,000 kilometers^2 of peninsular India.

Décollement The feature of stratified rocks in which upper formations may become "unstuck" from lower formations, deform, and slide thousands of meters over underlying beds.

Diatoms Microscopic golden-brown algae (chrysophytes) that secrete a delicate siliceous frustule (shell).

Differentiation The process by which a planet becomes internally zoned, as when heavy materials sink toward its center and light materials accumulate near the surface.

Dinoflagellates Unicellular marine algae usually having two flagella and a cellulose wall.

Diploid cells Cells having two sets of chromosomes that form pairs, as in somatic cells.

Dipnoi An order of lungfishes with weak pectoral and pelvic fins; not considered ancestral to land vertebrates.

Disconformity A variety of unconformity in which bedding planes above and below the plane of erosion or nondeposition are parallel.

DNA (deoxyribonucleic acid) The nucleic acid found chiefly in the nucleus of a cell that functions in the transfer of genetic characteristics and in protein synthesis.

Docodonts A group of small, primitive Late Jurassic mammals possibly ancestral to the living monotremes.

Dolomite (or dolostone) A carbonate sedimentary rock of which more than 50 percent is the mineral dolomite $CaMg(CO_3)_2$.

Domain A major taxonomic division ranking higher than a kingdom. The three domains are the Archaea, Bacteria, and Eukarya.

Dome An upfold in rocks having the general configuration of an inverted bowl. Strata in a dome dip outward and downward in all directions from a central area. An example is the Ozark dome.

Dryopithecine In general, a group of lightly built primates that lived during the Miocene and Pliocene in mostly open savannah country and that includes Dryopithecus, a form considered to be in the line leading to apes.

Dynamothermal (regional) metamorphism Metamorphism that has occurred over a wide region, caused by deep burial and high temperatures associated with pressures resulting from overburden and orogeny.

Echinoderms The large group (phylum Echinodermata) of marine invertebrates characterized by prominent pentamerous symmetry and a skeleton frequently constructed of calcite elements and including spines. Cystoids, blastoids, crinoids, and echinoids are examples of echinoderms.

Edentata An order of placental mammals that includes extinct ground sloths and gyptodonts, as well as living armadillos, tree sloths, and South American anteaters.

Ediacaran fauna The late Proterozoic fauna of multicellular animals first discovered in Australia but subsequently found in rocks about 600 million years old in many continents.

Endemic population The native fauna of any particular region.

Entelodonts A group of extinct artiodactyls bearing a superficial resemblance to giant wild boars.

Eon A major division of the geologic time scale. The Phanerozoic Eon comprises all of the geologic periods from the Cambrian to the Holocene. The term is also sometimes used to denote a span of 1 billion years.

Epifaunal organisms Organisms living on, as distinct from in, a particular body of sediment or another organism.

Epoch A chronologic subdivision of a geologic period. Rocks deposited or emplaced during an epoch constitute the series for that epoch.

Era A major division of geologic time, divisible into geologic periods.

Eukaryote A cell containing a true nucleus, enclosed within a nuclear membrane, and having well-defined chromosomes and cell organelles.

Eurypterids Aquatic arthropods of the Paleozoic, superficially resembling scorpions and probably carnivorous.

Eustatic Pertaining to worldwide simultaneous changes in sea level, such as might result from change in the volume of continental glaciers.

Evaporites Sediments precipitated from a water solution as a result of the evaporation of that water. Evaporite minerals include anhydrite gypsum ($CaSO_4$) and halite ($NaCl$).

Evolution The continuous genetic adaptation of organisms or species to the environment.

Facies A particular aspect of sedimentary rocks that is a direct consequence of sedimentation in a particular depositional environment.

Fault A fracture in the Earth's crust along which rocks on one side have been displaced relative to rocks on the other side.

Felsic Term used to describe light colored igneous rocks having the general composition of granite.

Fermentation The partial breakdown of organic compounds by an organism in the absence of oxygen. The final product of fermentation is alcohol or lactic acid.

Filter-feeders Animals that obtain their food, which usually consists of small particles or organisms, by filtering it from the water.

Fissility That property of rocks that causes them to split into thin slabs parallel to bedding. Fissility is particularly characteristic of shale.

Flood basalts Regionally extensive layers of basalt that originated as low-viscosity lava pouring from fissure eruptions. The lavas of the Columbia Plateau and the Deccan Plateau are flood basalts.

Fluvial Pertaining to sediments or other geologic features formed by streams.

Flysch Thick sequences of rapidly deposited, poorly sorted marine clastics.

Focus (earthquake) The location at which rock rupture occurs that generates seismic waves.

Foliation A textural feature especially characteristic of metamorphic rocks in which laminae develop by growth or realignment of minerals in parallel orientation.

Foraminifera An order of mostly marine, unicellular protozoans that secrete tests (shells) that are usually composed of calcium carbonate.

Formation A mappable, lithologically distinct body of rock having recognizable contacts with adjacent rock units.

Fossil The remains or indications of an organism that lived in the geologic past.

Fractional crystallization The separation of components of a cooling magma by sequential formation of particular mineral crystals at progressively lower temperatures.

Fusulinids Primarily spindle-shaped foraminifers with calcareous, coiled tests divided into a complex of numerous chambers. Fusulinids were particularly abundant during the Pennsylvanian and Permian periods.

Galaxy An aggregate of stars and planets, separated from other such aggregates by distances greater than those between member stars.

Gamete Either of two cells (male or female) that must unite in sexual reproduction to initiate the development of a new individual.

Gamma rays Very high-frequency electromagnetic waves.

Garnet A family of aluminosilicates of iron and calcium that are particularly characteristic of metamorphic rocks.

Gene The unit of heredity transmitted in the chromosome.

Gene pool All the genes present in a species population.

Genus The major subdivision of a taxonomic family or subfamily of plants or animals, usually consisting of more than one species.

Geochronology The study of time as applied to Earth and planetary history.

Geologic range The geologic time span between the first and last appearance of an organism.

Geology The science of the Earth, including the materials that the planet is made of, the physical and chemical changes that occur on and within the Earth, and the history of the Earth and its inhabitants.

Glacier A large mass of ice, formed by the recrystallization of snow, that flows slowly under the influence of gravity.

Glauconite A green clay mineral frequently found in marine sandstones and believed to have formed at the site of deposition.

***Glossopteris* flora** An assemblage of fossil plants found in rocks of late Paleozoic and early Triassic age in South Africa, India, Australia, and South America. The flora takes its name from the seed fern *Glossopteris*.

Gondwanaland The great Permo-Carboniferous Southern Hemisphere continent, comprising the assembled present areas of South Africa, India, Australia, Africa-Arabia, and Antarctica.

Goniatites One of the three large groups of ammonoid cephalopods with sutures forming a pattern of simple lobes and saddles and thus not as complex as either the ceratites or the ammonites.

Gowganda Conglomerate An apparent tillite of the Canadian Shield. The Gowganda rests on a surface of older rock that appears to have been polished by glacial action.

Graptolite facies A Paleozoic sedimentary facies composed of dark shales and fine-grained clastics that contain the abundant remains of graptolites and that are associated with volcanic rocks.

Graptolites Extinct colonial marine animals considered to be hemichordates. Graptolites range from the late Cambrian to the Mississippian.

Gravity anomalies The differences between the observed value of gravity at any point on the Earth and the calculated theoretic value.

Greenhouse effect A process in which incoming solar radiation that is absorbed and reradiated cannot escape back into space because the Earth's atmosphere is not transparent to the reradiated energy, which is in the form of infrared radiation (heat).

Greenstones Low-grade metamorphic rocks containing abundant chlorite, epidote, and biotite and developed by metamorphism of basaltic extrusive igneous rocks. Great linear outcrops of greenstones are termed greenstone belts and are thought to mark the locations of former volcanic island arcs.

Guide fossil A fossil with a wide geographic distribution but narrow stratigraphic range and thus useful in correlating strata and for age determination.

Gutenberg discontinuity The boundary separating the mantle of the Earth from the core below. The Gutenberg discontinuity lies about 2900 kilometers below the surface.

Gymnosperms An informal designation for flowerless seed plants in which seeds are not enclosed (hence, "naked seeds"). Examples are conifers and cycads.

Hadrosaurs The ornithischian duck-billed dinosaurs of the Cretaceous.

Half-life The time in which one-half of an original amount of a radioactive species decays to daughter products.

Haploid cell A cell having a single set of chromosomes, as in gametes (see diploid cells).

Hercynian orogeny The major late Paleozoic orogenic episode in Europe that formed the ancient Hercynian mountains. Today, only the eroded stumps of these mountains are exposed in areas where the cover of Mesozoic and Cenozoic strata has been removed by erosion.

Heterotroph An organism that depends on an external source of organic substances for its nutrition and energy.

Holocene A term sometimes used to designate the period of time since the last major episode of glaciation. The term is equivalent to Recent.

Homologous organs Organs having structural and developmental similarities due to genetic relationship.

Hydrosphere The water and water vapor present at the surface of the Earth, including oceans, seas, lakes, and rivers.

Hylobatidae A group of persistently arboreal, small apes that are exemplified by the gibbons and siamangs.

Hyomandibular bone (in fishes) The modified upper bone of the hyoid arch, which functions as a connecting element between the jaws and the braincase in certain fishes.

Ichnology The study of trace fossils (tracks, trails, burrows, borings, castings, etc.).

Ichthyosaurs Highly specialized marine reptiles of the Mesozoic, recognized by their fishlike form.

Ictidosauria A group of extinct mammal-like reptiles or therapsids whose skeletal characteristics are considered to be very close to those of mammals.

Igneous rock A rock formed by the cooling and solidification of magma or lava.

Infaunal organisms Organisms that live and feed within bottom sediments.

Ion An atom that, because of electron transfers, has excess positive or negative charges.

Island arcs Chains of islands, arranged in arcuate trends on the surface of the Earth. The arcs are sites of volcanic and earthquake activity and are usually bordered by deep oceanic trenches on the convex side.

Isopachous map A map depicting the thickness of a sedimentary unit.

Isotopes Atoms of an element that have the same number of protons in the nucleus, the same atomic number, and the same chemical properties but different atomic masses because they have different numbers of neutrons in their nuclei.

Karoo system A sequence of Permian to Lower Jurassic rocks, primarily continental formations, which outcrop in Africa and are approximately equivalent to the Gondwana system of peninsular India.

Kaskaskia sequence A sequence of Devonian-Mississippian sediments, bounded above and below by regional unconformities and recording an episode of transgression followed by full flooding of a large part of the craton and by subsequent regression.

Komatiite An ultramafic volcanic rock prevalent in Archean terranes.

Lacustrine Pertaining to lakes, as in lacustrine sediments (lake sediments).

Lagomorph The order of small placental mammals that includes rabbits, hares, and pikas.

Laramide orogeny In general, those pulses of mountain building that were frequent in Late Cretaceous time and were in large part responsible for producing many of the structures of the Rocky Mountains.

Lateral fault A fault in which the movement is largely horizontal and in the direction of the trend of the fault plane. Sometimes called a strike-slip fault.

Laurasia A hypothetic supercontinent composed of what is now Europe, Asia, Greenland, and North America.

Limestone A sedimentary rock consisting mainly of calcium carbonate.

Lithofacies map A map that shows the areal variation in lithologic attributes of a stratigraphic unit.

Lithosphere The outer shell of the Earth, lying above the asthenosphere and comprising the crust and upper mantle.

Litopterns South American ungulates whose evolutionary history somewhat paralleled that of horses and camels.

Logan's line A zone of thrust-faulting produced during the Taconic orogeny that extends from the west coast of Newfoundland along the trend of the St. Lawrence River to near Quebec and southward along Vermont's western border. (Named after the pioneer Canadian geologist Sir William Logan.)

Lophophore An organ located adjacent to the mouth of brachiopods and bryozoans that bears ciliated tentacles and has as its primary function the capture of food particles.

Low-velocity zone The interior zone of the Earth, characterized by lower seismic wave velocities than the region immediately above it.

Lunar maria Low-lying, dark lunar plains filled with volcanic rocks rich in iron and magnesium.

Lycopsids Leafy plants with simple, closely spaced leaves bearing sporangia on their upper surfaces. They are represented by living club mosses and vast numbers of extinct late Paleozoic "scale trees."

Magnetic declination The horizontal angle between "true" (geographic) north and magnetic north, as indicated by the compass needle. Declination is the result of Earth's magnetic axis being inclined with respect to the Earth's rotational axis.

Magnetic inclination The angle between the magnetic lines of force for the Earth and the Earth's surface; sometimes called "dip." Magnetic inclination can be demonstrated by observing a freely suspended magnetic needle. The needle will lie parallel to the Earth's surface at the equator but is increasingly inclined toward the vertical as the needle is moved toward the magnetic poles.

Mammoth The name commonly applied to extinct elephants of the Pleistocene Epoch.

Marsupials Mammals of the Order Marsupialia. Female marsupials bear mammary glands and carry their immature young in a stomach pouch.

Mass spectrometer An instrument that separates ions of different mass but equal charge and measures their relative quantities.

Mastodonts The group of extinct proboscideans (elephantoids), early forms of which were characterized by long jaws, tusks in both jaws, and low-crowned teeth.

Medusa The free-swimming, umbrella-shaped jellyfish form of the phylum Cnidaria.

Meiosis That kind of nuclear division, usually involving two successive cell divisions, that results in daughter cells having one-half the number of chromosomes that were in the original cell.

Mélange A body of intricately folded, faulted, and severely metamorphosed rocks, examples of which can be seen in the Franciscan rocks of California.

Mesosphere The zone of the Earth's mantle where pressures are sufficient to impart greater strength and rigidity to the rock.

Metamorphic rock A rock formed from a previously existing rock by subjecting the parent rock to high temperature and pressure but without melting.

Metamorphism The transformation of previously existing rocks into new types by the action of heat, pressure, and chemical solutions. Metamorphism usually takes place at depth in the roots of mountain chains or adjacent to large intrusive igneous bodies.

Metazoa All multicellular animals whose cells become differentiated to form tissues (all animals except Protozoa).

Meteorites Metallic or stony bodies from interplanetary space that have passed through the Earth's atmosphere and hit the Earth's surface.

Meteors Generally small particles of solid material from interplanetary space that approach close enough to the Earth to be drawn into the Earth's atmosphere, where they are heated to incandescence. Sometimes called "shooting stars." Most disintegrate, but a few land on the surface of the Earth as meteorites.

Micrite Pertaining to a texture in carbonate rocks that, when viewed microscopically, appears as murky, fine-grained calcium carbonate. Micrite is believed to develop from fine carbonate mud or ooze.

Milankovitch effect The hypothetic long-term effect on world climate based on positional changes between the Earth and the Sun. The changes provide a possible explanation for repeated glacial to interglacial climatic swings.

Miliolids A group of foraminifers with smooth, imperforate test walls and chambers arranged in various planes around a vertical axis. Miliolids are common in shallow marine areas.

Mineral A naturally occurring element or compound formed by inorganic processes that has a definite chemical composition or range of compositions, as well as distinctive properties and form that reflect its characteristic atomic structure.

Mitosis The method of cell division in which each of the two daughter nuclei receives exactly the same complement of chromosomes as had existed in the parent nucleus.

Mobile belt An elongate region of the Earth's crust, characterized by exceptional earthquake and volcanic activity, tectonic instability, and periodic mountain building.

Mohorovičić discontinuity A plane that marks the boundary separating the crust of the Earth from the underlying mantle. The "moho," as it is sometimes called, is at a depth of about 70 kilometers below the surface of continents and 6 to 14 kilometers below the floor of the oceans.

Molasse Accumulations of primarily nonmarine, relatively light-colored, irregularly bedded conglomerates, shales, coal seams, and cross-bedded sandstones that are deposited subsequent to major orogenic events.

Mold An impression, or imprint, of an organism or part of an organism in the enclosing sediment.

Mollusk Any member of the invertebrate Phylum Mollusca, including bivalves (pelecypods), cephalopods, gastropods, scaphopods, and chitons.

Monoplacophorans Primitive marine molluscans with simple cap-shaped shells.

Monotremes The egg-laying mammals.

Morganucodonts Early mammals found in Triassic beds of Europe and Asia and characterized by their small size and their retention of certain reptilian osteologic traits.

Mosasaurs Large marine lizards of the Late Cretaceous.

Multituberculates An early group (Jurassic) of Mesozoic mammals with tooth cusps in longitudinal rows and other dental characteristics that suggest they may have been the earliest herbivorous mammals.

Mutation A stable and inheritable change in a gene.

Mytiloids Bivalvia having rather triangular shells that are most commonly identical but in some forms unequal. The edible mussel *Mytilus* is a representative form.

Nappe A large mass of rocks that have been moved a considerable distance over underlying formations by overthrusting, recumbent folding, or both.

Nektonic Pertaining to swimming organisms. (Some texts shorten the term to nektic.)

Neogene A subdivision of the Cenozoic that encompasses the Miocene and Pliocene.

Neutron An electrically neutral (uncharged) particle of matter existing along with protons in the atomic nucleus of all elements except the mass 1 isotope of hydrogen.

Nevadan orogeny In general, those pulses of mountain building, intrusion, and metamorphism that were most frequent during the Jurassic and Early Cretaceous along the western part of the Cordilleran orogenic belt.

Newark series A series of Upper Triassic, nonmarine red beds (shales, sandstones, and conglomerates), lava flows, and intrusions located within downfaulted basins from Nova Scotia to South Carolina.

New World monkeys Monkeys whose habitat is today confined to South America. They are thoroughly arboreal in habit and have prehensile tails, by which they hang and swing from tree limbs.

Niche The full range of physical and biologic conditions under which an organism can live and reproduce.

Nonconformity An unconformity developed between sedimentary rocks and older plutons or massive metamorphic rocks that had been exposed to erosion before the overlying sedimentary rocks were deposited.

Normal fault A fault in which the hanging wall appears to have moved downward relative to the footwall; normally occurring in areas of crustal tension.

Nothosaurs Relatively small early Mesozoic sauropterygians that were replaced during the Jurassic by the plesiosaurs.

Notochord A rod-shaped cord of cartilage cells forming the primary axial structure of the chordate body. In vertebrates, the notochord is present in the embryo and is later supplanted by the vertebral column.

Notungulates A group of ungulates that diversified in South America and persisted until Plio-Pleistocene time.

Novaculites A term applied originally to rocks suitable for whetstones and, in America, to white chert found in Arkansas. Now applied to very tough, uniformly grained cherts composed of microcrystalline quartz.

Nuclear fission tracks Submicroscopic "tunnels" in minerals produced when high-energy particles from the nucleus of uranium are forcibly ejected during spontaneous fission.

Nucleic acid Any of a group of organic acids that control hereditary processes within cells and make possible the manufacture of proteins from the amino acids ingested by the cells as food.

Nuclides The different weight configurations of an element caused by atoms of that element having differing numbers of neutrons. Nuclides or isotopes of an element differ in number of neutrons but not in chemical properties.

Nummulites Large, coin-shaped foraminifers, especially common in Tertiary limestones.

Offlap A sequence of sediments resulting from a marine regression and characterized by an upward progression from offshore marine sediments (often limestones) to shales and finally sandstones (above which will follow an unconformity).

Oil shale A dark-colored shale rich in organic material that can be heated to liberate gaseous hydrocarbons.

Old World monkeys Monkeys of Asia, Africa, and southern Europe that include macaques, guenons, langurs, baboons, and mandrills.

Onlap A sequence of sediments resulting from a marine transgression. Normally, the sequence begins with a conglomerate or sandstone deposited over an erosional unconformity and followed upward in the vertical section by progressively more offshore sediments.

Oölites Limestones composed largely of small, round or ovate calcium carbonate bodies called oöids.

Ophiolite suite An association of radiolarian cherts, pelagic muds, basaltic pillow and flow lavas, gabbros, and ultramafic rocks such as periodotite regarded as surviving masses of former oceanic crust largely destroyed in former subduction zones.

Oreodonts North American artiodactyls of the middle and late Tertiary.

Ornithischia An order of dinosaurs characterized by bird-like pelvic structures and including such herbivores as the ornithopods, stegosaurs, ankylosaurs, and ceratopsians.

Orogenic belt Great linear tracts of deformed rocks, primarily developed near continental margins by compressional forces accompanying mountain building.

Orogeny The process by which great systems of mountains are formed. (Orogenesis means mountain building.)

Ostracoderms Extinct jawless fishes of the early Paleozoic.

Ostracodes Small, bivalved, bean-shaped crustaceans.

Ostreoids The "oyster family" of bivalves (pelecypods).

Outcrop An area where specific rock units are exposed at the Earth's surface or occur at the surface but are covered by superficial deposits.

Palatoquadrate Upper jaw element of primitive fishes and chondrichthyes.

Paleoecology The study of the relationship of ancient organisms to their environment.

Paleogene A subdivision of the Cenozoic that encompasses the Paleocene, Eocene, and Oligocene epochs.

Paleogeography The geography as it existed at some time in the geologic past.

Paleolatitude The latitude that once existed across a particular region at a particular time in the geologic past.

Paleomagnetism The Earth's magnetic field and magnetic properties in the geologic past. Studies of paleomagnetism are helpful in determining positions of continents and magnetic poles.

Paleontology The study of all ancient forms of life, their interactions, and their evolution.

Pangea In Alfred Wegener's theory of continental drift, the supercontinent that included all present major continental masses.

Panthalassa The great universal ocean that surrounded the supercontinent Pangea prior to its break-up.

Paraconformity A rather obscure unconformity in which no erosional surface is discernible and in which beds above and below the break are parallel.

Partial melting The process by which a rock subjected to high temperature and pressure is partially melted, and the liquid fraction moved to another location. Partial melting results from the variations in melting points of different minerals in the original rock mass.

Pectenoids Bivalvia exemplified by the scallops. They have generally subcircular shells and straight hinge lines.

Pelycosaurs Early mammal-like reptiles exemplified by the sail-back animals of the Permian Period.

Period A subdivision of an era.

Perissodactyl Progressive, hoofed mammals characteristically having an odd number of toes on the hind feet and usually on the front feet as well.

Permineralization A manner of fossilization in which voids in an organic structure (such as bone) are filled with mineral matter.

Petrification The process of conversion of organic structures, such as bone, shell, or wood, into a stony substance, such as calcium carbonate or silica.

Phanerozoic The eon of geologic time during which the Earth has been populated by abundant and diverse life. The Phanerozoic Eon followed the Cryptozoic (or Precambrian) Eon and is divided into the Paleozoic, Mesozoic, and Cenozoic eras.

Phenotypic trait A trait that is the observable expression of an organism's genes.

Phosphorite A sediment composed largely of calcium phosphate.

Photosynthesis The process of synthesizing carbohydrates from carbon dioxide and water, utilizing the radiant energy of light captured by the chlorophyll in plant cells.

Phyletic gradualism Gradual evolutionary change of one species population into another.

Phytoplankton Microscopic marine planktonic plants, most of which are various forms of algae.

Phytosaurs Extinct aquatic, crocodile-like thecodonts of the Triassic.

Pillow lava Type of lava that is extruded under water and in which many pillow-shaped lobes break through the chilled surface of the flow and solidify. (Resembles a pile of pillows.)

Pinnipeds Marine carnivores such as seals, sea lions, and walruses.

Placer deposit An accumulation of sediment rich in a valuable mineral or metal that has been concentrated because of its greater density.

Placoderms Extinct primitive jawed fishes of the Paleozoic Era.

Placodonts Extinct walrus-like marine reptiles that fed principally on shellfish.

Plankton Minute, free-floating aquatic organisms.

Plate tectonics The theory that explains the tectonic behavior of the crust of the Earth in terms of several moving plates that are formed by volcanic activity at oceanic ridges and destroyed along great ocean trenches.

Platform That part of a craton covered thinly by layered sedimentary rocks and characterized by relatively stable tectonic conditions.

Plesiosaurs The group of extinct Mesozoic marine reptiles (sauropterygians) characterized by large, paddle-shaped limbs and broad bodies, with either very long or relatively short necks.

Pluvial lake A lake formed in an earlier climate when rainfall was greater than at present.

Polyp The hydra-like form of some cnidaria in which the mouth and tentacles are at the top of the body.

Population (species population) A group of individuals of the same species occupying a given area at the same time.

Porphyry A textural term used to describe an igneous rock in which some of the crystals, called phenocrysts, are distinctly larger than others.

Precambrian Pertaining to all of geologic time and its corresponding rocks before the beginning of the Paleozoic Era.

Primary earthquake waves Seismic waves that are propagated through solid rock as a train of compressions and dilations. Direction of vibration is parallel to direction of propagation.

Principle of cross-cutting relations The principle that states that geologic features such as faults, veins, and dikes must be younger than the rocks or features across which they cut.

Principle of biologic succession The principle that states that the observed sequence of life forms has changed continuously through time, so that the total aspect of life (as recognized by fossil evidence) for a particular segment of time is distinct and different from that of life of earlier and later times.

Principle of temporal transgression The principle that stipulates that sediments of advancing (transgressing) or retreating (regressing) seas are not necessarily of the same geologic age throughout their lateral extent.

Proboscidea The elephants and their progenitors.

Productid An articulate brachiopod usually with a flat or concave brachial valve and a convex pedicle valve. Productids were present from Devonian to Permian.

Prokaryotes Organisms that lack membrane-bounded nuclei and other membrane-bounded organelles.

Prosimii The less advanced primates, such as lemurs, tarsiers, and tree shrews.

Proteinoids Extra-large organic molecules containing most of the 20 amino acids of proteins and produced in laboratory conditions simulating those found in nature.

Proteins Giant molecules containing carbon, hydrogen, oxygen, nitrogen, and usually sulfur and phosphorus; composed of chains of amino acids and present in all living cells.

Proton An elemental particle found in the nuclei of all atoms that has a positive electric charge and a mass similar to that of a neutron.

Pterosaur A flying reptile of the Jurassic and Cretaceous.

Punctuate equilibrium The model of evolution proposing that long periods of little or no evolutionary change are punctuated by short episodes of rapid change.

Pyroclastics Fragments of volcanic debris that have usually been found fragmented during eruptions.

Pyroxene group A group of dark-colored iron and magnesium-rich silicate minerals.

Queenston delta A clastic wedge of red beds shed westward from highlands elevated in the course of the Taconic orogeny.

Radioactive decay The spontaneous emission of a particle from the atomic nucleus, thereby transforming the atom from one element to another.

Radiolaria Protozoa that secrete a delicate, often beautifully filigreed skeleton of opaline silica.

Recumbent fold A fold in which the axial plane is essentially horizontal; a fold that has been turned over by compressional forces so that it lies on its side.

Red beds Prevailing red, usually clastic sedimentary deposits.

Regoliths Any solid materials, such as rock fragments or soil, lying on top of bedrock.

Regression A general term signifying that a shoreline has moved toward the center of a marine basin. Regression may be caused by tectonic emergence of the land, eustatic lowering of sea level, or prograding of sediments, as in deltaic build-outs.

Relative geologic age The placing of an event in a time sequence without regard to the absolute age in years.

Replacement A fossilization process in which the original skeletal substance is replaced after burial by inorganically precipitated mineral matter.

Reverse fault A fault formed by compression in which the hanging wall appears to move up relative to the foot wall.

Rhynchonellids A group of brachiopods having pronounced beaks, accordion-like plications, and triangular outlines.

Rift valley A valley formed by faulting, usually involving a central fault block that moves downward in relation to adjacent blocks.

Rudists Peculiarly specialized Mesozoic bivalvia often having one valve in the shape of a horn coral, covered by the other valve in the form of a lid.

Rugosa The large group of solitary and colonial Paleozoic horn corals.

Ruminant A herbivorous, cud-chewing ungulate.

Salt dome A structural dome in sedimentary strata resulting from the upward flow of a large body of salt.

Sarcopterygii Lobe-finned bony fishes, including air-breathing crossopterygian fishes.

Sauk sequence A sequence of upper Precambrian to Ordovician sediments bounded both above and below by a regional unconformity and recording an episode of marine transgression, followed by full flooding of a large part of the craton and ending with a regression from the craton.

Saurischia An order of dinosaurs with triradiate pelvic structures, including both the gigantic herbivorous sauropods and the carnivorous theropods.

Scleractinid coral Coral belonging to the order Scleractinia, which includes most modern and post-Paleozoic corals.

Sea-floor spreading The process by which new sea-floor crust is produced along midoceanic ridges (divergence zones) and slowly conveyed away from the ridges.

Secondary earthquake wave A seismic wave in which the direction of vibration of wave energy is at right angles to the direction the wave travels.

Sedimentary rock A rock that has formed as a result of the consolidation (lithification) of accumulations of sediment.

Sessile Pertaining to the bottom-dwelling habit of aquatic animals that live continuously in one place.

Series The time-rock term representing the rocks deposited or emplaced during a geologic epoch. A series is a subdivision of a system.

Shelly facies In general, sedimentary deposits consisting primarily of carbonate rocks containing the abundant fossil remains of marine invertebrates.

Siliclastic sedimentary rock. A sedimentary rock such as a conglomerate, sandstone, or shale that is composed largely of particles of silicate minerals such as quartz, feldspar, and mica.

Silicoflagellates Unicellular, tiny, flagellate marine algae that secrete an internal skeleton composed of opaline silica.

Silicon tetrahedron An atomic structure in silicates consisting of a centrally located silicon atom linked to four oxygen atoms placed symmetrically around the silicon at the corners of a tetrahedron.

Sonoma orogeny Middle Permian orogenic movements, the structural effects of which are most evident in western Nevada.

Sorting A measure of the uniformity of the sizes of particles in a sediment or sedimentary rock.

Spar carbonate As viewed microscopically, the clear, crystalline carbonate that has been deposited in a carbonate rock as a cement between clasts or has developed by recrystallization of clasts.

Species A unit of taxonomic classification of organisms. In another sense, a species is a population of individuals that are similar in structural and functional characteristics and that in nature breed only with one another.

Sphenodon (tuatara) Large, lizard-like reptiles that have persisted from Triassic to the present and now inhabit islands off the coast of New Zealand.

Sphenopsids A group of sponge-bearing plants that were particularly common during the late Paleozoic and were characterized by articulated stems with leaves borne in whorls at nodes.

Spiracle In cartilaginous fishes, a modified gill opening through which water enters the pharynx.

Spontaneous fission Spontaneous fragmentation of an atom into two or more lighter atoms and nuclear particles.

Spore A usually asexual reproductive body, such as occurs in bacteria, ferns, and mosses.

Stage The time-rock unit equivalent to an age. A stage is a subdivision of a series.

Stapes The innermost of the small bones in the middle ear cavity of mammals; also recognized in amphibians and reptiles.

Stasis A long interval of little evolutionary change.

Strain Deformation of a rock mass in response to stress.

Stratification The layering in sedimentary rocks that results from changes in texture, color, or rock type from one bed to the next.

Stratified drift Deposits of glacial clastics that have been sorted and stratified by the action of meltwater.

Stratigraphy The study of rock strata, with emphasis on their succession, age, correlation, form, distribution, lithology, fossil content, and all other characteristics useful in interpreting their environment of origin and geologic history.

Stratophenic phylogeny The traditional model of evolution in which organisms are arranged in treelike fashion, with the most recently evolved groups on the upper branches and the older, ancestral groups on the lower branches and trunk.

Stromatolites Distinctly laminated accumulations of calcium carbonate having rounded, branching, or frondose shape and believed to form as a result of the metabolic activity of marine algae. They are usually found in the high intertidal to low supratidal zones.

Stromatoporoids An extinct group of reef-building organisms now believed to have affinities with the Porifera and noted for the large, often laminated masses constructed by the colonies.

Subaerial Formed or existing at or near a sediment surface significantly above sea level.

Subduction zone An inclined planar zone, defined by high frequency of earthquakes, that is thought to locate the descending leading edge of a moving oceanic plate.

Sublittoral zone The marine bottom environment that extends from low tide seaward to the edge of the continental shelf.

Surface earthquake waves Seismic or earthquake waves that move only about the surface of the Earth.

Symmetrodonts A group of primitive Mesozoic mammals characterized by a symmetric triangular arrangement of cusps on cheek teeth.

Syncline A geologic structure in which strata are bent into a downfold.

System The time-rock unit representing rock deposited or emplaced during a geologic period.

Taconic orogeny A major episode of orogeny that affected the Appalachian region in Ordovician time. The northern and Newfoundland Appalachians were the most severely deformed during this orogeny.

Taxon (pl. taxa) Any unit in the taxonomic classification, such as a phylum, class, order, or family.

Taxonomy The science of naming, describing, and classifying organisms.

Tectonics The structural behavior of a region of the Earth's crust.

Teleosts The most advanced of the bony fishes, characterized by thin, rounded scales, completely bony internal skeleton, and symmetric tail. Teleosts range from Cretaceous to Recent.

Terebratulids A group of Silurian to Recent, mostly smooth-shelled brachiopods having a loop-shaped attachment for the lophomore. Terebratulids were most abundant during the Jurassic and Cenozoic.

Terrane A three-dimensional block of crust having a distinctive assemblage of rocks (as opposed to terrain, which implies topography, such as rolling hills or rugged mountains).

Tethys seaway A great east-west trending seaway lying between Laurasia and Gondwanaland during Paleozoic and Mesozoic time and from which arose the Alpine-Himalayan Mountain ranges.

Thecodonts An order of primarily Triassic reptiles considered to be the ancestral archosaurians.

Therapsids An order of advanced, mammal-like reptiles.

Thermal plume A "hot spot" in the upper mantle believed to exist where a huge column of upwelling magma lies in a fixed position under the lithosphere. Thermal plumes are thought to cause volcanism in the overlying lithosphere.

Thermoremanent magnetism Permanent magnetization acquired by igneous rocks as they cool past the Curie point while in the Earth's magnetic field.

Theropods The carnivorous saurischian dinosaurs.

Thrust fault A low-angle reverse fault, with inclination of fault plane generally less than 45A7.

Till Unconsolidated, unsorted, unstratified glacial debris.

Tillite Unsorted glacial drift (till) that has been converted into solid rock.

Time-stratigraphic unit (chronostratigraphic unit) The rocks formed during a particular unit of geologic time. Also called time-rock unit.

Tippecanoe sequence A sequence of Ordovician to Lower Devonian sediments bounded above and below by regional unconformities and recording an episode of marine transgression, followed by full flooding of a large region of the craton and subsequent regression.

Titanotheres Large, extinct perissodactyls (odd-toed ungulates) that attained the peak of their evolutionary development during the Early Eocene. Brontotherium is a widely known titanothere.

Trace fossils Tracks, trails, burrows, and other markings made in now-lithified sediments by ancient animals.

Transcontinental arch An elongate, uplifted region extending from Arizona northeastward toward Lake Superior. During the Cambrian, the arch was emergent, as indicated by the observation that Cambrian marine sediments are located on either side of the arch but are missing above it. (Along the crest of the arch, post-Cambrian rocks rest on Precambrian basement.)

Transform fault A strike-slip fault bounded at each end by an area of crustal spreading that tends to be more or less perpendicular to the trace of the fault.

Triconodonts A group of primitive Mesozoic mammals recognized primarily by the arrangement of three principal cheek tooth cusps in a longitudinal row.

Trilobites Paleozoic marine arthropods of the class Crustacea, characterized by longitudinal and transverse division of the carapace into three parts, or lobes. Trilobites were especially abundant during the early Paleozoic.

Tuff Volcanic ash that has become consolidated into rock.

Turbidites Sediment deposited from a turbidity current and characterized by graded bedding and moderate to poor sorting.

Turbidity current A mass of moving water that is denser than surrounding water because of its content of suspended sediment and that flows along slopes of the sea floor as a result of that higher density.

Tylopod The artiodactyl group to which camels and llamas belong.

Unconformity A surface separating an overlying younger rock formation from an underlying formation and representing an episode of erosion or nondeposition. Because unconformities represent a lack of continuity in deposition, they are gaps in the geologic record.

Ungulate Pertaining to four-legged mammals whose toes bear hoofs.

Uniformitarianism A principle that suggests that the past history of the Earth can be interpreted and deciphered in terms of what is known about present natural laws.

Vagile Pertaining to the bottom-dwelling habit of aquatic animals capable of locomotion.

Varve A thin sedimentary layer or pair of layers that represent the depositional record of a single year.

Vascular plants Plants, including all higher land plants, that have a system of vessels and ducts for distributing moisture and nutrients.

Veneroids A group of bivalves exemplified by the common clam, *Mercenaria*.

Vestigial organ An organ that is useless, small, or degenerate but representing a structure that was more fully developed or functional in an ancestral organism.

Vindelician arch A highland area believed to have formed a barrier separating the Germanic and Alpine depositional areas during the Triassic and Jurassic.

Volatile element A chemical element that generally occurs as a gas at moderate to high temperatures.

Williston basin A large structural basin extending from South Dakota and Montana northward into Canada; well known for the petroliferous Devonian formations deposited therein.

Zechstein sea An arm of the Atlantic that extended across part of northern Europe during the late Permian and in which were deposited several hundred meters of evaporites, including the well-known potassium salts of Germany.

Zircon A zirconium silicate mineral that is useful in uranium-lead isotopic dating.

Zone A bed or group of beds distinguished by a particular fossil content and frequently named after the fossil or fossils it contains. More formally known as a biozone.

Zygote The cell formed by the union of two gametes. Thus, a zygote is a fertilized egg.

INDEX

In this index, an italic *f* following the page number indicates a *figure*, and an italic *t* after the page number indicates a *table*. Entries from the appendices are preceded by a capital A.

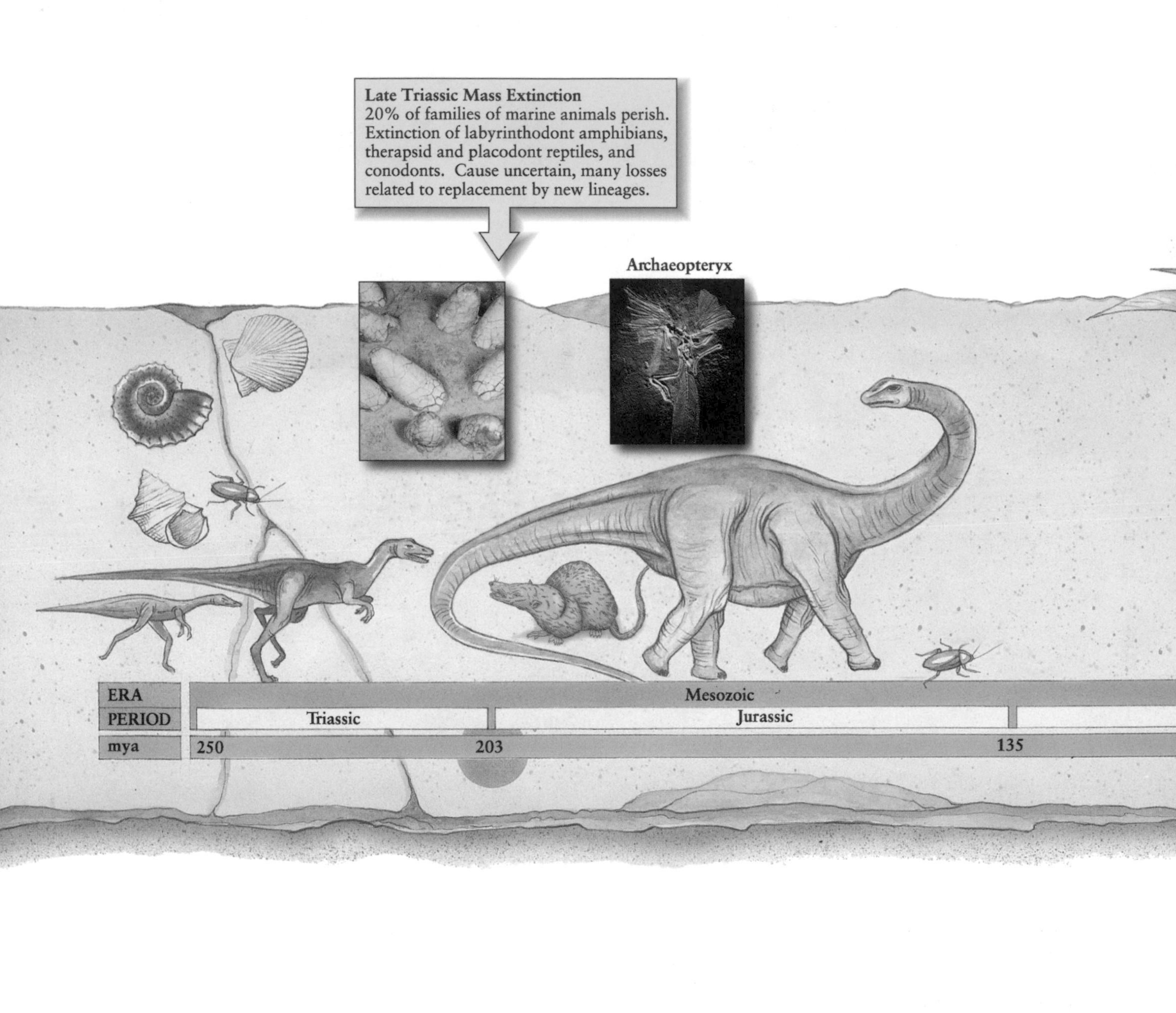

Late Triassic Mass Extinction
20% of families of marine animals perish. Extinction of labyrinthodont amphibians, therapsid and placodont reptiles, and conodonts. Cause uncertain, many losses related to replacement by new lineages.
Archaeopteryx
ERA
PERIOD
mya
Mesozoic
Triassic
Jurassic
250
203
135